Gravity from the ground up

Gravity
from the ground up

Bernard Schutz

*Max Planck Institute for
Gravitational Physics
(The Albert Einstein Institute)
Golm, Germany*
and
*Department of Physics and Astronomy
Cardiff University, UK*

PUBLISHED BY THE PRESS SYNDICATE OF THE UNIVERSITY OF CAMBRIDGE
The Pitt Building, Trumpington Street, Cambridge, United Kingdom

CAMBRIDGE UNIVERSITY PRESS
The Edinburgh Building, Cambridge, CB2 2RU, UK
40 West 20th Street, New York, NY 10011-4211, USA
477 Williamstown Road, Port Melbourne, VIC 3207, Australia
Ruiz de Alarcón 13, 28014 Madrid, Spain
Dock House, The Waterfront, Cape Town 8001, South Africa

http://www.cambridge.org

First published 2003

Printed in the United Kingdom at the University Press, Cambridge

This book was designed, composed and typest by the author using LATEX 2$_\varepsilon$. The author wishes
to acknowledge his indebtedness to the developers of TEX, LATEX, and the many packages that
were used in this project. The main text is set in Adobe Aldus supplemented by Adobe
Palatino; the text in boxes is set in Bigelow & Holmes' Lucida Bright Sans Serif; mathematics
uses in addition other Lucida Bright fonts and the Euler font designed by H. Zapf.

A catalogue record for this book is available from the British Library

ISBN 0 521 45506 5 hardback

To my children

Rachel, Catherine, and Annalie

who have not known a time when I was
not writing "the book"!

Contents

Preface

From the author to the reader

Why this book is about gravity

During the 35 years that I have done research in gravitation, I have watched with amazement and delight as my colleagues in astronomy have, step-by-step, opened up almost the entire Universe to our view. And what a view! There are punctures in space called black holes that capture gas and stars with a relentless and unbreakable grip; there are 10 km balls called neutron stars that are immense overgrown atomic nuclei with more mass than our Sun, that spin about their axes hundreds of times per second while emitting intense beams of radiation; there are bursts of gamma-rays from the most remote regions of the Universe that are so intense that they outshine the rest of the Universe for a short time; and most strikingly of all there was the beginning of time itself in an explosion of pure energy, driven by a force we do not understand, in which matter as we know it did not exist, in which even the laws of Nature themselves were mutable.

This astonishing Universe has captured the imagination of many people, among them many scientists. Physicists trained in a number of disciplines have applied themselves to explaining these and many more less spectacular but equally important phenomena, such as: how the chemical elements were made; where stars come from and how they evolve and die; how the vast systems of stars called galaxies formed and why they have grouped themselves into clumps and long chains; why a Universe filled with bright stars seems to contain even more matter that cannot form stars – and so remains dark.

From all this scientific activity has come a great deal of understanding. We know not just *what* happens, but in many cases *how* and *why*. Physicists, astrophysicists, and astronomers have been able to put together a coherent story of how our Universe began and of how its immense variety evolved.

> The central theme of the story of the Universe turns out to be *gravity*.

Gravity is the one force of Nature that operates everywhere; it controls the effects of all the other forces wherever they act; it regulates countless natural clocks, from the orbits of planets to the lifetimes of stars. Gravity rules the most violent places in the Universe – quasars, pulsars, gamma-ray bursters, supernovae – and the most quiet – black holes, molecular clouds, the cosmic microwave background radiation. Today gravity binds stars and galaxies and clusters of galaxies together, but much earlier it pushed the Universe violently apart. Gravity explains the uniformity of the Universe on very large scales and its incredible variety on small scales. Gravity even laid the path toward the evolution of life itself. If we understand how gravity works, then we begin to understand the Universe.

Rich as our understanding of the workings of gravity in the Universe has become, it is far from complete. The gaps are not just hidden regions, phenomena yet to be discovered, although when such discoveries occur they are sure to bring more

▷ The link between gravity and the wonders of astronomy goes right back to Galileo, who founded the science of gravity. Using a telescope for the first time, Galileo became the first person to understand that the Milky Way is composed of stars, that Venus shines by light reflected from the Sun, that the Sun is plagued by spots, that Jupiter holds its own satellites in orbit around itself in imitation of the Solar System. Our amazement at astronomers' discoveries today helps us to appreciate what Galileo's contemporaries must have felt at his.

amazement and delight. The most exciting gaps are those in our understanding of the laws of Nature.

The enormous advances in astronomy that I have witnessed in my working life have brought us to the threshold of a profound revolution in our understanding of gravity itself. Many physicists today are working to unify gravity with the other forces of Nature, which will lead to what is called a quantum theory of gravity. There are aspects of the Universe that will not be explained without this new theory, and there are clues to the new theory in many of the currently unexplained puzzles of the Universe.

▷ The word "revolution" gets used so much these days in discussions of progress in the sciences, that I hesitate to use it here. But I know no better word.

This book is about gravity at the threshold of this revolution. We will take a tour of the Universe from the ground up. We will start at the surface of the Earth and move outwards through the Solar System, the Galaxy, and beyond to a scale where our Galaxy is the merest atom in the corpus of the Universe.

We will learn about gravity and the other laws that govern the Universe, first as understood by Newton and his successors, then as understood by Einstein and modern physicists. We will use these laws to see how the parts of the Universe work, how they relate to one another, and how they may have come to be. By the end of our tour we will see the Universe and its physical laws, not merely as a collection of fascinating but separate phenomena, but rather as a unity.

▷ Our tour will be thought-provoking, sometimes demanding, even laborious. But we will not leave you, the reader, behind. If you start with high-school mathematics skills – see the section "How this book uses mathematics" beginning on the next page – then you will be able to follow the discussion all the way. And if you put in the effort to study and run the computer programs that allow you to study areas where simple mathematics does not suffice, then you can reach real expertise in some areas. Our goal is ambitious, but I hope you find it worth the effort!

Our goal is not just to wonder and marvel at our Universe, nor simply to admire the cleverness of the scientists who have made the Universe at least partially understandable. Instead, *our goal is to understand how the Universe works, to begin to think about the Universe in the same way that these scientists themselves do.*

How gravity evolves

Gravity, the oldest force known to mankind, is in many ways also the youngest. It is understood well enough to explain stars, black holes and the Big Bang, and yet in some ways it is not understood at all. Explaining gravity required the two greatest scientific minds of modern history, Isaac Newton and Albert Einstein; and now hundreds of the brightest theoretical physicists are working to invent it once again. Each time gravity has been re-invented, it has sparked a revolution. Newton's theory of gravity stimulated huge advances in mathematics and astronomy; indeed, it was the beginning of modern theoretical physics. Einstein's theory of gravity, which he called general relativity, opened up completely unexpected phenomena to investigation: black holes, gravitational waves, the Big Bang. When, sometime in the future, gravity changes into quantum gravity, possibly becoming just one of many faces of a unified theory of all the physical forces, the ensuing revolution may be even more far-reaching.

Each of these revolutions has built on the previous one, without undermining it. Newton's gravity is just as important today for explaining the motions of the planets as in Newton's time. It is used to predict the trajectories of spacecraft and to understand the structure of galaxies. Yet general relativity underpins all of this, because Newton's gravity is only an approximation to the real thing. We need only Newton to help us understand how a star is born and evolves; but when the star's evolution leads to gravitational collapse and a supernova, then we have to ask Einstein's help to understand the neutron star or black hole that is left behind. When we have a theory of quantum gravity, it won't stop us from using general relativity to explain how the Universe expanded after the Big Bang; but if we want to know where the Big Bang came from, and why (or whether) time itself started just then, we will need to ask the quantum theory.

There is a deeper reason for this continuity from one revolution to the next. As an example, consider the fact that two of the fundamental ideas in Einstein's general relativity, called the principle of relativity and the principle of equivalence, originated with Galileo. Einstein's revolution brought a complete change in the mathematical form of the theory, added new ideas, and opened up new phenomena to investigation. But there was a profound continuity in physical ideas, and these were as important to Einstein as the mathematical form of the theory. The coming quantum revolution will surely likewise be grounded firmly in concepts that physicists today use to understand gravity. These physical ideas are the subject of this book.

Of course, this book deals mainly with what we already know about the role that gravity plays in the Universe, which is the result of the first two revolutions. You, the reader, will learn what Newton's gravity is, and how it regulates planets, stars, and galaxies. You will learn what relativity means, and how general relativity leads to black holes and the Big Bang. But I want you also to be able to follow the continuity of ideas, to see for example how Newtonian gravity prepared the way for relativity. In the earliest chapters we will see that Newtonian gravity already contained half of relativity, that it contained the equivalence principle that ▷ Chapter 1
guided Einstein to general relativity, even that it foresaw the existence of black holes ▷ Chapter 2
and gravitational lenses. In the same way, general relativity contains seeds that will ▷ Chapter 4
blossom only when the third revolution arrives, such as why the theory allows the cosmological constant. I will try to point out some of these seeds as we go along, ▷ Chapter 27
usually by asking questions that general relativity or modern astronomy suggests but does not answer. The final chapter is devoted entirely to such questions. ▷ Chapter 27

Why this book is about more than gravity

Because gravity is the dominant force anywhere outside of the surface of the Earth, this book covers a lot of astronomy. But instead of just touring randomly around the Universe, we have a theme: gravity as the engine that makes things happen everywhere. This theme unifies and simplifies the study of astronomy. If we understand gravity on the Earth, then it is easier to understand it in the Solar System. If we ▷ Chapters 1–3
understand it in the Solar System, then we have an easier time grasping how it acts ▷ Chapters 4–8
in stars and black holes. And so it goes, right up to the largest scales, to the Universe ▷ Chapters 9–13, Chapter 21
as a whole. ▷ Chapters 24–27

Because gravity usually acts in concert with other forces of physics, studying gravity this way also gives us the opportunity to investigate much of the rest of physics along the way. For example, quantum theory and gas dynamics play important roles in stars, and so we study them in their own right where we need them. ▷ Beginning in Chapter 7
Even if you have studied these subjects before, you may be surprised at some of the connections to other parts of physics that you will discover by looking at them in the context of explaining a star.

How this book uses mathematics

You may already have guessed that this book is not a "gee-whizz" tour of the Universe: this is a book for people who are not afraid to think, who want to understand what gravity is, who want to go beyond the superficial level of understanding that many popular books settle for. But this is also not an advanced textbook. We shall steer a careful middle course between the over-simplification of some popular treatments and the dense complexity of many advanced mathematical texts.

This book has equations, but the equations use algebra and (a little) trigonometry, not advanced university mathematics. What is required in place of advanced mathematics is thought: readers are asked to reason carefully, to follow the links between subjects. You will find that you can climb the ladder from gravity on the Earth to gravity (and even anti-gravity) in the Universe if you go one step at a time, making sure you place each foot securely and carefully on the rungs as you climb. In return for putting in the thought that this book asks, you can get much further than you might have expected in understanding gravity and its manifestations in astronomy. School students and university undergraduates will find that this book offers them an early avenue into subjects that are usually regarded as much too advanced for them.

There is no calculus in this book, despite the fact that calculus is the workhorse mathematical tool of physics. Wherever possible I have tried to present a physical argument as a substitute for the mathematical one that physicists are used to. This has the great advantage that it makes connections between different parts of physics clearer and the logical reasoning more direct. It has the disadvantage, of course, that it is not always possible to do this: there are places where using more advanced mathematics really is necessary for a pen-and-paper treatment. In such cases I have often turned instead to a computer program. These programs are not "black boxes": their construction is discussed in detail. See the next section for a discussion of why they are good substitutes for advanced mathematics.

Sometimes I have had to resort to that awful phrase "it can be shown with more advanced mathematics" or something like it. I have avoided this whenever I could, but there are times when it seemed to me that any argument I could give for a

▷ See Figure 19.1 on page 242.

particular result would be over-simplified, it would hide or corrupt the truth. It is best in such situations to be honest and accept that our mathematical tools at this level are not always sufficient. Our aim is not to cut corners, but always to remain true to the physics.

In fact, it is possible to read this book while avoiding most of the equations, if you want to. All the extended algebraic calculations are placed in special boxes, called investigations. These are set aside on a light-gray background. Skipping these boxes might be a good strategy if you are short of time, or on your first reading. If you skip them you will just have to take on faith some of their results, which are then used in the main text. Many of the investigations contain exercises, which offer

▷ Solutions to the exercises can be found on the book's website. See the next page.

you a chance to test your understanding. I believe strongly that doing exercises is the most effective way to get comfortable with an important result. If you are using this book as a textbook for a course, then I hope your teacher will expect you to do the exercises!

How to go beyond this book by using computers

For those of you who have access to computers and want to use them, I have provided a way to reach the results of some very advanced mathematics by using computer programs that only require the mathematical level of the rest of this book. This is your best way to get to some of the results that algebra alone cannot reach. The programs can be downloaded from the website (see the next section) and used right away – just run them and look at the results. You can then change some of the numbers they work with, for example to compute the orbit of Jupiter instead of Mercury, without looking inside the program. But the way to get the most out of them is to study the investigations in which they are described, look inside the programs, and even experiment with changing the code.

As an example of the power of computer programs, consider the motion of a planet around the Sun. Newton's law of gravity giving the forces that govern the motion of the planet is not hard to write down or to understand using pure algebra.

Using it – solving it – to find the planet's orbit is not so easy with pen and paper. The usual way that university physics students learn how to show, for example, that the orbit is an ellipse is by using some rather sophisticated calculus. They may have to wait until their second or third year to get to this important result. In Chapter 4 we achieve the same thing by writing a simple computer program that moves the particle along in its orbit step-by-step. The law of gravity translates directly into a prescription for algebraic calculations that the computer can readily do. The orbit that comes out is clearly an ellipse. The calculation can be done as accurately as one wishes by simply telling the computer to take smaller steps. Repetitious computations like this are what computers do best.

Even better, a computer program can be modified and used again in situations that would require even more sophisticated calculus to make progress. We modify the orbit program slightly in Chapter 13 to explore what happens when three stars interact with one another, a situation that often results in one of them being expelled from the system at high speed. We modify the orbit program yet again in Chapter 21 to show that orbits of bodies around black holes are not ellipses, but rather make rosette patterns. And finally in Chapter 24 we modify it another time to calculate the expansion of the Universe itself.

▷This so-called three-body problem is a classic problem that cannot be solved by analytic calculus alone. All scientists who study it use computer programs.

The programs are available from the website of this book (see below). They are written in the free and widely available language Java®. Since the popularity of this language is steadily increasing, you may find that following the computer programs here will help you to learn a language that will be useful to you later. In any case, the Java language is not very different from the most popular language of all, C. If you have never programmed before, the package you can download from the website provides an easy way of starting.

Using the glossary and website

There are two aids for you to get more information than is in the text: the glossary, which begins on page 421, and the website. The glossary contains definitions of many of the terms used in this book. Some of the terms in the glossary are not defined in the book because I have assumed that most readers will know what they mean, or because they are not central to the subject matter. Other terms in the glossary are important words that are explained somewhere in the book but which are placed in the glossary so that you can conveniently look them up when you encounter them again. All terms in the glossary are printed in **boldface** type when they are first used in the book. I don't use boldface for any other reason, so whenever you see it you will know that it marks an entry in the glossary.

▷Terms printed in **boldface type** in the text are contained in the glossary.

The website for this book is

> http://www.gravityfromthegroundup.org

It contains

- the Java programs for you to download;

- a free version of the Triana® software environment for running the programs and displaying their results graphically;

- solutions of all the exercises;

- links to allow you to download and install Java and other programs needed for your computer;

- additional illustrations for some of the chapters;

- a way of submitting comments, misprints you have found, or suggestions that could be incorporated in future editions; and

- links to useful websites where you can follow up some of the material covered in the book.

Visit the website: it is a valuable addition to this book, and it is completely free.

From the author to his colleagues

Teaching gravitation

Although this book is aimed at beginners, many readers may be my colleagues, professional physicists who may be using the book in a course they are teaching, supervising a student who is using the book for a self-study program, or just looking for a different point of view on the subject. For such readers, this section enlarges on the pedagogical side of my approach to this subject.

The aim of this book is to introduce gravitation theory as a unified subject, and especially to show the key role that gravitation plays in the phenomena of the Universe. An associated pedagogical goal is to develop the reader's ability to think physically by using physical reasoning rather than advanced mathematics to move through the subject. The restriction to elementary mathematics presents real challenges in presentation, but it allows one to treat the entire theory in a unified way, from Newton to Einstein and beyond ... from the ground up, in other words.

Mathematics is not just a powerful tool in physics, it is the *reference language* of science: it is the language in which the fundamental theories are written, the medium which is used to deduce the predictions from a theory. But physicists generally supplement mathematical deduction with physical reasoning. Indeed, physicists who can do this reliably are widely admired for their great "physical intuition". When physicists are inventing new theories, searching for new physics, or trying to explain phenomena not previously encountered, physical reasoning often leads and mathematical reasoning follows: new ideas suggested first by physical arguments are then put into mathematical form and tested for consistency and suitability. Yet when physicists teach known physics to newcomers, the balance more often falls heavily toward mathematical reasoning, the reference form of the theory. Students need of course to master the mathematical form of a theory in order to be able to work seriously with it or to go beyond it, and physics teaching generally focuses on that requirement.

But I believe that this focus is often too narrow. It is important to remind ourselves that there is usually a line of physical reasoning that moves along parallel to the mathematical. Ideally, each way of thinking supports the other. But for students with unsophisticated mathematical tools, it should be possible to make significant progress using mainly physical reasoning. After all, if the principal theories of physics were invented by using physical arguments as guides, then it should be possible to teach important things about those theories in the same way.

Putting mathematical presentation first has another undesirable pedagogical side-effect: it is customary to teach some physical theories in discontinuous segments in order to allow students time to learn more sophisticated mathematics in between. Nowhere is this more arresting than for gravitation theory. Newton's law of gravity is presented in high school, but even using it to find the simple elliptical orbit of a planet must wait until the student masters integral calculus, in the first or second year of a university course. Because of its use of tensors and differential geometry, general relativity has to wait until the final undergraduate year at the earliest; most physicists encounter it first as graduate students, if at all.

Yet there are very good reasons for teaching gravitation theory as a unified subject. The continuity of physical ideas and phenomena is strong. Consider the following sampling, which is by no means exhaustive.

- The equivalence principle and the principle of relativity – so important to Einstein – originated with Galileo.

- Most physicists find it remarkable that black holes and the gravitational deflection of light were discussed by scientists more than a century before Einstein (see Chapter 2). Yet surely this simply means that the links between Newtonian and Einsteinian gravity go deeper than most of us assume.

- There are more similarities than differences between Newtonian and relativistic stars. Even the gravitational effects caused by the spins of stars and black holes have their roots in Newtonian gravity (Chapter 19).

- If we want to trace the histories of the objects in Newton's universe, such as planets and people, all the way back to their ultimate roots, we inevitably encounter the hot, slightly lumpy plasma that we call the Big Bang. The inevitability of the Big Bang has as much to do with the argument of the eighteenth-century physicist Olbers, who understood how strange it is that the sky is dark at night, as it has with Einstein.

- And even Einstein's theory of general relativity itself, for all its mathematical complexity, is arguably as close in physical content to Newton's as it was possible for Einstein to make it while still respecting special relativity.

To teach the broad sweep of gravity as a unified whole to an audience that normally only gets taught about circular planetary orbits, I have followed the pedagogical philosphy outlined earlier: using the minimum level of mathematical sophistication, I have tried to progress through gravitation theory as much as possible by using physical arguments. This started out, quite frankly, as an experiment, a challenge to myself, and I have learned much from it, especially about the connections between and continuity of ideas in this subject. For example, it is satisfying that it is natural to introduce both the principles of relativity and of equivalence in Chapter 1, followed immediately by the gravitational effect on time in Chapter 2, without ever leaving the vicinity of the Earth, before even considering Newton's law of gravitation. When these principles turn up again in special and general relativity, they are old friends. When I explain in Chapter 19 that gravity in Einstein's picture is found mainly in the curvature of time, it is not hard to justify this from the discussion in Chapter 2. It is equally satisfying to calculate the Newtonian gravitational force exerted by a spherical body (Chapter 4) – using one of the computer programs to do the integral calculus – and then to find that one needs nothing more than this to calculate the evolution of a homogeneous and isotropic cosmological model (Chapter 24). It is fascinating to calculate the fundamental normal mode frequency of the Sun (Chapter 8) in Newtonian gravity and then to find that the same formula comes within a factor of two of the right answer for the pulsations of a disturbed black hole (Chapter 21). It is equally fascinating to discover that one can derive the Lense–Thirring effect quantitatively from Newtonian gravity and special relativity only if one uses the Einstein form of the active gravitational mass as the source of gravity (Chapter 19), and thereby to establish deep links between Newtonian gravity, the spinning black holes (Chapter 21), and the inflationary universe (Chapter 24). The list could be much longer.

This approach to teaching gravitation will surely not appeal to everyone, but I hope that especially my scientific colleagues in relativity will find it amusing to see how many threads continue from Newtonian gravity to relativity, how many apparently abstract and mathematical properties of relativistic gravity have clear and simple physical derivations, and how much easier it is to introduce general relativity if Newtonian gravity has been taught in a way that emphasizes the ideas that continue into relativity.

Guiding students through this book

The book can be used by teachers for guided self-study, as a textbook for a course on physics or astronomy for non-scientists, or as a main or supplementary text in conventional university physics and astronomy courses.

Courses for non-scientists need to excite and challenge students without overwhelming them with mathematics. With its emphasis on developing physical intuition, this book aims directly at what is probably the most important goal of such a course: students should learn what it means to think like a physicist. The exercises can play an important role. Depending on the length of such a course and the background of the students in it, the teacher may want to be selective in what material to focus on. I would welcome feedback (via the website) from anyone teaching such a course.

When using this book with physics or astronomy undergraduates, the obvious problem is that the book has a "vertical" integration: it covers material that is usually treated in different courses in different years. Indeed, that is why I have written it. It can be helpful for beginners to expose them to some of these advanced ideas and to let them explore them with the aid of the computer programs before they reach the mathematical level needed to treat them in the conventional way, later in their education. Again I would welcome feedback via the website from lecturers who use this book in such courses, either as the main or as a supplementary text.

The computer programs deserve special attention from the teacher. They fill the gaps between algebra and calculus for beginners, while for students who continue to study physics they are good preparation for later analytical attacks on problems.

Let me give an example. Consider the computer program for finding the motion of a planet around the Sun, to which I referred earlier. The mathematical way that undergraduate physics students learn that its orbit is an ellipse is by writing Newton's law of motion as a differential equation and solving it using fairly sophisticated calculus. They often have to wait a year or two in their undergraduate course before they have the skill to do this. The solution can instead be found using a computer if we replace the differential equations with finite difference equations. By formulating Newton's law from the start in terms of finite differences – the change in velocity in a small but finite time-interval is approximately the acceleration at the beginning of the time-interval times the time-interval – we have an immediate entry into the computer simulation. The formulation is obvious to students, and just as obvious is the idea that if one makes the time-step smaller and smaller then the computer solution becomes a better and better approximation to the real thing. This is calculus in practice, and if students meet calculus later in their mathematics education, then they know they have already been doing it on the computer.

Computers are already used in this way in many introductory physics courses. Some of the best use spreadsheets, because they are widely available, they contain all the required mathematical operations, and they can display results as graphs. For this book, however, I have chosen instead to use the programming language Java®. It is also widely available, free, and mathematically complete. The Triana®environment that can be downloaded for free from the website provides the ability to run pro-

▷Besides the four equations-of-motion computer problems mentioned earlier, the book applies computers to a variety of other problems. Students can prove that the gravitational field outside a spherical body is the same as if all its mass were concentrated at its center, by adding up the forces from small elements of the body. They can make a computer model of the Earth's atmosphere, and later use the same program to model the Sun and a neutron star. In each case the problems are formulated from the start in terms of small differences rather than derivatives. And there are no compromises: we don't have to over-simplify these problems in order to put them on the computer.

grams as black boxes and get graphical output. I prefer the fact that Java programs are closer in structure to those that students may write later in their careers (in Fortran, C, or even Java). But lecturers who already use spreadsheets with their students should have little trouble transferring the programs to that format.

Acknowledgements

This book has taken many years to write, and in that time I have learned much from many colleagues in astronomy, physics, and relativity. Some have contributed directly to this book with constructive criticism, creative input, or tracking down resources and historical material; others have simply taught me things I did not know. I would especially like to acknowledge my indebtedness to Robert Beig, Werner Benger, Jiří Bičák, Curt Cutler, Thibault Damour, Karsten Danzmann, Mike Edmunds, Jürgen Ehlers, Jim Hartle, Günther Hasinger, Jim Hough, Klaus Fricke, Matthew Griffiths, Geraint Lewis, Elke Müller, Charlie Misner, Jürgen Renn, Rachel Schutz, Ed Seidel, Kip Thorne, Joachim Wambsganss, Ant Whitworth, Chandra Wickramasinghe, and Cliff Will. None of them, of course, bears responsibility for any errors that remain in the book. My employers, the Max Planck Society and Cardiff University, have kindly made it possible for me on a number of occasions to to get away from my normal duties and write. My editors at Cambridge University Press – Simon Mitton, Rufus Neal, Simon Capelin, Tamsin van Essen – deserve special thanks for their patience and encouragement while waiting for a manuscript that must have seemed like it might never arrive, and which kept growing well beyond its original planned length. Fiona Chapman of Cambridge University Press helped enormously with the presentation. And last but by no means least, I want to thank my family, especially Siân, for putting up with my hiding away to write on countless evenings, weekends, and holidays, and for their unwavering belief that the final result would be worth it.

Ready to start

This preface is long enough. Beginners, or rusty old-timers, should check the review of background material that follows next; others should jump straight to Chapter 1. I wish the reader a satisfying and enlightening journey through the universe of gravity, from the ground up.

Bernard Schutz
Golm, Germany
16 February 2003

Background: what you need to know before you start

As explained in the preface, I have used high-school mathematics to present some of the material in this book. If you want to know what that means, if you want to learn whether you have the background necessary to do the mathematics, then scan through this introductory material. But remember, it is not necessary to follow all the derivations, particularly the ones in the boxes, if you just want to learn what the main ideas in modern gravity and astronomy are. So if you find your mathematics too old or rusty, then see how you get along without it.

High-school mathematics

The mathematics used is basic numeracy, algebra, and a tiny bit of trigonometry (which you can skip).

It is essential to understand scientific notation for numbers, that is how to write numbers in the form 3.2×10^6 and know what the factor 10^6 means. Scientists use this notation all the time, because otherwise they would be writing out long confusing strings of zeros. The number 3.2×10^6 means $3\,200\,000$, obtained by moving the decimal point in 3.2 six places to the right. Similarly, the number 5.9×10^{-3} is 0.0059, obtained by moving the decimal point three places to the left.

Scientific notation also allows scientists to hint at the accuracy with which they know a number, or at least intend to use it. So if a scientist measures a brief lapse of time to the nearest thousandth of a second and gets 0.021 seconds, then he can write it as 2.1×10^{-2} s. If the measurement accuracy were greater, say it came out to be 0.0210 accurate to one ten-thousandth of a second, then the scientist could write 2.10×10^{-2} s. The number of figures quoted before the power of ten is called the number of *significant figures* in the expression. Generally, in working out calculations, you should aim to keep only as many significant figures in your answer as there were in the least accurately known number you used in your calculation. There is no point using π to ten figures to compute the area of a circle if you have measured its radius to only 10% accuracy.

Problems in physics often involve numbers that have *units*, like seconds or meters. We will use SI units in this book, and the values of important physical numbers in these units are given in the Appendix. Some readers may be more comfortable with American or British Imperial units, like feet, miles, pounds. But all fundamental scientific work is done these days in SI units, so I will assume that you know the conversions.

I also assume you know the basics of algebra. For example, you know how to manipulate and solve simple equations, for example:

$$a + b = c \Rightarrow a = c - b.$$

I often make remarks in the book that we will "solve for a variable" in an equation, as we have done for a above. This means adding, subtracting, dividing, multiplying, or whatever is needed in order to isolate the variable on the left-hand side of the equation. But we will not need to use advanced solution techniques, like solving quadratic or cubic equations.

When dealing with numbers that have units in algebraic expressions, treat the units as if they were variables. That is, if you square a length of 10 m to get the area of a square, then you square not only the number but also the unit, so that the result is 10^2 m^2 = 100 m^2, or 100 square meters. We shall always use negative exponents on units to denote division, where in words we would use "per": a speed of 15 meters per second is written as 15 m s^{-1}, and an acceleration of 9.8 meters per second per second is written as 9.8 m s^{-2}. Remember as well that, while you can freely multiply numbers that have different units (and get a result whose units are the product of the units of all the factors), you can't add numbers with different units at all. To add 2 seconds to 3 meters is nonsense. If you find yourself doing this in a calculation, you have made a mistake! Go back and find it.

Angles are always interesting. The everyday unit for angles is the degree, of which there are 360 in a circle. Scientists occasionally use this too, if they have an angle with a convenient size, like 90°. But the standard measure for angles in more advanced mathematics is the radian, and this pops up occasionally in this book as well. The radian measures essentially the fraction of a full circle that the angle represent, except that it expresses this as a fraction of 2π. So a full circle has 360° or 2π radians. One radian is therefore $360/2\pi$ degrees, or about 57.3°. The measure is convenient because it is the ratio of the length of the arc of a circle that the angle intercepts to the radius of the circle.

Being the ratio of two lengths, the radian is a *dimensionless* number. Dimensionless (or pure) numbers have a particular importance in mathematics. There are certain mathematical operations that can be done only with dimensionless numbers, such as evaluating the expression $x + x^2$. We will meet some situations like this in this book, especially when we introduce the exponential function.

Physics

This book is about introducing advanced ideas in physics in a simple way, so it helps if the basis for these ideas is already present. I assume that you have had an elementary introduction to physical science, so that we have some common language and ideas. Remember to check the glossary for any terms or words that you are unsure of, especially when you first encounter them written in **boldface** type.

▷Terms in **boldface** are defined in the glossary.

For example, the study of mechanics is the science of how things move under forces. The basis of mechanics is the set of three laws of motion of Newton. We will discuss these, but it helps if you review them beforehand. The first law says that objects move in straight lines if there are no forces acting on them. The second is usually expressed as the equation $F = ma$, which relates the acceleration a of a body to the force applied to it F and its mass m. The third law is the one that causes most grief in first discussions of mechanics: to every force there is an equal and opposite reaction. Since this is discussed in some detail in Chapter 2, don't worry too much about it before.

It helps if you have looked at Hooke's law for springs, which states that the force with which a spring pulls back is proportional to the length by which it has been stretched. It also helps if you have encountered the definitions of momentum and, less critically, of angular momentum. But none of these is a show-stopper for this book.

I also assume that the basics of electricity and magnetism are familiar. For example, electric charges come in two types: opposite charges attract each other and similar charges repel. Magnetism similarly has two poles, and is created by moving electric charges.

We shall learn a lot in this book about the details of atoms and quantum theory, so it helps to review some basic facts. All matter is composed of atoms, and each atom has a dense nucleus (made of protons and neutrons) surrounded by orbiting electrons. The electrons are responsible for chemistry: atoms bind to one another to form molecules by the forces that attract the electrons of one atom to the nucleus of another. The number of electrons equals the number of protons in a normal atom, but if electrons are removed it is called an ion. The number of protons determines the kind of element that the atom belongs to, while the number of neutrons can vary. Two nuclei that have the same number of protons but different numbers of neutrons are called different isotopes of the same element. We will use the notation ^{238}U to represent this information: the symbol U tells us we are dealing with uranium, and the prefix 238 represents the total number of protons and neutrons in the nucleus, so it tells us which isotope we have.

Light plays a central role in astronomical obsevations, so its properties occupy much of this book. By passing light through a prism you can split it into its component colors. This is called its spectrum. Color corresponds to the wavelength or frequency of the light. White light, such as is produced by a light bulb, has a smooth distribution of intensity across the colors of its spectrum. But sometimes the intensity is concentrated at particular colors. Light from a fluorescent light is like this. These "spectral lines" are signatures of the atoms emitting the light. We will go into this in some detail in the book.

Other kinds of radiation are related to light; they are generically known as electromagnetic waves. Radio waves and microwaves are the same thing, only with longer wavelengths. They are usually produced by making electrons accelerate in an antenna; accelerating charges emit electromagnetic radiation. X-rays and gamma-rays have very short wavelengths compared to visible light.

Naturally, it helps if you have some background in gravity. Although we will introduce almost everything we need, it makes things easier if you have seen some things before. For example, the history of gravity is an important issue in the history of ideas: Aristotle's view that objects fall to the ground because they are seeking their natural place was a big obstacle when Galileo began to formulate laws of motion and gravity mathematically. And Newton's law of gravity, that the force between two objects was proportional to the product of their masses and inversely proportional to the square of the distance between them, was a revolutionary step whose importance we shall discuss in the book. Finally, everyone knows that Einstein invented relativity and the formula $E = mc^2$ (which you do not need to know the meaning of before you read this book). But he also invented general relativity, which is the theory of gravity that physicists and astronomers use now, replacing Newton's. The last half of the book is devoted to studying general relativity and its applications in astronomy and fundamental physics.

This book takes you on a tour of the Universe, but again it helps if you have some idea of where we are going. The Earth is one small planet orbiting a modest star called the Sun. The Sun is one of 10^{11} stars that make up the Milky Way galaxy. This is a spiral galaxy, one of perhaps 10^{11} galaxies that are within the reach of our telescopes. The immensity of it all is astonishing.

With modern telescopes astronomers can see extremely far away. Since they use light, which has been traveling at the speed $c = 3 \times 10^8$ m s^{-1}, this means they are seeing distant objects as they were at an earlier time. Today astronomers can look back in time most of the way to the beginning of time in the Big Bang. The Big Bang led to the Universe that we see, full of galaxies rushing away from one another. But don't worry about this: we will deal with it thoroughly in the text.

Computing

The computer programs and exercises are optional, and everything you need to know about them can be downloaded from the website. You don't need to be a computer programming whiz, but you need to have had some experience with computer languages to understand the programs. However, to run the programs and see the results, play with the different possibilities – anyone can do this.

If you are comfortable with these things, then let's not waste any more time thinking about how to learn about gravity. Let's do it.

Gravity on Earth:
the inescapable force

Gravity is everywhere. No matter where you go, you can't seem to escape it. Pick up a stone and feel its weight. Then carry it inside a building and feel its weight again: there won't be any difference. Take the stone into a car and speed along at 100 miles per hour on a smooth road: again there won't be any noticeable change in the stone's weight. Take the stone into the gondola of a hot-air balloon that is hovering above the Earth. The balloon may be lighter than air, but the stone weighs just as much as before.

This inescapability of gravity makes it different from all other forces of nature. Try taking a portable radio into a metal enclosure, like a car, and see what happens to its ability to pick up radio stations: it gets seriously worse. Radio waves are one aspect of the *electromagnetic force*, which in other guises gives us static electricity and **magnetic fields**. This force does not penetrate everywhere. It can be excluded from regions if we choose the right material for the walls. Not so for gravity. We could build a room with walls as thick as an Egyptian pyramid and made of any exotic material we choose, and yet the Earth's gravity would be right there inside, as strong as ever. *Gravity acts on everything the same way.*

Every body falls *toward* the ground, regardless of its composition. We know of no substance that accelerates *upwards* because of the Earth's gravity. Again this distinguishes gravity from all the other fundamental forces of Nature. **Electric charges** come in two different signs, the "+" and "-" signs on a battery. A negative **electron** attracts a positive **proton** but repels other electrons.

> There is a simple home experiment that will show this. If you have a clothes dryer, find a shirt to which a couple of socks are clinging after they have been dried. Pulling the socks off separates some of the charges of the molecules of the fabric, so that the charges on the sock will attract their opposites on the shirt if they are held near enough. But the socks have the same charge and repel each other when brought together.

The existence of *two* signs of electric charge is responsible for the shape of our everyday world. For example, the balance between attraction and repulsion among the different charges that make up, say, a piece of wood gives it rigidity: try to stretch it and the electrons resist being pulled away from the protons; try to compress it and the electrons resist being squashed up against other electrons. Gravity allows no such fine balances, and we shall see that this means that bodies in which gravity plays a dominant role cannot be rigid. Instead of achieving equilibrium, they have a strong tendency to collapse, sometimes even to **black holes**.

These two facts about gravity, that it is ever-present and always attractive, might make it easy to take it for granted. It seems to be just part of the background, a constant and rather boring feature of our world. But nothing could be further from the truth. Precisely because it penetrates everywhere and cannot be cancelled out, it

In this chapter: the simplest observations about gravity – it is universal and attractive, and it affects all bodies in the same way – have the deepest consequences. Galileo, the first modern physicist, founded the equivalence principle on them; this will guide us throughout the book, including to black holes. Galileo also introduced the principle of relativity, used later by Einstein. We begin here our use of computer programs for solving the equations for moving bodies.

▷ Remember, terms in **boldface** are in the glossary.

▷ The picture underlying the text on this page is of the famous bell tower at Pisa, where Galileo is said to have demonstrated the key to understanding gravity, that all bodies fall at the same rate. We will discuss this below. Photo by the author.

is the engine of the Universe. All the unexpected and exciting discoveries of modern astronomy – **quasars**, **pulsars**, **neutron stars**, black holes – owe their existence to gravity. It binds together the gases of a **star**, the stars of a **galaxy**, and even galaxies into **galaxy clusters**. It has governed the formation of stars and it regulates the way stars create **chemical elements** of which we are made. On a grand scale, it controls the **expansion of the Universe**. Nearer to home, it holds planets in orbit about the Sun and satellites about the Earth.

The study of gravity, therefore, is in a very real sense the study of practically everything from the surface of the Earth out to the edge of the Universe. But it is even more: it is the study of our own history and evolution right back to the **Big Bang**. Because gravity is everywhere, our study of gravity in this book will take us everywhere, as far away in distance and as far back in time as we have scientific evidence to guide us.

Galileo: the beginnings of the science of gravity

We will begin our study of gravity with our feet firmly on the ground, by meeting a man who might fairly be called the founder of modern science: Galileo Galilei (1564–1642).

In Galileo's time there was a strong interest in the trajectories of cannonballs. It was, after all, a matter of life and death: an army that could judge how far gravity would allow a cannonball to fly would be better equipped to win a battle over a less well-informed enemy. Galileo's studies of the trajectory problem went far beyond those of any previous investigator. He made observations in the field and then performed careful experiments in the laboratory. These experiments are a model of care and attention to detail. He found out two things that startled many people in his day and that remain cornerstones of the science of gravity.

> First, Galileo found that the rate at which a body falls does not depend upon its weight. Second, he measured the rate at which bodies fall and found that their acceleration is constant, independent of time.

After Galileo, gravity suddenly wasn't boring any more. Let's look at these two discoveries to find out why.

The story goes that Galileo took two iron balls, one much heavier than the other, to the top of the bell tower of Pisa and dropped them simultaneously. Most people of the day (and even many people today!) would probably have expected the heavier ball to have fallen much faster than the lighter one, but no: both balls reached the ground together.

The equality of the two balls' rates of fall went against the intuition and much of the common experience of the day. Doesn't a brick fall faster than a feather? Galileo pointed out that air resistance can't be neglected in the fall of a feather, and that to discover the properties of gravity alone we must experiment with dense bodies like stones or cannonballs, where the effects of air resistance are small. For such objects we find that speed is independent of weight.

But surely, one might object, we have to do much more work to lift a heavy stone than a light one, so doesn't this mean that a heavy stone "wants" to fall more than a light one and will do so faster, given the chance? No, said Galileo: weight has nothing to do with the speed of fall. We can prove that by measuring it. We have to accept the world the way we find it. This was the first step towards what we now call the *principle of equivalence*, which essentially asserts that gravity is indistinguishable from uniform accleration. We shall see that this principle has a remarkable number of consequences, from the weightlessness of astronauts to the possibility of black holes.

In this section: Galileo laid the foundations for the scientific study of gravity. His demonstration that the speed of fall is independent of the weight of an object was the first statement of the principle of equivalence, which will lead us later to the idea of black holes.

Figure 1.1. Galileo Galilei moved science away from speculation and philosophy and toward its modern form, insisting on the pre-eminence of careful experiment and observation. He also introduced the idea of describing the laws of nature mathematically. Meeting strong religious opposition in his native Italy, his ideas stimulated the growth of science in northern Europe in the decades after his death. Image reproduced courtesy of Mary Evans Picture Library.

Investigation 1.1. Faster and faster: the meaning of uniform acceleration

In this investigation, we work out what Galileo's law of constant acceleration means for the speed of a falling body. The calculation is short, and it introduces us to the way we will use some mathematical symbols through the rest of the book.

We shall denote time by the letter t and the speed of the falling body by v (for *velocity*). The speed at time t will be written $v(t)$. The acceleration of the body is g, and it is constant in time.

Suppose the body is dropped from rest at time $t = 0$. Then its initial speed is $v = 0$ at time $t = 0$, in other words $v(0) = 0$. What will be its speed a short time later?

Let us call this later time Δt. Here we meet an important new notation: the symbol Δ will always mean "a change in" whatever symbol follows it. Thus, a change in time is Δt. Similarly, we shall call the change in speed produced by gravity Δv. Normally we shall use this notation to denote small changes; here, for example, I have defined Δt to be "a short time later". We shall ask below how small Δt has to be in order to be "short".

The acceleration g is the change in the speed per unit time. This definition can be written algebraically as

$$g = \frac{\Delta v}{\Delta t}. \qquad (1.1)$$

By multiplying through by the denominator of the fraction, we can solve for the change in speed:

$$\Delta v = g\Delta t. \qquad (1.2)$$

Equation 1.1 basically defines g to be the *average* acceleration during the time Δt. If we take Δt to be very small, then this gives what we generally call the *instantaneous* acceleration. In this sense, "small" effectively means "as small as we can measure". If I have a clock which can reliably measure time accurate to a millisecond, then I would take Δt to be 1 ms if I wanted the instantaneous acceleration.

Now, Galileo tells us that the acceleration of a falling body does not in fact change with time. That means that the average acceleration during *any* period of time is the same as the instantaneous acceleration g. So in this particular case, it does not actually matter if Δt is small or not: Equation 1.1 is exactly true for any size of Δt. If we let t be any time, then we can rewrite Equation 1.2 as

$$\Delta v = gt.$$

We assumed above that the body was dropped from rest at time $t = 0$. This means that the initial speed is zero, and so the speed at a later time is just equal to Δv as given above. But if the body has an initial downward speed $v(0) = v_0$, then its subsequent acceleration only adds to the speed. This means that

$$v(t) = v(0) + \Delta v,$$

or

$$v(t) = gt + v_0. \qquad (1.3)$$

Exercise 1.1.1: *Speed of a falling body*

Using the fact that the acceleration of gravity on Earth is $g = 9.8 \text{ m s}^{-2}$, calculate the speed a ball would have after falling for two seconds, if dropped from rest. Calculate its speed if it were thrown downwards with an initial speed of 10 m s^{-1}. Calculate its speed if it were initially thrown *upwards* with a speed of 10 m s^{-1}. Is it falling or still rising after 2 s?

The acceleration of gravity is uniform

Galileo performed a number of ingenious experiments with the rather crude clocks available in his day to demonstrate that the acceleration of falling objects is constant. Now, the acceleration of an object is the rate of change of its speed, so if the acceleration is constant then the speed changes at a constant rate; during any given single second of time, the speed increases by a fixed amount. We call this constant the **acceleration of gravity**, and denote it by g (for gravity). Its value is roughly 9.8 meters per second per second. The units, meters per second per second, should be understood as "(meters per second) per second", giving the amount of speed (meters per second) picked up per second. These units may be abbreviated as m/s/s, but it is more conventional (and avoids the ambiguous[†] ordering of division signs) to write them as m s^{-2}.

As with any physical law, there is no reason "why" the world had to be this way: the experiment might have shown that the speed increased uniformly with the distance fallen. But that is not how our world is made. What Galileo found was that speed increased uniformly with time of fall.

We can find out what Galileo's law says about the distance fallen by doing our first calculations, Investigation 1.1 and 1.2. These calculations show that uniform acceleration implies that the speed a falling body gains is proportional to time and that the distance it falls increases as the square of the time. The calculation also has another purpose: it introduces the basic ideas and notation that we will use in later investigations to construct computer calculations of more complicated phe-

In this section: near the Earth, bodies accelerate downwards at a uniform rate.

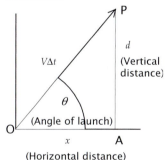

Figure 1.2. For the calculation in Investigation 1.3 on page 5, the vertical and horizontal distances traveled by a cannonball launched at an angle θ are the sides of a right triangle whose hypotenuse is the total distance $V\Delta t$.

[†]Ambiguity: does m/s/s mean (m/s)/s or m/(s/s)? Either would be a valid interpretation of m/s/s, but in the second form the units for seconds cancel, which is not at all what is wanted.

Investigation 1.2. *How the distance fallen grows with time*

Here we shall calculate the distance $d(t)$ through which a falling body moves in the time t. Again we shall do this with simple algebra, but the ideas we use will lay the foundations for the computer programs we will write to solve harder problems later. Accordingly, much of the reasoning used below will be more general than is strictly necessary for the simple problem of a falling body.

We follow similar reasoning to that in Investigation 1.1 on the previous page. We are interested in the distance $d(t)$ fallen by the body by the time t. During the first small interval of time Δt, the body falls a distance Δd. (Our Δ notation again.) The *average* speed in this time is, therefore,

$$v_{avg} = \frac{\Delta d}{\Delta t}.$$

Solving this for Δd gives

$$\Delta d = v_{avg} \Delta t. \qquad (1.4)$$

Now, we saw in Investigation 1.1 that the speed of the body changes during this interval of time. It starts out as zero (in the simplest case we considered) and increases to $g\Delta t$. So it seems to be an obvious guess that the average speed to use in Equation 1.4 is the average of these two numbers:

$$v_{avg} = \tfrac{1}{2}(g\Delta t + 0) = \tfrac{1}{2}g\Delta t.$$

If we put this into Equation 1.4, we find

$$\Delta d = (\tfrac{1}{2}g\Delta t)(\Delta t) = \tfrac{1}{2}g(\Delta t)^2. \qquad (1.5)$$

I have said "an obvious guess" because it might not be right. If the acceleration of the body were a very complicated changing function of time, then its average speed over a time Δt might not be the average of its speeds at the beginning and end of the time-interval. For example, for some kinds of non-uniform acceleration it might happen that the body was at rest at the beginning and end of the interval, but not in between. Then its average speed might be positive, even though our guess would give zero.

Our guess is really only a good approximation in general if we choose the time-interval Δt small enough that the body's acceleration does not change by much during the interval. This gives a new insight into what is meant by a short time-interval: it must be short enough that the body's acceleration does not change by very much.

Of course, in the case of a falling body, the acceleration is constant, so we can expect Equation 1.5 to be *exact* for any time-interval, no matter how long. So if we replace Δt by t and Δd by $d(t)$, we find

$$d(t) = \tfrac{1}{2}gt^2. \qquad (1.6)$$

Now suppose the body initially had a speed v_0. Then the average speed during the time Δt would be $v_0 + g\Delta t/2$, so Equation 1.5 would become

$$\Delta d = (v_0 + \tfrac{1}{2}g\Delta t)\Delta t = v_0\Delta t + \tfrac{1}{2}g(\Delta t)^2.$$

Then, if the body does not start at distance $d = 0$ but rather at distance $d(0) = d_0$, we have that $d(t)$ at a later time is

$$d(t) = \tfrac{1}{2}gt^2 + v_0 t + d_0. \qquad (1.7)$$

This is the full law of distance for a uniformly accelerating body.

The calculation we have just done may seem long-winded, especially to readers who are comfortable with calculus, because the operations I have gone through may seem like a beginner's introduction to calculus. This is not my aim, however. It will become clear in future examples that what we have actually met here is a method of doing calculations by **finite differences**; this method is at the heart of most computer calculations of the predictions of physical laws, and we will see that it will help us to solve much more difficult problems involving the motion of bodies under the influence of gravity. We can use finite differences reliably provided we use intervals of time that are short enough that the acceleration of a body does not change by much during the interval.

Exercise 1.2.1: *Distance fallen by a body*

For the falling ball in Exercise 1.1.1 on the preceding page, calculate the distance the ball falls in each of the cases posed in that exercise.

---

nomena. Anyone who can do algebra can follow these investigations.

Trajectories of cannonballs

In this section: Galileo introduced the idea that the horizontal and vertical motions of a body can be treated separately: the vertical acceleration of gravity does not change the horizontal speed of a body.

We can now take up one of the subjects that contributed to the Renaissance interest in gravity, namely the motion of a cannonball. We have discovered that the vertical motion of the ball is governed by the law of constant acceleration. What about its horizontal motion? Here, too, Galileo had a fundamental insight. He argued that the two motions are *independent*.

Consider dropping a rubber ball in an airplane moving with a large horizontal speed. The rate at which the ball falls does not depend on how fast the plane is moving. Moreover, imagine an observer on the ground capable of watching the ball: it keeps moving horizontally at the same speed as the plane even though it is free of any horizontal forces. That is, while it falls "straight down" relative to the passengers in the plane, it falls in an arc relative to the observer on the ground.

Let us transfer this reasoning to the example of a cannonball launched at an angle to the vertical so that its vertical speed is v_0 and its horizontal speed is u_0. Since there are no horizontal forces acting on the ball if we neglect air resistance, it will keep its horizontal speed as it climbs and falls, and the time it spends in the air will be the same as that of a ball launched vertically with the same speed v_0. Galileo showed that the trajectory that results from this is a parabola.

This would be easy for us to show, as well, by doing a little algebra.

Investigation 1.3. The flight of the cannonball

Here we show how the finite-differences reasoning of the two previous investigations allows us to construct a computer program to calculate the flight of a cannonball, at least within the approximation that the ball is not affected by air resistance.

From this book's website you can download listing of the Java program CannonTrajectory. If you download the Triana software as well you can run the program and compute the trajectory of a cannonball fired at any given initial speed and at any angle. Figure 1.3 on the next page displays the result of the computer calculation for three trajectories, all launched with the same speed at three different angles. (The Triana software will produce plots of these trajectories. The figures produced for this book have, however, been produced by more sophisticated scientific graphics software.)

Here is how the program is designed. The idea is to calculate the body's horizontal and vertical position and speed at successive times spaced Δt apart. Let d be the vertical position and x the horizontal one, both zero to start. If the ball is launched with speed V at an angle θ with the horizontal (as in Figure 1.2 on page 3), then our first job is to deduce the vertical and horizontal speeds, which Galileo showed behaved independently of one another after launch.

Suppose that we turn off gravity for a moment and just watch a cannonball launched with speed V at an angle θ to the ground. Then after a small time Δt, it has moved a distance $V\Delta t$ in its launch direction. This is the distance OP in Figure 1.2. Simple trigonometry tells us that this distance is the **hypotenuse** of a right triangle whose other sides are the lines PA (the vertical distance d it has traveled) and AO (the horizontal distance x it has traveled). Then by definition we have

$$\sin \theta = \frac{d}{V\Delta t} \quad \Rightarrow \quad d = V\Delta t \sin \theta,$$
$$\cos \theta = \frac{x}{V\Delta t} \quad \Rightarrow \quad x = V\Delta t \cos \theta.$$

The vertical speed is the vertical distance d divided by the time Δt, and similarly for the horizontal speed. We therefore find that the initial vertical speed is

$$v_0 = V \sin \theta$$

and the initial horizontal speed is

$$u_0 = V \cos \theta.$$

The horizontal speed remains fixed, so the horizontal distance increases by $u_0 \times \Delta t$ each time step. The vertical speed decreases by $g\Delta t$ each time step, and we calculate the vertical distance using the *average* vertical speed in each time step. (In vertical motion, upward speeds are positive and downward ones negative.) The program sets up a **loop** to calculate the variables at successive time-steps separated by a small amount of time.

Normally one would expect a program like this to become more accurate for smaller time-steps, because of the remark we made in Investigation 1.2: our method of taking finite steps in time is better if the acceleration is nearly constant over a time step. In this case the relevant time step is Δt. By making Δt sufficiently small, one can always insure that the acceleration changes by very little during that time, and therefore that the accuracy of the program will increase. But in the present case that does not happen because our method of using the average speed over the time-step gives the *exact* result for uniform acceleration.

Let us look at the results of the three calculations in Figure 1.3 on the following page. Of these, the trajectory with the largest range for a given initial speed is the one that leaves the ground at a 45° angle. In fact it is not hard to show that this trajectory has the largest range of all possible ones. What is this range? We could calculate it from the results of Investigation 1.2, but in the spirit of our approach we shall try to guess it from the numerical calculation.

Given that the initial angle will be 45°, the range can only depend on the initial speed V and the acceleration g. The range is measured in meters, and the only combination of V and g that has the units of length is $V^2/g : (m\,s^{-1})^2/(m\,s^{-2}) = m$. We therefore can conclude that there is some constant number b for which *range* $= bV^2/g$. (This reasoning is an example of a powerful technique called **dimensional analysis**, because one is trying to learn as much as possible from the units, or **dimensions**, of the quantities involved in the problem.)

The numerical results let us determine b. Since the calculation used $V = 100\,m\,s^{-1}$, it follows that V^2/g is 1020 m. From the graph the range looks like 1020 m as well, as nearly as I can estimate it. Since the value of b is likely to be simple, it almost surely equals 1. An algebraic calculation shows this to be correct:

$$\text{maximum range} = V^2/g.$$

The reader is encouraged to re-run the program with various initial values of V to check this result.

Exercise 1.3.1: *Small steps in speed and distance*

Suppose that at the nth time-step t_n, the vertical speed is v_n and the vertical distance above the ground is h_n. Show that at the next time-step $t_{n+1} = t_n + \Delta t$, the vertical speed is $v_{n+1} = v_n - g\Delta t$. Using our method of approximating the distance traveled by using the average speed over the interval, show that at the next time-step the height will be

$$h_{n+1} = h_n + \tfrac{1}{2}(v_n + v_{n+1})\Delta t = h_n + v_n\Delta t - \tfrac{1}{2}g(\Delta t)^2.$$

Exercise 1.3.2: *Suicide shot*

What is the *minimum* range of a cannonball fired with a given speed V, and at what angle should it be aimed in order to achieve this minimum?

Exercise 1.3.3: *Maximum range by algebra*

For readers interested in verifying the guess we made above from the numerical data, here is how to calculate the range at 45° algebraically. The range is limited by the amount of time the cannonball stays in the air. Fired at 45° with speed V, how long does it take to reach its maximum height, which is where its vertical speed goes to zero? Then how long does it take to return to the ground? What is the total time in the air? How far does it go horizontally during this time? This is the maximum range.

Exercise 1.3.4: *Best angle of fire*

Prove that 45° is the firing angle that gives the longest range by calculating the range for any angle and then finding what angle makes it a maximum. Use the same method as in Exercise 1.3.3.

But instead we show in Investigation 1.3 on the previous page how to use a personal computer to calculate the actual trajectory of a cannonball. These computer techniques will form the foundation of computer programs later in this book that will calculate other trajectories, such as planets around the Sun, stars in collision with one another, and particles falling into black holes.

Galileo: the first relativist

In this section: Galileo introduced what we now call the principle of relativity, which Einstein used as a cornerstone of his own revolutionary theories of motion and gravity almost 300 years later.

It would be hard to overstate Galileo's influence on science and therefore on the development of human society in general. He founded the science of **mechanics**; his experiments led the English scientist Isaac Newton (1642–1726) to discover his famous laws of motion, which provided the foundations for almost all of physics for 200 years. And almost 300 years after his death his influence was just as strong on Albert Einstein. The German–Swiss physicist Einstein (1879–1955) replaced Newton's laws of motion and of gravity with new ones, based on his theory of relativity. Einstein's revolutionary theories led to black holes, the Big Bang, and many other profound predictions that we will study in the course of this book. Yet Einstein, too, kept remarkably close to Galileo's vision.

The main reason for Galileo's influence on Einstein is that he gave us the first version of what we now call the **principle of relativity**. We have already encountered Galileo's version: the vertical motion of a ball does not depend on its horizontal speed, and its horizontal speed will not change unless a horizontal force is applied.

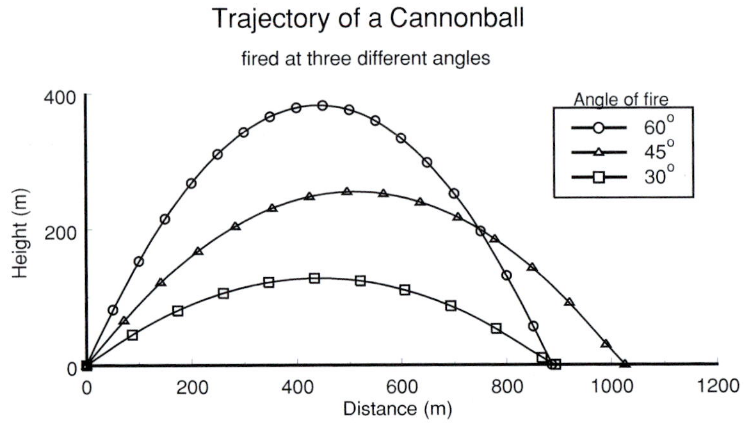

Where we used a fast-flying airplane to justify this, Galileo imagined a sailing ship on a smooth sea, but the conclusion was the same: an experimenter moving horizontally will measure the same acceleration of gravity in the vertical direction as he would if he were at rest.

Galileo took this idea and drew a much more profound conclusion from it. The radical proposal made half a century earlier by the Polish priest and astronomer Nicolas Copernicus (1473–1543), that the Earth and other planets actually moved around the Sun (see Figure 1.4), was still far from being accepted by most intellectuals in Galileo's time. Although the proposal explained the apparent motions of the planets in a simple way, it was open to an important objection: if the Earth is moving at such a rapid rate, why don't we *feel* it? Why isn't the air left behind, why doesn't a ball thrown vertically fall behind the moving Earth?

Figure 1.3. *Trajectories computed by the program developed in Investigation 1.3 on the previous page, for three angles of firing, each at the same initial speed of 100 m s⁻¹. The trajectory at 45° goes furthest.*

Galileo used the independence of different motions to dispose of this objection. Galileo's answer is that a traveler in the cabin of a ship on a smooth sea also does not feel his ship's motion: all the objects in the cabin move along with it at constant speed, even if they are just resting on a table and not tied down. Anything that falls will fall vertically in the cabin, giving no hint of the ship's speed. So it is on the Earth, according to Galileo: the air, clouds, birds, trees, and all other objects all have the same speed, and this motion continues until something interferes with it. There is, in other words, no way to tell that the Earth is moving through space except to look at things far away, like the stars, and see that it is.

Figure 1.4. The Copernican view of the planets known in Galileo's day as they orbit the Sun.

Today we re-phrase and enlarge this idea to say that *all* the laws of physics are just the same to an experimenter who moves with a uniform motion in a straight line as they are to one who remains at rest, and we call this the *principle of relativity*. We shall encounter many of its consequences as we explore more of the faces of gravity.

Unfortunately for Galileo, his clear reasoning and his observations with one of the first telescopes made him so dangerous to the established view of the Roman Catholic Church that in his old age he was punished for his views, and forced to deny them publicly. Privately he continued to believe that the planets went around the Sun, because he had discovered with his telescope that the moons of Jupiter orbit Jupiter in the same way that the planets orbit the Sun.

> Today we recognize Galileo as the person who, more than anyone else, established the Copernican picture of the Solar System.

Figure 1.5. Part of a sketch by Galileo of the positions of Jupiter (open circles) and its moons (stars) on a sequence of nights (dates given by the numbers). The big changes from night to night puzzled Galileo. At first he believed that Jupiter itself was moving erratically, but after a few observations he realized that the "stars" were moons orbiting Jupiter in the same way that the planets orbit the Sun.

2

And then came Newton:
gravity takes center stage

B orn in the same year, 1642, as Galileo died, Isaac Newton revolutionized the study of what we now call physics. Part of his importance comes from the wide range of subjects in which he made fundamental advances – mechanics (the study of motion), optics, astronomy, mathematics (he invented calculus), ... – and part from his ability to put physical laws into mathematical form and, if necessary, to invent the mathematics he required. Although other brilliant thinkers made key contributions in his day – most notably the German scientist Gottfried Leibniz (1646–1716), who independently invented calculus – no physicist living between Galileo and Einstein rivals Newton's impact on the study of the natural world.

Nevertheless, it is hard to imagine that Newton could have made such progress in the study of motion and gravity if he had not had Galileo before him. Newton proposed three fundamental **laws of motion**. The first two are developed from ideas of Galileo that we have already looked at:

> The *first law* is that, once a body is set in motion, it will remain moving at constant speed in a straight line unless a force acts on it. This is just like the rubber ball dropped inside the airplane of Chapter 1.

This is basically Galileo's idea, which led him to his principle of relativity, that motions in different directions could be treated independently. Notice that, since particles travel in *straight* lines unless disturbed, the directions along which motion is independent must also be along straight lines.

> Newton's *second law* is that, when a force is applied to a body, the resulting *acceleration* depends only on the force and on the mass of the body: the larger the force, the larger the acceleration; and the larger the mass of a body, the smaller its acceleration.

This dependence of the acceleration a on the force F and the mass m can be written as an equation:

$$a = F/m.$$

It is more conventional to write it in the equivalent form

$$F = ma. \tag{2.1}$$

The second law: weight and mass

The second law fits our everyday experience of what happens when we push something. If we have a heavy object on wheels (to allow us to ignore friction for a moment) and we give it a push, its speed increases (it accelerates) as long as we continue to push it. Then it moves along at a constant speed after we stop pushing. (Friction eventually slows it down, but that is just another force exerted by the surface it is moving across.) If we push it harder, it accelerates faster, so it is not

In this chapter: we learn about Newton's postulate, that a single law of gravity, in which all bodies attract all others, could explain all the planetary motions known in Newton's day. We also learn about Newton's systematic explanation of the relationship between force and motion. When we couple this with Galileo's equivalence principle, we learn how gravity makes time slow down.

Figure 2.1. Brilliant and demanding, Isaac Newton created theoretical physics. Besides devising the laws of gravity and mechanics, he invented calculus, still the central mathematical tool of physicists today. (Original engraving by unknown artist, courtesy AIP Emilio Segrè Visual Archives, Physics Today Collection.)

In this section: how force, mass, and acceleration are related to one another, and the difference between weight and mass.

unreasonable to guess that the acceleration might be proportional to the force we exert on it. Moreover, if we load more things on top of the object we are pushing, then to get the same acceleration, we need to push harder, so again we might guess that the force required would be proportional to the mass. Newton not only made these guesses, but he assumed (or hoped!) that the force would depend on nothing else besides the mass of the body and the acceleration produced.

Newton made an important distinction between two concepts that are often used interchangeably in everyday language: **mass** and **weight**. The mass of an object is, as we have just seen, the way it "resists" being accelerated. (Physicists sometimes call this its **inertia**.) The weight of an object is the force of gravity on it. When we step on our bathroom scales, we measure our weight; if we were to put the scales on the Moon, where gravity is weaker, we would get a lower reading. Our mass would not have changed, however. It would take the same force to accelerate us on a smooth horizontal track on the Moon as it would on the Earth.

▷A good illustration of how everyday language uses such terms differently from the way we use them in physics is provided by dieting. About twice a year, when I go on a diet, I tell my friends that I am trying to lose weight. Mercifully, none of them has yet pointed out that the surest way to lose weight is to go to the Moon! That would not really help, of course, since what I am really trying to do, for the sake of my health, is to lose *mass*. If I stay on the Earth, then of course losing weight implies losing mass.

The second law leads to a remarkable insight when combined with Galileo's discovery that bodies of different masses accelerate under gravity at the same rate: it tells us that a body's *weight* must be proportional to its *mass*. Here is the reasoning.

Suppose we lift a heavy body off the floor and hold it. What we *feel* as its weight is really the sensation of exerting an upwards force upon it to hold it against the force of gravity. When we exert this force on a body, it remains at rest in our hands. By the first law, we conclude that the total force on it is zero: our upwards force just cancels the downwards force of gravity on the object. Therefore the weight of the body *is* the force of gravity on it. If we now release the body, the force of gravity on it is the same but is no longer balanced by our hands' force. The body accelerates downwards: it falls.

But what *is* its acceleration in response to this force of gravity? Galileo observed that the acceleration of the body does not depend on its weight: it is the *same* for everything. Now, in Equation 2.1 on the previous page, the only way that we can change the force F (the weight) without changing the acceleration a is if we change the mass m in proportion to F: the force of gravity on a body is proportional to its mass.

Newton's reasoning here leads to an experimentally verifiable conclusion. Both mass and weight could be measured independently, the weight using scales and the mass by measuring the acceleration of the body in response to a given *horizontal* force and then using Equation 2.1 on the preceding page to infer its mass. If we divide the weight by the mass, we should get the acceleration of gravity. Put mathematically, this says that the force of gravity F_{grav} on any object equals its mass m times the acceleration of gravity g:

$$F_{grav} = mg. \tag{2.2}$$

In honor of Newton, scientists have agreed to measure force in units called *newtons*: one newton (N) of force equals 1 kg times an acceleration of $1\,\mathrm{m\,s^{-2}}$. The weight of a body in newtons is then just its mass in kilograms times the acceleration of gravity, $9.8\,\mathrm{m\,s^{-2}}$.

As we have noted, Equation 2.2 is experimentally verifiable: if it holds for any body, then this experiment serves as a test of the second law itself.

Once this law of motion was checked experimentally, Newton's argument led to a reformulation of Galileo's *principle of equivalence*: the mass of a body (ratio of force to acceleration) is proportional to its weight. From this statement and Equation 2.1 on the preceding page,

Galileo's original observation that all bodies fall with the same acceleration follows.

This was the way two centuries of physicists thought of the principle of equivalence. It was strikingly confirmed by the Hungarian physicist Baron Roland von Eötvös (1848–1919) in 1889 and in 1908. In one of the most accurate physics experiments of his time, von Eötvös showed that many different materials fell with the same acceleration, to within a few parts in a billion!

We now know that this form of the equivalence principle applies not just to ordinary bodies, but also to bodies with the strongest of gravitational fields in general relativity, even to black holes. However, Einstein's general theory of relativity did change Newtonian mechanics in some respects, so the way that modern physicists think about the principle of equivalence is also rather different from the Newtonian form. We will have to wait a few pages before we take a look at the modern reformulation, because we have not by any means finished with Newton's work yet. He made two further landmark contributions to our subject: his third law of motion and his law of gravitation.

The third law, and its loophole

Newton added another law, not explicit in Galileo's work, but which he needed to make the whole science of mechanics self-consistent.

> Newton's *third law* states that if I exert a force on an object, then it exerts a force back on me that is exactly equal in magnitude and opposite in direction to the one I have applied. This law is often paraphrased as "action equals reaction".

This law often strikes newcomers to the subject as contradictory: if there are two equal and opposite forces, don't they cancel? If the object I push on moves and I don't, doesn't that mean that I pushed harder on it than it pushed on me? These difficulties are always the result of mis-applying Newton's second law. Only the forces acting *on* an object contribute to its acceleration. The equal and opposite forces in the third law act on different objects, and so there is no way that they can cancel each other.

To see our way past such doubts with a concrete example, consider again the feeling of a weight in the hand. Suppose I hold an apple. I have to exert a force on the apple, equal to its weight, to keep it in one place. This force is exerted through my hand, but the hand doesn't stay where it is all by itself: it is kept there by the force exerted on it by my arm. (The tired feeling I eventually get in the muscles of my arm leaves me no room to doubt this!) Since the hand isn't moving but the arm is exerting a force on it, there must be a balancing force on it as well, and this can only come from the apple. So as I exert a force on the apple, it exerts a force back on me. How much of a force? Newton argued that this "reaction" by the apple must be equal to the force I exert on it.

There are several ways of seeing that this is reasonable. Suppose, for instance, that the force exerted by the apple on my hand was only half of its weight. What makes the hand special, that it gets back only half the force it gives out? Why wouldn't it be the other way around, that the apple should receive from my hand only half the force it exerts on my hand? This lack of symmetry, where the hand gets only half the force back that it exerts, while the apple gets twice its force back, makes the "half-reaction" law illogical.

> The third law has an important practical consequence: it is responsible for almost all propulsion. We walk by pushing backwards with our feet

on the ground; the ground then pushes forwards on us, and that is why we move. Similarly, a rocket pushes hot gas out of its nozzles in the backwards direction; the gas exerts its reaction to push the rocket forwards. Jet planes, swimming fish, and flying birds all use the third law to get around. What about sailing ships and downhill skiers?

Logical as the third law is, there is a loophole in it which we will find important when we discuss general relativity. The argument we have just gone through applies strictly only to forces exerted by bodies in constant contact; that is, by bodies that are touching and not moving relative to each other. But some bodies can exert forces without being in contact.

Electric charges, for example, exert forces over considerable distances, and these forces weaken as the distance increases. Now, the laws governing electric and magnetic forces (which together are called **electromagnetism**) tell us that these forces can only be transmitted at the speed of light, no faster. For example, if two electric charges move relative to one another, then the forces they exert on one another have to change as the distance between them changes. These changes travel through space at the speed of light. It thus takes some time for the force to be transmitted over a distance, so there will be a delay between the time when charge "A" exerts its force on charge "B" and the time when it receives the reaction force from "B". The electric charges will have changed their relative distances in this time, and then the two forces will not be equal.

In fact, this imbalance of forces leads to what is called **radiation reaction** on moving, accelerating charges: an accelerating charge experiences a force that depends on the rate of change of its acceleration. This force usually opposes its motion and acts as a kind of friction. When we send out radio waves from a radio transmitter, we have to drive the electrons in the transmitter into exactly the right sort of motions to produce the desired radio waves. The power required to transmit from such antennas goes primarily into overcoming the radiation reaction force on the charges. Newton knew nothing about this and did not allow for it in his formulation of the third law. We shall return to the subject of radiation reaction in more detail in Chapter 22 when we discuss gravitational radiation.

> We would therefore be safer paraphrasing the third law as: "action equals reaction for bodies in constant contact".

Preview: Newton's gravity

In this section: with a leap of imagination that even Galileo had not attempted, Newton showed that the same gravity that made apples fall to Earth makes the Moon stay near the Earth and the Earth near the Sun. A simple mathematical formula was consistent with all known data for planetary orbits.

Now we come to another of Newton's jewels, his law for the gravitational attraction between the Sun and its planets. We will look much more closely at its consequences when we study planetary motion in Chapter 4, but we describe it briefly here, not only because it is closely linked to his laws of motion and the equivalence principle, but also because in the law of gravity we see Newton's imagination at its boldest.

Believing that the laws of motion ought to apply everywhere, not just on the surface of the Earth, Newton realized that the orbital motion of the Moon about the Earth and of the planets about the Sun implied that the heavenly bodies were under the influence of *forces*, forces which moreover had to be exerted over considerable distances, since telescopes didn't reveal any horse carts pushing the planets around. (Greek and Roman mythology didn't stand up to the test of observation!)

The Earth already exerted the force of gravity on objects (falling apples, etc.) that were not in direct contact with it, so could the Earth's gravity extend very much farther away? Could gravity be the force responsible for keeping the Moon in its orbit?

To answer "yes" to this question, something we take so much for granted today, was in Newton's day a brilliant and even courageous leap. Newton realized that he would be ignored and even ridiculed unless he could extend the law of gravity to the heavens in a simple and convincing way. His courage was rewarded: in Galileo's time the German astronomer Johannes Kepler (1571–1630) had shown from painstaking calculations that the planets followed elliptical orbits with the Sun always at one focus of the ellipse, and Newton was able to find a force law that predicted exactly that.

The law of gravity *was* simple and, in the end, utterly convincing: the gravitational force between two bodies is proportional to the product of their masses and inversely proportional to the square of the distance between them.

Figure 2.2. Johannes Kepler was the foremost mathematical astronomer of his time. Working in Prague, his detailed studies of the motions of the planets made it possible for him to show that orbits were ellipses long before Newton explained this. The calculation was immense: more than a thousand sheets of arithmetic for his calculation of the orbit of Mars survive. Kepler was also an accomplished mathematician, proving the close packing theorem for spheres and explaining why logarithms worked. Image by permission of AIP Emilio Segrè Visual Archives.

This can be expressed as an equation. If the two bodies have masses M_1 and M_2, respectively, and are separated by a distance r, then the force of gravity exerted by each on the other is

$$F_{\text{grav}} = \frac{GM_1 M_2}{r^2},$$
(2.3)

where G is a constant of proportionality, called *Newton's gravitational constant*. Its value, as measured by the best modern experiments, is 6.6720×10^{-11} m^3 s^{-2} kg^{-1}. The dependence of the force on the masses M_1 and M_2 of the bodies follows directly from our discussion of the equivalence principle: the force of gravity on a body is its weight, and this must be proportional to its mass. By the third law, the other body will experience the same force, so the force must be proportional to its mass, too. The only new element in this law is the way it depends on distance r. We call it an **inverse-square** law.

How Newton decided that the dependence on r should be as $1/r^2$ is a story that shows that Newton had more than just courage: he had the capacity for immense hard work and persistence. For in order to prove to himself that the inverse-square-law force gave elliptical orbits, he had to invent the calculus. And even after proving the law to himself, he still refrained from publishing it for many years until he could iron out a particularly difficult detail that he felt might otherwise have proved his undoing. We will return to this difficulty when we examine the orbits predicted by this force, in Chapter 4. Newton's delay in publishing his work also led to bitter disputes with some contemporaries, especially with Robert Hooke (1635–1703), who claimed to have invented the inverse-square law himself.

Action at a distance

Newton's gravitational force between heavenly bodies was an *instantaneous* force: no matter how far apart two bodies were, the force between them would respond instantly to any change in their separation. Newton did not try to invent any mechanical way of describing the force, say by hypothetical particles traveling between one body and another – such intermediaries would travel at a finite speed, not instantly.

The way Newton's gravitational force behaved was called **action at a distance**. It was one of the most unpalatable aspects of his theory to many of his contemporaries, and Newton was forced to defend it vigorously. One can see why he needed it, for otherwise he could not have argued that his third law of motion (action-reaction) would apply in the heavens: the argument we gave above – that forces that are transmitted with a delay don't necessarily have equality of action and reaction – would have been devastating. In order to preserve his third law, and (very importantly)

In this section: Newton's law was audacious: gravity could act across distances apparently without anything in between. Einstein removed that and returned to a model for gravity that more resembles other influences: gravitational effects move through space at the speed of light. But Newton's law was the right one for his time, a single simple assumption that could explain a wealth of data.

in the absence of any evidence to the contrary, Newton held firm to instantaneous forces.

Einstein's general relativity replaced Newton's law of gravity with a more complicated theory that does have a finite speed of transmission of gravitational influences (the speed of light), so modern gravity does not involve action at a distance. But it would be hard to find a physicist today who would argue that Newton was "wrong" to take the position he did. Newton's sense of how to make progress with gravitational theory was unerring: by keeping it simple he provided physicists with a law which satisfactorily explained Solar System motions for two hundred years. And when some tiny discrepancies with his theory were finally observed (in details of the orbit of Mercury), the instantaneous aspect of his theory was not to blame. We will see in Chapter 21 that these discrepancies were finally explained by the curvature of space in general relativity.

Newton's theory of gravity did not immediately affect the study of gravity on the Earth, but in common with his three laws of mechanics it shows how he relied on a sense of simplicity to formulate his physical laws. The world is a complicated place, but within its complexity Newton tried to formulate his laws as simply as experience would allow. The third law, relying on symmetry between bodies, is an example: there was no experimental hint of the exceptional case we pointed out above (bodies in relative motion), so he did not allow for it.

Newton's law of gravity was likewise much simpler than the data he was trying to explain. It had a previously unknown constant of proportionality in it (G), and a perhaps surprising exponent (*inverse-square* law), but with just those two numbers he could explain the huge number of detailed observations of the Moon and the five known planets. Moreover, by uniting the theories of planetary motion and of terrestrial gravity, we can justly give him credit for having devised the first **unified field theory**. The theme of simplicity in the face of complex phenomena, especially of simplification by unification, has ever since been a dominant one in physicists' attempts to explain the world. We shall see that it applies particularly to Einstein's relativity, despite its exaggerated reputation for difficulty.

The new equivalence principle

In this section: Galileo's equivalence principle is reformulated into its modern version. An experimenter who falls freely in a gravitational field measures no gravity at all nearby. Weightlessness of astronauts is the classic example.

As I mentioned above, the modern view of the equivalence principle is somewhat different from both Galileo's and Newton's. This is because experiments have taught us that light has some special properties that make it hard to fit into Newtonian gravity. Newton couldn't have known this, for it was not possible to study light accurately with the technology of his day. But by the late nineteenth century, experiments began to force physicists to realize that light was somehow special. Today we know that it is fundamental to Einstein's **special relativity** theory that light has a *fixed* speed, called c, about 3×10^8 m s^{-1}. (I shall deliberately leave this statement a little vague here, and explain what a "fixed speed" means in Chapter 15.) We have also learned that light has *zero* inertial mass.

It is clear that the motion of light will not be described by the simple law $F = ma$, and since our formulation of the equivalence principle involves this equation, we might feel that light could violate the principle. However, since even the meaning of acceleration is unclear in the case of light, it is more sensible to reformulate the equivalence principle without mentioning acceleration or inertial mass. We will then see in the next section that, once we have done this, we can actually make striking predictions about the effect of gravity on light, without ever referring to $F = ma$.

One thing Galileo might have observed, had he been inclined to think this way about the problem, was that if he himself had fallen off the Leaning Tower of Pisa

---

Investigation 2.1. The effect of motion on light: the Doppler effect

One of the most important effects in the study of light, or indeed of any wave, is the Doppler effect. For sound waves, the Doppler effect causes the whistle of an approaching train to sound at a higher pitch than if the train were at rest. And as the train moves away, the pitch falls to a lower note. Analogous things happen to light.

In outer space, far from any gravitational fields, light travels in a straight line at the constant speed c and with unchanging frequency (color). If two experimenters moving relative to one another look at the same beam of light, they will generally see it to have different colors: this is called the Doppler shift of light. In particular, if one experimenter measures the light to have frequency f_0, and if for simplicity the second experimenter is moving in the same direction as the light is going, only with a speed v that is small compared to c, then the second experimenter will measure the frequency to be

$$f_1 = f_0(1 - v/c). \qquad (2.4)$$

This is smaller than f_0, and it means that, for example, blue light will be shifted toward the red end of the spectrum, and other colors will be shifted in the same sense. This is therefore called a *redshift* of light. In Figure 2.3 there is a visual derivation of the Doppler effect, leading to Equation 2.4.

If, on the other hand, the second experimenter were moving directly towards the source of the light, then we could use Equation 2.4 with v replaced by $-v$. This would increase the frequency of the light and result in a *blueshift*.

The important lesson to understand is that the frequency of a light beam is not an intrinsic property of the beam; rather it depends also on the motion of the experimenter who measures it. When we discuss the redshift of light produced by a gravitational field, we shall have to be careful to define who the experimenter is who measures the redshift, since other experimenters could see the same light with a blueshift.

---

along with the two heavy balls, then on the way down the balls would simply have stayed beside him: they would have behaved relative to him as if they had had no forces on them at all. (Again we are neglecting air resistance here, for clarity.)

Precisely, this means that if he had given one of the balls a push in any direction, even downwards, its subsequent motion relative to him would have been with uniform speed in a straight line. Both he and the ball would have been accelerating downwards relative to the Earth, but when we talk about their motion relative to one another, this common acceleration subtracts away exactly, leaving only uniform relative motion.

This happens only because, for gravity, the acceleration of every body is the same. Such statements would not be true of other forces, like electromagnetism. This lack of relative acceleration, even in a gravitational field, allows us the following formulation of the equivalence principle.

> In a gravitational field, all objects behave in such a manner that they appear to be completely free of any gravitational forces when observed by a freely-falling experimenter.

To a freely-falling Galileo, the laws of physics would be the same as they would be in outer space, far from any gravitating bodies. This is a formulation that Galileo and Newton would have accepted, even if they would have found it strange, and it is particularly suitable for us because all mention of mass and acceleration has disappeared from it.

It is possible to perform a home experiment that directly illustrates this version of the equivalence principle very well. It is a "toy" that was given to Einstein on his 76th birthday, described in Figure 2.5 on page 17.

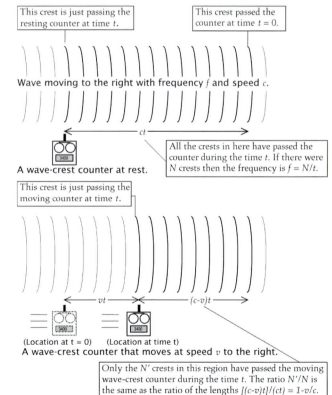

This crest is just passing the resting counter at time t.

This crest passed the counter at time $t = 0$.

Wave moving to the right with frequency f and speed c.

ct

A wave-crest counter at rest.

All the crests in here have passed the counter during the time t. If there were N crests then the frequency is $f = N/t$.

This crest is just passing the moving counter at time t.

vt $(c-v)t$

(Location at $t = 0$) (Location at time t)
A wave-crest counter that moves at speed v to the right.

Only the N' crests in this region have passed the moving wave-crest counter during the time t. The ratio N'/N is the same as the ratio of the lengths $[(c-v)t]/(ct) = 1-v/c$. Therefore the moving counter measures a frequency $f' = N'/t = (1-v/c)N/t = (1-v/c)f$.

Figure 2.3. A visual derivation of the Doppler effect. The wave runs from left to right. The counter counts the number of crests of the wave that pass it. The frequency of the wave is the number of crests per unit time. This depends on whether the counter is moving or not.

Investigation 2.2. The effect of gravity on light: the redshift

To understand how gravity affects light, we imagine a beam of light of a particular frequency f_{bottom} that is shining upwards from a source on the ground. An experimenter stands on a tower of height h directly over the source, and he measures the frequency of the light when it reaches him. He calls this frequency f_{top}. What is the relation between f_{bottom} and f_{top}? We will use the equivalence principle, which means introducing another experimenter who is freely-falling.

Suppose, then, that the experimenter at the top has a companion who falls off the tower at the moment that the beam of light leaves the ground. (Fortunately this is only a thought experiment!) As a freely-falling experimenter, the companion finds that light moves as if it were in outer space; in particular, the frequency that he measures does not change with time. At the instant he leaps off, his is still at rest with respect to the ground, so he measures the same frequency f_{bottom} as an experimenter on the ground would measure. By the equivalence principle, this is also the frequency he (the companion) measures when the light reaches the top of the tower a moment later.

But in this brief moment, the companion has begun to fall. Relative to the companion, the experimenter at the top is moving *away* from the light source at the time of reception of the light at the top, and so the fixed experimenter's frequency f_{top} will be *redshifted* with respect to the companion's frequency f_{bottom}. If our version of the equivalence principle is right, then light is redshifted as it climbs out of a gravitational field.

We can find out how much this redshift is by calculating the speed of the falling companion when the light arrives at the top. To travel a distance h, the light takes a time h/c. In this time the companion falls with the acceleration of gravity, g. His final speed v is therefore just g times the time of fall, or gh/c. From Equation 2.4 on the previous page we then have

$$f_{top} = f_{bottom}(1 - gh/c^2). \qquad (2.5)$$

The magnitude of the effect is very small, but not too small to be measured. For example, if the tower were 100 m high, then in Equation 2.5 gh/c^2 would be only 1.1×10^{-14}, so the change in the frequency of light would have to be measured to an accuracy of a few parts per 10^{15}. Two very high precision experiments by R V Pound, G A Rebka, and J L Snider in the 1960s confirmed the effect with good accuracy. (See Figure 2.4.) Today it is checked every day when routine corrections for the redshift are put into the time-signals of the GPS satellite system, as described in the text.

It is important to understand that, as we remarked at the end of Investigation 2.1 on the preceding page, the redshift is a property of the experimenters as well as of light. Thus, light at any height does not have a unique frequency. If we measure its frequency using experimenters at rest relative to the ground, then there will be a redshift. If we measure its frequency by using freely-falling experimenters, there will be none.

Exercise 2.2.1: *Redshift to a satellite*

Calculate the redshift gh/c^2 if h is the distance from the ground to a satellite in low-Earth orbit, 300 km. Suppose the "light" is actually a radio wave with a frequency of 10^{11} Hz. How many cycles would the transmitter emit if it ran for one day? How many fewer would be received in one day by the satellite? How long did it take the transmitter to generate these "extra" cycles?

The gravitational redshift of light

In this section: a direct consequence of the equivalence principle is that light changes its frequency as it climbs out of a gravitational field. It shifts toward the red end of the spectrum.

The effect of gravity on light can now be found by demanding that it should behave as if there were no gravity when it is observed by a freely-falling experimenter. This means, in particular, that it should follow a straight line with no change in its frequency. In Investigation 2.1 on the previous page we show that the frequency of light is affected by the motion of an experimenter. This is called the *Doppler effect*. Light can experience a **redshift** or a **blueshift**, depending on the motion of the source relative to the experimenter. Then in Investigation 2.2 we show how the Doppler effect and the new equivalence principle combine to tell us how gravity affects light as it moves upwards from the ground.

The result is that the frequency of a beam of light climbing up out of a gravitational field is Doppler-shifted towards lower frequencies. Blue light becomes more red, so this is called the *gravitational redshift*.

It is important to remember that the frequency of light depends on the state of motion of the experimenter who measures it, so that when we say that light suffers a gravitational redshift, we mean that the experimenters must both be at rest relative to the Earth; a freely-falling experimenter, for example, should see no redshift at all.

When general relativity was first proposed, physicists soon realized that it predicted the gravitational redshift of light, and observations of the redshift of light leaving the Sun and other stars were regarded as an important test of the validity of general relativity. But we can see from our discussion that *any* theory of gravity can be expected to predict the effect if it respects the principle of equivalence. The modern view is that observations of the gravitational redshift test the equivalence principle.

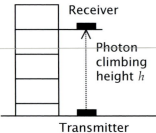

Figure 2.4. *A sketch of the experimental arrangements for the Pound–Rebka–Snider experiment, which first detected the gravitational redshift on Earth.*

Gravity slows time

The gravitational redshift leads us to a very profound conclusion about time itself: gravity makes it run slower. Suppose we build two identical clocks, in such a way that the clocks tick once in every period of the oscillation of the electromagnetic wave that we use in the redshift experiment. We place clock one on top of the tower in the experiment and leave the other on the ground. By design, the one on the ground ticks at the same rate as the frequency of the light signal that we emit there. Suppose we keep it there for, say, 10^{20} ticks. (Since visible light oscillates at about 10^{15} times a second, this would be about one day.) Now, the clock at the top of the tower receives the light redshifted, so the light frequency at the top of the tower is less by one part in 10^{14} (see Equation 2.5). The clock at the top is therefore ticking faster than the arriving light by that same factor.

Now, the light going up the tower is just a wave; one oscillation corresponds to the arrival of one "crest" of the wave. Crests don't disappear on the way up, so exactly as many oscillations of light arrive at the top during the experiment as were emitted at the bottom: 10^{20} in this case. But during the experiment, the clock at the top has ticked more times, by one part in 10^{14}. That means it has ticked 10^6 times more than the one on the ground. When the experiment finishes, we immediately bring the clock from the top of the tower down to the ground and compare it to the one on the ground. The one that has been sitting on the tower is ahead of the one on the ground, by these 10^6 ticks. This is only one nanosecond (1 ns = 10^{-9} s), but it is measurable.

Let us take stock of what we have learned. Given two identical clocks, if we place one for a while higher up in a gravitational field and then bring it down to the other one, we will find it has gone faster. This conclusion applies to any clock, biological or physical, regardless of how it is made: the workings of the clock did not come into the argument above.

> Since all clocks run faster higher up, we conclude that time itself runs faster higher up in the gravitational field: after all, time is only what we measure using clocks.

This is not just an abstract point. Today there are in orbit around the Earth a number of satellites that form the Global Positioning System (GPS). Launched by the US Air Force, they constantly send radio signals down to Earth that can be used in navigation: with a GPS receiver one can pinpoint one's location to within 10 m, an extraordinary accuracy. The satellites carry precise atomic clocks, the most accurate clocks that can be made. Because of the effect of gravity on time, these tick faster than do clocks on the ground; the difference is about three microseconds per day. (A microsecond or μs is 10^{-6} s.) Yet to give a position that is accurate to 10 m requires clocks that are accurate to the time it takes the radio waves from the satellites to travel 10 m, which is 0.03 μs. Therefore, this redshift correction *must* be taken into account in order for the system to function. (Actually, there is also a velocity correction that we will go into in Chapter 15, and this has to be taken into account as well.)

The routine use of the GPS by airplanes, ships, long-distance trucks, and even private cars confirms the gravitational redshift and the effect of gravity on time to a much higher accuracy than the original Pound–Rebka–Snider experiment.

In this section: from the redshift of light it follows that time itself slows down when gravity is strong. The GPS navigational satellite system, which relies on highly accurate clocks, must take this effect into account in order to maintain its accuracy.

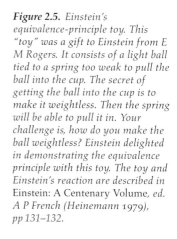

Figure 2.5. *Einstein's equivalence-principle toy. This "toy" was a gift to Einstein from E M Rogers. It consists of a light ball tied to a spring too weak to pull the ball into the cup. The secret of getting the ball into the cup is to make it weightless. Then the spring will be able to pull it in. Your challenge is, how do you make the ball weightless? Einstein delighted in demonstrating the equivalence principle with this toy. The toy and Einstein's reaction are described in* Einstein: A Centenary Volume, *ed. A P French (Heinemann 1979), pp 131–132.*

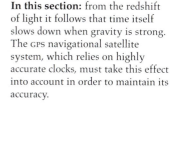

Light-weight ball — Weak spring

In this section: the seemingly simple consideration of gravity on Earth has given us the tools we need to study most of the Universe that modern astronomy reveals to us. But, in addition, it has brought us to deep conclusions: the effect of gravity on the color of light, the slowing of time by gravity. These ideas form the foundation on which we will build the modern theory of gravity later in this book.

Summing up

Although we have confined ourselves in these first two chapters mainly to gravity on the Earth's surface, and we have explored its properties with simple experiments and straightforward reasoning, we have uncovered a rich treasure of different ideas. We have been led to the principle of equivalence, the principle of relativity, Newton's laws of motion, the gravitational redshift of light, and the fact that gravity affects time itself.

We also now have a more complete answer to Copernicus' critics than Galileo could have given: we don't feel the motion of the Earth around the Sun because the whole Earth is in free fall, so there is nothing to feel.

An important practical idea which we have seen in Galileo's work is that motion in different directions is *independent*: the vertical acceleration tells us the change in the speed in the vertical direction, and similarly for other directions.

In the next twelve chapters we will extend our exploration of gravity into the Solar System, to stars and to galaxies, but apart from Newton's law of gravitation we will not have to introduce any new ideas that do not already arise in terrestrial experiments. This is surely one of the most satisfying aspects of physics, that by drawing conclusions from experiments on the Earth we can make sense of what is happening in distant parts of the Universe. Not until Chapter 15, when we begin to get into special and general relativity, will we need some essentially new ideas.

3

Satellites:
what goes up doesn't always come down

In this chapter: we use the equivalence principle to explain how satellites stay in orbit. We generalize the computer program of Chapter 1 to compute orbits of satellites.

▷Communications and many weather satellites, which must be in "geostationary" orbits, are an important exception, being in distant orbits. We will return to these orbits in Chapter 4.

▷The picture behind the words on this page is the Hubble Space Telescope (HST), a satellite launched by the National Aeronautics and Space Administration (NASA), with participation from the European Space Agency (ESA) as well. The HST has opened the most distant reaches of the Universe to view by making observations above the atmosphere of the Earth. Its low orbit makes it accessible to astronauts for repair and for upgrading its instruments. Image courtesy ESA.

Many people assume that satellites orbit the Earth far above its surface, but the numbers tell a different story. Most satellites orbit at less than 300 km above the ground. Compared with the radius of the Earth, 6400 km, this is very small. Their orbits just skim the top of the atmosphere. We can expect, therefore, that the acceleration of gravity on such a satellite will not be very different from what it is near the ground. How then can it happen that the satellite doesn't fall to the ground like our cannonballs in the first chapter?

The answer is that it tries to, but the ground falls away as well. Imagine firing a cannon over a cliff. Eventually the ball will fall back to the height from which it was fired, but the ground is no longer there. The Earth has been cut away at the cliff, so the ball must fall further in order to reach the ground.

If we kept cutting the Earth away, the ball would just keep falling without ever hitting the ground. Now, the Earth is spherical, so it is already "cut away". Moreover, gravity attracts bodies toward the center of the Earth, so as the body moves around the Earth, the direction in which it is trying to fall keeps changing. Therefore, if we fire the cannonball fast enough, it might just keep falling toward the Earth without ever reaching the ground. It would then be a satellite.

It is not hard to get a rough idea of how fast a satellite has to be going in order to stay in orbit. In the first chapter, in Investigation 1.3 on page 5, we saw that the maximum range of a cannonball fired with speed V is V^2/g, where g is the acceleration of gravity, $9.8\,\text{m s}^{-2}$. If we set this range equal to the radius of the Earth, $R = 6400\,\text{km}$, and solve for V, then we must get a number which has about the right size. The result is that, as a first guess, V must be $(gR)^{1/2}$.

The calculation in Investigation 3.1 on page 22 shows that this guess is exactly right: the orbital speed of a satellite near the Earth's surface is

$$V_{\text{orbit}} = (gR)^{1/2}, \tag{3.1}$$

which is $7.9\,\text{km s}^{-1}$.

Given that the circumference of the satellite's orbit is 40 000 km, one orbit at this speed will take 5100 s, or 84 minutes. Since we have used values of g and R appropriate to the surface of the Earth, the true period of a typical near-Earth orbit at 300 km altitude is a bit longer, more like 91 minutes. (Geostationary orbits are much higher: see Chapter 4.)

We can solve Equation 3.1 for the acceleration g by squaring and dividing by R. If we change the symbol for acceleration to a (which will denote any acceleration – we reserve g to mean the Earth's surface acceleration) then we get

$$a = V_{\text{orbit}}^2/R. \tag{3.2}$$

This is the general expression for the acceleration of a particle that follows a circular orbit with radius R and speed V_{orbit}.

Taking motion apart

In this section: we show how to do what Galileo did, namely to consider the motion and forces separately in vertical and horizontal directions. We distinguish the idea of speed from that of velocity.

Before looking at Investigation 3.1 on page 22, we will find it helpful to make a clear distinction between the words "speed" and "velocity". I have used the two words virtually interchangeably up to now, but from now we should keep them separate. Let us illustrate the difference with an example, the cannonball of Chapter 1, which moves both vertically and horizontally.

First, let us be clear on what we mean by the vertical and horizontal motions. We could measure the cannonball's horizontal motion, for example, by firing the cannon when the Sun is directly overhead, and then watching the ball's shadow. By our discussion in Chapter 1, the shadow will move at a constant speed. Similarly, we could discover the vertical motion by shining a bright light at it from behind, and watching the cannonball's shadow on a screen directly in front of it. This motion would be indistinguishable from that of any other ball fired with the same vertical speed, regardless of its horizontal speed.

Together, the two motions completely describe the body's trajectory, and it is usual to give the *set* consisting of the horizontal speed and the vertical speed the name *velocity*. The word "speed" then refers to the rate of change of a distance with time, but the word "velocity" contains directional information: knowing the speed in both the vertical and horizontal directions, we can figure out the actual direction in which the cannonball is moving.

Mathematically, this set is called a **vector**. The vertical speed is called its vertical **component**. Thus, the components are just the usual numbers we have been considering. Mathematicians and physicists reserve the word "velocity" for the vector and "speed" for numbers associated with specific directions, and we shall do the same.

An important number associated with the velocity is the particle's total speed, which means the number of meters per second the particle goes in its own direction, rather than its projection along one of the two directions. When I use the word *speed* on its own, rather than as horizontal speed or something like that, then I will mean total speed. By the **Pythagorean theorem**, the total speed is the square-root of the sum of the squares of the components of the velocity.

Acceleration, and how to change your weight

In this section: weightlessness in free fall, and weighing much more when a rocket takes off, are ways in which the equivalence principle changes the weight of astronauts.

When astronauts orbit the Earth they are weightless: they float across their cabin, they release pencils or balls from their hands and watch them float too, they perform space-walks outside their spacecraft and do not require a tether in order to stay up there with the craft. It is tempting to conclude from their weightlessness that gravity is very weak so far from the Earth, but that can't be right. We have seen that they are not particularly far away; and in any case, if gravity is so weak, why do they stay in orbit instead of flying off in a straight line? The real explanation is to be found in the equivalence principle.

> The astronauts and their spacecraft are in free fall, so all the objects near them behave as if there were no gravity. Without gravity, things are weightless.

Figure 3.1. *An Airbus research aircraft about to go into free fall, which it can sustain for up to 22 s. Image courtesy Novespace/Airbus.*

Being free of gravitational forces means having no weight. We shall see below that astronauts in orbit around the Earth are in free fall, constantly falling towards the Earth. Therefore, they are perfect examples of the hypothetical falling Galileo. When pens float alongside them, we see direct proof of the equivalence principle. A weightless environment can also be created for a limited time by allowing an airplane to fall freely. Research organizations use such aircraft regularly. (See Figure 3.1.)

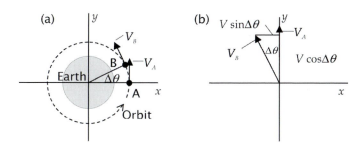

Figure 3.2. *Velocities at two nearby points A and B on a circular orbit, as required for the calculation in Investigation 3.1 on the next page. The angle between the points is Δθ. Velocities are drawn as arrows, whose length is equal to the total speed V. In panel (a) we see the circular orbit, the two points A and B, and the velocity arrow (tangent to the circle) at each point. The arrows have the same length but different directions.*

In (b) we see the same velocity arrows, moved to the origin and magnified so they can be compared. The angle between them is the same as the orbital angle Δθ between points A and B. At point A the speed is only in the y-direction. At B it has components in both directions, which are marked in the diagram. Remember that both velocity arrows have length V.

To fill out this explanation, let us ask what the sensation of weight really is. How do we *experience* our own weight? What we feel are the effects of the floor pushing up on us to support us against gravity. This force is transmitted to our bones and joints, which are somewhat compressed in our legs and extended in our arms as we stand. Our internal organs hang from their attachment points inside our chest and abdomen, and these supporting tissues are also stressed. All through our bodies there are nerve endings picking up these stresses and telling us that we are not weightless. If instead we find ourselves in free fall, these stresses disappear: no forces need be exerted by the supporting tissues to keep our lungs in place or our elbows together. The *sensation* of weight disappears completely.

> The floor, not the force of gravity, is responsible for our weight.

This argument can be turned around to explain why astronauts feel heavier when their rockets are firing, or indeed why we feel heavier when an ascending elevator starts up. Since gravity works only by inducing accelerations, any other steady acceleration mimics gravity. If a rocket accelerates at five times the acceleration of gravity (even in everyday language this is called "5 g's"), its occupants will feel five times as heavy. Astronauts can be trained for this on a large centrifuge. Using Equation 3.3 on the next page for circular acceleration, we find that a centrifuge of 5 m radius creates an acceleration of 5g when its speed is 16 m s⁻¹, which corresponds to one revolution every two seconds. Anyone who has ridden a roller coaster or who has been unlucky enough to have been caught in strong turbulence in a high-flying airliner will have experienced the discomfort of such changes in weight first-hand, generated by accelerations that are only roughly 1g in size.

Notice that it is normally only possible to eliminate the effects of gravity in a small region. Two experimenters in free fall at different places on the Earth are falling on radial lines through the center of the Earth, so they are accelerating relative to one another. Indeed, when we begin to study Einstein's point of view on gravity, we will see that the physical part of gravity is its non-uniformity.

If gravity were everywhere uniform we could not distinguish it from acceleration. This is the sense of the word *equivalence* in *equivalence principle*. Therefore, the changes in gravity over distances tell us that we are really dealing with gravity and not simply a uniform acceleration. We will elaborate on this subject in Chapter 5.

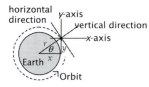

(a) The relationship between the vertical and the x-axis depends on the angular position θ of the point on the orbit. The orbital radius r and the x- and y-coordinates shown here are used below.

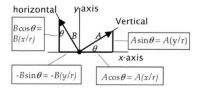

(b) Taking a vertical vector (A) and a horizontal one (B) and finding their components in the x-y coordinates.

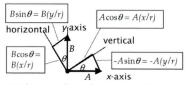

(c) Taking an x-directed vector (A) and a y-directed vector (B) and finding their horizontal and vertical components.

Figure 3.3. *Coordinates for the satellite-orbit calculation in Investigation 3.2 on page 23. The diagrams show how to find the components of vectors from vertical/horizontal coordinates to the x–y system. The orientation of the vertical and horizontal coordinates is as in Figure 3.2. Expressions for components are given in boxes.*

Investigation 3.1. Keeping a satellite in a circular orbit

Here we examine the geometry of a circular orbit in order to find the acceleration required to keep a satellite in uniform circular motion. Then we can set this equal to g to find the speed of a satellite orbiting near the Earth. In a later investigation we will test this result with a computer simulation of a satellite trying to get into orbit. But first we should look a little more at vectors, which were introduced in the text to describe velocity.

Other quantities may also be vectors. The position of a particle is given by a set of two numbers, which are its coordinates; this vector is called the **position vector**, or sometimes the **displacement** of the particle from the origin of coordinates. The acceleration is also a vector, but there are no special names for distinguishing between the acceleration vector and its components.

Now, suppose that the satellite's speed is V and its orbital radius is R. A small arc of the orbit is shown in panel (a) of Figure 3.2 on the preceding page. When the satellite is at point A its speed is in the direction shown by the arrow, whose length is V. When it gets to point B its speed hasn't changed, but the direction it is traveling in has, as shown. This change requires an acceleration, since its speed in both the x- and y-directions has changed.

In Figure 3.2 I show the geometry of the situation. Suppose for simplicity that the orbit lies in the plane of the Earth's equator. This is the plane of the diagram in the first panel of the figure. The x- and y-axes are rectangular coordinates in this plane, whose origin is at the center of the Earth. The orbit is a circle in this plane. The points A and B between which we wish the calculate the acceleration are indicated by dots on the orbit, and the velocity of the particle at each of these points is shown as an arrow. The length of the arrow is the same at the two points, but its direction has changed. The points are separated by an angle of $\Delta\theta$, which we eventually will assume is small. The velocity arrow is rotated by the same angle from A to B.

In the second panel, the velocities themselves are shown, magnified by a factor of two. What we want is to find the change in the speeds in the x- and y-directions. The velocity at A has an x-speed of zero and a y-speed of V. The parts of the velocity at B can be deduced by using the trigonometry of the right triangle whose hypotenuse is the velocity arrow of B, which has length V. Then the x-speed of this velocity is $-V \sin \Delta\theta$, and its y-speed is $V \cos \Delta\theta$. Thus, the change in the x-speed from A to B is $-V \sin \Delta\theta$, while the change in the y-speed is $V(\cos \Delta\theta - 1)$.

From this figure it is also apparent that for very small $\Delta\theta$ the largest piece of the change of velocity is in the x-speed, which is directed perpendicular to the direction of motion at A, i.e. toward the center of the circle. This means that the *instantaneous acceleration of a circular orbit is directed towards the center of the orbit*. What we have to do is to figure out how big this acceleration is.

Since the satellite travels at (total) speed V, it moves once around the Earth in a time equal to the circumference of its orbit divided by its speed. This is the period of its orbit, P:

$$P = 2\pi R/V.$$

In a small time Δt, it will travel a fraction $\Delta t/P$ of the orbit, so the corresponding angle $\Delta\theta$ will be (in degrees)

$$\Delta\theta = 360° \frac{\Delta t}{P}.$$

If this angle is very small, then it is clear from Figure 3.2 on the previous page that the change in the speed in the y-direction is very nearly equal to the arc of a circle of radius V subtended by the (small) angle $\Delta\theta$. This is simply the circumference of such a circle times the fraction $(\Delta\theta/360°)$. The result is

$$\text{change in } y\text{-speed} = 2\pi V \frac{\Delta\theta}{360°} = 2\pi V \frac{\Delta t}{P}$$

$$= 2\pi V \frac{\Delta t}{2\pi R/V} = \frac{V^2}{R} \Delta t.$$

The acceleration is this change divided by Δt,

$$a = \frac{V^2}{R}. \tag{3.3}$$

This is the desired formula. Notice that it does not contain Δt or $\Delta\theta$. This is because we made various approximations that were valid only if these were sufficiently small. This means that Equation 3.3 is the *exact* expression for the *instantaneous* acceleration.

For a satellite orbiting near the Earth, this acceleration will be just g. Putting this into Equation 3.3 and solving for V gives Equation 3.1 on page 19.

Exercise 3.1.1: *Vectors*

Quantities that have a value but no direction are called **scalars**. Decide whether the following physical quantities should be described mathematically by scalars or by vectors: (a) the mass of a rock; (b) the electric force on a charged particle; (c) the temperature of a room; (d) the slope of a hill.

Exercise 3.1.2: *Period of a satellite orbiting near the Earth's surface*

Use Equation 3.1 on page 19 to calculate the orbital period of a satellite near the Earth. Assume that the acceleration of gravity at the height of the satellite is the same as on the ground, $g = 9.8 \text{ m s}^{-2}$. Take the radius of the orbit to be the radius of the Earth, 6400 km, plus the height of the satellite above the Earth, 300 km.

Getting into orbit

In this section: how satellites get into and stay in orbit.

I remarked in Chapter 2 that the orbits of planets are ellipses. This property also applies to satellites of the Earth, the circular orbit being a special case of an ellipse. An important feature is that the orbit is *closed*: a satellite will always return to the place where it was set into orbit. More than that, when it returns to that spot it will be moving in the same direction as before. Therefore if we were to try to launch a cannonball into orbit just by firing it at a sufficiently high speed, we would not succeed: its elliptical orbit would make it want to return to the launching point from *below*. It would necessarily hit the ground somewhere else as it tried to pursue its orbit into the interior of the Earth.

To get something into orbit, one has to give it at least two pushes: one to get it off the ground, and a second to put it into an orbit that does

---

Investigation 3.2. Achieving orbit

Here we ask the computer to calculate for us the orbit of a satellite that is launched from a height of 300 m (say from the top of a cliff), and given a perfectly horizontal velocity. Does the satellite hit the Earth, or does the Earth fall away faster than the orbit does?

The calculation is basically the same as the trajectory program in Chapter 1, with the one difference that the acceleration of the body depends on where it is. As we saw in Investigation 3.1, the acceleration must always point towards the center of the Earth. We will make it slightly easier to calculate the orbit by assuming that the total magnitude of the acceleration is always g, the same as the acceleration on the surface of the Earth. This is not quite right, since gravity gets weaker as we get further from the surface, but we will not take that into account until we do the orbit program for the Solar System in Chapter 4. By taking the acceleration to be constant, we will not get the right orbits for a particle launched faster than the circular velocity, since that particle will move further away from the Earth.

Let us call a_x and a_y the accelerations in the x- and y-directions, respectively, at any time t. (In the program EarthOrbit they are called ax and ay.) From Figure 3.3 on page 21, we can see that, at a point of the orbit given by the angle θ, a line of total length g that points towards the center of the circle has x-length $g\cos\theta$. Since the cosine of this angle is, by the first part of the figure, $\cos\theta = x/r$, this is gx/r. But the acceleration is directed *towards* the center, so it has to be given a negative value: the change of x-speed produced by this acceleration is negative because it is directed in the negative-x sense. Putting all this together gives

$$a_x = -gx/r.$$

Similarly, for the acceleration in the y-direction, the geometry shown in Figure 3.3 on page 21 shows that

$$a_y = -gy/r.$$

We take $g = 9.8\ \mathrm{m\ s^{-2}}$.

As we did in Chapter 1, we will find approximate changes in the x- and y-speeds during a small interval of time Δt by multiplying the accelerations in these directions at time t by Δt. If we let the x-speed be v (v in the program) and the y-speed be u (u in the program), then the approximate change in v is

$$\Delta v = a_x(t)\Delta t, \quad \text{and} \quad \Delta u = a_y(t)\Delta t. \qquad (3.4)$$

In the program, we denote Δv by deltaV, and similarly for other variables. We also need the calculate the changes in positions, which are determined by the speeds. We have

$$\Delta x = v(t)\Delta t, \quad \text{and} \quad \Delta y = u(t)\Delta t. \qquad (3.5)$$

Recall that in Investigation 1.3 on page 5, we saw that this method gave the exact result for a cannonball trajectory. Unfortunately, we cannot expect this to be true here, as well, because in our case the acceleration changes from point to point. Therefore, as we have discussed before, we have to take a small enough time-step to give an accurate result. The program on the website uses a time-step of 0.4 s, and gets a fairly good circular orbit when we take the initial speed to be exactly the orbital speed given by Equation 3.1 on page 19. By changing the time-step you can explore the question of how accurate the calculation is.

Another important feature of this program (and an essential element of any computer program!) is how it stops. We have given it a way of knowing when the orbit has gone once around, so it stops if the orbit hits the Earth or after it goes around once. The logic is explained on the website.

An important point about this calculation is that we perform all our calculations with the x-speed and the y-speed, not the vertical and horizontal parts of the velocity. The reason is Galileo's principle of relativity, which was incorporated into Newton's first law, as we discussed in Chapter 2. The independent motions of a body are those that would continue unchanged in the absence of external forces, namely those along straight lines. Since the vertical and horizontal directions change with time, they cannot be treated separately. But the x- and y-motions are separate, and they depend only on the accelerations in those directions. All our later computer programs for the motion of bodies will consistently use these rectangular coordinates.

The conversion of x-y speeds and positions to vertical-horizontal ones depends on where in the orbit one is. From Figure 3.3 on page 21 one can deduce the following conversions for the components of any of the vector quantities (distance, speed, acceleration):

$$(x) = (vert)\cos\theta - (horiz)\sin\theta$$
$$(y) = (vert)\sin\theta + (horiz)\cos\theta$$
$$(vert) = (x)\cos\theta + (y)\sin\theta \qquad (3.6)$$
$$(horiz) = -(x)\sin\theta + (y)\cos\theta.$$

Here the notation is that (x) stands for the x-component of the quantity involved, $(vert)$ for its vertical component, and so on. As the boxes in Figure 3.3 on page 21 show, the trigonometric functions in the above equations may be replaced by expressions using the coordinates of the point on the orbit:

$$\sin\theta = \frac{y}{r}, \quad \text{and} \quad \cos\theta = \frac{x}{r}.$$

---

not collide with the Earth. That is why the "boost phase" of a satellite launch is always followed by a crucial "orbit insertion" event in which rockets are fired again. We cannot launch satellites just by shooting them out of a cannon.

Most satellite launches use more phases of acceleration than just the boost and insertion. Usually the boost is divided into two or three phases, contributed by different stages of the rocket. This is not required by the properties of orbits in a gravitational field. Rather, it is useful for keeping the amount of fuel used to a minimum. Once a certain amount of fuel has been used, the empty fuel tanks are a useless weight for the rocket, so it is more efficient to drop them off and use the remaining fuel to propel less weight into orbit.

In order to illustrate the expression for the circular orbital speed that we derived above, and to develop the computer techniques and programs that we will need in later chapters to look at orbits around the Sun, binary and multiple star systems, and orbits around black holes, we discuss a simple satellite-launching problem in Investigation 3.2. We imagine that the satellite is launched from a certain height in

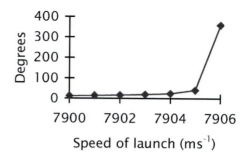

Figure 3.4. *A few attempts at getting into orbit, as calculated by the program* EarthOrbit. *The "satellite" is fired horizontally from a height of 300 m. The trajectory stops where it hits the Earth, which is taken to be a perfect sphere of radius 6 378 200 m. We do not show a picture of the trajectory, since it is so close to the Earth that different trajectories would all come out superimposed. Instead, we plot the angle through which the orbit turns before the trajectory hits the Earth, as a function of the launch speed. Note how much difference small changes in this speed make. The final speed, 7906 m s^{-1}, is the one that attains orbit.*

a horizontal direction. With too little speed, it falls and hits the Earth. The angular distance around the Earth that the satellite travels for a given launch speed is shown in Figure 3.4. Reaching orbit is remarkably sensitive to the exact speed. With a launch speed of 7900 m s^{-1} the satellite hardly gets anywhere. If its launch speed rises to only 7906 m s^{-1}, the satellite goes into orbit.

The Solar System:
a triumph for Newtonian gravity

As children of our age, we find it natural to think of the planets as cousins of the Earth: remote and taciturn, perhaps, but cousins nevertheless. To visit them is not a trip lightly undertaken, but we and our robots have done it. Men have walked on the Moon; live television pictures from Mars, Jupiter, Saturn, Uranus, and Neptune have graced millions of television screens around the world; and we know now that there are no little green men on Mars (although little green bacteria are not completely ruled out).

Among all the exotic discoveries have been some very familiar sights: ice, dust storms, weather, lightning, erosion, rift valleys, even volcanos. Against this background, it may be hard for us to understand how special and mysterious the planets were to the ancients. Looking like bright stars, but moving against the background of "fixed" stars, they inspired awe and worship. The Greeks and Romans associated gods with them, and they played nearly as large a part in astrology as did the Moon.

It was a giant step forward for ancient astronomers, culminating in the great Greek astronomer Ptolemy, to show that what was then known about planetary motions could be described by a set of circular motions superimposed on one another. These were called **cycles** and **epicycles**. It was an even greater leap for Copernicus to argue that everything looks simpler if the main circular orbits go around the Sun instead of the Earth. Not only was this simpler, but it was also a revelation. If the Earth and the planets all circle the Sun, and if the Earth is simply in the third orbit out, then probably the planets are not stars at all, and the Earth might be a planet, too.

This probability became a virtual certainty when Galileo trained his first telescope on the night sky. Not only did he discover that Jupiter had moons, just like the Earth, only more of them, he also saw craters and mountains on the Moon, indicating that it was rocky like the Earth; and he saw the phases of Venus, the shadow that creeps across Venus as it does the Moon as these bodies change their position relative to the Sun. This meant that Venus was a body whose size could be measured: it was not a mere point-like star. When Kepler (whom we met in Chapter 2) showed, by extraordinarily painstaking calculations, that the planetary orbits were actually ellipses, the modern *description* of the motion of the planets was essentially complete.

But Galileo's study of motion on Earth soon raised an even bigger problem. Since bodies travel in straight lines as a rule, and since the planets do not, what agency forces them to stay in their orbits? Newton saw this problem clearly, and he had the courage to say that it needed no extra-terrestrial solution: gravity, the same gravity that makes apples fall from trees, also makes the planets fall toward the Sun. But what was the *law* of gravity? What rule enables one to calculate the acceleration of the planets?

In this chapter: applied to the Solar System, Newton's new theory of gravity explained all the available data, and continued to do so for 200 years. What is more, early physicists understood that the theory made two curious but apparently unobservable predictions: that some stars could be so compact that light could not escape from them, and that light would change direction on passing near the Sun. Einstein returned the attention of astronomers to these ideas, and now both black holes and gravitational lenses are commonplace.

▷This name is pronounced "Tolemy". His full name was Claudius Ptolomæus, and he lived in Alexandria during the second century AD. Little else is known of him.

▷The image behind the text on this page is from a beautiful photograph of the Moon taken by the Portugese amateur astronomer A Cidadao on 1 March, 1999. Used with permission.

Figure 4.1. Both the Moon (left) and Mars (right) appear to be desolate, uninhabited deserts. But Mars appears to have experienced erosion, probably by flowing water, at some time in its past. Left image courtesy NASA; right courtesy NASA, the Jet Propulsion Laboratory (JPL), and Malin Space Science Systems.

How to invent Newton's law for the acceleration of gravity

In this section: we learn how knowledge of the Moon's distance, which was available to Newton, makes the law of gravity that he invented very plausible.

Let us look first at the nearest "planet": the Moon. If the laws of mechanics postulated by Newton are to apply in the heavens as well, then we should be able to deduce from the motion of the Moon what the force on it is. Suppose that the Moon is in a circular orbit. This is a good first approximation, but we shall have to return to the question of elliptical orbits later. We have seen in Chapter 3 that, for circular motion at speed V and radius R, the acceleration a is V^2/R. We can eliminate V in terms of the radius of the orbit R and the period P, because the speed is just the distance traveled (circumference of the orbit) $2\pi R$ divided by the time taken, P. This means that the acceleration is $(2\pi/P)^2 R$ towards the Earth.

Now, Newton knew the distance R to the Moon; even the Greeks had a value for it, by measuring its **parallax** from different points on the Earth, as shown in Figure 4.2. Since Newton also knew the period P of the Moon's orbit, he could work out its acceleration.

If we consult Table 4.1 for the modern values of these numbers, we can calculate from Equation 3.1 on page 19 or Equation 3.3 on page 22 that the acceleration of the

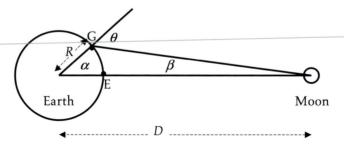

Figure 4.2. The direction to the Moon is different at different places on the Earth; this is called the Moon's parallax. One can use the parallax to determine the Moon's distance D. Suppose the Moon is directly overhead at a point E on the Earth, and suppose one measures its position in the sky from point G at the same moment, obtaining that it lies at an angle θ down from the vertical there. The point G is known to be more northerly on the Earth than E, by a latitude angle α. Simple geometry then says then that the angle between E and G as seen from the Moon is β = θ - α, and the angle of the triangle opposite the desired distance D is 180° - θ. By the **law of sines** for triangles, we have sin(180° - θ)/D = sin β/R, where R is the radius of the Earth. This can be solved for D. Ptolemy performed essentially this measurement of θ, and knew α and R reasonably accurately. He deduced that the Moon's distance was 59 times the Earth's radius. The right value is just over 60.

Table 4.1. *Data for the Moon.*

Average distance from the Earth, R (km)	Period of orbit, P (s)	P^2/R^3 (s^2 km^{-3})	Average speed (km s^{-1})	Mass (kg)
3.84×10^5	2.36×10^6	9.84×10^{-5}	1.02	7.3×10^{22}

Moon is 0.0027 m s^{-2}. Newton reasoned that if this was due to the Earth's gravity, then (like any gravitational acceleration) it could not depend on the Moon's mass, so it could depend only on how far the Moon was from the Earth. In particular, he guessed that it might depend only on how far it was from the *center* of the Earth. Compared with the acceleration of gravity on the Earth's surface, 9.8 m s^{-2}, that of the Moon is smaller. How much smaller? The ratio of the two accelerations, 9.8/0.0027, is 3600. The ratio of the radius R of the Moon's orbit, 384 000 km, to the radius of the Earth itself, 6380 km, is 60.

> Clearly, the ratio of the accelerations is the *square* of the ratio of the distances taken in the opposite sense: the acceleration produced by the Earth is inversely proportional to the square of the distance from the Earth's center.

We have already seen in Equation 2.3 on page 13 that the simplest form of such an inverse-square law of gravity that obeys the equivalence principle is

$$F_{\text{grav}} = \frac{GM_1 M_2}{r^2},\qquad(4.1)$$

where M_1 in this case is the mass of the Earth and M_2 that of the Moon.

One can imagine how Newton might have reacted to this result: such a simple relation cannot be coincidence! Surely it must also apply to the planets in their orbits around the Sun. But Newton knew that the orbits of the planets were ellipses, not simple circles. Could that also be a consequence of the inverse-square law? That is what we turn to next.

The orbits of the planets described by Newton's law of gravity

In Table 4.2 on the following page I have listed the main properties of the planets and their orbits. Here we encounter for the first time the astronomer's unit of distance in the Solar System, the **astronomical unit**, denoted AU. It is defined as the average distance of the Earth from the Sun, 1.496×10^{10} m. In these units, distances become easier to comprehend: 2 AU is just twice the radius of the Earth's orbit, but what is 3×10^{10} m? (It is just 2 AU again.)

Now look at column 4, where I have calculated the ratio of the square of the period to the cube of the average distance from the Sun. Kepler had noticed that these values were remarkably similar for all the known planets (out to Saturn), and we now call this *Kepler's third law*. (He didn't know the absolute distances between planets very well, but could deduce their ratios from observations, and that was enough to deduce the constancy of this number.)

Newton recognized that this strange relation provides the crucial evidence that the gravitational force does indeed fall off as the square of the distance. Again we consider an idealized circular orbit, for simplicity. From Equation 3.2 on page 19, the quantity in column 4 is, in terms of the acceleration a and the radius R,

$$\frac{P^2}{R^3} = \frac{4\pi^2}{R^2 a}.\qquad(4.2)$$

If this is to be constant, then $R^2 a$ must be constant, or a must be proportional to $1/R^2$, exactly as we inferred from the Moon's orbit.

In this section: given Newton's law of gravity and the distances to the planets, the law must predict all the details of their orbits. This is the critical test that the law had to pass before Newton would believe it.

Notice that in this calculation we had no "adjustable" parameters: if the data hadn't fit our proposed law we would have had to throw the law away. But they *did* fit: surely we are on the right track to an understanding of the planetary orbits.

Encouraging as this argument is, it does rely on the idealization of circular motion. Moreover, the numbers in column 4 are not perfectly constant: there are small but measurable deviations. Before he could convince himself of his law of gravity, Newton needed to show that a $1/R^2$ acceleration also produced the elliptical orbits that Kepler had observed, and that the deviations in column 4 could be predicted from the ellipticity of the orbit, something much more complicated. To do this Newton had to invent the calculus.

With a personal computer we can come close to providing a demonstration of this result in a matter of minutes, without calculus. The website contains a program which calculates planetary orbits. It is based on the trajectory program of Chapter 1, adapted to the planetary case by the calculations in Investigation 4.1. It produced the orbit of Mercury shown in Figure 4.3.

Our computed orbit closes smoothly on itself, and it "looks" elliptical. The **eccentricity** of the ellipse may be calculated by estimating the ratio, q, of the maximum to the minimum distance from the Sun:

▷Remember that the Sun sits at a focus of the ellipse, not its center. So the ratio q is not the same as the ratio s of the short axis to the long axis. $e = (1 - s^2)^{1/2}$.

$$e = \frac{1 - q}{1 + q}. \qquad (4.3)$$

My rather crude estimate of q from the graph gives $e = 0.21$, which is acceptably close to the observed eccentricity, 0.206 (i.e. the ratio of Mercury's short axis to its long one is 0.98). Apart from Pluto, Mercury has the most elliptical orbit of all the planets.

I started the computer calculation of Mercury's orbit at its minimum distance from the Sun (4.6×10^7 km), which is called its *perihelion* distance. To compute the orbits of other planets you may use the data in Tables 4.2 and 4.3. A nice thing to try is the orbits of Neptune and Pluto, because, as the tables make clear, they cross!

Of course, a numerical calculation cannot *prove* that the orbits are perfect ellipses, because there is always some inaccuracy in the fact that we take finite time-steps; instead of changing smoothly, the acceleration changes in discrete steps. This calculation cannot be a substitute for Newton's proof using calculus. Nevertheless,

Table 4.2. Planetary data. One year is 3.1557×10^7 s. The number of known satellites of most of the outer planets is likely to go up with further exploration. For comparison, the mass of the Sun is 1.989×10^{30} kg, and the radius of the Sun is 6.9599×10^5 km. Astronomers are discovering an increasing number of small planetary bodies outside the orbit of Pluto, but none as large as Pluto. They are not usually called planets, and are not in this table.

	1	2	3	4	5	6	7
	Average distance from the Sun, R		Period, P	P^2/R^3	Average orbital speed	Mass	Known satellites
	(10^6 km)	(AU)	(y)	(10^{-10} s^2 km^{-3})	(km s^{-1})	(10^{24} kg)	
Mercury	57.9	0.387	0.241	2.98	47.9	0.33	0
Venus	108.2	0.72	0.615	2.97	35.1	4.87	0
Earth	149.6	1.00	1.00	2.97	29.8	5.97	1
Mars	228.0	1.52	1.88	2.97	24.1	0.642	2
Jupiter	778.3	5.20	11.86	2.97	13.1	1900	39
Saturn	1429.4	9.54	29.46	2.99	9.65	568.41	22
Uranus	2871.0	19.22	84.01	2.98	6.80	86.83	21
Neptune	4504	30.06	164.1	2.97	5.43	102.47	8
Pluto	5913.5	39.5	247.0	2.97	4.74	0.0127	1
Moon	0.384	(from Earth)	0.0748	9.84×10^5	1.02	0.073	0

Investigation 4.1. How to follow the orbit of a planet

The calculation of the orbit of a planet is very similar to the one for the satellite around the Earth. The only essential change is in the law for the acceleration. Here we adopt the law that the acceleration is towards the Sun (which we place at the origin of coordinates) and has magnitude

$$a = k/r^2. \qquad (4.4)$$

The distance to the Sun is given by r. The constant k can be inferred from Kepler's third law, the constancy of the values in column 4 of Table 4.2. The way to do this is suggested in Exercise 4.1.1.

Newton was able to estimate k for the Earth-Moon system rather crudely from the then available estimate of the distance to the Moon.

The x- and y-accelerations are similar to those used in Investigation 3.2 on page 23, but with g replaced by k/r^2:

$$a_x = -k\frac{x}{r^3}, \qquad (4.5)$$

$$a_y = -k\frac{y}{r^3}. \qquad (4.6)$$

By changing the program EarthOrbit to use this as the acceleration, we could obtain a working program for planetary orbits. Giving it initial data from Table 4.3 on page 32 produces good orbits, if the time-step is small enough. Exercise 4.1.3 suggests that you do this and compare the periods and eccentricities of the orbits you compute with the data in table.

In Exercise 4.1.4 you will see that the program is not terribly accurate unless you take small time-steps. There are two important further changes we will make to turn EarthOrbit into the more accurate Orbit. For readers who are interested in how good computer programs are constructed, Investigation 4.2 contains a description of these refinements: a predictor–corrector to make each time-step of the calculation more accurate, and a time-step halver to maintain the accuracy of the finite-difference method at each time-step as the orbital conditions change. These changes are not strictly necessary for planetary orbits, but they will allow us to use the program later for the more demanding problems of binary stars and multiple stars. The resulting program also does very well for extreme Solar System orbits, such as those followed by comets.

Exercise 4.1.1: *Inverse-square-law constant*

From Equation 4.2 on page 27, show that the constant k in Equation 4.4 is $(2\pi)^2/(P^2/R^3)$. Evaluate this to give 1.327×10^{11} km^3 s^{-2}.

Exercise 4.1.2: *Measuring the mass of the Sun and the Earth*

The Newtonian law of gravity, Equation 4.1 on page 27, tells us the force on a body of mass M_2 exerted by the Sun (mass M_1 in the equation). Combine this with Newton's second law, $F = ma$, to show that the acceleration of the body of mass M_2 is $a = GM_1/r^2$. Use this to show that the force-law constant k in Equation 4.4 is $k = GM_1$. Convert this value of k to more conventional units using meters to find $k = 1.327 \times 10^{20}$ m^3 s^{-2}. (Hint: since 1 km $= 10^3$ m, it follows that $1 = 10^3$ m km^{-1}. Multiply by the cube of this form of the number 1 to convert the units for k.) Now use the value of $G = 6.6725 \times 10^{-11}$ m^3 s^{-2} kg^{-1} to find the mass of the Sun. Do a similar calculation for the value of Kepler's constant for the Moon, given in Table 4.2, to find the mass of the Earth.

Exercise 4.1.3: *Simple orbit simulations*

Run the orbit program as modified in this investigation for the planets Mercury, Jupiter, Neptune and Pluto. Start with the perihelion distance and speed given in Table 4.3 on page 32 and infer the value of k for each planet using Equation 4.2 on page 27 and the data in Table 4.2. Compute at least one full orbit. You will have to choose a time-step that allows the planet to move only a small distance at each step, and a number of time-steps that allow the planet to go all the way around. Use Equation 4.3 to calculate the eccentricty of the orbit. See how close you come to the eccentricity given in Table 4.3. Compare your orbit for Mercury with that shown in Figure 4.3.

Exercise 4.1.4: *Assessing the accuracy of the simple orbit program*

For the planet Mercury compute ten successive orbits and see if they lie on top of each other. They should do this, so the extent to which they do not reflects the inaccuracy of the approximations in the computer program. Reduce the time-step size, increasing the number of time-steps accordingly. Do the successive orbits lie more accurately on top of one another? As an estimate of the error, estimate the angle that the perihelion position of the orbit has moved after ten orbits. Calculate this error for different time-steps. Plot the error against the time-step. Is there a simple relation between them?

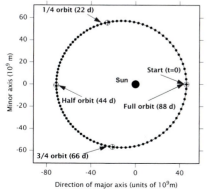

Figure 4.3. *Simulation of the orbit of Mercury using the computer program* Orbit *from the website.*

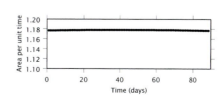

Figure 4.4. *Kepler's area law computed for the orbit of Mercury. Each dot is the area of the triangle between the Sun and two successive dots on Mercury's orbit, divided by the time it takes Mercury to move between them, in units of 10^{20} m^2 per day.*

Investigation 4.2. *A more sophisticated and accurate orbit program*

This investigation is for readers who want to learn some of the ingenious methods by which computer experts improve the accuracy of computer programs. We will explore two improvements. One is to improve the accuracy of each time-step, and the other is to control the accuracy of each time-step.

(a) Improving accuracy. Recall that in Chapter 1, we found that the most sensible approximation to the change in the position of a body during a small interval of time Δt is obtained by *averaging* the speeds at t and $t+\Delta t$ and multiplying by Δt (Equation 1.5 on page 4). By the same reasoning, it would be sensible to find the change in a body's velocity by averaging its acceleration. We did not need to do this in Chapter 1, because there the acceleration g was constant. We did not do it in Chapter 3, even though for the program EarthOrbit the acceleration did not have a constant direction. There we just took the change in velocity to be the acceleration at time t times Δt (Equation 3.4 on page 23). This kept the program simple, but not as accurate as it could be. So here we would like to use

$$\Delta v = \tfrac{1}{2}[a_x(t) + a_x(t + \Delta t)]\Delta t, \qquad (4.7)$$

and

$$\Delta u = \tfrac{1}{2}[a_y(t) + a_y(t + \Delta t)]\Delta t. \qquad (4.8)$$

Similarly, the best way of computing the changes in the position of the body is to find the average of the speeds over the interval:

$$\Delta x = \tfrac{1}{2}[v(t) + v(t + \Delta t)]\Delta t, \qquad (4.9)$$

and

$$\Delta y = \tfrac{1}{2}[u(t) + u(t + \Delta t)]\Delta t. \qquad (4.10)$$

Although we would *like* to use these equations, there is a difficulty that we did not have to face in Chapter 1: here the acceleration depends upon the position, so we cannot calculate, say, $a(t+\Delta t)$ without knowing $x(t + \Delta t)$, but we cannot calculate $x(t + \Delta t)$ without knowing $v(t + \Delta t)$, for which we need the acceleration $a_x(t + \Delta t)$. Are we trapped in a circle with no exit? Not if we remember that with a computer we are only solving the equations with a certain accuracy, not exactly.

The method we will use is called a predictor–corrector technique. The idea is to guess, or "predict", the values of, say, x and y at the later time by just multiplying the speeds at time t by Δt; these are the positions that we used in EarthOrbit. Then we use this (admittedly somewhat incorrect) position to calculate a_x and a_y at $t + \Delta t$. Although they are not exactly right, they should be better than not correcting for the change of the acceleration with position. These values can then be used in Equation 4.7–Equation 4.8 to find v and u at $t + \Delta t$. Now comes the beautiful step: these can in turn be used in Equation 4.9–Equation 4.10 to find *new* values for x and y at $t+\Delta t$. These are the "corrections" to the first "predictions". Since the corrected positions are calculated from better values of the acceleration and velocity, they should be better than the predicted positions.

We need not stop here. We can use the corrected positions as *new* predictions to give better accelerations, thence better velocities, and thence even better corrected positions. For a person using a hand-calculator, predictor–corrector is tedious and time-consuming. But a computer is good at doing things repetitively. It is easy to program the computer to repeat this procedure as often as we wish. We just have to tell the computer when to stop making new corrections. We tell the computer to compare the prediction and correction at each stage, and when their difference is smaller than some predetermined accuracy level that we have given to the computer, then the process stops. This insures that Equation 4.7–Equation 4.10 are satisfied to whatever accuracy we desire. The program Orbit includes this feature. Mathematicians call this technique "iteration". If successive changes in the predicted position become smaller and smaller, we say the method "converges" to the right answer. Note that it only converges to a solution of Equation 4.7–Equation 4.10. We are still left with the possibility that the time-step we chose in these equations was too large for these equations to give a good approximation to the real orbit. Fixing this problem is the aim of the next improvement.

(b) Uniform accuracy. The second new feature is the adjustment of the time-step to maintain a uniform accuracy. Consider what happens on a very eccentric orbit, such as that of a comet. Its speed is slow when it is far from the Sun, but it begins to move much more rapidly as it gets closer in. As we saw in Investigation 1.3 on page 5, we can only trust our finite-difference methods for computer programs if important physical quantities do not change much during the time-step. If we have a fixed time-step Δt, we expect much greater accuracy in our calculation of the orbit far away than near the Sun, where the position changes much more during Δt. Similar remarks apply to the accuracy of the velocity calculation if the acceleration changes by a large amount: if this happens, then even Equation 4.7 will not give accurate results.

Since the predictor–corrector method already looks ahead at the predicted position at time $t + \Delta t$ of the planet, we can look at the acceleration there to see if it is very different from the value at time t. If the difference between the accelerations at the original and the predicted positions is more than some preset fraction of the original value of the acceleration, then we should take a smaller time-step. I have written the program so that it goes back to the time t and cuts the time-step Δt in half. The test for the change in the acceleration is then applied again. The halving goes on and on until a satisfactorily small time-step is reached. This ought to give uniform accuracy over the whole orbit.

However, the way I have implemented this idea is crude, because at the next time-step its size reverts to the original one given as part of the data for the calculation. Since halving the time-step takes computing time, the program may run slowly for a highly eccentric orbit. You might like to see if you could improve the halving routine to make the program Orbit run faster.

Exercise 4.2.1: *Assessing the accuracy of the improved method*

Repeat Exercise 4.1.4 on the preceding page with the improved program Orbit. In particular, does the error depend in a different manner on the *average number* of time-steps?

by taking a small time-step and setting the accuracy parameters in the computer program to be small, we can make ourselves very confident of the result.

What is the value of G?

Newton's law of gravity contains the constant of proportionality that we now call G, Newton's constant of gravitation. In fact, Newton did not actually know the value of G. Remember that the acceleration of the Moon is the force on it divided by its mass. If the Moon has mass m and the Earth mass M, then the force between them will be GmM/r^2, so the Moon's acceleration will be GM/r^2. Newton could deduce the product GM from observations of gravitational accelerations. (We can see how to do this in detail in Investigation 4.1 on page 29.) He could only deduce G if he could independently estimate M, the Earth's mass.

Newton actually tried to do this by assuming that the Earth's density was five times that of water, which he felt was a reasonable guess based on the density of rocks on the Earth's surface. Multiplying this density by the Earth's volume gave him a value for G of about $10^{-10}\,\mathrm{m^3\,s^{-2}\,kg^{-1}}$. This is not bad, considering the data available to him. The value of Newton's gravitational constant is now measured to be 0.667×10^{-10} in these units. We can turn his argument around and use it to deduce the mass of the Earth from our values for g and the modern value of G. If M is the Earth's mass and R its radius, then Newton's law of gravity says that the downward acceleration of an object at the surface of the Earth is

$$g = \frac{GM}{R^2},$$

which means that

$$M = \frac{gR^2}{G}.$$

Using $g = 9.8\,\mathrm{m\,s^{-2}}$ and $R = 6400$ km, we find $M = 6.0 \times 10^{24}$ kg. (We use essentially the same method to deduce the Earth's mass from the Moon's acceleration in Exercise 4.1.2 on page 29.)

This is all very well if we know G, but how is G measured? The only way to separate G from M in the gravitational acceleration is to measure the gravitational acceleration produced by a body of known mass. Early attempts at this used mountains as the "known" mass: the direction that a plumb bob hangs will be slightly affected by the gravitational pull of a nearby mountain, and this is measurable by comparing the directions of plumb bobs on either side of the mountain. This isn't very accurate, however, and the modern method is due to the Englishman Henry Cavendish (1731–1810) (later Lord Cavendish), who succeeded in 1798 in measuring the mutual gravitational attraction of two balls in the laboratory. The force is very small, and his experiment was a marvel of precision physics for its day.

> Even today, the measurements of G are accurate to only slightly better than one part in a thousand. By contrast, the product GM for the Sun is known to one part in one hundred million, from accurate tracking of interplanetary space probes' orbits. So the limit on the accuracy with which we know the mass of the Sun or any of the planets is the inaccuracy of G.

The reason for the relative imprecision of measurements of G directly is the weakness of the gravitational force between laboratory-sized objects, compounded by the difficulty of knowing exactly what the mass M of the laboratory mass is, and how it is distributed within the body. Real metal balls, for example, are never really

In this section: Newton introduced the proportionality constant G, but astronomical measurements could not determine its value. Only Earth-based experiments with objects of known mass could tell Newton the value of G.

Table 4.3. *More data on the planets. The eccentricity is defined in terms of the ratio s of the minor and major axes of the ellipse by $e = (1 - s^2)^{1/2}$. The **perihelion** distance is the closest a planet gets to the Sun. Its maximum speed occurs at perihelion, where it is traveling perpendicularly to the direction to the Sun. Note that at perihelion, Pluto is closer to the Sun than Neptune, but its greater speed there carries it on a more eccentric orbit that is mostly outside Neptune's orbit Pluto's peculiar orbit may have something to do with Neptune. Notice that its orbital period is exactly 3/2 that of Neptune.*

	Eccentricity of the orbit e	Perihelion distance (10^6 km)	Maximum speed (km s^{-1})
Mercury	0.206	46.0	59.22
Venus	0.007	107.5	35.34
Earth	0.017	147.1	30.27
Mars	0.093	206.8	26.22
Jupiter	0.049	740.3	13.52
Saturn	0.053	1349.0	10.15
Uranus	0.046	2735.0	7.105
Neptune	0.012	4432.0	5.506
Pluto	0.249	4423.0	6.17

uniform and **homogeneous** inside, and as we will see later in this chapter, any non-sphericity will affect the force they exert.

Kepler's laws

In this section: Kepler's laws for planetary motion follow from Newton's. We can prove that to ourselves using computer programs to follow the orbits of planets.

We have come across Kepler's third law. What about his first and second?

> *Kepler's second law* is the observation that planets follow orbits that are ellipses with the Sun at one focus.

We have already seen in Figure 4.3 on page 29 that Mercury's orbit is an ellipse, so it is not particularly surprising that this extends to all planets.

But *Kepler's first law* is something new. The statement of it is that the line from the Sun to the planet sweeps out equal areas in equal times.

> Put another way, Kepler's first law tells us that, if we look at any triangle whose corners are the Sun and two points on the orbit near to each other, then the area of the triangle divided by the time it takes the planet to go from one point to the other will be the same, no matter where the points on the orbit are.

I have calculated this ratio for all pairs of adjacent points on the orbit shown in Figure 4.3. The resulting values are graphed in Figure 4.4. The values are constant, verifying Kepler's first law.

The first law contains important information. Given that we know the orbit of the planet, the law of areas allows us to calculate where the planet is after any time. It is the law that determines the speed of the planet in its orbit. Physicists and astronomers today have a different name for this law. It is called the law of **conservation of angular momentum**. The quantity physicists call the **angular momentum** of the body is just twice the mass of the body times Kepler's area sweeping *rate* (area swept out per unit time). The constancy of Kepler's area rate implies that the angular momentum is constant (which, in this case, is what physicists mean by "conserved").

The Sun has a little orbit of its own

In this section: the masses of the planets move the Sun.

As we remarked above, the equality of action and reaction means that if the Sun exerts forces on the planets, then the planets exert forces on the Sun, and the Sun must move in response to them. This motion turns out to be fairly small, but not insignificant. Because Jupiter is so massive, it exerts the largest force on the Sun.

Just as Jupiter moves on (roughly) a circle, so will the Sun. These circles should have a common center on the line joining the two bodies. The Sun's acceleration will be smaller than Jupiter's by the ratio of their masses (since the forces on the two are the same), and it must go around its circle with the same period as Jupiter's orbit. We saw earlier that the acceleration of circular motion is proportional to the radius divided by the square of the period, so we can conclude that the radius of the Sun's circle is smaller than Jupiter's by the ratio of their masses. This ratio is 0.0009547, so the radius of the Sun's orbit is 743 100 km. For perspective, compare this to the Sun's own physical radius of 695 990 km.

> Therefore the Sun executes an orbit about a point just outside itself with a period of 11.86 years. On top of this are smaller motions due to the other planets, particularly Saturn.

Geostationary satellites

Communications satellites relaying telephone and television signals from place to place on the Earth have to stay above their receiving and transmitting stations all the time in order to be effective. This means they have to be in an orbit which has a 24 hour period, so that the Earth will turn at just the right rate to keep their stations under them. Let us find out how this can be arranged.

In this section: to arrange for an orbiting satellite to hover over the same location on the Earth, the satellite must be much further away than most.

We have seen in Chapter 3 that an orbit at an altitude of 300 km (a radius of 6700 km) has a period of about 90 min. By Kepler's third law, the square of the period is proportional to the cube of the radius of the orbit. Since we want a period that is roughly 16 times as long, we want the square of the period to be 256 times as large. The radius will depend on the cube root of this, which is 6.35: the radius needs to be 6.35 times as large as the radius of the 90-minute satellite. This is 6.6 times the radius of the Earth itself, or 42 500 km. This places such a satellite 36 100 km above the ground.

The gravitational attraction of spherical objects

There is one point in our deduction of the law of gravity that we have glossed over, but which caused Newton the most difficulty of all. The gravitational attraction exerted by a body falls off as the square of the distance to it, but if the body is not very small, what do we mean by the distance to it? When we calculated the Moon's acceleration, we assumed (with Newton) that the important distance was that to the Earth's *center*. The reason that Newton worried about this was that it wasn't just a matter of definition in his law, but rather a question of the self-consistency of his theory.

In this section: a key to keeping orbit calculations simple is that spherical bodies in Newton's theory of gravity exert the same force on distant objects as they would if all their mass were concentrated at a point. The size of the Sun, for example, does not need to be taken into account when finding the orbits of planets. We prove this property using a computer program.

To understand what this means, consider the Earth not as a single body, but as a composite made up of tiny particles distributed throughout its whole volume. Each of these particles exerts a gravitational attraction on the Moon that is directed towards the particle itself, not towards the center of the Earth. Thus, the center of the Earth must be some *average* center of attraction. This is something Newton felt he would have to be able to deduce from his theory, rather than simply postulating it.

When he finally solved the problem, he found that the force of gravity does indeed vary as $1/R^2$ outside an exactly spherical body, where R is the distance to its center, but that the gravitational attraction of bodies with other shapes was more complicated. Now, since the Earth and all the other bodies in the Solar System are roughly spherical, his law of gravity could be used without significant error. It is an illustration of Newton's intellectual thoroughness and honesty that he was nevertheless unwilling to publish his theory of gravity until he had solved this subtle and difficult point.

The importance of this result goes far beyond the fact that it makes orbital calculations easier. It means that if I am at a given distance r from the Earth, then the *size* of the Earth does not affect the gravitational force I feel, provided that in changing the radius R of the Earth I do not change its mass, and provided that R

stays smaller than r. So it I were to imagine the Earth shrinking for some reason, it would not affect me if I stayed at the given distance r outside it. On the other hand, if I attach myself to the shrinking Earth's surface, then the gravitational acceleration at its surface will increase as $1/R^2$.

> This difference between the behavior of gravity at different places is crucial to an understanding of *black holes*: when a star shrinks to form a black hole, gravity on its surface gets stronger and stronger, but at a fixed point outside the gravitational field does not change. We will return to black holes later on in this chapter.

We shall show that spherical bodies have the acceleration assumed above by the same method that Newton used, except that where he did his calculation using calculus, we shall do it on a computer.

Rather than deal with a whole sphere, it is sufficient to consider only a very thin spherical shell of matter, because the whole sphere can be built up out of such shells. Our shell will be subdivided into many tiny parts, each of them effectively a **point mass** at a different distance from the place where we want to compute the attraction. We shall show that by adding up all these separate attractions, we get a result which is proportional to the total mass of the shell and inversely proportional to the square of the distance to its center. The calculation is in Investigation 4.3.

The result of the computer calculation of Newton's result is shown in Table 4.4. In the first column is the number of zones into which I have divided the shell. Since each zone is treated as a point mass, we should expect that the calculation will get more accurate as the size of a zone shrinks, that is as the number of zones increases. The table bears this out, since the difference between the numerical result and Newton's gets smaller as the number of zones increases.

Other features of Table 4.4 are also worth noticing. Consider how the accuracy of the computer calculation with a fixed number of zones gets worse as the place where the force of gravity acts nearer the surface of the shell. This is an effect of the zoning: those zones nearest the point where we calculate the force make a big contribution to the force, since the force is proportional to the reciprocal of the distance squared. The fact that these nearby zones are treated as point particles when they really are not is more important if these zones are nearby. Nevertheless, when the number of zones is increased, the accuracy improves.

> Another feature that Table 4.4 reveals is that the force of gravity *inside* the hollow sphere is zero! Just outside the shell the force is large, but after we cross inside the shell it drops to zero.

Newton was also able to show this. It means that if we dig a deep hole into the Earth (which we idealize as spherical for this discussion), the force of gravity that we feel depends only on the mass of the part of the Earth that is *inside* the radius that we have reached. The material outside our radius exerts no net gravitational pull on us. If we reach the center of the Earth (hypothetically!), the force of gravity will vanish entirely.

Playing with the orbit program

In this section: we experiment with different kinds of orbits.

Having constructed the orbit calculator, we don't have to stop after just calculating a few planetary orbits. We shall use it below to calculate the Newtonian prediction of the deflection of light as it passes the Sun, which is responsible for the phenomenon of **gravitational lensing**. Another interesting question to ask is, what is the speed a planet needs in order to escape from the Sun, i.e. to get into an orbit that never comes back?

Investigation 4.3. *Spheres are just as attractive as point masses*

A sphere can be decomposed into thin concentric spherical shells. If we can show the result for a thin shell then it will be true for a sphere, since all the shells have the same center. Suppose we have a shell made of a material with density ρ and of thickness ϵ. (The Greek letter ρ (rho) is the usual symbol physicists use for density. Physicists and mathematicians also use ϵ (epsilon) to represent something that is taken to be very small.) Choose any point on it as a North pole and put down lines of **latitude** θ and **longitude** ϕ. We shall use these lines to form a grid on the sphere. We shall take any small section of the sphere thus marked out and idealize it as a point mass. By adding up the gravitational forces of these point masses we will get Newton's result.

Suppose the angles θ and ϕ are measured in degrees, some of which are shown in Figure 4.5 on the following page. Suppose further that the grid of lines of latitude and longitude are spaced apart by the small angles $\delta\theta$ and $\delta\phi$, respectively. If the radius of the Earth is R, then the circumference of the circle of constant latitude shown in Figure 4.5 is $2\pi R\cos\theta$. Any small segment of it of angular length $\delta\theta$ will have a length in proportion: $2\pi R\cos\theta(\delta\theta/360)$. A line of constant longitude ϕ is a great circle of circumference $2\pi R$, so a small segment of angular length $\delta\phi$ has length $2\pi R(\delta\phi/360)$. So any small region of the sphere enclosed by pairs of adjacent grid lines, as in Figure 4.6 on the next page, has an area equal to the product of these,

$$\text{area} = \left(\frac{2\pi R}{360}\right)^2 \cos\theta \, \delta\theta \, \delta\phi. \qquad (4.11)$$

This is really only an approximate answer for the area, because we have used the formula for the area of a rectangle in a plane, and the region is really part of a sphere. However, provided $\delta\theta$ and $\delta\phi$ are small enough, the error won't be large.

Now, since the shell has thickness ϵ, the volume of the tiny region we will approximate as a point mass is ϵ times the area, and so its

mass is ρ times this,

$$\text{mass} = \epsilon\rho \times \text{area}.$$

The computer program to do the calculation is called Sphere–Gravity, on the in website. It has no special tricks. After choosing both a location at which we want to calculate the acceleration of gravity and a radius for the shell, we just calculate the acceleration due to each piece of the shell and add them all up. For convenience we take the point at which we want to compute the acceleration to be at the origin of our x–y–z coordinate system, and we take the sphere to be centerd at the point a distance d from the origin on the x-axis. (See Figure 4.6 on the following page.)

If the North pole is also on the x-axis, a distance $d + R$ from the origin, then a point on the shell at latitude θ is a distance $R\cos\theta$ from the x-axis and a distance $d + R\sin\theta$ along the x-axis, so it is at a distance $r = (d^2 + R^2 + 2Rd\sin\theta)^{1/2}$ from the origin. (Note that we call θ here the latitude angle that was called β in Figure 4.5 on the next page.) The acceleration this small section of the shell produces in the x-direction is the mass of the small piece of the shell divided by r^2 times the cosine of the angle α in the diagram. This is just $(d + R\sin\theta)/r$. The computer program simply multiplies the mass of the piece by the cosine factor and divides by the square of the distance. It then adds up all these contributions from each of the little patches on the sphere. (Notice that I have left out the constant G. It is not important for this calculation: we want to show that the sum of the accelerations of the small pieces equals Newton's acceleration, and this will still be true if we divide both by G.)

Each piece of the shell also produces accelerations perpendicular to the x-axis, but these must sum to zero because of the symmetry of the problem. Since each direction perpendicular to x is equivalent to every other one, the net attraction could not point along any one perpendicular direction, so it points along none: its perpendicular component must be zero. So we need not bother to add these up.

Exercise 4.3.1: *Area of Colorado*

The American state of Colorado is a spherical rectangle of the kind we have just described. Its northern and southern boundaries have latitude $41°$ and $37°$, respectively. Its eastern and western boundaries have longitude $102°$ and approximately $109.1°$, respectively. Given that the radius of the Earth is 6.3782×10^6 m, what is the area of Colorado?

We shall show in Chapter 6 that the escape speed is $(2GM/R)^{1/2}$, where R is the radial distance from the Sun at which the planet starts out. This is just $\sqrt{2}$ times the circular orbital speed, in other words 41% larger. However, this does not depend on the initial direction the planet takes, as long as it doesn't actually crash into the Sun. You might like to try to use the orbit program to test this equation and its lack of directional dependence.

Here again the test cannot be perfect, not only due to the numerical errors but also because no one can allow the program to calculate forever in order to verify that

Number of patches	Ratio of radius of shell to distance to its center					
	0.01	0.1	0.5	0.9	1.5[a]	5.0[a]
400	0.10	0.11	0.23	6.26	0.20	0.002
2500	0.016	0.01	0.037	0.93	0.032	0.0003
40 000				0.052		

Table 4.4. Gravitational attraction of a spherical shell. Figures given are percentage deviation of the numerical result from the exact Newtonian result. In the first four columns, increasing the number of divisions of the shell clearly brings the numerical answer closer to the Newtonian one. In the last two columns the acceleration is calculated at a point inside the shell, where the Newtonian answer is zero.

[a] The figures given in the last two columns are the computed acceleration as a percentage of the acceleration that there would be at that point if the shell were small enough to be inside that radius.

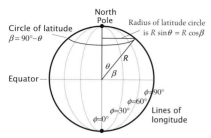

Figure 4.5. *A grid of lines of longitude and latitude form a coordinate system for a sphere, as used for the calculation in Investigation 4.3 on the previous page. A circle of constant latitude has a radius that depends on where it is. Circles of constant longitude all have the same radius, that of the sphere itself.*

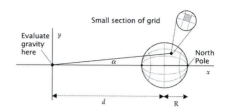

Figure 4.6. *Geometry for the calculation of the force of a spherical shell in Investigation 4.3 on the previous page. The force is calculated at a location a distance d from the center of the sphere, whose radius is R. Each small section of the sphere, like the one shown, makes a contribution.*

the planet doesn't really turn around at some large distance. Nevertheless, it should be possible to get a reasonably good demonstration with only a few calculations. Naturally, one can also calculate the escape speed from the Earth if one uses the mass of the Earth in the formula. For a space probe launched from the Earth's surface, this speed is 11 km s^{-1}.

The orbit program might tempt those with lots of computer time to try to evolve the whole Solar System for a long period of time, but this is not recommended! Inaccuracies build up after a few dozen or a few hundred orbits and render results over longer spans meaningless. Faster computers can run more accurate programs, but still the results cannot be trusted for more than a few tens of millions of years. This is one reason that, despite our knowledge of the "exact" laws of planetary motion, we have a very incomplete knowledge of what the early Solar System was like.

Black holes before 1800

We now have enough knowledge about gravity to take a look at one of the most remarkable speculations of eighteenth century physics: what we now call the black hole. In the late 1700s the British physicist John Michell (1724–1793) and the French mathematician and physicist Pierre Laplace (1749–1827), both of whom were well-acquainted with Newton's laws of motion and gravity and with the equivalence principle, independently put together two simple facts:

1. No object can escape from a body if its speed is less than $(2GM/R)^{1/2}$.

2. Light travels at a finite speed c. This had been proven by the Danish astronomer Olaf Roemer (1644–1710) in Newton's time, but the value of c was not well-known.

They then reasoned that light cannot escape from a body whose escape speed exceeds c. This inequality can be solved for the radius R of the body to give

$$R \leq R_g \quad , \qquad \text{where} \quad R_g = \frac{2GM}{c^2}. \tag{4.12}$$

So if it were possible to shrink a body of a fixed mass M down to a size smaller than R_g then it would appear black to the outside world.

Notice that the limiting radius R_g depends only on M and on the constants of nature c and G. Today we call this the *gravitational radius*

In this section: the equivalence principle and the realization that light has a finite speed led 18th century physicists to the possibility of "dark stars", stars that exert gravity but cannot emit light. These are the Newtonian versions of Einstein's black holes. Not until long after Einstein did scientists realize that Nature creates these objects abundantly.

▷Although Laplace is well-known to most modern physicists and mathematicians, Michell has sunk into obscurity. This is somewhat unfair, since in his day he was regarded as one of the premier scientists in Britain. It was he who suggested to his friend Cavendish the experiment to measure G that we referred to above.

▷Today we know c has the value 2.998×10^8 m s^{-1}, quoting only the first four figures.

of a body of mass M. The remarkable thing is that it gives exactly the radius of what general relativity calls a black hole, which is something from which no light can escape.

We shall look in more detail at the modern notion of a black hole in Chapter 21, but here it is worth saying that the Michell–Laplace black hole is not identical to its modern counterpart. In particular, Michell and Laplace envisioned light *starting off* with speed c as it leaves the body and gradually slowing down as it gets further away, eventually actually turning around and falling back in if the inequality was satisfied. So someone close to the body might still see a few "tired" beams of light before they turned around. Today, however, we believe that light always travels at speed c, and that if it leaves a body it cannot then turn around and fall back in. The modern idea of a black hole is that if a body has a radius smaller than its gravitational radius, then light never leaves it at all. No observer anywhere outside it would see any light from it at all.

Nevertheless, Michell's and Laplace's fundamental instincts here were sound. They had the courage to extend the equivalence principle to light, to say that light was affected by gravity just the same as anything else. The equivalence principle is fundamental to Einstein's general relativity, and underlies the modern black hole as well. Michell and Laplace could not have anticipated the development of relativity more than a century later, but within their own perspective they showed remarkable vision.

Light is deflected by the Sun's gravity

In this section: another "modern" phenomenon that was anticipated long ago by physicists working in Newtonian gravity is the fact that light rays, when passing close to the Sun, alter their direction. Einstein's theory was not new in making this prediction, but it predicts twice as large an effect as Newtonian theory. This was first verified in 1919.

Another consequence of the equivalence principle that nineteenth century physicists were perceptive enough to work out is that light will change direction as it passes the Sun. The reason is the same as the one underlying black holes: the effect of gravity on a particle depends only on the particle's speed, so if we set that speed to c then we will find out what happens to light itself.

We can do this by simply adapting our computer program `Orbit` that calculates Solar System orbits. Instead of using initial conditions appropriate to Mercury or another planet, we use initial conditions for a light ray coming from a distant star. The light will initially be traveling on a straight line at speed c. Its speed is much larger than the escape speed of the Sun, so that after passing the Sun, it is again moving on a straight line; but its direction will be different. Running the computer program `Orbit` with three different sets of initial conditions of this kind leads to the trajectories shown in Figure 4.7 on the following page.

To understand the diagram, we need to discuss the size of the expected deflection. Suppose, if the light were not affected by gravity, that the line would pass a minimum distance d from the center of the Sun. This is called the *impact parameter* of the light ray. In Investigation 4.4 on the following page we show that the deflection angle, measured in **radians**, is roughly $2GM/c^2 d$, where M is the mass of the Sun. A light ray just grazing the surface of the Sun would have an impact parameter approximately equal to the radius of the Sun, $d = 7 \times 10^8$ m, from which we would deduce a deflection of less than one second of arc (less than 10^{-6} radians). This would not be noticeable if plotted on a graph like Figure 4.7, so to do the figure I artificially shrunk the Sun to a point and allowed the light to have a very small impact parameter. I chose three values of d: 10 km, 15 km, and 20 km. The deflections measured from the graph are, respectively, $16.9°$, $11.3°$, and $8.4°$. It is easy to check that they agree with the prediction of the above formula, which we calculated only roughly in Investigation 4.4 on the next page.

This formula for the deflection was first derived by Cavendish in 1784 and inde-

---

Investigation 4.4. The Newtonian deflection of light

We shall derive the deflection by using the principle of equivalence, in much the same style as we derived the gravitational redshift in Investigation 2.2 on page 16. Consider light passing a star. Since light must travel on a straight line with respect to a local freely-falling observer, and since these observers all fall towards the center of the star, the light must continually bend its direction of travel in order to go on a straight line with respect to each observer it happens to pass. We can estimate the size of the effect, at least roughly, by the following argument within Newtonian gravity.

Let us consider just one freely-falling observer, who is at rest with respect to a star of mass M at the point where the light beam makes its closest approach to the star as it passes it by. Let this closest distance be d, the impact parameter. (This is not quite how we defined the impact parameter, but it is close enough for our approximate argument.) The observer's acceleration towards the star is $g = GM/d^2$.

Traveling at speed c, the beam of light will experience most of its deflection in a time of order d/c, the time it takes for the light to move significantly further away from the star. During this time, the observer has acquired a speed $v = gd/c = GM/cd$ perpendicular to the motion of the light. By the equivalence principle, the light must also have acquired roughly this same speed transverse to its original direction. Since its speed in the original direction is very little changed, we can calculate the angle of deflection by simple geometry: the tangent of the deflection angle is v/c. For small angles, the tangent of an angle is equal to the angle as measured in radians. This leads to the estimate that the angle of deflection will be

$\alpha = v/c = GM/c^2 d$ radians.

The total deflection should be double this, since the light will experience the same deflection coming in to the point of nearest approach as going out:

Newtonian prediction of the deflection angle in radians = $\dfrac{2GM}{c^2 d}$.
(4.13)

This turns out to be *exactly* the answer that a very careful calculation would give. But we do not need to do that calculation to check Equation 4.13. We only need to use the computer program Orbit with the right initial data and measure the results.

How would one measure this? If we look at the position of a star just once, as its light is passing near the Sun, we won't know how much deflection it is suffering, since we don't know its true position. The way to do it is to measure the position of a star when its light is passing nowhere near the Sun. Then, perhaps six months later, when the Sun is near the position of the star, measure the position again. It is clear from Figure 4.7 that the apparent position of the star moves *outwards*, away from the Sun. The only difficulty is in seeing the star when the Sun is near. But during an eclipse of the Sun, all the light from the Sun is blocked by the Moon, and so it is possible to see stars very close to the Sun's position. This is how the effect was eventually measured. The observed result is twice the number given by Equation 4.13, consistent with general relativity.

Exercise 4.4.1: *Light deflection by other bodies*

Any gravitating body will deflect light. Estimate, using Equation 4.13 above, the amount of deflection experienced by a light ray just grazing the surface of the following bodies: (a) Jupiter, whose radius is 7.1×10^4 km; (b) the Earth; (c) a black hole of any mass; and (d) you.

---

pendently by the German astronomer Johann G von Soldner (1776–1833) in 1801. They regarded it as a mere curiosity, since measuring the apparent positions of stars to an accuracy of an arcsecond or so was impossible in their time. Einstein himself independently re-derived the formula using the equivalence principle in 1909, and he pointed out that the expected deflection might well be observable with the telescopes of his day.

Figure 4.7. The deflection of light by a point-like mass as calculated in Newtonian gravity. The mass has the mass of the Sun, but to show the effect, it has been made almost as compact as a black hole. This allows trajectories to experience strong gravity and exhibit large deflections.

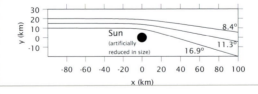

However, before a suitable opportunity arose for observing the effect, Einstein moved on to devise the theory of general relativity (1915). In this theory, there is an extra effect that causes light to deflect twice as much, so that the new prediction would be $4GM/c^2 d$. (We shall calculate this effect in Chapter 18.) A deflection of this size was indeed measured in an eclipse expedi-

▷Eddington was one of the first true *astrophysicists*, a scientist who used the theories of physics to understand the nature of astronomical objects. He was an early champion of Einstein's general relativity. Dyson was the British Astronomer Royal from 1910 to 1933.

tion in 1919 led by the British astronomers Sir Arthur Eddington (1882–1944) and Frank W Dyson (1868–1939). The accuracy was enough to distinguish between the old Newtonian deflection and the new general relativistic one. The verification of this prediction of general relativity did more than anything else to make Einstein a celebrity, a household name.

Tides and tidal forces:
the real signature of gravity

The tides wash the margins of all the great oceans, regulate the lives of sea urchins and fishermen, power the great **bore waves** on rivers like the St. John, the Amazon, and the Severn. For most of us the tides are romantic, primeval, poetic. Standing on an ocean beach, we might be impressed by this tangible manifestation of the gravity of the distant rock we call the Moon, but few of us would be led to reflect on how fundamental the tides are to an understanding of gravity itself. But *fundamental* is the right word. In the modern view, the *real* signature of gravity, the part of gravity that can't be removed by going into free fall, is the *tidal force*, whose most spectacular effect on Earth is to raise the ocean tides. In this chapter we will examine this aspect of gravity, starting with the simplest effects first and working our way up to ocean tides and then to tides elsewhere in the Solar System and beyond. We will return again and again in later chapters to the fundamental role of tides. Indeed, many astronomical systems transmit tidal forces as signals right across the Universe, signals that we call **gravitational waves**.

Tidal forces in free fall

When we formulated the modern version of the equivalence principle in Chapter 2, we talked about experiments performed in free fall. The simplest such experiment is just to carry a stone with us in free fall and then release it at rest. The principle of equivalence says that nothing happens: it just stays alongside us as we fall. A little thought will convince us that this is only true if the body is very near to us, and if we limit the duration of the experiment.

Consider two stones falling freely towards the Earth, one just above the other. The lower stone, being closer to the Earth, experiences a slightly larger acceleration of gravity than the higher stone, since gravity gets weaker at larger distances. This means that even if the two stones start out falling with the same speed, the lower one gradually acquires a slightly larger speed than the higher one, and the distance between them increases, as in Figure 5.1 on the following page.

This is due entirely to the fact that the Earth's gravitational field is *non-uniform*: it pulls with different accelerations in different places. If the acceleration of gravity were strictly the same everywhere, then the two stones would stay at the same separation forever. The non-uniformity (we sometimes say **inhomogeneity**) of the Earth's gravitational field has the effect of pushing the stones apart if they are placed one above the other.

Next consider two stones falling side by side. If they start from rest, they will both fall on radial lines directly toward the center of the Earth. This means they will not quite keep their initial sideways separation: they will approach each other as they approach the Earth. Again this is an effect of the non-uniformity of the Earth's gravitational field, which pulls in different directions at different locations.

To an experimenter falling freely with the stones, the overall acceleration of gravity disappears (the equivalence principle), but these residual

In this chapter: we study tidal gravitational forces. These are the forces that are not removed in free fall, because they come from non-uniformities in the gravitational acceleration. Their effects are visible all over the Universe, from the ocean tides on the Earth to the disruption of whole galaxies when they get too near to one another. The precise calculation of the tidal effects on Mercury's orbit left a tiny part of Mercury's motion unexplained by Newtonian gravity, its first failure. Einstein's general relativity explained the discrepancy.

In this section: tides arise from non-uniformities in gravitational accelerations. They are the part of the gravitational field that cannot be eliminated by going into free fall.

▷The figure on this page shows a volcanic eruption on Io, imaged by the Voyager 2 spacecraft in July 1979. Such eruptions are frequent, and are the result of the heating of the moon as it is deformed by changing tidal forces (see later in this chapter). Image courtesy of NASA/JPL/Caltech.

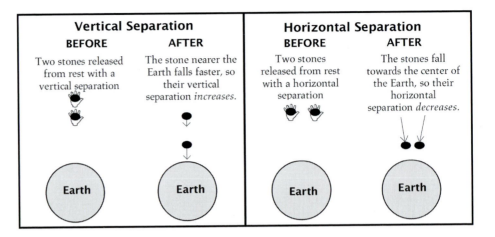

Figure 5.1. Tidal effects are most easily seen in the motion of freely-falling objects.

tidal effects of gravity remain: stones placed one above the other are pushed apart, while stones placed side by side are pulled together.

Notice that the tidal effect is proportional to the distance separating the stones, at least for relatively small separations. Horizontally separated stones accelerate toward each other in proportion to their separation: the larger their distance, the greater their **tidal acceleration**. The same is true for the vertically separated stones. This aspect of the tidal force suggests that it could be a significant force on scales of the diameter of the Earth, even though we don't notice it in local experiments. We shall also find that the increase of tidal forces with separation will be important to our discussion of the detection of gravitational waves, in Chapter 22.

The discussion of tidal forces we have just given would certainly have been acceptable, even obvious, to a nineteenth century physicist. But he might not have been prepared to place the tidal forces on a pedestal and call them the "real" gravitational force, the way we do today. It was principally Einstein who stressed the fundamental importance of the equivalence principle, who made uniform gravitational fields seem trivial, and who gave the tidal forces a special mathematical place in his theory of gravity, general relativity. We shall adopt this modern perspective here, and pay due respect to tidal forces by devoting this chapter to them.

But what about the equivalence principle? Are tidal forces its downfall? No, but only if we keep it *local*: given an experimenter in free fall who can measure things (such as distances, speeds, etc.) only to a certain accuracy, then there will be a certain region of space around him and a maximum duration of time for experiments in which he will not be able to detect the effects of the tidal forces. In this region the equivalence principle will be valid. Since no measurement is perfectly accurate, there is a real sense in which the equivalence principle applies in a small region but not everywhere.

Physicists use the words **local** for things that apply in small regions and **global** for things that apply everywhere. We therefore say that the equivalence principle is valid locally but not globally.

Ocean tides

In this section: how tides work: the way the Moon raises tides in the Earth's oceans.

Let us now see how tidal forces actually raise the tides. The Moon exerts a gravitational force on the Earth, equal and opposite to that which the Earth exerts on it. In response to this force, the Earth executes a small circular motion about its average orbital motion as it circles the Sun, just as the Sun orbits a point near it because of

Jupiter's force on it. (Recall the discussion of this motion in Chapter 4.) But this small circular motion is not the whole story, because the force which any piece of the Earth feels from the Moon's gravity depends on how close it is to the Moon.

The Earth is a large body, so the acceleration of the Moon's gravity on the side of it nearest the Moon can be substantially larger than on its far side. If the Earth were made of tissue paper, this difference in acceleration would probably tear it apart. But the Earth is tougher than that: its internal forces (both the mechanical forces that make rocks rigid and its own gravitational force on things on its surface, like oceans and people) are more than strong enough to resist this difference in acceleration.

Because of these internal forces, the parts of the Earth nearest the Moon cannot fall freely in the Moon's gravitational field, so they do not accelerate with the full acceleration of the Moon's gravity there. Instead, they stay attached to the Earth and accelerate at the same rate as the rest of the Earth. Similarly, parts of the Earth furthest from the Moon would, if they were free of the Earth's forces, fall toward the Moon with a smaller acceleration than the rest of the Earth, but they are not free to do so: they stay attached to the Earth, too. The net effect is that the Earth accelerates toward the Moon with the *average* of the acceleration of the Moon's gravity across it. Only the *center* of the Earth is truly freely-falling. This is illustrated in Figure 5.2.

Now consider an experimenter sitting at the center of the Earth. From his freely-falling point of view, the vertical stretching action of the tidal force of the Moon's gravity will try to pull the side of the Earth nearest the Moon away from the center. Even though the Earth's internal forces hold it together against this pull, the Earth is not perfectly rigid, and it will bulge slightly toward the Moon. The tidal force has an even more drastic effect on the oceans, which are not rigidly connected to the surface. An ocean whose center is on the side facing the Moon will be raised in elevation by this force, causing it to pull away from its shores, giving a low tide. An ocean whose edge is on the side nearest the Moon will find its water drawn towards this edge, giving a high tide.

But the part of the tidal force that might be more unexpected is that this *same* thing happens on the side of the Earth furthest from the Moon as well. Again as seen by the freely-falling experimenter at the center of the Earth, the vertically stretching tidal force of the Moon pushes the far side of the Earth *away*. What is really happening here of course is that the more weakly accelerated far side of the Earth is being left behind as the Earth accelerates toward the Moon: it is pulled toward the Moon, but not as strongly as the Earth as a whole, and so, *relative to the center of the Earth*, it is pushed away. An ocean centered on the far side will bulge out, too, and its shores will experience a low tide as well.

In one day any given ocean will experience *two* low tides, once because of its nearest approach to the Moon and the other because of its farthest recession away from it. Tidal effects have this characteristic behavior under rotations: places where the tidal effect is similar are separated by a rotation of only 180°.

We will encounter this symmetry again when we discuss gravitational waves in

Acceleration of the Moon's gravity on Earth.
Length of arrow indicates size of acceleration.

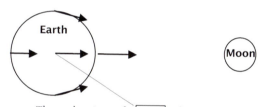

The acceleration at the center is the mean acceleration with which the solid Earth will fall. The acceleration of gravity due to the Moon is larger near the Moon and smaller further away.

Residual acceleration of the Moon's gravity,
after subtracting the mean acceleration of the Earth.

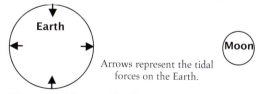

Arrows represent the tidal forces on the Earth.

Figure 5.2. *Tides raised by the Moon on the Earth arise from the residual acceleration of gravity left when the solid Earth falls at the average of the acceleration of the Moon's gravity across the Earth.*

Chapter 22.

Actually, the first high tides of consecutive days do not occur at exactly the same time of day, because the Moon has moved along in its orbit during the intervening day. Since the Moon orbits the Earth in the same direction as the Earth turns, any place on Earth has to wait somewhat more than 24 hours before it is again closest to the Moon. The Moon takes 27.3 days to orbit the Earth, so in one day it moves a fraction 1/27.3 of an orbit. Since the Earth takes 24 hours to turn full circle, it takes a fraction 24/27.3 of an hour to turn through the same angle as the Moon goes through in a day. This amount of time, about 53 min, is the amount by which a high tide is delayed past the time it arrived on the previous day. Similarly, the time between successive high (or low) tides is half of the full day-to-day period, about 12 hours 26 min.

Tides from the Sun

In this section: the Sun raises tides almost as high as the Moon does.

The Moon is not the only body strong enough to raise tides on the Earth. The Sun, though much further away, is also much more massive, and its tidal forces on the Earth are very similar in size to those of the Moon. The other planets have a negligible effect.

> The fact that the Sun and Moon happen to exert similar tidal forces on the Earth is deeply related to another "accidental" fact that might at first seem to be completely unconnected, namely that **eclipses** occur, i.e. that the Moon and the Sun are of similar angular size on the sky.

The reason for the relation between these two facts is explored in Investigation 5.1.

Spring and neap tides

In this section: the action of the Sun and Moon together is responsible for the seasonal variations in tides, from spring to neap and back again.

Investigation 5.1 shows that the Sun has a tidal effect on the Earth that is about 42% of that of the Moon. Thus, when these two effects reinforce each other, the tidal forces are 1.42 times the Moon's alone, while when they work against each other they are only 0.56 times the Moon's. Thus, the ratio of the maximum to the minimum tidal force is about 2.5.

Figure 5.3. Orbital alignments that give spring tides and neap tides. When the Moon is at the location of the darkly shaded circles, the tidal forces on the Earth are at their maximum. When the Moon is at the location of the lightly shaded circles, the tidal forces are at their minimum.

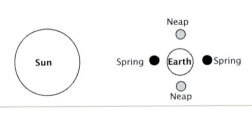

When do the two forces add? Since the tidal effects are the same on opposite sides of the Earth, the tidal forces of the Sun and the Moon reinforce each other if the three bodies all lie on a straight line, either with the Moon between the Earth and the Sun or on the other side of the Earth, as in Figure 5.3. This tide is called a "spring" tide. Similarly, the minimum tidal effects occur when the three bodies form a right triangle. This is a "neap" tide. (The word "neap" comes from Old English, where it meant helpless or weak.)

> Therefore, there are two spring tides per month and two neap tides, and the spring tides are associated with the full and new Moons.

> For the same reason that the interval between successive high tides in a day is slightly more than 12 hours, so too the interval between successive spring tides is slightly more than half of the Moon's orbital period, which is 13.7 days. Since the line joining the Sun and the Earth has rotated in this time because of the orbital motion of the Earth, the actual time it takes for the Moon to rejoin this line is a little more than a day longer, 14.8 days.

Investigation 5.1. Tides and eclipses

We want to calculate the relative strength of the Sun's and the Moon's tidal forces on the Earth, and to do this we have to discover how the tidal force depends on the distance r of the Earth from the body producing the tides. The overall gravitational force falls off as $1/r^2$, but the tidal force, which is the difference between the forces at two nearby points, falls off faster, as $1/r^3$.

To see this, we introduce an important algebraic expression, called the binomial theorem. This gives the value of $a + b$ raised to any power n, where a and b are any two numbers:

$$(a + b)^n = a^n + nba^{n-1} + \tfrac{1}{2}n(n-1)b^2 a^{n-2} + \cdots, \quad (5.1)$$

where I have only given the first three terms. If n is an integer, only the first $n+1$ terms are non-zero. If $n = 1$ the third term is zero, and we have the simple identity $(a + b)^1 = a + b$. If $n = 2$ then we have the quadratic formula $(a + b)^2 = a^2 + 2ab + b^2$, and the terms represented by the $\cdots$ in Equation 5.1 all vanish. If $n \geq 3$, Equation 5.1 only gives the first three terms. However, there is an important special case where the extra terms left out don't affect the answer very much. Consider the form of Equation 5.1 when $a = 1$:

$$(1 + b)^n = 1 + nb + \tfrac{1}{2}n(n-1)b^2 + \cdots.$$

If in addition b is very small compared to 1, then b^2 is even smaller, and higher powers of b get smaller and smaller, so that even if they were considered in this equation they would not make much contribution. We have arrived at a very useful conclusion: letting ϵ denote the very small number, we have that if ϵ is sufficiently small, then

$$(1 + \epsilon)^n \approx 1 + n\epsilon. \quad (5.2)$$

(The symbol "$\approx$" stands for "is approximately equal to".)

Consider two points a distance r and $r + h$ from the gravitating body, on the same radial line. The acceleration at distance r is k/r^2, where k is a constant. The acceleration at the distance $r + h$ is

$$\frac{k}{(r + h)^2} = \frac{k}{r^2(1 + h/r)^2},$$

where I have factored r out of the term in the denominator. By the binomial approximation, this is approximately given by

$$\frac{k}{r^2}(1 + h/r)^{-2}. \quad (5.3)$$

We can now use Equation 5.2 to evaluate the factor containing h/r in this expression. Since h is small compared to r (for the Moon and

Earth we have seen that h/r is about 1/60), terms containing $(h/r)^2$ (or higher powers of h/r) are small compared to the term involving h/r itself, and we will neglect them. The result is that

$$(1 + h/r)^{-2} = 1 - 2h/r + \cdots,$$

so expression (5.3) becomes

$$k/r^2 - 2kh/r^3 + \cdots. \quad (5.4)$$

This is the acceleration at $r + h$. The tidal acceleration is the difference between this and the acceleration at r itself, which just means subtracting off the first term of Equation 5.4. The result is

$$\text{tidal acceleration} = -2kh/r^3. \quad (5.5)$$

This establishes the $1/r^3$ fall-off of the tidal effects.

Let us now get rid of the constant k in the above expressions and replace it by what we know it to be, $-GM$, where M is the mass of the body producing the gravity. This tells us a crucial fact, that the tidal force is proportional to M/r^3. Now, the average density ρ of a body is its mass divided by its volume. Since the volume of a sphere is proportional to the cube of its radius R, its density is proportional to M/R^3. Turning this around, we find that its mass is proportional to ρR^3.

This in turn means that the tidal forces it produces are proportional to $\rho(R/r)^3$. Now, the ratio $2R/r$ is the *angular diameter* of the sphere on the sky, measured in radians. Put another way, the number of degrees of arc that a sphere spans on the sky is $360°$ times the fraction of a full circle that its diameter occupies, which is proportional to R/r. *Therefore the tidal acceleration is proportional just to* $\rho \times$ (angular diameter of the body)^3, with a constant of proportionality that depends only on pure numbers (like π), Newton's constant G, and of course the difference in the positions of the two points whose tidal effects are being examined.

Since the Moon and the Sun have almost the same angular size as seen from the Earth (which is why eclipses are so spectacular), the tidal accelerations they produce across the diameter of the Earth are proportional to their densities. The Moon, made of rocks, has an average density of 3300 kg m^{-3}, which is about 2.4 times as dense as the Sun. Therefore the Moon's tidal effects on the Earth are 2.4 times as large as the Sun's. This makes the Sun less important than the Moon, but not of negligible influence. Other planets have similar densities to the Sun and Moon, but very much smaller angular diameters, so they do not exert significant tidal effects on the Earth.

Exercise 5.1.1: *Testing the binomial approximation*

Use a pocket calculator to verify that Equation 5.2 gives a good approximation for small ϵ. For the following values of ϵ and n, evaluate the approximate value $1 + n\epsilon$, the exact value $(1 + \epsilon)^n$, the error (their difference) and the *relative error* of the approximation, which is defined as the error divided by the exact value: (a) $n = 2$, $\epsilon = 0.01, 0.1, 1.0$; (b) $n = 3.5$, $\epsilon = 0.01, 0.1$; and (c) $n = -2$, $\epsilon = 0.01, 0.1$. (Recall that negative powers indicate the reciprocal, so that $(1 + \epsilon)^{-2} = 1/(1 + \epsilon)^2$.)

In the above discussion of how the two tidal forces add up, we have made the unspoken assumption that the orbit of the Moon around the Earth is in the same plane as the orbit of the Earth around the Sun, so that the three planets can actually form a straight line when they are in their best alignment. This is very nearly the case, but not quite. The Moon's orbit is inclined at an angle of 6° to the Earth's orbital plane, tilted in a direction that rotates with time (a period of roughly 20 years) because of the Sun's gravity. This means that twice a year the best alignment of the three bodies is as much as 6° away from a straight line, while three months later, when the line they form is nearly parallel to the intersection of the two planes, their alignment can be nearly perfect. This gives a small seasonal variation in the strength of the spring and neap tides.

What the tidal forces do to the oceans, the Earth, and the Moon

Once Newton understood the universal nature of gravity and its variation with distance, he realized that he could explain the tides in the manner which we have just described. This was a significant piece of experimental support for his theory of gravity. But if we try to go further than this and actually predict the time of arrival of a tide at a particular place, the tidal range (difference in the height at high and low tides), and the variations of these with the day of the month, we find that the problem is hopelessly complex.

The reason is not hard to understand. Although the tidal *forces* are easy enough to describe, the response of the oceans to them depends on a large number of variables: the depth of the ocean in various places, the shape of the coastline on which the tides are measured, the density of the ocean (which depends on how much salt it contains), and even such day-to-day irregular conditions as the local atmospheric pressure and the strength of the winds. Even in the present age of modern computers, the prediction of tides along complicated coastlines is largely a matter of judgment based on their behavior in the past.

In certain places, such as the Bay of Fundy in Canada or the Bristol Channel in Britain, where the tidal surge of a large body of water is funneled into a narrow end at just the right distance from the opening, the tidal range can exceed 20 m and the pressure of this water can force a spectacular cresting wave that travels upstream in rivers that empty into this end. This wave, called a tidal bore (Figure 5.4), is very sensitive to changes in the tidal response of the body of water, so that even fluctuations in atmospheric pressure due to storms at sea can have a marked effect on the wave. The largest in the world is at Ch'ient'ang'kian (Hang-chou-fe) in China, where the wave can exceed 7 m in height.

Figure 5.4. The bore traveling upstream (to the left) along the River Severn in England in September 1976. The wave is breaking gently. Note the difference in water level before and after the wave. Enthusiasts often surf this wave. Photo copyright by the author.

The ocean tides are the most obvious manifestation of tidal forces, but what is the effect of the tides on the rocky body of the Earth and the Moon? Tides raised by the Earth have dramatically affected the Moon, causing it to show the same face to the Earth at all times. It is interesting to see how this happened.

At one time the Moon was spinning much faster on its axis than it is now. The tides raised by the Earth would make the rocks of the Moon bulge toward and away from the Earth, but the rotation of the Moon tended to carry this bulge around with it. Because friction slows down the response of any system to the forces on it, friction in the Moon caused the bulge in any group of rocks to lag behind the tidal force driving it. This meant that the bulge actually reached its maximum in the rocks shortly after they had rotated *past* the point of closest distance (and, on the far side, the point of furthest distance) to the Earth. The Moon therefore presented a bulge to the Earth that was not quite pointing towards the Earth. This bulge is illustrated in Figure 5.5, where its size and lag angle have been exaggerated for clarity.

The tidal force of the Earth naturally tried to align this bulge with the Earth–Moon line, which in this case meant pulling on the bulge *against* the direction of rotation (see Figure 5.5). The result was that the tidal force of the Earth slowed down the rotation of the Moon. As the rotation got slower and any group of rocks took more time to pass through the region nearest the Earth, the effect of the tidal

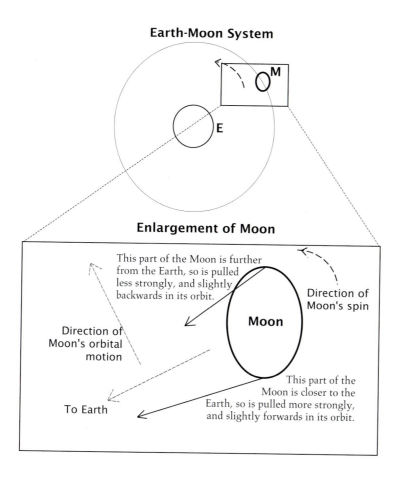

Earth-Moon System

Enlargement of Moon

This part of the Moon is further from the Earth, so is pulled less strongly, and slightly backwards in its orbit.

Direction of Moon's spin

Moon

Direction of Moon's orbital motion

To Earth

This part of the Moon is closer to the Earth, so is pulled more strongly, and slightly forwards in its orbit.

Figure 5.5. How tidal forces led the Moon to show the same face to the Earth at all times. The Moon's bulge lagged behind the tidal forces when it was spinning rapidly, leading to an elliptical shape that did not point towards the Earth. All rotations in this diagram are counterclockwise: the spin of the Moon and its orbital motion. The diagram illustrates the Earth's gravitational forces on the Moon at two places, on the near-side bulge and on the far-side bulge. These forces both point towards the center of the Earth, so they are not parallel to each other. The force on the near side of the Moon tends to twist the Moon against its rotation, while the force on the far side tends to increase the rotation. But the force on the near side is stronger, because it is nearer the Earth, so the net effect is to slow the Moon's rotation down. At the same time, the near-side force also has a small component pushing the Moon forward in its orbit, while the far-side force does the opposite. Again, the near-side force wins, and the net push is along the orbital motion. This forces the Moon further from the Earth.

forces got larger: with more time to act on any region of the Moon, the tidal forces were able to raise larger tides in the rocks, and this accelerated the slowing down of the Moon. Eventually, the Moon came to present the same face to the Earth at all times, so the bulge was able to align exactly with the tidal force, and now the Moon no longer loses rotation. We say that the Moon is now in **synchronous rotation** with its orbit, since it rotates on its axis once each orbit.

The Moon's tidal effect on the Earth similarly tends to decrease the rotation rate of the Earth, but the Moon's effect is much weaker on the more massive Earth, so it has not yet brought the planet into synchronism with its orbit. A billion years ago, the day was only about 18 hours long.

The loss of spin by the Moon and the Earth has had another effect on the two: it has driven them further apart. The same tidal forces that have aligned the Moon's figure with the direction towards the Earth have also tended to give the center of the Moon a slight push in the same direction as it is orbiting the Earth, with the result that it has been flung away from the Earth. (You can see this if you study Figure 5.5 carefully.) The radius of the Moon's orbit has grown, and its orbital period increased.

In addition to the effects of the Earth on the Moon's orbit, there are small effects due to the Sun and the other planets. I have mentioned above how the plane of the Moon's orbit rotates because of the Sun. So too does the location of the perigee of its orbit, from the same cause. These effects are similar to those produced by Jupiter in the orbits of the other planets, which we shall discuss below.

▷The increase in the radius of the Moon's orbit can also be understood as a consequence of the conservation of angular momentum, which we met in Chapter 4. The loss of spin by the two bodies increases the angular momentum of the orbit, forcing them apart.

▷The *perigee* is the point of closest approach to the Earth. We saw in Chapter 4 that the nearest approach of a planet to the Sun is its perihelion, the suffix "-helion" referring to the Sun. The Moon or an Earth satellite has a perigee, "-gee" being a modification of "geo", referring to the Earth.

In this section: tidal phenomena can be seen everywhere in astronomy. Mercury is tidally locked to its orbit, Jupiter's moon Io is heated so much by tides that it has volcanos, the asteroids are remnants of a failed attempt to form a planet too close to the tidal influence of Jupiter, and galaxies – whole systems of stars – crash together and disrupt one another tidally.

Tides elsewhere in astronomy

Whenever two extended bodies are sufficiently near to one another, one can expect tidal forces to operate. Mercury is so close to the Sun that the tidal force of the Sun across it is nearly three times as large as those the Earth experiences from the Moon. Mercury orbits the Sun once every 88 days, and it turns on its axis once every 58.6 days, exactly 2/3 of its orbital period. This ratio isn't an accident: it is due to the tidal effects of the Sun.

Let us look at this another way, from the point of view of someone standing on Mercury. In two orbits Mercury spins three times. But this spin is with respect to the distant stars: during this time the person on Mercury has seen the stars go around three times. On the other hand, the Sun has also moved through Mercury's sky. Since Mercury has made two orbits of the Sun, the Sun has appeared to move twice through the sphere of stars, as seen from the ground, but in the opposite direction. So it has actually gone through Mercury's sky only once.

Mercury has a "tidally locked" day that lasts twice as long as its year.

(If it always presented the same face to the Sun, as the Moon does to the Earth, then its day would be infinitely long.) Calculations show that this arrangement can be stable if Mercury is not perfectly symmetrical about its rotation axis: Mercury's spin is probably not slowing down any more.

A spectacular example of the effects of tides is Io, the nearest to Jupiter of the four moons discovered by Galileo with his first telescope. When the Voyager 1 spacecraft flew by Jupiter in 1979 it observed no less than *eight* volcanic eruptions (see the image on page 39). These eruptions and the smooth, craterless surface of Io suggest that this volcanism has been going on at a steady rate for a very long time.

▷ The heat that volcanism on Io requires is far in excess of what can be being liberated by natural radioactivity in Io's interior. Radioactivity is thought to be the ultimate source of the heat that drives the Earth's volcanic and tectonic activity. All Earth rocks contain trace amounts of radioactivity, but when added up over the volume of the Earth's interior, the source of heat is enough to keep the interior molten. Volcanos burst out when hot molten rock manages to puncture the crust of the Earth at a weak point. Radioactivity does not force the molten rock out; it simply provides the heat that keeps it liquid.

Io is heated by friction caused by the tidal forces of Jupiter on its moon. These are 250 times as strong as the Earth's forces on the Moon, chiefly because Jupiter is 300 times as massive as the Earth. Io is tidally locked to Jupiter, presenting the same face to it all the time. But it rocks back and forth about this position because of the tidal effects of Jupiter's other large moons, Europa and Ganymede. These two moons have orbital periods that are tidally locked to Io's: mutual gravitational forces between the moons have arranged that Io's period is half of Europa's and Europa's is half of Ganymede's. The regular tidal "bumping" of Io by these moons has built up a significant wobble, and the distortion of Io's tidal bulge during its wobble generates heat through friction inside Io. The distortion is not a small effect: parts of Io's surface can go up and down by as much as 100 m. It is not surprising that such large motions can lead to volcanism on this moon.

The asteroid belt, a system of large planetesimals orbiting the Sun between Mars and Jupiter, looks like the remnants of the formation of a planet that was stopped prematurely. The most likely cause is Jupiter's and the Sun's tidal forces: the weak binding forces holding together a pair of planetesimals was no match for the tidal forces. We shall discuss how this happens in more detail in Chapter 13.

Outside the Solar System tidal effects are also common. Most stars seem to be in binary systems, in which two or more stars orbit one another. Sometimes they can be quite close, closer than the Earth is to the Sun, or even in contact, so that their outer surfaces actually touch. In such situations, the shapes of the stars can become very distorted, and gas may even flow from one star to the other, sometimes with spectacular results. We will discuss these sorts of stars in Chapter 13.

Galaxies, too, can exhibit tidal effects. A galaxy is a collection of anywhere from 10^9 to 10^{12} stars, bound together by their mutual gravitational attractions. When two such galaxies get too close, the tidal forces of one can strip stars away from the

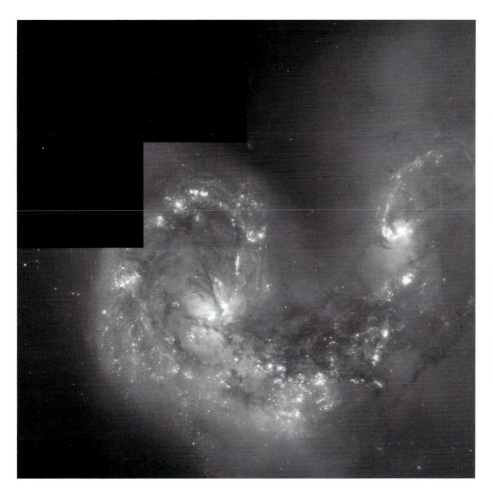

Figure 5.6. *A photo-mosaic the Antennae galaxies, taken by the Hubble Space Telescope (HST). These are two galaxies which are undergoing a collision. The non-uniformity of the gravitational accelerations of each galaxy on the other disrupt the normal orbital motion of the stars. The galaxies will eventually merge into a galaxy with a much smoother appearance. This is a snapshot of a collision that will take hundreds of millions of years to complete. The Milky Way and the great galaxy in Andromeda, our Galaxy's closest large neighbor, are similarly falling towards one another, and may collide like this in a billion years or so. (Courtesy NASA and its National Space Science Data Center (NSSDC).)*

other, producing chaotic streams of stars. Pairs or groups of such interacting galaxies are a common sight in photographs taken by the biggest telescopes (Figure 5.6).

> Our own galaxy, the Milky Way, may well now be showing the after-effects of such a tidal encounter. The Magellanic Clouds, bright patches of stars well away from the Milky Way in the sky visible from the Southern Hemisphere, are now known to be the brightest parts of a whole stream of stars extending right down to the Milky Way. The origin of this Magellanic Stream is still a matter of debate among astro-physicists, but one idea is that it may have been torn from the Milky Way by the tidal forces of another galaxy, or vice versa.

In fact, astronomers have discovered a region in the Milky Way, the other side of the center of the Galaxy from the Sun's location, where there is a large group of stars all traveling together with a different speed from most other stars. These stars may be the remnants of a small galaxy that is currently being torn apart and swallowed by the Milky Way. This may have happened many times in the history of the Milky Way. When such clumps get absorbed by the Milky Way, they go into orbits that retain a "memory" of how they fell in; they do not randomize their motions rapidly, because individual stars do not often come close enough together to deflect each other from their orbits. European astronomers are preparing a space mission called GAIA that could measure the speeds of stars all over the Milky Way

▷ The GAIA mission is one of the most ambitious space missions yet designed. For its wide range of scientific goals, see its website, http://www.estec.esa.nl /spdwww/future/html/gaia.htm.

so accurately that it would be able to identify such fossil "streams" of stars in the Milky Way, and thereby open a window into the past history of our own Galaxy.

Jupiter gives Mercury's story another twist

In this section: the effect of the tidal forces of Jupiter on Mercury's orbit is to push the ellipse around the Sun. This is called the precession of the orbit.

Nineteenth century astronomical observations of the motions of the planets became so precise that it was easy to see that the orbits of planets were not the perfect ellipses that one would expect if the Sun were the only gravitating body: the effects of the gravitational pull of the other planets caused slight but measurable deviations from ellipses. Many of these effects can be thought of as tidal effects.

Consider, for example, the effect that Jupiter has on Mercury's orbit about the Sun. Mercury will go around the Sun several times while Jupiter changes its position only slightly. We have seen in Chapter 4 that the Sun executes a small orbit because of Jupiter's gravitational pull on it. Mercury will follow the Sun as it does this, keeping the Sun at the focus of its elliptical orbit. So Mercury's orbit will not remain a perfect ellipse relative to the stars: it is best described as an ellipse that changes its location gradually, as the Sun moves. The direction of the major axis of the ellipse does not change during this motion: the ellipse keeps its orientation.

All this is because the Mercury–Sun system falls freely in Jupiter's gravitational field. But Jupiter's tidal forces will have a further effect on the ellipse of Mercury's orbit. We can see what to expect by thinking of Mercury's orbit, not as empty space, but as a line along which Mercury's mass is spread out. This is an acceptable approximation because Mercury orbits so much faster than Jupiter: it executes almost 50 orbits during one of Jupiter's, so the mean gravitational effect of Jupiter is indeed spread out along Mercury's orbit.

Now, when Jupiter is near one of the "bulges" in the ellipse of Mercury's orbit, it will tend to pull that bulge toward it, just as the Earth tries to align the Moon's bulge in Figure 5.5 on page 45. But Jupiter moves, while the direction of the bulge stays fixed in space. When Jupiter is approaching the bulge from behind it, it will tend to pull the bulge backwards. After it passes the bulge, it will tend to pull the bulge forward. But the first situation will last a little less time than the second: when approaching the bulge, the fact that Jupiter pulls it towards itself makes it reach the bulge more quickly than if it had not pulled the bulge, while the opposite happens after Jupiter passes the bulge.

The net effect, therefore, is to give the bulge a net pull in the direction of Jupiter's orbit. The perihelion of Mercury's orbit – its point of closest approach to the Sun – will move forward, i.e. in the same sense as Jupiter is moving. We say that Mercury's ellipse *precesses* forwards. All the planets have similar effects on one another's orbits. In the nineteenth century, mathematical physicists developed powerful approximation methods to calculate these precession effects; they needed them, because they did not have electronic computers! Much of modern mathematics has its roots in these calculations.

Triumph of Newtonian gravity: the prediction of Neptune

In this section: Neptune was discovered when it was shown that small unexplained motions of Uranus could be explained as tidal perturbations by an unseen planet.

Nineteenth century mathematicians found that they had to take into account all these small perturbations on the orbits in order to reconcile observations with Newtonian theory, and thereby to discover if Newtonian gravity was really an accurate description of gravity. Without the aid of electronic computers, the calculations involved were mammoth, and the mathematical techniques these scientists invented to simplify their job founded the modern branch of mathematics known as perturbation theory. The triumph of their calculational proficiency, and of Newtonian gravity, was the prediction by John C Adams (1819–1892) and Urbain Le Verrier (1811–1877) that certain perturbations of the orbit of Uranus could be explained if there were another planet, further away from the Sun. The planet was indeed discovered near the predicted location in 1846, and named Neptune. Newtonian gravity seemed unassailable.

Tiny flaw of Newtonian gravity: Mercury's perihelion motion

The very calculations that gave Newtonian gravity its triumph also brought about its one failure in planetary theory: its inability to account for the full precession of Mercury's perihelion. Observations showed that Mercury's orbit precesses by about 574 arcseconds per century, about 0.16 degrees. Le Verrier, who had predicted Neptune, calculated in the 1850s that the effects of all the planets on Mercury could not account for the whole precession. By the 1880s, astronomers knew how to explain only 531 arcseconds per century, leaving 43 arcseconds per century unexplained.

Naturally, scientists tried the same route as for Neptune: postulate an extra planet or other sort of matter. But none was discovered. The problem became so serious that a modification of Newtonian gravity was proposed, to change the exponent 2 in the inverse-square law to something slightly larger than 2. Readers who play with the orbit program may experiment with other exponents, and should observe that orbits precess forwards if the exponent is increased a little, and backwards if it is decreased. Since the amount of precession was slight, the required change in the exponent was small; but even this turned out to be inconsistent with better and better observations of the Moon's orbit. Not until Einstein's theory of general relativity appeared was there a satisfactory explanation of this precession. We will return to this story in Chapter 18.

In this section: the tidal effects of all the planets together cannot account for the total precession of the orbit of Mercury. A tiny residual amount went unexplained until Einstein showed that his theory of general relativity predicted the effect exactly.

Interplanetary travel:
the cosmic roller-coaster

S ome of the most exciting moments in the exploration of space in the last thirty years have been provided by a succession of unmanned spacecraft that have explored more and more remote reaches of the Solar System. The early Moon-orbiters, scouts for later Moon landers, were succeeded by spacecraft that visited Mercury Venus, Mars, Jupiter, Saturn, Uranus, Neptune, various comets, and the Sun itself.

But to explore the Solar System in this way requires stronger and stronger rockets, much stronger than are required simply to get a spacecraft away from the Earth's gravitational pull. In order to do the most with the rockets available to them, planetary scientists have used a remarkable trick, called the **gravitational slingshot**: they have used the gravitational pull of another planet, such as Jupiter, to give their spacecraft an extra kick in the direction they want it to go. In this chapter we will try to understand how this works, not only for getting spacecraft into the outer parts of the Solar System, but also for getting them very close to the Sun.

Getting away from the Earth

We remarked in Chapter 4 that the *escape speed* from the Earth is 11.2 km s^{-1}, which is $\sqrt{2} = 1.414$ times the orbital speed at the Earth's surface. We shall prove this in Investigation 6.1 on page 53, which readers should read in connection with the next section. We will see there that, when launched with the escape speed, a spacecraft will just barely get away: if the Earth were the only gravitating body around, it would coast away at an ever-decreasing speed that would tend towards zero as it got far away. In the context of the Solar System, "far away" is still relatively near to the Earth. A spacecraft launched with the speed of 11.2 km s^{-1} in any direction from the Earth would soon find itself roughly stationary with respect to the Earth, i.e. orbiting the Sun with the same speed and therefore in roughly the same orbit as the Earth itself.

Getting away from the Earth to another planet therefore must require a launch with a speed greater than 11.2 km s^{-1}, but the result of such a launch will depend on the direction the spacecraft goes, relative to the Earth's motion around the Sun. If it is shot out in the forward direction, then its excess speed will add to the Earth's own orbital speed, and the result will be an orbit that carries the spacecraft farther from the Sun, in an orbit with a perihelion of 1 AU. This orbit will take the spacecraft outwards to other planets. If the spacecraft is shot in the backward direction, its excess speed will *subtract* from the Earth's speed, resulting in an orbit that falls in closer to the Sun.

Let us imagine an extreme case: trying to get a spacecraft completely out of the Solar System. Since the Earth's orbit is roughly circular, the escape speed from the Sun is 1.414 times the Earth's orbital speed of 29.8 km s^{-1} (see Table 4.2 on page 28), which makes 42.1 km s^{-1}. This is the speed the spacecraft must have, relative to the Sun. When we launch the spacecraft, it already has the same speed as the Earth

In this chapter: mastering interplanetary navigation has opened up the planets to exploration in the last 50 years. The discoveries have been astonishing. The motion of spacecraft teach us much about mechanics: about energy and the way it changes, about momentum and angular momentum, and deepest of all about the role that *invariance* plays in modern physics.

In this section: we learn how to get enough speed to reach other planets, and shows that it is harder to get to the Sun than to escape from the Solar System.

while it sits on the launch pad. We can use this fact to take maximum advantage of the Earth's orbital speed, by shooting the spacecraft directly forward in the Earth's orbit, so that we only need to "top up" the speed by another 0.414 times the Earth's speed, or 12.3 km s^{-1}. This is, of course, the speed *after escaping from the Earth's gravity*. If we send the spacecraft out in any other direction, the Earth's speed will not contribute so much to its final speed in that direction, so it will require more of a boost to get it away from the Solar System.

To get the launch speed from this we have to add to the final speed the Earth's escape speed, 11.2 km s^{-1}. This gives a minimum launch speed that is more than twice as large as the escape speed from the Earth itself. This is rather large, and would require a powerful rocket and a great deal of fuel. We will see that it is cheaper to use the rocket to get as far as Jupiter and then to use the slingshot mechanism to get further.

Surprisingly, it is even harder to send a spacecraft very close to the Sun than to escape from the Sun altogether: for this we must insure that after it escapes from the Earth the spacecraft stops nearly dead in its orbit *relative to the Sun*, so that it can fall straight in towards it. This means that it must be shot out in a backwards direction so that its excess velocity relative to the Earth after escaping from the Earth is nearly equal to the Earth's orbital speed around the Sun, 29.8 km s^{-1}. When added to the escape speed, this requires more than three times the Earth's escape speed, so it follows that it would require a much bigger rocket to reach the Sun than that required to get away from the Solar System entirely. Here, too, we shall see that it is better to send such a spacecraft to *Jupiter* first, and let Jupiter direct it towards the Sun!

Plain old momentum, and how rockets use it

In this section: we learn about ordinary momentum and use it to explain how rockets work.

We have met, in this chapter as well as in Chapter 4, the idea of the conservation of angular momentum, which governs orbits around the Sun. We shall now add to this the law of conservation of (ordinary) **momentum**, which will help us understand how rockets move around in the Solar System.

The momentum of a body, say a rocket, is defined as the product of its mass m and its velocity.

$$\text{momentum} = \text{mass} \times \text{velocity}. \tag{6.1}$$

It is important to distinguish between the rocket's speed and its velocity, because the velocity depends on the direction the rocket is going in. One of the deep laws of physics is:

> the total momentum of a collection of bodies is constant in time if there are no forces acting on it from outside.

This law is called the law of *conservation of momentum*. This can help us understand how rockets propel themselves. Rockets carry their own fuel. They burn it at a controlled rate and expel the exhaust gases out the back. This accelerates the rocket forwards. How does this happen: how does having a hole at the back help the rocket move forward?

The gases that come out of the rocket nozzle have a small mass compared to that of the rocket, but they have a large speed. So they carry a lot of momentum. There are no forces acting on the rocket from outside (let's forget the small effect of gravity for the purposes of this discussion), so the total momentum of the rocket plus gases is constant. Therefore, the momentum carried away by the gases is lost by the rocket. But this momentum is directed backwards, so it is negative: the velocity of the exhaust gases is a negative number. The rocket loses this negative number from

---

Investigation 6.1. Escaping – you can get away from it all if you have enough energy

Energy holds the key to deciding whether a satellite will be able to escape from the Sun or not. Let us look at how much energy an escaping orbit has.

We look at the detailed form of the definition of the total energy, from Equations 6.8 and 6.9:

$$E = \frac{1}{2}mv^2 - \frac{GmM_\odot}{r}. \tag{6.2}$$

A body has escaped from the Sun if it can get arbitrarily far away, i.e. if we can make r as large as we want. This means that the second term in this equation can be made as small as we like, so that eventually the body is coasting with a constant speed v_{far} given by

$$E = \frac{1}{2}v_{far}^2.$$

A body just barely escapes if its final speed is zero, which means that the total energy on its trajectory is zero: *the trajectory that only just escapes has zero total energy*. It turns out that this is true as well if we turn the sentence around: if a trajectory has total energy zero, then it is the path of a body that will get arbitrarily far from the Sun, but whose speed goes to zero as the distance gets larger.

Now, suppose the body starts out at a distance R from the Sun. In order to follow a trajectory of zero energy, it must have a speed v_{escape} given by setting E to zero in Equation 6.2:

$$\frac{1}{2}mv_{escape}^2 = \frac{GmM_\odot}{R} \Rightarrow v_{escape} = \left(\frac{2GM_\odot}{R}\right)^{1/2}. \tag{6.3}$$

This is the escape speed from the Sun. For other bodies, we just replace $M_\odot$ with the mass of the body from which we are escaping. We quoted this result in Chapter 4.

It is interesting to compare the escaping orbit with a circular one. Since the acceleration of gravity by the Sun at this distance is $g = GM_\odot/R^2$, it follows from Equation 3.1 on page 19 that the circular orbital speed is

$$v_{circ} = \left(\frac{GM_\odot}{R}\right)^{1/2}, \tag{6.4}$$

which means that

$$v_{escape} = \sqrt{2}\,v_{circ}. \tag{6.5}$$

The total energy of a circular orbit is simple, as well:

$$E_{circ} = -\frac{GmM_\odot}{2R} = \frac{1}{2}V. \tag{6.6}$$

Put another way, the energy equation for a circular orbit implies

$$E = K + V = \frac{1}{2}V \Rightarrow 2K + V = 0. \tag{6.7}$$

Do not be worried by the fact that the total energy is negative. The only thing that is ever measurable is the change in the total energy from place to place or from orbit to orbit. If we were to add some huge constant energy to all energies to turn them into positive quantities, we would still have the same differences between energies, and the same physics.

Exercise 6.1.1: *Escaping from anywhere*

Calculate the escape speed from the Solar System for a satellite starting at the average distance from the Sun of each of the planets listed in Table 4.2 on page 28. In each case, find the ratio of this speed to the average speed of the planet (column 5 of the table).

---

its momentum, and this means it actually *increases* its momentum. The equation looks something like this:

> momentum gained by rocket = -momentum carried away by gases.

The minus sign cancels the minus sign in the momentum of the gases and leads to an increase in the momentum of the rocket. This is called the rocket equation.

There is a deep relation between this equation and Newton's laws of motion. There has to be: we should equally well be able to explain the acceleration of the rocket by the fact that the exhaust gases exert a force on it, propelling it forwards. Then we would use $F = ma$ to calculate the acceleration a. However, here we have to be careful: when Newton wrote down this equation he assumed that bodies would keep the same mass as they accelerate. But the rocket does not: its mass is always changing. We can see that Newton's second law has a slightly different form in this case from the following argument.

The acceleration a is the change in velocity divided by the time-interval during which the velocity changes. If we multiply both sides of the equation $F = ma$ by this time-interval, we get something that reads:

> Force × time-interval = mass × change in velocity.

For a body with constant mass, the right-hand side is the same as the change in the product of mass and the velocity, or the momentum. So for such a body, an equivalent expression is

> Force × time-interval = change in momentum.

These two equations are not equivalent for a body whose mass changes during the time-interval; that is, their right-hand sides are not the same. In an extreme case, a rocket could change its momentum by losing mass without changing its speed (say, by just pushing one of the astronauts gently out the door!), so the right-hand side of the second equation would be non-zero but that of the first would vanish. They can't both equal the force times the time-interval. So which one is right?

Newton's third law (Chapter 2) tells us the answer. Let us think about the rocket again. The exhaust gases exert a certain force F on the rocket, during whatever time-interval we want to consider. By the third law, the rocket exerts the equal and opposite force, $-F$, on the gases. So the left-hand sides of the two equations above are equal and opposite for the rocket and the gases. If the second equation is the correct one, then we will conclude that the change in the momentum of the gases is equal and opposite to the change in the momentum of the rocket, which is another way of saying that the total momentum is constant: momentum is conserved. The first equation is not equivalent to this, and so would not give conservation of momentum. Therefore, the second equation above is the correct version of Newton's second law to use when the mass of a body is changing. We will come back to this in Chapter 15 when we discuss the acceleration of bodies that go close to the speed of light.

Energy, and how planets never lose it

In this section: we define the total energy of an orbiting planet and show that it is constant.

Our discussion of orbits in the Solar System will be simpler if we first verify a third conservation law, the law of **conservation of energy** during the motion of a planet or spacecraft around the Sun. This is one of the most remarkable and profound ideas in all of physics.

In everyday language, "energy" is a measure of activity, or at least readiness for activity; and after a long period of activity we usually use up our energy and get tired. In physics, **energy** has a more precise definition, but one that is not unrelated to the everyday one. Here we will only discuss the physicists' definition of the energy of a planet; in later chapters we will begin to see its relation to our personal version of energy.

A planet of mass m moving in orbit around the Sun, has two kinds of energy. It has energy associated with its total speed v, called its *kinetic energy*,

$$\text{kinetic energy } K = \tfrac{1}{2}mv^2, \tag{6.8}$$

and it has another kind of energy by virtue of its distance r from the Sun, called its *gravitational potential energy*,

$$\text{potential energy } V = -\frac{GmM_\odot}{r}. \tag{6.9}$$

The symbol $M_\odot$ is the symbol astronomers always use for the mass of the Sun, which is about 2×10^{30} kg. The standard (SI) unit that scientists use for energy is the **joule**, abbreviated J. In Equation 6.8, the units on the right-hand side work out to be kg m^2 s^{-2}; one kg m^2 s^{-2} is, by definition, equal to one joule.

A more familiar concept in everyday life is the related unit for **power**, which is defined to be the rate at which energy is used: energy per second. This unit is the **watt** (W), equal to 1 J s^{-1}. Thus, a 100 W light bulb consumes 100 J of electrical energy every second. We have not yet made the connection between the energies of a body's motion and other energies, like electrical. We will not do that here, but we will return to the subject and develop it further at several points in later chapters.

We shall not try to justify the precise forms of the definitions given above of kinetic and potential energy; rather we shall investigate them "experimentally", by

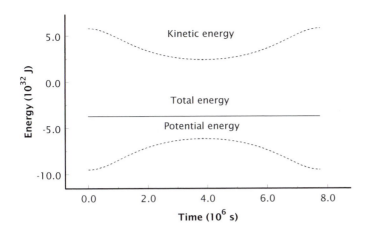

Figure 6.1. *The law of conservation of energy is illustrated for Mercury's orbit by using data from the computer calculation of Mercury's orbit. Although the kinetic and potential energies, as defined in Equations 6.8 and 6.9, respectively, change a good deal during the planet's orbit, their sum remains constant.*

using our computer program to show that the total energy, the sum of the two, is constant.

Before that, let us insure that we see that the individual definitions make sense. The kinetic energy increases as the speed increases, so it certainly measures how "energetic" the motion is. Moreover, bodies that are more massive also have more kinetic energy at any given speed. The potential energy at first looks strange, since it is negative. Far from the Sun it is essentially zero, because of its dependence on $1/r$. As r gets smaller, so the planet gets closer to the Sun, the potential energy gets larger in absolute value, and hence more negative. So the potential energy *decreases* as we get nearer the Sun. This energy is called "potential" in the sense that a planet far from the Sun has more potential to increase its speed and hence its kinetic energy by falling in towards the Sun than does a planet that is already close to the Sun, where the potential energy is less. We will return to why scientists use the word "potential" below, after defining the conservation law for energy.

> Now we can formulate the law of conservation of energy: for a planet orbiting the Sun, the value of the total energy $E = K + V$ is constant everywhere along the orbit.

As in Chapter 4, we verify this conservation law by using the orbit program to calculate the values of the kinetic, potential, and total energies for the orbit of Mercury. These are displayed in Figure 6.1. It is clear that the total energy remains constant even as the kinetic and potential energies change. Readers are encouraged to verify this for any other orbits they may have calculated. We shall take this as a sufficient verification of the law, so that we can use it in our discussions of interplanetary travel below.

We can now understand the name "potential energy" a little better. Since the sum $K + V$ is constant, an orbit can go from a region in which K is small to one where K is large. But K cannot grow indefinitely large: the difference in V in the two places gives the change in K, so that V really does contain the potential increase in K along the orbit.

Conservation of energy has important uses when we try to understand the orbits of bodies around the Sun. The most important of these is to imply that there is a single *escape speed* that any body has to reach in order to get away from the Sun, regardless of the direction it takes. This speed depends only on how strong gravity is where the body starts out. If the body is a distance R from the Sun, then its escape speed is

$$v_{escape} = \left(\frac{2GM_\odot}{R} \right)^{1/2}. \qquad (6.10)$$

If it reaches this speed it will be able to move into interstellar space.

Getting to another planet

In this section: how to use conservation of energy to discover the limits on a space probe's motion among the planets.

Before we see how the slingshot works, we have to see how we can get to other planets in the first place. Here our orbit program should help us. Suppose, for example, we want to reach Mars. In principle all we would have to do is to run the program a few times with various values of the initial velocity at the Earth's position, to see what is the minimum initial speed relative to the Earth that will just get us to the planet.

In practice, this trial-and-error method would be painfully slow unless one had a very fast computer. Instead, we shall show in Investigation 6.2 on page 58 how to calculate the right speed from the law of conservation of energy and Kepler's first law. After that we can use the computer simply to verify that the answer we get really does work.

The best strategy for launching a spacecraft so that it just barely reaches one of the outer planets is to take as much advantage as possible of the velocity of the Earth and shoot the spacecraft forwards in the Earth's orbit. Since the spacecraft already has the Earth's velocity when it sits on the launch pad, this strategy means that we have only to supply the excess speed required beyond the Earth's speed. This keeps the launch cost of the spacecraft as small as possible. Such an orbit has a perihelion at the Earth and it reaches its maximum distance from the Sun at Mars' orbit.

Suppose we launch from the Earth, at a distance R_1 from the Sun, at a target planet a distance R_2 from the Sun. Let r denote the ratio R_2/R_1; then r is just the orbital radius of the target planet expressed in astronomical units (AU). We show in Investigation 6.2 on page 58 that, in order just to reach the outer planet, the spacecraft is required to have a speed relative to the *Sun* after escaping the Earth of

$$v = \left(\frac{2r}{r+1} \right)^{1/2} v_\oplus, \qquad (6.11)$$

where $v_\oplus$ is the Earth's orbital speed, 29.8 km s⁻¹. For a trip to Mars, where $r = 1.52$, we find that v must be 32.7 km s⁻¹ (Exercise 6.2.2 on page 58). To reach Jupiter, we require 38.6 km s⁻¹.

These are speeds relative to the Sun after escaping the Earth. To get the launch speed in each case, we have to subtract the Earth's orbital speed of 29.8 km s⁻¹ and to add the 11.2 km s⁻¹ Earth escape speed. For Jupiter, this works out to a launch speed of exactly 20 km s⁻¹, which is significantly smaller than the 23.5 km s⁻¹ that is required to escape the Solar System.

It is easy to use the program Orbit on the website to verify these numbers, and therefore to provide an "empirical" verification of the above formula. For example, I ran the orbit program for a trajectory that just reaches Jupiter, using an initial position of 1 AU from the Sun, an initial velocity of 38.6 km s⁻¹ parallel to the Earth's orbit, and a time-step of 8000 s. The calculated orbit reached a maximum distance from the Sun of 5.24 AU, close enough for our purposes to Jupiter's actual distance of 5.2 AU. When it reached Jupiter its speed was 7.36 km s⁻¹, which is again close enough to the value of 7.42 km s⁻¹ predicted by Equation 6.13 or Equation 6.15 on page 58. The spacecraft took 2.9 years to reach Jupiter's orbit. Naturally, to encounter Jupiter, the launch must be timed correctly, so that Jupiter is in the right position in its orbit when the spacecraft arrives. The orbit program can give the information required for this as well. The reader is encouraged to try the program for some other planets, to see how long it would take to reach them.

As just noted, reaching the orbit of another planet is fruitless unless the planet is there to encounter the spacecraft at the right time. This means that there has to be a favorable disposition of the planets to allow a launch, and this may happen only once a year for the direct trajectories we have discussed so far, and less frequently for the slingshot effect. If one is unable to use the best "launch window", then reaching the target planet will require a greater launch speed. The penalty can be severe. For example, if one tried to reach Jupiter on a trajectory that went purely radially outwards from the Earth's orbit, it turns out that this would require a speed relative to the Earth after escape of 48.2 km s^{-1}. (You could try to verify this using the orbit program as well. If you do, bear in mind that the program requires the speed of the spacecraft relative to the *Sun*, which for this case is 37.9 km s^{-1} directly away from the Sun.) Thus, it is important for interplanetary missions that the launches take place on schedule!

The principle of the slingshot

Now that we know how to get a spacecraft to Jupiter, we can start thinking about how Jupiter can give it a further push to go somewhere else. The basic idea of this gravitational slingshot is that Jupiter (or another planet) supplies the extra energy that we could not get from our rocket engines. The energy comes from Jupiter's motion, but Jupiter is so massive and has so much energy of motion (its kinetic energy) that these encounters make no significant change in Jupiter's own orbital motion.

In this section: we see how Jupiter can accelerate a space probe even though the energy of the space probe's orbit around Jupiter is conserved.

However, underneath this simple statement — that Jupiter will give our spacecraft a push — is a subtlety that we encounter immediately if we try to understand how it works. Is it really possible for Jupiter to give a spacecraft a push at all? After all, our orbit program shows no such effect: if we send a spacecraft on a trajectory around the Sun, and we follow its orbit as it falls towards the Sun and then comes back out, we always find that it returns to the same place as we started it with exactly the same speed. We made a point of showing this, because it is required by the fact that orbits in Newtonian gravity are closed. So the Sun doesn't give our spacecraft any extra push: it doesn't have any more speed after its encounter with the Sun than it had before. The same would be true if we did an orbit calculation for a spacecraft that does not stay in orbit about the Sun, but instead falls towards the Sun with a large initial speed; when the spacecraft returns to the radial distance from which it started, it has the same speed as it started with, although in a different direction. This is just a consequence of the conservation of energy: since its potential energy *V* depends only on the distance of the spacecraft from the Sun, it follows that its kinetic energy, and therefore its speed, also depends only on the distance from the Sun. So if the Sun doesn't give anything an extra push, then how will Jupiter be able to do it?

The resolution of this apparent contradiction is to remember that all speeds are meaningful only when referred to some standard of rest. In the orbit program, we take the Sun as our standard of rest, and so the correct statement is that a spacecraft that encounters the Sun will have the same speed *relative to the Sun* after the encounter as it had before. The same will be true of encounters with Jupiter: it will have the same speed *relative to Jupiter* after the encounter as it had before. But for the spacecraft traveling through the Solar System, the important speed is not its speed relative to Jupiter, which is unchanged by the encounter, but its speed *relative to the Sun*, which can indeed change.

Let us look at a simple example of using Jupiter as a slingshot. Suppose a space-

---

Investigation 6.2. The reach of an orbit

The range of distances that a spacecraft can explore is governed by its energy and by Kepler's first law, which we met in Chapter 4. In this investigation, we will use these laws to get the speed we have to give a spacecraft to move it from one place to another.

We consider the energy of an orbit that is required to go from a minimum distance R_1 to a maximum distance R_2 from the Sun. Suppose it has a speed v_1 at R_1, its perihelion. As we know, it follows an ellipse, and arrives with a smaller speed, v_2, at R_2, which we call its **aphelion**, its furthest distance from the Sun. The equation of energy conservation is one equation that relates these four quantities. For any given orbit, the quantity E in Equation 6.2 on page 53 must be the same wherever it is calculated. This implies

$$\frac{1}{2}mv_1^2 - \frac{GmM_\odot}{r_1} = \frac{1}{2}mv_2^2 - \frac{GmM_\odot}{r_2}. \qquad (6.12)$$

We can get another relation from Kepler's first law. Recall that this says that the area swept out in any fixed time Δt by the line from the Sun to the planet is the same anywhere along the orbit. (This is also called the law of conservation of angular momentum.) Although this could be difficult to work out at a general position along an elliptical orbit, it is not hard at the perihelion and aphelion, where the velocity is momentarily perpendicular to the direction to the Sun. If we consider a small time Δt just as the planet is passing perihelion, the planet will move a distance $v_1\Delta t$ in this time, and the small triangle in Kepler's law will have equal sides of length R_1. (At other points of the orbit, these sides would not be equal and the calculation of the triangle's area would be more difficult.) The area of a triangle is one-half its base times its height. This triangle's base is $v_1\Delta t$, and its height is R_1, to an excellent approximation. So it has area

$$\text{area at perihelion} = \tfrac{1}{2}R_1 v_1 \Delta t$$

A similar calculation at aphelion gives

$$\text{area at aphelion} = \tfrac{1}{2}R_2 v_2 \Delta t.$$

Kepler's first law says these are equal, so we have

$$\tfrac{1}{2}R_1 v_1 \Delta t = \tfrac{1}{2}R_2 v_2 \Delta t.$$

Cancelling out the factors of $\tfrac{1}{2}$ and Δt, we find a second and very simple relation among the distances and speeds at perihelion and aphelion:

$$R_1 v_1 = R_2 v_2. \qquad (6.13)$$

If we solve Equation 6.13 for v_2 and substitute the result into Equation 6.12, we get an equation that we can solve for R_2 in terms of R_1 and v_1: in other words, we can predict the aphelion of an orbit if we are given its perihelion distance and speed. After multiplying by $2/m$, this equation can be put into the form

$$v_1^2 - \frac{2GM_\odot}{R_1} = \frac{R_1^2 v_1^2}{R_2^2} - \frac{2GM_\odot}{R_2}.$$

This can be simplified by introducing the symbol L_1 to denote the ratio

$$L_1 = \frac{GM_\odot}{v_1^2},$$

which is nothing more than the radius of a *circular* orbit about the Sun that has orbital speed v_1. (This will lie somewhere between R_1 and R_2.) The equation for R_2 now becomes, after dividing by v_1^2, multiplying by R_2^2, and arranging terms,

$$\left(\frac{2L_1}{R_1} - 1 \right) R_2^2 - 2L_1 R_2 + R_1^2 = 0.$$

This is a **quadratic equation** for R_2. It would be easy to use the general solution for such an equation (Exercise 6.2.1), but we can do something even simpler by observing that one solution of this equation must be R_1 itself: R_1 is a place where the velocity is perpendicular to the radius, so both of our original Equations 6.12 and 6.13 apply. This means that $R_2 - R_1$ is a *factor* of the above equation, which can in fact be written

$$(R_2 - R_1)\left[\left(2\frac{L_1}{R_1} - 1 \right) R_2 - R_1 \right] = 0.$$

The second factor provides the other solution, for the aphelion:

$$\left[\left(2\frac{L_1}{R_1} - 1 \right) R_2 - R_1 \right] = 0,$$

which gives finally

$$R_2 = R_1 \left(\frac{2GM_\odot}{R_1 v_1^2} - 1 \right)^{-1}. \qquad (6.14)$$

This is the desired expression for the aphelion distance in terms of the perihelion distance and speed. We can put this back into Equation 6.13 to solve for v_2, the aphelion speed:

$$v_2 = \left(\frac{2GM_\odot}{R_1 v_1^2} - 1 \right) v_1. \qquad (6.15)$$

We can now ask the question we had in mind from the beginning, which is to find the speed v_1 that we need to give to a space probe to get it from the Earth's orbit at R_1 to Jupiter's at R_2, provided it is fired straight ahead along the Earth's orbit. It is convenient in Equation 6.14 to replace $GM_\odot/R_1$ by $v_\oplus^2$, the square of the Earth's (circular) orbital speed. Then solving for v_1 gives

$$v_1^2 = \frac{2r}{r+1}v_\oplus^2, \qquad \text{where } r = R_2/R_1. \qquad (6.16)$$

This is equivalent to Equation 6.11 on page 56.

Exercise 6.2.1: *Solving the quadratic equation*

The general solution of the quadratic equation $ax^2 + bx + c = 0$ for x is

$$x = -\frac{b}{2a} \pm \frac{1}{2a}\left(b^2 - 4ac \right)^{1/2}, \qquad (6.17)$$

where the $\pm$ sign indicates that there are two solutions, found by taking either sign in the expression. Apply this formula to solve the quadratic equation above for R_2. Show that the two roots are R_1 and the root given by Equation 6.14.

Exercise 6.2.2: *Getting from the Earth to other planets*

Use Equation 6.16 to calculate the speed needed to go from the Earth's orbit to the orbits of Mars, Jupiter, and Saturn. The derivation of this formula actually did not need to assume that $r > 1$, so use it for Venus, too.

craft approaches Jupiter along Jupiter's own orbit, catching up with it from behind. Jupiter's orbital speed is 13.1 km s^{-1}, and let us assume for this illustration that the spacecraft is going at 15.1 km s^{-1} relative to the Sun. Then it is approaching Jupiter at a relative speed of 2 km s^{-1}. Again to make the illustration simple, let us assume that the encounter turns the spacecraft completely around, so that afterwards it leaves Jupiter going back toward where it came from. It will leave Jupiter with same the relative speed of 2 km s^{-1}, but now this is directed backwards along Jupiter's orbit, so that the resulting speed relative to the Sun is only 11.1 km s^{-1}. The encounter has slowed the spacecraft down *relative to the Sun*.

This is the sort of trajectory one would look for in order to send a spacecraft close to the Sun. Alternatively, we could have arranged for the spacecraft to approach Jupiter from the other direction, and the result would have been to speed it up and send it further out in the Solar System.

So have we partly lost conservation of energy? Is energy conserved when we measure it relative to Jupiter but not relative to the Sun? No; if that were the case then the law would not be a law at all. If we go back to measuring speeds relative to the Sun, then we have to take into account all the energies, both that of the spacecraft and of Jupiter. If the encounter speeds up the spacecraft, then it must slow down Jupiter. But because Jupiter's mass is so large, the change in its speed is too small to notice. Conservation of energy is fine, but the kinetic energy of a planet is so large that it is an essentially infinite reservoir on which we can draw for our planetary explorations.

Using Jupiter to reach the outer planets

Are the numbers we have quoted earlier realistic? If not, how effective could Jupiter be in a real situation? If we want to reach, say, Saturn from Jupiter's orbit, then we can use our previous formula to tell us the minimum speed we need to have when we leave Jupiter's orbit. Taking $r = 1.83$, which is the ratio of the orbital radii of Saturn and Jupiter, and using Jupiter's speed of 13.1 km s^{-1} in place of $v_{\oplus}$ in Equation 6.11 on page 56, we find that we need a minimum speed relative to the Sun of 14.9 km s^{-1} to reach Saturn from Jupiter. This means we need to leave Jupiter with a speed of at least 1.8 km s^{-1} relative to it, in the forward direction in its orbit.

If we could get an encounter with Jupiter that turned the spacecraft entirely around, we therefore would need to have reached Jupiter's orbit at a point slightly *in front* of Jupiter with a speed 1.8 km s^{-1} *lower* than the speed of Jupiter in its orbit, so that the spacecraft effectively approaches Jupiter from the front. This is an orbital speed of 11.3 km s^{-1}. This is the *maximum* speed we could allow in Jupiter's orbit for the slingshot mechanism to work: a higher speed would mean a lower speed of approach between the spacecraft and Jupiter and consequently a smaller boost from Jupiter of the spacecraft's speed.

The actual speed of the spacecraft when it reaches Jupiter is, as we have seen above, about 7.4 km s^{-1}, directed along the orbit of course. This is considerably below the maximum allowable for the mechanism to work, which means that we have plenty of leeway for playing with such things as the trajectory of the orbit to Saturn.

In fact, this margin allows us the flexibility to cope with another effect that we have so far ignored: a real encounter does not usually turn the spacecraft around by 180°. The angle by which the incoming and outgoing directions of the spacecraft relative to Jupiter differ is determined by the spacecraft's speed approaching Jupiter and by how close it actually approaches Jupiter. For safety reasons, the spacecraft must be kept well away from the planet's surface. So we cannot expect to get the full boost from Jupiter that our simple arguments suggest.

Even given this limitation, we could in principle use the slingshot to boost us

In this section: we examine the details of using Jupiter to boost the speed of a space probe.

Figure 6.2. The trajectories of Pioneers 10 and 11 and Voyagers 1 and 2, showing encounters with planets that have sent all four spacecraft out of the Solar System. (Courtesy JPL/NASA/Caltech.)

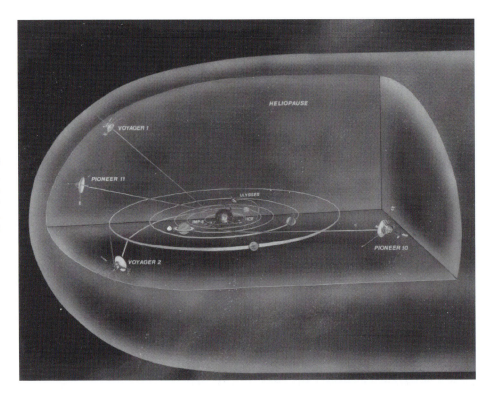

far beyond Saturn. Our spacecraft is closing with Jupiter at a speed of 5.7 km s^{-1}, so that (again, if it could be turned around and sent forward by Jupiter) its speed leaving Jupiter would be 5.7 + 13.1 = 18.8 km s^{-1}. The escape speed from the Sun from Jupiter's orbit is 13.1$\sqrt{2}$ = 18.5 km s^{-1}, so a slingshot from Jupiter might propel the spacecraft out of the Solar System entirely. This is in fact what has happened to both Voyager spacecraft, although their trajectories are complicated by encounters with several planets (Figure 6.2).

Interestingly, the Earth itself can be used to provide a slingshot to propel a spacecraft out to Jupiter and beyond. In Figure 6.3 we illustrate the trajectory of the Galileo spacecraft, which is a Jupiter probe. Its orbit first fell towards the Sun from the Earth, was boosted by Venus, then encountered the Earth twice more, propelling it out to Jupiter.

Slinging towards the Sun

In this section: we consider other possibilities for the slingshot mechanism, such as reaching the Sun or using the inner planets. Comets reach the inner Solar System by the slingshot mechanism, mainly using Jupiter.

What about using Jupiter to put a spacecraft near the Sun? Here we ask Jupiter to slow the spacecraft down. To reach within, say, 0.1 AU of the Sun, a spacecraft must have an orbital speed at Jupiter's orbit of no more than 2.5 km s^{-1}, which we can again obtain from Equation 6.11 on page 56 with r = 0.1/5.2 and $v_\oplus$ replaced by Jupiter's speed of 13.1 km s^{-1}. This orbital speed represents a speed relative to Jupiter of 10.6 km s^{-1} in the backwards direction. Before its encounter with Jupiter the spacecraft had to have the same speed of approach relative to Jupiter.

This situation is more difficult to analyze, because here one would want to take advantage of the fact that the spacecraft's trajectory is not deflected through 180° by Jupiter. One can see roughly how to make it work by considering two extreme cases that both result in trajectories with a speed of 2.5 km s^{-1}, one in which the spacecraft is indeed turned completely around and one in which there is no encounter at all. In the first case we approach Jupiter from behind going faster than it, and in the second we approach from the front. The first requires an initial spacecraft speed of 13.1 + 10.6 = 23.7 km s^{-1} relative to the Sun, while the second requires the same 2.5 km s^{-1} initially as the craft will have finally.

Neither speed is available to us, the first because it requires too much energy and the second because it is smaller than our spacecraft will have when it reaches Jupiter. But somewhere in between these two extremes is a trajectory that approaches Jupiter from the side with a speed relative to the Sun more like 6 km s^{-1}, and which will leave Jupiter in exactly the backwards direction provided we arrange the angle of deflection of the orbit correctly. (This is determined by the distance of closest approach to Jupiter.) Given that the two extreme speeds are 23.7 and 2.5 km s^{-1}, it seems clear that

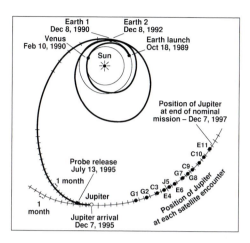

Figure 6.3. The trajectory of the Galileo spacecraft shows how the Earth itself can be used as a slingshot. (Courtesy JPL/NASA/Caltech.)

the real case will be closer to the second case, where we approach Jupiter nearly head-on from in front. Only a relatively small deflection will be required to remove a few kilometers per second of speed to get the spacecraft down from 6 to 2.5 km s^{-1}. We will not do this calculation in any greater detail here.

Interested readers may wish to consider other ways of reaching the inner Solar System, such as using an encounter with Venus to reach Mercury. Another option would be to approach Jupiter partly from below the plane of its orbit; diving under it this way could produce an orbit that is out of the plane of the planetary orbits, and which would then pass over the poles of the Sun. All of these tricks have been used or proposed for interplanetary exploration.

Artificial spacecraft are not the only objects that experience the slingshot mechanism. Outside the orbit of Neptune, stretching over many hundreds of AU, is the **Kuiper Belt**, a zone full of planetesimals that never formed into planets. It is named after the Dutch astronomer Gerard Peter Kuiper (1905–1973). Astronomers have only recently discovered how abundant these asteroids are, and how various their sizes are. In fact, Pluto seems more aptly described as a giant asteroid from the Kuiper Belt rather than a planet; its properties are very different from the gas giants, but very similar to the asteroids. Sometimes asteroids from the Kuiper Belt reach the orbit of Jupiter or Saturn, either because they have elliptical orbits or because they have collided with other asteroids. They can be slung by one of these planets into an orbit that takes them much closer to the Sun, and then they present a danger to the Earth. The collision that is thought to have assisted in the extinction of the dinosaurs was probably with one of these objects.

Outside the Kuiper Belt is the **Oort Cloud**, home of the comets. This region, extending as much as 10^5 AU from the Sun, has been named after another Dutch astronomer, Jan H Oort (1900–1992). The comets are believed to resemble the building blocks out of which the asteroids and from them the planets were formed. At distances so far from the Sun that they cannot collide often enough to build planets, these objects remain museum pieces of the earliest stage of planetary formation. There is great interest among astronomers in studying comets to learn about this period. If it were not for Jupiter, and to some extent Saturn, we would not see any in the inner Solar System. Although comets appear to have very eccentric orbits, they rarely fall close to the Sun. Instead, when they reach the orbit of Jupiter they are slung, like Kuiper-Belt asteroids, sometimes deeper into the Solar System. They too present a danger to the Earth.

Force and energy: how to change the energy of a body

In this section: we show how the application of a force changes the kinetic energy of a moving body.

The law of conservation of energy strikes us as remarkable when we first meet it because the total energy of an orbiting planet is composed of two parts that each can change, the kinetic and potential energies; only their sum remains constant. This cannot be a magical coincidence: it must reveal something deeper. In this section we see that there is a simple way to see how a force applied to a body (like a planet) changes its kinetic energy.

Let us consider the energy of a planet a little more carefully. Since the potential energy depends only on r, the distance of the planet from the Sun, it follows that the potential energy of a planet on a perfectly circular orbit remains constant in time. From this it follows that the kinetic energy also remains constant. This is the same as saying the speed of the planet is constant, and of course that is what we expect in a simple circular orbit. But the conclusion we are interested in is that the kinetic energy of a planet changes only as its distance from the Sun changes.

Let us see what the change is. Suppose an orbiting planet of mass m moves inwards from a distance r by a very small amount δr to a distance $r - \delta r$. Then in Investigation 6.3 we see that the potential energy changes from $-GmM_\odot/r$ to approximately $-GmM_\odot/r - (GmM_\odot/r^2)\delta r$. Since the total energy is constant, the change in the kinetic energy is the opposite (negative) of the change in the potential energy:

$$\text{change in kinetic energy} = \frac{GmM_\odot}{r^2}\delta r. \tag{6.18}$$

Notice that the force of gravity (Equation 2.3 on page 13) appears in this expression. In other words we can write this as

$$\text{change in kinetic energy} = F_{\text{grav}}\delta r. \tag{6.19}$$

In other words, the change in energy is the product of the force that moves the planet and the distance the planet moves. In this case, the force is directed inwards, towards the Sun, and we have assumed that the planet also moves inwards, so both the force and change in position are in the same direction. The result is an increase in the kinetic energy. This equation works only for small steps δr in radius. To find the total change in kinetic energy when there is a large change in radius, one must add up successive small changes, in the same spirit as our computer program for the orbit moves the planet in small steps.

Equation 6.19 is quite general, and works no matter what force is applied to a particle. Quite generally, the change in kinetic energy of any particle equals the distance the particle is displaced times the force acting in the same direction as the displacement. If the force acts in the direction *opposite* to the motion of the body, as happens for example with the force of friction as a body slides along a surface, then the change in kinetic energy is negative: we must put a minus sign into Equation 6.19. This corresponds to our expectations: friction reduces the speed and hence the kinetic energy of the body.

Physicists define the right-hand side of Equation 6.19 as the **work** done by the force of gravity. In general the definition is

$$\text{work done by a force } = \text{force } \times \text{distance through which the force acts.} \tag{6.20}$$

As with many common words that physicists use, *work* is not quite the same in physics as in everyday life. Sitting at her desk, a physicist does no work, according

---

Investigation 6.3. The change in the potential energy

Here we perform a short calculation to find the way the potential energy changes when there is a small change in the distance of a planet from the Sun. The potential energy at the new position $r - \delta r$ can be manipulated with a little algebra into a form that makes it easy to approximate:

$$-\frac{GmM_\odot}{r - \delta r} = -\frac{GmM_\odot}{r(1 - \delta r/r)} = -\frac{GmM_\odot}{r}\left(1 - \frac{\delta r}{r}\right)^{-1}.$$

In this last form of the expression we can use Equation 5.2 on page 43 to approximate the last factor:

$$\left(1 - \frac{\delta r}{r}\right)^{-1} \approx 1 + \frac{\delta r}{r}.$$

When we put this into the expression for the potential energy in the previous equation we find

$$-\frac{GmM_\odot}{r - \delta r} \approx -\frac{GmM_\odot}{r} - \frac{GmM_\odot}{r^2}\delta r.$$

This is the expression we use in Equation 6.18.

Exercise 6.3.1: *Changes in potential energy*

Justify (or fill in) the algebraic steps that lead from one term to the next in the first equation in this investigation.

---

to this equation, since she does not change her position. However, no doubt she still expects to get paid for what she does at the desk!

Time and energy

We began this chapter by learning how important the law of conservation of energy is. Then we seemed almost to lose the law, in the gravitational slingshot. Of course, energy conservation does still hold in the slingshot, as long as we add together the energies of both bodies. This is not surprising: we should expect that the spacecraft and the planet could exchange energy with each other. But there is another lesson we can draw from this chapter, and that is about the deep relationship between energy and time.

We looked at two kinds of problems: the motion of a body (planet or spacecraft) around the Sun, and the motion of a body around Jupiter. Both were motions under the action of gravity, and in both cases the body that created the gravitational field was too large to be affected by the body. Yet in one case (the Sun), the total energy of the body was constant, and in the other the total energy changed. The only significant difference between the two problems is that in the first case, the Sun was standing still, and in the second Jupiter was moving. That is, in the first case the gravitational field was time-independent, while in the second the field at any given location depended on time (as Jupiter moved past).

We have here a glimpse into one of the most profound relationships in physics: when there is some underlying time-independence in a physical situation, there is usually a conserved energy, and vice versa. The single body moving past Jupiter does not have a conserved energy because it experiences a force field that is time-dependent. But if we consider the body and Jupiter together, then they move in the background field of the Sun, which is time-independent, so their total energy is conserved.

> All the fundamental forces in physics, such as the electric force, gravity, and the nuclear force, work in such a way that the total energy of a collection of bodies is conserved provided that any forces on the bodies from outside the collection are independent of time at any one location.

Essentially, energy is conserved for these bodies if all the rest of the Universe is time-independent. When we come to consider **cosmology** – the study of the Universe as a whole – and the observed expansion of the Universe, we will see how we lose the law of energy conservation: as the Universe expands, its energy simply disappears.

In this section: we make a fundamental and deep connection between energy conservation and time-invariance of the laws of physics.

What about other conservation laws? We have also met conservation of angular momentum and of ordinary momentum. Both of these, as well, are associated with some kind of "independence". In the case of angular momentum, it is angular independence: the angular momentum of a planet is constant on its trajectory because the gravitational field of the Sun is independent of the planet's angular position around the Sun. The Sun is spherically symmetric, and this leads to conservation of angular momentum. Ordinary momentum is sometimes called linear momentum because it is conserved in situations where the external forces are constant along straight lines. The ordinary momentum of a planet is certainly not constant along its orbit, and this is because the Sun's gravitational field is not constant in any fixed direction. But when, for example, billiard balls collide on a billiard table, the effect of the Sun's (or indeed the Earth's) gravity is unimportant, and the table itself is flat in any horizontal direction, so the total momentum of the balls in the collision is constant.

Physicists and mathematicians have a name for the general concept that includes time-independence and angular independence. They call it **invariance**. They say that the gravitational field of the Sun is invariant under a change of time (from, say, now to tomorrow), and it is also invariant under a change of angular position. The relation between conservation laws and invariance is something that physicists believe is built into the laws of physics at their deepest level. In fact, physicists today who work on discovering the laws of physics at the highest energies imitate this principle by looking for more abstract kinds of invariances. The approach has been successful so far. Such theories of physics are called *gauge theories*, and all theories of the twentieth century that have unified the nuclear, electromagnetic, and weak forces are gauge theories (see Chapter 27). The principle of invariance is one of the deepest in physics.

Atmospheres:
keeping planets covered

7

There would be no life as we know it on Earth without the atmosphere. Even life in the oceans would not exist: without the atmosphere's thermal "blanket", the oceans would freeze. Yet in the beginning, the Earth probably had a very different atmosphere from its present one. The other planets, with their different masses and different distances from the Sun, all have vastly different atmospheres from the Earth's. In the retention of the atmosphere, and in the subsequent evolution of the atmosphere and of life itself, gravity has played a crucial role.

In this chapter, as we look at the role that gravity has played in this story, we shall encounter fundamental ideas about the nature of matter itself: how temperature and pressure can be explained by the random motions of atoms, why there is an absolute zero to the temperature, and even why atoms cannot quite settle down even at absolute zero. We shall also construct a computer program that builds atmospheres, and we will use it to model not only the Earth's atmosphere, but those of other bodies in the Solar System.

In this chapter: we study the way the atmospheres of the Earth and other planets have developed. We learn how to calculate their structure, and we meet some of the fundamental physical ideas of gases, such as the absolute zero of temperature. We discover the ideal gas law, and we see how pressure and temperature really come from random motions and collisions of atoms. Finally, we look more closely at what happens in a gas at absolute zero, and have our first encounter with quantum theory.

In the beginning ...

The Sun and planets formed some 4.5 billion (4.5×10^9) years ago. We know this from studies of radioactive elements in old rocks, whose decays provide us with a number of natural clocks. The oldest rocks known are older than 4.1 billion years. From theoretical studies of the Sun, which we will describe in Chapter 11, we know that it takes about 4.5 billion years for a star of the Sun's mass to evolve into one that looks like the Sun. It is clear, therefore, that if we want to understand where out atmosphere came from, then we must take a big leap in time-scale, from the orbital periods of planets and space probes we discussed in the last chapter, which are measured in years, to the long perspective of several billions of years.

In this section: how the planets formed, and where their atmospheres came from.

We do not know a lot about the formation of the Solar System, apart from when it happened. Most likely the planets formed from the same cloud of gas that formed the Sun, material that was not incorporated into the shrinking star, perhaps because it was rotating too fast. How the planets formed from this gas has been made much clearer to us by recent planetary exploration, and particularly by the exploration of the Moon by the astronauts on board the **Apollo** missions.

The first task of explaining the formation of the planets is to account for the great differences between them. The Sun is composed mostly of hydrogen, with some 20–25% helium, and traces of other elements. The giant planets, like Jupiter and Saturn, are also dominated by hydrogen. How, then, is the Earth so solid, with plenty of silicon, iron, oxygen, nitrogen, and other "heavy" elements, but comparatively little hydrogen and helium? Why are all the inner planets rocky and the outer planets gaseous?

All the planets seem to have formed from the hard **dust grains** that pepper the giant **interstellar clouds** of gas. Interstellar clouds are the places where stars form, so their overall composition is like that of the Sun. But, unlike the Sun,

they are very cold, cold enough to allow carbon and heavier elements to form tiny condensations that astronomers call dust. These whisker-like grains, only fractions of a millimeter long, are very common in gas clouds, where they are good at blocking the light traveling to us from more distant stars. We shall learn more about where grains come from in Chapter 12.

The gas from which the planets formed contained its share of dust. As a result of random collisions, grains began sticking to each other through molecular forces and building up large lumps. Eventually a number of lumps grew so large that they exerted a significant gravitational pull on their surroundings, pulling in nearby smaller lumps. These are called **planetesimals**. The outer planets (Jupiter, Saturn, Uranus and Neptune) seem to have retained much of the original gas that was in the disk, although they probably started out with rocky cores. Pluto is an exception, and its equally exceptional orbit suggests that it was formed in a different way. Pluto may simply be the nearest large object in the Kuiper Belt.

The **terrestrial planets** (Mercury, Venus, Earth, and Mars) probably trapped gas around themselves initially. But they were close enough to the Sun for this gas to get hot. Since these planets are relatively small, their gravity was not strong enough to hold onto the gas, and the planets lost their initial atmospheres.

In their process of merging from larger and larger bodies, these planets probably experienced their largest collisions last. This helps to account for the fact that many of them spin about an axis that is not perpendicular to the plane of their orbits, and in fact Venus spins in the opposite sense to all the others. The spin is the "memory" of the orbital plane of the last big fragment to merge into the planet. This is also thought to explain our Moon: after the Earth was formed, it was hit by a rogue planet the size of Mars, expelling enough material to form the small "planet" that now orbits the Earth in a repetition in miniature of the formation of the planetary system around the Sun.

The distant planets – Jupiter, Saturn, Uranus, and Neptune – formed in regions where there was more material in the initial cloud and where the temperature was too low to boil off their atmospheres. They may have solid cores but these are hidden from view. They all have moons, some of which may well have been small planets that formed nearby and were captured by three-body collisions. (We will study these in Chapter 13.)

The gas near the terrestrial planets was lost as the planets formed. Only elements trapped in dust grains did not escape. And here is the clue to where the present atmospheres came from: trapped inside the minerals of the grains were not only solid elements, but also some gases, primarily carbon dioxide and nitrogen, with a significant amount of water vapor. These gases were released from the grains when they were heated by the high pressures deep in the interior of the planets. They gradually leaked out of the planets to form the raw material of their present atmospheres. This process is called *outgassing*.

Mercury is hot and small, its gravity too weak to retain an atmosphere at the high temperature to which the Sun heats the planet. So the gases that leaked out simply drifted away, and the planet has no atmosphere. Venus, Earth, and Mars managed to retain small atmospheres. Considering that they all began with similar composition, the fact that these planets have radically different atmospheres now is a striking testimony to the fact that planets *evolve*. The most important factors in their evolution have been geological activity (such as volcanos) and the control of the temperature of the atmosphere by the **greenhouse effect**.

... was the greenhouse ...

Water vapor, carbon dioxide, and methane are **greenhouse gases**: they allow sunlight to pass through to the surface of the planet, but they are opaque to the **infrared** (heat) radiation that the planet radiates back into space. (We will find out more about the greenhouse mechanism in Chapter 10.) They trap this energy near the planet's surface, raising its temperature to well above what it would be on a planet with no atmosphere. Greenhouse gases in the Earth's primitive atmosphere kept it warm enough to have liquid oceans, despite the possibility that the Sun when it first formed was perhaps 25% dimmer than it is today.

On the Earth, we currently worry about the greenhouse effect that might accompany an increase in the small concentration of CO_2 present in today's atmosphere: there is the possibility that human activity will raise the planet's temperature uncomfortably high. Venus, whose atmosphere is more than 95% carbon dioxide, provides an example of an extreme greenhouse effect: Venus is far hotter than the Earth.

The contrast between Venus and the Earth is particularly striking, because they are nearly the same size and at similar distances from the Sun. Venus and the Earth should therefore have started with similar atmospheres. Moreover, both planets have had many volcanos, which release large amounts of carbon dioxide into the atmosphere. Perhaps it is not surprising, then, that Venus has a lot of carbon dioxide. The question is, how has the Earth managed to keep carbon dioxide concentrations low and thereby moderate its greenhouse effect?

In fact, the Earth has just as much carbon dioxide as Venus, but it is no longer in the atmosphere: it is almost all bound up in limestone rocks, which are made of calcium carbonate ($CaCO_3$). These rocks are made from the deposits of shelled animals, laid down over long periods of time. This shows that, on the Earth, the removal of atmospheric CO_2 has been helped by life itself.

How does this work? When rain falls through the air, some carbon dioxide dissolves in it, and forms in fact *carbonic acid*, H_2CO_3, which is just the combination of a water molecule (H_2O) and one of carbon dioxide (CO_2). When this acid enters the oceans, the bicarbonate ion HCO_3^- is freely available to shelled animals, which combine it with calcium that has also been dissolved in the oceans to make more shell, basically more calcium carbonate, $CaCO_3$. When the shelled animal dies, its shell eventually gets compressed into limestone rocks.

But animals cannot have been involved in this process on the early Earth. Even before shelled animals evolved, the Earth was removing the carbon dioxide released by volcanos, converting carbon dioxide into minerals by chemical means. The bicarbonate ions would form minerals and precipitate out of the oceans if their concentrations were high enough.

What I have described here is a balance that is called the *carbon dioxide cycle:* volcanos put CO_2 into the atmosphere and chemical reactions and living organisms take it out. This cycle plays the crucial role in stabilizing the small amount of carbon dioxide remaining in the Earth's atmosphere, and thus in keeping the greenhouse effect under control. Scientists generally agree on its importance, but the details of how it works are still not clear.

It is remarkable that, at least today, animals help to control the Earth's temperature and keep it fit for life. The attractive possibility that living creatures are actively modifying the Earth's climate in order to make the Earth a suitable place for life is known as the Gaia hypothesis: it is hotly debated among scientists today.

This debate is of more than academic importance. The balance of the carbon

In this section: the sizes of the atmospheres of the rocky planets are very different today. This may reflect the combined effects of differences in the planets' masses and distances from the Sun.

dioxide cycle is now being altered in a possibly dangerous way by human activities that release carbon that had been stored up in coal and oil reservoirs and in forest wood. The total carbon in these stores is small compared to the carbon in rocks, but we are releasing it at a high rate, and (probably not coincidentally) the concentration of carbon dioxide in the atmosphere is rising rapidly. A better understanding of the history of the Earth's atmosphere is now important for making predictions about how the Earth's greenhouse will respond to these human activities.

Equally relevant is the radically different history of Venus' atmosphere. Its original temperature may have allowed liquid water to form oceans, as it did on Earth. But Venus is closer to the Sun than the Earth is, and the carbon dioxide in its original outgassed atmosphere seems to have raised the temperature on Venus high enough to evaporate its oceans. Then the water vapor at the top of the atmosphere was broken up by the Sun's **ultraviolet radiation** into hydrogen and oxygen gas. The hydrogen, being light, escaped. The oxygen, being highly reactive, combined with surface rocks. This gradual depletion of water and the continued release of carbon dioxide by volcanos (see Figure 7.1) resulted in an atmosphere dominated by CO_2, with very little of the original water in it.

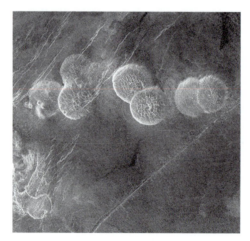

Figure 7.1. Venus shows abundant evidence for volcanic activity, including these unusual "pancake" volcano structures imaged by the NASA Magellan misson's radar. Courtesy of NASA/JPL/Caltech.

After Venus lost its water, there was no rain to remove carbon dioxide from its atmosphere, and that sealed its fate. It is intriguing to speculate that perhaps life did actually evolve in the oceans of Venus in its early days, since the planet was not very different from the Earth, and since geological evidence on the Earth suggests that life arose here no later than 1 billion years after the Earth formed, and perhaps even earlier. Life would presumably have been destroyed on Venus when the greenhouse effect evaporated the oceans.

In the end, the difference between Venus and Earth may just be that Venus is a little closer to the Sun, and its carbon dioxide cycle could not cope with the extra solar energy coming in. If that is the case, what will happen to the Earth if the Sun continues to warm up over the next billion years?

Mars also retains an atmosphere, although it is much thinner than the Earth's. Like Venus, Mars' atmosphere is primarily carbon dioxide. Unlike Venus, where all the CO_2 released by the planet has remained in the atmosphere, most of it on Mars seems to have become locked up in rocks on the Martian surface.

As on the Earth, rain would have removed CO_2 from the atmosphere. Mars shows many geological features that suggest that it had oceans and rivers at one time, for example the channels in the right-hand image in Figure 4.1 on page 26. The conversion of CO_2 to minerals probably occurred through natural chemical reactions, although we cannot exclude the possibility that life evolved on Mars early in its existence and also assisted the removal of carbon dioxide. If this did happen, it was suicide: the big difference between the Earth and Mars is that Mars seems not to have had enough volcanos to replenish the carbon dioxide that was being removed, and thereby to stabilize the greenhouse effect by maintaining a small concentration of CO_2 in the atmosphere.

Volcanos are driven by the release of heat in the interior of a planet. Planets are hot when they form, because of the impacts of all the planetesimals. But planets the size of the Earth and Mars cool quickly, so volcanic activity would go away if there were no further source of heat. For the Earth and Venus, this source is radioactivity.

The decay of naturally-occurring uranium and other elements inside the Earth provides the heat – and the Earth's gravity provides the pressure – that keeps the interior molten. The outward flow of this heat energy drives **plate tectonics**, the drifting and collisions of continents. Plate tectonics keeps volcano activity going. Mars is smaller, so it loses the heat it generates more quickly, leaving its interior temperature too low to drive tectonics and volcanos. Therefore, the size of its atmosphere has steadily declined. This has cooled the planet off: with less CO_2, there is less of a greenhouse effect, and the temperatures decreases.

What happened to the water on Mars? It is probably still all there, frozen into a permafrost. What about life on Mars? If it had time to evolve before Mars cooled off (perhaps 1 billion years) then it froze, or it became starved of CO_2 before that.

...and then came Darwin

As we have seen, the Earth's atmosphere has changed radically since it was first formed. In a combination unique at least to the Solar System, geological, biological, and physical forces have changed the original water–carbon dioxide atmosphere into the oxygen–nitrogen atmosphere we breathe today.

The original life forms evolved in the oceans and were adapted to living with and using the dissolved nitrogen and carbon dioxide from the early atmosphere of the Earth. They used carbon dioxide as their food, and produced free oxygen as a waste product. Gradually, the free oxygen built up to levels that were poisonous to these original organisms, with the result that their closest descendants today, such as anerobic bacteria, can be found hiding only in rare, oxygen-poor habitats. But new forms of life evolved that actually liked this world-wide pollutant, and from them all the present oxygen-using plants and animals are descended.

The present atmosphere is the result of a balance among a number of forces. I mentioned earlier that the amount of carbon dioxide in the atmosphere has been maintained by a cycle in which chemistry and living things deplete it and natural processes replace it. In all of this, gravity plays a quiet role, regulating the rate of volcanic activity and the rate of atmospheric circulation that leads to weathering. It would be difficult to make an Earth with the same balance of effects if gravity were half its strength or if the planet were twice as massive. Life as we know it can only exist on a geologically unstable planet, and the Earth's size and gravity are perfect for this.

In this section: the evolution of life changed the atmosphere of the Earth, adding oxygen. Life as we know it requires a certain balance of geological activity and atmospheric chemistry. Because of the role gravity plays in these balances, life might not survive on a planet with a very different mass.

The ones that get away

We saw earlier that hydrogen escaped from Venus' atmosphere because it is light. This has been happening to all the light gases in all the atmospheres of the terrestrial planets: hydrogen and helium have been outgassed, although in much smaller concentrations than in the original cloud of gas, and they have escaped into space.

Does this not contradict Galileo? If gravity makes all things fall at the same rate, regardless of their mass, how can it selectively keep heavier elements in an atmosphere? The explanation is that the *temperature* of the gas is the crucial selecting factor.

Consider a gas with a given temperature, made up of a variety of **atoms** and **molecules**, all with different masses. According to laws first discovered by the Austrian scientist Ludwig Boltzmann (1844–1906), the *kinetic energy* of a typical atom of mass m depends only on the temperature of the gas, and on *nothing* else.

In this section: the composition of an atmosphere depends on how many atoms of a given type reach escape velocity. The mean speed of heavier atoms is smaller than lighter atoms, so smaller planets tend to have atmospheres with heavier atoms.

Since the kinetic energy is $\frac{1}{2}mv^2$ (see Equation 6.8 on page 54), the typical speed of an atom in a gas of a given temperature is smaller if the atom's mass is larger.

We shall look more closely at Boltzmann's description of gases later, but it should be clear that not all atoms of a given mass could have exactly the same speed: there is a random distribution of speeds, some much faster, some much slower. It is the *average* speed of each type of atom that will be determined by the gas temperature and the mass of the atom. When a gas is a mixture of several types of atoms and molecules, each type will have the same temperature and hence the same average kinetic energy. Then heavier atoms and molecules in the mixture will have smaller average speeds.

Now we can see how gravity affects an atmosphere. The deciding factor for whether or not any body is bound to another by gravity is whether its speed exceeds the *escape speed* for that particular body. Since light atoms will have a higher speed than heavier atoms, they will have a greater tendency to escape. Those atoms which have by chance a random speed much higher than the average speed for the prevailing temperature will escape sooner. Although some atoms of any mass will always escape, there will be a critical mass above which so few atoms have enough speed to escape that atoms and molecules of that type and those that are more massive will be effectively bound to the planet.

Since more massive planets have larger escape speeds, they will retain more of their lighter atoms and molecules than less massive bodies can. Massive planets far from the Sun at low temperatures have retained all their light gases, like hydrogen. Smaller planets near the Sun have atmospheres that consist primarily of heavier atoms and molecules.

Atoms that are light enough to escape from a planet can be present in its atmosphere only if they are constantly replenished. Helium, for example, exists in the Earth's atmosphere mainly because it it generated by radioactivity in rocks: some radioactive elements produce alpha-particles, which are nuclei of helium. When these particles slow down and come to rest in rocks, they acquire two electrons and become helium atoms. They are not chemically bound to the rocks, so eventually they escape into the atmosphere, where they stay for a geologically short time before escaping into space.

The Earth's atmosphere

In this section: to understand even the simplest aspects of the structure of the Earth's atmosphere, we must learn about how pressure and gravity balance.

The Earth's atmosphere is an extremely complex system. It has many distinct layers; it is subject to continual mixing caused by weather systems and by **convection** of hot air from the ground; it is dragged along with the rotation of the Earth; it receives large amounts of water vapor over the oceans and dumps much of the water on the continents; it is heated by the Sun at a number of different altitudes, depending on which radiation-absorbing gases are where; and its composition is constantly evolving from the actions of natural forces and of man.

A full discussion of these complexities is well beyond our scope here, and in fact is well beyond the scope of any present computer model of the atmosphere. The vast number of effects that need to be taken into account would overwhelm the speed and memory of even the biggest supercomputers today, and in many cases the complexity of the physics and chemistry of the atmosphere defies understanding at present. Nevertheless, computer models of the atmosphere are helpful and informative if they are interpreted with due care.

For us, the computer can again be an aid to solving even the simplest equations that describe atmospheres. Our aim here is to understand enough of what affects an atmosphere to be able to use a computer to model the Earth's static atmosphere, where we neglect any changes due to weather, circulation, and so on.

The fundamental point is that the structure of the atmosphere is essentially a balance between gas pressure, which pushes the atmosphere up, and gravity, which keeps it down. We already know enough about gravity to write our program. But we do not know enough about pressure yet. We need to discuss two points about pressure that are crucial to the construction of a simple computer program. The first is how pressure manages to push things up; the second is the way that temperature affects pressure. Only after we discuss these will we be in a position to see how gravity and gas pressure balance each other to determine the structure of the Earth's atmosphere.

Pressure beats gravity: Archimedes buoys up balloons

A good way to understand how pressure acts to keep an atmosphere up is to see how it pushes a helium-filled balloon up through the atmosphere. We all know that the balloon rises because helium is lighter than air: at a given pressure and temperature, helium atoms weigh considerably less than the nitrogen and oxygen molecules of the surrounding air. But how, in detail, does the balloon rise? Where are these **buoyancy** forces acting, and where do they come from?

It is clear that the forces must come from the pressure forces in the surrounding air: the balloon is not in contact with anything else that can exert forces on it. Scientists define the pressure on any surface, such as a spherical balloon, to be the *force per unit area, acting perpendicular to the surface.* So on each little patch of balloon surface, there is a pressure force pointing inwards toward the center of the balloon. The *pressure force* on the patch equals the *pressure* at that place on the balloon times the area of the little patch.

Now I will pose an apparent paradox. Pressure is an **isotropic** force. This means that at any point inside a gas, pressure pushes with the same force in all directions. In other words, it doesn't matter whether I position the balloon just above the point or just below it, just to the left or to the right of the point in question, the pressure force will be exactly the same.

How, then, can the balloon rise? After all, a balloon has weight, so gravity is pulling it down. For it to rise, there must be another force pushing it up even more strongly. How can pressure provide this force? Doesn't the pressure of the gas above it push it down with the a force just equal to the pressure of the gas below it pushing up? If this is so, don't the two forces exactly cancel?

The answer is, of course, no. The reason is that the pressure above the balloon is not acting *at the same point* as the pressure below the balloon: it is acting on the other side of the balloon. Therefore the isotropy of the pressure is no reason to expect the two forces to cancel each other exactly, provided the pressure changes from one *place* to another. We say that pressure is isotropic, but it is not homogeneous: it is not the same at all different locations in the atmosphere.

In fact, pressure goes down as we go higher in altitude. This is the reason for the familiar sensation in our ears if we change altitude too quickly: the air pressure inside our heads does not change as quickly as that outside, and the inequality of pressure on the two sides of the eardrum causes a painful strain on that delicate membrane. So the balloon has less pressure on its top than its bottom, and the net effect of pressure is to push upwards on the balloon.

> It is only the *non-uniformity* of pressure across an object that provides a net pressure force on it.

But of course this is only part of the story. The same thing happens to, say, a rock held in the air: the pressure above it is lower than the pressure below it. Yet it

In this section: we learn exactly what the combination of forces is that lifts balloons into the air, and what needs to be balanced for the atmosphere to be in a steady state.

will fall when let go, not rise. For the balloon to rise, we have to have that the net upwards pressure force is actually larger than its weight.

Where is the dividing line? What is the critical weight that an object must have so that gravity and the pressure difference across it just balance and allow it to remain at rest? To answer this, let us try to find something that does remain at rest when it is let go, something that is said to be *neutrally buoyant*. The simplest example of such a thing is air itself. The air that is displaced by the balloon when we place it in the atmosphere, the air that would otherwise occupy the same place as we have put the balloon: this air would not move. So it has just the right weight to remain at rest. Therefore, if the contents of the balloon are *lighter than air*, like helium, the balloon will rise. If the object is heavier, like a rock, it will fall.

> The motion of an object in an atmosphere is the result of the net force that results when the pressure force down on it from above, the pressure force up on it from below, and the downward force of gravity on it are all added together.

Now we can return to the question we started with in this section: how does pressure support the atmosphere? Our example of neutral buoyancy shows how: if we replace our helium-filled balloon with the parcel of air again, then we see that it is held up against gravity by the pressure difference across it:

> For the atmosphere to be in equilibrium (in other words, perfectly at rest) the pressure force at the bottom of any parcel of air must exceed that at the top by the weight of the parcel.

▷ The word "hydrostatic" reveals how scientists began thinking about buoyancy: by studying things floating on water. The prefix *hydro-* is from the Greek word for water.

The mathematical expression for this rule is Equation 7.1 of Investigation 7.1, which is called *the equation of hydrostatic equilibrium*. The neutral-buoyancy argument of this section, which tells us that objects immersed in a fluid (such as air or water) will sink if they weigh more than the fluid they displace, was first worked out by the Greek scientist and engineer Archimedes (about 287–212 BC).

A very simple application of the law of hydrostatic equilibrium is to the whole atmosphere. Consider a slender column of air, stretching from the ground to the top of the atmosphere. The pressure at the top is zero, so the pressure force difference from top to bottom is the pressure force at ground level, and this is the pressure there times the cross-sectional area of the column. This must exactly balance its weight.

▷ The fact that pressure is isotropic, exerting forces in all directions, is what makes hydraulic machines work: the tubes that hold the brake fluid in an automobile braking system transmit the pressure from the brake pedal to the wheels regardless of twists and turns in the line. Pascal invented the hydraulic press, so we owe our automobile braking systems to his insight. He also founded the mathematical theory of probability, so we owe the calculations of our automobile insurance premiums to his insight, too!

> The atmospheric pressure on any area of the ground must be large enough to support the entire weight of the column of air above that area. Since atmospheric pressure is measured to be $1.013 \times 10^5 \, \mathrm{N \, m^{-2}}$, the weight of the air above a balloon of radius 10 cm and cross-section $0.0314 \, \mathrm{m^2}$ is 3140 N, which by Equation 2.2 on page 10 is the weight of a mass of about 320 kg, or four heavy men.

The units for pressure, $\mathrm{N \, m^{-2}}$, are given a standard name, the pascal, denoted Pa. This is named after the French physicist and mathematician Blaise Pascal (1623–1662), who first pointed out the isotropy of pressure.

Pressure beats gravity again: Bernoulli lifts airplanes

Given that atmospheric pressure can support four heavy men, it is perhaps not so surprising that the atmospheric pressure difference across a small helium balloon can be enough to give the balloon a small push upwards. But balloons are small-fry: atmospheric pressure really shows its strength when it lifts airplanes.

In this section: we explain how even objects that are heavier than air can use air pressure to keep them aloft.

Investigation 7.1. The balance of pressure and gravity in an atmosphere

Here we shall translate into equations the words about hydrostatic equilibrium at the end of the section on buoyancy. We fix our attention on a certain cubical parcel of air in the middle of the atmosphere, whose imaginary boundaries are drawn simply to distinguish it from the rest of the atmosphere. To remain at rest, the net force on the gas in it must vanish, regardless of the fact that its sides are imaginary. Newton's laws apply to any collection of particles, even if they do not form a solid body.

We take the length h of the sides of the cube to be small enough that there is only a small change of pressure across the cube, and (consequently) that the density of the air inside the cube is essentially uniform.

The three forces that affect the vertical motion of the cube's contents are the pressure at the bottom, the pressure at the top, and the force of gravity (the weight of the parcel of air). Let us consider them in turn.

Pressure force on the bottom of the parcel. If we denote the pressure at the bottom by p_{bottom}, then the force on the bottom is the pressure times the area of the bottom, which is a square of side h:

$$F_{bottom} = p_{bottom} h^2.$$

Pressure force on the top. Let the change in the pressure going from the bottom to the top be Δp; since the pressure falls with height, we expect this will come out to be a negative number. Then the pressure at the top is $p_{top} = p_{bottom} + \Delta p$, and the pressure force on the top of the cube is

$$F_{top} = -(p_{bottom} + \Delta p)h^2$$
$$= -p_{bottom} h^2 - \Delta p\, h^2.$$

Here the overall minus sign is needed because this force points *downwards*, in the direction opposite to the force on the bottom.

Force of gravity. The weight of the parcel is its mass times the acceleration of gravity, g. The mass, in turn, is the volume h^3 of the parcel times the density of air inside the parcel, which we call ρ.

Since this force also points downwards, we have

$$F_{gravity} = -g\rho h^3.$$

Net force. These three forces must add up to zero for the atmosphere to be in equilibrium. The equation is

$$0 = F_{bottom} + F_{top} + F_{gravity}$$
$$= p_{bottom} h^2 - p_{bottom} h^2 - \Delta p\, h^2 - g\rho h^3$$
$$= -\Delta p\, h^2 - g\rho h^3$$

This can be solved for the pressure step going from bottom to top:

$$\Delta p = -g\rho h. \tag{7.1}$$

This is called the *equation of hydrostatic equilibrium*, written in terms of *differences* between the top and bottom of the parcel. It is accurate only as long as h is sufficiently small, since the density ρ used in the equation is the average density over the parcel, and there would be errors if the density changed much across the parcel.

Readers who know calculus will recognize the beginnings of a standard limit argument. (Others should ignore this paragraph and the next!) As h tends to zero, so does Δp. Since h is really the change of altitude z in the atmosphere, it is convenient to change notation and replace h by Δz. Then dividing Equation 7.1 by Δz and taking the limit gives

$$\lim_{\Delta z \to 0} \left(\frac{\Delta p}{\Delta z} \right) = \frac{dp}{dz} = -g\rho.$$

This is the calculus version of the equation of hydrostatic equilibrium, the version that most physicists would think of if asked to write down the equation. But our finite-difference version, Equation 7.1 above, is the starting point for physicists when they actually solve the equation on a computer. We shall build our computer program for constructing model atmospheres on the finite-difference version, which contains the same physics as the differential equation.

Exercise 7.1.1: *How fast does a helium balloon rise?*

The density of the helium in a balloon filled at, say, a fairground is 0.18 kg m⁻³, while the density of the air around it is 1.3 kg m⁻³. Using Equation 7.1, compute the pressure difference across the air that the balloon will displace, assuming for simplicity that it is a *cube* of side 20 cm. Then compute from this the net pressure force on the balloon itself when it is inflated and takes the place of the air. Next, compute the weight of the balloon (neglecting the rubber of the balloon itself) and calculate its initial acceleration (upwards). What multiple of g is this? Will it keep this acceleration as it moves upwards? How many balloons would be required to lift a 60 kg woman?

Simple buoyancy is not enough to lift a airplane. Being much heavier than the air it displaces, an airplane will not float upwards the way a dirigible does. To fly, an airplane creates an artificial kind of buoyancy: it uses its *speed* to create a greater pressure difference between the upper and lower surfaces of its wings than there is in the resting atmosphere. We have learned enough about Newton's laws now to be able to understand how this works.

The basic effect was first observed by the Swiss physicist Daniel Bernoulli (1700–1782): if a fluid speeds up to get past an obstacle, its pressure goes *down*. The reason is simply Newton's second law. To speed the gas up, there must be a net force on it in its direction of motion. A small piece of the fluid that moves around the obstacle does not touch the obstacle, so the only forces it experiences are pressure forces from the surrounding fluid. The acceleration must, therefore, ultimately come from the net pressure forces in the fluid. For pressure to give a net force in the direction of motion, the pressure must be larger behind the bit of the fluid we are considering, and smaller ahead of it. Therefore, when the fluid accelerates to pass around an obstruction, its moves from a region of higher pressure to one of lower: its pressure drops. When it passes the obstacle, it rejoins the flow of the rest of the fluid, so its

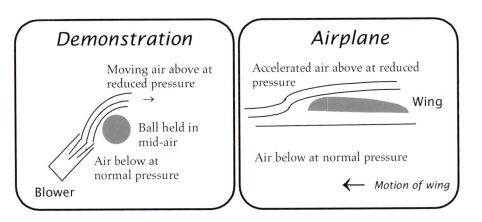

Figure 7.2. Illustrating the Bernoulli effect as it is often demonstrated in science exhibitions (left) and as it works to lift an airplane (right).

speed drops. To slow it down, the pressure must rise as it moves further forward. The minimum pressure will therefore be found at the place where the speed of the fluid is greatest.

Many readers will have seen a popular demonstration of the Bernoulli effect in a science museum or exhibit. It consists of a jet of air directed at a large inflated ball, which causes the ball just to hang in mid-air (see Figure 7.2). The ball hangs where the air from the jet will strike it just above its center and pass over it. This reduces the pressure of the air above the ball, increasing its buoyancy and stopping it from falling. This artificial buoyancy is called *lift*.

The wing of an airplane is shaped to make use of the same effect. It is rounded on top and flat below. As the airplane's engines push the airplane through the air, the air that passes above the wing has further to go, since the curved surface is longer than the flat one. The air passing above the wing accelerates, and its pressure drops. This provides the lift that the airplane needs to fly.

▷By extending wing flaps before landing, the pilot insures that air passing over the wing has further to travel, and therefore that it goes faster and provides more lift.

Pilots change the lift on a wing by modifying its shape with wing flaps. In this way they can provide enough lift at low speeds, while taking off or landing. Nature knew about the Bernoulli effect, of course, long before scientists did. Just like pilots, gliding birds shape their wings to take advantage of the same physics.

Helium balloons and the equivalence principle

In this section: how to use the equivalence principle to solve a difficult physics problem involving balloons.

Our discussion of buoyancy has a lovely link with our earlier study of the equivalence principle. The buoyancy of a helium-filled balloon gives rise to a classic "trick" physics problem, which we have all the preparation necessary to solve correctly. Try this experiment with your friends, but be sure the road is empty when you do!

Suppose you are in a car moving at a constant speed with the windows shut and the vents closed, and you are holding a helium-filled balloon by its string. The balloon keeps the string vertical. Now the car brakes hard. Which way does the balloon move: towards the front or the back of the car?

Are you ready for the answer? Consider the equivalence principle, which in its simplest form says that a uniform acceleration produces the same effects as a gravitational field in the opposite direction. So the environment inside the car when it brakes would be the same if we took the uniformly moving car and added a gravitational field pointing towards the front. Then the buoyancy of the balloon causes it to rise against gravity: in this case, to move towards the *back* of the car.

If you don't like to use the equivalence principle, you can work it out in terms of the pressure forces we discussed earlier, since the pressure of the air in the car will be higher at the front than at the back as the car decelerates. But you will still have to find something to play the role of the "weight" of the balloon when you work out the net horizontal force, since a rock would certainly not move towards the back of the car, despite the pressure difference. You are welcome to elaborate this argument (and of course a good physicist should be able to do so), but for my taste the equivalence principle provides a much more elegant approach to the solution.

Absolute zero: the coldest temperature of all

Buoyancy is only half of the story of how an atmosphere is supported. Now we turn to the other half. The temperature of an atmosphere is important because hotter gases tend to expand and have a lower density at a given pressure, so they are more buoyant.

In this section: by considering how the volume of a gas changes with its temperature, we are led to conclude that there must be an absolute zero of temperature.

What is the relation between temperature and volume? If a gas is cooled under fixed pressure, experiment shows that its volume decreases in direct proportion to its temperature; if we make twice as large a change in its temperature, its volume decreases by twice as much. Expressed as a word equation, this is

$$\text{change in volume} \ = \ \alpha \ \times \ \text{change in temperature,} \qquad (7.2)$$

where α is a constant that will depend on the pressure. This is called *Charles' Law*, named for the French physicist and mathematician Jacques-Alexandre-César Charles (1746–1823), and it has a remarkable consequence.

▷Charles also understood buoyancy and was prepared to stake his life on it: he was the first person to ascend in a hydrogen balloon, reaching heights well over 1 km.

Consider a finite volume V of the gas. If we keep reducing the temperature, eventually the decrease in volume in Equation 7.2 will equal V: the volume will have shrunk to zero! By Equation 7.2, this has required only a *finite* lowering of the temperature. The temperature at this point is surely the coldest one can make that gas. Therefore Charles' law implies that every gas has a *coldest* temperature.

Even more remarkable is that we can easily show that this coldest temperature is the *same* for every gas, no matter what pressure or volume we start with. Recall an everyday experience with temperature: if we place ice cubes into a warm drink, we get a cooler drink in which the ice has melted. In general, if we put two bodies that have different temperatures into contact, they gradually approach a common temperature somewhere in between the two. The warmer one cools and the colder one warms.

Now suppose the two bodies are two gases that have both been cooled to their own coldest-possible temperatures. If these temperatures are different, and we bring them into contact, then the warmer one will cool further. But this is a contradiction: it is already as cold as it can be. Therefore, our assumption that the coldest temperatures were different was wrong: the coldest temperature must be universal.

Charles' Law implies that there is a universal lowest temperature. Experiment reveals that this is about -273°C or -460°F. It is natural to define a new *absolute* temperature scale for which this temperature is defined to be zero. We call the coldest temperature *absolute zero*.

The absolute temperature scale (whose degree size is the same as the **celsius** scale) is called the **kelvin** scale, and temperatures are denoted by "K". The freezing point of water is about 273 K, and the boiling point of water is about 373 K. The advantage of using this scale is that, at constant pressure, the volume of a gas is zero when the temperature is zero, so Charles' law can be re-phrased to say that the volume is directly proportional to the absolute temperature. Thus, Charles law tells us that, if a given gas has volume V at atmospheric pressure and at absolute

▷It is conventional in scientific writing today just to write "K", not the old-fashioned "°K", after a temperature measurement.

temperature T, then at absolute temperature $2T$ and the same pressure it will have volume $2V$.

Why there is a coldest temperature: the random nature of heat

In this section: we learn about one of the fundamental achievements of physics, namely the understanding that temperature measures the energy of the random motions of atoms. This insight is fundamental to understanding atmospheres and, in later chapters, stars.

Figure 7.3. Ludwig Boltzmann was one of the giants of physics at the transition from the nineteenth to the twentieth century, when the foundations of modern physics were laid. In Boltzmann's time, not all physicists accepted that matter was composed of atoms. Many of the ideas Boltzmann used go back to Bernoulli, who argued that pressure could be caused by the random motion of small particles, and that this would naturally explain the increase of pressure with temperature. But Boltzmann put mathematics to these ideas, and turned them into a testable physical theory. By doing this, Boltzmann made a great step toward establishing the reality of atoms. Unfortunately, Boltzmann's story is a good deal more tragic than that of most other physicists: in his middle age, ill and depressed, at least partly by the resistance his ideas had met among older physicists, he committed suicide. He never understood that he had completely converted the younger generation of physicists to his point of view, and he did not know that his theories were actually on the verge of experimental verification. Image courtesy Österreichische Zentralbibliothek für Physik.

Although the universal nature of the absolute zero of temperature has been verified over and over again in laboratory experiments, there might be something unsatisfying about our approach to it so far: we have no real explanation, no real understanding of how this can be. A more satisfying explanation was provided by Boltzmann, whom we mentioned earlier in this chapter.

Boltzmann was the principal founder and exponent of the branch of physics that we now call **statistical mechanics**. (Other important contributions were made, independently, by James Clerk Maxwell – whom we will meet in Chapter 15 – and by the American physicist Willard Gibbs, 1839–1903.) Boltzmann showed that *all* the known properties of simple gases could be explained if one took the view that a gas was composed of atoms that moved randomly about inside a container, frequently colliding with each other and with the walls of the container. He showed that pressure was the result of the forces of all the small atoms hitting the walls randomly. To make his calculation work, he needed to make only one simple assumption about the relationship between the average kinetic energy of an atom in the gas[†] $\langle K \rangle_{\mathrm{avg}}$ and the absolute temperature:

$$\langle K \rangle_{\mathrm{avg}} = \left\langle \tfrac{1}{2}mv^2 \right\rangle_{\mathrm{avg}} = \tfrac{3}{2}kT, \tag{7.3}$$

where we have used Equation 6.8 on page 54, the definition of the kinetic energy for an atom of mass m. The constant k is called *Boltzmann's constant*, and it has the value

$$k = 1.38 \times 10^{-23} \ \mathrm{kg \ m^2 \ s^{-2} \ K^{-1}}.$$

Equation 7.3 is the quantitative form of the relation between temperature and kinetic energy that I referred to at the beginning of this chapter. We study Boltzmann's argument in more detail in Investigation 7.2 on page 78.

The idea that kinetic energy should be proportional to temperature was not just an arbitrary assumption. What Boltzmann showed was that when a large collection of atoms move and collide randomly, they tend to share out their kinetic energy equally: when a rapidly moving and a slowly moving atom collide, they usually both bounce off with speeds somewhere in between. This is so similar to what happens when hot and cold bodies are placed into contact, that Boltzmann drew what was to him an obvious conclusion: temperature essentially *is* the kinetic energy of a typical atom of the gas.

This leads, of course, to a simple explanation of *why* bodies in contact tend to approach the same temperature: their atoms at the point of contact tend to share energy, and when they collide with atoms behind them inside their respective bodies, this sharing tends to make all energies – hence both temperatures – the same. Moreover, and this is where our real interest is in this section, Boltzmann gives us a natural explanation for absolute zero: absolute zero is the temperature at which there is no longer any random kinetic energy inside the body. At absolute zero, all the atoms are perfectly at rest with respect to each other. The fact that this lowest temperature should be the same for all bodies is obvious in this picture.

Why does absolute zero lead to zero volume? Remember that in Charles' law the pressure is held constant, so there is always some pressure from outside on the

---

[†]The use of angle brackets $\langle \cdots \rangle_{\mathrm{avg}}$ is a conventional notation for a statistical average (also called the **mean**) over a large number of random events. In this case the average is over random motions of molecules.

gas. As its temperature decreases, the random motions of its atoms get slower, and their ability to resist compression decreases, so the volume decreases. Ultimately, when the atoms stop moving, they have no resistance to compression at all, and the volume goes to zero.

Notice that temperature is related to the *random* kinetic energy of the atoms. If we take a body at absolute zero and make it move at a constant speed, each atom will have a kinetic energy, but there will be no random motion: all the atoms are at rest with respect to one another. So the temperature will still be zero.

Although we have characterized absolute zero as a state in which the atoms of the gas stop moving, this state cannot actually be reached: no matter how one tries to remove kinetic energy from a gas, one will always do something to disturb it a little and leave a small amount behind. This may be very small, so one may try to get as close to zero as one likes; but absolute zero is unattainable. To date, temperatures below 0.001 K have been reached in small samples, and a metal bar weighing more than a ton has been cooled to below 0.1 K. (This bar has another connection with gravity: it is used in a gravitational wave detector, which we will discuss in Chapter 22.)

> These low temperatures may be the lowest ever seen anywhere in the Universe. We will see later that the Universe began as a hot gas (the Big Bang), and has been cooling off ever since. But there is a background of stray radiation left from the Big Bang that keeps the temperature of all natural objects above a minimum of about 2.7 K. To get colder than that probably requires some deliberate intervention. If the Earth contains the only intelligent life in the Universe, then cold temperatures may have existed only here.

The ideal gas

We have studied the way the volume of a gas depends on its temperature, but in doing so we held the pressure constant. We must now ask about changes in pressure.

In Boltzmann's picture, it is clear that, if we fix the volume of a gas and reduce its temperature to absolute zero, then the random motions of atoms go to zero, and the pressure (which results from the impacts of gas atoms on the walls of the container) must also go to zero. Conversely, as we raise the temperature at constant volume, we should expect the pressure to rise. Boltzmann showed by calculations what experiment had already confirmed: that the pressure is directly proportional to the temperature in these circumstances.

> These two laws can be combined into a single relation, which is called *the ideal gas equation of state*: the absolute temperature of a gas is proportional to the product of pressure and volume. This is expressed mathematically as:

$$pV \propto T. \tag{7.8}$$

This is the key relation for seeing how the density and pressure of the atmosphere change as we go up in altitude (see Figure 7.4 on page 79). In Investigation 7.2 on the next page we derive this relation from Boltzmann's point of view. We find there that the constant of proportionality in this equation is just Nk, where N is the total number of atoms in the gas and k is Boltzmann's constant.

Our work in Investigation 7.2 also gives us another important relation, namely that the typical velocity v of atoms in the gas can be found just from its density ρ and pressure p. Since sound waves in a gas are nothing more than some atoms

In this section: the simplest gas consists of independent atoms that collide with one another but do not stick together or lose energy through collisions. We learn that the pressure, volume, and temperature of such a gas have a simple relationship to one another, and that the sound speed depends on the ratio of pressure to density.

▷The symbol "$\propto$" stands for "is proportional to".

Investigation 7.2. *The ideal gas according to Boltzmann*

Before Boltzmann, scientists understood two simple relations among the pressure p, volume V, and absolute temperature T of a gas: $V \propto T$ with p fixed (Charles' law), and $p \propto T$ with V fixed. By multiplying these two equations, one gets a single relation among all three quantities:

$$pV = \beta T, \qquad (7.4)$$

where β is a constant, independent of p, V, or T. Boltzmann showed how to find β in terms of the atoms that make up the gas. We sketch his argument here.

Boltzmann observed that pressure represented the force that the gas exerts on the walls of its container, and by Newton's third law (Chapter 2) this is the same as the force exerted by the walls back on the gas. How does this force act? If the gas consists of atoms, then any wall will exert a force only while atoms are in contact with it, bouncing off it. The result of the force on the atom is to turn it around, to reverse the component of its velocity that is perpendicular to the wall. If we had to know the details of this process, such as how long it took the atom to turn around, how hard or soft the wall was, and so on, then the calculation would be hopelessly complicated.

Luckily, we are only interested in the *average* force exerted by the wall, so we can make a simplification: if the typical time between the collision of one atom with the wall and the collision of the next atom is Δt, then the average force exerted by the wall is the same as would be required to turn a single atom around during the time Δt. The reason is that, if the time it takes an atom to turn around is longer than Δt, then the force exerted on each atom will be less than we calculate, but at any time there will be several atoms being turned around, so the total force exerted by the wall will be the same as if one atom were turned around in the time Δt. A similar argument shows that this is the right *average* force when the time to turn an atom around is shorter than Δt, as well.

We shall now calculate how the pressure force depends on the atoms' average speed. If the average speed of an atom is v, then the component of a random atom's velocity perpendicular to the wall is proportional to v. The acceleration experienced by an atom turned around in time Δt will be the change in its velocity divided by Δt, so this will be *proportional to* $v/\Delta t$. If the atoms have mass m, then the pressure exerted by the wall will satisfy

$$p \propto mv/\Delta t.$$

Now we need Δt. The time between successive collisions of atoms with a wall will certainly depend on v: the faster the atoms travel, the smaller will be the interval between collisions. It also depends on the number of atoms per unit volume: the more atoms there are within a given distance of the wall, the more collisions there will be. The number per unit volume is the total number in the container, N, divided by its volume, V. We have thus argued that Δt is proportional to $1/v$ and to $1/(N/V)$, or that $\Delta t \propto V/Nv$.

When we put this together with the previous equation, we find

$$p \propto mv^2 \, N/V. \qquad (7.5)$$

Multiplying by V and using Equation 7.3 on page 76 to replace the typical value of v^2 by something proportional to the temperature T, we find

$$pV \propto NkT.$$

This is equivalent to Equation 7.4, and it tells us one more thing, namely that the constant β in that equation contains Nk, the total number of atoms in the gas and Boltzmann's constant.

Now, Boltzmann was able to do the calculation better than we did, because he was careful to do the averages over all the directions of the actual velocities of the atoms in the gas, so he could calculate the constants of proportionality in each of the steps. For his definition of k, as given in Equation 7.3 on page 76, he showed that the constant of proportionality in the above equation was just one:

$$pV = NkT. \qquad (7.6)$$

The argument we have given shows that this equation holds for all gases: atomic hydrogen, molecular oxygen, and inert helium all obey Equation 7.6. This is called the *ideal gas equation of state*.

This equation is "ideal" because we have made an oversimplification by assuming that the atoms interact with one another and with the walls only when they actually collide. In real gases, there can be various electric forces between atoms even when they are well separated, that make slight modifications in this equation. The real exceptions come when the gas becomes a liquid (or worse, a solid), as most gases do at low enough temperatures. Then the interactions between atoms become very strong, and the system cannot even be approximated as one composed of free atoms colliding occasionally.

One consequence of our derivation is an expression for how the pressure depends on the average kinetic energy of the atoms. By replacing kT in Equation 7.6 by $2\langle KE \rangle_{\text{avg}}/3$, we find

$$p = \frac{2}{3} \frac{N}{V} \langle KE \rangle_{\text{avg}}. \qquad (7.7)$$

We will use this equation when we study stars.

Another look at Equation 7.5 will show how, by making measurements on a gas, we can deduce the speed of its atoms. The quantity mN/V in the right-hand side of this equation is just the density ρ of the gas, the total mass per unit volume. So we learn that $v^2 \propto p/\rho$.

The ideal gas equation of state gives another perspective on "why" the helium-filled balloon rises. Both the balloon and the air it replaced had to have the same pressure, since they were surrounded by air with that pressure. They had the same volume and temperature, too, so they must therefore have had the same number N of atoms (or molecules, in the case of air). The force of gravity on the helium balloon is less because each atom of helium is so much lighter than an average molecule of air (which is a mixture of molecules of nitrogen, N_2, oxygen, O_2, and other gases).

Exercise 7.2.1: *How many atoms in a balloon?*

Consider the cubical helium-filled balloon of Exercise 7.1.1 on page 73. If the pressure inside the balloon is atmospheric pressure, $p = 10^5 \, \text{N m}^{-2}$, and the temperature is $T = 300 \, \text{K}$ (about 81 F), then use Equation 7.6 to calculate the number N of helium atoms in the balloon. The size of this answer justifies the approximation that we can average over large numbers of randomly moving atoms.

Exercise 7.2.2: *What is the mass of a helium atom?*

Use the answer to the previous exercise and the density of helium given in Exercise 7.1.1 on page 73 to calculate the mass of each helium atom. Use the density given for air to calculate the *average* mass of an air molecule. (Since air is a mixture of gases, we only obtain the *average* mass this way.)

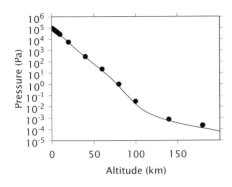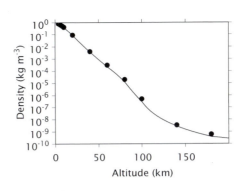

Figure 7.4. Comparison of predictions of the computer model of the Earth's atmosphere (solid lines) with the measured data (points).

bumping into others and making them bump into ones further away, the speed of sound is also given just by the pressure and density:

$$v^2_{sound} \approx p/\rho. \tag{7.9}$$

In our discussion of Boltzmann's picture of a gas as composed of atoms bouncing around, we did not really talk about what the atoms are. For Boltzmann, they were just little particles that somehow characterized the gas. In his gas laws, the "atoms" are whatever fundamental units the gas is composed of. Thus, if the gas consists of single atoms, as in helium gas, then the particles are the helium atoms. But if the gas is a molecular gas, such as oxygen, which normally exists as O_2, then Boltzmann's laws apply to the molecules. For example, the number N in the constant of proportionality in Equation 7.8 on page 77 would be the number of O_2 molecules, not the number of oxygen atoms.

An atmosphere at constant temperature

Imagine a column of air above a square drawn on the Earth. Let us go up from the Earth a small height, perhaps a few centimeters, so that there are N air molecules above the square to that height. Now imagine marking off successively higher steps, each of which makes a volume that contains the same number N of molecules. (These steps are not generally equally spaced in altitude, of course, because the density is decreasing – the air is getting "thinner".) If the air is still or moving slowly, then the forces on it must be in balance. What are these forces?

First, what is the gravitational force? We shall assume that the atmosphere does not extend very far from Earth, so that the acceleration of gravity g is constant everywhere inside. This is not a bad assumption, since the top of the atmosphere is certainly within 300 km of the ground, which is the altitude where many satellites orbit. This is less than 5% of the radius of the Earth, so to a reasonable approximation we can neglect the weakening of gravity as we go up. Then the gravitational force will be the mass of each volume times g.

Next we need to calculate the pressure forces. In order to be in equilibrium, the pressure force on the bottom of each volume must exceed the pressure force on its top by the weight of the molecules in the volume. Since each volume contains the same number of molecules, this weight is the same for each volume; and since we have constructed our volumes to have equal areas, the pressure change from one step to the next must be the same, all the way up. When the pressure falls to zero, we are at the top of the atmosphere.

To calculate the pressure changes, we have to have information about the way the temperature changes with height. In this section, we make the simplest assumption: we consider only the constant-temperature, or **isothermal**, atmosphere.

In this section: the simplest atmosphere to study is one with a uniform temperature. We show that it is of infinite extent: it cannot have a top boundary. Therefore, real atmospheres must have non-uniform temperatures that fall to zero at the top.

▷We assume, of course, that the temperature is not absolute zero!

Although not an entirely realistic representation of what happens on the Earth, it is nevertheless an instructive first example to think about, because it is easy to see what happens. Besides, as we shall see, this situation does arise in portions of other planetary atmospheres.

For a constant temperature, the volume of a gas is inversely proportional to its pressure. As the pressure goes down from one step to the next, the volume must go up in proportion. Now, since the cross-sectional area of each volume is the same, this means the height of each step increases in inverse proportion to the decreasing pressure. How far up do we have to go to get to the top of this atmosphere?

The answer is infinitely far. As the pressure drops towards zero, the height of each step increases to infinity, and we never quite reach the top.

If this seems strange, consider it from another point of view. The molecules in a gas at a non-zero temperature have a non-zero speed; recall that their average kinetic energy is proportional to the temperature. If we reach the top of the atmosphere with a non-zero temperature, then the molecules near the top will still have non-zero speeds. They will therefore not stay below the point we have taken to be the top: some of them will shoot above it, so that there will be gas above the point we thought was the top. This is a contradiction. Therefore, for a gas in equilibrium, the top of the atmosphere is not only a place of zero pressure; it is also a place where the temperature must fall to zero. Although the isothermal atmosphere is easy to calculate, it is not a good approximation to whole planetary atmospheres.

Figure 7.5. Measurements of temperature of the Earth's atmosphere (dots). The solid curve is simply a fit to the dots.

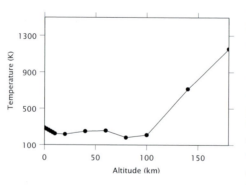

Before leaving this section, we pose a question we ignored earlier: why *is* it that the top of the atmosphere is marked by zero pressure? Why not by a finite positive pressure? (It can't be a negative pressure, because a gas cannot maintain a negative pressure. The physicists' term for negative pressure is **tension**.) If the atmosphere came to an end at a finite pressure, the gas just at the edge of the atmosphere would have a finite pressure below it, hence a finite force on it, with no force above it holding it down. We could consider a very thin layer of such gas at the edge, with tiny mass, so that the finite pressure below it would blow it away. Such an atmosphere could not remain in equilibrium. Therefore, the pressure in an atmosphere that is in equilibrium must go to zero at the top, and indeed it must do so in such a way that a small sliver of air at the edge feels a small pressure below it just sufficient to balance its small weight.

The Earth's atmosphere

In this section: we use a simple computer program to make an excellent numerical model of the Earth's atmosphere up to 250 km. Beyond that, the physics gets very complicated and the atmosphere cannot be represented as an ideal gas.

We can now look at the Earth's atmosphere in detail. The atmosphere is subject to many influences: physical forces from the Earth's rotation, heating from the Sun, and a huge number of chemical effects from chemical reactions and such processes as evaporation and precipitation. All of these affect the way the temperature behaves as one goes higher in altitude.

To try to understand this system in detail is a challenging research field. Our present understanding of some parts of the problem, such as the behavior of **ozone**, is seriously incomplete. One of the most important sets of data that researchers use in trying to understand the atmosphere is its *temperature profile*: the way the temperature behaves with altitude. If we use the measured values of the tempera-

---

Investigation 7.3. Making an atmosphere

Our purpose here is to explore the ideas that go into our computer program Atmosphere that makes a realistic model for the atmosphere of the Earth. The program can be found on the website. It relies on the discussion in the text of this chapter up to and including the section on the constant-temperature atmosphere. We will find that this program will be easy to modify to construct models of the Sun and other stars in later chapters.

Our technique is to use the equation of hydrostatic equilibrium, Equation 7.1 on page 73, to move upwards in the atmosphere with constant steps in altitude h.

Suppose that we let the variable z stand for altitude. At the ith step, we call the altitude z_i and the pressure and density p_i and ρ_i, respectively. Then at the next step the new altitude will be

$$z_{i+1} = z_i + h,$$

and Equation 7.1 on page 73 tells us that the new pressure will be

$$p_{i+1} = p_i - g\rho_i h. \tag{7.10}$$

To repeat this at the new height we need the new density ρ_{i+1}. We can get this from p_{i+1} if we know the temperature at this height, since then the ideal gas law, Equation 7.6 on page 78,

$$pV = NkT,$$

will give us the necessary information. Here is how to get the density from this.

The mass m of a parcel of air of a given volume V is the product of V and its density ρ. It is also equal to the number N of molecules it contains times the mass of each molecule. Since air is a mixture of gases, what we want is the average mass of a molecule. Now, molecules are basically composed of a few protons, an equal number of electrons, and a few neutrons. Moreover, the mass of a **neutron** is about equal to that of a proton, and is nearly two thousand times the mass of an electron. It follows that the mass of any molecule will essentially be the number of protons and neutrons times the mass of a proton m_p. The number of protons plus neutrons in a molecule is called its *molecular weight*. (For a single atom, scientists use the name *atomic weight* for the same thing.) The average mass of the molecules in a mixture of gases will be the average number of protons and neutrons times m_p. We shall call this average of molecular weights the *mean molecular weight* of the gas, and we use the symbol μ for it. Then we can write the *average* mass of a molecule of a gas as μm_p. For air, the mean molecular weight at sea level is $\mu = 29.0$. The two ways of calculating the mass of our parcel of air thus give

$$\rho V = N\mu m_p.$$

We can solve this for the ratio

$$\frac{N}{V} = \frac{\rho}{\mu m_p}. \tag{7.11}$$

From this and the ideal gas law we obtain

$$\rho = \left(\frac{\mu m_p}{k}\right)\frac{p}{T}. \tag{7.12}$$

This equation determines the density at each height, if we are given the pressure and temperature.

We have seen how we get the pressure by using the equation of hydrostatic equilibrium. How do we get the temperature?

As explained in the text, the temperature of the atmosphere is determined by a very complicated set of influences, and it would

be hopeless to try to model them in a simple computer program. Instead, we shall rely on the fact that the temperature of the atmosphere can be *measured* at different altitudes. These values are then put into the computer program, which uses them in the density calculation.

Our computer program, available on the website, handles this in a straightforward manner. Built into it are values for the fundamental constants, for μ, and for p_0, the pressure at the bottom of the atmosphere. It also has the values of the altitude at which the temperature is measured, followed by the values of the temperature at those altitudes. These constitute a *temperature table*. As the comments in the program explain, the temperature at altitudes where it is not measured is approximated by taking the measured temperature at the two nearest altitudes and assuming it behaves as a straight line between them. Thus, if the temperature is known to have the values T_1 and T_2 at heights z_1 and z_2, respectively, then its value at height z somewhere between is

$$T(z) = \frac{T_2 - T_1}{z_2 - z_1}(z - z_1) + T_1.$$

The program gets underway by computing a few useful numbers. In particular, it is useful to compute the so-called **scale-height** h_{scale}, which is a *rough* guide to the eventual height of the atmosphere. This is obtained by taking Δp in Equation 7.1 on page 73 to equal the pressure difference between the bottom and top of the atmosphere, and solving for h. Since at the top of the atmosphere $p = 0$, this gives

$$h_{\text{scale}} = p(0)/g\rho(0). \tag{7.13}$$

The program uses this as a guide for choosing the step size h. This is useful, for otherwise too small a value of h would waste computer time and too large an h would be inaccurate. The program ignores h_{scale} after this, so it does not assume that the atmosphere actually terminates there.

Then the program enters its main loop, stepping upwards in altitude until the pressure goes negative. There are some built-in safety measures to prevent the program taking too many steps.

There is one place where the program deliberately departs from realism, and that is at the top of the atmosphere. As explained in the text, the Earth's atmosphere becomes isothermal at high altitude, and so does not fit our notion of a finite atmosphere: it would in principle go on forever. In practice, the temperature is so high that the atmosphere is an **ionized** gas, and magnetic fields play an important role in what happens at these altitudes. The amount of material out there is so small, however, that we make little error at lower altitudes if we simply substitute an artificial cutoff at a high altitude. We do this by assuming that, above the highest altitude at which the temperature is supplied in the temperature table, the temperature is determined by the density by a relation of the form

$$T \propto p^{1/2}. \tag{7.14}$$

This is artificial, but it brings the atmosphere neatly to a termination. We will see in the next chapter that the relation between temperature and pressure inside stars is not unlike this equation.

The results of the calculation are plotted in Figure 7.4 on page 79. They are remarkably good, especially considering that our computational method is in essence very simple. We have not even introduced the predictor–corrector orbit program Orbit, and yet we still have been able to follow the density and pressure as they decrease by a factor of more than 10^5.

Exercise 7.3.1: *Finding values between measured points*

Show that the equation used above for finding the temperature between measured points, $T(z) = (T_2 - T_1)(z - z_1)/(z_2 - z_1) + T_1$, does in fact describe a straight-line relationship between the height z and the temperature $T(z)$. Show that the line passes through the measured points (z_1, T_1) and (z_2, T_2).

ture, as shown in Figure 7.5 on page 80, then we can construct an accurate model of the equilibrium atmosphere without understanding all the forces that shape the temperature profile. Data values from Figure 7.5 are used in the computer program on the website that we use to construct this model.

This temperature profile shows dramatic changes of temperature; the changes mark the boundaries of different layers of the atmosphere, where different physical processes take place. The temperature initially falls slowly with height, as anyone who has traveled in mountains would expect, until one reaches an altitude of about 15 km, where the stratosphere begins. In the stratosphere, the temperature gradually rises with height. This is mainly caused by the absorption of ultraviolet light from the Sun by the *ozone*. Then comes the mesosphere at about 50 km, where T falls again, until one reaches the thermosphere at about 100 km. From there the temperature rises dramatically and reaches a roughly constant value of 1500 K out to very great distances. So the outer regions of the atmosphere are in fact isothermal!

▷Ozone is O_3, and is formed by a number of chemical processes. It is an excellent absorber of ultraviolet light, shielding us from this damaging solar radiation. Man-made chemicals have eroded levels of ozone, especially in polar regions, increasing the exposure of people in those areas to ultraviolet radiation.

We have seen in the previous section that our simple atmospheres do not have a top if they are isothermal. In the case of the Earth, the outer regions are anything but simple! The gas there is ionized, magnetic fields are important, and the influence of the solar wind (see Chapter 8) begins to be important. Fortunately, the amount of mass in the outer reaches of the atmosphere is so small that whatever happens there has little effect on the structure of the lower atmosphere, and our computer program makes a very good model of this region, up to some 250 km (see Figure 7.4 on page 79).

Figure 7.4 shows that there are slight deviations between our model and the real atmosphere. These are mostly due to two effects we have left out: first, the composition of the atmosphere changes as one goes up; and second, the strength of the Earth's gravity decreases slowly as one goes higher. Despite these small differences, it is remarkable that we have been able to model the atmosphere so accurately with so little sophisticated mathematics.

If we want to go much beyond this, however, we are quickly humbled by the complexity of the physics. To explain the temperature profile would require an enormous computer program that contains not only the solar heating and the chemistry at different altitudes but also the dynamics of the atmosphere: convection of gas from one layer to the next, the influence of weather and storms, the many other time-dependent effects. To calculate the weather in any detail also requires complex programs, to take into account the variation of atmospheric properties from place to place, the effect of geography and variations in ocean temperature, the heating by the Sun, the transfers of energy and water between oceans and air, and so on. These areas are among the most active and challenging research areas in all of science today, and they make computing demands that exceed the capacity of the biggest available computers.

The atmospheres of other planets

In this section: temperature measurements on some other planets allow us to build models of their atmospheres too.

We can adapt the computer program to give us models of the atmospheres of other planets and indeed of some of the moons in the Solar System, just by replacing appropriate numbers. In Table 7.1 I have gathered the necessary data for three bodies: Venus, Mars, and Saturn's moon Titan. The output of the revised computer model for Venus is displayed in Figure 7.6.

Notice that all the temperature profiles include a temperature inversion: a place where the temperature begins rising again with altitude. This effect, which occurs because of the absorption of sunlight by the atmosphere, is seen in all Solar System atmospheres. Regarding Mars, the structure of its atmosphere is variable with time,

	Venus				Mars				Titan			
	$p_0 = 9.4 \times 10^6$ Pa, $\mu = 43.2$, $g = 8.6$ m s^{-2}				$p_0 = 730$ Pa, $\mu = 43.5$, $g = 3.74$ m s^{-2}				$p_0 = 1.6 \times 10^5$ Pa, $\mu = 28$, $g = 1.44$ m s^{-2}			
h (km)	0	57	90	135	0	10	70	100	0	40	60	600
T (K)	730	290	170	200	230	205	140	140	96	74	160	175

Table 7.1. Atmospheric data (altitude h and temperature T) for some Solar System bodies.

being strongly affected by dust storms, the seasons, and the latitude. Our model gives only an average structure. Titan, Saturn's largest satellite, is bigger than Mercury and retains a significant atmosphere. This is composed primarily of molecular nitrogen, but its history and evolution are shrouded in mystery. The temperature profile in Table 7.1 bears a striking resemblance to that of the Earth.

Quantum theory and absolute zero

To end this chapter, we turn to a matter of considerable importance to some topics we will discuss later in the book. I introduce it here in order to correct a misimpression that I deliberately allowed earlier, regarding what happens as the temperature of a gas is lowered to zero. According to Boltzmann, the motions of molecules will also go to zero as T goes to zero, so that a gas at absolute zero consists of molecules completely at rest (or at least it would if we could actually attain it). This turns out not to be quite what happens in the real world, because of the principles of the aspect of physics that we call **quantum theory**. We cannot go into quantum theory in any detail in this book, but we will use one of its most important principles in various places throughout the book: the *Heisenberg uncertainty principle*.

The brilliant German physicist Werner K Heisenberg (1901–1976) (see Figure 7.7 on the next page) was one of the founders of quantum theory. The principle that bears his name states that all measurable properties of any physical system come in certain pairs, which have a special relationship to one another: as one measures one member of the pair more accurately (say, with a better experiment), the other member inevitably becomes harder to measure accurately. Even given perfect measuring instruments, there is a minimum uncertainty in the measurement of the second member that is inversely proportional to the uncertainty in the measurement of the first. The constant of proportionality is known as *Planck's constant h*, and it has the value

$$h = 6.626 \times 10^{-34} \text{ kg m}^2 \text{ s}^{-1}.$$

We shall meet the man after whom the constant is named, Max Planck, in Chapter 10.

The value of h is so small that the uncertainties that are the subject of quantum theory rarely intrude into everyday measurements: our measuring instruments are typically too crude to be limited by quantum uncertainties. But it is not just actual measurements that are limited by these uncertainties: the limits apply to anything that *could in principle be measured*. This is how the uncertainty principle affects the behavior of matter.

A particularly important pair of measurable quantities consists of the position and the momentum of a particle. The

In this section: why atoms continue to vibrate even as as gas approaches absolute zero. These "zero-point" vibrations, a consequence of quantum theory, are responsible for many phenomena in the Universe that we will study later, involving neutron stars, black holes, and the Big Bang.

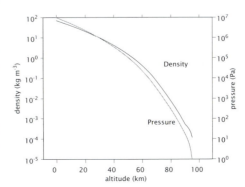

Figure 7.6. The atmosphere of Venus.

Figure 7.7. Werner Heisenberg formulated the theory now known as quantum mechanics when he was only 24 years old, during a holiday on an island in the North Sea where he had gone to escape hay fever. A year later a very different but equivalent formulation was achieved by the Austrian physicist Erwin Schrödinger (1887–1961). Heisenberg was a complex figure. Rather than leave Germany when Hitler began attacking Jewish physicists and even his own work on quantum mechanics, he remained and led Germany's unsuccessful program to build atomic weapons during the Second World War. After the war he continued to play a distinguished role in German physics. His wartime activities and his later attempts to portray himself as having been fundamentally opposed to creating nuclear weapons for Hitler have led to a continuing debate among historians and physicists about Heisenberg's character. His secret wartime visit to Nils Bohr (whom we will meet in Chapter 10) became the subject of the thought-provoking play Copenhagen, *by Michael Frayne. Photograph courtesy Mary Evans Picture Library.*

momentum is defined as the mass of the particle times its velocity, so the uncertainty here is basically between position and velocity. If the position is known (or could in principle be known) to great precision, then the particle's velocity would be very uncertain. This would have the consequence that a moment later, the position would be completely unknown: pinning the particle down at one moment forces it to squirt off in some completely random direction the next moment. For this reason, an accurate position measurement is not repeatable.

Now consider the gas at absolute zero. Boltzmann, who worked before the invention of quantum theory, would have expected that the molecules would be completely at rest. Thus, at least in principle, it would be possible to measure the position and (zero) velocity of each particle with arbitrary precision. Boltzmann would not have given this a second thought. But quantum theory forbids it.

Instead, what happens at absolute zero temperature is that each atom tries to reduce its velocity (hence its momentum) to as small a value as possible. This requires it to have as large an uncertainty in position as possible. If the gas is confined to a certain container, then the size of the container determines the maximum position uncertainty, and the minimum velocity uncertainty. Each particle retains a small average kinetic energy, whose size then depends on the volume of the container. In solids cooled to near zero, this is called the **zero-point energy** of vibration.

This illustrates the deep property of quantum theory, that the uncertainty is not just a value that is imprecisely known. The quantity that is uncertain can be thought of as undergoing random **quantum fluctuations** within this uncertainty. The energy of a cold particle is not just uncertain: the particle vibrates with this minimum energy even near the absolute zero of temperature.

In the Earth's atmosphere, the effect is completely negligible. But in some of the things we will study later, in systems as diverse as neutron stars, gravitational wave detectors, black holes, and the Universe itself, the uncertainty and its associated quantum fluctuations and quantum zero-point energy will be of critical importance.

8

Gravity in the Sun:
keeping the heat on

We have seen how the Sun's gravity holds the planets in their orbits. The Sun's gravity also holds itself together. Like all stars, the Sun is a seething cauldron, its center a huge continuous hydrogen bomb trying to blow itself apart, restrained only by the immense force of its own gravity. In this chapter, we will see how the Sun has managed to maintain an impressively steady balance for billions of years. In the course of our study, we will learn about how light carries energy and we will build a computer model of the Sun.

Sunburn shows that light comes in packets, called photons

The Sun glows so brightly because it is hot. We can infer just how hot it is from its color. The color and temperature of the Sun are related to each other in just the same way as for hot objects on the Earth. For example, watch the burner of an electric stove as it gets hotter; it changes in color from black to red. It won't get any hotter than red-hot. But if you watch objects in a really hot fire, such as a blacksmith uses, you will see them change from red to a blueish white as they heat up.

> As the temperature of an object increases, the radiation it emits moves toward shorter wavelengths, i.e. from red toward blue.

This change in color comes about in the following way. We saw in Chapter 7 that in hotter objects the molecules and atoms move faster. This means that when they collide and emit radiation, the radiation usually has higher energy. Now, it is a remarkable fact, which we will explore here, that higher-energy radiation has shorter wavelength: blue light is more energetic than red. It follows from these two observations that hotter objects tend to be bluer.

That light carries energy is obvious to everyone: the warmth of sunlight is caused by the conversion of the energy carried by the light into thermal energy (random kinetic energy) in our bodies. The fact that light of a certain *color* carries a *specific amount of energy* is a deeper property of physics, but it can be illustrated with an equally commonplace event: getting sunburned.

On a clear hot day, if you have sensitive skin, it does not take long to get a good red sunburn. But if you apply a blocking sunscreen lotion, you can remain in the same sunlight for hours without a burn. The lotion acts like a "filter" that prevents light of wavelength shorter than a certain ultraviolet wavelength from reaching your skin. No matter how much light of other colors reaches the skin, no matter how much energy in total the sunlight transfers to your skin, if it does not have a short enough wavelength it will not do the damage. There is clearly something different about the longer wavelengths of light. We will see that the difference is that the longer wavelengths of light do not carry enough energy to set off the chemical reactions in the skin that lead to sunburn.

The relation between the energy and the wavelength of electromagnetic radiation was discovered by Einstein. It was part of his explanation of the **photoelectric**

In this chapter: we learn how the Sun holds itself up. The key is another discovery of Einstein, that light actually comes in packets called photons. These form a gas that helps support the Sun. Photons move randomly in the Sun, taking millions of years to get out. We compute the structure of the Sun, and learn why stars and planets are round, while asteroids and comets are lumpy. Finally we study the vibrations of the Sun, which reveal the details of the Sun's interior to astronomers.

In this section: to understand stars, and in particular the Sun, we first learn about photons: packets of light whose energy is proportional to their frequency. The simple phenomenon of sunburn illustrates the way photons behave. The idea of a photon was first introduced by Einstein.

▷ The image beneath the text on this page is a picture of the Sun taken by the SOHO spacecraft on 14 September 1999, through a special filter. It shows a *superprominence*, the large loop of hot gas streaming out of the Sun. When such a prominence moves towards the Earth it can disrupt communication and electricity supplies, and cause aurora. The Sun is a turbulent, violent ball of gas that is only kept together by the strong force of its self-gravity. Image courtesy NASA/ESA.

effect, which is a metallic version of sunburn. It had been observed that light falling on certain metals can eject electrons, but only if the light has a short enough wavelength. This threshold wavelength depended upon the metal. As the wavelength of the light decreased further, the electrons came out with more and more kinetic energy.

Einstein proposed that light actually comes in discrete packets, which we now call **photons** or **quanta**. Each photon carries a fixed amount of energy that can be transferred to an electron or other particle if the photon collides with it. This energy can then be converted into kinetic energy of the electron. Einstein suggested that the energy of a photon is determined entirely by its wavelength: the shorter the wavelength, the more energy. He then proposed that each metal has what is effectively an "escape speed" caused by the attraction of molecular forces inside the metal, so that if the kinetic energy given to an electron by a photon were too small, it would not attain this speed and would therefore not be ejected. Once the wavelength of the photon was short enough to give the electron its escape speed, the electron would use up a certain amount of its kinetic energy escaping, and the rest would turn up as kinetic energy of the ejected electron. This is analogous to what happens when spacecraft escape from the Earth.

▷Physicists call the minimum energy for escape the *work function* of the metal.

This neatly explained all the experiments on the photoelectric effect, but it was nevertheless a revolutionary step in physics. Physicists had been used to thinking of light as a wave. A water wave's energy depends on its height, not its wavelength: we avoid swimming in the sea if the waves are large, not if they have very short spacing! The idea that light waves carried energy in discrete amounts, which depended on the wavelength, meant that scientists had to start thinking about light as if it were a particle. This took some getting used to.

> But the experimental evidence in favor of Einstein's proposal is overwhelming, and this so-called wave–particle duality of light is something that modern physics has come to embrace, even if it is a little hard to visualize in concrete terms. It is a fundamental aspect of quantum theory. Light behaves like a wave in some respects, for example when it refracts or interferes, and like a particle in other respects, such as by carrying fixed amounts of energy.

We shall more often refer to photons in the rest of this book than to light waves. Photons make a host of astronomical facts easier to understand.

The relation between wavelength and energy that Einstein proposed is remarkable because Einstein did not need to introduce a new constant of Nature to make the theory fit the observations: he only needed to use ones that were already known to be important for the physics of light: the speed of light, c, and Planck's constant h. Planck's constant had only recently been introduced by Max Planck to describe the spectrum of the radiation emitted by hot bodies. We have already encountered it in Chapter 7, where we saw how it plays a fundamental role in the uncertainty principle. We shall introduce its importance for light here, but defer a discussion of Planck's original use for it until we study the colors of stars in general in Chapter 10.

▷The Greek letter λ, called *lambda* and pronounced "lam-da", is standard physics notation for the wavelength of a wave.

Einstein showed that the energy carried by a photon of wavelength λ is inversely proportional to λ, the constant of proportionality being h times the speed of light c:

$$\text{energy } E \text{ of a photon} = hc/\lambda. \tag{8.1}$$

This relation is described more fully in Investigation 8.1.

Investigation 8.1. The colors of energy

Here we learn how to find the energy carried by a photon of a given color. Einstein's postulate for the photoelectric effect led, with other developments, to the quantum theory. In quantum theory, light (or any other electromagnetic radiation) is really composed of *photons*, which can be thought of as little packets of energy. The amount of energy E carried by each photon packet is directly proportional to the *frequency* of the light, and the proportionality constant is h, *Planck's constant*, which we met in Chapter 7:

$$E = hf, \tag{8.2}$$

where f is the frequency of the light (measured in units of cycles per second, which scientists call Hertz, denoted Hz). Notice that this equation is in fact a further illustration of the close relationship between energy and time that we first met in Chapter 6.

Because we often think in terms of the *wavelength* λ of light rather than its frequency f, we shall convert this equation using the relation between wavelength and frequency for a wave whose wave speed is c (which in our case is the speed of light):

$$f = c/\lambda, \tag{8.3}$$

where the speed of light has the value $c = 2.998 \times 10^8 \text{ m s}^{-1}$. Then we find that the energy of a photon is

$$E = hc/\lambda. \tag{8.4}$$

Visible light has a wavelength in the range 0.4–0.7 μm. (One μm is 10^{-6} m, and is sometimes called a **micron**. Readers who are used to old-style units may prefer **Ångstroms**; one micron is 10^4 Å.)

If we insert the values of h and c into the previous equation and then multiply it by 1 μm/1 μm (which equals 1, of course), we get

$$E = \left(\frac{1.986 \times 10^{-25} \text{ J m}}{\lambda} \right) \left(\frac{10^{-6} \text{ m}}{10^{-6} \text{ m}} \right)$$

$$\approx 2 \times 10^{-19} \left(\frac{10^{-6} \text{ m}}{\lambda} \right) \text{ J}. \tag{8.5}$$

This is a handy way of writing Equation 8.4 in a way that shows the scale of energies involved, and allows one to do the arithmetic more easily in one's head. A 1 μm photon (infrared light) carries about 2×10^{-19} J of energy. A green photon with a wavelength of 0.5 μm has twice this energy.

These energies are very small in everyday terms, as Exercise 8.1.2 shows. From the result of that exercise, it is not surprising that the eye is unaware of the discrete nature of the packets of energy that keep striking it: so many arrive per second that they merge into a continuous stream of energy. But these packets, or quanta, of energy do play an important role in a huge variety of situations, from the workings of individual atoms to the structure of stars.

Exercise 8.1.1: *Frequency of light*

Find the frequency (in Hz) of light whose wavelength is 0.5 μm.

Exercise 8.1.2: *Photons from a light-bulb*

Show that a 100 W light bulb (which emits 100 J of energy each second) must be giving off something like 10^{21} photons per second.

Exercise 8.1.3: *Sunburn*

The DNA molecules that carry genetic information in the nuclei of living cells are very sensitive to light with a wavelength of 0.26 μm, which breaks up DNA molecules. Deduce from this the binding energy of the chemical bonds within DNA. Ultraviolet light of wavelength 0.28 μm is the most effective for inducing sunburn. What is the threshold energy required to stimulate the chemical reactions that lead to sunburn?

Exercise 8.1.4: *Gamma-rays*

When some elementary particles decay, they give off so-called **gamma-rays**, which are really high-energy photons. A typical energy released in this way is 10^{-12} J. What is the wavelength of such a gamma-ray? What is its frequency?

A gas made of photons

If photons behave like particles, colliding with electrons and exchanging energy with them, then there can be circumstances in which it would make sense to speak of a *photon gas* mixed with an ordinary gas of electrons and ions, in which collisions between photons and gas particles would be as common as between the gas particles themselves. This happens inside stars, where the density of gas is so great that, as we explain later in this chapter, the photons bounce off atoms of the gas a fantastic number of times before they reach the surface. At each bounce they exchange energy with the gas.

This has two important effects. First, the collisions with photons exert *pressure* on the gas particles; and second, the exchange of energy with gas particles means that the photons in any part of the Sun come into *equilibrium* with the gas: the typical energy of a photon is roughly the same as that of a gas particle. Thus, the photon gas really behaves like a gas: it has a temperature and a pressure.

Now, the energy of a gas particle is determined by the temperature of the gas, and this in turn must equal the energy of a typical photon if collisions occur often enough to insure that energy is frequently exchanged between particles.

There is thus a characteristic photon wavelength associated with any

In this section: radiation can form a gas of its own, which provides some of the pressure that holds up stars. We estimate the temperature of the radiation from the Sun from a its color. We also introduce a new unit for energy, the electron volt.

Figure 8.1. Albert Einstein in 1905, when he was still working at the patent office in Bern, Switzerland. He revolutionized research in three different fields with his scientific papers that year. Photo courtesy ETH Library Picture Archive, Zürich.

▷ The reason for the name *electron volt* is that 1 eV is the energy acquired by an electron when it is accelerated by an electric field corresponding to a difference in potential of 1 V. Electrons inside television monitors are accelerated by a difference of several thousand volts before they are directed at the screen.

In this section: Einstein published five extraordinary papers in a single year, 1905, making breakthroughs in three different problems.

▷ This was the first discussion in physics of the *random walk*, to which we will return later in this chapter when we discuss the diffusion of light through the Sun.

temperature. The wavelength of the photons emitted by the Sun is an indication of the temperature of the gas near the Sun's surface.

Now we can come back to sunburn and use it to estimate the Sun's temperature. The fact that the Sun emits significant amounts of ultraviolet light means that we will get what physicists call an "order-of-magnitude" answer if we just set the thermal kinetic energy of particles in the Sun, $3kT/2$, equal to the energy of an ultraviolet photon, which we computed in Exercise 8.1.3 on the previous page to be about 7×10^{-19} J. Solving for T gives $T = 34\,000$ K. We should expect this to overestimate the temperature of the Sun, perhaps by as much as a factor of 5 or 10, since most of the Sun's light comes out in the visible region, not the ultraviolet. But at least it tells us that the surface of the Sun is at least several thousands of degrees, but not as high as several million. Often in physics such order-of-magnitude estimates are all one needs to get a reasonable understanding of a physical phenomenon.

Here is the appropriate place to introduce a new unit of measure for energy, one that is better suited to the tiny energies of photons and atomic interactions than the joule: it is not very convenient to keep writing numbers like 10^{-19} J. Physicists have introduced the unit called the **electron volt**, abbreviated eV, which is equal to

$$1\,\text{eV} = 1.602 \times 10^{-19}\,\text{J}. \tag{8.6}$$

The energy carried by one-micron (infrared) light is thus about 1.24 eV.

We do a better job of estimating the Sun's temperature in Investigation 8.2, by using the wavelength at which the Sun is brightest, and we improve on this method even further in the next chapter. The temperature of the surface of the Sun is actually about 5800 K.

At this temperature, ordinary materials cannot exist in the solid or liquid state: the Sun is a ball of hot gas consisting of electrons and ions, called a **plasma**. It is important to remember that the temperature we measure from the color of the Sun is only its *surface* temperature, since the light we see comes only from a thin surface layer called the **photosphere**. Inside the Sun temperatures rise sharply, to around 10^7 K in the center. We shall see why later in this chapter. We don't see the light from this region directly because of all the collisions that photons undergo inside the Sun. We only see the photons that have finally escaped after their last collision.

Einstein in 1905

Einstein's paper on the photoelectric effect was one of *five* landmark papers that he published in one extraordinary year, 1905. Besides introducing the quantum nature of light in order to explain the photoelectric effect, his other papers were equally revolutionary. Two of them established the theory of special relativity, to which we will return in Chapter 15. The other two explained the so-called **Brownian motion**, in which microscopic specks of dust floating on the surface of water had been observed to execute completely random motions. Einstein showed that huge numbers of random collisions with molecules, each making an unobservably tiny change in the motion of the dust, would add up to the observed motions. The Brownian-motion papers helped to establish the correctness of Boltzmann's atomic theory, and Einstein was even able to calculate for the first time the average masses of the molecules.

Einstein's paper on the photoelectric effect was one of the papers that founded quantum theory. The fact that Einstein founded three fields of physics in a single year (while holding down a full-time job in the patent office in Bern, Switzerland) is as impressive an indication of his genius as is his monumental work on general relativity ten years later, which occupies the second half of this book.

Investigation 8.2. How hot is the Sun?

Here we want to infer the temperature of the Sun from the wavelength of the radiation it emits. We can rewrite Equation 8.5 on page 87 in terms of electron volts to give

$$E = 1.24 \left(\frac{1 \, \mu m}{\lambda} \right) \text{eV}. \tag{8.7}$$

As we mentioned in the text, there is a characteristic photon wavelength associated with any temperature. This is given by setting E in Equation 8.7 equal to $\frac{3}{2}kT$ from Equation 7.3 on page 76, where k

is Boltzmann's constant. The result is that

$$\lambda \approx \left(\frac{9600 \, \text{K}}{T} \right) \mu m. \tag{8.8}$$

In the center of the Sun, where the temperature reaches 10^7 K, photons have typical equilibrium wavelengths of 0.001 μm and energies exceeding 1000 eV. Throughout the Sun, the energies of photons are enough to ionize hydrogen, and to strip electrons from other atoms too. Such a gas of electrons and ions is called a *plasma*.

Exercise 8.2.1: *Temperature of the Sun*

If one analyzes the colors of the Sun, one finds that the greatest amount of light is emitted in the *blue-green* region of the spectrum, around 0.5 μm. Show that this gives an *estimate* of the Sun's temperature of $T = 19\,000$ K. This is closer to the real temperature (5600 K) than our estimate in the text, but we will get a much better estimate by refining this technique in the next chapter. (The eye sees the Sun as yellow, not blue-green, partly because it has greater sensitivity to yellow light and partly because blue light is scattered by the atmosphere.)

Einstein was also the right man at the right time. Physicists were just learning how to probe the world of atoms and particles, and this world did not behave the way they expected, based on their experience with **macroscopic** objects. Einstein had an extraordinary ability to free his mind from prejudices and begin thinking in the new ways that atomic physics demanded. And not just to think, but to calculate, to make predictions that could be tested by experiment.

Interestingly, when Einstein received the 1921 Nobel Prize for physics, it was for the photoelectric effect. His work on relativity was explicitly excluded, since in the eyes of the awarding committee, it had not yet been sufficiently confirmed.

Gravity keeps the Sun round

We know the mass (1.99×10^{30} kg) and radius (6.96×10^8 m) of the Sun, and from them we can work out that the mean density (mass divided by volume) of the Sun is about 1400 kg m^{-3}, or 1.4 times the density of water. To compress a gas whose interior temperature is several million degrees to beyond the density of water requires a great deal of force.

What is this force in the Sun? The answer can only be gravity, the gravitational attraction of one part of the Sun for another. This mutual attraction would, if unresisted, simply pull the material of the Sun inward towards a single point. The resistance to this collapse is provided mainly by gas pressure, and secondarily by the pressure provided by all the photons that are produced in the center of the Sun and gradually make their way outwards, scattering off electrons and nuclei in the Sun countless times as they go. The Sun exists in a state of balance between the outward push of gas and radiation pressure (the pressure of the photon gas) and the inward pull of gravity.

The photons produced in the center come from **nuclear reactions**, which are processes that change nuclei of some atoms into other nuclei. These reactions release a great deal of energy, and are the chief source of the energy that makes the Sun (and all other stars) shine. The energy from these reactions leaves the Sun in two main forms: as photons and as **neutrino**s, which are very light particles produced in many nuclear reactions. We will look at how nuclear reactions work in Chapter 11. For now, we just assume that there is an energy source in a small region around the center of the Sun.

The shape of the Sun is also determined by gravity. Gravity is, like pressure, an *isotropic* force, that is a force that has no preferred direction: the gravitational attraction exerted by any particle is the same in all directions. As long as the Sun is

In this section: gravity singles out no special direction, nor does pressure, so stars and other large bodies are basically round. However, smaller bodies can be irregular in shape if chemical forces are significant. We calculate that any body with more mass than 1/1000th of the mass of the Earth should be round: gravity should dominate chemistry. This fits well with observations in the Solar System.

a balance between pressure and gravity, it can't help but form a ball that is round. If there were corners or other special places on its surface, then if we stood at the center of the Sun and looked outwards, there would be some directions different from other directions. Since there is nothing in gravity or pressure to single out these directions, they cannot exist: the Sun should be a sphere.

We have left out of this discussion three extra influences that *can* single out directions: rotation, a magnetic field, and the presence of a nearby gravitating body, such as a companion star in a binary system. We will discuss rotation below, since it does affect the Sun. Many stars have magnetic fields similar to that of the Earth, with a North and South magnetic pole. Pulsars, which we shall study in Chapter 20, have fields an incredible 10^{12} times stronger than the Earth's. If the field is strong enough, it can cause the star to have a distorted shape, particularly near the poles. The Sun's field is not that strong. Even the superprominence illustrated in the figure on page 85 contains a negligible amount of mass. Jupiter acts like a companion "star", distorting the shape of the Sun by tidal effects (see Chapter 5), but the effect is too small to measure. We will return to a more detailed discussion of the distorting effects of companions when we meet binary stars in Chapter 13. All in all, the Sun has no choice but to be spherical.

The arguments of the last paragraphs apply in fact to any astronomical bodies for which gas pressure and gravity are the main forces. But there are many astronomical bodies that are not round, because other effects dominate. Dust grains are whisker-shaped because of chemical forces; **asteroids** are irregular because the chemical forces that shape their rocks are as important as gravity; and on a very large scale, galaxies can be disk-shaped or cigar-shaped because the motions of large numbers of individual stars define their outlines.

In the case of asteroids, we can combine Boltzmann's understanding of the kinetic energy of a molecule (Chapter 7) with what we learned in Chapter 6 about escape speeds to answer the following elementary question.

> Why are planets and moons round, while ordinary rocks and even asteroids and the cores of comets have corners?

Figure 8.2. *Phobos is one of the two moons of Mars, and has a mass smaller than the number we calculate in Investigation 8.3 to be the minimum mass for a rocky body to be forced by gravity to be round. Its irregular shape is consistent with our calculation. Image courtesy* NASA.

The answer has to do with melting. If, when a body was formed in empty space, temperatures got high enough to melt it, then gravity would, as we have just argued, make it spherical, as long as rotation, magnetic fields, and tidal effects were not too important. Since the atoms and molecules that form planets heat one another by colliding as they fall together, their kinetic energy when they collide must be comparable to their gravitational potential energy. Given a molecule of mass m, falling onto a planetary body of mass M and radius R, its kinetic energy when it arrives will be something like GMm/R. The collisions randomize the direction of this energy, turning it into heat. The temperature, according to Boltzmann, will be given by setting $3kT/2$ equal to this energy. In Investigation 8.3 we put these expressions together and find that a rocky body in our Solar System should be round if its mass exceeds 3×10^{21} kg. This number depends on some assumptions, especially that the body is composed of silicate

---

Investigation 8.3. Why the Moon is round

We do a little algebra here to find the minimum mass M a body would need to have in order to melt as it forms. Once molten, gravity will shape it into a sphere. But if it does not melt, then it can have any irregular shape.

A molecule of mass m falling onto the body will have a kinetic energy approximately equal to GMm/R, where R is the size of the body (its radius, if it is spherical). This is an approximation, and in fact the energy could be more, but we are only interested in a rough answer. The collision transforms this into random kinetic energy, or heat, with a temperature T given by Boltzmann's relation

$$\frac{3}{2}kT \approx \frac{GMm}{R}.$$

We want to find the mass M required to make T high enough to melt the material. So we will assume that we know T, and set it equal to the melting point of rocks when we start doing numbers below. We shall also take m to be the mass of a typical molecule in a rock crystal when we do the numerical calculation. So we would like to solve this equation for M in terms of T and m, but we don't yet know R, the size of the body. To get a sensible value for R let us assume that we know the density ρ of the body, which we will take later to be the density of rock. Knowing ρ gives us a further approximation (again assuming a roughly spherical body)

$$\rho \approx M/\left(\tfrac{4}{3}\pi R^3\right),$$

which can be solved for R to give

$$R \approx M^{1/3}\rho^{-1/3}.$$

From now on I will ignore factors like the $4\pi/3$ in the density equation, since our answers are only going to be rough order-of-magnitude approximations anyway. If we put this into our first equation and solve for M we find

$$M \approx \left(\frac{kT}{Gm}\right)^{3/2} \rho^{-1/2}, \tag{8.9}$$

again dropping simple numerical factors.

Now, Solar System bodies typically have the density of rocks, about $\rho = 6000 \text{ kg m}^{-3}$. Let us take the molecule to be SiO_2, the main constituent of sand. The silicon **nucleus** contains 14 protons and 14 neutrons, and each oxygen nucleus contains 8 protons and 8 neutrons. Altogether, there are 60 protons and neutrons in one molecule. The mass of the molecule is about 60 times the mass m_p of a proton. (We neglect the small mass difference between protons and neutrons, and we neglect the mass of the electrons, which are a fraction of a percent of the mass of the nuclear particles.) Looking up the mass m_p in the Appendix, we find that the mass of the molecule is $m = 1 \times 10^{-25}$ kg. Finally, the melting temperature of silicon dioxide is about 2000 K. Putting all these into Equation 8.9, we find the minimum mass of a round body in the Solar System to be about

$$M \approx 3 \times 10^{21} \text{ kg.} \tag{8.10}$$

This is about 5% of the mass of the Moon, and much larger than the mass of any known asteroid or comet.

Exercise 8.3.1: *Rounding off the Moon*

Do the algebra that leads to Equation 8.9 from the two equations that precede it. Then put the given numbers into the formula to arrive at Equation 8.10.

---

rocks. For icy bodies, the mass would be a bit smaller. We should treat this as an order-of-magnitude estimate of the smallest mass of a round astronomical body.

For comparison, the mass of the Moon is 7.3×10^{22} kg, so its round shape is no surprise. Our minimum "round" mass is much larger than the mass of any known asteroid or comet, so we should expect asteroids and comets to have irregular shapes, as indeed they all do. In fact, many planetary moons in the Solar System are of smaller mass. For example, Mars' moon Phobos has a mass of 10^{16} kg and is very irregular, as Figure 8.2 shows. In fact, the largest irregular body in the Solar System is Saturn's moon Hyperion, whose mass is 1.8×10^{19} kg. So our rough calculation is not bad.

The Sun is one big atmosphere

Because the Sun is a balance between gravity and pressure, it is like one giant atmosphere. All the discussion of Chapter 7 can be directly applied here to help us understand the Sun's structure. We will extend the computer program of that chapter to help us make a numerical model of the Sun, and in the next chapter we will apply it to building other stars as well.

What changes do we need to make to apply our atmosphere program to the Sun? One obvious one is that the atmosphere program assumed we were dealing with the gravity of the Earth, not of the Sun. Changing this is not just a matter of changing the value of g, the acceleration due to gravity. For an atmosphere, which is a thin layer sitting on top of a big planet, one can assume without losing too much accuracy that the acceleration due to gravity is the same everywhere in the atmosphere. This assumption does not work for the Sun. For example, at the center of the Sun the acceleration must be zero, since particles are being pulled by the different parts of

In this section: the structure of the Sun is described by the same basic equations as we used to determine the structure of the Earth's atmosphere.

Investigation 8.4. How the gas in the Sun behaves

As mentioned in the text, we do not have direct measurements of temperature inside the Sun. We instead assume that the physics inside the Sun can be summarized by a relatively simple *equation of state*: a relation between pressure, temperature, and density. We shall use what physicists call a **power-law** relationship between density and pressure, that is a relationship where one variable is proportional to the other raised to a constant power (exponent). The usual way physicists write this is:

$$p = C\rho^\gamma. \qquad (8.11)$$

Astrophysicists call this a *polytropic* equation of state, which is another word for "power-law", but they adopt a somewhat strange way of writing the exponent. They use a *polytropic index n* in place of the polytropic exponent γ, defined by

$$n = \frac{1}{\gamma - 1}, \quad \text{or} \quad \gamma = 1 + \frac{1}{n}. \qquad (8.12)$$

A star with a polytropic equation of state is called a polytrope.

Given the ideal gas law (Equation 7.12 on page 81) for a gas with mean molecular weight μ, we can solve it for the pressure to obtain

$$p = \frac{k}{\mu m_p}\rho T. \qquad (8.13)$$

It follows that the temperature can be expressed in terms of the pressure by the power-law given in Equation 8.15,

$$T = A\,p^\beta,$$

with the constants

$$\beta = 1 - \frac{1}{\gamma} = \frac{1}{n+1}, \quad \text{and} \quad A = \frac{\mu m_p}{k}C^{1/\gamma}, \qquad (8.14)$$

where C is the proportionality constant in Equation 8.11. The constant C can be determined if the pressure and temperature are given in one place, say at the center of the Sun.

Assuming the Sun to be a polytrope is in fact not as arbitrary as it might seem. In regions that are dominated by convection of heat from the interior of the Sun, and in regions where most of the pressure is provided by the radiation making its way outwards through the Sun, the equation of state does indeed follow such power-laws. The physics hidden behind this statement is a little beyond our scope here. We shall simply adopt the polytropic equation of state and look at the models it produces.

---

the Sun equally in all directions.

The acceleration due to gravity inside the Sun is not hard to compute, however. Recall that in Chapter 4 we saw that inside a spherical shell, the force of gravity due to the shell is zero, whereas outside it the force is the same as if all the mass were concentrated at the central point. If we consider a point inside the Sun, then if we draw a sphere about the Sun's center through the point in question, the sphere divides the Sun into an "inside" and an "outside". The material outside the sphere does not contribute to the gravity, while that inside acts from the center. The acceleration due to gravity at the point is, therefore, just the acceleration due to that part of the mass of the Sun that is within the radius in question. This means that the computer has to keep track of how the mass of the Sun is increasing as we go out in radius, but computers are good at doing such things.

The other part of the program that needs to be modified is the computation of the temperature. This is not quite as easy to handle as gravity, because it requires a discussion of gases that goes beyond our treatment in Chapter 7. This is the subject of the next section.

Figure 8.3. *Comparing our solar model with the Standard Solar Model. Our computer model assumes that the relation between pressure p and density ρ is $p \propto \rho^{1.357}$ everywhere, and then adjusts the constant of proportionality and the central pressure to set the model's total mass and radius to those of the Sun. Of course, this relation is only an approximation to the real physics inside the Sun, so our model cannot describe all the details of the Standard Model. In particular, it overestimates the temperature everywhere.*

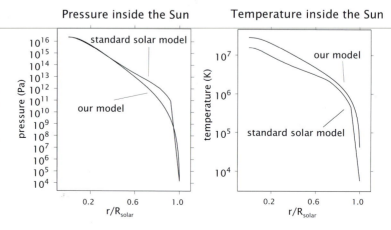

The Standard Model of the Sun

The biggest difference between modeling the Sun and modeling the Earth's atmosphere is that the interior of the Sun is not directly observable: we can't send a spacecraft into it to measure the temperature down in the center! In Chapter 7 we were able simply to specify the temperature of the Earth's atmosphere as a function of height, allowing direct observation to replace detailed modeling of the chemistry and physics that lead to the specific temperatures observed. For the Sun, we have no choice: we must put in all the physics to make the best model. When all this is done – including the nuclear energy generation in the center, the pressure of radiation flowing outwards through the gas, and the convective motions of the gas as it is heated from below – and when all the unknown variables are adjusted so that the model matches whatever we can observe – such as the radius, mass, composition, age, neutrino radiation, and total energy output of the Sun – the resulting model is called "the Standard Model" of the Sun.

In this section: we discuss how to modify the computer program we used for the Earth's atmosphere to make a model of the Sun. When astrophysicists put in all the physics that they believe to be relevant in the Sun, they obtain what is called the Standard Model, their best guess about what the Sun is like deep inside.

We can't attempt here to recreate the Standard Solar Model: many of the best astrophysicists in the world spend their working lives trying to make good models of the Sun and other stars! Instead, we will employ some approximations that are well grounded both in the science of gases – thermodynamics – and in experience in astrophysics. We have seen in Investigation 8.2 on page 89 that the trapping of photons in the Sun brings the photons into temperature equilibrium with the gas. It is therefore possible as a first approximation to ignore the *flow* of energy outwards and to treat the Sun as a static ball of gas and radiation. Then the next simplifying approximation is to assume that the pressure and temperature are related to one another by the following simple expression, called a *power-law*:

▷The Sun's age must be greater than that of the oldest rocks on the Earth, but presumably not much greater.

$$T = A \, p^{\beta}, \tag{8.15}$$

where A and β are constants.

▷A power-law is just a relationship between two quantities in which one of them is proportional to the other raised to a fixed power.

Although this may look like a great over-simplification, we will see that if we choose the power β appropriately, we can make a fair approximation to the Standard Solar Model. A star that has such a power-law equation of state relating pressure, density, and temperature is called a **polytrope**. We explore these ideas further in Investigation 8.4.

▷Of course, all the complicated physics is hidden away in determining the right values of β and the constant of proportionality. We will simply choose the values that give us the best approximation.

The structure of the Sun

In Investigation 8.5 on the next page we assemble the various elements of our discussion above into an outline of the computer program we use to construct our model of the Sun. In this section, we can take a look at the output from this program to see what we can learn about the Sun itself. We display the output in the form of the two graphs in Figure 8.3. These graphs show the pressure and temperature of both our model and the Standard Model, as a function of radius within the Sun. The radius is expressed as a fraction of the full solar radius $R_{\odot}$, which is $R_{\odot} = 6.96 \times 10^8$ m.

In this section: we discuss the features of our computer model of the Sun. Although we have left out much of the physics, the model is a remarkably good approximation to the Standard Model.

In these graphs we compare the output of our computer program with one of the most recent versions of the Standard Model. To arrive at the model displayed in this figure, I have experimented with a few choices of the value of the index n. For each n, I found the values of the central pressure and temperature that gave the model the same total radius and mass as the Sun. The value of n that seemed to give the best overall approximation to the Standard Model was $n = 2.8$, whose model is shown in Figure 8.3. (Pressure is shown in the first graph in Figure 8.3, but density is not displayed.) The appropriate values of central pressure and temperature are given in the computer program Star on the website.

Investigation 8.5. A computer model of the Sun

The changes that we need to make to our planetary atmospheres computer program Atmosphere on the website to adapt it to the Sun are relatively minor, and they make the program considerably *simpler*. The result will be a new computer program called Star.

The first change is to recognize that the acceleration due to gravity changes as one goes outwards through the star. As we noted in the text, the local gravity at any distance r from the center of the Sun depends only on the mass inside the sphere of radius r. We therefore define a variable called $m(r)$ (the array M in the program) whose value is the mass inside a sphere of radius r. If we move outwards from the sphere of radius r to one of radius $r + h$, with h very small compared to r, then the difference between $m(r)$ and $m(r + h)$ will be the mass inside the thin shell of thickness h. If the shell is thin enough, then the density inside it will be essentially the same everywhere, and equal to $\rho(r)$. The volume of such a shell is its thickness h times its area $4\pi r^2$, so we have the new equation for the increase of mass as one goes outwards:

$$m(r + h) = m(r) + 4\pi r^2 \rho(r)\, h. \qquad (8.16)$$

Thus, as we go outwards, we need not only to decrease the pressure according to the equation of hydrostatic equilibrium, Equation 7.1 on page 73, but also to increase the mass at each step. This leads to other minor changes in the program, such as (1) the fact that in place of the acceleration of gravity g in the pressure equation we have to use Newton's law,

$$g = \frac{Gm(r)}{r^2}, \qquad (8.17)$$

and (2) in the computer program the variable G no longer stands for g but instead for Newton's constant G.

The second change is the treatment of temperature. Since we adopt the polytropic law in Equation 8.15 on the previous page, we can get rid of all the steps in the program Atmosphere that had to do with reading in the temperature profile and using it to calculate the temperature at any height. Instead, once the pressure is known at any radius r, the temperature and density can be calculated immediately from it.

These are the only two changes in the program that come from the physics. We need also, however, to consider a technical point about how to get the computer to solve the equations. This is because there is one danger lurking in Equation 8.17. Since the radius of the Sun starts out at zero at the center, a careless programmer could wind up asking the computer to divide by zero to calculate the local acceleration of gravity. Computers don't like to do this, and they usually crash, stopping the program with an error.

First of all, we must be sure that there is no real problem with Equation 8.17 at the center of the Sun: $m(r)$ is zero there too, so we must see what happens near the center, as r gets smaller. Mathematically, we say we are looking at the *limit* $r \to 0$ to see whether the ratio $m(r)/r^2$ is in fact well-behaved.

The density reaches a maximum at the center: let us call it ρ_c (the variable rhoC in the program). Consider a tiny sphere of radius r about the origin. Within it, the density will not change much from place to place, so its mass will be the density times the volume of the sphere,

$$m(r) \approx \tfrac{4}{3}\pi r^3 \rho_c \qquad \text{for small enough } r.$$

Then the acceleration of gravity at such a small radius r is

$$g(r) \approx \tfrac{4}{3}\pi G \rho_c r.$$

In the limit as r gets smaller and smaller, this equation becomes exact (the approximation sign $\approx$ is replaced by equality), and we have

$$\lim_{r \to 0} g(r) = 0.$$

It should not be surprising that the acceleration of gravity at the center is zero: the Sun's gravity is pulling on the center equally in all directions, thus cancelling itself out and leaving no net pull at all.

The problem is therefore only to avoid asking the computer to perform a division at zero. We take the easiest way around this: we start the computer at radius h, the first step away from the center. We set the values at the center to the obvious ones: $m = 0$, $p = p_c$ (given as initial data), $T = T_c$. This allows one to find the constant C and therefore to find $\rho = \rho_c$ from the polytropic equation Equation 8.11 on page 92. Then at $r = h$ we set $m = \tfrac{4}{3}\pi h^3 \rho_c$ and we approximate $p \approx p_c$ and $\rho \approx \rho_c$. (Because the pressure and density reach a maximum at the center, they do not change much from $r = 0$ to $r = h$, so these approximations are reasonably good if we take h small.)

Then we step outwards as we have done in Atmosphere. There are more accurate ways of starting out at $r = h$, but provided h is small enough our method is good enough. The interested reader is encouraged to try to invent better ways and to test them.

The program on the website incorporates the changes I have described, and with its comments it should be relatively straightforward to understand.

One sees from the figure that the model's pressure fits reasonably well overall, but that it is too low in the central region. A graph of density would show a similar trend. The temperature is not so good, being overestimated everywhere in our model by a factor of about two.

These inaccuracies are caused by two things. First, the composition of the Sun changes from inside to outside, because nuclear reactions are generating much heavier elements in the center; in the model we have assumed that the mean molecular weight μ in Equation 8.13 on page 92 is everywhere equal to its value at the surface. The surface value is 1.285, while the Standard Model takes $\mu = 1.997$ at the center. The second complication is that the effective value of the index n should be allowed to change with radius, partly because we have left out radiation pressure from the photon gas, and partly because the gas in the Sun outside $0.74R_\odot$ is in steady convection, bubbling outwards and then sinking downwards in a long, slow rolling motion.

Despite the differences, the agreement between our simple model and the very much more elaborate Standard Model is gratifying: considering that the Sun's central temperature is more than 1000 times larger than the surface temperature, our overestimate by a factor of two is a relatively small error. Using only the most el-

---

Investigation 8.6. Using the computer to model the Sun

In using the computer program Star described in Investigation 8.5, one has to face a big difference from the planetary atmospheres case: we have no direct observations of the interior of the Sun, so we do not know directly what values to adopt for the pressure and temperature there. These are needed to start the calculation off. All we know is what the real radius and mass of the Sun are, and these are the *results* of the computer program. It might seem that we are stuck with a trial-and-error approach to finding the structure of the Sun: try a set (p_c, T_c) and use the program to find what they predict about M and R. Then choose different starting values to see if the results are closer to or further from the true values $M_\odot$ and $R_\odot$.

In fact, there is a more systematic way to guide one's choices of new values for p_c and T_c. If we look at the equation of hydrostatic equilibrium for the Sun,

$$\Delta p = -\frac{Gm(r)\rho(r)}{r^2}h,$$

then we may ask what happens if we take h to be R itself: take one giant step from the center to the surface. Then Δp will be the difference between the surface pressure, which is zero, and the central pressure p_c: $\Delta p = -p_c$. If we take the mass term $m(r)$ to be the total mass M, the density $\rho(r)$ to be ρ_c, and r to be R, then although the equation is not accurate, it gives us a starting approximation for p_c:

$$p_c = \frac{GM\rho_c}{R}.$$

Next, we treat the mass equation, Equation 8.16, the same way. If we imagine that the central density is the density everywhere, then one jump from center to surface gives

$$M = \frac{4}{3}\pi R^3 \rho_c. \tag{8.18}$$

By solving this for ρ_c and substituting it into the equation for p_c we get

$$p_c = \frac{3GM^2}{4\pi R^4} \propto \frac{M^2}{R^4}. \tag{8.19}$$

I have dropped all the constants in the second form and simply written it as a proportionality, because with all of our approximations we cannot trust the actual value this equation will produce. But it does tell us something extremely important: given two stars with the same polytropic equation of state, we can expect that their central pressures will scale with mass and radius in approximately the way given by Equation 8.19.

The ideal gas law tells us how the central temperature behaves, and with the same approximations it gives

$$T_c \propto \frac{M}{R}. \tag{8.20}$$

This very simple equation helps us treat the central pressure in the same way as the central density.

Now we can see how we can correct erroneous values of the central pressure and temperature. Suppose we start with central values p_1

and T_1, and suppose they give a model with mass M_1 and radius R_1. We want to find starting values p_2 and T_2 that will give us the right values $M_\odot$ and $R_\odot$. We write down the relevant proportionalities:

$$p_1 \propto \frac{M_1^2}{R_1^4}, \quad T_1 \propto \frac{M_1}{R_1}, \quad p_2 \propto \frac{M_\odot^2}{R_\odot^4}, \quad T_2 \propto \frac{M_\odot}{R_\odot}.$$

Assuming that the constants of proportionality are the same in each case (a big assumption: we will come back to this), we can divide the second set of equations by the first to obtain

$$p_2 = p_1 \left(\frac{M_\odot}{M_1}\right)^2 \left(\frac{R_1}{R_\odot}\right)^4, \quad T_2 = T_1 \left(\frac{M_\odot}{M_1}\right)\left(\frac{R_1}{R_\odot}\right). \tag{8.21}$$

This allows an *intelligent* correction of the first guesses for p_c and T_c, and can be used over and over again until the values of M_1 and R_1 converge to $M_\odot$ and $R_\odot$, respectively. But will it work? After all, the approximations don't seem very convincing: maybe the "constants" of proportionality in Equations 8.19 and 8.20 will depend on the structure of the star in some way that will make them change with changes in the central values of p and T, thus invalidating Equation 8.21.

The general answer is that if the first values p_1 and T_1 are pretty close to the right ones, then the constants of proportionality can't change much, and the corrected values of p_c and T_c will be better than the old ones. The procedure will close in on the right model if we repeat the corrections often enough, provided we start close enough to the right answer in the first place.

The proof of a pudding is in the eating. Try the method on the computer model. (Don't use the values of p_c and T_c supplied in the computer program, since they are the "right" ones.) You will be pleasantly surprised: you will find you need only *one* correction to get very close to the right mass and radius, no matter how far from the correct values you start! The method actually works better than we should expect. The reason is an "accident" that we have not made use of: for polytropes, the proportionalities in Equations 8.19 and 8.20 are in fact strict proportionalities, provided one keeps the polytropic index fixed.

The last open question is, what is the right index for the polytropic equation of state? We cannot answer this here: the Sun is not really a polytrope anyway. All we can do is find a polytropic index that comes close to the structure of the Standard Model. After some experimentation, I have settled on the value of 2.8 for the variable called index. For a model with the solar mass and radius, this gives a pressure curve that is quite close to that of the Standard Model, and a temperature curve that is usually within a factor of two of the standard one. This is about the best one can do with a single polytropic equation of state valid everywhere. The graphs in Figure 8.3 on page 92 show that our model is not a bad representation of the Sun, but it is clear that we would have to put in more physics to get all the detail right.

---

ementary techniques, and only observed data at the surface of the Sun (its radius, mass, and composition), we have learned quite a bit about the unseen regions of the Sun.

How photons randomly 'walk' through the Sun

Let us now put together two facts of common experience to learn a little about what happens to photons inside the Sun. The first fact is that the light that comes from the Sun's surface is visible light, with some ultraviolet. It can burn our skin, but it isn't strong enough to get inside our bodies, like X-rays do. The second fact is that radioactivity commonly produces *gamma-rays*, which are light waves that have more than enough energy to damage the insides of our bodies. Now, if the Sun is powered by nuclear reactions, then it seems sensible to assume that the nuclear reactions in the Sun's center also produce gamma-rays. Why then do we get visible

In this section: remarkably, it is not easy for a photon generated deep inside the Sun to get out: it takes millions of years. We show that this is due to random scattering from gas particles. We construct a computer program to describe what mathematicians call a random walk.

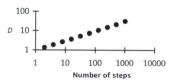

NET DISTANCE TRAVELED

Figure 8.4. *The net distance traveled in a random walk, as a function of the number of steps in the walk. The horizontal scale shows the number of steps, and the vertical scale is the average distance from the origin that the walk finishes, in units of the average length of each step of the walk. This is the output of the computer program* Random *from the website.*

▷Compton scattering is named after its discoverer, the American physicist Arthur Holly Compton (1892–1962). The phenomenon of scattering, in which a photon can lose or gain energy, is a further demonstration that light behaves like particles as well as like a wave.

light from the surface, and not gamma-rays?

The answer has to do with the photons' energy: the energetic gamma-rays produced inside are somehow losing energy before they reach the surface. The only way this can happen is through scattering: photons scattering from electrons and getting into temperature equilibrium with them. One of the biggest over-simplifications of our computer model is that we have neglected the transport of photon energy outwards through the Sun. In fact, getting the photon energy to the surface to be radiated away is a surprisingly long and tortuous process.

We will come back to the nuclear reactions that generate the Sun's energy in Chapter 11, where we discuss them in the context of all stars. Here we simply assume that energy in the form of photons is being released in the center: what happens then?

The radiation at the center of the Sun is very energetic, having been produced as gamma-rays. Each photon has millions of electron volts of energy when it is produced. But the gas in the Sun is really a plasma of individual charged particles, and photons scatter off charged particles very easily. This scattering is called **Compton scattering**, and the resulting exchanges of energy between electrons and photons lead to thermal equilibrium between matter and radiation: both have the same temperature.

If we want to decide how important scattering is inside the Sun, the key question is, how far can a photon travel between one scattering and another? The answer is surprisingly easy to work out from some simple basic numbers, and we do this in Investigation 8.7. We see there that the typical distance that a photon can travel in the Sun before it Compton scatters off another particle is no more than about 3.6 cm, which is a fraction 5×10^{-11} of the radius of the Sun!

If the photon were to travel on a straight line from the center to the surface, it would scatter 2×10^{10} times before emerging. This is certainly a lot of scatterings, so it is not surprising that the photon loses energy as it goes along. But in fact it cannot move on a straight line, since every scattering changes its direction of travel. The photon executes what mathematicians call a *random walk*, moving in random directions with steps of average length 3.6 cm. We show, using the simple computer program Random, described in Investigation 8.7, how a photon makes gradual progress outwards in this random, aimless way.

In fact the photon must scatter about 2×10^{10} *squared* times – 4×10^{20} times! – before it reaches the surface of the Sun. In doing so it will travel a total distance equal to 2×10^{10} times the radius of the Sun in order to get out, which at the speed of light takes more than a thousand years! Photons don't stream outwards; they diffuse very gradually.

Of course, it over-simplifies matters to imagine that a photon retains its identity all the way along this walk. In fact, photons are often absorbed by ions, and new ones are sometimes generated when charged particles collide. But the calculation still tells us how long it takes the energy carried by photons to diffuse outwards.

One effect of all this scattering is that, wherever the photon finds itself, it will be part of a photon gas that is in temperature equilibrium with the gas of the Sun. All the initial energy of the photon is lost quickly, and it adopts the energy of the particles (the free electrons and ions) that it is scattering from. Only at the very surface of the Sun does the probability that the photon will escape without a further scattering become large. This surface of last scattering is called the *photosphere* of the Sun, and the photons that come to us from it have the energy of the gas there, not the energy they started with at the center.

Investigation 8.7. The aimless walk of a photon through the Sun

The Sun is a dense cloud of electrons and ions, so it is not an easy place to be a photon. Photons scatter from charged particles, and ignore electrically neutral particles. Photons will also scatter from neutral *atoms*, because they actually encounter the electrons orbiting around the nuclei of the atoms, and they scatter from these.

It is not hard to estimate how far a photon can go before it scatters. To do so, we need two numbers: how many scatterers there are in a given volume of the Sun, and how "big" a scatterer is. The photon's problem is a bit like that of the ball in a pinball machine: if the scatterers are big enough, and if there are enough of them, then the photon can't go far without running into one.

We shall get a minimum estimate of how much scattering takes place by assuming that the photons scatter only from electrons and protons. In fact, in the Sun (as in most stars), ions of other elements contribute a very large amount to the scattering. Scientists use the word *opacity* to describe the amount of scattering, and ions of elements heavier than hydrogen and helium provide most of the opacity in the Sun. So our calculation here sets a lower limit on the opacity.

Assuming our scatterers are just electrons, what is their "size"? This size really refers to a kind of sphere of influence: how close a photon can get to the electron before it has to scatter. In quantum theory, the electron is not a solid particle of fixed size, but it does have a well-defined range of influence on photons, which is given roughly by what physicists call the "classical electron radius". Its value is about $r_e = 2.8 \times 10^{-15}$ m. (See the Appendix for a more accurate value.)

This means that an electron presents an area to the photon equal to the cross-sectional area of a sphere of the same radius, which is $2\pi r_e^2$. (This underestimates the actual effective cross-sectional area for scattering by about 2/3, but this is close enough for our calculation.) If the photon comes within this "target" area around the position of the electron, it will scatter strongly. If it passes further away, it may still scatter more weakly, but we will not make a huge mistake if we just treat the electron as a solid target of radius r_e.

The number of electrons per unit volume in the Sun is easy to calculate: the Sun is mainly hydrogen, which has just one electron and one proton per atom, and almost all of these atoms are actually ionized: the electrons and protons are separated. Therefore, the number of electrons equals the number of protons. Essentially all the mass of the Sun is in its protons, since the mass of an electron is only about 1/2000th of the mass of a proton. The number of protons in the Sun is then roughly the mass of the Sun divided by the mass of a proton: $N_p = M_\odot/m_p$. Looking up these numbers in the Appendix, we find $N_p = 1.2 \times 10^{57}$. This is also then the number of electrons in the Sun. The average number per unit volume is this divided by the volume of the Sun. The Sun's radius is $R_\odot = 7 \times 10^8$ m, so its volume is $V_\odot = 4\pi R_\odot^3/3 = 1.4 \times 10^{27}$ m^3. The average number of electrons per unit volume, n_e, is the ratio: $n_e = N_e/V_\odot = 8.4 \times 10^{29}$ electrons per cubic meter.

We now calculate the average distance a photon can travel before it encounters an electron. Imagine the electrons as being solid balls of radius r_e, distributed randomly around the present location of our photon. If the photon moves in some directions, it will immediately run into an electron. In other directions, it will miss the nearby ones and travel a larger distance before hitting one. Imagine drawing a sphere around the photon's present location. If this sphere is sufficiently small, the photon will have a pretty good chance of reaching it without hitting an electron. If the sphere is large, the photon will almost certainly hit an electron before it reaches the sphere. The sphere which the photon has roughly a 50–50 chance of reaching before it encounters an electron must be the one which contains just enough electrons that their cross-sectional areas equal the area of the sphere. We could just cover the inside of such a sphere if we arranged the electrons uniformly. In fact, they are arranged randomly, overlapping in places and leaving gaps elsewhere, so all we can say here is that this sphere is about the right size for a ray randomly directed outwards from the center to have a good chance of hitting an electron no matter what direction it takes.

Now, protons also scatter photons, and since the proton charge is the same as the electron charge, except for sign, protons scatter photons just as well as electrons. It follows that the number of scatterers inside the sphere is actually twice the number of electrons. (Remember, we are ignoring complications due to ions, which in fact provide much more opacity than electrons and protons.)

Our argument tells us that the average distance the photon can go before scattering, ℓ, is roughly the radius of this sphere. The number of scatterers inside the sphere, E, satisfies $4\pi\ell^2 = 2\pi r_e^2 E$, or $E = 2(\ell/r_e)^2$. On the other hand, we know how to find E from the volume of the sphere and the number of electrons per unit volume: $E = 2n_e \times 4\pi\ell^3/3$. Setting these equal gives an expression that can be solved for ℓ:

$$\ell = \frac{3}{4\pi n_e r_e^2} = 3.6 \text{ cm.} \qquad (8.22)$$

This is called the *mean free path* of a photon in the Sun. It is a fraction $\ell/R_\odot = 3.6 \times 10^{-2}/7 \times 10^8 = 5 \times 10^{-11}$ of the radius of the Sun.

Now assume that the photon moves outwards by taking steps of this size, but in random directions: after each scattering its direction is different in a random way. We can find out the effect of this by writing a simple computer program to simulate such a random walk.

The program Random on the website uses the fact that computers are good at generating random numbers. (In fact, since nothing inside a computer is really random, computers use clever tricks to generate what are called pseudo-random numbers, which for most purposes can be used as if they were truly randomly chosen.) By choosing random steps in each of the three coordinate directions, we can simulate the aimless motion of a photon in the Sun.

We want to know how far from the center a photon will get after a certain number of random steps. The program selects the number of steps and then performs a large number of trials for "walks" of this number of steps, calculating the average distance from the center at the end of each walk. It also calculates the average distance the photon goes in one step. In Figure 8.4 we plot the results. The axes of the graph are both logarithmic, which means that the linear distance increases uniformly for each factor of ten increase in the variable. This is ideal for showing relationships where one variable is a power of another. In this case, the graph shows that the average net distance of the walk, D, after N steps is $N^{1/2}$ times the average length of one step:

average finishing distance of a random walk of N steps

$$= \sqrt{N} \times \text{average length of one step.}$$

To see this, try a few cases. When the horizontal variable is 100 (a walk of 100 steps), a line through the points would give a vertical value of 10 (a net distance of 10 steps). When the horizontal variable is 1000, the vertical value is about 30, which is as close to the square-root of 1000 (31.6) as one can estimate from this graph.

So we have learned how the photon makes its way out: if it wants to reach the surface, which we have seen is equivalent to 2×10^{10} steps distant from the center, then it will have to make this number *squared* of actual steps to do so. This is 4×10^{20} steps of 3.6 cm each, a total walk of 1.4×10^{19} m, or 2×10^{10} times the radius of the Sun. At the speed of light, this takes 1500 years.

In reality, scattering from other ions makes the mean free path of a photon much smaller, so that the time it takes photons to emerge from the Sun is more like a million years!

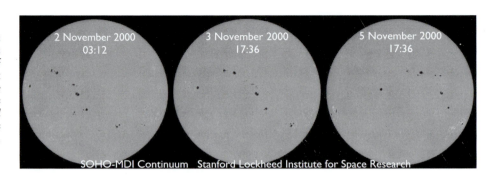

Figure 8.5. *The rotation of the Sun can clearly be seen from the motion of sunspots. Here, a sequence of images made by the* MDI *instrument on the* SOHO *satellite shows that the pattern of sunspots rotates nearly uniformly around the Sun. Aadapted from images courtesy* NASA/ESA.

The surface of the Sun is by no means as simple as our solar models would suggest, with the pressure and density going smoothly to zero. The outward streaming of radiation, the presence of magnetic fields, and the outward flow of pressure waves generated inside the Sun (see the section on solar seismology below) all conspire to produce a very complex region. The solar corona is a kind of atmosphere for the Sun. It extends far outside the Sun and, while being very rarefied, is also very hot. (Yet another place in the Solar System where the temperature begins to increase outwards!) Leaving the Sun is a constant stream of particles, called the *solar wind*. They flow outwards through the Solar System, disturbing the environments of all the planets. On the Earth, the very energetic particles produced by solar magnetic storms produce the aurora borealis and aurora australis phenomena.

Rotation keeps the Sun going around

In this section: we learn that the Sun rotates, which causes it to be slightly non-spherical. Sunspots give the evidence.

In an earlier section I said that the Sun would collapse toward a single point if it were not for the pressure that holds it up against gravity. How, then, do the planets stay "up" in their orbits against the gravity of the Sun? After all, they are not affected by any significant pressure from the Sun. The answer is, of course, easy: they rotate about the Sun. Evidently, rotation can be an important source of support against gravity, so it is time we discussed it in the context of the Sun.

▷ Sunspots are places where the tangled magnetic field of the Sun pokes out of its surface. They come in pairs, like poles of a magnet. Every 11 years the Sun's magnetic field reverses, with North becoming South and vice versa. Sunspot numbers wax and wane on the same 11-year cycle.

The evidence that the Sun does rotate is dramatic: **sunspots** migrate across the face of the Sun and often circle completely around it, returning for a second time before disappearing. This is illustrated in Figure 8.5. The motion of the spots is always the same, and the period of their rotation is always the same, regardless of whether the spots are large or small. Its cause is therefore not to be found in the spots themselves, but in the rotation of the Sun. Its rotation period is about 30 Earth days at the equator. The pole of the rotation axis is well aligned with the pole of the rotational motions of the planets, and the Sun's rotation is in the same sense as that of the planets.

Rotation should in principle change the shape of the Sun. At the solar equator, rotation will contribute a **centrifugal effect** that will bulge the equator outwards. At the pole, there is no such effect. We would therefore expect the Sun to have an elliptical shape, with the long axis in the equator. Such a three-dimensional elliptical shape – like a jelly doughnut – is called an oblate ellipsoid.

▷ The other way of making an ellipsoid, with the long axis joining the poles, produces a football or egg shape, called a prolate ellipsoid. Some galaxies are thought to be prolate, and prolate shapes can arise briefly when stars collapse, but the oblate form is much more common in the Universe.

The Sun's rotation period is very long, however, in terms of the amount of rotation that would be needed to make a significant distortion. If there were a hypothetical planet orbiting the Sun immediately above its equator, it would have an orbital period of only 2.8 h. Put another way, if the Sun spun with a rotation period of 2.8 h, then it would begin throwing material off from its equator. Since its actual period is around 30 days, it is rotating very slowly.

How much distortion would we expect from this rotation? The shape of the Sun can be measured by, say, the ratio of the minor to the major axis of its elliptical shape. This is a dimensionless number, so we might expect this to depend on the rotation rate of the Sun through another dimensionless ratio. The only one available is the ratio of the actual rotation rate to the maximum possible rotation rate. When the ratio is zero (no rotation) the distortion is also zero. So one might guess that the distortion would be proportional to the ratio raised to some exponent.

There is a simple argument that the exponent cannot be one: if it were, then it would change sign if the Sun rotated the other way, because this would give the Sun a negative period relative to the planets. But changing the sense of the Sun's rotation cannot change its shape, so the shape must depend on the second power of the ratio.

▷This kind of argument, based on changing signs, may seem surprising when you first encounter it, but it can be a powerful guide to understanding the solutions to many kinds of problems.

The next simplest guess is that the distortion will be proportional to the *square* of the ratio of these two rotation speeds, and more detailed calculations show that this is in fact correct. Since the ratio of speeds is of order 1/300, we should expect the expansion of the equator of the Sun to be of the order of 10^{-5} of its radius. This is exceedingly difficult to measure, since the edge of the Sun is not very well-defined: its brightness decreases gradually near its edge, not sharply. But recent observations confirm that the distortion is of this order.

Solar seismology: the ringing Sun

Observations of the Sun that we have encountered so far tell us about either the surface of the Sun – its brightness, size, composition, temperature, rotation, and surface magnetic field – or the very center where nuclear reactions are taking place. What about the vast, relatively inactive region in between? Until recently, we had no observational information about this region. The science of solar seismology, called **helioseismology**, is changing all that.

In this section: the Sun has characteristic frequencies of vibration, just like any other body. The turbulent flow of energy out from the center excites these vibrations, and astrophysicists measure their frequencies. These are the best data we have about the nature of the interior of the Sun.

The Sun isn't simply the quiet ball of gas that our equilibrium model calculates. The energy generated in the center and the convection of that energy outwards produces, at some depths, a slow rolling of gas out and back again. This motion disturbs the surface of the Sun, producing small motions that can be detected by specially constructed solar telescopes. These observations have the potential to tell us as much about the interior of the Sun as studies of seismic waves have told us about the interior of the Earth.

The key to grasping the importance of these observations is to understand that the Sun has certain **characteristic frequencies** of vibration, just as does any other physical system, such as a violin string, a drum, an organ pipe, a bell, or a half-filled soft-drink bottle. It will repay us to think a little about the characteristic frequencies of such systems.

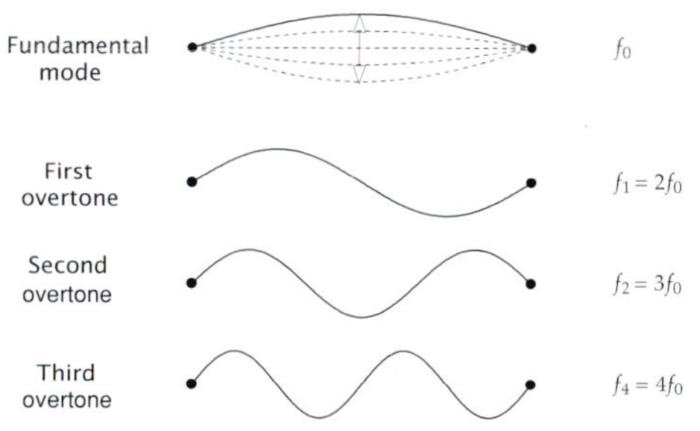

Fundamental mode	f_0
First overtone	$f_1 = 2f_0$
Second overtone	$f_2 = 3f_0$
Third overtone	$f_4 = 4f_0$

Figure 8.6. The first few characteristic frequencies of vibration of a string and the vibration patterns – normal modes – associated with them. The string is held fixed at its end points, as on a violin. When the string is set in motion with exactly the pattern shown, then it will move up and down with that same pattern, as shown for the fundamental mode. For the other modes, only the shape at one moment is shown. The fundamental frequency, f_0, depends on the length, thickness, and tension of the string.

The set of frequencies of vibration of a violin string (the *acoustic spectrum* of the string) is relatively simple, consisting of the **fundamental frequency** (lowest frequency) and its **overtones**, which are just simple multiples of the fundamental frequency. Associated with each frequency is a pattern of vibration: at the fundamental frequency, the string vibrates as a whole. At the first overtone, which has twice the frequency, the string vibrates in two halves; if one half is moving upwards, the other is moving downwards, so that the midpoint of the string does not move at all. Such patterns are called the **normal modes** of the string.

When the string is bowed or plucked, the resulting motion is usually a combination of many normal modes, so that the sound produced includes several frequencies. For the ear, frequencies that are an octave apart (i.e. where one frequency is just twice the other) sound harmonious when heard together. Therefore, the string is ideal for making musical instruments: its fundamental and first overtone are separated by an octave. The fundamental and overtones of open and closed pipes have a similarly pleasant sound when heard together, so such pipes form the basis of organs and wind instruments.

Membranes, from which drums are made, do not have harmonious overtones. Therefore, drums are made so that either the normal modes damp out (decay) rapidly, producing a dull "thud" as in a typical bass drum, or the membrane is stretched over a "kettle" that resonates with and amplifies the fundamental frequency more than the higher overtones, as in orchestral tympani drums.

Figure 8.7. The classic example of the importance of normal modes in engineering is the Tacoma Narrows bridge disaster. The bridge was built across a waterway in the American state of Washington. Wind blowing across the bridge excited a normal mode whose pattern was a torsional oscillation of the roadway. The bridge became a tourist attraction on windy days, until a particularly strong wind drove the oscillations to a larger size than the structure could cope with. The roadway broke up on 7 December 1940. Reproduced with permission of the Smithsonian Institution.

The analysis of normal modes is an important part of many other aspects of modern life. Engineers must routinely assess the frequencies and vibration patterns of all sorts of structures, including skyscrapers, road bridges, automobile chassis, aircraft bodies and components, and so on. All must be checked to see that the patterns of vibration are acceptable; that the characteristic frequencies are not likely to lead to the mode being amplified by external forces on the object (see Figure 8.7); and indeed that the structure is *stable*, which means that small disturbances of the structure will not spontaneously grow larger and larger.

The Sun vibrates in a way that is similar to other mechanical systems. We can deduce its typical vibration frequency from a simple argument. We saw in Equation 7.9 on page 79 that the square of the speed of sound in a body, including the Sun, is proportional to its pressure divided by its density. We also argued that this is similar to the square of the random speed of atoms in the Sun. But we know that this is also approximately kT/m, where m is the mass of the atoms (or ions). We also know that the average temperature of the Sun is not high enough to give the atoms the escape velocity, but it is not much lower either, so we can put this chain of argument together and roughly say that

$$v_{\text{sound}}^2 \propto v_{\text{escape}}^2 \propto GM_\odot/R_\odot.$$

Now, the frequency of vibration of the Sun must have to do with sound waves crossing the Sun back and forth in a regular pattern. The time it takes to cross is $R_\odot/v_{\text{sound}}$, and the frequency f is the reciprocal of this. This leads to the very important relation

$$f^2 \propto GM_\odot/R_\odot^3 \propto G\rho_\odot. \tag{8.23}$$

The fundamental frequency of vibration of the Sun is proportional to the square-root of its average density. When one puts numbers from the Appendix into this, one finds $f \approx 0.6\,\text{mHz}$.

Notice that the fundamental frequency of vibration is similar to the orbital frequency of a satellite at the surface of the Sun. The orbital speed is given by $v_{\text{orb}}^2 = GM_\odot/R_\odot$. The orbital frequency f_{orb} equals the circumference of the orbit divided by v_{orb}, which implies

$$f_{\text{orb}} \propto GM_\odot/R_\odot^3. \tag{8.24}$$

Thus, both the orbital frequency and the vibration frequency depend just on ρ. This is not a coincidence. Gravity is at work in both, fixing not only the orbital speed but also the pressure required to hold up the Sun, and therefore the sound speed.

Investigation 8.8. Making sure the Sun lasts a long time

We know from the age of the Earth that the Sun has been around a long time. It must therefore be stable, in other words resilient in its response to disturbances. Here we shall show that the great age of the Sun tells us that its polytropic index should be less than three.

The argument, like many that we have seen, is remarkably simple. We start with the rough solution of the structure equations for a star, Equation 8.19 on page 95, the most important part of which we reproduce here:

$$p_c \propto \frac{M^2}{R^4}.$$

We have already noted that this proportionality is strictly true if the star is a polytrope. Now consider a *sequence* of stellar models, all of which have the same composition and equation of state (so the constant of proportionality in this equation is the same) and the same mass. The members of the sequence will differ from one another in their central pressure and temperature, and their radii. If we fix the central pressure, we get a unique model, with a well-defined temperature and radius.

Since the masses of all the stars on our sequence are the same, we can write Equation 8.18 on page 95 as

$$M \propto \rho_c R^3 = \text{const.} \quad \Rightarrow \quad R \propto \rho_c^{-1/3}.$$

Replacing R in the equation for p_c by this, we find

$$p_c \propto \rho_c^{4/3}. \qquad \text{density?} \qquad (8.25)$$

Along a sequence of stars of the same mass, the central pressure will be proportional to the 4/3 power of the central pressure.

What does this have to do with the ability of the star to resist a slight compression or expansion? The answer is that the balance between pressure and gravity that determines the structure of the star is the same one that determines how the star will respond to a slight compression. The compression changes the gravitational field of the star by making the star more compact: gravity gets stronger. Compression of any gas also increases its pressure. If the compression produces more than enough extra pressure to resist the extra gravity, then the pressure will push the star out again. Such a star is said to be *stable*, and it will simply oscillate in and out, in one or more of its modes of oscillation. If on the other hand the star produces less extra pressure that is needed to resist the extra gravity, the star will continue to contract. Such a star is *unstable*, and even a very small compression will lead to its collapse.

Now consider a star just between these two cases: a star for which the pressure builds up exactly as much as is needed to compensate the extra gravity, and so the star remains just in equilibrium, neither bouncing back nor contracting further. Since the compression has not changed the mass of the star, compression must make it follow a sequence of equilibrium models of constant mass. We have seen above that along such a sequence the central density and central pressure are related by $p_c \propto \rho_c^{4/3}$. However, the equation of state of the star is, by hypothesis, a polytrope of the form $p \propto \rho^\gamma$. It follows that *if the equation of state has a polytropic exponent $\gamma = 4/3$, the star will remain in equilibrium when compressed.*

What about other stars? If the polytropic exponent γ exceeds 4/3, then the pressure increases faster for a given compression (a given change in ρ) than for the case of 4/3. Such a star will bounce back from compression, and so is stable. Conversely, if γ is less than 4/3, the star is unstable. If we re-express these results in terms of the astronomers' polytropic index n, as defined by Equation 8.12 on page 92, then the case of marginal stability ($\gamma = 4/3$) is $n = 3$. Models with $n < 3$ are stable, those with $n > 3$ unstable. Our solar model, with $n = 2.8$, is stable, as we expect from its long life.

Like musical instruments, the Sun will have an acoustic spectrum of characteristic frequencies, but these will in fact be much more complicated than those of any musical instrument. This is due to two factors: first, the Sun vibrates as a three-dimensional object, whereas most musical instruments use either one-dimensional vibrations (strings and air columns) or two-dimensional vibrations (drums, gongs, bells). Second, the Sun is held together by gravity.

In musical instruments and other everyday objects, there is usually a fundamental mode whose frequency is the lowest of all, and whose pattern of vibration involves the structure vibrating as a whole. Other modes have more complex patterns and higher frequencies.

The spherical oscillations of the Sun follow the same pattern: a lowest frequency and an ascending series above it. In Investigation 8.8 we discuss the forces that drive the mode with the lowest frequency. There we show that if the Sun is a polytrope with an exponent γ in the equation of state $p = K\rho^\gamma$ that is larger than 4/3, then it is stable: a spherical disturbance will make it oscillate rather than collapse or explode.

But when we look at modes that are not spherical, where, for example, the Sun is dimpling in somewhere and out somewhere else, the story is more complex. The fundamental mode still exists and has a pattern in which all layers of the Sun move together, but its frequency is actually in the middle of the spectrum. Above the fundamental is a whole series of modes (called pressure modes, or *p-modes*) of ascending frequency that resemble the pattern in terrestrial objects, but in addition there is a second series of modes *descending* in frequency, which are associated with buoyancy motions of different layers of gas in the Sun. These are called gravity modes, or *g-modes*. In both sequences, the frequencies are not "harmoniously" related; the exact values of the frequencies depend on the detailed structure of the Sun. In addition, the rotation of the Sun changes the frequencies of those modes whose patterns of vibration involve waves rotating one way or another around the Sun.

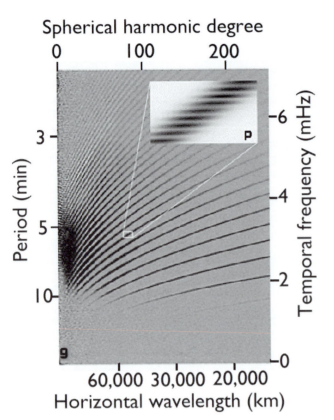

Spherical harmonic degree

This complexity of the spectrum, and its sensitivity to the exact details of solar structure, explain why scientists are interested in measuring the frequencies of the Sun. By first constructing a numerical model of the Sun in great detail, then computing from it the frequencies of vibration that one would expect, and finally comparing those frequencies to the observed ones, one can test the accuracy of one's model. If the modes do not compare well, then the model can be changed until it reproduces the observed frequencies. The modes provide us with essentially the only way that we can "see" into the vast portion of the interior of the Sun in which nuclear reactions are not taking place. Figure 8.8 shows an example of the large amount of data that scientists have been able to gather.

The science of solar seismology is still young, but it has already provided corrections to the way the Standard Model of the Sun treats the flow of photons outwards and it has severely constrained solutions to the solar neutrino problem, which we will discuss in Chapter 11. Observations are continuing from satellites, like SOHO (see Figure 8.8), and from a number of ground-based observatories.

Figure 8.8. *A chart of the frequencies of the p-modes of the Sun, as observed by the* MDI *instrument on the* SOHO *satellite. The horizontal axis is, as indicated, the horizontal wavelength of the waves, increasing to the* left. *The vertical axis is the frequency (increasing upwards) or the period (increasing downwards). The modes evidently form families. The g-modes, near the bottom of the chart, have not yet been observed. Image courtesy* NASA/ESA.

9

Reaching for the stars:
the emptiness of outer space

In this chapter: how astronomers measure the brightness and distances of stars.

With this chapter, we let gravity lead us out of the familiar territory of the Solar System and into the arena of the stars. This is a tremendous leap: the furthest planet, Pluto, is never more than 50 AU away from the Earth, while the nearest stars to the Sun – the αCentauri system – are 270 000 AU away! In between is almost nothing. Yet, just as gravity determines the structure of the Sun, so also it governs the stars.

Stars are the workplaces of the Universe. Stars made the rich variety of chemical elements of which we are made; they created the conditions from which our Solar System and life itself evolved; our local star – the Sun – sustains life and, as we shall see, will ultimately extinguish it from the Earth.

Leaping out of the Solar System

In this section: the huge number and variety of stars.

▷The biggest stars are called giants, and the smallest are neutron stars.

The huge variety of kinds of stars gives a clue to why they can do so many different things. There are stars that are 20 times larger than the whole Solar System, and others that are smaller than New York City. Big stars can blow up in huge **supernova** explosions; small ones can convert mass into energy more efficiently than a nuclear reactor. The material of which stars are made can take the form of a rarified gas thinner than the air at the top of Mt Everest. Or it can be so dense that ordinary atoms are squashed down into pure nuclei, so that a thimbleful of such material would contain more mass than a ball of solid steel 600 m across.

Stars affect each other in many ways. As they form together out of vast clouds of gas, most of them form pairs and triplets circling one another. The disturbances they produce in each other can lead to a range of fascinating phenomena, from **nova** outbursts to intense emissions of **X-rays**. Exploding stars can dump their debris into gas clouds that eventually form other stars. Some stars collide; a few even fall into black holes, with spectacular consequences. In all of these processes, gravity plays an organizing role; and all of them are important for understanding the Universe and indeed the origins of life on Earth.

▷The image under the text on this page is a photograph of the sky showing the constellation of Orion, one of the easiest to recognize. In the sword, the group of three objects arranged in a roughly vertical line, the central fuzzy one is the Orion Nebula, which is a nursery where new stars are being formed. (See Chapter 12.) (Image copyright Till Credner, AlltheSky.com, used with permission.)

▷As a rule of thumb, a typical galaxy has 10^{11}–10^{12} stars.

Most important of all is that there are a lot of stars. The number we can see in the sky on the darkest of nights is a mere handful compared to the hundred billion stars that make up the collection that we call the Milky Way. There are so many stars that their mutual gravitational forces are strong enough to hold them together in a single spectacular **spiral galaxy** like that shown in Figure 9.1 on the following page. And there are perhaps a hundred billion such galaxies in the part of the Universe that we can study with our telescopes. That adds up to a lot of stars!

It is impossible to understand stars and the ways they affect one another unless one first comes to grips with the enormous distances between them. Ancient astronomers had some idea of the size of the Earth and the distance to the Moon. But all they knew about stars was that they were far away, very far away, too far to measure. Actually measuring the distance to the stars was one of the greatest steps in the development of modern astronomy. In this chapter, we shall concentrate on

Figure 9.1. The spiral galaxy called the great galaxy in Andromeda, the Andromeda galaxy or M31, which is similar to our own Milky Way in size and shape. Like our own, it has several small satellite galaxies. If this were a photo of the Milky Way, the Sun would be about three-quarters of the way out in the disk. The view from that location would reveal an arc of stars circling the sky, with fewer stars in other directions. This is just what we see in our sky, and we call it the Milky Way. The Andromeda spiral is the nearest large galaxy to our own, and we are bound together gravitationally. In fact, we are approaching one another and will collide within a few billion years! Use of this image is courtesy of the Palomar Observatory and Digitized Sky Survey created by the Space Telescope Science Institute, operated by AURA, Inc. for NASA and is reproduced here with permission from AURA/STSCI.

how this is done, and on what we can learn about stars as a direct result of measuring their distances. In the next chapter, we shall look at what actually goes on inside stars and how they are born and die.

How far away are the stars?

In this section: how astronomers know the distances to stars, and what they are. The parallax method is the most direct, but is only the first step on a complex distance ladder.

▷Compare this use of parallax with that described in Figure 4.2 on page 26.

▷An arcsecond is 1/3600th of a degree, or 4.85×10^{-6} rad.

Astronomers measure the distance to the nearest stars by the same method that the ancient Greeks used to measure the distance to the Moon: triangulation. Where the Greeks took sightings on the Moon from different places on the Earth, modern astronomers take sightings on stars from different places on the Earth's orbit around the Sun. The change in the apparent position of a star when it is viewed from different places is called its parallax, and it is described in Figure 9.2.

Astronomers can measure parallax angles at least as small as one-tenth of an arcsecond. A star with a parallax of 1 arcsecond would be 206 000 AU distant. (This can be obtained from the formula in the caption of Figure 9.2 by setting α to 4.85×10^{-6}.) In practice, the nearest stars have parallaxes a bit smaller than this, but nevertheless this distance is fairly typical of the distances between individual stars. For this reason it has become a fundamental unit of length, and astronomers have given it a name: the **parsec**, the distance that gives a *parallax* of one arc *second*. It is abbreviated as pc, and its value is

$$1 \, \text{pc} = 2.06 \times 10^{5} \, \text{AU} = 3.08 \times 10^{16} \, \text{m} = 1.91 \times 10^{13} \, \text{mi} = 3.25 \, \text{light years}.$$

Astronomers express all cosmic distances in parsecs or in larger units derived from the parsec, such as the **kiloparsec** (abbreviated kpc) – about one-tenth the size of a typical galaxy – or the **megaparsec** (abbreviated Mpc) – typical of the distances between galaxies. We shall use these units too, since using meters or miles or other

Figure 9.2. The distance to a star can be determined from the change in the direction to the star as the Earth moves from one side of the Sun to the other. The different directions are indicated by the arrows from the Earth to the star in January, and then six months later in July. The parallax is defined to be the angle α indicated in the diagram. If α is measured in radians, then if it is small enough (always the case in practice) it is related to the distance d to the star to an excellent approximation by the equation $\alpha = (1 \, AU)/d$.

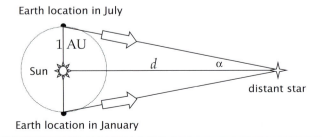

human-scale measures would only confuse us with the large powers of ten we would always have to use.

These distances are truly vast. The mass of the Solar System is almost all concentrated in the Sun, a ball of radius 7×10^8 m. The separations between stars are typically 10^8 times larger than this. The intervening space is largely empty: if we draw a box around the Sun whose sides are halfway to the nearest stars, then the Sun occupies only about one part in 10^{24} – one million-million-millionth – of the volume of this cube! In this vast space there is some diffuse gas (adding up to perhaps 10% of the mass of the Sun), but there may be as much as ten times the mass of the Sun in **dark matter**. This unseen substance is one of the great puzzles of modern astronomy. In Chapter 14 we will deduce its existence from its gravitational effects, but astronomers so far have not directly observed it. Yet even this dark matter is spread out over such a large volume that its density is unimaginably small.

In trying to measure stellar distances by parallax, astronomers have a problem: the twinkling of the stars. The Earth's atmosphere is turbulent, and light from a star has to pass through it before it reaches our telescopes. This turbulence causes the image of the star to jump around randomly. To the eye this causes the familiar and rather pleasant effect of twinkling. But to the astronomer, this is a nuisance, because it limits the accuracy of any single measurement of the position of a star from the ground to typically about one arcsecond.

▷Astronomers call this twinkling effect "seeing", and they build telescopes in places where the seeing is very good – the twinkling is small.

The way to get around this problem is to make many measurements of the position of a star. By patiently performing hundreds of such measurements on the same star, astronomers can estimate the average position of the star and remove much of the confusion caused by twinkling. In this way, astronomers using telescopes on the ground have measured parallaxes smaller than 0.1 arcseconds, and so measured the distance to stars more than 10 pc away.

The obvious way to remove the twinkling problem completely is to make these measurements from a telescope in orbit about the Earth, above the disturbing effects of the atmosphere. A specially designed satellite called Hipparcos has done just that. Built and launched by the European Space Agency (ESA), it has measured parallaxes of thousands of stars to an accuracy of about 0.01 arcseconds or better. These data have greatly improved astronomers' understanding of many aspects of stars and their evolution. We will see an example in Figure 12.4 on page 140.

For stars that are more distant than we can reach even with Hipparcos, our estimates of distance are decidedly less accurate. There are many complications, but almost all methods use the brightness of a star. We can estimate the distance to a star by comparing how bright it appears with how bright we think it really is. The method is described in the next section.

How bright are stars?

The everyday term *brightness* has two different uses, so physicists use different words for them. The first sense of brightness is the total energy given off in a unit time. For example, a light bulb may be rated at 100 W. This means it gives off an energy of 100 J in each second. (For an ordinary tungsten incandescent bulb, most of that energy is in heat, and only a small part comes out as light; but all the energy comes out in one way or another.) Physicists call this the **luminosity** of the object, its total emission of energy in a unit time.

In this section: we learn about the difference between the apparent brightness of a star and its absolute brightness, also called respectively the flux and luminosity of the star.

The other meaning is really the *apparent brightness*, which describes how a given star looks dimmer and dimmer as it gets further away. This happens because the total energy being given out in radiation is spread out over a larger and larger area as it moves out from the star. Any given observer gathers light from a fixed area, perhaps the area of the pupil of the eye, or the aperture of the telescope. The further away the star is, the smaller will be the fraction of the star's energy that will fall on a given detecting area, so the dimmer the star will seem to be.

Since the light from the star is spread over the area of a sphere surrounding the star at any distance, and since the area of the sphere is proportional to the square of the distance r to the star, the energy falling on a given detecting area decreases as $1/r^2$. Because less energy falls on the area as it moves further and further away, the star becomes dimmer and dimmer. If one imagines using a detector of a given unit area (say $1\,\text{m}^2$), then one can measure the energy falling on that unit area in a unit time. This is what the physicist calls **flux**: energy per unit time per unit area. So the apparent brightness of a star is called its flux, measured in joules per second per square meter, or W m^{-2}. For a "point" source of light, like a star, the apparent brightness is proportional to $1/r^2$.

Let us use these ideas to discover how luminous the nearest star to the Sun is: αCentauri. Its flux can be measured and turns out to be $2.6 \times 10^{-8}\,\text{W m}^{-2}$. We saw earlier that this star is at a distance of 2.7×10^5 AU, or 4.1×10^{16} m. If we multiply the flux by the area of the sphere over which the star's light is being spread at this distance ($4\pi r^2 = 2.1 \times 10^{34}\,\text{m}^2$), we get the star's luminosity: 5.5×10^{26} W.

What happens if we do the same calculation for the Sun, using its measured energy flux (usually called the **solar constant**), $1355\,\text{W m}^{-2}$, and its distance, 1 AU or 1.5×10^{11} m? We obtain the solar luminosity,

$$\text{solar luminosity} = 1\text{L}_\odot = 3.83 \times 10^{26}\,\text{W}. \tag{9.1}$$

The luminosity of αCen turned out to be 1.4 times the luminosity of our Sun. This is a reassuring result: we are not finding anything wildly different for our nearest star.

▷Given that an ordinary tungsten bulb converts only about 2% of its energy into visible light, αCen is as bright as a 100 W light bulb shining 2.5 km away. That may sound dim, but αCen is just about the brightest star in the night sky, easily visible to the naked eye.

Astronomers' units for brightness

In this section: astronomers measure brightness in magnitudes, a logarithmic scale that goes backwards, so that the brightest stars have the lowest numbers.

Astronomers have evolved their own way of describing the luminosity and apparent brightness of stars. In astronomy books one does not find the conventional units of watts (W) or watts per square meter (W m^{-2}). Instead, one finds that the apparent luminosity of a star is described as its **apparent magnitude**, or simply its **magnitude**, and is called m. One also finds the total luminosity described as the **absolute magnitude** M.

The details of the definitions of these magnitudes are given in two analyses, first Investigation 9.1, then Investigation 9.2 on page 108, but there are two features that we should note here. First, magnitudes run the "wrong" way: stars with larger magnitudes are *dimmer* than stars of smaller magnitudes! And second, magnitudes run on what we call a **logarithmic scale**, which means that if the magnitude of one star is larger by one than the magnitude of another star, then the first star is dimmer than the second by a fixed *factor*, in this case about 2.512. A difference of two magnitudes implies a brightness ratio of $(2.512)^2$, or 6.31. The magnitude scale is arranged so that a difference of 2.5 magnitudes corresponds to a ratio of exactly 10 in brightness. As is described in Investigation 9.1, these peculiar brightness scales came about through a combination of historical practice and adaptations to the physical properties of the human eye.

On these scales, the star αCen has an absolute magnitude M of 4.3 and an apparent magnitude m of -0.08. The Sun has an absolute magnitude M of 4.7 and an apparent magnitude m of -27. These numbers for absolute magnitude reflect the fact that the Sun is less luminous than αCen by a factor of 1.4. The apparent magnitude numbers are dominated by the fact that the Sun is so much closer to the Earth than is αCen, so it has a much greater apparent brightness.

Investigation 9.1. How ancient astronomers constructed their magnitude scale

In the days before there were any systematic units for physical measurements of energy or brightness, ancient astronomers needed to classify stars according to their relative brightness. It probably seemed natural to them to use a scale a bit like one they might have used for important people: the brightest stars were stars of the *first magnitude*, somewhat dimmer stars were of the second magnitude, and so on. Their brightness scale thus was the reverse of what would seem natural to us today: the larger the magnitude, the dimmer (less important) the star.

Ancient astronomers took the steps on the magnitude scale to correspond to levels of brightness that the eye could clearly distinguish. Moreover, in trying to keep the steps from one magnitude to another uniform, they devised a scale where given magnitude changes represent given *ratios* between brightnesses. Basically, this is because the eye and the brain are much better at saying "star X is twice as bright as star Y" than at saying "star X is brighter than star Y by the brightness of star Z". Here is how these two statements would get translated into measurement scales.

"Star X is brighter than star Y by the brightness of star Z." Let us consider the second approach first. It is the one that the eye is not able to do well and hence does not correspond to the scale adopted by ancient astronomers, so we will be able to discard it. But it is important to understand it, since it would have led astronomers to the sort of scale that a modern physicist would try to devise.

Suppose the eye could in fact sense fixed brightness differences. Then ancient astronomers would have constructed their scale by deciding upon some chosen star as their brightness standard. Let us call this star Z, and denote its brightness by B_Z. (Since ancient astronomers would have had no way of relating this brightness to other brightnesses on the Earth, they might just have called this brightness 1.) They would then have found another star, say Y, that was brighter than Z by the brightness of Z itself, and they would have assigned this a brightness of $2B_Z$. They would have looked for brighter stars,

and found one, say X, that was brighter than Y by the brightness of Z: this would make $B_X = 3B_Z$. This is what we call a *linear* scale: the quantity used to describe the brightness changes from one star to the next in a way that is proportional to the change in the brightness itself.

Unfortunately, physical properties of the eye prevented ancient astronomers from using this sort of scale. If star X is 100 times the brightness of star Z, while star Y is 99 times as bright as Z, then the eye can see all three stars at once but it cannot even tell that X is brighter than Y, let alone that the difference is just the brightness of Z. So it is impossible to describe a linear scale for stellar brightness using the eye as a measuring instrument. Let us therefore consider the alternative.

"Star X is twice as bright as star Y." This is the sort of statement that the eye *can* make fairly accurately. It can tell that, say, Y is twice as bright as Z, and that X is twice as bright as Y. It does not need to measure brightness differences to do this, only brightness *ratios*. Actually, what the eye is good at is telling that the ratio of brightness between Z and Y is the same as between Y and X: it is not so good at telling whether this ratio is exactly 2 or maybe 2.5 or 1.6 or something in between.

Ancient astronomers used this property to devise their magnitude scale. They took a given star, say X, as their standard and called it a star of the first magnitude (in modern language, $m = 1$). Then they found one, called Y, that seemed roughly half as bright as X and called it a star of the second magnitude ($m = 2$). They further found another star, Z, that had the same ratio to Y as Y had to X, and called that a star of the third magnitude ($m = 3$).

We call this a logarithmic scale because, as the table below shows, changes in our brightness scale (the magnitude) are proportional to changes in the *logarithm* of the brightness itself. Since the magnitude decreases as the brightness increases, the scale runs in reverse.

The ancient astronomers' magnitude scale

Hypothetical star	Brightness	Magnitude	Logarithm of brightness
Z	B_Z	3	$\log(B_Z)$
Y	$2B_Z$	2	$\log 2 + \log(B_Z)$
X	$4B_Z$	1	$2\log 2 + \log(B_Z)$

Standard candles: using brightness to measure distance

Now, αCen has other properties that astronomers can measure. For example, they can obtain its spectrum by dispersing its light through a prism or a diffraction grating. Although the general shape of the spectrum will depend mainly on the star's temperature (we will look at how this happens below), the details of the spectrum are a very sensitive "signature" of the star. Suppose an astronomer measures the spectrum of another star and finds that it is almost identical to that of αCen. Experience has shown astronomers that the two stars will also be very similar in their other properties, such as their mass, radius, and total luminosity. If it turns out that the second star has a flux (apparent luminosity) that is only one quarter that of αCen, then it is likely that this is because the star is twice as far away, so that its light is spread over a sphere whose area is four times as large as that over which αCen's light is spread when it reaches us.

This is how astronomers estimate distances to stars further away than a few tens or hundreds of parsecs. It is almost the only way that distances can be measured until we reach the enormous distances that separate the giant clusters of galaxies in the Universe, where we can use the expansion of the Universe itself to measure distances. Obviously, the accuracy of the apparent-brightness method of estimating distances depends on how well we can estimate the total luminosity of the distant

In this section: astronomers estimate the distances to most objects from their apparent brightness. This works well if the instrinsic luminosity of the object is known. The search for such "standard candles" is one of the most fundamental activities in astronomy.

Investigation 9.2. How modern astronomers construct their magnitude scale

The modern definition of *apparent magnitude* is taken from the ancient one with only minor changes. The most significant change is that the ancient astronomers' estimate of a factor of 2 decrease in brightness for one step in magnitude was a bit low, and one can make a reasonable fit to the ancient magnitudes by taking the ratio to be nearer 2.5. In order to make things simple, astronomers define a ratio of brightness of 10 to be a change of magnitude of exactly 2.5.

In equations, this is fairly straightforward. Take two stars of brightness B_Z and $B_Y = 10B_Z$. The logarithms of their brightnesses are $\log(B_Z)$ and $1 + \log(B_Z)$. If the magnitudes are to decrease by 2.5 then we have

$$m_Y - m_Z = -2.5[\log(B_Y) - \log(B_Z)].$$

If we combine the logarithms into a single term, then we get

$$m_Y - m_Z = -2.5 \log\left(\frac{B_Y}{B_Z}\right). \tag{9.2}$$

This shows clearly that magnitude differences depend on the *ratios* of brightnesses. If two stars have a magnitude difference of 1, then Equation 9.2 shows that the logarithm of their brightness ratio will be $1/2.5 = 0.4$ and so their brightness ratio will be $10^{0.4} = 2.512$. This is close enough to the ratio of 2.5 mentioned at the beginning of this section.

The apparent magnitude scale is fixed if we adopt one star as a standard. Here we find the second difference between modern and ancient astronomers. Because the three brightest stars are in fact Southern Hemisphere stars, while the magnitude system was invented by Northern Hemisphere astronomers, the modern scale has to assign some stars to negative apparent magnitudes. The modern scale is chosen so that our old friend αCen is 7.6% brighter than a standard *zero*-magnitude star. This gives us an alternative to Equation 9.2,

$$m_{star} = -2.5 \log\left(\frac{\text{Flux from star}}{\text{Flux from } \alpha\text{Cen}/1.076}\right)$$

$$= -2.5 \log\left(\frac{\text{Flux from star}}{2.4 \times 10^{-8} \text{ W m}^{-2}}\right). \tag{9.3}$$

(See Table 10.1 on page 110 for a list of magnitudes of the brightest stars.)

Modern astronomers also know that stars are at different distances from the Earth, so that in order to understand them physically we need to measure their intrinsic brightness, or luminosity, and not just their apparent brightness. Astronomers have adopted a scale for the absolute magnitude M based on the apparent magnitude scale. The *absolute magnitude* of a star is numerically equal to the apparent magnitude it would have if it were 10 pc from the Earth.

Since αCen is 1.33 pc away, we would have to place it 7.5 times further away to make its apparent magnitude equal its absolute magnitude. This would reduce its flux by $(7.5)^2 = 56.25$ and increase its apparent magnitude by $2.5 \log(56.25) = 4.38$. So the absolute magnitude of αCen is $-0.08 + 4.38 = 4.3$. A star with absolute magnitude 0 would have a luminosity larger than that of αCen by a factor of $10^{(4.3/2.5)} = 52$. This would give a luminosity 2.9×10^{28} W, so that the absolute magnitude can also be written as

$$M = -2.5 \log\left(\frac{\text{Luminosity of star}}{2.9 \times 10^{28} \text{ W}}\right)$$

$$= -2.5 \log\left(\frac{\text{Luminosity of star}}{75 \text{ L}_\odot}\right). \tag{9.4}$$

Why have modern astronomers stuck to such an ancient and inconvenient system of magnitudes when most physicists have adopted more modern measuring scales for other quantities? The answer is at least partially that some astronomers are conservative and are reluctant to break the continuous tradition of astronomy that makes it the oldest of the mathematical sciences. But a much more important reason is simply that the logarithmic magnitude scale is useful. Stars and other objects come in a huge variety of luminosities and apparent brightnesses, and it is useful to have a scale where a change of magnitude of, say, 100 implies a brightness ratio of 10^{40}. This is big enough to span even the huge variations encountered in modern astronomy. Since it is useful to have a logarithmic scale for brightness, one might as well continue to use the ancient one!

Exercise 9.2.1: *Magnitude of the Sun*

Use the solar constant, given just before Equation 9.1 on page 106, to compute the apparent magnitude m of the Sun, using Equation 9.2. Use Equation 9.4 to calculate the absolute magnitude M of the Sun.

Exercise 9.2.2: *Stellar magnitudes*

A particular star is known to be ten times further away than αCen and five times more luminous. Compute its apparent and absolute magnitudes.

---

star from other things we can measure about it. This depends on our finding similar stars whose luminosity is known, or can at least be calculated from some theoretical ideas about the object. Astronomers call such objects **standard candles**: objects whose intrinsic brightness is known.

In the past, astronomers have often changed their estimates of interstellar distances. However, in the last two decades of the twentieth century, some painstaking work with space observatories, coupled with a better understanding of important standard candles, has made astronomers' distance-scales much more accurate. Most astronomers now feel that their distance estimates, even over very large cosmological reaches, have errors no larger than 10%, and probably smaller.

10

The colors of stars:
why they are black (bodies)

Gravity is the engine that drives the Universe. But it does not work alone, of course. In fact, one of the most satisfying aspects of studying astronomy is that there is a role for essentially every branch of physics when one tries to explain the huge variety of phenomena that the Universe displays. One branch of physics, however, stands out from the rest because of its absolutely central place in helping us to learn about the Universe, and that is the study of the way hot bodies give off light.

Almost all of the information we have from astronomical bodies is carried to us by light, and almost all the light originates as radiation from some sort of hot region. The great breakthrough in physicists' understanding of such thermal radiation was made by the German physicist Max Planck (1858–1947) at the start of the twentieth century. (See Figure 10.2 on page 112.) The story of this breakthrough is the story of physicists' first steps toward quantum theory. It is also the story of the beginnings of a real understanding of the heavens.

We take a look at this story, which will lead us to two fundamental ideas that together will unlock the secrets of a great deal of astronomy. These ideas are **black-body radiation** and the ionization of hydrogen. We will put them to work for us repeatedly in the next few chapters.

The colors of stars

We will start with the color of light. The different colors are, of course, just manifestations of the different wavelengths of light. The overall color of a body depends on the amounts of light of the various colors present in its emissions. Stars vary in color from red to blue, as you can easily see by using a pair of binoculars on a dark night.

Table 10.1 on the following page lists the magnitudes and distances of the five most prominent stars. Notice that the magnitude of αCen is not quite what I quoted in the last chapter. This is not an error: it is because in this table I have used only the brightness of the stars in *visible* light. This is the so-called **visual magnitude** V of the star, and it is the most important measure of brightness for observations performed with the eye. Our previous discussion of brightness assumed we were dealing with all the radiated energy from the star, even radiation that comes out in the infrared or ultraviolet parts of the spectrum. This total emission is measured by the so-called **bolometric magnitude**, called M_b or m_b. The discrepancy in the magnitude of αCen is due to the fact that some of its energy comes out in the ultraviolet and infrared: our earlier value was its bolometric magnitude, and Table 10.1 contains only its visual magnitude.

Stars emit more light at some wavelengths than at others. If a star is brighter at the blue end of the spectrum (short wavelengths) than at the yellow end (long wavelengths), then it will look blue, and if another star is brighter in the yellow than in the blue it will appear yellow.

Table 10.1. *Magnitudes of the five brightest stars. The first three are in Southern Hemisphere constellations. Most are nearby, but notice how far away, and how intrinsically bright, Canopus is!*

Rank	Name	Constellation	m	M	d (pc)
1	Sirius	Canis Major	−1.46	1.4	2.7
2	Canopus	Carina	−0.72	−8.5	360
3	Rigil Kentaurus	Centaurus	−0.27	4.4	1.3
4	Arcturus	Boötes	−0.04	−0.2	11
5	Vega	Lyra	0.03	0.5	8.1

Astronomers have made this idea precise by defining a number they call the **color of a star**. This is based on measuring the magnitude of a star in different parts of the spectrum. If we filter the light through a blue filter before we measure the magnitude, we obtain the blue magnitude of the star, called B. If we filter through a filter in the central part of the spectrum, we get the visual magnitude V. These colors have become international standards, so that any astronomer wanting to measure B will use a filter that passes exactly the same range of wavelengths of light as any other astronomer would use. There are at least 11 such standard filters, ranging from the infrared nearly to the ultraviolet parts of the spectrum.

The nice thing about using a logarithmic scale for magnitudes is that the difference between any two magnitudes depends on the *ratio* of the brightnesses that the magnitudes measure – see Equation 9.2 on page 108. Therefore, if one takes the difference between any two filtered magnitudes, say $B - V$, one gets a number that measures the ratio of the blue brightness to the visual brightness. Now, since both brightnesses diminish with distance in the same way, their ratio is independent of how far away from the star we are.

> Astronomers define the **color index** of a star to be the difference $B - V$. They measure the color index using only apparent magnitudes for V and B, not even needing to know how far away the star is.

Why stars are black bodies

In this section: black bodies absorb all light that falls on them, so stars are black bodies. Hot black bodies also radiate light, and they play a key role in quantum physics.

We saw in the last chapter that the color of the Sun tells us how hot it is, and that holds just as well for other stars. In fact, despite all the possible complications of stars – their size, their pulsations, their varying composition – there is a remarkably consistent relationship between the color and the temperature of a star. This is because stars are excellent examples of what physicists call **black bodies**! At first this seems like an outrageous abuse of common-sense language: how can a brilliant star be a black body?

The explanation is that the words "black" and "bright" actually refer to different physical processes. Physicists adopt the very reasonable definition that a body will be called a black body if it absorbs all light that falls on it. This applies not only to everyday blackness, such as black cats, black ink, or the black of the night, but also to stars. If we were to shine a light at the Sun, for example, the Sun would just swallow it up: no light would be reflected, and none would be transmitted through to the other side. So the Sun is, by this definition, black.

Our difficulty with this is that we are used to thinking that black objects are also *dark:* they do not shine. This is because in everyday circumstances, bodies either absorb or reflect light, and their color is determined by what wavelengths they reflect. If they are black (absorbing) they are also dark (sending no light back to us). However, bodies can also emit light all by themselves, and if they are sufficiently hot, we will see the emission. The burner of an electric stove starts out black, but when heated it glows red. For a physicist, it is still a black body: the black covering will still absorb any light that hits it. But in addition, it glows. Stars are the same.

Investigation 10.1. Black bodies

The spectrum of light emitted by a body is a measure of how much light comes out at different wavelengths. This is not quite so easy to define as it may at first seem, so here is how physicists do it.

- First of all, "how much light comes out" means the *rate* at which energy is being emitted in light: it is an energy per unit time.

- Next, the energy is radiated by the surface of the black body, and since each piece of the surface is independent of every other piece, the energy radiated will be proportional to the area. So physicists speak of the energy radiated per unit time and *per unit area*. Recall that in Chapter 9 we called this the *energy flux*.

- Of course, the energy comes out in various directions, so to avoid complications we will consider only the total energy that a piece of the surface radiates towards the outside of the body.

- Finally, the energy radiated may depend on the wavelength of the light. Since photons come out with a whole range of wavelengths, the chance of finding a photon with *exactly* some given value of the wavelength λ is essentially zero; it is more correct to speak of the energy carried away by photons whose wavelengths fall in some given range.

The result of all this is that we shall characterize the spectrum as the *radiated energy flux between two wavelengths*.

Let us consider, then, two wavelengths λ and $\lambda + \Delta\lambda$, where $\Delta\lambda$ is meant to be very small. If $\Delta\lambda$ is small enough, the flux that comes out in this range will simply be proportional to $\Delta\lambda$: if we take half the wavelength range, the energy coming out will be half. (This only works if the spectrum is essentially constant within the range, which will always be true for a small enough range.) We shall call this flux per unit wavelength F_λ:

$$\text{flux between } \lambda \text{ and } \lambda + \Delta\lambda = F_\lambda \Delta\lambda.$$

This spectrum F_λ for a black body is illustrated in Figure 10.1 on the next page for a few temperatures. Its shape was known from experiment, but nineteenth century physicists could explain only the falling part of the spectrum at long wavelengths. The fact that the curve reached a maximum and turned over was first explained by Planck, who also derived from his arguments on quantized energy levels (see the main text) the famous formula that describes the curve:

$$F_\lambda(T) = \frac{2\pi h c^2}{\lambda^5} \frac{1}{\left(e^{hc/\lambda kT} - 1\right)}, \qquad (10.1)$$

where T is the temperature, c is the speed of light, k is Boltzmann's constant (see Chapter 7), and $h = 6.626 \times 10^{-34}$ J s is *Planck's constant*, which we met in Chapter 7. For readers who have never encountered it before, the symbol e represents a famous and important number in mathematics. It arises in many problems of calculus, as often as π arises in geometry. Its value is approximately $e = 2.71828$. It is a pure number, just like π, and so it carries no units. Raising e to a power, say x, is such a common operation in some parts of mathematics that it is given a special name: $e^x = \exp(x)$. This is called the **exponential function**, and it can be found on scientific calculators (where it is usually called e^x) and as a built-in function in computer languages like Java (where it is called exp). It may be used mathematically just like the functions $\sin$ or $\cos$.

It is clear from the curves in Figure 10.1 that the brightness of a body depends on the wavelength and temperature. The body at 5000 K is brightest in the optical part of the spectrum. The body at 100 K is brightest in the infrared. And the body at 10^6 K is brightest in the X-ray region. The color of the body will clearly depend on its temperature: the cooler the body, the longer the wavelength of most of the light it emits.

Since almost all of our understanding of the Universe can be traced back to the Planck function, we should take some time to understand its properties. This is the subject of Investigation 10.2 on page 117.

So the reason we think the two ideas are contradictory is that most everyday objects are simply not hot enough to be bright and black at the same time. But stars are. See Investigation 10.1 for more details.

The color of a black body

Now, the color of the glow of a hot body depends on its temperature. Bodies hotter than the electric burner may glow white hot. Cool bodies, such as the stove's resting burner, emit radiation too, but we don't see it because it is in the infrared region of the spectrum, to which our eyes are not sensitive.

Nineteenth century physicists found experimentally that the color or spectrum of the glow emitted by a "perfect" black body depends *only* on how hot it is, independently of what the body is made of or what shape it has. (An example of this spectrum is given in Figure 10.1 on the next page.) Physicists of the time were able to explain satisfactorily why the spectrum depended only on the temperature, essentially by showing that if two black bodies of the same temperature emitted different kinds of radiation and yet absorbed everything (because they were black), then one could use them to construct a **perpetual motion** machine. This fascinating style of argument is common in the branch of physics called **thermodynamics**, but we would be going too far from the theme of this book if we tried to give its details here.

The thermodynamic arguments also incidentally proved that a black body is *more efficient* at giving off light than a body of any other color: this is why, in fact, stove burners are manufactured black. But one thing the nineteenth-century physicists could not explain was the *shape* of the spectrum, or in other words the

In this section: Max Planck founded quantum physics by explaining the spectrum of light emitted by a black body. He postulated that light was always emitted with an energy proportional to its frequency. The proportionality constan, called Planck's constant h, is one of the most fundamental numbers of physics.

▷A perpetual motion machine is a machine that will run by itself forever, without requiring any supply of energy. Since real machines always have a little friction or other losses, perpetual motion requires the creation of energy from nothing. Since this is not possible, a circumstance that would lead to perpetual motion is also not possible.

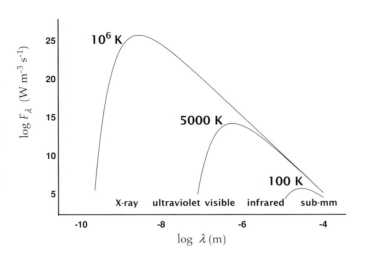

Figure 10.1. The black-body spectrum for the three temperatures 100 K, 5000 K, and 10⁶ K. Notice how the wavelength at which the maximum occurs decreases with increasing temperature. The function plotted is the energy emitted by the body per unit surface area, per unit time, and per unit wavelength. This is defined in Investigation 10.1 on the previous page.

Figure 10.2. Max Planck was one of the leading physicists of the early 20th century. A pioneer of quantum theory, he used his considerable influence in German science to nurture the development of both quantum mechanics and relativity. Near the end of his life he tried in vain to moderate Hitler's attacks on Jewish scientists, including Einstein. He lived to see one of his sons executed by Hitler for treason. Germany's network of pure-science research institutes is now named after this complex and profoundly influential scientist. Photo courtesy Mary Evans/Weimar Archive.

actual color associated with any given temperature. What is so special about the wavelengths where the curves in Figure 10.1 turn over? Why should a given temperature determine a certain wavelength of light?

The explanation finally came, just at the turn of the century, from Max Planck. With his explanation he made the first tentative step toward quantum theory. Since all previous attempts to explain the spectrum had failed, it was almost inevitable that any explanation would involve a new and strange hypothesis about matter. Planck's new hypothesis was strange indeed. Many physicists believed that the atoms of a black body (or of any other body) emitted light by vibrating. A vibration with a frequency f would emit light at frequency f. When an atom absorbed this light, it would be set into vibration with frequency f. So equilibrium between the radiation and the walls of the black body involved countless events in which light energy was interchanged with vibration energy.

Planck postulated that all the atoms vibrating with a given frequency f could exchange energy only in discrete amounts, only in energy "parcels" of size then its energy of vibration could have only certain values, namely:

$$E = hf, \qquad (10.2)$$

where $h = 6.626 \times 10^{-34}$ J s is *Planck's constant*, which we met in Chapter 7.

This is the same formula that Einstein used to explain the photoelectric effect, as we saw in Chapter 8. Einstein worked after Planck, and he gave this formula a more radical interpretation: he assumed that the reason that energy exchanges had to involve only quanta of energy of this size is that light itself can carry only such quanta of energy. For Planck, light was still a continuous electromagnetic wave, and the quantization was something to do with the way atoms behaved. Einstein put the focus onto light itself.

Planck had no theory from which he could predict the value of h. However, when combined with Boltzmann's statistical mechanics of the atoms, Planck's new hypothesis led exactly to the prediction of the shape of the curves in Figure 10.1. Planck could measure the value of h by finding the value that best made his theoretical curve fit experimental measurements, such as those in Figure 10.1. Then, once he had obtained the value of h from the curve for one temperature, he found that the he could exactly predict the measurements for all other temperatures. This was the triumph of his theory, and the sign that the constant h was a new and fundamental physical quantity.

Here is why Planck's hypothesis determines a relation between temperature and the wavelength of light: by introducing a new constant h with dimensions of energy times time, it is possible to start with a temperature T, find from it a "typical" energy kT, and from it deduce a number with the dimensions of time, h/kT. This number can be turned into a wavelength λ by multiplying by the speed of light: $\lambda = hc/kT$. Although it may seem that we have just played a mathematical game devoid of any physical reasoning, we have in fact learned one important thing: if the theory does associate a special wavelength with a temperature T, it will probably be roughly the same size as the number hc/kT, since that is the only number with the dimensions of length that we can find in the theory. It was the absence of a special constant like h in the theories of physics before Planck that prevented physicists from finding any special wavelengths associated with light.

> This reasoning is another example of dimensional analysis, which we first used in Investigation 1.3 on page 5. Used properly, it can be a powerful first step towards understanding difficult problems.

Why does quantization of light energy lead to the black-body formula? Although the details are well beyond our scope here, the outline is not hard to grasp. Suppose we are given a hollow box whose rough interior walls are at a temperature T. A tiny hole in the wall of the box will be a black body: shine any light onto the hole and it will go in, with almost no probability of its re-emerging from the hole directly. But the radiation that comes out of the hole will be the same as that inside the cavity, so the cavity contains black-body radiation.

Then the typical vibration energy of the atoms in the walls will be about kT. Changes in these energies can occur only in multiples of hf, because (following Einstein) the properties of light force this. Now, since there is only a finite amount of energy in the walls, and this will be shared among all the atoms, there will be very few with very high frequencies of vibration, because this would involve very large energy exchanges, so there will likewise be little light at very short wavelengths. There will also be few atoms vibrating with nearly zero energy, so there will be little light at long wavelengths. The light distribution must therefore peak at some intermediate wavelength, and that will be proportional to hc/kT. Although Planck's argument went somewhat differently, since he did not then have the benefit of the insight of Einstein, it nevertheless led to the same result: a curve that fit the experimental observations perfectly.

Planck was the first to suggest that energies might in some way be quantized. Planck's hypothesis, bold as it was, was carefully restricted only to the energies of the atoms. We now know that all measurable quantities are quantized: the angular momentum of a spinning particle, the linear momentum of a particle moving around in a box, and so on. This is the province of the quantum theory.

Relation between color and temperature: greenhouses again

How does this relate to stars? A star is black because it absorbs essentially all the light that strikes it. So the spectrum of the light it emits will depend only on its temperature, and this will be the same spectrum as is emitted by hot bodies in the laboratory. In turn, the spectrum uniquely determines the color of the black body, so that hotter bodies emit more energy in the blue, while cooler ones emit more in the yellow. Measuring the color allows one to determine the temperature. In this way we know that stars have surface temperatures ranging from about 2000 K to about 30 000 K.

In this section: greenhouse gases trap heat on the Earth because the Earth is colder than the Sun, so the spectrum of light it emits is different from the one it absorbs.

Not only stars, of course, but any body emits radiation with a color or characteristic wavelength that depends on its temperature. This explains the greenhouse effect that we met in Chapter 7. The energy arriving on the Earth from the Sun comes with visible wavelengths, since the temperature of the Sun is about 5000 K. The Earth, with a temperature of only 300 K, re-radiates this energy at much longer wavelengths, in the infrared. This means that it is possible for a gas to allow solar

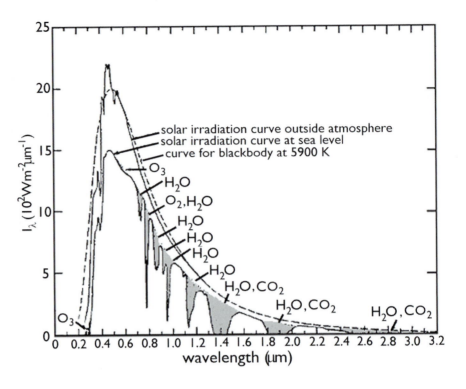

Figure 10.3. The solar spectrum at the top of the atmosphere is similar to the black-body curve for 5900 K. The solar spectrum reaching the ground is also illustrated, to show how much of it is absorbed by the atmosphere. The molecules responsible for the absorption are indicated. Figure based on illustration in the CEOS CD-ROM, (http://ceos.cnes.fr:8100/-cdrom-98/astart.htm).

radiation to hit the Earth (i.e. to be transparent at visible wavelengths) and still to block radiation leaving the Earth (to be opaque at infrared wavelengths). Greenhouse gases do just this.

Spectral lines: the fingerprint of a star

In this section: the spectrum of light from a star contains features, called lines, that arise in the photosphere, the layer from which photons leave a star. The spectrum contains detailed information about the star itself.

But does this not contradict another thing that we mentioned earlier, namely that the spectrum of a star is closely related to its other properties, such as its mass and size? Yes, in fact, there is a contradiction if we take the black body model too absolutely. In fact, every star is almost, but not quite, black. One reason is that, although stars are mainly made of a hot gas of individual protons and electrons, they contain other elements. If an incoming photon with just the right energy (frequency) strikes an atom of one of these elements in the outermost layers of a star, it can be reflected out of the star. So elements prevent stars from being perfect black bodies. Instead, one finds that the spectrum of a star has an overall shape similar to the black-body curve, but superimposed on this shape are narrow features, called **spectral lines**, that are caused by the absorption or emission of light of particular wavelengths by the atoms in the outer parts of the star. It is these features that are unique to each type of star, and indeed, if we go to enough detail, unique to each individual star.

In Figure 10.3 we can see what the spectrum of the Sun looks like, compared to the spectrum of a black body of 5900 K, which seems to be the temperature that gives the best approximation to the spectrum. The general shape follows the spectrum fairly well, but the spectral lines make noticeable diversions. If we see a star somewhere else whose detailed spectrum matches this, then we may be quite sure that it is similar in size, mass, and composition to our Sun.

Readers who remember how we modeled the Sun in Chapter 8 may wonder

Figure 10.4. *The high-temperature corona of the Sun can be seen only when the dominant light from the photosphere is blocked out, such as in this photograph of the 1999 eclipse, taken in Hungary. (Copyright Pavel Cagas, Zlin Astronomical Society.)*

what we mean here by the surface temperature of the Sun. After all, the Sun's temperature drops rapidly near the surface from very large values inside to almost zero, and then rises again to millions of degrees in the very outer regions. Where does the number 5900 K fit into this? The key to understanding this is to recall that photons scatter an enormous number of times as they make their way out from the central regions to the surface. At some point, they become free: the probability of a further scattering before they leave the Sun becomes very small.

These photons are the ones we see with telescopes. Because of the huge number of scatterings that a photon experiences on its way out from the center of the Sun, the region in which this last scattering takes place is localized to a very thin shell, and we call this shell the photosphere. The temperature of the gas at the photosphere is the surface temperature of the Sun. The higher temperatures outside this do not affect most of the light leaving the photosphere, because there is so little gas outside this point that few photons ever get scattered by this gas. It is only visible during an eclipse, when the photons of the photosphere are blocked out (see Figure 10.4).

Notice that all the spectral lines of the Sun in Figure 10.3 are dips in the spectrum rather than rises. This means that light from inside the Sun is being absorbed by the elements responsible for the lines. We call this an **absorption spectrum**. There are occasions, especially in the more exotic objects like quasars (Chapter 14), where spectral lines are seen in emission: there is more light coming to us from the lines than from the black body background.

Why are there lines in the first place? The reason is the modern version of Planck's great insight: atoms can exist in only certain energy states. When they emit or absorb a photon they can make a transition only to another of the allowed energy states, so the photon can have an energy that must similarly follow Equation 10.2 on page 112. In the gas that forms the black body, all vibration frequencies are represented, so the photons come with all wavelengths. But if there are traces of specific elements, which have vibration frequencies characteristic of that element,

Table 10.2. Radii of the five brightest stars, inferred from measured luminosities and temperatures.

Rank	Name	Luminosity (W)	Temperature (K)	Radius (m)	$R/R_\odot$
1	Sirius	8.0×10^{27}	8000	1.6×10^9	2.3
2	Canopus	7.3×10^{31}	15 000	4.4×10^{10}	63
3	Rigil Kentaurus	5.0×10^{26}	6000	7.1×10^8	1.02
4	Arcturus	3.5×10^{28}	4470	1.1×10^{10}	16
5	Vega	1.8×10^{28}	9500	1.7×10^9	2.5

then they can absorb preferentially at certain wavelengths and produce distinct features. This is what happens to give stars their unique fingerprints.

Now, Planck postulated that the allowed energy states were evenly spaced, the difference in energy being the same from each one to the next. But this does not fit the spectra physicists observe. The next great step toward quantum theory after the work of Planck and Einstein was taken by the Danish physicist Niels Bohr (1885–1962). He devised a more complicated rule in which the spacing in energy decreased as the energy went up, and this rule agreed with the simplest spectra, such as that of hydrogen. Further refinements to the rule allowed it to match more complicated spectra. The different spacing of the energy levels did not undermine Planck's derivation of the black-body spectrum, because Einstein had already shown that the black-body spectrum only needed the allowed energies of *photons* to be evenly spaced, and not those of the atoms.

How big stars are: color and distance tell us the size

In this section: the Stefan–Boltzmann law says that the luminosity of a star is proportional to the fourth power of its temperature and to its surface area. From this we find that stars range in size from hundreds of times the size of the Sun down to smaller than the Earth itself.

We come now to one of the most interesting consequences of being able to measure the distance to stars, which is being able to say how big they are. How is it that we can say with confidence that stars range in size from many times the size of the Earth's orbit down to sizes much smaller than the Earth itself?

The key lies in the black-body spectrum again. The spectrum (color) of the light emitted by a black body depends only on its temperature. But the *amount* of light emitted depends also on the size of the black body: the total light emitted by the black body is just the sum of the light emitted by each patch, and it must therefore be proportional to the *surface area* of the black body.

This is a remarkably simple conclusion: if we measure in the laboratory the total emission of a black body at some temperature T, and then if we find another black body of the same temperature that emits twice as much light, it must have twice the area. For stars, which are basically spherical in shape, if we know the area then we know the radius. We described earlier how knowing the distance to a star tells us the total energy emitted by it (its absolute magnitude). We also saw that measuring the color of the star tells us its temperature. These two together then tell us how big the star is.

In Investigation 10.2 we shall see that the law relating luminosity, area, and temperature is remarkably simple. The luminosity is proportional to the area and to the fourth power of the temperature:

$$L = \sigma A T^4, \tag{10.3}$$

where σ is the constant of proportionality called the *Stefan–Boltzmann constant*. (Yes, Boltzmann turns up here too!) Its value is

$$\sigma = \frac{2\pi^5 k^4}{15 c^2 h^3} = 5.67 \times 10^{-8} \text{ W m}^{-2} \text{ K}^{-4}, \tag{10.4}$$

where the units are watts per square meter of surface area per degree kelvin to the fourth power. This is called the Stefan–Boltzmann law.

Investigation 10.2. Exploring the Planck Function

There are two very important properties of the Planck spectrum that are not hard to understand, but which are central to the way astronomers use the function to learn more about stars.

- The first property is that the wavelength where the peak of the spectrum occurs is inversely proportional to temperature. (This is called Wien's law.)

- The second is that the total luminosity of a black body is proportional to the fourth power of its temperature. (This is called the Stefan–Boltzmann law.)

In fact, both laws were known to physicists before Planck deduced the full theory of thermal radiation, but it is easier for us to understand them now as a consequence of his theory.

Although the expression for F_λ may seem so complicated that these results might be difficult to prove, the situation is actually rather simpler than that. The key to simplifying F_λ is to give a simple name to the exponent of e in Equation 10.1 on page 111. We define a new variable x to be:

$$x = hc/\lambda kT. \qquad (10.5)$$

The various quantities in this expression all have complicated units of measurement (especially h), but we now show that the units must all cancel out to give a number with *no* units at all, a **dimensionless number**. This is because x enters F_λ as an exponent, a power: Equation 10.1 on page 111 contains

$$e^{hc/\lambda kT} = e^x.$$

Now, one can raise a number, say 3, to a power that is a pure number, like 5, to get $3^5 = 243$; this means that we multiply together five factors of 3. But if the power has dimensions, the expression has no meaning: what are 5 km factors of 3? Interested readers should check that x is indeed dimensionless, but it must work out that way for F_λ to make any sense at all. (Readers may also like to verify that x is just proportional to the *frequency* of the light, in fact that x is just the ratio of the frequency f to what one might call the "thermal frequency" kT/h, which is the frequency a photon would have if its energy were equal to the typical thermal energy kT.)

Let us now take x to be a variable in F_λ instead of λ itself. If we solve Equation 10.5 for λ we get

$$\lambda = hc/xkT. \qquad (10.6)$$

We now substitute this into Equation 10.1 on page 111 to get

$$F_\lambda = \frac{2\pi k^5 T^5}{h^4 c^3} \frac{x^5}{e^x - 1}. \qquad (10.7)$$

Apart from a coefficient out front that is constant for a given temperature, the function at the heart of this is the dimensionless function

$$f(x) = x^5/(e^x - 1), \qquad (10.8)$$

whose properties depend only on x.

Now we can look at the laws we wish to establish, the Wien and Stefan–Boltzmann laws. The peak of the spectrum for a given temperature occurs where $f(x)$ reaches a maximum. Let us call this value x_{max}. We don't need to know its value. All we need is the relation above for λ as a function of x: the maximum will occur at the value of λ given by

$$\lambda_{max} = hc/x_{max} kT.$$

This proves that the peak wavelength is inversely proportional to T. In Investigation 10.3 on page 119 we will see how to calculate the value of x_{max}. The result gives the Wien law

$$\lambda_{max} = 0.29/T \text{ cm}, \qquad (10.9)$$

where T is given in degrees kelvin. For example, if a spectrum peaks in the visible region, say at 0.5 microns (5×10^{-7} m), then its blackbody temperature is 5900 K, just like the Sun.

The Stefan–Boltzmann law has a similar foundation. The energy radiated between λ and $\lambda + \Delta\lambda$ is, for sufficiently small $\Delta\lambda$, just equal to

$$\text{flux} = F_\lambda \Delta\lambda.$$

Now we need to convert $\Delta\lambda$ to an equivalent range of the variable x. The wavelengths λ and $\lambda + \Delta\lambda$ correspond to two values of x. Since x decreases as λ increases, we call these values x and $x - \Delta x$, respectively:

$$x = \frac{hc}{kT} \frac{1}{\lambda}, \quad \text{and}$$

$$x - \Delta x = \frac{hc}{kT} \frac{1}{\lambda + \Delta\lambda}.$$

Their difference gives Δx:

$$x - (x - \Delta x) = \Delta x$$
$$= \frac{hc}{kT} \left(\frac{1}{\lambda} - \frac{1}{\lambda + \Delta\lambda} \right),$$
$$= \frac{hc}{kT} \left(\frac{\Delta\lambda}{\lambda(\lambda + \Delta\lambda)} \right)$$
$$\approx \frac{hc}{kT} \left(\frac{\Delta\lambda}{\lambda^2} \right),$$

where the last step is an approximation that gets better and better as we make $\Delta\lambda$ smaller and smaller. Next we replace the factor λ^2 with its equivalent in terms of x to get

$$\Delta x = \frac{kT}{hc} x^2 \Delta\lambda.$$

We can now solve this for $\Delta\lambda$ to get

$$\Delta\lambda = \frac{hc}{kTx^2} \Delta x.$$

If we put this into the flux expression and use Equation 10.7 for F_λ, we find

$$\text{flux} = \frac{2\pi k^4 T^4}{h^3 c^2} \frac{x^3}{e^x - 1} \Delta x. \qquad (10.10)$$

This is proportional to a pure number depending on x and to T^4. If we ask for the total flux of light from the body, we have to add up contributions like this from all ranges of wavelengths, from $x = 0$ (the longest wavelengths) to $x = \infty$. Then even the dependence on x goes away, and the result is that the total flux of energy radiated from the surface of a black-body is proportional to T^4. The full equation for this is justified in Investigation 10.3 on page 119:

$$F = \frac{2\pi^5 k^4}{15 c^2 h^3} T^4. \qquad (10.11)$$

The dependence on k, T, h, and c is as in Equation 10.10 above. The pure numbers (such as π^5) come from finding the area under the curve $x^3/(e^x - 1)$, which we do in Investigation 10.3 on page 119.

Now we can at last test whether our ideas about black bodies have any relation to the real stars, in particular to the Sun. Each square meter of the surface of a black body at a temperature of 5900 K, like the Sun, shines with a power of 6.42×10^7 W. Since from Equation 9.1 on page 106 the Sun's luminosity is 3.83×10^{26} W, it must have a surface area of $3.83 \times 10^{26}/6.42 \times 10^7 = 5.97 \times 10^{18}$ m². From the formula for the area of a sphere, $A = 4\pi r^2$, it follows that the solar radius is 6.89×10^8 m.

This is very close to the accepted value of 6.96×10^8 m, which is obtained by direct measurements by spacecraft. The closeness of these two numbers is a triumph for the black-body model of the Sun! We can expect to use it with confidence on other stars.

In Table 10.2 on page 116 we look at the same stars as in Table 10.1 on page 110, only this time we list their luminosities and temperatures and the radii we infer by the method we have just applied to the Sun. There is a huge range of sizes, from about the size of the Sun to more than 60 times its size. This small selection of stars illustrates an important point: most stars are either about the same size as the Sun or they are big. Astronomers call the normal stars **main sequence stars** for reasons that will become clear in the next chapter. The big stars are called **giants**.

In fact, the range of size is even greater than we have shown with this selection of stars. For example, the star Sirius is actually two stars, one of which is very dim. Called Sirius B, its luminosity is 9.6×10^{23} W, but its color is very blue and its temperature is a very high 14 500 K. The only way it can have such a high temperature and yet be so dim is for it to be small. The radius we infer is only 5.5×10^6 m, or 0.86 times the radius of the Earth! This star is truly remarkable, because we can see the gravitational effect it has on its much brighter companion, Sirius A, and infer from this that its mass is actually 1.05 times the mass of the Sun. (We will see in Chapter 13 how to do this.) It has more than the mass of the Sun squeezed into a volume smaller than the Earth! Such a star is called a **white dwarf**, and we will find out in Chapter 12 how such extraordinary stars can exist.

Even this is not the end of the scale of sizes. When we come to study pulsars in Chapter 20, we will see that they are neutron stars. They have masses greater than the mass of the Sun, yet their typical radii are only 10 km, smaller than a good-sized city! Because their sizes are inferred by means other than those we have employed here, we shall reserve a full discussion of these incredible objects to the later chapter.

But why are stars as hot as they are, and no hotter?

In this section: the surface temperatures of most stars are not very different, and this comes from the way photons free themselves from the star.

We have come a long way in our understanding of stars just by learning that they are black bodies. We have used that knowledge to measure their sizes. But we can do even better: with a little thought, we can actually predict the temperatures of the stars. Notice a rather remarkable feature of Table 10.2 on page 116. The range of temperatures of the stars in the table is not large. While their radii range over a factor of about 60, and their luminosities over more than 10^5, their temperatures differ by less than a factor of four. Is there something, then, that fixes the temperature of the star?

To answer this we must remind ourselves that we are looking at the surfaces of the stars, not their interiors. And then we meet a puzzle: when we solved for the structure of the Sun in Chapter 8, we found that we predicted that the temperature of the Sun should fall smoothly to zero at its surface! (In fact, as we noted in Chapter 8, complex dynamical processes heat the outer corona of the Sun to a very high temperature, but so little mass is out there that it has no influence on the normal visible properties of the Sun.) What, then, do we mean by the temperature of a star: is not its surface temperature zero?

We can solve this puzzle by thinking about what happens to photons trying to get out of the star into space, eventually to hit our eye. Photons have a special affinity for charged particles, like electrons and protons. Radiation is given off by any moving, accelerating charge: the radio waves coming from the radio transmitting tower near your home come from electrons that race up and down the tower at the right frequency. Neutral atoms accelerating give off no radiation. The time-reverse of emitting radiation is absorbing it, and the same considerations apply: a light beam

---

Investigation 10.3. Computing the Planck function

Readers who have access to a computer and who already understand the exponential function e^x and its inverse, the natural logarithm $\ln(x)$, may wish to verify the graphs in Figure 10.1 on page 112 and the results of Investigation 10.2 on page 117 by using the computer program Planck on the website. In order to understand how the program is constructed, we have to look at a difficulty that arises in calculating functions like $f(x)$ that depend on the exponential function e^x.

If one simply programs the expression for $f(x)$ directly, one soon finds a difficulty: the exponent on e can get very large and overflow the limits that ordinary computers set on such things ($|x| < 200$ or so). The way to get around this problem is to calculate the natural logarithm of $f(x)$,

$$\ln[f(x)] = 5 \ln x - \ln(e^x - 1), \qquad (10.12)$$

and then to make separate approximations to the exponential function for three different ranges of the variable x. These are as follows.

- If we get into trouble because x is too large an exponent for the computer to be happy with, then of course we will also have $e^x \gg 1$ so we may neglect the 1 subtracted from e^x in the second term of Equation 10.12. This means we have

$$\ln(e^x - 1) \approx \ln(e^x) = x,$$

  the last equality following from the inverse property of the exponential and the natural logarithm.

- If x is neither large nor small, say $0.01 < x < 100$, there is no problem evaluating Equation 10.12 directly in the computer, so no approximation is needed.

- If x is smaller than, say, 0.01, we can make use of a remarkable (and very profound) property of the exponential function, namely that for small x we can approximate

$$e^x \approx 1 + x, \qquad |x| \ll 1. \qquad (10.13)$$

If this seems surprising, remember that any number raised to the power of *zero* is equal to 1. So it is not surprising that when x is nearly zero then e^x is nearly 1. What is remarkable is that e^x differs from 1 just by x itself! Experiment with this on your pocket calculator. For example, if $x = 0.1$ I find that $e^x = 1.10517$. This is pretty good: the error of the approximation is 0.005, or one-half of one percent of the answer. The approximation works if x is negative as well, and it gets better as $|x|$ gets smaller. This allows us to make the following approximation in Equation 10.12:

$$\ln(e^x - 1) \approx \ln(1 + x - 1) = \ln x.$$

These three cases are the basis of the program Planck given on the website, which calculates $f(x)$.

The program finds the maximum of $f(x)$ in a simple way: as it steps through its range of values of x, it tests each computed value of $f(x)$ to see if it is larger than the largest previous value. If it is, then this value becomes the new maximum and its value of x the new x_{max}. After all values of x have been tested, we have found the "global" x_{max}. The program finds the maximum to be $f_{max} = 21.20$ at $x_{max} = 4.95$. This value of the maximum leads to Wien's law, Equation 10.9 on page 117, when we note that $x = hc/\lambda kT$. Solving for λ_{max} gives

$$\lambda_{max} = \frac{hc}{x_{max} kT} = \frac{0.29 \, \text{cm}}{T},$$

when T is measured in degrees kelvin. This is just what we quoted in Equation 10.9 on page 117.

The program also finds the coefficient in the Stefan–Boltzmann law by simultaneously computing the area under the curve $x^3/(e^x - 1) = f(x)/x^2$. This is equivalent to summing up the fluxes from all the little wavelength ranges, each given by Equation 10.10 on page 117. The method for estimating the area under a curve using a computer is described in Figure 10.5. The program Planck gives 6.494. More sophisticated calculations using the tools of the calculus give an exact value of $\pi^4/15$. This evaluates to 6.494, keeping the same number of places. Our simple computer program has given us the right answer for Wien's law and the Stefan–Boltzmann law to several places accuracy, and we did not have to use any advanced mathematical techniques.

---

has an easy time making charged particles move, and thereby being partly absorbed and partly scattered as a result. But light passing through a medium with no free electrons is little affected by it. Indeed, metals are good at reflecting light because the electrons in them are free to move around, whereas in transparent materials like glass the electrons are firmly held to their atoms.

Because the Sun is a highly ionized gas, composed of bare protons and electrons, light has a hard time passing through it. Photons in the hot inner regions of the star do not reach us directly: they scatter off the electrons and protons too readily. However, as one goes outwards from the center of the star to its surface, the temperature of the gas decreases, until eventually it falls low enough to permit hydrogen (its main constituent) to remain in its neutral atomic state. At this point, photons are free to escape and reach us. So on this argument, we would expect the surface temperature of the star to be about the temperature required to ionize hydrogen.

This temperature is not hard to calculate. The energy required to pull an electron

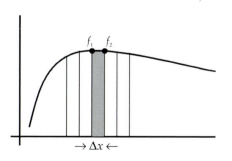

Figure 10.5. *For Investigation 10.3, we approximate the area under the curve $f(x)$ by dividing it into many small sections of width Δx and replacing the curve by a straight line across the top of each section. The approximation to the area of the shaded trapezoidal section shown here is then $(f_1 + f_2) \times \Delta x/2$. The approximation is good if Δx is small enough that the curve is practically straight across the section.*

▷Recall the definition of the electron volt (eV) given in Chapter 8: 1 eV = 1.602×10^{-19} J.

away from the proton in a hydrogen atom is about 13 eV. In a hot gas, this is provided by collisions: if the energy transferred by one atom to another in a collision exceeds this amount, the result is likely to be that an electron gets knocked off an atom. The energy transferred in a collision should be comparable to the energy of the moving gas atoms, which is about $\frac{3}{2}kT$. Setting this equal to 13 eV gives a temperature of $T = 10^5$ K. This is about a factor of 2 larger than the Sun's surface temperature, which means that our reasoning is pretty good but not perfect. The main error is simply that we don't need *all* the photons to be energetic enough to ionize the hydrogen atoms. If the temperature were lower, there would still be photons with energies above 13 eV, but there would just be fewer. That would be acceptable if there are still enough of them to ionize any hydrogen atoms that happen to form by the recombination of a proton and electron. Exactly what this temperature should be depends on a balance that involves the density of the gas (which affects how often hydrogen atoms form by recombination) and the details of the distribution of energies of different photons at any given temperature. We need not worry about these details here.

Looking ahead

In this section: we have learned how to determine the overall properties of a star. This forms our platform for investigating what goes on inside in the next chapters.

What we have learned here and in the previous chapter sets the stage for the next four chapters. Now that we have seen how astronomers have learned how far away, how bright, how hot, and how big stars are, it is not difficult to guess that we will be able to discover much more about what goes on inside them. In the next chapter we see how the balance between the inward pull of gravity and the outward pressure of hot gas is maintained by the steady nuclear reactions that make stars shine, and from which life itself ultimately derives. Then we will look at what happens when this balance fails, and stars "die", turning themselves into white dwarfs, neutron stars, or even black holes. After that we will look at stars in pairs, orbiting one another, sometimes so closely that they exchange gas and even feed black holes. Then in Chapter 14 we will look at the ways that stars form larger groups held together again by gravity: star clusters, galaxies, clusters of galaxies. We shall again have to calibrate our rulers and change our perspective on size: the distances between galaxies will make the distances between stars seem miniscule.

Our study of galaxies will conclude the first half of the book. After that, we have to widen our ideas about fundamental physics to include relativity. Once we do that, the second half of the book opens up to us: black holes, gravitational waves, cosmology, quantum gravity, and the beginning of time itself.

11

Stars at work:
factories for the Universe

In this chapter we open the door to our own history. Surely one of the most satisfying discoveries of modern astronomy is how the natural processes of the Universe led to the conditions in which a small planet could condense around an obscure star in an ordinary looking galaxy, and life could evolve on that planet.

The evolution of life seems to have required many keys, but one of them is that the basic building blocks had to be there: carbon, oxygen, calcium, nitrogen, and all the other elements of living matter. The Universe did not start out with these elements. The Big Bang, which we shall learn more about in Chapter 24 to Chapter 27, gave us only hydrogen and helium, the two lightest elements. All the rest were made by the stars. Every atom of oxygen in our bodies was made in a star. It was then expelled from that star and eventually found its way into the hydrogen cloud that condensed into the Sun and the Solar System.

Indeed, the atoms in our bodies have a durability that makes the time they spend being part of us seem minute. All the elements participate in vast recycling schemes on Earth: the carbon dioxide cycle (which we mentioned in Chapter 7), the calcium cycle, and so on, in which they move in and out of living organisms. But these cycles are mere epicycles on a grander cycle, in which the atoms are made in stars, pushed out into giant interstellar clouds of gas, and incorporated into new stars. In their new stars they might actually be torn apart and the pieces – neutrons and protons – re-assembled into new elements. They might then be expelled to go around the grand cycle once more, or they might become trapped in a dying star and be lost forever.

This grand cycle of stellar birth, death, and re-birth is the main business of the Universe. The balance between gravity and nuclear reactions drives this cycle. We shall look more closely at the processes of birth and death in the next chapter. Here we look at what keeps the star going during its normal lifetime.

Star light, star bright ...

If we want to understand what the stars are doing, let us go straight to their centers and ask: where does the energy come from that makes the stars shine? The short answer is: nuclear reactions. The conversion of hydrogen and helium into other elements gives off energy, and that comes out as light from the star's surface. Physicists only learned about nuclear physics in the twentieth century. Nineteenth century astronomers speculated about the subject, but they did not have enough understanding of physics to know what made the stars shine. We shall first see why the mechanisms they knew about could not work, and then see why nuclear reactions do.

If you are an astronomer speculating about why the stars shine, but you don't know about nuclear physics or you want to look for other sources for the energy radiated by the Sun and stars, then two possibilities should come to mind. One is that perhaps there are some chemical reactions going on inside. After all, chemical reactions are the source of most of our heat on Earth: burning wood, gas or coal

In this chapter: we look at the way stars have created the chemical elements out of which the Earth, and our bodies, are formed. The nuclear reactions in generations of stars that burned out before our Sun was formed produced these elements. But the physics is subtle, and nearly does not allow it. We examine this issue, and also show how the study of a by-product of nuclear energy generation in the Sun, neutrinos, has revealed new fundamental physics.

In this section: we ask where the energy of the Sun could be coming from. Chemical reactions or gravitational contraction cannot supply enough energy. Nuclear reactions can, because they convert mass into energy. Moreover, they have a characteristic signature: the Sun should be emitting the elusive particles called neutrinos.

releases energy that heats our houses and, after conversion into electricity, lights our rooms. Chemical reactions differ from nuclear ones in that they do not convert one kind of atom into another; they just rearrange the ways that atoms combine to form molecules. Maybe stars are forming certain kinds of molecules and releasing heat that way.

The other possibility is that stars are contracting, getting smaller, falling gradually inwards upon themselves. This would release energy. This energy is the same sort that would be released if a small mass fell to the Earth into, say, a bucket of honey: the energy of its fall is dissipated by friction in the honey, and the result is that the honey is heated slightly. If enough energy were dissipated this way, the honey would be hot enough to shine (if it didn't boil away first!).

The problem with both of these explanations is that neither chemical nor contraction energy can last long enough. Let us take the Sun in particular, since it is the star we know most about. The Earth is known to be about 4.54 billion years old, from studies of the radioactive decay of elements in its rocks, and the geological evidence is that the Sun has had pretty much the same luminosity for all that time. We show in Investigation 11.1 how to calculate how much energy could come from either process.

> Our calculation shows that the Sun could not derive its luminosity from gravitational energy for more than 1% of its present age, and that chemical energy is even less effective. Gravitational contraction could not have released enough energy to have kept the Sun shining for more than a few tens of millions of years. Nineteenth century astronomers nevertheless believed that contraction powered the Sun, and this led them into great conflicts with geologists, who knew that the Earth was older, and with Darwinian evolutionists, who needed much longer time-scales for evolution to work.

The puzzle of the energy source for stars began to clarify the moment Einstein discovered his famous equation $E = mc^2$, about which we will have much more to say in Chapter 15. For now we only need to know that Einstein predicted that mass and energy can be converted into each other. We see in Investigation 11.2 on page 124 that, if one could find a process that converts even a small fraction of the mass of a hydrogen atom into its equivalent energy, there would easily be enough energy to power the Sun. However, physicists did not know what the details of such a process might be until the discovery of the neutron in the 1930s.

The key advance was the realization that one could *make* helium out of hydrogen by a nuclear reaction that essentially converts four protons (which are the nuclei of hydrogen) and two electrons into a single nucleus of helium, which consists of two protons and two neutrons. This nucleus is also called an alpha particle. This reaction gives off energy, converting about 0.8% of the mass of each hydrogen atom into energy. As Investigation 11.2 shows, this is more than enough to power the Sun for the required time. The reaction does not happen easily, though. Nuclear reactions only occur if particles get very close, and this does not happen often because the protons have electric charges that repel each other. To get them close enough together, they must have a large enough speed to overcome the electric repulsion. In stars, they get this speed from the random thermal motion of gas particles at the temperature of the core of the star. This means that nuclear reactions only occur if the core is hot and dense enough. They do not occur in the core of the Earth, for example. The details of the reactions involved are explored in Investigation 11.4 on page 127.

Investigation 11.1. What **doesn't** *make the stars shine*

Since we know the Sun is at least 4.54 billion years old, we can ask whether any proposed source of energy could last that long. Suppose every hydrogen atom in the Sun did something once during the lifetime of the Sun that released energy: it engaged in a chemical reaction, or it simply fell into the Sun and released its energy of fall. How much energy would it have to release on that occasion to be able to account for its share of the steady luminosity of the Sun?

The Sun has a mass of $M_\odot = 2 \times 10^{30}$ kg. Each hydrogen atom has a mass of $1 m_p = 1.67 \times 10^{-27}$ kg. Therefore there are $N_p = M_\odot / m_p = 1.2 \times 10^{57}$ protons (hydrogen atoms) in the Sun. (The electrons are so light that they don't affect this calculation. We are ignoring the other elements, since we will see that errors of 10 or 20% won't be important to our conclusions.) Geologists find evidence that the Earth has had liquid water (oceans) on its surface for its whole history, so we can conclude that the Sun has been shining with roughly its present luminosity of $L_\odot = 3.8 \times 10^{26}$ W for at least $t_\odot = 4.5 \times 10^9$ y $= 1.4 \times 10^{17}$ s. Therefore it has given off a total energy of $E_\odot = L_\odot \times t_\odot = 5.4 \times 10^{43}$ J so far. This is an energy of $E_\odot / N_p = 4.5 \times 10^{-14}$ J per hydrogen atom. Let us call this the "energy duty" on atoms in the Sun: in order to join the Sun, the average hydrogen atom is required to "pay" on average 5×10^{-14} J once during its time in the Sun.

Now what of our two candidate sources of energy – are they capable of paying this duty?

- *Chemical energy.* There are so many possible chemical reactions that one might think it would be impossible to decide whether chemical reactions could do the job. After all, could there not be some mysterious chemical at work whose reactions give off enormous amounts of energy? The answer is no: all chemical reactions involve separating electrons from atoms and placing them around other atoms, and so chemical reactions involving any given atom cannot give off more

energy than it would take to remove an electron completely from that atom. This energy is called the ionization energy of the atom, and for all atoms it is less than or about equal to 10 eV. (Recall the definition of the electron volt (eV) given in Chapter 8: 1 eV $= 1.602 \times 10^{-19}$ J.) Thus, an atom of hydrogen engaged in a chemical reaction could only pay an "energy duty" of about 10^{-18} J, too small by a factor of more than 10^4. (Multiple chemical reactions are no solution: once the electron has been given up, it can't be given up again, and hydrogen has only one to give.) Chemistry is not the answer: it could not make the Sun shine for perhaps a million years, not ten billion.

- *Gravitational energy of infall.* The Sun was formed from a cloud of gas that contracted, heating itself up. Could the Sun still be contracting and generating heat? The energy it would generate per hydrogen atom would be roughly the same as one would get if one allowed an atom of hydrogen to fall onto the present Sun from far away. (Again, errors in our estimates of factors of 2 or even 10 will turn out not to matter to our conclusions.) This is just the reverse of "launching" an atom from the surface of the Sun and letting it escape from the Sun's gravity. The speed the infalling particle would have when it hit the Sun is then exactly the same as the escape velocity of the Sun. In Chapter 4 we saw that the escape velocity from the Sun is $v_{escape} = (2GM_\odot/R_\odot)^{1/2}$. The kinetic energy of the atom when it reaches the Sun will thus be $\frac{1}{2} m_p v_{escape}^2 = GM_\odot m_p/R_\odot = 3 \times 10^{-16}$ J. This is better than for chemical reactions, but still inadequate: the Sun might shine by contraction for 100 million years, but not 5 or 10 billion.

Some other energy source is needed. Only nuclear reactions seem to be able to do the job.

Exercise 11.1.1: *Chemical bangs*

Can it really be true that chemical reactions all give off the same energy per atom, to within a factor of, say, 10 or 100? Don't the chemical reactions that make a TNT bomb explode give off far more energy than the chemical reactions that heat up a smoldering rubbish dump? Explain why the answer to this question is no.

Exercise 11.1.2: *Turning on the lights*

Once a cloud of gas begins to contract to form a star, roughly how long does it take before nuclear reactions begin to power the star? Will the star shine before this?

The conversion of hydrogen to helium is not quite as simple as suggested in the last paragraph. It turns out that one of the particles produced as a by-product of the reaction is the neutrino. One of Nature's most elusive particles, the existence of the neutrino was guessed at in the 1930s, by theoretical arguments that we will describe in Chapter 16. But it was not directly detected until experiments during the period 1953–1956, by the American physicists Clyde Cowan (1919–1974) and Frederick Reines (1918–1998).

Despite its elusiveness, the neutrino plays a key role in many problems in astronomy, in particular powering supernova explosions (Chapter 12). Neutrino astronomy is in its infancy, but we will see below that it has already completely revolutionized our understanding of the physics of neutrinos. One reason for the neutrino's importance is that it is emitted in abundance whenever nuclear reactions occur in astronomy. For example, two neutrinos are released every time a helium nucleus is created. Another reason is its very elusiveness: it travels fast (nearly at the speed of light), and it hardly interacts with anything at all. This means it can go through matter, such at the outer layers of a star, with little hindrance. Only gravitational waves, which we will meet in Chapter 22, have greater penetrating power than neutrinos.

▷ This confirmation of the neutrino won belatedly for Reines a share of the 1995 Nobel Prize for physics, but this honor came too late for Cowan, who had already died.

Investigation 11.2. What does make the stars shine

Mass and energy are related. We will see in Chapter 15 that relativity tells us that if a system gives off an amount of energy E, then its mass must decrease by E/c^2, where c is the speed of light. For chemical reactions, the release of, say, 1 eV from a reaction involving, say, a pair of oxygen atoms makes a negligible change in the mass of the system. The lost mass is $1\,eV/c^2 = 2 \times 10^{-36}$ kg, while the masses of the atoms together amount to about 32 proton masses, or 5×10^{-26} kg. So the system loses less than one part in 10^{10} of its mass in a chemical reaction.

When nuclear reactions take place, on the other hand, the mass changes are significant, so that we can deduce the energy released just by looking at the masses of the particles involved. If four protons combine to form a helium nucleus, the starting mass is 6.69×10^{-27} kg, while the final mass is 6.64×10^{-27} kg, a change of 5×10^{-29} kg, or almost 1%. This is more than 10^7 times as much energy as is released in a chemical reaction. (Other particles, usually electrons or **positrons**, are also involved in these reactions in order to keep the total electric charge the same before and after, but the electron mass of 9×10^{-31} kg is negligible here.) In energy units, it amounts to 4.5×10^{-12} J released for every helium nucleus formed.

By comparison with the energy duty of 5×10^{-14} J that we saw in Investigation 11.1 was required of each proton in the Sun, this is huge. Nuclear reactions have more than enough energy to power the Sun. In fact, is this excess too much of a good thing? Does it mean that the Sun ought to be either more luminous or more long-lived than it is? The answer is no: the nuclear reactions only take place at the very center of the star, where the temperature and density are high enough. So only about 1% of the mass of the star actually gets to be involved in the nuclear reactions that convert hydrogen to helium. We will see in Chapter 12 what happens when these reactions have exhausted the hydrogen fuel in the central part of the star.

Exercise 11.2.1: *Water power*

If all the hydrogen in a teaspoonful of water were converted into helium, how long would that water power a 100 W light bulb? Take a teaspoon to contain 5 g of water.

...first star I see tonight

In this section: what does the emission of light by the Sun and other stars tell us about the history of the Universe itself? We argue that the Universe cannot be infinitely old: it must have had a beginning.

Let us step back from the details of the source of the Sun's energy to think for a moment about where our investigation is leading us. The Sun began its life as a star roughly 5 billion years ago. What was the Universe like before that? Were there other stars, which were formed earlier, and may by now have died? If so, then what was the Universe like before them? Another generation of stars, perhaps? But earlier again – could the process of forming stars have been going on forever? Is the Universe infinitely old?

The answer is that it can't be infinitely old. In any region of space, say the space occupied by our Galaxy, there is a finite amount of hydrogen. Every generation of stars converts some amount of it into helium. If this had been going on for too long in the past, we would not have much hydrogen in stars today. Yet all stars we see are primarily composed of hydrogen. Observations leave room for maybe three or four earlier generations of stars, but not for an infinite number.

Could there be a way out of this: maybe the helium is somehow re-converted back into hydrogen by some mysterious process? Any such process would need to replace the energy that the original reaction gave off, and where would the process get such energy? The original energy has simply gone away, carried into empty space by the light given off by the stars. Any energy for making hydrogen again would have to come from other nuclear reactions, and they would in turn lead to the same argument.

Figure 11.1. Fred Hoyle was one of the most influential astrophysicists of the twentieth century. We will discuss his fundamental contributions to the understanding of how elements are made in stars below. He is also credited with having coined the phrase "Big Bang" for the beginning of the Universe. He seemed to relish taking controversial scientific positions, of which the C-field was an example. In his later years he became embroiled in heated debates about panspermia, also mentioned below. (Reproduced courtesy of N C Wickramasinghe.)

Time marches on. The Universe *does* get older. Conversely, there was a time when the Universe was young. There *was* a first star.

It is important to understand that this is not in itself an argument for what we call the "Big Bang", although it is a step toward it. The Big Bang is now the almost universally accepted model for how the Universe began. Our arguments above from nuclear physics do not tell us how the Universe began. All they tell us is that, before a certain time not too long ago (as measured in stellar lifetimes), the Universe was doing essentially nothing. The theory of the Big Bang makes a further step beyond this to say that the Universe expanded from a point a few stellar lifetimes ago, and it probably didn't exist at all before that time. We will look at the reasons we believe this, and what it actually means, in Chapter 24.

The only way to avoid the conclusion that the Universe has a finite age is to postulate that somehow energy is simply created in order to replace that which is lost as stars shine. The British astrophysicist Sir Fred Hoyle (1915–2001) and the Indian astrophysicist Jayant Narlikar (b. 1938) made such a postulate, called the **C-field** ("C" for "creation"). They invented this field in order to support the so-called **Steady-State model of the Universe**.

The C-field makes matter to fill in the empty spaces of the Universe left behind as it expands, so that it can be "steady" while expanding. We will return to this subject in Chapter 24. The weight of observational evidence today is heavily against the Steady-State model and its C-field.

Cooking up the elements

Once we accept that nuclear reactions are changing hydrogen into helium, it is natural to ask what other reactions are happening. In particular, if the amount of helium in the Universe is increasing with time, is that also happening to other elements?

The answer is yes: just as the lightest element, hydrogen, can make helium, so can several helium nuclei react to form heavier elements, such as carbon, oxygen, and silicon.

> Much of the story of the formation of the elements depends, of course, on the details of the nuclear physics, but one fact is of overriding importance: the most tightly bound collection of protons and neutrons that it is possible to form is the nucleus of iron.

This means that the sequence of reactions forming elements heavier than helium stops with iron. If iron reacts with anything else to form a heavier nucleus, energy must be added to make the reaction go.

We do of course find other elements on the Earth. One of the most important heavy elements is uranium, more than four times heavier than iron. But there is not much of it around, and that means it was formed in unusual circumstances, when energy from some source other than nuclear reactions was available to drive the reactions past iron. The fact that uranium is formed by the addition of energy to other nuclei partly explains why it is good as a nuclear fuel: by splitting it apart into smaller nuclei, one liberates some of the energy that was put in originally.

We believe that most uranium was formed in the giant stellar explosions that we call *supernovae*. We shall explain in the next chapter what supernova explosions are and where the extra energy comes from that goes into making uranium and other heavy elements. It is interesting to reflect on the fact that the ultimate source of our nuclear energy on Earth is not our Sun.

> The Sun is responsible for most of the forms of energy we use, such as chemical (burning coal, oil, gas, or wood) or environmental (wind, hydroelectric, and direct solar power). But nuclear power plants release energy that was stored up from an ancient supernova, the death of some massive anonymous star long before the Sun was born.

It is even possible to estimate how long ago that supernova event took place, or in other words to estimate the age of the heaviest elements in our bodies. This is about 6.6 billion years. The estimate comes from measuring how abundant two particular **isotope**s of uranium are today. Since they decay at different rates, one can work backwards in time to determine how long ago they were equally abundant. The original supernova would have made them in roughly equal quantities, so this gives the estimate of their age. The details of the calculation are in Investigation 11.3 on the following page.

In this section: stars are the factories where elements are made. All elements heavier than hydrogen and helium are made by stars. The process is not straightforward, and would not happen at all if it were not for a strange coincidence of nuclear physics. We owe the evolution of life to the fact that a certain nuclear reaction proceeds much more easily than it should, allowing carbon and oxygen to be formed in stars.

▷The plural of the word *nova* is *novae* in most languages that use this scientific term. Increasingly in English authors use *novas*. I will remain with the international form in this book.

Investigation 11.3. Finding out how long ago "our" supernova occurred

The supernova that gave birth to the heavier elements in the cloud of gas that eventually condensed to become the Solar System made the heavy elements in our bodies. It is interesting that we can actually work out approximately how long ago that supernova event occurred.

The key is the fact that the supernova made these elements very quickly, within the space of a few minutes. Many of the nuclei that were made then were radioactive. Those that were very unstable decayed into other elements rapidly and disappeared completely. But many radioactive elements decay very slowly, and these are still present today. Two such nuclei are ^{235}U and ^{238}U, two isotopes of uranium. The notation used for identifying nuclei is described in the Introduction.

These two nuclei are not very different, and the processes that formed them were not sensitive to the small effects that make them decay over long time-scales. From the point of view of the physics that dominated the supernova explosion, these two isotopes were stable end-products, produced in essentially the same way. One nucleus got, at random, three more neutrons than the other. Therefore, they were produced in roughly equal numbers. Detailed calculations of the physics of the supernova explosion suggest that the number of ^{235}U nuclei produced was just 1.7 times the number of ^{238}U nuclei. The explosion produced, of course, comparable numbers of ^{236}U and ^{237}U as well, but these are more unstable and have since decayed away.

Since their production, ^{235}U has been decaying slowly with a **half-life** of 7.0×10^8 y, and ^{238}U has been decaying with a half-life of 4.5×10^9 y. The half-life is the time it takes half of the nuclei in a sample to decay. Thus, if there are N_0 nuclei of a certain isotope present at the beginning, then after a time equal to a half-life there are $N_0/2$. After two half-lives, the number is $(N_0/2)/2 = N_0/4$. After k half-lives there are

$$N = N_0/2^k$$

nuclei left. Now, this equation is true even if k is not an integer. So

after, for example, 1.5 half-lives, the number of nuclei has decreased by a factor of $2^{1.5} = 2.83$. The exponent k here is the number of half-lives, or in other words the ratio of the actual time to the half-life. So if we use the letter[a] τ to denote the half-life of the isotope and we look at the sample after a time t has elapsed, then we are looking at it after $k = t/\tau$ half-lives. The general equation for the number of nuclei remaining is then our previous equation with k replaced by t/τ:

$$N(t) = N_0/2^{t/\tau}. \tag{11.1}$$

At the present time, the ratio of the number of ^{235}U nuclei to the number of ^{238}U nuclei is about 0.007, as determined from Moon rocks brought back to Earth for analysis by the Apollo astronauts. Let us suppose that we began at time $t = 0$ (the time of the supernova event) with a sample that had 10^{20} nuclei of ^{238}U and 1.7×10^{20} nuclei of ^{235}U. These are in the correct ratio for elements just after the original explosion, but the overall number of 10^{20} is arbitrary. All we will be interested in is ratios of numbers, not the overall number remaining.

We can simply calculate the number remaining and their ratio after successive steps of 10^9 y, using the right value of τ for each element. For ^{235}U the number will be $10^{20}/2^k$, where $k = t/7 \times 10^8$ y is the number of half-life steps in the time t. For ^{238}U it will be the same formula but with $k = t/4.5 \times 10^9$ y. The results are summarized in the table below.

What the table shows is that sometime between 6 and 7 billion years after the supernova, the ratio of the two isotopes falls to 0.007, its present value. Therefore, the supernova that gave birth to all our heavy elements occurred between 6 and 7 billion years ago. Since we believe our Sun is only about 5 billion years old, this means that the formation of the Sun did not follow immediately after the supernova: it was apparently not triggered by the supernova. Radioactive dating of other elements shows that the Earth's rocks themselves were formed about 4.5 billion years ago.

	Time after supernova (y)						
	1×10^9	2×10^9	3×10^9	4×10^9	5×10^9	6×10^9	7×10^9
Number of ^{235}U remaining	6.3×10^{19}	2.3×10^{19}	8.7×10^{18}	3.2×10^{18}	1.2×10^{18}	4.5×10^{17}	1.7×10^{17}
Number of ^{238}U remaining	8.6×10^{19}	7.3×10^{19}	6.3×10^{19}	5.4×10^{19}	4.6×10^{19}	4.0×10^{19}	3.4×10^{19}
Ratio ^{235}U/^{238}U	0.74	0.32	0.14	0.059	0.033	0.011	0.0049

Exercise 11.3.1: *Getting the age of uranium right*

Perform the calculations to fill in the above table, using Equation 11.1 with the two given half-lives. Then take time-steps of 0.1 billion years within the last interval to show that the supernova occurred about 6.6 billion years ago.

---

[a]The Greek letter τ is frequently used in mathematical notation to represent particular values of time. It is normally pronounced like "out" backwards, but some speakers say "taw".

If the stars and their explosions are continually making more and more heavy elements, what was the Universe like when the first generation of stars was just forming? By looking for old stars, stars of the first generation that were small enough that their nuclear reactions have not yet run their course (why small stars age slowly will be explained in Chapter 12), astronomers have learned that the gas the first stars formed from was not pure hydrogen: about 20% or so by weight was already helium.

This initial helium must have been made from hydrogen in the early stages of the Big Bang. The bang was hot, very hot, and the same nuclear reactions that now make helium in stars also made helium then. However, the expansion of the Universe cooled the reacting gas off very quickly, freezing out a certain concentration of helium, but quenching the production of heavier elements. Most of the elements of which we are made were left to be cooked up in early generations of stars.

Investigation 11.4. How to make helium

The key nuclear reaction runs something like (but not exactly, as we will see later) the following:

$$p^+ + p^+ + p^+ + p^+ \longrightarrow {}^4He^{+2} + 2e^+ + \quad \text{energy.} \qquad (11.2)$$

In this equation we use some of the conventional notation of nuclear physics: p^+ denotes a proton, e^+ a positron (a positively charged electron, also called an anti-electron), and ${}^4He^{+2}$ a nucleus of helium-4. (Helium-4 is the most common form of helium, having a nucleus consisting of two protons and two neutrons.) As is indicated in the equation, the reaction gives off energy, which comes out as the kinetic energy of the products.

A word on the notation in equations like Equation 11.2: we write the name of a nucleus by giving not only the symbol of its element, such as Fe for iron, but also a preceding number to indicate the total number of neutrons and protons in the nucleus. The most common form of iron has 28 protons and 28 neutrons, and so is called ${}^{56}Fe$. When it is important to indicate the electric charges involved, which for ${}^{56}Fe$ is 28, we write it as a following number: ${}^{56}Fe^{+28}$. The charge on the nucleus determines the element it belongs to, so that explicitly showing the charge is redundant; we only do it when necessary for clarity. On the other hand, the number of neutrons in the nucleus is not important for the chemical reactions of the element; two nuclei with the same number of protons but different numbers of neutrons are called *isotopes* of the same element. The preceding number attached to the symbol for the element allows one to distinguish one isotope from another. Some nuclei have alternative names and symbols: the nucleus of ordinary hydrogen is a single proton, also called p or p^+.

The main thing to watch for in equations like Equation 11.2 is that certain quantities must balance on both sides. For example, the total electric charge of the particle going into the reaction must equal that of the particles coming out, since charge can be neither created not destroyed. In the above equation, the four protons on the left carry four positive charges. On the right, the helium nucleus contains two charges and the two positrons (called e^+) each carry one, balancing the charge. A similar *conservation law* applies to the total number of protons and neutrons. Protons and neutrons are called **baryon**s, and we say that the total baryonic "charge" must balance. The helium nucleus has two protons and two neutrons, each of which has a positive baryonic charge, so the total of protons and neutrons balances as well. This, we shall see, is significant.

Evidently, the nuclear reaction listed above proceeds by changing two protons into neutrons. The balance of electric charge is taken care of by creating two positrons. However, there is another balance law that must be obeyed: **lepton** conservation. Electrons and positrons belong to a class of particles called leptons, and in any reaction the total *lepton* number must not change. Now, electrons have leptonic charge +1, while positrons have leptonic charge -1. The *neutrino* ν is also a lepton, and has leptonic charge +1. Therefore, the creation of two positrons (leptonic charge -2) can be balanced if two neutrinos (leptonic charge +2) are created.

Physicists have discovered that there are in fact *three* families of leptons, each with its own separate balance law. We have been discussing reactions involving the electron family here, so the neutrinos that are created are called **electron neutrinos**, denoted by ν_e. We can now rewrite the reaction in Equation 11.2 in its correct form:

$$p^+ + p^+ + p^+ + p^+ \rightarrow {}^4He^{+2} + 2e^+ + 2\nu_e + \text{energy.} \qquad (11.3)$$

Although it was clear in the 1930s that in principle the Sun could shine this way, the details were not worked out until the 1950s, and they may still be in need of refinement, as we shall see below. The basic problem is that it is exceedingly rare for four protons to collide all at once, so that the direct reaction given in Equation 11.3 hardly ever happens. Instead, there is a rather complex path through many intermediate reactions, whose end effect is the same as Equation 11.3.

To follow this path, we need to introduce a few more nuclear particles into our discussion. These are **deuterium** – the heavy form of hydrogen that is used to make heavy water – and ${}^3He$, the light isotope of helium. Deuterium (${}^2H$) is made from one proton and one neutron. Add one more proton and one gets ${}^3He$.

There are many ways to formulate a more realistic path to the formation of ${}^4He$. Pairs of protons collide to form deuterium, emitting a positron and a neutrino. Two deuterium nuclei may then collide to form ${}^4He$. Or a proton may collide with deuterium to form ${}^3He$. A fourth proton hits this and forms ${}^4He$, with the emission of a further positron and neutrino. This network of reactions is called the *p–p chain*.

There are other ways to do the same thing, the most important of which is the so-called carbon cycle, in which nuclear reactions involving carbon nuclei have the net effect of facilitating Equation 11.3, without changing the total number of carbon nuclei; carbon acts as a **catalyst**. The carbon cycle dominates the energy output of stars somewhat more massive than our Sun.

The p–p chain has some side-chains, as well. Nuclei collide and transmute at random, so that some reactions occur in the Sun involving lithium, boron, and other light elements. These are not significant in terms of the energy they contribute to the Sun, but they have assumed great significance in what has become known as the solar neutrino problem. We shall look at this later in this chapter.

The various conserved charges that we have met here all have deep connections with the fundamental forces of physics. Electric charge is responsible for the electromagnetic interactions of particles. Likewise, baryonic charge is associated with the so-called **strong interaction**, which is the glue that binds protons and neutrons together in nuclei, overwhelming the electric repulsion of the protons for one another. The leptonic charges are associated with the weak interaction, which is responsible for beta decay.

There is a strong feeling among modern physicists that in fact all these forces are aspects of a single force, described by a **grand unified theory** of fundamental physics. We already know that the weak interaction and electromagnetic interaction are aspects of the so-called **electroweak** force, and there is strong evidence for unification of this with the strong force. If the forces are not completely separate, then there is the possibility that the conservation laws will not be completely separate, either, so that it may occasionally happen that a single proton will decay into a combination of leptons. Experimental evidence against this happening puts a lower bound on the half-life of the proton of 10^{33} y, so the violation of the conservation law for baryons will be a rare event indeed! Although these considerations do not matter much to the nuclear physics inside stars, they will matter deeply when we come to the most profound questions raised by the modern study of cosmology in Chapter 27.

The discovery that nuclear reactions power the Sun opened up vast new territory in astrophysics. Why should the transmutations stop at helium? Why not carry on and make carbon, oxygen, iron, indeed all the heavier elements? This is indeed what has happened: all the principal elements in our bodies were manufactured in stars that lived and died before the Sun and Earth were formed.

The solar neutrino problem

The main thrust of this chapter is to explore how stars made the elements of which we are composed, and thereby made life possible. We are a side-effect of the nuclear physics that is going on inside stars. The main evidence supporting the picture of nuclear energy generation that we have developed so far is indirect: theoretical calculations based on nuclear physics experiments on the Earth predict models for stars that agree well with their observed properties, such as their luminosities and

In this section: fewer neutrinos from the Sun's nuclear reactions are detected than expected. The reason is becoming clear: neutrinos transform themselves as they move through space between the Sun and the Earth.

temperatures. But direct evidence about the nuclear furnaces in stars is hard to obtain, because we can't see directly into them. The one exception is our Sun. Because the Sun is so near to us, we have the opportunity to look for the neutrinos that the nuclear reactions emit. Elusive as they are, they can be detected. The good news is that they *have* been detected. The bad news is that there aren't as many of them as physicists had expected. This is called the solar neutrino problem.

Earlier in this chapter we met the neutrino, and we learned more about it in Investigation 11.4. A key property of neutrinos is that they have a characteristic called the **lepton number**. This is analogous to electric charge, in that it is *conserved:* the total lepton number of all the particles in a nuclear reaction is the same before and after the reaction.

▷Lepton number has some similarities to electric charge. For example, the total electric charge in a reaction must not change even if charged particles are created or destroyed. Similarly, the total lepton number must not change: leptons must be created or destroyed in pairs with equal and opposite lepton number. Lepton number is therefore a kind of "leptonic charge".

This leptonic charge is responsible for the nuclear force that physicists call the **weak interaction**, or weak force. When particles carrying a non-zero lepton number collide with other such particles, they can scatter because of the weak force between them. The amount of scattering is much smaller than that which would occur if the particles also had electric charge, which is why the force is called "weak". Neutrinos do not have electric charge, so they can scatter from other particles *only* via the weak force. The result is that neutrinos produced in the center of the Sun stand little chance of ever scattering off anything else on their way out. Clearly, they will pass right through the Earth almost unhindered as well. This makes them excellent probes of the central conditions in the Sun: when we detect solar neutrinos on the Earth, we "see" directly into the core of the Sun itself.

The problem with detecting solar neutrinos is the other side of the same coin: they don't interact much with matter. The flux of neutrinos itself is huge. Every second something like ten billion solar neutrinos (10^{10}) pass through your body. But they just pass through, leaving almost no energy behind, inducing almost no nuclear reactions. So they will also pass through any detector we might build almost without noticing it. Physicists must use extremely sensitive techniques to find solar neutrinos at all.

▷Contrast the easy ride that neutrinos have with the plight of photons in the Sun. We saw in Chapter 8 that a photon scatters more than 10^{20} times on its way out from the center! This illustrates the feebleness of the weak interaction compared with the electric forces.

The first solar neutrino observations were made by the American physicist Raymond Davis (b. 1914) in the Homestake Gold Mine in South Dakota. He went below ground into the mine in order to use the rocks above him to screen out **cosmic rays** (high-speed particles hitting the Earth from space), which might otherwise have obscured the signal from the neutrinos. The basis of the experiment is a reaction in which a solar neutrino transforms a chlorine nucleus into one of a radioactive isotope of argon. The chlorine is in liquid form inside the experimental chamber, while the argon is a gas. By extracting the gas periodically and counting the number of radioactive argon decays, and by using the known rates at which neutrinos will interact with chlorine, Davis is able to infer the neutrino flux falling on the detector and hence the neutrino luminosity of the Sun. Davis has created a new unit for measuring the flux of neutrinos, the **Solar Neutrino Unit**, abbreviated SNU. One SNU represents one captured neutrino for every 10^{36} target chlorine atom in each second.

▷Cosmic rays are fast-moving protons that fly around the Milky Way. Most seem to be particles ejected from supernova explosions long ago, which follow random paths in and around the Milky Way, guided by our Galaxy's weak magnetic field. Some cosmic rays have energies above 10^{20} eV, which makes them the most energetic particles scientists have ever dealt with. The origin of these particles is a deep mystery, which we will explore in Chapter 27.

This is an experiment of extraordinary delicacy, but over more than 20 years of collecting data, Davis has consistently measured a capture rate of about 2.5 SNU, which is a factor of two or three below the predictions of the Standard Model of the Sun's interior. The Standard Model starts from considerations like those in Investigation 8.5 on page 94, where we built a model of the Sun. To this rather simple approach, physicists add the best information available about the nuclear reactions that can take place, and carefully take account of the radiation pressure provided by the outgoing radiation. When they build a model that has the mass, luminosity, size, and composition of the Sun, it predicts a neutrino luminosity about three times larger than Davis sees.

Such a discrepancy is very serious, and indicates that there is something we don't understand about either the Sun or neutrinos. The interpretation of Davis' experiment is made more complex by the fact that he does not detect neutrinos from the p–p chain, which supplies almost all the energy radiated by the Sun, but from a minor side reaction. This is because the nuclear reaction Davis uses in his detector requires the incoming neutrino to have a relatively large energy, and the p–p neutrinos are not energetic enough. The only neutrinos produced in the Sun that have enough energy are produced as a result of one of the inconsequential side reactions that are going on all the time. The most important one is:

$$^{7}\text{Be} + \text{p}^{+} \longrightarrow {}^{8}\text{B}. \qquad (11.4)$$

In turn, the boron nucleus ^{8}B is unstable to a form of **beta decay**, emitting a positron and a particularly energetic neutrino when it decays back to ^{8}Be. This reaction does not contribute significantly to the energy output of the Sun, but the rate at which it proceeds, and hence the flux of neutrinos it produces at the Earth, depends sensitively on the local conditions in the solar core, especially the temperature and density.

For some time, Davis' experiment was the only one able to detect solar neutrinos, so there was always a possibility that there was some flaw in the experiment or in the nuclear physics calculations that are required to interpret it. Then an independent measurement of the neutrino flux was made by the Kamiokande neutrino detector in Japan. This detector was built to look for possible spontaneous decays of the proton, which we refer to in Investigation 11.4 on page 127. The detector entered the realm of astrophysics when it registered neutrinos from the supernova explosion called sn1987a that occurred in the Large Magellanic Cloud in February 1987. I shall have more to say about the neutrinos from sn1987a and the Kamiokande detector in Chapter 12. For now, the important thing is that this detector is also sensitive only to high-energy neutrinos, so it directly tests Davis' experiment. Within the experimental errors, it has confirmed the shortfall in the neutrino flux.

What could be the cause of this deficit? Physicists first tried to modify the Standard Model of the Sun, reasoning that we know less about the solar interior than about the nuclear physics that can be tested in the laboratory. But they have found that changing the flux expected from the Sun by such a large amount is hard to do. In particular, the evidence from helioseismology (Chapter 8) places very strong constraints on the temperature and density profiles inside the Sun, and made it unlikely that the solution to the problem could be purely astrophysical.

A second possibility is that the rates of some of the minor nuclear reactions inside the Sun have been overestimated because we don't understand the nuclear physics itself as accurately as we think. Nuclear experiments are underway in order to test this understanding. There have recently been revisions in some reaction rates, and these have helped reduce the gap between theory and observation. But a substantial difference still remains.

However, there is a third, very interesting, class of solutions that has been proposed, and which has very recently gained strong observational support. The idea is that something happens to the neutrinos during their flight from the Sun to the Earth. At first one might try out the idea that they could simply decay into something else. Then the Sun might be producing the right number of them, but they don't reach us. Unfortunately, this is not consistent with the fact that Kamiokande saw just about as many neutrinos from the supernova in 1987 as scientists expected, assuming they do not decay. If half of them might decay just going from the Sun to the Earth, then essentially all of them would have disappeared coming from the supernova, which occurred 10^{10} times further away! So we have learned from the supernova that neutrinos do not simply disappear even as they travel over the vast distances between stars.

A more subtle variation on the decay idea, however, does seem to be happening: neutrinos change into other neutrinos and back again, in an oscillating manner, as they move through space. In Investigation 11.4 on page 127 we pointed out that there are actually *three* types of neutrino, each associated with one of the three charged leptons. However, only the electron neutrino will trigger the nuclear reactions in Davis' detector. If electron neutrinos transmute into other varieties between the center of the Sun and here, then this would explain the low observed flux. If they oscillate back again over a longer journey, and if each type of neutrino does this at a different rate, then there would be no contradiction with the neutrino observations of SN1987A. All it would mean is that the observed neutrinos were about one third of the original number, and since scientists have only a very approximate idea of the number of neutrinos to expect from a supernova, this would be an acceptable solution.

The neutrino oscillation solution violates the separate laws of lepton number conservation, but we would have to accept this; at least the *total* lepton number would still be conserved. The issue is one for experimental physics, and theory will have to follow. A number of detectors have recently been built to look for solar neutrinos with different energies, from different parts of the nuclear reaction chain. Many of them use gallium as a target rather than chlorine. They have measured capture rates about half of the prediction of the Standard Solar Model. The Kamiokande detector has been replaced by a much larger one, called SuperKamiokande. It has measured a solar neutrino flux of 2.32×10^6 cm^{-2} s^{-1}, again about half of the prediction of the Standard Solar Model. These results are all consistent with the Davis measurement, since they measure neutrinos from different reactions. They all reinforce the view that neutrino physics is responsible for the puzzle.

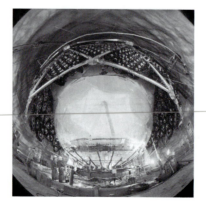

Figure 11.2. A fish-eye-lens view of the SNO detector. The inner white vessel holds 1000 tonnes of heavy water (in which deuterium replaces normal hydrogen), with which the neutrinos interact. The metal cage is the top part of the structure holding phototubes that register light emitted by particles produced by the neutrinos. Note the scale of the people in the photo. (Photo courtesy Ernest Orlando Lawrence Berkeley National Laboratory.)

At the time this book is being written (2002), scientists have just had first results from a new heavy-water detector. Called the Sudbury Neutrino Observatory (SNO), it is located in a mine owned by the Inco Mining Company in Sudbury, Ontario. This detector has the ability to distinguish between different types of neutrinos, and it should finally point the way to a solution. The first data from SNO are a measurement of the electron neutrino flux from the Sun. The flux is less than SuperKamiokande, only 1.75×10^6 cm^{-2} s^{-1}. Since SuperKamiokande is sensitive to some of the non-electron neutrinos, and since the Sun should be producing only electron neutrinos, this implies that some of the produced neutrinos have changed into other types by the time they get to the Earth. Physicists have estimated from the details of the two experiments that the original flux of electron neutrinos leaving the Sun should be 5.44×10^6 cm^{-2} s^{-1}, in excellent agreement with the predictions of the Standard Solar Model.

Further observations expected from SNO, including its own direct measurements of the total neutrino flux, should settle the matter and point the way to measuring values for the different oscillation wavelengths. The implications for theories of fundamental physics are only just beginning to be assessed. One implication is already clear, however: the neutrino must have a small mass. It cannot be a massless particle. This is another consequence of special relativity. We will see in Chapter 15 that massless particles travel at the speed of light, and that particles moving at the speed of light experience no lapse of time: time stands still for them, and if they had an internal clock it would not advance at all. No dynamical process, like oscillation from one type of neutrino into another, could happen; nothing at all could change for a massless neutrino. Other experiments have already shown strong evidence for tiny neutrino masses, less than one millionth of the mass of an electron. Neutrino oscillations also require masses of this order.

▷ As this book was being finished, the 2002 Nobel Prize for Physics was announced. It was shared by Davis and Masatoshi Koshiba (b. 1926), founder of the Kamiokanda experiment, with Riccardo Giacconi (b. 1931), a pioneer of X-ray astronomy.

Life came from the stars, but would you have bet on it?

Viewed from different perspectives, the evolution of life on Earth can seem either almost inevitable or wildly improbable. Stellar astronomy forms the background against which the story of life unfolds. Consider the astronomical ingredients. Stars had to form before the Sun, make elements heavier than helium, and return a good fraction of them back to the interstellar clouds of gas from which the Sun's generation of stars would form. The Sun had to condense from the cloud of gas, leaving a sufficiently massive disk of gas around it from which the planets formed. The Sun needed to be hot enough to warm the planets, but not so hot that it would exhaust its nuclear fuel before life had time to evolve.

In this section: the improbability of life is a subject of intense interest to many people. The stars provided the raw materials, the Sun provided the nursery. But the laws of physics seem to have a lot of fine-tuning that allowed just the right conditions to be present for life as we understand it.

None of these ingredients is very unusual. Star formation will require many generations to exhaust the hydrogen supply, so there will be a long time in which stars are forming from clouds seeded with heavy elements. There is increasing evidence that many, perhaps most, stars like the Sun have left disks behind, from which planets might form. Indeed, surveys using specially designed telescopes are now turning up many planets around nearby stars. And the Sun is a very ordinary star: there are many like it that will shine steadily for the billions of years apparently required for life to evolve. These minimal conditions for life probably exist in billions of places just in our own Milky Way galaxy.

Life as we know it also probably required that the Earth have just the right distance from the Sun, and just the right mass and composition to allow things like volcanism and plate tectonics (continental drift) to continue over billions of years. This is somewhat more special, but it still might not be very surprising if there were millions of places in the Milky Way where life could evolve, and possibly is doing so right now. It is certainly not possible to estimate accurately the likelihood of this. The most famous attempt to do so, by Frank Drake (b. 1930), resulted in an equation full of undetermined factors (see Figure 11.3 on the next page).

Yet there is much that seems less probable if we look further back in time. When we end our study of cosmology in Chapter 27, we will see that we have no real explanation for why the Universe is the age it is. It might have happened that the Big Bang was not quite so big; then the expanding material of the early Universe might have turned around and re-contracted after only, say, a few million years. This would not have allowed enough time for life to evolve anywhere. We will also see, more dramatically, that the very laws of physics as we know them today took shape in the very early Universe. The mass of the proton and electron, the exact relative strengths of the nuclear and electromagnetic forces, the sizes of the density irregularities that eventually led to the formation of stars and galaxies in the expanding Universe: all of these things may have been determined essentially

Figure 11.3. *The Drake equation for estimating the probability that life could have evolved in our Galaxy could also be applied to equally difficult problems! Reprinted with kind permission of Mark Heath.*

at random just after the Big Bang.

This is relevant, because the evolution of life seems to depend on a little "fine-tuning" of the fundamental physics here and there. For example, we noted above that heavier elements form in stars when helium nuclei combine to form carbon, oxygen, and so on. There is a deep mystery, an incredible coincidence, underneath that bland statement. Carbon has 12 particles in its nucleus: six protons and six neutrons. It therefore requires three helium nuclei as building blocks. Now, if one relies on random collisions to drive nuclear reactions, it is very improbable that three alpha particles will converge at the same place at the same time. Ordinarily, one would expect that two would collide, forming a nucleus with eight particles in it, and then some time later a third helium nucleus would collide with the eight-particle nucleus to form carbon. Unfortunately, *there are no stable nuclei with eight particles.* Any such nucleus formed from two alpha particles will immediately disintegrate. In the conditions inside stars, such objects do not last long enough for a third alpha particle to come by. Nor was it any more likely in the Big Bang, which we will study in Chapter 25. When physicists first began to study these things, it seemed there was no way to explain elements heavier than helium.

This bottleneck would have prevented the formation of planets and the evolution of life, if it were not for an apparent accident. In a brilliant flash of insight, Hoyle realized that there was another possibility. *If* three alpha particles had a bit more attraction for one another than one would expect on general grounds, then the rate of three-particle reactions would be higher: the particles would not have to come quite so close to one another to get drawn into the reaction that forms carbon. He calculated that if there was a sufficiently strong extra attraction, then it would show up as what physicists call a long-lived **excited state** of the carbon nucleus. This means that if one were to fire another particle (maybe another alpha particle) with

---

Investigation 11.5. *Where do the photons come from?*

The Sun radiates light to us, yet in the nuclear reactions of Investigation 11.4 on page 127 we have not seen any that produce photons (gamma-rays). Where does the light come from?

There are two aspects to the question: how the photons are generated deep inside the Sun, and then how they get to the surface. In the core of the Sun, there are two places where photons can be made. The first is as a result of the production of positrons, the e^+ particles in, for example, Equation 11.3 on page 127. The positrons quickly meet electrons, which are their anti-particles, and annihilate via the reaction

$$e^+ e^- \longrightarrow 2\gamma. \qquad (11.5)$$

Each of the two photons is very energetic, having about 0.5 MeV of energy. (The symbol **MeV** denotes "million electron volts", which is 10^6 eV $= 1.6 \times 10^{-13}$ J.) These photons quickly lose their energy, however, in the Compton scattering process we shall describe below.

The second way that photons are generated in the central core is simply by the collision of charged particles. For example, most of the time that two protons collide they do not produce deuterium; instead, they simply deflect each other's motion and change some of their energy into a photon:

$$p + p \rightarrow p + p + \gamma. \qquad (11.6)$$

This slowing down of the protons gives rise to the physicists' name for this reaction, **bremsstrahlung**, from the German *bremsen* (to brake) and *Strahlung* (radiation).

In addition, photons are constantly running into protons and scattering off them, exchanging a little energy each time. This is called *Compton scattering*. It can, of course, occur between photons and electrons or other charged particles as well. Compton scattering and bremsstrahlung keep the photon gas at the same temperature as the particles, i.e. they insure that the typical energy of a photon equals the typical energy of a proton or electron. In particular, the gamma-ray photons produced by positron annihilation quickly lose their excess energy this way. They dissolve into the general background gas of photons.

In this manner, we say that the interior of the Sun generates a photon gas that is in thermal equilibrium with the particles. The overall temperature of this equilibrium is determined by gravity, by the pressure required to hold up the whole mass of the star against its gravitational self-attraction.

Exercise 11.5.1: *Entropy of the Sun*

We have seen in Chapter 8 that a typical photon in the Sun takes 10^6 y to randomly "walk" out of the Sun. That means that the Sun contains all the photon energy it generated by nuclear reactions in the last million years. This must all be in the form of photons, since the particles in the Sun have the same total energy today as they had a million years ago. (a) Calculate from the solar luminosity how much energy the Sun contains in photons. (b) If the average temperature inside the Sun is 10^5 K, calculate mean energy of each photon. (c) From these two results estimate the number of photons inside the Sun. (d) From the mass of the Sun, assuming for simplicity that it is composed entirely of hydrogen, calculate the number of protons (hydrogen nuclei) in the Sun. (e) Find the ratio of the number of photons to the number of protons in the Sun. This is a measure of what physicists call the **entropy** of the Sun.

---

just the right energy at a carbon nucleus, the impact would cause it to split up into three alphas, but these would not fly apart; the extra attraction would hold them together, oscillating about one another, until they eventually emitted a gamma-ray photon and settled back down to a normal carbon nucleus.

Experiments soon found the predicted effect. Carbon *does* have such an excited state, three alpha particles *do* attract one another more than one might at first expect, and carbon *can* form inside stars by three-alpha collisions. The synthesis of elements beyond carbon then needs only a succession of two-particle reactions: add one helium nucleus to carbon and one gets stable oxygen; combine oxygen and carbon nuclei and one gets stable silicon, and so on. So life seems to hinge on the existence of this one excited state of carbon. This seems to be a small detail of the laws of physics. If certain fundamental numbers, like the mass of the electron, the unit of electric charge, or the strength of the nuclear force were to have been just slightly different, then life would simply not have been possible: the elemental building blocks would not have been there.

Another example of the special nature of the laws of physics is given in the next chapter, where we consider the death of stars by supernova explosions. Some of these are triggered by the collapse of the core of a massive star when it runs out of nuclear fuel. The collapse releases energy, and this blows the rest of the star apart. These explosions are one of the ways that elements heavier than helium are placed into clouds of interstellar gas, ready to act as seeds for planets and for life in the next generation of stars. It appears that our own Solar System formed from gas enriched by such an explosion, without which we would not be here.

It seems, from theoretical calculations, that it is not easy to blow such stars apart. The energy released by such a collapse is carried away by neutrinos, which leave only just barely enough energy to blow the envelope of the star away. Now, we shall see that this collapse only occurs when the mass of the exhausted core reaches about one solar mass. This mass can be calculated, quite remarkably, from simple fundamental constants of nature: Newton's constant of gravitation G, Planck's constant h, the speed of light c, and the mass of the proton m_p. If the proton were a bit more massive, then (as we shall see) the core would collapse when it had less than one solar mass. Such a collapse would release less energy, and perhaps not enough would be available to blow apart the star. What is more, the collapse causes an explosion only because something halts the collapse, causing a rebound of the infalling material. This something is the formation of a neutron star, which we will also study in Chapter 12. We will see there that the existence of neutron stars depends on the exact strength of the nuclear forces. A small weakening of these forces would have led the collapse to form black holes, with little or no rebound, and the interstellar medium would not have been enriched as much as it has been.

This sort of fine-tuning can be found elsewhere, as well, and we will give more examples in the final chapter. This has given rise to a point of view about the history of the Universe that is called the **Anthropic Principle**. The mildest form of this principle holds that, since we are part of life, any universe in which the fine-tuning prevented life from forming would not have had us in it to puzzle over the fine-tuning. Therefore, the fine-tuning is no puzzle; it must be taken for granted from the simple fact that we are here to discover it. This point of view seems plausible if one believes that there might be many "universes", so that the cosmological experiments can be repeated many times, and there is nothing special about ones that happen to produce life on obscure planets. One might imagine a repeating Big Bang, endlessly cycling through expansion and re-collapse (the "Big Crunch"). Or one might imagine the Universe to be extremely large, and that different regions of it have different values of such things as the proton mass. We would then not see such regions because light has not had time to reach us from them since the Big Bang.

A more radical version of the Anthropic Principle is more metaphysical: the Universe is fine-tuned to produce life because its *purpose* is to produce life. Scientifically, this could be a dead end, since it discourages further questions about the fine-tuning, except possibly to try to find relations between examples of it that seem rather distant from one another. I prefer the first version of the principle, especially since, as we will see in Chapter 27, many physicists are working hard now to arrive at a theory of quantum gravity, and this may well predict that our Universe is not the only one, that the values of the fundamental constants are not the only ones that have occurred or will occur.

I have not addressed here the other ingredients necessary for life, especially the enormous chain of chemical reactions building upon one another to make the complex molecules of life on the present Earth. It is usually assumed that this happened spontaneously on the Earth, using a combination of catalysts and natural selection to guide the chemistry. Some astronomers, including Hoyle, have revived the idea of **panspermia**, which is that life could have spread from one star to another, perhaps propelled by strong mass flows that astronomers observe from some kinds of giant stars, so that it need not have arisen on Earth. That still requires that the complex chemistry take place somewhere in the Galaxy, but it allows biologists to look in environments very different from the early Earth for the very first steps toward evolution.

12

Birth to death:
the life cycle of the stars

The cycle of birth, aging, death, and re-birth of stars dominates the activity of ordinary galaxies like our own Milky Way. The cycle generates the elements of which our own bodies are made, produces spectacular explosions called supernovae, and leaves behind "cinders": remnants of stars that will usually no longer participate in the cycle. We call these white dwarfs, neutron stars, and black holes.

Governing this cycle is, as everywhere, gravity. An imbalance between gravity and heat in a transparent gas cloud leads to star formation. The long stable life of a star is a robust balance between nuclear energy generation and gravity. This balance is finally lost when the star runs out of nuclear fuel, leading to a quiet death as a white dwarf or to a violent death as a supernova.

Even the cinders, all unusual objects, can be understood from simple calculations based on elementary physical ideas. We have already met black holes. White dwarfs and neutron stars exist in a balance between gravity and quantum effects: they illustrate the deepest principles of quantum theory. Exotic as these may seem, life on Earth would not exist without the neutron stars and white dwarfs of our Galaxy.

Starbirth

The Milky Way is filled not only with stars but with giant clouds of gas that have not yet formed stars. We call these clouds "molecular clouds" because they are dense enough for chemical reactions to take place to form molecules. The chemical elements in the clouds were put there by earlier generations of stars, by processes we will study later in this chapter. Such clouds contain many simple molecules, mostly molecular hydrogen (H_2, with traces of carbon monoxide and formaldehyde), and they also contain solid grains of interstellar dust.

We met dust grains in Chapter 7; they are the raw material of which the planets were made. Dust also obscures the astronomer's view of distant stars. There is a small amount of dust in any direction we look, scattering light and making stars seem redder than they really are, just as dust in the Earth's atmosphere makes the Sun seem red at sunrise and sunset. In some molecular clouds, the dust is so thick that it obscures everything beyond it, sometimes sculpting spectacular shapes in the night sky (see Figure 12.1 on the following page).

The chemistry of clouds is interesting in its own right, and sometimes leads to spectacular consequences. One of the most extraordinary ones is the interstellar **maser**. A maser is like a **laser**, but the emission comes out in radio waves and not in light. Because clouds are rarified, collisions among molecules are rare. More common are collisions with cosmic rays. These act as the "pump" which puts energy into the maser; in the right circumstances, the result can be that clouds continuously emit pencil-thin, intense beams of radio waves in random directions. Scientists developed masers and lasers in the laboratory only relatively recently, but Nature has been producing them for billions of years!

In this chapter: stars form in molecular clouds and die when they burn up their fuel. Small stars die quietly as white dwarfs, larger stars explode as supernovae. In both cases, they return some of their material to the interstellar medium so that new stars and planets can form. White dwarfs, and the neutron stars that usually form in supernova explosions, are remarkable objects. They are supported against gravity by purely quantum effects, so they do not need nuclear reactions or heat to keep their structure. We learn about the quantum principles involved and use them to calculate the size and maximum mass of white dwarfs.

In this section: stars form when portions of gas clouds collapse. The criterion governing collapse is the Jeans criterion, which we derive.

▷The figure underlying the text on this page is from a computer simulation of a supernova explosion. The gas, rotating about a horizontal axis in this image, has "bounced" from the neutron star core (center left) and is moving outwards through the envelope with turbulent convection. From a paper by K Kifondis, T Plewa, and E Müller in AIP Conf. Proc. 561: Symposium on Nuclear Physics IV (New York, 2001). Used with permission of the authors.

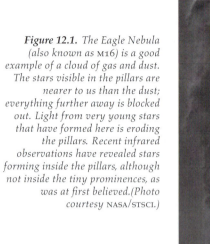

Figure 12.1. The Eagle Nebula (also known as M16) is a good example of a cloud of gas and dust. The stars visible in the pillars are nearer to us than the dust; everything further away is blocked out. Light from very young stars that have formed here is eroding the pillars. Recent infrared observations have revealed stars forming inside the pillars, although not inside the tiny prominences, as was at first believed.(Photo courtesy NASA/STSCI.)

If a cloud is sufficiently dense or sufficiently cold, parts of it can begin to contract to form stars. There is a minimum size for a region that will contract: it has to have enough mass and enough gravity to overcome the pressure in the cloud. This minimum size is called the **Jeans length**, because it was first determined by the British astrophysicist Sir James Jeans (1877–1946). It is given by

$$\lambda_{\text{Jeans}} = \left(\frac{\pi k T}{G \rho m} \right)^{1/2},$$

(12.1)

We work out this size in Investigation 12.1.

▷Actually, this equation is a special case of the Jeans formula, where we make the assumption that the temperature T of the gas does not change while it contracts. This is the case for clouds that are unable to trap heat until they get much denser.

Once parts of a dense molecular cloud start to contract, starbirth gets underway. Astronomers call the contracting gas a **protostar**. Astronomers see many clouds where stars are forming. One is illustrated in Figure 12.1. There are probably several things that can provide the initial disturbance that triggers the contraction of a region in a Jeans-unstable cloud: explosions of supernovae of stars already in or near the cloud, collisions between two clouds, or the compression of a cloud as it moves into a region of stronger gravitational field in the Milky Way (i.e. through the plane of our spiral galaxy). Simulations of star formation using supercomputers are beginning to shed light on these mechanisms and on what happens when stars begin to form (Figure 12.2 on page 138). These simulations are, unfortunately, far beyond the scope of our Java programs!

As the protostar contracts, it eventually becomes dense enough to trap radiation and begin to behave like a black body. At this point its temperature rises sharply

---

Investigation 12.1. Forming stars by the Jeans instability

The key insight into how stars form from molecular clouds is simply to understand that the cloud is *not* a star! This may seem obvious, but in physical terms it means that the cloud is not a black body: it does not trap radiation, but instead is transparent to photons. In Chapter 8 we discussed the stability of a star, and we found that the pressure in a star could balance gravity when the star is compressed if the polytropic index was larger than 4/3. Here we study the same balancing act, only in a transparent cloud.

In a long-lived molecular cloud, the balance between gravity and gas pressure has to be maintained without internal energy generation. Thin clouds have low temperatures, typically 20 K, where energy input from cosmic rays coming into the cloud or heating by nearby hot stars can balance the energy losses from emitting photons. These photons are released when molecules of the cloud vibrate, say after encountering a cosmic ray or colliding with another molecule. The molecules emit low-energy photons in the **microwave** or **sub-millimeter** parts of the spectrum. These low-energy photons are not scattered by the dust in the cloud, so the cloud is truly transparent at these wavelengths, despite being opaque to optical light, as in Figure 12.1. In this way a cloud can be kept at a fixed temperature despite any disturbances, unlike what would happen if the cloud trapped its photons like a black-body star.

In such a cloud, the key balance is between the random kinetic energy of the molecules and their gravitational potential energy. If the cloud contains a region of size R and mass M that is large enough that molecules moving with their thermal motion (at the cloud's temperature T) do not have the escape speed from that region, then roughly speaking they are trapped by gravity, and the region will begin to collapse if any small disturbance gives it an inward push. On the other hand, if molecules can escape from the region then a disturbance will not have time to collapse before the molecules diffuse away. To find approximately the size R at which a spherical region becomes unstable to collapse we simple set the average kinetic energy $3kT/2$ of a molecule of mass m equal to (the absolute value of) its gravitational potential energy GMm/R, and then replace the mass M of the region by its expression in terms of its density ρ, $M = 4\pi\rho R^3/3$. This gives

$$\tfrac{3}{2}kT = \tfrac{4}{3}\pi G\rho R^2.$$

If we solve for R we get

$$R = \left(\frac{9kT}{8\pi G\rho m}\right)^{1/2}.$$

Our argument is a bit crude, since molecules near the surface of the region can leave the region even with smaller speeds, which will diffuse the contraction. A more sophisticated mathematical analysis first performed by Jeans leads to the slightly larger *Jeans length* λ_J:

$$\lambda_J = \left(\frac{\pi kT}{G\rho m}\right)^{1/2}. \tag{12.2}$$

This is about three times larger than our estimate, but it has the same dependence on temperature, density, and molecular mass, so this indicates that our physical analysis is correct.

Any part of the cloud larger than this size has enough self-gravity to collapse. The mass of this region is $4\pi\rho\lambda_J^3/3$, which is called the Jeans mass

$$M_J = \frac{4\pi}{3}\rho\lambda_J^3 = \frac{4\pi^{5/2}}{3}\left(\frac{kT}{Gm}\right)^{3/2}\rho^{-1/2}. \tag{12.3}$$

The important part of this formula is that M_J decreases as ρ increases. Thus, as the unstable region of initial size λ_J collapses, its density rises. This lowers the Jeans mass and makes smaller parts of the cloud unstable to collapse. If, realistically, the density is not uniform in the cloud, then a collapsing region is likely to *fragment* into many smaller regions, and this fragmentation could occur on many scales. It only stops happening when the cloud cannot maintain its original temperature, either because it becomes less transparent or because it is heated by the contraction faster than the molecules can radiate energy away. At this point the temperature can rise, and the conditions for further instability and fragmentation become more like those we discussed for stars in Chapter 8.

Exercise 12.1.1: *The conditions for star formation*

A typical molecular cloud has a temperature $T = 20\,\mathrm{K}$, a composition mainly of molecules of H_2 (molecular hydrogen), and a density that corresponds to having only 10^9 molecules of H_2 per cubic meter. Calculate the Jeans length and Jeans mass of this cloud. Compare the mass you get with the mass of the Sun.

---

and it begins to shine with visible light. The energy released by contraction, which we calculated in Investigation 11.1 on page 123, continues to be radiated away so the star can continue to contract. Although this energy is not sufficient to power the Sun for billions of years, the protostar can shine for a million years or so on it. (See Exercise 11.1.2 on page 123 for the calculation.)

The gravitational thermostat

Once a star is formed, it will live in a fairly steady fashion for a very long time. We can't understand the death of the star until we understand how it manages to live quietly for as long as it does.

The nuclear reactions that take place in the Sun also take place in thermonuclear explosions, commonly called hydrogen bombs. Why is it then that the Sun has not blown itself up: how can it burn hydrogen in such a steady way? The answer is in the way gravity holds the black-body star together.

Suppose something happened inside a star to make the reactions proceed faster, such as a small increase in the temperature. Then the extra energy released (as photons and in fast particles) would immediately tend to expand the central part of the star, reducing the temperature and density, and thereby reducing the rate of

In this section: the luminosity of a star remains steady, despite all the turbulence inside, because gravity and gas pressure strike a cooperative balance.

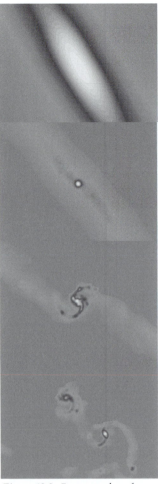

Figure 12.2. Four snapshots from a computer simulation of the formation of stars as a result of the collision of two clouds of gas. The stars form along filaments, and often form binary and triple pairs. Images courtesy A P Whitworth, Cardiff University.

In this section: most of the lifetime of a star is spent on the main sequence, which means that it resembles the Sun.

reactions. If the opposite happened, say a temperature decrease, then the reactions would put out less energy, there would be less pressure in the gas, gravity would make the star contract and heat up, and the reactions would go faster again.

> The self-corrections provided by gravity keep the nuclear reactions in a star in a steady state.

This kind of self-correction does not happen in a molecular cloud, which is transparent to radiation, and that is why the Jeans instability grows. Eventually it leads to the formation of stars that *are* able to maintain this balance.

This kind of self-correction is also much harder if one does not have self-gravity. In the thermonuclear bomb, there is no attempt to sustain the reactions. Instead, the material is arranged so that nearly all the desired reactions take place before the released energy has a chance to scatter the material. In a nuclear fusion reactor, the only way to keep the material together for long enough is by using magnetic fields. This is very difficult to do, and experimenters have found that there are many situations in which a small change in the density or reaction rate does not get corrected naturally.

> The goal of fusion research is to find a way to sustain the reactions at a high enough temperature and density to get more energy out than has been put in to make the magnetic field and heat the gas. This goal is getting nearer, but progress is slow, principally because fusion reactors do not have the natural thermostat provided by self-gravity.

The main sequence

We have seen in the previous chapter that stars "burn" hydrogen to make helium. The rate at which they use the hydrogen, and hence the luminosity of the star, depends on how massive the star is. More massive stars are hotter in the center, which means that nuclear reactions will proceed faster, and the star will shine more brightly. Such stars also have more hydrogen to burn, of course, before they end their normal life, but it turns out that their luminosity increases so strongly with their mass that their lifetime actually shortens.

> The more massive the star, the shorter will be its life before it reaches the end of its hydrogen-burning time.

In Figure 12.3 I have plotted the result of theoretical calculations of both the luminosity and lifetimes of normal stars. Notice that the lifetime of massive stars is as short as 10 million years, only 1/1000th of the Sun's lifetime. This figure shows another interesting curve: the amount of energy released by an average kilogram of mass of each star over the life of the star is roughly the same for all stars.

> Stars differ in the rate at which they extract energy, but all of them get about the same amount out of each kilogram in the end. This shows us that stars live as ordinary stars until they have exhausted their nuclear fuel. It is the end of nuclear burning and not some other cause that brings about the death of a star.

In Investigation 12.2 on page 141 we extract some numerical information from this figure.

Although stars spend long periods of time in a fairly steady state, burning with the luminosity shown in Figure 12.3, they do evolve slowly as their interior composition changes and as they lose mass. The Sun is losing mass at a very slow rate at the moment, but it may have lost a significant fraction of its original mass in the first billion years after it formed, and it is likely to lose almost half of its mass as it evolves into a red giant, some 5 billion years from now. Composition changes in the Sun may also affect its luminosity, which is probably very slowly increasing.

We show in Investigation 12.2 on page 141 that the luminosity L of a star is a very sensitive function of its surface temperature T: $L \propto T^8$. This can be seen in direct observations. Recall that astronomers can measure the surface temperature of a star by measuring its color. They cannot directly measure its absolute luminosity because the distance to a star is usually unknown, but if we consider only stars that are in a given cluster of stars, all at the same distance from us, then the ratios of their absolute luminosities will be the same as that of their apparent brightnesses. Since the luminosity–temperature relation is only a proportionality, it should then be true that for stars in a given **star cluster**, (apparent brightness) $\propto$ (surface temperature)^8.

In Figure 12.4 on the next page we display such a plot for the stars of a single cluster. Astronomers call this plot the **Hertzsprung–Russell diagram**, after the two astronomers who independently devised it, the Dane Ejnar Hertzsprung (1873–1967) and the American Henry Norris Russell (1877–1957). The normal stars that we have been describing lie in a diagonal band with a slope that is consistent with the proportionality we have deduced. This band of stars is called the *main sequence*. Notice, however, that there are well-defined bands off the main sequence that also contain a large number of stars. These must be stars which are not in the normal stage of their life cycle.

Some of these are newly-formed stars (pre-main-sequence stars) that are shining by gravitational contraction, but most are stars which have finished hydrogen burning and have moved off the main sequence. The behavior of stars as they leave the main sequence is very complex, and even now is not fully understood, although it can be modeled quite well on a computer. (This sort of computer program would also be beyond our scope here!) But again gravity plays a crucial regulatory role.

Giants

Consider the stars above and to the right of the main sequence. They have generally lower temperatures yet larger luminosities than main sequence stars. This means they must be much larger in size: they are called *giants*. Here is how stars become giants.

When a star exhausts the hydrogen in its center, there is little energy being produced to hold up the weight of the outer parts of the star (which contain, after all, 90% or more of its mass). The inner part of the star contracts and heats up. What happens next depends sensitively on how much mass the star has and what its exact composition is: it is rather sensitive to how much carbon, oxygen, and so on were in the initial cloud that contracted to form the star. Various things can happen in various combinations, as follows.

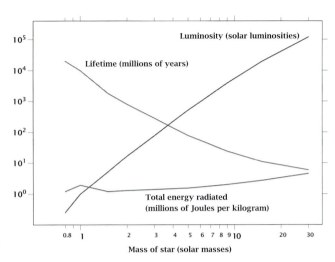

Figure 12.3. *Luminosity and lifetime of normal stars. Stars of larger mass have higher luminosities and shorter lives. The energy released by an average kilogram of the star is, however, essentially the same for all stars. All three curves are referred to the scale on the left, using the units indicated for each curve.*

In this section: when stars run out of hydrogen to burn, they change their interior structure, and become giants. We examine the causes.

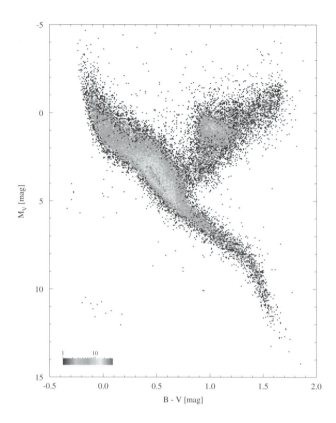

Figure 12.4. The temperature-luminosity diagram of stars measured by the Hipparcos satellite, mentioned in Chapter 9. The vertical axis is the absolute visual magnitude, which measures the luminosity of the star. The horizontal axis is the color of the star, measured by B - V, the difference between the blue and visual magnitudes. The Hipparcos survey parallaxes gave distances to stars accurately enough to deduce good values for their absolute magnitudes. Image courtesy ESA.

▷Astronomers use the word **nebula** for any diffuse cloud of gas around a star.

In this section: once a star can't support itself against gravity, it will collapse either to a white dwarf or a neutron star. Both objects are supported by quantum-mechanical forces.

1. The contraction heats the core enough to ignite hydrogen on its boundary, so that a hydrogen-burning shell develops. This usually is accompanied by an increase in luminosity due to the increased temperature.

2. The contracting core can reach a high enough temperature to make another reaction happen: the conversion of ^{4}He into ^{12}C. This reaction releases energy, and the star can settle for a while as it uses the helium as fuel. When this is exhausted, the core may contract, heat up further, and ignite carbon burning, converting ^{12}C into ^{16}O by the addition of ^{4}He. Reactions can go further, up to ^{56}Fe, but each step releases less energy and consequently lasts less time.

3. The release of an increasing amount of radiation from a contracting core actually makes the outer part of the star expand. The expansion can be very dramatic. Its radius can get so large that, despite its greatly increased luminosity, its surface temperature actually decreases, and it becomes distinctly red. Such a star is called a *red giant*. If the mass of the star is large (more than about eight times the mass of the Sun), we call the star a **supergiant**.

4. The star can get so large that gravity is very weak at its surface, and the pressure of the radiation leaving the star can blow off a steady strong "wind" of gas. Our Sun has such a wind, but giants have winds on much larger scales. They can lose large amounts of mass this way, up to 10^{-5} of a solar mass per year. On an astronomical time-scale, it does not take long for such winds to change the mass of the star significantly, or to transform the star into a shell of gas with a small hot star in the middle. Such shells are called **planetary nebulae**. Figure 12.5 on page 142 shows such a star. If a massive star loses enough mass, its hotter interior regions become exposed and it can become a blue giant.

The end of the main sequence lifetime of a star inevitably leads to great changes. These take place on relatively short time-scales, typically several million years. During this period, they are unusually bright, so that a large fraction of the stars visible to the naked eye are giants, even though they represent a small fraction of the total population of stars. The ultimate end of their post-main-sequence life is decided by how much mass the star has. That is the subject of the rest of this chapter.

Degenerate stars: what happens when the nuclear fire goes out

When a star's nuclear fuel runs out, which must eventually happen, then big change is unavoidable. Since gravity itself never "runs out", what happens next is dominated by gravity. In fact, it is easier to make calculations about the fate of a star than about its normal life. We shall see that relatively simple ideas lead us to white dwarfs, neutron stars and supernova explosions. We will continue the theme in Chapter 20 on neutron stars and Chapter 21 on black holes.

Investigation 12.2. Stars on the main sequence

This investigation is an exercise in extracting information from graphs, such as the curves in Figure 12.3 on page 139. This figure is plotted on what we call *logarithmic scales*. By that we mean that the main tick marks on, say, the vertical scale are not evenly spaced, but go up in powers of ten. The *logarithm* (the power of ten itself) of the vertical axis increases in uniform steps from one main tick mark to the next.

In this figure the horizontal axis is plotted logarithmically as well, although this is not so noticeable since the range is smaller. Whenever data span many powers of ten, and one is interested as much in the small values as in the big ones, it is useful to plot data this way.

What we see in Figure 12.3 is that all the curves are fairly straight. This implies something simple about the relations between quantities. A straight line has the general form of

$$y = mx + b, \qquad (12.4)$$

where m is the slope of the line and b its y-intercept, the value of y where the line passes through the y-axis. If we consider the relation between luminosity L and stellar mass M in Figure 12.3, then the approximately straight line is a relation between their logarithms:

$$\log L = m \log M + b. \qquad (12.5)$$

(Don't get confused between the slope m and the mass M: we use similar letters just in order to follow familiar notation for each.) We estimate m as follows. As M increases from 1 to 30, we see from the graph that L increases from 1 to about 1.5×10^5. In terms of logarithms, $\log M$ increases from 0 to about 1.5 while $\log L$ increases from 0 to about 5.2. The slope is the change in $\log L$ divided by the change in $\log M$, or about 3.5.

Now, if we raise both sides of Equation 12.5 to the power of 10, we get a relation between L and M themselves. Recall the properties of logarithms:

$$10^{\log x} = x, \qquad a \log x = \log(x^a), \qquad \log x + \log y = \log(xy),$$

so that we get from Equation 12.5

$$L = \beta M^{3.5}, \qquad (12.6)$$

where we define β by $\log \beta = b$. This is the way the y-intercept appears in the final result. We are more interested in the exponent here (the slope of the curve in the figure) than in the intercept. We have learned that, for normal stars, luminosity is proportional to the 3.5-power of the star's mass.

In a similar way, we can deduce that the lifetime τ of a star is related to its mass by $\tau \propto M^{-2.5}$.

Since luminosity is energy released per unit time, the product of luminosity and lifetime is the total energy released over the star's life. If we then divide this by the mass of the star, we get the average energy released per kilogram. Given the proportionalities, we find

$$\frac{L\tau}{M} \propto \frac{M^{3.5} M^{-2.5}}{M} = \text{const}. \qquad (12.7)$$

This is interesting because it is directly related to the nuclear reactions going on in the star. No matter what the total mass of the star is, each kilogram in it gives up on average about one million joules over the lifetime of the star.

This is what we called the "energy duty" in Investigation 11.1 on page 123 when we discussed where the Sun's energy came from. We concluded there that this energy was so large that it could only come from nuclear reactions. We now see that this is true for all stars, not just the Sun. Notice that this is the average energy given up by each kilogram of the star. Since most of the mass of any star is too far from the hot center for nuclear reactions to take place, the kilograms that actually do undergo nuclear reactions give up much more energy than this.

For the normal stars described in Figure 12.3 on page 139, the rapid increase in luminosity with mass has significant effects on the structure. Although there is more mass to generate gravity, there is much more pressure from the radiation flowing outwards through the star. The result is that more massive stars are larger. The radius of a normal star turns out to be roughly proportional to its mass to the 0.7 power: $R \propto M^{0.7}$.

Now, the surface temperature of a star is determined by its radius and luminosity: as we saw in Chapter 10, the luminosity is proportional to the fourth power of the temperature T and the square of the radius. Inverting this gives $T \propto L^{1/4} R^{-1/2}$. If we take the luminosity and radius to depend on the mass as above, we find two interesting relations: $T \propto M^{1/2}$ and $L \propto T^8$, where I have rounded off the exponents to simple integers and fractions.

Exercise 12.2.1: *Inverting a logarithmic equation*

Go through the steps leading from Equation 12.5 to Equation 12.6. First put the definition of β into Equation 12.5. Then write $m \log M$ as $\log(M^m)$. Finally combine this term with the β term on the right-hand side of Equation 12.5 to get the logarithm of Equation 12.6. Justify each step you make in terms of the rules given above for the use of logarithms.

Exercise 12.2.2: *Dependence of stellar lifetime and luminosity on mass*

In the same way as we estimated the exponent in the relationship between luminosity and mass, estimate the exponent in the relationship between lifetime and mass. Do you get -2.5, as given above? The curves for luminosity and lifetime are not perfect straight lines, so representing them by a single constant exponent is an approximation. Estimate the error in this approximation by giving a range of values for both exponents that would acceptably represent the graphs.

Exercise 12.2.3: *Energy radiated per kilogram: is it a constant?*

Test the assertion that the energy radiated per kilogram is constant, independent of the mass, by estimating the exponent from the graph in the same way as the lifetime and luminosity exponents were estimated. Is the exponent really zero? If not, can you explain this in terms of the uncertainty in the other two exponents that you arrived at in the previous exercise? In other words, is the exponent you get for the radiated energy within the range of exponents you would get if you selected various values for the other exponents and put them into the relation in Equation 12.7?

It would be natural to assume that, when the nuclear fuel runs out, the star will just begin to contract, and it will keep contracting until, perhaps, it shrinks to a point. After all, what can halt the contraction? What can provide as much resistance to gravity as had the now-exhausted nuclear reactions?

Remarkably, there *is* something, at least for small enough stars. This is the pressure of what we call a **degenerate gas**. To understand what a degenerate gas is, we must go back to what we learned about the Heisenberg uncertainty principle in Chapter 7. If we try to pin down the location of, say, an electron to some great accuracy, then the *momentum* of the electron will be very uncertain. If the electron is confined in a certain region of space, say the volume of a star, then its position will be uncertain to no more than the size of the star. That creates a minimum uncertainty in its momentum, or equivalently in its velocity.

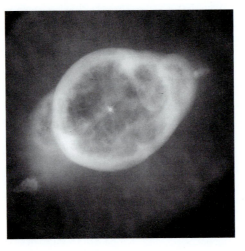

Figure 12.5. The planetary nebula NGC3242: the glowing gas has been ejected by the star visible at the center of the nebula, and is now reflecting light from the star towards us. The complex and beautiful structure may result from the magnetic field of the star. Image courtesy NASA/STSCI.

For ordinary stars, this is not important. But as a star contracts after the end of nuclear burning, the space in which its electrons are confined gets smaller, and so their minimum velocity goes up. This randomly oriented velocity provides an effective pressure that can act against gravity. In Investigation 12.3 we see that this can indeed counteract gravity, but on its own it would lead to a "star" of impossibly tiny dimensions, about 10^{-31} m in size! This is certainly not what actually happens, because there is another quantum principle at work: the *Pauli exclusion principle*, named for the Swiss nuclear physicist Wolfgang Pauli (1900–1958). This says that electrons are individualistic: any electron will resist all attempts to force it to behave in an identical way to any other electron.

Now, electrons are intrinsically identical to one another. They all have exactly the same mass and electric charge. They also have another property, which comes purely from quantum theory: they have a small amount of angular momentum, called **spin**. You can think of each electron as spinning about some axis. The amount of spin is tiny, just $h/4\pi$. All particles in quantum theory have a spin (for some it is zero), and it always comes in multiples or half-multiples of the value $h/2\pi$: spin is quantized. The spin of an electron is half of this, and so electrons are said to be spin-½ particles.

Under the exclusion principle, no two electrons can be identical in all their properties. The only ways in which they can differ from one another is in their motion and spin. The remarkable aspect of spin-½ particles is that they can have only two *independent* orientations for their spin. Since no two electrons are allowed to have the same momentum and the spin, at most two electrons can have the same momentum.

The effect of the exclusion principle is that, as the star cools, only *two* electrons can have the minimum momentum allowed by the uncertainty principle. The rest of the 10^{57} electrons in the star have to have larger momentum, all of them different from each other by at least the minimum uncertainty in momentum. Such a gas is called a *degenerate electron gas*.

The collapsing star will therefore have a population of electrons, some of which have quite a large momentum, hence quite a large energy. This population is called the *Fermi sea of electrons*, after the Italian nuclear physicist Enrico Fermi (1901–1954) who first described it. As we see in Investigation 12.4 on page 144, this Fermi sea will support the star against gravity in the manner described before, but because the electrons have much larger average momentum and energy, the balance between degeneracy pressure and gravity occurs when the star is much larger than if the ex-

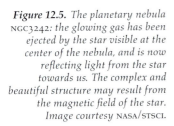

▷Planck's constant h has dimensions of angular momentum.

▷Spin is one of the deep mysteries of quantum physics. While it is possible to measure how an electron spins about any axis, there is a sense in which the axis of spin is not defined until it is measured. If two electrons have identical momentum and different spin, and if one of them is measured to be spinning about a certain axis in a clockwise sense, then the other one will always be measured to spin about the same direction in the counterclockwise sense. The second one acquires a direction of spin as soon at the first one is measured. We have no room in this book for a discussion of the fascinating subject of how measurements are made in quantum physics. That would require a book on its own!

Investigation 12.3. Degenerate matter, part 1: collapsing too far

Here we explore the physics of degenerate stars, where rather simple calculations lead us to striking conclusions. Our aim is to calculate how much pressure a degenerate star has. This pressure will be present even when a star is so cool that the ordinary thermal pressure does not hold it up. We will base our calculation on our study of the structure of the Sun and stars that we began in Chapter 8, and on the two fundamental principles of quantum theory that we have met so far: the Heisenberg uncertainty principle (Chapter 7) and the Pauli exclusion principle, discussed in the text of the present chapter.

In the uncertainty principle, the two related quantities we will use are the position of an electron on, say, the x-axis, and its x-momentum. Let an electron have speed v and momentum $m_e v$, where m_e is the mass of the electron. Suppose it is confined within a star of radius R. Since pressure comes from the momentum of particles, and since we want to find the pressure that is there even when thermal momentum has gone away, we want to find the *least* momentum allowed by the uncertainty principle. This will always be there.

The minimum momentum is associated with the maximum uncertainty in the position of an electron, which is the size of the star itself, $2R$. Assuming this, the x-momentum is at least

$$\Delta m_e v = h/2R, \tag{12.8}$$

where h is Planck's constant. This contributes a kinetic energy of $(\Delta m_e v)^2 / 2 m_e$, just from the x-motion. The three directions combine to give a *minimum kinetic energy* of

$$\langle KE \rangle_{min} = 3h^2/8 m_e R^2. \tag{12.9}$$

This minimum random kinetic energy immediately leads us to the gas pressure. Recall our discussion of an ideal gas in Investigation 7.2 on page 78. The pressure of such a gas is given in terms of the random kinetic energy of the gas particles by Equation 7.7 on page 78. Substituting the above minimum kinetic energy into this gives a minimum pressure of

$$p_{min} = \frac{h^2 N_e}{4 m_e V R^2}, \tag{12.10}$$

where V is the volume of the star and N_e is the number of electrons in the star.

Now, the structure of the star is a balance between pressure and gravity. We shall treat the structure equations approximately and see what we can learn from them about the degenerate star. We will then not expect the numerical values of quantities we deduce to be exact, but our results should represent at least roughly the relations between different physical quantities.

First we need the relation between the mass of the star and the number of electrons. The mass is determined by the number of protons and neutrons, since the electrons have negligible mass. If the star were composed of hydrogen, there would be lonly one proton per electron, and the mass of the star would be $m_p N_e$. More realistically, the star will be made of helium, carbon, and other elements up to iron. These typically have equal numbers of protons and neutrons

in their nuclei, so that the mass of the star is about $M = 2 m_p N_e$. Although details of the composition of the star could change the factor of two by a small amount, our other approximations make more of a difference in the final answer than this.

Recall the equation giving the approximate way the pressure scales with the mass and size of a star, Equation 8.19 on page 95: $p = 3GM^2/4\pi R^4$. If we put the minimum pressure in here from Equation 12.10, use the mass we have just deduced, set the volume V to $4\pi R^3/3$, and solve for R, then we find the relatively simple relation

$$R = \frac{h^2}{4G\mu m_e m_p M}. \tag{12.11}$$

If we take a solar mass for M (2×10^{30} kg) and use $\mu = 2$, then we have $R = 2.7 \times 10^{-31}$ m.

This is rather small! White dwarfs are indeed small stars: Sirius B is the size of the Earth. But here we have a radius much smaller than an atom! It is also much smaller than the radius we calculated for a black hole of a solar mass in Chapter 4, so we would not expect matter that behaves like this to form stars at all, but rather just to disappear into a black hole. So what has gone wrong with our calculation? The answer is that we have left something out of our considerations so far: the *Pauli exclusion principle*.

The Pauli exclusion principle applies to neutrons, protons, and electrons, which are the main constituents of matter. Such particles are called **fermions**. We will see in Investigation 12.4 on the following page that the exclusion principle requires fermions to form degenerate stars of the expected size.

We should note here, however, that the Pauli exclusion principle does not apply to all kinds of particles. Particles called **bosons** do not "exclude" one another. Photons are bosons, but they do not form stars by themselves because they do not have mass. Most other known bosons are unstable elementary particles, which would not last long enough to make a star. But there is speculation among particle physicists today that there may be a stable boson with a very small mass, and if it exists then it could in principle form **boson stars**, with radii smaller than white dwarfs. If bosons do not interact with one another, then the boson star would have a size given only by the uncertainty principle, as we have calculated above. The mass of the boson would then determine the maximum mass that a boson star could have without collapsing into a black hole. Conversely, in order to have a boson star of a certain mass, say a mass comparable to that of a white dwarf, then there is a maximum allowed mass for the bosons themselves. We examine this in the exercise below.

Larger boson stars are possible for a given boson mass if bosons repel one another, so that the uncertainty principle does not determine the size of the star. At the present time (2002) there is no compelling experimental evidence that such particles or stars might exist, but the Universe has proved itself to be full of surprises. Such stars could be detected by observing the gravitational radiation emitted by binary pairs of boson stars. As we shall see in Chapter 22, gravitational wave detectors will routinely conduct searches that could reveal such systems.

Exercise 12.3.1: *Boson stars*

(a) In Equation 12.11, replace the proton and electron masses by a single boson mass m_b and assume that $\mu = 1$. This gives the formula for a star composed of just one type a particle, the boson of mass m_b. For such a star of total mass M, calculate the following ratio

$$2GM/Rc^2 = 8M^2 m_b^2/m_{Pl}^4,$$

where m_{Pl} is the Planck mass defined in Equation 12.20 on page 146.

(b) The ratio above is the ratio of the size of a black hole, $2GM/c^2$, as given in Equation 4.12 on page 36, to the size of the star. We will see in Chapter 21 that the star cannot be smaller than a black hole, so this ratio must be less than one. Show that this sets a maximum mass on a boson star made from bosons of mass m_b:

$$M_{max} = 8^{-1/2} m_{Pl}^2/m_b.$$

(c) Find the largest mass m_b that the boson could have in order to allow boson stars of a solar mass to exist. Find the ratio of this mass to the mass of a proton. You should find that the boson needs to have very much less mass than a proton.

Investigation 12.4. Degenerate matter, part 2: white dwarfs

According to the Pauli exclusion principle, if there are N_e electrons, they all must form pairs that have different values of the momentum separated by at least $\Delta m_e v$. It follows that, in three dimensions, the largest momentum in any direction must be at least $N_e^{1/3}\Delta m_e v$. (We omit factors of 2 and π and so on here, since our treatment of the structure of the star is approximate anyway. If you are unsure of where the factor of $N_e^{1/3}$ comes from, see Exercise 12.4.1 below for a derivation.) The *average* momentum is within a factor of two of this, so we will take this to be the typical momentum of an electron in a degenerate Fermi gas.

Going through the steps leading to Equation 12.11 on the previous page again gives this time a radius larger by a factor of $N_e^{2/3}$:

$$R = N_e^{2/3}\frac{h^2}{4G\mu m_e m_p M}.$$

Using the fact that $N_e = M/\mu m_p$, we have

$$R = \frac{h^2}{4Gm_e(\mu m_p)^{5/3}M^{1/3}} = 10^7 \text{ m, for } M = 1M_\odot. \quad (12.12)$$

This radius is slightly greater than the radius of the Earth. As we noted above, we can trust it to be correct to perhaps a factor of two or so, but not better than that. In fact, more detailed calculations show that a white dwarf of one solar mass has a radius about 90% of that of the Earth. Our answer is quite close to the right one, considering the simplicity of the mathematical approximations we have made. This tells us that we have not left out important physics.

What is the *equation of state* of a degenerate Fermi gas? Recall our use in Investigation 8.4 on page 92 of a power-law (polytropic) relation between pressure and density to make a simple model of the Sun. Here we will find that the degenerate Fermi gas obeys such a law.

The pressure given in Equation 12.10 on the previous page for a gas without the exclusion principle needs to be multiplied by $N_e^{2/3}$, the *square* of the factor by which the average momentum goes up when we take account of the exclusion principle. That gives

$$p_{\text{Fermi}} = \frac{h^2 N_e^{5/3}}{4m_e VR^2}. \quad (12.13)$$

If we note that $R \propto V^{1/3}$, we see that the pressure depends only on the ratio N_e/V, the number of electrons per unit volume. Since this is itself proportional to the mass density $\rho = \mu m_p N_e/V$, we arrive at the *Fermi equation of state*

$$p_{\text{Fermi}} = \beta\rho^{5/3}, \quad (12.14)$$

where β is a constant that depends on h, m_e, μ and m_p. Our value for β is not exact because our calculations are approximate in places, but what *is* exact is the exponent: the degenerate Fermi gas is a polytrope whose polytropic index, as defined by Equation 8.12 on page 92, is $n = 3/2$.

Exercise 12.4.1: *Momentum in the Fermi sea*

Here is where the factor of $N_e^{1/3}$ comes from. First we consider the easier case of electrons confined in a one-dimensional "box", say along a string of finite length. We return to the three-dimensional star later. If each pair of electrons has a distinct momentum, separated by $\Delta m_e v$ from its neighbors, then we could mark out a line on a piece of paper, start with the smallest momentum allowed ($\Delta m_e v$), and make a mark each step of $\Delta m_e v$. Each mark represents the momentum of one pair of electrons. If we have N_e electrons, then there will be a total of $N_e/2$ marks. We would have to make marks in the negative direction too (electrons moving to the left), so the largest momentum will be $(N_e/4)\Delta m_e v$. Now suppose the electrons are confined to a two-dimensional square sheet of paper. Show that, leaving out factors of order unity, their maximum momentum is $N_e^{1/2}\Delta m_e v$. (Hint: each pair of electrons occupies a *square* of momentum uncertainty.) Similarly, show for three dimensions that the result is $N_e^{1/3}\Delta m_e v$.

---

clusion principle were not operating. A star supported by electron degeneracy is called a *white dwarf*. We show in Investigation 12.4 that a white dwarf's radius depends on its mass and on a simple combination of fundamental constants of physics (specifically, on G, m_e, m_p, and h). A star with the mass of the Sun should be about the same size as the Earth. Its density is, therefore, huge: about one million times the density of water!

The protons in the star should also be subject to the same uncertainty and exclusion principles, and therefore to the same minimum momentum. This will create a degenerate proton gas and an associated degeneracy pressure. But the kinetic energy of a proton that has the same momentum as an electron is much less than that of the electron: kinetic energy is $\frac{1}{2}mv^2 = (mv)^2/2m$, and since momentum is just mv, it follows that the kinetic energy of two particles that have the same momentum is inversely proportional to their masses. The proton is nearly 2000 times more massive than the electron, so the degenerate proton gas would have only $1/2000^{\text{th}}$ of the energy of the degenerate electron gas. For this reason, the electrons provide essentially all the pressure in such a situation.

Not all particles are subject to the exclusion principle. In fact, the exclusion principle only applies to particles with half-integer spin, such as spin-½. This includes all the ordinary particles of matter: electrons, protons, and neutrons. But some particles have whole-integer spin, either spin-0 (pions), spin-1 (photons), or − as we will see in Chapter 27 − spin-2 (gravitons). These do not exclude one another; in fact they have some preference for ganging up together in the same state! Photons,

▷Pions are elementary particles that are made in particle accelerator experiments. Gravitons are the quantized form of gravitational waves.

for example, would never form the kind of electromagnetic waves that radios and mobile phones use if no two of them could be the same. Lasers, which essentially emit strong beams of light with all the photons in exactly the same wavelength and state of oscillation, exist only because photons like being the same as one another. Particles that obey the exclusion principle are called fermions, after Fermi. Particles that like being together are called bosons, after the Indian physicist Satyendra Nath Bose (1854–1948).

Although degenerate matter has strange properties, the natural evolution of a star brings it to the point where degeneracy becomes important. Astronomers see white dwarfs in their telescopes, and they have just the size we have calculated. These huge objects depend for their very existence on the strange physics of quantum theory. Because they are composed of matter whose structure cannot be described in the conventional language of forces, they would have been incomprehensible to Newton. Yet they are abundant: one in ten stars is a white dwarf; and, as we noted in Chapter 10, one of the brightest stars in the night sky, Sirius, has a white dwarf in orbit about it.

The Chandrasekhar mass: white dwarfs can't get too heavy

Unfortunately for some stars, but fortunately for the evolution of life on Earth, the story of degeneracy does not stop here. The problem is that, if the star is very massive, then gravity will force the degenerate electrons into such a small volume that their typical speed becomes close to the speed of light. In this case, we have to treat the electrons by the rules of special relativity.

We will study special relativity beginning in Chapter 15, but here we need only one new fact that is explained there: the momentum carried by a photon is just proportional to its energy, in fact is E/c. This is very different from the situation for low-speed electrons, where the momentum is twice the kinetic energy divided by the speed of the particle. The difference is in part due to the fact that in special relativity, mass has energy, so the relationship between energy and momentum must include the total mass-energy of a particle. Now, since at a speed close enough to the speed of light, any particle behaves more and more like a photon, the momentum carried by a fast electron is also just its energy divided by c.

Now, the energy of the gas particles is directly responsible for the pressure of the gas, so when the electrons become relativistic, the pressure starts to increase only in proportion to the uncertainty momentum, rather than to its square. This significantly weakens the degeneracy pressure that electrons can exert, and in fact we show in Investigation 12.5 on the next page that it leads to a universal *maximum mass of a white dwarf*, the maximum mass that can be supported by degenerate electrons. This mass is called the *Chandrasekhar mass* M_{Ch}, after the Indian astrophysicist Subrahmanyan Chandrasekhar (1910–1995) who discovered it. Its value is in the range 1.2 to 1.4 times the mass of the Sun, depending on the exact composition of the star when it reaches the density of the white dwarf.

> The Chandrasekhar mass is one of the most remarkable numbers in all of physics. As we show in Investigation 12.5 on the following page, it depends mainly on some fundamental constants of nature, not on fine details of atomic or nuclear physics.

It seems to be an accident of our Universe that this particular combination of the constants of nature gives a mass for a relativistic degenerate white dwarf that is similar to the masses of ordinary stars. If this mass had come out to be much larger or much smaller than a solar mass, the death of most stars would be radically

In this section: the kind of quantum-mechanical support that white dwarfs use can only support a little more than the mass of the Sun. With more mass, the star will collapse.

Figure 12.6. Subrahmanyan Chandrasekhar made many important contributions to astrophysics, sharing the 1983 Nobel Prize for Physics. His discovery as a very young man of the limiting mass for white dwarfs led to a bitter conflict with Eddington (see Chapter 4), who felt that Nature simply could not behave in such a way! Chandrasekhar lost this battle, escaping from Cambridge to Chicago. Subsequent research, of course, vindicated his work completely. Chandrasekhar's modest manner, his devotion to science, and his erudition won him the respect and affection of generations of scientists. Image courtesy University of Chicago.

Investigation 12.5. Deriving the Chandrasekhar Mass

For stars with relativistic electrons, we need to re-calculate the structure and equation of state from the beginning, since Equation 12.9 on page 143 is wrong for this case. As we will see in Chapter 15, the energy of a photon is just the speed of light times its momentum. Therefore, this must be almost true even for ordinary particles moving at close to the speed of light:

$$E = pc. \tag{12.15}$$

Given the same uncertainty in momentum, $\Delta m_e v = h/2R$, using the exclusion principle to give a typical momentum that is a factor of $N_e^{1/3}$ larger than this, and then following the same steps as before for the ideal gas, we arrive at a pressure

$$p = \frac{hcN_e^{4/3}}{3RV}. \tag{12.16}$$

Setting this equal to $3GM^2/4\pi R^4$ as in Investigation 12.3 on page 143, and replacing the volume by $4\pi R^3/3$, we get

$$\frac{hcN_e^{4/3}}{4\pi R^4} = \frac{3GM^2}{4\pi R^4}. \tag{12.17}$$

Here we notice a remarkable and unexpected thing: the radius of the star drops out, and we are left with an equation that determines a single mass! This mass is the *unique* mass of a fully relativistic white dwarf. For non-relativistic white dwarfs we could choose a mass and find a radius, or vice versa. Here, we have no choice about the mass, and presumably the radius can be anything at all! This rather remarkable discovery was made by Chandrasekhar, and so we name the unique mass after him. Our expression for it is, from the previous equation,

$$M_{Ch} = \left(\frac{hc}{3G} \right)^{3/2} \left(\frac{1}{\mu m_p} \right)^2. \tag{12.18}$$

This evaluates to about 1.4 solar masses. Of course, our calculation is only approximate, but it turns out that we have got the right value almost exactly.

Since real electrons don't exactly obey Equation 12.15, but come closer and closer to it the more relativistic they get, we should regard the Chandrasekhar mass also not as the exact mass of any particular white dwarf but rather as an upper bound on the mass of all white dwarfs. Less massive stars have fewer relativistic electrons. More massive stars simply cannot be supported by electron degeneracy pressure at all.

Is the fact that we have not determined the radius of this star a worry? Not really: again, in a real star, the electrons are not fully relativistic, the electron gas is not perfectly ideal, and there is some pressure support from the protons or other nuclei. All these make small corrections, but they are enough to guarantee that any real star's radius will be determined by its mass. The radius will be about the radius of the Earth, as before.

Importantly, the equation of state of the relativistic white dwarf is also a polytrope, but this time with a different power. Steps similar to those used in Investigation 12.3 on page 143 give the relation

$$p = \beta \rho^{4/3}, \tag{12.19}$$

so that the polytropic index is 3.

Now we remind ourselves of the calculation we did for the stability of the Sun, in Investigation 8.8 on page 101. A polytrope of index 3 is only marginally stable against collapse. Any small correction to the properties of white dwarfs could cause them to be unstable. One correction is that, on compression, some electrons and protons tend to combine into neutrons, removing electrons from the degenerate sea and reducing its pressure. This makes collapse more likely. A second correction is general relativity: in Einstein's theory of gravity, the critical polytropic index actually needs to be somewhat smaller than 3, so that an $n = 3$ gas is actually unstable. Both of these effects become important for highly relativistic white dwarfs, and lead them to be unstable to gravitational collapse a bit before they reach the Chandrasekhar mass.

It is interesting to note that the Chandrasekhar mass can be expressed in terms of two simpler masses: the proton mass m_p and a number with the dimensions of mass that is built only out of the fundamental constants of physics, h, c, and G:

$$m_{Pl} = \left(\frac{hc}{G} \right)^{1/2} = 5.5 \times 10^{-8} \, \text{kg}. \tag{12.20}$$

This mass is called the **Planck mass**, hence the symbol m_{Pl}. In terms of these simple masses we have

$$M_{Ch} = \left(\frac{1}{3^{3/2}\mu^2} \right) \frac{m_{Pl}^3}{m_p^2}. \tag{12.21}$$

The Planck mass was first discussed by Planck himself. He noticed, soon after introducing his constant h, that from the fundamental constants h, c, and G one could build numbers with any dimensions one wanted: a mass, a length, a time, and so on. These are now called the Planck mass, the Planck length, etc. We do not yet know exactly what role these quantities play in physics, but we expect it to be fundamentally associated with the quantization of gravity, since they involve both h and G. We will return to this in Chapter 21.

Exercise 12.5.1: *Deriving the Chandrasekhar mass*

Derive the expression in Equation 12.18 by the indicated method.

Exercise 12.5.2: *Relativistic degenerate gas equation of state*

Find the constant β in Equation 12.19.

different. Since stellar death provided our Solar System with the raw ingredients of life, we owe much to the Chandrasekhar mass!

Neutron stars

In this section: neutron stars are also supported by degeneracy pressure, but here it is the neutrons which form the supporting distribution. Their maximum mass exceeds $2M_\odot$.

What, then, happens to a contracting star if its mass exceeds the Chandrasekhar mass? Electron degeneracy fails because the electrons have become relativistic, but the protons are still available. Because the proton mass is much larger than that of the electron, protons do not become relativistic until the star is much smaller. When the star is the size of a white dwarf the proton degeneracy pressure is negligible, but as the star contracts further this pressure grows until it can support the star.

The calculations in Investigation 12.4 on page 144 show that a degenerate star's radius is inversely proportional to the mass of the particle that is responsible for the degeneracy pressure, and it also depends somewhat on the composition. As we show in Investigation 12.6 on the next page, the result is that the star will continue to contract until it reaches a radius about 1/600th of the radius of a white dwarf. This size is about 10–20 km.

> Can there really be stars with the mass of the Sun that are only 10 km in size?? Remarkably, the predictions of our simple calculations are borne out by observations: more than 1000 such *neutron stars* have now been observed, and possibly about 0.1% of all stars are this size!

In order to understand how we can identify such neutron stars in astronomical observations, we need to study them in more detail.

While the contraction from the white dwarf stage is occurring, a crucial change takes place within the material of the star. As the density increases, the energy of a typical degenerate electron gets to be so large that it exceeds the energy equivalent of the difference between the mass of the proton and the neutron. (The neutron is slightly more massive than the proton.) The result is that it costs less energy to combine a proton and an electron into a neutron than it does to keep the electron in the Fermi sea. The electrons then almost all combine with protons to form neutrons. The contracting material is then a *degenerate neutron gas*. The energy that is released by this nuclear transformation is carried away by the neutrinos that are given off when this happens.

This transformation makes no difference to the degeneracy pressure, since neutrons have essentially the same mass as protons, and they both, like electrons, obey the Pauli exclusion principle. Moreover, as we see in Investigation 12.6 on the following page, the maximum mass that can be supported by neutron degeneracy pressure is much larger than the Chandrasekhar mass for white dwarfs, so, unless the contracting star is very massive, the contraction of the star can be halted when neutrons become degenerate.

This is why we call this a *neutron star*. As we noted, its radius is about 10–20 km. We shall study the properties of these ultradense objects in some detail in Chapter 20, and learn there that they are associated with *pulsars*, of which some 1000 have now been identified.

It is worth noting here, however, that neutron stars take us into the province of relativity, where we should not trust Newtonian gravity too much. We can see this by calculating the escape velocity from a neutron star whose mass is $M = 1M_\odot$ and whose radius is $R = 10$ km. The Newtonian formula for the escape velocity is $v^2 = 2GM/R$. A little arithmetic gives $v = 1.6 \times 10^8$ m s^{-1}, or about half the speed of light! Newtonian gravity can only indicate the general features of neutron stars, but it cannot give a good quantitative description of them. That will have to wait until Chapter 20.

Fire or ice: supernova or white dwarf

We now have enough understanding of the possible forms of equilibrium stars to look at what happens to giant stars at the end of their nuclear lives. Some die away quietly as white dwarfs, and others explode spectacularly as supernovae.

We saw above that giants have cores made of the waste products of nuclear reactions. These waste products are nuclei of moderate mass, like carbon and oxygen, mixed with helium. There is no hydrogen in the core, of course. The core is surrounded by a shell in which nuclear reactions still take place, but as the material in the shell is exhausted, the shell moves outwards and the core increases in mass. This burned-out core is supported by electron degeneracy pressure – there is no

In this section: when a star runs out of nuclear fuel, it can end in a supernova explosion or in a quieter contraction to a white dwarf. *Some say the world will end in fire, some say in ice, ...* (Robert Frost)

Investigation 12.6. Neutron stars as degenerate stars

Now, the degeneracy calculations we have performed in previous investigations have not used any special properties of the electrons: they are just the particles that supply the degeneracy pressure. As we note in the text, a *neutron star* is a star where neutrons supply the degeneracy pressure. All the formulas before apply then to neutron stars, if we replace m_e by m_n, which to our accuracy is the same as m_p. We can also set $\mu = 1$, since practically all the particles in a neutron star are neutrons. This means that the radius of the star from Equation 12.12 on page 144 will be smaller by the ratio m_e/m_p but larger by the fact that the white dwarf factor of $\mu^{5/3} = 3.2$ is set to one for neutron stars. This means the radius of a neutron star

should be about 1/600th of that of a white dwarf, or about 17 km. This is within the range of what the more detailed calculations give, even in general relativity (see Chapter 20).

Notice also that the Chandrasekhar mass in Investigation 12.5 on page 146 for neutron stars should be larger than that for white dwarfs only because of the factor of μ^2. This raises the maximum mass of a neutron star to about five or six solar masses. The effects of relativity, however, drastically reduce this number. We shall see in Chapter 20 that the extra strength of relativistic gravity reduces the maximum mass of a neutron star to about $2M_\odot$.

▷The wind from a giant is much stronger than the wind that comes from our Sun at present. Astrophysicists do not completely understand the complex processes that lead to the expulsion of such a large amount of material.

▷The situation is like that of Investigation 11.1 on page 123, where a star radiates its thermal energy away. White dwarfs have only 10^{-4} of the surface area of the Sun, making their luminosity is smaller by the same factor. It can therefore take tens of billions of years for the dwarf to cool off.

▷Why don't these reactions go rapidly? After all, they release energy, so they need no energy input to drive them. The barrier is the electric repulsion of nuclei for each other. At low temperatures the nuclei have low speeds and don't come near enough to one another to trigger nuclear reactions. Only when the temperature reaches a certain value do nuclei have enough speed to overcome their electric repulsion and begin to react. This temperature is reached as the core grows towards the Chandrasekhar mass inside a massive giant star.

other available form of support – and so it is just a small white dwarf inside the giant. The composition of its nuclei has little effect on the electron pressure, but it has a small effect on the size of the Chandrasekhar mass: stars with heavier nuclei have a larger critical mass. What happens as the core grows depends on the initial mass of the star. It is a complex process that astrophysicists must simulate on computers in order to understand. The outline of what happens is clear now, but many details are still poorly understood.

For stars of moderate mass, say less than about 8–10 solar masses, the growth of the core is slow enough, and the stellar wind at the surface of the giant is large enough, that before the mass of the core reaches the Chandrasekhar mass the rest of the star has been blown away by the steady wind. The core never exceeds the Chandrasekhar mass, and the endpoint is a white dwarf star that gradually cools off, a dead degenerate cinder. Our own Sun, which starts out with less than a Chandrasekhar mass, will end like this, as a white dwarf of perhaps 0.6 solar masses. For a time the expanding shell of expelled gas is be illuminated by the part of the star that still remains, forming a beautiful planetary nebula, as in Figure 12.5 on page 142.

This peaceful and silent end is not available to more massive stars. Above about 8–10 solar masses, the loss of mass during the giant phase is not fast enough to prevent the core reaching the Chandrasekhar mass. When this happens, it collapses.

The core that collapses does not have the original composition of the waste products. Typically by this time it has changed and become dominated by iron. We saw in the previous chapter that iron is the natural end-point of nuclear reactions, that reactions among carbon and oxygen and other nuclei release energy when they form iron. As the core of the giant accumulates mass, these reactions convert most of the nuclei to iron. So by the time the core reaches the Chandrasekhar mass and collapses, its composition is inert. It has no chance to release energy from further reactions during its collapse.

There is therefore nothing now to halt the collapse until the former white dwarf reaches neutron star density, where neutron degeneracy pressure can build up. During the collapse, the increasing density makes the protons in the iron nuclei combine with the free electrons to form neutrons. When the collapse reaches the density of a neutron star, the speed of infall will be nearly the escape velocity from a neutron star, since the starting point of the collapse is very large compared to the size of a neutron star. We estimated above that this is about half the speed of light! At this speed, the collapsing star shrinks from white dwarf size (6×10^6 m) to neutron star size in less than a tenth of a second. We call this free fall **gravitational collapse**.

What happens when neutron degeneracy pressure builds up to halt this incredible speed? If there were no friction, no dissipation of the energy of collapse, then the star would have to "bounce" and re-expand, just like a rubber ball hitting a brick wall. However, the conversion of electrons and protons into neutrons during the collapse has produced neutrinos that have already removed some energy, so the bounce will be a little weaker. What is more, at the point of maximum density something new happens. Neutrinos, which can pass through normal matter virtually without scattering, become trapped: the density is high enough to scatter them many times as they move through the star.

The effect of this is that neutrinos quickly come into thermal equilibrium with the hot neutron matter, and a neutrino gas builds up. Much of the kinetic energy of infall is converted into neutrino energy. When the bounce starts and the density goes down a little, these neutrinos can suddenly escape, carrying away a great deal of energy. This is a sort of shock absorber, which prevents the star from rebounding back to its original white dwarf size again. Most of the star is now trapped at the enormous density of the neutron star.

But the rest of the material of the star has also been falling in, and begins to hit the outer layers of the core, just as the neutrinos are beginning to expand away. The neutrino gas runs into the infalling envelope of the star, and what physicists call a "shock wave" develops. Familiar examples of shock waves are the sonic boom, the bow wave in the water in front of a fast-moving ship, and the tidal bore found on some rivers, as in Figure 5.4 on page 44.

▷ A shock develops when an object (here the expanding core) moves into a fluid (the envelope) with a speed faster than the local sound speed. The fluid cannot move out of the way fast enough and a large density difference develops just in front of the object.

What happens after the shock forms seems to depend sensitively on details of the nuclear physics, much of which is not yet fully understood. But computer simulations suggest that, at least in many cases, the neutrinos remain trapped long enough to help push the shock outwards into the infalling envelope, with enough energy to blow the envelope away. The expanding envelope is heated by the shock, so that nuclear reactions take place in it at a very rapid rate. When the shock reaches the outer boundary of the star, the star suddenly brightens up, and we see a *supernova*. Meanwhile, at the center, the collapsed core either settles down into a neutron star, or – if much further material from the envelope falls down onto it – collapses again to a black hole. We will discuss both possible outcomes in later chapters.

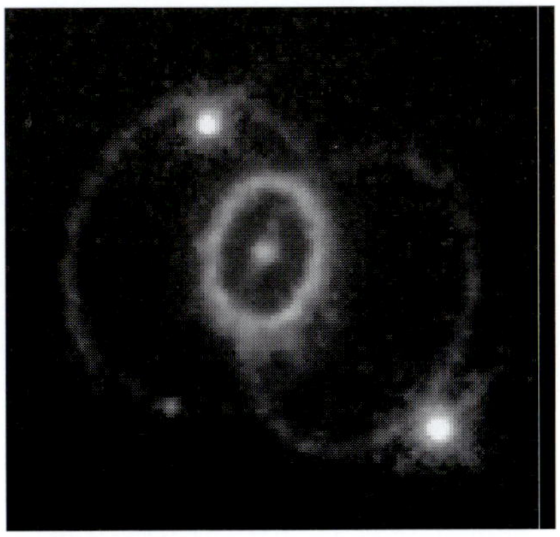

Figure 12.7. The supernova of 1987 in the Large Magellanic Cloud was accompanied by the formation of these extraordinary rings when light from the supernova hit shells of gas that the original giant star had expelled during a phase of mass-loss. The supernova light caused these shells to glow in fluorescence. Image courtesy of NASA/STSCI.

It should not be a surprise that the envelope can be blown away by the neutrinos. The energy released by the collapse of the core is enormous. When we study general relativity later in this book we will learn how gravity can convert mass into energy. Gravitational collapse to a neutron star converts a larger fraction of the mass of the core into energy than happens in a nuclear reactor or nuclear bomb, and much of this energy is carried away by the shock. The envelope has been sitting in a relatively weak Newtonian gravitational field, and it is no match for the thundering impact of the shock. Despite the fact that the envelope may contain ten or twenty times the mass of the core, it blows away at a high speed.

What we have described is called by astronomers a **supernova of Type II**. Supernovae of Type II are among the most spectacular events visible in optical telescopes. The most recent one visible to the naked eye from the Earth was the supernova of 1987, called SN1987A (see Figure 12.7). Located in the Large Magellanic Cloud, which is a small galaxy in orbit about our own Milky Way galaxy (see Chapter 14), it seems to have occurred in a blue giant star of about 20 solar masses. It

▷ We will discuss supernovae of Type I below.

Figure 12.8. The Crab Nebula is the result of a supernova explosion recorded by Chinese astronomers in 1054. The explosion left behind a neutron star, which today is seen as a pulsar (see Chapter 20 and especially Figure 20.4 on page 270). The pulsar is the lower of the two bright stars near the center of the nebula, oriented along a diagonal line from lower right to upper left. This nebula is also known as M1. Photo by Jay Gallagher (U. Wisconsin)/WIYN/NOAO/NSF.

was the brightest star in the southern hemisphere sky for a time.

Physicists were fortunate enough to have detected not only light but also the neutrinos from this supernova: when the trapped neutrinos escaped, about 11 of them induced nuclear reactions in the Kamiokande proton-decay experiment in Japan, and a few others registered in similar experiments in the USA and Russia. This number is about what one would have expected, and their detection provided a clear verification that the picture of the supernova mechanism described here is fundamentally right.

Supernova explosions are not commonplace events. The last one visible to the naked eye before SN1987A was recorded by Kepler in 1604. Once the supernova is triggered, the cloud of gas continues to expand for thousands of years. Astronomers see many such supernova remnants relatively near the Sun. The most spectacular is the Crab Nebula, shown in Figure 12.8.

We are particularly fortunate to be able to see in this remnant the neutron star that was formed by the explosion. This neutron star is a *pulsar,* and we shall discuss it in Chapter 20.

Death by disintegration

In this section: some stars end in a giant nuclear explosion. These are called Type I supernovae.

A supernova of Type II leaves a neutron star or black hole behind. Other explosions can be so violent that they leave nothing behind at all. These occur when a white dwarf star formed long before undergoes a long-delayed gravitational collapse. This can happen when, for example, the star is in a binary system and gas from the companion star falls onto the white dwarf. We will discuss such phenomena in the next chapter.

After a long period of accumulating mass, the old white dwarf could reach the Chandrasekhar mass. What happens next is very different from what happens inside a giant. The old white dwarf is still composed of carbon, oxygen, and other nuclei lighter than iron. It is hot, but not hot enough to allow the nuclear reactions that form iron to take place. The material falling on it is mostly hydrogen, which converts quickly to helium at the temperature of a white dwarf, but not to iron. So there is still plenty of nuclear energy available in the material of an old white dwarf.

> When its long-postponed collapse finally begins, the old white dwarf is an enormous nuclear bomb waiting to happen.

During the collapse the increasing density and pressure leads to a rapid increase in nuclear reactions, as the nuclei combine to form heavier nuclei. Since most of the original nuclei are lighter than iron, these reactions release energy. This energy is enough to stop the collapse well before it reaches the density of a neutron star and blow the white dwarf completely apart. This is what most astronomers believe is the mechanism of the explosion that they call a **supernova of Type Ia**. These explosions are the brightest of all supernovae, and astronomers have been able to detect them at huge distances. In fact, although Type Ia supernovae are even rarer than Type II, astronomers have been able to find enough of them at very great distances to demonstrate that the expansion of the Universe is apparently accelerating rather than slowing down. We will come back to this extraordinary and unexpected observation in Chapter 14 and in the final chapters on cosmology.

> ▷ Nuclear reactions do not release as much energy as gravitational collapse, so a Type II supernova is more energetic than a Type Ia. But most of this energy comes out as kinetic energy of the envelope, so the photon luminosity of a Type Ia is in fact larger than that of a Type II.

What is left behind: cinders and seeds

When almost all stars die, much of their material gets locked up forever in a stellar cinder: a white dwarf, a neutron star, or even a black hole. But at the same time, much of their material is returned to interstellar space to be recycled into further generations of stars. Winds that blow away the outer envelopes of massive stars return gas that is enriched in carbon, oxygen, and other elements vital to life. The gas ejected in a supernova explosion is different. It has been thoroughly processed by the shock wave that ejected it, and the material ejected is rich in very heavy elements, from iron to uranium. Not only are some of these elements vital for life, but as we have seen the uranium powers geological activity on the Earth, without which life could not have evolved. Without white dwarf cores to trigger supernovae and neutron stars to stop the collapse and generate the bounce, we would not be here!

> **In this section:** we remind ourselves that the existence of compact star remnants is essential for the formation of life.

Binary stars:
tidal forces on a huge scale

In this chapter: we look at a number of astronomical systems that are affected by tidal forces, inhomogeneity of the gravitational field. These systems include binary stars, interactions between planets, mass flows between stars, X-ray binaries, and the three-body problem. We use computer simulations to explore realistic examples of many of these systems.

Binary stars are stars bound in orbit about one another by their gravitational attraction. Most stars seem to form in binary systems or in systems containing more than two stars. This is not really surprising: stars form from condensations in giant clouds of gas, so where one star forms, others are likely, and they may form close enough to each other to be bound together forever. We saw this in the numerical simulation reproduced in Figure 12.2 on page 138.

We have already studied special cases of binaries: planets in motion around the Sun, and the Moon around the Earth. These orbits allow us to measure masses in the Solar System. We learn the Sun's mass once we know the radius and period of the Earth's orbit. Similarly, we measure the mass of the Earth by studying the motion of the Moon (and of artificial Earth satellites). In the same way, binary star orbits are used to measure the stars' masses. Binaries are often our best, indeed our only, way of measuring the masses of stars. When two compact stars (such as neutron stars or black holes) form a binary system, they can be used to test our ideas about gravity: one of the most stringent tests of general relativity is that it predicts perfectly the observed orbits in a certain binary neutron star system.

The nuclear physics in the core of a star in a binary system is normally not affected much by the companion. The core is fairly dense and small, so the tidal forces on it from the other star do not change the conditions there very much. But the gravitational field of the companion can have dramatic effects on the outer parts of the star, and these can eventually change completely the course of evolution of the star. When binaries consist of a main sequence star and a cinder (a neutron star, for example), the cinder can strip the ordinary star of gas, with spectacular results. Such binaries are sites where nova explosions occur; they can produce X-rays in abundance; and they sometimes shoot out extraordinarily narrow beams of particles traveling at nearly the speed of light. Binary evolution is one of the most intensively studied subjects in astronomy today.

Looking at binaries

Binary systems turn up in astronomical observations in many different ways. Some binaries are easy to observe directly: with a telescope one can watch, over a period of months or years, the positions of two stars change as they orbit one another. Such **visual binaries** are relatively rare, since they have to be close enough to us for our telescopes to be able to separate the two images despite the blurring caused by the Earth's atmosphere. Sirius A and B, described in Chapter 10, are a good example of a visual binary. The Hipparcos satellite, mentioned in Chapter 9, has greatly increased the number of such binaries known.

It is much more common to learn that a star is in a binary system by recording its *spectrum*. We saw in Figure 10.3 on page 114 that the spectrum of a star is peppered with sharp features called spectral lines. The wavelengths of these lines are characteristic of the atoms that emit or absorb the light. These wavelengths

▷The background picture on this page is an X-ray image of the binary stars Sirius A and B, taken by the Chandra satellite. The rays are optical effects produced by the telescope. The Sirius system was discussed in Chapter 10. Image courtesy NASA/SAO/CXC.

In this section: we learn how astronomers know which stars form binary systems, and what can be learned from observations.

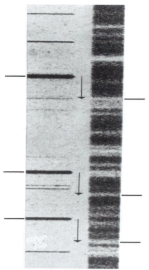

Figure 13.1. *The spectral lines of the observed spectrum (left, in negative) are shifted from those of the reference spectrum (right). The amount of the shift indicates the velocity of the star along the line-of-sight. The marked lines are all lines of neutral iron.*

can be measured in the laboratory and compared with the observed spectrum. If the star is moving away from us, then the Doppler shift, explained in Figure 2.3 on page 15, will shift the wavelengths of all the lines to the red (longer wavelengths). Similarly, if the star is moving towards us, the lines shift to the blue. So by comparing the positions of spectral lines in an observed spectrum with those in the lab, we can determine the speed of the star along the line-of-sight. This is illustrated in Figure 13.1. Importantly, motion of the star in directions perpendicular to the line-of-sight produces no Doppler effects and is not measurable this way.

If we look at the spectrum of a star in a binary system, then the Doppler shift of the lines should change with time, as the star's orbital velocity changes. Observed for long enough, the changes should be periodic, that is they should repeat after one orbital period of the binary. Binaries that are discovered this way are called **spectroscopic binaries**, and they constitute the overwhelming majority of known binaries.

If both stars are of comparable brightness, then it may be possible to see two sets of spectral lines, shifting in different ways but with the same period. But it usually happens that only one set of lines is visible, either because the second star is too dim for its lines to be seen in the light from the first, or because the second star is a "cinder" (a white dwarf or a neutron star) that does not have prominent spectral lines.

In a spectroscopic binary, we only learn about velocities along the line-of-sight. The orbital plane, on the other hand, will be oriented at random to the line-of-sight. So we only get partial information about the orbit. In an extreme case, if we happen to be looking at the orbit "face-on", directly down onto the orbital plane, there will be no motions along the line-of-sight, no Doppler shifts, and we might not recognize the system as a binary at all. At the other extreme, if our line-of-sight is in the plane of the orbit, we see the whole motion. Astronomers who wish to use binaries to measure the masses of stars need to try to unravel these uncertainties. We shall look at how they do this in the next section.

The orbit of a binary

We saw in Chapter 4 that the orbit of a planet around the Sun was a perfect ellipse, at least if the gravitational effects of other planets are ignored. It would be natural to expect that the situation would be more complicated when the "planet" is actually another star, whose mass is comparable to that of the star it is orbiting. After all, in the Solar System we idealized the Sun as being fixed at one point, undisturbed by the weak gravitational pull of the planets. In a binary star system, neither star will stand still, and so we might expect the orbits to be much more complex.

Remarkably, this is not the case.

In this section: the orbits of stars in a binary system are ellipses, just like planetary orbits.

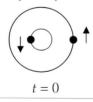

$t = 0$

$t = P/2$

Figure 13.2. *For the discussion in Investigation 13.1 on page 156, two stars in circular binary orbits shown at times half an orbital period apart.*

Provided the stars in a binary system are themselves spherical, *both* of their orbits will be ellipses, just as for planets in the Solar System.

We show this for the special case of circular orbits in Investigation 13.1 on page 156. For the general elliptical case, we turn to the computer program for orbits that we constructed in Chapter 4. A simple modification in Investigation 13.2 on page 157 is enough to demonstrate the elliptical nature of the orbits of both stars. The results of two computer runs are shown in Figure 13.3.

The restriction to stars that are spherical is the same as we had in the Solar System: the gravitational field of a spherical star is the same as if all its mass were concentrated at its center. This restriction is not always valid. When the stars are close to one another, the tidal forces of one deform the shape of the other, and the orbit can become much more complex.

Figure 13.3. *Results of computer simulations of the orbits of two binary systems. On the left is a system consisting of two 1M⊙ stars with initial speeds of 13 km s⁻¹ at their point of furthest separation, which is taken as twice the minimum separation of Mercury from the Sun. This should be compared to Mercury's speed of 59 km s⁻¹ at that point. The stars plunge together rapidly. The right-hand figure shows a system in which star A has mass 2M⊙ and B has 1M⊙. The initial speed of A is 20 km s⁻¹, while that of B is 40 km s⁻¹. In both diagrams, the positions of the stars at selected times are illustrated by dots on the orbits. The stars' positions always lie diametrically opposite each other on a line through the common focus of the orbital ellipses, which is shown as a "+".*

Computer Simulations of Binary Star Orbits

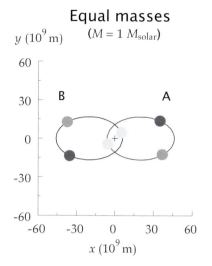

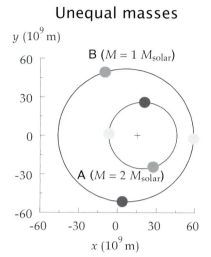

Planetary perturbations

To a good approximation, the Solar System can be regarded as a binary system involving the Sun and Jupiter. In this time-dependent gravitational field, the other planets make their orbits around the Sun. We have mentioned before that, although the orbits would be elliptical if the Sun were the only body creating the gravitational field, the orbits are not exactly elliptical when we take account of Jupiter's influence. To see this, I have again modified the computer program that was used to calculate Mercury's orbit in Chapter 4; the new program can follow Mercury in the gravitational field of the Sun and one other body. I assume the Sun and "Jupiter" follow circular binary orbits. The program MercPert is available on the website.

The effect of Jupiter on Mercury in our Solar System is rather small, although easily measurable if one uses observations of the orbit performed over a century or so. However, when the Solar System first formed, things may have been much more interesting. In particular, astronomers speculate that several Jupiter-sized planets may have formed and spiralled into the Sun rather quickly. In order to illustrate what might have happened to an inner planet in such a situation, on a short time-scale, I have re-shaped the present Solar System: it is nice how computers allow one to play with such ideas! This hypothetical planetary system has a planet 100 times as massive as Jupiter, lying in the orbit of Venus. Such a massive planet so near to Mercury makes the changes in Mercury's orbit easier to see, but similar things would happen if Jupiter were less massive. Astronomers have recently discovered planets around other stars that are not very different from this super-Jupiter.

The results of our simulation are illustrated in Figure 13.4 on page 158. In the left panel are shown the orbits of all three bodies. Notice the extraordinary event where Mercury actually moves out and loops around the planet. This was not something I set out to achieve: it happened on the first run with this configuration.

> Mercury in effect indulges itself in the gravitational slingshot here. Its orbit after the encounter goes much closer to the Sun than before, just as we noted would happen to an artificial space probe that meets Jupiter in this way (Chapter 6).

In this section: each planet is a kind of binary body with the Sun, but the planets also affect each other. We construct imaginary Solar Systems and use computer simulations to see what might have happened in the early Solar System.

Investigation 13.1. Two stars making circles around each other

It frequently happens that binary orbits are nearly circular, and in that case it is not hard to find out how to use observations to give us information about the masses of the stars.

First, we need to convince ourselves that circular orbits are possible. In Figure 13.2 on page 154 I illustrate the geometry. Two stars of different masses move on circles of different radii but with a common center. The orbits have the same period, so the stars are always on opposite sides of the center. Can we insure that the system can be set up to behave like this?

Suppose we first choose an arbitrary value for their common orbital period P. We next have to decide on the sizes of their orbits. Once we choose the orbital radii, we then know what speed we have to give the stars, which has to be just the right amount to get the stars around their orbits in the time P. The radii must be chosen so that the acceleration required to keep each star on its circular orbit equals the gravitational acceleration produced by the other star at the distance separating them (which is the sum of the orbital radii). Then each star will at least initially tend to move on its circle, and the stars will continue to lie diametrically opposite each other.

Since the speed is constant in circular motion, a short time later exactly the same conditions will continue to hold, and so the stars will continue to follow their circles. Therefore, circular motion of the type shown in Figure 13.2 on page 154 is possible, given the right initial conditions. In fact, even when the initial conditions are not right, friction in the system (such as results from tidal deformation of the stars) can circularize the orbit.

Suppose the stars, called "1" and "2", orbit on circles of radii R_1 and R_2, respectively. Let their masses be M_1 and M_2. Suppose their orbital period is P, the same for both stars. Let their total separation be called $R = R_1 + R_2$. Consider the acceleration of star 1. Traveling a circle, it has uniform acceleration towards the center of $4\pi^2 R_1/P^2$, from Chapter 4. The gravitational acceleration produced by the other star is GM_2/R^2, depending on the mass M_2 of the other star. Circular motion requires these to be equal:

$$\frac{GM_2}{R^2} = \left(\frac{2\pi}{P}\right)^2 R_1. \tag{13.1}$$

There is an analogous equation for the second star, obtained by exchanging the indices 1 and 2:

$$\frac{GM_1}{R^2} = \left(\frac{2\pi}{P}\right)^2 R_2. \tag{13.2}$$

Let us first divide these equations. That is, we divide the left-hand side of Equation 13.1 by the left-hand side of Equation 13.2 and similarly for the right-hand sides. The ratios remain equal, and most of the factors cancel out. We are left with the simple expression

$$\frac{M_2}{M_1} = \frac{R_1}{R_2}. \tag{13.3}$$

The sizes of the orbits are in inverse proportion to the masses of the stars. The heavier star executes the smaller orbit. We already knew this: we have used it in Chapter 4 to determine the radius of the circle the Sun moves on due to Jupiter's gravitational pull.

Next we add the two equations, by adding the left-hand sides to each other and the right-hand sides to each other. This gives

$$\frac{G(M_1 + M_2)}{R^2} = \frac{4\pi^2}{P^2} R. \tag{13.4}$$

This is interesting because it shows that simply measuring the period P of the binary gives us the ratio $(M_1 + M_2)/R^3$.

If we knew the stars' separation R we could then infer the total mass of the two stars. This is observable in a visual binary, but not in a spectroscopic binary. In fact, in a visual binary one can determine everything, since one can measure both orbital radii: these immediately give M_1 and M_2 from the original equations.

But visual binaries are not common. Normally astronomers observe a spectroscopic binary with only one set of lines, the other star being too dim to appear in the spectrum. Let us call this star 1. By replacing R in Equation 13.4 by $R_1 + R_2 = R_1(1 + R_2/R_1) = R_1(1 + M_1/M_2)$, we obtain the following useful equation:

$$\frac{M_2^3}{(M_1 + M_2)^2} = \frac{4\pi^2 R_1^3}{GP^2}. \tag{13.5}$$

Astronomers can measure the period P, and they can *almost* measure R_1. What they actually measure is the Doppler shift of the spectral line, as in Figure 13.1 on page 154, which tells them the speed of the star along the line-of-sight. Combined with the measured period P, this tells the observer how much distance the star moves toward and away from the Earth during its orbit. But it does not reveal how much the star moves in a plane perpendicular to the line-of-sight, which astronomers call the **plane of the sky**. So they do not measure R_1, but rather the projection of R_1 onto the line between the Earth and the star. It is easy to see that this differs from the true radius by the sine of the angle between the line-of-sight and a line perpendicular to the true plane of the orbit. Astronomers call this angle the **angle of inclination** of the orbit, and use the symbol i for it. Thus, observers measure $R_1 \sin i$. Multiplying Equation 13.5 by $\sin^3 i$ gives a right-hand side that is measurable, and therefore the left-hand side is a known quantity for spectroscopic binaries. This is called the **mass function** of the circular orbit:

$$f(M) = \frac{M_2^3 \sin^3 i}{(M_1 + M_2)^2}. \tag{13.6}$$

It has dimensions of mass, but its value only constrains the masses of the two stars. From observations of a single star in a spectroscopic binary, one cannot determine the individual masses without additional information.

Additional information is sometimes available. If the second star's spectrum is also visible, then its velocity determines the mass function with indices "1" and "2" reversed; this allows the mass ratio of the two stars to be determined. In a single-spectrum system, the visible star may have a standard and well-understood spectrum, so that theoretical calculations determine its mass; then the mass function can be used to constrain the value of the companion's mass. Alternatively, the companion star may pass right in front of the star that we see, eclipsing it; this requires that the angle i is nearly 90°.

We will see another example of extra information in Chapter 20, when we consider binaries containing pulsars. For very compact binaries, general relativity predicts extra effects that can be measured and used to determine the individual masses and the angle i.

In the right-hand panel of the figure, I show the orbit of Mercury relative to the Sun; that is, I plot the x- and y-distance of Mercury from the Sun. This is not identical to the path of Mercury in the left-hand panel, because the Sun moves. After the near encounter with the planet, Mercury's orbit is considerably more elliptical than it would have been with no planet there. It also does not keep its orientation in space: the (imperfect) "ellipse" it traces out turns counterclockwise. The likelihood exists that a further encounter with the massive planet would send Mercury plunging into the Sun.

Interested readers are encouraged to play with this program. There is an infinite

---

Investigation 13.2. Simulating the orbits of a binary pair

Binary orbits are generally not circular. To demonstrate to ourselves that they are actually elliptical, we do the same as we did before: we simulate their orbits using the computer, and then look at the shape of the orbits. This cannot be exact, but it can be made very convincing.

On the website is the program `Binary` that does this calculation. It is adapted from the program `Orbit`, which did the orbit of Mercury that was displayed in Figure 4.3 on page 29. The adaptation is straightforward. The main complication is that there are now two bodies to follow. I have called them A and B, and their coordinates, for example, are (xA, yA) and (xB, yB). Each statement in `Orbit` that refers to the coordinates, velocity, or acceleration of the body has become two statements, one referring to body A and the other to body B. Tests for, say, the appropriate time-step size are performed on both orbits, and both must pass the test.

The main change in the physics underlying this program is the fact that the gravitational acceleration no longer comes from the Sun, located at a fixed point which we took to be the origin of the coordinate system. Instead, the attraction on body A comes from body B, and vice versa. We must therefore insure that, no matter how the bodies move, the acceleration of gravity is calculated correctly.

This change is not hard to make. Let us write down again the equations we used for the acceleration produced by the Sun, Equations 4.5 and 4.6 on page 29:

$$x\text{-accel} = -k\frac{x}{r^3},$$

$$y\text{-accel} = -k\frac{y}{r^3}.$$

What do the terms in this equation mean? The denominator r is just the distance from the planet to the Sun. In the binary problem, we would replace this with r_{AB}, the distance between the two stars. The factors of x and y in the numerators of the planetary acceleration are the coordinates of the position of the planet. Now, the coordinates themselves have no absolute significance: they only show where the planet is in relation to the Sun, which in the earlier calculation was at the origin of the coordinate system. If the Sun, or another body exerting a gravitational force, were at a position with coordinates (x_S, y_S), then we would replace x by the *x-distance* from the Sun to the planet, which is $x - x_S$. Similarly we would use $y - y_S$ in place of y.

In the binary problem, the acceleration of body A produced by B is therefore found by replacing x by $x_A - x_B$ and y by $y_A - y_B$. Similarly, the acceleration of B produced by A is found by replacing x by $x_B - x_A$ and y by $y_B - y_A$. This is done in the program.

The output of the simulation, as displayed in Figure 13.3 on page 155, shows the elliptical nature of both orbits clearly, even when the stars have different masses. Notice that the ellipses share one focus, indicated by the "+" in the diagrams. The stars remain on opposite sides of this focus at all times. In our study of the circular orbit problem in Investigation 13.1, we saw that the bodies remained on opposite sides of the origin; this is the focus of the circular orbits. The distances of the stars from the center of the circle were inversely proportional to their masses: $M_1 r_1 = M_2 r_2$. The same is exactly true for the elliptical orbits: the distances of the stars at any time from the focus are in inverse proportion to their masses.

I must not omit one small but important point about using the binary orbit program. When choosing initial speeds for the stars, make sure that the total *momentum* vanishes:

$$M_A U_A + M_B U_B = 0, \qquad M_A V_A + M_B V_B = 0. \qquad (13.7)$$

If initial data are chosen that violate this, then the stars will still orbit one another, but the whole system will move as well! We will see an example of a moving binary in a later analysis.

---

variety of possible configurations to try, and interesting things will happen very often. In particular, try replacing Mercury with a comet in a very elongated orbit plunging toward the Sun. If it comes near the planet, it could encounter it and end up in a much more *circular* orbit.

We have met here a rather complicated situation, where there are three bodies in mutual gravitational fields. In fact, this is far beyond the reach of pen-and-paper mathematical calculations, and it would not normally be treated in, say, courses leading to an undergraduate astronomy degree. But, given the original orbit program, it is hardly any extra work to get our computers to show us what we might expect to happen here. Our computers open up whole new subjects for us to think about.

Of course, the complexity of the problem does still have an impact: one really has to sit down with the computer and run many versions with different initial data in order to develop a feeling for the variety of things that might happen. I encourage the reader to do this, since the variety is really very large and interesting.

Tidal forces in binary systems

The effect of "Jupiter" on Mercury is essentially the action of the tidal gravitational forces exerted by the large planet on the orbit of the smaller body. There are many other places in astronomy where such tidal forces are important. We met some in Chapter 5. Here are some more.

- *Mass transfer between binary stars.* One of the most dramatic examples of tidal forces occurs in binary star systems, where one star begins to pull mass off the other. We have idealized our binary systems above by assuming the stars are spherical. But stars are not rigid, so if they get too close, the tidal

▷Our simulated comet will not, however, be captured by the planet, as happened to comet Shoemaker–Levy 9, causing it to crash spectacularly into Jupiter in 1994. The reason is that capturing a comet requires that some of the comet's energy be given up to friction, as when a comet breaks apart. In our computer program, there is nothing to simulate friction.

In this section: tidal forces can make gas flow from one star to another, emitting X-rays. They can help one star to capture another into a binary. And they explain why the inner planets like the Earth became rocky.

Motion of the three bodies

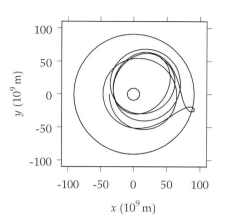

Orbit of perturbed "Mercury" relative to Sun

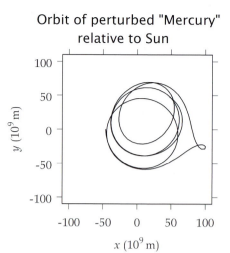

Figure 13.4. Computer simulation of the effect that an imaginary massive planet in Venus' orbit would have on Mercury. All bodies orbit counterclockwise. Mercury begins between the Sun (which executes the small circular orbit) and the planet (large circular orbit), and on its second orbit happens to interact strongly with the planet, looping around it.

forces of each star will deform the other. If one star evolves into a giant, the result can be that the tidal forces actually pull the weakly bound outer layers off the giant and onto the other star. The numerical simulation in Figure 13.4 shows how this can happen. Suppose that the body we have called "Mercury" is really just a bit of the outer parts of a giant star sitting where the Sun is. Then the companion star pulls this bit of the giant over to itself (Mercury loops around "Jupiter"). What will happen next is different for gas than for a planet like Mercury. Instead of just going once around "Jupiter" and heading back to the original star, as Mercury does, a parcel of gas will run into other gas that has been pulled off before. It will stop and sink into a disk around the companion. This is called an **accretion disk**. We will look at two examples of such systems in more detail below: **cataclysmic variable** stars and **X-ray binary** stars.

- *Tidal capture.* Suppose two stars that are not originally in orbit about one another happen to pass very close to each other. If the stars were point particles, they would deflect each other's direction of motion, but then they would separate again and fly off on their own trajectories. But real stars will affect each other tidally. The tidal forces may pull material off one or both stars; they will stretch each star and cause frictional heating; they could also change the intrinsic rotational motion of one another. All of these effects can remove energy from the stars' motion, with the result that they may not have quite enough to get away from each other afterwards. They then fall into a highly eccentric elliptical orbit, repeatedly passing as close to each other as before. On each pass, they lose a bit more energy, so the orbit becomes smaller and more circular. This is called tidal capture, and it is thought to happen to a significant number of stars in the centers of the **globular clusters** of stars that we will study in Chapter 14. It also operates in the star-forming simulations shown in Figure 12.2 on page 138, where long chains of protostars merge into one another.

- *Why the inner planets are rocky.* Although the formation of the planets is still shrouded in some uncertainty, one thing is clear: they did not simply form as big balls of diffuse gas that subsequently evaporated away and left rocky cores. Instead, they probably formed by the agglomeration of small

---

Investigation 13.3. Forming rocky planets

We saw in Investigation 5.1 on page 43 that the tidal force exerted by a body of mass M on another body at a distance r from it is proportional to M/r^3. This in turn is proportional to the *mean density* of the body, if its mass were spread over the entire region out to r.

From this it is easy to see what would happen to a cloud of gas condensing near the Sun. Consider the tidal force of the Sun on the outer part of the condensation, relative to the gravitational force of the condensation itself. This self-force is nothing more than the tidal force of the body on itself, so the two forces are in proportion to the two relevant densities. The tidal force of the Sun is proportional to the mean density of the Sun spread out over the whole region inside the orbit of the condensation. The force of the condensation on itself is proportional to its own density.

Now, in the original collapsing gas cloud from which the Sun and planets formed, the density must have been higher near the center than near the edge. That means that the mean density of the gas inside the orbit of the condensation must have been larger than the density at the condensation. The gas inside went on to form the Sun, but its tidal effect on the condensation didn't change. Since it had the larger density, its tidal effect dominated. The condensation would therefore have had a very hard time forming.

If the protoplanet formed from rocky condensations that stuck to one another when they collided, then the argument would be different. The density of the asteroids was much higher than the density of the gas as a whole, and they would not have been torn apart by the Sun. Whatever the details were, it is clear that gravitational tides would have stopped purely gravitational condensations. Some chemical processes in the early nebula were required to produce the asteroids, which could then form seeds for the planets.

---

rocky asteroids that formed from the interstellar dust grains in the cloud of gas that formed the Sun. The reason is the Sun's tidal effect on the disk in which the planets formed. The Sun would always have been strong enough to have torn apart any ball of gas that was condensing into a planet. This is explained in Investigation 13.3. For the inner planets (out to Mars), this meant that they grew with little gas present; their present atmospheres come from the release of gas that was trapped in the original rocky asteroids. The more gaseous outer planets managed to trap some of the gas of the original cloud, but only after they had developed massive cores from the asteroids. This was easier at a great distance from the Sun, possibly because the gas there was colder. The present-day asteroid belt is probably a planet that never formed: the combined tidal forces of the Sun and Jupiter prevented even the rocky asteroids from accumulating into a planet.

Accretion disks in binaries

Once accretion disks have formed around certain kinds of stars, the subsequent events can be dramatic. Material that falls into a disk quickly gets pushed into a circular orbit around the central star, since that minimizes friction with the material already there. But since circular orbits at different distances from the central star have different periods, there is inevitably some friction always present. The result is that the material slowly spirals into the center, and its orbital energy is converted into heat. Accretion disks around compact stars, such as white dwarfs and neutron stars, are very hot near the central star.

We show in Investigation 13.4 on page 161 that the random thermal kinetic energy of particles in the disk is determined by a balance between the rate at which mass moving through the disk releases gravitational energy and the rate at which that energy can be radiated away from the disk. We find that the temperature near the center of a disk around a white dwarf of mass $1M_\odot$ and radius 5×10^6 m that is accreting $10^{-10} M_\odot$ per year is 8×10^4 K. The typical thermal kinetic energy of the particles is about $7\,\text{eV}$. Such a disk would emit thermal radiation in the near-ultraviolet part of the spectrum, at a wavelength of $0.17\,\mu\text{m}$.

Accretion disks around white dwarfs are responsible for a wide range of observed phenomena. The material that falls on the dwarf from a normal companion comes from the companion's atmosphere, so it is mostly hydrogen. When it lands on the white dwarf, it is hot and much denser. After a certain amount has accumulated on the surface of the dwarf, a nuclear chain reaction can occur, converting the hydrogen to helium. This sudden release of energy causes an explosion, which we see as a nova. This is very different from a supernova, in which a whole star is disrupted. In

In this section: an accretion disk can emit X-rays, blow up as a nova, or funnel mass onto a central star until it becomes a supernova.

a nova, the disruption is temporary; accretion soon resumes, and it is only a matter of time until the next nova occurs in the same system.

Accretion disks can be responsible for supernovae, as well. If the accretion rate is high enough, then after a long period of accretion and despite many nova outbursts, the white dwarf's mass will have increased to the Chandrasekhar mass, and the star will collapse. As we saw in the previous chapter, this white dwarf typically contains much material that can still undergo nuclear reactions, and the result can be a giant nuclear reaction that incinerates the whole star. This is believed to produce Type Ia supernova explosions.

Between nova explosions, an accreting white dwarf can still appear to be unusual. The flow of material onto the star can be irregular for many reasons, and minor outbursts can occur with surprising regularity. Such binary systems are called *cataclysmic variables*.

Compact-object binaries

In this section: the most spectacular accretion phenomena occur when the central star is a neutron star or black hole. Observing the emitted X-rays is one of the main ways of identifying black holes.

If we took the same accretion disk as we placed around a white dwarf in the previous section and put it around a neutron star of radius $10\,\mathrm{km}$, it would have a central temperature of $8 \times 10^6\,\mathrm{K}$, and would radiate thermal X-rays. The typical thermal energy of a particle would be $0.7\,\mathrm{keV}$. (The notation "keV" means kiloelectron volts, or $10^3\,\mathrm{eV}$.) This is called a "soft" X-ray. We see such systems because they are strong emitters of X-rays. There are over 100 so-called X-ray binaries in our Milky Way galaxy.

If one can identify the "normal" star (usually a giant) that provides the gas for the accretion disk, then one can usually measure the Doppler shift of its spectrum as it orbits the neutron star. This gives us some information about the mass of the neutron star, as we have explained earlier. It is quite remarkable that all neutron stars identified in this way have masses of about $1.4M_\odot$.

The compact object at the center of an accretion disk could also be a black hole. Here the energy would be a little higher, but the system would still be an X-ray source. Since black holes are likely to be formed when the collapse to a neutron star involves more mass than the maximum for neutron stars, black holes are expected to be more massive than neutron stars. We now observe a number of systems where the mass of the compact object is likely to be $8M_\odot$ or more. Since this is much higher than the maximum mass of neutron stars, these are identified as black holes. We shall describe black holes in Chapter 21.

Fun with the three-body problem

In this section: we modify the computer program so it can simulate three stars of similar masses interacting with one another. Spectacular consequences follow. We simulate a "factory" for black hole binary systems.

We have studied above the problem of three bodies interacting by gravity, which astronomers call the "three-body problem". However, we have not made it the most general three-body problem we might imagine, because we have not let the gravitational field of Mercury affect the other two bodies. We have treated Mercury as if it were a "test particle", a probe of the field created by the other two bodies. Astronomers therefore call this the "restricted three-body problem".

This is all right if Mercury's mass is small, as it is here. But the general case is even more interesting, and we turn to it now. The full three-body problem opens up a new possibility that the restricted problem does not have. Although a system of three stars may start out mutually bound, in the sense that none of them has the escape velocity to get away from the other two, it is possible for two of them to form a very close binary pair, giving the third such a strong "kick" that it leaves the system altogether. In fact, such behavior is the norm rather than the exception.

I have modified the orbit program to calculate the full three-body problem. This means treating each star the same, and allowing its gravitational forces to act on

---

Investigation 13.4. Accretion disks

We have learned enough physics in previous chapters to predict fairly accurately the temperature and observable features of accretion disks. The essential feature of an accretion disk is that material flows through it, gradually falling onto the central star. This flow is driven by friction. The mechanism creating the friction could be complicated, but it seems to be the case that many accretion disks are in a fairly steady state, so that the mass flows through the disk at a constant rate.

What makes accretion disks hot? They are hot, of course, because gravitational energy is being released by the material falling in. Now, a disk without friction would consist just of circularly orbiting gas: no matter would flow through it onto the central star, no energy would be released, and the disk would have zero temperature. (Think of the rings of Saturn here.) An accretion disk can remain hot, with a steady luminosity, only if material flows through it. The reason is that it must constantly replace the energy it radiates away with new gravitational energy released by the flow of matter through it. The flowing gas must, of course, come from somewhere and go to somewhere. The gas that reaches the inner edge of the disk falls onto the central star or into the central black hole. If the disk is in a steady state, then this gas has to be replaced by new gas arriving at the outer edge of the disk. Normally this gas comes from a companion star in a binary system.

Let us look at what happens in a steady disk. Suppose that, in a small interval of time Δt, an amount of mass ΔM arrives at the outer edge of the disk, and the same amount leaves from the inner edge. For a steady situation, the mass arriving, ΔM, will be proportional to the time interval Δt. If we call the constant of proportionality M_t, then we can write[a]

$$\Delta M = M_t \Delta t.$$

Now, each small amount of mass m that reaches the inner edge of the accretion disk, whose radius is R, has had to release a total energy roughly equal to the energy it would need to escape from this radius, because escaping is just the time-reverse of falling in. (I say roughly, not exactly, because the gas still has orbital kinetic energy and maybe a little inward speed as well at the inner edge. If it encounters the central star there, then all this energy will be released as it hits the star. But if the central object is a black hole, then the kinetic energy will go into the hole and not be converted into radiation. This would roughly halve the energy released.) This escape energy is just the gravitational potential energy at the inner edge $E = GMm/R$. Then if ΔM amount of mass moves through the disk in the time Δt, it releases an energy $\Delta E = GM\Delta M/R$. This energy is released all over the disk, but more is released near the inner edge, where the gravitational potential energy is largest.

Now, the *luminosity* L of the disk equals the energy released per unit time, i.e. $L = \Delta E/\Delta t$. Using the fact that $\Delta M/\Delta t = M_t$ allows us to write

$$L = GMM_t/R. \tag{13.8}$$

To find the disk's *temperature*, we assume that the disk is a *black body*. Knowing the luminosity of the black body, we need to estimate its surface area in order to deduce its temperature. Although the accreting matter will release its energy over the whole of the disk as it gradually spirals through it, most of the energy is released in the central region, especially if the material finally accretes onto the surface of a central star. We won't be far wrong, therefore, if we assume that the area is πR^2. This is not exact, but it will give us a good idea of what temperatures to expect.

Equating the energy released to the energy radiated at temperature T gives

$$\frac{GMM_t}{R} = \sigma(\pi R^2)T^4. \tag{13.9}$$

Solving this for the temperature gives

$$T = \left(\frac{GMM_t}{\pi\sigma R^3}\right)^{1/4}. \tag{13.10}$$

This equation leads to the numbers quoted in the text.

Astronomers often use formulas like Equation 13.10 above that can be applied over a wide range of values of some of the parameters. For example, one astronomer might be dealing with accretion onto a white dwarf, another with accretion onto a neutron star; the difference in R is a factor of 600. Similarly, different sorts of binary systems could have values of M_t that differ by factors of 1000, depending on the mass and size of the companion star. It is useful in such situations to pick values of the parameters suited to one situation, calculate the desired numbers, and then show how the result would scale with changes in the values of the parameters. In this way, we do the arithmetic involving numbers that don't change (like G) once and for all. For example, if we take $M = 1\,M_\odot$, $M_t = 10^{-10}M_\odot\,\mathrm{y}^{-1}$, and $R = 5 \times 10^6$ km, then we get $T = 8 \times 10^4$ K. Having calculated this once, there is no need to go through the trouble of looking up values of G and σ again for a different central stellar radius. Instead, we express the general result in the following way:

$$T = 8 \times 10^4 \left(\frac{M}{1\,M_\odot}\right)^{1/4} \left(\frac{M_t}{10^{-10}M_\odot\,\mathrm{y}^{-1}}\right)^{1/4}$$
$$\times \left(\frac{R}{5 \times 10^6\,\mathrm{m}}\right)^{-3/4}. \tag{13.11}$$

The proof that this is equivalent to Equation 13.10 is that (1) when the values assumed above are used, the answer is right; and (2) the value of T in Equation 13.11 depends on the variables M, M_t, and R in the same way as in Equation 13.10.

Now when we change the parameters, we can do the calculation more easily. For example, take the same accretion rate onto a solar-mass neutron star with $R = 10$ km. All values are the same except for R, which is a factor of 500 smaller. Since the temperature depends on $R^{-3/4}$, we conclude immediately that the temperature is a factor of $500^{3/4} = 105$ larger. Recall that the wavelength of the emitted radiation depends inversely on the temperature. A temperature of just under 10^5 K, as in Equation 13.11, is cooler than the Sun and therefore emits predominantly infrared radiation. When we replace the white dwarf by a neutron star, the wavelength goes right down into the X-ray region of the spectrum. So when astronomers first began observing X-rays, they began to find neutron stars and black holes in binary systems, not white dwarfs.

Exercise 13.4.1: *Accretion disk temperatures*

Perform the arithmetic to get the temperature $T = 8 \times 10^4$ K in Equation 13.11. Then use this equation to calculate the other values of temperature given in the text.

Exercise 13.4.2: *Accretion disk luminosities*

Take the equation $L = GMM_t/R$ for the disk luminosity and write it in a similar normalized form to Equation 13.11, scaling the mass of the central object to $10^9 M_\odot$, the accretion rate M_t to $1\,M_\odot\,\mathrm{y}^{-1}$, and the radius to 10^{13} m. These values are appropriate to accretion disks around the giant black holes that power quasars (Chapter 14).

---

[a]Readers who are familiar with calculus will recognize M_t as the derivative of M with respect to t, dM/dt, which is constant for a steady flow. Other readers will not need to know this.

every other star. The program Multiple, available on the website, is written so that any number of stars can be used, so it can investigate the four-body, five-body, and in general the *n*-body problem. Of course, the more bodies that one uses, the slower the program runs: since each body's motion must take account of the forces from every other body, the number of calculations necessary to advance one time-step is roughly proportional to n^2 for n bodies. So the four-body problem takes nearly twice as long to run as the three-body problem. I would encourage only those readers with fast computers to go beyond three bodies.

The result of the first run I made with the three-body program is shown in Figure 13.5. I took three stars of masses 1, 2, and $3M_\odot$, and arranged them with separations comparable to the Mercury–Sun distance. I gave them small initial velocities, to insure the system was bound overall: no star had the escape speed from the gravitational field of the other two. All motion was in a single plane. The result, as is evident from the chart, is that the two smaller stars formed a bound binary pair, and expelled the third star. The pair runs off to the left in orbit around each other, and the single star moves the other way. The system is no longer bound: it breaks into two pieces.

Figure 13.5. Expulsion of a star from a three-body system. The result of the simulation described in the text is that two stars form a close binary and expel the third. The starting positions are marked "0", and their positions after 80 days are marked "80". The axes are calibrated in astronomical units (AU).

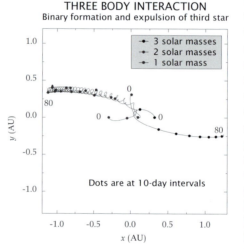

THREE BODY INTERACTION
Binary formation and expulsion of third star

Dots are at 10-day intervals

How can it happen that a system that is initially bound later becomes unbound? The answer is in gravitational energy. When two stars become more tightly bound, they release energy, just as a particle falling onto a star releases energy. This energy must go somewhere, and it goes into the motion of the third star. This is not by some magic. If the third star were not present, the first two could not form a tight binary pair: they would fall towards one another and then recede to the same distance. The forces exerted by the third star allow the outcome to be different: it acts as the "marriage broker". By Newton's third law, the force exerted on the pair by the third star is exerted back on the third star by the pair: the result in this case is to bind the pair more tightly and expel the third. Under other circumstances, the result could have been a different binary pair, but the remaining single star would still have been less tightly bound, and may have been expelled. I encourage the reader to experiment with different initial conditions, just to get a feeling for the frequency with which stars are expelled from triple systems.

Such events do happen in astronomical systems. We shall meet globular clusters – dense systems of millions of stars – in the next chapter. Black holes tend to settle into their centers, where they occasionally – over millions of years – undergo close three-body encounters. Astronomers speculate that such collisions could lead to the formation of numerous close binary black holes, and in the first decade of the twenty-first century astronomers will be searching for tell-tale gravitational waves from such systems. In the next chapter we will look at the globular clusters that produce these black holes, and at their galaxies, which contain even more massive central black holes that might have formed in similar ways; in Chapter 22 we will learn what these waves of gravity really are.

14

Galaxies:
atoms in the Universe

W̲e are now ready to make another step outwards in our exploration of the Universe: we change from looking at stars to looking at *galaxies*. As we saw in Chapter 9, galaxies are vast collections of stars. Our own Galaxy, the familiar Milky Way, contains about 10^{11} stars. Figure 14.1 on the following page shows photographs of two typical galaxies. Galaxies are held together by the mutual gravitational attraction of all the stars. It is remarkable indeed that Newton's force of gravity, which he devised in order to explain what held the planets in their orbits, turns out to explain just as well what holds the whole Milky Way together.

Galaxies are more than just collections of stars. The collective gravity of all the stars makes the centers of galaxies very unusual places. Stars and gas crowd together so densely that in some cases they can form immense black holes, with masses of millions or even billions of stars. These black holes then become the sites of intense activity: as gas falls towards them and heats up, it can shine more brightly than the whole remainder of the galaxy. Even more astonishing are the **jets**: two collimated streams of ionized gas, shooting in opposite directions out of the centers of some galaxies at nearly the speed of light, maintaining their intensity and direction for millions of years. We shall study below various ways that this activity shows itself: quasars, giant radio galaxies, Seyfert galaxies.

Galaxies also hide enormous amounts of dark matter. We don't yet know what this dark matter might be, but galaxies provide us with the evidence that it is there: it produces much more gravity than can be explained by the stars that we can see. The problem of the **missing mass** is one of the most intriguing in all of astrophysics.

Galaxies are, of course, very large. When we jumped from the Solar System to the nearest star, we increased the scale of distances by a factor of some 10^5. In this chapter we jump by an even larger factor. The size of a galaxy is already a factor of 10^4 larger than our previous distance-scale: the typical size of a galaxy is tens of kiloparsecs (kpc), compared to the typical distances between stars of 1 pc. And the separations between galaxies are larger by a further factor of 100, up to the megaparsec (Mpc) scale. Despite these enormous distances, the concentrating force of gravity in the centers of galaxies often leads to intense activity that takes place in regions the size of our Solar System.

Galaxies mean another kind of jump for us: a jump in time to modern astronomy. In earlier chapters, when we described the Solar System, we were on ground that would have been familiar to Newton and Galileo. When we studied stars, we dealt with issues that would not have surprised most nineteenth century astronomers. Even the first speculations about black holes and the gravitational deflection of light belong to the eighteenth century. Modern astronomy and astrophysics have made huge advances in our *understanding* of the planets and stars, but the objects themselves were part of the known Universe of the nineteenth century.

In this chapter: we finally reach the basic building blocks of the Universe: galaxies. Galaxies come in many shapes and sizes. They foster the formation of stars and harbor giant black holes in their centers. They contain only some of the mass in the Universe: much more is dark and unidentified. As beacons of light, they allow astronomers to measure how rapidly the Universe is expanding. Their first stages of formation are imprinted on the cosmic microwave background radiation.

▷ The image under the text on this page shows our Milky Way galaxy as seen in the radio waves called microwaves. It was compiled from observations by the COBE satellite. The view shows the full 360° sky around our location, flattened onto a single view. Courtesy NASA Goddard.

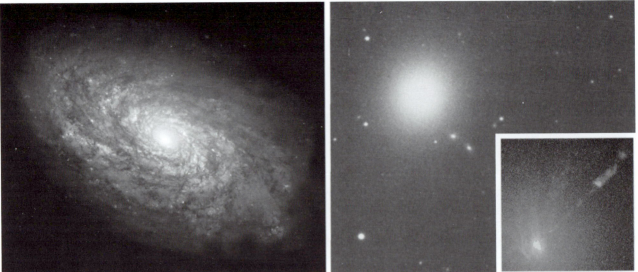

Figure 14.1. *Photographs of two galaxies that are representative of the two main types seen by astronomers. The galaxy on the left is a spiral galaxy known as* NGC4414. *It is very similar in form to the Milky Way. The galaxy on the right is an elliptical galaxy known as* M87. *It is one of the largest galaxies in the Virgo Cluster of galaxies. It has ten times as many stars as the Milky Way. The small fuzzy objects near it are globular clusters in orbit around it. Despite its smooth outer appearance,* M87*'s center contains a jet of gas moving outwards at close to the speed of light. This is illustrated in the inset photo. The jet indicates the presence of a massive black hole at the center. We shall see below that studies of the motion of gas near the center of the galaxy indicate that the black hole has a mass larger than* $10^9 M_\odot$! *All photographs courtesy* NASA/STSCI. *The photos of* NGC4414 *and the jet in* M87 *are* HST *images, courtesy* NASA/STSCI. *The photo of* M87 *is courtesy of the Palomar Observatory and Digitized Sky Survey created by the Space Telescope Science Institute, operated by* AURA, *Inc. for* NASA *and is reproduced here with permission from* AURA/STSCI.

Galaxies, on the other hand, were not recognized as being stellar systems outside the Milky Way until the twentieth century. And the discovery of their incredible activity had to await the opening of new windows on the Universe: radio astronomy, X-ray astronomy, and the use of Earth-orbiting astronomical observatories. The study of galaxies is quintessentially modern.

Globular clusters: minigalaxies within galaxies

In this section: globular clusters are small self-contained star systems in our Galaxy. They may have been building blocks from which the Milky Way was assembled. They are fragile and easily disrupted, but also seem to be rich factories of binary stars and black holes.

It is useful to begin our study of galaxies by looking at globular clusters, which share some of the properties of galaxies on a smaller scale. A representative globular cluster is illustrated in Figure 14.2. They are clusters of typically a million stars, all formed at about the same time, held together by their mutual gravitation.

Although the picture looks crowded, the distances between stars in a cluster are still very much larger than the sizes of stars, so direct collisions are rare. Distant encounters between stars can, however, still transfer small amounts of energy between them, and in globular clusters it generally takes less than a billion years for such encounters to share out the energy of the stars randomly. Scientists call this time-scale the **relaxation time**, and they say that globular clusters are **relaxed**.

This means that the velocities of stars at any point inside the cluster are fairly random, in direction as well as size. The stars form what is called a **collisionless gas**. Unlike the air in a room, where gas molecules collide very frequently, stars in globular clusters collide directly almost never. This is fortunate: gas molecules survive their collisions unharmed, but stars would be completely destroyed!

Globular clusters are prized by astronomers as museums of stars. Because they have held their "collections" intact since they were formed, they are excellent places to test ideas about stellar evolution. Hertzsprung–Russell diagrams (Chapter 12) are particularly interesting for

---

Investigation 14.1. Boiling away a globular cluster

We shall calculate here roughly how much energy is required to "boil away" the stars from a globular cluster. A typical cluster contains $M_{cl} = 10^6$ stars of average mass $1 M_\odot$, in a sphere of radius $R_{cl} = 50$ pc. The escape velocity from such a cluster is

$$v_{escape} = \left(\frac{2GM_{cl}}{R_{cl}} \right)^{1/2} = 13 \text{ km s}^{-1}. \qquad (14.1)$$

The energy a star of mass $m = 1 M_\odot$ needs to escape is then

$$E_{escape} = \tfrac{1}{2} m v_{escape}^2 = 2 \times 10^{38} \text{ J}. \qquad (14.2)$$

To boil off the whole cluster means adding roughly this energy to every star, which requires a total energy of

$$E_{boil} = 10^6 E_{escape} = 2 \times 10^{44} \text{ J}. \qquad (14.3)$$

This may seem like a large amount of energy, but we must compare it to other kinds of energy that may be available, such as the energy released when stars form close binary pairs.

The energy released by a binary pair when the pair is formed is the same as the energy required to split it up again. The calculation is similar to the previous one. If the stars have an orbital separation R, then the escape velocity of star 1 (mass M_1) from star 2 (mass M_2) satisfies $v_{escape}^2 = GM_2/R$. The energy of star 1 when it has this speed is

$$E_{binary} = \tfrac{1}{2} M_2 v_{escape}^2 = GM_1 M_2 / 2R. \qquad (14.4)$$

Suppose the stars each have one solar mass and are separated by the radius of a white dwarf, about 5×10^6 m. This could represent a very close white dwarf pair or a well-separated neutron star binary. Then this evaluates to 3×10^{43} J.

This is already about 10% of the binding energy of the cluster, and it is only one binary system. The formation of a handful of such systems could easily provide enough energy to expand the cluster or even disrupt it. And if a very close pair of neutron stars is formed, with a separation of, say, 100 km (still 10 times the neutron star radius), the energy released would be seven times as much as would be required to boil off all the stars from the cluster!

How would the energy released in this way get transferred to all the stars in the cluster? We saw in Figure 13.5 on page 162 that when a close pair is formed in a three-body encounter, the excess energy goes to the third one. If this happened in a cluster, the third body would have so much energy that it would simply shoot straight out of the cluster and leave the cluster essentially unchanged. But if instead the binary is first formed with a relatively wide orbit, perhaps highly eccentric, then it could shrink by a succession of much smaller energy transfers to other stars in its neighborhood. These stars would not receive enough energy to escape the cluster, and so eventually they would transfer their energy to the cluster stars generally, and the cluster as a whole would change.

Exercise 14.1.1: *Binding energy of a cluster*

Show that E_{boil} can be expressed as

$$E_{boil} = \frac{GM_{cl}^2}{R_{cl}}. \qquad (14.5)$$

In this form it is usually called the *binding energy* of the cluster. This is only an approximation, of course, accurate to a factor of two or so.

---

globular clusters: because all the stars have the same age, they show the relative rate of evolution of stars of different masses. The oldest globular clusters also tell us what the abundance of elements was in the gas that the first generation of stars formed from.

Despite the fact that some globular clusters have been around since the Milky Way formed, they are not robust structures. We show in Investigation 14.1 that they can in fact be disrupted completely by the energy that is released when a single close neutron star or white dwarf binary system is formed inside them, perhaps as a result of a chance three-body encounter like the one simulated in Figure 13.5 on page 162. Given their fragility, it is possible that most of the initial globular clusters of the Milky Way have already been disrupted, either by internal events as in Investigation 14.1 or by the tidal gravitational effects of other clusters or the Galaxy itself.

Globular clusters are, however, not merely museums. They process many of their stars in unusual ways. The most massive stars gradually sink towards the center, giving up energy to lighter stars by gravitational interactions. The most massive objects normally formed in globular clusters are black holes, so over a period of time the centers of globular clusters become rich in black holes. Three-body collisions among such holes (again as in Figure 13.5 on page 162) can form a binary black hole system, and some of these might be observed by gravitational wave detectors, as we shall discuss in Chapter 22.

Figure 14.2. *Photograph of the globular cluster* M3. *Photo courtesy of the Palomar Observatory and Digitized Sky Survey created by the Space Telescope Science Institute, operated by* AURA, *Inc. for* NASA *and is reproduced here with permission from* AURA/STSCI.

Describing galaxies

Like virtually everything else in astronomy, galaxies come in a wide variety of shapes and sizes. Figure 14.1 illustrates the two main types, spiral galaxies and **elliptical galaxies**. Spirals have a central bulge surrounded by a wide, thin disk.

In this section: most galaxies come in one of two basic shapes, spiral and elliptical. Ellipticals may result from mergers of spirals.

The bulge has a "hot" stellar distribution, in the sense that the random kinetic energy of a typical star is comparable to the kinetic energy it needs to orbit about the center of the galaxy. The disk is cold, since the stars have small random velocities compared to their orbital speeds. Besides its stars, the disk generally has lots of gas and dust.

Spiral galaxies take their name from the striking spiral patterns seen in many of them. It appears that the distribution of stars is actually much more symmetrical about the central axis than these patterns might suggest. The patterns instead trace out the places where new stars are being formed.

This is an interesting illustration of what astronomers call a **selection effect**. We will see below that the size and mass of a typical spiral are such that the orbital period of a star in the disk about the center of the galaxy is of the order of 10^8 years. But we saw in Chapter 12 that massive stars live only about 10^6 years. These massive stars form very bright giants at the end of their lives, which stand out much more in a photograph. An observation limited to a certain minimum brightness will inevitably *select* a much larger fraction of the bright stars than of the ordinary ones. So the spiral features are the locations of the brightest stars, not the vastly more numerous ordinary ones. And since these stars live only a short time, we are seeing them where they form.

Scientists understand some aspects of why the star-forming regions of many galaxies have such a regular spiral shape. They generally agree that this has to do with a **density wave** of some sort, which moves through the disk and compresses molecular clouds, triggering star formation. But how the wave is maintained, and what the triggering mechanism is, are still matters of debate. It should also be remarked that many galaxies do not have spiral star-forming regions: these regions can be much more irregular. **Irregular galaxies** are the third broad classification of galaxy appearance.

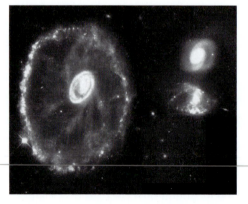

Figure 14.3. A head-on collision between two galaxies. The galaxy on the left was a spiral galaxy before one of the two galaxies on the right passed directly through its center, sending a shock wave outwards through the gas of the galaxy. As the wave travels out, it triggers star formation, resulting in a ring of bright young stars. The ring gives the galaxy its name: the Cartwheel Galaxy. Photograph by the HST *courtesy* NASA/STSCI.

Elliptical galaxies, by contrast, seem to be virtually free of gas and dust. They look more like globular clusters or the central bulges of spirals. Ellipticals can be much bigger than spirals. In Figure 14.1 on page 164, NGC4414 has a mass of about $10^{11} M_\odot$, while M87 has a mass ten times larger. Moreover, such giant ellipticals often seem to be the places where galactic activity prefers to occur. Quasars (see below) and giant radio galaxies tend to be ellipticals.

For a long time, the contrast between spirals and ellipticals was very puzzling to astronomers. Why should galaxies have formed in two such different ways? How could one explain how a relatively "clean" elliptical galaxy could harbor in its center a massive black hole, and how could it "feed" the hole in order to keep the activity going?

These questions are still open and much debated among astronomers, but the outline of a solution is emerging. The key is galaxy collisions. We see many examples of spiral galaxies colliding with one another, either head-on (Figure 14.3) or in a near miss (Figure 14.4). Computer simulations of what happens when two spiral galaxies collide have been able to reproduce observed galaxies, such as those in the photographs, with remarkable fidelity. The photographs show galaxies that have collided but have not (yet) merged. However, in a certain fraction of cases, galaxies merge completely. It is now believed that such mergers result in the formation of giant elliptical galaxies.

The tidal gravitational forces associated with the collision compress the gas clouds of the galaxies, triggering star formation on a huge scale, in the way we saw simulated in Figure 12.2 on page 138. Many of these stars evolve rapidly to the supernova stage, and if there are enough supernova explosions in a short time, the remaining gas and dust will simply be blown away. Because the collision usually involves galaxies whose spiral disks were not in the same plane, the final galaxy will not have a disk-like shape any more. It will simply be a highly disturbed collection of stars that eventually relaxes to an elliptical shape.

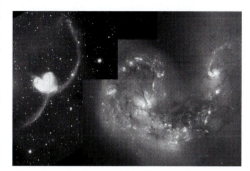

Figure 14.4. *A collision between two galaxies that completely changes their appearance. The image is a composite of a ground-based photo (left) and a high-resolution image of the central region taken by the Hubble Space Telescope (right). The galaxies, called* NGC4038 *and* NGC4039 *, were originally well-separated spirals. Tidal gravitational forces have expelled long streams of gas and stars and stimulated a huge burst of star formation, especially at the join between the two. Photomontage courtesy* NASA/STSCI.

It may be that in the center of the merged galaxy, the stronger gravity and the initial availability of gas and dust can lead to the formation of a black hole. Or it may be that the original spirals already have black holes in their centers – there is a growing body of evidence that most spirals do, including our own. Again, numerical calculations show that the two black holes can spiral into the center of the new galaxy and even merge together in a relatively short time, perhaps 10^9 years. Issues like this are at the heart of current research into active galaxies.

Galaxies are speeding apart

We can't learn much about galaxies until we establish their distances. Astronomers in the nineteenth century had observed galaxies, but most assumed that they were at the same distance as the ordinary stars of the Milky Way. This made them rather small and insignificant in the grand plan of things. But the American astronomer Edwin P Hubble (1889–1953) changed all that when he determined in 1929 that galaxies were well outside the Milky Way, and were in fact completely separate stellar systems.

Hubble showed more than that, in fact. It was already known from observations of the spectral lines of galaxies that they were receding from us at various speeds. This might, of course, have indicated that they were small systems expelled from the Milky Way by some mechanism, and this is what a large number of astronomers initially believed. But once Hubble had distances, he discovered that the speed of recession of a galaxy was always proportional to its distance from us. We write this *Hubble law* in the following way:

$$v = Hd, \qquad (14.6)$$

where v is the recession speed, d is the distance to the galaxy, and H is a constant of proportionality that we now call the **Hubble constant**. Its value, in units that make sense to astronomers, is about 70 km s^{-1} Mpc^{-1}. In conventional units, this is about 7×10^{-15} s^{-1}. The accuracy to which it is known is better than 10%.

The Hubble expansion law, plus the large distances involved, demolished the ejection theory and showed that the Universe as a whole is expanding. We shall explore this implication beginning in Chapter 24.

In this section: Hubble discovered that distant galaxies are moving away at speeds proportional to their distances from us. This indicates that the entire Universe is expanding.

▷Astronomers initially called galaxies "nebulae" because they did not know that they were composed of individual stars. Like true nebulae (clouds of gas around stars), galaxies appeared to be just smudges on the sky.

Measuring the Universe: the distances between galaxies

In this section: exploring the Universe depends on knowing the distances to galaxies. Astronomers use a complex hierarchy of distance measures. Only in recent years have they been able to measure the scale of the Universe accurately.

▷Of course, galaxies also have random velocities on top of the systematic expansion of the Universe. If the galaxy is too near, these will dominate v and make the Hubble method unreliable. In any distance determination, allowance has to be made for this uncertainty.

Figure 14.5. Edwin Hubble's patient measurements on hundreds of galaxies proved first that they were outside the Milky Way, and second that the Universe was expanding and consequently of finite age. Few astronomers have had as a profound an influence on human thought as he. The first Space Telescope was fittingly named after him (HST). Reproduced with permission of AIP Emilio Segrè Archive.

▷When astronomers refer to "the Galaxy" instead of just "the galaxy" they mean our own galaxy, the Milky Way!

▷Galaxy and nebula names come from catalog names. The brightest are in the list compiled by the French astronomer Charles Messier (1730–1817); these names begin with M. Many more are listed in the New General Catalog, compiled in Ireland by the Danish astronomer J L E Dryer (1852–1926); their names begin with NGC.

The Hubble Law gives an excellent way of measuring the distance to a galaxy, provided one knows the value of H. The spectrum of light from a galaxy reveals its redshift, from which one deduces its velocity v. Then one just divides v by H to get d. But the central problem of cosmology in the last half of the twentieth century was to determine the value of H. Only since about 1990 have astronomers begun to agree on its value.

Astronomers try to determine H by measuring the distances to some galaxies independently of the Hubble method, and then measuring v to determine H. We shall see below that getting reliable distances to enough galaxies is a very difficult job, and the astronomers' best-guess value of H has changed many times because of this. It is ironic that Hubble's own distances – and hence his own value for H – were systematically wrong. They were the best that could have been done at that time, but the distances came out a factor of five or ten smaller than our present estimates.

The reason for the difficulty is the complexity of the chain of argument that leads to the distances. We described some of the steps in this chain in Chapter 9. Each step requires a standard candle, a class of objects whose intrinsic brightness is known, so that their apparent brightness can be used to measure their distance. An important class of *variable stars* called Cepheid variables can be seen in nearby galaxies, and the orbiting Hubble Space Telescope (HST) has extended observations of them to more distant galaxies. These are useful because their intrinsic luminosity is correlated with their period of variability, so that by timing the regular variations in the star's brightness an astronomer can deduce its luminosity. Cepheids in turn can be used to calibrate other standard candles, such as supernovae of Type Ia, and certain kinds of ionized gaseous regions around hot stars ("HII regions"). As we mentioned in Chapter 12, Type Ia supernovae have become particularly important in recent years because they can be detected so far away that they not only can be used to determine the expansion rate H of the Universe, but also its rate of change, called the acceleration of the Universe. We will see in Chapter 24 and subsequent chapters that these measurements have given the completely unexpected result that the Universe is actually accelerating its expansion.

The size of the Galaxy is much better determined: the Sun orbits the center at a radius of some 8 kpc. The mass of the Galaxy is not so well determined, but it is about $10^{11} M_\odot$. Near the Galaxy are several small satellite galaxies, the most prominent of which are the Magellanic Clouds, visible from the Southern Hemisphere. The nearest large galaxy is M31, seen in Figure 9.1 on page 104, which is somewhat larger than the Milky Way. M31 has a prominent satellite elliptical galaxy called M32, whose mass is about $10^{10} M_\odot$. M32 is the nearest elliptical galaxy to us, so it has been intensively studied. Despite M32's relatively small size, observations by the Hubble Space Telescope indicate that it probably has a sizeable black hole in its center. M31 and M32 are about 0.5 Mpc from the Milky Way, and falling toward it.

Masses of galaxies and their luminosities are hard to determine. Only in the last two decades have astronomers had instruments, usually in space, that could observe galaxies at most wavelengths, finally getting an estimate of their total emission. The way astronomers measure the masses of the Sun, of planets, and of stars is to monitor objects in orbit about them, and infer their masses from Newton's law of gravity. This will not usually work for galaxies, because orbital times are too long: we can't wait 10^8 years to see what our exact orbital period around the center of the Galaxy is! Of course, if we could measure our acceleration towards the galactic center accurately enough, we could measure the mass of the Galaxy quickly. Only one such measurement has ever been performed, using what astronomers call the

Hulse–Taylor binary pulsar system, whose catalog name is PSR1913+16. We will study this very important system in Chapter 22. For the most part, astronomers use approximate measures of the gravitational accelerations produced by galaxies to measure their masses. We will discuss these when we consider the problem of missing mass in the next section.

When all else fails, astronomers estimate the mass of a galaxy from its brightness. Using a rule of thumb called the **mass-to-light ratio**, astronomers multiply the luminosity of a galaxy by a rather uncertain number to get its mass. The rule of thumb takes account of the fact that much of the mass of the galaxy does not radiate light. Astronomers use values of M/L between 10 and 40 solar masses per solar luminosity. But again, this number is uncertain because of the uncertainties in mass and luminosity of all galaxies.

Most of the Universe is missing!

Given that it is not possible to follow the orbit of a star or satellite galaxy around a galaxy whose mass we wish to measure, and that we cannot determine the instantaneous acceleration of the object, how are we to estimate the galaxy's mass? Gravity is the only reliable way to do it, but we need to make additional assumptions.

Within galaxies, the usual assumption is that stars follow circular orbits. If a star has a speed V in a circular orbit of radius R, then its acceleration towards the center of the circle is (see Investigation 3.1 on page 22) V^2/R. Equating this to the gravitational acceleration of the galaxy (assuming all its mass to be concentrated at its center, at least as a first approximation), GM/R^2, gives an expression for the mass of the galaxy:

$$M = \frac{V^2 R}{G}. \tag{14.7}$$

Let us try to measure the mass of our Galaxy this way. Astronomers know that the distance R to the galactic center is 8 kpc. But we don't know our orbital speed directly, again because the motion is too slow for us to watch it. Instead, we use a more indirect method. Stars slightly nearer the center should be going faster, because Equation 14.7 says that $V^2 R$ is a constant. By measuring the difference in speed between stars slightly nearer the center than the Sun and those slightly further away, it is possible to determine V itself from Equation 14.7.

Actually, the situation isn't quite so straightforward. For one thing, stars have random motions on top of their orbital motion, so one has to average over a suitable sample of stars in the two different positions. For another, one cannot completely neglect the fact that the mass of the Galaxy is not concentrated at the center: there is mass between the two positions that affects the difference in orbital speeds. The first astronomer to work out how to make a good estimate of the mass of the Galaxy this way was J Oort, whom we met in Chapter 6.

While this method is probably very good, it does rely on the untested assumption of circular motion. If for some reason the orbits of stars near the Sun all follow a single ellipse, on average, then the mass we estimate for the Galaxy will be systematically wrong.

When we apply this circular-orbit assumption to external galaxies, it sometimes is easier to use, and it gives us our first indication of missing mass. If a spiral galaxy is nearly edge-on, then its circular orbits will be moving directly towards us or away from us in certain places. These velocities produce Doppler shifts in spectral lines, and so by measuring these shifts one can deduce the speed of matter in the galaxy in different places. If we average the speeds measured at, say, 5 kpc to one side and to the other of the center, we should obtain the overall velocity of the galaxy away from (or towards) us. If we take the difference of these speeds, we should obtain

In this section: the central problem of understanding galaxies today is that there appears to be much more dark matter than luminous. This is not concentrated in regions where galaxies emit light, but is spread outside and between them.

2*V*. Knowing the distance to and size of the galaxy (not always easy!) gives us the orbital radius *R*, and hence the mass *M*.

If the gas or stars whose motion we measure in this way are far enough out from the center of the galaxy, then one would expect that the mass inside the orbit would be fairly constant. Then, by Equation 14.7, one would expect to see V^2R constant, or $V \propto R^{-1/2}$. What we actually see in almost every case where measurements can be made far from the galactic center is illustrated in Figure 14.6: *V* stays relatively constant or even increases as *R* increases.

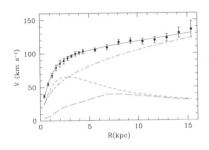

Figure 14.6. The orbital speed of gas in the spiral galaxy M33. If the mass of the galaxy were all in its center, then we would expect V to decrease. Instead, it is slowly increasing as far out as it can be measured. The two lower dashed lines are contributions to the velocity from the visible galaxy. The upper dashed line is the deficit that must be made up by invisible mass. Figure from E. Corbelli and P. Salucci, Mon. Not. Roy. astr. Soc., 311, 441 (2000), with permission of the authors.

What are we to make of the **rotation curve** shown in this figure? If *V* is constant, then Equation 14.7 on the previous page tells us that the mass of the galaxy inside a distance *R* is proportional to *R*. Now, the rotation curves obtained in this way use very weak radio waves from neutral hydrogen gas orbiting the galaxy. This is sometimes detectable two or three times as far from the center as any visible light from the galaxy, in other words in regions well outside the photographic image of the galaxy. They tell us that the mass is still increasing: there is a huge amount of dark matter out there, perhaps two or three times as much mass as one would infer from the photographic image.

What is this missing, or dark, matter? No one yet knows, despite years of investigation. We will return to this question after the next section, once we have seen that the spaces between galaxies hide even more missing mass than the galaxies themselves do.

Gangs of galaxies

The ever-attractive nature of gravity makes it inevitable that galaxies are not spread out uniformly through the Universe. Instead they tend to group together in what are called clusters of galaxies. Our own Local Group is a small, loose cluster consisting of 3 spiral galaxies (Andromeda, the Milky Way, and a smaller spiral called M33, illustrated in Figure 14.6), several minor galaxies, and many satellite galaxies such as M32 and the Magellanic Clouds.

In this section: galaxies often come in groups called clusters, with hundreds or thousands of members. These clusters provide additional evidence for missing matter, and they also give clues to how galaxies were formed in the very early Universe.

There are many clusters that are much more populous, having 100 to 1000 members. The nearest big cluster of galaxies is the Virgo Cluster, at a distance of about 18 Mpc, containing a few hundred galaxies. Figure 14.7 is a photograph of this cluster. The elliptical galaxy M87 shown in Figure 14.1 on page 164 is a giant elliptical in the center of the Virgo Cluster: it may well have been formed by the merger of two or more spirals that were brought to the center by the collective gravitational force of the whole cluster. Clusters also group into **superclusters**; we will discuss these in Chapter 25.

Figure 14.7. The central portion of the Virgo Cluster of galaxies. The elliptical galaxy M87 illustrated in Figure 14.1 on page 164 is at the center of this picture. Several hundred galaxies of various sizes may belong to this moderately rich cluster. Use of this image is courtesy of the Palomar Observatory and Digitized Sky Survey created by the Space Telescope Science Institute, operated by AURA, Inc. for NASA and is reproduced here with permission from AURA/STSCI.

When we try to estimate the masses of clusters, we find further evidence of even more missing mass. The mass of a cluster has to be inferred from its gravity, and there are several ways to do this. We shall consider three of them.

One way is to estimate the mass from the observed velocities of cluster galaxies. This is called the **virial method**, and it is a generalization of simple ideas we have seen before. A planet in a circular orbit around a central star has a kinetic energy that is exactly one-half of the energy it needs to escape from the star. The same factor of one-half applies in more complex systems involving many orbiting bodies, provided that the cluster of galaxies is relaxed, in the sense we defined for globular clusters above. In a cluster of galaxies, no single galaxy has a nice circular orbit about the center, but the *average* energy of all the galaxies at some position will be the same as the energy of a circular orbit there. And the total kinetic energy of all the galaxies will be just half of the "escape energy", which for a cluster is the energy required to break it up, its binding energy. This is the energy we looked at in Investigation 14.1 on page 165.

To use this, astronomers try to measure the kinetic energies of all the galaxies and then equate the total to half the binding energy. The binding energy depends somewhat on the distribution of galaxies (how concentrated towards the center they are), but it is proportional to GM^2/R, as in Investigation 14.1 on page 165. In principle, by measuring the distribution and velocities of galaxies and determining the distance to the cluster and hence its radius R, one can infer M. There are many uncertainties in doing this, not the least of which is the fact that it is hard to be sure that any given galaxy is a member of a cluster. On top of this, many clusters are not really relaxed. Nevertheless, the method indicates that the true mass of a cluster can be up to 20 times the visible mass of the galaxies. Interestingly, the first application of this method, and therefore the first demonstration of the missing mass, was by the Swiss physicist and astronomer Fritz Zwicky (1898–1974) in 1933. Since even the understanding that galaxies were external systems were only four years old at that time, astronomers did not know what to make of this result, and simply ignored it for decades! We will learn more about Zwicky's far-sighted research in Chapter 20 and Chapter 23.

Our second way of estimating cluster mass is another form of the virial argument, made possible by X-ray observations. Somewhat to the surprise of most astronomers, X-ray telescopes have detected strong X-ray emission from the spaces *between* galaxies in most dense clusters. This indicates that there is a hot gas in this space. This gas seems, from the observations, to be fairly smoothly distributed, and so it is almost certainly relaxed. The gas allows astronomers to estimate the mass of the cluster. Doing it crudely, we just equate half the mean kinetic energy $3kT/2$ of an atom of the gas (which is mainly hydrogen) to its gravitational potential energy, $GM_{cl}m_p/R_{cl}$. A rich cluster might contain 500 galaxies in a region of radius 3 Mpc, and its gas might emit X-rays with an energy that reveals that the gas temperature is 50 million degrees kelvin. Simple arithmetic allows us to solve for M, which comes out near to $10^{15}M_\odot$, about 20 times the visible mass of the 500 galaxies. This is consistent with the numbers found by the virial method on galaxy speeds. Note that this is an estimate of the total mass of the cluster, not just the mass of the gas generating the X-rays.

The third method, which is unrelated to virial or dynamical estimates, is to use gravitational lensing, which we will study in Chapter 23. According to general relativity, the direction light moves is deflected when it passes any gravitating body, and the consequences of this are often easily seen in astronomical photographs. When light from a distant quasar or galaxy passes by a galaxy cluster on its way to us, then astronomers can estimate the mass that the cluster must have in order to produce

Figure 14.8. *Fritz Zwicky was one of the most creative, but also one of the most idosyncratic, astronomers of the mid-twentieth century. Trained as a theoretical physicist, he moved to the California Insitute of Technology in 1925 and remained there until his retirement in 1968. Observing with some of the world's most powerful telescopes, he used his physics background to make deeply perceptive interpretations of his data. He was the first to recognize the problem of the missing mass, the first to suggest that supernovae produced neutron stars, the first to face the likelihood that black holes could form, and the first to suggest that gravitational lensing would be observable. All of these ideas became mainstream astronomy, but in most cases only after his death. Perhaps because his contributions were so often ignored or undervalued by his contemporaries, Zwicky developed a reputation of being short-tempered, impatient, and critical toward other astronomers, although to students he was frequently welcoming and encouraging. He also developed a theory of mental processes called morphology, which has a small following today. The photograph shows Zwicky in 1970 (reproduced courtesy of the Fritz-Zwicky-Stiftung, Glarus, Switzerland).*

the observed deflection. This usually leads to estimates a little larger than those of the virial and X-ray methods. Interestingly, this method is beginning to be used to detect dark clusters, regions that produce gravitational lensing but contain no visible galaxies or X-ray emission at all!

Some of the mass of clusters that is not in the galaxies is the X-ray emitting gas itself. This gas is presumably left over from the time the galaxies formed, so it might be thought to be "primordial", i.e. not processed through stars. But studies of the spectra of X-rays from the gas indicate that it contains many elements that can only have been formed in stars, such as iron. This indicates that the dynamics of clusters is complicated, and that gas moves into and out of galaxies over time.

The intensity of the X-ray emission allows one to estimate how much gas there is. Compared to the mass in the galaxies, the cluster gas dominates: there is perhaps five times as much mass in the diffuse gas as in the galaxies. But this is not enough to account for the total mass of the clusters: the X-ray gas amounts to only about 25% of the total mass of a typical cluster. The rest is dark, emitting no visible light, no X-rays, no infrared light, no radio waves.

The missing mass

In this section: something must be between galaxies that is providing the background gravity but is not participating in the gas dynamics that leads to the emission of light, X-rays, radio waves, or infrared radiation. This is the dark matter. Its identity is not known. The puzzle of dark matter is one of the most important in all of astronomy and physics.

The missing cluster mass may be the same stuff as is missing in galaxies themselves; there is more of it because the spaces between galaxies are so much larger. What form the dark matter takes is as uncertain for clusters as it is for individual galaxies. Possible explanations fall into two groups.

First, it could consist of objects of astronomical size, such as very small stars or large black holes. The mini-stars are called **brown dwarfs**. They are not massive enough to ignite nuclear reactions, so they just quietly contract, shining weakly by radiating their gravitational potential energy away. We know that such a star must be dim, since we showed in Investigation 11.1 on page 123 that gravitational potential energy would not sustain the Sun's luminosity for more than about 0.1% of its lifetime. Therefore, the average luminosity of a brown dwarf over several billion years must be less than $0.001L_\odot$. Such stars are not easy to detect.

Black holes in the dark matter might be left over from a hypothetical early burst of star formation and collapse that occurred as the galaxy formed. Many scientists believe that the first generation of stars, made of pure hydrogen and helium, would contain more massive stars than formed later, and many of these might have formed black holes.

These two populations would be difficult to observe directly, but might show up in studies of gravitational lensing within the Galaxy. Teams of astronomers are systematically monitoring millions of stars in the hope of finding chance moments when one of these hypothetical dark objects passes close enough to the line-of-sight to one of the stars to magnify its intensity briefly. The good news is that these stud-

▷MACHO stands for MAssive Compact Halo Object.

ies have uncovered a large population of dark stars with a mass apparently between 0.1 and $0.5M_\odot$, which are called MACHOs. But the bad news is that they are not plentiful enough to account for most of the missing mass of our own Galaxy, let alone that of most clusters.

A second possibility, and the one that is favored by most physicists, is that the missing mass could consist of a smooth distribution of some kind of elementary particle. If so, the particles cannot carry electric charge, for otherwise in the X-ray emitting gas in clusters the "dark" particles would collide with the ordinary gas and emit X-rays themselves, creating more intense emission than is observed. So the particles must be electrically neutral. The problem is that physicists don't know of any neutral particles that have enough mass to provide the extra gravity and are stable against radioactive decay. A free neutron, for example, decays into a proton, electron, and neutrino in only 11 minutes. Dark matter has to last for more than 10 billion years! Neutrinos have mass and, collectively, are stable, but their masses are too small. Not only would the dark matter require more neutrinos than physicists believe could ever have been produced, even in the early Universe, but considerations of how galaxies formed in the first place require particles that have at least hundreds of times the neutrino mass.

Studies of galaxy formation lend considerable support to the dark matter hypothesis, despite the fact that the dark particles are not known. We will come back to this issue below, and in much more detail in Chapter 25, where we can study the formation of galaxies in the context of the overall expansion of the Universe. For now, we will just note that most physicists believe that galaxies did not form spontaneously, but rather that they needed "seeds": strong centers of gravity to start pulling the gas in until it reached a high enough density to make the gas collapse inwards and begin forming large numbers of stars.

The seeds could have come from these dark matter particles, provided the particles had high enough mass and weak enough interaction with ordinary matter (with the protons and electrons) to cool off rapidly as the Universe expanded, to clump, and to pull in the ordinary gas. This hypothesis, called "cold dark matter", has been studied in numerical simulations on large supercomputers, and it seems to produce galaxies with the size, number, and distribution that astronomers observe. So the cold dark matter (CDM) hypothesis neatly explains both the missing mass and the formation of galaxies.

The elementary particles in this picture are sometimes called WIMPs. While they are not predicted by the standard theories of particle physics, they could plausibly emerge from unified theories of the nuclear, weak, and electromagnetic forces. All that is required now to turn CDM from theory into fact is to observe WIMPs directly. Millions of them must pass through us every second. Like neutrinos, they are very elusive and almost never collide with atoms of our bodies. Experimental searches for WIMPs are now underway; if enough exist to account for the missing mass, they ought to be detected and identified in the near future.

▷ WIMP stands for weakly interacting massive particle.

There are other possible seeds for galaxy formation. Among the best-studied alternatives are **cosmic strings**. These are long, incredibly thin concentrations of mass that grow in the early Universe in some theories of high-energy physics. They could also provide sites where galaxies form. Simulations of galaxy formation do not show such good agreement with the distribution of galaxies that we see today, and cosmic strings would not easily provide the missing mass within galaxies, since they are typically much longer than the spaces between galaxies. For these reasons, most astronomers today favor WIMPs over cosmic strings.

▷ The evidence at present favors the WIMPs over the MACHOs!

Although it is rather embarrassing for astrophysicists to have to admit that they do not know what most of the mass in the Universe is, the missing mass problem is one of the crown jewels of modern astronomy. Much of astrophysics deals with the application of known laws of physics to try to model and understand astronomical phenomena. But occasionally astronomy offers the only way for a new discovery to be made in fundamental physics. Newton used astronomy to determine his law of gravitation. Studies of solar neutrinos have revealed the striking phenomenon of neutrino oscillations. Gravity, by telling us what the masses of galaxies really are, appears now to be pointing the way to further new physics. The new particles do not fit into the known theory of the nuclear forces. They would be indicators of physics outside of this theory, and most physicists would expect them to be vital evidence for the theory that will unify all of the forces of physics. We will return to this issue in Chapter 27.

If there is new physics to be discovered here, it would not have been found except for the painstaking study of the dynamics of galaxies. We will see a further modern example of how astronomy can be used to discover new physical laws when we discuss the way elements were formed in the Big Bang (Chapter 25).

Radio galaxies: the monster is a giant black hole

In this section: radio galaxies emit radio waves from regions far outside the visible galaxy, powered by jets of gas emitted from their central regions. The sources seeem to be giant black holes formed from millions or even billions of solar masses.

When radio telescopes began systematic observations of the heavens in the late 1940s, they soon made completely unexpected discoveries. Astronomers knew that the Sun would emit radio waves, and radio emission from the Milky Way had already been detected. But radio astronomers discovered that many other galaxies were intense sources of radio waves, and that the radio emission was not coming from ordinary stars in these galaxies but from an unknown source associated with the galaxy as a whole.

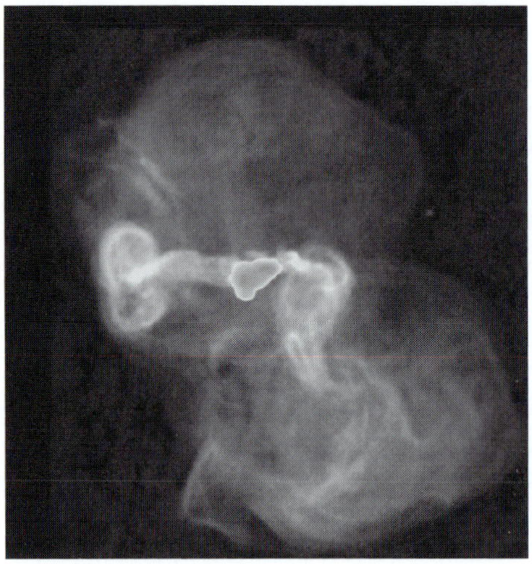

Figure 14.9. Radio emission from M87 is aligned with the jet in the inner region and then spreads out and changes direction. Since the galaxy is more than 10 kpc in radius, or roughly 30 light-years, the large region of emission outside the galaxy indicates that the jet has been active for hundreds of thousands of years. Radio image by Owen, Eilek, and Kassim at the Very Large Array in New Mexico.

Radio emission from galaxies is generally associated with jets of gas streaming outwards from a central black hole at nearly the speed of light. Figure 14.9 shows this for the elliptical galaxy M87, which we have seen in previous illustrations. The radio emission comes from a large region surrounding the galaxy, which is coincident with the brightest part of the radio image. In the inner regions it is aligned with the jet we saw in Figure 14.1 on page 164. Notice that the radio emission goes in both directions from the center, which means that there is probably a jet in both directions, even though only one is visible in Figure 14.1. As the jet leaves the galaxy, the radio emission pattern makes a turn. This could indicate that the jet is running into gas outside the galaxy and is being deflected. The size of the radio lobes indicates that the activity has been taking place for at least hundreds of thousands of years.

These features are absolutely typical of giant radio galaxies. Indeed, M87 is a baby among them: the most luminous ones are thousands of times as bright and ten times the size. Their activity has been going on for millions of years.

What are we to conclude from this? The only mechanism available to a galaxy for maintaining a single direction steady over such a long time is rotation: a rotating disk of gas and/or stars will define an axis of rotation that can normally remain fixed for very long times. Moreover, the dynamical studies of the inner region of M87 indicate that there is a black hole there. Presumably this is not a coincidence.

How does the black hole generate the jets? Where does the energy come from, for example? Nuclear energy is simply not adequate. Consider the numbers: many radio galaxies radiate 10^{38} J s^{-1} in radio waves, which is ten times as much as a typical galaxy radiates in optical light. Yet, as we see in the pictures, the jet originates in a tiny region in the center. No set of nuclear reactions such as we described in Chapter 11 for normal stars could produce this prodigious energy. One has to think of mechanisms for converting the mass of whole stars into energy. The radio luminosity above is the equivalent of converting 1/60th of the mass of the Sun into pure energy every year, using Einstein's famous equation $E = mc^2$, which we will study in the next chapters. And this conversion process must be sustained for millions of years.

Gravity in relativistic situations offers ways of doing this. A particle falling onto a neutron star reaches it with a speed equal to the escape velocity. We saw in Chapter 12 that the escape velocity from a neutron star is about half the speed of light. Its kinetic energy at this velocity, $mv^2/2$, is therefore a good fraction of its total rest mass energy mc^2. This energy is in principle available to any processes near the neutron star that could convert it into power for a jet.

In fact, neutron stars are much too small to act as centers for the jet phenomenon. If, say, 20% of the infalling mass is converted into jet energy, the remaining 80% of the mass has to stay near the neutron star, since to send it back out would take as much energy as the mass released by its falling in. If the jet requires the conversion of 1/60th of a solar mass in energy each year, then 1/15th of a solar mass per year must accumulate on the star. After only something like 15 years, this would push the neutron star over the upper mass limit and convert it into a black hole. Over a few million years, at least $10^5 M_\odot$ will have accumulated in the region where the jet originates. Therefore, the mechanism needs a massive central black hole for its relativistic gravitational field. Astronomers call this massive black hole "the monster".

Other possibilities have been proposed: supermassive relativistic stars, extremely dense clusters of neutron stars, and others. It seems, however, that even if one could somehow form such systems, they would not last long before collapsing to form a massive black hole. The conclusion that the monster is a massive black hole seems inescapable.

Quasars: feeding the monster

The discovery of the enormous luminosity of quasars in 1963 by the Dutch astronomer Maarten Schmidt (b. 1929) was a landmark in the development of modern astronomy. Radio astronomers had identified a class of unusual, intense radio sources that did not seem to be associated with galaxies. Optical observations revealed point-like images at the positions of some of the radio sources, but the images were not like ordinary stars. In particular, their spectra did not look like spectra of stars, and in fact no-one could identify any of the lines. They were called *quasistellar objects*, a name that has evolved into *quasar*, and is frequently abbreviated QSO.

Schmidt decided to see if he could interpret the spectrum of one of these objects, called 3C273, by applying a very large Doppler shift to some standard spectral lines of hydrogen. He found that he could indeed fit the spectrum of 3C273, provided he used a shift corresponding to a recessional velocity of 15% of the speed of light. This was far larger than any velocity that had by then been measured for galaxies. Interpreted as a Hubble velocity, it meant that 3C273 was one of the most distant objects known. Although it looked like a dim star on photographic plates, its great distance meant that it was actually one of the most luminous objects known.

The redshifts of other quasars were soon measured, and a number of features emerged. First, quasars put out much more light than an ordinary galaxy, so it is likely that they are associated with some phenomenon that takes place in the center of some galaxies, like that which produces radio galaxies. Second, most quasars are so far away that the light we see has been traveling to us from them for a good fraction of the time since the Big Bang. The Universe was younger then, and it is natural to conclude that quasars have something to do with the early stages of the formation of their "host" galaxies. Third, quasars were much more numerous in the early Universe than they are now. They were so numerous that a sizeable fraction of galaxies must have had quasars in them at one time, although in our Galaxy's

In this section: quasars also seem to contain black holes, and they give a clue to the source of the energy: gas falling towards the black hole.

▷Initially there was some skepticism that the enormous quasar velocities should be interpreted as part of the Hubble expansion. But very sensitive optical observations have revealed many quasars in clusters of normal-looking galaxies of the same redshift. There can therefore be no doubt about the enormous distances to these objects, and hence their enormous luminosities.

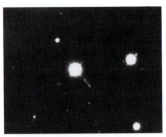

Figure 14.10. *The jet from* QSO *3C273. It comes straight out of the center of the object. The jet illustrates the close similarity between quasars and radio galaxies. Photo courtesy* AURA/NOAO/NSF.

neighborhood they seem to have died out completely.

The key observation that shows that quasars are related to radio galaxies is that they also display jets. In Figure 14.10 we see the jet from the original quasar, 3C273. It comes straight from the heart of the image.

Quasars give us, in addition, some very important information on how big the region emitting their radiation can be. The brightness of quasars is very variable, and some have been known to change their brightness by a factor of two in a few minutes! Whatever the mechanism for changing the luminosity may be, relativity tells us that it cannot involve influences that move faster than light, so the size of the emitting region must be smaller than the distance light can travel during the time that the luminosity changes. In this case, this is a few light-minutes, or less than 1 AU, less than the distance of the Earth from the Sun.

> All the light of a quasar, more than the normal emission from a whole galaxy, originates in a region smaller than our Solar System!

This is consistent with the conclusions we came to for radio galaxies. Even a huge black hole of $10^9 M_\odot$ has a size of $2GM/c^2 = 20$ AU (recall the formula for the size of a black hole, Equation 4.12 on page 36), so there is plenty of room to fit such a hole in the monster's chamber! However, there is not much room for any other kind of object that could produce the light from a quasar.

Once quasars were discovered, it became possible to look for them using optical images, even without radio positions. It has emerged that only about 20% of quasars actually emit detectable radio energy. Moreover, similar objects, on a somewhat smaller scale, have been found in the centers of ordinary galaxies. These *active galactic nuclei* come in a wide variety of forms, and classes of them have special names: Seyfert nuclei and BL LAC objects are two. Because quasars are so bright, it is difficult to see their surrounding galaxy, but a few relatively nearby ones have been detected (including that of 3C273). They are all giant ellipticals. The active galactic nuclei, however, can be present in both spirals and ellipticals.

In fact, it now seems that *most* galactic nuclei show some modest level of activity. This means that by observing nearby galaxies, we have a chance of seeing details of the phenomenon that we would never be able to resolve in distant quasars. Observations of M87 using the Hubble Space Telescope have shown that its jet originates in a region that contains a disk of orbiting gas that is no larger than 20 pc and whose orbital speed is at least 750 km s^{-1}. This speed is 25 times the orbital speed of the Earth around the Sun, yet the orbiting gas is 4 million times further from the center of its orbit than the Earth is. Since the orbital speed of a planet is $v = (GM/R)^{1/2}$, the central mass M is proportional to Rv^2. It follows that the central mass is 2.5×10^9 times as large as the Earth's central mass, which is of course the Sun. For an object of this mass to be smaller than 20 pc in size, it must be a black hole: no method of concentrating this much mass in that small a region could avoid gravitational collapse for long. This is the biggest monster for which astronomers have such conclusive evidence at present.

But we still have the question: what causes the phenomenon? Although the question cannot yet be answered in full, quasars and active galactic nuclei bring the answer closer. Why should the quasar phenomenon have died away with time? How can M87 have such a huge black hole and yet be only modestly active, compared to quasars? And how can active black holes be responsible for the production of jets: don't they trap everything that falls into them?

The final question holds the key. Black holes provide the gravitational attraction that allows matter falling towards them to release such huge amounts of energy, but it must be that the material is stopped, or at least slowed down, before it actually reaches the hole. If infalling material has some rotation, then it will form an accretion disk around the hole, and it will only gradually spiral in. Before it reaches the hole it will have released a lot of energy. When it falls into the hole, it makes the hole rotate.

How the jet is produced is very unclear at present. It might be that the accretion disk is very thick near the hole and only allows a small opening along the rotation axis, through which matter is expelled by the complicated pressure forces in the disk. Alternatively it is also possible that magnetic fields generated by the matter in the disk could interact with a rotating black hole to generate a jet.

Given that the monster has to be fed by gas falling into the accretion disk, the decay in quasar activity with time could be explained by famine: in the original galaxy there are only a certain number of stars that are in orbits that take them close enough to the black hole to be disrupted by its tidal forces and end up in the disk. Once these have been eaten, the hole becomes quiet. When galaxies merge, stellar orbits become disturbed and the quiet holes suddenly have much more to eat again, and activity can start up again.

Galaxy formation: how did it all start? Did it all start?

Quasars are associated with galaxies that are young. Astronomers have been conducting intensive searches to find galaxies at an even earlier stage, when they are forming, presumably by the contraction of a diffuse cloud of gas in the very early Universe. They have found very interesting objects, very distant, very young. Figure 14.11 shows a sample of these objects found by the Hubble Space Telescope. They are unlike any galaxies we see today. They are mere fragments. It seems clear from studies like these, as well as from numerical simulations using supercomputers, that galaxies form from multiple mergings of such fragments.

In this section: astronomers are beginning to probe the time when galaxies formed, and they see them arising from mergers of smaller objects. We have known for a long time that there was a time when galaxies formed. Otherwise the sky would not be dark at night. This is Olbers' Paradox.

Observations of the **cosmic microwave background radiation**, which we will study in Chapter 24, are also beginning to yield information about the mechanisms of galaxy formation (Chapter 25). The initial density irregularities that grew to form galaxies have left imprints on the background radiation, and measurements during the first decade of the twenty-first century should reveal much of the detail of the earliest phase of structure formation in the Universe.

To understand this phase, we need to study cosmology, the history of the Universe in the large. Galaxy formation is but one step in a long chain of events that led to the Universe we observe: normal matter (protons, electrons, neutrons) first took form, hydrogen and helium were made from these building blocks, the dark matter began to clump, hydrogen and helium were

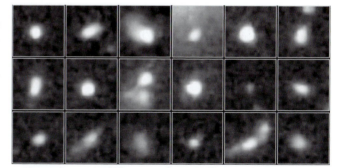

Figure 14.11. *A Hubble Space Telescope collection of images of very distant, very young clumps of stars that are in the process of combining to form galaxies as we know them. Taken from* STSCI PRC96-29B, *images by R Windhorst,* NASA/STSCI.

drawn in and began forming stars, galaxies, and clusters. After that, we have already drawn the outline of the remaining steps: stars made the heavier elements, the Sun formed from a cloud of gas with plenty of heavy elements, the Earth formed from these heavy elements, and (leaving out a few more steps) here we are!

But wait – before going down this road, how can we be sure that there was a time when galaxies were young, when stars were just beginning to form? We saw in Chapter 11 that the Universe cannot be infinitely old, because stars systematically use up the hydrogen and make more and more heavy elements. There is an even more dramatic way of seeing that the Universe had to have a beginning, a way that was already discussed in Newton's day. It is what we now call **Olbers' Paradox**.

> Look up at the night sky on a clear night. It is dark. That is enough to show that the Universe had a beginning!

The argument is as follows. Let us assume a model of the Universe that would have seemed very reasonable to eighteenth century physicists: the Universe is filled with a uniformly distributed collection of stars, and is infinite in extent, and infinitely old. This was, in fact, the model favored by Newton. Then the problem is that the light radiated by stars would also be uniformly distributed in such a Universe. Moreover, since energy is conserved, the light radiated by stars long long ago is still running around the Universe, and since this has been going on for an infinite amount of time, the amount of light at any location in the Universe would be infinite. The night sky would not only be bright, it would fry us!

▷ Put mathematically, the light from any star at a large distance r from us is dimmer than that from a star at the nearer distance R by the ratio $(R/r)^2$, but a spherical shell surrounding us with thickness δr has a larger volume if it is at a distance r than if at R, by the inverse factor $(r/R)^2$. Thus, the number of stars (in such a uniform model) is exactly large enough at r to provide the same amount of light here as the stars at R. Since there are an infinite number of such shells, there is an infinite amount of light.

This argument was put to Newton by the English physician William Stukeley (1687–1765), but Newton ignored it. The German astronomer Wilhelm Olbers (1758–1840) took the issue seriously, and tried to find a way out. He argued that the light might be absorbed by dust or by other stars. But this does not work: it just leads over time to the heating of the absorber, which will then re-radiate the light. Another way out would be to assume that the part of the Universe containing stars is of finite extent, surrounded by empty space, so that the light reaching us is finite as well; the light radiated by stars long ago has left our part of the Universe and is streaming out through the empty space around it. But this is no solution either: a finite Universe must collapse in on itself through Newtonian gravity, and so it would have only a finite lifetime, as well as a finite history. Newton himself apparently believed that God would intervene periodically to stop the Universe collapsing on itself, but he did not postulate a Universe of finite extent.

That no-one before Einstein seriously discussed a Universe of finite age as a resolution of Olbers' Paradox illustrates the fact that, until Einstein, few scientists seriously thought that cosmology was a province for purely scientific investigation. All that has changed, as we will see in Chapter 24. Today, cosmology is the deepest application of gravity to science. But to understand it we need to understand Einstein's gravity, general relativity. So we shall defer our final investigation of galaxies, how they formed, and how they are distributed in the Universe, until after we have learned how Einstein re-formulated Newton's concept of gravity.

15

Physics at speed:
Einstein stands on Galileo's shoulders

We have allowed gravity to take us on a tour of the Universe in the first half of this book. It has taken us from the planet Earth to the rest of the Solar System, then to other stars, and from there to galaxies. Gravity wants to lead us further, because we have not yet come to understand its most profound consequences. These include black holes, which we met briefly in Chapter 4, and the Big Bang, which is the beginning of time itself.

These matters require *strong gravity*: gravity that is strong enough to trap light in a black hole or to arrest the expansion of the entire Universe. Studying strong gravity takes us beyond the limits where we can trust Newton's theory of gravity and his laws of motion.

The reason is speed: if gravity is strong, then speeds get large. If we shrink a star until it is compact enough to turn it into a black hole, then the escape speed from its surface gravitational pull will exceed the speed of light; all other speeds near it, such as the speed of a nearby orbiting planet, will be close to the speed of light. But when we try to understand phenomena that involve speeds close to that of light, we need relativity. We need to improve our theory of gravity to make it a *relativistic* theory.

Einstein showed physicists and astronomers how to do that. He did it in two steps. First he showed how physics gets modified when things move at close to the speed of light; this came to be called the special theory of relativity, or special relativity. In this work, he basically brought Galileo's principle of relativity (recall Chapter 1) up to date. The second step came ten years later, when he brought Newton's gravity up to date with relativity. Einstein's relativistic theory of gravity came to be called the general theory of relativity, or general relativity.

"If I have seen farther," Newton once said of his relationship to scientists of an earlier age, "it is by standing on the shoulders of Giants." The same could be said of Einstein: his special relativity rests squarely on Galileo's shoulders, and he made his general theory of relativity as close to Newton's theory as he could.

Fast motion means relativity

Unfortunately, special and general relativity are probably the worst-named theories of modern physics. Their names convey little meaning, and this sometimes causes confusion right from the start. Here are thumbnail definitions of what the theories are really about.

Special relativity is Einstein's description of how some of the basic measurable quantities of physics – time, distance, mass, energy – depend on the speed of the measuring apparatus relative to the object being studied. It shows how they must change in order to guarantee that Galileo's principle of relativity (that the laws of physics should be the same for

In this chapter: we embark on relativity. We present the fundamental ideas of special relativity. Einstein based it partly on Galileo's relativity, partly on a new principle about the speed of light. We discover the main consequences of the theory, which we require for the development of general relativity in the rest of this book.

▷ The picture behind the text on this page represents the products of the head-on collision of two protons in the CDF experiment at Fermilab, a major particle-physics accelerator laboratory in Illinois. Hundreds of sub-atomic particles are produced in each collision. Such collisions illustrate many of the predictions of special relativity detailed in this chapter. The protons collide at nearly the speed of light, with mass-energy nearly 1000 times larger than their rest-masses. This mass-energy is converted into the rest-masses and energies of the product particles. Accelerator experiments like this test special relativity stringently millions of times per day. This image is captured from the live display of results on the experiment website. See http://www.fnal.gov.

In this section: relativity theory is required when motions can be a significant fraction of the speed of light. Gravitational fields that are strong enough to accelerate bodies to such speeds must be described by a theory of gravity that is compatible with the principles of relativity.

every experimenter, regardless of speed – recall the discussion in Chapter 1) should hold even at speeds near that of light.

Because it deals with the general properties of measurements, special relativity is not really a theory about any particular physical system. Rather, it is a set of general principles that all the other theories of physics have to obey to deal correctly with fast-moving bodies. All the theories of physical phenomena – for example, mechanics (the theory of forces and motion), electromagnetism (the theory of electricity and magnetism), and quantum theory (the theory of the sub-microscopic physics of electrons, protons, and other particles) – have relativistic versions that physicists use when they need to understand a situation where speeds get close to that of light.

Gravity is no exception: it must also follow the principles of special relativity.

General relativity is Einstein's relativistic theory of gravity.

General relativity replaces Newton's theory, which works well for slow motion and is still good for most purposes when describing Solar System orbits and the structures of stars and galaxies. But when gravity is strong enough to accelerate bodies to nearly the speed of light, then we have to turn to general relativity to find out what really happens. In Chapter 19 we will see by explicit calculation that Newtonian gravity cannot predict correctly the gravitational field of a moving body.

In what situations do we require a relativistic theory of gravity? One answer – Einstein's answer – is that we need it everywhere, because it is simply unacceptable to have theories of physics that are inconsistent with one another. In Einstein's day this was the *only* answer, since there were no big observational problems with Newtonian gravity. But today we can see much more of the Universe, and there are places where Newtonian gravity simply will not work satisfactorily. So today there is the astronomer's answer, which is that we need to do better than Newton whenever gravity is strong enough so that speeds within the system being observed are near the speed of light. These speeds could be escape speeds or speeds of random motion of gas particles; in the latter case the pressure becomes large, so that p is of the same order of magnitude as ρc^2.

We saw in Chapter 4 and Chapter 6 that, in Newtonian gravity, the escape speed from a body of mass M and size R is $(2GM/R)^{1/2}$. We can make this large by either increasing M or reducing R, or both. Black holes and neutron stars normally form by reducing R: the nonrelativistic inner core of a star collapses, raising the escape speed until we need relativistic gravity to describe it. Cosmology, the study of the Universe as a whole, needs a relativistic description because it involves a large mass M. If we imagine a region of the Universe with a uniform average mass density ρ, then the mass in a sphere of radius R is $M = 4\pi\rho R^3/3$, so the escape speed $(2GM/R)^{1/2}$ from that region is

▷This ρ is obtained by spreading the mass of stars and galaxies smoothly over the entire region; it is therefore much less than the density within a star.

$$\text{escape speed from region of size } R \text{ and density } \rho = (8\pi G\rho R^2/3)^{1/2}. \qquad (15.1)$$

This increases in proportion to R, so if we take a big enough region of the Universe we are bound to reach a point where the Newtonian escape speed would be c, and therefore the Newtonian description of gravity would fail. It is precisely on this length-scale that we need general relativity to provide us with a consistent model of the Universe.

Before we can run, we must walk; before we can understand the Universe, we must learn relativity. The present chapter opens the door to relativity. It introduces the basics of special relativity, covering all the essential ideas and the few formulas that we will need when we go on to general relativity, black holes, and cosmology.

Special relativity is both fascinating and – let us admit it right away – worrying. It insists that we change the notions of time and length that we have taken for granted all our lives. This insistence fascinates some people who meet the theory for the first time, and it raises resistance in others: "How can that *be*?" is the frequent question. For readers who want to go deeper into the theory, or who really need to find out how it can *be!*, I have developed the themes covered in the last part of the present chapter more fully in the next, Chapter 16. Reading the next chapter is not essential: if all you want to do is to get on to black holes, then you can jump to Chapter 17 at the end of this chapter.

Relativity is special

Although the name "special relativity" is not particularly informative, it does remind us that Einstein built it on the foundations of Galileo's principle of relativity, which we met in Chapter 1. Relativity is a thread that has run through physical thinking for centuries, and Einstein reminded his generation of physicists how important it was. In his paper on special relativity in 1905, he built a revolution in physics on the very traditional foundation of the principle of relativity.

Recall the principle of relativity as we phrased it in Chapter 1:

> *Relativity:* All the laws of physics are just the same to an experimenter who moves with a uniform motion in a straight line as they are to one who remains at rest.

Einstein insisted that this was one of *only two* guiding principles that all theories of physics had to follow. He was firmly in Galileo's tradition here, but it happened that nineteenth-century physicists had by and large begun to doubt that this principle was correct. Einstein rescued the principle of relativity from oblivion.

Einstein's second guiding principle was far from traditional. In fact, it was so radical that even today beginners have difficulty believing that it can be true.

> *Speed of light:* It is a fundamental law of physics that the speed of light has a particular, fixed value, which we usually call *c*. Because this is a law of physics, it follows from the principle of relativity that every experimenter who measures the speed of light will get the *same* value for it. This is true *even if two different experimenters moving relative to each other measure the speed of the* same *beam of light.*

We are used to the speed of any object relative to ourselves changing if we change our own speed. Thus, if I drop a ball while traveling on a train, the ball acquires a small downwards speed before it hits the floor. But someone watching from the platform of a station that my train speeds through will decide that the ball has a large horizontal speed as well, equal to the speed of the train. There is nothing surprising in all this, because we are used to the idea that all speeds are relative. Galileo taught us this.

But Einstein said no: light is different. The speed of light is not relative. If, on the train, I shine a flashlight forwards, then the photons travel away from me at the speed of light *c*, about $3 \times 10^8 \, \mathrm{m \, s^{-1}}$. The person on the station platform ought, if Newton and Galileo were right, to measure the speed of the photons to be larger, equal to *c* plus the speed of the train. But in fact, said Einstein, that person will measure the speed of the *same photons* to be exactly *c* relative to the platform as well. *They do not gain anything from the speed of the train!*

At first this seems impossible, since it seems to conflict with everyday experience. But we must be cautious here, and not try to shape Nature to our own preconceptions. Our experience of speeds and how they change is entirely confined

In this section: Einstein's theory of special relativity is based on two fundamental principles. One is Galileo's relativity principle. The other was introduced by Einstein: the speed of light will be the same to all experimenters, no matter what their state of motion. This radical departure from the way all speeds had previously been assumed to behave gives special relativity all if its surprising and hard-to-accept results. Despite the difficulty we may have in accepting it, it is well verified by experiment. This *is* how light behaves.

▷Here we refer only to the speed of light in vacuum, i.e. to the speed of free photons. When light travels through a material, like glass, it moves more slowly. Although physicists often say that the speed of light in glass is slower than the speed in vacuum, this is shorthand for what is really going on. In glass, what travels is a complicated interaction between the electric and magnetic fields of the atoms. The interaction begins when light is absorbed on one side of the glass pane, and results in light being emitted at the other side, but what is inside the glass is more complicated than just light. The fact that this interaction moves through the glass at a slower speed does not contradict Einstein's second postulate.

to slow speeds: a train going at 100 mph (160 kph) is only moving at $1.5 \times 10^{-7} c$. The ball that falls from my hand in the train reaches the floor at a speed of only $6\,\mathrm{m\,s^{-1}}$, or $2 \times 10^{-8} c$. We have no direct experience of combining large speeds: we might indeed shine light inside a train, but our brains and nerves are not quick enough to sense how fast the light travels down the train, or whether it is slower or faster when viewed from the platform.

Instead of relying on our low-speed experience, Einstein worked out mathematically the consequences of his two guiding principles. He found that two speeds (that of the train and that of the projectile, in our example) combine by a more complicated rule than Galileo's. For small speeds, this rule gives the usual addition of speeds in the way that we expect, to an excellent approximation, so it is consistent with our own experience. But, for large speeds, Einstein's rule tells us that the speeds combine only partially; in the extreme case, when one of the speeds is the speed of light, then two speeds always combine to give the speed of light exactly again, regardless of what the other speed is.

This rule, which we write down in Investigation 15.1, undermines our objection that Einstein's prediction is contrary to our experience. Instead, his prediction is fully consistent with our experience, which deals only with the way speeds combine for small speeds. Only for large speeds, where we have no experience, does the law begin to deviate from our expectations.

Einstein had a good reason for introducing this second guiding principle, even though it seemed to contradict experience. The problem that he was trying to solve had been around since the Scottish physicist James Clerk Maxwell (1831–1879) had shown in the middle of the 1800s that electricity and magnetism are really two special cases of the general force called electromagnetism. By unifying electricity and magnetism in this manner, Maxwell had been able to show that the electromagnetic field should have waves, and he could even make a numerical prediction for the wave speed. This was close enough to the measured speed of light for physicists to realize that Maxwell's equations had explained what light was.

However, this result had a puzzling side, because the laws said that light would travel with this speed *regardless* of the speed of the system that emitted the light. The speed of the system would affect the frequency of the light through the Doppler shift (Chapter 2), but not the wave speed. Light therefore does not behave like a projectile, something thrown out by its source, whose speed depends on the speed of the source. Instead, light behaves like a wave in a medium, having a speed that depends on the medium carrying the wave but not on the speed of the source.

We will see in the next section that most physicists of Maxwell's time interpreted this to mean that there was a material substance that carried light vibrations at a fixed speed, regardless of the speed of the source of the light. They called this medium the ether. But nobody could find any direct evidence for the ether, and Einstein therefore decided to explore the alternative: if the laws of electromagnetism said that the speed of light was always the same number, then maybe we should just accept that as a law of physics itself. Maybe there was not an ether for it to have a speed in; maybe it just had this speed in all circumstances.

Physicists have a shorthand word for Einstein's second guiding principle, that the speed of light is the same to all experimenters. They say that the speed of light is *invariant* under a change in the speed of the apparatus that measures it. The invariance of the speed of light is not something we can decide just by thinking about it. It is a matter for experiment. Only experiment can tell us whether to follow Einstein in his second hypothesis. And experiments on special relativity are effectively performed every day: from nuclear power generators to giant particle

Figure 15.1. *James Clerk Maxwell was one of the giants of nineteenth century physics. By unifying the apparently separate phenomena of magnetism and electicity, Maxwell opened the modern age: telecommunication and power generation technology require the cooperation of magnetism and electicity. On a more subtle level, his electromagnetism was the first unified field theory, and the first of what physicists now call "gauge theories". This set the direction for all of fundamental physics, a direction still followed today. If this wasn't accomplishment enough, he also made fundamental contributions to statistical mechanics, thermodynamics, and the theory of colors. He died at the young age of 48, a few months after Einstein was born. Reprinted with permission of American Institute of Physics (Emilio Segrè Archives).*

▷We met the word invariant in Chapter 6, where we used it to describe situations that do not change with time or position. Here we discuss changes of experimenter.

Investigation 15.1. (Much) Faster than a speeding bullet ...

Consider two experimenters, one at rest and the other moving with speed u in the x-direction. Let the moving experimenter release a projectile moving with speed v *relative to him*, again along the x-axis. What is the speed V of the projectile relative to the experimenter at rest?

To Newton and Galileo, the answer was obvious:

$$V = u + v. \qquad (15.2)$$

The speed of the moving experimenter adds to the speed of the projectile to give the total speed relative to the experimenter who is at rest.

But this is not consistent with the principle that the speed of light is the same to every experimenter. In particular, if $v = c$, then relativity insists that we must get $V = c$, since both speeds are the speed of a photon. Equation 15.2 does not give this result.

In other words, in special relativity speeds don't add. Physicists use the word **compose**: the speeds u and v compose to get V. The Einstein velocity-composition law is

$$V = \frac{u + v}{1 + uv/c^2}. \qquad (15.3)$$

Let us try a few special cases to see what this law implies. Suppose, as before, that $v = c$. Then the numerator is $u + c$ and the

denominator is $1 + u/c = (u + c)/c$. When we divide the fraction, we get $V = c$, just as required. So this law is consistent with the principle that the speed of light is the same to all experimenters.

Another important special case of the Einstein law is when the speeds u and v are both very small compared to c. Then the fraction uv/c^2 in the denominator, which is a pure dimensionless number, is very small, and can be neglected compared with the one in the denominator. Then the Einstein law reduces to the Galilean law, Equation 15.2. This is an important consistency check: the predictions of relativity must reduce to those of Newtonian mechanics when speeds are very small compared to light.

This also explains why our intuition, based on everyday experience, leads us to expect that the Galilean law is right. Our senses and our everyday measuring devices can only deal with very small speeds, and for these the simple addition law is fine. Although we can see light, we can't sense its speed, so we don't have experience with seeing how the speed of light changed when we changed our own state of motion. If we had such sharp senses, then Galileo would have written down the full theory of special relativity from the start! Further properties of Equation 15.3 are explored in the exercises below.

Exercise 15.1.1: *More photon velocities*

Let $v = -c$ in Equation 15.3, corresponding to a photon moving backwards relative to the one we tested above. Show that again $V = -c$: the speed of the photon does not depend on the observer.

Exercise 15.1.2: *Computing the graph*

In Figure 15.2 on the following page we plot the composition law Equation 15.3 for the special case $u = 0.4c$. Compute V/c for the set of values $v = \{0.1c, 0.4c, 0.9c\}$. Compare them with the points plotted on the curve in the right-hand panel of the figure.

Exercise 15.1.3: *How fast is relativistic?*

If both u and v are $0.1c$, what is the fractional error in using the Galilean addition law? [The fractional error is the difference between the Einstein and Galilean results (the error), divided by the Einstein result (the correct answer).] If $u = v = 0.3$, what is the fractional error? Suppose V can be measured to an accuracy of $\pm 5\%$. What is the largest speed (again assuming $u = v$) for which one can use the Galilean formula and make errors too small to be measured?

Exercise 15.1.4: *Zero is still zero*

The Einstein composition law still has some features that we expect from everyday life (and logical consistency). Show that, if the projectile remains at rest with respect to the moving experimenter (so $v = 0$), then its speed relative to the experimenter at rest is the same as the speed of the moving experimenter, $V = u$. Show further that if the moving experimenter shoots the projectile backwards with a speed of $v = -u$, then it will be at rest with respect to the resting experimenter ($V = 0$).

---

accelerators to the GPS navigational satellite system that we mentioned in Chapter 2, many of today's high-technology devices would not function correctly if special relativity were wrong. Special relativity is one of the best-tested aspects of all of fundamental physics.

We should note here that physicists tend to use other words for what we call **experimenters** in this book. When you read other books on relativity you may read about **observers** or **frames**. Observers are the same as experimenters, and we will sometimes use the two terms interchangeably. A frame is the coordinate system used by the observer or experimenter to locate events in space and time, and we will have much to say about coordinates in these chapters on relativity. But we will not use the word "frame" in this book, except in Chapter 24, where we use it to describe a coordinate system spanning the whole Universe, something that is not really easy to envision as a single experimenter or observer. Otherwise we avoid the term "frame", because it is impersonal, and it might lead you to think that the results of special relativity are somehow to do with bad definitions of coordinates. The words "experimenters" and "observers" are more appropriate, because they should make you think of careful scientists who make measurements with the best techniques,

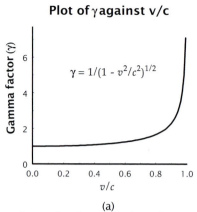
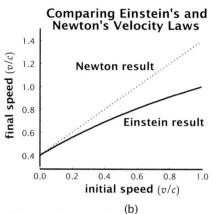

(a)   **(b)**

Figure 15.2. The two figures show how much the predictions of Einstein differ from those of Galileo and Newton. (a) The left-hand panel shows the Einstein γ-factor, $\gamma = 1/(1 - v^2/c^2)^{1/2}$. For a given speed v (shown as a fraction of c on the horizontal scale), the height of the graph gives the factor by which time stretches, masses increase, and lengths contract at that speed. This factor is always bigger than one, and it gets infinitely large as v approaches c. (b) The right-hand panel compares one example of the Einstein law for combining speeds with that which Newton and Galileo would have expected (and which our low-velocity intuition leads us to expect, too). Suppose a rocket ship is moving at the fixed speed 0.4c relative to us, and suppose it fires a particle forwards with speed v relative to the rocket. Then the two curves show the two predictions of the speed of this particle relative to us (vertical scale) as a function of its speed relative to the rocket (horizontal scale). For the Einstein law (solid curve), the final speed never gets bigger than c, and only reaches c when the particle's speed relative to the rocket is c. This means that a photon fired from the rocket will have speed c relative to the rocket and to us, in accordance with one of the founding principles of special relativity. By contrast, the dotted line shows the speed that Newton and Galileo would have expected, namely 0.4c + v, which can exceed c. Simple as this formula is, Nature follows the Einstein law instead.

and who really do measure all the unexpected results of special relativity.

The Michelson-Morley experiment: light presents a puzzle

In this section: the experimental foundations of special relativity go back to the Michelson–Morley experiment. It gives direct evidence that the speed of light is an invariant. Until Einstein, physicists could not make sense of the result of this experiment, and most of them simply ignored it. It is not clear whether Einstein himself realized its importance until after he was led to his postulates by examining Maxwell's equations.

▷Interferometry is a classic example of a phenomenon seen repeatedly over the centuries, where a key technology is invented by a scientist simply as a tool for the investigation of a deep question in "pure" science. Interferometry is used today in industrial machining, geology, pollution control, astronomy, and countless other areas.

In fact, the earliest experiment that supported Einstein's guiding principle on the speed of light was actually done well before Einstein's 1905 paper on special relativity. It is the famous Michelson–Morley experiment, performed in the 1880s. It was designed to test how the speed of light changed in certain circumstances. When it found that there was no measurable change, it became one of the puzzles that physicists of the day could not understand. Only after Einstein's work did scientists accept that the Michelson–Morley experiment had been telling them all along about the invariance of the speed of light.

It is worthwhile spending a little of our time looking at the Michelson–Morley experiment. Not only will it show us that Nature really does adhere to Einstein's second guiding principle, but it is also a chance to look at a remarkable invention, the Michelson **interferometer**. This instrument, which the American scientist Albert A Michelson (1852–1931) devised in order to measure changes in the speed of light, has developed into one of the most important high-precision measuring instruments of modern science and technology. And it is being developed further today to detect one of the most significant predictions of general relativity: gravitational waves. We will look closely at this instrument in the next section. This discussion will prepare us for studying gravitational waves and their detection, in Chapter 22.

Michelson invented the interferometer to perform his first experiment on light in 1881. This produced such an unexpected result that he repeated it with greater precision with his American collaborator Edward W Morley (1838–1923) in 1887. The aim was to show that light had a different speed relative to the laboratory when it traveled in the direction of the Earth's motion than when it moved in a perpendicular direction.

In fact, Michelson expected it to be *slower* along the Earth's motion than across it. From Maxwell's theory, physicists knew that light was a wave, and that it was just a short-wavelength version of radio or other electromagnetic waves. For most physicists, if something was a wave then it had to be a vibration in *something*, and they called this medium the *ether*. This was a hypothetical substance whose vibrations were light waves.

The problem with the ether was that it had to be everywhere, in order to carry light to us from the distant stars, and yet there was no independent evidence for it. For example, if the planets were moving through the ether on their orbits around the Sun, why did it not slow them down? Why did the planets follow Newton's laws so exactly? To get around this, physicists had to assume that the ether was frictionless, unlike any other substance known. Many physicists were uncomfortable with such implausible properties, and in fact it was Einstein's own discomfort that led him to throw out the idea of the ether and embrace the invariance of the speed of light. But in the 1880s physicists were not ready for this. Instead, they felt that they had to find direct evidence for the ether.

Michelson hit on a way to do this. He reasoned that, if light had a fixed speed relative to this medium, then as the Earth traveled through the medium, the speed of light relative to the Earth would be slower in the direction of the Earth's motion than perpendicular to it. So Michelson expected that by comparing the speed of light in the direction of motion of the Earth as it orbits the Sun with the speed of light in the perpendicular direction, he would be able to measure the speed of the Earth relative to the ether, and thereby give a strong demonstration that the ether was really there.

Instead, what he found must have been very frustrating to him, at least at first: he could detect no difference between the speeds, so it seemed that the Earth was always at rest with respect to this ether, regardless of its motion. The ether was undetectable in this experiment.

It is interesting to look at the way physicists in 1887 reacted to this experiment. The cleanest thing to do, after this, would have been to abandon the ether idea: if it is not measurable, maybe it does not exist. This would have required a radical reformulation of physics, however, and no-one was able to do this until Einstein, 18 years later. In fact, most physicists found the experiment difficult to incorporate into their thinking. Many recognized the importance of the experiment, but could not fit it into the rest of what they knew about physics. Some went to extremes to defend the ether, such as postulating that the ether, far from being frictionless, was dragged around by the Earth; but such "fixes" raised other problems and were clearly contrived. Other physicists simply put the problem aside as being too difficult, and they worked on something else, where they knew they could make progress.

The most important attempts actually to find a plausible way to explain the Michelson–Morley result were by the Dutch physicist Hendrik A Lorentz (1853–1928) and the Irish physicist George F Fitzgerald (1851–1901). They pointed out, independently of each other, that if the interferometer actually *contracted* by a certain amount in the direction of its motion, then this would compensate the expected smaller speed of light in that direction and allow the photon going this way to return at exactly the same time as the photon moving across the motion. The mathematical content of their work was elaborated by French physicist Henri Poincaré (1854–1912).

We shall see below that a length contraction is indeed a prediction of Einstein's theory, and we call it today the **Lorentz–Fitzgerald contraction**. But the contraction in Einstein's theory does not take place in the same circumstances as Lorentz and Fitzgerald predicted, and Einstein's explanation of the Michelson–Morley ex-

▷ When Einstein derived the Lorentz–Fitzgerald contraction from his two fundamental principles, Lorentz and Poincaré were among the first physicists to recognize the significance of his new approach, to applaud the genius of this young patent clerk in Switzerland, and to help open the doors of the academic world to him.

periment is very different from that of Lorentz and Fitzgerald. We will return to this difference below.

Michelson's interferometer: the relativity instrument

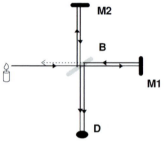

Figure 15.3. A sketch to show the principle of the Michelson interferometer

Michelson's interferometer, shown in Figure 15.3, is basically a device for measuring the tiny differences in the time it takes light to travel along two perpendicular paths. This can't be done using a stopwatch, because light travels too fast. Instead, the device only compares the two times, measuring the *difference* in light-travel-time along the two directions, but not telling us what the travel time along either direction actually is. To see if light travels at different speeds in two different directions, this is all one needs.

Here is the simple idea behind the device. Suppose it can be arranged that two photons leave the light source (the candle) at exactly the same time, so they reach the point B at the same time. Suppose further that one photon then travels to mirror M1 and the other to mirror M2. If the lengths B–M1 and B–M2 are the same, then when reflected, they will return to B at the same time if and only if they travel at the same speed in the two directions. If one of the speeds is larger, then the photon traveling that way will return before the other one. If one can measure the *difference* in arrival times, one can measure the difference in the speeds. In Investigation 15.2 we look at how, by using the interference of light in the two arms, the interferometer can measure tiny differences in arrival times.

In the Michelson–Morley experiment, the instrument is carried through space by the motion of the Earth. By aligning one "arm", say B–M1, in the direction of motion of the Earth, and the other arm (B–M2) across that motion, the experimenters expected to see a difference in speeds along the two arms. *They saw no difference.* No matter how the instrument was oriented, no matter what direction it was carried by the Earth, the two photons both arrived back at the point B at exactly the same time. The accuracy of their experiment was good enough to see a difference even if the speed of the Earth through the ether were only 1% of the speed of the Earth around the Sun. But they saw nothing.

> Following Einstein, we now interpret Michelson and Morley's result as a direct demonstration that the speed of light does not depend on the speed of the instrument measuring it.

It is not hard to see how an interferometer could be used for other high-precision experiments today. Since we know that the speed of light is the same in each arm, any difference in the arrival times of light after traveling in the two arms must be caused by a difference in the *lengths* of the arms. Interferometers today are used to make sensitive length measurements. We will see that this is exactly what is needed when looking for gravitational waves, which can make minute changes in the arm-lengths of a suitably constructed interferometer.

Now, this description of how the interferometer works is clearly somewhat over-simplified. The principle is correct, but there are many impractical aspects of the description. It is not practical to get just two photons to leave the light source at the same time, and it is not practical to measure the difference of their arrival times back at point B directly. It is even rather difficult to insure that the two arms B-M1 and B-M2 are exactly the same length. Readers who want to find out how Michelson actually did it will find a more realistic description in Investigation 15.2. But readers who do not consult this investigation will not miss anything essential. In particular, our discussion of how modern astronomers expect to use Michelson's interferometer to detect gravitational waves will require only the ideas just described.

Investigation 15.2. How an interferometer works, and why it got its name

In the Michelson experiment, the light source is a continuous beam of light, not just an emitter of two photons. The difference in arrival times at the detector D is detected by allowing the two beams to *interfere* with one another.

Interference should be a familiar phenomenon to anyone who has watched water waves in a harbor or in a bathtub. When waves pass through each other, there are places where the height of the water goes up and other places where it is cancelled out. The high places are places where the peaks of the two individual waves coincide. The low places are places where a peak of one wave coincides with a trough of the other. If the two waves have the same wavelength, then the pattern of peaks and valleys where they interfere can stay in one place for a relatively long time.

Light behaves in the same way, as an oscillating electromagnetic wave. When two light beams interfere, they make a pattern of light and dark spots or stripes. These are called fringes. In order to get a good interference pattern, it helps to use light of a single wavelength. Broad-band light can be made to interfere in one place, but if the two beams have similar wavelengths then they will make an interference pattern over a wider region. This makes it easier to set up the interferometer and to live with small differences in arm-length. Michelson filtered his light to a narrow band of colors; today scientists typically use monochromatic (single-color) lasers as the light source.

Consider, therefore, a single-color beam of light leaving the light source (the candle in the diagram) and reaching the beam splitter B, drawn as a gray diagonal element in the diagram. This is a half-silvered mirror, which means that half the light goes through it and half is reflected. These two beams of light leave the beam splitter with their oscillation peaks and valleys locked in step together, since they came from the same original beam.

After reflecting from the mirrors M1 and M2, they arrive back at the beam splitter. Both beams are split again, with half the wave reflecting and half transmitting. Here the where the interference phenomenon takes place. If the arms are exactly the same length, and the speed of the light in the arms is the same, then the beam coming from M1 that goes through the beam splitter back towards the light source, and the beam from M2 that is reflected by the beam splitter toward the light source, will be exactly in phase with each other, and they will add together to make a strong beam leaving the interferometer in this direction. At the same time, the light beams that head toward the detector D will be exactly out of phase with each other, and they will nullify one another so that *no* light goes to the detector. One could make other configurations: if the two arms differ, for example, by one-quarter of a wavelength, then the light will all go to D, with nothing going back toward the source.

What Michelson expected was that, during the course of the experiment, as the Earth turned and changed the direction of the arms of his interferometer, the speed of light in one arm would change relative to the other, and this would change the interference arrangement. So he looked for changes in the amount of light falling on the detector D that had a period of 12 h, the time it takes the interferometer to rotate to an equivalent configuration. He could detect changes induced by differences of speed between the two arms as small as 1% of the expected difference in speed (the speed of the Earth around the Sun), but he saw none.

Special relativity: general consequences

The Michelson–Morley experiment shows us that Einstein's principle that the speed of light should be the same for all experimenters is correct, even though it is radically different from our expectations. A theory founded on such a radical idea is bound to have consequences that are equally radical. In this section I shall list the important consequences of special relativity that we will need to know about in order to go on and study black holes and cosmology.

The list in this section will be brief but comprehensive. It will cover all the effects of special relativity that we will need in our discussions of general relativity and its astronomical implications. Readers who seek a deeper understanding of special relativity itself will find a section on each of the following points in the next chapter, Chapter 16. These sections will give more derivations where necessary, discuss the interrelations between these points, treat the experimental support for various effects, and dispose of worries that there might be internal contradictions (paradoxes) within special relativity.

1. *Nothing can travel faster than light.* No matter what forces one uses to accelerate a particle (or a rocket) to higher speeds, the object will never reach the speed of light. This comes from Einstein's formula for the combination of speeds, as illustrated in Figure 15.2 on page 184.

2. *Light cannot be made to stand still.* Obviously, since light has the constant speed c, it cannot be brought to rest. This principle applies to light that is free to move in empty space. In many circumstances, light interacts with other things: it bounces off a mirror or travels through a piece of glass. In these circumstances the "light" wave can travel at different speeds (or even come to rest instantaneously as it is being reflected from the mirror). But in this case the speed refers, not to pure light, but to a property of the interaction of light with other matter. In a vacuum, light travels at one speed only.

In this section: we list the most important consequences of the principles of special relativity, along with brief explanations and key formulas. Each consequence is treated in more detail in the next chapter. The list includes:
● nothing can travel faster than light;
● light cannot stand still;
● time slows for moving bodies;
● moving objects contract in length;
● simultaneity depends on the observer;
● mass depends on speed;
● energy and mass are equivalent;
● photons have zero rest-mass;
● the Doppler effect is changed.

▷The speed of light is the limit on all speeds.

▷Anything that travels at the speed of light cannot be made to come to rest.

▷ Time runs slower for moving bodies. It stands still for light. This is called time dilation.

3. *Clocks run slower when they move.* It is not possible to lose Galileo's simple formula for adding speeds together without also losing his simple notions of space and time. In order that the speed of light should be the same to two different experimenters, something unexpected must happen to the way they measure time and space, since a speed is simply the ratio of a distance to a time. What happens to time is that it slows down at high speed. For example, if an unstable elementary particle decays in a time t as measured by an experimenter at rest with respect to the particle, then it will decay in a time

$$t' = \frac{t}{\sqrt{1 - v^2/c^2}},$$

as measured by an experimenter moving past the particle at speed v. Here we meet for the first time an expression that is so important in equations in special relativity that it is given its own symbol, γ:

$$\gamma = \frac{1}{\sqrt{1 - v^2/c^2}}. \tag{15.4}$$

In the left-hand panel of Figure 15.2 on page 184, I have drawn γ as a function of v/c. For a speed v that is small compared to that of light, this is nearly equal to 1, and the two times differ by very little. Only when v becomes close to c do relativistic effects become important. The slowing of time is called **time dilation**. Notice that as v approaches c, t' gets longer and longer. In other words, for a photon (going at speed $v = c$), time stands still. Photons do not age. For this reason, they also do not decay: the only way a photon can change is to interact with something outside it. We referred to this in Chapter 11, to show that if neutrinos change their type as they move, then they must have mass.

▷ The size of a moving object contracts along its motion. This is called the Lorentz–Fitzgerald contraction.

4. *The length of an object contracts along the direction of its motion.* Not only time is affected: democratically, lengths also depend on speed. If an experimenter at rest with respect to, say, a length of pipe, measures its length to be L, then an experimenter moving past with speed v will measure the length L' to be shorter, by the same factor as time got longer:

$$L' = L\sqrt{1 - v^2/c^2} = L/\gamma. \tag{15.5}$$

This is the *Lorentz–Fitzgerald contraction*, and is the same formula originally written down by Lorentz and Fitzgerald as they tried to explain what happened in the Michelson–Morley experiment.

In special relativity, this is really just a counterpart to the time dilation: the two are two faces of a single coin. To see this, consider a rocket ship crossing the Galaxy at such a speed that $\gamma = 10^{12}$. Now, the Galaxy has a diameter of about 100 000 light-years, and the rocket is traveling at very close to the speed c, so the clocks on Earth tick about 100 000 years while the rocket makes the trip. But the clocks on the rocket tick at only $1/10^{12}$ of this time, which is 3 s! However, on the rocket the astronauts do not feel any different: by the principle of relativity, they consider themselves to be at rest, and every clock on board is ticking at its normal rate. How, then, can they cross the Galaxy in only three of their seconds?

Length contraction is the answer: the Galaxy has, as they measure it, a diameter of 100 000 light-years divided by 10^{12}, or only 10^9 m. Since the Galaxy is flashing past them at nearly the speed of light, it takes only 3 s to go completely past. So length contraction can be derived from time dilation, and vice versa.

This illustrates a fundamental aspect of relativity, that different observers (in this case the astronauts and an observer at rest in the Galaxy) must always make the same prediction for the outcome of an experiment (in this case the number of seconds ticked on an astronaut's wristwatch as he crosses the Galaxy), but they may explain the outcome of the experiment in different ways (time dilation or length contraction).

5. *There is no universal definition of time and simultaneity*. If two events that happen in different places are measured to occur at the same time by one experimenter, they may not be measured to occur at the same time by other experimenters that are moving with respect to the first. We call this the **loss of simultaneity**. To Newton and Galileo, *before* and *after* had invariant meanings: everyone would agree that event A happened before event B. This seemed only logical, since event A might have been part of the cause of event B, and it would be contradictory if some else determined that B occurred first. In relativity, this logic only requires that the notion of before and after is required to apply only to events that can influence one another. Thus, if A could cause B, then everyone must agree that A was earlier. But A can only cause B if light can travel from A to B: no influences travel faster than light. Therefore, if B is too far away for light to get there from A by the time B happens, then there can be no cause-and-effect relation and there is no logical need for different experimenters to agree on which one occurred first.

> ▷ Simultaneity is not something that all experimenters will agree on. This disagreement is closely tied to the time dilation and length contraction.

Events that occur at the same time in different places as measured by one experimenter are exactly of this type: neither can be the cause of the other. So it happens that relativity does not give them a unique order: to one experimenter they are simultaneous, to another A may occur first, and to a third B may occur first. But all three experimenters will agree that light cannot make it from one to the other, so they can have no causal effect on each other.

On the other hand, if light *can* travel from A to B, then all experimenters will also agree on this, and all will place B later than A (though by differing amounts of time, depending on the time dilation effect). So relativity preserves a notion of before and after, of *future* and *past*, but it does not apply that relation to all possible pairs of events.

This means that it is not possible to maintain Newton's idea of a three-dimensional absolute space, for which time is just a parameter: in Newton's world everyone would agree on what space looked like at a given time. In Einstein's world, there is just **spacetime**, the four-dimensional continuum of all events that occur anywhere at any time. Notice that **events** are the "points" of spacetime: an event is something that occurs in a particular location at a particular time, so it is a "dot" in spacetime. One experimenter will group a particular set of events into 3D space at a particular time, but a different experimenter could equally validly decide that a very different set of events constituted space at that particular time.

Two events that cannot have a cause-and-effect relationship with one another are said to have a **spacelike** separation in spacetime. Two events that can be connected by something traveling at less than the speed of light are said to have a **timelike** separation in spacetime. Two events that can be connected by a single photon are said to have a **lightlike** separation.

Relativity mixes notions of time and space. If we change point of view from one experimenter to another in relative motion, then there is a transformation in how we distinguish space from time, in how we reckon the passage of time, and in how we measure distances. This whole change in point of view is called the **Lorentz–Fitzgerald transformation**. It has a mathematical expression, but we need not deal with that. The main thing is that any one experimenter can use his or her own conventions on time and space consistently, but there is nothing absolute about them. Another experimenter's conventions will do just as well, even though they are different. This mixing of time and space will be discussed thoroughly in Chapter 17, where it will form the basis of our study of general relativity.

▷ As an object moves faster, its inertial mass increases, so it is harder to accelerate it. This enforces the speed of light as a limiting speed: as the object gets closer to the speed of light, its mass increases without bound.

6. *The mass of an object increases with its speed.* We noted above that no force, no matter how strong, could accelerate a particle to the speed of light. Does this mean that Newton's second law, $F = ma$, is wrong? After all, if I take F to be large enough, I should be able to make the acceleration large enough to beat the speed limit of c. No: relativity has a way out of this potential contradiction. The mass m in Newton's equation gets unboundedly large as the particle gets near to the speed of light, again by the ubiquitous γ factor:

$$m = \gamma m_0, \tag{15.6}$$

where m_0 is the mass that an experimenter at rest with respect to the particle would measure. Physicists call m_0 the particle's **rest-mass**. In fact, they define the rest-mass of any object to be its inertial mass when it is at rest. If the object is complex, like a gas with lots of random internal motions, then the rest-mass is the mass when the average momentum of all the particles is zero.

▷ Einstein's most famous equation expresses the *equivalence* of energy and mass, not just the ability to convert between them. The kinetic energy of a moving body accounts for its increased inertial mass. Any object that gains energy, say from heat, is also harder to accelerate because it has more mass.

7. *Energy is equivalent to mass.* Here we meet the most famous equation associated with Einstein:

$$E = mc^2.$$

To see what it means, and why it makes sense, consider again what happens if we try to accelerate a particle up to the speed of light. We keep applying an immense force to it, but since its mass is increasing rapidly, its speed hardly changes. Nevertheless, the force is doing work, by Equation 6.20 on page 62, and we have to keep supplying energy to keep the force going. What is happening to this energy? In conventional Newtonian language, we would at least expect the kinetic energy of the particle to be increasing. This was introduced in Equation 6.8 on page 54. Einstein showed that the energy accounting comes out right if the total energy of the particle is just its total mass times c^2. This includes its kinetic energy and, of course, a new energy: the rest-mass energy m_0c^2 that the particle has even when it is at rest.

This new concept had many important implications for physics. For one, it meant that energy has inertia: the more energy one puts into a system, the more mass it has and therefore the harder it is to accelerate. For another, it became possible to imagine the conversion going the other way: reducing the mass of an object and releasing the corresponding energy in another form. This is the implication that is most familiar to us: nuclear reactors and nuclear explosives work in this way. But it is important to understand that this conversion happens in everyday life, too, although on a scale that we don't notice.

If an automobile has a rest-mass $m_0 = 1000\,\text{kg}$ (its mass when it is standing still), then when it is moving at speed $v = 100\,\text{km hr}^{-1}$ (about 60 mph), so that it has $v/c = 9 \times 10^{-8}$, its total mass is larger by about $4 \times 10^{-12}\,\text{kg}$. (See Exercise 15.3.2 on the next page for the details of this calculation.) If two such cars collide head-on and come to rest, then the rest-mass of the wreck is larger than the sum of the two original rest-masses by twice this amount, or about 8 picograms (less, of course, the mass of the hubcaps and other pieces that roll away, the mass-equivalent of the sound energy radiated by the collision, the mass left in tire skid-marks on the road, and so on). This 8 pg of mass takes the form mainly of extra chemical energy in the deformed structures of the cars. It is a real mass: it would show up in a precision weighing of the wreck, and it would contribute to the inertia of the wreck if the rescue vehicles try to push the mess off the road.

8. *Photons have zero rest-mass; their momentum is proportional to their energy.* When a particle is accelerated to nearly the speed of light, its mass increases without bound. How, then, do photons get to the speed of light with finite energy? The only consistent answer is that their rest-mass should be zero. This is not really a well-defined notion, since photons cannot be brought to rest in order to measure their rest-mass. It is really only a convenient way of speaking about photons to explain why they do not fit into the rest of mechanics. But it is also a new perspective that allows us to speculate that perhaps there are other particles that have zero rest-mass as well. They, too, would travel at the speed of light. From this point of view, the "speed of light" is more fundamental than light itself: it is the speed of all zero-rest-mass particles. The neutrino (see Chapter 11) was at first thought to be massless and to travel at speed c, although observations today suggest that it has a very small mass. When gravity is turned into a quantum theory, some physicists expect that there may be a particle associated with gravitational waves, called the **graviton**, which will also be massless. But, as we shall discuss in Chapter 27, it may be very different from the photon.

▷ Traveling at the speed of light, photons are special. They have no rest-mass and they carry a momentum that is proportional to their energy, a very different relationship from the one that governs non-relativistic particles.

We have seen that the momentum carried by a photon is important in astronomy. The momentum of any particle is its mass times its speed. In relativity, the mass is its total mass m. This is equivalent to its total energy E divided by c^2. For a photon, whose speed is always c, the momentum is therefore

$$\text{photon momentum} = (E/c^2) \times c = E/c.$$

9. *The Doppler redshift formula changes slightly.* The changed notion of time in special relativity leads to a simple modification of the formula for the redshift of a photon. Remember how we derived the formula, by a visual method using Figure 2.3 on page 15. We counted the number of wave crests that passed by a moving wave-crest counter, and compared that with the number that passed one at rest. The number of crests passing per unit time is the frequency of the wave. Now we have to take into account that the moving counter's clock is running a bit more slowly than the one at rest. So if the counter at rest counts a number N crests in a time t, the moving counter counts a number $N' = N(1 - v/c)$ crests (from Figure 2.3 on page 15) in a time $t' = t/\gamma$ (Einstein's time dilation). When we divide the number of crests by the time, the counter at rest measures a frequency $f = N/t$, while the moving

▷ Because of the time dilation effect, even velocities across the line-of-sight to a body will slow time down and therefore change the apparent frequency of light it emits. So the Doppler formula must be modified to take account of time dilation.

Investigation 15.3. *Relativity at small speeds: making Galileo happy*

The formulas of special relativity look rather complicated when one first meets them, with all those factors of v^2/c^2 and $(1 - v^2/c^2)^{-1/2}$. All these factors are necessary to deal with phenomena at or near the speed of light. But when speeds are small, we must expect to get the Galilean and Newtonian results as well. We saw how this worked with the velocity composition law in Investigation 15.1 on page 183. We want to look at it more systematically here, since there are so many formulas with these factors in them.

One can use the binomial theorem Equation 5.1 on page 43 to show that, for small speeds v,

$$\left[1 - \left(\frac{v}{c}\right)^2\right]^{-1/2} \approx 1 + \frac{1}{2}\frac{v^2}{c^2}. \qquad (15.8)$$

The term $v^2/2c^2$ is an estimate of the size of the relativistic correction to a Galilean or Newtonian formula. For example, the mass of a particle with rest-mass m is larger than the rest-mass by

$$\Delta m = m\left[1 - \left(\frac{v}{c}\right)^2\right]^{-1/2} - m \approx \frac{1}{2}\frac{v^2}{c^2}m.$$

The equivalent excess energy is $\Delta mc^2 \approx \frac{1}{2}mv^2$. This is what is called the kinetic energy in non-relativistic physics. We see that it is a low-velocity approximation to the correct value of the energy associated with the motion of the particle.

Similarly, in the Lorentz–Fitzgerald contraction, the change of length is

$$\Delta L = L\left[1 - \left(\frac{v}{c}\right)^2\right]^{1/2} - L \approx -\frac{1}{2}\frac{v^2}{c^2}L.$$

If the speed is 1% of the speed of light, for example, then the contraction is only 0.005% of the original length.

If we consider the volume of a rectangular box which is made to move parallel to one of its sides, then it will contract along that length and not along the other two, so its volume will decrease in direct proportion to the length of the contracting side. The above formula then has the consequence that for small speeds the change in volume is approximately

$$\Delta V = -\frac{1}{2}\frac{v^2}{c^2}V. \qquad (15.9)$$

We will need to use this below when we discuss the dynamics of moving fluids.

Exercise 15.3.1: *Slow-velocity expansion*

Use the binomial expansion Equation 5.1 on page 43 to show that the expansion of $(1 - v^2/c^2)^{1/2}$ for small v/c is

$$\left[1 - \left(\frac{v}{c}\right)^2\right]^{1/2} = 1 - \frac{1}{2}\left(\frac{v}{c}\right)^2 + \cdots.$$

Exercise 15.3.2: *How much mass is in kinetic energy?*

Consider the example given in the text, of an automobile with a rest-mass of 1000 kg. Show that its kinetic energy at a speed of 100 km hr^{-1} has a mass equivalent of 4 pg (4 picograms, or 4×10^{-12} g).

---

counter gets $f' = N'/t'$, which works out to be

$$f' = (1 - v/c)\gamma f = \frac{1 - v/c}{\sqrt{1 - v^2/c^2}}f. \qquad (15.7)$$

This is the formula when the moving counter is going away from the source of light, as seen by the counter at rest. This produces a decrease in the frequency, or a redshift. If the moving counter is approaching the source of light, there is a blueshift, an increase in the frequency, because the sign of v changes and the numerator in this equation is then bigger than one. Because the denominator is always smaller than one, the redshift and blueshift are larger than one gets from the non-relativistic Doppler formula. Notice that there is even a Doppler shift if the moving counter is moving *perpendicular* to the direction to the source of light. In this case, the non-relativistic Doppler shift is zero, because the motion of the counter does not add or subtract any wave crests from the number counted by a counter at rest. But there is still the time dilation, which reduces the amount of time that the moving counter measures while it counts the crests. This produces a blueshift in relativity where there is none in the Newtonian Doppler formula. This is called the transverse Doppler shift.

The extra inertia of pressure

In this section: we single out an unexpected consequence of special relativity: the more pressure a gas has, the harder it is to accelerate.

So far we have looked at how special relativity affects isolated bodies as they go faster: clocks (which go slower), rulers (which get shorter), accelerated particles (whose masses increase), and so on. But the Universe is not just composed of iso-

lated bodies. In our tour of the Universe we have studied gases in order to understand stars, and at the end of our tour we will again need to understand gases in
order to understand the Universe as a whole. In this section we will discover, perhaps rather unexpectedly, that in special relativity the pressure inside a gas plays
an important part in the inertia of the gas: the more pressure the gas has, the more
difficult it is to accelerate.

While at first sight this might seem like a mere curiosity, it has very far-reaching
consequences. When we study neutron stars in Chapter 20 we will find that this effect makes the neutron gas "weigh" more, and this in turn forces the star to have a
higher pressure, which only makes it weigh even more, and so on. This pressure-
feedback effect eventually makes it impossible for the star to support itself: the inertia of pressure opens the door to the black hole.

The inertia of pressure can be traced to the Lorentz–Fitzgerald contraction. In
Investigation 15.4 on the next page we show how to calculate the extra inertia, but
even without much algebra it is not hard to see why the effect is there. Consider
what happens when we accelerate a box filled with gas. We have to expend a certain amount of energy to accelerate the box, to create and maintain the force of
acceleration. In Newtonian mechanics, this energy goes into the kinetic energy of
the box: as its speed increases so does its kinetic energy. This happens in relativity
too, of course, but in addition we have to spend some extra energy because the box
contracts.

The Lorentz–Fitzgerald contraction is inevitable: the faster the box goes, the
shorter it gets. But this shortening does not come for free. The box is filled with
gas, and if we shorten the box we reduce the volume occupied by the gas. This
compression is resisted by pressure, and the energy required to compress the gas
has to come from somewhere. It can only come from the energy exerted by the
applied force. This means the force has to be larger (for the same increase in speed)
than it would be in Newtonian mechanics, and this in turn means that the box has
a higher inertia, by an amount proportional to the pressure in the box.

In fact, the formula for the extra inertia is simple. If the box has a mass-density
ρ (which, in relativity, includes the mass associated with all the different forms of
energy in the gas) and pressure p, then the density of inertial mass is

$$\text{inertial mass density} = \rho + p/c^2. \qquad (15.10)$$

This equation is derived in Investigation 15.4 on the following page. It is simple to
use. If the box has a volume V, then the total inertial mass in the box is $(\rho+p/c^2)V$, in
the sense that force F required to produce an acceleration a is just $F = [(\rho + p/c^2)V]a$.
This is simply a consequence of special relativity. We will find that inertial mass
density useful in our study of cosmology: it is a key to understanding Einstein's
cosmological constant and the theory of inflation.

Conclusions

The ideas we have discussed and the formulas we have derived in this chapter will
lead us naturally into relativistic gravity and its consequences in later chapters.
However, some readers may want a more detailed discussion of the points described
here. For example, why is the time dilation not an internal contradiction in special
relativity: if experimenter A measures that the moving clocks of experimenter B are
going slowly, how can relativity be preserved? After all, the principle of relativity
says that an experiment should not single out a preferred speed. Thus, if experimenter B measures the rate of A's clocks, which after all are moving with respect
to B, then B should find that they are going more slowly. But how can the clocks

*In this section: our survey of
special relativity will be sufficient
for the rest of the book, but readers
can find more depth in the next
chapter. For general relativity, skip
to Chapter 17.*

Investigation 15.4. How pressure resists acceleration

We shall look at this only for slow motions, where the effects of special relativity are small. Suppose we have a box at rest that is filled with a uniform gas. We denote the volume by V, the mass density by ρ, and the pressure by p. Suppose next that we apply a small force to the box and accelerate it until it has a speed v that is small compared to c. The key question is, how much energy did we have to put in to get the gas up to speed v? For simplicity, we will only ask about the gas, not the container: in astronomy we usually don't have containers: one part of the gas of a star is held in place by gravity and the pressure of other parts of the gas.

Once it is at speed v, the gas in the box has acquired a kinetic energy, so one might think that the total energy that we had to add to the box in order to accelerate the gas in it would have been equal to this kinetic energy, $\frac{1}{2}mv^2 = \frac{1}{2}\rho V v^2$, where in the second expression we have used the fact that the mass m in the box is ρV. But this is not the whole story, because the Lorentz–Fitzgerald contraction has shortened the length of the box and therefore changed its volume. Making a box smaller when it contains a fluid with pressure p requires one to do work on it, in other words to put some energy into the gas. This extra energy represents the extra inertia of the gas: it is harder to accelerate the gas because it takes work not only to accelerate the existing energy but also to compress the gas as the Lorentz–Fitzgerald contraction demands.

We only need to work out this extra energy in order to see why. The energy one has to put into it is just $-p\Delta V$, where we denote the change in volume by ΔV; the minus sign is needed so that when the box contracts (ΔV negative) then the energy put into the box is positive. Using Equation 15.9 on page 192 to get the change in volume, we find that the extra energy we put in is

$$\frac{1}{2}\frac{v^2}{c^2}pV.$$

This energy does not just disappear; it goes into the internal energy of the gas in one form or another, depending on the details of the gas molecules. At least some of the energy goes into raising the temperature of the gas (the random kinetic energy of the molecules).

The total energy required to accelerate the gas-filled box can be written in a simple way:

$$E = \frac{1}{2}mv^2 - p\Delta V$$

$$= \frac{1}{2}\rho V v^2 + \frac{1}{2}\frac{v^2}{c^2}pV \qquad (15.11)$$

$$= \frac{1}{2}\left(\rho + \frac{p}{c^2}\right)v^2 V.$$

The last expression is the one we need to examine. The energy required to accelerate the box is proportional to the sum $\rho + p/c^2$. This energy comes from the work done by the force we must use to accelerate the box, so the force had to be larger than we might have expected. Put another way, for a given applied force, the box accelerates less than we would have expected by measuring its mass, since some of the energy we put in goes into the internal energy of the gas instead of the kinetic energy of the box. Scientists therefore say that the inertia of the box is larger than just its rest-mass, and in particular they call the quantity $\rho + p/c^2$ the *inertial mass density* of the gas. If we want to know how much force is required to accelerate a fluid we have to know the inertial mass density, not just the rest-mass density. This is purely a consequence of special relativity.

---

of B go more slowly than those of A, and at the same time the clocks of A are going more slowly than those of B? The same worry arises for the length contraction. In fact, these apparent contradictions are not real: they result from not considering carefully enough what is being measured. To allow us more space for a discussion of these deep and profound issues, and also to give readers a glimpse of the enormous body of experimental evidence that now supports special relativity, each of the points discussed in the list earlier in this chapter is given a separate section in the next chapter. The interested reader can use these sections to become much more deeply acquainted with special relativity.

This extra material is not essential for our investigations of relativistic gravity in later chapters, so readers who want to stick to the main line of development can safely leave them out and go straight to the first chapter on general relativity, Chapter 17.

16

Relating to Einstein:
logic and experiment in relativity

Our introduction to special relativity in the last chapter covered the basics, but it may have raised more questions for you than it answered. Before reading the chapter, you may have been very happy with the simple idea that everyone would agree on the length of a car, or the time it takes for the hands on a clock to go around once. If so, you have now learned to question these assumptions, that Nature does not really behave like that. If you want to fit these ideas together into a more logical framework, and if you want to learn something about why scientists are so sure that Nature really follows the principles of special relativity, then this chapter is for you. Read on.

In the previous chapter I listed some important effects of special relativity and gave a brief description of each, such as time dilation and the equivalence of mass and energy. In most cases, I left out the derivations, the algebra that linked one result to another. In this chapter I will fill in some of these gaps. Each of the points in Chapter 15 has its own section here, in which I give an argument to derive it from basic principles. I shall use the style that Einstein himself favored, that of a "thought experiment", an idealized physical situation where it is easy to work out what must happen. I shall then back this up with a description of a real experiment, where the same basic feature is tested and verified. This set of experiments illustrates why physicists have such confidence in special relativity; indeed, special relativity is one of the best-tested theories in all of physics.

Many of the results are surprisingly simple to derive algebraically from Einstein's basic principles. For readers who want to see how this works and how they relate to one another, Investigation 16.1 on page 201 contains the essential algebra and arguments.

Nothing can travel faster than light
Thought experiment

Imagine you are an experimenter using a linear particle accelerator, which employs strong electric fields to push a charged particle, like an electron, faster and faster in a straight line. In your experiment, you send a photon down the accelerator at the same time as you start accelerating an electron. Suppose that, a few moments later, the electron has reached the speed $0.999c$ with respect to you. (You are at rest on the ground.) From your point of view, the photon is now some distance ahead of the electron and still pulling away, but only gradually. Imagine now another experimenter who is flying past you with the same speed as that of the electron, and who measures the speed of the electron and of the photon. The electron's speed is momentarily zero with respect to the flying experimenter, of course. And, by the invariance of the speed of light, this experimenter measures the photon's speed to be c. So the electron, despite its enormous acceleration, has from this point of view not come one bit closer catching up with the photon! If you wait another few moments, until the electron reaches the speed of $0.999999c$, there will still be

In this chapter: we examine the foundations of special relativity in detail, deriving all the unusual effects from the fundamental postulates, examining the experimental evidence in favor of each one, and showing that the theory is self-consistent even if at first sight it seems not to be.

▷The image under the text on this page illustrates length contraction. The top figure is after Leonardo da Vinci's famous drawing. The bottom figure has the dimensions that an experimenter would measure if the experimenter were flying across the original drawing at a speed of $0.9c$.

In this section: how to understand that the invariance of the speed of light prevents anything going faster than it, or indeed even catching up with a photon.

▷Interestingly, this argument on not accelerating up to the speed of light does not exclude from Nature the possibility that there are particles that simply start out at speeds faster than light. These logical possibilities are called **tachyons**. There is no evidence that they exist, and because they create problems with causality (see the section on the loss of simultaneity below) most physicists do not expect them to exist.

▷The *synch*rotron gets it name from having to *synch*ronize its forces with the exact position of the particle as it goes around faster and faster.

In this section: the invariance of the speed of light also means that photons can never come to rest.

another (imagined) experimenter who would measure this electron to be at rest and the photon still to be moving at the full speed c. If you think about this, you will see that, no matter how much you accelerate it, the electron won't reach the speed of the photon, because there is always a perfectly good experimenter for whom the two speeds are not even close! The only possible conclusion is that the electron simply cannot travel at the speed of light. And since it can't get to the speed of light, it can't go faster than light.

Real experiment

Linear accelerators like this one, and circular accelerators (called synchrotrons) that push electrons (or protons) around a circle, operate successfully every day. While they do not directly measure the speed of the electron relative to the hypothetical photon, they do something just as good. They need to anticipate exactly where the electron is at any moment so they can give it just the right push to keep accelerating it. If the electron did not turn up in the right place, as predicted by special relativity, then these enormous machines would simply not work. If an electron ever moved faster than light in such a machine, the experimenters would soon know it!

Light cannot be made to stand still

Thought experiment

Since every experimenter must measure light to have speed c, there is no experimenter for whom light can stand still.

Real experiment

Strictly speaking, what we are saying is that, if a photon or any other particle travels exactly at speed c, then it cannot go any slower, and so it cannot be made to stand still. This is established indirectly by the successful operation of particle accelerators, as just described. It is a separate question to ask if photons are such particles: do real photons travel with speed c? This is equivalent to asking if they have a non-zero rest-mass. This is an experimental question about the nature of photons, and so far there is no evidence for any rest-mass. It is important to understand that, if experiments did show that photons had a small but non-zero rest-mass, then this would not upset special relativity. It would mean that we would no longer want to call c the speed of *light*, so we would call it something else, like the Einstein speed. But it would still be fundamental even if light happened not to follow it, and it would still be a barrier to the speed of all objects, including sluggish photons.

It is also important to understand that Einstein's principle applies to the speed that light travels only if the light is free to move without disturbance. Obviously, if a photon is reflected backwards by a mirror, then at the moment of reflection one might be tempted to say that it is "standing still". But in such a case we are not dealing with a photon on its own. It would not reflect if it were not interacting with the electrons and protons of the mirror. What actually happens on a microscopic level is that the incoming photon is absorbed by the electrons of the mirror, which are set into oscillation by the photon's oscillating electric field. The result is, for some materials (shiny ones), that the electrons' oscillation creates a new photon that moves away from the mirror in the opposite direction. The incoming and outgoing photons are free and move at speed c, but they are not the same photon, because at the "moment" of reflection (which actually lasts no more than a few oscillation periods, perhaps 10^{-14} s) there is no independent photon at all.

Not only mirrors, but also transparent materials undergo complex interactions with light. When "light" moves through water or glass, what actually moves is a composite wave in the electric fields of photons and of the atoms of the material. The incoming photon causes atoms to oscillate, which then disturb other atoms further along, which oscillate, and so on through the material. This wave of disturbance moves at speed less than c, a speed that is usually called the "speed of light" in the material. But it is not the speed of a free photon, which is always c (provided photons are massless). If the frequency of the photon is high enough, the electrons of the material won't be able to respond to it, and the photon will be able to travel freely through it. This is why X-rays penetrate most materials, and why the speed of light in any material always gets closer to c as the frequency of the photon gets higher.

Clocks run slower when they move

Thought experiment

We shall show this by considering a very simple kind of clock, one that just reflects a photon between two mirrors and "ticks" once for every round-trip that a photon makes. This is a good clock for us to use in a thought experiment, even if it is rather impractical to make, because it directly involves light, whose simple properties we completely understand. Suppose that the time it takes for light to go up and back when the mirrors are at rest is τ, the time for one tick of the clock. Now imagine the clock is moving with a certain speed v in a direction perpendicular to the line joining the mirrors. Now the light has to travel further on its round-trip. As Figure 16.1 on the next page shows, the light travels on the two sides of a triangle, whose base is the distance the clock travels during the round-trip travel time. Since the side of the triangle is longer than the distance between the mirrors, the total distance traveled by light in the moving clock is larger than in the clock at rest. By Einstein's principle of the invariance of the speed of light, the photon travels at the same speed in each case, so it must take longer to go up and back in the moving clock.

The calculation in Investigation 16.1 on page 201 shows that the time it takes for the clock to tick if it moves at speed v is longer by the factor $\gamma = (1 - v^2/c^2)^{1/2}$ than the time it takes if it is at rest:

$$\Delta t_{\text{moving clock}} = \gamma \Delta t_{\text{clock at rest}}. \tag{16.1}$$

This is called the *time dilation* effect: clocks that move run more slowly, so time is stretched out (dilated). Note that this must happen to all clocks, not just ones based on light. If we have built a sufficiently accurate mechanical clock that keeps time with our light clock, and we place the two side-by-side and at rest, then they will remain synchronized for as long as we wish. We can arrange, for example, for them to give off flashes of light once every second, simultaneously. If another experimenter travels past us on a train, then our light clock has a non-zero speed with respect to the experimenter, and so it will run slowly as measured by the experimenter. This must then also be true of the mechanical clock, since we know it will emit a flash of light every time the light clock does; because these flashes happen right next to each other, the moving experimenter must see the two clocks flash at the same time, just as we do. The mechanical clock, which could be any other clock, therefore runs just as slowly as the light clock. A particularly important clock is our biological clock: a moving experimenter ages more slowly than one at rest!

Real experiment

Time dilation is easy to measure directly. Nature provides us with a number of natural clocks in the form of unstable elementary particles and nuclei. Particles decay at random: some decay rapidly and some take a long time. And they have no memory: a particle that has by chance lived 100 years is just as likely to decay in the next second as a particle of the same type that was created one millisecond ago. For this reason, the decay is fully characterized by one number, called the *half-life* of the particle type, a concept that we introduced in Chapter 11. It follows that, in a sample of N identical particles, there will be on average $N/2$ decays in the time τ.

In this section: the only way that light can have the same speed to all observers is if observers disagree on time and space measurements. Here we see that moving clocks must run slowly.

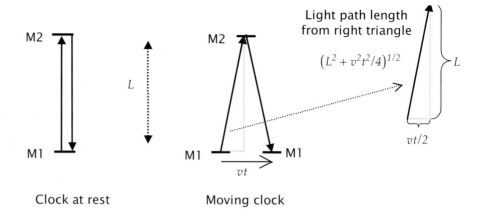

Figure 16.1. A simple clock based on reflecting light. This clock ticks once when a photon makes a round-trip from mirror M1 to mirror M2. When the clock is at rest (shown at the left in the figure) the photon travels a distance 2L for this. When the clock is moving (shown at the center), light must travel a longer distance in order to return to the mirror M1, which has moved since the photon left it. The geometry used to calculate this distance is shown at upper right.

So if we measure the number of decays per unit time in a sample of particles, we have a natural clock, a natural way to measure τ. Every time the number of atoms reduces by one-half, this natural clock ticks a time τ.

An observation of time dilation using such a clock was first made for particles called **muons**, which are produced abundantly in the upper atmosphere when high-energy cosmic rays strike oxygen or nitrogen nuclei. The half-life τ of a collection of muons at rest is 2.2×10^{-6} s. Even if the muons are produced moving at close to the speed of light, they could travel no more than 660 m in 2.2×10^{-6} s before losing substantial numbers to decay. Yet muons are detected easily at ground level, many tens of kilometers below where they are produced. And at the top of a 3000 m mountain, experiments see only a few more than at ground level, rather than the factor of $2^{(3000\,\mathrm{m}/660\,\mathrm{m})} \approx 23$ more that would be expected if half of them decayed every 660 m. Time dilation explains this: since they travel at nearly the speed of light, their internal clocks slow down dramatically, and they live much longer according to our clocks. Similar experiments can be done with unstable particles produced at high speeds in accelerators, and the predictions of special relativity are confirmed to a high accuracy.

A more practical application of time dilation today involves the Global Positioning System (GPS) that we discussed in Chapter 2 as an illustration of the gravitational redshift. This redshift produces considerable differences between the rates at which clocks on the ground and in orbit run, and these differences have to be corrected often in order for the navigation system to work. What we did not explain in Chapter 2 is that time dilation produces differences of a similar size, so that it too has to be calculated and removed with the gravitational redshift: an annoying but unavoidable nuisance for the navigation system! If special relativity were not right, we would quickly learn about it from the GPS.

Novices to special relativity often worry that the time dilation effect is inherently self-contradictory, and that this should show up in experiments. The worry goes as follows: if experimenter A measures experimenter B's clocks to run slowly, simply because B has a speed v relative to A, then the principle of relativity implies that B will also measure A's clocks to run slowly, since the speed of A relative to B is also v. But this seems to be a contradiction: how could B be slower than A and A be slower than B? This is an important question, and one that goes to the heart of understanding special relativity. I shall give considerable attention to it in a separate section on the so-called twin paradox, at the end of the chapter. But there is a brief answer that we can look at here and see that the appearance of a contradiction comes from comes from comparing what are in fact two different measurements.

Let us look carefully at how each experimenter performs the measurement. For example, when A measures the rate of ticking of one of B's clocks, A must effectively use *two* of his own clocks: one to record the time where B's clock first ticked, and the second to record the time where B's clock next ticks. A needs two clocks because B's clock is moving. Thus, A's measurement involves a comparison of three clocks in total: two of A's clocks, which must run at the same rate (must be synchronized), and one of B's clocks. By the same reasoning, the experiment performed by B involves two of B's clocks and only one of A's. So although both experimenters describe their experiments as a comparison of one set of clocks with another, they actually compare different sets of clocks, so they are not doing the same experiment, and they do not need to get "consistent" results.

The length of an object contracts along its motion

Thought experiment

How can the muon result be explained to an experimenter traveling with the muons? Since they are at rest with respect to this experimenter, then half of them decay after only 2.2 µs. The Earth has been approaching the experimenter at nearly the speed of light during this time, but even so it cannot have traveled more than 660 m. Yet most of the muons have reached the ground. The inescapable conclusion is that the ground is less than 660 m from the top of the atmosphere, as measured by the experimenter moving with the muons. A length that is more than 20 km as measured by an experimenter at rest on the Earth has contracted to less than 660 m when measured by an experimenter moving at nearly the speed of light.

In this section: just as time slows down with speed, so also lengths contract along the direction of motion.

This effect is called the *Lorentz–Fitzgerald contraction*, because Lorentz and Fitzgerald were the first to propose that it and time dilation actually occurred. The formula is, following the pattern of earlier ones,

$$L_{\text{moving object}} = \sqrt{1 - v^2/c^2}\, L_{\text{object at rest}}. \tag{16.2}$$

But there is a crucial difference between what Lorentz and Fitzgerald predicted and what Einstein showed really happens. For Lorentz and Fitzgerald, the speed v in this formula was the speed of the object through the ether. Thus, in the Michelson–Morley experiment, they expected that the length of the arm of the interferometer that lay along the direction of motion of the Earth was physically shorter than the other arm, but that this was unfortunately unmeasurable: as soon as Michelson held a ruler up against this arm to measure its length, the rule would contract by the same amount, so the arm would appear to have its rest-length. Nevertheless, to Lorentz and Fitzgerald, the arm "really" was shorter. For Einstein, the length *is* what the experimenter measures, and the contraction occurs only when the object moves relative to the experimenter who makes the length measurement. This fundamental difference in interpretation is the main reason that Einstein gets the credit for discovering the contraction effect, even though the mathematical expression is the same as for Lorentz and Fitzgerald, and even though we honor their contribution by naming the effect after them.

Real experiment

We have already noted above that unstable muons must see the Earth's atmosphere greatly contracted, in order for them to reach the ground in their decay lifetime. The same effect occurs for elementary particles in accelerators; from their point of view, the accelerator tube must be very short, so that they don't decay in the middle of it. It must not be thought that the Lorentz–Fitzgerald contraction is somehow an illusion produced by a problem measuring time or by clocks that don't behave

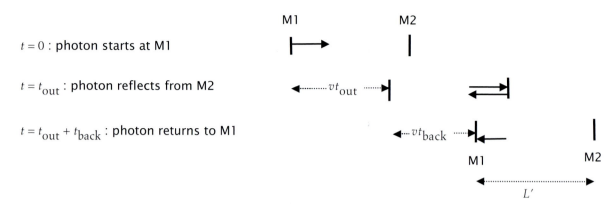

Figure 16.2. *The simple clock of Figure 16.1 on page 198, now moving along its length. Three snapshots are shown, at the moments when the photon starts out from mirror M1, reflects from M2, and returns to M1. The distance between the mirrors is called L′, to allow it to be different from the distance L when they are at rest. On its way out, the photon travels much further than L′, because the end mirror is moving away from it. On the way back, it travels less than L′. Its total round-trip time must be the same as for the moving clock in the previous figure, since the two are at rest with respect to one another and so must stay in synchronization.*

well: it is a real contraction. For example, if one tries to accelerate a solid body, then its contraction requires an extra input of energy to squeeze the atoms of the body closer together. This extra energy goes into the total mass of the moving body. Without it, the mass would not increase in proportion to γ, as we saw above that it must.

Loss of simultaneity

In this section: the disagreements over time measurements lead to a breakdown of the idea that there is a universal time. Any single experimenter can measure that a given pair of events occur at the same time; but to another experimenter, one event will occur before the other. Neither experimenter is "wrong".

According to relativity, we simply have to give up the idea that it is possible to say that two things happened at the same time and expect everyone to agree. If they happen at the same time *and* at the same place, then everyone will agree about this. (We used this for our two flashing clocks above.) But if the events are separated, the notion of simultaneity is not universal.

Thought experiment

A simple thought experiment involving two of our light-clocks will give us a direct example of how the notion of simultaneity depends on the experimenter. Let us put the two clocks next to each other on a table, at rest, but oriented so that their arms are perpendicular to one another. Let the ends of the clocks where the photons start out be in the same place. Now suppose the photons in each clock start out at the same time. The clocks are identical, except for orientation: the photons set out in perpendicular directions. Because the arms have the same length, the photons reach the end mirrors at exactly the same time. These reflections are *simultaneous* to us. But of course they occur in different places.

Now we shall see that they cannot be simultaneous to an experimenter who watches the same thing happen as the clocks move. Suppose our laboratory is in a train, and this is traveling at speed v along the direction of one of the clocks. Then we know that the experimenter on the ground will measure our clocks to be running slowly, but still both go at the same rate: the total time it takes a photon to go out and back in both clocks is the same. Here, however, we are interested only in the first part of the photon's journey, going out. For the clock perpendicular to the motion of the train, the journeys out and back are identical, so the reflection event occurs at exactly one-half of the time of a "tick". (This is the situation shown in Figure 16.1 on page 198.) For the clock oriented along the direction of motion, the outward and return journeys of the photon are not identical. On the outward leg, the distant

Investigation 16.1. The light-clock shows how time and space warp at speed.

The time dilation follows directly from the principle that the speed of light is the same to all observers. We will use the light-clock illustrated in Figure 16.1 on page 198.

First, let us agree that the clock at rest, on the left in the figure, is a good clock, and once calibrated it will run at the same rate as any other clock we could construct, held at rest with respect to us. So if, as we shall show, the clock runs at a different rate when moving, then this will apply to all other clocks, including the psychological perception of time.

The clock ticks once each time the photon returns to the bottom mirror M1. While the clock is at rest, this takes a time $\tau = 2L/c$. Now let the clock move, or equivalently let a moving experimenter measure the time it takes for the same clock to tick, using an identical light-clock at rest in his own laboratory. All we need to compute is the distance light travels in one tick. From the figure, it is clear that it travels along a triangular path because the mirror M1 is now moving and it has to meet it after being reflected at M2. If the return trip takes time t, the figure shows that the path has total length $2(L^2 + v^2 t^2/4)^{1/2}$.

Now, the photon traveled at speed c, so the time t it took is this distance divided by c. This gives us an equation to solve for t:

$$t = 2\sqrt{L^2 + v^2 t^2/4}/c.$$

Squaring this gives

$$t^2 = 4L^2/c^2 + v^2 t^2/c^2,$$

which can be solved for t. We shall call this value of t the ticking time of the moving clock, τ':

$$\tau' = \frac{2L}{c}\sqrt{1 - v^2/c^2} = \gamma\tau,$$

where γ was defined in Equation 15.4 on page 188. Since γ is larger than 1, the tick time τ' of the moving clock is larger than that of the clock at rest, so the moving clock is running slowly.

From this we can also derive the Lorentz–Fitzgerald contraction: it is the other face of time dilation. Consider the muons created in the upper atmosphere and heading for the Earth, as described in the text. Because of time dilation, they live much longer than their half-life as measured when they are at rest (2.2 μs), allowing them to travel much further, so that many reach the ground.

Now, how can this be explained to an experimenter traveling with the muons? Since they are at rest with respect to this experimenter, then half of them decay after only 2.2 μs. The Earth has been approaching the experimenter at nearly the speed of light during this time, but even so it cannot have traveled more than 660 m in this time. Yet the muons have reached the ground. It follows that the ground must be less than 660 m from the top of the atmosphere,

as measured by the experimenter moving with the muons. A length that is more than 20 km to an observer at rest on the Earth has contracted to less than 660 m by moving at nearly the speed of light.

We can deduce what the size of this contraction should be by simply turning the clock we used for the time dilation thought experiment on its side, so it is now moving at speed v along its length. This cannot change its ticking rate, since the clock is at rest with respect to the moving clock we computed above, just turned in a different direction. An observer at rest with respect to them would expect them to maintain synchronization. From this we will see that its length must change. We can see the situation in Figure 16.2.

On its way from mirror M1 to M2, the photon takes a time t_{out}. In this time it has to travel the length of the clock, L' (not necessarily equal to L), plus the extra distance that M2 has moved in the time t_{out}, which is vt_{out}. It travels at speed c along this path (Einstein again), and so we again have two expressions for the distance it traveled, which must be equal:

$$ct_{out} = L' + vt_{out},$$

from which we can deduce that

$$t_{out} = L'/(c - v). \tag{16.3}$$

Similarly, the return journey is shorter because the mirror M1 is now catching up with the photon. So a similar expression for the distance traveled in terms of the time t_{back} for this journey is

$$ct_{back} = L' - vt_{back} \quad \Rightarrow \quad t_{back} = L'/(c + v).$$

Now, these two times must add up to the round-trip time we had when the clock was oriented perpendicular to its motion, which was $2L/c(1 - v^2/c^2)^{1/2}$. This determines L' in terms of L:

$$\frac{L'}{c - v} + \frac{L'}{c + v} = \frac{2L}{c}\frac{1}{\sqrt{1 - v^2/c^2}}$$

$$\Rightarrow L' = L\sqrt{1 - v^2/c^2}. \tag{16.4}$$

We see that the length of the clock when it moves along its length is shorter than its resting length, by the factor $(1 - v^2/c^2)^{1/2}$. This is the only way to keep the clock ticking at the same rate, independent of orientation, *and* to have the speed of light the same for all experimenters, regardless of their state of motion relative to one another. It is important to keep this last fact in mind: if Newton had been doing this calculation he would have obtained a very different result, because for him the speed of light would have been different in different circumstances; where we always simply used c, he would have used different and more complicated expressions. But we know from experiment that Einstein was right and Newton wrong on this.

Exercise 16.1.1: *Practical time dilation*

An airline pilot spends 20 h per week flying at a speed of 800 km h^{-1}. Over a career of 30 years, how much younger is he than if he had never flown?

Exercise 16.1.2: *Exploring the Galaxy*

A future civilization manages to construct a rocket that will move through the Galaxy with $\gamma = 10^4$. What speed does it go? How long will it take to cross the Galaxy, some 20 kpc? How wide is the Galaxy as measured by the voyagers?

Exercise 16.1.3: *The dangers of space exploration*

Show from Equation 15.7 on page 192 that a photon with an energy E as measured by an experimenter at rest in the Galaxy, and which is approaching the spaceship of the previous exercise head-on, has an energy relative to the spaceship of $E' = 2\gamma E$. Suppose the spaceship is approaching a normal star like our Sun. Use Wien's law (Equation 10.9 on page 117) to infer that the radiation reaching the spacecraft from the star will have an effective temperature of 50 million K! That means it would be a source of gamma-rays, and the inhabitants of the spacecraft would have to shield themselves carefully.

mirror is running away from the photon, while on the return journey the mirror at the back of the clock is approaching the photon. The outward journey must, therefore, take longer than the return journey, which means it must take longer than half a "tick" of the clocks. So the reflection event in this clock will be later than that in the other clock. *Two events that are simultaneous to one experimenter are not necessarily simultaneous to another.*

Simultaneity is therefore not a universal property of pairs of events. Two events that are separated in space may be simultaneous to one experimenter and not to another. In fact, we see in this example that the event that is more to the rear of the train, which was the reflection along the line perpendicular to the motion of the clocks, is the one that occurred first. This is a general property: if two events are simultaneous to one experimenter, and that experiment is viewed by another experimenter who is moving, then the event toward the rear will happen first. Notice that events separated perpendicular to the direction the train is moving remain simultaneous: this effect applies only in the direction of motion.

This is an inescapable consequence of the way the speed of light behaves: if the speed of light were to combine with other speeds in the way that Galileo and Newton expected, then the experimenter on the ground would agree with the one on the train that the two events were simultaneous. Because light does not change speed when viewed by the different experimenters, simultaneity depends on who measures it.

But the same property of light implies that some pairs of events will have an invariant time-ordering. These are events that can be connected by a photon or, indeed, by something traveling slower than light. Here every experimenter can observe the event where the photon started and the event where it finished, and clearly (since light always travels at speed c) the time between the two events will be non-zero. Pairs of events that can be connected by a single photon are said to be lightlike-separated. If the events occur even closer to one another, so that a particle traveling at less than c can go from one to the other, we say they are timelike-separated. Events that occur too far apart to be connected by a photon are said to be spacelike-separated.

All the effects of special relativity that we have studied – time dilation, Lorentz–Fitzgerald contraction, the loss of simultaneity – are related to each other. We saw above how to relate the Lorentz–Fitzgerald contraction to the time dilation effect. Let us here derive the Lorentz–Fitzgerald contraction from the loss of simultaneity. The first question to answer, and in one sense the deepest, is: how can we, at least in principle, measure the length of something that moves, say the train of the previous example? This requires us to compare its length with the length of a ruler or other length standard that is at rest with respect to us. How do we compare a moving length with one at rest?

One acceptable way is to imagine the train moving past a wall. If we mark the locations of the front and rear of the train on the wall, we can then measure the distance between the marks at leisure to get the length of the train after it has passed. Of course, if we adopt this method, we must insure that we mark the front and rear *at the same time*. It would not make sense to mark the location of the rear at one time and then wait a minute to mark the location of the front: the train would have moved in the meantime, and the marks would not be separated by the true length of the train. They have to be marked at the same time.

We see, therefore, that measuring the length of a moving train in this way requires us to use the notion of *simultaneity*. Since simultaneity depends on the experimenter, so does the length of the train.

It will reward us to look a bit more closely at how this works. We have just seen that a clock at the back end of the train runs ahead of the one at the front, as far as we (experimenters on the ground) are concerned. Therefore, we do not make the marks when these two clocks read the same time, say 10:00. Instead, if we happen to mark the front of the train when its clock reads 10:00 , then we must mark the rear when its clock reads a bit *later* than 10:00. From the point of view of the experimenter on the train, we have not done it right: we have waited too long to mark the rear, and therefore we have obtained too short a length. The rear has caught up a little with the front during the extra time we allowed at the back. But from our point of view, we have done the right thing: we have marked the two ends at the same time, as we measure time. And this leads, of course, to a length that is smaller than the length measured by the experimenter on the train: the Lorentz–Fitzgerald contraction.

The relativity of simultaneity therefore leads directly to the relativity of lengths. Our argument shows that it happens only to lengths oriented along the direction of motion: because simultaneity holds in perpendicular directions, perpendicular lengths do not contract.

The notions of timelike and spacelike separations, which we introduced above, are related to some simple ideas about where and when events occur. If two events are timelike-separated, then by definition there is a particle moving at less than the speed of light that goes from one event to the other. Suppose we jump on a spaceship traveling at this speed. If this particle is at one place next to the spaceship when the first event occurs, then it will be at the same place (in relation to the spaceship) when the second event occurs, because the particle is not moving relative to the spaceship. If follows that, if two events have a timelike separation, there is a class of experimenters who will measure the positions of the two events in space to be the same: they occur in the same place at different times for these experimenters.

By analogy, it is not hard to see that, for any pair of *spacelike*-separated events, there is a class of experimenters who will see them as simultaneous: they occur at the same time in different places.

This has implications for the particle called the tachyon, which we mentioned earlier. If it travels faster than the speed of light, then it can travel from one event A to another B that is spacelike-separated from the first. There will be some experimenters who will measure these two events to occur at the same time, and for these experimenters the tachyon travels infinitely fast. For some other experimenters, the event A at which the tachyon started occurs *after* the event B at which the tachyon finished (the ordering of spacelike-separated events in time depends on the experimenter's state of motion), so for these experimenters the tachyon actually travels backwards in time. For this reason, it is hard for most physicists to believe that tachyons exist: they would seem to upset all our notions of cause and effect.

Real experiment

Simultaneity is important for the GPS that we have already mentioned several times. In order to provide a receiver on the ground with the correct information from which to deduce its position, the clocks on the satellites to be coordinated in some way. Suppose we were setting up this system and tried to arrange for the clocks in the satellites to be synchronized with each other. Consider a pair of satellites, one following the other in orbit around the Earth. If they synchonize with each other, then from the point of view of an observer on the ground, the clock in the leading satellite will be behind that in the following one. This is not acceptable: they should be synchronized as measured by ground-based experimenters. This means that, when the satellites pass time signals among themselves, they must be aware of the fact that their own clocks are not synchronized with one another. If the clocks

▷The situation for the GPS is slightly more complicated, since the satellites orbit in a gravitational field, and this means that gravitational effects on clocks, like the gravitational redshift, must also be taken into account. But the principle of the discussion here is not changed by these complications.

did not behave as relativity requires, then we would soon measure this. The GPS system is, therefore, a continuous demonstration of the relativity of simultaneity.

The mass of an object increases with its speed

As we described earlier when we introduced this idea, by "mass" we mean the usual symbol m that appears in Newton's second law, $F = ma$. This is called the inertial mass of the object. But because the mass changes with speed, this form of Newton's law is not adequate. We saw in Chapter 6 that when the mass of an object changes as it moves, we need to replace the simple version, $F = ma$, with the equation that we called the rocket equation: the product of the applied force with the time it acts equals the change in the momentum of the particle. Since the speed of the particle hardly changes when it is already near c, the increase in the momentum must be almost entirely due to an increase in the mass of the particle.

Thought experiment

It is not hard to derive the formula for the mass of a moving particle from the formula for the relativistic Doppler shift, Equation 15.7 on page 192 above, and the equivalence of energy and mass, $E = mc^2$. The derivation is instructive because it shows that the increase of mass with speed is a direct consequence of time dilation. Since mass and energy are proportional, this is another illustration of the deep connection between time and energy that we first met in Chapter 6.

Imagine a simple physical event, where a particle of rest-mass m_0 suddenly decays into two photons of equal energy. (A particle called the π^0 does this.) The particle disappears altogether, and the photons travel away from it in opposite directions. Seen by an experimenter who was at rest with respect to the particle before it decayed, the two photons have equal energy, $m_0 c^2/2$. That is all we need to know about this event.

Now suppose an experimenter watches the same event while speeding past it at speed v, and suppose that this speed is in the direction that one of the photons takes after the decay. Then the initial energy of the particle depends on its total mass m_1. Let us forget that we have a formula for m_1, and try to derive it by conservation of energy, using the energy of the two photons as measured by this experimenter. The two photons now come off with different energies. The one that is going in the same direction as the experimenter is redshifted. Its frequency as measured by the experimenter at rest was its energy divided by Planck's constant h, $f_0 = m_0 c^2/(2h)$. Its frequency as measured by the moving experimenter is, by the relativistic redshift formula Equation 15.7 on page 192, $f_0 \gamma (1 - v/c)$. Its energy is h times this.

The other photon is going in the other direction, so it is blueshifted. Since it has the same frequency f_0 with respect to the first experimenter, its frequency with respect to the moving experimenter is $f_0 \gamma (1 + v/c)$. Again, its energy it h times this. The total energy, therefore, as measured by the moving experimenter, is $2 h f_0 \gamma$: the Doppler factors of v/c have just cancelled. Putting back f_0 into this gives us a total energy of $\gamma m_0 c^2$. By energy conservation, this must be the total mass-energy of the particle before it decayed, or $m_1 c^2$. We find from this the formula quoted earlier without proof, that the inertial mass is $m_1 = m_0 (1 - v^2/c^2)^{-1/2}$.

Where has the factor of γ come from? It is the new factor of γ in the relativistic Doppler formula. This arose, as we saw above, from time dilation. It is the same factor that causes the transverse Doppler effect. This shows again the deep relationship between time and energy that we first mentioned at the end of Chapter 6.

Real experiment

Verification of the increase of inertial mass with speed comes again from the synchrotron accelerator. Keeping the accelerated electron on the circular track requires the machine to produce precisely the right acceleration, even as it goes faster and faster. This is done by calculating the force needed to produce that acceleration in a particle whose mass depends on speed in just the way Einstein's theory predicts. If this prediction were wrong, accelerators would simply not work.

The record for a particle moving close to the speed of light is not held by a manmade particle in an accelerator, but rather by a cosmic ray. Protons regularly hit the upper atmosphere at high speeds, and are called cosmic rays. They produce the showers of muons that we used to illustrate time dilation. By measuring the muons and other particles produced by the collisions of cosmic rays with oxygen and other nuclei in the atmosphere, astronomers can infer the speed and mass of the incoming proton. The largest energy so far measured is about 10^{21} eV, or about 160 J. This one elementary particle carried as much energy as your body would extract from eating two spoonfuls of sugar! The mass equivalent to this energy (see the next section) is about 10^{12} times the rest-mass of the proton. That means that the proton was traveling at a speed of $0.99999999999999999999999995\,c$ when it hit the Earth!

Energy is equivalent to mass

Thought experiment

Remember that energy is conserved, but it can be converted between different forms. It follows that *any* form of energy put into a particle contributes to its inertial mass. For example, if I begin with two protons, and squeeze them together against their mutual electric repulsion, then the force I have exerted to push them together has done work and put energy into the system. If I then hold them together somehow, I have a system with more energy than I started with. Its mass must be larger than the masses of the two protons alone. If the system is just sitting at rest somewhere, then this mass is its rest-mass. By allowing the two protons to fly apart, I convert some of this rest-mass back into energy, the kinetic energies of the particles. Therefore, even rest-mass, at least for composite particles, is convertible into other forms of energy. Moreover, this is not a mysterious process: one is just releasing energy that was put in when the composite object was assembled. The law of conservation of energy must include rest-mass energy. This is m_0c^2 for a particle of rest-mass m_0.

Notice that, apart from the factor of c^2, mass *is* energy. Rest mass is one part of the energy of a system, but its total mass is its total energy divided by c^2. All forms of energy in a composite system contribute to its rest-mass, and a moving system has a mass that is greater than its rest-mass. This extra energy can be called its kinetic energy, but it is *not* given (except for slowly moving particles) by the Newtonian formula $mv^2/2$. We will find the correct formula below.

Real experiment

When the electron accelerated in a real synchrotron smashes into a target, which is what high-energy-physics experiments usually require, very sensitive and fast measuring machines measure the energy released, in terms of the rest-masses and kinetic energies of all the particles produced in the target. This always totals the energy put into the electron's total mass by the forces that accelerated it.

This law holds to such accuracy that it can allow physicists to discover new particles. The neutrino, which we learned about in Chapter 11, was first noticed this way: the energy (and momentum) in the particles that were identified as having come from the decay (splitting up) of a particular initial particle did not add up to the energy and momentum of the initial particle, even once allowance had been made for the energy in its rest-mass. In 1934, following an earlier suggestion by Pauli, Enrico Fermi showed that the deficits in decays of similar particles could always be made up by postulating that there was an undetected particle that traveled

In this section: the close relationship between energy and time leads in relativity to a link between energy and time dilation, so that the faster an object goes, the more energy it has. And this energy is exactly proportional to its inertial mass.

▷We met both Pauli and Fermi in Chapter 12.

at the speed of light and had no electric charge. He named the particle the "neutrino", which means, in Italian, a small neutral particle. As we have seen, it took roughly twenty years to develop the technology to make detectors sensitive enough to register neutrinos directly.

Nuclear reactors and nuclear weapons are, of course, the standard examples of the conversion of rest-mass into energy. In these devices, a composite particle – often a nucleus of uranium or plutonium – is split into two smaller particles whose rest-masses total less than that of the first. The excess rest-mass appears as the kinetic energy of the two smaller products, and this is the source of energy for the device. The hydrogen – or "thermonuclear" – bomb works the other way, by fusing hydrogen nuclei to form helium, which also has a lower rest-mass than that of the "raw material" nuclei. As we saw in Chapter 11, this is also the power source in stars.

Photons have zero rest-mass

Thought experiment

The fact that a photon must have momentum as well as energy follows from the same thought experiment that we used above to derive the dependence of the mass on the particle's speed. As measured by the experimenter at rest with respect to the particle that decays, there is zero total momentum because the particle was not moving. But now look at the decay of the same particle from the point of view of the moving experimenter. The particle has initial speed v, so initial momentum $m_1 v = \gamma m_0 v$. After the decay, where does this momentum go? If momentum is conserved, which is a fundamental principle we don't want to give up, then the photons must carry away the momentum as well as the energy of the particle. Einstein showed that the momentum p of a photon is related to its energy E by

$$p = E/c. \qquad (16.5)$$

It is easy to show that this formula is exactly what is needed in this case to give momentum conservation. The momentum of the forward-going (redshifted) photon is $(hf_0/c)\gamma(1 - v/c)$, and the blueshifted one has momentum $-(hf_0/c)\gamma(1 + v/c)$, which is negative because the photon is going backwards relative to this experimenter. Added together, these give a total momentum of $-2hf_0\gamma v/c^2$. Putting in $f_0 = m_0 c^2/2h$, we find that the total momentum is $-\gamma v m_0$. Since, before the decay, the particle was moving backwards with speed v with respect to this experimenter, and since its mass was γm_0, this total momentum is exactly the momentum the particle had before the decay. Einstein's formula, Equation 16.5, is just what is needed to insure conservation of momentum.

In fact, this formula is not very mysterious. It is a special case of the general expression for momentum, $p = mv$. This relation holds in relativity just as in Newtonian mechanics provided we use the total inertial mass of the particle. Now, since the inertial mass is just the total energy E divided by c^2, we also have that the momentum is $p = Ev/c^2$. Now, a photon has $v = c$, so for it we find $p = E/c$, as above.

Real experiment

Real accelerator experiments have to take into account both the energy and the momentum of any photons (gamma-rays) emitted in a reaction. Using the rules that the photon has no rest-mass and has a momentum proportional to its energy, such calculations always give consistent answers.

A more direct demonstration of the momentum of a photon is the *radiometer*, a device consisting of four vanes, each painted black on one side and white on the other, able to spin about its axis in a vacuum. When light strikes it, the black sides absorb the light and its momentum, but the white sides reflect the light and there-fore give the outgoing photons new momentum. This means that the force on the white sides exceeds that on the black, and the device spins appropriately.

Massive giant stars are also an example of the effects of the momentum carried by light: radiation pressure is their main support against gravity, and this is nothing more than the exchange of outward momentum from photons to gas particles, pre-venting the gas from falling inwards under gravity. And as a final example, recall our derivation of the Chandrasekhar mass in Chapter 12, which computed the mo-mentum of relativistic electrons from the formula for the momentum of the photon. Gravitational collapse, supernovae, pulsars, and black holes all owe their existence at least partly to the fact that a photon's momentum is proportional to its energy!

> ▷ Do not confuse the radiometer with a device sold in toy shops that looks similar to it, but which spins in air. There the situation is complicated by the heating of the air, which is stronger near a black side, and which can cause the device to spin in exactly the opposite direction from that which would result from light pressure alone!

Consistency of relativity: the twin paradox saves the world

In the section above on time dilation, I described why it is not a contradiction that both experimenters A and B measure each other's clocks to be going slowly. I pointed out that they actually perform different experiments. You might feel this is a little unsatisfying, because all I showed was that there does not have to be an inconsistency in relativity; I didn't really prove consistency.

There is in fact a much more subtle way to try to construct a situation in which special relativity looks self-contradictory. This is usually called the "twin paradox". Because it is clever and it really brings out our conceptual difficulties with relativity, I will describe it here. But note from the start: it fails. It is not a true paradox at all. Relativity comes through it perfectly consistently.

The idea is to get away from an experiment that has to use three clocks to mea-sure the rates of time of two different experimenters. Here is a version of the twin paradox that does not actually involve twins.

> **In this section:** to test the logical consistency of special relativity, we confront the apparent contradiction in time dilation. If one experimenter sees another's clocks to be running slowly, how can the other experimenter measure that the first one's clocks are also going slower?

By the year 2202, overpopulation so threatened the Earth that the government of the (united) planet decided on a radical, but humane, solution. The exploration of the nearby part of the Galaxy for planets similar to the Earth had been fruitless: no place was known where excess Earthlings could be sent to live. Food was becoming a serious problem: almost all the land was used for dwellings, and the oceans had been over-farmed.

Their solution was ingenious: all of Earth's 100 billion people were distributed among 100 million spacecraft (constructed by solar-powered robots that mined the Moon), and each such community of 1000 people was assigned a list of target stars to visit in an attempt to find a new home. Their instructions were to stay at any star that turned out to be suitable, and to broadcast their discovery around the Galaxy so that other communities could go there. (In fact, many communities decided in secret that if they found a good new planet they would never tell anyone else!) If a star had no suitable planets, the community was to return to Earth immediately, re-stock their food supply (which had meanwhile been grown, stored, and packaged by robots), take a one-year vacation, and then head out for their next target. On average, it would take 1000 trips by each community before every star in the Galaxy had been visited.

It was only because of special relativity that this solution could work. At the speed of light, a spacecraft would take up to 60 000 years to cross the Galaxy and return. Yet the people on board would experience such a large time dilation that they would age very little. In practice, for the technology then available, the spacecraft produced an average time dilation factor of $\gamma = 10^4$, so they aged no more than six years on their round-trip. Communities that had nearer target stars returned after even less on-board time. And the plan was made attractive by the one-year vacation between trips: the ratio of this vacation to the length of a trip was better than the ratio of vacation to work-time that most of the population were entitled to in their Earth jobs.

The beauty of this strategy was that it solved the Earth's food shortages *even if no community ever found another Earth-like planet!* The reason was that, while the various communities were away, the Earth had many years to grow enough food to give the travelers when they returned. Since time for the spacecraft population was slowed to 10^{-4} of its rate on Earth, they ate very slowly while away (as measured by Earth clocks)! Time dilation was the perfect appetite-suppressant!

The plan was accepted by the people of the Earth, in most cases reluctantly, and they began to organize their small communities for the first set of trips. But then the political consensus was threatened in a dangerous way. A demagogue who was a fiery and persuasive speaker but a poor physicist began to build up opposition to the plan. Here is what the demagogue claimed.

Consider, instead, time dilation from the point of view of the people in the rockets leaving Earth. Once they reach their steady cruising speed, they are perfectly good experimenters, and when they compare the rate of time on Earth with the rate in their spacecraft, they will see that time on Earth is going slowly. From their point of view, Earth is receding from them at nearly the speed of light and hence suffers an enormous time dilation. Instead of the Earth being the place where food would grow for thousands of years before the community returned, it was really the other way around: the community would eat up its six-year on-board food supply and return to an Earth that had been growing food for only about five hours!

More sober politicians replied that it obviously could not turn out both ways: either the Earth had aged more than the communities when they returned, or they had aged more than the Earth, but not both at once. Then, since the Earth just sat around while the communities zipped around the Galaxy, it was clear that the communities were the ones that were the travelers and suffered time dilation, and not the Earth.

The demagogue replied that relativity claimed that all experimenters were equal, none was better than any other. It was a democratic theory, and there should accordingly be a vote to decide which point of view was right. In any case, how could anyone believe in time dilation at all when it gave two conflicting results? Maybe the right answer was that there was no dilation: the communities would have aged just as much as the Earth on their return, and since that could be as much as 60 000 years, nobody would live long enough to return. The one-year vacation was a bad joke, and the whole plan was a conspiracy by high government officials to get rid of everyone and then turn their own communities' spacecraft around to come back to a depopulated Earthly paradise. Many people began to believe that the demagogue was right.

Despite the polemic style, the demagogue had a point: time dilation is reciprocal, so how can we tell whether the Earth or the small communities will have aged more by the time they meet up again? Does this really indicate a logical flaw in the theory, as the demagogue claimed?

The key to answering the demagogue is to demonstrate that the two "experimenters" are not really on the same footing. The Earth-based experimenter is fine: clocks on the Earth can be constructed, synchronized, and run for the thousands of years required to see the communities return. But the communities are not ideal experimenters. In particular, they have to turn around. This changes their way of measuring time.

A good way to see the effect of this is to imagine a community that has a schism just before reaching their target star. Half of them do not want to tell anyone else if they find a good planet, and the other half do. So only half of them (the honest ones) stop at the star, while the remainder (the secret ones) continue on.

Once the honest ones find that there is in fact no suitable planet at this particular star, they set off on their return journey. But now there is a huge time dilation between the two halves of the original community, since they are traveling at high speeds in opposite directions. Their relative speeds are now even larger than the speed of each relative to Earth. In fact, it is possible to show that the time dilation factor γ between the two groups is twice the *square* of the time dilation factor between each of them and the Earth, provided all speeds are close to the speed of light.

From the point of view of the secret group that continues without stopping (and which is therefore as good a set of experimenters as the people who never left Earth), time on the honest returners' clocks suddenly begins to go incredibly slowly, much more slowly even than time on Earth. From the secret group's point of view, Earth's clocks soon overtake the clocks of the honest returners, and by the time the returners get back, they find they have aged much less than the Earth. Of course, the returners may still not be expecting this, but by turning around they have changed their definition of time in such a way that they no longer agree with the secret group that they had agreed with before: the two groups begin at that point to have different ideas on simultaneity, for example, as well as on honesty.

So the returners are not good judges of what to expect about the behavior of time, and their expectations should be discounted. The demagogue was wrong to rely on their definition of time.

Rather than tell you right away how the political crisis on Earth turned out, I will give you the chance to decide your own ending to the story. Is the argument about the community that splits convincing enough to have defeated the demagogue? Correct physics does not necessarily win votes. My own ending to the story is upside down in Figure 16.3.

Because there is no contradiction between the time dilation measured by different experimenters, there is also no contradiction between consequences of time dilation, such as the Lorentz–Fitzgerald contraction. One can find books full of intellectual puzzles called "paradoxes" of special relativity, but in each case the challenge is to see how the wording of the puzzle leads one into thinking wrongly that the theory is self-contradictory. There are no real paradoxes in relativity.

Here is what happened between the demagogue and the government. The government felt that an abstract argument, such as the one given in the text, might not sway enough people. It therefore decided to demonstrate the appropriate form of time dilation experimentally. It constructed a circular particle accelerator and a source of muons. The source produced muons in bursts. Alternate bursts were fed into the accelerator, which took them to a gamma factor of about 10^4, or into a "muon cooler", which reduced their speeds without allowing them to come into contact with any container. By monitoring the products of muon decay, one could compare the lifetime of a muon at rest (in the cooler) with one in the accelerator. The accelerated muons were a very good analog of the communities of people going to other stars: they traveled at the same speed and kept returning to the same place. Ordinary people were invited to come to the laboratory and measure the rate of radioactive decay. When the muons went into the cooler, the initial radioactivity was high. When they went into the accelerator, the rate of decays went down dramatically. The government guessed correctly: the demonstration convinced the population that the demagogue had done the physics wrong. The communities formed and began their explorations of the Galaxy. Experimental particle physics had, at the last minute, saved the population of the Earth from starvation! The demagogue, however, was not convinced, or was too proud to admit to being wrong, so she persuaded the government to make an exception in her case and to allow her to stay behind on Earth. That suited everybody. She died, of course, before anyone returned.

Figure 16.3. This is my ending for the story of how the Earth saved its population from starvation.

Relativity and the real world

Relativity is not a mere intellectual game, stimulating though it may be. We have seen from the experiments described in this chapter that it plays a central role, not only in physics experiments like big particle accelerators, but also in practical navigation systems and in nuclear power generators.

In fact, our lives depend on special relativity: if the Sun could not convert some of its rest-mass into energy, we would simply not exist. Our evolution has also been critically affected by special relativity at many times in our cosmological history. The protons, neutrons, and electrons of which we are made themselves came into existence about three minutes after the Big Bang; before that the Universe contained only a hot plasma of material in which particles were constantly being converted into photons and photons back into particles – rest-mass into energy and back again, in equilibrium. But, as we will see in Chapter 25, the Big Bang gave us only a Universe of hydrogen and helium, and not much in the way of heavier elements. The important elements of our lives – oxygen, carbon, iron, silicon, nitrogen, phosphorus, sodium, sulfur, chlorine, aluminum, lead, copper, zinc, silver, gold, uranium, and more – were all formed in stars as by-products of the conversion of matter into energy, and in fact we have seen that many of them were formed in the explosion of a big star (a supernova explosion), which happened before our Sun formed and which mixed the new elements into the gas cloud from which our Sun and its planets condensed many years later. Supernova explosions are highly relativistic events, converting something like one percent of the mass of a star into explosive energy, a much higher fraction than in a nuclear bomb. This spectacle of relativity was an essential step on the road to creating life on Earth.

It is in astronomy that the most spectacular consequences of relativity are found. We see protons (cosmic rays) that hit the upper atmosphere of the Earth traveling at speeds incredibly close to c. We see jets of gas shooting out of quasars at nearly the speed of light. We see spinning stars (pulsars) that rotate so fast that their surfaces are moving at one-tenth of the speed of light. We see regions of space containing black holes, that trap light and therefore everything else. In all of these phenomena, gravity also plays a key role in making them happen. It is time, therefore, to make our next step forward, to learn about general relativity. When we make the union of gravity with special relativity, we will begin to understand the Universe.

Spacetime geometry:
finding out what is *not* relative

When Einstein began to develop his theory of gravity, he knew he had to build on special relativity, but he felt strongly that he also had to preserve Galileo's other great contribution to physics, the principle of equivalence (Chapter 1). As with special relativity, Einstein worked by blending the old and the new in equal proportions: special relativity combined the old principle of relativity with the new principle of the universality of the speed of light; in his new theory of gravity Einstein combined the old principle of equivalence with his new theory of special relativity.

Einstein required more than ten years, including six of intensive work, to bring these two principles together in a way that was also consistent with Newton's theory of gravity and with all the observational evidence. The resulting theory came to be called general relativity. Conceptually elegant but mathematically complex, it made a great number of new predictions, almost all of which are now verified by experiment or astronomical observation. General relativity turned Einstein into a household name, and justly so: it is one of the triumphs of theoretical physics.

The observational evidence that Einstein used was mainly the fact that Newtonian gravity was so successful in describing the motion of the planets. The one unexplained gravitational effect was the extra shift of the perihelion of Mercury's orbit, which we described in Chapter 5. Although Einstein knew about this problem, he did not use it to guide his development of general relativity; rather, he kept it to one side and used it as a test of the validity of his equations once he had arrived at them. As we describe in the next chapter, Chapter 18, his theory passed this test with flying colors.

Gravity in general relativity is ...

Let us repeat here the astonishing statement in the last paragraph: Einstein began his quest for a relativistic theory of gravity using essentially the *same* observational evidence about gravity that was available to Newton! The invention of general relativity was not driven by an urgent need to explain new experimental results. Einstein did have something that Newton did not, but it was a theory, not an observation: special relativity. Einstein's main objective was to achieve *theoretical consistency* between gravity and the rest of known physics. It is perhaps all the more amazing, therefore, that in the end Einstein devised a theory that made many new and completely unexpected predictions that could be tested by experiment and astronomical observation.

Our purpose for the rest of this book is to learn about general relativity and its applications. This will take us on a journey to some of the most interesting phenomena in astronomy. We will have to steer a careful course between the rocky shoals of too much mathematical complexity and the becalmed waters of over-simplification. There is a huge amount that can be understood well with the level of mathematics we use in this book, and readers will find that the phenomenology of relativistic

In this chapter: we take our first steps toward understanding general relativity by describing special relativity in terms of the geometry of four-dimensional spacetime. This geometry describes in an elegant and visual way the algebraic predictions of special relativity that we met in the previous chapters. The geometry of special relativity is flat, and we learn how the equivalence principle will allow us to curve it up and produce gravity.

▷Underneath the text on this page is the familiar Mercator projection map of the entire Earth. This map illustrates strikingly the fact that the surface of the Earth cannot be represented faithfully on flat paper. The Earth is curved, and mapping it flat distorts distances. In this case, the distances near the poles are exaggeratedly large.

In this section: we look ahead at the ways we will learn to use general relativity in the rest of this book.

gravity can be understood, not just learned about, from the few basic principles that we develop, carefully, in this and the next two chapters.

Here are some of the things we will learn how to do.

▷Chapter 18
- We shall learn how to reproduce the effects of a Newtonian gravitational field by using Einstein's geometric ideas.

▷Chapter 18
- We shall see how to work out the gravitational deflection of light, getting the correct relativistic value instead of the Newtonian one we found in Chapter 4.

▷Chapter 21
- We shall compute the orbit of a planet around a black hole, and show that the orbit is not a closed ellipse but rather a precessing ellipse, describing a rosette pattern over time.

▷Chapter 19
- We shall learn that the main differences between the predictions of general relativity and Newtonian gravity can be traced to a difference in the *source* of gravity, and in particular the way that pressure helps to create Einstein's gravity.

▷Chapter 19
- We shall deduce that rotating stars and black holes must produce gravitational accelerations that resemble the magnetic forces of electromagnetism, in that they depend on the *velocity* of the object being accelerated.

▷Chapter 20
- We shall compute the structure of a neutron star and see why stars that are too heavy must collapse to black holes.

▷Chapter 22
- We shall compute the effect of a gravitational wave on a detector, and so understand why the new gravitational wave astronomy is so interesting.

▷Chapter 23
- We shall see how gravity creates some of the most beautiful pictures in astronomy, multiplying and distorting images of distant galaxies and quasars as they pass through gravitational lenses.

▷Chapter 25
- We shall calculate the history of our expanding Universe back to the Big Bang, learn how the elements hydrogen and helium were made, and speculate on how the huge amount of dark matter in the Universe helped stars and galaxies to form.

▷Chapter 27
- We shall understand, from the way pressure creates Einstein's gravity, why cosmologists believe that the Universe underwent a period of very rapid expansion at the beginning, and why its expansion may even today be accelerating rather than slowing down.

▷Chapter 27
- We shall glimpse the links between gravitation theory and the theories of the other fundamental forces in physics, as some of the brightest theorists working in physics today struggle to produce a theory of physics containing all the forces in one unified whole.

This is a tantalizing menu for the remainder of our exploration of gravity, but it also an indication of the broad sweep of applications of general relativity in astronomy today. Einstein's invention, devised purely for mathematical consistency, has become essential for the interpretation of the world we see around us. Gravity, the same everyday gravity that Galileo probed with his inclined planes, is the key to understanding the modern Universe.

These predictions of general relativity are radical enough, but what is even more revolutionary about the theory is the *way* it describes gravity.

Until Einstein, gravity was thought of as simply a force, like the electric force. Einstein described gravity instead as geometry.

Rather than being a force exerted by one body directly on another, gravity was more indirect: one body would cause space and time to curve, and the other body would move in response to this **curvature**. This is unfamiliar language for us: we are used to the idea of a force, but what does it mean that gravity is geometry? The purpose of this chapter and the next is to help us to understand Einstein's way of thinking about gravity.

...geometry

Since Einstein describes gravity in terms of geometry, our natural first question is, what do we mean by the word *geometry*? Consider ordinary spaces we are familiar with, such as the surfaces of spheres, or the **Euclidean plane** as represented by a flat piece of paper. All such spaces are smooth and continuous, but when we speak of their geometry we mean something more: we mean their shape, the distances between points in the space, and so on. We calculate distances typically by using coordinates. For example, if I give you the latitude and longitude of both New York and London, you could in principle calculate the distance between them along a great circle on the Earth's surface. This sort of calculation is routine for airlines.

Now, the latitude and longitude of a city are coordinates that locate it on the Earth, just as the *x*- and *y*-coordinates locate points on a graph. We generally need coordinates in order to specify which points (cities) we are talking about, and then we use them to compute the distance. But we know that the distance is something that does not depend on the coordinate system we use. For example, we might use longitude measured, not from the Greenwich meridian, but from (say) a line passing through Disneyland California: we could call this the Mickey Mouse coordinate system for the Earth. Although this would change the values of the longitude coordinate we use to describe every city, it would not change the distances between cities.

We want to describe the geometry of relativity. We have already seen that time and space must both be involved, since both are distorted and even mixed by the Lorentz–Fitzgerald transformation. We must therefore explore the geometry of spacetime, the four-dimensional continuum with three spatial dimensions and one time dimension that is the arena for special and general relativity. The unification of space and time into spacetime is one of the most important conceptual advances that special relativity led physicists to. We define and explore it in the next section.

The geometry of a space, like the Earth's surface, is described by the distances between places, not the coordinates of the places. It is something that is a property only of the space itself. When we study the geometry of special relativity and then of spacetimes with gravity, we will of course have to use coordinates (such as *t*, *x*, *y*, and *z*) to describe events in the spacetime. But we have seen that in special relativity two different observers will use different coordinates. The geometry of the spacetime must not depend on which observer describes it. So we must find ways of describing the geometry using invariant distances between events.

This invariant will be called the **spacetime-interval**. This is a word we have used often in this book to represent a particular lapse of time. In relativity it is used in a very specific manner, to represent a measure of separation of events in time or space that is agreed by all experimenters, independent of the coordinates of the events. We will define it later in this chapter and then use it repeatedly through the rest of the book. The geometry of spacetime is determined by the spacetime-intervals between events. Spacetimes that describe gravitational fields

In this section: we learn what geometry is and why it can be used to explain gravity. The key is a distance measure in spacetime called the interval.

will differ from the spacetime of special relativity by having different spacetime-intervals between events.

Spacetime: time and space are inseparable

In this section: spacetime is the four-dimensional arena in which all things can be described. Einstein showed that we cannot separate space from time easily. We learn the language of spacetime and illustrate the entanglement of space and time with an example loosely based on the legend of William Tell. We will see how to describe the geometry even of special relativity.

Our study of special relativity in the previous chapter has already told us that the world is not constructed in the way we may have thought, and certainly not in the way that Galileo and Newton thought. Time is not absolute: different experimenters measure it differently, and no single experimenter has a better definition than another. Nor is space absolute: solid objects have different lengths when measured by experimenters moving at different speeds.

These ideas were just as troubling and counter-intuitive to physicists of Einstein's day as they are today to new students who encounter them for the first time. When physicists began to think more deeply about *why* the ideas were troubling, they found that it helped them to stop thinking about time and space as distinct and separate things, and instead to join them together. Since time is one-dimensional (the history of ancient Rome, for instance, can be ordered along a single line) and space has three dimensions, their combination is a four-dimensional realm. We call this spacetime.

A single point of spacetime occupies, therefore, both a particular location in space and a particular moment in time. Just as space is the collection of all "places", or *points*, spacetime is the collection of all "happenings", or events. A **spacetime diagram** is a graph that records the entire history of an experiment or of some other process.

We can clarify what this means by drawing a spacetime diagram, as in Figure 17.1. I will illustrate the idea by recording in this diagram, in a simplified way, the history of the legendary episode where the Swiss patriot William Tell was compelled to shoot an apple from the head of his son. The diagram can only show two of the four spacetime dimensions, so I have chosen to show time (vertically) and the x-direction of space (horizontally). I align my x-direction with the direction the arrow took when flying from Tell to his son.

In the left-hand panel of this figure, we imagine that Tell stands at the origin of the space coordinates ($x = 0$) and fires the arrow at time $t = 0$. The *event* of firing the arrow is the intersection of the time and space axes. The arrow does not remain at the origin, but instead moves to the right (positive x) as time increases. This motion is shown as a slanting dashed line. This is the history of the arrow's progress from Tell to the apple: at any time t one can find out where the arrow was by just looking at the point on the line where its t-coordinate value has the desired value. Such a line is called a **world line**.

Similarly, the apple has a world line. This is the vertical dashed line. It is vertical because the apple does not move while the arrow approaches: it stays at the same value of x all the time. The single *event* at which the arrow pierces the apple is then the intersection of the two world lines, shown as the gray dot. This event belongs to the histories of both the arrow and the apple. We see that *events* are represented as single points in the diagram: they have no extent in time or space. *Objects*, on the other hand, remain objects through time, so they are represented by continuous lines that go upward in time.

Notice that William Tell's son is not shown. Fortunately for him, his head's world line does not intersect that of the arrow! His head stayed at the same x-location as the apple, but it was at a different height (say, a different value of the coordinate z). Therefore we can't show it on this diagram: we have no room for the z-dimension here. Nor can we show Tell's subsequent escape from the scene, in which he rode off in the y-direction!

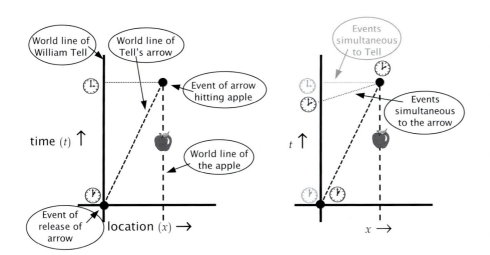

Figure 17.1. Spacetime diagram of what happened when William Tell shot the apple. The arrow starts at the spatial origin (x = 0) and travels to the right as time increases (slanting dashed line). The apple stays in one place (vertical dashed line). The intersection (gray dot) of the world lines of the arrow and apple represents the event at which the apple was pierced by the arrow. For an explanation of the way the diagram illustrates time dilation, see the text.

Relativity of time in the spacetime diagram

The idea of drawing this diagram might have occurred to Newton as easily as to us, so where does relativity come in? One way to introduce relativity is to ask about how much time it took the arrow to fly from Tell to the apple. Suppose, for example, that Tell was wearing a very accurate Swiss wristwatch, whose world line is the vertical axis because the watch (and Tell) stayed at the origin $x = 0$ during the flight of the arrow. Suppose, next, that Tell looked at his watch just when the arrow hit the apple. Importantly, we don't mean that Tell looked at his watch when he saw the arrow hit the apple: this would be later than the time at which the arrow actually hit, since it would take light some time to reach Tell so that he could see it. As a good experimenter, Tell would have to correct for the time it takes him to get the information that the arrow reached its goal. We assume that he does this, that he manages to look at his watch at exactly the moment (according to his measurements) when the arrow hits the apple. What time was it on the watch then? Since the two events did not occur at the same place (the watch was on Tell's wrist when he looked at it, while the arrow was somewhere else), relativity tells us we should be careful in answering this question.

Before we answer, let us look more carefully at the diagram itself. Who is the experimenter who recorded the time and position of the apple and arrow in order to draw the diagram? Assuming the experimenter was careful and accurate, the only important question is, what was his or her state of motion? Clearly, since the apple stays at the same x-position in this diagram, the experimenter was at rest with respect to the apple, and therefore at rest with respect to the ground. The diagram we have drawn is the natural way for such an experimenter (Tell himself, for example) to describe these events.

Tell's wristwatch is therefore a good recorder of this experimenter's definition of time. Since all events that occur at the same time are at the same height in this diagram, we only need to draw a horizontal line from the gray dot to the vertical time-axis in order to discover what time Tell thinks it is when the arrow strikes the apple. This line is shown as a dotted horizontal line in the left-hand panel.

But we know that a different experimenter, say one who was moving at the same speed as the arrow, would give a different answer to this. If Tell, in a demonstration of supreme self-confidence, had attached a similar Swiss watch to the arrow just before releasing it, then because of the time dilation effect, the flying clock would

In this section: we continue to work with William Tell, seeing here how to represent the effects of special relativity on time.

have ticked a little less time when it reached the apple than Tell would have seen when he looked at his wristwatch. Of course, for realistic arrow speeds, this would be an incredibly tiny time difference. Let's agree, for the fun of the story, that the watch is accurate enough to measure this difference!

Because the time on the flying watch was less than the time at which Tell looked at his own watch, the flying experimenter would have decided that the arrow hit the apple a little *earlier* than Tell believed: this experimenter believes that Tell looked at his watch too late. Put another way, the set of all events that the moving experimenter regards as being simultaneous with the piercing of the apple is a line that is not horizontal in this diagram: it tilts a little upwards to the right in order to intersect Tell's wristwatch's world line at a time that is earlier than the moment when Tell looks at the watch. This tilted line is the dotted line shown in the right-hand panel of the figure.

The two panels in this figure show, in a striking and graphical way, how the idea of time changed from Newton to Einstein. For Newton, time was universal: provided two experimenters synchronized their clocks and agreed to start them at the same time, they would always agree on what events occurred at what times. The left-hand panel of Figure 17.1 on the preceding page would represent Newton's concept of time correctly for any experimenter. Horizontal lines would connect events that were simultaneous, and all experimenters would agree on this.

But Einstein's time dilation, which leads to the loss of simultaneity, changes all that. For the experimenter at rest with respect to the apple, lines that are horizontal in this diagram are lines of constant time. For the experimenter flying with the arrow, lines of constant time are tilted with respect to the first set of lines.

The flying experimenter would of course not draw them tilted in his or her own spacetime diagram. In that diagram the arrow would be at rest, the moving apple would follow a world line slanted to the left, and Tell's line of constant time would be tilted upwards to the left. Readers may find it interesting to draw a spacetime diagram the way the flying experimenter would.

Time dethroned ...

In this section: the special role that time played in Galilean and Newtonian physics does not continue in relativity. Instead, we need to know about space and time together.

If time is not universal, do we have to forget completely our present notions of time? No, because relativity preserves the most important aspect of time, which is the separation of cause from effect. Relativistic time keeps a consistent direction: we saw in the previous chapter that the ideas of future and past are still well-defined. Different experimenters may disagree on the order in time of events that cannot have a causal relationship with one another, such as the event of piercing the apple and the event of Tell glancing at his watch. But they will always agree on causally related events, such as the fact that Tell fired the arrow *before* it hit the apple.

In relativity, however, there is not just one time. Time is best regarded as a coordinate in spacetime, just as the value of x is a coordinate. We know what it means that x is "just" a coordinate: two different experimenters could orient their x-axes in different directions. For example, there was no need for me to have put the flight of Tell's arrow on the x-axis; the y-axis or halfway between would have done just as well. We are used to thinking that there is no special, or preferred, orientation for the x-, y-, and z-axes. Since Einstein, time is the same kind of thing: one experimenter draws the line of simultaneity horizontally in the spacetime diagram in Figure 17.1 on the previous page, while another draws it slightly tilted.

This tilting of the lines of constant time is the main reason that physicists find it so useful to think of the four-dimensional spacetime as a single continuous entity. Three-dimensional space could be represented by a horizontal slice through spacetime, just as we represent the x-axis by a horizontal line in the figure. But how horizontal: which experimenter do we use to define the slice? There is no unique way of identifying three-dimensional space within spacetime. Spacetime is a single continuum; space and time can be separated from one another only by choosing the coordinates of a single experimenter, and the separation will depend on which experimenter we choose.

So far, we have only described spacetime as a kind of four-dimensional map that charts not simply places but entire histories. This is how we will look at the Universe through the rest of the book. But to describe gravitation, we need to put some real geometry into spacetime.

> It is not enough just to have a map. We have to know whether the map is flat, curved, crumpled, perforated, whatever: gravity is in the crinkles in the map of spacetime!

...and the metric reigns supreme!

The fundamental reason that relativity merges time and space into spacetime is that time and space are separately not invariant: different experimenters get different results for the length of time something takes, or the distance between them. But spacetime itself is invariant, and it is one of the most remarkable facts about special relativity that it provides us with a new, invariant, unified measure of distance in *spacetime* itself. This measure, which is called the spacetime-interval, is a combination of distances in space and in time. It measures the spacetime distance between *events*. And it is invariant: all experimenters who measure times and distances carefully will get the *same* value for the spacetime-interval between any two events, even if they get different individual values for the time difference and distance between the events.

Here is the definition of the spacetime-interval. Suppose, as measured by a certain experimenter, two events are separated by a time t and a spatial distance x. Then in terms of these numbers the spacetime-interval between the two events is the quantity

$$s^2 = x^2 - c^2 t^2. \tag{17.1}$$

Notice that this is written as the square of a number s. The spacetime-interval is the quantity s^2, not s. In fact, we will not often deal with s itself. The reason is that s^2 is not always positive, unlike distance in space. If ct is larger than x in Equation 17.1 then s^2 will be negative. In order to avoid taking the square-root of a negative number, physicists usually just calculate s^2 and leave it at that. You should just regard s^2 as a single symbol, rather than as the square of something.

This quantity is important *because* it is invariant. Two different experimenters can calculate it and will get the same answer. Let us see how this happens by first calculating a spacetime-interval from William Tell's measurements, and then from those of the flying experimenter. We will compute the spacetime-interval between the following two events: the first event is the firing of the arrow, and the second is the piercing of the apple.

Suppose that, as measured by Tell, the distance of the shot was ℓ and the time of flight of the arrow was t. Then the spacetime-interval, as measured by Tell, is

$$(s^2)_{\text{Tell}} = \ell^2 - c^2 t^2.$$

Notice that this will come out to be negative. The distance light would travel during the flight of the arrow is ct, and this must of course be much larger than the actual distance the arrow traveled, ℓ. Therefore the spacetime-interval between these events is negative.

In this section: the rule for calculating intervals in spacetime is given by the metric. The intervals are experimentally measurable and will be agreed by all experimenters, just as distances computed from the theorem of Pythagorus are agreed by all measurers in Euclidean geometry. This is the key to an invariant idea of the geometry of spacetime.

▷There are a number of different variations in the definition of the spacetime-interval. Some scientists multiply it by –1 or divide it by c^2 before calling it the spacetime-interval. Since these multipliers are constants, these differences are simply matters of convention, like measuring spatial distances in miles or kilometers.

▷If this sounds strange, place yourself on a moving train going into a tunnel: from your perspective, first the entrance to the tunnel rushes over you with a whoosh, then there is the noise of the tunnel, and then the end of the tunnel passes over you and the sky re-appears. You have not moved from your seat during all of this.

The flying experimenter, moving with the arrow, sees space and time differently. For one thing, the arrow stays in one place relative to this experimenter, so the distance x between the two events is zero. In this view, first William Tell and then the apple fly past the experimenter. Tell releases the arrow just as he passes our experimenter, the arrow stands still, and then the apple smashes into it! As we have seen above, the time τ between the two events, as measured by the flying experimenter (or equivalently by the Swiss watch attached to the arrow), is shorter than that which clocks on the ground measure. By the time dilation formula, the time τ is

$$\tau = t/\gamma.$$

From this we can easily compute the spacetime-interval measured by the flying experimenter, bearing in mind the definition $\gamma = (1 - v^2/c^2)^{-1/2}$. Since the distance is zero, we have simply

$$(s^2)_{\text{flier}} = -c^2\tau^2 = -c^2t^2/\gamma^2 = -c^2t^2(1 - v^2/c^2) = -c^2t^2 + v^2t^2.$$

Now, the product vt is just the distance ℓ that the arrow flies as measured by Tell, which shows that this spacetime-interval is exactly the same as the spacetime-interval as computed by Tell. Even though the two experimenters measure different values for distance and time, the differences are just what is needed to insure that the spacetime-interval they compute is the same for both.

> The spacetime-interval between two events does not depend on the experimenter who defines time and space. A different experimenter may, because of Lorentz–Fitzgerald contraction and time dilation, assign different values to Δt and Δx, but the spacetime-interval between two given events, calculated using Equation 17.1 on the previous page, will be the same for all experimenters.

The spacetime-interval is the most fundamental number associated with pairs of events. It gives a distance measure on spacetime. You may already have noticed that it has a certain similarity to the Pythagorean theorem, which defines distances in the ordinary two-dimensional plane:

$$\ell^2 = x^2 + y^2. \tag{17.2}$$

The spacetime-interval differs from this equation by using a minus sign in front of one of the terms. This is an important difference. If two events occur at the same time (as measured by some experimenter) then the spacetime-interval is the same as the square of their Pythagorean distance, which is positive. But if they occur at the same place at different times then their spacetime-interval will be the negative of the square of the time between them. This change of sign is what keeps time distinct from space in relativity: although neither is absolute, and they will be separated from one another in different ways by different experimenters, time and space are not identical, and the different sign in the spacetime-interval is the way the distance measure on spacetime respects that difference.

The formula for the spacetime-interval also contains the constant c, which can be thought of as a weighting factor between space and time, telling us how much a given time difference contributes to the spacetime-interval in relation to distances. Since the weighting is a constant, it is not particularly important here. But we will see when we discuss curved spacetimes below that *curvature* comes about essentially when the relative weightings of terms in the spacetime-interval change from place to place in the spacetime. Gravity is represented mathematically by the weighting factors in the spacetime-interval. Because the weighting factors in special relativity are constant, we say that the spacetime of special relativity is *flat.*.

The spacetime-interval has another name: the **spacetime metric**. The word *metric* is used in geometry for distance measures, particularly those that involve the squares of coordinate separations. Our discussion of relativity shows us that the metric takes over the role that, in Newtonian mechanics, was shared by separate measures of time and distance.

The geometry of relativity

Now we are ready to talk about geometry: we have a space (called spacetime) and a distance measure (the spacetime-interval). Einstein himself did not at first seem to think geometrically about spacetime. It was his former mathematics professor in Switzerland, Hermann Minkowski (1864–1909), who pointed out how important the geometry of spacetime was. Because of his contribution, physicists refer to the spacetime of special relativity by the name **Minkowski spacetime**.

Figure 17.1 on page 215 is a depiction of part of Minkowski spacetime. I have not marked time and distance units along the axes. If I had, and if I had used conventional units like seconds for time and meters for length, then the world line of a photon would be so tilted over that one would not be able to distinguish it from a line parallel to the x-axis. Its slope, $\Delta t/\Delta x = 1/c \approx 3 \times 10^{-9}$ s m^{-1}, would be too shallow to draw. Put another way, in one second of time (the vertical direction in the diagram), a photon would travel such a large distance in x (the horizontal direction) far that it would not only be off the page, it would be off the Earth!

This would therefore not be a good way to draw a diagram of a part of spacetime in which we want to record relativistic effects. Since relativity only differs from Galilean and Newtonian physics when things move at speeds close to c, it is better to use different *units* in a spacetime diagram to keep a photon's world line at a reasonable angle. The units that many physicists use for spacetime diagrams involve a re-scaling of the time coordinate to one we shall call T:

$$T = ct. \tag{17.3}$$

This is shown in Figure 17.2. The time coordinate has been expanded so much by this re-scaling that now a photon world line has a slope of one. This new time coordinate has dimensions of distance: we measure time by the distance light travels in that time. One meter of time is the time it takes light to go one meter, or 3×10^{-9} s. This is similar to something that most people are familiar with, measuring distances in light-years. One light-year is the distance light travels in a year. If we had re-scaled the x-axis to light-seconds, and kept time in seconds, we would similarly have produced a photon world line with a slope of one. But it is more conventional in relativity these days to re-scale time, so that time is measured in "light-meters".

In terms of these new units, we can write the spacetime-interval in the simpler way:

$$s^2 = x^2 - T^2. \tag{17.4}$$

The striking thing about this distance measure is that it can be either positive or negative. This is very different from the distance measure in ordinary space. This means that the geometry of spacetime will have something that is not familiar from ordinary **Euclidean geometry**: the spacetime-interval between two well-separated events can actually be zero. Let us face this squarely: what is a zero spacetime-interval?

In this section: the spacetime of special relativity is called Minkowski spacetime. Although there is no universal way to separate it into time and space, there are invariant divisions of spacetime. They are given by light-cones.

▷When we learn about spacetimes that represent gravitational fields – black holes, cosmologies – we will see that many are named after their discoverers: Schwarzschild, Kerr, Friedmann, and so on. Minkowski has the distinction of having been the first to describe an important spacetime in relativity, the spacetime that has no gravitation!

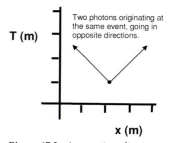

Figure 17.2. A spacetime diagram in natural units, showing the world lines of two photons, one traveling to the right and the other to the left.

If I set $s^2 = 0$ in the previous equation, I find $x = \pm T$. Remembering the definition of the re-scaled time T in Equation 17.3 on the preceding page, this implies

$$x = \pm ct. \tag{17.5}$$

This is just the equation for something moving at the speed of light. The positive sign is for a photon going to the right, the negative sign to the left. The two lines that are drawn in Figure 17.2 on the previous page are the lines that go through all the events that have zero spacetime-interval from the event at the origin of the diagram. We call these lines the **light-cone**, because of what it would look like if we added a further spatial dimension to the diagram. If we include the y-axis, say pointing out of the page, then there are world lines of light that move out from the origin in all directions in space, always moving forward in time at the speed of light. These lines, taken together, form a cone whose apex is at the origin. This is the light-cone of the origin.

Other events have light-cones too: the set of all light world lines that pass through a given event is the light-cone of that event. Any event on this light-cone will have a zero spacetime-interval from the original event. Since the spacetime-interval is independent of which experimenter measures it, the light-cone of any event is an *invariant*: all experimenters will assign the same events to the light-cone of any given event. If you think about this, you will see that this is nothing more than one of the fundamental principles of special relativity: all experimenters measure the speed of light to be the same value c.

Events have light-cones going into the past as well. These consist of all light world lines that converge on the given event from the past. We speak of the past light-cone and the future light-cone of any event.

The invariance of the light-cone has other consequences. It divides spacetime into separate regions, relative to a given event. The interior of the future light-cone consists of events that are separated from the given event more by time than by space, so they have negative values of their spacetime-interval from the event. They are called the timelike future of the event. Similarly, the timelike past is the interior of the past light-cone. The exterior of the light-cone is a single region whose events are separated from the given event (the one at the apex of the cone) more by space than by time, so this region is the spacelike "elsewhere" of the given event. All experimenters will agree on this division of spacetime relative to a given event.

Proper measures of time and distance

In this section: the spacetime-intervals lead to definitions of proper time and proper distance that all experimenters will agree on.

Just as the Pythagorean distance in space is the true distance that someone would measure if they walked along the line, so is the spacetime-interval a measure of the true distance, or **proper distance** in spacetime. If I want to measure the length of something, even say a moving train, then as we saw in the last chapter I must make the measurement at a given time: I have to take the distance between the locations of the ends at the same time, according to my own clocks. This means that when I compute the spacetime-interval between the events that I used (the events that correspond to the locations of the two ends at the given measurement time), then the time-difference is zero and the spacetime-interval will be exactly the square of the distance that I measure. We say that the spacetime-interval gives the proper distance between two events that have a spacelike separation; in other words, it is the distance that an experimenter would measure between them if the events were simultaneous to the experimenter.

The same holds for timelike spacetime-intervals. For example, we saw above that the watch that William Tell attached to the arrow ticked a time whose square was just (in our units) the negative of the spacetime-interval between the events. So the spacetime-interval along a timelike world line tells us what the rate of ticking is of a clock that moves along that world line. All we need to do is to take the square-root of its absolute value, and divide by c if we want time measured in time units instead of distance units. This measure of time, the ticking of a clock that moves along a world line, is called the **proper time** along the world line, and is usually denoted by τ:

$$\tau = \frac{1}{c}\left|s^2\right|^{1/2}. \tag{17.6}$$

It is interesting to ask what are the events in spacetime that have a given proper distance or proper time from the origin, say in Figure 17.2 on page 219. In ordinary space, the points that are at the same distance from the origin form a circle (in two dimensions) or a sphere (in three). What is the analog in spacetime?

In Minkowski spacetime, the points at the same *spacetime-interval* from the origin satisfy an equation of the form

$$x^2 - T^2 = \text{const.} \tag{17.7}$$

This is the equation of a hyperbola in the x–T space. We call this the **invariant hyperbola**.

One such hyperbola is shown in the diagram in Figure 17.3. Since the spacetime-interval tells us the time measured by a clock traveling on a world line (the proper time), a clock that moves on a straight world line from the origin to any point on the hyperbola ticks the same total proper time, regardless of which world line it moved on. Since different straight world lines describe clocks moving at different speeds, the points on the hyperbola are the events that can be reached from the origin in a fixed given proper time (in the case shown, this is 5 m of time, or 1.6×10^{-8} s) by traveling at different speeds.

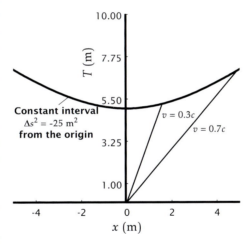

Figure 17.3. The hyperbola connects all the points that can be reached from the origin in a proper time of 5 m (measuring time in light-meters) by clocks traveling at different speeds. Two such clock world lines are shown.

Because of time dilation, the clocks that travel faster take longer to tick out the given proper time, so one can see in Figure 17.3 that they go further than one would have expected without special relativity. As an extreme case, it would be possible in principle to send a team of astronauts across the Galaxy and back in a proper time as short as, say, a year, if we could accelerate their rocket to a speed sufficiently close to c. This was the basis of the time dilation fantasy at the end of the last chapter.

Equivalence principle: the road to curvature . . .

The spacetime of special relativity is the simplest one to describe, which is why we have spent half of this chapter on it. But it is also the most boring of all the spacetimes we will meet: it has no gravity, it is flat and the same everywhere, it is just a static background arena for the events that happen in it. Spacetimes that have gravity are more interesting: they participate in the physics that happens in them, they affect the physical systems that they contain, and they make possible all kinds of new phenomena, such as black holes (Chapter 21) and gravitational waves (Chapter 22). They do this through their curvature.

In this section: the spacetime of special relativity is flat. Curvature is needed to describe gravity. The key to understanding what curvature means in spacetime is the equivalence principle.

It is the equivalence principle that nudges us along the road to the picture of gravity as curvature of spacetime. According to Galileo's version of the principle of equivalence (Chapter 1), the effect of gravity on a body depends only on its state of motion: given an initial position and velocity, the subsequent motion of the body is fully determined. The body's color, number of baryons, charge, and so on do not affect its gravitational acceleration.

Now, the location and velocity of a body are both properties of the body's world line. Consider the world line of a body going at a constant speed in the x-direction, shown in Figure 17.4. The speed can be inferred from the slope of the line. This would also be true of an accelerated body, whose world line would not be straight: its slope at any point gives its speed at that time.

▷This means the slope of its tangent at that point.

Figure 17.4. The world line of a body contains the information needed to compute its speed. Unlike the path in ordinary space, the world line tells us where the body is at any time, so one can read off its speed. The diagram shows that the speed between two events is the inverse of the slope of the world line joining the events. Since the diagram uses the time coordinate $T = ct$, the inverse of the slope is the body's speed relative to c, in other words v/c.

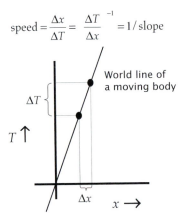

$$\text{speed} = \frac{\Delta x}{\Delta T} = \frac{\Delta T}{\Delta x}^{-1} = 1/\text{slope}$$

Now, according to the equivalence principle, if we know the location and the velocity (the speed and its direction) of a body, then gravity completely determines what its subsequent motion will be. Put another way, if we know that the world line of a body passes through a certain event with a certain slope, then that information completely determines the rest of the body's world line, in a given gravitational field. Every particle that starts out from that event with that slope (velocity) will follow the same world line, provided of course that gravity is the only influence on it.

This means that gravity determines *preferred* world lines, the ones which all particles follow, regardless of their mass, color, etc. The world lines of particles as they move through spacetime under the influence of gravity can therefore be thought of as properties of the spacetime itself, not of the particles: it is irrelevant whether it is a proton or a piano in orbit around the Sun, the orbit will be the same. Einstein reasoned that we should focus our attention on these world lines, and find a description of gravity that showed why certain world lines were special.

Notice that it is crucial that we be in *spacetime* for this geometrical argument to work: it would not work in ordinary three-dimensional space. The path of a body in ordinary space traces out where it has been, but does not tell us when it was there, so it does not tell us its speed. William Tell's arrow, once released, follows the world line of a free body until it encounters the apple. But its spatial path, which we have drawn as a portion of the x-axis, could equally well be followed by a bird, supported by the aerodynamic forces on its wings. The spatial path by itself does not contain enough information to tell us about the effect of gravity. There is thus no possibility that gravity could be related just to the geometry of ordinary space. The equivalence principle tells us to think about the geometry of the full *spacetime*.

... is a geodesic

In this section: freely falling bodies follow world lines in spacetime that are as straight as possible. These are called geodesics. Orbits are curved lines when projected just into ordinary three-dimensional space, but in spacetime they are locally straight.

Now, mathematicians know that curved spaces have special sets of lines. Consider one of the simplest examples of a curved space, the ordinary sphere. The special lines on the sphere are the great circles. These are the curves of "least effort": if you walk on a large sphere (like an idealized Earth), and if you just keep walking straight ahead and never bother to make up your mind to change direction, then you will walk along a great circle.

Suppose you meet a little dimple in the sphere, a shallow basin rather like a crater on the Moon. If you keep "following your nose" in this way, your path will generally emerge from the dimple going along a different great circle. Your path has been deflected, but you never decided to change your direction. You followed your nose, putting one foot in front of the other, and the geometry decided your direction for you.

By following your nose, you are imitating Newton's first law of motion: once a body is set in motion, it will continue on a straight line unless acted upon by a force. Einstein's new idea was that gravity should *not* count as an external force acting on the body, but rather that bodies affected by gravity are really just obeying Newton's first law. They keep going as straight as they can. If their world lines curve, that is because spacetime itself is curved.

These special paths are called **geodesics**. Geodesics are defined as lines that go as straight as possible on a surface.

> If the geodesics of a surface are straight, so that geodesics that start out parallel remain parallel and don't intersect one another, then mathematicians call the surface a **flat space**.

In the spacetime of special relativity, particles move on straight world lines. The spacetime of special relativity, Minkowski spacetime, is therefore by this definition a flat spacetime. So this definition of flat and curved is consistent with our earlier, rather vaguer, notion that Minkowski spacetime is flat because it is boring!

The equivalence principle: spacetime is smooth

Everyone has experience of curvature by handling curved surfaces, so it is helpful to try to understand geodesics on such surfaces first. One of the first distinctions we make about surfaces is whether they are smooth or have a corner. A curved surface that is smooth can still bend sharply, but it does not have an edge where a lot of bending occurs all at once. The difference is the following. If you look closely at any point on a smooth curved surface, then in a small enough area around it, the surface is effectively flat: it is not much different from a flat piece of paper. If you look closely at a point on a true corner, it is still a corner: there is always the same amount of bending, no matter how small an area around the point we look at.

> In general relativity, we always assume spacetime is smooth in the same way that a curved surface is smooth. Gravity may be strong, but it does not concentrate its curvature in sudden jumps.

Now, consider the geometry of a very small patch of a smooth surface. If we stay near enough to the central point, it is hard to tell that the surface is curved at all. Think, for example, about the Earth. On a large scale, the Earth is curved: we all know that it is impossible to draw a map of the entire Earth on a flat piece of paper and expect it to represent faithfully the distances between all the points. Maps of the Earth are generally cut in a number of places to keep their distortions to a minimum, and even these maps are not perfect. However, maps of cities do not have such problems. To the accuracy with which we need to represent the distances between places in a city, the curvature of the Earth is not important. Of course, local curvature caused by hills might still prevent a map from being faithful. In that case, faithful flat maps could only be made for smaller areas, such as a section of the hillside. Mathematicians say that the Earth (and any other smoothly curved body) is **locally flat** because we can draw a local map that is flat and is as accurate as we want it to be, provided we cover a small enough part of the space with it.

In this section: the equivalence principle links the spacetime of special relativity with that of general relativity: curvature is noticeable only over large regions.

▷ This is almost always true, but like most rules it can have exceptions. The cosmic strings that we will discuss in Chapter 19 are exceptions.

For gravity, the local flatness of spacetime means that, in a small part of space-time, gravity does not matter: spacetime looks like a portion of Minkowski space-time. We have actually met this before, in Chapter 2. There we saw that the modern way of phrasing the equivalence principle is that a freely-falling experimenter does not see any effects of gravity, at least in a small enough region. We now see that we can re-phrase this in geometrical language:

> The curvature of spacetime is smooth, and the locally flat observers are the freely-falling experimenters. Experimenters who fall freely with exactly the Newtonian acceleration of gravity see no gravity; photons and other particles move on straight lines in their local coordinates. The equivalence principle tells us that the geodesics of a spacetime with gravity are the paths of freely-falling objects.

Local flatness gives us a way of drawing a geodesic on a curved surface, at least in principle. If you are drawing a great circle on the sphere and then hit that dimple, how should you continue? Just make a small map of the locally flat region around your present location, near the (smooth) dip into the dimple. Make the region covered by the map much smaller than the dimple. Then follow the straight line on the map that is going in the same direction you have been going up to now. When you get to the edge of this map, draw a new one, also very small, around the new location and do the same thing. Always go straight according to the map. In the end you will have changed direction, because all the little maps can't be joined into one large flat map. But it is by following the little maps that you can determine your geodesic path.

Einstein's great insight was to understand that the equivalence principle lead naturally to a picture in which the effects of gravity could be represented by a curved spacetime whose geodesics, constructed in the way we have just described, are the paths that particles follow in the gravitational field. What is more, two locally straight paths that start out near one another will not remain exactly parallel; the may diverge or cross as they move through regions of slightly different curvature. This is a perfect representation of what we called "tidal forces" in Chapter 5: the forces that make nearby freely-falling particles diverge or converge. So the curvature of spacetime describes the tidal forces. Since the tidal forces are the part of gravity that can't be removed by going to a freely-falling observer, the curvature of spacetime *is* gravitation.

Einstein came to his insight early in his search for the right theory of gravity, and in fact there is no reason why other scientists could not have come to the same conclusion much earlier. After all, there is nothing relativistic about this picture of gravity as curvature. In Newtonian physics there is also a spacetime, which Newton did not talk about, but which we drew in Figure 17.1 on page 215. Newtonian gravity incorporates the equivalence principle, so it would have been possible for anyone who knew geometry to have reformulated Newtonian gravitational mechanics in this way. Scientists have done this since, but nobody did this before Einstein.

Since we are comfortable with Newtonian gravity, the best way for us to get a feeling of what gravitational curvature means is to find a curved-spacetime description of Newton's theory. We want to blend the equivalence principle with special relativity, as Einstein did. In the next chapter we will start with the spacetime of special relativity and follow what we have learned here to add in enough to get a spacetime with Newtonian gravity. This will enable us to get a good idea of what Einsteinian gravity is all about.

Einstein's gravity:
Einstein climbs onto Newton's shoulders

Geometry is at the heart of Einstein's picture of gravity. The best place to see how gravity as curvature works is in the Solar System, where the predictions must be very close to the description given by Newton. In this familiar arena, we can compare the old and new ways of looking at gravity. In this arena, too, general relativity meets and passes its first two crucial tests: explaining the anomalous advance in the perihelion of the planet Mercury (which we puzzled over in Chapter 5), and predicting that light should be deflected as it passes the Sun by twice the amount that would be calculated from Newtonian gravity (see Chapter 4).

We saw in the last chapter that the equivalence principle tells us that it is not possible to represent gravity just by the curvature of space; the curvature of space-time must include the curvature of time as well.

At this point, you may ask (indeed, you *should* ask) "What does curved time *mean*? What does it *look* like?"

Curved time sounds at first like a formidably abstract idea. But it is not nearly as abstract as it may seem. In fact, we will see below that we already know it by a different name: Newtonian gravity. The link is the way that gravity affects time, the gravitational redshift. The curvature of time is just the fact that the gravitational redshift is different in different places, i.e. that the gravitational field is not uniform. The way this leads to Newtonian gravity is already contained in our earlier discussions of the equivalence principle and of the deflection of light as it passes the Sun: we will just have to look at those discussions in a new way below in order to see Einstein's gravity in its simplest form.

Since the curvature of space must be a new form of gravity. It was the first really new element that Einstein added to gravity in the Solar System. We shall see how space curvature makes the new predictions that we mentioned above and that established the correctness of Einstein's theory, the deflection of light by the Sun and the precession of the perihelion of Mercury.

Our first step towards relativistic gravity must be to learn how to describe a curved spacetime. The best way to do that is first to describe a curved surface, such as the surface of a sphere, or of a more irregular object. The main idea always is to describe distances: you know what a surface is like if you can calculate distances along it.

Driving from Atlanta to Alaska, or from Cape Town to Cairo

A good way to develop an understanding of how to measure distance on a curved surface is to think about driving a car on a long journey. To guide you on this trip you need maps. Maps come in a variety of kinds: you can get large-scale maps of whole countries, regions, even continents; or you can get fine-scale city maps showing each small street.

In this chapter: we use Einstein's geometrical picture of gravity to study the motion of planets and light in the Solar System. We learn how to understand the curvature of time, and why Newtonian gravity is fully described by this curvature. We work out how the curvature of space changes the Newtonian deflection of light and makes Mercury's orbit precess. Since the extra deflection of light has been measured, we know what Solar System curvature Einstein's equations must predict when we encounter them in the next chapter.

▷Under the text on this page is an image of Mercury, the only astronomical object known in the nineteenth century whose motion could not be explained by Newtonian gravity. Einstein's theory, constructed without reference to Mercury's motion and without any freedom to adjust the theory to explain the motion, nevertheless exactly predicted the anomalous extra motions that had been observed. This triumph convinced Einstein and many other astronomers that general relativity had to be correct. The image is a photomosaic recently processed from images taken by the Mariner 10 mission. Courtesy NASA and the Astrogeology Team, US Geological Survey.

In this section: coordinates are familiar to anyone who has navigated by using maps. Estimating distances on a map requires re-scaling the map reference coordinates.

Figure 18.1. *This map of the area around New York City in 1970 used grid lines spaced by half a degree in latitude and longitude. Since this is not near the equator, these steps have different lengths. To calculate distances from references to these grid lines, one would have to use the weighted form of the Pythagorean law, Equation 18.1. The weighting factors are given in fact by Equation 18.2. Image from the U.S. National Atlas, courtesy of The General Libraries, The University of Texas at Austin.*

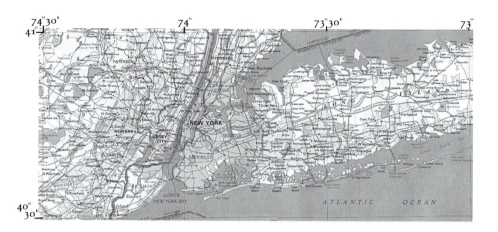

The city maps are usually drawn as if the city were flat and the geometry were Euclidean, and they often place North at the top of the map. A good map will contain reference numbers, such as squares labeled by numbers going across the map and letters going down the side, so that a given region of the city is denoted by, say, reference 3E or 1P. These references are just *coordinates* for the map.

The map should give you a distance-scale, and you could use this to convert the coordinates to standard Cartesian coordinates, so that you could use a distance x instead of a location "1", y instead of "D", so that reference location 1D is the same as (x, y). Then you could measure diagonal distances across the map using the Pythagorean rule given in Equation 17.2 on page 218,

$$\ell^2 = x^2 + y^2.$$

It could happen that the grid reference lines have different spacing in the two directions. That is, the width of a single reference block in the horizontal direction (going from 1 to 2, say) is a distance A in kilometers, while the height of a vertical reference block (from C to D) is B kilometers. An example of such a map is shown in Figure 18.1. To convert this back to an equation giving the true distance between map reference points one would have to use the distances A and B as weighting factors on the changes in the reference locations:

$$(\Delta\ell)^2 = A^2 \times (\text{change in horizontal reference})^2 +$$
$$B^2 \times (\text{change in vertical reference})^2. \quad (18.1)$$

Notice we have introduced a subtle variation here: the second equation deals with *changes* in coordinates and the distance between them. We will find below that the best way to describe distances along surfaces is to take very small steps, using the Pythagorean rule in a form like Equation 18.1, and add up the steps. This equation allows us to calculate what we called proper distance in the last chapter: the true distance independent of coordinate system.

This is fine for getting around the city or its metropolitan area, but if your journey is long, you will soon leave the domain of this map. What then?

Suppose that all you have to guide you is a big collection of local maps of all the cities, towns, and rural regions. You go from one map to another as you drive. You could navigate this way, but you would forever be puzzling over the map edges. How does the road we took on the last map match up to one we are on now? Our road left the edge of that map at map reference 1E; is it the same as the road entering the edge of the new map at 8A? How far did we travel to get across that last map? Is the scale on the new map the same? Is it oriented so that North is at the top?

This is messy, but it has the advantage that you are always looking at a Cartesian representation of distances. Once you get used to the scale of each map, you can quickly estimate distances from the Pythagorean rule as given in Equation 18.1. But you constantly have to work out how the maps join. There is a better way to navigate, and that is to buy a single map, drawn in a smooth way, with a single map reference system (a single coordinate system) that unifies the whole region through which you travel. You then use the smaller maps only if you need fine detail somewhere.

This makes navigation vastly simpler, but it introduces some distortions. For example, if the region through which you journey is large enough, you may find that the vertical map reference lines are only approximately North–South lines, and that the true northerly direction changes relative to these lines as you move across the map. If the journey takes you far from the equator, then the further you go the smaller are the East–West distances for a given change in the map reference: on a flat map, Canada and Scandinavia look much bigger than they actually are. (See the Mercator projection of the Earth on page 225 for an extreme example.) In other words, it is not possible to keep the *scale* of the map the same everywhere.

You can live with these distortions, of course, if you are aware of them. Suppose, on your large map covering the entire journey, you wanted to write down a rule for calculating distances anywhere on the map, in terms of the map's reference system. You want to extend Equation 18.1 to be valid everywhere, not just in one local map. The extension is not difficult. We just have to recognize that the numbers A and B convert from coordinates to real distances, and that this conversion may be different in different places as the scale changes gradually across the map. Your distance rule would look the same as Equation 18.1, but now A and B would be *functions of position*. Their values would depend on where you are on the map.

Dimpled and wiggly: describing any surface

Let us look at a concrete example of what we have just described. Suppose my map has reference coordinates that I call x and y, but which are *not* Cartesian: they just stand for whatever reference numbers go across and down the map, respectively. Suppose that I find that distances on my journey are well-described by the following form of the Pythagorean rule:

$$(\Delta\ell)^2 = (\Delta x)^2 + \sin^2 x (\Delta y)^2. \tag{18.2}$$

In this section: how to compute distances along a curved surface in any kind of coordinate system.

This is a special case of Equation 18.1, where we have re-named the coordinates (map references) and taken $A = 1$ and $B = \sin x$. Now, what surface am I describing? What does it look like?

> Don't assume that, just because I have called the coordinates x and y, this is a flat plane! Even if we used "Martha" and "Fred" for the map reference coordinates, the geometry would be the same.

Take a guess at the answer for the geometry before you read the next paragraph.

The answer is that the space described by Equation 18.2 is actually the *surface of a sphere of radius 1*. It describes the entire Earth, with map reference coordinates that are the usual spherical polar coordinates: the coordinate x is the polar coordinate angle that is usually called θ and the coordinate y is the polar coordinate angle ϕ, both angles measured in radians (see Figure 18.2). Put another way, Equation 18.2

gives the distance between nearby points on the Earth if we identify the coordinate y with the *longitude* and the coordinate x with the **co-latitude** (90° minus the latitude, so that the North Pole is at $x = 0°$ and the South Pole at 180°). We also need to set the radius of the Earth to 1, so that means we are using a distance-scale in which one unit of distance equals 6400 km.

▷Q: What is the longest distance you have ever driven, measured in Earth radii?

To see why this equation describes a sphere, consider latitude and longitude on the Earth. Forget Equation 18.2 on the preceding page for the moment. Just think about moving all the way around the Earth on a circle of constant latitude, i.e. keeping a constant distance from the Pole. How far have you gone? You can't answer this without knowing your latitude. A journey around the Earth along the equator is longer than a journey around the Earth at a high latitude.

Equation 18.2 answers this question mathematically. Remembering that x is co-latitude and y longitude, we see that the effect of the factor $\sin^2 x$ is to change a given difference in longitude Δy into the corresponding distance in the correct way. (Actually, of course, it changes the squared longitude difference into the correct squared distance.) The closer one goes to the pole, the smaller is the co-latitude x, the smaller is $\sin x$, and so the smaller is the actual distance associated with the journey. The factor of $\sin x$ is just the radius of the circle of constant co-latitude x, as measured in units of the Earth's radius.

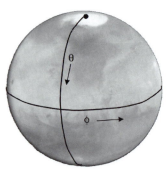

Figure 18.2. The way the latitude and longitude coordinates θ and ϕ run on a sphere, illustrated using a picture of Mars taken by the Hubble Space Telescope. (Photo courtesy NASA/HST/STScI.)

For use in later chapters we should write the same equation using the more conventional names for the coordinates, θ for the co-latitude and ϕ for the longitude. Then the distance relations on a unit sphere (sphere of radius 1) are given by

$$(\Delta\ell)^2 = (\Delta\theta)^2 + \sin^2\theta(\Delta\phi)^2. \tag{18.3}$$

If the sphere has radius r, then all distances scale in proportion to r. Since this expression gives us the squared distances on the sphere, the appropriate relation for a sphere of arbitrary radius r is

$$(\Delta\ell)^2 = r^2(\Delta\theta)^2 + r^2\sin^2\theta(\Delta\phi)^2. \tag{18.4}$$

Notice that there is no term here involving Δr. All distances are measured along the sphere, at fixed r, so there are no changes in r to take into account.

Equation 18.1 on page 226 therefore gives us lots of flexibility to describe curved surfaces, but it does not have the most general form for calculating distances. It can be extended even further by putting in a term that is a product of Δx and Δy. This leads to the *most general possible* expression for the distance between nearby points on a curved surface in a given coordinate system (x, y):

$$(\Delta\ell)^2 = A^2(\Delta x)^2 + B^2(\Delta y)^2 + 2C\Delta x\Delta y, \tag{18.5}$$

where A, B, and C depend on position on the surface, i.e. they are functions of the coordinates x and y.

The only difference with Equation 18.1, apart from calling the coordinates by their conventional names x and y, is the extra term with the coefficient C (which, in this context, has nothing to do with the speed of light). This coefficient corrects for the fact that the coordinate lines of x and y may not always be perpendicular to each other on a general map. This kind of thing is inevitable if the geometry is not perfectly smooth. If the map contains a mountain, for example, then there would be no way to draw the horizontal map reference lines in such a way that — if they were painted on the mountain and not just drawn on the map — they would cross the vertical map reference lines at right angles everywhere. Therefore the Pythagorean rule cannot be used in the simple form of Equation 18.1 on page 226.

It is not hard to see why this problem can be cured with the term involving C, containing a product of Δx and Δy. Remember the "cosine rule" for the length ℓ of the side of a triangle opposite to an angle θ, formed from sides of length x and y:

$$\ell^2 = x^2 + y^2 - 2xy \cos \theta.$$

When the triangle's two sides are perpendicular, then $\theta = 90°$, so that the last term is zero, leaving the usual Pythagorean theorem. But in general one needs the cosine term to compensate for the fact that the distances x and y are measured in directions that are not perpendicular to each other. We see that the cosine formula is a special case of the formula Equation 18.5 with $C = -2 \cos \theta$ and $A = B = 1$. The cosine formula can be thought of as the distance formula in flat space with straight coordinates that are skewed, so that there is an angle θ between the coordinate axes, an angle that is not necessarily a right angle.

Of course, the cosine rule is a formula that is correct only for a triangle in a flat two-dimensional plane. But every smooth curved space is locally flat (i.e. flat if we look at a small enough piece of it), so if the differences Δx and Δy are small enough, we can interpret the distance formula Equation 18.5 exactly as a version of the cosine rule. Therefore, C at any point just measures the angle θ between the directions of the coordinates at that point: $\cos \theta = -C/2AB$.

> This shows that an ordinary space (not spacetime), where squared distances must be positive, cannot have a distance formula with arbitrary coefficients: C^2 must be smaller than or equal to $4AB$ (in order that $\cos \theta$ should be less than or equal to one) and A and B themselves must be positive.

Figure 18.3 illustrates a coordinate system for a two-dimensional surface that is curved. It shows that, even when the coordinates are drawn in a very regular and smooth way, they stretch and turn to follow the surface. If we choose any grid line in the diagram and move along it, we see that the distances (measured along the surface, of course) between successive intersections with other grid lines do not remain a constant length: the grid is stretched and compressed. We also see that grid lines do not always intersect at right angles. The distance formula for this surface, in this coordinate system, will have non-zero functions A, B, and C.

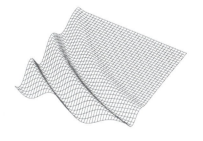

Figure 18.3. This drawing shows a simple, nearly-rectangular coordinate system drawn on a curved (wavy) two-dimensional surface. It is impossible to keep the coordinates smooth without stretching them. Generally the coordinate lines also cannot be made to intersect at right angles.

It is important to understand that the functions A, B, and C depend on the coordinates we have chosen, as well as on the curvature of the surface. There are many different ways I can draw coordinates on a surface, and the amount of stretching and squashing of the coordinates I need to do will depend on how I draw them, even though the surface will remain the same.

> Therefore, while the functions A, B, and C contain information about the curvature of the surface, they are not uniquely determined by the curvature: they depend on the coordinate system as well.

▷ In fact, it is possible to have complicated functions even on the flat plane, just by choosing the coordinates differently from the usual Cartesian coordinates x and y. For example, the Pythagorean theorem for small distances using polar coordinates r and θ in the plane is

$$(\Delta \ell)^2 = (\Delta r)^2 + r^2 (\Delta \theta)^2.$$

Newtonian gravity as the curvature of time

How do we use distance measures in curved spaces to describe Newtonian gravity? We discussed curved two-dimensional surfaces because we could visualize them, but

In this section: we learn how to understand the curvature of time.

Figure 18.4. How a time coordinate could be assigned to events on Earth by a distant experimenter using his own clock. The distant experimenter (the "master"), at rest, sends a photon to his "slave" on Earth (1), who reflects it back (2). The master notes the time he receives it (3). Since light takes the same time to travel to the Earth as to return, the master assigns the average of the times of (1) and (3) to the reflection event (2) as its time-coordinate value. He tells his slave this much later, in a letter (4), but that is okay: the event has been given a unique time. If the master does this again, then the slave will notice that the elapsed coordinate time is different (longer) than the time elapsed on a clock on the Earth. Being a slave, he is in no position to object to this! Nor should he: the master determines the coordinate time to be assigned to things, even if they are not the true (proper) times. That is what a time-coordinate is.

it is harder to visualize a four-dimensional curved spacetime! The remarkable thing is that the mathematics that we have developed for measuring distances changes very little when we adapt it to describing gravity.

One big change is that in spacetime, the distance measure is the *spacetime-interval*, not the Euclidean distance, so we expect the coefficient of the time term to be negative.

The only other change is that in principle we need to use all three dimensions for space: gravity is a property of our real three-dimensional world. In the last chapter we saw how to write the spacetime-interval in just one space dimension, Equation 17.1 on page 217. To put in the other two dimensions is simple:

$$\Delta s^2 = -(c\Delta t)^2 + (\Delta x)^2 + (\Delta y)^2 + (\Delta z)^2. \qquad (18.6)$$

The extra two space dimensions y and z have the same footing as x, and for purely spatial spacetime-intervals ($\Delta t = 0$) this is the standard Pythagorean rule in three dimensions. It follows from our discussion above that a general curved spacetime is described by modifying the spacetime-interval in Equation 18.6 to add in variable coefficients and "mixed" terms. Since this could get extremely messy, we'll only just note that it must be done to be fully general, but we won't need to do it for our discussions! Instead, we focus here on putting a variable coefficient in front of $(\Delta t)^2$.

The key to linking the notion of curvature in time with Newtonian gravity is the gravitational redshift. We have already seen in Chapter 2 that the gravitational redshift affects the rate at which clocks run. Imagine that we establish a time coordinate in the Solar System so that the time assigned to any event is the time that is recorded by a clock far from the Sun when the event occurs. It is worth thinking a little about how this time coordinate could be set up. Let's do it for the Earth, as a concrete example.

As we have remarked before, a clock on the surface of the Earth runs slower than one far away. How do we measure this? Let the clock on the Earth send out a radio signal each time it ticks. Then this signal will take a while to reach the distant clock, but the signals from both ticks take the same amount of time to travel, so the time between their arrivals at the distant clock will depend only on the time between their emissions: the time between ticks of the Earth-bound clock.

> We *define* our time coordinate t even at the Earth-bound clock to be the time elapsed on the distant clock between Earth-bound ticks. This will be longer than time on the Earth-bound clock, because of the gravitational redshift. But since clocks at different altitudes on the Earth will also run at different rates, there is nothing special about the Earth-bound clock. Our global time coordinate t has the advantage that it is possible to define it anywhere.

Of course, this time is not the time that the Earth-clock ticks. Our t is just a coordinate, a way of locating events in time. It is not meant to be directly a physical measurable. The proper time, given by the spacetime-interval, is the time on the local clock.

The gravitational redshift forces us to be careful about defining time. We saw in our discussion of the GPS in Chapter 2 that we now have to take into account the differences in redshifts of different clocks in our daily timekeeping on the Earth. The gravitational redshift causes local clock time (the proper time) to be different from our time-coordinate time t, so it is exactly the factor that converts from coordinate

---

Investigation 18.1. *The gravitational redshift tells us how time curves*

In Chapter 2 we saw that the effect of a Newtonian gravitational field was to change the rate at which clocks ticked. Now, the proper time given by the spacetime-interval is the time on clocks. The coordinate time is rather arbitrary, but it is helpful to take it to be the same as the proper time of clocks that are far from the gravity of the star or black hole that we are considering. These are "our" clocks, the clocks that we as astronomers far away from the system use to measure time.

Suppose that after a given amount of coordinate time Δt, a clock at rest in the gravitational field has ticked a proper time $\Delta \tau$. The relation between these depends on the clock's position in the gravitational field. For a clock at a distance r from a Newtonian body of mass M, a simple extension of the redshift calculation of Investigation 2.2 on page 16 shows that this relation is

$$\Delta \tau = \left(1 - \frac{GM}{c^2 r}\right) \Delta t. \tag{18.8}$$

Notice that proper time and coordinate time are equal when we are far away from the star or black hole ($r \rightarrow \infty$), which is how we defined the time coordinate t. This equation is only valid if the Newtonian field is weak, i.e. if $GM/c^2 r \ll 1$.

Now, along the world line of a clock that is at rest, the spatial coordinates don't change, so if we use the spacetime-interval to calculate the proper time, we can set $\Delta x = \Delta y = \Delta z = 0$. The negative of the spacetime-interval Δs^2 is the square of the proper time, $\Delta \tau^2$, times c^2, so we are led to

$$\Delta s^2 = -\left(1 - \frac{GM}{c^2 r}\right)^2 (c\Delta t)^2. \tag{18.9}$$

The term that is squared can be simplified by expanding the square:

$$\left(1 - \frac{GM}{c^2 r}\right)^2 = 1 - 2\frac{GM}{c^2 r} + \left(\frac{GM}{c^2 r}\right)^2.$$

The last term on the right-hand side is very small, since we are assuming the gravitational field is weak. For example, on the surface of the Earth we have $GM/c^2 r \approx 10^{-8}$, so the last term is about 10^{-8} as large as the second term. We will neglect it now. We must not throw away the second term, however, since that is where all the deviations from special relativity occur! We get

$$\Delta s^2 = -\left(1 - \frac{2GM}{c^2 r}\right)(c\Delta t)^2. \tag{18.10}$$

The time part of the spacetime-interval is therefore determined by the gravitational redshift effect. For the spatial coefficients, we make the simplest assumption: keep them the same as in special relativity. This gives the spacetime-interval Equation 18.7. We will see that this works perfectly for geodesics that represent particles going at non-relativistic speeds.

By the way, don't think that Equation 18.9 is more accurate than Equation 18.10, just because we have dropped the squared term to get the result. In fact, Equation 18.9 is itself not fully correct when gravitational fields are strong, so it can have errors of the same general size as the term we have neglected. Ironically, it turns out that Equation 18.10 is actually the form that is correct for strong fields as well as weak ones.

Exercise 18.1.1: *Redshift near the Sun*

Derive Equation 18.8, starting from Investigation 2.2 on page 16. Calculate the redshift experienced by a photon with a wavelength of 0.5 μm as it travels from the surface of the Sun to a very distant observer. Calculate the redshift of the same photon if it is observed by a space observatory in the Earth's orbit but far from the Earth. Finally, calculate the redshift if the same photon is observed by an astronomer on the surface of the Earth.

---

time to proper time. The gravitational redshift is therefore precisely the "squashing" factor that we are looking for in the spacetime-interval formula for the Solar System. The details of how to put the redshift factor into the spacetime-interval are worked out in Investigation 18.1. The result is a spacetime-interval where only the time-coefficient is variable:

$$\Delta s^2 = -\left(1 - \frac{2GM}{c^2 r}\right)(c\Delta t)^2 + (\Delta x)^2 + (\Delta y)^2 + (\Delta z)^2, \tag{18.7}$$

where M is the mass of the Newtonian star (the Sun could be replaced in this argument by any star) whose gravity we represent by this curved spacetime, and where $r^2 = x^2 + y^2 + z^2$ is the distance from the star to the point (x, y, z) in space where the clock is.

▷ Note how different a time coordinate is from a time measurement. To *measure* the time passing on Earth, the clock must be on the Earth too. To assign an arbitrary *time-coordinate* one can use any clock, and it is particularly convenient to use one so far away that the Earth's gravity does not slow it down.

Do the planets follow the geodesics of this time-curvature?

The test of whether our expression Equation 18.7 for the spacetime-interval represents the real world is whether this spacetime has geodesics that are the Newtonian trajectories of particles orbiting the mass M. This may again sound like a hard thing to show, but it is not. In fact, we have essentially done all the work we need to show this. We just have to assemble the various components of the argument.

The key is to realize that the locally flat coordinates in a spacetime are the coordinates of observers who fall freely with the acceleration of gravity. These observers can, by the equivalence principle, expect to do local experiments and have them come out exactly as in special relativity, as if there were no gravity. Now, one of the ex-

In this section: we show that the motion of any object acted on by a Newtonian gravitational force can be fully described instead as a free motion along a geodesic of a geometry with curved time.

periments they can do is to watch another nearby freely-falling particle. Since there is no gravity in their local (freely-falling) spacetime coordinate system, this particle will move on a straight line through their coordinates.

But this is the definition of a geodesic: a geodesic is a straight line in a locally flat coordinate system. Therefore, the geodesics of a spacetime in which the locally flat coordinates are those of experimenters falling freely in a gravitational field are the trajectories of freely-falling particles in the same gravitational field.

▷ Fundamentally, we have turned our old derivation of the gravitational redshift completely around. In Chapter 2 we derived the redshift from the Newtonian gravitational force. In the present chapter, we have derived the Newtonian "force" (really, the equivalent spacetime geometry) from the gravitational redshift. From Einstein's point of view, the redshift is the more fundamental of the two, since it directly measures the geometry of spacetime. The motion of particles follows almost incidentally from that geometry.

This proves that Equation 18.7 on the preceding page describes a spacetime geometry in which particles that follow geodesics will move on exactly the same trajectories as particles would do in a flat spacetime with the Newtonian gravitational force acting. We have therefore found a curved-spacetime picture of Newtonian gravity. The curvature here is *only* in the time-direction. Curvature in time is nothing more than the gravitational redshift: time advances at different rates in different places, so time is curved. We have found that the gravitational redshift fully determines the trajectories of particles in the gravitational field.

We have arrived at this goal with a minimum of calculation. We did not have to do any calculus or solve any differential equations. Yet we now know what it means physically when we say that time is curved: it means that the rate at which clocks run changes from place to place, even when the clocks are at rest with respect to one another. The curvature of time is in the gravitational redshift, and the gravitational redshift is enough to insure that freely-falling bodies follow their Newtonian trajectories.

All of Newtonian gravitation is simply the curvature of time.

How to define the conserved energy of a particle

In this section: since the gravitational redshift changes the energy of a photon, it tells us how to compute energy changes for all particles as they move through curved time. We learn how to define an energy that is conserved along geodesics.

The frequency of photons changes because of the redshift, so their energy also changes. Nevertheless, it is possible to define a conserved energy in relativity, just as it was in Newtonian gravity. This should not be surprising, especially considering that energy depends not only on the particle but also on who is measuring it.

Let us identify the experimenters who measure the redshifted energy of a photon. They are local experimenters who are at rest with respect to the star. They each perform a local experiment to measure the frequency of the light, and they find that it decreases as the photon climbs away from the Earth. If they had been observing a freely-falling particle they would have found a similar result: the speed, and hence the kinetic energy, of a particle gets lower and lower as it gets further and further from the star.

▷ Actually, although we are far from other astronomical bodies, we do sit deep in the Earth's gravitational field. Astronomers agree to use a universal time coordinate that matches proper time on the Earth, not in interstellar space.

▷ The conservation of the energy we have defined is not actually trivial, as explained below.

An observer very far away, so far that GM/rc^2 is too small a correction to measure, is in a special position: this is where we are when we observe almost all astronomical systems outside the Solar System. In an ideal case, where we consider only the gravity of a single star, this distant observer lives for all practical purposes in the spacetime of special relativity. In special relativity, the energy of a particle or photon is constant as it moves. Although this is not true of the photon that moves near the star, it becomes true when that photon moves far from the star: as it leaves the gravitational influence of the star, its energy (frequency) becomes constant, regardless of where it is going.

This energy, as measured by a distant observer, is called the *conserved energy* of the photon. It is conserved in what might seem to the reader to be a trivial sense, in the sense that it is a number that is *defined* to be a constant equal to the energy when the photon is far away. If we associate this number with the photon even when it is near the star, this does not mean that it is measurable near the star. It is, rather, a property of the photon (or particle) and the overall spacetime, which corresponds to the result of doing a direct measurement on the photon when it is far away.

Now, this might seem an odd thing to define, but let us see how it works out. The way to define this energy is to apply the redshift effect, to multiply the energy as measured by any local experimenter near the star by the right factor to get the energy as seen far away. Since the redshift is just a change in clock rates, and since we now have a spacetime-interval in Equation 18.7 on page 231 that describes clock rates, the definition of the conserved energy is straightforward:

$$\text{conserved total energy} = \text{local measured energy} \times \frac{\sqrt{|\Delta s^2|}}{c\Delta t}.$$

Let us write E_{local} for the locally measured energy and $E_{\text{conserved}}$ for the conserved total energy. Then using the spacetime-interval we get

$$E_{\text{conserved}} = \sqrt{1 - \frac{2GM}{c^2 r}}\, E_{\text{local}}. \tag{18.11}$$

Since we are assuming that $GM/c^2 r$ is a very small number, we can use the binomial approximation in Equation 5.2 on page 43 to replace the square-root in this equation by $1 - GM/c^2 r$. It is not hard for photons to put in the energy $E = h\nu$ and get back the gravitational redshift formula that we started with.

More interesting is what this implies for particles that do not travel at the speed of light. Their local energy is just the energy that a special-relativistic experimenter would measure, so if they have speed v at some location, then their local energy is given, as we have seen in Chapter 15, by their rest-mass mc^2 times the relativistic gamma-factor $\gamma = (1 - v^2/c^2)^{-1/2}$. If the velocity is slow, then using the binomial approximation again gives $\gamma = 1 + v^2/2c^2$, and the total energy is

$$E_{\text{local}} = mc^2 + \tfrac{1}{2}mv^2.$$

Here we see clearly that the local energy is just the kinetic energy of motion, plus of course the rest-mass energy. The total energy is this times $1 - GM/c^2 r$. This multiplication involves two expressions, each containing two terms. Therefore the result has four terms. However, one of them is the product of the two very small quantities, and is of the same size at the terms we dropped in using the binomial approximation. So the final result is, to the accuracy we have been working, just

$$E_{\text{conserved}} = mc^2 + \frac{1}{2}mv^2 - \frac{GMm}{r}. \tag{18.12}$$

This is, apart from the constant rest-mass term, exactly the form of the conserved energy that we had in Chapter 6 by adding up Equations 6.8 and 6.9 on page 54.

We have learned something deep here. The constant energy of an object orbiting the Sun is the locally measured energy it has when its trajectory takes it far from the Sun. The relativistic total energy is the generalization of this concept to relativity. However, some orbits in Newtonian systems cannot get far away; they are said to be bound to the central star. These include the normal circular orbits of planets. They have negative Newtonian energy, which means their total relativistic energy is less than their rest-mass energy. This proves that they cannot escape to distant regions on these trajectories: no object could have a locally measured energy less than its rest-mass. But the total conserved energy can nevertheless be defined by asking what extra energy would be required to get the planet far away with zero speed, so that its locally measured energy far away is its rest-mass energy. Then the actual energy of the object on the bound orbit is its rest-mass energy less this escape energy. In this way, all conserved energies are measured with respect to a distant experimenter.

In relativity, when speeds are not slow and gravitational fields are not weak, the full energy will be much more complicated than Equation 18.12 on the previous page. However, as long as the geometry of spacetime does not change with time, it is possible to define the conserved energy.

▷ If the geometry were to change with time, then the energy of the particle when it arrives at the distant observer will depend on just what wiggles and changes in the geometry it has encountered on the way. The energy will not be a constant since a different particle starting out the same way could finish with a different energy. We saw in our discussion of the gravitational slingshot in Chapter 6 that energy can only be conserved if the forces are time-independent. This applies to the geometry of spacetime as well.

The conserved total energy of a particle is especially important because it is not only constant along the trajectory of the particle, provided it falls freely (follows a geodesic), but it is also conserved for a collection of particles. This follows from the local conservation of energy. Consider two particles that collide. Before their collision they each have a conserved energy and it is constant, so their total conserved energy is constant. When they collide, their local energies change, but only in such a way that the total local energy is constant. The local and conserved total energies differ from one another only by terms that depend on position, like GMm/c^2r. Since the collision takes place at a particular position, these terms are the same for the two particles, so that their total conserved energies after the collision add up to the same value as before. This argument can clearly be generalized to many particles or to a continuous body.

> If the geometry is time-independent, then the total energy of a material system, as measured by an experimenter very far away, is conserved, independent of time.

The deflection of light: space has to be curved, too

In this section: we calculate, with simple algebra, the correction to the Newtonian deflection of light that is caused by the curvature of space. The measured deflection tells us how space must be curved near the Sun.

We made remarkable progress in fashioning gravity as geometry by our discussion of the curvature of time. Had Einstein wanted just to describe Newtonian gravity, he could have stopped there. But Einstein already knew from special relativity that there is no unique way to distinguish time from space. If time is curved, as measured by one experimenter, then space will be curved to another. What physical effects follow from the spatial curvature?

In Equation 18.7 on page 231 the spatial coordinates obey the usual three-dimensional Pythagorean theorem everywhere, so they are coordinates of flat Euclidean space. To introduce spatial curvature, we must put other coefficients in front of $(\Delta x)^2$ and its relatives.

Why did we not go to this complication already in Equation 18.7? Did we make an over-simplification? After all, maybe the spatial coefficients are similar to the one for time. Would they not affect the geodesics so that they do not reproduce the Newtonian orbits? The answer, profoundly, is no.

> We can neglect the coefficients in the spatial part of the spacetime-interval because they actually have little effect on the motion of a planet.

The reason for this is the slow speed v of a Newtonian gravitational orbit. In a given time ΔT (as measured in meters), the distance traveled by the orbiting body is $(v/c)\Delta T$, which is much less than ΔT. Therefore, the time coefficient in the spacetime-interval dominates the value of the interval for planetary motion: the other terms in the spacetime-interval are small, and therefore the effects produced by their slight deviations from one can have only a slight effect on the Newtonian motion.

When we come to look at the motion of photons, however, the rest of the metric does matter, because the contributions to the spacetime-interval of the terms involving Δx are comparable to those from ΔT.

> The motion of photons is sensitive to the curvature of space, and measuring what photons do can tell us what corrections we have to put into the spatial part of the spacetime-interval.

Since the spacetime-interval in Equation 18.7 on page 231 produces exactly the same geodesics as a Newtonian gravitational field, the geodesic of a photon that passes a distance d from the Sun *in this geometry* will be deflected by the Newtonian value of $2GM/c^2d$, as given in Equation 4.13 on page 38. But, as we saw in Chapter 4, observations of the 1919 eclipse showed that the deflection was actually twice this:

$$\text{angle of deflection of light by the Sun} = \frac{4GM_\odot}{c^2 d}. \qquad (18.13)$$

We must, therefore, find appropriate corrections to the spatial part of the spacetime-interval. The corrections must of course involve the quantity GM/c^2d, since it is the only physical quantity available.

Again, although the problem of determining what this correction should be sounds difficult at first, there is a plausible argument that will allow us to guess the result. Let us put an arbitrary correction term into Equation 18.7 on page 231 of the form we expect. This amounts to considering a metric of the form

$$\Delta s^2 = -\left(1 - \frac{2GM}{c^2 r}\right)(c\Delta t)^2 + \left(1 + \gamma \frac{2GM}{c^2 r}\right)\left[(\Delta x)^2 + (\Delta y)^2 + (\Delta z)^2\right], \qquad (18.14)$$

where γ is a constant whose value we have to determine.

This metric is already much simpler than a general one with spatial corrections would be: I have taken the coefficients of all three spatial coordinate changes to be the same, and I have not allowed any "mixed" terms. Since we are looking for a small correction to flat space, we take this function to be $1 + 2\gamma GM/c^2r$.

To determine γ, remember that, for a photon, the time difference (in distance units) $\Delta T = c\Delta t$ between two events will roughly equal the spatial distance. Therefore, the spatial corrections to the spacetime-interval will have the same "weight" as the corrections to the time part: when we add up distances to get the spacetime-interval, both corrections are multiplied by the same $(\Delta T)^2$ and then added in. Moreover, since the sign of the time part of the spacetime-interval is negative, the spatial correction will *increase* any physical effect caused by the time part if it has the *opposite* sign to the time correction. I have written the corrections in Equation 18.14 with opposite signs already, so we can conclude from this that, if the time part alone creates a certain deflection of light, then the time and space corrections together will make a deflection that is $1 + \gamma$ times as large. A full calculation of the trajectory of a photon confirms this conclusion.

We have seen that the observations indicate that the deflection of light is $4GM/c^2$. This is twice the Newtonian value, i.e. twice the value that one would get from only the time correction in the spacetime-interval. From this we conclude that $1 + \gamma = 2$, or $\gamma = 1$.

> The full spacetime-interval for the spacetime curvature created by a star like our Sun is, therefore,

$$\Delta s^2 = -\left(1 - \frac{2GM}{c^2 r}\right)(c\Delta t)^2 + \left(1 + \frac{2GM}{c^2 r}\right)\left[(\Delta x)^2 + (\Delta y)^2 + (\Delta z)^2\right]. \qquad (18.15)$$

▷ This use of the symbol γ has nothing to do with our other use of γ in the time dilation or Lorentz–Fitzgerald contraction formulas. I have used γ here because it has become the conventional symbol physicists use for this constant when they discuss measurements of the spatial curvature in the Solar System.

▷ Excluding mixed terms in the interval is justified in this case by the fact that the spatial geometry must be spherically symmetric: the Sun is round, and I should be able to rotate my x–y–z coordinate system in an arbitrary way about the Sun's position without changing the form of the spacetime-interval. This means that the spatial part of the spacetime-interval should look just like flat space except for being multiplied by an arbitrary function of the distance r from the Sun.

▷Our argument used only photons that passed by the Sun, so we cannot use it to deduce the geometry inside. That's good: we don't have to ask, at least not yet, what happens when r is small enough to change the sign of the coefficient of $(\Delta t)^2$. But we *will* ask this question in Chapter 21.

In this section: our estimate of the curvature of space sets a challenge to any geometrical theory of gravity that wants to replace Newton's. Different theories can be expected to predict different curvatures. Einstein found that his theory predicted the correct curvature automatically.

Apart from the approximation that we should be far enough from the star that the effects of gravity are weak enough to be represented by Newtonian gravity, this spacetime-interval is a full representation of the geometry outside the Sun. Its curvature determines the motion of any particle, from a slowly moving planet to a speedy photon. That we were able to derive it from the observed deflection of light past the Sun, again with only a minimum of mathematics, is testimony to the fundamental simplicity of Einstein's ideas on gravity.

Space curvature is a critical test of general relativity

We have calculated both the curvature of time and the curvature of space in the Solar System, but we should not go further without noticing that there was a big difference between the premises we used to get them. In this difference lies the explanation of why the measurement of the deflection of light by the Sun was so important for the acceptance of general relativity as the correct theory of gravity.

We derived the curvature of time from the gravitational redshift and the equivalence principle: it followed directly from Einstein's basic picture of gravity, and especially from the importance he gave to the equivalence principle. It is therefore the case that *any* relativistic theory of gravity that embraces the equivalence principle in this geometrical way will have the same curvature of time that we derived. The gravitational redshift is the key that opens the door to the geometrical description of gravity. But it leads to no new physical effects all by itself.

It is important to realize that Einstein's theory consists of more than just his picture of gravity as spacetime curvature. The key part of his theory is what are called the **Einstein field equations**, which tell physicists how to calculate the corrections to the spacetime-interval when they know what systems create gravity in any particular situation. Two different theories of gravity could both adopt the geometrical point of view, give the same curvature of time, and yet make very different predictions about the curvature of space.

So far, we have deduced the curvature of space, not from a general principle, but from the *observations* of the deflection of light by the Sun. If the observations had given a deflection three times the size of the Newtonian one, then we would have set γ to 2 and been just as happy with it. In fact, it is the job of the theory of gravity to *predict* this deflection, but this can be done only when we use the theory to calculate the curvature that the Sun should produce. This is what the Einstein equations are designed to do, and we will show in the next chapter that they do indeed predict $\gamma = 1$.

Other scientists, before and after Einstein, have suggested other equations for gravity, usually more complicated, and they usually produce different kinds of spatial curvature.

The spatial curvature, as measured by the deflection of light, is a key *test* of the particular theory of gravity, not just of the general geometrical framework in which we describe gravity.

The remarkable fact about general relativity is that – as we shall see – it predicts that the spatial curvature produced by the Sun will be exactly as given in Equation 18.15. Einstein did not know about the size of the effect when he devised general relativity: it was only measured three years later. The measurement of the deflection, showing that the spatial curvature was exactly as his theory predicted, coming on top of his earlier explanation of the precession of the orbit of Mercury (see the next section), was the event that propelled Einstein into superstardom, making his name a household word for "genius".

How Einstein knew he was right: Mercury's orbital precession

Einstein was convinced of the correctness of his theory even before the deflection of light was measured. The reason is that the theory had already explained the tiny anomaly in Mercury's orbit: the extra precession of its perihelion. .

We have seen that photons respond more to the curvature of the spatial part of spacetime than do planets, because they move faster: the contribution of Δx to the total spacetime-interval for a photon is similar in size to that of ΔT. Planets, on the other hand, have much smaller values of Δx for a given time spacetime-interval ΔT, so the effect of spatial curvature on their orbits is smaller. But it won't be exactly zero, and the effect it has will be something new, not contained in Newtonian gravity. Can we estimate what the spatial terms might do to an orbiting planet?

It seems reasonable to expect that the effect of spatial curvature on a planetary orbit would be similar to that on a photon's trajectory, at least qualitatively. For the photon, spatial curvature increased the deflection, drawing it further away from its original direction and toward the Sun.

So let us consider what might happen to a particle that is in an elliptical Newtonian orbit around the Sun. Its orbit should, like the photon's, turn a little bit more than the Newtonian orbit would. This would mean that the orbit would no longer be exactly a closed ellipse: in the time it takes the planet to go from one perihelion to the next, the orbit will turn by a little more than 2π. This is exactly what is observed for Mercury: the perihelion *advances* by a small amount each orbit, as we saw in Chapter 5.

The extra angle per orbit should be similar to the angle by which the photon is deflected by the spatial curvature, $2GM/c^2r$, essentially because this is the only physical number that is available in this problem. But it need not be exactly this value. For one thing, the photon experiences the deflection as it passes by the Sun on one side, while a planet goes all the way around. Moreover, there may be other effects besides spatial curvature that could induce a perihelion precession. For example, we derived the form of the spacetime-interval in Equation 18.15 from the Newtonian motion of the planets. Einstein's theory might add further, smaller correction terms. These could have the form, for example, of adding $(GM/c^2r)^2$ times some extra coefficient into the time coefficient in the spacetime interval.

Einstein's theory does indeed predict such extra terms (the "extra coefficient" of the previous sentence turns out to be two), just as it predicts the spatial curvature. We cannot calculate them here from what we know of the theory so far, but when all such terms are taken into account, Einstein's theory predicts that the perihelion position of a nearly circular orbit will advance by an angle of

$$\delta\phi_{\text{peri}} = \frac{6\pi GM}{c^2 r} \tag{18.16}$$

per orbit, a factor of $3\pi/2$ times the total angle by which light is deflected. If we evaluate this angle using numbers appropriate for Mercury's orbit (setting M to the mass of the Sun and r to the radius of Mercury's orbit, which we can obtain from Table 4.2 on page 28), we get an advance for each orbit of 4.8×10^{-7} radians (just under 0.1 arcseconds). Astronomers don't have the accuracy to detect this advance on each orbit, but fortunately they don't need to. Since the same advance occurs for each orbit, the successive advances just accumulate. In 100 years, Mercury makes about 415 orbits, so the accumulated perihelion advance is an easily measurable 41 seconds of arc per century.

This is almost exactly the extra perihelion advance of 43 arcseconds per century that had been known but unexplained for decades before Einstein arrived at general

In this section: even before the curvature of space was measured using light deflection, Einstein had shown that the spatial curvature predicted by his theory explained the anomalous precession of the orbit of Mercury.

relativity. (See the discussion in Chapter 5.) If we had used a more accurate formula that included the slight effect of the eccentricity of Mercury's orbit on the prediction of general relativity in Equation 18.16 on the preceding page, then agreement would be within the observational errors, which are less than one arcsecond per century. This is a very satisfactory agreement.

Einstein was well aware of the problem of the unexplained perihelion advance while he was working on general relativity, and when his development of the theory had gone far enough for him to calculate the effect on Mercury's orbit, he immediately found the value in Equation 18.16.

> By this time, his theory had no unknown constants in it, so there was nothing left to adjust to fit the observations: either the theory predicted the right amount of precession or it didn't. It did.

> The prediction of the correct result for the precession of the perihelion of Mercury, coming at the end of a long and painstaking calculation, gave Einstein palpitations of the heart. In his own words: "For a few days I was beside myself with joyous excitement." He knew then that his theory was the right one.

The perihelion advance and the extra non-Newtonian part of the deflection of light are only two of a number of new effects in the Solar System that Einstein's theory predicts. These are called **post-Newtonian** effects: small corrections to Newton's predictions. A number of post-Newtonian predictions of Einstein have been verified to accuracies of better than 1%. It is important to stress that Solar System tests are weak-field tests, so they leave open the possibility that the correct theory of gravity is one that differs from Einstein's only for very strong fields, like those near a black hole. Moreover, we expect that the theory will fail when quantum effects are important, since the theory does not take quantum gravity into account at all. We will return to this theme in later chapters.

Weak gravity, strong gravity

In this section: we look ahead at the task of the next chapter, to show how the curvature of spacetime is generated by the bodies in it.

The discussion in this chapter and the next, though based on weak gravitational fields, will explain a great many things about strong-field situations in general relativity. These situations, involving compact objects or cosmology, are the places where general relativity is most needed, and where it predicts and explains phenomena that even Einstein never dreamed of when he wrote down his equations. We will begin our study of these in Chapter 20. But first we need to complete our journey through relativity by learning how to work out the gravitational fields that different objects will create. We need to learn about Einstein's field equations.

Einstein's recipe:
fashioning the geometry of gravity

W e are now ready to go to the heart of general relativity, to learn how matter *generates* gravity. This subject is usually left out of discussions of general relativity below the level of an advanced university course. The reason is mathematics, not physics: Einstein formulated his field equations, his gravity-generating equations, using the language of differential geometry. This is the mathematical discipline that deals with curvature, and it is far from elementary. The physical ideas that Einstein expressed in this mathematical language are simply too important, however, to pass over. In this chapter we whittle down the mathematics to a form that is as close as possible to the algebra we used in our earlier chapters on Newton's gravity. This allows us to share in Einstein's thinking, to see what general relativity really predicts about the world we live in.

We are stepping here into a realm that is amazingly rich with new ideas. We will see how Einstein introduced a modest change in the way that the Newtonian part of the gravitational field is generated, a change that led, step-by-step, to modern concepts like cosmological **inflation** and the accelerating Universe. We will see how stars curve the space they live in, by just enough to explain the observed deflection of light as it passes the Sun. We shall see why gravity in special relativity implies that moving bodies create a new kind of gravitational effect, called **gravitomagnetism**. This effect is about to be tested in an experiment in orbit, a test that will (coming full circle) also probe the foundations of the theory of cosmological inflation. The phenomena we meet here form the basis of all the remaining chapters: why stars collapse to black holes, why rotating black holes can power quasars, how binary systems radiate gravitational waves, how the expansion of the Universe can be accelerating – all these and more are grounded in an understanding of how gravity is created.

Not that the mathematics of differential geometry is unimportant: far from it. Einstein's equations are quite beautiful when expressed this way. Not only are they compact (as we will see below) but they have a very deep symmetry: they have exactly the same form in any coordinate system, so they give equal status to any observer/experimenter, inertial or not. This extension of the principle of relativity is called the **principle of general covariance**. Where the principle of relativity placed all *experimenters* on the same footing, the principle of covariance says that the field equations must be able to be used in any *coordinate system*, no matter how peculiar. This is rightly regarded as a beautiful and powerful aspect of Einstein's theory.

To make the mathematics simpler, we have to set aside general covariance, at least temporarily. We shall analyze how matter creates gravity from the point of view of a particular observer, an observer who is essentially at rest with respect to the sources of gravity. We shall follow the pattern of the previous chapter, where we separated the curvature of time from the curvature of space. Here we will look

In this chapter: we study the equations that show how matter generates gravity in general relativity. We identify four properties of matter and gravity that act as sources of gravity, and we show how these different sources produce different gravitational effects. Using only a little algebra, we compute the curvature of space and get the observed deflection of light as it passes the Sun. We show how special relativity and the curvature of time lead to something called the dragging of inertial frames. We examine the special properties of the cosmological constant as a source of gravity.

▷The image beneath the text on this page is from a numerical simulation of the merger of two black holes. It shows a measure of the curvature of space that carries information about gravitational waves. In the center can be seen two small blobs surrounded by a larger one: these are the two original black holes after they have merged into a single one. The rest of the picture shows gravitational waves moving out. Image by W Benger, Zuse Institute Berlin (ZIB), from a simulation performed by scientists at the Max Planck Institute for Gravitational Physics (Albert Einstein Institute, AEI), Washington University (WASHU), and the National Center for Supercomputer Applications at the University of Illinois (NCSA).

▷Our picture of general relativity
builds on insights gained by
generations of physicists –
successors to Einstein – who took
apart the complex field equations
and painstakingly won the physical
insights and developed the physical
intuition that are necessary for
applying general relativity to the
real world of astronomy.

In this section: we list the four
distinct sources of gravity in
general relativity: active
gravitational mass, active curvature
mass, momentum, and gravity
itself.

▷When something physical can
help create itself, as in item 4 in the
list, physicists say that it is
non-linear. This refers here to the
relation between the gravitational
field and its source. In Newton's
gravity, if the density of mass
inside one star is twice that of
another, while their sizes are the
same, then the gravitational field of
the first will be twice as strong as
the second. The field will be
proportional to the density; a graph
of the field strength against the
density of the star will be a straight
line, which mathematicians describe
as a **linear** relationship.
That does not happen in general
relativity. The larger density will
create a larger field, of course, but
then this larger field will create
even more energy, and the result
will be that the field of the denser
star is not just twice as strong as
that of the other. The graph of the
field strength against the density
will not be straight line: it will be
non-linear. It is even possible in
general relativity to have a pure
gravitational field, with no matter
at all, acting as its own source!

separately at how matter generates the curvatures of time and of space. We shall learn how to compute the geometry of gravity when gravity is weak, and we will build on this insight to understand how strong gravity works.

Of course, if the insights we obtain by simplifying the mathematics in this way are valid, then we should expect that there must be a sense in which they are independent of observers. After all, if one observer predicts that a star should collapse and it does, then all other observers must have been able to make that prediction as well, based on their own measurements. So at the end of our discussion we will look again at the principle of covariance that guided Einstein, use it to draw further insights into the theory, and return to the very real beauty of the theory that has been called the greatest creation a single human mind has ever achieved.

Einstein's kitchen: the ingredients

You can think of this chapter as an excursion into the kitchen of general relativity, where we will see how the gravitational fields of general relativity are made. Creations from this kitchen, such as black holes and cosmology, are displayed in every popular book on relativity, but you don't always get to go into the back room and see how they are cooked up. In later sections we will study Einstein's recipe, how he begins with the fundamental ingredients and arrives at the finished product. The recipe is called Einstein's equations. We start here with the foundation of any good recipe: the ingredients.

The ingredients are what physicists call the *sources of gravity*: things, like mass, that are responsible for gravity. The richness of the predictions of general relativity comes directly from the rich variety of sources of gravity that Einstein uses.

Newton believed that only the masses of objects create gravity. This was a powerful principle, which worked well for two and a half centuries. But, with only one ingredient, the result was inevitably a little monotonous. By contrast, Einstein's gravity involves at least four main ingredients, four distinct kinds of sources. We list them here and explain them in subsequent sections:

1. The **active gravitational mass** plays the role in Einstein's gravity that the ordinary mass plays in Newton's: it produces the main gravitational effect, namely the curvature of time.

2. The **active curvature mass** generates the curvature of three-dimensional space, which is totally absent from Newtonian gravity.

3. The ordinary *momentum* of matter generates what physicists call *gravito-magnetism*, the part of gravity that acts on masses in a way that resembles the way magnetism affects charged particles.

4. *Gravity itself creates gravity*. This is inevitable, since energy has mass in relativity, and even in Newtonian gravity there is an energy associated with the gravitational field, which we called the gravitational potential energy. Thus, gravitational fields have energy and this feeds back into the gravitational field.

The fourth source is fundamental to Einstein's picture of gravity, but it makes the equations hard to solve. You think you have a solution, but you have to change it to take into account the way your solution acts as part of its source. Normally this kind of problem is solved using computers. The huge variety of possible gravitational fields is now being explored numerically by physicists who use the most powerful available supercomputers. Because of the non-linearity of general relativity, we will only explore in this chapter how Einstein's equations create relatively weak gravitational fields. For weak gravity, the contribution of the field itself to making more field can be ignored compared to other sources, and we are back to a linear problem.

These four sources are enough for us to gain deep insights into the way gravity works in general relativity, without using sophisticated mathematics. What we lose by ignoring the mathematics is the principle of general covariance, the equivalence of all observers. This is not just an abstract idea. It simplifies many of the concepts. It enabled Einstein actually to postulate only *one* non-gravitational source for the gravitational field, which we will return to later. The symmetry of the principle of covariance provides relationships among the first three of our sources, implying for example that if the energy density is a source, then the momentum density must also be a source. These three sources of gravity are not independent, not chosen arbitrarily. Einstein was led to them because they are different aspects of one mathematical object. So the price we pay for ignoring general covariance is that we don't see this deeper layer of unity in the theory.

In fact, the principle of general covariance insures that we are safe even if we ignore it! Since any observer, any coordinate system, is a valid system for describing gravity, the point of view of the observer we will use – one who is at rest with respect to the systems we study – is just as good as that of any other. And since any gravitational field is weak in the neighborhood of a freely-falling observer (the equivalence principle), even our assumption of weak fields is not as drastic as it might first have seemend. In fact, our four sources and our prescriptions for generating gravity from them will be excellent guides even to strong-field gravity, like black holes. Even when we study cosmology, where the gravitational fields are strong enough to control the whole Universe, we will be able to explain the expansion and acceleration of the Universe entirely in terms of weak-field gravity.

So let's get going. Welcome to Einstein's kitchen!

Einstein's kitchen: the active gravitational mass comes first

In Newton's theory, gravity is created by mass. So in general relativity, we should expect that the main source of the curvature of time will be mass. Indeed, in the last chapter we used the mass of the Sun to obtain the spacetime geometry of the Solar System. We were able to compute the effect of gravity on a relativistic particle, namely a photon as it passes the Sun, but the source of gravity – the Sun – was non-relativistic. What would gravity be like if its source were more relativistic?

Even the Sun can be made relativistic: just view the Solar System from a rocket ship moving past at $0.5c$. Would such an observer still be able to use Newton's prescription to calculate the gravitational redshift due to the mass of the Sun and to predict the motion of, say, the Earth around the Sun?

Immediately we see a problem: "mass" has no unique meaning in relativity. Rest mass is an invariant, but it is not really a suitable source of gravity. It is not even particularly well-defined for a composite object like the Sun; do we mean the total mass of the Sun as measured when it is at rest, or the sum of the rest-masses of all the particles, each as measured by an experimenter at rest with respect to it? These are different, because (as we saw in Chapter 16) the total mass of the Sun at rest includes the energy of the photons and the kinetic energies of all the particles, along with their rest-masses. So rest-mass is not a suitable source.

A better relativistic generalization of Newton's mass is the total mass-energy; this is at least conserved during nuclear reactions. But the mass-energy of the Sun is not an invariant in relativity, so the observer in the rocket will calculate a very different value for it. We might therefore expect that other properties of the Sun, which may also depend on its velocity, might be sources.

What can these properties be? When moving, the Sun will have a momentum,

In this section: the active gravitational mass generates the curvature of time, which is the most important part of the geometry of gravity. Its density is defined as the density of ordinary mass-energy, plus three times the average pressure divided by c^2.

▷ Rest mass has another problem. What would happen to a gravitational field created by rest-mass when rest-mass is turned into energy by nuclear reactions? Would gravity disappear? This seems unreasonable. Rest mass is a dead end.

EINSTEIN SIMPLIFIED

Figure 19.1. Although we simplify Einstein's equations in this chapter by presenting them only for weak gravitational fields and only for a particular kind of observer, we retain the essential way that they link curvature to the properties of the matter that creates it. The simple form helps us to understand how relativistic gravity works even when the field is strong. Simplify, but don't go beyond the point of recognition! Cartoon copyright S Harris, reprinted with permission.

so this would certainly be a candidate. On the microscopic scale, the random momentum of the gas particles inside the Sun gives it pressure, and in fact the overall motion of the Sun past the rocket observer gives it something that physicists call **ram pressure** (which we will define below), so we might expect that both momentum and pressure could contribute to the gravitational field. Now, momentum has a direction, so when we talk about a source for the curvature of time, which has no spatial direction, momentum is not a candidate. But pressure is. And in Einstein's theory, pressure does indeed play a key role in creating the curvature of time.

▷We shall keep momentum on the ingredient list, however, and use it later as a source for other parts of the gravitational field.

If we write Einstein's equations down from the point of view of an observer who is at rest with respect to the Sun (or another body that is creating gravity) then we find that the source of the curvature of time is the active gravitational mass:

density of active gravitational mass

$$= \text{density of total mass} + 3 \times \text{average pressure}/c^2. \quad (19.1)$$

Here the term "total mass" means all energies added together, and converted to their mass equivalent; and "average pressure" means the pressure averaged over the three directions in space. In a gas, pressure is the same in all three directions, so the average pressure is just the ordinary fluid pressure. We will later meet other situations where the averaging gives a different result because the pressure is not isotropic.

In Newtonian situations the contribution of the pressure is negligible, essentially because, as we saw in Investigation 7.2 on page 78, pressures of gases are typically of the same size as their densities times the square of the random velocities of gas molecules. When one divides this by c^2 as in the above equation, the term is much less than the density of mass itself. So the extra gravity produced by pressure was not noticed before Einstein.

Einstein's kitchen: the recipe for curving time

In this section: the active gravitational mass curves time in our approximation just as in Newtonian gravity.

Knowing that the source of time curvature is the active gravitational mass already gives us some insight into relativistic gravity, but in many cases one wants to be able to compute the curvature of time explicitly. Here we shall spell out the rule for computing the coefficient of $(\Delta T)^2$ in the spacetime-interval, when gravity is not too strong, so we don't have to worry about the gravitational field as an additional source.

The rule is similar to, but actually simpler than, the rule we used in Investigation 4.3 on page 35 to compute the Newtonian gravitational acceleration produced by a sphere. It is similar because we are dealing with a generalization of the Newtonian gravitational field. It is simpler because here we shall only calculate the rate of running of clocks (the gravitational redshift) in different places rather than the acceleration of planets that the non-uniform redshift (the curvature of time) leads to. The following steps lead to the coefficient of $(\Delta T)^2$ at any given point inside or outside the body that is the source of gravity, as long as the gravitational field is not very strong and the body is at rest or nearly so.

1. Divide the body into small pieces.

2. For each piece, multiply the density of active gravitational mass by the volume of the piece, divide by the distance to the point, and multiply by G/c^2.

3. Add up these numbers for all the pieces of the body. Call this sum Φ.

4. The coefficient of $(\Delta T)^2$ is $-1 + 2\Phi$.

There is a name for the gravitational effect produced by this redshift factor: it is the **gravitoelectric field** of general relativity. This is the part of the gravitational field that is most like the Newtonian gravitational acceleration. The term "gravitoelectric" is used because there is a strong analogy with the electric part of the electromagnetic field, which is called the Coulomb or electrostatic field. We will encounter similar terminology below when we investigate the part of the gravitational acceleration that resembles magnetism, called the gravitomagnetic field.

The construction of the gravitoelectric field is, of course, very similar to that of Newton's field, only with a different source. It follows that, if the source of gravity is perfectly spherical, then outside the source the field is independent of the size of the region occupied by the source and is independent of whether the source is moving in and out with time, as long as it remains spherical. We saw that this was true in Newtonian gravity in Chapter 4, and it is also true in general relativity.

The gravitoelectric part of Einstein's gravity governs the motion of slowly-moving bodies, even near highly relativistic sources. The inclusion of the pressure in the source for this part of gravity leads to the most dramatic differences between the predictions of Einstein's gravity and Newton's gravity. Here is a partial list, a preview of what we will study in this and later chapters.

- *Gravitational collapse to black holes.* The large pressure of gas in relativistic objects makes the gravitational field stronger. When we study neutron stars in the next chapter, we will see that this makes it impossible for neutron stars to exist above a certain mass. Supporting the extra mass requires more pressure; that just adds to gravity and requires, in turn, more pressure. For neutron stars above a certain mass, this feedback mechanism runs away: it is never possible to add enough pressure to support the star. Instead, the star will collapse to a black hole.

- *Gravity has a magnetic-like side to its effect on matter.* We will show below that the application of the principles of special relativity to the gravitoelectric Einstein field is enough to derive the *gravitomagnetic* part of the gravitational field. The inclusion of pressure in the active gravitational mass is crucial here; it will allow us to show that the gravitomagnetic part must be present, and to derive it quantitatively, from simple arguments based on special relativity.

▷The reason it is simpler is that the redshift is pnly one number at each point, while the acceleration, requiring a direction, is three numbers (a vector) at each point. If we wanted to calculate the acceleration in relativity, it would be even more complicated than the Newtonian calculation.

▷Remember that $T = ct$ is the time measured in distance units, i.e. in light-meters.

- *Zero gravity or even **anti-gravity**!* As we noted right at the beginning of this book, Newton taught us that gravity is always attractive. That is because its source is mass, and mass is always positive. Well, nearly always: in quantum field theory there is the possibility of negative energy, and we will look briefly at that in Chapter 27. But in everyday situations we have come to expect gravity to be attractive. However, there is nothing unusual about **negative pressure**. In physics this is called tension. Ordinary positive pressure pushes outwards. Negative pressure simply pulls inwards. If you wrap your newspaper with a stretched rubber band, you are handling negative pressure in the rubber band. By including pressure in the source for gravity, Einstein opened the possibility that a system with very large negative pressure could have zero or even negative active gravitational mass.

The last item leads to the most dramatic differences from Newton's gravity, yet it is not always emphasized in introductions to general relativity. We will meet examples of zero and negative gravity in this book. Cosmic strings, which we first mentioned in Chapter 14, have zero active gravitational mass, their only gravitational effect being to curve space but not time. We will see what peculiar effects this can have on matter near them in Chapter 25. Even more important, Einstein himself introduced a negative-pressure field into general relativity; he called it the **cosmological constant**. His purpose was to find ways to turn the active gravitational mass negative! The cosmological constant is so important in modern physics and astronomy that we will focus on it later in this chapter. In the final chapters on cosmology we will see how scientists use negative active gravitational mass to explain the observed acceleration of the expansion of our Universe and as the basis for the theory of cosmological inflation.

Einstein's kitchen: the recipe for curving space

In this section: the active curvature mass produces the curvature when fields are weak and the gravitational field is isotropic. It works in the same way as the active gravitational mass.

We saw in the last chapter that the motion of photons past the Sun showed an extra deflection caused by the spatial curvature, and that this was a key test of general relativity. Here we learn how in general relativity this curvature is generated.

For weak gravitational fields, the coefficient in the spacetime-interval of, say, $(\Delta x)^2$, will be almost one, with a small correction. This correction is determined by the Einstein field equations. Since there are six spatial terms of the form $(\Delta x)^2$, $(\Delta x)(\Delta y)$, and so on, there are six coefficients to determine. This means that the general case will be quite complicated, possibly involving six different sources of spatial curvature.

We will simplify to just one source by assuming that the matter that produces gravity has *isotropic pressure*, which is something we defined in Chapter 7. This means that the pressure is the same in all directions, so that the averaging over directions that we used in the previous section is not necessary. This is not a strong restriction: the pressure inside a star at rest, for example, is isotropic. The pressure of the hot gases in the early Universe was similarly isotropic. So in many of the cases we will be interested in, the assumption of isotropy is fine.

Now, for weak gravitational fields in general relativity, the spatial curvature produced by matter with isotropic pressure has only *one* source, which we shall call the density of active curvature mass:

$$\text{density of active curvature mass} = \text{density of total mass - pressure}/c^2. \quad (19.2)$$

The similarity to Equation 19.1 on page 242 is striking. Both contain the mass density and the pressure, but in the case of the active gravitational mass, the pressure is multiplied by three and added, while for the active curvature mass the pressure is subtracted.

The rule for computing the coefficient of, say, $(\Delta x)^2$ is similar to the rule for computing the coefficient of $(\Delta T)^2$ from the density of active gravitational mass. The following steps will evaluate this coefficient at any given point inside or outside the body that is the source of gravity, as long as the gravitational field is not very strong:

1. Perform steps 1-3 on page 243 for the active curvature mass; call the result Ψ.

2. The coefficient of $(\Delta x)^2$ is $1 + 2\Psi$.

Because we have assumed isotropy, the coefficients of $(\Delta y)^2$ and $(\Delta z)^2$ are the same as that of $(\Delta x)^2$, and the coefficients of the mixed terms like $(\Delta x)(\Delta y)$ are all zero.

Notice that this gives the following form for the spatial distance element of a curved spacetime whose sources have isotropic pressure:

$$\Delta \ell^2 = (1 + 2\Psi)\left[(\Delta x)^2 + (\Delta y)^2 + (\Delta z)^2\right].$$

This is exactly the form we assumed in our discussion of light deflection, Equation 18.15 on page 235, provided we identify $\Psi = GM/c^2 r$. This is equal to the Newtonian field Φ, as we noted there. Now we can see that Einstein's equations do predict that $\Psi = \Phi$. They are the same because, to a good first approximation, their sources are the same. Both the active gravitational mass and the active curvature mass are dominated by the mass density ρ for systems with weak gravitational fields and small pressures. They are therefore equal, to within the accuracy with which we needed them in order to calculate the curvature for a photon's trajectory.

> We have proved that Einstein's equations really do predict the observed deflection of light. We have established that general relativity passes a key observational test of its validity.

Although the active curvature mass creates the spatial curvature of Einstein's gravity, it does not lead to the kind of dramatic consequences that we saw for the active gravitational mass. Scientists generally believe that p/c^2 will not exceed in absolute value the mass-density of any relativistic field, so the active curvature mass should always be non-negative.

Einstein's kitchen: the recipe for gravitomagnetism

We are now in a position to understand one of the most remarkable features of general relativity: the existence of gravitational effects analogous to the magnetic effects of electromagnetism. When the gravitomagnetic gravitational field is present, the gravitational acceleration of a particle depends on its velocity as well as its location.

The existence of gravitomagnetism must be related to special relativity: since a particle being accelerated by gravity can be at rest with respect to one observer but moving with respect to another, the gravitomagnetic effects seen by the second observer must somehow be part of the gravitoelectric field seen by the first. In this section we will deduce the gravitomagnetic field in exactly this manner, by looking at the acceleration of a particle from the point of view of two different observers and insisting that the accelerations they predict should be the same.

It is well-known to theoretical physicists that one can deduce the existence of the magnetic field of electromagnetism from the electric field by such an argument. A particularly elegant demonstration of this was given by the brilliant American theoretical physicist Richard P Feynman (1918–1988), in which he showed how to calculate the force of magnetism by applying the rules of special relativity to the

In this section: by demanding that our theory of gravity predict the same things when used by two different observers, we show that there must be a third kind of gravitational effect, which is called gravitomagnetism. Its source is momentum and it affects only moving bodies. Our derivation follows a similar derivation of magnetism from electricity and special relativity by the physicist R P Feynman.

▷ Some scientists use the term *magnetogravity* instead of *gravitomagnetism*.

electric force. In this section we adapt Feynman's argument in order to derive gravitomagnetism from the gravitoelectric field. The outline and logic of the argument is presented in the main text of the section, and should be accessible to all readers. Some of the details of the calculation are reserved for Investigation 19.1 on page 250 for those who want to follow the whole calculation.

Consider the following system, illustrated in Figure 19.2. There are two streams of identical moving particles, each stream being perfectly straight and very long compared to their separation $2d$ from each other. They are also very thin (in cross-section) compared to their separation. The two streams are parallel to one another, and they have equal and opposite velocities (v and $-v$). To keep things simple, we suppose that all the particles move with the velocity of their stream: there are no random motions. The particles in the stream are so numerous that the stream is essentially a continuous string of matter.

▷Feynman's argument appears in his influential undergraduate physics textbook, *The Feynman Lectures on Physics*, R P Feynman, R B Leighton, & M Sands (Addison Wesley, Reading, Mass, 1964), vol 2.

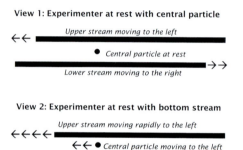

Figure 19.2. Two streams of particles moving in opposite directions leave a central particle undisturbed (top panel). When viewed by an experimenter at rest with respect to the lower stream (bottom panel), the top stream has more mass and should pull on the central particle. The particle can only remain at rest if there is a velocity-dependent gravitational repulsion from a moving stream.

In Newtonian gravity, if the stream has a very small cross-section, then the gravitational acceleration produced by the string will depend only on the amount of (rest) mass in the string per unit length. We call this number μ, and it has units of kilograms per meter..

Now we do the following simple idealized experiment.

We place a particle exactly in the middle between the two streams, at rest. Because of the symmetry of the situation, the net gravitational force on the particle must be zero. Regardless of what theory of gravity one uses to calculate the force, the fact that one stream is the mirror image of the other means that the influence of one will cancel that of the other.

The particle will remain at rest in this (unstable) equilibrium position.

Let us view the same system from a rocket ship (or other experimental laboratory) that is moving at the same speed as the bottom stream. From the point of view of this experimenter, the bottom stream consists of particles that are at rest; the top stream, on the other hand, is moving to the left at speed $2v$. (We will assume that the speed v is much less than the speed of light c, so that we don't have to worry about relativistic corrections to the velocity-addition rule.) The particle in the middle is moving at speed v to the left. But the physical system is the same, just viewed by a different observer. That means that the particle in the middle remains in the middle, moving at its constant speed v, but not falling toward one or the other stream.

However, when the moving experimenter tries to calculate the gravitational forces he expects on the particle, then it is not so clear that they will balance. Let us consider what this experimenter expects to happen if he believes Newtonian gravity is all he needs to know, so he does not use Einstein's active gravitational mass. He assumes that the total mass-energy of the stream creates gravity, and he (like the first experimenter) has a machine that can measure accurately the mass per unit length along each stream. We will compare the mass per unit length that the two experimenters measure for each stream.

The top stream is moving *faster* relative to the second (rocket) experimenter than to the first. This means that each particle will have a larger mass-energy as measured by the second experimenter. What is more, the Lorentz–Fitzgerald contraction of lengths pushes the density of mass-energy higher still. So the second experimenter measures a much *higher mass-energy* per unit length for the top stream than the first experimenter does.

The bottom stream, by contrast, is at rest with respect to this second experimenter, while it is moving relative to the first experimenter. The situation is the reverse of that for the top stream, so the same reasoning leads to the conclusion that the second experimenter measures a *smaller mass-energy* per unit length for the bottom stream than the first experimenter does.

The rocket experimenter, assuming as he does that Newtonian gravity is the whole story, expects that the gravitational force exerted by the top stream will be larger than that exerted by the bottom one, and that the particle in the middle will begin to fall towards the top stream.

> But the particle does not fall toward the top stream. So the experimenter in the rocket must conclude that there is more to gravity than the simple Newtonian force.

We knew already on general grounds that Newtonian gravity is not compatible with special relativity. Here we have an explicit demonstration of it.

Now, we already know about one correction to Newtonian gravity: the active gravitational mass in relativity includes the pressure term that is shown in Equation 19.1 on page 242. So maybe all we have to do is calculate what we have called the gravitoelectric gravitational acceleration in general relativity; maybe then the central particle will experience exactly balancing accelerations.

What, however, is the pressure term in our situation? Our streams have no conventional fluid pressure, since the particles have no random motions. However, the overall bulk motion of the particles creates what physicists call a "ram pressure", which is basically the pressure that the stream would exert if it were running up against (ramming into) a wall. This pressure does contribute to the gravitational field. We calculate the ram pressure in Investigation 19.1 on page 250, and we find it is proportional to μv^2. When we put that into the expression for the active gravitational mass, however, it does *not* correct the imbalance of the accelerations as computed by the rocket experimenter. In fact, it is easy to see that it makes it worse. The pressure is positive, and it is non-zero only in the top stream, where the gravitational attraction was already too high.

> Adding in pressure only makes the imbalance worse! We are driven to the conclusion that there has to be more to gravity in general relativity than just the gravitoelectric part of the field.

We clearly need a further acceleration, produced by the top stream, that *repels* the central particle, keeping it in equilibrium with the weaker gravitoelectric acceleration produced by the bottom stream. Since we did not need to invoke such an acceleration when we looked at these streams from the point of view of an experimenter at rest with respect to the central particle, it is natural to expect that this acceleration will turn out to be associated with the *motion* of the particle with respect to the second experimenter. In order to cancel out the excess active gravitational mass density of the upper stream, this velocity-dependent force must work in such a way that a moving stream *repels* a particle that is moving in the same direction.

▷ You might wonder what would happen if, despite our earlier objections, the moving experimenter assumed that *rest-mass* created Newtonian gravity, instead of total mass-energy. It should be easy for you to see that the Lorentz–Fitzgerald contraction still makes the rest-mass density higher in the top stream than the bottom, so this variant of the assumed Newtonian force would also draw the particle towards the top.

▷ In fact, there is an alternative relativistic theory of gravity in which the pressure enters the active gravitational mass with the opposite sign, and with the right coefficient so that the ram pressure cancels the higher density of the upper stream and leaves the particle in balance, with no further corrections. This theory has no gravitomagnetism. However, it predicts the wrong result for the anomalous perihelion shift of Mercury, even giving it the wrong sign, so it is not a viable theory. This illustrates the important point that the ultimate test of a theory is its agreement with experiment.

The situation is very close to that in magnetism: an electric current (stream of moving positive charges) will create a magnetic field that actually *attracts* a positively charged particle that is moving in the same direction as the current. Here we see that a moving stream of particles will repel a particle moving in the same direction. The sign of the effect is different (attraction in one case, repulsion in another), but this is just because in electromagnetism the sign of the electric part of the acceleration is also different from that in gravity: electric charges of the same type repel each other, while in gravity two masses attract. Apart from this sign, there is such a close analogy to magnetism that we call this velocity-dependent gravitational effect *gravitomagnetism*.

▷ It should be clear from our derivation that the words "gravitoelectric" and "gravitomagnetic" are used only to draw an analogy with electromagnetism. They are purely gravitational effects; they have their source in the mass and momentum of particles, not in electric charge or electric current.

The gravitomagnetic effect is created by the moving stream, so it is a gravitational effect whose source is the *momentum* of the particles. We have therefore found the field created by the third source of gravity in the list on page 240. In the first experimenter's view, both streams create gravitomagnetism, but the central particle is at rest, so it does not feel the effect. In the second experimenter's view, the bottom stream is at rest and therefore does not create this effect, but the top stream does, and it just compensates the extra gravitational attractiveness of the top stream to produce the same net gravitational attraction as the bottom stream exerts. We calculate the size of this effect in Investigation 19.1 on page 250. The argument gives exactly the gravitomagnetic effect that one could calculate from Einstein's field equations directly, with the mathematics of differential geometry! Our derivation is just as good, and uses only elementary algebra.

▷ If you have mastered the right-hand rule for magnetism, then this paragraph should make sense to you. If you have not met this sort of reasoning before, you may want to skip to the next paragraph!

By analogy with the magnetic field, the direction of the gravitomagnetic effect can be determined by something we might call the two-hand rule, as follows. Let the thumb of your right hand point in the direction of the momentum of the top stream. Then let your fingers curl up around this direction. The fingers follow lines of the gravitomagnetic part of the gravitational field, which are circles around the stream. Now, to calculate the effect on a passing particle, take your *left* hand and let the fingers curl in the following way. First point the fingers in the direction of the motion of the passing particle. Then curl them so that their tips point along the direction of the gravitomagnetic effect, which you just determined using your right hand. When your left hand is oriented so that the fingers can curl from the one direction to the other as described, then your left thumb points in the direction of the gravitomagnetic acceleration of the particle.

▷ We can use the Newtonian acceleration a_N here rather than the full gravitoelectric acceleration, because the extra pressure terms are already corrections to a_N of order v^2/c^2, so they become terms of order v^4/c^4 in a_M, and we have neglected such corrections in this argument.

The size of the gravitomagnetic effect, as calculated in Investigation 19.1 on page 250, has a simple formula, expressed as a correction to the gravitoelectric acceleration produced by any source. If the Newtonian gravitational field of a system would produce an acceleration that has magnitude a_N, and if the source moves with speed v_s and the particle with speed v_p, then the magnetic-type gravitational acceleration of the particle will have magnitude

$$a_M = a_N \frac{4v_s v_p}{c^2}. \tag{19.3}$$

The direction of the magnetic-type acceleration is given by the two-hand rule.

> This equation allows us to compute the gravitomagnetic acceleration produced by *any* moving system on a moving particle, if we know the two velocities and the Newtonian acceleration the system produces.

The idealized, infinitely long streams of particles have served their purpose, so we can forget them now and focus on more realistic systems.

For example, let us write down the magnitude of the gravitomagnetic acceleration due to a single particle source of mass M moving with speed v_s. If we look at the acceleration at a distance r from it, then the Newtonian acceleration is, in magnitude, $a_N = GM/r^2$, so the gravitomagnetic acceleration on a particle with speed v_p has magnitude

$$a_M = \frac{4GMv_s v_p}{c^2 r^2}. \tag{19.4}$$

Like the Newtonian acceleration, it falls off as $1/r^2$.

Notice that we can write the gravitomagnetic *force* on a particle of mass m, which is just ma_M, in the form

$$F_M = \frac{4G}{c^2 r^2}(Mv_s)(mv_p).$$

It is therefore possible to regard gravitomagnetism as a coupling between the *momentum* of the source and that of the particle.

In this way we see that momentum creates its own kind of gravity.

The geometry of gravitomagnetism

So far, we have talked in Newtonian language about gravitomagnetism, describing the way it acts on particles. It is natural to ask how it fits into the geometrical picture: where, in the calculation of the spacetime-interval, does gravitomagnetism come in?

In this section: gravitomagnetism comes from the mixed time–space coefficients in the general interval.

First we have to decide what we expect to find in the spacetime-interval. The spacetime-interval represents the gravitational field created by the source of gravity. It does not contain any properties of the particles that are affected by gravity: they move on geodesics of this spacetime geometry. So when we look for the gravitomagnetic acceleration terms in the spacetime-interval, we are looking only for the first factor in the following equation, which is just Equation 19.3 re-written in a convenient way:

$$a_M = \left(a_N \frac{4v_s}{c}\right) \frac{v_p}{c}.$$

We have factored out the part of the acceleration that depends on the particle, its (dimensionless) speed v_p/c, so that what is inside the large parentheses is the part due just to the source of the field. This is the *gravitomagnetic field*:

$$\text{gravitomagnetic field} = \frac{4v_s}{c} \times \text{Newtonian gravitational field.} \tag{19.5}$$

▷Just where we place the factors of c in defining these parts of Equation 19.3 is, of course, arbitrary, but it seems simplest to keep things dimensionless where possible, so that the gravitomagnetic field has the same dimensions as the Newtonian field.

The rule is that the magnitude of the gravitomagnetic effect on a particle is just the gravitomagnetic field times the dimensionless velocity v_p/c of the particle. The direction of the effect is given by the two-hand rule.

We expect to find the gravitomagnetic field of Equation 19.5 encoded somewhere in the spacetime-interval. We can discover where it is by the application of a symmetry argument. Consider what happens to our example when we reverse the sense of time, as if we took a video of the experiment and played it backwards. Then all the velocities would go the other way, and by the two-hand rule the sense of the gravitomagnetic field would reverse. The end effect, the acceleration of the particle, would not change: it would still be repelled from the top stream. But this would come about because of two compensating changes of sign. We would be multiplying the particle's own velocity, which has changed sign, by the gravitomagnetic field, which has also changed sign.

So the gravitomagnetic field itself must be contained in the spacetime-interval in a term that changes sign when we change the sign of T. Moreover, we get a similar change of sign if we reflect the experiment in a mirror perpendicular to the

Investigation 19.1. How big is gravitomagnetism?

Here we shall see how to calculate the gravitoelectric part of the gravitational attraction, which comes from the active gravitational mass in general relativity, and we will find the shortfall that must be made up by gravitomagnetism. The first experimenter sees a symmetrical situation, so it is not of interest to us to calculate the forces from his point of view: they will only cancel out completely and leave the particle at rest. So we will focus on the second experimenter, flying in a rocket that is moving at the same speed as the bottom stream.

The bottom stream produces the same gravitoelectric acceleration in general relativity as it would in Newtonian theory, because it has no pressure and no velocity (as measured by the second experimenter). Let us call the mass per unit length of this stream as measured by the second experimenter μ'. This differs from μ, which is measured by the first experimenter, but we won't need to find the relation between the two.

In Newtonian theory, an infinitely long line with a mass-per-unit-length of μ' will create a certain gravitational acceleration in a particle a distance d away. This acceleration is proportional to μ'. It also depends on d, but since in our situation d will not change, and it is the same for both streams and (importantly) for both experimenters, we won't need to know the dependence on d. We will just write the acceleration produced by the bottom stream as measured by the second experimenter as

$$a_N = \alpha\mu',$$

where the constant α contains all the things we don't want to bother with.

In general relativity, the gravitoelectric acceleration produced by the *upper* stream, as calculated by the second experimenter, will be different from the Newtonian acceleration we have just written down for three reasons. First, the second experimenter will measure the mass of each particle to be larger than its rest-mass, because of the extra kinetic energy. Second, the experimenter will measure a smaller separation between the particles, because of the Lorentz–Fitzgerald contraction. And third, general relativity tells us that the pressure has to be added into the expression for the active gravitational mass, as in Equation 19.1 on page 242. We need to work out all three corrections.

1. *The transformation of mass.* We learned from Equation 15.6 on page 190 how mass depends on speed. Given that the speed of the top stream is $2v$, the mass of each particle as measured by the second experimenter is a factor $\left[1-(2v)^2/c^2\right]^{-1/2}$ larger than the rest-mass.

2. *The Lorentz–Fitzgerald contraction.* By Equation 15.5 on page 188 for the Lorentz–Fitzgerald contraction, the particles in the top stream are closer to one another by the factor $\left[1-(2v)^2/c^2\right]^{-1/2}$, which further raises the density of mass along the stream. Thus, to the first experimenter, the mass density per unit length of each stream is

$$\text{mass density of moving stream } = \mu'/\left[1-(2v)^2/c^2\right].$$

We will work only with first corrections to the Newtonian formulas, so we can write (recall Equation 5.2 on page 43)

$$\left[1-(2v)^2/c^2\right]^{-1} - 1 \approx 1 + (2v)^2/c^2 = 1 + 4v^2/c^2.$$

This implies

$$\text{density of top stream } \approx \mu' + 4\mu'v^2/c^2.$$

3. *Pressure contribution to the active gravitational mass.* In our example there is no ordinary pressure inside the streams, but there is still an effect of the same type. If the stream were

to run into a solid wall, like spraying a jet of water from a hose against a wall, then there would be a large pressure on the wall, even if there were no internal pressure at all in the stream. This is called the "ram pressure" of the stream, and in general relativity, this kind of ram pressure will create gravity the way ordinary pressure does.

The ram pressure can be calculated by asking what sort of pressure the wall has to exert on the water stream in order to avoid being knocked over by the spray. If the density of the stream (this time we mean mass per unit *volume*) is ρ and its speed is u (which we will set to $2v$ later), and if the cross-sectional area of the hose is A, then in a small time t the mass of water that hits the wall will be $m = \rho \times A \times ut$. This water has momentum mu. To stop the water, the wall has to exert a force equal to the change it makes in the momentum of the water divided by the time, or $F = \rho A u^2$. The force per unit area, or in other words the pressure exerted by the wall as it continuously resists the water stream, is $p_{wall} = \rho u^2$. This is, by Newton's law of action and reaction, the same as the ram pressure of the water stream itself, or of any other stream of uniformly moving particles.

Now, in our example we are interested in the active gravitational mass per unit length of our streams, not their mass per unit volume. This is because we assume the streams are so thin that all the particles in a given cross-section of the stream are effectively the same distance from the central particle. Since the ram pressure is just u^2 times the mass density ρ, then by analogy the ram pressure contribution to the active gravitational mass per unit length will be $\mu'u^2/c^2 = \mu'(2v)^2/c^2 = 4\mu'v^2/c^2$.

We have only to worry about the requirement in Equation 19.1 on page 242 that we need the *average* pressure. The pressure in the stream in directions perpendicular to the stream is zero: there is no ordinary pressure and there is no velocity to make a ram pressure. So the average over the three directions of the pressure is $(4\mu'v^2+0+0)/3 = 4\mu'v^2/3$. Then the final correction is three times this divided by c^2:

pressure part of the active gravitational mass

$$= 4\mu'v^2/c^2.$$

This adds to the density of the top stream to give

total active gravitational mass

$$\approx \mu' + 4\mu'v^2/c^2 + 4\mu'v^2/c^2 = \mu' + 8\mu'v^2/c^2.$$

When all three corrections are added into the active gravitational mass, we find that the gravitoelectric Einstein gravitational acceleration of the particle due to the top stream, as measured by the experimenter at rest with respect to the particle, is

$$a_E = \alpha\mu'(1 + 8v^2/c^2).$$

We find, therefore, that the rocket experimenter calculates that the gravitoelectric gravitational attraction of the top stream exceeds that of the bottom by $8(v^2/c^2)\alpha\mu$. If there were no other gravitational effects, the central particle would move upwards with this acceleration.

Since the central particle does not move, there must be a magnetic-type acceleration, depending on velocities, that exactly compensates this. Since it should depend on both the speed of the particle, $v_p = v$, and the speed of the source, $v_s = 2v$, we can write this acceleration as a repulsion from the source of magnitude

$$a_M = 4(v_p/c)(v_s/c)\alpha\mu.$$

x-direction. This changes the direction of all the velocities in the same way that changing the sense of time did. So again, the gravitomagnetic effect must be in a term that changes sign when we replace *x* by -*x*. There is only one term that changes sign when we do either operation, and that is the mixed term containing the product $(\Delta T)(\Delta x)$.

> The coefficients of the mixed terms between time and space coordinates in the spacetime-interval, like $(\Delta T)(\Delta x)$, create the gravitomagnetic effects in the motion of particles following geodesics of a geometric gravitational field.

Gyroscopes, Lense, Thirring, and Mach

Once we realize that gravitomagnetism exists, we can find many situations where it can be seen. The most important are the effects caused by rotating masses. Long streams of particles, such as we treated in the last section, are rare in the Universe, but rotating stars and black holes are common. In this section we will see how to estimate the gravitomagnetic effect of a rotating star, how this is being measured today for the Earth, and how it is related to an old philosophical idea called Mach's principle.

From the two-hand rule, we can determine the effect of gravitomagnetism on bodies moving in other ways than the central particle of the example we studied first. In particular, suppose in Figure 19.2 on page 246 that a particle is moving directly *towards* the top stream, from above it. Then the gravitomagnetic effect will bend its path *in the direction of motion of the stream*. By symmetry, this will happen to any particle approaching the stream. This effect has acquired a rather dramatic name in general relativity. It is called the **dragging of inertial frames**. What this means is that the stream seems to change the local standard of rest. Particles that are at rest far away and then fall toward the stream are dragged along it a little, as if the stream were a jet of water pushing though still water, entraining some of its surroundings along it.

Let us see how this dragging can be important in realistic situations. Consider the gravitational field near the rotating Earth. The rotation of the Earth can be approximated, for the purposes of our little discussion, as a stream of matter moving around a loop. Then if we are near one side of the loop, we see gravitomagnetic forces from the near side, pushing us one way, and from the far side, pushing us the other way. These effects tend to cancel each other, but since the force depends on distance, the cancellation is not perfect, and the near side wins. The further we are from the Earth, however, the less significant is the difference between the distances to the two sides of the loop, so the better is their cancellation.

> The net result is that the gravitomagnetic force caused by the rotation of the Earth falls off with distance from the Earth faster than it does from a single moving particle, proportional to $1/r^3$ rather than the $1/r^2$ of the basic Newtonian force.

The rotation of the Earth also gives this dragging force a twisting character. This is most easily seen if we imagine placing a spinning gyroscope exactly in the *center* of the rotating loop that we take as a model for Earth (Figure 19.3 on the next page). This idealized experiment will help us understand what will happen in more realistic situations.

Suppose the gyroscope is oriented horizontally, with its axis parallel to the equatorial plane, pointing momentarily at longitude 0°. Suppose also that the gyro is spinning in a positive sense, which means it is spinning counterclockwise when looking down the axis from the longitude 0° point on the loop.

In this section: through gravitomagnetism, spinning bodies can cause particles near them to rotate in the same sense. Two experiments are trying to measure the effect using gyroscopes and satellites. The effect is as close as general relativity comes to suporting the ideas of Mach on where inertia comes from.

▷In relativity, a *frame* is the coordinate system of an observer, so the term "dragging of inertial frames" describes the way a freely-falling observer is swept along in the direction of rotation by gravitomagnetism.

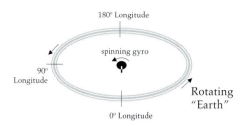

Figure 19.3. Idealization of the geometry of a spinning Earth dragging a gryroscope situated at its center. We represent the Earth as a loop of mass concentrated at the equator, and the gyroscope as a disk spinning about a horizontal axis pointing toward longitude 0°.

Figure 19.4. Drawing of the Gravity Probe B (GP-B) satellite, which will carry the most sensitive gyroscope ever constructed into orbit to measure the Lense–Thirring effect. The gyroscope will change its direction by only 42 milli-arcseconds in one year, and GP-B is designed to measure that to an accuracy of 1%. This angular precision, half a milli-arcsecond, is about the angular size of a medium-sized dog at the distance of the Moon! Drawing courtesy of Gravity Probe B.

Figure 19.5. This fused-quartz sphere coated with superconducting niobium is one of the gyroscopes carried on GP-B. The size of a table-tennis ball, it is so spherical that its irregularities are nowhere more than 40 atoms high. This is just one of many challenges that have been met in designing this extraordinarily sensitive satellite. Image courtesy of Gravity Probe B.

Now, the gyro is just a loop of mass moving in a circle about its own axis. On the top of the gyro the mass of this loop is moving towards longitude 90° W (just off the coast of Ecuador). The part of the Earth in western longitudes is moving eastward, and exerts a dragging force on this part of the gyro to pull it towards the direction of 0° longitude. The part of the Earth on the other side of the gyro, in eastern longitudes, is moving in the opposite direction, but the gyro is moving away from it, so it "anti-drags" the gyro, again pulling the top part of it towards 0° longitude. The bottom part of the gyro is moving in the opposite sense, so it must feel a force towards 180° longitude, in the South Pacific. The net result of these two forces is a torque (twisting force) trying to pull down on the part of the gyro's axis that points toward 0° and up on the opposite side.

Now, we know what happens when we try to do this to a gyro: it rebels. It simply turns to the side.

> The effect of the gravitomagnetic forces, therefore, will be to change the direction of the gyro's axis, causing it to rotate slowly in the same direction as the Earth spins. It is not hard to convince oneself that this will happen to any gyro oriented horizontally, even if it is not at the center of the Earth. This is called the Lense–Thirring effect, after the two scientists – Josef Lense (1890–1985) and Hans Thirring (1888–1976) – who discovered it only two years after Einstein published his general theory.

A gyro can be used to measure the Lense–Thirring effect. Two experiments are presently underway using very different kinds of gyros. One of them is called Gravity Probe B, illustrated in Figure 19.4. This is planned for launch by NASA in 2003; it will carry a very sensitive gyroscope (Figure 19.5) into orbit around the Earth to try to measure the predicted effect over a period of a year or more. The other experiment uses existing satellites, called LAGEOS and LAGEOS2, shown in Figure 19.6. If a satellite has an orbit that goes over the Poles, it is moving in the same way that the mass of the gyroscope was moving in our example above, so the orbit will get twisted by the Lense–Thirring effect in the same way: the orbit will gradually precess eastwards. By precise tracking of the orbits of these satellites, which have been specially designed to minimize the effects of atmospheric drag, and which have nearly polar orbits, the second group of scientists is presently measuring the dragging. Their initial results have confirmed the predictions of Einstein's theory, and higher accuracy is expected in the near future.

Astronomers are beginning to see the effects of frame-dragging near neutron stars and black holes. We will return to the evidence for this in Chapter 21, but it seems that the Lense–Thirring effect may soon be used to measure the spin of black holes in astronomical systems. Other astronomers are proposing a very high-accuracy astrometry satellite called GAIA, a successor to the Hipparcos mission described in Chapter 9. This would be able to see the extra deflection effects produced by dragging as light passes near the Sun. The Sun is spinning, so light that passes on one side of it will be affected by dragging differently than light on the other side, and it will be possible to measure the interior spin of the Sun for the first time in this way.

Suppose the experiments near the Earth are not able to confirm the details of frame-dragging: suppose gravitomagnetism is not as predicted by general relativity. What then? Our derivation makes it clear that only two assumptions are needed to derive the standard formulas: special relativity and the Einstein expression for the active gravitational mass. We would not like to give up special relativity, since it is tested in many other places. We would therefore have to look for a different expression for the active gravitational mass, and that would have all kinds of implications for gravity.

> Since Einstein's equations work so well in other situations, it seems very likely that the Gravity Probe B and LAGEOS experiments will verify the Lense–Thirring effect. But surprises are always possible!

The dragging effects of gravity are reminiscent of philosophical ideas that go back to the Moravian physicist Ernst Mach (1838–1916). Mach was intrigued by the question of why bodies have inertia. Why does it require a force to accelerate a mass? What is so special about the state of uniform motion that it requires no acceleration? Put simply, uniform with respect to what? Mach suggested that the Universe itself establishes what is meant by uniform velocity, that a velocity can be maintained without an external force if it is uniform with respect the Universe. He speculated that this condition had a real cause, that bodies exerted an influence on one another that resisted their relative acceleration. Although Mach did not turn these ideas into a successful theory, they appealed to Einstein and he gave them as one of the influences that shaped the way he searched for a relativistic theory of gravity.

The dragging of inertial frames seems Machian in spirit. It is a real influence that seems to try to bring one thing closer to the state of motion of another. But the fit to Mach's ideas is very imperfect. For one thing, gravitomagnetism depends on the direction of motion. A body falling towards a stream of matter is indeed pulled in the direction of its motion, but a body moving away from the stream is accelerated in a direction *opposite* to the motion of the stream! And a body at rest feels no influence from the motion of the stream at all.

> In fact, despite Einstein's interest in Mach's ideas, Einstein's own theory sheds no light on what creates inertia.

Figure 19.6. *The* LAGEOS *satellite is the complete opposite of* GP-B. *It is a passive satellite, with no working components. Covered in mirrors, its only job is to reflect laser beams back to Earth, which are used in range-finding to find its exact position. The relativity experiment is a spin-off from the satellite's main mission, which is to track continental drift by measuring the motion of the ground stations that track the satellites. From the tracking data, however, scientists are beginning to discern the gradual precession of the orbit of one of them induced by the Lense–Thirring effect. Image courtesy* GSFC.

The cosmological constant: making use of negative pressure

The history of the cosmological constant is one of the oddest chapters in the development of general relativity. Einstein reluctantly and belatedly introduced this new term into his theory, and then he later withdrew it. But now astronomers think they have measured it, cosmologists imitate it in their theory of cosmological inflation, and physicists find that it comes naturally out of their theories of high-energy physics. In this section we will simply describe the way the cosmological constant works, and what is special about it. We are thereby preparing ourselves for our discussion of the physics and astronomy of the cosmological constant in the last chapters.

When Einstein invented general relativity, astronomers did not yet know that the Universe was expanding. Einstein wanted to be able to make a mathematical model of the whole Universe that was static, neither expanding nor contracting. To do this he needed something that would counteract the attractive force of the matter in the Universe. Fortunately for him, his equations gave him the right loophole: negative pressure. If he could introduce enough negative pressure, then he could arrange for the total density of active gravitational mass to be zero.

In this section: the cosmological constant can be viewed as a physical fluid with a positive density and a negative pressure. We derive the remarkable and unique properties of this special fluid: it has no inertia, exerts no pressure forces, stays the same density when it expands or contracts, and creates a repulsive gravitational field: anti-gravity. These properties allowed Einstein to introduce it safely into his equations in order to stop the Universe from collapsing.

As we remarked above, negative pressure is called tension. But in ordinary materials, it normally is present only if the material is acted upon by outside forces, such as being stretched in one direction like a rubber band. Normal materials do not have tension in their resting state.

> No ordinary matter displays isotropic negative pressure in its normal state. Einstein's suggestion was something entirely new. He had no physical model for his cosmological negative pressure. It was a mathematical device to produce a Universe with zero gravitoelectric force on the large scale.

It is not hard to imagine why Einstein was never happy with this idea, despite the fact that it did what he wanted. For one thing, it was what scientists describe as *ad hoc*, something that has no other justification than to patch things up. The negative pressure had no foundation in observation, and was introduced simply to rescue the theory from the difficulty of the expanding or contracting Universe that it predicted. Einstein had no physical mechanism for producing the pressure.

Even worse, in order to make the Universe static, the amount of pressure had to be exactly right, just enough to cancel out the energy density of the universe in the active gravitational mass. If the pressure were not large enough to cancel the attraction of the energy density, then the Universe would slow down and perhaps re-collapse; if the pressure over-compensated for the energy of the Universe, then the Universe would expand in an accelerated way. Einstein's static universe model was *unstable* to small changes in its density.

But Einstein recognised that negative pressure was the only way he could get general relativity to give a static Universe, so he pursued the idea.

> Einstein found that there was one and only one way to introduce this negative pressure and still preserve his principle of relativity. His cosmological constant introduces an energy density and pressure into the Universe that are constant in time and in space, and that moreover are the same no matter which observer measures them. The cosmological "fluid" is completely invariant. This brilliant mathematical insight has consequences in modern cosmology well beyond anything Einstein could have imagined.

Let us see what we can make of this idea.

Einstein wanted to introduce negative pressure without giving up the most fundamental feature of general relativity, that it does not pick out any special observer, or place, or time. Since his cosmological negative pressure was to be fundamental, not tied to any accidental matter field or configuration, he needed the pressure to be constant in space and in time, so that an observer could not pick out any special place or time by measuring the pressure. It had to be a fundamental *constant* of nature. Einstein actually introduced, instead of a fundamental pressure, a fundamental constant Λ, which he called the *cosmological constant*. The uniform cosmological pressure he needed, p_Λ, is defined in terms of Λ by

$$p_\Lambda = -\frac{c^2 \Lambda}{8\pi G}. \tag{19.6}$$

The sign allows Λ to be positive to give the negative pressure required for a static universe. The other constants in the definition show that Λ itself has the dimensions of a frequency squared, or $1/(\text{time})^2$.

Figure 19.7. *Ernst Mach is mainly remembered today for his work on supersonic motion: the Mach number of a projectile or aircraft is the ratio of its speed to the speed of sound. But he also speculated about profound questions in physics, psychology and philosphy, often advocating his positions stubbornly. He was one of the most vocal (and last!) opponents of the atomic theory of matter, and strongly attacked Boltzmann's theories, despite the fact that he and Boltzmann were colleagues at the University of Vienna. Regarding inertia, Mach was dissatisfied that physicists since Newton had studied only the forces that were required to accelerate masses, but not why the masses had inertial mass in the first place. Mach hoped to find a deeper physical principle underlying Newton's laws. Image courtesy Charles University, Prague.*

Einstein found that he could make the pressure invariant if the cosmological constant also generated a mass density

$$\rho_\Lambda = -\frac{p_\Lambda}{c^2} = \frac{\Lambda}{8\pi G}. \qquad (19.7)$$

The cosmological fluid that has this pressure and density has remarkable properties. First, let us ask if there is any way we can detect this fluid, other than by observing its gravitational effects. What are its local properties?

- *Zero inertial mass density.* One might ask if this cosmological fluid could be felt in non-gravitational ways. For example, does it have inertia, so that it would make things harder to move? We calculated the inertial mass density in Chapter 15, and we saw in Equation 15.10 on page 193 that it is $\rho + p/c^2$. With $p_\Lambda = -\rho_\Lambda c^2$, this fluid has *zero* inertial mass! It can be accelerated with no cost, no effort. This property is the key, as we will see momentarily, to the invariance of the pressure and density against a change of observer.

- *Zero pressure force.* With a large negative pressure, surely this fluid would exert observable pressure forces on things in the Universe. But no, pressure forces act only through pressure *differences*, as we saw in Chapter 7. A uniform pressure, even a negative one, exerts no force.

The cosmological fluid is remarkable indeed:

> Einstein's cosmological constant is *undetectable* in non-gravitational experiments. It contributes nothing to local dynamics. It offers no resistance to objects moving through the vacuum. Its pressure is uniform, so it exerts no direct forces on objects. You can't *feel* the cosmological energy density or pressure. The vacuum is just as empty with it as without it, except for its gravitational effects.

This aspect of the cosmological constant was particularly repugnant to Einstein, who had only recently succeeded in getting rid of the nineteenth-century notion of the ether, as we saw in Chapter 15. Now apparently he was forced to introduce something just as strange.

Now let us see how this fluid could have these invariant properties. Normally, the density and pressure of a fluid depend on the observer, on the speed of the fluid relative to the observer. Consider, first, how the pressure of the cosmological fluid might depend on its motion. We saw in our earlier derivation of gravitomagnetism, in our discussion of the active gravitational mass of a stream of particles, that when a fluid moves, then in this direction the density contributes something to the pressure. We called this the ram pressure, and saw that it equals ρv^2. But that discussion assumed that the fluid was non-relativistic, in the sense that not only was the speed v small but also the pressure was small. When the pressure is large, so that p/c^2 is similar in size to ρ, then the ram pressure must be modified. We must use the inertial mass per unit volume, $\rho + p/c^2$, instead of just ρ. The inertial mass per unit volume is the quantity that measures the inertia of the fluid, which determines the pressure it would exert if it ran into a wall. So the ram pressure of a relativistic fluid moving at a small speed v is $(\rho + p/c^2)v^2$. Now, we have already seen that the inertial mass density of the cosmological-constant fluid is zero. Therefore, its pressure as measured by an observer who is moving with respect to the fluid is exactly the same as for an observer at rest. The pressure p_Λ of this fluid is an invariant.

In the same way, the density ρ_Λ is also an invariant. We leave the proof of this to Investigation 19.2 on page 257. And there is one other property that Einstein

▷ As if to rub salt into Einstein's wounds, modern physicists use the term **quintessence** for some new theories of physics that introduce fields with negative pressure. Quintessence was the name Aristotle used for the ether.

required. His equations of gravitation require that any matter of fluid in spacetime should obey the laws of conservation of energy and momentum in any small volume. His new cosmological fluid needs to pass this test too. We shall see, again in Investigation 19.2, that the law of conservation of energy insures that, as the Universe expands, the density and pressure of this fluid remain *constant*, just as Einstein required. With these properties, the cosmological constant provided Einstein with just what he needed: a force that could keep the Universe static and at the same time did not single out a preferred observer, place, or time.

> Then came Edwin Hubble. When Einstein learned of Hubble's discovery that the Universe was expanding, he bitterly regretted having invented the cosmological constant. He reasoned that, if he had had the courage to stay with his original theory, he would then have predicted the expansion before it was discovered, and his prediction might well have led astronomers to look for the expansion earlier than they did. The expanding universe would have been seen as a further experimental test of and triumph for general relativity. To Einstein, one of his greatest blunders was not having had confidence in his original equations in the cosmological arena.

Physicists today take a more generous view of Einstein's "blunder". Spurred on by theoretical considerations in fundamental physics, which suggest that this kind of cosmological fluid could be a natural consequence of theories of high-energy physics, physicists are looking for ways to predict a cosmological constant with a value that would account for recent astronomical observations that the expansion of the Universe appears to be accelerating. They actually use the word "field" instead of "fluid" to describe the cosmological constant, but that is nomenclature. We will return to a discussion of this in our final chapter, Chapter 27.

The big picture: all the field equations

We have seen the detail of Einstein's theory, we have used it to calculate the deflection of light by the Sun, and we have shown that the gravitational field includes interactions between momenta and between spins. We could go on to study the phenomena of relativistic gravity. But we have not yet asked where Einstein's theory came from, what led Einstein to his creation. We will spend the rest of this chapter trying to look at general relativity from Einstein's own perspective.

> Einstein was looking for equations that would generate the curvature that represents gravity and at the same time obey the principle of general covariance. We have seen that the way we describe geometry, for example the coefficients in the spacetime interval, depends on the coordinate system. Einstein had to find a way to allow the geometrical description and the sources to change when the observer changed, but to get the same geometry in the end. If things worked out correctly, the different parts of the gravitational field would fit together in just the right way to compensate for the different values measured by different observers for the individual sources.

This is easy to say, but hard to do. Einstein could make no assumptions to simplify the form of the spacetime interval; he had to work with the most general form. We have written such a spacetime interval for two dimensions in Equation 18.5 on page 228, and we saw it had three coefficients, which depended on three functions A, B, and C. In four dimensions, there are ten coefficients: four for the terms like $(\Delta T)^2$ and $(\Delta x)^2$, and another six for mixed terms like $(\Delta x)(\Delta y)$ and $(\Delta T)(\Delta z)$. That means

▷It is worth asking how Einstein arrived at this remarkable prescription for a cosmological fluid. He certainly did not follow the route I have taken in presenting it here; this method fits well with physicists' perspective on the constant today, but it was not Einstein's perspective. In fact, he was led to Λ by the principle of general covariance. Once one has studied the full mathematics of Einstein's equations, the cosmological constant actually seems like a rather natural modification of the theory. Einstein regarded Λ as a fundamental constant of Nature, which he introduced as a modification of the field equations (i.e. of the *recipe*). He did not think of the cosmological constant itself as a fluid, as a source of gravity, as a new *ingredient* for the old recipe. We shall see this in Equation 19.9 on page 258.

In this section: we meet the full field equations of general relativity and learn why Einstein was led to postulate them, and in what sense they are simple and elegant.

▷This is exactly the kind of compensation that we saw above when we used the gravitomagnetic force to balance the excess gravitoelectric force seen by a moving observer. We see from that example how well Einstein succeeded.

Investigation 19.2. The remarkable physical properties of the cosmological fluid

Einstein defined his cosmological constant in a very special way, ensuring that there was a strict relationship $p_\Lambda = -\rho_\Lambda c^2$. This can be thought of as the equation of state of the cosmological fluid. (We introduced the notion of an equation of state in Chapter 7.) This equation of state has a remarkable property. It guarantees that, as the Universe expands, the mass density of this cosmological "fluid" remains *constant* in time. Unlike all normal gases, a fluid with $p = -\rho$ does not get diluted by expansion. Nor does its density depend on its speed, so it does not pick out any preferred observer.

To see how the density remains constant, consider an isolated box filled with such a fluid. Suppose the box has initial volume V and then expands slowly to $2V$. Since the fluid in the box has tension, a force is required to expand the box. The force does work, as defined in Equation 6.20 on page 62, and this adds energy to the fluid. The mass equivalent to this energy adds to the mass already in the box. We shall show that when the pressure is that of this peculiar cosmological fluid, the added energy is just enough to insure that the density of the fluid in the larger volume is the same as in the smaller.

We can make this verbal explanation quantitative with a simple set of calculations. As our first step we will find out how the energy in any fluid changes when it expands and contracts. Consider a rectangular box of volume V with one movable side. The fluid in the box has pressure p. If p is positive, the pressure pushes outwards on the walls. Now apply a force F that moves the movable side inwards, like a piston. Let us suppose that the movement is very small, so that the force F just balances the pressure. Then we can calculate F from the fact that the pressure is the force of the gas per unit area on the wall. If the wall has area A then we simply have $F = pA$. Now, if the wall moves a small distance δx, then the work done by this force is

$$W = F\delta x = pA\delta x.$$

The product $A\delta x$ is the reduction in the *volume* of the box. So the change in volume is $\delta V = -A\delta x$, the minus sign indicating that the volume of the box has been reduced. The result of all this is a very general law about work on gases:

$$\text{work done on a gas to change its volume} = -p\delta V. \qquad (19.8)$$

Now, the work done by the external force F is work done against the atoms or molecules of the gas. The pressure is nothing more than the result of untold numbers of collisions between gas particles and the walls. So when the wall moves inwards, it pushes a little on each particle that it encounters, making it rebound from the wall

a little faster than if the wall had not been moving. So the work done by F equals the increase in kinetic energy of these particles, in close analogy with the situation we met when we first introduced the concept of work in Chapter 6. There the work done by the gravitational force increased the kinetic energy of a body orbiting the Sun, as in Equation 6.19 on page 62. In a fluid, however, collisions among the gas particles themselves quickly transfer this energy around the fluid, sharing it roughly equally among all the molecules. The result is that the energy of the fluid has increased by the amount of work done by the external force.

Now, let us look at our box full of cosmological fluid. Its initial energy content was $\rho_\Lambda c^2 V$. During the expansion, the pressure did not change, and the volume changed by V. The work done, and hence the change in the total energy in the fluid, is $-p_\Lambda V$. Since $p_\Lambda = -\rho_\Lambda c^2$, the energy in the box has increased by $\rho_\Lambda c^2 V$, so that the new total energy is $2\rho_\Lambda c^2 V$. But the volume is now $2V$, so the energy *density* is $\rho_\Lambda c^2$ and the mass density is ρ_Λ, unchanged from before the expansion. This demonstrates that, as the Universe expands, it is consistent with local conservation of energy (i.e. conservation of energy in any small region of the Universe) that the cosmological constant should not change.

The constancy of energy with volume also explains how the mass-energy density of the fluid is independent of the observer. To see this, we suppose that an observer at "rest" measures mass density ρ_Λ, and that, as above, this fluid is in a rectangular box. This time the box does not have a movable wall. Suppose another observer moves at a small speed v along one edge of the box. This observer will notice two things. First, the box is, of course, moving at speed v past him. The fluid in it therefore should have more mass-energy than when it is at rest, because it has the kinetic energy of its motion. However, in this particular case, as we have seen in the text, the inertial mass density of this fluid is *zero*: it can be accelerated for free, without any energy cost. (The energy of the box, made of ordinary matter, is not of interest to us here.) The second thing the observer will notice is that the box is shorter, because of the Lorentz–Fitzgerald contraction (Chapter 16). Normally this would raise the density of mass-energy in it, but we have seen above that for this kind of fluid, changing the volume of the fluid has no effect on its density. The net result is that the density of this fluid is invariant under a change of observer, just as the is pressure.

So the cosmological constant is quite remarkable: it provides an all-pervading energy density and negative pressure that are the same to all observers, at all places, and at all times in the history of any universe model, even expanding ones.

Exercise 19.2.1: *Upper bound on the cosmological constant*

The fact that Newtonian gravity describes the orbits of planets in the Solar System very well, using only one parameter (the mass of the Sun) for all planetary orbits, suggests that the cosmological constant must create a smaller mass density than the mean mass of the Solar System out to Pluto's orbit. (a) Calculate this mean density by dividing the mass of the Sun by the volume of a sphere whose radius is the radius of Pluto's orbit. (b) From this, calculate the value of the cosmological constant Λ that would give a mass density ρ_Λ of the same value. Use Equation 19.7 on page 255.

---

that a geometrical theory requires ten equations, in which the ten coefficients are determined by the properties of the source of the gravitational field: energy, pressure, and so on.

Einstein had another problem, though: as we have remarked earlier, the geometry of spacetime does not uniquely determine the values of the coefficients, since we are free to change the coordinates that we use to describe spacetime. Indeed, Einstein wanted to build into his theory this freedom to choose coordinates. But that meant that the equations of the theory could not possibly determine all ten metric coefficients in terms of the sources. If they did, that would amount to determining the coordinates too.

Einstein's breakthrough came when he found that he could write down ten equations that were not all independent. He could derive some of the equations from the others, *provided the sources of the gravitational field obeyed the laws of conservation of energy and momentum in any locally flat coordinate patch.* This was a very significant step. It meant that his geometrical gravity would respect the equivalence principle completely, so that not only would freely-falling particles "feel" no gravitational field, but also freely-falling *gases* would behave just as if there were no gravitational field. At the same time, Einstein's equations would only determine six combinations of the metric coefficients, the remaining being determined by coordinate choices.

This requirement, that gravity should be compatible with energy conservation in ordinary matter, almost fully determined the equations of the theory.

> The elegance and beauty that mathematicians and physicists find in general relativity comes partly from the way Einstein started with an apparently horrible prospect, namely trying to find ten equations that would work in any coordinate system and that would predict Newtonian gravitational effects when gravity was weak, and yet managed to arrive at a theory that does all this and can be written down in one line.

Here it is:

$$G_{\mu\nu} + \Lambda g_{\mu\nu} = (8\pi G/c^4)T_{\mu\nu}. \qquad (19.9)$$

We aren't going to work with these equations, of course. But it would be a shame to spend a lot of time in this book discussing the theory and repeatedly mentioning the equations, without ever writing them down in their most general form!

The fundamental unknown quantities here are the ten metric coefficients that describe the spacetime-interval, denoted here by $g_{\mu\nu}$, and called the **metric tensor**. The symbol $G_{\mu\nu}$ on the left-hand side is related to the curvature of spacetime, and is constructed from the ten metric coefficients; its name is the **Einstein curvature tensor**. The idea is to solve these equations for the metric coefficients as functions of position in space and time, given (as the source of gravity) the density, pressure, and momentum of any matter fields that are present: the first three of our sources are on page 240. These sources are all part of the object $T_{\mu\nu}$ on the right-hand side, which is called the **stress–energy tensor**. This is the single source to which we referred earlier in the chapter, when we described the symmetry of the principle of general covariance.

The constant Λ is Einstein's cosmological constant. He placed it on the left-hand side of Equation 19.9, as part of the equation to be solved, rather than on the right, with the sources. We noted why earlier.

What makes the Einstein equations mathematically challenging is not just that they use the language of tensors. More important is that the Einstein curvature tensor can only be constructed from the metric by using calculus. It is a function of the **derivatives** of the metric tensor. It is a non-linear function, so it is in the Einstein tensor that our fourth "source" on page 240 is to be found. The Einstein equations form a set of what mathematicians call **differential equations**, and their complexity is so great that they can be solved by algebraic methods only in special circumstances, such as when one is looking for solutions with a particular symmetry. Full solutions, for example those that represent collisions of black holes, must be solved on supercomputers.

The search for simplicity

The fact that Einstein's equations can be written in a single line, using only a few symbols, is a reflection of the fact that they are, conceptually, simple equations. The symbols $G_{\mu\nu}$ and $T_{\mu\nu}$ refer to meaningful combinations of mathematical entities

▷ The word **tensor** used in these names refers to a mathematical object that is a generalization of a **matrix**, or array. The symbols μ and ν are labels (called **indices**) that can together be taken in ten different combinations to make the full set of equations. The word **stress** is a physicists' word for things like pressure and tension. This is how the pressure contributes to the creation of gravity, as we have seen it does in general relativity.

In this section: we meet Occam's razor and show how Einstein used it when devising, and later revising, the field equations.

that were known to physicists and mathematicians even before Einstein; they are not just shorthand for long strings of algebra.

In arriving at his famous equations, Einstein followed a long-cherished principle in science, called **Occam's razor**: "It is vain to do with more what can be done with less."[†]

> Named for the Englishman William of Occam (1300–1349), who was putting into words what had already been practiced by Greek scientists long before, this principle is interpreted by physicists today to mean that, when trying to find a new theory to fit some observed facts, one should always aim for the simplest description.

Inevitably, there will be many theories that might fit the facts, including the trivial one that says, for example, "It is a law of Nature that the Earth should take one year to go around the Sun and it is another law of Nature that Venus should do so in 0.72 years." Such a "theory" merely re-states observed facts without offering explanations or relations between them, and so is unsatisfactory. Newton's law of gravity explains these facts and the other planetary periods, plus much more, using only one observed fact, the mass of the Sun. This is an illustration of the simplicity of the theories of physics.

Einstein's original field equations had no cosmological constant, so they had a simplicity similar to Newton's gravity, in that they introduced no new measurements or important numbers that are not already present in Newton's gravity and special relativity. The original equations just use the constants G and c plus the mathematics of spacetime curvature. In a very real sense, Einstein's theory is the simplest theory that makes Newton's gravity compatible with special relativity and the other laws of physics.

Later, when Einstein felt compelled by astronomical evidence to introduce the cosmological constant, he retreated a little from the initial simplicity of the theory. But he still used Occam's razor: he found a way of introducing a cosmological repulsion, or anti-gravity, that did not require any special observer or coordinate system. The very special and peculiar properties that this cosmological fluid possesses allowed Einstein to keep the principle of general covariance and avoid introducing anything new except Λ. This shows that Occam's razor is not so conservative that it prevents innovation and the modification of old and inadequate theories. Instead it imposes a form of discipline, keeping the innovations as simple as the new facts allow.

General relativity

Our approach in this chapter to the field equations – taking them apart and looking separately at the most important sources of gravity – has given us considerable insight into general relativity, but there are some aspects of the theory that this method does not directly illuminate. To fill in these gaps, here is a partial list to help us get ready for later chapters.

In this section: we step back and look again at the theory from Einstein's point of view. We mention some predictions, like gravitational waves and cosmology, that we have not looked at so far in this chapter.

- *Einstein's equations predict gravitational waves.* This is inevitable in a theory that obeys special relativity and embodies Newtonian gravity. Since no influence, not even a gravitational one, is allowed to travel faster than light, it follows that the changes in a gravitational field that are caused by changes in its source (such as the orbital motion of a pair of binary stars) must travel outward no faster than light. This outward motion of the changes of gravity is

---

[†]*Frustra fit per plura, quod fieri potest per pauciora.*

like a wave moving along the surface of a pond from the point where a stone falls into the water. We call this a gravitational wave. In general relativity, gravitational waves move at exactly the speed of light. Chapter 22 is devoted to gravitational waves.

- *General relativity can deal consistently with cosmology.* The idea that gravity is geometry rather than an extra force has this unexpected and useful side-effect. Newton's rule for computing gravity makes no sense if one tries to apply it to an infinite Universe, where one has to add up the effects of an infinite number of galaxies. General relativity avoids this problem because gravity is just geometry; it does not add up direct long-range forces. As long as the geometry of the Universe is smooth, then gravity evolves with time in a regular way, regardless of how big the Universe is. Moreover, since changes in gravity move at a finite speed (c), very distant parts of the Universe do not affect us. If they are so far away that light could not have reached us since the Big Bang, then they can have no influence on our local geometry. We shall see that, in fact, all we need is our local weak-field equations to compute the geometry of cosmology.

- *The Einstein equations are not the only ones that one might invent.* There are more ways than one to write down a generally covariant set of equations for a curved spacetime that satisfies the equivalence principle. But the success of general relativity in experiments and Solar System observations has shown that any changes to Einstein's equations need to be small. They are most likely to arise from the next item on our list, **quantum gravity**.

- *General relativity is not a quantum theory of gravity.* Planck's constant is conspicuously absent from Einstein's equations. There is therefore no uncertainty principle: all gravitational effects can, at least in principle, be measured with arbitrary accuracy. This can lead to logical contradictions with the rest of physics, if for example one imagines using gravitational means to measure the positions of elementary particles. Since we believe that the Universe is basically quantum in nature, we expect that general relativity will ultimately have to be replaced by a quantum version. It is likely that this will effectively change Einstein's equations by adding correction terms proportional to Planck's constant. We will return to this subject in Chapter 27.

Looking ahead

In this section: we are ready to apply the principles of general relativity to the systems discovered by astronomers.

We have now laid the foundations for the remainder of this book. We have opened the door to the rich and fascinating world of relativistic gravitation. We have had hints before, about black holes and neutron stars, about gravitational collapse and gravitational waves, about inflation and the Big Bang. But now we are in a position to understand these ideas and phenomena in a deeper way. We will start with neutron stars, progress to black holes, look at the new astronomy that gravitational wave detectors will soon open up, learn how the deflection of light is being used to discover dark masses, and then confront the ultimate: cosmology, the Universe as a whole. Almost every proton, neutron, and electron in our bodies has been in existence since about three minutes after the Big Bang. We are ready now to begin to understand the history of the matter of which we are made.

20

Neutron stars:
laboratories of strong gravity

I n previous chapters, we have seen how the new ideas in Einstein's gravity make small but striking corrections to the predictions of Newton's gravity, bending light more strongly as it passes the Sun and causing the orbits of planets to precess. Working out these corrections helped to ease us into the theory, to see that relativistic gravity is a natural development from Newtonian gravity. But the real excitement in modern astronomy and theoretical physics is in situations where Newtonian gravity doesn't even come close to being right. The Universe demands that astronomers use general relativity to explain what they see, and the deepest questions of fundamental physics demand that physicists even go beyond general relativity to find their answers. In this chapter we open the door on the richness of modern gravity by studying our first example of really strong gravitational fields: neutron stars.

Neutron stars are effectively giant nuclei, held together by gravity. If Isaac Newton had understood enough nuclear physics, he could have predicted their existence, and he could have given a rough description of them within his theory of gravity. We did this in Investigation 12.6 on page 148. When we look at this below, we will see that such a calculation merely shows us that Newtonian gravity cannot give a particularly accurate description of neutron stars: relativity cannot be ignored or relegated to a small correction.

What Newton also would not have been able to do, even with the best nuclear physics, is to have predicted how *abundant* neutron stars are. Possibly one star in every thousand in our Galaxy is a neutron star. Newton also could never have guessed how spectacularly they show themselves off, as pulsars and intense sources of X-rays. Containing more mass than the Sun, in a region smaller than a large city, a typical neutron star spins on its axis tens of times per *second*, nurtures a magnetic field billions of times stronger than the Earth's, and – with an interior temperature of millions of degrees or more – is the ultimate high-temperature superfluid and superconductor.

Nuclear pudding: the density of a neutron star

In Investigation 12.6 on page 148, we calculated roughly some of the properties of neutron stars from basic quantum theory and Newtonian gravity. Here we take a different point of view, and show that, without knowing much about quantum theory, it is easy to see that a neutron star should have the same density as an ordinary heavy nucleus, like that of uranium. It may seem strange to try to extrapolate from a tiny nucleus to a huge neutron star, but nuclei have one unusual property that makes this possible.

This property is that the nuclei of *all* heavy elements have very similar densities, about 2×10^{17} kg m^{-3}. This is a huge density by ordinary standards, some 2×10^{14} times the density of water. Since almost all the mass of an atom is concentrated in its nucleus, the nuclei occupy a very small part of the volume of an atom, smaller in

In this chapter: we study neutron stars, our first example of strong relativistic gravity. Neutron stars are known to astronomers as pulsars and X-ray sources, and they are at the heart of supernova explosions. They are giant nuclei containing extreme physics, including superstrong magnetic fields, superconductivity, and superfluidity. Neutron stars only exist because of a few coincidences among the strength of the nuclear, electric, and gravitational forces; without these coincidences, life would never have formed on Earth.

▷Underlying the text on this page is a sketch of a *pulsar*, which is a spinning magnetic neutron star. The magnetic field lines (arcs) converge on the magnetic poles, which are hot spots, emitting beams of radio, visible, X-ray, and gamma-radiation. The magnetic poles lie near the equator of the spinning star, whose spin axis might point vertically on this page. The poles sweep the sky like a lighthouse, so that if the Earth is in one or both beams, we see the star turn on and off. Figure 20.4 on page 270 shows a photographic record of the light from such a star flashing 30 times a second.

In this section: neutron stars are simply huge nuclei, held together by gravity rather than nuclear forces. Their existence depends on the push–pull nature of the nuclear forces, which stop gravitational collapse when the protons and neutrons get about as close to one another as in a normal nucleus.

size by the cube-root of 10^{14}, or about 5×10^4: the radius of the nucleus of an atom is roughly 50 000 times smaller than the orbital radii of its electrons. Put graphically, if the nucleus were magnified to the size of an apple, then its electrons would be 1.6 km (1 mile) away! All the space between the nucleus and its electrons is empty.

Now, because nuclei have this same high density no matter how many neutrons and protons go into the nucleus, every **nucleon** (every proton or neutron) occupies the same volume as every other one, and this volume is virtually the same, regardless of whether there are 10 nucleons or 100.

This is a rather unexpected behavior: one would normally expect that the nuclear forces that attract nucleons together would get stronger as more nucleons are added, and the density would increase. This happens when gravity provides the attractive force and the matter is ordinary gas: when more mass is added to a star, its density normally increases. So nuclear forces must be different from gravity somehow, to keep the nuclear density at its special value.

The difference can be understood by analogy: if we fill a box with plastic balls, the density of balls (number per unit volume) does not change with the number of balls we add. Balls keep piling up, but, as long as none of them gets crushed, the density is determined only by the size of each ball. A small box with 10 balls and a large box with 100 balls will have roughly the same density. The reason is that the balls are hard: when they get sufficiently close to one another, they resist being pushed any closer.

> The uniform density of nuclei means that the nuclear forces must have a *hard core* of repulsion that keeps nucleons a certain distance apart. When nucleons are further apart than the size of this hard core, the nuclear forces attract them together. This attraction holds nuclei together against the repulsive force of the positive electric charges on all the protons. But the nuclear attraction must change to repulsion when the nucleons get sufficiently close.

The nuclear density quoted above tells us that each nucleon of mass 1.67×10^{-27} kg occupies a mean volume of about 8×10^{-45} m^3, which is the volume of a cube of side 2×10^{-15} m. Now, the nucleons will be separated by the sum of both repulsive hard cores, so the radius of the hard core should be no more than half the side of this cube, 10^{-15} m. The core radius has been measured experimentally to be about 4×10^{-16} m. This is consistent with our estimate: one would expect nucleons to keep a little further apart than their minimum core radius, since in a nucleus they form a quantum Fermi gas (recall Chapter 12) in which the nucleons move around and have a quantum uncertainty in their positions.

Physicists do not clearly understand the forces between nucleons when they are pushed up against this core. Much research in modern nuclear physics is directed at understanding the attractions and repulsions between nucleons at short range, and some of the tools of that research are giant accelerators that can smash heavy nuclei together to form super-dense collections of hundreds of nucleons. But, if we are explaining the density of neutron stars, we can take the basic hard core as a starting point.

We saw in Chapter 12 that when white dwarf cores of giant stars collapse, the high densities force electrons and protons to combine into neutrons. Yet this in itself does not make a neutron star. The object can become a star in equilibrium only if it can support itself against gravity. Because the nuclear forces are attractive until they reach the hard core, this support can happen only if the density is close to the nuclear density.

Here is the argument in detail. Suppose we have a gas of neutrons whose density is much less than nuclear density. Then when neutrons collide with one another, as must happen at random all the time, they will have a tendency to stick together and form large nuclei; the extra pressure of neutrons from outside will make these condensations bigger than ordinary nuclei like uranium. So the gas will be a mixture of free neutrons and big nuclear lumps. Now consider what happens when the gas is compressed. Collisions become more likely, and the result will be that many neutrons will get stuck in lumps and not contribute to the gas pressure: remember from Chapter 7 that the pressure of a gas depends on its temperature and the number of particles in the gas, not on the masses of the particles. If free neutrons are lost to the lumps, then the pressure will not build up fast when the gas is compressed, and it will not be able to hold itself up against gravity. The collapsing core of a giant star in a supernova explosion therefore continues to collapse well after neutrons have been formed.

▷ Readers who have read Investigation 8.8 on page 101 will have already gone through this argument about pressure in detail.

> When the density of the collapsing core reaches nuclear density, the lumps all merge into a smooth "pudding", and further compression sees a rapid increase in pressure from the hard-core repulsion. Collapse stops, and a neutron star with the density of an ordinary nucleus is formed.

Now we see that the incredibly small size of neutron stars compared to the Sun is not so hard to explain. In normal matter, nuclei are separated by the huge distances occupied by the intervening electrons. In a neutron star, Nature has simply managed to remove all that wasted space, and put all the nuclei right up against one another.

It takes a whole star to do the work of 100 neutrons

This argument tells us what the density of a neutron star should be, but it does not tell us about the mass. It does not tell us whether this phenomenon should occur with stars or with basketballs. Why are there neutron stars, and not neutron basketballs? Or neutron galaxies?

The answer, of course, is gravity. To see why, let us imagine trying to make a neutron basketball. If one takes a heavy nucleus, say of uranium, and tries to build it up into a neutron basketball by adding one nucleon at a time, something goes wrong: as soon as a nucleon is added, the nucleus spits it out again, or worse still the whole nucleus divides in half. This is *radioactivity*.

Heavy nuclei decay through radioactivity because they are unstable. This happens basically because of the second feature of the nuclear force that did not come into our previous discussion but which is obvious if we look at everyday life from the right point of view: even the attractive part of the nuclear force is of very *short range*. We can see that it must be short range, since essentially all the properties of ordinary materials can be explained by using just the electric and magnetic forces that electrons and nuclei exert on one another through their electric charges. The nuclear forces are intrinsically strong, since they can hold all the protons in a nucleus together, despite their mutual electric repulsion. But they do not extend very far from the nucleus, since they do not influence chemistry. Unlike gravity and the electrostatic force, which fall off as $1/r^2$ as one goes away from the source, the nuclear force must fall off much more rapidly as one leaves the nucleus.

This means that, as one adds nucleons to a nucleus, there will come a point where nucleons on one side of a nucleus no longer feel the attraction exerted by those on the other side. Protons still feel the electrostatic repulsion of other protons, however, so if one adds protons to a sufficiently large nucleus, they will simply be pushed out

In this section: the neutrons in a neutron star are held together by gravity. We show here that gravity is only strong enough to replace the binding forces that hold nuclei together when there are as many nucleons as in a typical star. This coincidence is one of the deep mysteries of nature, because without neutron stars there would be no life on Earth.

again: the new proton feels a nuclear attraction from only a few nucleons, but a repulsion from all the existing protons.

If one adds neutrons, one avoids this repulsion, but one still runs into a problem: the Pauli exclusion principle (Chapter 12). As one adds more and more nucleons of either kind, the new ones cannot have the same low kinetic energies of the existing ones, since the existing ones have filled up all the low-energy quantum states. So new nucleons must have higher energies, and at some point these energies will be enough to escape from the attraction exerted by the nearby nucleons. At this point, the nucleus will accept no new nucleons. This happens at roughly an atomic number of about 210: nuclei with more than 210 neutrons and protons in total tend to be unstable. This is about the location of lead in the periodic table.

▷ Lead is an abundant mineral on Earth because it has been produced by the radioactive decay of heavier nuclei over the ages.

So we are frustrated in our attempts to build a nucleus with the mass of a basketball by the short range of the nuclear forces. To hold a bigger nucleus together requires a long-range force, and the only candidate is gravity. Electric forces won't do, since like charges repel, and an equal mixture of positive and negative charges will exert no net long-range force. So only gravity can stabilize nuclei bigger than lead.

Yet gravity is a weak force, and the attraction it exerts within an ordinary nucleus is tiny compared to the other forces. Gravity can only provide the glue to hold together a large nucleus if the self-gravitational force of the nucleus is comparable to the nuclear forces between nuclei. This is going to require a large amount of mass.

We can in fact compute just how much mass is required by a relatively simple argument. It is observed experimentally that the "escape energy" of a nucleon from a nucleus is about 8 MeV, the same for most nuclei. This is the energy that has to be supplied to a nucleon to get it away from the nucleus, and nuclei become unstable when the exclusion principle forces new nucleons to have this energy inside the nucleus. A nucleon that has a kinetic energy of 8 MeV inside a nucleus has just enough speed to escape. We show in Investigation 20.1 that this escape speed is about 13% of the speed of light.

▷ Recall that the symbol "MeV" represents a million electron volts, which is 10^6 eV $= 1.6 \times 10^{-13}$ J.

Now, gravity can prevent this escape if it raises the escape speed: if the escape speed from a large clump of neutrons exceeds this value, then the clump will be one big stable self-gravitating nucleus: it will be a neutron star. In Investigation 20.1 we show that a star with the density of a nucleus has an escape speed exceeding the nuclear escape speed if the mass of the star exceeds roughly $0.02 M_\odot$.

> An object with more than 2% of the mass of the Sun and the density of a nucleus has strong enough gravity to keep the nucleons bound together. This is our estimate of the minimum mass of a neutron star.

▷ Thoughtful readers will realize that our argument here is certainly an oversimplification, since nuclei are not electrically neutral, and the electric repulsion of the protons must affect their structure and in particular the escape energy. Neutron stars are neutral, so their escape energy will depend only on the nuclear forces. However, since the neutrons in a nucleus are not affected by the nuclear force, and still they have escape energies comparable to those of the protons, the argument here should be accurate to within an order of magnitude.

So there are no neutron basketballs.

Despite the simplicity of our argument, our estimated minimum mass is very close to the value of $0.1 M_\odot$ that full calculations give in general relativity, using more sophisticated nuclear physics. Considering that we have bridged a gap of a factor of 10^{53} from a nucleus of mass, say, 10^{-25} kg to the mass of a star, to have come within a factor of five of the right result is close indeed!

▷ Notice that our way of calculating the mass of the neutron star from the binding energy of a nucleus is essentially the same argument as we used in Investigation 8.3 on page 91 to calculate the minimum mass of an object in the Solar System that is round, from the binding energy of silicon dioxide.

What about the *maximum* mass of a neutron star? As for white dwarfs, the maximum mass for neutron stars is set by the balance between the inward pull of gravity and the amount of pressure the nuclear matter can sustain. In Investigation 12.6 on page 148 we calculated the maximum mass of neutron stars in the same way as we calculated the Chandrasekhar mass for white dwarfs, and we obtained the result that the maximum mass should be about five or six solar masses.

Investigation 20.1. Minimum mass of a neutron star: no basketballs

Experiments show that the energy required to remove a nucleon from an ordinary stable nucleus is about 8 MeV. From this energy it is possible to deduce an "escape speed" for a nucleon from the formula

$$K = \tfrac{1}{2}mv^2,$$

where K is the escape energy. Some arithmetic gives, using the mass of the proton for m (see Appendix A),

$$v_{escape} = 4 \times 10^7 \, \text{m s}^{-1},$$

or 13% of the speed of light.

Now, we want the escape speed from the neutron star to exceed this. This speed is (in Newtonian gravity) $(2GM/R)^{1/2}$, where M is the mass of the star and R its radius. The star must have nuclear density ρ_{nucl}, which means that we can deduce its radius from its mass. Writing down the equation for the average density,

$$\rho = \frac{M}{\tfrac{4}{3}\pi R^3},$$

and solving for R, we find

$$R = \left(\frac{M}{\tfrac{4}{3}\pi\rho} \right)^{1/3}.$$

This gives a gravitational escape speed for a "nucleus" of mass M:

$$v_{escape} = 1.8 G^{1/2} M^{1/3} \rho_{nucl}^{1/6}. \tag{20.1}$$

Using the value of the nuclear escape speed for v_{escape} here gives a minimum value for M, which is

$$M_{min} = 4 \times 10^{28} \, \text{kg} = 0.02 M_\odot. \tag{20.2}$$

Exercise 20.1.1: *How big is the nuclear hard core?*

Use the mass 1.67×10^{-27} kg of a nucleon and the density 2×10^{17} kg m^{-3} to calculate the volume occupied by each nucleon in a nucleus. If the nuclei are contained in cubical boxes, how big is each box? What is the size of the hard core, the irreducible radius of a nucleon?

Exercise 20.1.2: *Calculating the minimum neutron star mass*

Solve Equation 20.1 for M and use the value of ρ_{nucl} in the previous exercise to verify the minimum mass in Equation 20.2.

Exercise 20.1.3: *What does a neutron star look like?*

Taking the mass of a neutron star to be $1 M_\odot$, what is its radius? What is the escape speed of a projectile leaving its surface? What is the speed with which a projectile falling from rest far away reaches the surface? What fraction of the rest-mass of such a projectile is its kinetic energy when it arrives at the surface? What is the orbital speed of a particle in a circular orbit just above the surface of the star? What is its orbital period? Do all calculations using Newtonian gravity, even though the speeds are relativistic.

Exercise 20.1.4: *Thermal effects in neutron stars*

If the binding energy of a nucleon is 8 MeV, what temperature would the star have to have in order to boil off a nucleon? Since the pressure support for the star comes from the hard-core repulsion and not from random thermal motions of the star, it is possible for stars to cool off after formation without changing their properties. Give an argument that a star is "cold" (thermal effects are unimportant for its structure) if its temperature is smaller than the one you have just calculated. Assume the star has a temperature of 10^6K. What is its black-body luminosity? (See Equation 10.3 on page 116.) What is the wavelength at which it is brightest? (See Equation 10.9 on page 117.)

However, this is too simple an estimate, since Newtonian gravity is just not accurate enough for such compact stars. One needs to use general relativity to calculate their structure. We will do this in Investigation 20.3 on page 280, but for now we only point out that the effect of using relativity is to lower the maximum mass to somewhere between two and possibly three solar masses. Its actual value is not known: uncertainties in nuclear physics prevent reliable calculations.

We have learned that neutron stars can only exist in a rather restricted range of masses, between perhaps 0.1 and two solar masses. In fact, their lower limit in practice will normally be much larger, since a collapsing star will stop at the white dwarf stage if its mass is less than the Chandrasekhar mass. Neutron stars should only form if their masses are somewhat larger than $1M_\odot$. A collapsing star above the maximum mass will continue to collapse, and will form a black hole.

It is also interesting to ask what happens if we have a neutron star that subsequently gains or loses mass. If it gains mass, perhaps from a companion in a binary system, then it can be tipped over the maximum and it will collapse to a black hole. If it loses mass, again to a companion in a neutron star binary (see below), then when it reaches the minimum mass it will no longer be bound together and will undergo a catastrophic nuclear disintegration: it will *explode*.

The most remarkable and fortunate coincidence about these masses is that the maximum mass is larger than the Chandrasekhar mass. This coincidence allows neutron stars to form in the first place. The maximum mass is a property of nuclear physics and general relativity. The Chandrasekhar mass depends on Newtonian gravity and atomic physics. We could imagine a Universe in which the nuclear repulsive core was smaller, so that neutron stars were denser and the effects of general relativity correspondingly greater, leading to a maximum mass smaller than the Chandrasekhar mass, which is unaffected by the nuclear hard core. In such a Universe, collapsing stars bigger than white dwarfs would form black holes directly. And in such a Universe, people would not exist.

The reason is that the nuclear hard-core repulsion plays a key role in the chain of events that leads to life on Earth. We have seen that the elements of which we are made were formed in stars, and that the heavier elements are spread into interstellar clouds by supernovae. Our Sun and Earth formed from clouds seeded with oxygen, silicon, and many other elements essential for life, by a long-ago supernova. But that supernova could not have happened if neutron stars could not form. If the collapsing core of the giant star that became the supernova could simply have continued to collapse to a black hole, then there would have been no "bounce", no shock wave to blow off the envelope of the giant star. Instead, all the gas in the giant star would have fallen into the black hole. The vital elements carried by the supernova gases would never have left the star and found their way into our Solar System.

> We owe our existence to the existence of neutron stars, and in particular to the neutron star that formed in that long-ago supernova. We must be thankful that Nature has arranged for the Chandrasekhar mass to be smaller than the maximum mass of neutron stars.

What would a neutron star look like?

Let us ask a few questions about the typical properties of a neutron star, just using the numbers we have obtained so far and assuming we can use Newtonian gravity.

Let us suppose the star has a mass of $1M_\odot$. This can't be far wrong, since a star with a mass less than the Chandrasekhar mass (see Chapter 12) will support itself at the density of a white dwarf and not collapse to a neutron star. So by taking a mass of $1M_\odot$ we are probably underestimating a little, but it will suffice to give us an idea of what the star will be like. In Exercise 20.1.3 on the preceding page you have the opportunity to do the calculations leading to the numbers below.

With a density of 2×10^{17} kg m^{-3}, the star's radius will be about 13 km, or 8 miles: it would just cover Manhattan Island. The escape speed for a particle at its surface is 1.4×10^8 m s^{-1}, or about half the speed of light. The orbital speed of a satellite at its surface is about 10^8 m s^{-1}, one-third of the speed of light. The period of such an orbit is 0.8 ms. This also sets the maximum spin rate of a neutron star: it could in principle rotate about 1000 times per second without flying apart.

The compactness of the star tells us also that the gravitational redshift of light from its surface will be significant. By the equivalence principle, an observer falling freely from far away will see no change in the frequency of light. Such an observer reaches the surface with the escape speed, $c/2$. The redshift seen by observers that remain at rest with respect to the star, then, is the same as that seen by an observer who is receding from a source of light at this speed. This will produce a lengthening of the wavelength of the light by at least a factor of two. If astronomers could see spectral lines in the radiation from a neutron star, they should be strongly redshifted.

These numbers can only be approximately correct of real neutron stars, since they tell us that typical speeds associated with the star are good fractions of the

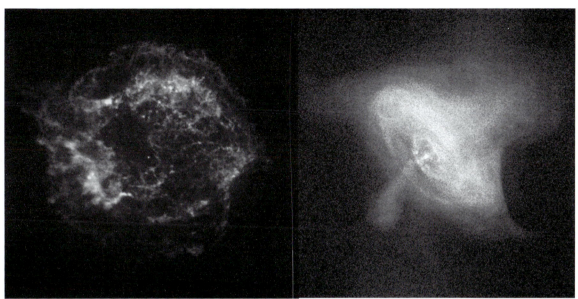

Figure 20.1. *The Chandra X-ray satellite has excellent sensitivity and imaging capability. It took these two photos of supernova remnants. The one on the left is called Cassiopeia A, which is a supernova that exploded about 300 years ago. The X-rays reveal the expanding cloud of gas and, for the first time, a bright spot at the center. Further studies will reveal whether this spot is thermal radiation from a neutron star or from a disk of gas falling into a black hole; either kind of object might have been created. The photo on the right is of the central part of the Crab Nebula; an optical photo of the entire nebula is shown in Figure 20.4 on page 270. This nebula was formed by a supernova explosion about 1000 years ago, and it contains a neutron star. The photo reveals a complicated cloud of gas around the neutron star, and very interestingly a jet of gas shooting out from the neutron star. This is a miniature form of the phenomenon seen in quasars. The Chandra satellite is named after the astrophysicist S Chandrasekhar, whose work was introduced in Chapter 12. Image courtesy of* NASA/CXC/SAO.

speed of light. This means that Newtonian theory is suspect, and we really have to use general relativity to get reliable numbers. We shall do that below. But first, let us consider what even these approximate numbers tell us about where we might expect neutron stars to be found and what they might look like.

One thing is clear; the ordinary thermal radiation from neutron stars should not be visible on ordinary photographic plates. Suppose a neutron star has a very high surface temperature, say as much as 10^6 K. Then the black-body luminosity (Equation 10.3 on page 116) is about one-third of the luminosity of the Sun. (See Exercise 20.1.4 on page 265.) Most of this energy comes out, however, near wavelengths of 3×10^{-9} m (Equation 10.9 on page 117), which is in the X-ray band of the spectrum. Only a tiny fraction emerges in the visible region, so we do not expect to see this black-body radiation in photographs, but we should hope to find it with X-ray telescopes. X-rays are the ideal means for studying neutron stars; see Figure 20.1. There are also spectral lines in the X-ray band, for example from ions of iron that have been stripped (by the high temperatures) of all but one electron. One might see these lines strongly redshifted.

> *Observable property 1:* neutron stars should emit thermal X-rays with redshifted spectral lines. The redshift is a *diagnostic*, separating neutron stars and black holes from other possible sources of X-rays.

Another observable property of neutron stars is their short time-scales: with their small size and large velocities, any dynamical process will happen very quickly. If astronomical phenomena are found that involve changes on millisecond time-scales, then one should consider whether a neutron star might be responsible. In fact, one should not be surprised to find a neutron star that is rotating rapidly, because any rotational speed of the original collapsing star should increase during collapse, just as an ice-skater spins faster when she pulls her arms in. Any spin rate, right up to the breakup speed, would in principle be possible.

> *Observable property 2:* neutron stars can exhibit variability on millisecond time-scales, either from pulsations or from rapid rotation. This is also a diagnostic feature, since only neutron stars or black holes can be compact enough to allow such rapid changes.

A third feature we might predict about neutron stars is that they could have strong magnetic fields. This is because, according to the laws of electromagnetism, when an electrical conductor changes size, the magnetic field it is carrying will change in proportion to the inverse square of the size of the conductor. This is the same proportionality as in the law of conservation of angular momentum. By the same calculation as we did above, a neutron star could have a magnetic field larger than that of its progenitor by 5×10^9. Ordinary stars have fields of a few gauss, so we might expect to find magnetic fields of billions of gauss on neutron stars.

Magnetic fields in astronomy are usually associated with radio emission: strong radio sources tend to have strong magnetic fields. The fields accelerate free electrons, and when they accelerate they give off electromagnetic radiation. So one might expect that any compact source of unusual radio radiation might be associated with a neutron star.

> *Observable property 3:* neutron stars might have strong magnetic fields, and these could make them strong radio sources. The existence of radio emission is not unique to neutron stars, but if the radio emission indicates a very strong magnetic field or exhibits very rapid time-variability, then this would also be diagnostic of a neutron star.

▷ Zwicky, introduced in Figure 14.8 on page 171, was led to the idea of neutron stars by his study of supernovae, which he was the first to identify as a special class of phenomena. It is remarkable that, only a couple of years after the discovery of the neutron, Zwicky was prepared to postulate whole stars made of neutrons! See Figure 20.2 and Figure 20.3 for brief introductions to Landau and Oppenheimer, respectively.

We have identified three observable properties of neutron stars that could help in finding them. This list could in principle have been made at any time since neutron stars were first predicted, independently in the 1930s by Zwicky (whom we met in Chapter 14) and the Russian physicist Lev Landau (1908-1968). The American physicist J Robert Oppenheimer (1904–1967), inspired by the work of Landau, and working with graduate students, showed convincingly within general relativity that the formation of neutron stars by gravitational collapse was possible and indeed in some circumstances inevitable. Thus many parts of our discussion were understood by the 1950s, and yet they were ignored by most astronomers. Neutron stars still proved elusive to observe, partly because no-one was looking, and partly because the technology available to astronomers was not what was needed.

Where should astronomers look for neutron stars?

In this section: here we follow up our predictions and suggest how one might design observations to find neutron stars.

If enough astronomers in the 1950s had taken the idea of neutron stars seriously, where might they have looked? Where might they look today? Since we expect neutron stars to be formed in supernova explosions of Type II (which are triggered by core collapse), the first place to look for them is in the clouds of gas that mark the supernova remnants, as in Figure 20.1 on the previous page and Figure 20.4 on page 270. These often – but not always – contain observable neutron stars.

Supernova remnants fade away after a few tens of thousands of years, while neutron stars have been produced in our Galaxy for billions of years. Therefore, most neutron stars should be scattered randomly around the Galaxy. An isolated, old neutron star would be very difficult to observe, being too cool even to give off much X-radiation. So the other place to look for neutron stars would be in binary systems, where any interactions between a neutron star and its companion might reveal the neutron star.

The problem with binaries is that the supernova explosion that produces the neutron star is likely to disrupt the binary system. This is not caused by the exploding gases themselves: they flow around the companion star and give it hardly any push. But the gases carry away mass, and this reduces the gravitational attraction between the companion and the neutron star that is left behind.

If the companion is a star of small mass, then it is easy to see what happens. The escape speed of the companion from its orbital position is only $\sqrt{2}$ times larger than the circular orbital speed, so if the supernova reduces the central mass enough to reduce the escape speed by the same factor, then the companion will find itself moving with enough speed to escape from the neutron star, and the binary will fly apart. Now, the orbital and escape speeds depend on the square-root of the central mass, so we conclude the following.

> If the supernova expels more than half of the mass of the star, then a low-mass binary companion will be expelled and the binary will not survive.

Most supernovae would be expected to expel far more than half of the original mass. If the system does survive, the excess speed of the companion will turn the initially circular orbits into elongated ellipses.

On the other hand, if the companion has much more mass than the original supernova star, then the binary will survive. To see why, turn the argument around and regard the supernova star as the small mass orbiting a large mass. Then if the small mass splits into any number of pieces, each will orbit the companion in the same orbit as the original. If then most of these pieces are sent away by the explosion, any piece left behind will not be affected: its orbit is determined by the companion, not by the other pieces.

> So the fate of a binary after a supernova depends very much on the ratio of the masses of the two stars: if the supernova occurs in the less massive of the two, it may remain as a binary, but otherwise probably not.

And even in those binary systems that survive, one would expect the initially circular orbits to have become elliptical. Since the statistics of binaries are poorly known, and the evolution of the stars in them is even less well-understood (particularly in terms of the all-important question of how much mass is transferred from one star to another, or is lost from the system entirely during the evolution of the stars), it is difficult to make reliable predictions at present of how many neutron stars should be in binaries.

One conclusion is safe to draw, however. Since most stars start out in binaries, most neutron stars will have been formed in binaries. Those which then leave the binary do so with a large speed, the escape speed from the orbit. This could be anything up to the escape speed from the surface of a star like the Sun, if the binary orbit had been small enough. This is some 600 km s^{-1}. Observations of pulsars (which are described in the next section) show that they have even higher velocities than this argument would suggest. Some are believed to be traveling faster than 1600 km s^{-1}, and their mean speed is around 400 km s^{-1}. These speeds are actually too high to be explained by orbital breakup, since most binary orbits are fairly widely separated and hence have fairly low velocities. The conclusion that astronomers have drawn is that neutron stars are given a "kick" when they are born,

Figure 20.2. *Lev Landau was the giant of Soviet physics from the 1930s onwards. His insistence on the highest standard of mathematical ability and his physical insight were legendary. His books and pedagogical legacy are still influential around the world today. Photo courtesy* AIP *Emilio Segrè Archive.*

Figure 20.3. *J Robert Oppenheimer's research in theoretical physics and relativity was interrupted by the Second World War, during which he led the American team that developed the fission bomb. After the war he worked to avoid an arms race between the USA and the Soviet Union, and he opposed the development of the fusion bomb. For these stands he was excluded from further advising the government of the USA, and he was treated like a traitor by some politicians and parts of the popular press. The photograph was taken around 1944, when he was leading the bomb project. Image courtesy U. S. National Archives,* ARC *picture 29-1233a.*

Figure 20.4. This composite shows that the Crab Nebula (Figure 12.8 on page 150) contains a pulsar. The series of images on the right show the central two stars, one of which is the pulsar and the other of which is a background star that happens to be in the image. The images were produced digitally by recording only the light arriving at the telescope during a particular millisecond of the pulsar's 33 ms period, and adding up many such images over a long exposure time, all from the same phase of the pulsar's period. Thus, the first image is the brightness during the first millisecond of every period, the second image (below the first) is the brightness during the second millisecond, and so on. The background star remains constant while the pulsar turns on and off twice, which occurs as the two pulsar beams sweep past the Earth. (See the drawing underneath the text on the first page of this chapter.) Most pulsars are not visible in optical photographs, but their radio emission and sometimes their X-ray emission behaves in a similar way. The images were made from a two-hour exposure taken at NOAO Kitt Peak in Arizona in October 1989. Image courtesy N A Sharp/AURA/NOAO/NSF.

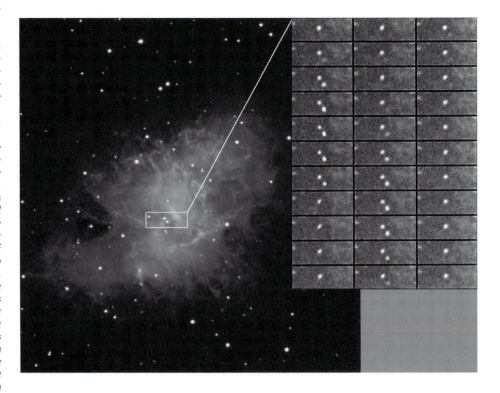

averaging around 400 km s^{-1}. This presumably has to do with the turbulent hydrodynamics in the gravitational collapse that forms the star, but there is no widely accepted model of such kicks yet. As we remarked above, kicks could also change the spin of the star.

The space velocity of these pulsars is large. The speed of the Sun in its orbit around the center of the Galaxy is only about 200 km s^{-1}, so some neutron stars will have enough speed to escape from the Galaxy entirely. Others should form a population of neutron stars of high speed, distributed broadly around the Galaxy, not confined to the disk of the Galaxy (the Milky Way) where they were produced.

Although we have not reached firm conclusions about where to find neutron stars, we can be comforted by one further conclusion: there must be a lot of them out there. This conclusion comes from some elementary reasoning. Supernova explosions of Type II are rare events, but from the statistics of supernovae in our and external galaxies, it seems that they have been occurring on average about once every 30–100 years in our Galaxy. Over the lifetime of the Galaxy, 10^{10} years, there have been more than 10^8 supernovas. A good fraction of them must have produced neutron stars, so the total number produced must be around 10^8. Given that the Galaxy has 10^{11} stars, it follows that one in a thousand stars may be a neutron star. There could be even more, if we have underestimated the supernova rate (hidden supernovae) or if neutron stars can be formed in other ways that are not so spectacular.

Pulsars: neutron stars that advertise themselves

A major reason why neutron stars were not discovered earlier is that astronomical technology was not suited to discovering them. For example, the first astronomical X-ray source to be identified (other than the Sun) was the object Sco X-1, seen in a rocket-borne X-ray experiment in 1962. It is now known to be a neutron star, but the first observations did not have the sensitivity to pick out any of the diagnostic features of a neutron star, such as rapid variability or redshifted spectral lines. Optical astronomers had little chance of identifying a neutron star, because even if it emitted enough light, the long exposures required for photographic plates would have prevented them from seeing any rapid variability. Indeed, the neutron star in the Crab Nebula was known for a long time to be a candidate for the object left behind by the explosion, since its spectrum did not resemble that of an ordinary star. But there was not enough information in the optical spectrum for a confident identification. Radio astronomers had, by our earlier discussion, the best chance, but for many years their telescopes were not designed to pick up rapid variability.

The absence of a positive identification of a neutron star, and the extreme properties expected of them, led most working astronomers to regard them as a theoretician's fantasy, if they thought about them at all! It was therefore a complete surprise when they turned up as *pulsars*.

The story of the discovery of pulsars is well worth remembering, for it shows that, despite scientists' attempts to pursue scientific research in a planned and orderly fashion, some of the most important discoveries arrive in ways that are virtually impossible to predict. Astronomers at Cambridge University in England, led by Anthony Hewish (b. 1924), had constructed a special telescope to look for rapid variations in the radio waves arriving at the Earth from distant radio sources. This was not because they were looking for neutron stars. Rather, the radio sources themselves were expected to be fairly constant, and the expected variations were produced by irregularities in the interstellar plasma and the solar wind that the radio waves pass through before reaching the Earth. The instrument was designed to give information about these plasmas by detecting fluctuations on time-scales shorter than a second. With hindsight, we can see that Hewish had built the first radio instrument capable of identifying neutron stars, but no-one realized this until later.

Most of the observations fit the expected pattern of irregular fluctuations, but a graduate student, S Jocelyn Bell (b. 1943), noticed fluctuations that seemed to be periodic, with a period of about 1 s. After checking that nothing was wrong with the telescope, Bell convinced Hewish and her other colleagues that the radiation was coming from an astronomical source, and in 1967 the result was announced. Soon many other such sources were found, with different periods. The flashing sources were named *pulsars*.

Although our discussion in the previous section makes neutron stars an obvious candidate for pulsars, astronomers at first had to consider many alternatives, such as white dwarfs oscillating in their fundamental mode of radial vibration (as we discussed for the Sun in Chapter 8).

> But one property of pulsars proved decisive: the pulsations kept time with remarkable stability over many years. The only motion in astronomy that can keep such good time is rotation: pulsars had to be associated with rotating stars.

The period of rotation then points to neutron stars. No star can rotate faster than the orbital period of a satellite at its surface, and this depends just on the average

In this section: when neutron stars were actually first observed, it was through their strong magnetic fields and rapid spin. With magnetic fields 10^{12} times stronger than the Earth's, spinning many times per second, and immensely strong gravitational fields, pulsars are laboratories of extreme physics.

▷ The pulsar phenomenon was so unexpected and puzzling that some astronomers seriously wondered at first if the signals were messages from intelligent beings far away across the Galaxy. But the regularity of the pulses, the power required to produce them, and the discovery of several other pulsars very quickly led to the conclusion that the phenomenon was a natural one. Nevertheless, during the first couple of years astronomers frequently – and only half-jokingly – used the acronym LGM for these objects, an abbreviation for Little Green Men!

density of the star. A period of 1 s requires a density of 6×10^{11} kg m^{-3}. This would just barely allow a model based on very rapidly rotating white dwarfs, but when pulsars were discovered with periods as small as 30 ms, the density limit went up to 6×10^{14} kg m^{-3}. Only neutron stars can reach these densities. Astronomers accepted that they had finally discovered rotating neutron stars.

Rotation provides the "clock" that keeps the pulsar ticking regularly, but what provides the pulses of radio waves? The simplest answer is that a pulsar is like a lighthouse: there is a beam shining in a single direction, which is swept past our telescopes by the rotation of the pulsar. Every time it passes us, we see a pulse. The beam emits continuously, but we see it only intermittently. Such a beam can in principle be created by a strong magnetic field, which we saw was a feature we could expect of pulsars. If the magnetic poles of the pulsar's field are not near the rotational poles, as they are on the Earth, but instead lie in the pulsar's rotational equator, then what we could be seeing is a view down onto the pulsar's magnetic pole every rotation.

In this picture, we would expect a second pulse as the other magnetic pole passes our view, and indeed we do see this in many cases. A good example is the Crab pulsar, shown in Figure 20.4 on page 270. However, if the pole lies somewhere between the rotational axis and the rotational equator, then we would see only one pole and not the other. Moreover, for every pulsar we see, there should be many more whose beams never pass over us, and which we therefore do not see.

Although neutron stars were expected by physicists, the idea that they would send out flashing beams to tell us where they are had never been dreamed of! For the discovery of pulsars, Hewish shared the 1974 Nobel Prize for Physics. Many scientists felt that Bell should also have had a share. We may never know why the Nobel committee neglected the key contribution of this graduate student to the project. The Nobel committee seems to have been more sensitive to the contribution of a graduate student when, in 1993, they awarded the prize for the second time to pulsar astronomers for the discovery of the binary pulsar system PSR1913+16. This story will be told in Chapter 22.

The mystery of the way pulsars emit radiation

Astronomers are in essentially universal agreement about the association between neutron stars and pulsars, and they agree that the beam is related to a magnetic field that is not aligned with the rotation axis. But astronomers agree about little else. In particular, the way in which the magnetic field produces the radiation in its beams is not at all understood.

It may be related to the terrestrial phenomenon of the aurora, in which charged particles from the Sun move along the Earth's magnetic field toward the magnetic poles, sometimes creating beautiful displays of light as they reach the Earth. In the case of pulsars, the strong magnetic fields (apparently of order 10^{12} G) are able to pull charges off the surface of the star and send them along the field to the poles, so the emission phenomenon is self-feeding.

Pulsars emit more than just radio waves. The beam can contain visible light and X-rays too. The Crab pulsar emits both pulsed X-rays (Figure 20.1 on page 267) and pulsed light (Figure 20.4 on page 270). These images dramatically illustrate just how unusual pulsars are.

The fact that astronomers see optical light from the Crab may seem to contradict our earlier estimate that the thermal radiation from a neutron star would be too weak to be visible. There is no contradiction, because this light is not produced by black-body radiation. It is produced by energetic particles moving at speeds near the speed of light near the magnetic poles of the star. In fact, the absence of an image of

▷The name "pulsar" is of course a misnomer: the stars are simply rotating, not pulsating. The name was created at a very early stage, when the first natural assumption was that astronomers were seeing pulsations of some kind of star. By the time it became accepted that they were spinning, it was too late to change the name!

In this section: the details of how a spinning magnetic neutron star emits radiation in its beams are still unknown. Somehow the magnetic field creates beams of radiation streaming out from the magnetic poles.

the Crab pulsar during times when the pulse is not arriving is proof that the thermal radiation from this star is very weak.

The rotation rate of pulsars and how it changes

The visibility of the Crab pulsar in so many wavelengths does tell us that there is an enormous amount of energy available for producing this "non-thermal" radiation. Where does this energy come from?

In the end, regardless of the detailed mechanism by which the radiation is produced, the main source of the radiated energy is probably the rotation of the star itself. So as the various forms of radiation carry energy away, the star must slow down. This is in fact seen: most pulsars that have been observed accurately have been observed to change their pulse period slightly over a number of years. The slowing down is very slight, sometimes gaining less than 10^{-15} of a period in each period. Such small changes are measurable only because the pulses themselves are so regular.

In this section: pulsars lose much more energy than they put into their beamed radiation, and this energy lights up the gas around them. By losing energy, they slow down, and the rate of slowing allows an estimate of their ages. Most are only a few million years old.

> Small as the slowing down is, the energy lost by the pulsar is enormous. In Investigation 20.2 on page 275 we show that the Crab loses about 1.3×10^{30} J of rotational kinetic energy in each period of rotation. This amounts to an energy loss rate about 10^4 times the luminosity of the Sun!

It is worth reminding ourselves where a pulsar's energy of rotation came from in the first place. The pulsar formed in collapse and, as we pointed out earlier, collapse leads in a natural way to a rapid spin. The energy that goes into the spin comes from the gravitational potential energy released when the star's core collapsed, so it is gravity that originally supplied the store of energy for this powerhouse.

> We should also note that the Crab pulsar is a *slow* rotator: even though it spins 30 times a second, its gravity is strong enough to hold it together even if it spun 1000 times a second. In particular, its shape should be nearly spherical: it won't bulge out much at the rotational equator.

More than 1000 pulsars are now known, and they have a wide range of periods, down to 1.6 ms, or 600 rotations per second! For longer periods, longer than about 20 ms, the emission seems to be weaker in pulsars with longer periods. It seems that whatever mechanism produces the pulses gets turned off once the star spins down to a certain rate. This accounts for why there are so few observed pulsars when we expect there to be so many neutron stars. Most neutron stars in the Galaxy may well be old, dead pulsars, still rotating at a modest rate but no longer pulsing.

Where does the energy lost from the spin of the pulsar go? Most of it is not going into the pulsed radiation that is seen by radio and optical telescopes: the Crab pulsar emits about 10^{24} J per rotation in each of these, only one millionth of its total rotational energy loss. Most of the energy probably goes into the acceleration of high-energy particles, some of which contribute to the pulses, and into low-frequency electromagnetic waves generated by its rotating magnetic field. Any rotating magnet will emit electromagnetic radiation with a frequency equal to the frequency of rotation of the magnet. For the Crab pulsar, this is 30 Hz.

Waves at such a low frequency get strongly absorbed by the thin plasma that surrounds the star, the remnant of the gas ejected from the supernova star that formed the Crab. So astronomers cannot hope to detect it on Earth. This is a pity, because if this radiation were detected, it would directly measure the magnetic field of the Crab and determine if this radiation really accounts for the energy loss. It is

possible that pulsars lose rotational energy in other ways, for example in gravitational radiation, to which we will return in Chapter 22.

If we do assume that the radiation of low-frequency radiation from the spinning
magnetic field accounts for the slowdown, then it is possible to calculate the magnetic fields of pulsars. Most pulsars, like the Crab, have field strengths of the order
of 10^{12} G. (Remember that the Earth's field is of order 1 G!) These fields are strong
enough to make the emission mechanism discussed above plausible. But as yet,
there is no independent evidence for the strength of the magnetic fields; scientists
can only estimate them from the spindown rate.

Pulsar magnetic fields fall into three groups. Normal pulsars have fields of order
10^{12} G. Faster-spinning **millisecond pulsars**, with periods below 10 ms, seem to
have much weaker fields, of order 10^9 G. And there is a small group of very slowly
rotating **magnetars** with fields of order 10^{15} G. These large differences in field presumably indicate different histories or modes of formation of stars in these classes.
We will have more to say about millisecond pulsars below. Magnetars are in fact not
normally seen in radio telescopes; they are found from the strongly pulsed X-ray
emission.

The slowing down of the pulsar also allows us to estimate its age. If the slowing
down were constant, so that the rate of change of the pulse period is constant for
the whole lifetime of the pulsar, then we would have a simple equation

period = original period + rate of change of period × time.

An estimate of the age of the pulsar is obtained by taking the original period to be
zero (a neutron star spinning infinitely fast!) and solving for the time it takes to
produce the present period with the observed rate of change of the period. If we
let the rate of change of the period be represented by the symbol $\dot{P}$, which is the
conventional symbol used by astronomers, then the age turns out to be $T = P/\dot{P}$.
This is of course just an estimate. It could be an overestimate, since the original
spin of the star was not infinitely fast. It could be an underestimate if the rate of
change of the period was not constant in time, but was lower at earlier times. In
fact, physicists expect that it was higher at earlier times, since the energy lost to
radiation from the spinning magnetic field increases as the fourth power of the spin
rate, so the spindown was much stronger when the pulsar was young. This leads to
the conventional definition of the so-called "spindown age" of the pulsar,

$$T_{\text{spin}} = \frac{P}{2\dot{P}} ;$$

this is still, of course, an estimate.

For the Crab pulsar, the observed $\dot{P}$ is 4.3×10^{-13}. (Notice that the rate of change
of the period, $\dot{P}$, is a dimensionless number, because it is formed by dividing a number with dimensions of time – the change in the period – by another number with
dimensions of time – the time in which the change took place.) From this, the
spindown age is 1200 years. Astronomers in China happen to have recorded the supernova that produced the Crab in the year 1054, some 950 years ago. This is good
agreement, considering the simplicity of the assumptions.

When astronomers use this method to compute the ages of other pulsars, they
find that the Crab is the youngest known, and the oldest are some 10^9 years old. In
fact, pulsar ages seem to correlate with their magnetic field strengths. The oldest
pulsars, with ages up to 10^9 y, have millisecond periods and weak magnetic fields.
Normal pulsars have ages up to about 10^7 y, although there are so many of them
that it is not surprising to find a few with ages of a few thousand years. The magnetars are much younger, on average, than normal pulsars.

Investigation 20.2. Pulsars lose enormous amounts of energy

In terms of the total energy that a pulsar is getting rid of, pulsars are much more luminous than ordinary stars. The energy they lose as they slow down is enormous.

We can estimate that energy as follows. The energy of rotation is basically kinetic energy. But different parts of the star rotate with different speeds: the surface is going fastest and the center doesn't move at all. We shall approximate the rotational kinetic energy by assuming an em average speed: we won't be far wrong if we calculate the kinetic energy of a body with the mass of the neutron star moving with a speed that is half of the surface speed of the pulsar.

Consider the Crab pulsar. We can only do a rough calculation to see how large some of the numbers can be. Suppose the pulsar has a mass of $1 M_\odot$ and a radius of 13 km, as our earlier calculations suggest is appropriate for neutron stars. Then the surface, rotating at 30 times per second, has a speed $v_{surf} = 2.5 \times 10^6$ m s^{-1}, less than 1% of the speed of light. We take the average speed of the material in the star to be $v_{avg} = v_{surf}/2$, and so the kinetic energy of rotation K is

$$K = \tfrac{1}{2} M_\odot v_{avg}^2 = \tfrac{1}{8} M_\odot v_{surf}^2 = 1.5 \times 10^{42} \text{ J}. \qquad (20.3)$$

Observations show that, in one period of rotation, the Crab pulsar's rotational speed decreases by a fraction 4.3×10^{-13}:

$$\frac{\Delta v_{surf}}{v_{surf}} = -4.3 \times 10^{-13}.$$

The energy K changes accordingly by the (negative) amount ΔK given by

$$K + \Delta K = \tfrac{1}{8} M_\odot \left(v_{surf} + \Delta v_{surf}\right)^2.$$

Squaring the speed term on the right-hand side and subtracting the original expression for K given in Equation 20.3 above, we obtain

$$\Delta K = \tfrac{1}{8} M_\odot \left[2 v_{surf} \, \Delta v_{surf} + (\Delta v_{surf})^2\right].$$

Now we divide by the original K to get

$$\frac{\Delta K}{K} = 2\frac{\Delta v_{surf}}{v_{surf}} + \left(\frac{\Delta v_{surf}}{v_{surf}}\right)^2.$$

Since $\Delta v_{surf}/v_{surf}$ is so small, the second term on the right-hand side is completely negligible, and we have the simple result that the fractional decrease in the kinetic energy of rotation is twice that of the rotational speed itself, or -8.6×10^{-13}.

Now, the total kinetic energy is 1.5×10^{42} J, so the loss of energy in one period of rotation is 1.3×10^{30} J. Since one period takes only 0.033 s, this amounts to a rate of energy loss of 4×10^{31} J s^{-1}. Compare this with the total luminosity of the Sun, about 4×10^{27} J s^{-1}. The Crab pulsar is losing energy at the same rate as 10 000 Sun-like stars put together!

Exercise 20.2.1: *Pulsar energy storehouse*

A pulsar stores its energy as rotation. Estimate how much energy was released when the neutron star was formed by calculating the approximate gravitational potential energy of the neutron star, $-GM^2/2R$. You should find that the rotational energy is a small fraction of what was available when the star formed. What happened to the rest of the energy?

Puzzles about the rotation of pulsars

Some pulsars, particularly the youngest ones, show sudden small jumps in their spin rates, which astronomers call **glitches**. These seem to arise from the structure of the neutron star, whose interior is mainly fluid but which is thought to have a jelly-like **crust** of material formed from the heavy nuclei that neutron matter condenses into when the densities are below that of nuclei. Perhaps this crust breaks once in a while as the slowing of the star reduces its equatorial bulge, leading to a "starquakes". Or, perhaps the crust slows down a little more than the interior and once in a while has to be brought back up to the spin rate of the interior by the forces that connect the crust to the interior. Glitches are regularly seen in the Crab and other young pulsars.

By contrast, the older millisecond pulsars have not been observed to glitch. In fact they are excellent time-keepers. Some of them may even be better than the best atomic clocks made on Earth. This extraordinary stability has made them useful tools for conducting extremely sensitive observations that have verified the existence of gravitational waves and placed strict upper bounds on how much gravitational radiation has been left over from the Big Bang. We shall return to this subject in Chapter 22.

It may seem surprising that millisecond pulsars are old (as inferred from their very large spindown age) and yet they spin faster than the known young pulsars. This would indeed pose a big problem if there were so many millisecond pulsars that it seemed that all pulsars must turn into millisecond pulsars when they get old. But this is not the case. There are very few millisecond pulsars, especially considering that they remain visible for 10^9 y or more, while normal pulsars seem to stop radiating after about 10^7 y. Most old pulsars are probably slow rotators. The millisecond pulsar minority are formed in a special way, involving pulsars in binary

In this section: old fast pulsars, young slow ones, glitching pulsars, vibrating pulsars: pulsars puzzle astronomers more and more.

systems. We will return to that subject a little later in this chapter.

Pulsars in binary systems

Some of the most interesting pulsars are those in binary systems. Not many are known, which is consistent with our earlier discussion of binary disruption. But those that are observed are important in a number of respects. Significantly, almost all of them are millisecond pulsars, which suggests that pulsars that remain in binaries after being formed go on to become millisecond pulsars. Since they probably form in the same way as isolated pulsars (indeed, most isolated pulsars probably formed in binaries), something in the binary system must spin them up later. This probably involves the transfer of gas from the companion star onto the pulsar. This gas swirls around the neutron star as it gets near it, so when the gas reaches the star it carries considerable angular momentum, and it spins the star up. During the mass-transfer phase, the swirling gas will be hot enough to emit X-rays, and many systems like this have been discovered by X-ray telescopes in orbit. We will discuss this process in more detail in the section on X-ray binaries below.

Among the most significant pieces of information astronomers get from binary pulsars are their masses. Recall that all masses in astronomy are measured by studying orbits: the Sun's mass from the orbits of planets, the Earth's mass from the orbits of satellites, and so on. When pulsars orbit other stars, their orbits can tell us about their masses. Astronomers learn not just about the masses of the companions, but also about the masses of the pulsars themselves, since the pulsars are not simple test particles but instead produce observable effects on the motion of their companions.

Almost all the radio pulsars known to be in binary systems have companions that are either white dwarfs or other neutron stars. It may be that neutron stars in orbit about main-sequence stars have difficulty becoming pulsars, possibly because of the effects of gas coming from the companion. After the companion has evolved to a white dwarf, or has become a neutron star without disrupting the system, the binary is "cleaner", and the neutron star becomes a pulsar.

> Pulsars give good information about the binary orbit because they are such stable clocks. The pulses are emitted by the pulsar at very regular intervals of time as measured by the pulsar itself, but their arrival times at the Earth change with the motion of the pulsar.

This is essentially the Doppler shift of the pulse rate, similar to any other Doppler shift. For binary systems, it is convenient to look at the Doppler effect in terms of the time interval between successive pulses, rather than in terms of the changes in the frequency of pulsation.

When the pulsar is in the part of its orbit nearest the Earth, the pulses take less time to reach the Earth than they do when the pulsar is on the other side of its orbit. The times between successive pulses therefore change in a periodic way as the pulsar orbits the companion. These changes are easily measured: the maximum delay between when a pulse might be expected (if there were no binary motion) and when it actually arrives is the light-travel time across the orbit, which can be several seconds. Measuring this effect tells us the size and period of the orbit, so it tells us something about the masses of the two stars.

The orbits of two-neutron star binaries are generally elliptical, again as expected from our binary disruption discussion. These are particularly fruitful, because, as we have seen in Chapter 18, elliptical orbits precess in general relativity: their orientation in space changes slowly. This can again be measured from the pulse train, since the exact pattern of delays depends on the orientation of the ellipse with respect to the line-of-sight to the binary system. This gives a further constraint on the masses of the two stars and their separation.

Moreover, as the pulsar orbits the companion in an elliptical orbit, it finds itself sometimes nearer the companion than at other times. Just like any other clock, its pulsation rate experiences a gravitational redshift, which makes the pulses slow down and speed up periodically. Finally, the radiation from the pulsar follows a curved path as it passes the companion because of the deflection of light by the companion (see Chapter 18; discussed further in Chapter 23), and this introduces an anomalous time-delay into the arrival times of pulses that adds to the one produced by the redshift. Again because pulsars are such good clocks, the combination of these two small but changing effects can be measured, and it provides further information about the mass of the companion star and the size of the orbit.

The information in the combination of these measurements is enough to determine all the characteristics of the binary: the stars' masses, their separation, and even the angle that the orbital plane makes with the line-of-sight to the system. There are now two pulsar systems where the eccentricity is large enough and observations have continued long enough to make these measurements, and the results have given astronomers a surprise:

> The masses of all four neutron stars in these two binaries are very close to $1.4M_\odot$.

There is mounting evidence from X-ray observations, which we will discuss below, that other neutron stars have masses about $1.4M_\odot$ as well.

It would seem that some mechanism operates to produce a fairly uniform mass, at least for neutron stars formed in binary systems. The mass of $1.4M_\odot$ is close to the Chandrasekhar mass for normal white dwarfs, as we saw in Investigation 12.5 on page 146, so it might be thought that the explanation is simple: only stellar cores above the Chandrasekhar mass will collapse, and they do so as soon as they reach that limit, so almost all neutron stars will have formed from cores of the same mass. The difficulty with this argument is that the mass is wrong: a white dwarf core that collapses with the Chandrasekhar mass of $1.4M_\odot$ will form a neutron star of about $1.2M_\odot$, because of the large amount of energy that is lost when it collapses. This energy loss is also a mass loss, by $E = mc^2$, and this can be as much as 10–20% of the original mass of the white dwarf. So a $1.4M_\odot$ neutron star must have come from the collapse of a core with a mass of about $1.6M_\odot$, or some extra mass must have been added to the star after the initial collapse. It is not yet clear what this means about the circumstances in which these stars form.

X-ray binary neutron stars

Not long after the discovery of pulsars, astronomers found other neutron stars in an equally unexpected place: X-ray binary systems. We saw above that one might expect to detect X-rays from neutron stars, because at least young ones can be very hot. But the first X-ray satellite observatories did not see the expected relatively weak black-body X-radiation from points located in the middle of supernova remnants. Instead, they found powerful X-ray sources scattered over the sky, and it soon became clear that at many of these locations there was also a visible giant star whose spectrum showed periodic Doppler shifts. These sources were evidently binary stars which somehow produced huge amounts of X-rays.

Assuming that the X-rays were black-body radiation, the sources had to have high temperatures. Moreover, luminosity of the sources required much larger surface areas than the area of a neutron star. The temperatures could not be supplied by the visible star, whose surface temperature could be measured from the observed

▷Once the masses of the stars in a binary system have been determined, it is possible to make one further prediction: how much energy should be radiated by the orbital motion in gravitational waves. We will see in Chapter 22 how to calculate the radiation emitted. But in general terms, it is clear that the effect of losing energy must be to gradually shrink the stars' orbits, bringing them closer together. As the orbit shrinks, the orbital period goes down as well. This should be noticeable in radio observations, and indeed in two systems this effect has now been seen. We will discuss the importance of this for understanding gravitational radiation in Einstein's theory in Chapter 22.

In this section: some binary systems with neutron stars are spectacular emitters of X-rays. These systems convert mass into energy far more efficiently than any nuclear power station or explosive made by man.

visible light. They had to be associated with its binary companion, which was not visible in optical light. It was significant that the phenomenon usually seemed to involve giant stars, stars that were at a point in their evolution where they had suddenly expanded greatly in size.

These facts soon led astronomers to the now-accepted **accretion** model of binary X-ray sources. The unseen companion that is the source of the X-rays is a neutron star or black hole, and it is close enough to the visible star that when the visible star expanded to become a giant, it began to dump mass onto the compact star. This happens in some cases because the giant star begins blowing off mass at a considerable rate (a giant version of the solar wind), and in other cases because the giant expands so far that the part of it nearest to the neutron star actually begins to be dominated by the gravity of the neutron star, so that a steady stream of gas falls towards the neutron star.

Figure 20.5. An artist's sketch of accretion onto a neutron star in a binary system. Because the neutron star is in orbit about its companion, the matter that falls onto it is rotating, and foms a flattened disk. Drawing courtesy NASA.

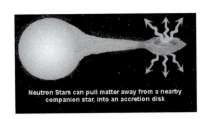

In either case, because of the orbital motion of the compact star, the gas does not fall onto it in a spherical manner. Instead, it first swirls into a disk in the plane of the orbit, and then the gas gradually spirals down onto the compact star. As it spirals towards the star, it releases its gravitational energy. The spiralling-in must happen due to friction within the gas, so the re-leased energy goes first into thermal motions in the gas, and then into black-body radiation. The radiation that we see is coming from the disk, not from the compact star: the disk has much larger surface area and therefore radiates much more energy. The disk is called an accretion disk, since it contains the gas that will accrete (accumulate) onto the central star.

The compact star must be at least as compact as a neutron star: if it were a white dwarf, the temperature of the disk would never get high enough to make X-rays. We show this when we study accretion disks in Chapter 21. Because the gravity of a neutron star is so strong, the amount of energy released as a particle moves through the disk is a large fraction of its rest-mass, up to 10–20%.

> Using strong gravity as a mediator, accretion disks convert a much larger fraction of their rest-mass into energy than do the nuclear reactions in-side stars!

With this much energy available, it is possible to power X-ray sources with very little mass: the amount of mass accreted by a neutron star can be as little as $10^{-9} M_\odot$ per year. Since the giant star may remain a giant for only a few million years, the total amount of mass dumped onto the neutron star need be only a small fraction of a solar mass. But this much mass is enough to change the spin of the neutron star substantially during this time: it is easily possible to spin stars up to 600 Hz as they accrete the rotating gas from the disk.

We shall return to accretion when we consider black holes. There we will put more numbers into the discussion above.

Gamma-ray bursts: deaths of neutron stars?

In this section: gamma-ray bursts, the most energetic events astronomers observe, may be associated with neutron stars that are on their way to becoming black holes.

In the 1960s, the United States Air Force launched the first of a series of Vela satel-lites designed to detect tests of nuclear weapons in space that had been banned by treaty. Later Vela satellites contained gamma-ray detectors and good clocks, so that the position of an explosion could be determined by differences in the arrival time of the gamma-rays at different satellites. These satellites never detected violations of the test-ban treaty, but in the early 1970s Air Force scientists realized that over the years a number of events had been detected that had come from astronomical sources.

Since astronomers could not find any other wavelengths of light associated with these **gamma-ray bursts**, they could not determine their source. It was not even known *where* they were coming from. But as data accumulated, especially from the 9-year mission of NASA's Compton satellite (launched in 1991), it became clear that the bursts were at least as far away as remote portions of the Galaxy, if not further, and that they must be coming from objects as compact as neutron stars. The principal reason for this was the time-scale of the bursts. Each burst was highly individual, but most lasted from about 1 s to several hundreds of seconds. Moreover, each burst was composed of sub-bursts where the intensity changed dramatically on times as short as a millisecond. As we noted above, this pins them down to neutron stars.

Finally in the mid-1990s astronomers, using other satellites, managed to find visible light emitted by the bursts. For a few hours to days, the region around the burst would brighten with what astronomers call an "afterglow". Bursts were always found to be located in distant galaxies. In fact, there seemed to be no limit on how far away they could be seen. They were bright enough to be seen as far away as any galaxy. During the burst the gamma-rays were thousands of times as luminous as the entire host galaxy. Since then, bursts have been seen that rival the luminosity of the entire Universe, for the brief seconds that they shine. Such an enormous energy can only be obtained by converting a good fraction of a solar mass into pure radiation. Given that it happens on short time-scales, this means that something catastrophic is happening to a neutron star or black hole. Bursts happen several times a day somewhere in the Universe.

Astrophysicists have come up with various proposals for what happens in a gamma-ray burst. As we have seen, neutron stars sometimes form binary systems, whose orbits shrink gradually because of the emission of gravitational radiation. What happens when an orbit shrinks so much that the stars are brought together? They cannot form a new neutron star, since the maximum mass is thought to be less than the $2.8 M_\odot$ that the two stars contain. (See the discussion of this below.) They might quietly collapse to a black hole, but this does not seem likely, given the speed with which they collide and the highly distorted shape that they have when they first encounter one another. It seems more lkely that they explode, at least partially. This explosion should produce visible X-radiation, and it might even produce a gamma-ray burst.

Another possibility is that the burst occurs at the end of the in-spiral of a binary system containing a neutron star and a black hole. While no such systems have yet been identified in radio pulsar observations, they should exist. In this case, the neutron star is disrupted by the tidal forces of the black hole and must form a thick ring close to the hole, which loses energy and spirals into the hole. In this process, it may well be possible to expel some of the energy in a jet of gas that gives rise to the burst.

Some astrophysicists have proposed that the event takes place inside a very massive star, as a kind of super-supernova, called a **hypernova**. It might still involve the formation of a neutron star at an intermediate stage, but this star does not live long before forming a black hole and releasing further energy.

Whichever of these models is correct, if any, the energy and timing of the burst strongly suggest that this is the catastrophic death of a neutron star. The energy comes, in the final analysis, from the gravitational energy released when matter forms objects as compact as neutron stars or black holes, as we calculated in Exer-

▷The Compton satellite was named after Arthur Compton, discoverer of the scattering of photons by electrons, whom we met in Chapter 8. It used Compton scattering of gamma-rays to detect them.

Investigation 20.3. Building neutron stars in general relativity

Despite the complexity of general relativity, it is possible to cast the equations for the structure of a star in a form that is very similar to the Newtonian equations. It then becomes simple to modify our program Star to produce such models with little effort.

Let us remind ourselves of the Newtonian equations. We imagine stepping outwards in radius with steps of size h. The mass of the star increases according to Equation 8.16 on page 94:

$$m(r + h) = m(r) + 4\pi r^2 \rho(r) \, h.$$

The pressure decreases by the equation of hydrostatic equilibrium, a combination of Equation 8.17 on page 94 and Equation 7.1 on page 73:

$$\Delta p = -G \frac{\rho \, m(r)}{r^2} \, h.$$

In relativity, it turns out that the mass equation above remains exactly the same. The relativistic corrections to the hydrostatic equilibrium equation all involve terms proportional to $1/c^2$, which would be very small in the non-relativistic limit, where c is large compared to any physical speed. The new equilibrium equation must be calculated using the full mathematical framework of general relativity, but when that is done the result is remarkably simple. There are only three changes, and each of them can plausibly be related to a property of relativity that we have already encountered. They are as follows.

1. Replace

$$\rho \quad \rightarrow \quad \rho + p/c^2.$$

   This is the inertial mass density, defined in Equation 15.10 on page 193, which governs how a fluid accelerates under an applied force.

2. Replace

$$m(r) \quad \rightarrow \quad m(r) + 4\pi r^3 p/c^2.$$

   This is analogous to the active gravitational mass. In Newton's theory, the density of the active gravitational mass is just ρ, and $m(r)$ is the total mass inside a radius r. In relativity, the active mass density is $\rho + 3p/c^2$ (for a fluid with isotropic pressure, as we have assumed here). If we multiply the extra term, $3p/c^2$, by the volume inside radius r, $4\pi r^3/3$, then we just get the extra term in the relativistic structure equation. Now, at one level this "derivation" is just a coincidence, because the actual volume of the star inside the radius r is not given by the flat-space formula, since space is curved. And the total pressure "inside" radius r would not be obtained by multiplying the value of p at r by this volume, because p is larger at smaller radii. But the coincidence is nevertheless interesting, and it does make it clear that this term does give the role played by the active gravitational mass in the equation governing the structure of the star.

3. Replace

$$r^2 \quad \rightarrow \quad r^2(1 - 2Gm(r)/c^2 \, r).$$

   This is related to the curvature of the three-dimensional space of the star. Recall from Chapter 19 that, at least for weak gravitational fields, the coefficient of $(\Delta x)^2$ in the spacetime-interval is $1 + 2\Psi$, where, to our accuracy, Ψ is the same as the Newtonian field $\Phi = -Gm(r)/c^2 \, r$. So the square of the change in the proper distance $(\Delta \ell)^2$ between two points is $(\Delta x)^2(1 - 2Gm(r)/c^2 \, r)$ near a point at radius r. This is close to the form of the new term in the equation, but not exact. Again, the analogy is enough to help us to see that this term in the equation is an effect of spatial curvature.

The result of these replacements is the relativistic equation of hydrostatic equilibrium, called the *Oppenheimer–Volkov equation*,

$$\Delta p = -G \frac{(\rho + p/c^2)(m(r) + 4\pi r^3 p/c^2)}{r \, (r - 2Gm(r)/c^2)} \, h. \qquad (20.4)$$

We have used this equation in the program Neutron.

A few cautionary words are in order about the interpretation of some of the symbols. We have said in the text of this chapter that physical variables such as p and ρ are measured by a locally freely-falling experimenter at rest in the fluid. But r and $m(r)$ are not defined this way, since they are not locally measurable. The radius r has a well-defined geometrical meaning, even in the curved spacetime of a star. Since the spacetime is spherically symmetric, there are surfaces around the center of the star that are perfectly spherical. The coordinate r assigned to any surface is the circumference of the sphere divided by 2π, just as in flat space. This may not seem worth commenting on, until one realizes that the definition makes no mention of the distance to the center. In a spherical but curved space, this distance will generally not be equal to r. We have already noted this in our discussion of the way the denominator of the equilibrium equation changes. We shall see an example of this in Chapter 21.

The mass $m(r)$ is best regarded as an auxiliary variable. It cannot be interpreted as the mass inside the sphere whose radius is r, as it would be in Newtonian gravity, since (1) mass is not easy to localize in relativity, especially because gravitational potential energy has to be counted in the total mass but does not reside in any particular place, and (2) the volume inside radius r will not in general be the same as in flat space, so the factor of $4\pi r^2 h$ in the mass equation is not the actual volume of the shell of thickness h. In practice, this does not cause a problem: one never needs to know how much mass is within a certain radius of the star.

Finally we need to discuss the equation of state. We shall only use the non-relativistic neutron-gas equation of state, despite the fact that the neutrons do become relativistic when the density goes much beyond the normal density of nuclear matter. The reason for neglecting relativistic corrections is that, at the point where they become important, other aspects of nuclear physics are so poorly understood that the relativistic corrections will by themselves not be very accurate. It is important to understand, however, that the results of our models at high densities are only indicative, and cannot be relied on quantitatively.

The non-relativistic equation of state is of the form given by Equation 12.14 on page 144:

$$p_{Fermi} = \beta \rho^{5/3}, \qquad (20.5)$$

where the argument in Investigation 12.4 on page 144 does not determine the constant β accurately. We merely quote the correct coefficient, which can be derived by careful calculations like those outlined in Investigation 12.4:

$$\beta = \left(\frac{9h^6}{320\pi^2 m_n^8} \right)^{1/3},$$

where in this equation h is Planck's constant. In SI units this evaluates to

$$\beta = 5.3802 \times 10^3 \text{ kg}^{-2/3} \text{ m}^4 \text{ s}^{-2}.$$

This value is used in the program Neutron on the website.

The program to evaluate the models is essentially the same as that used for the Sun, the program Star, also on the website. Only one difference needs to be commented on here: the irrelevance of temperature. The equation of state we use has a fixed coefficient β, given by the physics of a degenerate neutron gas. It is not affected by the temperature, since in a degenerate gas the pressure comes from quantum effects and not the random motions of the atoms. So not only is temperature not important to the structure of the star, it is also wrong to calculate it from the degenerate pressure using the ideal gas law. So we leave it out of the program entirely.

Other details of the program are explained on the website.

cise 20.2.1 on page 275. This energy is so large, and it is released so quickly, that it makes the nuclear reactions going on in ordinary stars seem insignificant. Fortunately for the Earth, our Sun relies only on the tame nuclear physics!

The relativistic structure of a neutron star

Now we return to the theory of neutron stars. Neutron stars of a solar mass are highly relativistic, as we have seen. It is not reasonable to expect that Newtonian stellar models, of the type we made for ordinary stars and white dwarfs, would be very accurate here. In fact, we shall see that they are more than inaccurate: Newtonian models completely miss some very important features of neutron stars.

The structure of a neutron star in general relativity is affected by a number of things. First, of course, there is the curvature of spacetime, and particularly that of space itself. Then there are the effects of special relativity in its equation of state. And finally, one has always to be careful about how one defines the quantities one deals with, since different observers may see them differently.

A good example of the problem of definition is given by the mass density ρ. First of all, it must now include all energies, since all energies have an equivalent mass (the energy divided by c^2) and they all contribute to the inertia of an element of the fluid. Second, this mass(-energy) density will depend on the observer. Different observers measure different energies for any body, and the fluid in the star will be no exception. On top of that there is the Lorentz–Fitzgerald contraction (see Chapter 15), which means that an observer moving past a little piece of the fluid will measure its volume to be smaller, and hence its density to be larger. So what do we mean by the *density* of the gas in a neutron star?

In order to avoid confusion over a multitude of possible definitions, physicists have settled on the following convention. When they use a word in relativity that is the same as one in Newtonian physics that refers to a property of a gas, such as density or temperature, they mean the analogous quantity *as measured by a freely-falling experimenter who is at least momentarily at rest in the part of the fluid which is being measured.* Being freely-falling, the observer can make a measurement as if there were no gravity, as if the whole system were governed just by special relativity. Since the experimenter is at rest with respect to the fluid, the effects of relativistic speeds do not come into it.

It must be stressed that this is a convention on the definition of symbols like ρ and T. There is of course real physics in the interactions between fluid streams that move with respect to one another, and between the gravitational fields they produce. An observer moving through a fluid might try to measure the density, and would not be likely to get ρ, since that is the density measured by an observer at rest. It is the job of general relativity to deal with these things, and it does it very well. But when we want to describe the state of a fluid, we will always use the measurements that an experimenter at rest inside it would make.

In this section: we use general relativity to make realistic computer models of neutron stars. The equations governing their structure differ from those for Newtonian stars in ways that correspond directly to the new features of relativistic gravity that we learned about in previous chapters.

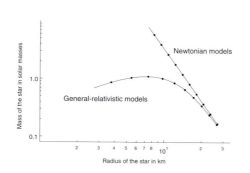

Figure 20.6. The mass–radius relation for neutron stars constructed in both general relativity (using Neutron*) and Newtonian gravity (using* Star*). In both cases the equation of state is the non-relativistic form of the neutron-gas equation of state that was derived in Chapter 12. The data show that general relativity predicts a maximum mass for neutron stars even when Newtonian stars using the same internal physics have no maximum mass.*

We cannot derive here all the differences between the equations that govern the structure of a star in relativity and those in Newtonian gravity. Despite the complexity of Einstein's equations, it is remarkable that the equation of hydrostatic equilibrium can be written in a form that is so similar to the one we have in Newtonian mechanics. We describe this equation in Investigation 20.3 on page 280, where we also go on to construct models of neutron stars using the computer program Neutron on the website, which is a modification (and in some ways a simplification) of the program Star.

The results of several runs of Neutron and Star are shown in Figure 20.6 on the previous page. This figure shows the main features of how relativity affects neutron stars, so we shall discuss it here at some length.

The relation of mass to radius for neutron stars

In this section: general relativity sets a maximum on the mass of neutron stars. This comes from the fact that pressure creates gravity. Fortunately for life on Earth, this maximum mass is *larger* than the Chandrasekhar mass.

Figure 20.6 on the preceding page compares the mass and radii of models constructed within relativity and Newtonian gravity, starting with the same central density and using the same equation of state. The differences between the curves are striking, and they are due entirely to the relativistic corrections to the Newtonian structure equations. The differences range from 2% for low-density stars up to more than 70% for the very dense models.

The most obvious difference between the two sequences is that the relativistic stars have a *maximum mass*. The value of this maximum is too low, a little lower than the measured mass of binary neutron stars. This simply means that we are not using the right equation of state, and we know that already: we have not put in relativistic corrections for the neutron gas, and on top of that we are not modeling the hard-core repulsive part of the nucleon–nucleon force at all. This hard core should increase the pressure at high densities and raise the maximum mass considerably. So the figure shown here is instructive more for what it tells us about general relativity than about real neutron stars.

And it tells us that general relativity will place a maximum on the mass of a sequence of stars, even when there is no such maximum in Newtonian sequences. In fact, the Newtonian sequence moves along a perfectly straight line in this logarithmic plot, which means that the relation between mass and radius for Newtonian stars is a power-law: $M \propto R^\alpha$. This is an illustration of a fact that we used but did not prove in Chapter 8, that in Newtonian gravity a polytropic model has a perfect scaling: if one knows its structure for a certain central density and temperature, then one can deduce its structure for other central values by simple scaling.

The relativistic equations do not obey this scaling: every model is different from every other one, even for a polytropic equation of state.

> The maximum of the mass has a fundamentally important consequence. Any collapsing cloud of gas with a mass larger than this maximum cannot stop at neutron star density; it must continue collapsing, presumably to a black hole. This makes the formation of black holes all but inevitable in astronomy.

The only way to avoid black holes is for nature to insure, magically, that collapsing clouds do not exceed this mass. But this does not seem plausible. The maximum mass is a property of the very dense neutron-matter equation of state, and should not be related to processes that lead to collapse in giant stars. Moreover, giants extend in mass up to many tens of solar masses. It seems unlikely that in all cases the star can insure that no more than, say, two solar masses of material actually collapses.

The existence of a maximum has another important but less obvious consequence: it tells us that all the neutron star models with a mass smaller than the maximum but with a central density that is higher than that of the maximum-mass star are *unstable* and will themselves collapse if disturbed.

The reasoning is similar to that used in Investigation 8.8 on page 101. Imagine starting with the star which has exactly the maximum mass, and then constructing another model with a slightly larger central density. Since the mass is a maximum, the new model has essentially the same mass as the previous one, but it has a larger density and therefore a smaller radius. Being an equilibrium model, it represents a fluid configuration that will remain static indefinitely. Physically, the structure of the star is the same as if we had taken a single star and compressed it: its mass would not change but its central density would go up and its radius would decrease. Since our calculations tell us that after these changes we still have an equilibrium model, we learn that the model with maximum mass is *neutrally stable*: if it is compressed it will neither bounce back out or collapse – it will just remain in equilibrium at the smaller radius.

Now, this neutral stability is a very special property: it indicates a point of transition between stability and instability along a sequence of equilibrium models. Consider a hypothetical sequence of stars which makes such a transition. On one side of the transition point, the stars are stable: they respond to a compression by bouncing back. On the other side they are unstable; they respond by collapsing further. The transition point must therefore be a point where the response is to remain exactly in equilibrium. Therefore, when we build a sequence of stars, we need only make a mass–radius plot and look for places of maximum (or minimum) mass: these are places where stability changes along the sequence.

So we only need ask which side of the point of maximum mass is the stable side and which is the unstable one. The models to the right have lower density, and the very low density ones are very similar to the Newtonian models. The polytropic exponent γ that we have used here is 5/3, which we saw in Chapter 8 means that the Newtonian models are all stable. It follows that the relativistic ones far to the right of the transition point are also stable. Moving from these to higher central densities, the stability cannot change until the mass reaches a maximum or minimum, so we conclude the following.

▷This argument only tells us about stability against *spherical* disturbances. We shall not consider non-spherical ones here.

> All the relativistic models in Figure 20.6 on page 281 to the right of the maximum neutron star mass are stable, and all the high-density models to the left are unstable.

If we had used a better equation of state at high densities we would have seen the same general feature, with the mass reaching a maximum, but the mass would be higher and the peak at maximum might be sharper. I have not bothered to introduce a more realistic equation of state because at present there is still considerable uncertainty about the details at high densities.

Neutron stars as physics labs

The uncertainty we presently have in the equation of state is one motivation for studying neutron stars: observations have the potential to tell us things about nuclear physics. Good measurements of the radius as well as the mass of a neutron star, or some indication of what the actual maximum mass is, would bring a rich reward in nuclear physics.

The physics of the interior of a neutron star has many more challenges than just computing the correct equation of state. Here are a few.

Neutron stars clearly have strong magnetic fields, which lead to pulsar emission, but physicists do not know what creates and maintains the magnetic field. It may be similar to the processes that maintain the Earth's field and that of the Sun (which are also not very well understood), or it might be totally different. Any explanation

In this section: neutron stars contain matter in extreme conditions. Their interiors are as yet veiled from observation, but they probably contain superfluid and superconducting material, even at temperatures of millions of degress. They may have jelly-like crusts and solid cores. Future observations of neutron stars may reveal even more exotic physics.

must also explain why the magnetic poles lie near the equator of a neutron star, whereas for the Earth and the Sun the poles are near the rotation poles.

The neutrons in the deep interior are thought to form a what physicists call a superfluid. This is a fluid that moves with no **viscosity**, no friction. Whether or not this happens depends on details of the nuclear physics that are not testable in laboratory experiments. Superfluidity is a peculiarly quantum phenomenon; it arises from the fact that the neutrons are all identical to one another. Under some conditions, they all move in exactly the same way, and it takes a lot of energy to make them scatter from each other or from other particles, because to do so would place the scattered particles into a quantum state where they can be distinguished from the others. The indistinguishability of the particles also means that they cannot rotate about one another, because again this would mean that the particles at the center of rotation would be distinguishable from the others. Superfluids can only rotate about special lines called **vortices**, within which the fluid is not a superfluid. But neutron stars do rotate, so they must be threaded with enormous numbers of these vortices. Physicists understand little about these vortices, or the way they interact with the magnetic field whose axis runs perpendicular to them.

The protons in the interior may also form an analgous fluid, a superconductor. This has no electrical resistance. This might or might not be related to the mechanisms that create the magnetic field.

At the very center of the star the density may be high enough to allow some even more exotic physical processes. These may involve elementary particles that are hard to study in the lab, such as **quarks**, which are the building blocks from which protons, neutrons, and other strongly interacting particles are made. The central core is thought to be liquid, but under some equations of state it may be solid.

Our understanding of nuclear physics is not good enough today to exclude the possibility that there is another stable state of matter even denser than nuclear matter. This would be called **quark matter** or **strange matter**, and it is possible that this is the most stable state of matter, leading to **strange stars**.

The outer parts of the star, where the density is low enough that not all matter has turned into neutrons, consist of layers of very unusual neutron-rich isotopes of familiar elements. The properties of the semi-solid outer crust – how it wobbles, breaks, re-forms, etc. – depend on these nuclei and how they fit together into regular patterns. The vortices of the interior rotation terminate on the inside of the crust, which is not superfluid, so it rotates normally. Slippage of these vortices along the crust is one model for the glitches in young pulsars.

Scientists today are only able to give relatively superficial descriptions of this physics; they do not have enough experimental data to help them to understand the details of these fascinating objects. It is not possible simply to see inside the stars. Data from supernovae that form neutron stars will help. More helpful will be the observation of gravitational waves from neutron star vibrations (Chapter 22), because then neutron star seismology could begin to reveal the inner structure in the same way that helioseismology has shown us the inside of the Sun (Chapter 8). However, it will probably be many decades before astronomers can gather enough data to unravel the mysteries of the structure of a neutron star. Until then, neutron stars will remain the most mysterious stars in the Universe. Only the interiors of black holes are more effectively hidden from our view.

21

Black holes:
gravity's one-way street

B lack holes. No term evokes the mystery of modern gravity more than this one. The mystery of black holes is more than an invention of popularizers of astronomy and relativity. Black holes were certainly a mystery to Einstein and his contemporaries. Yet today black holes are everywhere: in X-ray binaries, in the centers of galaxies, and of course in books, like this one, on relativity and gravity!

Theorists attacked the problem of understanding black holes, not by using astronomical evidence, but by using lessons they had learned from quantum mechanics. Quantum thinking demanded that physicists ask only questions about things that could be measured, not about what is hidden from experiment. Thus, they can measure that light behaves sometimes as a particle (the photon) and sometimes as a wave, but they find it useless to ask what *is* a wave–particle.

> In quantum thinking an object *is* only what it *does*.

Theorists found this disciplined way of thinking helpful for relativity too. It meant that one should only ask what one can measure about black hole, and not worry about the rest. The rest includes, for example, strange things that might happen to coordinates. If a distant observer or an experimenter falling into a black hole can measure something, then it is real and important. If not, then forget it.

> This rule is a good one for learners of relativity too. Always frame questions in terms of what an observer could measure. If it is impossible to frame the question that way, then it is not a question that has any real meaning.

Many physicists contributed to developing this point of view, but none has been a stronger advocate of it than the American physicist and teacher John Archibald Wheeler (b. 1911, see Figure 21.2 on page 288), who, incidentally, was the person who coined the name "black hole".

The first black hole

Remarkably, the black hole was the very first exact solution of Einstein's equations that physicists found! Einstein had contented himself at first with approximate solutions that made key predictions, like the bending of light by the Sun and the precession of the perihelion of Mercury (Chapter 18). But the German astronomer and theoretical physicist Karl Schwarzschild (1873–1916) almost immediately found the solution describing exactly the gravitational field outside a spherical star.

Here is the spacetime-interval Schwarzschild found:

$$\Delta s^2 = -\left(1 - \frac{2GM}{c^2 r}\right)(c\Delta t)^2 + \left(1 - \frac{2GM}{c^2 r}\right)^{-1}(\Delta r)^2 + r^2\left[(\Delta\theta)^2 + \sin^2\theta(\Delta\phi)^2\right].$$

(21.1)

In this chapter: we study general relativity's most intriguing prediction: black holes. We look at the central place they have in Einstein's theory, their role in astronomy today, and the direction they are giving to efforts to unify gravity and quantum theory. We calculate orbits around black hole, examine the astronomical evidence for black holes, and learn about wormholes, the Hawking radiation, and black hole entropy.

▷The image under the text on this page is a spacetime drawing of the light-cones near a black hole. It is a thin slice of time, showing in which directions light can travel after being emitted from different locations in space near the black hole. Far from the center, the cones are vertical, and they open 90°, as in flat Minkowski spacetime. But near the center, they tilt more and more inwards and get narrower. Image courtesy W Benger, ZIB and AEI.

In this section: the Schwarzschild black hole was the first solution found for Einstein's full equations, but one of the last to be understood. Besides describing a black hole, it gives the gravitational field outside any spherical (non-rotating) star.

Figure 21.1. Karl Schwarzschild was one of the most brilliant astrophysicists of his time. A friend of Einstein and the director of the astronomical observatory in Potsdam, near Berlin, he was serving with the German artillery on the Russian front in the First World War when he received copies of Einstein's papers announcing general relativity. Two months later, in January 1916, Schwarzschild sent Einstein the solution, which Einstein immediately had published. Tragically, four months later Schwarzschild was dead, killed by disease.
One can only speculate on what he might have accomplished had he survived the war. In particular, he would likely have grasped the importance of Chandrasekhar's maximum mass for white dwarfs, and this could have led to a much earlier recognition of the astrophysical importance of neutron stars and black holes. Schwarzschild's son Martin (1912–1997) later fled Hitler, settled in the United States, and became one of the pioneer developers of the theory of stellar evolution. Image reprinted with permission of the Universitäts-Sternwarte Göttingen.

In this equation there is a constant named M, which is the mass of the black hole. Notice that this constant enters in combination with G and c, so that the equation actually contains only the length GM/c^2. We first met this length in Equation 4.12 on page 36, where we called it the *gravitational radius* R_g associated with the mass M.

Although the whole expression may at first look complicated – it is probably the most complicated equation we will write down in this book – it can be understood if one approaches it by asking questions about measurables. We take it apart in Investigation 21.1 and see what its pieces say. But one piece is so important that we should treat it here: the piece governing the curvature of time.

We know that the time part of the spacetime-interval gives the rate at which clocks at a fixed place in space run. If a clock is at rest, so that along its world line $\Delta r = \Delta\theta = \Delta\phi = 0$, then the Schwarzschild spacetime-interval tells us that its proper time lapse $\Delta\tau$ associated with a coordinate time lapse of Δt is given by

$$(\Delta\tau)^2 = -\frac{1}{c^2}\Delta s^2 = \left(1 - \frac{2GM}{c^2 r}\right)(\Delta t)^2. \tag{21.2}$$

This determines the curvature of time, as we saw in Chapter 18.

Remarkably, the time part of the spacetime-interval is exactly the same as in Equation 18.7 on page 231. This leads to three conclusions. First, if we are far away from the center of the Schwarzschild geometry (i.e. if r is much larger than the gravitational radius GM/c^2), the curvature of time is the same as it is far from a nearly Newtonian star whose mass is M. We saw in Chapter 18 that the curvature of time determines the orbits of planets and other slowly moving objects, so we are led to the following important conclusion.

> Gravity far from a black hole is just like gravity outside an ordinary star: you can't tell you are in orbit around a black hole just by measuring the orbits far away from it.

Second, we are right to call the constant M in the Schwarzschild geometry the "mass" of the black hole. It is the mass that an ordinary star would have if it produced the same Newtonian gravitational effects far away. Third, we have learned that the coordinate time t is the time as measured by experimenters far away, just as in Chapter 18.

At first, no-one knew that the Schwarzschild solution described a black hole. As a description of the geometry outside an ordinary star, it is just a generalization of Equation 18.15 on page 235. But suppose that the star is smaller than $2R_g$, so that the point $r = 2R_g$ is outside the star. Then we can presumably use the interval in Equation 21.1 on the preceding page at this value of r, but of course something strange happens there. The coefficient of $(\Delta r)^2$ in the spacetime-interval in this equation gets infinitely large as r approaches $2R_g$, so that it seems that proper distances stretch out infinitely far. And as Equation 21.2 shows, time seems to stand still there. We call this special radius $2R_g$ the **Schwarzschild radius**. To find out what happens at this radius, we have to ask about measurable things: what do black holes *do*?

What black holes can do -- to photons

In this section: the Schwarzschild solution exhibits strong gravitational redshifts. A photon can stand still at the horizon of the black hole.

The bad behavior at the Schwarzschild radius was the problem that took such a long time for physicists to solve. As we noted before, the way to understand these things is to ask questions strictly about measurable things. For example, how shall we understand that time seems to stand still at $r = 2R_g$?

Investigation 21.1. The Schwarzschild geometry

The Schwarzschild geometry is so important that it is useful to have a look at it. It is not hard to understand it, after the spacetime-intervals we have already studied.

Here, copied from Equation 21.1 on page 285, is the Schwarzschild spacetime-interval:

$$\Delta s^2 = -\left(1 - \frac{2GM}{c^2 r}\right)(c\Delta t)^2 + \left(1 - \frac{2GM}{c^2 r}\right)^{-1}(\Delta r)^2$$
$$+ r^2\left[(\Delta\theta)^2 + \sin^2\theta(\Delta\phi)^2\right]$$

The first things to notice are what is *not* in the spacetime-interval, i.e. how simple it is. One "not" is time: the coefficients are all independent of time. Time appears in the spacetime-interval as Δt, of course, but the *coefficients* of the coordinate changes are all independent of t. This tells us that the geometry described here is time-independent. It represents the gravitational field outside a star that is simply sitting there, doing nothing!

Another thing that is missing is any mixed term, such as a term with $\Delta t\Delta\phi$. From our discussion of the dragging of inertial frames in Chapter 19, we would expect such terms if the star were rotating. Therefore, this geometry is that outside a static star, one that has no internal fluid motions.

Now let us look at what *is* in the spacetime-interval. The last term in square brackets is just the distance relation on a sphere of radius r, as we worked out in Equation 18.4 on page 228, so this expresses the fact that the geometry is spherical. More precisely, the two-dimensional surfaces on which t and r are constant (obtained by setting Δt and Δr to zero in the spacetime-interval above) have the same geometry as a sphere of radius r in flat space.

Now, in flat space r would be the distance to the center of the sphere. In this spacetime-interval, however, the term in square brackets has nothing to do with the distance to the center of the sphere, which is measured by the Δr part of the spacetime-interval. What it tells us is that, if we stay on the sphere, then little steps in angle require the same distance as little steps on a sphere in flat space

with a radius of r. In particular, if we add up a lot of little steps and go all the way around the sphere, we will measure a circumference of $2\pi r$, just as in flat space.

So the coordinate r is a "circumferential radius", defined by how broad the sphere is, not how far it is to the center. In a curved space the circumference does not have to equal 2π times the radial proper distance. So when writing down the spacetime-interval, we must always make choices about what coordinates to use; this is a point we made earlier. Here we have a coordinate system whose radial coordinate is defined by circumferences of spheres. Later we will meet a different coordinate system for this geometry.

The spherical part tells us how the radial coordinate is defined and how it can be measured (by measuring the circumference of a sphere), but it is not a surprise that it is there: we know Schwarzschild assumed a spherical geometry from the start. More interesting are the coefficients of the time part of the spacetime-interval and the radial part. We discuss the time part in the main body of this chapter. So let us turn to the spatial part.

The spatial curvature is determined by the coefficient of $(\Delta r)^2$. Looking at the expression, it appears that it is everywhere larger than one. This measures proper distance in the radial direction, so one can say that radial distances are bigger than we would find in flat space: two spheres with circumferences $2\pi r_1$ and $2\pi r_2$ are separated by a proper distance that is *larger* than $r_1 - r_2$. This is the effect of the curvature.

Why not go all the way to the center of the sphere: what is the radial distance? There is a problem with this question, because the coefficient of $(\Delta r)^2$ goes to infinity at the finite radius $2R_g = 2GM/c^2$. We explore the meaning of this difficulty in the main body of this chapter.

One of the points of genius of Schwarzschild's solution is his choice of the radial coordinate. He saw clearly how useful it would be to have a definition of the radius of the spheres that was not tied to a notion of radial distance. His r can be defined without reference to the center of the space, which could be buried in the middle of a star.

Recall that the coordinate time t is the time on a distant clock. We saw as early as Chapter 2 that time on a clock slows down relative to distant clocks as the clock goes deeper into a gravitational field. This is the effect of the gravitational redshift. So the fact that proper time on a clock should be smaller than coordinate time t, as in Equation 21.2, is to be expected. It just means that, if the clock sends out one photon every second by its own proper time, then the deeper into the gravitational field it goes, the longer it takes the photon to get out.

> What is unexpected here is that the proper time goes to zero at the Schwarzschild radius. This means, effectively, that if a clock at that radius emits a photon, the photon will never get out!

How can this be? Recall that photons move in such a way that the spacetime-interval along their world lines is zero. Look at Equation 21.1 on page 285. If we are at $r = 2GM/c^2$, then the coefficient of $(\Delta t)^2$ vanishes. So a world line which remains at one spatial location,[†] i.e. with $\Delta r = \Delta\theta = \Delta\phi = 0$, has zero spacetime-interval, regardless of Δt. Therefore a photon can just sit on the surface $r = 2R_g$ forever, never getting out.

> It follows that any photon emitted from *inside* $2R_g$ is trapped: not only will it not cross $2R_g$, it will in fact be pulled inwards, to smaller and smaller radii. Because light cannot come to us from inside $2R_g$ physicists call this surface the **horizon** of the spacetime.

▷ Remember how coordinate time was defined, illustrated in Figure 18.4 on page 230.

▷ Notice that the Schwarzschild radius is exactly the same radius that Michell and Laplace had identified as the place from which light could not escape, as in Equation 4.12 on page 36. The main difference between the Schwarzschild black hole and the older Michell–Laplace black hole is that in relativity, photons never cross $2R_g$ from the inside. For Michell and Laplace, photons were like balls that were shot outwards at a fixed speed. If a ball started from inside $r = 2R_g$, it would simply move outwards and then fall back.

[†] You might worry that at $r = R_g$ the coefficient of $(\Delta r)^2$ goes to infinity. To get around that, do this for r slightly larger than R_g and let r get smaller, always with Δr strictly equal to zero.

Figure 21.2. John Wheeler, one of the twentieth century's most imaginative and influential theoretical physicists. His many PhD students include Feynman (Chapter 19) and Thorne (Chapter 22). Wheeler was one of the physicists who helped clarify general relativity by focussing on what experimenters can measure. In this he followed the discipline of quantum physics, to which he had previously made key contributions. Wheeler coined the phrase "black hole". Photo by K Thorne, reprinted with permission.

The gravitational redshift

In Investigation 2.2 on page 16 we found that a photon climbing a distance h in a gravitational field whose acceleration was g would have a lower frequency at the top than when it started, by the ratio

$$\frac{f_{\text{top}}}{f_{\text{bottom}}} = 1 - \frac{gh}{c^2}. \tag{21.3}$$

This is adequate for small distances h near the Earth, but in general relativity we need a more general form. We can deduce this for the Schwarzschild geometry from the clock equation, Equation 21.2 on page 286. Let us examine the relationship between the slowing of clocks and the gravitational redshift in the Schwarzschild geometry.

Suppose the clock near the place the photon comes from, at rest in the coordinates at radius r_0, ticks once for each cycle of the photon, and the time between ticks on the clock is τ. Then the photon's frequency near the clock will be $f_0 = 1/\Delta\tau$. When the photon arrives at an experimenter far from the hole, where spacetime is effectively flat and the coordinate time t is proper time on clocks, then the distant observer can use the cycles of the photon to measure the ticking rate of the clock: the clock ticks once for every cycle of the photon. So the frequency f_{far} of the photon far away is the frequency of the clock ticks as measured by a distant experimenter. This redshift is, for our case,

$$\frac{f_{\text{far}}}{f_0} = \left(1 - \frac{2GM}{r_0 c^2}\right)^{1/2}. \tag{21.4}$$

Notice that as the starting radius r_0 gets near $2R_{\text{g}}$, the frequency far away goes to zero. As we have seen earlier, these photons never leave the Schwarzschild radius at all.

Danger: horizon!

Now, are we in danger of a contradiction? Have we not learned in Chapter 15 that a photon cannot stand still? So how can a photon emitted at the horizon simply stay there? The answer is that the curvature of spacetime only makes the photon stand still with respect to an observer *far away*. Remember that general relativity incorporates special relativity only *locally*: the physics as observed by a freely-falling experimenter, looking only over a small region of spacetime, must be the same as in special relativity. If a freely-falling experimenter were to cross the horizon and observe the photon, it would not be standing still with respect to the experimenter: like any photon, it would be traveling with the speed of light relative to a local freely-falling observer. But time curves so much between the Schwarzschild radius and the distant observer that a photon traveling radially outwards at the speed of light only just manages to keep its distance from him.

Let us think about the freely-falling experimenter that we have just mentioned. Suppose she has taken the precaution of falling while strapped into a rocket ship with powerful motors. Her plan is to fall freely for a while, doing some important physics experiments, and then turn on the rocket and get away from the black hole. If she falls across the horizon before turning on the rockets, it will be too late: since nothing can move faster than light, not even a powerful rocket, and since inside the horizon *all* photons, no matter what their initial direction, fall inwards, the experimenter will also inexorably fall inwards. We shall consider what terrible things will happen to her in a later section.

The falling experimenter's risk is made worse by the fact that spacetime contains no signposts telling unwary travelers where the horizon is.

There is no sign saying, like the title of this section, "Danger: horizon!" Spacetime is smooth, empty, and locally flat there.

One can calculate the curvature of spacetime there, and it is unremarkable. There is curvature, of course, but it is not necessarily very large. In fact, the curvature of space and time at the horizon is proportional to $1/M^2$, so the larger the black hole, the smaller is the curvature at the horizon. The horizon only marks the boundary between where photons can get out and where they cannot, but that is a property of the large-scale structure of the spacetime, not something that can be sensed locally.

The nature of the horizon is illustrated by the light-cones drawn in the image under the text on page 285. They gradually tilt inwards as one goes towards the center, and the difference from one cone to its neighbor is small. Only by accumulating these small changes does the big difference between the cones far away and those at the horizon arise.

> ▷ Recall that the density of material forming a black hole also decreases with the mass of the hole. Large black holes exert locally rather weak gravitational forces.

Getting away from it all

> In this section: how to use time dilation near a black hole.

The slowing of time near a black hole has measurable effects besides the gravitational redshift. For example, it would in principle be possible to put something into a slow-time storage locker near a black hole. Suppose you don't like your government, which is authoritarian and makes life in your country miserable, but you are not optimistic about outliving it. Just get into a rocket and spend a few days very close to a black hole. From the point of view of the government you left behind, your time near the black hole goes so slowly that your few days there take many years back home. When you come back the government may have changed and you will have only lost a few days. Of course, you may have lost all your family, friends and possessions, as well, but at least you are young enough to start over again!

> ▷ Before signing up to a trip like this, you might think twice! Not only does everyone at home age much faster than you, but there are risks. You have to remain at rest very close to the horizon, which is not easy to see locally. If you make a tiny navigational error, you might wind up trying to park on the wrong side of the horizon, and your holiday would become your funeral!

There is nothing inconsistent in this. It certainly conflicts with our ordinary sense of how time behaves, but once we accept that gravity curves *time*, as we argued it must in Chapter 17, then such things become possible. This example may seem fanciful, but it is simply an exaggeration of what happens every day to the signals that go back and forth to the GPS satellites. And it is nearly duplicated by the gas orbiting black holes in galaxies: later in this chapter we will see the evidence for this.

Singularities, naked or otherwise

> In this section: the most disturbing aspect of black holes is that they contain singularities: places where general relativity breaks down. At least they are hidden inside the hole: if they were outside the hole (naked) then gravitation theory would be in trouble. The cosmic censorship hypothesis expresses physicists' hope that this does not happen.

Let us return to consider what happens to the photon or observer that finds itself inside the horizon. Gravity forces it inwards, and here there is a real problem: at $r = 0$ there is what physicists call a **singularity** of spacetime: unlike at the horizon, the curvature of the spacetime does get infinitely large as one approaches $r = 0$, and so the tidal forces are arbitrarily large. Nothing could survive an encounter with such a singularity.

The existence of such a singularity is taken by many physicists to imply that general relativity cannot be a complete theory of gravity by itself: it does not predict what should happen to particles that encounter that singularity. There is good reason for not trusting general relativity near such singularities. They seem to imply the confinement of particles into very small regions, with very high energies. This may not be consistent with the Heisenberg uncertainty principle of quantum theory. It may be that when we have a consistent quantum theory of gravity, then we will find that the behavior of gravity near singularities is very different from what general relativity predicts.

Does this undermine our confidence in other aspects of black holes? Should we fear that quantum corrections will prevent black holes from forming in the centers of galaxies? The answer is no: the singularity is inside the horizon, and the horizon itself is not a place of strong curvature or any other effects that might make

Figure 21.3. Roger Penrose has helped establish the modern mathematical framework for general relativity. Working with Hawking (see below) he showed that gravitational collapse generally produces singularities, which led him to the Cosmic Censorship Conjecture (see text). He has also stimulated a wide debate outside physics on the origins of consciousness. Photo courtesy Oxford University and Roger Penrose.

In this section: we modify the orbit program to compute orbits around black holes. There are new features, including orbits that are trapped by the hole. The motion of equal-mass objects under their mutual attraction can only be studied with supercomputers.

▷Of course, just as in Newtonian gravity, once one goes inside the star the field will be different from that of a black hole, but outside they are identical.

quantum corrections important there. Most physicists believe that any changes to this picture that quantum gravity may bring will be confined to a region near $r = 0$, and will not even be visible from outside the hole. The horizon not only hides the singularity; it also hides our ignorance!

This point of view would be radically altered if it were possible to encounter singularities that are not hidden behind horizons. Such things are called **bnaked singularities**, and they would be a much more serious problem for general relativity than the singularities inside black holes. So far, no one has succeeded in establishing that very strong singularities can form naked. The contrary view is called the **cosmic censorship hypothesis**: framed by the British mathematical physicist Roger Penrose (b. 1931), it postulates that serious singularities will never be naked: Nature censors such sights from our eyes! The effort to prove (under suitably general conditions) or to find a counter-example to this hypothesis is one of the most interesting areas of mathematical research in relativity today.

The stakes are high. If naked singularities can form, then general relativity cannot be a complete theory: it cannot predict the future (even in its non-quantum form) even of systems that start out in a non-singular state. If this turns out to be true, then perhaps physicists will have to look toward a quantum theory of gravity (Chapter 27) to rescue the consistency of the entire theory, not just to make it compatible with the rest of physics!

What black holes can do ... to orbits

Let us turn once more to the effect of black holes on things near them, and specifically on the orbits of nearby particles. This is of more interest than just for black holes, since the gravitational field outside of a spherical star of mass M is identical to that of a black hole of the same mass. The study of orbits around black holes is then applicable to orbits around stars.

This is because, as we mentioned in Chapter 19, general relativity obeys a theorem similar to the one we proved for Newtonian gravity in Chapter 4: the gravitational field outside of a spherical body does not depend on the radius of the body, only on its mass, even if the radius is changing in time. Therefore, if a spherical star were to collapse in a perfectly spherical manner, in such a way that all its mass went to form a black hole, then the gravitational field outside the original radius of the star would not change at all. Planets outside would not even notice, provided the collapse was spherical or nearly so.

We saw this theorem in action earlier, where we observed that the gravitational field far outside a black hole is the same as that of a Newtonian star. Here we learn that this identity is true even close to the hole, where the field would be produced by a relativistic star of the kind we studied in Chapter 20. It turns out that, for neutron stars, the mass M of the star is the value of the quantity $m(r)$ at its surface. We already used this for the star's mass in plotting Figure 20.6 on page 281.

If orbits far from the hole are Newtonian, those nearby are certainly not. In Investigation 21.2 we develop the orbit equations for particles near black holes and relativistic stars. The modification from the Newtonian orbit equations is small, so it is an easy job to turn the program Orbit into RelativisticOrbit, which generates orbits around black holes. The details can be found on the website, and two example orbits are shown in Figure 21.4.

These orbits show important new features that are absent from Newtonian orbits. Both orbits begin close to the hole, at just five times the Schwarzschild radius, to insure that they show the effects without too much computer time. One of the orbits, which starts with a speed of $c/3$, spirals closer to the horizon and finally plunges across it. Such behavior is unknown in Newtonian gravity: no matter how near the

---

Investigation 21.2. Orbiting a black hole

In Investigation 4.1 on page 29 we developed the theory behind the program Orbit, which computed orbits in Newtonian gravity. Here we give the modifications that are necessary to do orbits around black holes and relativistic stars.

The acceleration of gravity produced by a Newtonian star of mass M has magnitude

$$a_{\text{Newtonian}} = \frac{GM}{r^2},$$

and it is directed towards the star. When we compute the acceleration components at an arbitrary location (x, y) (with the star at the origin), then we get Equations 4.5 and 4.6 on page 29:

$$a_x = -a_{\text{Newtonian}}\frac{x}{r} \quad \text{and} \quad a_y = -a_{\text{Newtonian}}\frac{y}{r}.$$

The only change that needs to be made to convert these equations into the ones that exactly describe motion around black holes is to change the magnitude of the acceleration $a_{\text{Newtonian}}$ to

$$a_{\text{relativistic}} = \frac{GM}{r^2}\left(1 + \frac{12K^2}{c^2 r^2}\right), \qquad (21.5)$$

where K is the "Kepler constant" of the orbit, which is defined just as it was for Kepler's first law (see Chapter 4). That is, K is the area sweeping rate of a Newtonian orbit with the same starting radius and velocity as the relativistic orbit has. As we remarked in Chapter 4, this rate is one-half of the angular momentum divided by the mass of the particle[a].

This extra term must be handled with care, because it involves the constant K that is not a property of the black hole but rather of the orbit. Once the starting position and velocity for the orbit are given,

one must calculate the area sweep rate and use it for K. The relativistic orbit equations given by this acceleration insure that this rate is constant, just as in Newtonian gravity.

Calculating the sweep rate for general starting values is difficult, so I have written the program RelativisticOrbits to start with values where it is simple. If the initial position is at a distance r along the x-axis and the initial velocity is v directed in the y-direction, then after a small time Δt the particle will have defined a right-angled triangle with one side of length r (the height of the triangle) and a perpendicular side of length $v\,\Delta t$ (the base). The area of the triangle is then $rv\,\Delta t/2$, and the area sweep rate is this divided by the time it took to generate the area, Δt. Therefore, the Kepler constant for this configuration is just $rv/2$.

Once calculated from the initial data, there is no need to recalculate the number: it remains the same. It just has to be used to compute the acceleration.

The last remark we need to make about our equations is the meaning of the time variable t when we use the given acceleration. It is the proper time of the orbiting particle, the time as kept on its own clock. So if you run the program and discover an orbital period, this is the time that elapses on the particle's own clocks during one orbit, not the time as seen by an observer watching the orbit from far away. To convert from one to another requires two steps: first use the special-relativistic time dilation to change from the particle's proper time to the time on clocks at rest along the particle's orbit, and then use the gravitational redshift given by Equation 21.4 on page 288 to convert to the time of the distant observer. This is not hard to do for an orbit that is perfectly circular, where the conversion factors are constant. It is rather more difficult for the orbits we have displayed.

[a]Experts in relativity who have not seen this form of the orbit equations before may want to know more detail. The Cartesian coordinates used here are in the equatorial plane of the Schwarzschild solution, and are $x = r\cos\phi$, $y = r\sin\phi$, where r and ϕ are the usual Schwarzschild coordinates. All speeds and accelerations are changes in these coordinates with respect to the proper time of the orbiting particle. The angular momentum referred to is the usual conserved p_ϕ, so the Kepler constant is one-half of the specific angular momentum.

---

center an orbit gets, it will always come out again. But this particle is doomed in relativity by having too small an initial speed: it gets gobbled up by the hole.

The other orbit begins with 20% larger initial speed, and this is enough to keep it out of the hole. It follows a roughly elliptical orbit, but the orbit moves around. This is an extreme example of the precession we calculated in Chapter 17.

The plunge orbit illustrates the danger of the hole to particles near it. There are many orbits that start off innocuously, but which wind up being trapped. In fact, it can be shown that there are no stable circular orbits at all around a Schwarzschild black hole for particles nearer than three times the Schwarzschild radius! This radius is called the **innermost stable circular orbit**. You could try showing this with the orbit program. Start at, say, 2.5 times the Schwarzschild radius and vary the initial speed to see if you get any orbits that stay even roughly circular (such as a mildly elliptical precessing orbit). You will find none.

Although the acceleration of a single small particle near a black hole is a simple modification of the Newtonian acceleration, more complicated systems in relativity

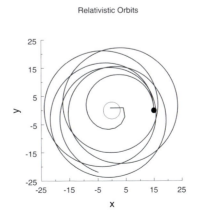

Relativistic Orbits

Figure 21.4. Two orbits near a black hole of one solar mass, as calculated by the program RelativisticOrbits. *The horizon of the black hole is shown as the circle in the center. Both orbits begin at five times the horizon radius, shown by the heavy dot. One orbit does not have enough initial speed to stay away from the hole, and is captured. (The kink in the orbit at the end is due to the small number of points output by the computer program – the orbit is calculated accurately enough.) The other orbit stays near the hole but precesses by about half a circle on each orbit.*

▷In the terminology we used in Chapter 19, Einstein's equations are non-linear, so it is not possible to describe complex systems by adding together simpler ones.

are not generalizations of similar Newtonian systems. In relativity it is not possible to write down simple equations even for a binary system, as we did for Newtonian gravity in Investigation 13.1 on page 156. Even the geodesic solutions for orbits that we have just found are only approximations to what real particles will do near a black hole, because real particles will radiate gravitational waves (see the next chapter), lose energy, and gradually spiral towards the hole. The result is that realistic solutions for binaries or more complicated systems must be found on a computer.

Figure 21.5. *This image illustrates the merger of two black holes of unequal mass, from an in-spiralling orbit, as calculated in a supercomputer simulation. The outer surface is the horizon that forms around them both, indicating that by this stage of the simulation they have merged into a single black hole. The ghosts of the old individual horizons are still shown inside the new horizon, even though they do not have any significance at this point. Image adapted from one rendered by W Benger (ZIB) from a simulation performed by the AEI-WASHU-NCSA numerical relativity collaboration; used with permission.*

Such computer solutions are unfortunately not just simple generalizations of the computer work we do in this book. The solution of Einstein's equations must keep track of not just the positions of a few particles, but also the changing coefficients in the spacetime-interval everywhere. The whole of space and time is dynamical. Calculations like this require enormous computer memories and very fast processors. Supercomputers today (2002) are just becoming fast enough to do calculations like this, and scientists working in the field of numerical relativity hope soon to be able to provide physicists' first real insight into the behavior of strong gravitational fields in general relativity. Figure 21.5 shows an illustration of the output from a recent simulation of the merging of two black holes.

Making a black hole: the bigger, the easier

Several times in this book we have hinted that there are limits on what kinds of systems can remain as ordinary stars, and that if they cross those limits they will become black holes. General relativity alone does not tell us what those limits are, since the solution for the Schwarzschild black hole scales with the mass M, and this can have any arbitrary value. General relativity can contemplate black holes of 1 g or of one galaxy in mass, and is equally happy with both. The sizes of real black holes are determined by the astrophysical objects that make them.

One constraint on the physics of black hole formation is the density of material when it forms a black hole. A uniform cloud of gas of mass M just forming a black hole has a radius of $2R_g$. Taking its volume to be $4\pi(2R_g)^3/3$ then the density ρ_{bh} is roughly $M/(32\pi R_g^3/3)$. Putting in GM/c^2 for R_g gives, after some simplification,

▷The real volume of the gas cloud is not the volume of a sphere in Euclidean space, because space near the black hole is not flat. But we are not interested here in getting all the numbers exactly right, just in seeing approximately what numbers come out.

In this section: the density of material just as it forms a black hole depends on the mass of the hole being formed. Holes smaller than a solar mass require densities higher than nuclear, which stops them being formed. Supermassive holes in galaxies can form from material less dense than water.

$$\rho_{bh} \approx \frac{3c^6}{32\pi G^3 M^2} = 2 \times 10^{19} \left(\frac{M_\odot}{M}\right)^2 \text{ kg m}^{-3}. \qquad (21.6)$$

The middle expression is not very informative, so I have evaluated it for the mass of the Sun, and shown how it scales from this value for any mass M of the black hole.

> The density required to form a black hole scales inversely with the square of the mass of the black hole. Objects of small mass have to be compressed to high densities to form a black hole, while objects of very large mass can form black holes at low densities.

We can learn much from Equation 21.6. For example, if Nature were to try to form a black hole of the mass of the Sun, then the collapsing matter would have to compress to a density of about 2×10^{19} kg m^{-3}. This is a factor of 100 higher than the typical nuclear density of 2×10^{17} kg m^{-3}, at which we saw in Chapter 20 the nuclear repulsion is strong enough to stop the collapse. We should not expect to see black holes of $1 M_\odot$ in the Galaxy, and indeed none are known.

However, a black hole with a mass of $10 M_\odot$ forms exactly at nuclear density. It is no surprise, then, that the black holes observed in X-ray binaries (as described later in this chapter) are typically this size.

The giant black holes in galaxies clearly form at low densities. Some are known with masses of $10^9 M_\odot$ or more. Such a black hole would form with a density of 20 kg m^{-3}, less than the density of water! We don't know if such objects formed in one grand collapse of a cloud of this density, or if they started out smaller and gobbled up smaller stars and black holes to reach their present size. But if they did form by a single collapse, then clearly nothing would have stopped the collapse: the density was not even high enough to have triggered nuclear reactions.

At the other extreme, black holes of small mass would be very dense when they formed. A black hole of 10^{12} kg, which we will see later is an interesting mass, would require a density of almost 10^{56} kg m^{-3}. A density this large was only ever seen once in the history of the Universe: shortly after the Big Bang. If such black holes exist, they could only have formed in the Big Bang. They are not forming now.

> ▷We should be cautious because our estimate of the volume was rough and because gravity is very strong even before the collapsing cloud reaches the Schwarzschild radius. We saw in the last chapter that detailed calculations show that in fact gravity overwhelms nuclear physics when the star has a mass of no more than about $3 M_\odot$, and perhaps even as low as $2 M_\odot$. Our simple calculation here is consistent with this.

Inside the black hole

What about after the collapse? What happens to the material inside the black hole? Does it continue to collapse or does it stop when it finally reaches nuclear densities? The answer is that nothing can stop the collapse inside the hole, not even the hard-core nuclear repulsion.

The reason is that light cannot get out or even stand still in this region. Light is forced to fall inwards, even if we start it out in an outward direction. Therefore all freely-falling experimenters must also fall inwards.

> **In this section:** whatever falls inside the hole must continue moving to smaller radii. Nothing can hold itself up against such strong gravity. Everything reaches the singularity in a finite proper time.

> If a collapsing star managed to stop collapsing and keep a fixed size inside the black hole, then it would be traveling outwards relative to all freely-falling experimenters at a speed faster than light. This is not possible. Once inside, the star must continue moving to smaller values of the radius r, reaching the singularity in a finite proper time.

So once matter has fallen in, it cannot stop collapsing. We will return later in this chapter to the natural next question, of what happens when it reaches the center.

Disturbed black holes

We have seen in Chapter 8 that a star has a natural vibration frequency that depends just on its density. Having calculated the density of a black hole, we can therefore compute its expected natural frequency from Equation 8.23 on page 100:

> **In this section:** black holes have natural frequencies of vibration, just as other objects do. These will give a characteristic imprint on gravitational radiation from events where holes are formed or collide.

$$f_{\text{bh}} \approx \left(\frac{3}{32\pi} \right)^{1/2} \frac{c^3}{GM} = 35 \left(\frac{M}{1 M_\odot} \right)^{-1} \text{ kHz.} \qquad (21.7)$$

This is fairly close to the result of a full calculation of the ringing frequency of a black hole.

But how can a black hole oscillate? We have emphasized that the horizon is not a material surface, not a place which one can find by local experiments. Nevertheless, it does represent the boundary between what gets out and what is trapped, and the location of this boundary can change with time. If a small star falls into a black hole, it will disturb the location of this boundary, and the boundary will oscillate for a short time before settling down. Typically, it will execute only two or three oscillations in this way. The same happens when a black hole is formed by collapse, if the collapse is not perfectly spherical. The hole is formed in a non-spherical shape; the non-sphericity oscillates and dies away.

The oscillations produce a time-dependent gravitational field outside the hole, and as we will see in the next chapter this must lead to the emission of gravitational waves. So black holes can be dynamical and can emit gravitational radiation. Scientists are building detectors with the hope of finding this radiation, and numerical physicists are doing simulations to try to determine precisely the characteristics of this radiation. We will describe this more fully in the next chapter.

Limits on the possible

In this section: we look at some more simple numbers, the Einstein luminosity and the Planck scales, which lie at the boundaries of what we know about the physical world. Our style of discussion here is called dimensional analysis. We met it first in Investigation 1.3 on page 5.

We have drawn some important conclusions in the last sections from rather simple calculations. Here we do some equally elementary calculations that take us to the limits of physics. Black holes, by their nature, fix boundaries between the possible and the impossible, and these boundaries depend on combinations of the most fundamental constants of Nature, c, G, and h.

We start with a simple observation, that it is possible to form a number with the dimensions of *luminosity* by using only c and G:

$$L_E = \frac{c^5}{G} = 3.63 \times 10^{52} \text{ W.} \qquad (21.8)$$

This is called the **Einstein luminosity**. It is huge compared to the luminosity of most objects in the Universe. The luminosity of the Sun, for example, is only 3.8×10^{26} W, that of the Galaxy about 10^{40} W, and that of gamma-ray bursts around 10^{43} W for a few seconds. What significance can the very much larger Einstein luminosity have?

We will show here that L_E is an upper bound on all luminosities: anything that tried to have a larger luminosity would collapse to a black hole, pulled in by the self-gravity of the very energy it was trying to radiate away. To see this, consider the following extreme example. Suppose we have a system of mass M and radius R, which suddenly turns itself entirely into light, which then radiates away. We won't ask how this could happen; we will see whether a more realistic scenario teaches us anything after we follow this one through. So we have an energy of Mc^2, and it leaves the region R at the speed of light, so it takes a time R/c to do it. That means the object briefly has the luminosity

$$L = Mc^2/(R/c) = Mc^3/R.$$

Now, the region's size can't be smaller than its Schwarzschild radius, or it would be a black hole and the radiation would not get out. So we can take $R \geq 2GM/c^2$. In turn this implies

$$L \leq c^5/2G = 0.5L_E.$$

More realistically, if the radiation escaping was only a small fraction f of the mass of the object, then the luminosity would be $0.5fL_E$.

> The Einstein luminosity is the effective upper bound on the luminosity of any process.

Another universal number built out of fundamental constants is the Planck mass, which we met in Chapter 12. We shall see that we can also define the **Planck length** and **Planck time**. We reproduce Equation 12.20 on page 146 here:

$$m_{Pl} = (hc/G)^{1/2} = 5.5 \times 10^{-8} \text{ kg.} \qquad (21.9)$$

Associated with this mass is its gravitational radius Gm_{Pl}/c^2, called the Planck length:

$$r_{Pl} = (hG/c^3)^{1/2} = 4.0 \times 10^{-35} \, \text{m}. \tag{21.10}$$

And then there is the time it takes light to travel the Planck distance, r_{Pl}/c, called the Planck time:

$$t_{Pl} = (hG/c^5)^{1/2} = 1.4 \times 10^{-43} \, \text{s}. \tag{21.11}$$

These are all made only of fundamental constants. Since they include Planck's constant and Newton's constant, many physicists believe they must have something to do with quantum gravity. Perhaps r_{Pl}, for example, is the smallest length scale on which we could use general relativity to describe gravity.

To test this idea, let us introduce quantum ideas by using the Heisenberg uncertainty principle. Recall that a precise measurement of the size of something is accompanied by an uncertainty in its momentum, $\Delta x \Delta p \geq h$. Consider a black hole of mass M. For general relativity to be a valid description of Nature, it should be possible to localize the mass of the hole to within its gravitational radius, so that $\Delta x \leq GM/c^2$. It follows that the momentum of the hole cannot be defined to more precision than $\Delta p \geq hc^2/GM$. We can express this as a velocity uncertainty by dividing by the mass M to get $\Delta v \geq hc^2/GM^2$.

Now, for ordinary black holes this is really small. For a $10M_\odot$ black hole, the velocity uncertainty is $2 \times 10^{-67} \, \text{m s}^{-1}$. We won't notice this! But if the velocity uncertainty is of the same order as the speed of light, then it will not be possible to talk about a black hole without using quantum theory, since it will be an object with no well-defined velocity or position. So if we put $\Delta v = c$ into the above expression and solve for M we get

$$M \geq m_{Pl}.$$

> Physicists use the word **classical** to refer to non-quantum theories of physics, including general relativity. A classical black hole must have at least the Planck mass.

What might quantum black holes smaller than the Planck mass be like? Some scientists suggest that quantum fluctuations (see Chapter 7) might produce, temporarily, black holes of the Planck mass or smaller, which live for a time allowed by the uncertainty principle (h divided by their energy, $m_{Pl}c^2$, which gives the Planck time) and then disappear. This picture is called **spacetime foam**. The tiny black holes distort spacetime on the length-scale of the Planck length, so that it is not the smooth empty spacetime we think it is when we probe only on larger distance-scales. These are interesting ideas, but we will probably only know for sure what the Planck scales mean when we have a good theory of quantum gravity.

Other combinations of constants are possible. For example, the Planck density is m_{Pl} divided by r_{Pl}^3,

$$\rho_{Pl} = c^5/hG^2 = 8 \times 10^{95} \, \text{kg m}^{-3}. \tag{21.12}$$

This is not a density we can think of achieving in laboratory experiments! But one would not trust a model of the Big Bang, for example, at times so early that the density of matter was higher than this: one would want a quantum theory of the Big Bang. We will take these issues up again in Chapter 27.

The uniqueness of the black hole

So far, we have treated only spherical black holes, as described by the Schwarzschild geometry. Since Schwarzschild was the first person to discover any exact solution of Einstein's equations, one might expect that further research would have revealed

In this section: black holes have mass, spin, and charge: that's all!

lots of other geometries describing black holes. Remarkably, this is not at all the case.

> In fact, general relativity allows only one family of time-independent black hole solutions, of which the Schwarzschild geometry is one member (the one with zero spin and zero charge). In this family, the black hole is completely determined by giving only three numbers: its mass, its spin, and its electric charge.

When black holes are formed or disturbed, they can take on a variety of shapes temporarily. But after a few oscillations they settle down into a member of this single family by radiating away the disturbance in gravitational waves, as we saw earlier.

▷Imagine trying to describe everything about a *person* by giving just three measurements, for example the height, weight, and girth. It would be ridiculous to expect that any two people with the same height, weight, and girth would be identical: same hair color, same blood pressure, same sex, same dreams, same taste in food, same performance on a physics exam! It might seem even more ridiculous to expect this of black holes, which are formed by the collapse of enough matter to make at least 10^{29} human beings! Yet it is true.

The idea that just three numbers fully describe a black hole is astounding. No other macroscopic object is as simple. So where has all the variety gone – the variety of all the stars with their sunspots and eruptions, the planets with their mountains and red spots, the countless individual grains of interstellar dust, whatever fell into the hole – that squashed together to form the black hole? It has, of course, disappeared into the hole. Once the material is inside the horizon, it can send no information back out, so this variety leaves *no trace* in the gravitational field outside the hole. The three numbers that remain are very special sums over what went in: the mass, spin, and charge of the hole as measured by an observer at rest with respect to the hole and far from it.

Why these three? Why not the total number of baryons, or some extra numbers that might describe the detailed shape of the horizon, a few extra ripples or corners?

> Mass, spin, and charge characterize the black hole because they are the only three properties of a body that are conserved *and* that can be measured from far outside the body.

We learned in Chapter 19 that Einstein's equations demand the conservation of energy and momentum. For a body at rest, the linear momentum is zero but the spin (angular momentum) does not have to be. Both the total energy (mass) and angular momentum are conserved. Moreover, electric charge is also a conserved quantity. No chemical or nuclear reaction has ever been observed that changed the total electric charge.

▷Many physicists believe, however, that baryon and lepton number may not always be conserved. We will look at this in Chapter 25.

Physics has other conservation laws. Nuclear reactions seem to preserve baryon and lepton number, for example. However, the nuclear forces are short range: we saw in Chapter 20 that they influence only other nearby baryons. Outside a nucleus one cannot measure directly how many baryons are in it, whereas by using the Lense–Thirring effect of gravity (Chapter 19) one could in principle measure the spin of the nucleus even from far away. Quantities conserved by short-range forces cannot be felt outside of a black hole: we will never know how many baryons fell into any given hole. Only three quantities are conserved and measurable from far away: mass, spin, and charge.

Forming a black hole results in a huge loss of information. In thermodynamics, this information is associated with a concept called *entropy*. We will see later in this chapter that the entropy of a black hole is immense.

Spinning black holes drag everything with them

In this section: why spin shrinks the hole and makes a region outside it where nothing can stand still.

The Schwarzschild black hole has no spin, but a realistic cloud of gas collapsing to a black hole should rotate, and should therefore form a spinning black hole. The solution of Einstein's equations that describes such a hole was not found until 1963, by the New Zealand physicist Roy Kerr (b. 1934). The Kerr geometry is the unique family describing a time-independent black hole with spin and no electric charge.

The Kerr geometry brings gravitomagnetism to the black hole. Orbits that rotate in the same sense as the black hole experience the repulsion due to the gravitomagnetic effect that we calculated in Chapter 19. This allows them to approach closer to the hole, while remaining in circular orbits, than they could around a non-rotating hole. These closer orbits have shorter orbital periods. Astronomers believe they can now measure such orbits and verify that some known black holes are very rapidly rotating. (See Figure 21.7 on page 302.)

What is the horizon of a rotating black hole like? For the Schwarzschild black hole, we saw earlier that the horizon consists of light paths that stay at constant r, θ, and ϕ. For the Kerr horizon, the spin of the hole gives an extra repulsion to photons that orbit with the hole's rotation, so these are the ones best placed to resist the inward pull of gravity.

> It seems unlikely that astrophysical black holes will form with charge, so we will not describe the full family in this book.

> The horizon consists of photon world lines that rotate around the hole with the speed of the dragging of inertial frames on the horizon. The horizon is also smaller than the Schwarzschild horizon, since the repulsion allows some photons to come nearer the hole and still escape.

If the last photon to resist the inward pull of the black hole is rotating around it, then the light-cones near the hole are not only tilted inwards, as in the figure under the text on page 285, but also tilted in the direction of rotation. This must then be true for light-cones just outside the horizon, which means that photons and particles near the horizon must also rotate around the hole. Anything that stands still sufficiently close to the horizon must be following a world line that moves outside the light-cones, and is therefore going at faster than the speed of light relative to local freely-falling experimenters.

> Remember we want always to describe geometry in terms of measurables. For the horizon, the measurable is its area. The radius of the horizon is just a coordinate, but the surface area is a geometrical quantity. The area of a spinning Kerr horizon is less than that of a Schwarzschild horizon of the same mass.

> The horizon of a rotating black hole is surrounded by a region of finite size in which all particles and photons must move around the hole in the same direction as the hole rotates. It is impossible in this region to move backwards. The dragging of inertial frames near the horizon of a rotating black hole is irresistible.

The region in which dragging is so dominant has only a finite thickness. There is therefore another surface outside the horizon where it is possible for a photon to remain at rest with respect to a distant observer. This surface is called the **stationary limit**. Unlike the case for a non-rotating black hole, the stationary limit is not the horizon, because photons that rotate with the hole can escape from inside this surface.

The naked truth about fast black holes

Imagine letting particles fall into a black hole from orbits that have angular momentum in the same sense as the hole's. (This happens in black hole X-ray sources, as we will see below.) Then the hole's spin will increase, and the gravitomagnetic repulsion on co-rotating orbits will increase. By doing this it is possible to make the hole spin so fast that the gravitomagnetic repulsion is strong enough to allow photons to escape from any location: there is no longer a horizon!

In this section: with enough spin, a Kerr hole can have a naked singularity. But most physicists believe that these holes cannot form.

The Kerr black hole that has just enough spin to annul the horizon is one whose angular momentum is related to its mass M by exactly $J = GM^2/c$. This is called the "extremal Kerr" black hole. However, the Kerr solution allows, at least mathematically, much larger values of J. In these, the horizon that normally conceals the inner singularity is gone, but the singularity remains: they are examples of *naked singularities*. Physicists have calculated that the repulsive effect of gravitomagnetism on material falling toward a Kerr black hole will prevent any real hole from gaining this much angular momentum, so that cosmic censorship will prevail. Observations of X-ray sources, as described below, can test this belief.

Mining the energy reservoir of a spinning black hole

In this section: the rotational energy of a spinning black hole can be extracted using negative-energy orbits outside the horizon. We show why these orbits exist and suggest how Nature is using them to power jets from quasars.

Like a massive spinning flywheel, the rotation of a spinning black hole represents a reservoir of usable energy. Physicists have learned in principle how to tap this reservoir, and it also appears that Nature has learned how to as well. It all has to do with the remarkable properties of particles and photons inside the stationary limit, where particles can have *negative* conserved energy.

We have met negative energy before, when we defined the gravitational potential energy of a particle in a Newtonian gravitational field in Equation 6.9 on page 54. It is part of the total energy of a particle on an orbit, the other part being its kinetic energy. It is negative because it is the energy given up to the kinetic energy as the particle falls. The sum of these two energies, the particle's total energy, is constant, and it equals the kinetic energy that the particle would have if it could reach a very distant experimenter.

The simplest particles to discuss in the Kerr geometry are photons. The conserved total energy of a photon is defined by analogy to the Newtonian case simply as the energy the photon has when it reaches a distant observer. In relativity the total energy includes any rest-mass the particle's may have.

> For a photon, its *conserved* total energy is the gravitationally *redshifted* energy it has when it gets far away, regardless of what energy it began with.

Now, consider the photon that we mentioned earlier, which sits exactly on the stationary limit surface and is at rest with respect to a distant observer. A neighboring photon, just outside the stationary limit, would get out with a very large redshift, and therefore has very small positive total energy. The photon that is standing still does not get out at all, and therefore has *zero* total energy. It follows that a photon just inside the stationary limit could actually have *negative* total conserved energy. And if there are negative-energy photon world lines, then there must be particles with negative conserved energy inside the stationary limit, too, since particles can move as close as we like to the speed of light.

▷We will actually make use of the negative-energy photon orbits in the Schwarzschild geometry when we discuss the Hawking radiation below. The quantum uncertainty in their locations makes them accessible to some extent outside.
▷Because of the negative-energy orbits, the stationary limit surface is sometimes called the **ergosphere**. This name uses the prefix *ergo-*, which indicates energy.

Of course, such negative-energy orbits are found inside the stationary limit in the Schwarzschild solution as well, but in the Schwarzschild case the stationary limit is the horizon, so these orbits are inside the black hole and of no relevance or use to us outside. However, some of these negative-energy orbits in the rotating Kerr geometry are outside the horizon, so they can participate in physical processes.

> Particles in such negative-energy orbits can exchange energy with other particles. Penrose pointed out that this can be used to extract energy from the spinning black hole.

Imagine the following process. A distant astro-engineer drops two balls connected by a compressed spring toward the hole. Once inside the stationary limit, but still outside the horizon, the spring releases and the balls separate in such a way that one of them goes into an orbit that has negative total energy with respect to the distant engineer, although of course it has positive energy as measured by any local experimenter. Then, by conservation of energy (which still holds, even if some energies are negative), the other ball must be in an orbit whose total energy is more than the rest-mass energy of the original balls together. When this ball emerges from the stationary limit and reaches the distant engineer, it will have this large energy locally. The engineer has got back more energy than she put in: she is mining the hole!

> The energy comes from the rotation of the hole, because the negative-energy orbits all have negative angular momentum, so that when the ball that is left behind inside the stationary limit falls across the horizon, the spin of the hole decreases. Reassuringly, there is no perpetual motion here: eventually the engineer will extract all the spin and the hole will become a useless Schwarzschild hole!

▷ Notice that the engineer also extracts angular momentum from the hole. To mine the energy over a long period of time, she must also deal with the angular momentum.

Our astro-engineer is just a fantasy at present, but Nature may already be working the **Penrose process**. The giant black holes that power quasars and active galaxies are likely to be very rapidly rotating, because all the matter that they accrete comes in with high angular momentum. (We will look at the evidence for this in the next sections.) Moreover, as we saw in Chapter 14, these holes produce enormously powerful, highly collimated jets of gas, whose direction is sometimes stable over millions of years. More and more astronomers are beginning to believe that the source of the jets' energy is the rotation of the hole. It seems possible that magnetic fields generated in the accretion disk can penetrate the stationary limit and extract the rotational energy of the hole. Further research and many more observations will be required to discover the source of the energy in the jets.

Accretion onto black holes

Since most stars are members of binary systems, stars that form black holes usually start out in binaries. Unlike their cousins that form neutron stars, these events are much less likely to disrupt their binaries. Since much of the original star's mass stays in the black hole, the gravitational attraction between the binary stars does not weaken so much, and more of the systems will survive. So astronomers are not surprised to find that several percent of binary X-ray sources contain black holes.

As for the neutron star binaries we discussed in the previous chapter, it is often possible to estimate the mass of the accreting object from spectroscopic observations. Astronomers have studies systems that appear to contain black holes, which they call black hole candidates, to try to obtain at least a lower bound on the mass of the accreting object. When that lower bound exceeds, say, $5M_\odot$, then astronomers are confident that the object is not a neutron star: it must be a black hole.

In this section: some X-ray sources contain accreting black holes. Although the main features of the X-rays are the same as for neutron stars, there are distinctive features that provide strong evidence that there is a hole in the center.

Table 21.1. Black hole candidates in stellar systems. The first column contains two names, the first being the modern style (catalog name, position on the sky) and the second (in round brackets) being the older name by which the object was known, often as a variable star or nova. The second column is the mass with uncertainties, or the mass range allowed by the observations. Data assembled by K Menou, E Quataert, and R Narayan (1998).

Names	Estimated mass ($M_\odot$)
GRO J0422+32 (XN PER 92)	≥ 9
A0620-00 (XN MON 75)	4.9–10
GRS 1124-683 (XN MUS 91)	5–7.5
4U1543-47	1.2–7.9
GRO J1655-40 (XN SCO 94)	7.02 ± 0.22
H1705-250 (XN OPH 77)	4.9 ± 1.3
GS 2000+250 (XN VUL 88)	8.5 ± 1.5
GS 2023+338 (V404 CYG)	12.3 ± 0.3
0538-641 (LMC X-3)	7–14
1956+350 (CYG X-1)	7–20

The observations in fact usually indicate a mass around $10M_\odot$ for the compact object. This is why astronomers normally assume that black holes formed in stellar systems will typically have this mass. Table 21.1 gives a list of the best candidate black holes in stellar binary systems. Notice that they are all in our Galaxy or in the Magellanic Clouds. That is because X-ray telescopes are not yet sensitive enough to see many X-ray binaries in distant galaxies.

In Investigation 21.3 on the next page we look at the phenomenon of X-ray emission from accretion disks more closely, and learn why the compact objects at

---

Investigation 21.3. X-rays from gas near black holes

When X-ray telescopes above the atmosphere began making observations at photon energies of 1–10 keV, they discovered a host of sources that had not been known at visible wavelengths. To emit substantially at these energies, the temperature must be of the order of this energy divided by Boltzmann's constant k, which gives a temperature of about 10^7 K. Compared to the surface temperature of the Sun (perhaps 5000 K) or of giant stars (cooler still), this is very hot.

When observations revealed that these sources had companion stars, then from optical observations of the stars a distance could be estimated. This in turn allowed estimates of the total luminosity

in X-rays. Typical values (for CYG X-1, for example) are 10^{30} W.

We studied accretion disks in Investigation 13.4 on page 161. Equation 13.9 on page 161 gives the luminosity, and by putting a temperature of 10^7 K into its right-hand side we can estimate that the area of the emitting region is that of a disk of radius only 24 km. This is an extraordinarily small region to be visible when it is so far away! Using Equation 13.8 on page 161 for mass falling onto a black hole of mass $15 M_\odot$ (as in Table 21.1 on the previous page), we find an accretion rate of only 10^{-9} solar masses per year. Clearly this kind of source could last for a very long time.

Exercise 21.3.1: *Accretion disks*

(a) If the spectrum of an X-ray source looks like a black-body spectrum that peaks around 1 keV, show that the associated temperature of the body should be near 10^7 K. (b) If the luminosity of the X-ray source is 10^{30} W, estimate the surface area and effective radius of the region emitting the X-rays. (c) Find the rate at which mass accretes onto the compact object, assuming that its mass is $15 M_\odot$. Express the result in units of solar masses per year.

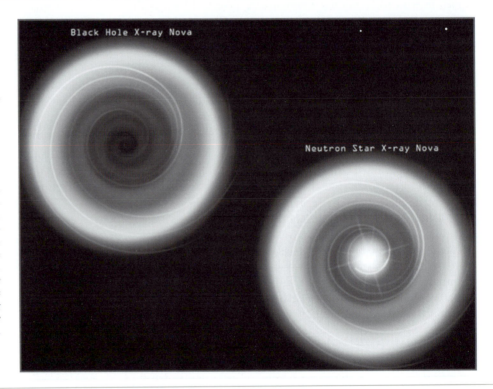

Figure 21.6. *This drawing illustrates how astronomers can tell whether there is a black hole or a neutron star at the center of an accretion disk. If the spectrum of the radiation reveals a particularly hot component, it is likely to be coming from the surface of the neutron star, which is heated by the impact of the accreting matter. If that component is not there, then it is likely that there is no surface at all: the gas is falling across the horizon. The figure refers to an X-ray nova, which is like its white dwarf counterpart discussed in Chapter 13, except that the mass overflows onto a compact object. These systems are ideal for the observations needed here, since it is possible to compare them during outbursts and quiet periods. Figure courtesy NASA/CXC/M. Weiss.*

the centers of these disks must be either neutron stars or black holes.

The signature of the supermassive black hole in MCG-6-30-15

In this section: We look at the evidence that one particular galaxy has a massive black hole in the center. In this case, it is possible to use the spectrum of X-rays to show that there is gas within a radius that is only 1.3 times the gravitational radius. Not only does this establish that the object is a black hole, but it also shows that the hole must be very rapidly rotating.

Most black holes are known because they accrete. The argument that they are actually black holes is usually somewhat indirect. For massive black holes, the argument is often that the object is so compact that it can't be anything but a black hole. If they are of modest mass, they can be distinguished from neutron stars by looking for radiation for the surface of a neutron star, as in Figure 21.6. Increasingly, however, astronomical instruments are getting to be so good that better diagnostics are becoming available. It is even becoming possible to measure the spin of the black hole. As an example, we look here at how astronomers have detected the spin of the supermassive black hole in the galaxy called MCG-6-30-15.

This galaxy has an active nucleus, a kind of mini-quasar, and astronomers have long believed it to contain a massive black hole. One argument for a black hole is that the X-ray emission is variable on very short time-scales. Recently, the XMM-Newton satellite took an X-ray spectrum of this object. An X-ray spectrum is just like a spectrum in visible light: the X-rays are sorted by wavelength, and the spectrum records the intensity of X-radiation over a range of X-ray wavelengths. The spectrum of MCG-6-30-15 is shown in Figure 21.7 on the next page. It looks very different from the spectrum of visible light from a star, such as Figure 10.3 on page 114.

The spectrum contains a number of emission lines, which are wavelengths where the intensity is higher in a small region than elsewhere. This is expected in an accretion disk. These lines arise in this case from ions of heavy elements that have been stripped by collisions of all their electrons, which is normal at the high temperatures in an accretion disk. Occasionally an electron is captured by such an ion and drops into the lowest-energy state, emitting an X-ray photon that carries away the released energy. If the captured electron had no kinetic energy before it was captured, then the emitted photon' energy will be exactly the energy required to ionize the atom. In practice, the electron has some extra kinetic energy, so the emitted photons have a spread of energies above this value. In turn, the emitted wavelengths are spread over a range below a fixed value.

The wavelengths just described are measured in the rest frame of the ion. We would expect to see (at least) the three following different modifications to the emission line when the radiation comes from an accretion disk around a black hole.

1. The lines should experience a gravitational redshift. In the spectrum of MCG-6-30-15 , the lines are strongly redshifted. The redshift is the same for all the measured lines, so it is unlikely that the lines themselves have been misidentified.

2. The lines should be spread out more by the Doppler effect of thermal motion. The gas is hot, so the ions are moving a high random speeds. This is also seen in this spectrum, but not as strongly as the next effect.

3. The lines should be skewed because of the Doppler effect of the rotation of the accretion disk. If we are not looking straight down on the disk, then X-rays from the side of the disk in which material is moving away from us should be redshifted beyond the gravitational redshift, and those from the other side should be blueshifted relative to the gravitational redshift. Without relativity this would be a symmetrical effect, and would just broaden the line. But in relativity there is an effect called **beaming**. Radiation that is emitted isotropically in the rest frame of a particle will come out preferentially in the forward direction when it is moving. So the gas coming toward us in the accretion disk will emit more X-rays toward us than the gas moving away from us does. The result is that the line will be more intense at the shorter wavelengths.

In the case of MCG-6-30-15 the redshift is huge. For the line in Figure 21.7 on the next page labeled O VIII (an oxygen ion), the rest wavelength is about 14.2 Å, or 1.4×10^{-9} m. The center of the line is at about 2.1×10^{-9} m, and this should be roughly where the pure gravitational redshift can be estimated. That is a change by a factor of 1.5. If we use the relativistic redshift formula from the interval for the Schwarzshild black hole, Equation 21.4 on page 288, and remember that the ratio of wavelengths is the reciprocal of the ratio of the frequencies, we find that $(1 - 2GM/rc^2)^{-1/2} = 1.5$, which we can solve to find that $r = 3.6GM/c^2$. But we

▷The interpretation of the X-ray spectrum of MCG-6-30-15 is an excellent example of modern astronomy at work. It shows how astronomers make use of both observations and theory (which they call modeling) to draw conclusions about such exotic objects. It also shows the key role played by technology: these observations were not possible before the satellite that made them was launched. Each new satellite and each advance in ground-based telescopes brings more data and leads to a more secure understanding of the Universe.

Figure 21.7. *X-ray spectrum of radiation from the black hole in the center of the galaxy known as* MCG-6-30-15 *, as measured by the satellite* XMM-*Newton. The dark jagged line represents the measured intensity of the X-rays (photons per second per unit wavelength interval) at the wavelength shown on the horizontal axis (in units of Å, or 10^{-10} m). The lighter line is a fit to a model in which the elements responsible for the broad features are fully ionized oxygen, nitrogen, and carbon. These data indicate that the inner edge of the accretion disk may be at $1.24GM/c^2$. As we discuss in the text, this black hole must therefore be spinning very rapidly. Figure from a paper by G Branduardi-Raymont and collaborators given at the Johns Hopkins University Workshop on X-ray Accretion onto Black Holes, proceedings at the website* `http://www.pha.jhu.edu/groups/astro/workshop2001.` *Used with permission of the authors.*

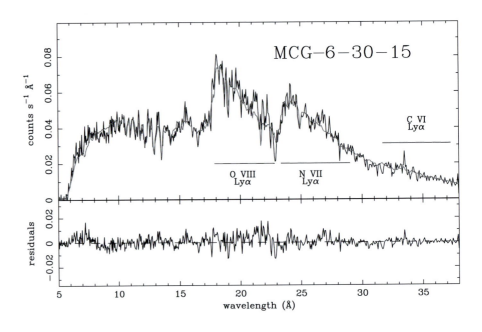

have already seen that accretion disks around a Schwarzschild black hole cannot extend within the last stable orbit at $6GM/c^2$. It can only go further in if helped by the repulsive effect of magnetogravity. So even from a crude inspection of the spectrum, we see that the black hole must be rapidly rotating. We are dealing with a Kerr black hole.

The authors of the paper from which the spectrum shown in Figure 21.7 was taken have done a careful calculation of what radiation to expect from a disk around a black hole with any spin. Their best fit to the data is the smooth line in Figure 21.7. Their estimate is that the inner edge of the accretion disk is at just $1.24GM/c^2$!

▷The astronomers have made a model of the accretion disk that fully uses the Kerr geometry, accounting for all three effects listed above for X-rays emitted from everywhere in the disk. The model even makes a correction for the curved orbits of the photons that reach us after leaving the disk.

This is an astonishingly small radius. If the hole were Schwarzschild, it would be inside the horizon! This black hole must be rotating almost as fast as the extremal Kerr hole. And it must be a black hole: we are, after all, seeing radiation from just outside the horizon.

Wormholes: space and time tubes

In this section: wormholes are present in black hole solutions in general relativity, but they close off faster than anyone can get through them. Physicists speculate on how to keep them open with negative energy, and how to use them for time travel. The object of the speculation is to test the limits of general relativity.

One of the most intriguing aspects of black holes is that there are black hole solutions in general relativity that appear to involve "bridges" to other places. These bridges are really tubes that connect one part of space to another, and have acquired the name **wormholes**. The simplest Schwarzschild black hole involves such a bridge, connecting our space to a totally different three-dimensional space. The character of the Schwarzschild wormhole is illustrated in Figure 21.8.

Unfortunately, this wormhole does not provide a way of communicating with or traveling to the other region of space to which it is connected. Any particle or photon falling into the black hole will reach the singularity at $r = 0$, not the other end of the wormhole. What happens is that the wormhole is dynamical, and it pinches off before anything can get through. The only way to pass through it is to go faster than light.

There is an even more important reason not to get too excited by the Schwarzschild wormhole: it is not present inside black holes that form by gravitational collapse. The reason is that the Schwarzschild solution only describes the *exterior* of the collapsing gas, and the wormhole is a feature of the interior. Inside the gas, there is no reason to expect connections to other parts of the universe to form spontaneously. They are only present in the mathematical solution that describes a black hole that is not formed by collapsing gas, but has existed forever.

Nevertheless, some scientists today are indeed excited by the prospects that wormholes may have some reality. The reason is quantum theory. It appears to be possible to keep a wormhole open long enough to allow a particle to pass through it if one can make *locally negative energy*. No ordinary physical systems have *locally* negative total energy – except in quantum theory.

Quantum theory allows uncertainties and fluctuations that are not allowed in non-quantum physics. Temporary fluctuations can produce photons of negative energy. In order to preserve the total energy, negative-energy photons form in pairs with positive-energy partners. These pairs almost immediately re-combine and disappear, since the quantum theory has to get rid of the negative-energy photon quickly in order to produce macroscopic physics of positive energy. But negative energy does exist for short times, in these quantum fluctuations. Like Planck-mass black holes, local wormholes may be an ingredient of spacetime foam.

More excitingly, physicists speculate that it might be possible to manipulate negative energy to build macroscopic wormholes. If they are able to open up and sustain a wormhole for a short time, then maybe an intrepid astronaut would be able to zoom through it into a different region of space!

Space travel is in fact only the second-most attractive possibility associated with wormholes: **time travel** is the first! Suppose the two ends of the wormhole emerge into the same space, next to each other. By accelerating one of them away from the other, it is possible to use the time dilation of special relativity (Chapter 15) to arrange that a particle that falls into the wormhole would emerge from the other end much earlier, in time to come back to the starting point just as it is about to fall into the wormhole the first time.

Ideas like these were until a few years ago just the province of science fiction. But even though they are now in the realm of serious scientific study, the expectations that scientists have about the importance of the ideas is very different from those of science fiction writers.

Some scientists feel strongly that time travel is a logical contradiction, and will somehow be ruled out by the laws of physics. The possibilities offered by wormholes are not, after all, solutions of the laws of physics, since we don't yet have the correct law of quantum gravity. Instead, they are speculations about the shape that such solutions can take. By studying the features of physical theories that may exclude

▷Actually, the Schwarzschild solution without gas starts at an infinite time in the past with a **white hole**, a time-reversed black hole. This is another reason that the pure-vacuum solution is unphysical; after all, our Universe appears not to have existed for all time in the past!

▷Gravitational potential energy can be negative, but that is not a locally defined energy: it requires a reference to a distant observer. Any local freely-falling experimenter measuring the energy of a particle will get just its rest-mass and kinetic energy, which are positive.

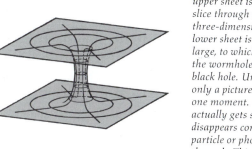

Figure 21.8. The wormhole of the Schwarzschild black hole. The upper sheet is a two-dimensional slice through our own three-dimensional space, and the lower sheet is another space, just as large, to which we are connected by the wormhole in the middle of the black hole. Unfortunately, this is only a picture of the situation at one moment. The wormhole actually gets smaller with time, and disappears completely before any particle or photon has time to get through. This wormhole is not a channel for communication.

such phenomena, scientists hope to learn what some of the basic features of quantum gravity may be.

Hawking radiation: black holes are truly black bodies

Negative-energy fluctuations may be speculative when it comes to making wormholes, but they are well-established in another aspect of gravity: they turn black holes into black bodies.

Recall our discussion of black bodies in Chapter 10. There we saw that any body that absorbs all the light that falls on it is a black body, and when heated to a given temperature T it will give off a characteristic spectrum of radiation. Now, a black hole certainly absorbs all the light that falls on it, so it is a black body. But we have seen that nothing from inside can get out of a black hole, so it would appear that it cannot be a source of radiation. Black holes therefore don't seem to fit comfortably into thermal physics.

However, black-body radiation is a quantum phenomenon: Planck invented his constant in order to describe it. Fittingly, therefore, when the British physicist Stephen Hawking (b. 1942) studied the quantum theory of electromagnetism near black holes, he found that black holes actually emit radiation, that in fact has a black-body spectrum.

How can black holes emit radiation? It should be no surprize that the answer lies in quantum uncertainty. All over spacetime the quantum electromagnetic field is undergoing the little negative-energy quantum fluctuations that we considered above. Normally they are harmless and invisible, because the negative-energy photons disappear as quickly as they form. But near the horizon of a black hole, it is possible for such a photon to form outside the hole and cross into it.

Once inside, it is actually viable: as we remarked earlier, it is possible to find trajectories for photons inside the horizon that have negative total energy. So such a photon can just stay inside, and that leaves its positive-energy partner outside on its own. It has no choice but to continue moving outwards. It becomes one of the photons of the **Hawking radiation**.

> In this picture, nothing actually crosses the horizon from inside to out. Instead, the negative-energy photon falls in, freeing the positive-energy photon. The net result of this is that the hole loses mass: the negative-energy photon makes a negative contribution to the mass of the hole when it goes in.

Once we accept that black holes can radiate, then it is not hard to estimate the wavelength of the radiation that they emit. The only length-scale in the problem is the size of the horizon. A photon with a wavelength λ equal to the radius of the black hole has (ignoring the curvature of spacetime in this simple argument) an energy equal to

$$E = h\nu = h\frac{c}{\lambda} = hc\frac{c^2}{2GM} = \frac{hc^3}{2GM}.$$

If black holes are indeed black bodies, absorbing everything that falls on them and emitting light, then their temperature T should be at least approximately related to this energy by setting $E = kT$, leading to the following estimate of the temperature of a black hole,

$$T = \frac{hc^3}{2kGM}.$$

Now, our argument cannot be expected to be exact, since we had no reason to take the wavelength equal to the radius of the hole rather than, say, its diameter or circumference, and since we must expect that the details of quantum theory and spacetime curvature will not be encapsulated in such a simple dimensional argument.

In this section: remarkably, like other perfect absorbers, black holes actually radiate a black-body spectrum. This is a purely quantum effect. The radiation from astrophysical black holes is undetectably small, but in the early Universe small black holes might have formed that would explode today.

Figure 21.9. Stephen Hawking is one of the most well-known physicists of our time. He has been able to reach the general public with deep questions that are at the current limits of physical theory. He has strongly influenced the development of theoretical physics, deepening the understanding of black holes and making a major step toward quantum gravity with his discovery black holes must emit thermal radiation. Photo courtesy S W Hawking.

▷If you worry about our choice of wavelength here, consider that the uncertainty in the location of a photon is about one wavelength. Very short wavelengths are localized outside the hole and fluctuations can't cross in time. Very long wavelengths hardly notice the hole and have little chance of finding their way inside.

Investigation 21.4. The decay of a black hole

Here we study how long it takes a black hole to lose a significant amount of mass because of Hawking radiation. The temperature of a Schwarzschild black hole given in Equation 21.13 allows us to calculate the luminosity from the standard formula for a black body, Equation 10.3 on page 116,

$$L = \sigma A T^4,$$

where σ is the Stefan–Boltzmann constant, defined in Equation 10.4 on page 116, and A is the area of the surface that radiates. The surface in this case is the horizon, so the area is the area of a sphere with radius $2R_g$. (Recall that the radial coordinate that Schwarzschild used is the one that measures the circumference of the sphere, not the distance to its center. Therefore, we can be confident that the area of the sphere that is the horizon is given by the usual formula for spheres, even if we don't know what space is like inside the horizon.) This gives

$$A = 4\pi(2R_g)^2 = 4\pi(2GM/c^2)^2 = (16\pi G^2/c^4)M^2.$$

Combining this with all the other quantities gives the luminosity of the black hole, and grouping terms in a special way, gives

$$L_{bh} = \frac{1}{30720\pi^2} \frac{ch/G}{M^2} \frac{c^5}{G}.$$

It is instructive to take this expression apart. The first factor is, of course, just a pure number. The second contains, in its numerator,

the quantity ch/G. It is the square of the Planck mass m_{Pl}, defined in Equation 12.20 on page 146, which we have discussed elsewhere in this chapter. So the second factor is dimensionless, being the ratio of the squares of two masses. The third factor is the Einstein luminosity, also discussed in the body of this chapter.

The Einstein luminosity is large, but the black hole only approaches this luminosity when its mass is as small as the Planck mass. For an ordinary hole, the factor in $1/M^2$ reduces the luminosity drastically. For example, a $10M_\odot$ black hole radiates 10^{-30} W!

The lifetime of a black hole can be estimated to be Mc^2/L_{bh}; this is an overestimate, since it assumes the luminosity will be constant in time, whereas it increases. But the increase is gradual, and so the estimate will be accurate to a factor of something like two. (A detailed calculation shows that the true lifetime is one-third of this estimate, which is not much error when we are dealing with such huge times.) For the $10M_\odot$ black hole, this estimate gives 2×10^{78} s, an unimaginably long time!

What is the mass of the hole that will just decay in the age of the Universe, about 10^{10} y, so that if these were formed in the early Universe, we would be seeing their explosions now? Just set the lifetime, Mc^2/L_{bh}, to this value and solve for M. The answer is that the hole should have a mass of about 10^{12} kg. This hole is too small to form today or at any time since galaxies formed, but perhaps in the very early universe conditions were different. There is no observational evidence for such holes, however.

Exercise 21.4.1: *Hawking radiation*

Perform the computations indicated in this investigation. Then find out how much time the hole has remaining when its temperature is high enough to produce electrons in its radiation (this will require kT to exceed $m_e c^2$).

Nevertheless, our answer is only a factor of $8\pi^2$ larger than the one that Hawking found, which is now called the **Hawking temperature** T_H:

$$T_H = \frac{hc^3}{16\pi^2 kGM} = 6 \times 10^{-8} \left(\frac{M}{M_\odot}\right)^{-1} \text{ K.} \tag{21.13}$$

This is so small for stellar-mass and supermassive black holes that it has little relevance to astrophysics. But Hawking's discovery is widely regarded as one of the first real steps toward a quantum theory of gravity. Although we have no such theory, many physicists expect that it must predict the Hawking radiation.

Through this radiation, black holes gradually lose mass. The smaller they get, the higher their temperature goes (by Equation 21.13), so the loss of mass accelerates. In Investigation 21.4 we use our knowledge about black-body radiation to calculate the lifetime of a black hole. For a one-solar-mass black hole, it is about 10^{67} y!

But smaller holes have shorter lifetimes. The mass of a hole that has a lifetime equal to the age of the Universe, about 10^{10} y, is 10^{12} kg. (See Investigation 21.4.) We have seen earlier that holes of this mass cannot form today, but it is conceivable that such **primordial black holes** did form by random fluctuations in the very early Universe.

These primordial black holes would be ending their lives today in an explosion. The amount of energy released in the last second of a black hole's life equals the energy equivalent of the mass of a black hole whose lifetime equals one second. This is a hole of mass about 10^6 kg, which converted into energy gives about 10^{23} J.

The Hawking radiation has linked black hole physics to two other, very different branches of physics: thermal physics and quantum gravity. When an unexpected result makes such links, they must be fundamental. In the next sections we will

▷ The release of this much energy in one second might be observable: it is only a fraction of a percent of the solar luminosity, but it would come out in gamma-rays; this does not explain the observed gamma-ray bursts. They have a luminosity that is up to 10^{22} times larger than this!

look at them further and learn why physicists find the Hawking radiation such a deeply satisfying result.

Black hole entropy: a link to nineteenth century physics

In this book we have discussed many aspects of gas dynamics in astronomy, but we have not yet studied the fundamental concept of entropy. Our study of black holes has led us to the point were it is now time to fill in this gap. The entropy of black holes is a remarkable illustration of the unity of the fundamental concepts of physics across different disciplines.

Entropy fundamentally has to do with measuring how much *information* a system contains. Information is related to order. An ordinary gas is highly disordered, its atoms moving in a random manner that is well described by only a few numbers, such as the density, composition, and temperature of the gas. A crystal lattice, on the other hand, has more structure, and correspondingly requires more information to describe it: the spatial arrangement of the atoms, their separations, the locations of any impurities, and so on. If a system is ordered, then it requires more information to describe it than if it is disordered. Entropy measures disorder. A highly ordered system has *low* entropy, and a messy system has *high* entropy.

Entropy was first introduced into gas dynamics by the German physicist Rudolf Clausius (1822–1888), but he did not associate it with disorder. This fundamental step was the greatest triumph of Boltzmann, whom we met in Chapter 7. He was able to show that his statistical mechanics, from which he could derive the pressure–density relation for gases, could also give a deeply satisfying definition of Clausius' entropy. Basically, Boltzmann showed that one could compute the entropy by counting the number of different ways that the molecules of a gas could be arranged to produce the same overall state of the gas: the same pressure, temperature, and density. This number is huge, of course, and the entropy is essentially the logarithm of it times the Boltzmann constant k.

Clausius had introduced entropy in order to describe heat flow. We have not needed to discuss it before because most of the fluid dynamics we have discussed in this book has been without heat conduction. In astronomy, heat flow is usually a secondary effect. But in engines and other technological systems, heat conduction is central to the function. Clausius originally defined the change in entropy of a system as the heat energy it absorbed divided by the temperature at which it absorbed the heat. When a system does things without losing heat, such as a gas expanding a piston in an idealized non-conducting environment, then the entropy of the gas did not change.

Since systems can gain or lose heat, their entropy can increase or decrease.

> The remarkable discovery of Clausius was that – essentially because heat always moved from high-temperature regions to low-temperature ones – the total change in entropy, summed over all the parts of a system that were exchanging heat with one another, was always *positive*. The entropy of the universe is always increasing.

This could be shown mathematically, but early physicists had no fundamental explanation for it.

Boltzmann showed that this was to do with disorder. Individual systems can get more ordered – I can clean up my desk once in a while, maybe – but the universe as a whole gets more disordered. When I clean my desk I expend so much energy that the entropy of the air in the room and of the chemicals in my body dramatically increase. (That's why I do it so rarely!)

> It is one of the deep mysteries of the world that entropy increases, *disorder* increases, as time goes on. This so-called thermodynamic **arrow**

of time has intrigued physicists for a long time. The universe seems to be continually losing information.

As soon as physicists came to understand black holes, they realized that black holes have an interesting link to entropy. Black holes swallow up lots of information. They are universal wastebaskets. Since they refuse to tell us what has fallen in, they are systems which have the same external state for lots and lots of possible internal states. This suggests that they may have a large entropy. But how big is it?

The first step toward a measure of entropy was a theorem by Hawking that the area of a black hole must always increase, provided energy is always positive. For a Kerr black hole, which is not spherical, the area of the horizon depends on both the mass and the angular momentum. Hawking showed that, even for Penrose-type processes, which extract mass from the hole, they do it in such a way that the area still increases. The Israeli physicist Jacob Bekenstein (b. 1947), then working in the USA, recognized that the area was a kind of entropy function. But was the entropy a function of the area? A multiple of it? Could it be exchanged with other entropies? And what about information and disorder? The Hawking area theorem gave physicists a hint of entropy at the level of Clausius: something had to increase with time, but what did it mean?

The answer came with Hawking's later discovery of the thermal radiation from the hole. This gave physicists a chance to calculate the entropy, since they could then use the classical physics result that the decrease in the energy of a hole through its Hawking radiation, divided by the Hawking temperature, was the decrease in its entropy. The result gave the remarkably simple result that the entropy is proportional to the horizon area A:

$$S_{bh} = \frac{\pi k c^3}{2Gh} A. \tag{21.14}$$

This is a huge entropy compared to that of ordinary objects. When a gas falls into a black hole, we really lose all information about it, and the entropy goes up enormously. When a black hole radiates some energy back into the outside world, the radiation carries its own entropy, so there can indeed be an exchange of entropy between black holes and other physical systems.

> The study of the temperature and entropy of black holes is called black hole thermodynamics. It is remarkable that such exotic macroscopic objects as black holes can fit into the microscopic physics of Boltzmann in such a direct way.

Why should the value of the entropy depend on Planck's constant h, i.e. involve quantum physics? Hawking has offered an explanation: that in a classical gravitational collapse, an outside observer never sees anything cross the horizon because time slows down near the horizon. Classically everything could always be observed, so no information would be lost. However, in a quantum world the hanging material could not be observed forever. Photons are quantized, so that eventually the material falling into the hole would send out its last photon, and then the outside observer would really have lost the information. So information is only lost because of quantum effects.

Black hole entropy: a link to twenty-first century physics

Despite their satisfaction with the unification of black holes with other thermal systems, physicists know that they are still working at the level of Clausius, able to define the entropy of a black hole, but not yet able to describe the link between black hole entropy and information or disorder. Most physicists believe that this link will help them towards another unification, that of gravity and quantum theory.

▷ Here is the let-out for the Hawking radiation, which he discovered some years after his area theorem. We saw above that black holes lose energy and therefore shrink in area because of negative-energy photons falling into them.

In this section: Hawking entropy is seen by most physicists as a key beacon on the obscure road to a quantum theory of gravity. We give a plausible derivation of it.

Let us write Equation 21.14 on the previous page in another way that is suggestive of how quantum gravity might define the entropy. We replace Planck's constant by the appropriate function of Planck length given in Equation 21.10 on page 295. Then the entropy equation can be re-written in the simple form

$$S_{\text{bh}}/k = \frac{\pi}{2} \frac{A}{m_{\text{Pl}}^2}. \tag{21.15}$$

The entropy of a black hole is proportional to the number of "Planck areas" m_{pl}^2 that would cover the area of its horizon. The proportionality is almost unity ($\pi/2$), apart from the requisite factor of k.

Now, we saw that in Boltzmann's statistical mechanics, the entropy of a system is k times the logarithm of the number of different microscopic ways that the given macroscopic system can be constructed. Taking this as a starting point, we might look for a way to count the number of ways a black hole can be made, as a way of calculating its entropy independently of the Hawking radiation. Maybe this number is the number of ways things can fall into the hole, or maybe it is the number of microscopic (quantum) states that would look like the same macroscopic black hole.

To see how this might work, consider the following rather simple approach, along lines originally suggested by Wheeler, whom we met earlier in this chapter. Imagine that in some quantum description of a black hole, the horizon is composed of an ensemble of "gravitons". (We shall discuss gravitons in Chapter 27.) These are presumably massless particles that travel at the speed of light and stay on the horizon. Suppose that the horizon is actually made up of such particles. Each might have an energy comparable to the Hawking energy, i.e. proportional to $1/M$. To make up the black hole mass M, the number N of such particles must be proportional to M^2, or in other words proportional to the area of the horizon.

The entropy of the hole could be the logarithm of the number of ways the horizon could be constructed from such particles. In quantum theory, the particles are not distinguishable from one another unless they have different spin states. Now, it turns out that gravitons can have two different spin states, which correspond to the two independent polarizations of classical gravitational waves that we will learn about in Chapter 22. The number of different ways to build the horizon would be roughly 2^N, with some correction to make the total spin equal to zero. The logarithm of this is proportional to N and hence to the area of the hole.

▷Physicists' attempts to find a fundamental derivation of black hole entropy have met with some success recently in **string theory**, which is one of the strongest candidates today for a method of unifying gravitation theory and quantum theory. Many take this as evidence in support of string theory. This illustrates the guiding role of Hawking radiation in physics today.

> Physicists are trying to make arguments like these more exact, and to ground them better in quantum gravity. For most such scientists, the calculation of the Hawking temperature and entropy is one of the acid tests of any proposed quantum theory of gravity.

This discussion has taken us to one of the frontiers of theoretical research in gravitation theory. We will take up these questions again in Chapter 27, but in order to discuss them adequately we need first examine three further frontiers of research: gravitational waves, gravitational lensing, and cosmology. These are introduced in the next three chapters.

Gravitational waves:
gravity speaks

One of the most radical changes in the behavior of gravity in going from Newton's theory to Einstein's is that Einstein's gravity has waves. When two stars orbit one another in a binary system, the gravitational field they create is constantly changing, responding to the changes in the positions of the stars. In any theory of gravity that respects special relativity, the information about these changes cannot reach distant experimenters faster than light. In general relativity, these changes in gravity ripple outwards at exactly the speed of light.

These gravitational waves offer a new way of observing astronomical systems whose gravity is changing. They are an attractive form of radiation to observe, because they are not scattered or absorbed by dust or plasma between the radiating system and the Earth: as we saw in Chapter 1, gravity always gets through. Unfortunately, the weakness of gravity, which we also noted in Chapter 1, poses a severe problem. Gravitational waves affect laboratory equipment so little that only recently has it become possible to build instruments sensitive enough to register them.

In this chapter we will learn what gravitational waves are, why scientists are confident that general relativity describes them correctly, how they are emitted by astronomical bodies, and what efforts are underway to detect them.

In our tour of the Universe so far, we have repeatedly seen that gravity is the engine at the heart of things, the force that dominates all others and controls stars, galaxies, and (as we will see in the final four chapters) the Universe itself. But gravity does not normally show itself to us directly. Scientists know what it does only because they can observe the photons (and sometimes other particles) emitted by the systems controlled by gravity. From these photons they infer, sometimes after long chains of deduction, what may really be going on inside the systems.

So far, gravity has been a silent engine. Scientists have never yet directly measured gravity from systems outside the Solar System. When gravitational wave detection becomes part of astronomy, astronomers will record in their laboratories the changing gravitational fields produced by some very distant bodies. Gravity will no longer be silent. It will tell us its story directly. Gravity will speak to us.

At this point scientists can only guess what it will say. This chapter looks at the best guesses they make today.

Gravitational waves are inevitable

From our explanation that gravitational waves simply arise from the restriction that no influence can travel faster than light, it is clear that gravitational waves will be a feature of any relativistic theory of gravity. Different theories may differ in the details of the waves, but all theories will have them. In this gravity is not unusual. All physical systems sustain waves: water waves, sound waves, pressure and buoyancy waves in stars, electromagnetic waves, and so on.

In this chapter: we meet the dynamical part of gravity. Gravitational waves are generated by mass-energy motions, carry energy, and act transversely as they pass through matter. Binary systems, involving compact stars or black holes, are the most important sources of detectable waves. The first detections are likely to be made by interferometers now under construction. The low-frequency observing window will be opened after 2010 by the planned international space-based LISA detector.

▷The drawing on this page is of the LISA detector, which is described later in this chapter. LISA is being prepared by ESA and NASA for launch into an independent orbit around the Sun in 2011. It consists of three independent spacecraft using laser beams to track the changes in their separations. It will observe low-frequency gravitational waves from massive black holes in the centers of galaxies and from binary systems in our own Galaxy. From an image provided through the courtesy of the Jet Propulsion Laboratory, California Institute of Technology, Pasadena, California.

In this section: special relativity forces any theory of gravity to have waves, but Laplace speculated about them two hundred years ago.

In fact, it is Newtonian gravity that is strange in this respect, because it does not have waves: when two stars in a binary system move around, their gravitational fields change instantaneously everywhere. So even if an experimenter is millions of light-years away, she could in principle feel the effect of the changing positions immediately, without any delay. This was called action at a distance, and some of Newton's contemporaries found this aspect of his theory disturbing.

Newton may have secretly shared this disquiet, but he was sensible enough not to let it deflect him from developing his theory. If he had tried to include wave effects, he could have hopelessly muddled the theory: experimental physics and astronomy in his day were simply not up to the job of measuring wave effects in gravity, and the whole theory might have been in trouble. Newton kept things as simple as he could.

The first physicist who seriously tried to work out the consequences of assuming that gravity might act with a delay and carry waves seems to have been Laplace. But his idea was that gravity was a kind of fluid, which emanated outwards from its source (such as the Sun) at a finite speed.

Laplace calculated that friction between the fluid and a planet would cause the planet's orbit to shrink. But he decided that, since no such shrinkage had been observed, the speed of gravity had to be large, in fact much larger than the speed of light. Laplace went no further with his speculations.

From the modern perspective, Laplace's speculations were impressive. He was on the right track: he wanted a finite speed and he looked for the right physical effect, orbital decay. What led him to the wrong conclusion is that he had the wrong model for gravity. The evidence, particularly electromagnetic theory and special relativity, that led Einstein to general relativity, was simply not available to Laplace. Given what he knew, he took a very modern point of view.

Knowing a little more than Laplace knew, we can already guess some of the properties of real gravitational waves. For example, in ordinary materials, the stiffer the material, the faster the wave speed. Since gravitational waves will travel with the fastest possible speed – the speed of light – it follows that space itself is effectively the stiffest possible "material". In stiff materials, it takes a lot of force and energy to make a small disturbance, so we can expect that gravitational waves will have small amplitudes even when created by major events, like supernova explosions, and that they will carry large energies in their small amplitudes. We will see in this chapter how all of these guesses work out in Einstein's theory.

Transverse waves of tidal acceleration

Just as the gravity of the Moon can be detected directly on the Earth through its time-dependent tidal forces (Chapter 5), so too are gravitational waves detectable through the time-dependent tidal accelerations they produce. The difference is that the force of the Moon comes from the curvature of time, whereas we will see that gravitational waves carry time-dependent spatial curvature.

> The only part of the gravitational field of a wave that we can measure directly is the non-uniform part, which acts in such a way that one section of an apparatus is affected by gravity differently than another. We can therefore only register the *differences* in gravitational acceleration across the region occupied by our experiment.

In Chapter 5 we saw that the effect of the Moon on the Earth was to deform it from a sphere into an ellipse. Gravitational waves act in a similar way on objects they encounter, but relativity changes some of the details.

▷ Laplace is the same physicist who was among the first to suggest that there could be black holes (Chapter 4)!

▷ The faster the waves travel, the weaker would be their effect on planetary orbits. In the limit of infinite speed, we get back to Newton's theory, where there is no orbit shrinkage.

▷ We will see below that the measurement of orbital decay in a different system has proved that gravitational waves exist.

In this section: gravitational waves act in the plane perpendicular to their direction of travel.

Gravitational waves produce tidal accelerations only in directions *perpendicular* to the direction they are traveling in. In general in physics, waves come with two types of action, producing motions either along or across the direction of motion.

Sound waves move air molecules back and forth in the same direction as the wave travels: this produces the compression and rarefaction of the air that constitutes sound. Physicists call sound waves **longitudinal**: they act a*long* the wave direction. By contrast, waves on the surface of water are **transverse**: the water moves up and down as the wave moves across the surface. Electromagnetic waves are also transverse.

Gravitational waves similarly act transversely. We shall show that this is a consequence of the property of general relativity (inherited from Newtonian gravity) that spherical motions produce no changes in the gravitational field (Chapter 19). Suppose that a gravitational wave that encounters a slab of material acts on the material by producing alternating compression and rarefaction (by its tidal forces) along the direction of motion, just as does a sound wave. Then imagine taking a film of this and running it backwards in time. Any physical process run backwards also satisfies the basic equations of physics, so it is a possible (if unlikely) event. In this time-reversed film, the density of the material oscillates and *produces* gravitational waves.

> This is a key concept: whatever action a gravitational wave has on matter is also the motion by which matter produces gravitational waves.

Now let us shape the material that produces waves into a perfect spherical shell. We arrange in some way that the shell oscillates in thickness, so that the density oscillates with time. Then this motion will produce gravitational waves, in our hypothesis. The waves must be spherical and will go outwards away from the shell, as well as inwards. But the mathematical theorem in general relativity that was quoted in Chapter 19 does not allow this: any spherical source produces a time-independent gravitational field outside it. Therefore, gravitational waves cannot act longitudinally. They must act transversely, like electromagnetic waves.

How gravitational waves act on matter

The analogy with electromagnetic waves breaks down when we consider the geometry of the way in which gravitational waves act on matter. Electromagnetic waves carry oscillating electric fields that make electrons move back and forth along a line, and the direction of the line is called the direction of the **polarization** of the wave. Gravitational waves are different. They produce deformations in the transverse plane that turn circles into ellipses, qualitatively similar to those we saw in Figure 5.2 on page 41. However, the deformation produced by the Moon is partly directed towards the Moon (the longitudinal direction), whereas gravitational waves are transverse.

Their action is illustrated in the top line of Figure 22.1 on the following page. The deformation ellipse that is produced by a wave has the same area as the original circle, so we say that gravitational waves in general relativity are *area-preserving*. Only two polarizations are illustrated, because only two are needed. The second is obtained by rotating the first by 45°. Any other action of a gravitational wave in the same plane can be described by combining these two.

Since the accelerations are *tidal*, the shape of the deformation is independent of the size of the original circle. If the circle were twice as large, the tidal accelerations across it would be twice as large, and the displacement these forces produce would be twice as large, leading to an ellipse with exactly the same shape, the same ratio of major to minor axes. Therefore, the measure of the strength of the gravitational

▷ You can prove that light is a transverse wave by using Polaroid, the semi-transparent material that is used in some sunglasses. If you take two pieces of Polaroid and place them over one another, then if they are oriented correctly they will pass about half the light through that falls on them. But if you rotate one piece by 90°, then the two pieces together will completely block all the light. This proves that light acts *across* its motion, because the rotation of one piece of Polaroid does not change anything in the longitudinal direction.

▷ The proof given here also works in electromagnetism to show that electromagnetic waves are transverse. This is because there is the same theorem for electric fields: a time-dependent spherical distribution of a given amount of electric charge has a static electric field outside it.

In this section: gravitational waves produce transverse tidal accelerations that deform circles into ellipses of the same area.

▷ To convince yourself that light acts along a line, look again at the experiment with Polaroid described in the margin earlier. When you rotate the second Polaroid sheet by 90°, then no light gets through. A further rotation by 90° restores transmission. The kind of geometrical object that is turned into itself by a 180° rotation is a line.

▷ In the case of light, there are also just two polarizations as well, but as the experiment shows they are obtained by rotating the line of action of light by 90°.

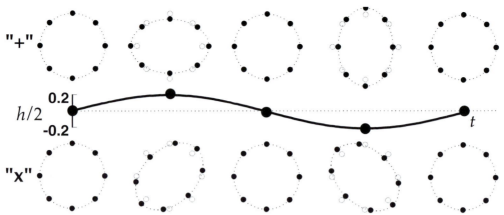

Figure 22.1. *The two lines of circles show how gravitational waves act in general relativity by producing relative changes in proper distances between nearby free particles. They turn circles into ellipses of equal area, and there are two independent ways of doing this. These are the two different polarizations that a gravitational wave can have in general relativity, and they are labeled by their conventional names, "+" for the top line and "×" for the bottom; these represent the orientation of the axes of the ellipses. The middle line of the figure shows the deformation of the ellipse, which is half the gravitational wave amplitude h, so the axis is labeled h/2. It shows the wave as a function of time, and the ellipses top and bottom correspond to the times shown by the dots on the curve. The relative distortions shown are of the order of 20%. This is greatly exaggerated compared to what we expect from real waves, the strongest of which may produce relative deformations of order 10^{-21}. We will discuss the technology of achieving this below.*

wave is this *relative* deformation. If the radius of the circle is called ℓ, and if the maximum displacement along an axis of the ellipse is called $\delta\ell$, then scientists define the *amplitude h* of the gravitational wave to be

$$h = 2\frac{\delta\ell}{\ell}. \qquad (22.1)$$

The factor of two is part of the definition, but we need not worry about why: this is just the definition that physicists have adopted. The relative deformation $\delta\ell/\ell$ itself is called the **strain** induced by the gravitational wave.

This figure shows the way the tidal accelerations produced by a gravitational wave deform a circle of particles if they are free to follow geodesics as the wave passes. (Remember of course that, because these are tidal accelerations, the whole assembly of particles may also have an overall free fall motion.) If the particles are part of a solid body, however, then the resulting deformation will be a result of all the forces, the tidal accelerations and the internal stresses of the material.

The fact that gravitational waves are transverse and do not act like the Moon does on the Earth implies that they are not part of the curvature of time, since that is where the Newtonian forces originate. They are purely a part of the curvature of space. When gravitational waves move through a region they do not induce differences between the rates of nearby clocks. Instead, they deform proper distances according to the pattern in Figure 22.1

Gravitational waves in other theories of gravity can act differently. Waves in theories called scalar theories of gravity are transverse but not area-preserving: the circle changes into a bigger or smaller circle. Some physicists expect that when general relativity is turned into a quantum theory of gravity (Chapter 27), it could become a scalar–tensor theory of gravity, in which case scalar gravitational waves might be present in real observations. So gravitational wave detectors will be trying to measure the pattern of action of any waves they detect.

Early confusion: are gravitational waves real?

The equivalence principle led to considerable misunderstanding and doubt about gravitational waves in the early development of general relativity. Physicists tended to think about waves acting like electromagnetic waves, which accelerate single particles relative to local inertial observers, and gravitational waves do not do that.

Many relativists, including at times Einstein himself, believed that gravitational waves were a mathematical illusion. The complexity of the mathematical formulation of general relativity prevented physicists from working out easily what the energy and momentum carried by the waves was. Some scientists believed that they could carry no energy, that gravitational waves were somehow not the same as waves in the rest of physics.

Fortunately, careful work by many physicists in the 1950s and 1960s clarified both the mathematics and the physics of gravitational waves. The picture I present in this chapter is the result of that work. Waves do carry energy, and we will see later in this chapter that astronomers observe the effects of the wave energy in certain astronomical systems. Just as with black holes, which were also not fully understood until the 1970s, astronomical observation has helped to clarify the physics of Einstein's equations.

I mention the early confusions here but will ingore them from now on. I have tried in the chapters on relativity and its consequences to distill the modern understanding of the theory down to the simplest principles and equations, to help you to see its logic and its physical content. But this understanding is the product of the work of dozens of the twentieth century's best physicists, who took Einstein's amazing baby and grew it to maturity. We are all today standing on the shoulders of those giants.

How gravitational waves are created

Imagine a gravitational wave emitted by a system somewhere, traveling through space, reaching the ring of particles drawn in Figure 22.1, and distorting them. Now imagine the whole process run backwards in time, as if you had taken a film of the wave and you now ran it backwards. As we have noted before, if something happens in physics, then its time-reverse is also possible, in the sense that it does not violate the laws of physics.

In the time-reversed film, the particles in Figure 22.1 move in and out in their elliptical pattern, the wave travels from the particles to what used to be its source, and the "source" moves in some way in response to the arrival of the wave. Now, what sort of motion is possible in the time-reversed "source"? Clearly, it can only move in the way shown in Figure 22.1, since it is responding to a gravitational wave in the time-reversed film. When we go back to the "real" situation, running forward in time, these motions are the ones that create the wave.

> The kinds of motions that give rise to gravitational waves are similar to the motions in Figure 22.1. A source must deform in some kind of irregular way to emit radiation.

In particular, a spherical star that collapses but remains spherical only deforms circles into smaller circles, and this motion will emit *no* gravitational radiation in general relativity.

More particularly, here is how to judge whether and in what direction a source will radiate gravitational waves.

> Look at the source from the desired direction. Since the waves act only in directions transverse to their motion, project the source's (perhaps

In this section: it took decades for physicists to cut through the mathematical complexity of general relativity and establish the physical reality of waves. They were helped by observations performed on the Hulse–Taylor binary pulsar PSR1913+16.

▷ This confusion is a good example of how physics develops. Although Einstein's theory emerged essentially complete in his 1915 papers, questions concerning the physical interpretation of the theory were not fully resolved until the 1970s. Physicists trying to discover the meaning of the theory were handicapped by its complex mathematics, and especially by the absence of any experiments or observations that could tell scientists how gravitational waves behaved. It was even difficult for physicists to understand just how confused they were: some physicists took passionate positions on the subject of gravitational waves, based on what we now know to have been flawed mathematical calculations.

In this section: we use a time-reversal argument to show that the motions in a detector mimic the motions of the source that creates the waves. Therefore the sources of waves are motions such as those in Figure 22.1.

▷ Not all physics has this time-reversibility property. It seems that certain reactions involving elementary particle do not conform: the time-reversal of some pheneomena just do not occur. But in the macroscopic world of astronomy, all the theories of physics allow time-reversed behavior, even if it is wildly *improbable* (like the spontaneous heating of a cup of coffee, due to the cooling of its already cooler surroundings).

complicated) internal motions onto the "plane of the sky", which means onto a plane perpendicular to the line-of-sight to the source. Then only the motions in that plane that are some combination of the motions in Figure 22.1 on page 312 will generate gravitational waves. Moreover, the detector will respond with exactly the same combination of motions: detectors simply mimic the tidal distortions of the source.

We will use this rule when we come to the point in this chapter where we discuss radiation from various astrophysical sources.

▷ Light waves have a very high frequency. For visible light it is about 10^{15} Hz.

The frequency of a gravitational wave is determined by the typical time-scale for things to happen in its source. If the masses radiating the waves move in and out, say, in 1 s, then the waves will have periods near one second and frequencies near 1 Hz. The upper bound on expected frequencies is about 10^4 Hz, because it is difficult to get large astronomical bodies, with masses comparable to the Sun or larger, to do anything on time-scales shorter than a tenth of millisecond or so. There is no lower bound, and in fact scientists are planning detectors that reach down below 10^{-3} Hz, also written as 1 mHz.

The set of all frequencies in a given gravitational wave is called its spectrum. Gravitational wave spectra are like sound spectra. There are some musical instruments that emit single notes, and others that emit thuds, bangs, or crashes. These instruments have counterparts in gravitational wave astronomy. Generally, a wave (either in sound or in gravity) will have a sharply defined frequency (its spectrum will contain a "line") if the motion of the source is regular and periodic or almost-periodic over a long time. Orbiting stars or stars vibrating in their normal modes (as in Chapter 8) emit narrow-line gravitational radiation. By contrast, a system that behaves in an irregular way, or whose radiation is so short-lived that there is time for perhaps one cycle of vibration or motion, emits a broad spectrum of waves, not concentrated sharply near any one frequency. The crashing gravitational collapse of the core of a giant star, which precedes a supernova explosion, could emit a broad spectrum of gravitational waves.

Strength of gravitational waves

In this section: we meet the quadrupole formula for the creation of gravitational waves.

For sources of gravitational waves that are not extremely relativistic, their Newtonian field effectively sets an upper limit on the amplitude of the emitted gravitational waves. If this were not the case, the tidal accelerations produced by the wave in its own source would exceed the source's self-gravitation, and tear the source apart; then we would have no source! For a source of mass M and size R, we should therefore expect that, as the wave leaves the source, $h \leq 2GM/Rc^2$, at least approximately.

Now, in all theories of gravity that have gravitational waves, the strength of the wave decreases as it moves away from the source, and this decrease is proportional to $1/r$, where r is the distance back to the source.

▷ This decrease in proportion to $1/r$ also happens to the amplitude of electromagnetic waves as they move away from a radiating antenna. We will see later that the energy flux carried by a wave is proportional to the *square* of its amplitude. The flux should therefore fall off as $1/r^2$, which is consistent with what we learned in Chapter 9 about light from a star.

Thus, the wave amplitude remains a constant fraction of the overall distortion in the metric produced by the mass of the source.

Combining this with the limit above, we get

$$ h \leq \frac{2GM}{rc^2} \qquad \text{outside the source.} \qquad (22.2) $$

This is already enough to tell us that realistic values of h should be small. For example, suppose we consider a gravitational wave coming from a neutron star in the Virgo Cluster of galaxies. Using a mass of $1.4 M_\odot$ and a distance of 50 million

---

Investigation 22.1. *Energy flux of a gravitational wave*

The flux of energy in a wave can be estimated from a physical argument. The only thing that a full calculation in general relativity is needed for is to get the coefficient in front right.

The energy flux must depend on the amplitude h of the wave, but it must not be simply proportional to h. Since h could be either positive or negative as part of the normal oscillation of the wave, anything that is proportional to h could also be positive to negative. The energy of the wave should not be affected by these regular oscillations. To make sure it carries positive energy, the flux should be proportional to h^2. (It could be h^4 or any other even power of h, but we make the simplest guess.)

Next, the energy flux must depend on the frequency f of the wave. If it were independent of frequency, then a wave with *zero* frequency, which is just a value of h that is constant in time, would have the same energy as a wave of high frequency. But a zero-frequency "wave" should have zero energy, since nothing is changing, no energy is being transported. As with h, the flux must be proportional to an even power of f, since frequencies, like amplitudes, can be negative.[a] Again for simplicity, we guess f^2.

So far we have guessed $F = \alpha f^2 h^2$, where α is a proportionality constant that does not depend on the properties of the wave. What can it be? Surely it will contain some simple numbers, like 2 and π, but it can also contain some fundamental constants of physics. We can expect it to depend on c and G. But we should not expect other numbers, like Planck's constant or the mass of a proton to come into this, since such things are irrelevant to the energy carried by a pure gravitational wave in empty space.

Now we apply dimensional analysis. We know that the flux has units of energy per unit area per unit time, which is $J\,m^{-2}\,s^{-1}$. Since a joule is one $kg\,m^2\,s^{-2}$, the dimensions of energy flux can be written as $kg\,s^{-3}$. Therefore αf^2 must have these units, since h is dimensionless. The dimension of the frequency f, since it is Hz, or oscillations per second, is s^{-1}. So we conclude that our unknown constant α must have dimensions $kg\,s^{-1}$. In particular it does not depend on meters.

These units for α must come from a combination of c and G. The dimensions of c are $m\,s^{-1}$, of G $m^3\,kg^{-1}\,s^{-2}$. The only combination of

them that cancels the dimension of meters is c^3/G, or some power of it. It is easy to see that the dimensions of c^3/G are $kg\,s^{-1}$. This is exactly what we want for α! So we have learned that α is a pure number (dimensionless) times c^3/G.

This is as far as our guessing method takes us. We have no way to guess the pure number. A full calculation in general relativity is required to get it right. The right value is $\pi/4$, which we include in Equation 22.4 on the following page. So even without a full calculation we came very close!

What have we learned from this analysis? We will answer that first by asking what could have gone wrong. Suppose we had not found a combination of c and G that gave the dimensions needed for α. If this had happened, then we would have had to go back to the beginning: we would have had to find a different dependence of F on f (possibly f^4 or f^{-2}?), or we would have had to include some other physical property of the wave (but what is there besides its amplitude h and frequency f?), or we would have had to include Planck's constant h (but that would have forced us to explain why defining energy in classical general relativity needs quantum theory).

Conversely, we might have erred by starting with a different guess for the dependence of F on f: we might have guessed $F = \alpha f^{-2} h^2$, for example. In this case, we would have stopped when we did the dimensional analysis: no combination of c and G would have given something with the units of energy flux. This would have been a clear signal to change the formula.

Altogether, we learn from this that, *if* gravitational waves carry energy, then the flux must be proportional to f^2. Unfortunately, the dependence on h is not constrained by dimensional arguments. There we must rely on the reasonableness of our original assumption.

Ultimately, no guessing argument like this is fully satisfactory. Physicists have other, more deductive ways of defining energy, starting from the fundamental equations of general relativity. Fortunately for us, they give the same answer, and of course they pin down the dimensionless factor $\pi/4$. But for our purposes in this book, it is sufficient to be able to see that the expression for the flux that one gets from these more advanced methods is reasonable.

Exercise 22.1.1: *Dimensional analysis*

Fill in the missing steps above that show that the dimensions of energy flux are $kg\,s^{-3}$. Then show similarly that the dimensions of c^3/G times the square of the frequency are the same.

Exercise 22.1.2: *Size of gravitational wave flux*

We saw that a gravitational wave arriving at the Earth might have an amplitude h as large as 3×10^{-21}. If its frequency is $1000\,Hz$, then calculate the energy flux from such a wave. Compare this with the flux of energy in the light reaching us from a full Moon, $1.5 \times 10^{-3}\,W\,m^{-2}$. Use Equation 9.2 on page 108 to compute the apparent magnitude of the source. Naturally, the source is not visible in light, so this magnitude does not mean a telescope could see it, but it gives an idea of how much energy is transported by the wave, compared to the energy we receive from other astronomical objects.

---

[a] If you are puzzled by the idea of a negative frequency, remember that frequency is the number of cycles of the wave per unit time. If we run time backwards, such as by making a film of the wave and running it backwards, then the number of cycles per unit time also goes backwards, and the wave has a negative frequency. But the backwards-running film shows a normal wave, one that you could have created in the forward direction of time with the right starting conditions, so it must also have a positive energy.

---

light-years, or $r = 4.6 \times 10^{23}$ m, we get $h \leq 6 \times 10^{-21}$. Our argument gives this as an *upper bound* on the strength of waves from such a source, and therefore on the distortions in shape that the wave produces in a detector.

How far below this upper bound do realistic wave amplitudes lie? Clearly this depends on the source. But when motions are not highly relativistic, it is possible in general relativity to make a simple approximation that works very well. The source must be the mass of the system radiating, since both the active gravitational mass and the active curvature mass are dominated by the ordinary mass-energy. But the overall mass of the system is constant and gives rise to the spherical Newtonian field, not to waves. We are looking for the part of the mass-energy that can follow

▷ This upper limit on realistic gravitational waves has set a target for detector developers since the 1960s.

the patterns in Figure 22.1 on page 312. It should not be surprising, therefore, that in general relativity:

> gravitational radiation is produced only by the mass-equivalent of that part of the *kinetic energy* of the source that has the elliptical pattern of Figure 22.1 on page 312 as seen from the direction of the observer of the gravitational radiation.

Written as an equation, the prediction of general relativity is called the **quadrupole formula**. It is similar to the expression for the corrections to the coefficients of the interval that we computed in Chapter 18:

$$h = \frac{8G}{rc^2}\left(\frac{K}{c^2}\right)_{\text{projected elliptical part}}.$$ (22.3)

▷ The notation "projected elliptical part" here means that only the part of the kinetic energy that contributes to source motions similar to those of the test particles in Figure 22.1 on page 312 contributes to the radiation. Each polarization must be treated separately. The factor of eight is not something we can derive here; we must just accept that a full calculation in general relativity justifies it. It takes into account both the mass-energy and pressure parts of the source as well as any gravitational potential energy (the non-linearity of Einstein's equations).

This gives a good approximation for radiation from systems where the velocities are small compared to c.

Einstein was the first to derive the quadrupole formula and yet, as I remarked earlier, he did not always have confidence in it. It took decades for physicists to be sure that it represented a good approximation, especially for realistic systems where gravitational potential energy was comparable with the kinetic energy. There were important contributions from Landau (whom we met in Chapter 20) and his Soviet colleague Yvgeny Lifshitz (1915–1985), and from Chandrasekhar (see Chapter 12), among many others. The subject is still an important area of research today, though not a controversial one. Physicists are developing better and better approximations to the radiation by refining Equation 22.3, in order to be able to recognize and interpret gravitational waves in the observations made by the detectors that we will describe below.

Gravitational waves carry energy, lots of energy

In this section: with the help of an analysis we calcualte the energy carried by a gravitational wave. We see that even weak waves carry huge energies.

Gravitational waves clearly can transfer energy from one system to another. For example, if the particles in Figure 22.1 on page 312 are embedded in a viscous fluid, then their motion will transfer energy to the fluid, and long after the wave is gone the energy will remain. The energy transferred should be small, because we know the waves have great penetrating power.

To find out what energy is carried by the waves requires a small calculation, so it is reserved to Investigation 22.1 on the preceding page. The result, however, is important enough to write down here. Let us consider a **plane wave**. This is a wave from a source that is so far away that the wave passes us with a flat wave front, all parts of the wave traveling in the same direction with the same amplitude h. Suppose in addition that the gravitational wave is a simple sine-wave oscillation with a frequency f (measured in hertz). The appropriate measure of the energy carried by the wave is its *energy flux*, the energy carried by the wave through a unit area per unit time. The formula derived in Investigation 22.1 on the previous page is

▷ We introduced the idea of energy flux in Chapter 9, where we discussed the apparent brightness of stars. The apparent magnitude of a star is a measure of the flux of light energy we receive from it. By analogy, we have here the formula for the energy flux carried by a gravitational wave.

$$F = \frac{\pi c^3}{4G}f^2 h^2.$$ (22.4)

The key point about this formula is that the energy is proportional to the squares of the amplitude and of the frequency. Each of the two polarizations of the wave contributes its own energy, so this formula must be used separately for the "+" and "×" amplitudes.

The constant c^3/G is a very large number, so that even when h is as small as we have found it to be, the flux can be large. In Exercise 22.1.2 on the previous page we find that the flux of energy carried by a gravitational wave can easily be larger than the flux of light energy we receive from a full Moon. Considering that the source of the wave could be in the Virgo Cluster of galaxies, while the Moon is by comparison right next door, it is clear that the emission of gravitational radiation by an astronomical object can be a catastrophic event, carrying away huge amounts of energy.

Because the equivalence principle allows us to wipe out any gravitational field locally, even a gravitational wave, the energy of a wave is really only well-defined as an *average* over a region of space whose size is larger than the wavelength of the wave, and over a time longer than the period of the wave. Extended bodies can therefore only extract the energy if they interact with the wave over a long enough time or a large enough distance.

In the present case, the geometry of spacetime is constantly changing because of the gravitational wave, so energy conservation needs to be treated carefully. Indeed, if we consider just a matter system (such as a detector for gravitational waves), then the waves are an external time-dependent influence on it, and we do not expect its energy to be constant. That is good: one hopes a wave will disturb the detector enough to allow us to measure it! To arrive at a conserved energy that can be exchanged between the detector and the wave, we have to treat the wave and detector together. This is not so easy in general relativity, because it is not easy to define the wave separately from the rest of the geometry.

To see the reason for this, consider water waves. Drop a rubber duck into the still water in a bathtub. Waves ripple out from the place where it lands. We have no trouble distinguishing the waves from the rest of the water, and eventually the waves disappear and we return to the same still water surface as before. By contrast, look at a stormy ocean during, say, a hurricane. Near the beach, what are the waves? Sometimes there is water, sometimes beach. The whole ocean is moving. There is no way to define waves as a disturbance *on* the water.

Strong, time-dependent gravitational fields must be treated with more care in general relativity than we are able to do here. Recall that we learned in Chapter 6 that energy is only conserved in situations where external forces are independent of time. For weak waves, it is possible to define their energy with reference to the "background" or undisturbed geometry, which is there before the wave arrives and after it passes. But if the geometry is strongly distorted, the distinction between wave and background has little meaning. In such cases, physicists do not speak about waves. They only speak of the time-dependent geometry. But normally such regions are small, and outside of them the waves take shape as they move away.

The Binary Pulsar: a Nobel-Prize laboratory

In 1974, two astronomers made a discovery that was finally to give gravitational radiation theory an experimental foundation. The American radio astronomer Joseph H Taylor (b. 1941) had sent his graduate student, Russell Hulse (b. 1950), to observe pulsars with the largest radio telescope in the world, the Arecibo telescope in Puerto Rico. Hulse noticed a signal that appeared to be a pulsar, but strangely its pulse frequency kept changing. He told Taylor, who soon joined him at Arecibo, and together they determined that the pulsar was changing its frequency in a periodic way, coming back to its original frequency every eight hours or so. For a star like a neutron star to change its rotation speed that rapidly seemed impossible, like trying to slow a thundering train. Something else had to be making the frequency change. The conclusion was inescapable: the pulsar was in orbit around another star, with a period of eight hours, and the change in the frequency was simply the Doppler effect as the pulsar went away and came back again in its orbit.

▷The weakness of the influence of the gravitational wave on the Earth shows that little of the energy carried by the wave is left in a detector. This is due to the weakness of gravity itself, not to any lack of energy in the waves.

Figure 22.2. *Joseph Taylor. (Photograph by Robert Matthew provided courtesy Princeton University.)*

Figure 22.3. *Russell Hulse. (Photograph provided courtesy Princeton Plasma Physics Laboratory.)*

In this section: the discovery of the first pulsar in a binary system provided the first experimental confirmation of the theory of gravitational radiation. It has become a test of extraordinary accuracy.

But an 8 h period is extraordinarily short. No binary had ever before been observed with such a short period. Mercury goes around the Sun in 88 days. A satellite of the Sun would have to be just skimming its surface in order to have an orbital period as small as 8 h. But the pulsar was not skimming a star: there was no evidence of friction making the orbit change quickly, and later optical observations of the pulsar's position did not reveal any star, not even a white dwarf. The pulsar, therefore, was orbiting another neutron star or a black hole. Whatever is there radiates nothing we can see. Because this was the first pulsar discovered in a binary system, astronomers began to call it *the* Binary Pulsar. Radio astronomers have subsequently discovered many other pulsars in binaries, so the name is no longer a good one. We shall call it the Hulse–Taylor binary pulsar or simply PSR1913+16.

The orbit of PSR1913+16 is highly relativistic, its speed being about 0.1% of the speed of light. The orbit is, fortunately, a rather eccentric ellipse, so the precession of the perihelion (in this case, it is called the **periastron**) is easy to measure because it is 4° per year. (Compare this to Mercury, where one waits a century or so for the effect to build up enough to measure it accurately!) As we saw in Chapter 18, the precession depends on the mass of the companion, but when (as is the case here) the satellite's mass is not negligible compared to the companion, it is not possible to determine each mass individually from the precession alone.

But Taylor, by repeated careful observations spread over many months, was able to extract another relativistic effect. He could see the change in the pulsar's spin rate as it moved closer to and further from its companion. As we discussed in Chapter 20, this is caused by two effects that act together. The first is the changing gravitational redshift as the pulsar moves in and out in the companion's gravitational field; this redshift affects the spin rate in the same way it would any other clock. The second is the bending of the path of the radio waves as they pass near the companion, which introduces a changing time-delay that adds to the gravitational redshift. The combination of these two effects and the precession allowed Taylor to deduce the masses of both stars, as shown in Figure 22.4. Remarkably, they are both of mass about 1.4$M_\odot$. Today the masses are known to an accuracy of better than 0.1%, the best mass determinations of any objects outside the Solar System.

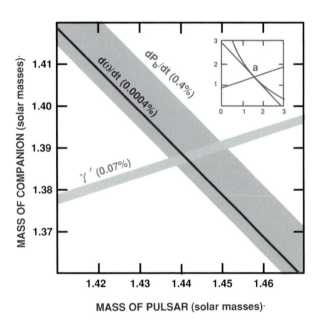

Figure 22.4. *This figure shows the way the masses of the stars in the Hulse–Taylor pulsar system are determined, and how the observed period decrease is consistent with them. The axes are the masses of the two stars, and the lines show how the observed properties of the system depend on the masses. The line labeled γ' is the combined redshift and time-delay term. Any combination of stellar masses on this line would give the observed delay. The width of the line indicates the spread of values allowed by the observations. The extremely narrow line labeled $d\varpi/dt$ is the region allowed by measurements of the periastron shift of the elliptical orbit. The narrowness of this line shows how well this is determined. The broader area around this line, labeled dP_b/dt, is the region allowed by the observed shortening of the orbital period. The fact that all three strips overlap in one region (at masses about 1.39 and 1.44 times the mass of the Sun) is a strong test of general relativity. In another theory of gravity, they need not coincide. The inset figure is the same figure drawn with a larger range of masses. This shows that the curve for the orbital period bends away from the periastron curve over a larger region; if general relativity were not correct then these two curves might not touch at all. Figure courtesy of C M Will.*

Because the companion has a mass in the range of masses of neutron stars, it seems unlikely it could be a black hole: pressure would have halted its collapse. So it is assumed to be another neutron star, but there are no direct observations of it, no pulses of radiation or faint glow of X-rays that might confirm this.

The Hulse–Taylor pulsar is a laboratory for relativity. It confirms the perihelion precession calculated by Einstein to much higher accuracy than Mercury does. It demonstrates the gravitational redshift of a huge clock, showing that the equivalence principle works even for timekeeping by the spin of relativistic stars. All this information is enough to

Example system	Component mass M	Orbit radius R	Distance r	f_{gw} (Hz)	t_{gw}	h	L_{gw} $(L_\odot)$
Hulse–Taylor	$1.4 M_\odot$	1×10^6 km	8 kpc	6.9×10^{-5}	7.4×10^9 y	3.5×10^{-23}	1.5×10^{-3}
NS-NS	$1.4 M_\odot$	50 km	200 Mpc	190	1.5 s	2.8×10^{-23}	4.7×10^{18}
MBH–MBH	$1.4 \times 10^6 M_\odot$	5×10^7 km	4 Gpc	1.9×10^{-4}	1.5×10^6 s	1.4×10^{-18}	4.7×10^{18}

Table 22.1. *Three binary systems of the type that could be detected by ground-based or space-based gravitational wave detectors. For simplicity the systems are assumed to contain equal-mass components in a circular orbit around one another. For each example we specify the masses of the stars, the orbital radius, and the system's distance from us; then we calculate the frequency of the gravitational waves f_{gw} from Equation 22.6 on the following page, the chirp time t_{gw} (orbital shrinking time-scale due to gravitational waves) from Equation 22.12 on page 321, the maximum gravitational wave amplitude h at the Earth from Equation 22.7 on the following page, and the gravitational wave luminosity L_{gw} from Equation 22.10 on the next page. The latter is given in units of the solar luminosity $L_\odot$. For the system in the first line, which is a circular-orbit version of the Hulse–Taylor binary pulsar system, the calculated chirp time is longer than the observed one by a factor of about 12, because of the eccentricity of the real orbit. This brings the stars closer together for a fraction of their orbits, and so the average value of the luminosity is larger. The second and third systems are binaries that have the same compactness, as measured by GM/Rc^2. Notice that they have the same luminosity, despite having very different masses. The more massive system (third line) has a longer lifetime, allowing it to radiate more energy in total. The third system also has the strongest amplitude despite being at a very great distance, where the cosmological expansion redshift is about one.*

tell us everything we would want to know about the orbit.

And on top of all of this, the orbit shrinks. As gravitational waves carry energy away from the orbit, the stars get closer together, and the orbital period decreases. This is exactly the effect Laplace looked for in planetary orbits. General relativity of course provides a prediction for the rate of shrinking, and it has no adjustable numbers in it. Since physicists know the masses and separations of the stars from the other relativistic effects, they can use general relativity to predict exactly how rapidly the period should decrease. We make an estimate of the energy radiated by the system in Investigation 22.2 on the next page, and from it the expected rate of change of the period in Investigation 22.3 on page 321. The prediction is that the period should lose $(2.4427 \pm 0.00005) \times 10^{-12}$ seconds per second. The uncertainty of $\pm 0.00005 \times 10^{-12}$ seconds per second comes from the uncertainties in the deduced masses of the stars. The measurement is that the system is losing $(2.4349 \pm 0.010) \times 10^{-12}$ seconds per second. The uncertainty here is the observational accuracy. The two numbers agree within the uncertainties, as is shown in Figure 22.4.

This is a stringent test of general relativity and a striking confirmation of the predictions of the theory regarding gravitational radiation. For their discovery of this immensely important system, Hulse and Taylor received the Nobel Prize for Physics in 1993. Unlike the case of Jocelyn Bell, to which we referred in Chapter 20, in this case the Nobel committee included the graduate student who first recognized the phenomenon. Perhaps the controversy over Bell's omission was a lesson learned by that committee.

▷The shrinking of the orbit happens because general relativity creates a small *gravitational radiation reaction* force, so named because it is the reaction of the orbit to the loss of energy to gravitational waves. We mentioned this in Chapter 2.

Gravitational waves from binary systems

Although the Hulse–Taylor binary system is radiating gravitational waves with a strength that physicists can compute exactly, there is little hope of directly detecting them in the near future: their frequency (given in Table 22.1) is too low for detectors now being planned, as we discuss later. Nevertheless, other binary systems are the most important gravitational wave sources that the detectors now planned or under construction will search for.

Astronomers now know that there are many other binaries with even shorter periods than the Hulse–Taylor system. A few systems that are known from optical or X-ray observations in our Galaxy have periods that will be detectable by the space-based detector LISA, which we will describe at the end of this chapter. Even

In this section: there is a wide variety of binary systems that could be radiating detectable gravitational waves. Coalescing neutron star and black hole binaries are among the most important targets of ground-based detectors, and a detector in space could obtain important information about a large variety of massive binaries.

Investigation 22.2. *Gravitational waves from the Hulse-Taylor binary pulsar system*

We will calculate here the wave amplitude that we expect from the Hulse–Taylor binary pulsar system PSR1913+16 and the energy it is radiating. The sizes of these numbers may surprise you!

For simplicity, we will consider here only binaries in which the masses of the stars are equal (call this M) and their orbits circular. Then because the two stars have the same mass, they also follow the same orbit, always lying opposite each other on a circle whose radius we will call R.

For a binary system with a circular orbit, the frequency of the gravitational waves is twice that of the orbit. The factor of two arises from the simple fact that, after half an orbit, the stars have replaced each other, and the gravitational field of the system is basically back to where it was at the beginning. So half an orbital period is a full gravitational wave period. This is true even if the stars do not have exactly the same mass, because the source of the gravitational waves is the elliptical asymmetry in the mass distribution, which is the same if one exchanges the two stars.

When we studied circular orbits in Investigation 3.1 on page 22, we found Equation 3.3 on page 22, that the acceleration of a body following a circular orbit of radius R with speed V is $a = V^2/R$. In the binary, this acceleration is produced by the gravity of the other star, which is a distance $2R$ away, so it is $a = GM/(2R)^2$. Setting these two expressions for a equal to one another tells us that the orbital speed is given by

$$V = \left(\frac{GM}{4R} \right)^{1/2}. \tag{22.5}$$

One gravitational wave period is the time it takes for one star to go halfway around the orbit, which is a distance of πR. At the speed V, this takes a time

$$P_{gw} = \frac{\pi R}{V} = \left(\frac{4\pi^2 R^3}{GM} \right)^{1/2}.$$

The gravitational wave frequency is the reciprocal of this:

$$f_{gw} = \frac{1}{2\pi} \left(\frac{GM}{R^3} \right)^{1/2}. \tag{22.6}$$

The amplitude of the radiated gravitational waves depends on the elliptical part of the kinetic energy of the system, projected onto the ellipses in Figure 22.1 on page 312. If we look down the axis of rotation of the orbit, then all the kinetic energy is in the plane of the sky. At one moment in the orbit the stars are moving in the x-direction in opposite senses, and a quarter of an orbital period later they are moving in the y-direction again in opposite senses. This is exactly what the test particles of the "+" pattern (top row) in Figure 22.1 on page 312 do, so all the kinetic energy of the stars contributes to the amplitude for this polarization. For two stars the total kinetic energy is $K = MV^2$, and we can use Equation 22.5 for V. Then we get for the amplitude along the rotation axis:

$$h_+^{axis} = 2 \frac{GM}{Rc^2} \frac{GM}{rc^2}. \tag{22.7}$$

The "×" polarization has the same amplitude up the rotation axis. It must, because the system is executing circular motion, and a simple rotation of 45° changes the "+" pattern into the "×" pattern:

$$h_\times^{axis} = h_+^{axis}. \tag{22.8}$$

If we look at the system from a direction in the equatorial plane, then on average only half of the kinetic energy survives projection onto the plane of the sky, the rest being along the line-of-sight. And that half is only in the plane of the orbit: there is no circular symmetry from this viewing direction. So if we orient the plane along the x-axis in the viewer's coordinates on the plane of the sky, then the "+" amplitude will be half of its value on the axis, and the "×" amplitude will be zero:

$$h_+^{plane} = \frac{GM}{Rc^2} \frac{GM}{rc^2}, \qquad h_\times^{plane} = 0. \tag{22.9}$$

The amplitude expressions are based on a simple product of two terms. One of them, GM/Rc^2, measures how relativistic the system is: how large the gravitational radius is compared to the orbital radius. The second, GM/rc^2, is proportional to the Newtonian correction to the geometry of flat spacetime that produces the curvature of time and space for the Schwarzschild geometry.

The energy flux radiated by the system is given by Equation 22.4 on page 316, into which we can substitute the expressions above for the frequency and amplitude of the radiation to get (along the axis of rotation of the binary)

$$F_{axis} = \frac{\pi c^3}{4G} f_{gw}^2 \left(h_+^{axis} \right)^2 + \left(h_\times^{axis} \right)^2 = \frac{c^5}{2\pi G} \left(\frac{GM}{Rc^2} \right)^5 r^{-2}.$$

In the equatorial plane this is reduced by a factor of eight.

What is of most interest normally is, how much energy is the system losing to gravitational waves? We can find its gravitational wave luminosity L_{gw} by adding up the flux radiated in all directions. If the flux were uniform in all directions, then it would be radiating the same energy per unit area per unit time across all parts of any sphere surrounding the binary. Taking the sphere to have radius r, we would just have to multiply the total area of this sphere, $4\pi r^2$, times the flux to get the luminosity.

The binary is not quite this simple, since the flux varies with direction. So we need to multiply the area of the sphere by the *average* flux. The flux in the equator is only one-eighth of the flux at the pole, but there is much more area near the equator than at the pole, so we might guess that the average flux should certainly be larger than one-eighth times the above expression, but possibly not as much as one-quarter times it. A full mathematical calculation shows that the correct factor is one-fifth. With this factor we get a formula that is actually the correct result for a binary whose orbit is basically governed by Newtonian gravity, despite the roughness of our derivation:

$$L_{gw} = \frac{2}{5} \left(\frac{GM}{Rc^2} \right)^5 L_E, \tag{22.10}$$

where $L_E = c^5/G$ is the Einstein luminosity, introduced in Chapter 21. It is striking how sensitive the binary's luminosity is to how relativistic the system is: the "relativity factor" GM/Rc^2 is raised to the 5th power, so a binary with just twice the orbital radius of another will radiate only $1/32^{nd}$ (about 3%) of the energy. We expect to detect radiation only from the most compact systems.

Exercise 22.2.1: *Working out the algebra*

Fill in the algebraic steps that lead to all the numbered equations in this investigation.

---

more exotic are binaries in which two neutron stars are about to spiral together and form a single object. Two neutron stars will orbit one another hundreds of times a second in the last stages before coalescence, so the radiation will be observable by instruments built on Earth, if they are sensitive enough.

Such coalescing binaries are rare. The Hulse–Taylor system will spiral together about 100 million years from now. It is believed that there are a handful of other

---

Investigation 22.3. *The shrinking orbit of the Hulse–Taylor binary pulsar system*

Where does the energy radiated in gravitational waves come from? The stars themselves are not significantly affected: they retain their mass and size. The energy has to come from the orbital energy. We saw in Chapter 6 that the total energy of the orbit is conserved in Newtonian gravity. But gravitational radiation is not part of Newtonian gravity, and so the energy carried away by the waves results in a slow change in the orbital energy.

The orbital energy consists of two parts, kinetic and potential. They are actually closely related for the binary we are working with. We have seen that for the two stars together, $K = MV^2$. Using Equation 22.5, we get

$$K = GM^2/4R.$$

The potential energy of the two stars is the same as in Equation 6.9 on page 54, with m replaced by the mass of one of the stars and $M_\odot$ by the mass of the other (both of which of course are M), and the radius r replaced by the distance between the two stars in the binary, $2R$:

$$V = -GM^2/2R.$$

The result is that the total binary energy is

$$E = -GM^2/4R. \tag{22.11}$$

This depends on the stars' masses and their orbital radius. The masses don't change as the binary emits gravitational radiation, so that the only thing that can change is R. Since the energy of the orbit must decrease by the amount that is radiated, it must become more negative, or in other words its absolute value must become larger. That means that R must become smaller. As R shrinks, the gravitational wave frequency f_{gw} given in Equation 22.6 increases. The signal is a whistle of gradually ascending pitch, which physicists call a **chirp**.

We can use these equations to deduce a characteristic time for the orbit to change. Let us ask how long it takes to cut the orbital radius R in half, doubling the absolute value of the energy. This means that the energy radiated must be equal to the absolute value of the energy at the beginning of this time. If the luminosity (the energy radiated per unit time) were constant in time, then the orbit-halving time t_{gw} would satisfy the equation

$$L_{gw} t_{gw} = GM^2/4R.$$

Of course, the luminosity is not constant, so this is not exact, but it should still indicate how long we have to wait for a substantial change in the orbit. Using Equation 22.10 for the luminosity, we can solve for this characteristic time:

$$t_{gw} = \frac{GM^2}{4R} \frac{1}{L_{gw}} = \frac{5R}{8c} \left(\frac{GM}{Rc^2} \right)^{-3}. \tag{22.12}$$

This is called the **chirp time** of the binary. It is given by the light-crossing time of the orbit, $2R/c$, times a factor that is a sensitive function of how relativistic the system is. As the system shrinks, the chirp time gets shorter. This means that the chirp time is not very different from the full lifetime of the system: after the system has shrunk by a factor of two, it takes much less than the same amount of time again to shrink another factor of two, and so on until the stars coalesce.

In table Table 22.1 on page 319 we put some flesh on the abstract "bones" of all these formulas and evaluate the important numbers for three different equal-mass circular binary systems: a binary similar to the Hulse–Taylor binary today (but with a circular orbit), in our Galaxy at the distance astronomers calculate for the Hulse–Taylor system; a binary like the Hulse–Taylor binary at the time in the future when it is very near to coalescence, only placed at a distance of 200 Mpc from us, which is a distance where astronomers expect one such coalescence per year; and a binary consisting of two $10^6 M_\odot$ black holes at the center of a galaxy at a distance of 4 Gpc (which corresponds to a cosmological redshift of about $z = 1$). The implications of this table are discussed in the main text.

Now, when observing the gravitational waves, it is not usually possible to measure R or M directly: always measurable are h and f_{gw}, and if the system has a small enough orbital radius then t_{gw} may also be measurable. The properties of the system that determine these numbers are just the values of R, M, and the distance to the system r. These three unknown properties can be calculated if one can measure all three observables, since the three observables depend on R, M, and r in different ways. This leads to a profound result: one can measure the distance to a chirping binary just from the properties of the gravitational wave signal. *Chirping binaries are standard candles.* The distance can be estimated for any system whose orbit changes; it is not necessary to follow it all the way to the point where the stars coalesce.

We have demonstrated only that binaries consisting of equal-mass components in circular orbits are standard candles, but this important property actually extends to all binaries. Observed for long enough, the gravitational waves from any binary contain enough information to tell us how far away it is.

Exercise 22.3.1: *Radiation from example binaries*

Do the calculations that lead to the values in Table 22.1 on page 319 for the orbital numbers and chirp times from the values of M, R, and r given in the table.

Exercise 22.3.2: *Chirp times*

From the chirp time for the system that resembles the Hulse–Taylor pulsar that was calculated in Exercise 22.3.1, work out the rate of change of the period: what fraction of a second does the orbital period lose each second? Compare this with the measured number quoted in the text. Explain the difference. (See the caption for Table 22.1 on page 319.)

---

systems like the Hulse–Taylor system in our Galaxy that are too far away to be seen by today's radio telescopes, but if astronomers want to detect a few such events per year they must survey tens of millions of galaxies. This is a goal of present detector development.

What amplitude of radiation would we expect? From Table 22.1 on page 319 we find that we need to detect waves with amplitudes of a few times 10^{-23}. By comparison, the first laser-interferometer detectors, which are expected to begin operation in 2003, will have an initial sensitivity of around 10^{-21}.

Fortunately, physicists do not have to build detectors that are 100 times more sensitive than the generation now beginning to operate. The American physicist

▷ Remember from Equation 22.1 on page 312 that the amplitude h of a gravitational wave is a dimensionless number.

Figure 22.5. Kip Thorne has had a major influence on the development of astronomers' understanding of black holes and gravitational waves, and he has been a driving force behind the development of interferometric detectors, which we will consider later. In particular he helped to found the LIGO project, which we will discuss later in this chapter. Drawing by Glen Edwards, used with permission.

Kip Thorne (b. 1940) was the first to understand how to benefit from the fact that binary radiation lasts for many cycles and is highly predictable. During the time the waves are in the observable frequency band, detectors will register tens of thousands of cycles of radiation. This will allow scientists to do **pattern matching** on the gravitational wave data, i.e. to look for a weak signal that matches the exact pattern of cycles that are predicted. Events will be reliably identified when they are only about ten times more sensitive than the first ones being built now (2002). This improvement in sensitivity is expected by about 2007, and frequent observations of coalescences of neutron stars can be expected soon after that.

Signals from binary black holes could be five to ten times stronger. But it is much harder to estimate the number of binary black hole coalescences that might occur. As we mentioned in Chapter 14, globular clusters may be efficient factories for binary black holes. It is possible that detectors might see many more coalescences of black holes than of neutron stars, or indeed that the first event detected will be a black hole coalescence. We will have more to say about this kind of observation below.

Merging black holes of larger mass are targets for the space-based detector LISA. We saw in Chapter 21 that the mean density of a black hole goes down as its mass goes up, so the orbital frequency near the horizon also goes down. Waves expected from holes between 1000 and $10^7 M_\odot$ are in the LISA frequency window.

The lower end of this range represents very interesting objects. Computer simulations have suggested that the first generation of stars to form, which were composed purely of hydrogen and helium, had much larger masses than we see in stars today, up to perhaps $1000 M_\odot$. Many or even most of these may have formed black holes, and surely left behind a population of binaries. LISA will be sensitive enough to see any systems in its frequency window anywhere in the Universe.

More massive black hole binaries may form from black holes that are in the centers of galaxies, as a result of galaxy mergers, as we noted in Chapter 14. LISA could again see any merger involving holes smaller than $10^7 M_\odot$ anywhere in the Universe. When one of the holes is much smaller, say $10 M_\odot$, LISA might be able to follow thousands of orbits before the smaller object crosses the horizon of the larger. These orbits would contain detailed information about the gravitational field outside the black hole, and from that information physicists could not only measure the mass and spin of the big black hole but even test the theorem that all black holes must have the Kerr geometry (Chapter 21).

Astronomers have another reason for searching for binary signals from systems whose orbits shrink during an observation. Such systems are said to "chirp", because the frequency of their radiation increases with time. We show in Investigation 22.3 on the preceding page that such systems are *standard candles*: their waveforms encode their distance. If the radiation from chirping binary systems, including coalescing binaries, is observed by enough detectors to deduce the polarization and hence the intrinsic amplitude, then the systems will reveal their distance. This will be particularly interesting for, say, the $1000 M_\odot$ binaries that are seen at the time of the formation of the first stars.

Listening to black holes

Although astronomers are confident that many black holes can be identified using the techniques of optical, radio, and X-ray astronomy, all such identifications are indirect. They rely on electromagnetic radiation emitted by gas near the black hole. Apart from the impossibly weak Hawking radiation (Chapter 21), the only radiation that black holes themselves emit is gravitational radiation. When their horizons are distorted from their normally smooth shape by an interaction with another black

Figure 22.6. *A snapshot of the gravitational waves emerging in the equatorial plane of a computer simulation of the merger of two black holes that have fallen together from a nearly circular orbit. The spiral nature of the waves reflects the orbital in-spiral of the two holes. The tightness of the spiral indicates how rapidly the black holes themselves are moving: since the waves move outwards at the speed of light, the holes must be moving at nearly that speed to wind the spiral so tightly. The waves carry away about 3% of the total mass-energy of the holes. Adapted from an image by W Benger (ZIB), simulation by the Lazarus Project (AEI).*

hole or a star, then the horizon wobbles for a short time, emitting gravitational waves until it settles into its quiescent state. Detecting this radiation, which has a recognizable signature, will be the first direct positive observation of a black hole. Astronomers will be *listening* to the holes themselves.

Detection of such events will not just be awe-inspiring. It will also test general relativity more stringently than any of its tests have done so far. The merger of two black holes to form a single one, with the emission of enormous amounts of gravitational radiation, is about as far from Newtonian gravity as one can get. But to perform this test, scientists have to make independent calculations of what radiation general relativity predicts such a merger to emit.

Such calculations cannot be done with pen and paper! In fact, they can't yet be done as accurately as will be needed, even with the fastest supercomputers available today (2002). But the next generation of computers may be big enough and fast enough to perform accurate simulations of the coalescence of two black holes. Teams of scientists around the world are working intensively towards this goal, and in many ways the work is as difficult and time-consuming as is the effort to build sensitive gravitational wave detectors. There are many teams of scientists collaborating on this problem around the world. Figure 22.6 illustrates a recent computation by one team of the gravitational radiation emitted in the equatorial plane by two in-spiralling black holes of equal mass.

Black hole collisions are pure Einsteinian gravity in action. No matter what kind of a star collapsed to form the hole in the first place, the matter from that star is trapped inside (and probably turned into something like the state of matter at the beginning of the Universe!) and it cannot influence what happens to the outside of the hole any more. So when two holes merge, the result is independent of how they originally formed, and indeed it does not involve any matter at all. It is pure gravity in a vacuum. The merger is pure dynamical geometry.

> It is eerie to think that thousands of stellar black hole mergers take place every year in the Universe, yet every event happens in complete silence apart from the whispers of the emitted gravitational radiation. These waves carry huge energies; a single stellar-mass black hole merger event has a graviational wave luminosity greater than the luminosity in light of thousands of galaxies. Yet no-one has yet seen (or more properly, felt) a single event, and even the stars nearest the merging holes are hardly affected by the changes in gravity. Gravitational waves truly probe what Thorne has called the dark side of the Universe.

Gravitational collapse and pulsars

In this section: other potential sources of gravitational waves include gravitational collapse supernovae and pulsars with irregular mass distributions.

While binaries provide a huge variety of targets for gravitational wave detectors, there are other systems that could also be detected. The two that astronomers discuss most often are gravitational collapse and spinning, irregular neutron stars. Predictions about both are beset by many uncertainties.

Supernovae of Type II are rare and unpredictable events, occurring once in perhaps 50 years in any galaxy. Equally unpredictable is the radiation they will emit, because optical observations tell us little about how non-spherical the collapse and re-explosion will be. The best remedy for this uncertainty is to build detectors with great sensitivity. We have seen that an upper bound on the amplitude of this kind of radiation would be about 10^{-21}, and that is the sensitivity level of the first interferometric detectors, which will begin taking data in 2002. It seems likely, therefore, that first detections of supernovae will have to wait for the second-generation detectors.

However, gravitational wave astronomers must remain alert for such events. Gravitational waves are the form of radiation that will arrive first at the Earth from a supernova. If they can be recognized, then they will provide early warning to other astronomers that a supernova has occurred at a particular position, which should immediately be observed with other telescopes. The optical brightening of a supernova occurs several hours after the interior collapse, and has never been seen from the beginning.

The supermassive black holes in galactic centers may also have formed by gravitational collapse, whose radiation could be detected by lisa. This would, of course, help solve a number of mysteries about the origin of these ubiquitous objects.

Some spinning neutron stars could also be detectable sources. Unlike the narrow pencils of radio waves and light emitted by pulsars, any gravitational waves they emit would not be beamed. But the pulsar can nevertheless give off gravitational radiation if it is non-symmetrical about its axis of rotation.

Imagine a neutron star with a small lump on it somewhere. This could be a crack or deformation in the semi-solid crust of the star. Then as the star spins, the lump executes a circular motion not unlike the motion of the binary stars we examined above, and the radiation coming out will be similar. As for binaries, this radiation carries away energy. The effect would be to slow the pulsar down. Now, all pulsars are observed to be slowing down, but we have seen in Chapter 20 that this would be expected just from the electromagnetic radiation and relativistic particles they emit. Astronomers have no way of estimating how much of the slowdown to attribute to gravitational waves. For any pulsar, the measured slowdown therefore gives an upper bound on the possible radiation from that pulsar. If the Crab pulsar radiates only 1% of its spindown energy loss into gravitational waves, it will be detected by the first generation of detectors within the first month or so.

Such a gravitational wave signal will be steady over long periods of time, so one's ability to find it increases with time, just as for the coalescing binary. However, the time-scale for achieving a detectable amplitude by pattern matching is many months rather than a few seconds. This makes extra demands on a computer-based analysis of the data. The reason for this is that the motion of the Earth, carrying the gravitational wave detector, introduces Doppler shifts into the observed frequency of the gravitational waves, and this exact pattern will have to be matched if the full sensitivity of the detector is to be realized.

One of the more exciting possibilities is the discovery of a previously unknown neutron star, just by the gravitational radiation it emits. For this reason, scientists want to make gravitational wave surveys of the entire sky. To do this, scientists will have to remove the Doppler shifts as for known pulsars. However, they won't know ahead of time the position of the neutron star, so they won't know what pattern to look for. A survey involves looking for all possible patterns. In a data set covering several months, this is such a complex job that it will require very fast supercomputers.

Neutron stars may also emit short-lived bursts of radiation from their normal modes of vibration, for example in the second or so after they are created, before they settle down into a quiescent state. This brief burst of radiation would be rich in information about the interior structure of the neutron star, in much the same way that the normal modes of the Sun have told us much about the solar interior. But these modes would be likely to radiate only weakly, so observations of this type are a long-term goal for more sensitive gravitational wave detector development.

Gravitational waves from the Big Bang: the Big Prize

To my mind the most exciting possibility of all for the new detectors is that they may be able to detect a background of gravitational radiation in space that originated in the Big Bang, the event that started the expansion of the part of the Universe that we can see.

We will learn, beginning in Chapter 24, that the Big Bang also produced a background of electromagnetic waves, with microwave frequencies: the cosmic microwave background radiation. By studying this radiation, scientists have been able to learn an immense amount about the early Universe, about the formation of the elements of which we are made, and about the formation of the galaxies that fill the observable Universe. Most important of all, these observations have given solid support to the idea of a Big Bang in the first place. Detecting gravitational waves from the Big Bang would offer fundamental information of a different kind.

The reason is that the gravitational waves would have been emitted much earlier than even the microwaves. Because gravity penetrates everywhere, the Universe is transparent to gravitational radiation, and has been so from the first moment. Electromagnetic radiation, on the other hand, was trapped and could not move freely in space in the earliest phase of the Big Bang, when matter was so hot that the whole Universe was an ionized plasma. Not until about 300 000 years after the Big Bang did radiation become free. So, the microwave radiation tells us about the Universe when it was a few times 10^5 years old. This is, of course, tiny compared to the present age of the Universe, about 10^{10} years, but it is still much later than the time at which some of the most interesting things happened, as we will see in the final chapters of this book.

▷ Of course, there is no lower bound: there could be no radiation at all!

In this section: the most fundamental observation that gravitational wave detectors could make is to measure random gravitational waves left over from the Big Bang. These would contain the imprint of the laws of physics at energies much higher than scientists can reach in Earth-bound accelerators. However, the task of detecting these waves is not simple.

Gravitational waves, on the other hand, were emitted just when all the interesting and poorly understood physics was happening, within a tiny fraction of a second after the Big Bang. To detect this radiation is to look at the Big Bang itself, and to get our first glimpse of physics at energies higher than physicists can ever hope to reach with Earth-based particle accelerators.

Unfortunately, scientists understand the physics of this very early period so poorly today that they have no strong predictions about the amount of radiation there is. Physicists have come up with very interesting mechanisms that could produce this radiation, and if it were found it would provide insight into physics at very high energies. But these models are speculative. What can be said with confidence is that the radiation today would just be a random background, looking like some extra noise in any one detector. Two detectors would be able to identify it, however, by looking to see if the noisy behavior of one is keeping step with the noisy behavior of the other. But physicists don't know yet whether they can expect this noise to be detectable by the instruments beginning their work now. The goal of detecting this radiation is likely to become more and more important in the design of detectors as they increase in sensitivity.

▷ This is called measuring a correlation between the noisy outputs of the two detectors.

Catching the waves

The last few sections have been a kind of menu of what the Universe might be offering us in gravitational radiation. Of course, any good menu also has the "chef's special", which is a dish you didn't expect when you went to the restaurant. Astronomers have to be alert for such things: in fact, most of the interesting discoveries of the last four decades in astronomy have been things that were not predicted or expected on the basis of prior knowledge. It would be strange if that did not also happen with gravitational waves.

In this section: gravitational wave detectors have been under development since the 1960s, and early claims of detections have been rejected. But the field today benefits from the ground-breaking work done by early researchers.

So there is plenty of motivation for building detectors. But there is also plenty of reason to run away and do something else: trying to measure distortions in any man-made object at the level of one part in 10^{21} or even smaller is not a job to be undertaken lightly. In fact, it is a job that has taken many decades of work by a number of dedicated scientists, building detector prototypes that had only a small chance of detecting anything (such as rare supernovae in our Galaxy), gradually improving the technology until it was ready for the first generation of highly sensitive detectors, the ones that are under construction now.

The first gravitational wave detectors were developed by Joseph Weber (1919–2000) at the University of Maryland in the early 1960s. He considered a number of possible designs, and settled on what was the most practical for the technology of the time: a massive cylinder of metal, isolated as far as possible from external vibrations. When a gravitational wave hits this **bar detector** from its side, it induces a stretching and contraction along its length: just imagine the bar sitting along the horizontal diameter of the top row in Figure 22.1 on page 312. By instrumenting the bar to sense this stretching, Weber hoped to detect gravitational waves.

Weber's bar was one of the most sensitive instruments that had ever been built up to that time, but it was nowhere near the requirement of sensing a relative stretching of one part in 10^{21}. Probably his first bar did not do better than about one part in 10^{14} or so, although his subsequent instruments improved on this by perhaps a factor of 100. However, by the end of the 1960s Weber believed he was actually detecting gravitational waves, which were exciting two of his detectors simultaneously and frequently.

Many physicists responded to this extraordinary situation by building similar detectors. But detector after detector failed to confirm the observation. No other detector group found any significant excess of "events" over what they expected from normal thermal and vibrational noise. Eventually the scientific community concluded that, whatever Weber had seen, they were not gravitational waves.

Many of the groups that built bars like Weber's decided to stay in the field, even though the rewards might be a long time coming. Some of them improved bar detectors by isolating them from external disturbances better and especially by cooling them to liquid helium temperatures (about 4 K), reducing their random vibrations and making weaker gravitational waves easier to detect. Today (2002) there are two bars, both in Italy, that are cooled below 0.1 K. These are called NAUTILUS and AURIGA. Since each bar weighs several tons, it is fair to say that these could be the largest such cold objects that have ever existed, even since the Big Bang! As their instrumentation improves, they should be able to detect broad-spectrum bursts near 1 kHz with an amplitude of perhaps 10^{-20}. Larger solid detectors, with a spherical shape, are now on the drawing boards. They could go a factor of ten or more better than this, breaking through the 10^{-21} barrier.

Such detectors will be expensive to make, and will not be ready until at least the second decade of the twenty-first century. The 10^{-21} barrier will be broken first, not by a bar, but by an interferometer. That is what we will look at next.

Michelson returns: the relativity instrument searches for waves

We have seen in Chapter 15 how interferometers work. An interferometer is designed to compare two lengths and to detect tiny changes in the difference between the lengths. To see how this can be used to detect gravitational waves, look again at our favorite figure in this chapter, Figure 22.1 on page 312. In the top row, place the center of the interferometer at the center of the circle, let the mirror at the end of one of the interferometer arms sit on the circle along the x-axis, and place the other arm's mirror on the circle at the y-axis. Now follow what happens to the lengths of the arms (the distances between the mirrors and the center) as the wave passes. Each arm changes length, but as one is expanding the other is shrinking. So the gravitational wave induces a change in the *difference* between the arms, and this is exactly what the interferometer is designed to sense.

The present interferometer projects grew largely out of decisions by some of the first bar-detector groups to explore interferometry as an alternative. These groups built interferometer prototypes and proved the technology would work. This led to the funding of much larger instruments, which could reach their initial goal of 10^{-21} by 2003. Large size is important for these detectors, to take advantage of the action of tidal forces, which produce length changes proportional to size. Since an interferometer measures changes in length by comparing them to the wavelength of the light it uses, a larger detector will produce a larger signal more easily than a smaller detector.

The most ambitious project is the Laser Interferometer Gravitational-wave Observatory (LIGO) in the USA, which is building two interferometers with arms 4 km long. One is at Hanford, Washington (see Figure 22.7), and the other at Livingston, Louisiana. At Hanford there will be an additional 2 km interferometer within the same system. LIGO could begin taking data of good sensitivity in 2003. On a similar timetable, but smaller in size, is GEO600, a detector being built near Hannover, Germany, by a German–

▷ Weber himself continued to maintain until he died that something interesting had been exciting the detector, but he convinced few other scientists of this. The history of physics has other episodes of a similar nature, where results that are accepted for a time are later discarded, often unexplained. What counts to the majority of physicists is not whether each experiment can be explained: only if its results can be duplicated by other scientists does it demand to be accepted.

In this section: the first detections may be made by interferometers, descendents of Michelson's original instrument. We describe the principle of using these as detectors, the major projects that are building them – LIGO, VIRGO GEO600, TAMA300– and the kinds of challenges involved.

Figure 22.7. The LIGO detector at Hanford, in the US state of Washington. Each arm stretches 4 km, with mid-stations at 2 km. The arms house the world's largest vacuum system, inside of which intense laser beams monitor the separations between mirrors that are suspended in such a way that they are free to move along the direction of the arm in response to a gravitational wave. The system is able to sense motions as small as 4×10^{-18} m if they occur in 10 ms. Image courtesy of LIGO.

British collaboration of scientists. GEO600 is developing more advanced optical technology to achieve a sensitivity similar to that of LIGO with arms that are only 600 m long (see Figure 22.8). This technology, once proved in GEO600, will assist LIGO and the other larger detectors to upgrade their sensitivity later in the decade.

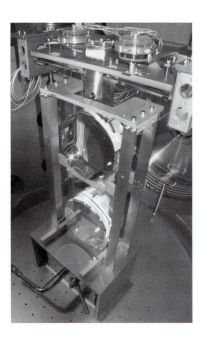

Figure 22.8. One of the mirror assemblies for the GEO600 detector. The bottom piece is the mirror, the upper disk a balancing part of the suspension. The mirrors are made of high-quality fused silica (quartz) and are suspended from fibers made of the same material. The quality of the suspension is such that if the mirror pendulum were set swinging and left alone in the vacuum system, it would swing for many months before the amplitude reduced by half. Image courtesy of GEO600.

Only slightly smaller than LIGO is the single 3 km Italian–French detector VIRGO, being built near Pisa. VIRGO should begin operations about one year after LIGO and GEO600. In Japan the TAMA300 detector in Tokyo, with 300 m arms, is a prototype for a later Japanese detector with 3 km arms, which is planned to leapfrog the initial LIGO and VIRGO detectors and go straight to second-generation sensitivity. There is a proposal in Australia for a detector called AIGO that would have even longer arms than LIGO, but as of 2002 this had not been funded.

Even over 4 km, a disturbance of 10^{-21} translates into a mirror motion of 4×10^{-18} m, less than the size of a proton. Of course, such instruments do not measure the positions of single protons; rather they sense the average position of the surface of a mirror, which can be defined to a high accuracy even though individual atoms move around by much more. However, to sense even an average displacement of the surface of a macroscopic object to this accuracy requires excellent mirrors and high-power continuous wave lasers. It also requires that the light move in a very good vacuum. The LIGO detectors have a vacuum whose volume is larger than any constructed before, even for the big particle physics accelerators. Finally, projects like these require money. LIGO is the largest single scientific project that the National Science Foundation in the USA has ever undertaken.

Developers of these instruments have had to learn how to cope with the same basic sources of instrumental noise as the bar detectors fight against: external vibration and internal thermal motions. These sources of noise set a lower limit on the frequency window at which they can observe. In this window, the sensitivity of an interferometer is limited mainly by how much light is used in the interferometer. This is because the mirror displacements caused by the gravitational wave are much smaller than a wavelength of light (most detectors use infrared light with a wavelength of about 1 μm), and the precision with which they can use light to pinpoint small displacements depends on how many photons are used. Each photon arrives, by the uncertainty principle, with a randomness that makes the interference slightly "fuzzy". The more photons one uses, the less the average randomness is. This fuzziness, called **shot noise**, is the third limiting factor on the sensitivity of interferometers.

The fact that several interferometers are under construction is not redundancy. As we noted above, the data from three or four interferometers observing a given source simultaneously are necessary to locate the position of the source in the sky and determine the polarization of the incoming waves.

This is a field that is developing rapidly. By the time you read this, the major instruments may already be operating. You should keep your eyes on the scientific press for further developments.

LISA: catching gravitational waves in space

We have seen that the frequency range around 1 mHz is very interesting for gravitational-wave astronomy. But it is not possible to observe this low-frequency band from the Earth, no matter how much scientists improve the technology. The reason is that the Earth itself creates gravitational noise that is stronger than the signals astronomers expect from these sources. A gravitational wave detector is simply an instrument that responds to tidal gravitational forces, of whatever origin. When a heavy truck drives past a laboratory, its small gravitational field can be much larger than a weak signal from space. If the vehicle takes 30 s to drive past, then its effect will confuse wave observations in a frequency band of $1/(30\,\text{s}) = 33\,\text{mHz}$.

This has nothing to do with mechanical vibrations: it is gravity itself, and it can't be screened (Chapter 1). The only remedy is to get far from the Earth, because these disturbances get weak very rapidly as one goes further away, while the size of gravitational waves from very distant astronomical sources essentially doesn't change.

Space-based searches for gravitational waves have already been made using communication signals between the Earth and interplanetary space probes. To track spacecraft in the Solar System, space engineers continually send out radio signals to them and receive signals back. If the radio waves were simply reflected from the spacecraft, as from a mirror, they would be too weak to detect when they returned to the Earth. Instead, space probes carry **transponders**, which are systems that receive a radio signal from the Earth and re-transmit back to Earth an identical signal; they effectively act as amplifying mirrors. Now, a passing gravitational wave would affect the time it takes radio waves to travel out to the probes and back; the signature of a gravitational wave in such data is unique, and sensitive searches can be made at very low frequencies. No positive detections have been reported so far, and it is unlikely that sensitivity will be improved in the near future to levels below $h = 10^{-16}$. This is not good enough, as we can see from Table 22.1 on page 319.

In this section: the most spectacular gravitational wave detector being planned is LISA, a space mission to detect low-frequency gravitational waves. Three spacecraft will orbit the Sun and measure tiny changes in their separations. The launch is planned for around 2011.

The LISA detector is designed to do better than this by roughly a factor of 10^6. Developed first for ESA by a team of European and American physicists, the mission is now a joint project of ESA and NASA, scheduled for launch in 2011. LISA will consist of three spacecraft orbiting the Sun in the Earth's orbit, about 20° behind the Earth. They will form a roughly equilateral triangle with arm lengths of about 5×10^6 km, many times larger than the the Sun. Of course the arms would be empty: space is already a good vacuum. But laser light would be transmitted from one spacecraft to another and back, between all three pairs. The three would form two essentially independent interferometers, which would together extract all the desired information from the waves. The arrangement chosen for the spacecraft insures that, as they follow their individual elliptical orbits around the Sun, their pattern remains an equilateral triangle that rotates once per orbit (see Figure 22.9). The mission could make observations for a

Figure 22.9. An artist's view of how LISA would look in orbit around the Sun, about 20° behind the Earth. The three spacecraft follow free orbits around the Sun, chosen so that they remain in formation, always facing the Sun and lying in a plane tilted at 60° to their orbits. The sizes of the spacecraft, Sun, and Earth are, of course, not drawn to scale. Image by Chris Osland and Jonathon Copeland, Rutherford Appleton Laboratory, used with permission.

decade or more.

The technology of LISA is fascinating in itself. The spacecraft will contain small cubes (called proof masses) that are the reference points for the interferometry. These must remain undisturbed so that they respond only to gravitational waves. Accordingly, they will fly freely inside the spacecraft, which will continually sense their position and fire tiny retro-rockets to counteract any external forces on the spacecraft (such as from variations in the radiation pressure of sunlight). The job of the spacecraft is to shield the proof masses so that it does not bump into them!

LISA also does its interferometry in a different way. Its "mirrors" are not reflectors, as in ground-based interferometers. Instead, they are transponders, just like the systems carried by interplanetary space probes to allow them to be tracked from the Earth. On LISA, however, they transpond laser light rather than radio waves.

LISA will make observations that are complementary to those made by ground-based detectors. In fact, LISA has been designed mostly by scientists who also work in ground-based projects. It will have sufficient sensitivity to see waves from black hole collisions wherever they occur in the Universe. It will also be able to do a census of binaries in the galaxy in its frequency range. Moreover, it will also look for a cosmological background of gravitational radiation in its frequency range, with a sensitivity comparable to that of the ground-based detectors, which operate at a much higher frequency. Even if it does not see the background, scientists and the agencies are already studying LISA follow-on missions dedicated to detecting the Big Bang. Gravitational wave astronomy is a field that can be expected to have a long future in space.

23

Gravitational lenses:
bringing the Universe into focus

As we have progressed through the story set out in this book, we have met and begun to understand many of the objects that astronomers regularly photograph: planets, stars, galaxies, supernovae. Astronomical photographs show, in fact, the astonishing variety of objects that make up our Universe. But, to my eye, the most spectacular and entertaining astronomical photographs are fashioned by the objects we will study in this chapter: gravitational lenses. Let's start this chapter with two, shown in Figure 23.1 on the following page. Gravitational lenses are a spectacular illustration of the working of general relativity in the Universe. And besides entertaining us with pictures of eerie beauty, they have become an important tool of astronomy, a way of probing the distribution of mass (and in particular the dark matter) in galaxies and clusters of galaxies.

Lensing can also be a nuisance, of course. Imagine trying to count the number of galaxies that existed at some early time in the evolution of the Universe, to try to pin down the details of galaxy formation, and being confronted with the second photograph in Figure 23.1. How do you know how many images of distant galaxies correspond to only a single galaxy? How do you know if the images are brightened by the lens, so that if the lens had not been there you would not have seen the galaxy in your survey at all? (In fact, the photograph in Figure 23.1 came from just such a project!) Whether lensing is a nuisance or a tool, whether an astronomer wants to remove its effects or use them to study other things, gravitational lensing is important, and it is one of the big research areas in astronomy today.

Pretty obvious, really, ...

Gravitational lensing is a direct consequence of the fact that light is deflected by gravity. The Cavendish–Soldner–Einstein derivation of light deflection, using only the equivalence principle (see the discussion in Chapter 4), is itself enough to establish the principle, although to do quantitative studies it helps to know (Chapter 18) that the deflection predicted by general relativity is twice that due just to the equivalence principle alone. That was established by the eclipse expedition of Eddington and Dyson in 1919.

Nevertheless, gravitational lensing did not become a serious part of astronomy until 1979, when two images of a distant quasar were first identified using radio observations at the Jodrell Bank radio observatory in England. During the intervening years, much of the basic theory had been worked out, but the chances of observing lensing were thought by most astronomers to be small. This was partly because astronomers had at that time no clear idea of how much mass there is in the Universe, and they had no idea how clumpy it would turn out to be. Clumps make better lenses than smooth distributions.

The lack of interest in lensing also reflected the technology of the day. In many lenses the image separation is less than one arcsecond, which is difficult for ground-based optical telescopes to resolve. Moreover, to see lensing of this angular size,

In this chapter: gravitational lensing has become one of the most important tools astronomers have for investigating the true distribution of mass in the Universe, and for measuring the Hubble expansion rate. We study how lensed images form, why lenses produce multiple images (always an odd number), why some are magnified, and how lensing and microlensing are used by astronomers.

▷The image under the text on this page is of the gravitational lens known as B1359+154, which is remarkable for showing six different images of the same distant galaxy. The circled images are the three galaxies that create the lens, at a cosmological redshift of about one. The remaining images represent the same very distant galaxy, which has a redshift of 3.2. (HST image courtesy STScI/NASA.)

In this section: lensing could have been predicted hundreds of years ago. When it was finally observed it came as a surprise.

Figure 23.1. *Two photographs of gravitationally lensed images. The first image is known as the Einstein Cross, or G2237+305: four perfectly symmetrically placed images of a single distant quasar around the central core of a spiral galaxy. The second is a complex and rich cluster of galaxies (the fuzzy objects) called CL0024+1654, that form a lens that produces at least five different images of a single much more distant spiral galaxy. (Left image courtesy Goddard Space Flight Center (GSFC), HST, and NASA. Right image taken by W Colley, E Turner, and T Tyson, HST/STSCI/NASA.)*

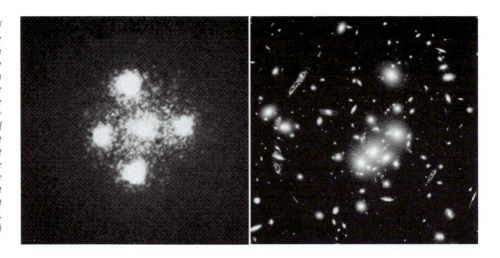

the lensed object should have a very small angular size as well, or all the effects will get washed out in its large image. Until the discovery of quasars in the 1960s, astronomers thought that the only objects distant enough to be lensed by galaxies would be other galaxies, and their large angular size would make lensing hard to identify. But quasars are different (as we saw in Chapter 14): bright enough to be seen even at great distances, they are nevertheless small enough that their appearance is point-like.

It is not surprising, therefore, that the first lensed object to be discovered was a quasar, and that it was discovered with a radio interferometer, which is an array of several radio telescopes that, when observing the same source together, have excellent angular resolution. The two images of the quasar were separated by only 6 seconds of arc. It is also not surprising that the discovery was completely by chance: the observers were just doing a survey of quasars. When they found two quasars unexpectedly close together, they went to an optical telescope and took spectra of both objects. Quasar spectra are like fingerprints: although they all share some general features, no two are exactly the same. In this case, however, the two spectra were identical. Astronomers had their first gravitational lens.

Since that time, hundreds of gravitational lenses have been discovered and studied. Lensing by huge clusters of galaxies and by individual stars smaller than the Sun has been detected. Gravitational lensing is teaching scientists about the dark matter on all length-scales, from within our Galaxy to within clusters of galaxies.

...but not always easy

In this section: the observation and interpretation of gravitational lenses present real challenges.

Although the basic ingredient of an analysis of lensing is just the equation for the relativistic deflection of light in a Newtonian gravitational field, Equation 18.13 on page 235, the use of this formula to understand realistic lenses requires to elaborate calculations on computers. The main problem is that Nature provides us with very complicated lenses to interpret. A telescope maker on Earth spends enormous effort to make the mirrors and lenses of a modern telescope smooth and perfectly shaped to small fractions of the wavelength of light, and then the telescope is used to observe light that has come to us through a bumpy, astigmatic, partly absorbing natural gravitational lens!

The principles of lensing are not difficult, however, and in this chapter we will concentrate on understanding simple lenses. We will discover the peculiar nature of the gravitational lens, divergent in some regions and convergent in others; we will see why lenses magnify objects and make them brighter; we will see why there are

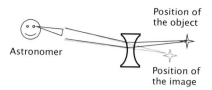

Figure 23.2. *A diverging (concave) lens is placed between an astronomical source and the observer (with telescope). Because the rays are spread apart as well as being deflected, when the astronomer receives them they seem to be coming from a nearer source, in a different direction. The location of this image is drawn with lighter lines. It is the direction the astronomer has to point the telescope in order to see the source.*

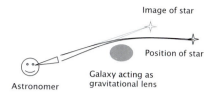

Figure 23.3. *A galaxy deflects light as it passes it on its way from a star to an astronomer. The angular position of the image is further from the lens than the true position of the star. Moreover, because the light ray nearer the galaxy has bent more than the other one, they are diverging faster when they reach the telescope than they would have been if the galaxy had not been in the way. This makes the image appear closer.*

in principle always an odd number of images of any lensed object, although not all of them will necessarily be bright enough to be detected; and we will see how lensing can be used to measure the mass of the lens itself and, possibly, the expansion rate of the Universe.

How a gravitational lens works

The most spectacular photographs of gravitational lenses are of systems with multiple images, as in Figure 23.1, but lensing also operates on single images, and it is easiest to start with that. We will see that a gravitational lens works, most of the time, as a **diverging lens**. To see what that means, we first look at the way a diverging lens made of glass works. In Figure 23.2, we imagine a glass lens between a star and an astronomer. The lens is concave. When a parallel beam of light passes through such a lens, it spreads out. This is why we call it a diverging lens. The spreading of the beam has a very interesting consequence: it leads to magnification. Consider a pencil of light rays emanating from one point on the star in the figure, and passing through the lens before arriving at the telescope. The pencil is diverging anyway, and in principle it would be possible to measure the distance to the star from the angle the diverging rays make. You just trace them back to where they intersect, and that must be where they came from. This is how the parallax method of determining astronomical distances works, as we saw in Chapter 4. This is also how our brains judge distances, using this divergence information as recorded by our binocular eyes.

When the rays pass through the glass lens, they get a bit of extra divergence, as well as an overall change of direction. When they arrive at the telescope, the astronomer infers where they came from by tracing straight lines back along the incoming rays. How this works is shown in the figure. The rays intersect in the "wrong place", of course. This is the *image location*, the place where the lens has fooled us into thinking the object is. Because the lens has made the divergence stronger, the image location is closer than the real location. The object appears closer than it really is. It follows that it looks larger and brighter than it would without the lens: it is magnified.

Now, the same thing happens with a gravitational lens, but with a rather different geometry. In Figure 23.3 we see a single galaxy acting as a lens. Instead of pushing the light away from the axis (the line between the astronomer and the center of the lens), as in the case of the glass lens, gravity pulls the light towards the axis. But the pull of gravity is stronger on the rays that are closer to the axis, closer

In this section: how gravitational lenses form images. How they work as diverging lenses, and why the divergence makes the image brighter.

to the galaxy. So these get bent more, and the result is that the whole beam is given a little extra divergence.

The net effect is the same as with a glass lens: the extra divergence makes the image appear closer, and therefore brighter. The figure shows that the overall bending of the path of the light moves the image *away* from the galaxy that acts as the lens.

Why images get brighter

In this section: we explain why a diverging lens makes images brighter. This applies to glass lenses as well as gravitational lenses.

It is not difficult to see from Figure 23.3 on the previous page that the image should seem closer, but why should it be brighter? After all, the light rays are diverging faster when they reach the astronomer; should not that make the image dimmer?

This apparent paradox is present in the theory of the glass lens, too. Its resolution is to realize that the brightness of the image is not represented by the rays in Figure 23.3. They show the light from one point on the star as it reaches many places in the telescope. The brightness of the image, on the other hand, is determined by the rays that reach a given place on the telescope from different places on the star: how much light do they bring from the star?

As a first step in understanding what happens, we discuss what happens when the observer looks through the telescope at a distant galaxy, rather than a tiny star. Consider a small pencil of rays from the observer that reach the galaxy, as in Figure 23.4. Suppose in fact that the pencil is so narrow that when it reaches the galaxy it covers only a small part of the surface of the galaxy. (These are the rays that, say, will bring the light to one **pixel** of the image of the galaxy on the observer's photographic plate.) Since the lens has made the pencil diverge more than it would have if the lens were not there, these rays intersect more of the surface of the galaxy than if the lens were not there. This tends to bring *more* light into the observer's eye. In fact, it exactly compensates the divergence we noted in the first paragraph of this section: the light from the surface of the galaxy is indeed being spread out more by the lens, so less of it reaches us from any part of the galaxy. But the lens allows us to fit more of the surface of the star into our pencil of rays, with the following net result.

Figure 23.4. The brightness of the image depends on how much light from the source arrives at a point in the astronomer's telescope. Therefore we draw light rays that originate at the astronomer, the opposite of the rays we drew for working out where the lensed image was, in Figure 23.3 on the previous page. The divergence of the rays means that they cover more of the surface of the object (a galaxy in this diagram) than they would have covered if the lens had not been there. This compensates the divergence of the rays from the galaxy, so that the brightness of any small angular part it is the same as if the lens were absent.

> A pencil of rays with a given angular width receives the *same* amount of light from the galaxy regardless of whether the lens is present or absent, provided that the pencil is smaller than the angular size of the galaxy.

Naturally, this is true only if the lens is transparent; we don't worry here about absorption or scattering of the light by the lensing objects.

Therefore in Figure 23.4 we draw the same situation, but we trace rays back from the astronomer to the star. They pass the galaxy and are lensed in exactly the same way, which means they are given a little extra divergence. The effect of this is that when they reach the star, they occupy *more area* on the star than they would have if the galaxy had not been there.

> The extra brightness of the image of the star comes from the fact that more of the star is contributing light to this point at the entrance to the telescope: the image of the star is brighter because more light from it arrives at the telescope than if the lens were not there.

The word that astronomers use for the amount of light received from a piece of the surface of an object into a given angle at the telescope (into a given pixel) is **surface brightness**. We have shown that the surface brightness of an object is unchanged by the lens. This applies to lenses in ordinary optics as well as to gravitational lensing.

But still, why do stars look brighter though a gravitational lens? Have we not just proved that they will be the same brightness? No, we have proved that a piece of the star, covering a given angular size in our observation, will be just as bright as before. But when we consider the whole star, we need to take account of the fact that the size of the image of the star is larger than without the lens, because the diverging lens has made the star appear closer. The star occupies a larger angular size on the sky. Since it has the same surface brightness, we get more light in total from it.

This effect is particularly important for stars, whose angular sizes are so small that astronomers cannot resolve individual parts of their surfaces. In a photograph, it is not possible to tell that the star is in fact larger, since its size is still too small for the telescope to resolve. All we see in the photograph is that there is more light from the star: it is brighter. Galaxies that are big enough to resolve in a photograph, on the other hand, are no brighter in a given area of the photograph than without the lens. They are simply bigger.

You might now ask, what happens to conservation of energy? If there is more light arriving at the telescope from the star, where is this energy coming from? Clearly, the star is not making any extra light, nor is the lens. The extra light in the telescope is light that would have gone elsewhere but is being re-directed by the galaxy into the telescope. Therefore, some other astronomer must be losing the light that should have arrived. Where is he?

The position of the astronomer in Figure 23.4 is not particularly special. Any astronomer will get a little extra light if the light passes the galaxy. The astronomers who lose out are on the other side of the star. The galaxy's gravitational attraction has pulled a little of the light that is meant to go to the right in the diagram and is sending it to the left, and this allows the re-distribution of light that makes the image brighter.

If the lens were more complicated, like the one we are about to look at in Figure 23.6 on the next page, then the situation is also more complicated. Different parts of an image can brighten up at the expense of neighboring parts, as well as at the expense of the unfortunate astronomers in the other half of the Universe.

Making multiple images: getting caustic about light

If you bought a new camera and found, when you had developed the first roll of film, that there were two images in the same photo of your grandmother holding her pet dog, you would feel cheated and you would demand your money back. But when gravitational lenses do this, we all get excited! In fact, we will see that gravitational lenses typically give you *three* images of your grandmother, and in one of them she is left-handed!

In this section we explore one of the extra images, called the second direct image. Its existence is easy to understand and of most interest for applications of lensing. In the next section we will investigate the Einstein ring, an important radius around the lens, which is the key to understand microlensing and the discovery of MACHOS (see page 172). After that we will thread our way through the subtle reasoning that shows that if there is a second image then there is also a third, in which left and right are reversed.

What normally happens is that there is only one image, a little brighter. Other rays, that pass on the other side of the lensing galaxy, do not deflect enough to reach the astronomer. This is illustrated in Figure 23.5 on the following page. Notice,

In this section: we learn that the number of images is related to the way the light rays from the object intersect one another in caustics.

▷Of course, if your grandmother really is left-handed, then in the third image she will be right-handed!

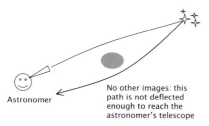

Figure 23.5. *Every galaxy (or any other mass) produces a small deflection of light rays, but for multiple images to form there has to be another path that brings light from the star to the astronomer. In this case, light going around the other side of the galaxy is not deflected enough to reach the astronomer.*

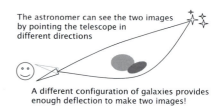

Figure 23.6. *If another galaxy is put into the lens, gravity may now be strong enough to deflect light on a second path from the star to the astronomer. This light arrives from a different direction, so the image looks like a distinct object. But it is just a different view of the same star.*

however, that the rays that pass on either side are approaching each other. They just don't have time to meet before one of them meets the astronomer. If the astronomer goes much further away, then she can find a location where both rays arrive at the same location, and she will see two images. It is important to realize that for gravitational lenses, as for any other kind of optics, the location of the observer is as important for determining what is seen as are the locations of the source and the lens.

Suppose the observer is very close to the lens, so close that there is only one image of a distant star, as in Figure 23.5. The other way to get two images is to modify the lens. In Figure 23.6 we have added another galaxy with a lot more mass to the lens. This bends the light from the star much more on that side of the lens, and directs it toward the astronomer. The astronomer can now see the two images by pointing her telescope in the two different directions.

For the astronomer to see multiple images, she must be sitting in a special region, where light rays emitted from the source in different directions intersect one another. The nice smooth light-cones of special relativity have no self-intersections. But even a small amount of matter in spacetime will distort light-cones enough to make them fold over on themselves. The boundaries of regions in which self-intersections occur are called **caustics**. To see multiple images, there needs to be a caustic between the observer and the lens.

> Figure 23.6 is highly exaggerated, of course. Normally astronomers see images with only small separations, and both images are typically in the same field of view of the telescope. But it is better to draw these exaggerated views when trying to understand the phenomenon, to avoid confusion between the rays.

Then let the observer move further away. When she encounters the first caustic, she will see three images instead of one. In Figure 23.6 it is clear that there are at least two images. We will see why the number actually goes up to three later in this chapter.

If the lens has a complicated shape, she might encounter another caustic if she moves even further away, or to the right or left. At this caustic, two more images appear, giving a total of five. Even more are possible.

The Einstein ring

In this section: we learn about the characteristic size of the lens, called the Einstein radius. For symmetrical lenses, an object directly behind it is lensed into a ring, called the Einstein ring. In more general lenses, all secondary images appear within this radius.

There is an important length-scale in gravitational lensing, called The **Einstein radius**. It is defined by assuming, just for the purpose of the definition, that all the mass of the lensing object is arranged in a perfectly spherical way, and the object being lensed is point-like, is directly behind the lens from the astronomer's point of view, and is at just the right distance for its light to bend enough to reach the observer. This is illustrated in Figure 23.7 on page 338. With this simplified geometry, the light from the point-like "star" is not focused into a single image, but is spread

---

Investigation 23.1. The Einstein radius and ring

Our aim here is to use the geometry of the triangle drawn in Figure 23.8 on the following page to calculate the distance of closest approach b to the lens. The trajectories of the light rays are approximated by straight lines in this calculation, but this is not a bad approximation for rays that have to travel vast distances between star, lens, and astronomer.

The key to the calculation is that light is deflected by the Einstein formula, and this gives the angle θ between the direction the light was going before it encountered the lens and the direction it is going after leaving the lens, given in Equation 18.13 on page 235:

$$\theta = 4GM/c^2 b.$$

This angle is, by the geometry of triangles, the same as the sum of the angles $\alpha + \beta$. Each of these is simply expressed in terms of the distances shown. Since in any reasonable case, the angles will be very small and so b will be very small compared to the other distances, we can use the small-angle approximation to the tangent function and write

$$\alpha = b/D_L, \qquad \beta = b/D_{LS}.$$

Then the equation $\theta = \alpha + \beta$ is

$$\frac{4GM}{c^2 b} = \frac{b}{D_L} + \frac{b}{D_{LS}}.$$

Solving this for b gives the result quoted in the body of the chapter, Equation 23.1. Dividing it again by D_L gives the quoted result for the angle α that the astronomer measures for the ring.

Actually, our calculation is valid only if the lens and source are nearby. If the distances D_L and D_{LS} are large on a cosmological scale (see the next chapter), then the curvature of spacetime means that they can't simply be added together in Equations 23.1 and 23.2 to give the overall distance from the observer to the source of light (the star). Instead astronomers use the symbol D_S for this.

Exercise 23.1.1: *Einstein ring*

Perform the indicated algebra to derive the Einstein radius and its angular size.

---

out into a ring, coming to the astronomer from all directions around the lens. This ring is called the **Einstein ring**. Even though real lenses are not usually so symmetrical, the size of the ring is an important measure for the lens. It is roughly the place where one can expect strong magnification of images.

Nature comes surprisingly close to producing an Einstein ring image once in a while. The Einstein Cross in Figure 23.1 on page 332 is an image produced by a galaxy with a roughly elliptical shape exactly in front of the true position of the quasar. The elliptical shape of the galaxy pushes the light into four images, rather than allowing it to spread evenly over the ring. The radius of the circle on which the images lie is the Einstein radius. In Investigation 23.1 we show, using simple trigonometry, that its radius is, for a lens of mass M,

$$b = \left(\frac{4GM}{c^2} \frac{D_L D_{LS}}{D_L + D_{LS}} \right)^{1/2}. \tag{23.1}$$

All the symbols in this equation are defined in Figure 23.8 on the next page. From the astronomer's point of view, what is important is the angular size of the ring, which (when measured in radians) is this distance b divided by D_L:

$$\alpha = \frac{b}{D_L} = \left(\frac{4GM}{c^2} \frac{D_{LS}}{D_L(D_L + D_{LS})} \right)^{1/2}. \tag{23.2}$$

In this configuration, a great deal of the light from the star is directed toward the astronomer. It is as if there were images of the star all around the ring instead of just in one place. If the astronomer is able to resolve the ring, as in the Einstein Cross, then she will see a ring or, more realistically, a series of images or beads arranged around the ring. Each will be as bright as a single lensed image of the star. If the ring is too small to resolve, then the astronomer will just see a very bright single image.

Now, perfect alignments of the lens with the source and astronomer are rare. So let us ask what might happen if the distant star or quasar is not inside the Einstein radius around the galaxy. Will there be any lensing at all?

Yes, because it is clear that light can approach closer to the lens than the Einstein radius. So if a star is displaced to the side, then to form a second image the light has to bend more on going past the galaxy. It can in principle do this by approaching closer to the galaxy. So stars outside the Einstein radius can still form secondary images if the lens is compact enough. However, because the light ray comes in closer to the galaxy, the apparent position of the *image* will be inside the Einstein radius. Thus, we have the following important result.

> All the secondary images produced by a spherical gravitational lens will be within the Einstein radius, even if the position of the object itself is outside this radius.

Now, of course real lenses are not perfectly spherical, and so our definition of the Einstein radius cannot be exact for more realistic lenses. But it serves as a good guide to the region in which one should look for images. And if you see a number of images of an object, then it is possible to estimate the Einstein radius from them. If these images are distributed around a circle, then you can be confident that the circle is the Einstein ring and the object itself is inside that ring too.

MACHOs grab the light

In this section: microlensing studies, where one looks for lensing by small objects, indicate that the halo of the Galaxy is composed partly of dark, compact objects.

One of the most interesting recent applications of lensing is to the search for the dark matter in the Galaxy. It is possible that at least some of this matter is in large objects, of the mass of Jupiter or larger, but which are not massive enough to initiate nuclear reactions and give off light. These are called brown dwarfs: *brown* because they don't radiate much light, and *dwarfs* because they are small stars. It is also possible that there are other small, dark objects of an unexpected nature, such as boson stars (Chapter 12). The generic, and somewhat whimsical, name that astronomers use for all of these is MACHOs, which we first met in Chapter 14.

Although dark, MACHOs could be detected if they act as a gravitational lens on a more distant star. The idea is that the MACHO would be moving in the halo of our Galaxy, and would pass in front of a distant star by random. The star would brighten up temporarily, and then as the MACHO moved on it would return to its original brightness. To detect MACHOs this way requires extensive monitoring of millions of background stars, waiting for chance events. It also requires painstaking investigation of each candidate event, to insure that it was not caused by something else, such as a variable star.

Let us first look at some typical numbers. If a MACHO is in the halo of our Galaxy, it might be a typical distance of 5 kpc away: $R_L = 1.5 \times 10^{20}$ m. The stars that astronomers are monitoring are either in the central bulge of our galaxy (10 kpc

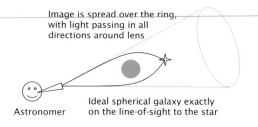

Image is spread over the ring, with light passing in all directions around lens

Astronomer

Ideal spherical galaxy exactly on the line-of-sight to the star

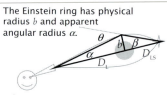

The Einstein ring has physical radius b and apparent angular radius α.

Astronomer

Figure 23.7. *If the star, lens, and astronomer are lined up perfectly, and if the lens is a sphere, then the image will be a ring: light will reach the astronomer equally well traveling around the lens in any direction.*

Figure 23.8. *We can solve, to good accuracy, for the Einstein radius b (the impact parameter of the light) by working with the right triangles shown. See Equation 23.1 and Investigation 23.1 on the preceding page.*

away) or in the Large Magellanic Cloud (50 kpc away), both of which offer large numbers of background stars in a small angular region. If we consider a hypothetic MACHO with a mass of $0.1 M_\odot$, then the Einstein ring has an angular radius of about 1.4×10^{-9} radians, which translates to 0.3 milliarcseconds. (In linear size it is about 80 AU, about the size of the Solar System. This is much larger than we would expect the MACHO itself to be.) This is far too small an angle to resolve with telescopes, so one would not look for the deflection of the image due to the lens. Instead, one would look for the brightening caused by magnification if the Einstein ring of the lens happens to pass over the star. This phenomenon is called **microlensing**.

The mass of the lensing object can be estimated from the duration of the brightening event. The event lasts as long as it takes the Einstein ring to pass over the star, so if we can estimate the speed of the lens then we can calculate the size of the ring and hence the mass of the object. The velocity of the lens can't be inferred directly, but it can be estimated statistically by assuming that the population of MACHOS has a random velocity sufficient to keep them in a halo around the Galaxy, i.e. that they have roughly the circular orbital speed of an orbit at 5 kpc around an object with the mass of the Galaxy.

Recent results from two groups of astronomers who have constructed automatic, computer-controlled telescopes to do these repetitive observations are puzzling. They find MACHOS, but with masses that seem more consistent with $0.5 M_\odot$ than with something below $0.1 M_\odot$. On theoretical grounds, it is hard to understand how an object with as much as half a solar mass could not be a normal star, radiating with enough light to detect. But the lenses are definitely dark. Intensive follow-up observations with large telescopes have not shown any conventional stars where the lenses should be.

The number of detected MACHOS is not large enough yet for the statistics to be good enough to establish the case for the large-mass MACHOS beyond a doubt. But if an explanation based on lenses in the Large Magellanic Cloud does not stand the test of time, then these measurements have revolutionary potential. Physicists are already talking about possible new stable states of matter, boson stars, ways of transforming small white dwarfs into neutron stars or black holes, or something not yet thought of that will produce dark objects with half the mass of the Sun.

The third image: the ghost in a mirror

We are now ready to look at the third image that must be present if there is a second image of the type illustrated in Figure 23.6 on page 336. The existence of this image becomes evident if we ask what happens to rays of light that go *through* the lensing galaxy rather than around it. We shall assume for the moment that the light does go through without being absorbed or scattered. We are interested here in the effect of the geometry of spacetime on the propagation of light.

Because gravity is attractive, light rays will still be deflected in the same direction if they pass inside as if they passed outside. But the acceleration due to gravity begins to get smaller as one approaches the center of the galaxy: at the center it is exactly zero. So the amount of deflection is *smaller* for a ray passing near the center than it is for one further away. The effect of this is that the interior of the galaxy acts like a converging lens rather than an diverging lens.

This is illustrated in Figure 23.9 on the following page, which shows a situation in which the initial direct image is formed by rays that pass through the galaxy. The rays enter the galaxy diverging in the normal way. Their passage through the galaxy reduces their divergence. The image they form at the telescope therefore seems further away than the true distance of the source object.

The more interesting effect is what happens when there are two direct images,

▷The calculation of the MACHO mass depends on assuming that the lensing MACHO is in the halo of our galaxy. But if in fact the lens is nearer to the star, say both in the Large Magellanic Cloud, then the mass required to produce the magnification goes down, because relative to the lens–star separation, the observer is much further away and will therefore require a much smaller deflection angle. So one possible explanation of this observation is that there are plenty of $0.1 M_\odot$ objects in the Large Magellanic Cloud, but not many in our own Galaxy's halo.

In this section: we explain why smooth gravitational lenses always produce an odd number of images, why nearly half of them are mirror images, and why these are usually not seen in astronomical observations.

as in Figure 23.10. Here we see the same situation as earlier, with two direct images formed by a pair of galaxies. Let us first ask a simple question. If there are other stars in the field of view, besides the one we calculated the lensing for, where are they in the image? It is clear from the way the rays pass through the diverging (exterior) part of the lens that the ray coming from the star to the left of the main star arrives at the telescope also displaced to the left of the ray from the main star, and similarly for the star on the right. In other words, the left–right order of the sources is preserved in the image. This is what we mean when we say that an image is a **direct image**.

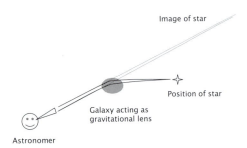

Figure 23.9. *This is the same situation as in Figure 23.3 on page 333, but now the position of the galaxy is slightly different, so the lensed rays have to go through the galaxy to be seen by the astronomer. The ray closer to the center is deflected less than the ray further from the center, so the initial divergence of the rays is reduced. This puts the image location further away.*

Now, it may seem that this is too obvious to spend time on. How could it be otherwise? The original image is also direct: a little experimenting with rays from the other stars will show that it also preserves the left–right ordering. And that is exactly the problem. To see why having two direct images creates a problem, consider sweeping the telescope slowly from the first image position to the second. This is a sweep to the right. For each position of the telescope, trace back the ray that enters it straight in. When the telescope points toward the first image, the ray that goes back to the star follows the upper curve. When the telescope points toward the second image, the ray goes back to the star again, this time taking the bottom path. At other positions, the ray will generally miss the star and go somewhere else. But what we will now show is that it has to hit the star again *somewhere in between the two direct image positions.*

The matter distribution in the galaxies is continuous and smooth, so we should not expect any big jumps in the direction this ray takes far away. It should either pass to the left of the star or to the right of the star, but it won't jump from one side to the other without passing through the star.

So let us begin the sweep. The telescope begins by pointing at the first (main) image. As it sweeps to the right, the ray from the telescope encounters the star just to the right of the main star, because the first image is a direct image. This star appears in the telescope field of view. The sweep continues until the direction of the telescope is near the second image. Just before it points to the second image, the telescope is to the left of that position. As we can see from Figure 23.10, it will be

The astronomer scans the sky near one image.

Figure 23.10. *When the astronomer looks at one of the images, she sees not just the central star but other stars as well if they are near enough. This figure illustrates the fact that a star to the left of the central star has an image that is also to the left, and similarly for a star on the right. The image is to the left because the diverging nature of the lens guarantees that the light from the star on the left will travel closer to the lens, and so it will appear in the image to have come from slightly to the left.*

The astronomer looks at the third image, which goes through the converging part of the lens.

Figure 23.11. *The third image must be formed by light that goes through the converging part of the lens, the mass of the galaxy. This figure shows how a ray from each of the stars reaches the telescope, pointing now at the third image. The rays cross, and the star on the left seems in the image to lie on the right. Because of the convergence, the third image also appears further away and dimmer (not illustrated in this diagram).*

pointing at the star just to the left of the main star, again because the image is a direct one.

But how did this ray get to the left of the main star? After it left the neighborhood of the first image, it was pointing to the right of the main star. To get to the left of the main star again, *it had to pass through the main star somewhere in between the two images.* But if this ray passes through the main star, then the telescope sees the star in that position: there is a third image between the first two.

What is even more remarkable is that, as the telescope scans across the main star in this third image, *the image position is moving to the left as the telescope moves to the right!* This is, after all, how we deduced that there had to be a third image: we had to get the image position over to the left again. This happens as we sweep across the third image.

So the third image is inverted, with left turned to right. If the image were really of your grandmother cuddling her dog, and if she is right-handed, then in this image she is left-handed.

Where does this image form? The only possibility is that it forms in rays that pass through the galaxies. As long as the rays are on one side or the other of the galaxies, they will behave in the direct-image way as the telescope turns. What happens is that, in the galaxies, the convergence of the lens actually makes the rays *cross*, so that the ray from the left winds up on the right. The resulting caustic allows the inversion to happen. This is illustrated in Figure 23.11.

Because it is an image through a converging part of the lens, the third image seems further away, and hence dimmer. Moreover, in real galaxies there is a good chance that the light will be scattered or absorbed as it passes through dust clouds. So the third image will generally be hard to see. We should not be surprised if double images are more common than triple images in observations.

There is no reason to stop at three images. A really complicated lens could produce three or four direct images, maybe more. But an extension of the reasoning we have used here shows that, for every extra direct image, there must be another inverted image. It won't always be easy to detect, but it must in principle be there. As long as the gravitational field of the lensing galaxy is smooth, there will always be an odd number of images in principle.

You may have worried by now that we have only worked in a single plane: our diagrams and the reasoning from them have stayed in the plane defined by the telescope, galaxy, and star. But in reality, the galaxy will be an extended object with complicated structure out of the plane, so might that not have an effect? The answer is, of course, yes. Light rays can be deflected out of the plane, and image forming becomes more complicated.

For this reason, rather sophisticated computer programs are used to study realistic lenses and make maps of the image and lens structures. But the principles are the same as we have discussed here, and no fundamentally new kinds of images arise. In particular, there is a general mathematical theorem that the number of images must be odd, and that, after the first, they come in pairs: one direct, the other inverted.

The only exception to this theorem is if there is a gravitational field without a smooth galaxy or other smooth mass distribution. In Newtonian physics, this would be the field of a "point mass", a particle with zero size but finite mass. This is a mathematical device, but a fiction, so in all realistic Newtonian gravitational fields the theorem about an odd number of images holds. However, in relativity, black holes create a field with no smooth center. Lensing by a black hole does not need to create an odd number of images; the light that would form the odd images gets trapped by the hole instead of passing through to make the image. In fact, lensing by black holes is a fascinating subject, since the light that passes near the horizon can circle the hole more than once before emerging. Images in such circumstances become very distorted!

Lensing shows us the true size of quasars

Gravitational lensing has its main application in astronomy today in the field of cosmology. This is partly because of the numbers: if we take the equation for the Einstein radius, Equation 23.1 on page 337, and put in numbers that are appropriate for cosmological distances, then the angles look rather different. For example, suppose the lensing mass is a dense cluster of galaxies with $M = 10^{14}M_\odot$, at a distance of 100 Mpc, and suppose the lensed object is a quasar twice as far away. (This is nearer than typical lensing systems, but it will serve to illustrate the point.) Then the Einstein radius works out to be about 30 kpc, about three times the radius of our Galaxy. The angular size of this ring in an astronomer's telescope is about 1 minute of arc. This is much bigger than the microlensing rings, and it is an angle that modern telescopes can easily resolve, even from the ground. For this kind of lensing, finding images will be as important as finding brightening.

▷The astronomer Zwicky, whom we first met in Chapter 14, was the first to suggest, in 1937, that lensing by galaxies could be observable. As with his other work on neutron stars and dark matter, he was well ahead of his time.

A picture of lensing, such as the image on the right-hand side in Figure 23.1 on page 332, can therefore yield quantitative information about the system. From the size of the Einstein radius, it is possible to estimate the mass of the lens. By comparing this mass to the mass that one would deduce from the brightness of the galaxies, one can estimate the amount of dark matter in galaxies and clusters. The amount found in this way is consistent with our other estimates, as described in Chapter 14.

The positions of the images are not the only information that astronomers can gather about the lens. If the lensed object is a quasar, it can be expected to vary in brightness by significant amounts on time-scales of days or weeks. These changes will not be seen at the same time in the different images, because the light rays of these images follow different paths. The paths have different lengths and, importantly, they experience different propagation time-delays due to the different gravitational redshifts they experience in the gravitational field of the lens. If one can guess the mass of the lens, say from photographs of the galaxies, then one can calculate the time-delay if one knows how far away the galaxies are. Since the time-delay is measured, one can use it to infer the distance to the lens. By measuring the redshift of the lensing galaxies, one can finally infer the value of the Hubble constant (Chapter 14). This is one of the key methods astronomers use today to measure the expansion rate of the Universe.

Alternatively, if one assumes a value for the Hubble constant, then one can use the measured time-delays to constrain models of the lens and pin down the overall mass and size of the lensing cluster. This is being done to determine the amount of dark matter in lenses, or indeed to discover regions in which there are condensations of dark matter with no visible galaxies at all.

One of the important side-effects of looking for correlated changes in brightness in the Einstein Cross lens system was the discovery that some brightness changes occur only in one image and not in the others. This does not mean that the images come from different sources, i.e. that the lensing model is wrong; the spectra of the four images are too similar for them to come from different objects. Rather, it indicates that another, short-lived lensing phenomenon may be acting. In particular, individual stars in the lensing galaxy will occasionally pass across the image and produce microlensing, just as we have described above. Scientists have shown that the number of observed events is consistent with what one would expect if the lensing galaxy has a population of stars similar to that of our Galaxy. More importantly, the duration of a microlensing event is determined by the time it takes the lensing star to pass across the image, and this in turn depends on the size of the bundle of rays that form the image and the speed of the lensing star. Taking a speed consistent with a star orbiting within the lensing galaxy, scientists have calculated the size of the bundle of rays in the image, and they have then worked backward to calculate the size of the quasar itself.

> Microlensing shows that the quasar in the Einstein Cross has a size no larger than 10^{13} m, or about 70 AU. This is an independent confirmation of the black hole model for quasars, since nothing else can be that small and yet have the required mass.

Weak lensing reveals strong gravity

As telescopes get larger and more sensitive, they can see and resolve the shapes of galaxies that are very far away. They can therefore also see the distortions in shape that are produced by intervening masses. While it is not possible to tell whether any single galaxy image has been distorted or just shows the true and irregular shape of the galaxy, it is possible to do this statistically if one can observe a large number of galaxy images in a single area of the sky.

The reason is that the distortions produced by a gravitational lens have a systematic orientation on the sky. Look again at the right-hand panel in Figure 23.1 on page 332. The distant lensed galaxy is always stretched in the direction along a circle surrounding the lens, and compressed in the radial direction. By contrast, any irregularities in the real shapes of galaxies should be randomly oriented on the sky. So if an image reveals slight distortions of the type seen in Figure 23.1, systematically arranged along circles surrounding a given center, then this is evidence of an intervening lens.

It might seem odd that images get compressed radially and stretched along the circle, since a gravitational lens is a diverging lens, and especially if we recall Figure 5.1 on page 40, which shows that the trajectories of falling objects are stretched apart in the radial direction and squeezed together in the horizontal direction. The situation is no different for light, but the effect of this is to compress the *image* radially, not stretch it. This is clear from Figure 23.4 on page 334, which displays the effect only of the radial compression of the geodesics. The figure shows that the light arriving into a given angle at the telescope covers more of the star than if the lens had not been there. This means that without the lens the image of the star would be larger in this direction. In the horizontal direction the effect is just the opposite.

As this chapter is being written (2002), astronomers are just beginning to employ the power of this statistical method to make estimates of the masses of the dark matter halos of clusters of galaxies. These estimates appear to be consistent with other measures of the dark matter, which is reassuring. Over the next few years, astronomers may be able to produce a complete map of the dark matter distribution in the nearby Universe. This is one more tool for discovering the way galaxies formed as the Universe expanded from its initial singularity. It is time, therefore, for us to turn our attention to the Universe as a whole, and begin to study cosmology.

In this section: by studying the very small distortions of hundreds of images, astronomers are beginning to detect the clumpiness of the dark matter distribution in the Universe.

24

Cosmology:
the study of *everything*

Cosmology is the study of the Universe as a whole. A century ago, scientists had only a vague idea about what even the Milky Way galaxy was like, and they were only able to make guesses about the Universe beyond. Most educated people believed that the nature and history of the Universe were simply matters for religious belief. The word "cosmology" referred to the set of beliefs one had about the whole world: Earth, God, Universe, Creation.

It is one of the most remarkable achievements of modern astronomy that it has turned cosmology into a scientific discipline. In fact, cosmology is one of the most active and productive areas of scientific research today. Stunning pictures taken by the world's most powerful telescopes have shown us what the Universe looks like at very great distances and very early times (Figure 24.5 on page 353), and they have allowed us glimpses of its very early history (Figure 24.6 on page 354). Indeed, as we shall see below, they have brought us the most startling revelation of all: the Universe had a beginning.

Every year new observations bring a greater understanding of how the Universe began and how it evolved, of where galaxies and stars came from and how they led to the evolution of life. Scientists are even beginning to trace changes in the very laws that govern the behavior of the elementary particles, from the beginning of time to the present day. And woven through it all, regulating every step, is gravity.

One of the things that makes the science of cosmology so very interesting is its tendency to stir up controversy. Because the "study of everything" brings astronomy into areas that have been the preserve of philosophy and religion, many questions in cosmology arouse emotional debates that are far more intense than in any other branch of physics.

> The discovery that there appears to have been a beginning to time itself – the Big Bang – has made many people question what their religions teach them about Eternity and Creation. Some scientists try to build bridges between religion and modern cosmology, but this turns out to be a personal exercise, and one scientist's explanation may offer little comfort to another. Some non-scientists seem to have felt so threatened by the notion of the Big Bang that they have rejected the validity of the scientific approach itself.

So cosmology is important to everyday life in a way that other interesting subjects in astronomy, such as the study of the giant black holes in the centers of galaxies, are not. Although its practical applications and commercial "spinoffs" are negligible (see Figure 24.1 on the next page), cosmology has nevertheless become one of the most important branches of modern physics.

Cosmologists believe that they can make progress toward understanding deep cosmological questions – such as the Big Bang and the nature of time – in a scientific

In this chapter: we introduce our study of cosmology. We focus on the measurements that astronomers can make about the Universe as a whole: the Hubble expansion and the acceleration of the Universe. We learn about homogeneity and the Copernican principle, about what the expansion does to space and what is in it, and how to compute the evolution of the Universe.

▷The image under the text on this page reminds us that creation myths and cosmologies were central parts of the belief systems of ancient peoples. It is remarkable how many cultures believed in a beginning of time, a moment of creation. The ancient Egyptians had several creation myths. The Hebrew creation story even orders the events in much the same way that modern science would, although on a vastly different time-scale. More than any other branch of physics, the scientific study of cosmology raises religious sensitivities and addresses questions that have long been regarded the domain of philosophy and belief. Image from a photograph copyright Jon Peck, www.ancient-mysteries.com, used with permission.

way, by the same methods that scientists use when trying to discover what is going on in the centers of galaxies. Like any scientific study, cosmology cannot answer every question, and indeed the understanding of many issues is very incomplete. This leaves plenty of room for religious belief and lively philosophical debate. Also like any other science, cosmology almost inevitably suggests new and exciting ideas, such as the possibility that by studying the Universe as a whole physicists can learn about the laws that govern the innermost workings of protons and electrons. And sometimes the new ideas raise troubling questions, such as what came "before" the Big Bang, and even whether the notion of "before" can make any sense.

Figure 24.1. "My big mistake was going into cosmology just for the money."
This is the wrong way to think about cosmology! Copyright by S Harris, used with permission.

I will return to these issues in the last of our four cosmology chapters, Chapter 27, after we have studied the scientific foundations of cosmology in astronomical observations and in the theory of general relativity. These four chapters on cosmology will have a different character from earlier chapters. Here we confront the limits of our knowledge, the limits of what our theoretical understanding of physics tells us. Studies of cosmology are even today calling fundamental assumptions into question and revealing places where new physical theories are needed. These chapters are a snapshot, at a time of rapid change, of a field in which some of the deepest questions in physics are being asked and at least partially answered. Cosmology is at the sharp end of physics, and inevitably some of what we discuss here is uncertain, tentative, speculative, paradoxical . . . and wonderfully exciting!

What is ''everything''?

Some cosmologists like to think that they are studying the whole Universe, possibly infinite in extent, from the beginning of time to the present. They will tell you that the entire Universe is like this or like that. Don't believe them! There are parts of the Universe that are too far away to see, and all we can do is speculate about them. Let us start our study by being careful about what it is we can and cannot investigate.

As in any other scientific study, we are only able to describe what we can observe, and in cosmology we can only study the part of the Universe that is *near enough* for us to observe. If we accept for the moment that the Universe began a finite amount of time ago, then there is a limit on how far away we can see anything. The most distant parts of the Universe which we can have any hope at all of observing are those regions that, at the time of the Big Bang, emitted photons or gravitational waves which are just reaching us today, having traveled at the speed of light ever since the Big Bang. The past light-cone that stretches backwards in time from our present moment is the boundary of this region, and it is called our **particle horizon**, or simply our horizon. This is illustrated in Figure 24.2. Any region further away (outside the particle horizon) has not yet had time to send us any signals, so we cannot even in principle say anything scientific about it at all.

This is then the province of *scientific* cosmology: the observable Uni-

In this section: cosmology studies the observable part of the Universe, inside the *particle horizon.*

▷The particle horizon is very different from the *event* horizon of a black hole. The event horizon prevents us from learning about what is inside, even in the future. The particle horizon simply tells us how far we can see at the present moment.

verse. Its size increases daily: every new observation could in principle bring information from a region of the Universe that until that moment was too far away to see. In practice, astronomical instruments are not yet good enough to see to these extreme theoretical limits, but in the next couple of decades gravitational wave detectors might achieve this. (See Chapter 22.)

Importantly, it is always possible that new information from a previously unobserved part of the Universe could upset all of our old theories. Tomorrow's observation might reveal a more distant region of the Universe which is, say, infinitely old, which did not experience a Big Bang and a beginning of time. Cosmology is just like all other sciences: new data can destroy old theories.

But cosmologists have a reason for not expecting anything as radical as this to happen to their picture of the Big Bang. They call this reason the **Copernican principle**, or more plainly the **principle of mediocrity**, and it can be tested by observations.

Copernican principle: "everything" is the same "everywhere"

Copernicus simplified our picture of the planets by dropping the assumption that the Earth was specially located at the center of the Solar System. He realized that the Earth was just one of the planets moving around the Sun. Later scientists came to understand that the Sun is just an ordinary star, moving about the center of our Galaxy, and that the Milky Way is a fairly ordinary galaxy, a member of an unremarkable small cluster of galaxies called the Local Group.

> We are, in effect, *mediocre:* we are not at the center of anything, and anyone looking at the Universe from a different location would not immediately be drawn to look at us by any special features of our neighborhood. What we see around us is typical of the Universe as a whole, and it seems reasonable to assume that this is true even of its unseen parts. This assumption is called the *Copernican principle.*

This does not mean that every part of the Universe is as boring as our immediate neighborhood; far from it! The Universe contains quasars, active galaxies, gamma-ray bursters, black holes, and a host of other exotic objects. But there are no more of them anywhere else in the Universe at the present time than there are around us.

The Copernican principle also reassures us that we shouldn't live in fear (or excited anticipation!) of seeing a completely different kind of Universe crossing over our particle horizon the next time we look in our telescopes. If this were to happen, then our moment in *time* would also be very special. This is because we can already see enough of our Universe to know that hypothetical cosmologists at some earlier time (say, a billion years ago on another planet elsewhere in the Milky Way) would have had no chance of seeing big changes to their picture of the Universe in short times.

These expectations are *not* scientific deductions. The Copernican principle is basically a philosophical assumption, but this does not make it unscientific. Scientists often use such assumptions as important guides to the framing of scientific theories, helping them to choose among a multitude of possible avenues to explore. The Copernican principle has the great advantage of simplicity: it tells us that we don't need to waste endless time speculating about what distant regions of the Universe look like. Unless we are forced by new evidence to think otherwise – and no philosophical principle should blind us to hard evidence – we shall assume they look the

In this section: the observable Universe is homogeneous on large scales. Our location in space is not particularly special.

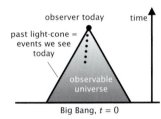

Figure 24.2. The part of the Universe that we can observe, at least in principle, is limited to the regions that have had time to send us light since the Big Bang. The past light-cone of the observer is the boundary of the observable region, called the particle horizon. Tomorrow, when the observer makes another observation, this boundary will have moved upwards and outwards. So the region outside the boundary contains events that could someday become observable. Using light, the observer sees only the events on the light-cone, but the events inside could have sent information at slower speeds. The observer's own history (that of the Earth and the elements from which it was made) is shown as the dotted line. This diagram over-simplifies the structure of the Big Bang itself, but it conveys the correct idea about the boundary of the observable region.

Investigation 24.1. *The rubber-band universe*

Here we look in some detail at how the rubber-band universe behaves, how it produces a Hubble law for people in it. The main difference between this "toy" universe and the real one is dimension: the rubber band is one-dimensional, whereas the real world is three-dimensional. But if we look in a fixed direction in our three-dimensional world, we will see a one-dimensional strip that behaves just like our rubber band.

The first point, and the most crucial, is that the rubber band is the whole universe to its inhabitants. They cannot leave it, nor can light signals or any other kind of physical means of transporting information. When one dot on the band looks at another dot, it looks along the band, not in a short-cut across it. That would take it into another dimension, outside of its one-dimensional universe. While this is not a hard rule to visualize in the case of a rubber band, it applies equally strongly to us in our real Universe. When we come to discuss the geometry of possible universe models in Chapter 26, it is important to bear in mind that we can in principle only measure within the three dimensions of the space, not into any extra dimensions that we may have to use for visualizing the model.

Now suppose that the circumference C of the band behaves in some arbitrary way as a function of time, denoted by $C(t)$. Then all separations between dots will change in exact proportion. So if dots 1 and 2 are separated by a distance $d_{1,2}(0)$ at time $t = 0$ (when the rubber band is relaxed), then at any other time t their separation will be

$$d_{1,2}(t) = \left(\frac{C(t)}{C(0)} \right) d_{1,2}(0).$$

Let us call the ratio of circumferences the *scale-factor* $R(t)$ of this universe:

$$R(t) = \left(\frac{C(t)}{C(0)} \right).$$

Then the equation governing the separation of dots along the band is

$$d_{1,2}(t) = R(t)\, d_{1,2}(0). \qquad (24.1)$$

Now, the average speed of separation of the dots, $v_{1,2}$, as measured along the band by either of them, will be the change of distance, $d_{1,2}(t) - d_{1,2}(0)$, divided by the time t. This gives

$$v_{1,2} = \frac{d_{1,2}(t) - d_{1,2}(0)}{t} = \frac{R(t) - 1}{t} d_{1,2}(0). \qquad (24.2)$$

We have derived Hubble's law here: the speed of separation is proportional to the original separation $d_{1,2}(0)$. The rubber band is a good model of an expanding or contracting Universe.

It is important that we have not imposed any particular stretching law: $R(t)$ is an arbitrary function of time. But we have insisted that the change of length takes place uniformly along the band, so that the relative distances remain the same. This is the equivalent of assuming the Universe is homogeneous.

Now suppose that the change is actually a contraction at constant speed v, so that

$$C(t) = C(0) - vt. \qquad (24.3)$$

Then the rubber band will contract to a point (reach zero circumference) at a time

$$T = C(0)/v.$$

For this contraction law, the scale-factor is $R(t) = 1 - [vt/C(0)]$, and the Hubble law is (from Equation 24.2)

$$v_{1,2} = -v\frac{d_{1,2}(0)}{C(0)}.$$

This is reasonable: the speed of approach of any two dots is the same fraction of the overall contraction rate v as their separation is of the whole circumference. In this case, the Hubble constant we infer is

$$H = -v/C(0),$$

which, not surprisingly, is constant for all time. The key result we have been looking for is that this is just the reciprocal of the time to contract to zero (apart from a sign):

$$T = |1/H|. \qquad (24.4)$$

Now, the same formula gives the age of a uniformly expanding rubber band as well. If the rubber band started at zero circumference and expanded, then a movie of it run backwards would show a rubber band contracting to zero circumference with constant speed. The age of the original expanding band is the same as the time it takes the contracting band in the movie to reach zero size. We have therefore learned that if a rubber band (or a universe) expands at a constant rate, then the reciprocal of the Hubble constant gives its age. We show in Exercise 24.1.1 that the approximate the age of the Universe is 14 billion years.

Of course, because of gravity, the expansion of the Universe is not constant: it may slow down or even speed up, as we will see later. That means that the reciprocal of the Hubble constant is only an approximation to the age. The Hubble constant is a constant in space if the Universe is homogeneous, but it is not constant in time.

Exercise 24.1.1: *Age of the Universe*

Assume that the Hubble constant has a value 70 km s^{-1} Mpc^{-1}. First convert this value to more standard units (s^{-1}) by converting megaparsecs to kilometers. Then take its reciprocal to find that the approximate age of the Universe is 14 billion years. If the Hubble constant is larger, what does this do to the approximate age?

---

same as the region we can see. This idea is related to Occam's razor, which we met in Chapter 19. The importance of simplicity is deeply ingrained in physicists' thinking.

The Copernican principle must be tested, of course: we cannot accept any guiding principle if it predicts things that conflict with observations. It is supported by a number of remarkable observations. Everywhere we look in the Universe, we observe that its appearance in one region is very similar to that in another with the same cosmological age. Astronomers have measured properties like the number of galaxies in a given volume; the shapes and colors of galaxies in different places; the speed of the Hubble expansion (see Chapter 14); and even some of the most fundamental numbers in physics, such as the ratio of the mass of the proton to the mass of the electron. In each case, the properties are the same everywhere they can be measured. This fact has a name: the Universe is *homogeneous*.

On the other hand, modern observations show that the Copernican principle does not hold in the time direction: the Universe was decidedly different long ago, filled with quasars expelling enormous jets of gas, with early generations of very blue and very hot stars, with gas poor in the heavier elements that are needed for life, with galaxies much nearer to one another than they are today, and hence with many colliding and merging galaxies. Today, the expansion of the Universe has turned it into a quieter place. We shall now see how we can understand this expansion.

The Hubble expansion and the Big Bang

We saw in Chapter 14 that Hubble measured the recession speeds of galaxies, and found that the speed of a galaxy was proportional to its distance from us:

$$v = Hd. \qquad (24.5)$$

In this section: we see how a homogeneous expansion of a homogeneous space leads automatically to the Hubble law.

What does this mean about the Universe? Can we understand the Hubble law in a simple way? The answer is yes.

> There is a simple way to demonstrate how Hubble's law arises in an expanding Universe. Let a rubber band represent a one-dimensional "universe": only points actually on the band represent points in this one-dimensional universe. Draw dots on it to represent galaxies, as in Figure 24.3. Arrange it as a circle, and then stretch it to two, three, and four times its original radius, retaining its circular shape as far as possible. Then the length of the arc of the circle between points separated by, say, a quarter of a circle increases at a uniform rate, which is half the rate that the length of the arc joining points separated by a half-circle does. The more distant are any two points, the more rapidly they move apart. This is just Hubble's law.

The rubber-band universe is explored more quantitatively in Investigation 24.1.

Notice that the rubber-band universe really is homogeneous: no dot occupies a special position, there is no natural "central" dot on the band. (The circle does have a center, but that is not part of the band. In our one-dimensional universe, only points along the band are part of the universe.) Because every dot is like every other

Measurements must follow the rubber band

Measurements across the band are unphysical

Figure 24.3. The rubber-band model of the Universe.

one, all dots see the same Hubble law. Every observer attached to a dot sees the universe expanding away from the "home" dot. This is a good model for what our Universe looks like, except we have to extend the model to three dimensions.

Did our rubber-band universe have to be a closed loop? No: if it were a long straight piece of rubber and we drew dots on it and then stretched it, we would have seen the same thing. The Hubble law would still apply, as measured by an observer on any dot. And still no dot would be in a special position.

Cosmologists make a distinction between *local* and *global* properties of the universe. Local properties are measurable directly. Global ones are properties of the Universe as a whole; as we emphasized earlier, these are usually more hypothetical. The Hubble expansion, for example, does not define the *global* structure of our rubber-band universe: it cannot tell us whether the Universe looks like a straight piece of rubber or a circular rubber band. It only tells us how it stretches, locally. In Chapter 25 and Chapter 26 we will take up the subject of what the Universe looks like in the large.

Our picture of the Universe as a homogeneous, expanding "gas" of galaxies is really very simple. What does this tell us about the Universe at earlier times?

If we add to the picture our expectation that gravity is attractive, so that the different parts of the Universe will be pulling back on each other, then we expect that the expansion should be slowing down. If so, then the expansion rate in the past would actually have been *faster* than it is today, and at some finite time in the past all the gas and galaxies in the Universe would have been squeezed together to an infinitely high density. This moment of infinite density is another example of a *singularity*, just as we found inside a black hole. Where the singularity in the black hole is the end of time for any particle that encounters it, the cosmological singularity is the *beginning* of time for all particles in the Universe. The expansion of the Universe away from this singularity is what we call the Big Bang.

▷ The homogeneity of the Universe makes this infinite density inevitable: everything in the Universe came together at exactly the same moment.

The accelerating Universe

In this section: astronomers have discovered that the expansion of the Universe is getting faster with time. We look at their methods, and especially the evidence from the supernovae that they use as standard candles to measure distances.

However, our simple and apparently obvious assumption that gravity is attractive has recently been called into question by astronomical observations. Astronomers have been able to use automated techniques to find large numbers of very distant Type Ia supernovae, and their measurements are providing a strong indication that the expansion of the Universe is actually accelerating today, that the Hubble constant was smaller in the recent past than it is today. If this is the case, then a singularity at the beginning of time is not an inevitable consequence of physical laws. It is a matter for observation, measurement, and physical theory to decide if it really took place.

We have often discussed supernovae of Type II in this book: they are triggered by the collapse of the core of a giant star, and they result in the formation of a neutron star or a black hole. But not all stellar explosions are triggered in this way. Another way to make an explosion is by accreting sufficient matter onto a white dwarf.

If a white dwarf star finds itself in a binary system with a star that is shedding matter, then some of that matter will accrete onto the white dwarf, just as in binary systems containing neutron stars or black holes. If accretion goes on long enough at a high enough rate, then the mass of the dwarf will increase to the Chandrasekhar mass (Chapter 12), the maximum mass allowed for the dwarf. When this happens, the star must collapse.

But the star does not usually collapse to a neutron star. Instead, if the nuclei in the white dwarf are relatively light, not having been processed as far as iron, then nuclear reactions begin to take place during the collapse. The star becomes an enormous nuclear bomb. The energy released in the runaway reactions disintegrates the star, and the result is a huge stellar explosion, a different kind of supernova.

The interesting thing about such explosions, from the point of view of measuring the cosmological constant, is that they are not very dependent on how the white dwarf was formed or on what kind of companion star is shedding material onto the dwarf. The Chandrasekhar mass depends only on fundamental constants of physics, so it should have been the same for the first stars as it is today. This mass largely determines the properties of the explosion. Within tight bounds, there is therefore not much variation in the explosion from one event to the next: the peak luminosity, for example, is fairly constant. Astronomers call such a system a "standard candle", a term we introduced in Chapter 9.

Here is how astronomers use Type Ia supernovae to determine the value of the cosmological constant. When a supernova in a very distant galaxy is detected, it is observed carefully, night after night, in order to find the maximum brightness. Since astronomers already know the intrinsic luminosity of the supernova at maximum brightness, they can then compute how far away the source is, and in particular what its redshift should be if we assume that the Universe has a particular Hubble constant and deceleration parameter.

The astronomers then directly measure the redshift of the galaxy containing the supernova and compare this with the predicted redshift. They adjust the Hubble constant and deceleration parameter until the predicted redshift matches the observed redshift, and in this way they determine these parameters.

This can only be done in a statistical way, of course. All measurements have some uncertainties, and so to get a good result astronomers use many tens of supernovae. These are searched for using automated telescopes that survey small regions of the sky to look for very dim and distant galaxies. When the brightness of a galaxy changes, the astronomers must investigate further to determine if it was a supernova, and then if it was of Type Ia.

At the time of writing (2002), two independent groups of astronomers had measured these parameters. The data of the two teams are illustrated in Figure 24.4 on the next page. The value of the Hubble constant indicated here is near to other estimates, around $70\,\mathrm{km\,s^{-1}\,Mpc^{-1}}$. But the deceleration parameter they measure is *negative*: the expansion of the Universe is accelerating!

These observations are very recent, and not all alternative explanations of the observations have been fully explored and ruled out. The scatter of the points in Figure 24.4 is large, and this raises the possibility that a small systematic effect, not so far taken into account, could change the conclusion, but intense work on this question has so far not found any reason to doubt the result. More data are needed, and astronomers are building new instruments and even satellites to gather it. But for now, the evidence for acceleration is strong, and it has led to a thorough re-examination of the physics and astrophysics of cosmology. Even if the acceleration proves in the end to be an illusion, the stimulus it has given to physicists and astronomers to come up with new ideas and justify old ones has been a positive result.

Was there a Big Bang?

The acceleration of the Universe was unexpected to most physicists, and there is as yet no accepted theoretical explanation of it. Evidently the Universe is filled with a physical field that exerts a repulsive gravitational effect that is larger than the attraction of all the normal matter in the Universe. We saw in Chapter 19 that it is possible to make the active gravitational mass negative, i.e. to achieve repulsive gravity, if the physical field has a large negative pressure. Most explanations of the accelerating Universe propose some such field. Einstein's cosmological constant, introduced in Chapter 19, is such a field. It may well be the explanation of the acceleration, but there are other alternatives. Astronomers are beginning to call this field **dark energy**, to draw a parallel with dark matter. Just like dark matter, dark energy is invisible except through its gravitational effects. But unlike dark matter, dark energy creates anti-gravity effects. We will see in Chapter 27 that it is likely that the eventual explanation of dark energy will have deep implications for theoretical physics.

Even without a theory we can ask what implications the observation of acceleration has for the Big Bang. Going backwards in time, the normal matter of the Universe gets denser, and so its contribution to the net force of gravity increases. If the dark energy behaves like a cosmological constant, then (as we saw in Chapter 19) its repulsive effect will remain roughly constant in strength. Given the fact that the observed acceleration is of roughly the same size as the expected *deceleration*, so that the gravitational effect of the dark energy is comparable to that of the matter

In this section: the Big Bang must have occurred if gravity was always attractive in the past; but dark energy might have prevented it.

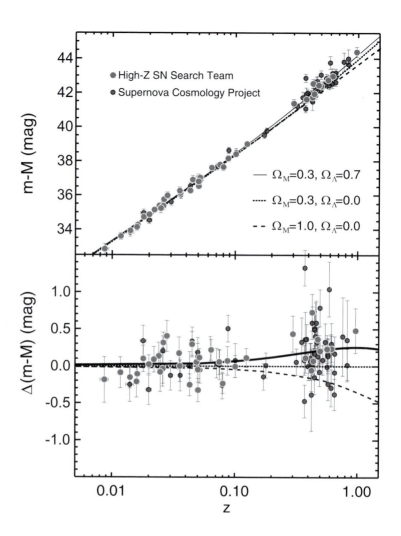

Figure 24.4. *Trend of distance against redshift for distant supernovae of Type Ia. This figure shows the evidence for the acceleration of the Universe at the time this book was completed (2002). In the upper plot, the horizontal axis gives the redshift of the supernova, as inferred from its spectrum and that of the galaxy containing it. This measures the speed of the expansion of the Universe (Chapter 14). The vertical axis gives the difference between the apparent magnitude m and the (standard) absolute magnitude M of the supernova, which (see Chapter 9) is a logarithmic measure of the distance to the object. In a simple cosmology with constant expansion speed (no acceleration or deceleration), the points would all fall on a straight line. Accordingly, the lower plot shows only the deviation of the points from that straight line. The expected relations for three model universes are drawn as lines. Don't worry at this point about the notations Ω_Λ and Ω_M, which we will define in later chapters. The line of long dashes is a universe that is decelerating and whose speed will approach zero as it gets larger and larger; the line of short dashes is a model that is decelerating less rapidly, so that it eventually will expand at a constant speed; and the solid line is a Universe that is accelerating. The points have a large scatter because of observational uncertainties, but the solid line gives the best fit: the deviation of the points from this line is significantly less than that from the other lines. The data come from two independent teams of astronomers, measuring different supernovae. Figure from High-Z Supernova Search, based on data from Riess, A. G., et al (1998) Astronomical Journal **116**, 1009, and Perlmutter, S., et al (1999) Astrophysical Journal **517**, 565.*

today, it follows that when the Universe was even one-half of its present age, the attractive (matter-created) part of gravity dominated, and the Universe was slowing down. On this hypothesis, the Universe had attractive gravity as long as it was smaller than this.

The net attractiveness of gravity at early times would still have made the Big Bang unavoidable. If we take our rubber-band universe and follow it back in time, assuming that it was always expanding at a constant rate, then at a *finite* time in the past, it was a single point. We show in Investigation 24.1 on page 348 that this time is just the reciprocal of the Hubble constant: if the speed increases with distance as $v = Hd$, then a universe with a constant expansion rate would have begun as a point at a time $T = 1/H$ ago. This is called the **Hubble time**, and in Investigation 24.1 on page 348 we show that its value, as measured by astronomers today, is around 14 billion years. If the net effect of gravity was actually attractive at early times, then the universe would have been expanding faster in the past than now, so that it would have actually taken less time to arrive at its present size than if the expansion had been constant. Therefore, the time $1/H$ would in fact be an *overestimate* of the age of the universe.

However, if the dark energy could have increased in density for some reason in the past, so that it always maintains its advantage over the matter (including dark matter), then the Big Bang is not inevitable. The Universe might have started its expansion from rest at a small but finite size, propelled to expand by the repulsive field. Since we don't know the properties of this repulsive field, we can't use theory alone to decide which of these two alternatives really happened. Instead, we have to appeal to observations.

We will see in this chapter that the observations strongly favor the Big Bang, so they suggest that the dark energy was weak at early times. However, no matter how far back in time our observations allow us to go in testing the existence of the Big Bang, it would still be possible to postulate that the repulsion at times earlier than we can directly observe was much stronger than at later times.

Indeed, the concept of inflation (Chapter 27) relies on just such a repulsion at very early times, and it is possible that inflation itself actually reversed an earlier contraction phase, preventing a true singularity. In this case, the Universe would have been small enough and dense enough to have looked like the conventional Big Bang in all of its measurable consequences, but it would have avoided a singularity.

In such theories (which seem unlikely to at present) one needs to distinguish the Big Bang, which refers to the expansion from a hot, dense state, from the singularity, which is the very beginning of time. Even if the singularity did not occur, the hot dense Big Bang is necessary to explain the formation of elements, of the cosmic microwave background, and of galaxies, all of which we study in the next chapter.

Only further research will tell what the nature of the dark energy is. It is very possible that the answers will come from a future theory of quantum gravity, which we will consider in the final chapter. Meanwhile, we will focus on today's "standard" model of cosmology, in which the repulsive field we see today was not important at early times, and the Big Bang is the expansion of the Universe away from a genuine singularity.

▷ Interestingly, the observed near-balance between repulsion and attraction today means that we apparently live in a special time in the history of the Universe. At much earlier times, the repulsion of the dark energy was not noticeable. At much later times, the attractive pull of the matter will probably also be negligible. It seems that the Universe has two comparable effects nearly in balance only at roughly the stage in the history of the Universe where people are measuring things. This is very non-Copernican. It seems hard to apply the Copernican principle at all to our place in time, even if it applies to our place in space.

Figure 24.5. The Hubble Deep Field South is one of the longest-exposure pictures ever taken with the Hubble Space Telescope. It reveals galaxies as they looked when the Universe was only one-fifth of its present age. A photo of the North field, in the opposite direction, shows very similar galaxy forms and numbers, even though there has not been enough time since the Big Bang for the more distant regions to have affected each other in any way. Photo courtesy HDF-S *Team and* NASA/STSCI.

Looking back nearly to the beginning

The most remarkable tests of the Copernican principle are a number of observations that look at things that are very far away indeed, and therefore very far in the past. Consider, for example, the galaxies in Figure 24.5. The light reaching us now from those objects has been traveling toward us for more than 80% of the time since the Big Bang. When astronomers look at the most distant galaxies in any particular direction, they are looking back in time most of the way to the Big Bang.

Now suppose we look at two groups of very distant galaxies that are in opposite directions from us. Light from a galaxy in one group has taken 80% of the age of the Universe to reach us; light has surely not had enough time to reach the galaxies in the other group that we can see in the opposite direction. That means that the galaxies in one group have not yet entered the observable universe of the other group.

In this section: the Big Bang seems to have started in the same way everywhere we can see.

Intelligent beings living today near the site of one group of galaxies can see our locality (as it was when the Universe was only 20% of its present age), but they can not yet see the galaxies in the other group. Yet *we* can see both groups, and we can see that the Universe around one group looks very similar indeed to the Universe around the other. So if people today living near the site of one galaxy are assuming the Copernican principle about the unseen region of the Universe in the direction of the other set of galaxies, then we can say that they are right.

Even more striking support for the Copernican principle comes from observations of the cosmic microwave background radiation. We turn to a discussion of this, which is the most important piece of evidence for the Big Bang at this time.

Cosmic microwave background: echo of the Big Bang

In this section: our earliest direct signal from the Big Bang is the microwave background. We learn how it was accidentally discovered and why it has such fundamental importance.

The early Universe was unimaginably dense and hot. Ordinary matter as we know it did not exist. At early enough times the temperature was so high that matter formed a plasma: the atoms of the gas were moving so fast at this temperature that when they collided they ionized one another, stripping the electrons away from the nuclei.

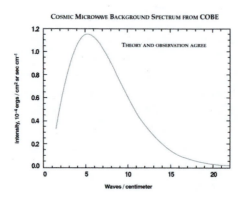

Figure 24.6. The spectrum of the cosmic microwave background radiation, as measured by the COBE spacecraft. It follows the expected black-body curve perfectly. The uncertainties in the measurements are smaller than the width of the line. This radiation was emitted when the Universe was only 3×10^5 y old, a fraction 2×10^{-5} of its present age. Courtesy COBE team and NASA/GSFC.

We met plasmas in Chapter 8, where we learned that the interiors of stars are plasmas. Plasmas are created on Earth in fusion reactors, where the goal is to collide the nuclei at high enough speeds to generate nuclear reactions and release energy that can be used to make electricity. Plasmas contain not just the nuclei and the free electrons. They also contain lots of electromagnetic radiation, in the form of gamma-rays or X-rays, depending on the temperature of the plasma. This radiation is normally thermal: in a dense enough plasma, the photons scatter frequently from the charged particles and quickly come to have the same average energy. They form a photon gas.

Now, dense plasmas are good "black bodies", as described in Chapter 10: any radiation that falls onto such a plasma will instantly be scattered by the charged particles and trapped within the plasma. It will not get through to the other side or be reflected. Because of Planck's remarkable law of black bodies, the radiation in a dense plasma will have a black-body spectrum. This was the case when the Universe was young and hot.

▷The epoch of decoupling is also frequently called "recombination", to indicate what is happening to the electrons and protons. However, this is an odd name for this event, since at earlier times the Universe was so hot that electrons and nuclei had never previously been combined. A better name for this epoch would be "first combination", but it seems too late to change this usage. Along with many other physicists, I prefer to use the term *decoupling*.

As the Universe expanded, its plasma cooled, and eventually the matter became cool enough for the electrons to combine with the nuclei to form a neutral gas. At this point, the radiation in the plasma suddenly became free. Photons prefer to scatter from charged particles, not neutral atoms, so once the particles in the Universe had become neutral, the typical photon would never be scattered again. For this reason the moment when electrons and nuclei combine is called the epoch of **decoupling**, when photons effectively decoupled from the rest of the matter.

At that moment, the photons had the spectrum of a black body with a temperature hot enough to ionize hydrogen. As the Universe expanded further, this photon gas also expanded, and like any expanding gas it cooled off. Since these photons have been moving freely through the Universe ever since, not scattering from anything, they have experienced a cosmological redshift, just as the light emitted later by galaxies has. (See Chapter 14.) This redshift applies to every photon, so the net result is that the photons still have a black-body spectrum, but its temperature has been redshifted to a much lower value. The details of how this happened are studied in Investigation 24.2 on the next page.

These photons have been observed, and they are in the microwave part of the spectrum. Their temperature, about 2.7 K (Figure 24.6), is a factor of about 1000 lower than that of a plasma that can ionize hydrogen. Investigation 24.2 on the next page shows us that this implies that the Universe has expanded by the same factor of 1000 since decoupling. Allowing for the overall slowing of the Hubble expansion, the Universe was about 0.003% of its present age when this radiation was emitted, roughly 300 000 y old.

> The cosmic microwave background radiation was discovered by accident by the American physicists Arno A Penzias (b. 1933) and Robert W Wilson (b. 1936) in 1965 at Bell Laboratories, because it was an unexpected and annoying source of noise in microwave transmissions of telephone conversations! For recognizing the importance of their discovery, Penzias and Wilson received the Nobel Prize for Physics in 1978.

Since the plasma of the Universe scattered photons easily before decoupling, the Universe was essentially opaque to photons then. Therefore, we have no hope of ever detecting any electromagnetic radiation that originated at a time earlier than the cosmic microwave background.

> When we look at the microwave background, we are seeing the Universe at a very early age. It is very significant, therefore, that the microwave background has the same property as galaxies: when we look at it in one direction and in the opposite direction, we see exactly the same sort of radiation, with the same temperature and the same degree of uniformity.

In fact, we don't need to look in opposite directions: because the radiation originates so much closer to the time of the Big Bang than does the light given off by galaxies, the microwave radiation coming to us from directions separated by only a couple of degrees on the sky is coming from regions that would have had no knowledge of one another at the time the radiation was emitted.

The rest frame of the Universe

Because it brings us our earliest glimpse of the Universe, the cosmic microwave background has been studied intensively by astronomers. It has a number of lessons to teach us, and they all suggest that the Copernican principle is correct to a very high degree of accuracy. One of the most remarkable is that it determines the mean rest frame of the Universe.

The microwave background does this because it enables us to measure our own speed relative to this rest frame. In each direction that astronomers look, the spectrum of the cosmic microwave background is a black-body spectrum, characterized by a single temperature. If the Universe is genuinely homogeneous, then when the radiation was last scattered, that temperature had to be the same everywhere. So the temperature of the radiation reaching us should be essentially the same in all directions.

However, because the Earth and the Sun have a random motion with respect to other galaxies, and therefore with respect to the mean rest frame of the galaxies, we

▷Interestingly, *gravitational waves* are not scattered significantly by anything, and certainly do not care whether matter is ionized or neutral. Therefore, gravitational radiation *can* come to us from a much earlier time, and searching for a cosmological background of gravitational waves is one of the most important observations that gravitational wave detectors plan to make. If it exists, such a background would have originated from very much closer to the Big Bang than the cosmic microwave background, at a time when the Universe was less than 10^{-30} s old!

In this section: the microwave background defines a standard of rest. Astronomers have measured the Sun's velocity relative to it.

▷Recall that a frame is an observer's coordinate system. So the rest frame of the Universe describes the preferred observer who is at rest in the Universe.

Investigation 24.2. Cosmic microwave background radiation in an expanding Universe

The cosmic microwave background radiation is a window into the early Universe. In this investigation, we will work out the two simple rules that govern how it changes when the Universe expands: (1) the wavelength of any photon remains proportional to the size of the Universe; and (2) the temperature of the photon gas is inversely proportional to this size. We will learn what the energy-density of the radiation is, and how to use it to measure our own speed relative to the cosmic rest frame.

We begin by asking how much energy the black-body radiation contains. We have seen in Chapter 10 that the total flux of energy emitted by the surface of a black body is given by the Stefan–Boltzmann law (Equation 10.11 on page 117):

$$F = \frac{2\pi^5 k^4}{15c^2 h^3} T^4. \qquad (24.6)$$

This is the energy emitted per unit area per unit time. We want to use this to deduce the density of energy, the energy per unit volume.

Suppose that part of the surface is a flat plane with area A. If we look at the photons that are radiated by this area during a very short time-interval Δt, then they must all have come from the region just inside the black body and very near the surface. In particular, no emitted photon could have traveled further than $c\Delta t$ during the time Δt, so all the emitted photons originated inside a thin volume behind the surface of width $c\Delta t$. However, not all the photons that were in this volume at the beginning of the time-interval actually emerged from the surface. Half of them, in fact, had velocities directed *away* from the surface, back towards the interior, so they did not contribute to the flux. Of the rest, most were moving at an angle to the surface, so in the time Δt not all of them were able to reach the surface and contribute to the flux. It turns out (see Exercise 24.2.3) that the photons moving toward the surface have an *average* speed towards the surface of only $c/2$; the rest of their motion is parallel to the surface.

It follows from this that, of the photons that are in the thin volume, only one-quarter reach the surface and contribute to the flux. The energy that emerges is, therefore, only one-quarter of the energy in the volume. If the energy density of the radiation is denoted by ϵ_{bb}, then we have

energy flux $\times$ area of surface $\times$ time-interval =

¼ energy density $\times$ volume of thin region,

which translates into the equation

$$FA\Delta t = \tfrac{1}{4}\epsilon_{bb}A(c\Delta t),$$

because the volume of the thin region is its surface area A times its depth $c\Delta t$. This can be solved for the energy density of black-body radiation:

$$\epsilon_{bb} = 4F/c = \frac{8\pi^5 k^4}{15c^3 h^3} T^4. \qquad (24.7)$$

Like the flux, this is proportional to T^4.

Our next step is to consider a box containing black-body radiation of a given temperature T. The interior walls of the box are perfectly reflecting mirrors. Recall that, if this box has a tiny peep-hole, then the hole itself is a black body, since radiation falling onto the hole from outside will pass in, be reflected around the inside the box, and have negligible probability of re-emerging. The hole is a perfect absorber.

Now let us gradually expand the box until it is twice its original size in all dimensions. The hole in the box is still a black body, so the radiation inside still has a black-body spectrum. But what is its temperature?

As the box expands, the photons cannot keep their original wavelengths. A photon that encounters a moving wall will reflect in such a way that its energy is constant as measured by an experimenter at rest with respect to the wall, not by the experimenter at rest relative to the box. This results in a decrease in the energy of the photon at each reflection. This energy is lost in doing work to expand the box. The loss of energy produces a redshift of the wavelength of the light.

The redshift is proportional to the energy of the photon itself and to the speed of the wall. The redshift formula tells us that, for a single encounter with a wall, $(\Delta\lambda/\lambda)_{single\ bounce} \propto v/c$. In a very small time Δt, the photon will have many encounters, proportional to $c\Delta t/d$, where d is the size of the box. Its wavelength will increase according to the rule $(\Delta\lambda/\lambda)_{during\ \Delta t} \propto v\Delta t/d$. But the product $v\Delta t$ is the change in the size of the box during the expansion, Δd. So we have deduced the simple relation

$$\Delta\lambda/\lambda \propto \Delta d/d.$$

A more detailed calculation of the statistics of the photon velocities would show that the constant of proportionality here is just one. In other words, the fractional change in wavelength is the same as the fractional change in the size of the box. *If the box doubles in size, each photon gets stretched to twice its original wavelength.*

Since the spectrum remains black-body, and since the wavelength at which the spectrum reaches a maximum is inversely proportional to the black-body temperature, it follows that *the temperature of a black-body photon gas scales inversely with the size of the container. Therefore the energy density decreases as the inverse of the fourth power of the size of the container.* This scaling of the energy density is physically reasonable: the number of photons per unit volume is going down as the inverse cube of the size of the container, and the energy per photon has been redshifted by another factor of the size.

All of this happens to photons in the Universe. As the Universe expands, photons get redshifted. If the mean distance between galaxies expands by a factor of two then the cosmic background photons stretch to twice their original wavelength, and the energy density of the radiation decreases by a factor of 16.

Exercise 24.2.1: *Energy density of the cosmic microwave background*

(a) Use Equation 24.7 to calculate the energy density of the cosmic microwave background, given its temperature of 2.7 K. (b) Show from this that the equivalent mass-density of the microwave background is 4.5×10^{-31} kg m^{-3}.

Exercise 24.2.2: *Motion through the cosmic background*

According to measurements by COBE, the temperature of the cosmic microwave background has a maximum value on the sky that is 3.15 mK warmer than the average, and it has a minimum in a diametrically opposite direction that is 3.15 mK cooler than the average, after correcting for the motion of the satellite around the Earth and that of the Earth around the Sun. (The abbreviation mK stands for *millikelvin*.) Give an argument to show that the observed radiation should be black-body at a red- or blueshifted temperature. Then show that the speed of the Sun relative to the cosmic rest frame is 3.5×10^5 m s^{-1}.

Exercise 24.2.3: *Random photons*

Devise a computer program, based on RANDOM, which allows you to calculate the mean speed toward the wall of the photons in the thin volume in our derivation of the energy density of a photon gas. Generate many random cases by choosing random directions for each photon (three random Cartesian coordinates) and calculating the speed toward the wall on the assumption that the total speed of the photon is c. Show that the mean speed toward the wall of those that move toward the wall is $c/2$.

would not expect to be exactly at rest with respect to the **surface of last scattering**: the distant sphere around us where the microwave photons finally became free. Instead, we would expect to be approaching that sphere in one direction and receding from it in the opposite direction. We would therefore expect to see a Doppler effect. The photons from one direction should be blueshifted, and therefore be hotter, and those from the opposite direction redshifted. This redshift preserves the form of the black-body spectrum but changes its temperature, as shown in Investigation 24.2.

This is indeed what is observed. The spectrum displayed in Figure 24.6 on page 354 is what results when the Doppler effects are removed. Scientists use the word **anisotropy** to describe the deviations from isotropy in the radiation. Measuring the anisotropy allows astronomers to calculate the Sun's velocity, as we describe in Exercise 24.2.2.

> Our Sun's velocity through the Universe is measured by this means to be 350 km s^{-1}.

It is interesting to reflect that relativity began with the assumption that we could never measure our own velocity, not in an absolute sense. We discussed this in Chapter 15. Has relativity now led to the opposite, a measurement of our velocity in the Universe? There is no real contradiction here, but the issue is intriguing nevertheless. The velocity we measure from the cosmic microwave background is, like every other velocity in relativity, a velocity *relative* to something. In this case, it is relative to the mean rest frame of matter at the time the universe became transparent to photons. We only determined this velocity by making a measurement on something outside of ourselves: the radiation itself. Velocities are still relative, even in cosmology.

But what is intriguing about this is that there is only one Universe, so the Universe really does have a preferred rest frame. This is the frame in which the mass in the Universe is at rest, on average. And since the random velocities of galaxies are small compared with the expansion speed of the Universe over distances we can easily observe, this rest frame really is the best frame for describing the physics of the Universe. When we consider how nuclear physics determined the creation of elements after the Big Bang, or how galaxies formed from small density irregularities, or how the microwave background itself was formed: for all of these questions it is very helpful to do the physics in the preferred rest frame of the Universe.

Are there other variations in the temperature of the microwave background in different directions? There should be, since the Universe is irregular on small scales. But the variations are incredibly small.

> After correcting for our motion, the temperature of the radiation in any given direction differs from that shown in Figure 24.6 on page 354 by an amount of order one part in 100 000.

This is an extraordinary degree of homogeneity. Once we take away the Doppler effect of our own motion, the radiation has the same temperature even when coming from regions that apparently could not have communicated with one another between the Big Bang and the time they emitted the radiation. The Copernican principle for observers at those remote locations seems very good indeed!

Big Crunch or Big Freeze: what happens next?

The engine at the heart of the Universe is gravity. Gravity is what makes the Big Bang slow down, gravity is what is making it speed up again, and gravity will decide its long-term future. If the Universe is indeed homogeneous and isotropic, then *gravity alone* will determine its future development. No other forces can act on

In this section: the future of the Universe depends on the way the dark energy behaves in time. It is likely to expand forever, but it might re-collapse.

masses in a homogeneous and isotropic Universe. If a force exists that pushes a bit of the Universe in a certain direction, then by isotropy there must be an opposing force of equal size pushing in the other direction.

> Gravity is the only force that determines what happens to the Universe as a whole, and indeed what has happened in the past.

However, although local pressure *forces* are not important, pressure itself helps to create gravity by contributing to the *active gravitational mass* (Chapter 19). If the Universe is filled with matter that has negative pressure, then this can make the active gravitational mass negative, and then gravity will turn repulsive. This appears to be happening now, with a field of unknown origin contributing enough negative pressure to accelerate the Universe.

If we want to predict the future of the Universe, we have to have certain information about its present state. We need the density and pressure of the Universe, so we can calculate the density of the active gravitational mass and therefore the gravitational acceleration. And we need the present Hubble expansion speed of the Universe, from which we predict future expansion speeds using the acceleration. And finally, we need to know the physics of how the density and pressure change when the Universe expands.

Since we assume that the Universe behaves the same way everywhere, there are only two main types of futures, always expanding or eventually re-contracting; and there is the marginal class in between, where the Universe just manages to keep expanding, but with its expansion speed approaching zero. These are illustrated in Figure 24.8 on page 361. Can we say which evolution will be followed by our Universe?

Yes, we can, at least in principle. We will see below how we can use what we learned about escape speeds in Chapter 6 to calculate the Universe's escape speed, the expansion speed it needs now to go on expanding forever.

This analogy between orbits and cosmologies isn't perfect, of course; the Universe is not the same as a small satellite in the gravitational field of a planet. One difference is that the Universe provides its own gravity. Because of this, the bound universe does not cycle in and out the way an elliptical orbit does. Once it starts to re-collapse, the bound Universe shrinks toward infinite density, which is the time-reverse of its behavior at the Big Bang. Scientists have come to call this possible future the **Big Crunch**. Gravity is so strong at the Big Crunch that the Universe encounters a singularity: we cannot use the known laws of physics to predict what will happen after that. Many scientists hope that, if the Big Crunch happens, quantum gravity will get the Universe through the singularity and into another phase of expansion, but this is pure speculation at present.

The evidence today, however, is that the Big Crunch may not happen. Instead, the measured acceleration of the Universe suggests that the Universe is actually supplying its own *anti-gravity!* This makes it behave in a way that is unlike anything we saw with planetary or cometary motion. The acceleration, if it continues forever, will make the Universe bigger and bigger, colder and colder. The stars will eventually burn out, and the Universe will be cold and dark forever. This is the **Big Freeze**.

Cosmology according to Newton

If we want to know what might happen to our Universe in the future, we must study the laws of motion of an expanding universe, find the balance between expansion speed and gravity. Let us first see how this works using Newton's law of gravity.

> ▷A good example of a local force that has no net result is the pressure *force*. Pressure forces act through non-uniformities, as we saw in Chapter 7. Therefore, they cannot directly accelerate the Universe's expansion, because there is no *net* pressure force on any part of the Universe.

> ▷Interestingly, these three possible outcomes are similar to the three types of orbit around the Sun that we met in Chapter 4: the particle in an elliptical orbit moves outwards, and then comes back in; the hyperbolic orbit moves outwards forever, limiting to a non-zero speed; and the marginal case of the parabolic orbit, which moves outwards forever, but with a limiting speed of zero.

> ▷An alternative to the Big Crunch and Big Freeze is the steady-state theory of Hoyle and Narlikar (Chapter 11). This gained support in the 1950s through 1970s, but the discovery of the cosmic microwave background and of evidence for the creation of elements in the Big Bang (see the next chapter) made the theory go out of fashion. The accelerating Universe is a further severe problem for this theory.

In this section: Newtonian theory is incomplete when describing cosmology.

The homogeneity of the Universe makes Newtonian cosmology almost as simple as studying the motion of a planet around the Sun. Homogeneity allows us to place ourselves at the "center" of the Universe and to calculate the forces on other galaxies that attract them towards us. All we want is to find the relative acceleration of the galaxy and ourselves; we do this by placing ourselves at the middle, so we have no acceleration, and calculating the acceleration of the galaxy due to its location. The key result that makes this calculation simple is as follows.

▷In one respect cosmology is even simpler than planetary motion: bodies in the Solar System move around the Sun as well as toward and away from it, while in cosmology a homogeneous Universe only moves in and out.

> The net gravitational force on a galaxy that is a distance d away from us is produced only by the part of the Universe that is within a distance d of us. This force is the same as if all the mass within this distance d were concentrated at a point at our own location.

The rest of the Universe, further away from us than the galaxy, *has no net gravitational effect on it!*

The argument for this key result is basically the theorem of Newton that we proved using the computer in Chapter 4. As illustrated in Table 4.4 on page 35, the Newtonian gravitational force on an object inside a *spherical* shell is zero. Let us consider the part of the Universe that is further away from us than the galaxy we are considering.

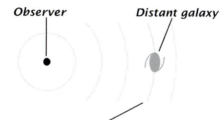

Observer   **Distant galaxy**

Beyond this shell, the outer shells exert no net force on the galaxy. All the force comes from inside this shell.

Figure 24.7. Dividing the Universe into concentric spherical shells, so that the gravitational force attracting a distant galaxy to us depends only on the mass closer to us than it.

If we can divide it up into a series of concentric spherical shells, we will have proved that it can exert no net gravitational force on the galaxy. If the Universe is truly homogeneous and infinte, then we can surely do this, so the result would follow. This idea is illustrated in Figure 24.7.

Unfortunately, this argument has a big problem. We appear to have to invoke the nature of the Universe arbitrarily far away: what happens if, in a part of the Universe so far away that we cannot yet see it, the mass distribution in the Universe actually comes to an end, and the edge is not spherically symmetric about our location? That would upset the argument, which required that we split the Universe up into exactly spherical shells centered on us. The ragged edge is far away but has a lot of mass, being distributed over a huge surface area, so it would have a measurable effect on gravity here. Is it acceptable that the evolution of the Universe in our neighborhood should depend on the detailed structure of the Universe outside the particle horizon? The problem here is clearly that we have pushed Newtonian gravity too far. Only in relativity, where regions very far away have not had time to affect us gravitationally yet, can this paradox be resolved.

▷The problem of the influence of very distant regions is a serious one in Newtonian gravity, and is one of the reasons that cosmology did not become a serious study until Einstein provided a better theory of gravity.

Cosmology according to Einstein

General relativity is essential if we want to describe the Universe in the large. There are in fact at least two reasons for this. The first is the finite age of the Universe, which implies, as we have seen above, that there is only a finite portion of the Universe (inside our particle horizon) that can have influenced us. Gravitational influences in general relativity cannot travel faster than light. Indeed, gravitational waves travel at exactly the speed of light. This tells us that, in the argument above where we divided the Universe into spherical shells, we need only go out as far as the distance that light can have traveled since the Universe began. Anything further away has had no influence yet on gravity here. So the gravitational force on a relatively nearby galaxy cannot depend on whether the Universe is spherical in a

In this section: Einstein's theory allowed the first consistent physical theory of cosmology. But to calculate its evolution one can get away with equations from Newtonian gravity.

region we can't observe; it only depends on the Universe being homogeneous out to as far as we can see. The observations of the cosmic microwave background radiation reassure us that this is indeed the case.

The second reason for needing general relativity is that, if we consider galaxies so far away that their Hubble recession speed approaches the speed of light, then Newtonian gravity must fail to be valid. Let us remind ourselves of Hubble's law, that the speed of recession of a galaxy is proportional to its distance,

$$v = Hd, \tag{24.8}$$

where H is Hubble's constant. In this expression, if we can make d big enough, then we can make v bigger than the speed of light. Clearly the expression in Equation 24.8 must change if we go far enough away from our Galaxy, even in a homogeneous Universe. This expression can only be a *local* approximation to the recessions speeds we measure.

Extending the Hubble law to large distances might seem to be a big complication, but it is something we can postpone worrying about until after we have studied the basic dynamics of the Universe. The reason is as follows.

> The expansion and contraction of a homogeneous Universe is a *local* phenomenon: as long as we can calculate the motion of *nearby* galaxies relative to us, then the homogeneity of the Universe guarantees that all other galaxies will behave the same way.

The key point, which we shall now argue, is that we can calculate the acceleration of nearby galaxies relative to us in Einstein's theory in the same *local* way, using concentric shells of matter, as we sketched in Newton's theory.

We have already noted in Chapter 21 that the gravitational field outside a spherical star in general relativity is identical to the field outside a black hole of the same mass. The inverse of this is also true: if we take a spherical distribution of mass in general relativity and cut a spherical hole in the middle, leaving the hole empty, then in the hole there will be no gravitational acceleration: spacetime will be perfectly flat, just as in special relativity.

▷ The inverse should be no great surprise: we saw in Chapter 19 that at least the dominant change in the way gravity is created in general relativity is that the Newtonian mass density is replaced by a combination of density and pressure, the active gravitational mass. Therefore, properties of the gravitational field that depend on symmetries, such as the way matter is distributed, should be the same in relativity as in Newtonian gravity.

So even in relativity, the gravitational acceleration of a galaxy relative to us depends only on the part of the Universe closer to us than it is. Now, if we consider only galaxies so near by that their recession speed is very much less than the speed of light, then we should be able to use Newtonian gravity to describe their motion. It follows that the dynamics of the Universe can be described by Newtonian gravity, provided that we use the correct relativistic source of gravity, which is the active gravitational mass.

> Despite its logical inconsistencies, a spherical Newtonian cosmology is an accurate approximation to the relativistic cosmology if we use the correct relativistic form for the source of gravity.

Evolving the Universe

In this section: we define and calculate key numbers, like the critical density and the density and deceleration parameters.

Here we shall see how to use Newtonian dynamics to find the dynamics of the relativistic Universe. First we will make a simplifying assumption and neglect all pressure, supposing that the Universe is dominated by the observed matter and the inferred dark matter. This is probably not accurate today, but it was a good approximation over much of the early evolution of the Universe, and especially at the time galaxies were formed. We shall therefore begin our study of the evolving Universe with these assumptions, and come back to the effects of pressure later. We call this the **matter-dominated** Universe.

Let us use the symbol ρ to represent the mass-density of the Universe averaged over volumes that today are about 100 Mpc in size. This will be the same everywhere, by our assumption of homogeneity. A galaxy at a distance d from us lies on a sphere whose volume is $4\pi d^3/3$. The mass closer to us than the galaxy is then $M = 4\pi\rho d^3/3$. The part of the Universe further away contributes nothing to the net force, and the net force on the galaxy is the same as it would be if all this mass were concentrated at the center (our location).

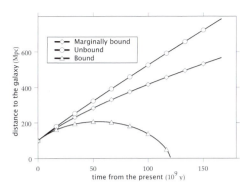

Figure 24.8. Three possible future developments of a simple matter-dominated universe that has a Hubble constant of $50\,km\,s^{-1}\,Mpc^{-1}$. The re-collapsing universe model has a density that is twice the critical density. The other models shown have density equal to and half of the critical density, respectively. For definiteness, we plot the position at any time of a galaxy that starts at a distance of 100 Mpc from the origin.

Now, the key idea that keeps matter-dominated cosmology simple is that any particular galaxy at the distance d will *always* feel only the gravitational force from the same mass M, as long as pressure is negligible and the matter is non-relativistic. All the mass closer to us expands less rapidly, thus never going further away than that galaxy; all the mass further from us expands more rapidly, thus never coming closer to us than the galaxy; and none of the mass in the volume disappears. It follows that the escape speed of the galaxy is exactly the same as if the galaxy were escaping from a fixed point mass M. This speed is, as we saw in Equation 6.10 on page 56,

$$v_{escape} = \left(\frac{2GM}{d}\right)^{1/2}.$$

Substituting $4\pi\rho d^3/3$ for M in this equation, we find

$$v_{escape} = \left(\frac{8}{3}\pi G\rho\right)^{1/2} d. \tag{24.9}$$

This has the same form as the Hubble law, Equation 24.8, namely $v \propto d$. It tells us that if the proportionality factor H in the Hubble law exceeds the proportionality factor $(\frac{8}{3}\pi G\rho)^{1/2}$ in this equation, then *every* galaxy – regardless of d – will be going faster than the escape speed, and so the Universe will continue to expand forever. Put another way, if the Universe is matter-dominated with a Hubble constant whose value is H_0, then there is a **critical density** ρ_c for which the Universe is just marginally bound. We find this value by setting the coefficient in Equation 24.9 equal to H_0, squaring, and solving for ρ:

$$\rho_c = \frac{3H_0^2}{8\pi G}. \tag{24.10}$$

Taking $H_0 = 70\,\text{km s}^{-1}\,\text{Mpc}^{-1}$, we find that the critical density today is about $\rho_c = 10^{-26}\,\text{kg m}^{-3}$.

Let us compare this number with astronomers' estimates of the mean density of the Universe. The smallest possible value for ρ is the density we obtain if we spread out the luminous mass (not the unseen dark matter) of observed galaxies. This is estimated by astronomers to be no more than $5 \times 10^{-29}\,\text{kg m}^{-3}$ if $H_0 = 70\,\text{km s}^{-1}\,\text{Mpc}^{-1}$. This is 200 times smaller than the critical density. Cosmologists often prefer to couch their discussions of the mass density in terms of a dimensionless quantity, the ratio Ω of the true mass density to the critical density, called the **density parameter**:

$$\Omega = \frac{\rho}{\rho_c}. \tag{24.11}$$

▷The fact that the mass in the volume is always the same is the key simplification of the matter-dominated assumption: when we consider a photon gas, whose energy redshifts away during the expansion, or a cosmological constant, whose energy density is constant with time, then the active gravitational mass inside the volume changes with time.

▷The critical density defined this way is important even for cosmologies that are not matter-dominated. We will see in Chapter 26 that it determines the curvature of space.

▷The symbol Ω is the capitalized version of the Greek letter omega, the last letter of the Greek alphabet.

The visible mass density contributes a fraction $\Omega_{vis} = 0.005$ of the critical density. By itself, the visible mass density could not stop the expansion of the Universe.

We saw in Chapter 14 that there is a lot of missing mass. We shall see in Chapter 25 that some of it – perhaps four times as much as the visible mass – is in the form of hydrogen and helium gas that has never formed stars. Much more than this is hidden dark matter, in a form that astronomers have not yet identified. The best estimates of the amount of dark matter on the cosmological scale give densities a factor of three lower than the critical density: $\Omega_M = 0.3$.

If the Universe at the present time were matter-dominated, then it would have less than the critical density, and it would expand forever. We calculate the actual deceleration of such a model in Investigation 24.3, where we address an important detail of principle: we show that the deceleration is also proportional to distance, so that the Hubble law (Equation 24.8 on page 360) remains true for all time. The result is the important equation:

$$a_{cosmol} = -\tfrac{4}{3}\pi G\rho d. \tag{24.12}$$

The fact that the inward acceleration increases in proportion to d implies that the Hubble law will hold for all time. Cosmologists usually write this equation in a slightly different way, defining the dimensionless **deceleration parameter** q by

$$q = \frac{4\pi G\rho}{3H^2}, \tag{24.13}$$

so that

$$a = -qH^2 d.$$

The value of the deceleration parameter today, q_0, is related to the dimensionless density parameter Ω of the Universe, defined in Equation 24.11 on the preceding page, by

$$\Omega = 2q_0. \tag{24.14}$$

The expansion of the Universe at present appears to be accelerating, so that means it cannot be matter-dominated. To include pressure, one simply replaces ρ by the density of active gravitational mass, $\rho + 3p/c^2$, in Equation 24.12. In particular, the dark energy must be added to other mass densities when comparing with the critical density of the Universe. Since the dark energy behaves like a cosmological constant, astronomers denote its density relative to the critical density by the symbol Ω_Λ.

As Figure 24.4 on page 352 shows, the density associated with the dark energy is just enough, when added to the density of the dark matter, to give the critical density, within the observational errors. This result is borne out by measurements of the cosmic microwave background radiation as well, as illustrated in Figure 27.2 on page 403.

This is an unexpected result that many physicists want to see explained. As we will see in Chapter 27, it is predicted by the theory of inflation.

However, when pressure is important the evolution of the Universe will of course be different. In particular, the work done by the pressure as the Universe expands will affect the mass-energy density ρ. Moreover, the pressure itself can change. Normally, to decide whether a particular model Universe will expand forever or re-collapse requires a computer calculation.

Investigation 24.3. Cosmological gravity

A key question is, does the expansion of the Universe maintain the Hubble law? Hubble discovered the expansion in the first place by finding that the speed of recession of a galaxy is proportional to its distance,

$$v = Hd, \qquad (24.15)$$

where H is Hubble's constant. If this describes the velocity of matter in the Universe now, then will the expansion of the Universe change it? After billions of years, will the expansion law look different, say with speed depending on the square of the distance? We don't expect it to, since the Hubble law is the only one that a homogeneous Universe can satisfy. But we need to check that the law of gravity does indeed maintain this. Otherwise, we have a logical inconsistency in our model of the Universe.

Now, the expansion of the Universe must be slowing down or speeding up, due to gravity, so Hubble's "constant" is generally not constant in time. If the Hubble law is preserved, then it follows that the acceleration (or deceleration) must also be proportional to the distance. Therefore, we expect to find, at least near to our Galaxy,

$$a = Kd, \qquad (24.16)$$

where K is a different "constant", a number that is independent of position but can change with time. Because, as we have seen in the text, the acceleration depends only on the mass closer to us than d, we can calculate this in the same manner as we calculated the escape speed. The mass closer to us than the galaxy is $M = 4\pi\rho d^3/3$, and the acceleration it produces is $-GM/d^2$. This gives the cosmological acceleration:

$$a_{\text{cosmol}} = -\frac{GM}{d^2} = -\frac{4\pi G\rho d^3}{3d^2} = -\frac{4\pi G\rho}{3}d. \qquad (24.17)$$

Reassuringly, the acceleration increases in proportion to d, just as we expected: *our model of an expanding homogeneous universe governed by the known laws of gravity is self-consistent.* The constant of proportionality in Equation 24.16 is then

$$K = -4\pi G\rho/3.$$

Cosmologists do not usually deal with K directly. Instead, they define a dimensionless measure of the deceleration, called the *deceleration parameter.* Here is how it is defined.

We have already noted that the Hubble constant has dimensions of $1/\text{time}$, and that $1/H$ is a measure of the age of the Universe. Now look at the proportionality constant K. Its dimensions are those of acceleration divided by distance, which works out to be $1/(\text{time})^2$. So the ratio K/H^2 is *dimensionless.* It is, to within a sign, what cosmologists call the deceleration parameter q:

$$q = -\frac{K}{H^2} = \frac{4\pi G\rho}{3H^2}. \qquad (24.18)$$

We can thus write the cosmological acceleration in Equation 24.12 as

$$a = -qH^2 d.$$

The present values of all these "constants" are denoted by a subscript "0": H_0, q_0, ρ_0, and p_0.

Starting from Equation 24.17, it is possible to calculate the expected evolution of the Universe. We show how to do this with the help of a computer in Investigation 24.4 on the next page.

If pressure is not negligible, say because of radiation in the early Universe or because of a cosmological constant, then we just replace ρ by the relativistic $\rho + 3p/c^2$ everywhere in this calculation.

Exercise 24.3.1: *The emptiness of the Universe*

Since the luminous mass in galaxies is primarily in hydrogen, what would be the mean volume occupied by a single hydrogen atom if the mass in galaxies were smoothed out over the entire Universe? (Use the mean density of visible matter given in the text, 5×10^{-29} kg m^{-3}.)

Exercise 24.3.2: *Local accelerations*

The nearest large galaxy to us is the Andromeda galaxy (also called M31), which is about 0.5 Mpc away and is falling towards our Galaxy, not receding from it. Take the mass of our Galaxy to be $10^{11} M_\odot$ and calculate the gravitational acceleration produced by our galaxy on the Andromeda galaxy, using the formula $a = -GM/r^2$. Calculate the cosmological acceleration given by Equation 24.17 at a distance of 0.5 Mpc, using the critical density ρ_c. Compare the two accelerations. Are motions within the local group of galaxies (those dominated by Andromeda and ourselves) strongly affected by the expansion of the Universe?

Exercise 24.3.3: *Relation between Ω and q*

Derive Equation 24.14 from Equations 24.11, 24.10, and 24.18.

The cosmological scale-factor

Astronomers describe how a cosmological model expands by using the cosmological scale-factor. We introduced this for the rubber-band universe in Investigation 24.1 on page 348. The idea is a way to describe changes in the size of a cosmological model, even if the model is infinitely large. No matter how large the model is overall, if the distance between two typical galaxies doubles over some period of time, then the "size" of the cosmology has effectively doubled. These relative size changes are the important aspects of cosmological expansion, not the overall size of the Universe. The **cosmological scale-factor** tracks this relative expansion.

Consider two galaxies at some particular reference time, say at the present moment t_0. Let their separation be d_0. At another time t they are separated by a distance $d(t)$. We define the scale-factor R to be the ratio of these distances

$$R = \frac{d(t)}{d_0}. \qquad (24.19)$$

In this section: the appropriate indicator of the expansion of the Universe is the change in relative distances between nearby points. This is directly measured by the cosmological redshift.

Investigation 24.4. Making the Universe grow

In this investigation, we shall see how to use a computer to evolve the equations for the matter-dominated Universe.

We start with an extension of the reasoning in Investigation 24.3 on the preceding page that led us to the critical mass density ρ_c. Every spherical shell surrounding us is affected only by the mass inside it. Since shells never cross, the mass inside it remains constant, so the fact that the portion of the Universe inside the shell is also expanding is irrelevant: it simply contributes a gravitational pull on the shell which is exactly like the pull on, say, a spacecraft launched from a planet of constant mass. Since the shell consists just of independent galaxies, all moving radially outwards, they all move on exactly the same trajectories that free particles would take if they were launched radially outwards from the same position with the same mass inside.

This makes it very easy to calculate the future development of the Universe: we just use our orbit program Orbit on the website, with initial numbers adapted to the present problem. For example, let us consider a galaxy that is 100 Mpc away from us at present, and let us take Hubble's constant to be 50 km s^{-1} Mpc^{-1}. Then the galaxy

has an initial speed of 5000 km s^{-1}. We can orient our x-axis for this problem to be along the direction from us to the galaxy. Then it will have an initial x-speed of 5×10^6 m s^{-1} and an initial y-speed of 0. Its initial x-distance is 100 Mpc, or 3×10^{24} m. Its initial y-distance is 0. If we take, as an example, a universe model which has half the critical density for this H_0, then the mass closer to us than the galaxy is $4\pi(\rho_c/2)x^3/3 = 3 \times 10^{47}$ kg.

The program requires the variable K to be GM, which in this case is 1.9×10^{37} in SI units. Since we will want to follow the Universe as it expands for something like the Hubble time of 6×10^{17} s, we take a time-step that is much smaller than this, 5×10^{14} s.

The curves in Figure 24.8 on page 361 show the results of this and two other simulations, one for a universe that has only half the critical density, and one for twice the density. The three graphs show how the position of the initial galaxy changes with time in each of three possible futures: re-collapse to a Big Crunch, expanding forever but slowing to zero speed, or expanding forever at a constant speed.

Clearly, R is a function of the time t, and of the reference time t_0. But it is *not* a function of the galaxies we chose. The reason is Hubble's law: if the two galaxies had initially been twice as far apart, their expansion speed would have been twice as large, so the distance between them would have increased by twice as much, and the ratio of $d(t)$ to the original d_0 would turn out to be exactly the same.

So the scale-factor tells us how the Universe is expanding. We can express other physical quantities in terms of it. For example, the number of galaxies in our expanding Universe is not changing, but the distances between them are increasing in proportion to the scale-factor. The mean density ρ_M of matter in the Universe is decreasing in inverse proportion to the volume containing any given collection of galaxies. Since the volume is the product of three lengths, all of which are increasing in proportion to R, the volume is proportional to R^3, and the density is inversely proportional to this:

$$\rho_M \propto 1/R^3.$$

The scale-factor is directly measurable in the cosmological redshift, and this is one of the most important relations in cosmology. The redshift is a Doppler shift, which we described in Figure 2.3 and Investigation 2.1 on page 15. If the light comes to us from a galaxy at a distance d, then the galaxy is receding with its Hubble speed $v = Hd$, and the redshift is $z = v/c = Hd/c$. The ratio of the wavelength the light has when we receive it, λ, to its original wavelength, λ_0, is

$$\frac{\lambda}{\lambda_0} = 1 + z = 1 + \frac{Hd}{c}. \tag{24.20}$$

Now consider what has happened to the Universe during the time the photon was moving from its source galaxy to us. The photon took a time $\tau = d/c$ to travel to us at the speed of light. In this time, its source galaxy moved from its original distance d to a further distance $d + v\tau$. Therefore, the scale of the Universe has increased by the factor

$$R = \frac{d + v\tau}{d} = \frac{d + (Hd)(d/c)}{d} = 1 + \frac{Hd}{c}.$$

▷ We won't see the galaxy at its new location until the photons it emits "now" have had time to reach us, of course.

This is identical to the wavelength ratio in Equation 24.20 above. We have therefore verified, by a different method, and for photons that are not part of the cosmic microwave background, the same remarkable and simple result as we had for the expanding photon gas in Investigation 24.2 on page 356:

the wavelength of a photon that moves freely through the Universe increases in direct proportion to the cosmological scale-factor R.

Expressed as an equation, this is

$$\lambda \propto R. \tag{24.21}$$

This allows us to look back in time and make conclusions about what the early Universe was like. For example, if we examine a quasar whose redshift is four, then the scale-factor of the Universe was only one-fifth (because $1 + z = 5$) of what it is today. The average distances between galaxies were only 20% of what they are today, and the temperature of the cosmic microwave background radiation was five times higher than today, or around 13 K.

What is the cosmological expansion: does space itself expand?

Because of the cosmological redshift, photons behave as if they were being stretched between the galaxies: their wavelengths increase exactly as the distances between galaxies increase. It might be tempting to conclude from this that the cosmological expansion stretches space itself, just as our rubber-band universe was built out of a stretching material. But this could be quite misleading from a physical point of view.

In particular, one must not think that *space* itself is everywhere enlarging, as if extra "points" were somehow being created among the old ones all the time, and everything was getting bigger in proportion to the Universe: wavelengths, sizes of atoms, sizes of people. If the sizes of atoms and the spacings between atoms in molecules were getting bigger, then the expansion of the Universe would be un-measurable. If our rulers were enlarging at the same rate as the wavelengths of photons from distant quasars, then we would not notice the redshift, since the incoming light would occupy the same fraction of the standard meter as it did when it left the quasar billions of years ago. The expansion of the Universe is an observable fact precisely *because* ordinary matter does not expand with it.

It is simpler from a physical point of view to think of the expanding Universe as a simple collection of particles (called galaxies) that are rushing away from one another. The redshift of light is a Doppler shift caused by the motion of the source galaxy away from us.

Notice how this looks from the point of view of a photon. Let us take its source galaxy as the standard of rest, the (arbitrary) center of the Universe. As it moves away from the source, it passes other galaxies. They are all moving away, but not as fast as the photon, which moves at the speed of light. The further the photon travels through the Universe, the faster is the speed of the galaxies it passes, since the faster galaxies have traveled further from the center since the Big Bang. Suppose that it happens to be detected by an astronomer in one of these galaxies. Then, the longer it has traveled, the faster will be the speed of the detecting galaxy relative to the source galaxy, and the bigger will be the detected redshift. The redshift increases with time, but this increase has nothing to do with a metaphysical stretching of space: it is simply the way the Doppler shift works in a homogeneous expanding Universe.

Why does ordinary matter not expand with the Universe? After all, each proton and electron starts out from the Big Bang participating in the cosmological expansion. The answer is that the particles would continue to expand if they remained free particles, influenced only by the smooth cosmological gravitational field. But they don't remain free.

The other forces of Nature, such as electromagnetism, disturb the cosmological expansion in small regions, binding individual particles to one

In this section: the expansion of the Universe is not a mystical expansion of space, but rather the expansion of distances between ordinary objects.

another, wiping out the initial relative velocity between them. Once the "memory" of the initial expansion is lost, atoms are governed by forces in their neighborhood, not by cosmological gravity.

▷Physicists who use supercomputers to simulate the formation of galaxy clusters – as a way of testing the cold dark matter hypothesis – use Newtonian gravity in their simulations, because it is so much easier to use than general relativity, and it is a perfectly adequate approximation. See Figure 25.3 on page 380 for the results of one such simulation.

This remark even applies to irregularities in gravity. Galaxies form from the expanding gas of the Universe because of some random irregularity in the density of the expanding gas, which causes a local increase in gravity that slows the expansion of the nearby gas. Eventually, if the initial irregularity is big enough, the local self-gravity is strong enough to reverse the cosmological expansion in the gas, and the gas becomes a gravitationally bound object, perhaps a galaxy or a cluster of galaxies. After that it can be described perfectly adequately by Newtonian gravity, ignoring the rest of the Universe. The *average* motion of the cluster of galaxies can't be wiped out by the forces between the particles in the cluster, so it still participates in the cosmological expansion, as a whole cluster, relative to other clusters and galaxies. But within the cluster, the expansion has been forgotten.

The age of the Universe

In this section: the computed age of the Universe depends on the value of the acceleration today. The age is probably between 12 and 13 billion years.

Let us again look backwards in time, to the Big Bang. We saw that the Hubble constant gives us an upper bound on the age of the Universe provided it was matter-dominated over most of its past. With a more realistic model for the Universe, one can get a better estimate of its age.

The early history of the Universe was very complicated, but this period was short: the Universe became matter-dominated after about 400 000 y. In recent times, however, things have become complicated again: the acceleration of the expansion has dominated for about the last half of the age of the Universe. This makes the estimate of the age very sensitive to the assumed value of the acceleration. The best estimate today is that it is between 12 and 13 billion years old. This is consistent with the estimated ages of all known stars and clusters.

But is this enough time to produce the Universe as we see it? Can stars and galaxies form in this amount of time, do galaxies have enough time to clump into rich clusters? This is not an easy question to answer. In fact, without dark matter, the answer would be no: not nearly enough time. We will see in the next chapter how the clumpiness of dark matter helps accelerate the formation of structure in the evolving Universe.

The Big Bang:
the seed from which we grew

We have now pushed our model of the history of our Universe back just about as far as we could hope to go: the Universe had a beginning, and that beginning was the source of all that happened afterwards – all matter, all stars, all galaxies, even life itself. Big Bang cosmology has placed modern physics in the remarkable position of being able in principle to trace back to the beginning every aspect of the world we live in, to say "This is where X came from, and this is how Y started".

Physicists have grasped this opportunity with enthusiasm. The study of what is often called **physical cosmology** – the evolution of the matter in the Universe after the Big Bang – is one of the most active and exciting branches of astrophysics today. Helped by powerful computers, physicists can now explain how the elements hydrogen and helium were made, where the cosmic background radiation came from, and (at least in outline) how galaxies and clusters of galaxies might have formed. Within these galaxies, we have already seen in Chapter 12 how stars arose and turned hydrogen and helium into all the heavier elements, and in Chapter 7 we speculated on how a tiny portion of these heavier elements became the planet Earth and produced the conditions that allowed an even tinier portion to become living things.

> We can, still very imperfectly, trace our own origins as humans right back to the beginning of the Universe.

This chapter is about *physical cosmology:* what happened in great arena of the Universe from the beginning to now.

Physical cosmology: everything but the first nanosecond

Let us imagine running the movie of the Universe's expansion backwards, so we observe it getting denser and denser. How far can we go and still claim to understand what is going on? The answer is startling: physicists can go back with confidence to within 10^{-10} s of the Big Bang. And they can even make some shrewd guesses about what happened as early as 10^{-35} s.

The reason for this remarkable success is the homogeneity of the Universe. Going back in time is just like compressing the matter in the Universe into smaller and smaller volumes. This makes the matter hotter, and the particles in the matter become more energetic. As long as the typical particle energy is less than the limits of our understanding of particle physics from man-made accelerators (about 1 TeV), we can be confident of our description of what happens at these energies and densities. By carefully solving the Einstein equations for the evolving Universe, generalizing what we did in Investigation 24.4 on page 364, scientists find that the average particle energy fell to about 1 TeV at about 10^{10} s after the Big Bang. This epoch marks a watershed in physicists' models of the Universe: before this time, they speculate; after this time they speak with fair confidence. We will look in Chapter 27 at the most interesting current speculations about the earliest phase of the Universe. In this chapter we deal with what is known with some confidence.

In this chapter: we study physical cosmology: how physics worked in the expanding Universe. This includes the formation of the elements hydrogen and helium, the role of dark matter, and the formation of galaxies and clusters. Physicists have achieved a remarkable understanding of the Universe after its first nanosecond.

▷ The background image on this page is a plot of the spatial distribution of galaxies in a thin wedge of space centered on our position, from the CfA Redshift Survey. Measured in 1985, this distribution gave astronomers their first indication that galaxies were grouped into chains as well as clumps. The human-like pattern (horizontal in this view, with the legs to the left) became a celebrity in its own right. Data from M Geller and J Huchra, image copyright South African Observatory (SAO).

In this section: physicists are fairly confident they understand the Universe after its first nanosecond!

▷ We met the electron volt, eV, as a unit of energy in Chapter 8. High-energy physicists usually describe energies as multiples of this unit: keV (10^3 eV), MeV (10^6 eV), GeV (10^9 eV), and TeV (10^{12} eV).

The expansion of the quark soup and its radiation

In this section: the earliest stage that is well-understood is the quark soup: when the Universe was still too hot and dense to allow protons and neutrons to exist individually.

▷This is nothing more than the kinetic temperature: express the energy E of 1 TeV in ordinary units (joules) and set it equal to $3kT/2$ to find T. You will get about 10^{16} K.

What was the Universe like at 10^{-10} s? If the Universe is highly homogeneous today, it must have been more so at these early times, since all the structure we see today (galaxies, and so on) developed at a later time as the Universe cooled down. We will describe the way structure developed later in this chapter. But little of it was present when the Universe was a hot ball of gas with a typical particle energy of 1 TeV and a corresponding temperature of 10^{16} K.

Ordinary matter as we know it could not exist at these temperatures. If there were any ordinary protons around, then their random collisions would break them up into their constituents, which physicists call quarks. Quarks are the oddest particles in physics: in groups of three they make protons or neutrons, and in groups of two they make π-mesons and other lighter particles. Yet one never sees them alone: single quarks cannot be peeled off from particles in accelerators. In the early Universe, the particles were packed so closely together that quarks were never alone. Instead, they blended together in a sea that physicists call the **quark soup**.

Besides quarks, there were many other particles in the early Universe. Whatever particles now constitute the dark matter were already there, but their density irregularities, which would be important for galaxy formation later on, were not significant at this time. The dark matter particles were neutral and had stopped interacting with the quarks or the photons by this time. They were already just a provider of a gravitational background.

And there were photons. With energies typical of the thermal energy, they had enough energy to form new quarks in reactions where two photons collide and two quarks emerge. By mechanisms like this and the reverse, the numbers of quarks and photons were maintained in a steady balance.

When particles like quarks or protons are produced by photons that collide with one another, they emerge with equal numbers of particles and anti-particles. The anti-particle of any particle has the opposite sign of the charge. So if electrons are produced, one is a normal electron and the other is a positron, or positively charged electron. If a proton is produced, an **anti-proton** (with a negative charge) is also produced. The anti-particle of a photon is just another photon. So in this way, no net charge is produced: the two photons initially have zero charge, and the two particles that emerge have zero total charge.

When an anti-particle and a particle of the same type collide, the result is often to produce a pair of photons, which is the time-reverse of the reaction described in the previous paragraph. Thus, a proton and anti-proton will annihilate each other to produce two photons. Similarly, a positron and an electron annihilate to two photons. Physicists refer to positrons, anti-protons, anti-neutrons, anti-quarks, and so on collectively as **anti-matter**.

The laws of physics prefer matter over anti-matter

In this section: if the laws of physics were perfectly symmetrical between matter and anti-matter, all matter would have been annihilated and we would not be here. We owe our existence to a small preference in the laws of physics for matter over anti-matter.

As the Universe expanded, the mean distance between quarks grew until they began to get too isolated. When this happened, they started to clump into twos and threes, forming ordinary protons, neutrons, π-mesons, and other particles. The corresponding anti-quarks also clumped to form anti-protons, anti-neutrons, anti-π-mesons, and so on.

As the Universe cooled further, the photons, whose gas stays at the same temperature as the particles because they collide frequently, no longer have enough energy to create proton–anti-proton pairs when they collide. At this point, there are still lots of collisions where protons and anti-protons annihilate to form photons, but the photons get redshifted by the expansion of the Universe and, by the time they meet other photons, no longer have enough energy to create protons and anti-protons again. This also applies to neutrons and anti-neutrons. The number of particles decreases steeply at this point because annihilations are dominant over creations.

In principle, all the protons should have annihilated against all the anti-protons. But in fact, it is obvious that they did not: we are all made of protons that survived this era. It is natural, then, to expect that some anti-protons also survived, but this apparently did not happen. The two were so well mixed that we should see anti-protons everywhere, and we don't. Instead, it appears that there were simply more protons than anti-protons, by a small amount. This can only reflect a fundamental asymmetry in the laws of physics, a preference for one kind of matter against its opposite.

As the Universe expanded and cooled, the same thing happened later for the electrons: when the temperature was too small to create electron–positron pairs, then the electrons and positrons annihilated. The asymmetry at this point was exactly the same: the same laws of physics allowed the same fraction of excess electrons to survive as for protons.

We can learn how slight the excess of protons was by counting photons today. The microwave background radiation has on average 10^9 times more photons in any region of space than there are protons and neutrons. This number has not changed much since the separation of photons and electrons and the subsequent annihilations took place. The number of electrons has not changed, and the number of photons has changed only by a factor of two or so by the processes we describe below. It follows that the excess of protons/electrons over anti-protons/positrons in the very early Universe was about 10^{-9}. This is a small clue to the nature of laws of physics that physicists do not yet understand. We will return to this point in Chapter 27.

> We owe our existence to this slight asymmetry in the laws of physics. If the laws were perfectly symmetrical between matter and anti-matter, then all the protons would have been annihilated in the early Universe, and there would have been nothing left to build stars, planets, and people from. The Universe would instead have been filled with pure radiation, cooling as it expanded.

It is interesting to reflect that we are formed from the waste that resulted from a slight imperfection in the laws of physics!

The Universe becomes ordinary

The annihilation of protons and neutrons stopped when the thermal energy kT fell to about the rest-mass-energy of a proton $m_p c^2$. It is easy to calculate that this temperature is about 10^{13} K. Since this temperature is smaller by a factor of 100 than the quark-soup temperature we quoted above, it follows that the mean photon energy had gone down by a similar amount, and therefore that the Universe had expanded by the same factor of 100. By Equation 25.1 on the next page, the time since the Big Bang had therefore increased by a factor of 10^4, to 10^{-6} s.

The electrons annihilated at a temperature of about 6×10^9 K, when the Universe was a further factor of 1600 larger, and a factor of 3×10^6 older. It was now 3 s old. This is the epoch at which ordinary matter appears.

> After the first microsecond, nuclear matter was already the normal material of which the nuclei of all elements is made. After the first three seconds, all the remaining exotic particles had disappeared, and the Universe was made of familiar stuff.

Notice how much physics takes place in times that seem short to us: everything really exotic is finished in the first three seconds! When one deals with the early

In this section: the excess protons and neutrons eventually dominated the composition of the early Universe, accompanied by electrons and neutrinos. Most other particles had gone away after the first few seconds.

Universe, one quickly begins to think, not in terms of time-intervals, but in terms of time *ratios*.

> The time-interval between the end of the anti-proton era at 10^{-6} s and its beginning at 10^{-10} s should not be thought of as 0.9999 µs; rather, it should be thought of as a ratio: time since the Big Bang has increased by a factor of 10^4, and the scale-factor by a factor of 100.

This way of thinking about cosmological time is shown in Figure 25.1 on page 375. This displays time logarithmically, which insures that time-intervals that have the same ratio are separated by the same distance along the line.

After the annihilation of the anti-protons, the total energy density of the Universe was dominated by the photon gas. The remaining particles had little total energy, and by colliding frequently with photons they kept the same temperature as the photons. We say that the Universe was radiation-dominated.

One can show (see Investigation 25.1) that a radiation-dominated cosmological model, at least soon after the Big Bang, expands in such a way that its scale-factor R is proportional to the square-root of the time t since the Big Bang:

$$R(t) \propto t^{1/2}. \tag{25.1}$$

What happened to the energy in the photon gas as the Universe expanded? The number of photons remained approximately constant, so the number of photons per unit volume was determined only by the volume of the clump, which increased as R^3. Just as we saw for galaxies in the previous chapter, the number of photons per unit volume was proportional to R^{-3}.

But we also saw there that the wavelength of a freely moving photon in an expanding cosmology increases with R. Therefore its frequency ν is inversely proportional to R, and its *energy* $h\nu$ is inversely proportional to R. So the *energy density* of the photon gas (energy per photon times number of photons per unit volume) was proportional to R^{-4}. Expressed as an equation, this is:

$$\text{energy density of a photon gas} \propto R^{-4}. \tag{25.2}$$

This is a deep result, and it applies to any period of time when photons are either dominant over matter or move freely without scattering from matter. In particular, it holds for the cosmic microwave background radiation today.

Besides the photons there were also neutrinos in the early Universe. Observations of neutrinos from the Sun now strongly suggest that the neutrino has a non-zero rest-mass, but they don't tell us yet what it is. But the rest-mass energy equivalent will be less than a few electron volts, very small compared to the average neutrino total energy in the early quark soup. So the neutrinos at this epoch would all have been traveling essentially at the speed of light. Moreover, the density of the soup was so large that the neutrinos scattered frequently, keeping them in equilibrium with the quarks and hence with the photons. So the neutrinos at this time formed a gas with a temperature and energy density similar to the photons, and there were likewise billions of neutrinos for every quark.

Making helium: first steps toward life

In this section: nuclear reactions stopped as the Universe expanded and cooled. Their main product was helium. Most of the helium in existence today was formed in the Big Bang.

We, as human beings, have a real interest in what happened in the early Universe: if things had been very different, and the right conditions for life had not emerged, then we would not be here. One of the indisputable essentials for life is the existence of heavy elements, such as carbon and oxygen. These are made in stars from the basic building blocks of helium nuclei. But the helium itself is not made in stars

---

Investigation 25.1. *Exact solutions for marginally bound model universes*

Here we determine how the scale-factor R of the Universe depends on time for different kinds of assumptions about the composition of the Universe. Our derivation will use rather simple ideas, but we will arrive at accurate answers.

We begin with the fundamental equation for the acceleration (or deceleration) of a galaxy at a distance d away from us in a matter-dominated Universe, Equation 24.12 on page 362, which we reprint here:

$$a_{cosmol} = -\tfrac{4}{3}\pi G \rho d.$$

In such a Universe, the density ρ is proportional to $1/R^3$, because as the Universe expands the particles simply get spread out over a larger volume. The location d of any particular galaxy is also proportional to R, so the right-hand side of this equation is proportional to R^{-2}.

The left-hand side is harder to work out. First, as we have just noted, the galaxy's distance from us is proportional to R. Therefore its velocity away from us is the rate of change of R. Without knowing how R depends on time yet (that is the goal of this calculation!), let us make the crude approximation that the speed of the galaxy is proportional to R/t. This approximation is exact if the speed is constant and R increases in direct proportion to t. If R is proportional to, say, t^2, then the approximation suggests that the speed is proportional to t. This is in fact correct, as we saw in the discussion of gravity on the Earth in Investigation 1.2 on page 4. Going another step further, we can guess that the acceleration of the galaxy will be proportional to R/t^2, again exactly as in Investigation 1.2.

The result of these guesses is that the acceleration equation implies

$$R/t^2 \propto 1/R^2.$$

We can multiply by R^2 and t^2 and solve:

$$R^3 \propto t^2, \qquad \Rightarrow \qquad R \propto t^{2/3}.$$

This turns out to be exactly correct: the scale-factor of a matter-dominated Universe is proportional to the -2/3 power of the time. You might worry that the method is crude, and indeed it is. In a moment we will show how we would be able to tell whether the method was giving a wrong answer. In this case, we are getting the right answer for a *particular* solution to the equation.

Our solution is not the most general one possible. We have explored the general solutions with a computer, and the results are in Figure 24.8 on page 361. In fact, since according to our present solution, R increases unboundedly with time, this solution cannot represent the bound Universe. Also, since the rate of change of R, which is proportional to R/t, goes to zero as t gets larger and larger, it also does not represent the unbound Universe. We are left with just one possibility: we have found the law governing the marginally bound solution in Figure 24.8 on page 361. Notice, however, that in this figure all three solutions behave very like one another in the early Universe. Therefore, we have also found a good approximation to any matter-dominated cosmology in its early phase.

If we want to examine other kinds of Universe evolutions, we must replace ρ by $\rho + 3p/c^2$ in the acceleration law:

$$a_{cosmol} = - \left(\frac{4\pi G}{3}(\rho + 3p/c^2) \right) d.$$

For example, the radiation-dominated Universe is a Universe filled with a photon gas. The pressure of such a gas is proportional to its density, in fact $p = \rho c^2/3$. So the right-hand side of this equation is proportional to ρ, just as in the matter-dominated case. However, as the Universe expands, the energy density of the photons decreases more rapidly. As we can infer from Equation 24.21 on page 365, the energy per photon decreases as $1/R$, while the number of photons per unit volume decreases in the same way as the number of

particles in a matter-dominated Universe, by $1/R^3$. This means that the right-hand side of this equation is proportional to $1/R^4$ for the radiation-dominated Universe. In Exercise 25.1.1 below, we show that this leads to the law $R \propto t^2$. Again this is the solution representing the marginally bound case, but it is also a good approximation to all the solutions in the very early Universe. This is particularly relevant, since the very early Universe (after any period of inflation) was radiation-dominated.

Let us try this reasoning on the case where the Universe is dominated by a cosmological constant, for which $p_\Lambda = -\rho_\Lambda c^2$. Then the right-hand side of the acceleration equation is proportional to ρ_Λ, which is a *constant*. The minus sign on the right-hand side has been cancelled by the minus sign that comes from $\rho_\Lambda + 3p_\Lambda/c^2 = -2\rho_\Lambda$, so that the acceleration is *positive*. This is how the cosmological constant produces an accelerating Universe. If we follow the same method for finding how R depends on time, we get

$$R/t^2 \propto R,$$

since the only dependence on R on the right-hand side is now the factor d. If we try to solve this for R, we get into trouble: R divides out, leaving us with the non-sensical result $t^2 \propto 1$.

So our method fails for this type of Universe. The approximation that the acceleration is proportional to R/t^2 is not consistent with the acceleration equation. We shall see how to find the right answer in a moment. But let us point out in passing that the fact that the method did not fail for the matter-dominated and radiation-dominated cases tells us that we can believe the answers that we got, however crude our initial guess was. Essentially, our initial guess did not have to be right, but if the equation gives a consistent result after making the guess, then this retrospectively confirms that the guess was correct.

So how do we proceed when the guess is wrong? We make another guess, of course. To see what new guess might be reasonable, let us first ask a simpler question. What if our equation was not for acceleration but for velocity? What if it said that the rate of change of R was proportional to R? This is a familiar situation in lots of physical problems. It happens in radioactivity, for example: the number of nuclei that decay in a given time is proportional to the number that are there. It happens to populations of rabbits, as well, if they are not limited by availability of water or food or by predators or disease: the number of new rabbits in a given year is proportional to the overall number of rabbits.

Such problems are known as *exponential* problems, and they have solutions in which the number of things (nuclei, rabbits) is proportional to e^{kt}, where k is a constant that is not determined by these arguments. We met the exponential function when studying the black-body law in Chapter 10.

Now let us go back to our problem with acceleration. Do we still have exponential behavior? The answer is yes. Again, let us assume that the scale-factor of the Universe increases exponentially in time. Then we have just seen that its rate of change is proportional to itself and therefore increases exponentially in time. This in turn means that the acceleration, which is the rate of change of the expansion speed, is also proportional to the scale-factor. This is exactly what the acceleration equation gives for the cosmological constant. Therefore, we have guessed a consistent solution this time, and it is the right one: *the Universe expands exponentially when dominated by the cosmological constant.*

The theory of inflation postulates that the Universe went through a phase of exponential expansion at a very early time. Recent observations of supernovae suggest that our Universe has recently entered this kind of phase again. If the laws of physics give us a cosmological constant that really is constant for all time (see Chapter 27), then our Universe will expand exponentially. Exponential expansion is very rapid: the bigger it gets, the faster it goes!

Exercise 25.1.1: *Radiation-dominated universe*

Find the dependence of the scale-factor on time for the radiation-dominated Universe. The analysis is similar to our derivation for the matter-dominated Universe above. The only difference is that, as explained above, the factor $\rho + 3p/c^2$ is proportional to R^{-4}. Show from this that $R \propto t^{-2}$.

in great quantities – the Big Bang supplied this basic ingredient. Here is how it happened.

We have seen that after the first microsecond, the matter in the Universe consisted mainly of protons, neutrons, neutrinos, photons, and the electron–positron gas. The positrons had little effect on nuclear reactions, so we ignore them here. Now, neutrons on their own are unstable particles: as we observed in Chapter 12, the neutron is slightly more massive than a proton and an electron, so it can decay into them and release some energy, which is carried away by a neutrino. But in the early Universe, electrons had so much energy that when they collided with protons there was more than enough to turn the pair into a neutron. So after 1 μs, neutrons and protons were more or less in equilibrium with one another, and with the electrons and neutrinos as well.

▷ The fact that we *are* here tells us something already about the early Universe: it had to make some amount of helium, for example. Reasoning of this kind is related to the Anthropic Principle, which we mentioned in Chapter 11 and to which we will return in Chapter 27.

This happy situation could not last forever, because the expansion of the Universe was relentlessly cooling things off. Eventually the density became low enough so that the neutrinos stopped scattering from the electrons and protons. This occurred at about 10^{-2} s. From that time until now, the cosmological background of neutrinos has been cooling off. Its temperature now is about 2 K, but since neutrinos are so much harder to detect than photons, it has never been directly observed.

The photons were still scattering off protons and electrons, of course, and insuring that all the particles stayed at the same temperature. All kinds of other collisions were also happening. Protons and neutrons would collide to form a deuterium nucleus, for example, but soon afterwards a photon would collide with the nucleus and break it apart. So some of the protons and neutrons were to be found in light nuclei at any time, but not many.

However, once the expansion had cooled off the photons sufficiently, they no longer had enough energy to break up the light nuclei that were constantly forming briefly by the collisions of protons and neutrons. The energy required to split up a deuterium nucleus is about 2 MeV. The equivalent temperature to this energy ($E = kT$) is 2×10^{10} K. Therefore, once the temperature of the photon gas had fallen below about 2×10^{10} K, there were not many photons around that could break up deuterium, so the random collisions of protons and neutrons quickly began to build up a density of deuterium. The deuterium nuclei occasionally suffered further collisions, and this led to the formation of helium nuclei, primarily ^{4}He, which consists of two protons and two neutrons.

Other light elements were formed at this time, up to ^{7}Li. But the expansion of the Universe reduced the density of these elements as rapidly as they could form, so nuclear reactions did not go beyond lithium in any quantity. The neutrons that were left free at this time (not inside deuterium or helium nuclei) subsequently decayed.

All of these nuclear reactions took place while the positrons were still present in large numbers. So at the end of the positron era at 3 s, the Universe contained protons, electrons, a decreasing population of free neutrons, and some nuclei. This was the gas out of which the first stars were made.

Does it correspond to reality?

In this section: we review the observational evidence supporting this picture of the early Big Bang.

Our story of the Big Bang so far has been based mainly on two sets of facts: astronomical observations of the expansion and homogeneity of the Universe, and our knowledge of laboratory nuclear and high-energy physics. But the calculations of helium formation make detailed predictions about the amounts of these elements that we should see around us, and these lead to independent checks on the theory. If the Big Bang predictions were seriously wrong, we would have to throw out at least some parts of the theory. In fact, as we shall see, the predictions are so good that they led astronomers to new results in high-energy particle physics that were later verified by accelerator experiments. The result has been an enormous strengthening of our confidence that the Big Bang model provides a good description of the history of the Universe, at least back to 1 s after the Big Bang.

Detailed calculations of the synthesis of the light elements at around $t = 1\,s$ are done by extensive computer calculations, but they also require one to make assumptions about a few numbers that astronomers do not have direct evidence for. One of these numbers is the density of protons and neutrons (nucleons) at the time of helium formation. Since most matter today seems to be dark, we can't simply trace the present density backwards in time. Instead, scientists calculate the predictions of the amounts of these elements produced by the Big Bang for a number of different values of the density of nucleons at 1 s, and compare the predictions with observational evidence about the amount of each element that the first generation of stars contained. These observations are done by taking spectra of stars and gas and measuring the strength of the lines in the spectra that are characteristic of the elements we are looking for.

This comparison has to be done carefully, since it is not easy to identify which stars and gas clouds (if any) belong to the first generation and which ones formed later from material "contaminated" by the waste products of the first generation. The problem is made easier by a key fact: deuterium and lithium are not created in the nuclear reactions that take place inside stars or when stars explode. So if astronomers observe deuterium and lithium in stars today, their abundances set a *lower bound* on the amount that was produced by the Big Bang.

When observations of lithium and deuterium are combined with observations of helium in the oldest stars known, the result is a fairly tight constraint on the amount of nucleons that were available to make these elements at about 1 s after the Big Bang. It tells us that *today* the cosmological density of nucleons is

$$\rho_{\text{nucleon}} = 0.14 \pm 0.03 \text{ nucleons per cubic meter.} \tag{25.3}$$

So if we were to spread the nucleons in the Universe smoothly out over its volume, there would be about one particle in every seven cubic meters! An average-sized room of $30\,m^3$ would contain just 4 hydrogen atoms. Multiplying by the mass of a proton, the nucleon mass density of the present Universe is $2 \times 10^{-28}\,kg\,m^{-3}$. This is only 2% of the critical density needed to turn the Universe around, if the present Hubble constant is $70\,km\,s^{-1}\,Mpc^{-1}$. In the language of the previous chapter,

$$\Omega_{\text{nucleon}} = 0.02.$$

Small as this is, it is much larger than the density of the luminous matter that astronomers observe directly, which is around $5 \times 10^{-29}\,kg\,m^{-3}$ for this value of the Hubble constant. So the production of He in the Big Bang tells us that perhaps 80% of the nucleons in the Universe are dark. This is remarkably consistent with the numbers that we get independently from studies of galaxies and clusters of galaxies, as described in Chapter 14. This is a further strong argument that the Big Bang is a good description of the early Universe.

Three and only three neutrinos: a triumph for Big Bang physics

These helium-formation studies have produced an even more remarkable test of their validity: astrophysicists were able to determine from them how many different kinds of neutrinos there are in nature. In order to produce the observed amounts of light elements, one does not just need exactly the right amounts of protons and neutrons to be available for the nuclear reactions. One also has to have the right

In this section: cosmologists predicted from the helium abundance that there could be only three kinds of neutrinos. Experimental particle physicists subsequently proved it.

expansion speed of the Universe at that time. If the Universe expands too fast, it produces less helium because the nuclear reactions turn off too quickly. In that case, there would be more deuterium left over. So the balance among the different elements today also tells us the expansion rate at the time of nucleosynthesis.

Now, one might think that the expansion rate today (Hubble constant) would tell us what the expansion rate was at the time of helium formation, but it is not so simple. The expansion rate since then has been slowing down because of the gravity of the Universe. To know the expansion rate at 1 s, we also have to know the self-gravity of the Universe. That depends on how much mass-energy there was at different times.

Today of course there is great uncertainty about the amount of mass, but fortunately that uncertainty does *not* affect the expansion rate at 1 s as much as one might expect. As we have seen earlier, the self-gravity of the Universe at 1 s was dominated by the *radiation* in it, not the particles. If the radiation was just photons, then by observing the microwave background today we could tell what the total self-gravity was at 1 s, and we could deduce the expansion rate of the Universe at that time. But we have left out the neutrinos in the Universe: they also behaved as a radiation gas at the time of helium formation, so their density contributes to our conclusions about the expansion rate then.

Now, at 1 s a neutrino gas would have been in equilibrium with the photon gas, so it would have had the same temperature and energy density. This would be true for each type of neutrino. Particle physicists have direct evidence for three kinds of neutrinos: electron neutrinos produced when an electron and a proton combine to form a neutron; mu neutrinos produced by the decay of a mu meson, or muon; and tau neutrinos produced by the decay of a tau meson. Each sort of neutrino would have formed a gas, so the density of the Universe at 1 s would have been at least four times the density of the photon gas itself.

But suppose there were a fourth kind of neutrino that particle physics experiments have not yet turned up. Particle physics theories allow this, and in fact some particle physicists have preferred theories with more than three kinds of neutrinos. Then the density of the Universe at 1 s would have been larger again, its self-gravity correspondingly larger, and the expansion speed it would have required at that time in order to reach the Hubble rate today would also have been larger. This would have quenched the helium production faster. The amount of helium decreases if there are more families of neutrinos.

> Even taking into account the uncertainties in the present total mass density of the Universe, astrophysicists found that the only way to fit the observed amounts of all the light elements today was in a Universe that had exactly *three* kinds of neutrinos and has a present density of nucleons given by Equation 25.3 on the preceding page.

More recent particle physics experiments have also shown that there are only three kinds of neutrinos. By observing the decay rate of the Z^0 particle, which can only decay into particles accompanied by neutrinos, and whose decay rate is therefore proportional to the number of different possible neutrino species available to it, particle physicists at CERN found that there were only three decay modes.

This confirmation of the conclusion that had already been drawn on the basis of Big Bang cosmology was a real triumph for the Big Bang model, and it has led to a great deal of collaboration since then between astrophysicists and high-energy physicists to see what further light cosmology can shed on the behavior of particles at very high energies. The Big Bang is one of the few places that energies above

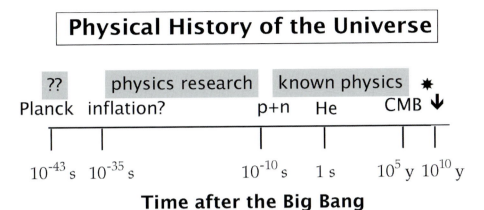

Figure 25.1. *A time-line encapsulating the main features of the evolution of the Universe. Starting at the Planck time of 10^{-43} s, physicists know very little about the laws of physics governing the Universe until about 10^{-35} s, where matter began to dominate over anti-matter in the expanding ball of energy. The physics of this process is a principal area for research in high-energy physics today, so the period up to about 10^{-10} s is one where physicists have some understanding of how the Universe behaved. At this time, electrons and neutrinos began to behave differently from one another, and we enter the realm of well-understood physics. Soon after this the protons and neutrons condensed from the quark soup, and by 1 μs the anti-protons and anti-neutrons had annihilated. Nuclear reactions formed helium and deuterium. Just before 1 s, neutrinos stopped scattering and became a free gas, and shortly thereafter the positrons annihilated. The cosmic microwave background (CMB) became a free photon gas at about 3×10^5 y, and the mass condensations that led to galaxy clusters began forming soon afterwards. The most distant quasars and galaxies that we see were shining and forming their earliest stars at the time *, and we live today near ↓.*

those reached by our present accelerators have ever been seen, so this interest is natural. Now that physicists have confidence in the basic correctness of the Big Bang model, they can use it to illuminate other branches of physics. We shall see more of how this works in the Chapter 27.

From nuclei to atoms: the Universe goes transparent

Once the nuclear reactions had finished, there followed a long period in which the Universe simply cooled off as it expanded further. Initially, of course, all the matter was fully ionized: the nuclei had formed, but the energy of the particles was too large to allow the electrons to be bound to the nuclei. As we noted in Chapter 10, it takes only about 13 eV of energy to remove the electron from a hydrogen atom. As long as the average energy of the particles in the Universe is bigger than this, any electron that does get trapped by a nucleus will be knocked off it almost immediately. So the electrons will remain free.

Now, photons only scatter off charges. Photons are basically little packets of oscillating electric and magnetic fields, and these fields are affected by charged particles but not by neutral particles. Neutral atoms can scatter photons, but only if the photon can get close enough to "see" the individual charges within them. So they are not nearly as effective as free electrons and nuclei. The early Universe was a black body – a perfect absorber of photons – for the same reason that we found stars to be black bodies in Chapter 10. The ionized plasma traps the photons. Since the photons were still the dominant source of gravity, their temperature decreased inversely with the scale-factor R, and the matter temperature followed suit. The neutrinos also cooled off in the same way, so that even though they had stopped exchanging energy with other forms of matter, they continued to have the same temperature.

But eventually the temperature had to fall to the point where the typical energy was about 13 eV. This is when we would expect atoms to start forming and staying bound. But in fact, this was not the temperature at which most of the matter be-

In this section: once the nuclear reactions had stopped, the next big event was the formation of atoms. This required the gas to cool to below the ionization temperature of hydrogen. After this time the gas of the Universe is largely neutral, so light propagates without scattering. Our observations of the cosmic microwave background therefore go back to this time, at which radiation and matter decoupled.

came neutral. Recall our discussion in Chapter 10 of the ionization of hydrogen in the outer layers of the Sun, and especially why the surface temperature of the Sun is lower than the temperature where the typical particle energy would be 13 eV. The situation is the same in the early Universe, only more extreme, because there is a huge imbalance between the number of photons and the number of nucleons. With 1 billion photons for each nucleus, if only a small fraction of the photons have energies above 13 eV they can keep the matter ionized. In any gas, there is a random spread of energies. Not until the temperature of the photon gas fell to about 0.6 eV were there too few photons to keep the nuclei ionized. This is the epoch of decoupling.

The temperature ratio from the helium formation epoch to this time is the same as the ratio of the characteristic energies, $(2\,\mathrm{MeV})/(0.6\,\mathrm{eV}) = 3 \times 10^6$. Because of the relation between the temperature of the photon gas and the scale-factor, this is the ratio by which the Universe expanded between these two times. Since the elapsed time is proportional to the square of this ratio in a radiation-dominated Universe, the time has increased by a factor of 9×10^{12}. This puts us at $t = 9 \times 10^{12}$ s, or about 3×10^5 years.

> Three hundred thousand years after the Big Bang, the Universe finally became neutral. After this time, it became largely transparent. The radiation that is now the microwave background was formed at this time and has been cooling off ever since.

The temperature at decoupling was the equivalent of 0.6 eV, which is about 4000 K. The microwave background temperature today is about 2.7 K. Therefore the photon gas has redshifted by a factor of about 1500 since decoupling. This is then the factor by which the Universe has expanded since then. The density of matter in the Universe has changed by the cube of this factor, about 8×10^9. If we take the density today that is indicated by the helium-formation arguments, then this density at decoupling was about $2 \times 10^{-18}\,\mathrm{kg\,m^{-3}}$. This is already a very low density compared to everyday densities on the Earth, which are 10^{21} times larger. Forming the Earth therefore required a great concentration of matter at later times.

Coincidentally, the end of the plasma era is accompanied by another change: the transition from radiation-dominated to matter-dominated evolution. As the Universe expands, it is inevitable that this transition will take place. The energy of each photon decreases as the Universe expands, while matter particles have a reservoir of energy that does not go away: their rest-mass energy. So eventually, no matter how many photons there are, their total mass-energy will drop below that of the matter. The rest-mass-energy of a nucleon is about 10^9 eV, and there are about 10^9 photons per nucleon. The cross-over, therefore, occurs when the photons have an average energy of 1 eV. By coincidence, this happens at about the same time as decoupling.

> After decoupling, the self-gravity of the Universe is dominated by matter. The background radiation of photons and neutrinos follows the expansion of the Universe but does not dominate it.

The evolution of structure

In this section: how the dark matter began the formation of galaxies.

Once matter becomes the dominant source of self-gravity, the details of the expansion and deceleration change somewhat, but the general trend was the same. The time since decoupling has been spent developing structure: this is the era during which clusters of galaxies, galaxies, and stars formed. This was a very complex physical process, which astrophysicists are now coming to understand, aided by simulations performed on the world's largest supercomputers. Despite this complexity, there are some key aspects of the problem that we can consider in this book and draw conclusions about.

The first issue for us is to try to understand how, in a homogeneous Universe, irregularities like galaxies could have formed at all. Fundamentally, galaxies, stars, and planets all owe their existence to the basic fact about gravity that we mentioned on the first page of this book: it is universally attractive. In a smooth, expanding gas, this leads to instabilities. Any small irregularity might grow through its self-gravity. What actually happens is that in a region of higher-than-average density, the cosmological expansion slows down; the region continues to expand and get less dense, but the density *contrast* with the average density of the Universe gets bigger. If the region containing the density contrast is large enough, or the contrast is big enough, the region can actually reverse its expansion and re-collapse. Then we have a potential galaxy cluster. We will study one way in which this might have happened below.

This leads directly to our second question: where did the density irregularities come from? Of course, nothing is perfectly homogeneous. The positions of atoms are random, and that inevitably leads to clumping. But, as we show in Investigation 25.2 on the next page, the random clumping of atoms is too small ever to explain how vast numbers of them could have come together into clusters of galaxies. Something had to provide larger-scale density irregularities. Fortunately, we know that the Universe also contains dark matter whose form is undetermined. It is natural to look to the dark matter to provide these irregularities, rather than try to invent yet another mechanism.

As we saw in Chapter 14, the dark matter could come in several different forms, all of which make different predictions about the nature of galaxy clustering. The leading dark matter candidate at present is a sea of heavy, electrically neutral particles: this has come to be called **cold dark matter** (CDM). We will shortly see why this is attractive. Another candidate is cosmic defects, left over from exotic particle physics processes in the early Universe. This includes cosmic strings and cosmic textures. We will discuss these in Chapter 27, but here we only need to note that they may have concentrated considerable energy into small regions, providing mechanisms to start the collapse of ordinary matter.

The third question we can answer concerns the time at which galaxies might have begun to form. It seems certain that ordinary matter – nucleons and electrons – did not participate in any structure formation until after decoupling. The reason is that the ionized matter was very closely tied to the photon gas, so any clumping would have had to involve the photons too. But photons do not stay in one place, so they don't clump for long. Once they diffused away, the particle clumping would similarly die out. So before decoupling, any clumping could only have involved dark, uncharged matter. This is in fact the great advantage of invoking dark matter to start galaxy formation; it can get started much earlier than ordinary matter can. Galaxies only began to form after decoupling, but they formed by falling onto clumps of dark matter that had formed long before.

This brings us to the fourth issue, which is one of the main uncertainties about galaxy formation: why did the dark matter clump, and how clumpy did it get before decoupling? We have shown in Investigation 25.2 on the following page that random fluctuations in particle positions cannot account for any sensible degree of clumping. The dark matter accounts for more mass than the ordinary matter, so is this problem not even harder for dark matter?

Investigation 25.2. *Can random clumping of particles lead to galaxy formation?*

The first reason one might offer as an explanation of the density inhomogeneities that led to galaxy formation is pure randomness. One would expect, even in a homogeneous Big Bang, that particles would have small irregularities in their locations, in the same way that the molecules of the atmosphere we breathe are not perfectly uniformly distributed even though the atmosphere when averaged over many particles is uniform.

However, random irregularities in the density due to such effects get small when there are a lot of particles. Essentially, for every particle that moves closer to another, there is likely to be another that moves further away. It is a bit like the random walk that we studied in Chapter 8, and it has the same statistics: if a given small region of space would have, on average, N particles, then random fluctuations in particle positions will change it typically by a number of order $N^{1/2}$. The density contrast produced by a fluctuation, which is the ratio of the density fluctuation to the average density, is of order $N^{1/2}/N = N^{-1/2}$, which gets smaller as N gets larger.

Now, the number of particles that one needs to make even a star, let alone a galaxy or a cluster of galaxies, is very large. It is the ratio of the mass of the Sun to the mass of a proton, something like 10^{57}, so random fluctuations in particle positions will provide such a col-

lection of particles with a typical over-density of no more than one part in about 10^{23}. In order to form a star, the slight excess gravity of this fluctuation has to amplify the density contrast to something of order one.

This is simply much too small an initial contrast for this to have happened by now. Cosmologically speaking, stars have not had a lot of time to form. Gravitational collapse could not have started until after the Universe cooled enough for its highly ionized matter to have become neutral. Before that, the electrons scattered off the photons easily, which kept them too hot to collapse. Only when they re-combined with protons, and the photons no longer had enough energy to scatter from them, did matter have a chance to start collapsing. This was the time of the formation of the cosmic microwave background. Since then, the Universe has expanded by only a factor of 1000 or so. This would not have been enough time for random inhomogeneities to grow by a factor of 10^{23}.

Since the 1950s scientists have recognized that they had no obvious explanation for the initial fluctuations that led to galaxy formation. One reason that the theory of inflation is so attractive to many cosmologists is that it offers a natural explanation for bigger density fluctuations.

Exercise 25.2.1: *Random clumping*

Experiment with random clumping using a tossed coin as your random-number generator. Use three successive tosses to generate a number between 0 and 7, using its binary representation. That is, if the coin comes up heads assign a 1 to a digit, and if tails a 0. With three tosses you get three digits, say 010, and that is the number 2. (The digits *abc* represent the number $4a + 2b + c$.) Record each such number you get. Generate a large set of them, say 80. Each number should come up on average ten times, but some will come up more often and some less, at random. The excess over the average should be, according to the argument above, $10^{1/2} \approx 3$. You should expect some numbers to come up at least 13 times, and others only 7. You might expect one bin to have twice as large a fluctuation, i.e. to reach 16 or 4. Now go on and do twice as many, 160 numbers. (You need 480 coin tosses to do this!) Then the average will be 20 and the expected fluctuation $20^{1/2} \approx 4.5$. Although the fluctuation is larger in this case, it is a smaller fraction of the average, so that the distribution of numbers among the bins is actually smoother. If you have the stamina, go to 320 numbers. Verify that the typical fluctuation is of order six.

The clumping mechanism for dark matter depends on what the dark matter consists of. For the most popular CDM model, random particle positions are no help, because there will be a similarly huge number of these particles as of baryons. Instead, physicists tie the CDM model to the idea of inflation, which we will study in Chapter 27. Inflation has the curious property that it amplifies density irregularities by a huge factor. Even a small density fluctuation from quantum uncertainties before the era of inflation can be amplified into a significant irregularity in the distribution of dark matter particles after inflation. We shall have to wait until Chapter 27 to see how this works.

If the dark matter is in cosmic strings, then the strings themselves are large-scale objects, so it might be thought that they would form points of attraction for ordinary matter easily. However, the situation is a little more subtle. We will see in Chapter 27 that cosmic strings have *zero* active gravitational mass: their huge density is exactly cancelled by the equally huge negative pressure, which acts in only one direction, so that the active gravitational mass is $\rho + p/c^2 = 0$. They do not curve time at all, so they do not directly form places where matter clumps. That is, a static string sitting in the middle of a cloud of gas does not pull the gas towards itself. Instead, strings curve space, and this can only be felt by matter that moves transversely across the string. If the string moves through space, as it would be expected to do, then matter flowing around it on one side is brought by this deflection into collision with matter flowing around it from the other side. These collisions could cause the over-densities that lead to the formation of galaxies. Cosmic strings could form galaxies in the wake they leave as they move through ordinary matter.

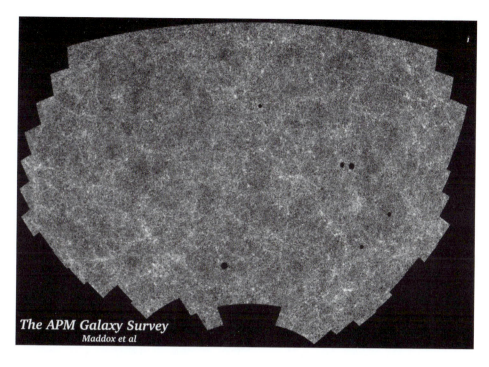

The APM Galaxy Survey
Maddox et al

Figure 25.2. *The galaxy distribution of the* APM *survey of a large region of the sky. The brightness of each point represents the number of galaxies there. The background is a smooth gray because of very distant galaxies, which are seen at an early time, so they do not clump very much. The brighter knots and filaments are made of nearby galaxies, which we see at a later stage of galaxy clustering. In fact these knots are superclusters, clusters of clusters, and the filaments between them are chains of clusters. Image courtesy of Steve Maddox, Will Sutherland, George Efstathiou and Jon Loveday, Astrophysics Department, Oxford University.*

Ghosts of the dark matter

Theories of how dark matter forms galaxies can be tested against observations in at least two ways, even without directly detecting dark matter particles in the laboratory. The first way is to look for special characteristics in the way galaxies clump. The second way is to look for traces of the dark matter's density irregularities in the cosmic microwave background. Both these tests currently strongly favor CDM .

Studies of large-scale statistics of galaxy positions have shown that, not only do galaxy clusters clump into larger superclusters, but the superclusters tend to be connected by linear filaments which are also locations of large numbers of galaxy clusters. That is, galaxy clusters form a kind of web-like distribution in space, with superclusters at the knots in the strings of the web. This is illustrated in Figure 25.2, which is a computer plot of the positions of millions of observed galaxies across a large portion of the sky.

Statistics of clustering that are inferred from this kind of survey can be compared with computer simulations of what kind of clumping one might expect in the CDM model. Figure 25.3 on the following page shows the results of four such simulations, viewed at three different times. Notice how similar the structures at the present time in the first row of the figure look to the structures in white in Figure 25.2. This is obtained with a CDM model in which the dark matter has a density equal to 30% of the critical density (as defined in the previous chapter), and a cosmological constant with 70% of the critical density, as is suggested by the microwave background studies we look at next. This simulation fits the observations better than ones with other assumptions.

The clumping of dark matter also has a small but measurable effect on the microwave background radiation. The temperature of the radiation is lower toward a strong clump than elsewhere. This is a gravitational redshift effect, but with a subtlety. If light falls into a gravitational field and then leaves it, we have seen that its path is deflected (gravitational lensing) but we did not mention any redshift. That is because there is no net redshift: the energy of the photon is conserved if the gravita-

In this section: observations today of the spatial distribution of galaxies and of the irregularities in the cosmic microwave background radiation give clues to the nature, density, and distribution of dark matter.

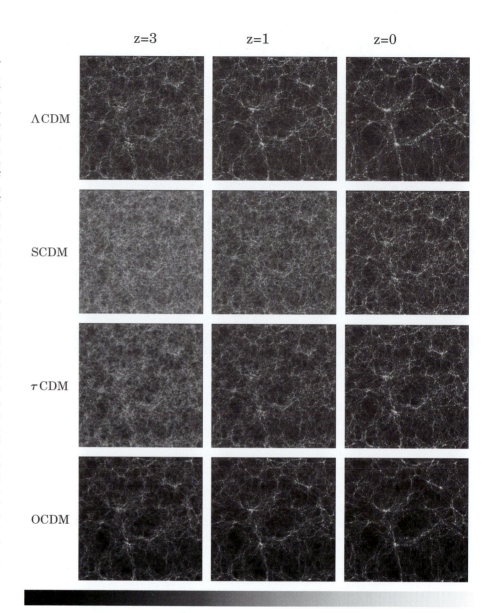

Figure 25.3. *Comparison of four simulations of galaxy clustering, with different assumptions about the cold dark matter. Each simulation is shown at three different times (redshifts) so that one can see how clustering gets stronger with time. Redshift z = 3 represents a time when the Universe was only one-quarter of its present size. Redshift z = 1 is when the Universe was one-half of its current size. Redshift z = 0 is the present time. Each image represents the same galaxies, so they are not shown to scale. In principle, one should reduce the images in the first column by a factor of four and those in the second by a factor of two, in order to see the expansion as well as the clustering. The top row is the preferred model, which produces clustering most closely like that in Figure 25.2 on the previous page and similar surveys. Its* CDM *has 30% of the critical mass, and it uses a cosmological constant with a further 70% of the critical mass density. The other models have different parameters. The second row, for example, is a model with no cosmological constant and a density of* CDM *equal to the critical density. The bottom row has the same* CDM *as the first but no cosmological constant. Figures made for the* VIRGO *consortium, by Joerg Colberg, published in Jenkins et al., 1998 Astrophysical Journal, 499, 20–40.*

tional lens is static. However, in the cosmological case, the gravitational field of the clump is getting stronger during the time that the photon passes through it, so the photon's energy is not conserved. (Recall our discussion of energy conservation and time-dependent gravitational fields in Chapter 6 in association with the slingshot mechanism.) The gravitational field is stronger when the photon leaves, so it loses energy and is redshifted.

This means that very precise measurements of temperature irregularities in the cosmic microwave background can give information on the density irregularities that were forming at the time of decoupling and later. The first such measurements were made in the early 1990s by the COBE satellite. We showed the overall spectrum measured by COBE in the last chapter in Figure 24.6 on page 354. COBE also measured the temperature fluctuations and showed that they have a size consistent with what

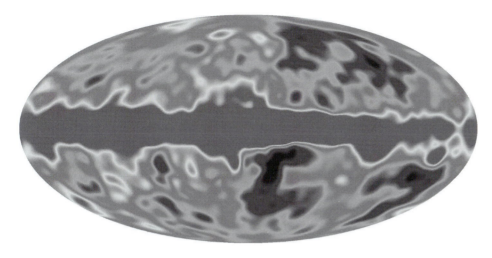

Figure 25.4. This map shows the irregularities in the temperature of the cosmic microwave background in different directions on the sky, as measured by the COBE satellite. The mean temperature has been subtracted; the fluctuations shown here are only of order 10^{-5} of the overall temperature. Courtesy COBE team and NASA/GSFC.

is needed to explain galaxy formation. The results, shown in Figure 25.4, give a striking visual representation of the CDM fluctuations before galaxies formed. Even better pictures are expected soon from the MAP satellite, which was launched in 2001. Later in the decade we can expect the launch of the most sophisticated cosmic microwave background satellite of all, called Planck.

Since COBE, more accurate measurements have been made by instruments that have been flown on high-altitude balloons. These experiments, called Boomerang and Maxima, examine fluctuations on much smaller angular scales, which are even more closely related to the galaxy count maps such as Figure 25.2 on page 379. These observations strongly favor the distribution of galaxies that would be produced by the CDM model with inflation, and are beginning to exclude cosmic strings as the main cause of galaxy formation. These experiments also have given independent information about the cosmological constant. We will see that evidence in Figure 27.2 on page 403.

What is the dark matter?

While galaxy clustering and the cosmic microwave background observations independently point toward the inflation with CDM model, they do not themselves tell us much about the CDM particle that is involved. All that is required to match observations is a neutral particle that does not easily scatter from baryons and whose mass is sufficiently large that the expansion of the Universe has cooled the random velocities of the particles enough to clump. As far as these results are concerned, the particles could be either as-yet undiscovered elementary particles (which we called WIMPs in Chapter 14) or, say, massive black holes.

The reason that scientists favor the former is that the element-formation studies that we described above require that the number of baryons should be a small fraction of the critical density, while the dark matter is a much larger fraction. So if the dark matter is in black holes, then either the black holes formed from some non-baryonic particles (in which case we still need unknown particles), or the black holes formed from baryons before the epoch of helium formation. Since the Universe was hot and smooth at that point, forming black holes would either require some exotic quantum process at the end of inflation, or it might require a very radically different model of the very early Universe in which there was a cold Big Bang. Neither of these options is simple, so physicists will continue to assume that CDM is an elementary particle until it is found in laboratory experiments or until experiments show somehow that these particles do not exist.

In this section: most astrophysicists favor the idea that the dark matter is made of uncharged elementary particles that do not feel the nuclear force. Other models are possible, and intensive searches are underway to identify the particles.

▷ Theoretical physicists would not be surprised to find WIMP particles. Modern theories suggest a large collection of so-called supersymmetric particles, many of which could have the right properties for CDM.

One of the components of the dark matter seems to be neutrinos. The accumulating evidence from studies of solar neutrinos is that neutrinos have a small mass, which is not zero but is smaller than 1 eV. Given, as we saw above, that neutrinos should be as abundant as microwave background photons, this is a significant amount of matter. Neutrinos could form up to 5–10% of the dark matter.

But neutrinos cannot be responsible for galaxy formation, because they were not "cold" at the time of decoupling. At that point, the neutrino temperature was of the same order as the photon temperature, an energy equivalent as we have seen above of a few electron volts. With rest-masses much smaller than this, neutrinos would have been moving at close to the speed of light at decoupling, and they could not have formed the stable, tight clumps needed to start galaxy formation. So neutrinos constitute **hot dark matter**: interesting, but not sufficient to complete our picture of the composition of the Universe.

The hunt for this elusive component of the Universe is one of the most interesting experimental activities today. Deep in underground laboratories, a number of groups of scientists are monitoring very sensitive equipment for evidence of unexpected particle events that cannot be explained by known particles. The labs are underground to screen out cosmic rays, which would otherwise create such a large background of events that the desired ones would be hard to identify. Some groups are looking for new kinds of nuclear reactions; others simply look for the tiny heat and sound waves generated by a collision between a dark matter particle and the material of the detector. Experiments are getting more sensitive all the time and have begun to put constraints on dark matter models. A direct detection of a dark matter particle, and a solution of the mystery of how they fit into our modern picture of particle physics, may be only a few years away.

Einstein's Universe:
the geometry of cosmology

I n the last two chapters we have made a lot of progress in exploring the future and past of the Universe, basically just by using local Newtonian gravity. We argued that the dynamics of an expanding, homogeneous and isotropic cosmology can be calculated from Newtonian gravity, at least if the pressure in the Universe is negligible, because all we need to look at is the local Universe, the part nearest us. The assumption that the Universe is homogeneous guarantees that the rest of the Universe will behave the same as our local region.

But this line of reasoning has its limitations. Even if we calculate the dynamics of the Universe this way, we don't learn what the distant parts of the Universe will *look like* in our telescopes. The curvature of space, which is not part of a Newtonian discussion, will affect the paths of photons as they move through the Universe. Moreover, if we want to ask deeper questions about the Universe, such as those we pose in the next chapter, then we should know something more about its the larger-scale structure. For this, we must turn to full general relativity. Only general relativity can provide a consistent picture over the vast scales we shall need to explore, out to where the cosmological speed of recession approaches the speed of light. So it is now time to learn about Einstein's description of cosmology.

Cosmology could be complicated …

As we have seen, Einstein's theory has the simplifying property that only matter within our past light-cone – matter that can send signals to us – can have influenced the evolution of the Universe we observe. This is logically much more satisfying than Newtonian gravity, where matter everywhere affects us with its gravity instantly. In fact, scientists did not study cosmology seriously before Einstein: the logical difficulty of applying Newtonian gravity to an infinite Universe, coupled with the fact that astronomers before the twentieth century had no idea how large the Universe was, left scientists with little to work with. When Einstein's theory showed how to treat gravity in a causal way and provided consistent cosmological models, scientists began to explore the subject.

The basically Newtonian view of cosmology we developed in the last two chapters was still based on relativity: we had to use the two facts that (1) only matter in our past light-cone affects our gravitational field, and (2) general relativity allows us to ignore the gravity due to spherical mass distributions further away from us than the galaxy whose motion we are computing. For homogeneous universe models, we were then able to ignore most of relativity and study the dynamics with essentially Newtonian equations. We will develop below the relativistic counterparts of these model universes, and we will see that in many situations they are remarkably similar.

But relativity is richer than Newtonian gravity. There are model universes that are not describable in Newtonian terms. Here is an example.

Imagine a homogeneous universe in which the expansion is different in different directions. For example, imagine that the Universe were expanding at twice the rate

In this chapter: we explore the three different geometries that a homogeneous and isotropic cosmology can assume. We see how to construct two-dimensional versions of these, which shows us why there are only three possibilities. We see how astronomical observations can measure this geometry directly.

▷The drawing under the text on this page illustrates how complicated three-dimensional solid objects could be. Why is the Universe apparently so simple?

In this section: the large-scale shape of the Universe could be very complex. Even if the Universe is homogeneous, it could be anisotropic: different in different directions.

▷Anisotropic expansion is different from what an observer would see if the extra speed of "expansion" were really caused by the observer's own motion, for then in one direction galaxies would be receding more rapidly than in perpendicular directions, while in the opposite direction galaxies would be receding more slowly.

▷Anisotropic expansion also challenges Newtonian cosmology: if the universe is infinite, should the gravitational field be calculated by dividing space into spheres rather than, say, into ellipsoids with the same shape as the expansion velocity? Newtonian gravity offers no unique resolution, while general relativity gives unique answers.

In this section: observations give no evidence that the Universe has any other large-scale geometry than the simplest: a homogeneous and isotropic space.

in a particular direction as in any perpendicular direction. Notice that, because this is an expansion, it looks the same if one looks in one direction or in exactly the opposite direction (which means turning 180° around). So the rapid expansion we are imagining occurs in both directions along a particular line.

Now, just having a higher expansion rate in one direction does not destroy the *homogeneity* of the universe: no matter where the observer is, the expansion in that particular direction would be twice as fast as in perpendicular directions. Just imagine a sheet of rubber as a two-dimensional universe model, and let the sheet be stretched in only one direction. It can remain homogeneous – the same everywhere – even as it expands. Because not all directions are the same, the expansion is not isotropic (recall the discussion of isotropy in Chapter 7). We say that this kind of universe model is homogeneous but *anisotropic*.

This sort of expansion could occur in a Newtonian universe too: if we start the universe off with such an expansion, then Newtonian gravity will keep it expanding in this anisotropic way. But in an Einstein universe model, the anisotropic expansion changes the gravitational field itself, and such models can differ dramatically from Newtonian ones.

So in principle, cosmology *could* be much more complicated than the Newtonian universe models we have studied so far.

...but in fact it is simple (fortunately!)

But the observational evidence all supports the simple cosmologies:

> When we look for evidence of this kind of anisotropy in the real Universe, we find none.

In particular, the cosmic microwave background radiation does not show any big systematic effects of this kind: once we have removed the Doppler effect of our own motion, its temperature is the same in all directions to a high accuracy. If the Universe were expanding anisotropically, we would expect to see one temperature along the direction of the more rapid expansion (in both directions along this line), and a different one in perpendicular directions. The deviations that we do measure seem to be random: there is no large effect along one line.

> This is the **homogeneity/isotropy problem**: out of all the possible kinds of universes we might have found ourselves in, it seems puzzling that ours is so nearly homogeneous and isotropic: why has Nature provided us with such a simple arena to play in?

The idea of inflation, which we shall study in Chapter 27, is an attempt to provide an answer to this question, among others.

Gravity is geometry: what is the geometry of the Universe?

In this section: we learn how to describe and measure the curvature of a homogeneous and isotropic space.

Einstein described gravity in terms of geometry. So when we look for model Einstein universes similar to the Newtonian ones we met in Chapter 24, we need to look for *geometries* for three-dimensional space that embody the remarkable homogeneity and isotropy that we see around us. When we do, we find that things are much simpler than we might have expected.

There are in fact only *three* possible kinds of homogeneous and isotropic models. They are commonly called the closed, open, and flat universe models. These three cosmological models were first discovered by the Russian physicist and meteorologist Aleksandr Friedmann (1888–1925). In Investigation 26.2 on page 389 we look at their geometry in some detail. But it is important here to understand why there are only three, and what they look like.

We are all familiar with at least one three-dimensional space that is isotropic and homogeneous: the standard Euclidean space that we grew up thinking we lived in! This flat model universe is one of Friedmann's geometries. What about curved spaces: are there any that are still homogeneous and isotropic?

Now, there are many ways to distinguish a flat space from one that is curved, but if the space is isotropic, so that nothing changes from one direction to another, then things are fairly simple. Notice first that a space that is fully isotropic has to be homogeneous. The reason is that, if it were *not* homogeneous, then there would be at least one place where things were different from other places (say, a "bump" somewhere). Now, if we were to stand anywhere else in the space and look around ourselves, we would see that space was different in one direction (looking towards the "bump") than in other directions: the space could not be isotropic about our location. For the space to be isotropic about all its points it must also be homogeneous.

Now, if a three-dimensional space is isotropic, then we can be sure that we can draw a perfect sphere about any point in it: since all directions are the same, then we just form a surface from all the points whose distance from the central point is constant. We call this distance the *radius* of the sphere. The sphere will be identical to a sphere in Euclidean space, and in particular we can draw a great circle on it, say its equator. The length of this circle is called the *circumference* of the sphere.

All this may seem trivial, just elementary geometry, but it turns out that such spheres give us a very simple measure of the curvature of the space we are in: just draw a sphere around any point and take the ratio of its circumference to its radius. In a Euclidean space, this ratio is of course just 2π. The converse is also true: if we are in an isotropic space and we find that this ratio is exactly 2π for one sphere drawn about one point, then it will be exactly 2π for a sphere of the same radius drawn about any other point, and by constructing all of these spheres we can build up the space and find that it must be Euclidean. How to do this is shown in Investigation 26.1 on the next page.

More interestingly, suppose we have an isotropic space and we measure the ratio of circumference to radius of one particular sphere and find that it is *smaller* than 2π. Then that space must be curved. Since the space is homogeneous, spheres of that size drawn around other points will have the same ratio, and (again as we show in Investigation 26.1) we can construct the space just from these spheres. In particular, the circumferences of other spheres of different sizes will all be determined by the properties of the first sphere we chose.

The sphere we chose first is not the only one that would generate this space: any other sphere in this space would have worked just as well. Spheres of different radii may have different ratios of circumference to radius, so this ratio for an arbitrary sphere does not characterize the curvature of the space. One way of defining the space and its curvature is to look at a *particular* sphere, say the one that has a radius of 1 m. Its circumference is a single number that completely defines the space: there is one and only one space for which a 1 m-radius sphere has that circumference.

> We have learned that a homogeneous and isotropic space is determined by giving just *one* number.

If that circumference is 2π m, then the space will be a three-dimensional Euclidean space.

But what if the circumference of the 1 m sphere is *less* than 2π m – is that really possible? How can one imagine a space that has circles whose radii are too large for their circumferences? The answer is that it is actually very easy to visualize such a space if we go down to two dimensions and try to find a two-dimensional surface

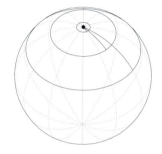

Figure 26.1. *When we do geometry on the sphere, we find that circles have smaller circumferences than they do in the flat Euclidean plane. Suppose that the sphere shown here has a radius of 1 m. Three circles centered on the pole are displayed. The smallest circle has a radius, measured along an arc drawn on the sphere down from the pole, of 0.1 m. A measurement of the length of its circumference would in fact give 0.62727 m, so their ratio is 6.2727, which is only 0.9983 times 2π. For the middle circle, whose radius is 0.5 m, the ratio is 0.9589. And for the largest circle shown, of radius 1 m, the ratio is 0.8415. The key point is that we are talking about the geometry of the sphere, so the radius of a circle must be measured along the sphere, not along a line that extends out into the surrounding three-dimensional Euclidean space.*

Investigation 26.1. The sphere, and only the sphere: a two-dimensional universe

Our goal here is to see why there is only one kind of two-dimensional homogeneous and isotropic space in which circles have smaller circumferences than 2π times their radii. We shall do this by construction, by making a sphere out of the circles.

We start from the premise that our space contains at least one circle whose circumference is a smaller multiple of its radius than it would be in flat space. We begin by choosing a size for this circle and drawing it in three-dimensional Euclidean space.

Our next step is to choose the location of its center, by which we mean the point on the two-dimensional sphere that will be its center. If we place this center in the plane defined by the circle, then the radius will be the circumference divided by 2π, which is exactly what we have in flat space because the plane containing the circle is flat. To make a sphere we need to give the circle a larger radius for its circumference. To do this we simply move the point that represents its center out of the plane of the circle, along a line perpendicular to that plane. Then our circle-plus-center looks like the largest circle in Figure 26.1 on the preceding page. Our circle is shown in the upper left in Figure 26.2. Its center is not shown in this diagram. We will see that choosing the first circle and the location of its center fully defines the 2-sphere.

What we cannot yet do is to draw the circle's radius in the 2-sphere, because that would require us to know the shape of the surface between the center and the circle. We have to "build" this space, so all we can do is start with a center displaced from the expected position. We can say something about the radius, however. The two directions along the circle must be equivalent, or the space would not be isotropic. Therefore, the radius must intersect the circle at a right angle. Any other angle would allow one to distinguish one direction from the other around the circle. Since this applies to smaller circles as well, the radial curve must lie in a plane perpendicular to the plane of the circle itself, and containing the displaced center.

The key to the construction of the sphere from this first step is a simple sequence of operations. Consider any point on the circle itself. This is a point of the space we are constructing, and that space is homogeneous. Therefore, if we move to this new point and look at the space, it must look the same. In particular, there must be another circle-plus-center, identical to the one we have just defined, in the space, with its *center* at this new point. The key to the construction is to determine how this second circle-plus-center fits with the first one. Its center is the new point, but how is it oriented?

The first part of the answer is that the second circle has to pass through the center of the first circle. This is because the points on the circle all have the same distance from the central point, and are indeed defined by this distance. Since the new point has this distance from the first center, then the first center must be on the equivalent circle drawn around the new point.

The second part of the answer follows from this: the new and old circles must *share* the same radial curve. The reason is that the radius must be the shortest distance within the space between the center and the circle, and so it must be the same curve whether we regard one point or the other as being at the center or on the circle. Now, we have already seen that this radial line must be perpendicular to the old circle. It must therefore also be perpendicular to the new one, and so both the old and new circles lie in planes perpendicular to the radius. This fixes the orientation of the new circle: given its center on the first circle, it must pass through the first center

and lie in a plane perpendicular to the radial curve. We construct the new circle by rigidly moving the first one until its center is on the new central point, and we tilt it until the new circle passes through the old center, keeping it perpendicular to the plane that we just identified.

The old and new circles are shown in the upper right in Figure 26.2. Notice that the two circles intersect at two points. This is good: we want them to form parts of the same surface, so they should certainly not pass over or under one another.

Now, the space we are constructing is also isotropic, so the same thing must happen at any other point on the first circle. This leads to a set of new circle-plus-center constructions distributed around the original circle, all passing through the first central point. We show in the bottom left in Figure 26.2 the members of this family that are separated by 1 radian around the first circle. By construction, these all intersect at the point that we chose as the center of our first circle, so the location of this point in our space is now clear from this diagram.

Then one can further build up the space by doing the same on each of these circles, allowing them to spawn more circles by taking their points to define new centers. The result of allowing one of the secondary circles to spawn more is shown in the lower right of Figure 26.2. It should be clear by now that we are filling in the ordinary 2-sphere. The radius of the sphere is determined by the displacement of the center above the plane of the circle that we adopted for the first circle: the smaller the displacement, the larger the sphere, and the closer to flat space the space is. But for any non-zero displacement, the construction will give a 2-sphere.

If by chance we had chosen a circle-plus-center object that turned out to span exactly $90°$ on the 2-sphere, then the circles would keep repeating and would not fill in the whole sphere, but with the exception of a set of such special objects, the repeating circle-plus-center objects will eventually pass through every point of the sphere.

This construction does not entirely rigorously show that the 2-sphere is the only two-dimensional homogeneous and isotropic space containing circles whose circumferences are smaller than 2π times their radii. Strictly speaking, it is only part of a rigorous proof. The rest of the proof must address the question of whether we were justified in assuming in the first place that our circle-plus-center should be drawn in three-dimensional Euclidean space. Maybe we would get another kind of surface, not a 2-sphere, if we started in a different space, say five-dimensional Euclidean space.

It is not hard to show that there is nothing new in five-dimensional Euclidean space. But there is something new if we start in the Minkowski spacetime of special relativity, as we describe in Investigation 26.2 on page 389. In such a space it is possible to draw circles whose radii are *shorter* than the circumference divided by 2π. By combining those circles in the way we have done here, one constructs, not a sphere, but a hyperboloid.

There are thus just three possibilities for constructing a homogeneous, isotropic, two-dimensional space: construct a sphere in Euclidean space out of circles with radii that are too large, construct a hyperboloid in Minkowski spacetime out of circles with radii that are too small, or construct a plane out of circles with radii that are just right.

Goldilocks would be pleased!

---

where the radii of circles are extra large compared to the circumferences. We have only to look at an ordinary sphere, as in Figure 26.1 on the preceding page. There we see three circles centered on the same point. Their radii are arcs drawn on the sphere, and from this it is easy to see that as the arcs get longer the circles do not grow in circumference as rapidly as they do in flat space.

Now, the sphere is a two-dimensional, homogeneous and isotropic space. Every such space in which circumferences are "too small" is a sphere of some size. The reason is that one tiny patch of the surface determines the whole surface, by homogeneity and isotropy. So if, as one increases the size of a circle from nothing to a tiny amount, the circumference begins to lag more and more behind 2π times the radius, and one only needs to know how serious the lag is: in one centimeter, does the circumference fall below $2\pi \times 1\,\mathrm{cm}$ by 1%, by 0.1%, by 0.01%, or ... ? This fractional lag determines the radius of the sphere: the smaller the lag, the more nearly flat the space is, so the larger the radius of the two-dimensional sphere. The uniqueness of the sphere can be shown by explicitly constructing the sphere from the circles. We show how to do this in Investigation 26.1 and Figure 26.2.

Friedmann's model universes

We have already seen that one of the model universes in general relativity will be three-dimensional Euclidean space, the obvious homogeneous and isotropic three-space. We have also seen that in two dimensions the sphere is likewise homogeneous and isotropic. The generalization of the two-dimensional spherical surface to 3 dimensions is an important geometry that we call the three-sphere.

The 3-sphere is defined as the set of all points that have the same distance from a central point in four-dimensional Euclidean space. Now, it is not easy to visualize four-dimensional Euclidean space, so I don't recommend trying. The properties of the three-sphere are very like those of the ordinary sphere (the 2-sphere). In particular, it is the only homogeneous and isotropic three-space in which spheres have circumferences that are "too small". Therefore, the three-sphere is also one of Friedmann's model universe geometries. It is usually called the because it has finite size.

Before asking about the third kind of model universe, I want to warn the reader about one pitfall in studying Figure 26.1 on page 385. It may seem that the excess radius is a cheat, that the circles would look like normal circles if we looked instead at their radii, not running along the sphere back to the pole, but in a plane slicing through the sphere and containing the circle. Since that plane is flat, the ratio of circumference to this kind of radius is still 2π. This is of course true, but irrelevant. Figure 26.1 is meant to illustrate a property of the *intrinsic* geometry of the sphere, i.e. what we would measure about circles if we were little ants confined to the surface of the sphere, only able to lay out lines and take measurements on the surface. We must confine ourselves to the surface in order to make a good analogy with the three-dimensional case that we are interested in: when we generalize to a three-sphere, then there is no physical way to slice through the circle to get it to be flat: one would have to slice into a fourth, unphysical, dimension to do this. In all three dimensions of a three-sphere, circumferences grow more slowly with radius than in Euclidean space.

And what about the opposite case, where circumferences grow more rapidly with radius than in Euclidean space? By the same reasoning, there is only one kind of three-dimensional geometry that has this property. It is harder to visualize this geometry, however, because an analogous two-dimensional surface cannot be drawn in

In this section: there are three types of homogeneous and isotropic Universe model: spherical, flat, and hyperboloidal. We learn how to construct them explicitly.

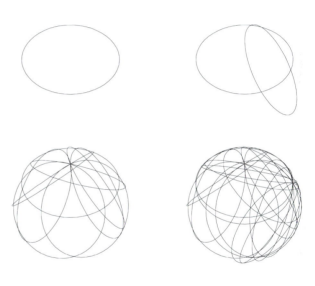

Figure 26.2. *A construction to show that the sphere is the only two-dimensional surface that is homogeneous, isotropic, and contains circles whose radii are larger than their circumferences divided by 2π. The four stages of construction are described in Investigation 26.1.*

▷This is, indeed, the situation we face if we do civil engineering on a large scale on the Earth: the way large road networks mesh is significantly non-Euclidean, and one can't make a faithful roadmap on a flat piece of paper if the area described is more than a few hundred kilometers in size.

three-dimensional Euclidean space like the 2-sphere can be. We describe in Investigation 26.2 how the geometry can be drawn as a surface in the Minkowski spacetime of special relativity. In this space, it looks like a **hyperboloid** rather than a sphere, and so we call it the hyperbolic model universe, or the *open model universe*. It is infinite in extent.

▷If the Universe we live in is really described by a three-sphere or 3-hyperboloid, is there any physical reality to the fourth dimension of a Euclidean or Minkowskian space in which the sphere is drawn? Is that dimension "really" there? There is no logical necessity for it to exist, since all our experiments are performed in three dimensions, but it might be philosophically pleasing if it did. Maybe this dimension might actually be one of the dimensions in the higher-dimensional spaces of string theory (see Chapter 27).

Friedmann did more than just characterize the geometries. He showed how each geometry is linked to the properties of the matter that create it. Recall our discussion of spatial curvature, Equation 19.2 on page 244, which asserts that in Einstein's equations, spatial curvature is produced by the combination $\rho - p/c^2$, at least for isotropic pressure. We wrote this down in the weak-field case, assuming the geometry was close to flat spacetime. In the case of cosmology the geometry has an added feature, namely the expansion. Friedman's geometries are spheres or hyperboloids at any one particular time, but they expand. The result of this is to modify the expression and replace the pressure by a term involving the Hubble expansion parameter:

$$\text{curvature of space} \quad \propto \quad \rho - \frac{3}{8\pi G}H^2. \tag{26.1}$$

Effectively, the pressure is replaced by an effective pressure equal to $3H^2/8\pi G$ when determining the spatial curvature. This applies even when there is real pressure in the cosmology: the sum of the pressure and any kinetic stresses (ram pressure as in Chapter 19) is replaced by the expansion term in H^2.

Now, if we use Equation 24.10 on page 361 to replace H^2 by the critical density ρ_c we get the simpler equation

$$\text{curvature of space} \quad \propto \quad \rho - \rho_c. \tag{26.2}$$

A universe that has just the critical density will have zero curvature, so it will be spatially flat; one that has more than the critical density will have positive curvature, which is to say that it will be the closed three-sphere; and one that has less than the critical density will be the open, hyperboloidal model.

The large-scale geometry of the three models is very different: the sphere is a finite space, the hyperboloid is infinite, and the flat model is, well, flat. It should not be a surprise that it is not possible for one model to transform itself into another at some time: the geometrical class is preserved during the evolution of the cosmology.

> If the density is critical at one moment (so that the Universe is flat) then it remains critical for all time.

This is not achieved by some mysterious mechanism; all that happens is that the acceleration equation (Equation 24.12 on page 362) insures that the Hubble constant changes with time at just the right rate to keep the critical density equal to the current mass density.

For matter-dominated cosmologies, the critical density is also related to the escape speed.

> Closed matter-dominated models will re-collapse, while open models expand forever.

If pressure is significant, however, this simple link between curvature and the future of the Universe is broken. A model could have more than the critical mass density, and so be closed; and yet in principle it could have enough pressure to make its active gravitational mass smaller, or even negative, so that it expands forever. We saw in Chapter 25 that a universe dominated by a cosmological constant will expand exponentially fast, so it never reaches a constant speed.

Investigation 26.2. The three Universes of Friedmann

In this investigation, we explore the nature of the three Friedmann geometries for the Universe. Since the shape of the Universe outside our past light-cone does not affect us, we can make any assumption we want about it. It simplifies the discussion, and fits the Copernican principle, if we assume that it is exactly like the part of the Universe we do observe. But we must bear in mind that this is only a matter of convenience.

We therefore ask what kinds of three-dimensional spaces are homogeneous and isotropic everywhere. Clearly, flat Euclidean space fits our requirement: no matter where we are, the geometry is the same as anywhere else, and in all directions. This is the simplest of the universe models.

The second kind of homogeneous and isotropic three-dimensional space is the three-sphere. It is a generalization of the usual sphere, which is called the 2-sphere because its surface is a two-dimensional (curved) space. (The 1-sphere is simply a circle, of course!) A perfect sphere is the same everywhere: the geometry has no bumps or defects to tell one where one is, and turning the sphere around does not change anything either. The 2-sphere is a two-dimensional isotropic homogeneous space.

The three-sphere is the same, extended to one more spatial dimension. Its formal definition is the set of all points in four-dimensional *Euclidean* space that are the same distance from a given point. Most people find it hard to visualize such a thing. I recommend not trying to, but instead trying to understand its properties by generalizing from 1- and 2-spheres.

Like a 2-sphere, which has a finite area, the three-sphere has a finite *volume*, but nevertheless a curve drawn in it never encounters an edge: curves just keep circling around and around the space. A three-sphere it is what mathematicians call a finite and *closed* space, which means it is finite but has no boundary.

Now, one can measure the radius of a sphere by measuring the way the circumferences of circles increase with radius, as in Figure 26.1 on page 385. We constructed a 2-sphere from this information in Investigation 26.1 on page 386. This is an important point: if we walk outwards in our own Universe and find that the circumference is not increasing as rapidly as we would expect in flat space, then it must be a spherical Friedmann universe, and we can even measure its radius! Once we have measured the size of the three-sphere by examining circles in some small region, the rest of the geometry has to wrap itself up into a three-sphere of this size, or else it must be inhomogeneous.

The other possible universe model is the one where the circumference of a circle increases *more* rapidly with radius than in flat space. This space can be described as a subset of points four-dimensional space *Minkowski spacetime*, which we met in Chapter 17. The third three-space has a definition that is closely analogous to that of the three-sphere: it has the same geometry as the set of all points that are at a constant *timelike spacetime-interval* from a given point in Minkowski spacetime.

What this means is the following. Choose any event in Minkowski spacetime to be the origin, $t = x = y = z = 0$. The timelike spacetime-interval between the origin and another event is just the time ticked on a clock that travels at a constant speed from the event at the origin to the other event. Another clock, traveling at a different speed and in a different direction, will reach a different event after the same time has elapsed on it. The set of all such events, reached by all possible clocks traveling at less than the speed of light, is the third space that is a possible Universe model. Because the spacetime-interval is given by Equation 18.6 on page 230, the equation for this space is

$$c^2 t^2 - x^2 - y^2 - z^2 = \text{const},$$

which is the equation of a hyperboloid in Minkowski spacetime.

This space is homogeneous because all observers agree that all possible clocks started at a particular event and arrived at the surface with the same *proper* time. This is observer-independent. Different observers regard different clocks as being at rest, so they would place the origin of coordinates in different places, but the space would look the same to them.

How do circles behave in this space? In this case, as a circle is enlarged, its radius increases along the hyperbola. Now, the hyperbola is getting closer and closer to a lightlike direction as we move outwards from the center of the coordinates. This means that the proper length of the radial curve does not increase very much as the circle is enlarged: lightlike lines have zero proper length. The result is just the opposite of the case for the sphere. As the circle increases in circumference, the radius fails to increase much, and the ratio of circumference to radius is *larger* than in flat space.

Just as for the sphere, even a small circle is enough to tell us the amount of curvature. Therefore, once we have determined how the curvature affects one circle, we have determined the geometry of the space. Mathematicians call this sort of curvature *negative*.

The hyperbolic three-space is still a true space, despite the fact that we have constructed it in a spacetime. Any two points within it are separated from one another by a spacelike spacetime-interval because they happen simultaneously to an observer whose velocity is the average of the velocities of the clocks that reach those points.

This space is not closed. With respect to a given experimenter, clocks that travel very close to the speed of light take longer and longer to tick the given time-interval. This is just the time dilation of special relativity, and it means that these clocks define points on the space that get further and further away from the origin. The hyperboloid just keeps going in Minkowski spacetime. It is an *open* space.

We have constructed three model universe spaces. And this is all there is: our construction leaves no room for any other homogeneous and isotropic spaces.

Cosmologists therefore speak of three possible Universe models: the flat, closed (spherical), and open (hyperboloidal) models. These names refer to their overall structure, which depends on the regions outside our past light-cone. However, their local geometry also reflects their shape, and this is measurable. We must always bear in mind that the only observable features of these models are their sections inside our past light-cone.

Notice that we can determine the geometry of the Universe by measuring the total mass density of the Universe in our neighborhood and the Hubble parameter. We will see in the next chapter that, when we do this for the observed Universe, we find that we appear to live in a nearly flat cosmology. The density of matter plus that of the dark energy is enough to reach the critical density, within observational uncertainties. Flatness is another prediction made by inflationary models of the early universe, as we will see in the next chapter.

However, it is important always to bear in mind that we can ever only know about a finite portion of the Universe, and our conclusions only apply to what we can see within our past light-cone. Beyond our particle horizon, the geometry might be very different. And if it is – so that the Universe is not homogeneous – then its future evolution will be different from what we would predict as well.

What the Universe looks like

How do we measure the curvature of the Universe? The simplest way is to try to measure the total mass density and the Hubble constant, and take the difference between the observed density and the inferred critical density, as in Equation 26.2 on page 388. This is an indirect determination of the curvature, since we are measuring the quantities that produce the curvature, rather than the curvature directly. But it has the great advantage that astronomers are already measuring the Hubble constant and the density for many reasons. The best estimates today suggest that the curvature is close to zero, that we live in a nearly flat cosmology. We will look more at these measurements in the next chapter.

Another approach might be to look for direct evidence of curvature. Distant galaxies and quasars are separated from us by a curved spacetime. There must be evidence of this curvature in observations of them.

The kind of evidence that one might look for is not hard to imagine. It follows directly from our way of defining the curvature of the Friedmann universe models: a geometry is curved if the circumference of a circle divided by its radius does not equal 2π. Now, if we observe a distant object whose size is known, then if we measure its *angular diameter*, then the circumference of a circle at the distance of an object is just its size divided by its angular diameter. We can measure the radius of the circle if we know how far away the object is, which is given by its redshift.

Figure 26.3 illustrates an extreme example of what can happen in a curved universe model. In the figure, the Universe is the closed three-sphere, and two objects (galaxies) of the same size are observed at different distances. Since the size of circles around the sphere begins to decrease as you go further and further away, the fraction of a full circle occupied by the more distant galaxy is much larger than that of the nearer, so its angular size is actually *larger* than that of the nearby galaxy, despite being further away. The general rule is that if the universe model is closed then angular diameters decrease less rapidly with distance than in a flat universe, and if the model is open they decrease more rapidly.

Astronomical observations that would reveal such an effect require a **standard meter stick**, an object whose physical size is the same in the early Universe as it is today. Like standard candles, a class of objects must be found whose size is not affected by evolution, or at least whose evolution in time is understood. A number of possible objects are currently the subject of investigation, including the angular sizes of the fluctuations in the microwave background radiation. It is possible that in the near future astronomers will have an independent check on the curvature of the Universe by measuring angular diameter evolution.

One theory actually predicts the curvature of the Universe: if the very early Universe underwent a period of inflation, as many astrophysicists now believe, then it should have stretched space nearly flat. Inflation is one of the subjects at the frontiers of physics and cosmology, to which we now turn.

In this section: the curvature of space produces larger angular diameters in distant objects than in flat space. This could be observed by astronomers.

▷The angular diameter is the angle on the sky that the object appears to occupy in our observation. This concept was first introduced in Chapter 5.

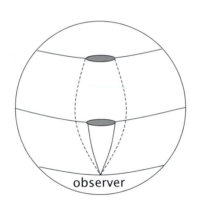

Figure 26.3. *The curvature of the Universe can dramatically affect the apparent angular size of an object. Two galaxies of identical physical size (shaded ovals) are at different distances from the observer in a closed spherical universe. Since we only want to see the effect of spatial curvature, we suppose the sphere is not expanding. Light travels along great circles, so the light rays from the more distant object to the observer (dashed lines) can arrive with a wider angle than those from the nearer. This would make the more distant object look larger than the nearer.*

Ask the Universe:
cosmic questions at the frontiers of gravity

The study of cosmology presents today's physicists with the biggest challenges to their understanding of gravity and of fundamental physics in general. Both on theoretical and on observational grounds, it seems that we will not be able to understand cosmology well until we understand physics better than we do today. But it also seems that cosmology could provide us with the keys to that deeper understanding of physics.

The biggest gap in physics is quantum gravity: we do not yet possess a consistent way of representing gravity as a quantum theory. There is no uncertainty principle in general relativity, no quantization of gravitational effects, no need to use probabilities in making predictions about the outcome of gravitational experiments. This seems inconsistent with the fact that all material systems that create gravity are quantum systems: if we can't say exactly where an electron is, how can we say exactly where its gravitational field is?

Many physicists believe that the way to quantize gravity is to unify it with the other forces of nature in a single theory in which electromagnetism and gravity are just different members of a single family of forces, and in which the unity among these forces only becomes apparent at very high energies, near the Planck energy. One would expect such a theory to predict new phenomena at these high energies. The only places where we know such energies have been met in the history of the Universe are (1) inside black holes, and (2) at the Big Bang. Phenomena inside black holes are hidden, but the Big Bang is very visible.

> By a combination of theory, experiment, and observation, physicists and astronomers hope to use the Universe as a laboratory to make big advances in physics. Cosmology is a hunting ground for clues to the ultimate unification of the physical forces.

Fortunately, there are many clues. We have met a number of them in passing during earlier discussions. Here is my personal list of cosmic puzzles whose solutions have the potential to revolutionize our understanding of physics.

- *Clue.* The Universe on the large scale is *homogeneous* and *isotropic*. Regions that, in the conventional Big Bang model, have not yet had enough time to have communicated with each other (see Figure 27.1 on page 393) are nevertheless very similar. They have the same density of matter, the same numbers and types of galaxies, the same degree of clustering; the microwave background radiation has the same temperature in all directions around us to a few parts in 10^6! How was this arranged?

- *Clue.* Galaxies could not have formed without *dark matter* seeds. Experimental searches for the dark matter may soon show what the cold component is. If it turns out to be a new elementary particle, then its identity will be a clue

In this chapter: we confront the limits of modern physics with puzzles and clues from cosmology. They have to do with the large-scale properties of the Universe, the formation of galaxies, and event the formation of life. The next big step in theoretical physics will be the unification of gravity with the other forces. The resulting theory should be able to address the questions we ask here, and go beyond them. It should clarify quantum theory, and even tell us something new about time itself.

▷ The unified theory could also predict new phenomena at lower energies, but none have been noticed in experiments. Some physicists have recently suggested that they might modify Newtonian gravity, making it stronger on distance-scales shorter than some characteristic length, which could be as large as 1 mm!

▷ The image under the text on this page is a *simulation* of the kind of data expected from NASA's MAP satellite, which began observation of the cosmic microwave background shortly before the completion of this chapter (2002). It will provide the most detailed map to date of the microwave background's irregularities. (Compare with Figure 25.4 on page 381.) These in turn will give physicists their best measures so far of the conditions in the very early Universe. Even higher-resolution data should come from the Planck mission, planned for launch by ESA by 2007. Courtesy MAP Science Team and NASA/GSFC.

to new kinds of physics. There are plenty of candidates for this new physics already, but scientists need experimental or observational data to tell them which ideas are right.

- *Clue.* When astronomers make their best estimate of the total mass density of the Universe, adding in the dark matter and dark energy densities, they find it *equals the critical density* as defined in Equation 24.10 on page 361. This is a very special value, because if the Universe is critical at one time, it remains critical for all time. Many physicists feel that this should have an explanation.

- *Clue.* Observations of the cosmic microwave background strongly support the idea of *inflation*, that the Universe underwent a very early phase of enormously rapid expansion, which was driven by dark energy with a negative pressure, like a temporarily large cosmological constant. The cause of inflation is shrouded in our ignorance about physics at the highest energies, but it is already clear that many fundamental processes can mimic a cosmological constant.

- *Clue.* Observations of the expansion of the Universe seem to show that the Universe has again entered a phase of *accelerated expansion* with a much smaller dark energy. This could be a remnant of the earlier inflationary phase, or a new physical field, or a permanent cosmological constant (or all three!).

- *Clue.* Theories of high-energy physics suggest that the Universe may contain *cosmic strings*, long concentrations of dark energy, thinner than any elementary particle. Cosmic strings do not curve time but they do curve space, and they could be detected by gravitational lensing.

▷For comparison with the flux of these cosmic rays, recall that in Chapter 11 we saw that in each second ten billion neutrinos of much lower energy pass through your body alone!

- *Clue.* Observations of the *highest-energy cosmic rays* have shown that the Earth is struck by about one cosmic ray with an energy larger than 10^{20} eV each second. This is a tiny flux of particles that have incredibly energy, some 10^8 times greater than physicists can produce in particle accelerators. The origin, and even the nature, of these cosmic-ray particles is a complete mystery. Maybe their sources are dark and represent new physics, or maybe the particles themselves are new.

▷Could the coincidences be explained by selection from a large set? For example, does the Universe have many Big Bangs in different places, each beginning with a different set of randomly chosen constants, so that some are guaranteed to allow people to evolve and ask these questions? (The British astrophysicist Martin Rees (b. 1942) has called such a universe a "multiverse".) Many physicists treat such speculations seriously, and they hope that quantum gravity will provide serious answers.

- *Clue.* The Universe would not have produced human beings if the laws of physics did not have some very special properties, including some apparent *coincidences among the fundamental constants of nature*. Some scientists see in these accidents a role for God, as the creator of the improbable machinery that led to life. But many others look for explanations within physics. We will discuss several mystifying coincidences below.

We will go through this list of puzzles in this chapter, weaving these challenges into a larger discussion of quantum gravity and the prospects for a unified theory of all the forces of Nature.

Unlike in previous chapters, here we are at the frontiers. Physicists' perspectives on what is puzzling, important, or fundamental change rapidly here. Even by the time you read this, some of the puzzles on our list may have been resolved; others may have been rendered irrelevant to fundamental physics; new ones might join the list. Progress, stimulated by new observations and new theoretical speculations, is sure to be rapid but unpredictable. But don't underestimate the difficulty of arriving at a full understanding of the physics of the Universe. It is work for a generation of physicists. Or more.

The puzzle of the slightly lumpy Universe

The Big Bang model of cosmology provides a framework for thinking about the history and future of the Universe, and we have seen that this framework provides simple cosmological models that seem to fit the observed facts very successfully. But the simplicity of these models raises two big questions which the Big Bang model does not answer.

In this section: the Universe is smooth on the large scale but lumpy enough to make galaxies. This combination is very special. Inflation provides explanations.

Why is the Universe so smooth on large scales, so homogeneous?

Given that it is smooth, why is it not even smoother? Why did it have enough initial irregularity on small distance-scales to form the stars and galaxies we see?

We have seen in earlier chapters that the homogeneous Friedmann model is a good model of the Universe. At the time of nucleosynthesis, when helium was being produced, the Universe was remarkably smooth on the very smallest scales: the ratios of the different isotopes show no evidence of the slight changes that would have been caused by significant inhomogeneity.

Dramatically, there are many measures that show us that the Universe was homogeneous at very early times, so early that the regions we compare would not have had time to communicate with each other in the standard Big Bang model. The helium abundance is the same in different directions; the numbers and types of galaxies at very high redshift are the same in opposite directions; and the microwave background temperature is the same to a few parts in 10^6 all over the sky!

Provided we believe we are not in a special place in the Universe from which it just happens to look homogeneous (this is the Copernican principle, introduced in Chapter 24), then this homogeneity poses a problem, a problem of *how*. The two regions producing helium could not have influenced one another physically in order to insure that they made the same amount of helium (by arranging to have the same density and temperature, for example), so *how* have they arranged to be so alike?

If the standard Big Bang is right, then the large-scale homogeneity we observe seems to be accidental. It requires that the initial conditions for different parts of the Universe were the same at the Big Bang. A messy, random initial start to the Universe could not have produced the Universe we see.

This conclusion is disturbing from the point of view of the Copernican principle, since it means that the Universe itself is very special: of all the kinds of universes that one could imagine, the one we have is exceptionally homogeneous.

Inflation modifies the standard Big Bang picture to offer an explanation of the homogeneity. Inflation proposes that there was a very early period dominated by dark energy that acted like a temporary cosmological "constant", driving a very rapid (exponential) expansion. The result is that large regions today have actually expanded from tiny regions just before the onset of the inflationary expansion, regions that were small enough to have become smooth in the very short time between the Big Bang and the beginning of inflation.

If inflation lasted long enough, with a strong enough acceleration, then everything we see today was once inside a tiny region. The smoothness we see on short and long scales could have been achieved by ordinary physical processes before inflation, even if the Universe initially had been very messy and random.

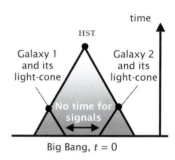

Figure 27.1. Two galaxies, observed by astronomers using the Hubble Space Telescope (HST), lie in opposite directions to one another, at such a great distance that they are seen when the Universe was much younger. They are so far apart that there has not been enough time for them to communicate, to exchange signals: their past light-cones reach the Big Bang before they intersect. This is the conventional picture of the Big Bang without inflation, and shows why the homogeneity of the Universe – the fact that the galaxies and their neighborhoods look so similar – is a difficulty for the standard Big Bang model. It requires that the Big Bang should have started in a very similar way in disconnected places.

But wait: this explains the homogeneity of the Universe, but what about its lumpiness? The uniformity certainly breaks down on small length-scales, because galaxies and stars have formed. Doesn't inflation smooth things out too much?

Remarkably, inflation offers an explanation of the lumpiness to. The expansion due to inflation had a side-effect: it created density irregularities through the amplification of quantum fluctuations that existed before inflation, the same kind of fluctuations that we described in our discussion of black-body radiation from black holes in Chapter 21. We will see how such tiny effects can give rise to the formation of massive galaxies in our discussion of inflation below.

> Inflation's explanation of galaxy formation is the critical argument that makes scientists take the idea seriously. It produces the right amount of inhomogeneity, on the right length-scales. Inflation promises to solve both the smoothness problem *and* the not-so-smoothness problem.

Inflation has given scientists the confidence that they can solve the mysteries of cosmology with scientific methods. They remain strong in their belief that the Universe is understandable.

> The combination of Einstein's law for gravity and modern thinking about the fundamental laws of physics may have enough power to penetrate through barriers that physicists formerly regarded as absolute, and to answer questions that physicists once would not have dared to ask. The answers, if they come, will come from the Universe itself.

It is time now to begin our study of inflation. The first step is to look at the cosmological constant in its cosmological setting.

Einstein's "big blunder"

When Einstein invented general relativity, Hubble had not yet discovered the expansion of the Universe, and the general opinion of astronomers of the day was that the Universe was static, that it had always existed in its present state. This was, of course, fundamentally because they were only looking at stars in our Galaxy. But it was also conditioned by the Judeo-Christian philosophical and religious beliefs shared by the scientists who developed physics in the nineteenth and early twentieth centuries.

Their society widely believed that the Universe was created at a finite time in the past, and in a "perfect" state, in just the same condition as they observed it. The idea that the Universe was dynamical – changing with time, evolving – would have been uncomfortable, in the same way that the idea that life was evolving was uncomfortable to many people at that time, scientists included. So, in the absence of evidence to the contrary, nineteenth century astronomers took the conservative route and assumed that the Universe was static. It would have been possible to have explored the assumption that the Universe was expanding, but in the absence of observational evidence this would have required a considerable leap of the imagination.

Einstein did not make this leap. He made many greater leaps than this in mathematical and theoretical physics, but he was not an astronomer, and he simply trusted the wisdom of the astronomers of his day.

> Einstein assumed that, since his theory of general relativity predicted a dynamical cosmology, it must be *wrong* when applied to the Universe as a whole!

▷ The absence of an explanation for homogeneity or clumpiness is, of course, fundamentally a philosophical difficulty rather than a strictly scientific one, since there is only one Universe, and a scientist can only measure what it is like, not how it might have been if things had been different at the Big Bang. Cosmology is not an experimental science: one cannot create new universes under controlled conditions and explore what happens to them! In fact, for some physicists homogeneity has a religious implication, proving that there was a design in the Big Bang, that the initial explosion of the Universe was carefully wrought by a Creator. But other physicists feel that it is now possible to look seriously for purely scientific explanations of the structure of the Universe. For them, the philosophical difficulty has been fruitful scientifically, because it has led them to the idea of inflation.

In this section: Einstein regretted introducing the cosmological constant, but today many scientists believe it exists in some form.

Nevertheless, Einstein was convinced that general relativity was basically sound: it was theoretically elegant and it made successful predictions of how gravity behaved in the Solar System. The question to him was, how could he fix it, change it a little so that it would work for cosmology too?

We have seen in Chapter 19 that Einstein solved this problem in a characteristically elegant manner, by finding a unique way to introduce a negative pressure into cosmology without giving up the principle of relativity. This pressure and its associated density would be independent of the observer, of position and of time.

Elegant or not, when Hubble discovered the expansion of the Universe, Einstein abandoned the constant. That made sense in his day. But today we can see that he may have been too hasty. We are not in his position, of trying to construct a static Universe. We have a worse problem: an accelerating Universe.

The cosmological constant in particle physics

The modern "rehabilitation" of the cosmological constant began, however, with completely unrelated developments in theoretical particle physics. Recall that Einstein had no physical model for the fluid that produces the negative pressure: the cosmological constant is *ad hoc*. Today particle physicists have a possible physical model. It seems that a cosmological constant may arise naturally in quantum theory.

Ironically, the story of the quantum justification for Λ starts with some other work by Einstein. Recall that we saw in Chapter 8 that Einstein had invented the concept of a photon. He had shown that light comes in discrete packets of energy, and the amount of energy depends on the wavelength of the light. He showed that this idea explained a number of experimental facts, but he did not work out a fully quantized theory of light to back it up.

In fact, it proved very difficult to find such a theory. Light is an electromagnetic wave, so physicists realized that they needed to invent a quantum theory of electromagnetism. Quantizing the atom, establishing the quantum theory that would predict correctly the spectral lines of different atoms: this was the work of Bohr, Heisenberg, and Schrödinger in the 1930s. But it took physicists another two decades to get a good quantum theory of electromagnetism, culminating in the independent work of he American physicists Richard P Feynman (mentioned in Chapter 21) and Julian Schwinger (b. 1918), and the Japanese physicist Sin-Itiro Tomonaga (1906–1979). Physicists finally had a theory that gave a precise meaning to the notion of a photon, almost 50 years after Einstein introduced it. The theory is called **quantum electrodynamics** (QED).

The reason for the long wait was that the photon presented physicists with some of the most difficult theoretical problems they had ever grappled with. We have learned enough about quantum physics in this book to be able to understand one of the difficulties. Einstein had shown that energy was quantized, that having more energy in the electromagnetic field meant having more photons. But unfortunately, there is the Heisenberg uncertainty principle. We met this principle in Chapter 7, where we saw that it forces atoms in a gas to have a certain zero-point energy, which can't be removed by cooling the gas. The wave oscillations of the electromagnetic field also have a minimum zero-point energy, which can't be removed. So although there may be *no* photons at all, quantum theory tells us that there is still some energy in the field. This purely quantum effect was something that Einstein could not have anticipated in his early work on photons, long before Heisenberg.

The zero-point energy of oscillation of atoms in a gas presents no real problems. The energy per atom is small, and there are only a finite number of atoms. But the problem gets more serious when one considers electromagnetism, the theory

In this section: one reason for expecting the cosmological constant to exist is that it comes out naturally from modern approaches to quantizing electromagnetism. In fact, the puzzle is that it should be very much larger than it is.

▷The three scientists shared the 1965 Nobel Prize for physics for QED.

of light. Photons are bundles of energy associated with electromagnetic waves of a particular wavelength, and the uncertainty principle requires a zero-point energy for the vibrations of *each* different wavelength. But there are an infinite number of possible wavelengths! Even if there were no photons around at all, there would be an infinite amount of energy associated with the zero-point vibrations. One of the successes of QED was showing how to deal with this zero-point energy – how to describe the way that charged particles affect one another and give off photons when accelerated, without being disturbed by the infinite energy that seems to pervade space.

The zero-point energy is controlled but not eliminated in QED. In fact, it leads to experimentally verified predictions. One of the most striking is the so-called Casimir effect, named for the Dutch physicist Hendrik Casimir (1909–2000). Imagine an idealized experiment, where two metal plates are placed parallel to one another a small distance apart. The plates are infinite in extent in both directions, and we imagine that they are perfect conductors of electricity, which means that they are shiny mirrors that reflect all photons. The electromagnetic waves between them now cannot have arbitrary wavelengths. The photons will reflect from the plates, bouncing between them. The allowed wavelengths are only those that fit exactly between the plates: one-half wavelength, one, one-and-a-half, two, and so on, just as a violin string (or a star, as in Chapter 8) has only certain allowed wavelengths or frequencies of vibration.

Now, in the idealized experiment, imagine bringing the plates closer together – to half of their original separation. Most of the originally allowed wavelengths of photons will still fit in the new separation, but one will not: the longest wavelength of the larger separation will not fit the new one. This means that its zero-point energy is no longer present in the space between the plates. By bringing the plates together, we have reduced the total zero-point energy of the space between the plates. This means that there should be an *attraction* between the plates: if by bringing them closer we liberate energy, then Nature will want to do this. This attraction exists even if we consider a more realistic case where the plates are of finite size, as long as the plates are very close together. It can be, and has been, measured. The only way to explain it is by ascribing reality to the zero-point energy of photons that are not even there!

The problem for gravity is that this energy ought to create gravity. The energy measured by the Casimir effect is only the difference between the total energy when the plates are in one position and that in another position: the difference between two infinitely large numbers, which in this case is a finite number. But for gravity, we expect that *all* the zero-point energy should make a gravitational field. If that were the case, space would curve up dramatically. So it appears not to be there, or at least not so much of it. Is there a way to get rid of this energy? Is it really there?

Let us ask in what way this energy would create gravity. There can be no special experimenter for measuring the zero-point energy, since it is a property of empty space.

> All experimenters must measure the same energy, regardless of their motion relative to one another. This means that the zero-point energy has to have exactly the same property that Einstein needed for the energy created by the cosmological constant. This zero-point energy is equivalent to a cosmological constant! The attraction between the conducting plates is due to the negative pressure associated with this energy.

Considering that the zero-point energy is potentially infinite, what limits are there on its size? The only limit that physicists generally agree on comes from the fact that it does not make sense to talk about photons with a wavelength smaller than the Planck length that we first met in Chapter 21, which is the smallest length that most physicists think can be used in theories that do not embody quantum gravity. If we add up the zero-point energies of all possible photons with wavelengths larger than this, the cosmological constant we get is huge, contributing a much larger energy density than any known matter in the Universe. In fact, since the calculation can only involve Planck's constant, the Planck length, and the speed of light, it must be a number made, like the Planck length itself, out of G, c, and h. And it must have the units of density, so it should be proportional to the Planck density, given in Equation 21.12 on page 295:

$$\text{natural zero-point mass density} \propto \rho_{\text{Pl}} = c^5/hG^2 = 8 \times 10^{95} \text{ kg m}^{-3}.$$

It would be unreasonable to expect the constant of proportionality to be so small that the natural mass density would be small compared to the critical density of the Universe, so although we have made the zero-point energy finite, we still have a big problem.

One way out is to postulate that the zero-point energy is really there, but that it is cancelled to a high accuracy by a cosmological constant of the opposite sign. The remainder would be the effective, observed cosmological constant. But this does not solve the problem. It just pushes the original question into a new one: why is the cosmological constant so large, so that when it cancels the zero-point energy the difference (the *effective* cosmological constant) is so small?

> So the question now facing physicists is not, does the cosmological constant exist? Rather, the question is, why is it so small? At the present time, physicists have no answer to this question.

It is ironic that what Einstein regarded as his biggest mistake might yet prove to be one of his most important contributions! And not just because the Universe may be accelerating today. Let us now look at what it may have been doing when it was just a baby.

Inflation: a concept waiting for a theory

Inflation is an idea, or a working hypothesis, about what happened in the early Universe to make it so homogeneous. It has other consequences, which are now well-supported by observation: that the Universe should be almost flat, that galaxies should have formed from initial density fluctuations of a certain size and distribution. Inflation is about what tricks the zero-point energy of particles in the early Universe might have played.

Inflation can't yet be called a physical theory, because we don't know its cause. Rather, it is a phenomenon that can occur in the very early Universe if the correct theory of high-energy physics has certain properties. The circumstantial evidence for it is strong. And inflation seems to be a feature of a large class of high-energy physics theories. Cosmological observations therefore have the possibility of guiding the development of these theories.

Here is a list of questions that inflation sets out to answer.

In this section: inflation answers many questions about cosmology, but it is not yet grounded in any fundamental theory of physics.

- *Q1.* How did the Universe get to be so homogeneous and isotropic on the large scale? We saw earlier in this chapter how difficult it is to understand the large-scale similarity between different regions of the Universe. Inflation offers an explanation.

- *Q2.* Why are there no **magnetic monopoles**? This is a serious problem for particle physics, but it is one that we have not yet come across in this book. A monopole is the magnetic analog of electric charge: it would be a purely North pole, or a purely South pole. The Universe has plenty of electrically charged particles, but apparently there are none that have a single magnetic charge. The only way we get magnetism is from moving electric charges, so that North poles are always accompanied by South poles on every magnet.

▷The word *mono*pole means one pole: a particle that is just a single magnetic pole. Inflation was originally invented in order to explain the absence of monopoles.

  But the laws of electromagnetism permit monopoles, which would create fields like charges do but with the electric and magnetic aspects exchanged. For example, a moving magnetic monopole would create an *electric* field. The standard theories of high-energy physics not only allow monopoles: they suggest that they should have been created in abundance in the early Universe. Inflation explains why they are absent today.

- *Q3.* Why is the density of the Universe so nearly critical? Dark matter observations (Chapter 14) and the theory of the creation of helium in the Big Bang (Chapter 25) tell us that the density of matter is within one-third of critical. The mass density associated with the dark energy carries a further two-thirds of the critical density. Considering all the possible values that the total density could have, why is it so near to critical? Inflation, at least in its simplest form, predicts that the Universe should be almost exactly critical.

- *Q4.* How did galaxies form: why did sufficiently large fluctuations in density occur in such an otherwise smooth Universe?

Inflation power: the active vacuum

In this section: we explain how inflation works. It relies on a change in the quantum state associated with the vacuum and a release of energy. This has analogies with phase changes in magnets.

Inflation relies on a form of the cosmological constant that arises temporarily in the early Universe. In this section we will see how such a temporary energy field can come out of the laws of physics. In the next section we will use our understanding of the active gravitational mass to show how this field drives the Universe into a rapid expansion.

Recall our earlier discussion of the Casimir effect. There we saw that, in quantum theory, the "vacuum" is a state in which particles are absent, but which still has plenty of energy, the zero-point or uncertainty energy.

> Many physicists now believe that it is possible that the laws of high-energy physics allow for there to be two or more different vacuum states, with different amounts of energy, and that the temperature of the Universe determines which vacuum it is in.

This may seem a contradiction in terms: how can there be two different ways in which particles can be absent? The difference is in the way the energy of the vacuum is determined. Two analogies may be helpful in seeing that this is possible.

The first analogy is with a waterfall. Imagine that you are boating on the Niagara River above Niagara Falls. The river moves placidly and you are so far from the falls that you can't see or hear them. You look around and feel, intuitively, that you are floating at "ground level". You have no sense that you are really high up on a plateau. Now suppose that on the following day you go fishing a long way downstream of the waterfall. Again you cannot see or hear the waterfall, so you sit on the riverbank, at "ground level", and you have no sense that you are lower than the level you were at the previous day.

The vacuum state in quantum theory is like the "ground level". It is not an absolute level, but just a state in which, under suitable conditions, there are no particles:

everything is quiet and placid. If the conditions change (near Niagara, you move from one place to another; in cosmology, the Universe changes its temperature), then the nature of the vacuum can change.

Let us pursue this analogy a little further. Suppose that you went from one ground level to the other by allowing yourself to drift up to the Falls and fall over them. If you manage to survive the drop over the waterfall, you will arrive at the lower ground level, but not right away, and you will not have made a smooth journey. For the Universe, the transition from one vacuum to another was also not smooth: it resulted in a huge release of energy, which created all the particles of which we are made today. But eventually, like the Niagara River, it settled down into its present placid state.

▷ This is just a thought experiment: don't try it yourself!

Here is the second analogy, which is actually quite a good one, since the physics and mathematics are similar: it is the formation of a "permanent" magnet. Many minerals acquire magnetism as they cool. For example, molten lava from a volcano has no magnetism. But as it cools, its molecules find that if they line up all their spins in a consistent way, then they will have a lower total energy than if their spins are randomly oriented. When the lava is hot, the kinetic energy of vibration is much larger than this spin orientation energy, so the lava does not have any systematic orientation. But when it is cooler, the random vibrations are weaker than the spin orientation effects, and the material prefers to align its spins. When all the molecules are oriented in a consistent way, their spins combine to create effectively a small electric current, and this is what creates the magnetic field of the object. Magnets that you buy in a store are made like this.

The final direction that the spin takes is essentially random, but it can be influenced. If the mineral is in a magnetic field when it cools, then the spins tend to line up with the external field. This external field just gives them a little nudge in the right direction; it does not force them to align. They "want" to align because there is an energy benefit to do so. The alignment process releases this energy, so for a brief time the mineral is reheated slightly.

▷ The magnetization of cooling lava provided one of the crucial pieces of evidence in favor of plate tectonics, the theory that the continents move around on the Earth. The Atlantic ocean is widening at the mid-Atlantic ridge, a deep furrow running North–South roughly midway in the ocean. Geologists found that the direction of the natural magnetism of the rocks on the ocean floor near the ridge alternates between North and South as one moves away from the ridge: on either side of the ridge there are alternating bands of magnetism. The explanation lies in the periodic reversals of the Earth's magnetic field. The alternating bands imply that the rocks are formed and then move away from the ridge, new ones being formed at the ridge to replace them as the Atlantic widens.

> Inflation in its simplest model is very similar to the process that produces magnetism in minerals. When the Universe was very hot, the laws of physics were simple. As we mentioned before, the standard view of physicists is that all the forces of Nature were then on an equal footing, all with the same strength. Then, as the Universe expanded and cooled, a different state became the preferred one. In this sense, there was an "alignment" in the abstract space of all possible strengths of forces and masses of particles. In this picture, some details of this alignment were random. This is called **spontaneous symmetry breaking**. In this process, a large amount of energy was released. Unlike the analogy with magnetism, where the energy difference is small, in this case the energy difference was huge. It is this energy release which created the Universe as we see it today.

The energy released is the zero-point energy of the initial vacuum state. At first it behaved like a cosmological constant, with no preferred rest frame. But eventually the energy was transferred to other fields, creating the photons, neutrinos and quarks of the very early Universe. During the cosmological constant phase, the Universe expanded rapidly. This was the epoch of inflation.

The switch from one vacuum state to another is called spontaneous symmetry breaking because the simplicity (symmetry) of the original vacuum is replaced by (broken into) the complexity we see in particle physics today. This is a kind of change of phase in the early Universe. Theories that unify the strong, weak, and electromagnetic interactions are called Grand Unified Theories (GUTs), and they all have to use spontaneous symmetry breaking to explain the fact that there is no unity today among these forces. In fact, there have been several epochs in the Universe where this happened, and each may have left its mark.

This is indeed the way many physicists think the Universe evolved, although study has shown that the inflationary phase may have been entered and left more gradually than the analogy with magnetism suggests. The most common guess is that inflation happened when the temperature of the Universe was about at an equivalent energy of $kT = 10^{16}$ GeV, which is about $\times 10^{-3}$ of the Planck mass-energy. This is called the GUT energy scale. This is below the Planck energy, but not by much, so inflation would have happened very early.

Inflating the Universe

In this section: inflation expands the scale-factor of the Universe exponentially, at a very early stage in the expansion of the Universe.

Now we can describe how inflation works. The dynamics of the inflationary Universe are remarkable.

To see what to expect, look at Equation 24.12 on page 362, but replace the density ρ with the active gravitational mass $\rho + 3p/c^2$, which gives the general expression for the acceleration of the expansion. This gives

$$a_{cosmol} = -\left(\frac{4\pi G}{3}(\rho + 3p/c^2)\right)d. \tag{27.1}$$

When the dominant forms of energy and pressure are given by the vacuum energy, which has the same properties as the cosmological constant, then $\rho + 3p/c^2$ is negative, equalling -2ρ. While the negative pressure exerts no local forces because it is uniform, it actually causes the Universe to expand. This would not happen if gravity were governed by Newton's law of gravity, where only the mass density creates gravity. But Einstein's theory allows inflation, and that is the crucial difference.

The expansion produced by this vacuum energy is particularly rapid. In Equation 27.1 we can do some simple dimensional analysis to get the time-scale. The left-hand side is an acceleration, which has dimensions of distance divided by the square of time, and so the right-hand side must have the same dimensions. The right-hand side contains the distance d, so the remaining factors must together have the dimensions of $1/\text{time}^2$. Thus, a characteristic time in the problem is obtained by taking the inverse square-root:

$$\tau = \left(\frac{3}{8\pi G|\rho + 3p/c^2|}\right)^{1/2} = \left(\frac{3}{8\pi G\rho_v}\right)^{1/2},$$

where ρ_v is the vacuum energy. This is the characteristic time-scale for the expansion. The expansion is exponential on this time-scale, as we noted in Chapter 25.

What was the density at the GUT scale when inflation may have happened? The density when the temperature was equivalent to the Planck mass-energy was presumably the Planck density, Equation 21.12 on page 295. At this time the Universe would have been radiation-dominated (rest-masses were probably unimportant in the energy density), so the density decreased as the fourth power of the temperature. Since the temperature went down by a factor of 1000, the density decreased by 10^{12} to a mere 10^{84} kg m^{-3}! With this density, the time-scale for exponential growth evaluates to about 10^{-38} s. The Universe roughly doubles its size every time the clock ticks 10^{-38} s! The full equation for the exponential time-dependence of the scale-factor is

$$R(t) \propto e^{t/\tau}.$$

A region the size of the Planck length at the beginning of inflation would reach a macroscopic size, say 1 mm, in only 73τ. So inflation does not need to last long to make a huge difference.

This exponential expansion can't go on forever, because the energy being released is converted into normal matter, whatever that is at the GUT scale! It presumably behaves like radiation, with a positive pressure. Without a big negative pressure, the exponential expansion ceases, and the Universe starts to decelerate. But it does so from the enormous initial expansion speed provided by inflation. This is the point where the standard Big Bang picture begins to take over. At a time later than 10^{-38} s but probably earlier than 10^{-30} s, the Universe is a hot gas of normal matter (the old vacuum energy) in a state of the present vacuum.

Inflation put to the test

Although inflation is not yet grounded in a theory of fundamental physics, scientists have actively explored various "scenarios", sets of assumptions about how inflation might behave in detail. These serve to restrict the possibilities for fundamental theories by eliminating variants that do not fit observed data.

In this section: observations confirm most of the predictions of inflation.

The earliest full version of inflation, proposed by the American physicist Alan Guth (b. 1947), turned out to be too simple: it produced too much density irregularity today by ending too quickly. Subsequent work has focused on "slow-roll" inflation, which ends more gradually and gives acceptable agreement with the density irregularities needed to explain galaxy clustering. One interesting variant, called chaotic inflation, works even if the initial conditions before inflation were highly variable from one place to another. In regions where the Universe was initially contracting, inflation never took place, and so human beings were never created. In this picture, we happen to be part of a patch that was initially expanding. This is relevant to our discussion of the Anthropic Principle, below.

In an earlier section we wrote down a list of problems that inflation tries to solve. Now that we know what inflation does, we can see how it produces solutions.

- *A1.* It is easy to see how inflation solves the homogeneity/isotropy problem. If the period between the onset of inflation and its cessation is long enough, the expansion would have inflated any small region into an enormous size. The Universe we see today could have come from something very small, so small that even at the early time of 10^{-38} s it would have had time to smooth itself out in the relatively quiescent period before inflation began. In this picture, the distant galaxies and the various regions of the Universe at decoupling were all part of the same original tiny domain.

- *A2.* This also shows how inflation solves the monopole problem. The reason that inflation is assumed to have occurred around the GUT energy is that the Universe reached this after forming monopoles. So even if monopoles were abundant before inflation, they will be dispersed so far apart that the chance of our encountering one now would be minimal.

- *A3.* Inflation also solves the problem of why the Universe is so close to its critical density. The reason is in the conditions at the end of inflation. The exponential expansion phase wiped out any memory of the initial expansion velocity before inflation set in. In the exponential expansion, the Hubble parameter is just the reciprocal of the time-scale τ. Its square is then

$$H^2 = \frac{1}{\tau^2} = \frac{8\pi G \rho_v}{3c^2}.$$

By Equation 26.1 on page 388, the universe has zero spatial curvature: it is flat! Of course, real inflation can only be approximately exponential, and it has to make a transition to ordinary expansion, so this equation will only be a first approximation. But it implies that after the Universe exits from the inflationary phase, it must remain nearly flat, with a density close to the critical density.

Notice that the inflation-dominated universe has a very special geometry. Because the cosmological constant fluid has the same density and pressure to all experimenters at all times, regardless of their motion, it follows that this universe model must look the same to all observers: unlike our present Universe, there is no preferred rest frame. This symmetry implies that the geometry of space is actually flat, with the galaxies flying apart from one another through it. The curvature of time in this model is produced entirely by the Doppler redshift of these galaxies relative to each other.

▷ The inflation-dominated universe model was discovered by the Dutch mathematician and astronomer Willem de Sitter (1872–1934) in 1917, immediately after Einstein introduced the cosmological constant. Nevertheless, de Sitter disliked the cosmological constant and argued (long before Hubble's observations) that general relativity implied that the Universe was expanding.

- *A4.* Finally, inflation provides an initial spectrum of density irregularities at an early time that can lead to galaxy formation. It does this by amplifying initial quantum fluctuations in the quantum fields that describe matter in the very early Universe. Although such fluctuations are initially tiny, they increase in size during the period of inflation.

Just at the beginning of inflation, the Universe is unstable: a small random fluctuation in density can initiate inflation in one place before another. Since inflation then changes the density exponentially with time, the density contrast between two places that start inflating at slightly different times gets larger and larger, amplifying by the cube of the factor by which the Universe expands. In this way a tiny quantum fluctuation can grow to the size needed to begin galaxy formation.

The details of how this density fluctuation now produces galaxies are very sensitive to assumptions one makes about the transition from inflation to the normal expansion. It also relies on the dark matter in the Universe, since this is free to start collapsing as a result of this overdensity, while the ordinary protons and electrons are tied to the photons. But numerical simulations give excellent agreement with observations so far, as in Figure 25.3 on page 380.

Is inflation still going on?

In this section: the acceleration of the Universe is evidence that some kind of weak inflation is happening today.

We have seen in Chapter 25 that observations of Type Ia supernovae have suggested that the Universe is accelerating even today, although at a much smaller rate than during the epoch of inflation. Detailed studies of the cosmic microwave background irregularities have independently given further evidence for this. This is shown in Figure 27.2.

While both inflation and the acceleration we see today are similar to the behavior of a universe model with a positive cosmological constant, it seems likely that something more complicated than a cosmological constant is driving this acceleration. Inflation, certainly, was not caused by a cosmological constant, simply because it was not constant: the epoch of inflation came to an end.

Scientists are therefore looking for a theory of a variable dark energy field, which can act for limited times, changing its strength in a natural way. As we noted in Chapter 19, some physicists call their proposals *quintessence*. The search for such a theory is in its infancy, and it may require much more data, both from physics experiments and from cosmological observation, before believable models can be found. But as long as astronomers continue to believe that the Universe had one or more periods of accelerating expansion, physicists will continue looking for an

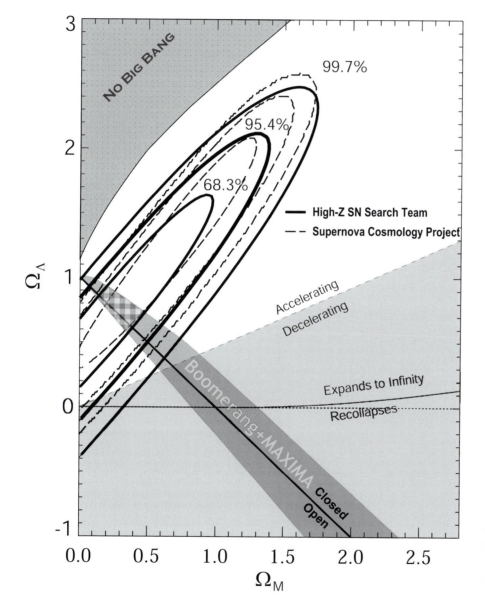

Figure 27.2. *This chart shows the implications for the large-scale nature of the Universe of the measurements of Type Ia supernovae and of the sizes of the fluctuations in the microwave background, according to the data available in 2001. The horizontal axis is the fraction of the critical density that is in matter, $\Omega_m = \rho/\rho_c$. The vertical axis is the fraction in dark energy, Ω_Λ, that is contributed by the cosmological constant Λ. If these add to one, then the Universe is flat: its total mass density is equal to the critical density. The downward sloping solid line is the line where these two numbers add to one. The region above the line has more mass, and so represents closed universes; the region below represents open universes. The various oval regions show the parts of the diagram consistent with the observations of Type Ia supernovae, as in Figure 24.4 on page 352. (The percentages drawn in the figure are confidence levels on the observational uncertainties: the observations tell us that the true Universe must lie within the outer ovals with a probability of 99.7%.) The dark wedge around the closed/open line is the region allowed by the observations of the microwave fluctuations. There is only a very small part of the diagram where the two observational constraints overlap, and this is shown as the hatched area. The center of this area is a model where the mass density of the Universe is 30% of the critical density and the density contributed by the cosmological constant is 70% of the critical density: a flat expanding Universe. Figure from the High-Z Supernova Search, based on data from de Bernardis, P., et al (2000) Nature **404**, 955, and Balbi, A., et al (2000), Astrophysical Journal **545**, 1.*

explanation. The acceleration could turn out to be the key clue pointing toward the next theory of fundamental physics.

Is Einstein's law of gravity simply wrong?

In this section: instead of dark matter and dark energy, maybe general relativity is wrong. This is possible, but an alternative theory is hard to find.

In trying to explain the many mysteries of the Universe, we have assumed that Einstein's gravity provides a good description of both the local dynamics of galaxies and the global Universe. From this it follows that puzzling observations – such as the high velocities of matter in galaxy clusters or the strong bending of light in gravitational lenses – tell us that there is hidden mass and dark energy.

However, general relativity is simply a theory of physics, and it must always be tested against observation. In the Solar System and in the Hulse–Taylor pulsar, it passes tests superbly well. It explains black holes and neutron stars. But it is hard to test the theory over the long distance-scales of galaxies and galaxy clusters. It is not surprising, therefore, that some scientists have posed an interesting question: could the missing-mass problem be solved essentially by changing the law of gravitation over long distances, without introducing a hidden form of matter? Is there really hidden mass, or just hidden gravity?

This is a natural question to ask, but it has not led to much progress in providing viable alternatives to the model of inflation with cold dark matter that the large majority of astrophysicists favor. There are two reasons for this. One is that it seems exceedingly difficult to modify general relativity on long length-scales without throwing out its successful predictions. The other reason is that the accumulating evidence for missing mass puts strong constraints on new theories.

This last point may seem surprising, but it comes about because any modification of the laws of physics must be *universal*: it must be the same everywhere in the Universe. If there is a new term in the law of gravitation, containing (say) a new fundamental constant of Nature that determines how strong it is and over what distance-scale it is going to be noticeable, then this must give a consistent explanation of the dynamics of every galaxy and galaxy cluster in which there is a missing-mass problem. There are dozens of well-studied clusters and galaxies, and there is sufficient variety among them to challenge any of the proposed modifications of gravity.

In particular, there seems to be no single length-scale on which missing gravity takes over from Newtonian or Einsteinian gravity. This is not surprising if one accepts that the missing gravity is created by missing mass: each galaxy or cluster condensed around an individual clump of dark matter, and since these clumps are random, the gravity they create will be different from galaxy to galaxy and cluster to cluster.

One kind of theory that does not suffer from this problem has gained much attention among physicists in the last few years. These are called "brane-world" theories, and they are inspired by string theory. We will look at them later in this chapter, and in particular we will see that some of them explain the dark matter as ordinary matter existing in other three-dimensional worlds separated from ours by a small distance in a fourth spatial dimension. In these theories gravity can bridge the gap between these worlds but the other forces of physics cannot, so we are unaware of their existence except for their gravitational influence on us.

The book is certainly not closed on new theories. If the dark matter particle is detected, much of the motivation to look at ideas like these will disappear. But if dark matter searches show that the required particles are not there, then scientists will take these theories much more seriously.

Cosmic defects

Inflation is only the most spectacular example of what can happen if modern theories of particle physics are applied to early cosmology. Whether or not inflation occurred, we might still find the Universe filled with what are called "defects", primarily cosmic strings and cosmic textures. These also arise from spontaneous symmetry breaking at lower energies.

In this section: cosmic strings could have been formed in phase transitions in the early Universe, and could produce detectable effects today.

When a symmetry breaks spontaneously, it breaks randomly. In different parts of the Universe, the values of certain fields in the theory will have random aspects that differ from one another from place to place. In some kinds of theories, these differences lead to cosmic strings, which are long thin condensations of trapped energy. Inside the string, space is still in the old "false" vacuum, which has lots of energy relative to the present "true" vacuum. If the string arose from symmetry breaking at an energy of 10^{16} GeV, then the energy can be enormous: the strings can have a mass per unit length of 10^{21} kg m^{-1}.

However, their gravitational behavior is not simply that of a massive piece of rope. Since the matter inside is trapped vacuum energy, its energy density ρ is accompanied by a negative pressure $p = -\rho c^2$. In this case, since the string is one-dimensional, the pressure acts in only one direction, and provides a tension along its length that keeps the string together. But the active gravitational mass is $\rho + 3\langle p \rangle / c^2$, where $\langle p \rangle$ is the average pressure. Since the pressure is zero in the two directions perpendicular to the string, the active gravitational mass is just $\rho + p/c^2$, and this vanishes!

> So a cosmic string does not curve time. Clocks are not redshifted by it and particles do not go in orbit around it, despite its immense mass.

How, then, can a string have any effect, and in particular how can it assist the formation of galaxies?

The primary effect of the string is to curve space. In Chapter 19 we defined the active curvature mass, $\rho - p/c^2$, but this was valid only for isotropic pressure. For strings, the result is more complex, and needs to be calculated from the details of Einstein's theory. The result is that the curvature in the direction along the string is zero: there is no curvature mass, and so proper distance along the string is the same whether measured inside or outside the string. On the other hand, curvature in a plane perpendicular to the string is non-zero, and it is generated by a density of curvature mass equal to 2ρ.

This has observational consequences. For example, we have seen that light deflection by the Sun depends on the spatial curvature as well as the curvature of time. Therefore, a cosmic string will deflect light that passes by it. This would lead to a kind of gravitational lensing in which one might get double images of a star, one from light that passes one side of the string and the other from light passing the other side. Astronomers have looked for this effect, but so far without success.

Our simple picture of the geometry around a string gets more complicated if two strings intersect, or if a string becomes dynamical and has oscillations, as it would be expected to do. Strings can break off when they intersect, and form closed loops. Dynamical strings give off gravitational waves (by shaking the curvature in the plane perpendicular to their length), and loops that do that will shrink and eventually disappear. Moving strings are also good seeds for galaxy formation: although time is not curved, the curvature of space causes the geodesics of particles near the string to deflect toward the loop. When this happens, densities increase in the string's wake, particles collide, and galaxies can form.

In fact, scientists have calculated that there is a simple relation between the amount of gravitational radiation that strings emit and their effectiveness as seeds for galaxy formation. Current limits on the amount of radiation are placing constraints on the number of strings, but they are not yet able to eliminate the theory that cosmic strings formed galaxies.

Observations of the microwave background are, however, placing much tighter constraints on cosmic strings. At short angular scales, strings produce a very different pattern of fluctuations than inflation. At this time (2002) it seems unlikely that strings were plentiful enough to have been the main cause of galaxy formation. But they could nevertheless still be there in large quantities, and could be observable through their gravitational effects, including gravitational radiation.

Cosmic rays

Cosmic rays are charged particles, mainly protons, that strike the Earth at very high energies. They are detected by looking for their collisions with atoms of the atmosphere. Historically, cosmic rays were the way physicists first studied what we now call "high-energy physics", before they could create high-speed particles in laboratory accelerators. The muon, one of the three types of leptons (see Chapter 25), was first discovered in the products of the collisions of cosmic rays with atoms of the atmosphere, and the time dilation of special relativity was convincingly verified by observing that fast-moving muons in cosmic rays lived much longer than ones at rest (Chapter 16).

Today cosmic rays are once again pushing physicists to the limits of what they can understand. The highest-energy cosmic rays are observed with energies above 10^{20} eV, which is far beyond any energy that can be produced in an accelerator. But what is challenging about them is not the physics of their collisions with other particles. What is puzzling is that there are so many of them. As we noted above, the Earth encounters roughly one such particle every second.

These high-energy particles do not seem to come from the Galaxy, because their arrival directions do not coincide with the plane of the Milky Way, and they are too energetic to have been deflected into their arrival directions by the small magnetic fields in interstellar space. On the other hand, they should not be able to move too far between galaxies, because they would lose energy to interactions with the cosmic microwave background radiation. When a low-energy photon of the microwave background collides with such a high-energy particle, many things can happen, including the production of other particles. This would cause a high-energy proton to lose its energy rapidly.

Figure 27.3. One detector of the Pierre Auger Observatory, an array of 1600 similar units being constructed in Argentina, able to observe high-energy cosmic rays striking anywhere in an area of 3000 km². It is hoped that this will provide enough data to solve the mystery of the ultrahigh-energy cosmic rays. Image courtesy Pierre Auger Observatory.

Calculations suggest that high-energy protons or other known elementary particles should not be able to move further than about 20 Mpc before their energy falls below about 10^{20} eV. Yet, within this distance (about the distance from the Earth to the Virgo Cluster), and in the directions from which the various particles have been seen to arrive, there are no visible sources. One would expect that any astrophysical object that could produce protons of such extraordinarily high energy would be doing something else as well, like producing light or X-rays or gamma-rays. But astronomers see no likely candidates.

The arrival directions of these particles are not known to high precision, so there are certainly galaxies from which they could have come. But the galaxies do not look like they contain anything special. The speed of these particles is so close to the speed of light that we would expect to see the event in which they were produced almost at the same time as we receive the particles. No supernovae or other spectacular events have been associated with observations of these particles.

And certainly there is no evidence for such events happening once every second!

Other events, much further away, would be more likely to produce particles at these energies. Gamma-ray bursts, for example, or perhaps quasars. But these do not happen so close to our Galaxy.

There is as yet no explanation of these high-energy cosmic rays. New instruments are under construction that will gather much more data and hopefully lead to a solution. The solution might be simple, such as a form of particle acceleration that scientists have overlooked and which is present in all galaxies. The particles might be very heavy nuclei, which would have the observed energy at a speed slow enough to avoid rapid energy loss. But it is also possible that we are being presented here with some completely new physics. Perhaps the sources are relatively nearby but dark (cosmic strings, magnetic monopoles, decaying massive dark matter particles, ...), or perhaps there is new physics in the cosmic-ray particles themselves. The resolution of this problem certainly has the potential to affect the other issues we are discussing in this chapter.

Quantum gravity: the end of general relativity

We have arrived now at the limits of general relativity. Almost everything in the earlier chapters has been standard physics. Even though black holes may seem exotic, they are well-understood theoretically and they have been identified observationally. They are not particularly controversial in physics today. Gravitational waves have not yet been detected directly, but there is little theoretical doubt about their existence and general properties. Our ignorance about the large-scale structure of the Universe is still high but the framework provided by Einstein's equations seems adequate to describe the evolution of the Universe that we see today.

We have also seen some more speculative physical ideas, especially in the first part of this chapter: inflation, the cosmological constant, cosmic strings. These are not so well-established, and some of them might either fall out of fashion tomorrow, or be turned into "standard physics" by a crucial astronomical observation next year. But all of these ideas are rooted in fundamental physics as we now understand it. The negative pressures and cosmic defects of these speculations are features that are expected from theories that describe the nuclear interactions. The speculative part is whether the correct theory will turn out to exhibit these features at just the right energy (or temperature) to explain the facts we observe.

Big as these speculations are, there is an even bigger hole in physicists' theories.

> The biggest incompleteness in physics has to do with gravity. Just as gravity drives the evolution of the Universe and of most things in it, gravity also drives the most fundamental and exciting theoretical research in physics today. Gravity is where the action is, if you are a fundamental theoretical physicist.

The reason is that Einstein's general relativity is not, cannot be, the last word on gravity. General relativity is what physicists call a classical theory of physics. It has none of the distinctive features of quantum theory, and that is a contradiction that must be fatal for general relativity.

The reason is simply the uncertainty principle. Consider the fact that in quantum systems, one cannot measure exactly how much energy the system has, or exactly where it is located. Nevertheless, gravitation theory tells us how to compute the gravitational field from the distribution of energy. So we could in principle measure the gravitational field far away with arbitrary precision (if it is a non-quantum field) and determine what the distribution of energy in the spacetime is with arbitrary precision, contradicting quantum theory.

In this section: the goal of theoretical physics is to unify gravity with the other forces and produce a quantum theory of gravitation. There are many situations where a quantum theory is needed.

The only way out of this contradiction seems to be to invent a theory of gravity that is a quantum theory, but which in appropriate circumstances makes predictions so close to those of general relativity that they also satisfy the observational evidence that supports general relativity. This is a standard situation when one wants a quantum theory of phenomena that are already well-described by a classical theory. The theory of electromagnetism does a good job on electric and magnetic fields that are used in everyday life, such as those that create and are created by electrical circuits in the home. But when one wants to describe electromagnetism on the level of single photons interacting with single electrons, then one needs the quantum version of the theory, QED.

In the same way, scientists need a quantum theory of gravity in order to study some phenomena with confidence. Here is a partial list of the places where quantum effects in gravity might make important changes from the predictions of general relativity.

1. *Singularities* (Chapter 21). The center of a black hole contains a place where, according to general relativity, time finishes. Any object that falls in and reaches this location will not progress further in time. Most scientists find this disturbing, and hope that the strong gravitational fields near a singularity will create substantial uncertainties in a quantum theory of gravity, and that these uncertainties will allow particles near the singularity to continue into the future indefinitely. However, it is also possible that a quantum theory of gravity will not remove singularities but embrace them in some way, so that they are no longer places where physical theory breaks down.

2. *Hawking radiation and naked singularities* (Chapter 21). There seems to be little doubt in the minds of most physicists working on these questions that black holes must emit radiation with a spectrum basically like that of a black body, but of course this cannot last forever. When all the mass of the hole has been radiated away, something must happen to the singularity inside. Does it become "naked", i.e. visible to the outside world? Does it disappear altogether? Or does it remain hidden behind a horizon that has shrunk to a point? In general, one might expect quantum gravity to say something about the cosmic censorship hypothesis, which we discussed in Chapter 21.

3. *The Big Bang* (Chapter 24). This was a singularity in our distant past. Perhaps a quantum theory will tell us how the Big Bang came about, whether there is any meaning to the notion of time *before* the Big Bang, whether there were many Big Bangs with different outcomes, whether the Big Bang was smooth or bumpy.

4. *Planck scales* (Chapter 21). The characteristic numbers that we call the Planck mass, length, time, and so on, are built from Planck's constant (quantum theory), the speed of light (relativity) and G (gravity). A quantum theory may make very different predictions about physics on these scales, which could only be attained in rare circumstances. In particular, on distance-scales shorter than the Planck length, spacetime might not even be continuous. Some physicists suggest that spacetime really has the structure of a tangle of disjoint loops, and only looks smooth when averaged over distances larger than a Planck length. Others suggest that it consists of tiny fluctuating wormholes (see below and Chapter 21), called spacetime foam. Still others think that spacetime might really have ten or eleven dimensions, the extra ones (above

the four of conventional general relativity) being visible to us only over distances of the order of the Planck length. (See the discussion of branes below.)

5. *Negative energy, wormholes, and time travel* (Chapter 21). We have seen that to keep wormholes open for travel one needs negative energy. It may be that quantum gravity can supply a source of such negative energy. If so, it is most likely to occur only on the Planck scale, leading to the idea of spacetime foam. But it is possible that quantum gravity will teach us how to make sustainable regions containing negative energy. The applications of this would be immense, and would come close to elements of the science fiction world of space travel. In particular, it would theoretically be possible to travel through a wormhole and emerge earlier, having traveled backwards in time.

6. *Shadow matter.* This is a long shot, but one that could cause enormous difficulties with gravitational experiments. The possibility exists that a theory that unifies all the forces of Nature will predict a class of matter that interacts with the rest of matter only gravitationally. Within this class there could be a complex series of interactions between its particles: they could have their own charges, strong forces, and weak interactions, but these would be insensitive to the ones that we know. In effect there could be a hidden Universe occupying the same space as we occupy, with its own structures and dynamics. It would be detectable through its gravitational effects, but would otherwise be dark and invisible. The missing mass could in principle come from this sector of the theory, and if dark matter experiments fail to detect weakly interacting particles then this may become a real possibility. But experimental confirmation would be exceedingly difficult to provide.

7. *Branes.* String theory, the leading candidate today for a unified theory of all the forces, and therefore the leading candidate for providing us with a quantized theory of gravity, requires that the real spacetime that we occupy have eleven dimensions, and that our four-dimensional world is just a subspace, a kind of membrane, in the larger space. This higher dimensionality is not arbitrary; the theory can only be made mathematically self-consistent in this number of dimensions. The way that our four dimensions fit into the larger spacetime is not known, but it is possible that they are like a twisted, kinked ribbon, called a **brane**. According to these models, the non-gravitational forces of physics act only in the brane but gravity can exert an influence over a region of the eleven-dimensional space near our brane. The initial assumption that this region would be as small as a Planck length has recently been challenged by some physicists, who suggest that it could be much larger, up to say 1 mm. If true, this so-called brane-world picture could have observational consequences: modifications of Newton's law of gravity at very small distances, generation of gravitational radiation in the very early Universe, creation of gravitational waves and gravitational attraction by matter that does not exist in our Universe but rather inhabits another nearby brane (referred to earlier in this chapter), explanation of the nature of the early Universe without invoking inflation, and perhaps more. It is interesting and exciting that even rather simple gravitational experiments, such as measuring the force of gravity over short distances, could in principle provide evidence for quantum gravity.

▷In fact, physicists working in string theory are excited about its prospects because it is the only theory that they have been able to make self-consistent. There is a feeling among some that there may be only one possible mathematical structure that can be made self-consistent, and it must therefore be the correct theory.

8. *Gravitons and the definition of energy.* It is tempting to expect that a quantized theory of gravity should involve quantized gravitational waves, which

physicists call gravitons. However, the concept has problems. The usual idea would be that they should be like photons, which carry quantized amounts of energy that depend on the photon's wavelength. The problem is that energy is not so easy to define in general relativity, and indeed even gravitational waves go away if one goes to a locally inertial frame and looks on a scale smaller than the wavelength of the wave. Gravitons need to be a little like Lewis Carroll's Cheshire Cat: if you look too closely they go away! This illustrates an aspect of quantum gravity that is not often discussed: how quantizing gravity will change conventional *quantum theory*. It seems likely that quantum gravity will inherit from classical general relativity its inability to define energy in an invariant way. Far from giving us a concrete picture of the graviton, quantum gravity might instead make our picture of the photon a little fuzzier!

In this list I have tried to anticipate how quantum gravity might affect predictions about phenomena that we can already study in classical general relativity. But quantum gravity need not, surely will not, confine itself to modifying what we already know. In the rest of this chapter I will speculate on areas where quantum gravity might solve entirely new problems or introduce entirely new ideas.

A Universe for life: the Anthropic Principle

In this section: we examine the fine-tuning of physical quantities that seems to have been required for the evolution of life. Many of these arguments can be found in two stimulating books, *The Antropic Cosmological Principle* by J D Barrow & F J Tipler (Oxford University Press 1986) and *The Life of the Cosmos* by L Smolin (Oxford University Press 1997).

We have already seen in Chapter 11 that the appearance of life on at least one planet in the Universe requires some special values of some fundamental physical quantities. If these values were substantially different, life could not have evolved. Quantum gravity, by unifying all the forces of Nature, could be in a position to explain how some or all of these values came to be. There might be no arbitrary parameters, no adjustable values. Quantities like the mass of the proton, for example, should either be predicted by the theory or be given a certain probability, from which we shall in principle be able to calculate what the probability was of having values that would have led to the evolution of life.

Here we expand the list of these numbers beyond those we mentioned in Chapter 11. We will take the view that the values of Planck's constant h, Newton's constant G, and the speed of light c are not numbers that have to be predicted by the theory. Their values depend on the human-based system of units in which we express them. In fact, we have seen in Chapter 21 that they just define a natural system units in which everything else can be measured. So we ask about how the evolution of life depended on certain dimensionless numbers, like the ratio of the mass of the proton to the Planck mass, which play a role in the physics that we have discussed in this book. We start with the ratio of the mass of the electron to that of the proton.

1. *Ratio of the mass of the electron to that of the proton.* If the mass of the electron were much larger, say comparable to the mass of the proton, then the structure of atoms would be very different. Electrons would orbit very close to the nucleus, and the energy required to ionize an atom by removing the electrons would be much larger. Chemistry would be totally different. While life might still be possible, it would be on terms that we would not recognize.

On an astronomical level, the structure of white dwarfs would be rather different, since the electrons would contribute considerably to their self-gravity. This would lower the Chandrasekhar mass (see Chapter 12) and perhaps make it impossible for supernova explosions to occur, since (as we also saw in Chapter 12) the process that produces the explosion is finely balanced, and a much lighter neutron star might not release enough energy to drive the outer layers of the giant star away. This would reduce the production of the heavy elements needed for life.

2. *Ratio of the mass of the proton to the Planck mass.* If the mass of the proton were larger, then again this would lower the Chandrasekhar mass, and supernovae might not occur. Even a few percent increase in the proton mass might have this effect, stifling the evolution of life.

   On the other hand, if the proton mass were smaller, the Chandrasekhar mass would be bigger, and the collapsing core might be too massive to stop at the neutron star stage: its self-gravity might overwhelm the nuclear forces and lead to the formation of a black hole directly. This would again deprive the collapse event of the strong rebound shock wave that blows the envelope away.

   > The evolution of life has been very sensitive to having exactly the right ratio of the proton mass to the Planck mass.

3. *Difference between the proton mass and the neutron mass.* The neutron has slightly more mass than the proton. This difference is large enough to allow a free neutron to decay into a proton, an electron, and an anti-neutrino. This decay is called beta decay. But in a nucleus, neutrons can be stable against beta decay: it takes a little more energy to make a proton in a nucleus than outside it, because of the presence of other nearby protons whose electrostatic repulsion of the new proton raises its total energy. This is not a large energy, so the existence of stable nuclei depends on the neutron having a mass very close to that of the proton. If it were much larger, all nuclei would decay by beta decay, and life would be impossible.

   If the neutron had a mass smaller than the proton by even a small amount, then electrons and protons would spontaneously combine to form neutrons, releasing the extra energy as a neutrino. Life would again be impossible, because there would be no chemistry: the Universe would consist only of collections of neutrons bound together by the nuclear attraction.

   > Like the proton, the neutron has only a small mass range available to it in which life can evolve.

4. *Energy levels of the carbon nucleus.* We discussed this in detail in Chapter 11: the synthesis of elements heavier than helium depends sensitively on the details of the energy levels of the carbon nucleus. A small change in these levels, caused by a small change in the strength of the nuclear force, would make life impossible.

5. *Strength of the hard-core nuclear repulsion.* Another way in which the nuclear forces affect the possibility of life is that the strength of the repulsion that takes place when two nucleons (protons or neutrons) come close to one another has an effect on the formation of neutron stars. If the repulsion is not strong enough, then when a white dwarf core of a giant star collapses (having the Chandrasekhar mass, which is determined only by the proton mass, not by the nuclear forces), the collapse will not stop before a black hole is formed. Recall that the neutron star has a radius that is only about three times larger than its Schwarzschild radius, so it only has to collapse to one-third of its radius in order to form a black hole. A modest decrease in the strength of the hard-core repulsion could allow this.

On the other hand, if the hard-core repulsion were to increase, then neutron stars would form with much larger radii and lower densities. The binding energy released would be smaller, and this could have the consequence (as we saw above) that the envelope of the giant star would not be blown off. In both cases, supernovae might not occur, some heavy elements would not form, and life as we know it would not be possible.

> Therefore – as with the proton mass, the neutron–proton mass difference, and the longer-range nuclear forces – there is only a small range of values of the strength of the nuclear hard-core repulsion which will allow life to evolve.

6. *The mass of the electron neutrino.* We have seen in Chapter 11 that there is strong evidence that the mass of the electron neutrino is not zero, but it is still very small. If it were much larger, then the neutron would be unable to decay, since it would not have enough mass to produce the masses of an electron, a proton, and an anti-neutrino (whose mass will be the same as that of the electron neutrino). This would distort the process of forming nuclei, since there would be no beta decay. The abundances of some of the elements that are used in the chemistry of life would be much less, and the chemistry of life would be very different, if not impossible.

7. *The mass of the dark matter particles.* As of this writing (2002), we do not yet know the nature of the dark matter. But if we assume that galaxies formed because of the dark matter, then a crucial requirement for the formation of life is that the dark matter particles had a mass heavy enough that they would be cold long before decoupling. The reason is that *galaxies themselves are essential for life.* The collapse of clouds of gas onto the dark matter condensations led to heating of the gas and the formation of stars. And it was crucial that stars formed in galaxies, not in isolation. Even if stars somehow had formed without galaxies, so that they were randomly sprinkled through the expanding Universe, then there would have been only one generation of stars. When they died, the matter they expelled through winds and explosions would have simply swirled through the spaces between stars, but would never have attained enough density to form another generation of stars. So the elements made by the first generation of stars would not have found their way into new stars, planets, and people. Galaxies therefore played a vital role in the evolution of life, encouraging the first generation of star formation and then retaining the gas released by those stars and mixing it into the giant molecular clouds from which second, third and subsequent generations of stars formed. These stars had the heavier elements that the first generation did not have, and they could form planets. The Earth is made almost entirely of atoms that were created in stars and held in place by the Galaxy until the Sun could form, and with it the planets of the Solar System. If the mass of the dark matter particles were too small, galaxies would not have formed at all.

8. *The violation of time-reversibility in fundamental physics.* We have seen in Chapter 25 that the violation of time-reversibility in the fundamental laws of physics led to the fact that the Universe had slightly more protons than anti-protons, slightly more neutrons than anti-neutrons. Since we are made of this excess, its existence is crucial to the evolution of life. If the Universe had been completely symmetrical between matter and anti-matter, then almost all the particles would have annihilated against one another, and the only particles

left in the Universe would be a few lucky ones that had avoided encounters with anti-particles. This particle–anti-particle mixture would have been fairly uniform, so that any clumping of the kind that led to stars in our Universe would just have led to further annihilations as particles and their anti-particles got closer together. Stars could not have formed in such circumstances, and of course neither could life.

It is sobering to reflect on the fact that this subtle and almost unobservable violation of time-reversal invariance, which is a feature of physics that even physics students generally do not learn much about until they reach graduate school, is one of the foundations of life itself! We can hope that quantum gravity will explain the violation of time-reversal invariance.

9. *Balance of dark matter and dark energy.* The Universe left the inflationary epoch with the critical mass density, but inflation does not prescribe how that density is shared between dark matter and dark energy. If there had been much less dark matter, then galaxy formation would not have occurred, and, as described above, life would not have been possible.

Biologists and chemists debate whether the evolution of life was inevitable, given the original conditions on the Earth. Astronomers try to estimate how many Earth-like planets the Galaxy might contain. These are arguments about probabilities, about how many places in the Universe might contain life like ours. However, if any of the physical parameters in the above list were substantially different, then probabilities would have been irrelevant: life as we understand it would simply have been impossible.

It may be that quantum gravity will tell us that these parameters were inevitable, and so there is no need for further discussion. But this seems to me to be unlikely. Instead, quantum gravity could predict the probability that the Universe had begun with the values of these parameters. Then we will be able to assign a probability that the Universe could have evolved life anywhere. This probability may come out rather small. If that happens, what will that mean?

As we mentioned in Chapter 11, the Anthropic Principle is relevant here. The existence of human beings implies that we live in a Universe in which the evolution of life was possible, so our measurements could not have come out any other way. But this is unsatisfying if there is one and only one Universe. If, however, quantum gravity tells us that there have been many universes, each beginning with random values of the constants, and the number of such cosmic experiments was unlimited, then the problem goes away: there will inevitably be universes in which the constants take the appropriate values, and we live in one of them.

Some physicists have speculated on how this might work out. Maybe the Universe is really bound and will re-collapse to a Big Crunch, and by quantum effects re-expand with another Big Bang with different fundamental constants. This seems unlikely to be a good explanation, since one of the fundamental constants is the expansion speed: if the speed of one such re-expansion is so large that the Universe never re-collapses, then the process comes to an end, and maybe not soon enough to have produced life with high probability. In fact, the process seems to be ending with our present Universe, which apparently will not re-collapse. It seems too much of a coincidence that the cyclic Universe would stop cycling just when life evolved on a tiny planet in an unremarkable corner of an anonymous Galaxy.

Another possibility is raised by inflation. If the Universe before inflation was very inhomogeneous, then different regions could have inflated in different ways. If quantum gravity tells us that these different regions had different values of the fundamental constants, then our problem might go away. We inhabit just one of many regions of the entire Universe, and in our region the parameters allowed life to evolve. In a neighboring region, the parameters might be very different. This region is far away now, too far for us to see. While this explanation could be correct, it would not be verifiable, and that is unsatisfactory.

Whatever the ultimate explanation turns out to be, the most important point of all is that explanations seem possible, that quantizing gravity could have much bigger implications than just increasing our understanding of gravity.

Causality in quantum gravity: we are all quantized

In this section: a quantum theory of gravity will have to address issues of time and causality beyond present quantum theory.

Another major conceptual change that quantum gravity is likely to bring about is to our notions of predictability in physics, of cause and effect. These were already modified by standard quantum physics, which removed the older, classical idea of Newton's era that one could at least in principle predict arbitrarily accurately what the outcome of any experiment would be, provided one had sufficiently good information about the physical conditions at the starting point. In quantum theory, one can only hope to predict a set of *probabilities* about the outcome of an experiment. If one does the experiment very many times with identical starting points, then one can test the prediction, verifying that the frequency with which any outcome appears is consistent with the probability predicted by quantum theory. In quantum gravity, even this amount of predictability may be eroded.

One source of trouble is that conventional quantum theory can assign probabilities to outcomes of the experiment only be making a sharp distinction between the experimenter and the experimental system; the experimenter remains classical and behaves with free will, setting up the system for the experiment as many times as desired, measuring the outcomes. But quantum gravity seems unlikely to be able to do this. For one thing, it must presumably give us a quantum cosmology, that is a quantum theory of everything, in which there is *no* outside observer to do experiments and to predict the frequency with which something happens. We are part of the cosmology! If quantum gravity can give us a logically consistent way of dealing with such a situation, then when it is applied to a laboratory experiment it is likely to merge experimenter and experiment into one, and remove the complete freedom that the conventional experimenter has to perform the same experiment over and over again.

By telling us how to merge the experimenter and the experiment into one system, quantum gravity may also help us to understand some existing experiments that are consistent with conventional quantum theory but which seem inconsistent with normal notions of causality. It is possible to create so-called entangled states, where for example two particles are created in such a way that their total spin is zero but the spin of each one is not fixed. Then measuring the spin of one determines the spin of the other. It is possible to show in these experiments that the spin of a particle is not determined at the time it is created and then carried along with it, in the way a classical particle would behave. The spin is not determined until it is measured. Up until that point, there are only probabilities of one spin value or another. However, once the first particle has been measured, the experiments show that measuring the spin of the second particle always gives the right value, the opposite spin to the first particle so that the two spins add to zero. So it appears that, by measuring one particle, one puts the other into a definite spin state too. But when the particles are so far apart by the time they are measured that it is possible to measure the second before any information has had time to travel from the first to "tell" it what spin state to assume, the second particle is still in the correct spin state! How can this information have traveled from one particle to the other? If it did not, how did physics conspire to give a correlated result? Knowing how to treat experimenters and their experiments as a single relativistic system might help resolve this paradoxical behavior, and quantum gravity ought to tell us just that.

Another puzzle for causality would be time travel through wormholes. It would be difficult to sustain a view that a person traveling back in time had the freedom to set up experiments or do anything else that we would call "free will", because then one falls into the "grandmother paradox": go back in time and murder your grandmother when she is still a little girl. Do you exist? Paradoxes like these seem to require that, if time travel is possible, the objects that travel back in time behave consistently in the past; quantum gravity must choose one kind of behavior in all of spacetime, and nothing is free to vary that. The grandmother paradox is no paradox because, just as you are about to run over your juvenile grandmother with a car, your car's tire bursts, you hit a tree, and you are killed! Your attempted assassination becomes one of your grandmother's favorite stories, one you heard when you were small, but which you did not realize was going to be about you! You have no free will, and the quantum state of the Universe is what it is and always has been: your time travel is just part of it.

A final source of difficulty for causality might be the evaporation of black holes. In conventional quantum theory, no information is in principle lost when a system evolves. The outcomes are only predicted with some probability, but if we do the experiment many times we will find that all aspects of the initial state will have some influence on the outcome. But with black holes this appears not to be the case. If we start with a star that is about to collapse to a black hole, there are some aspects of the distribution of mass inside the star that do not show up in the final state, which is an evaporating black hole. Much information has just been swallowed by the hole. When the hole evaporates away entirely, we will again have a smooth spacetime, but one that bears no information about some aspects of our starting spacetime. Black holes seem to be information destroyers of a fundamental kind. It may be that quantum gravity will rescue this situation, and will predict that the information we lack is actually hidden in tiny correlations among the photons emitted as the hole evaporates, or that the final spacetime is not as smooth as we have assumed. Or quantum gravity may just force us to live with a further erosion of our ability to predict things.

The quantization of time?

Playing with causality may seem like disturbance enough to our way of thinking about the world, but quantum gravity will go beyond this: it will play with *time* itself. In a way, time has been the main sub-theme of this book. Gravity expresses itself mainly through the changes it makes to time. The gravitational slowing down of time (the gravitational redshift), the curvature of time that is central to Einstein's explanation of how gravity works, the fact that time itself comes to an end for a particle that encounters a singularity, and the related fact that time itself began at the Big Bang – all of these are part of the fundamental connection between gravity and time.

Time is, however, among the most puzzling of concepts in modern physics. What *is* time? Why does it have its one-way character? There are many so-called *arrows of time*, ways of recognizing that time moves in one direction but not the other:

- *the psychological direction*, the fact that we have memories of the past but

In this section: time is one of the great mysteries of physics. A quantum theory of gravity must shed light on this, since time is just part of our geometry.

not of the future;

- *the statistical direction*, which means that heat always flows from a hotter body to a colder, that different gases can easily mix together but never spontaneously separate;

- *the wave direction*, the familiar observation that waves of any kind always move outwards from their source, never converging inwards on the system, as would happen in a film of a wave run backwards in time;

- *the quantum measurement direction*, where a measurement wipes out previous uncertainties in a quantum system, so that (for example) a particle can no longer reach a location after a measurement that it had a non-zero probability of reaching before it;

- *the cosmological direction*, the expansion of the Universe that appears to have the same direction everywhere, and which may not ever reverse; and

- *the fundamental-physics direction*, the tiny violation of time-reversibility in the laws of physics that is intimately connected with the fact that the early Universe had more matter than anti-matter.

Physicists and philosophers debate the relationships between these arrows of time, questioning whether only one is fundamental, and the others derivable from it, or alternatively whether there is a coincidence of two or more independent arrows that point in the same direction.

It is to be expected that quantum gravity will alter our notion of gravity and therefore of time. Perhaps it will allow us to give a meaning to "before" the Big Bang and to "after" a singularity. Perhaps it will illuminate the relationships between the different arrows of time. Perhaps quantum fluctuations in gravity will lead to quantum fluctuations in the sense of time. Most intriguingly, perhaps quantum gravity will give us a more radical notion of time, as an average over something more fundamental that happens on sub-Planck-scale dimensions.

We have seen that the Planck length is the smallest length-scale on which it is sensible to think of space in our conventional way, as a continuous background within which things happen. What is physics like on sub-Planck scales? Guesses and speculations abound. It could be a kind of spacetime foam, composed of Planck-mass black holes that fluctuate into and out of existence on Planck timescales. It could be a tangled web of strings or loops with no particular sense of dimension or direction. It could even be a set of discrete points linked together by mathematical relations that make them look continuous on larger scales. And if space has a messy structure on small scales, so too must time.

Time for the twenty-first century

In this section: we stand on the threshold of a revolution in gravitation theory as big as Einstein's. Most exciting of all is what the new theories may do to our concept of time.

The idea that time is not continuous allows physicists to begin to ask questions that would have been unthinkable in Eintein's time. How does time emerge from the tangled mess at sub-Planck scales? Why is there just one time dimension, out of many space dimensions? Is time a kind of *mistake*, a defect in a mathematical structure that normally would produce only space directions on scales larger than the Planck scale? Will quantum gravity teach us how to control time, to manipulate its direction or "flow"? Or, as hinted in the previous section, will quantum gravity simply take away our freedom to stand outside a physical system and manipulate it, even as it gives us the knowledge of what we could do with time if only we could do so?

In this book we have journeyed through the Universe, starting from the Earth and ending in the contemplation of the Universe on its largest and smallest scales. We have also made a journey through the world of scientific ideas. We started in Galileo's world, gaining deep and profound insights from simple experiments with cannonballs and pendula. We coupled his and Newton's insights into gravity with the understanding of atomic physics that scientists developed during the century from roughly 1850 to 1950, opening up the physics of planets, stars and galaxies. Then, halfway through our journey, we entered Einstein's conceptual universe, introduced at the beginning of the twentieth century. New ways of thinking about space, time, and matter have helped us appreciate the richness of the physical Universe revealed by the incredible blossoming of astronomy that took place in the second half of the twentieth century.

I write this just as we are beginning the twenty-first century, and the blossoming of astronomy shows no sign of abating. With new and more powerful instruments, astronomers are certain to continue to surprise us with new discoveries, at both astronomy's interface with particle physics and astronomy's interface with biology. At conferences, lectures, and press conferences, the excitement that astronomers feel for their subject is palpable: they are on the trail of some of the deepest secrets of the Universe, and they are getting new clues to these secrets every day.

Just as exciting is the promise the new century brings of a genuine revolution in our thinking: quantum gravity. This promise drives the work of thousands of the world's most talented theoretical and experimental physicists. When a good theory of quantum gravity finally arrives, it may or may not lead to technologically useful by-products, tools for society like microchips or nuclear energy. But it will surely create a new universe of ideas, one that may require as big a change in our thinking as the universe that Einstein created. It will re-define our understanding of cause and effect. It will illuminate the earliest steps on the long road that led to the evolution of our own lives.

Most profoundly – dwarfing all the other developments we may foresee in the realms of gravity, astronomy, cosmology – quantum gravity must re-define time itself. Time is at the heart of gravity. It slows and curves to make the planets orbit the Sun, it stops entirely inside a black hole, it doesn't even begin to advance for a photon or a gravitational wave. And for intervals shorter than the Planck time, it may not behave like time at all.

> To my thinking, the most profound result of quantizing gravity, the most important reason for encouraging – and joining! – the efforts of the thousands of physicists and astronomers who are trying to solve the puzzle of how to unite the visions of Albert Einstein and Max Planck, will be to discover how the *quantum nature of gravity* leads us to an understanding of the *quantum nature of time*.

A p p e n d i x :
values of useful constants

The following values are useful in the exercises and in evaluating equations in the text. All values are quoted in SI units, which are reduced to the fundamental units. Thus, the units (dimensions) of G are given as $\mathrm{m^3\,s^{-2}\,kg^{-1}}$ rather than the equivalent $\mathrm{N\,m^2\,kg^{-2}}$. The only exception is the Hubble constant, which is given in its conventional units.

Astronomers prefer to use their unit of a parsec (pc) rather than call it 30 exameters. In fact, astronomers are non-conformists in many ways. They still quote measurements in the older CGS system (centimeter-GRAM-seconds) rather than in SI units, so that astronomy publications are full of references to centimeters, grams, and ergs. In this book we stay with the SI units, except that we also use parsecs as a distance measure.

Symbol	Value	Units	Description
c	2.9979×10^8	$\mathrm{m\,s^{-1}}$	speed of light
G	6.6726×10^{-11}	$\mathrm{m^3\,s^{-2}\,kg^{-1}}$	constant of gravitation
g	9.807	$\mathrm{m\,s^{-2}}$	acceleration of gravity on Earth
m_e	9.1094×10^{-31}	kg	mass of the electron
m_p	1.6726×10^{-27}	kg	mass of the proton
m_n	1.6749×10^{-27}	kg	mass of the neutron
h	6.6261×10^{-34}	$\mathrm{kg\,m^2\,s^{-1}}$	Planck's constant
k	1.3807×10^{-23}	$\mathrm{kg\,m^2\,s^{-2}\,K^{-1}}$	Boltzmann's constant
eV	1.6022×10^{-19}	$\mathrm{kg\,m^2\,s^{-2}}$	electron volt
r_e	2.8179×10^{-15}	m	classical electron radius
σ	5.6705×10^{-8}	$\mathrm{kg\,s^{-3}\,K^{-4}}$	Stefan–Boltzmann constant
$M_\odot$	1.989×10^{30}	kg	mass of the Sun
$R_\odot$	6.9599×10^8	m	radius of the Sun
$L_\odot$	3.826×10^{26}	$\mathrm{kg\,m^2\,s^{-3}}$	luminosity of the Sun
$R_\oplus$	6.3782×10^6	m	equatorial radius of the Earth
$M_\oplus$	5.976×10^{24}	kg	mass of the Earth
y	3.1557×10^7	s	length of a year
AU	1.4960×10^{11}	m	radius of Earth's orbit about the Sun
p_{atm}	1.013×10^5	$\mathrm{kg\,m^{-1}\,s^{-2}}$	atmospheric pressure on the Earth
pc	3.0857×10^{16}	m	parsec
H_0	70	$\mathrm{km\,s^{-1}\,Mpc^{-1}}$	Hubble constant
H_0^{-1}	4.4×10^{17}	s	Hubble time

The SI units include a number of standard prefixes that indicate how a unit is changed by a power of ten. The table below lists the standard prefixes and their symbols (*e.g.* k for "kilo").

Size	Prefix	Symb	Size	Prefix	Symb	Size	Prefix	Symb
10^{-24}	yocto	y	10^{-3}	milli	m	10^9	giga	G
10^{-21}	zepto	z	10^{-2}	centi	c	10^{12}	tera	T
10^{-18}	atto	a	10^{-1}	deci	d	10^{15}	peta	P
10^{-15}	femto	f	10^{1}	deka	da	10^{18}	exa	E
10^{-12}	pico	p	10^{2}	hecto	h	10^{21}	zetta	Z
10^{-9}	nano	n	10^{3}	kilo	k	10^{24}	yotta	Y
10^{-6}	micro	μ	10^{6}	mega	M			

Glossary

Terms in the glossary are printed in **bold** where they first appear in the text.

absolute magnitude A measure of the intrinsic luminosity L of a star, defined as $M = -2.5 \log(L/2.9 \times 10^{28}\,\text{W})$. Because of the minus sign, stars with lower values of M are intrinsically brighter. A star that is five magnitudes brighter than another is 100 times brighter. *See also* apparent magnitude.

absorption spectrum The spectrum (color content) of the light from a star typically has dips in intensity at certain wavelengths, where elements in its outer atmosphere absorb the light preferentially. The details of the absorption are a fingerprint of the chemical composition of the outer part of the star and also carry information about its temperature, pressure, and density.

acceleration of gravity Near the surface of the Earth, this is the acceleration with which all bodies would fall to the ground if there were no resistance from air or other forces. Typically called g, it has the value $9.8\,\text{m s}^{-1}$. The acceleration at other distances from the center of the Earth or near other bodies will have different values.

accretion The process whereby gas falls onto an astronomical body. This can be via an accretion disk confined to a plane or in a more spherical way. The term is not used for the assembly of large amounts of gas to form a body in the first place; it is used when the body subsequently acquires smaller amounts of material. *See also* accretion disk.

accretion disk When gas falls from one star onto another in a binary system, it typically spirals around the second star before reaching it. The spiralling material forms a disk. If the second star is very compact, then the gas in the disk can reach high temperatures, where it will emit X-rays. Disks can also form around planets and individual stars during their formation. *See also* accretion.

action at a distance The term that describes the fact that the force of gravity in Newton's theory of gravity acts between separated bodies without any intermediary and without any delay. This is different from electromagnetism or general relativity, where waves of the field go from one body to another.

active curvature mass The term used in this book for the combination of density and pressure that is the source of the curvature of space in general relativity: density - pressure/c^2. *See also* active gravitational mass.

active gravitational mass The term used generally for the combination of density and pressure that is the source of the curvature of time in general relativity, which is responsible for most ordinary gravitational effects: density + 3pressure/c^2. *See also* active curvature mass.

angle of inclination Used to describe the orientation of an orbit from the perspective of a viewer on the Earth. It is defined as the angle between the line-of-sight and a line perpendicular to the plane of the orbit. A system with an inclination of $90°$ is one that is seen edge-on to the orbit.

Ångstrom A commonly used but non-SI unit of distance, defined as 10^{-10} m. This is typical of the sizes of individual atoms.

angular momentum The quantity that measures the amount of spin a body has. It is defined with reference to a particular point. In Newtonian physics and in special relativity, the angular momentum of a single particle of mass m moving on a circle of radius r about the reference point with speed v is the product mvr. For larger bodies the angular momentum is the sum of the angular momenta of all the particles in the body. When the motion is not circular, only the component of the velocity perpendicular to the radial direction from the reference point is used. In general relativity, the angular momentum can only be defined in certain circumstances, particularly when the spacetime is invariant under rotations about the reference point. *See also* momentum, component, conservation of angular momentum, invariance.

anisotropy The property that a system is not the same in all directions from a given point; the opposite of isotropic. *See also* isotropic, homogeneous, inhomogeneous.

Anthropic Principle Really a collection of several principles, variants on the theme that the Universe contains human beings because it was designed to contain them. The strong version assumes that the Universe was created with this intent; the weak version of the principle merely says that scientists cannot expect to observe a Universe that could not have created people, so that certain observed conditions are inevitable.

anti-gravity The term that describes a situation in which gravity is repulsive rather than attractive. This does not happen in Newton's theory, but can happen in general relativity when the pressure is so large and negative that the active gravitational mass is negative. This underlies the theory of inflation in the early Universe. *See also* inflation, cosmological constant, dark energy, quintessence.

anti-matter All elementary particles have a counterpart, called an anti-particle, that has the same (positive) mass, the same spin, and opposite electric charge; when a particle collides with one of its anti-particles, they annihilate each other and convert their mass-energy into photons or other particles. Anti-matter is the collection of all anti-particles of normal matter. The Universe began with an excess of matter over anti-matter;

otherwise the particles of which we are made would have been annihilated long ago. *See also* anti-proton, positron, photon.

anti-proton The anti-particle of a proton. *See also* anti-matter.

aphelion The point on an elliptical orbit around the Sun that is furthest from the Sun. If the central object is a star, the point is called the *apastron;* if the Earth, *apogee. See also* perihelion.

Apollo Name given to the US program begun in the 1960s to send men to the Moon. It consisted of a number of missions, at first in Earth orbit, then orbiting the Moon without landing, and finally a succession of landings. The first landing was Apollo 11 in 1969. The Apollo 13 mission nearly ended in disaster, but the astronauts successfully returned to Earth. The earlier Apollo 8 mission caught fire on the launch pad, killing its astronauts. The scientific legacy of the program is important. The astronauts returned with rock samples that have shown that the Moon was formed from the debris from an enormous collision between the Earth and a body the size of Mars. The astronauts left behind reflectors on the Moon that are still used to track the Moon's orbit to an extraordinary precision.

apparent magnitude A measure of the apparent brightness (energy flux) F of a star, defined as $m = -2.5 \log(F/2.4 \times 10^{-8} \text{ W m}^{-2})$. Because of the minus sign, stars with lower values of m appear brighter. A star that is five magnitudes brighter than another is 100 times brighter. *See also* absolute magnitude.

arrow of time The term that describes the perception that time advances in the same "direction", never reversing. The psychological perception of time is probably related to one or more different arrows of time that physicists have identified. These include the increase of entropy (disorder) with time, the spreading of radiation and waves outwards with time, and a tiny lack of time-symmetry in the fundamental laws of physics. *See also* entropy.

asteroids Small rocky bodies in the Solar System. They may be residues of a planet that did not quite form. Most orbit the Sun on roughly circular orbits, but sometimes encounters with one another place an asteroid on an orbit that plunges toward the Sun. The collision of one such body with the Earth about 60 million years ago is thought to have been responsible for the extinction of the dinosaurs. *See also* Kuiper Belt.

astronomical unit The mean distance of the Earth from the Sun, about 1.4960×10^{11} m.

atoms Basic units of matter, which combine to form the chemicals of which our world is made. Atoms consist of a small but massive positively charged nucleus composed of protons and neutrons, and a much lighter and larger cloud of electrons. *See also* proton, electron, nucleus, chemical elements, isotope.

bar detector A gravitational wave detector made of a metal cylinder, which is stretched into longitudinal oscillation by a gravitational wave. *See also* gravitational wave, interferometer.

baryon A collective name for protons, neutrons, and related unstable particles of larger mass. The total number of baryons (with anti-particles counted negatively) is conserved in nuclear reactions. Electrons are leptons, not baryons. *See also* lepton.

beaming A term used in special relativity to describe the effect of Lorentz–Fitzgerald contraction and time dilation on the direction of radiation from an accelerated charge. The faster the charge goes, the more its radiation is directed in the forward direction. *See also* Lorentz–Fitzgerald contraction, time dilation.

beta decay A form of radioactivity which is driven by leptons. Normally a particle decays to produce, among other particles, electrons or positrons, plus associated neutrinos. Because leptons interact with one another by weak interactions, beta decay usually happens on a longer time-scale than radioactivity involving rearrangements of baryons, such as nuclear fisson. *See also* baryon, lepton, neutrino, weak interaction.

Big Bang The name given to the beginning of the Universe, which seems to have occurred in a single explosion. The term was coined by F Hoyle.

Big Crunch The name given to the hypothetical end of the Universe, should it re-collapse to an infinite density. The evidence today is that it will not re-collapse, but instead progress to the Big Freeze. *See also* Big Bang, Big Freeze.

Big Freeze The name given to the hypothetical end of the Universe, should it continue to expand forever. The evidence is that this is likely to happen. *See also* Big Bang, Big Crunch.

black bodies A technical term for bodies that are perfect absorbers of radiation. Such bodies emit a characteristic spectrum of radiation that depends only on their temperature and not on their composition. *See also* black-body radiation.

black-body radiation The radiation emitted by a perfect black body; its spectrum depends only on the temperature, and its intensity on the surface area. Stars emit radiation that is a good approximation to the black-body spectrum, and black holes emit the Hawking radiation, which has a black-body spectrum. The larger the temperature, the shorter the typical wavelength of the emitted radiation. *See also* black bodies, Hawking radiation.

black holes Bodies that have such strong gravity that light cannot escape if it is emitted from within a certain region, whose boundary is called the horizon. Since nothing travels faster than light, black holes trap everything that gets within the horizon. *See also* horizon.

blueshift The shortening of the wavelength of radiation, which can be caused by motion or by gravity. *See also* redshift.

bolometric magnitude The brightness of a star, measured in magnitudes, using the light in a range of colors defined to span the visible spectrum. *See also* apparent magnitude, absolute magnitude.

bore waves A shock in water waves, which can build up into a high wall of water moving upstream. Seen on several rivers with strong tidal ranges.

boson Each elementary particle has spin, an intrinsic angular momentum that, because of quantum effects, always is either an integer or half-integer multiple of $h/2\pi$, where h is Planck's constant. Particles that have integer spin are called bosons, those with half-integer are fermions. Bosons have a preference for occupying the same quantum state, so that they bunch together. Photons are bosons, and laser light is an example of the way they try to conform to one another. *See also* fermion, quantum theory, photon, laser.

boson star A hypothetical star composed of a different and hypothetical form of matter: a boson field. If the mass and (repulsive) self-force of the field have suitable values, then it is possible to make stars of a mass and size similar to neutron stars. No such particles are known from experiment, but grand unified theories allow them. *See also* neutron star, boson, grand unified theories.

brane A technical term in string theory, which is the current leading contender for the way of unifying gravity with the other forces of Nature. In string theory, elementary particles are not point-like, but are instead represented by small closed loops (strings) in a space with ten dimensions. Subsurfaces of this space with more than one dimension generalize the notion of strings, and are called branes, from the word mem*brane*. *See also* string theory.

bremsstrahlung The electromagnetic radiation emitted by a rapidly moving charged particle when it is suddenly decelerated. If the initial speed is large and the deceleration great, the radiation is strongly beamed in the forward direction. *See also* beaming.

brown dwarfs Astronomical objects intermediate in mass between large planets and small stars. By definition, they do not have enough mass to raise their interior temperature to the point where nuclear reactions ignite; instead, they glow by radiating away their gravitational potential energy. Not many are known, but it is possible that there is a huge population of them that contributes substantially to the mass of our Galaxy.

Brownian motion The random movements of, for example, a speck of dust floating on the surface of water. Collisions with water molecules impart tiny changes in the speck's motion, which individually are invisible, but which happen so frequently that they randomly accumulate into apparently sharp changes in the motion of the speck. The speck executes a "random walk" across the water surface.

buoyancy The force that acts upwards on a body that is immersed in a medium of greater mean density, such as a hot-air balloon.

C-field A hypothetical field postulated by Hoyle and colleagues, which would be required in order for the Universe to obey the postulates of the Steady-State model of the Universe. *See also* Steady-State model of the Universe.

calculus The mathematical theory that deals with rates of change of functions. Invented by Newton and independently by Leibniz, it provides systematic ways to solve for the motions of bodies acted upon by forces, but goes well beyond this in being able to treat variations in anything, such as surfaces (curvature), areas, and much more. Calculus is the fundamental mathematical tool of physics: all the basic laws of physics are expressed fundamentally in the language of calculus.

cataclysmic variable A class of variable star in which there are large outbursts of visible light and X-rays caused by mass transfer onto a white dwarf from a giant star that is its companion in a binary system. *See also* white dwarf.

catalyst An agent that promotes a chemical reaction (or other process) without itself being changed by the end of the process. Normally the catalyst is modified during the process but is restored to its original state by the end. Most car exhausts today have catalytic converters, in which a catalyst like platinum helps to convert pollutants into harmless gases. Unless the catalyst is degraded by other chemicals, it will continue doing its job indefinitely.

caustics Places where light rays that start from the same source and that pass through a complicated optical system are made to intersect. You can easily see caustics by looking at light reflecting from a choppy water surface; the caustics are the edges of the bright regions that flicker past the eye.

celsius The standard temperature system in most of the world (apart from the USA) and in science. The zero is defined as the freezing point of water, and the boiling temperature of water is 100 C. Formerly widely known as the centigrade scale, since there are 100 degrees between freezing and boiling water. *See also* kelvin.

centrifugal effect The apparent outward force that a body experiences when executing circular motion. The circular motion is itself accelerated, and the centrifugal effect is actually caused by whatever force causes the body to move from a straight line. The centrifugal effect is an example of an "inertial force".

characteristic frequencies All material bodies vibrate when disturbed. Bodies of finite size normally vibrate freely with a set of frequencies that depend on their composition and shape. These are called the characteristic frequencies of the bodies.

chemical elements The building blocks of all the materials of our environment. The atoms corresponding to each element are identical except, possibly, for the number of neutrons in the nucleus, and their chemical behavior – the way they combine into compounds – is the same for all. *See also* atoms, isotope.

chirp A gravitational wave with increasing frequency and amplitude, emitted by a binary system whose orbit is shrinking because of the emission of gravitational radiation. As the orbit shrinks, the orbital period also goes down and the stars or black holes speed up. These effects make the gravitational radiation frequency and amplitude increase. *See also* chirp time, gravitational wave.

chirp time The time-scale on which a chirping binary system changes its frequency by a factor of two. A chirping binary is a binary system whose orbit is shrinking because of the emission of gravitational radiation. *See also* chirp.

classical A term used by physicists to describe theories of physics that do not incorporate quantum effects. *See also* quantum theory.

co-latitude Latitude is one of the coordinates we normally use for locating places on the Earth. It runs from -90° at the South Pole to +90° at the North Pole. The co-latitude measures the same angle but in a different way, starting at 0° at the North Pole and finishing at 180° at the South Pole. If the latitude is called β, then the co-latitude is $\theta = 90° - \beta$. When mathematicians discuss the geometry of abstract spheres, they normally use the co-latitude rather than the latitude as one of the so-called spherical coordinates.

cold dark matter If the dark matter inferred from astronomical observations consists of particles much more massive than normal atoms and carrying no electric charge, then it would have cooled off more rapidly as the Universe expanded, and could have formed massive clumps that later attracted normal matter gravitationally and began the process of galaxy formation. This cold dark matter is the standard model of galaxy formation. However, the hypothetical particles have not yet been identified or detected. *See also* dark matter, hot dark matter.

collisionless gas If a gas is so rarified that collisions between its particles are very rare, it is called collisionless. If the particle velocities are random the gas might still behave like a normal gas, with a definite pressure and temperature. This can happen if, for example, the gas is confined by walls with which the particles collide and exchange energy.

color index The difference of the blue magnitude and the visual magnitude of a star. Since these magnitudes are defined by logarithms of the brightness of the star, this difference depends on the ratio of the brightness of a star in the visual color band to that in the blue band. This ratio is independent of the distance to the star, since both brightnesses fall off with distance in the same way. It is therefore a measure of the intrinsic color of the star, and thus of its temperature. *See also* visual magnitude, color of a star.

color of a star The color of a star depends on the relative intensities of different colors in its light. Most stars look more-or-less white to the naked eye, because the black-and-white sensitivity of the eye is greater than its color sensitivity for weak light sources. But when the colors in the light are measured, stars turn out to have different balances. Some have much more red light than blue, some more blue than red. These differences reflect differences in the temperatures of stars: cooler stars are more red. *See also* color index, black-body radiation.

component Mathematicians use the word *vector* to describe a directed quantity, like velocity. The piece of the velocity along any particular direction is called the component along that direction. To describe a vector in three dimensions requires three components. The concept of a component is used also for tensors, where it refers to the elements of the matrix that rep-

resents the tensor in a particular coordinate system. *See also* vector, matrix, tensor.

compose In the context of this book, scientists use this term to describe how different velocities combine in special relativity. When body A measures the velocity of B and B measures that of C, the velocity of C as measured by A will not be simply the vector sum of the two previous velocities. If this were the case it would be easy to get velocities greater than the speed of light. Instead, relativity predicts a more complicated composition law, which never produces a speed exceeding that of light.

Compton scattering The scattering of photons from charged particles. Since photons carry electric fields, they interact with electric charges. The scattering, however, reveals the discrete nature of photons: they behave just like particles carrying a given energy and momentum and scatter from the charged particle into various directions. *See also* photon.

conservation of angular momentum In Newtonian physics and in special relativity the total angular momentum of any isolated system is constant in time. The parts of the system can exchange angular momentum, but the total is unchanged. In general relativity, this law holds only if the geometry of the spacetime is invariant under rotations about the reference point for the computation of the angular momentum. *See also* angular momentum, invariance, conservation of energy.

conservation of energy In Newtonian physics and in special relativity the total energy of any isolated system is constant in time. The parts of the system can exchange energy, but the total is unchanged. Each time physicists have uncovered a new force, a new branch of physics, they have found that there is an associated energy that can be defined in such a way that the total remains conserved. This is not arbitrary: it is only possible to define conserved energy if the geometry of spacetime is time-independent. In general relativity, this law can therefore only hold in certain circumstances: particles moving in time-independent geometries, or in the locally flat geometry of special relativity sufficiently near to any event, or in terms of total energy as measured by a distant experimenter sitting in the flat spacetime far away from an isolated star or black hole. *See also* energy, experimenter, invariance, locally flat.

convection When a fluid is heated from below too rapidly for ordinary conduction or for any radiative flux through the fluid to carry it away, then the fluid begins to flow in a roughly circular motion, absorbing heat at the bottom of the convection cell and releasing it at the top.

Copernican principle Copernicus argued that the Sun was at the center of the Solar System, not the Earth. This made the Earth an ordinary planetary body, not located in any special place in the Solar System. When this principle is extended to the Galaxy and the Universe, we would assume (unless there is evidence to the contrary) that our location is similarly not privileged. In particular, the Universe should look the same, statistically, to any other astronomer observing it from any other ordinary star in any other ordinary galaxy. This principle is sometimes called the *principle of mediocrity*.

cosmic censorship hypothesis Solutions of Einstein's equations for black holes contain singularities inside, locations where the predictive power of the laws of physics break down. These are regarded as a mild failure of general relativity, since the unpredictability is hidden from our view behind the black-hole horizon. However, if a singularity were to form outside a black hole, this would be a much more serious problem for physics. No robust examples of this are known, however, and Penrose suggested that perhaps there was a deep connection between singularities and horizons in general relativity. His cosmic censorship hypothesis is the proposal that there should exist a mathematical theorem to the effect that in generic situations singularities never appear outside a horizon: they are always "censored" by Nature. The conjecture is unresolved. *See also* singularity, naked singularities.

cosmic microwave background radiation The early Universe consisted of a dense, hot, expanding gas. When the gas cooled, it became transparent to radiation, and most photons released at that time have traveled freely through the Universe ever since. Astronomers detect this radiation as a black-body spectrum with a temperature such that the radiation is predominantly in the microwave part of the spectrum. Tiny irregularities in temperature from one direction to another are a snapshot of a very early phase of galaxy formation. *See also* decoupling, photon, microwave.

cosmic rays The Earth is bombarded by high-energy particles, mainly protons, from space. Most collide with gas in the upper atmosphere, so the radiation poses only a limited radiation risk at sea level. Most cosmic rays probably originate in supernova explosions, but there is a small flux of ultrahigh energy particles whose origin is a puzzle. *See also* supernova.

cosmic strings Certain theories of high-energy physics predict that, when spontaneous symmetry breaking occurs to form the laws of physics as we know them, there may be some locations where the original, symmetric form of the laws still holds. These regions must be concentrations of energy that was trapped because it could not decay. If these regions are one-dimensional, they are called cosmic strings. In principle such strings could be so plentiful and massive that they caused galaxies to form near them. However, the evidence today is against this mode of galaxy formation and in favor of cold dark matter. Nevertheless, lighter cosmic string are still possible and could be sources of gravitational radiation. *See also* spontaneous symmetry breaking, cold dark matter.

cosmological constant The term introduced by Einstein into his equations of general relativity, in order to insure that there could be solutions for static cosmologies, because astronomers at that time had not found evidence for an expanding Universe. The term created a repulsive force, a form of anti-gravity, that countered the attraction of normal matter. When the expansion of the Universe was discovered, Einstein rejected this term, but today astronomers have put it back in because they have found that the expansion of the Universe appears to be accelerating. Physicists have created alternative explanations of such effects, such as quintessence. The generic name for such effects in the

equations of general relativity is dark energy. *See also* anti-gravity, dark energy, inflation, quintessence.

cosmological scale-factor The mathematical quantity that tracks the expansion of the Universe. It can be defined to be any length at any particular initial time, but then it expands in proportion to the distances between galaxy clusters that are so far apart that their mutual gravitational attraction has a negligible effect on their velocities. The expansion speed and acceleration/deceleration of the Universe are defined by how this length-scale behaves with time relative to its initial value. *See also* cosmology, galaxy cluster.

cosmology The study of the Universe as a whole, its history, and the physical processes that led generally to the formation of galaxies and stars. *See also* physical cosmology, cosmological scale-factor.

critical density The mass density that the Universe would have if the gravitational attraction of the matter was just what would be required to reduce the expansion speed of the Universe to zero in the infinite future. When there is dark energy as well, the dynamics of the Universe will be more complicated, but the ratio of the total density of mass-energy to the critical value determines the overall spatial structure of the Universe: open (ratio smaller than one), closed (larger than one), or flat (equal to one). *See also* physical cosmology, dark energy.

crust The outer layer of a neutron star, where the density is not great enough for all the matter to exist as neutrons. The crust is composed of neutron-rich nuclei in equilibrium with free neutrons and electrons. The nuclei are thought to arrange themselves in a weak lattice, which resembles a jelly-like solid. The crust is not brittle, but rather pliant and yielding. As the only likely solid part of the star, it is responsible for many observed phenomena, and could be a source of gravitational radiation. *See also* neutron star.

curvature The property of a space that determines whether parallel lines can remain parallel when extended in as straight a manner as possible. A space has zero curvature (i.e. is flat) if parallel lines remain parallel. If they approach one another the space has positive curvature, if they diverge then it is negative. Einstein used the curvature of spacetime to describe the action of gravity in general relativity. The locally straight lines are the paths that free particles follow through spacetime. The curvature of time represents, to a first approximation, Newtonian-like gravity. *See also* spacetime, spacetime-interval, spacetime metric, flat space.

cycles Ancient astronomers tried to describe the apparent motion of the planets in the sky by complicated motions superposed. If the planets are assumed to go around the Earth, then they do not always go in the same direction. Sometimes they turn around and go backwards, an effect which is easily explained in the Copernican model of the Solar System when one takes into account the motion of both the Earth and the planet. But ancient astronomers, thinking the Earth was fixed, described the motion of a planet as a basic *cycle* around the Earth, added to a smaller *epicycle* that was a kind of circular motion back and forth along the orbit, sometimes having the net effect of moving the planet backwards.

dark energy Observations suggest that the expansion of the Universe is presently accelerating, and theories of inflation also require a period of rapidly accelerating expansion. To produce the acceleration one needs a physical field that has a negative pressure, large enough to make the active gravitational mass negative and produce an anti-gravity effect. Fields that have positive energy but large negative pressure are called dark energy. *See also* cosmological scale-factor, active gravitational mass, anti-gravity.

dark matter Astronomers have determined that the gravity that seems to bind together galaxies and galaxy clusters, and the overall gravitational field of the Universe, can only be explained if there is hidden matter, called dark matter. It is not known what it is composed of. *See also* cold dark matter.

deceleration parameter A dimensionless number that represents the acceleration or deceleration of the Universe today.

decoupling At first the Universe was hot enough to ionize hydrogen, so that its gas was a plasma of charged particles, through which photons could not move far without scattering. As the Universe cooled off, its gas became neutral, and photons could propagate. The transition is called the *decoupling* of photons from matter. It is sometimes called *recombination*. *See also* plasma, cosmic microwave background, photon.

degenerate gas If fermions are cooled and compressed to a sufficient density, then because no two identical fermions can occupy the same quantum state, not all of them can slow down to small velocities, as would classical particles. So they retain a residual pressure even as their temperature goes to zero kelvin. This is a degenerate gas, and the pressure is enough to hold up a neutron star against gravity. *See also* fermion, classical, quantum theory, kelvin, neutron star.

density parameter The ratio of the density of the Universe (or of one component of the Universe) to the critical density. *See also* critical density.

density wave In the theory of galactic structure, it is believed that a compression wave in the density of the stars and gas in a spiral galaxy is responsible for the observed spiral pattern. Most of the mass in such a galaxy is distributed symmetrically around the center, but the brightest stars are not: they are concentrated in the spiral arms. Since the time it takes a density wave to travel once around the galaxy is typically long compared to the lifetime of massive bright stars, it is believed that the spiral arms trace the location of the compression region of the wave. *See also* spiral galaxy.

derivatives In calculus, the functions that represent the rates of change of other functions. The velocity is the derivative of the position of a particle with respect to time. *See also* calculus, differential equations.

deuterium The nucleus of the isotope of hydrogen that has one proton and one neutron. *See also* isotope, nucleus, proton.

differential equations Equations consisting of functions and their derivatives. All the basic laws of fundamental physics are expressed as differential equations. *See also* calculus, derivatives.

dimensional analysis The technique of examining the consistency of an equation with the dimensions (basic units of mass, length, and time) of the quantities in it. It can be a powerful technique if, on physical grounds, the equation must contain only a few quantities with known dimensions. Then dimensional analysis can actually point the way to inventing the correct equation that relates them with one another. *See also* dimensions.

dimensionless number A number that has no dimensions (units). Arguments of non-linear mathematical functions, like the sine or exponential, must be dimensionless even if they are composed of quantities that individually have dimensions. *See also* dimensions.

dimensions The *type* of units carried by a physical quantity. For example, distances have dimensions of length. A distance can be given in units of meters, miles, microns, and so on, and its value will depend on the unit. But in each case, the quantity has the dimension of length.

direct image In optics, an image created by a system of mirrors and lenses which preserves the sense of left and right in the original object.

displacement A technical term for the vector position of an object from the origin of the coordinates. *See also* vector.

diverging lens A lens that causes initially parallel light rays to diverge (separate) after they pass through it.

dragging of inertial frames A colorful name for the effects of gravitomagnetism in which the trajectories of freely-falling bodies are dragged in the same sense as the motion of the source of the gravitational field. *See also* gravitomagnetism, stationary limit.

dust grains Interstellar gas clouds contain not only the basic gas from which stars form, but also dust: solid particles of ice and carbon compounds, which scatter light and obscure distant objects. These particles contribute much of the heavy elements when stars form.

eccentricity A measure of how non-circular an ellipse is. If the ratio of the minor axis (shortest diameter) of the ellipse to its major axis (longest diameter) is r, then the eccentricity is $e = (1 - r^2)^{1/2}$.

eclipse The blocking of the view of one astronomical body by another. Since the Moon and Sun have nearly the same angular size on the sky as seen from the Earth, there are times when the Moon totally blocks the light from the Sun at some locations on the Earth: total eclipses. Similarly, the Earth can come between the Sun and Moon, stopping sunlight from reaching the Moon: a lunar eclipse.

Einstein curvature tensor The mathematical construction used by Einstein to describe the part of the curvature of spacetime that directly equals the densities, momenta, and stresses of the matter producing gravity. The tensor depends on derivatives of the metric of spacetime. *See also* curvature, spacetime, spacetime metric, stress energy tensor, Einstein field equations, derivatives, tensor.

Einstein field equations The fundamental equations of general relativity. On one side of the equation is the Einstein curvature tensor, on the other the stress energy tensor of matter. The equations are differential equations because they contain derivatives of the spacetime metric. There are ten equations in all that must be solved for the ten components of the metric. The equations are non-linear and interlinked, so that in most cases a realistic solution can only be obtained by computer simulation. *See also* Einstein curvature tensor, stress energy tensor, tensor, spacetime metric, component.

Einstein luminosity A number with the dimensions of luminosity (energy per unit time) that is composed only of fundamental constants, c^5/G. It is the maximum luminosity that any physical object can radiate. *See also* luminosity.

Einstein radius The characteristic distance from a gravitational lens at which bright images might appear. If the lens is strictly spherical and the source is directly behind its center, then the image of the source will be a ring at this radius. The ring is called the Einstein ring. *See also* gravitational lensing.

Einstein ring *See* Einstein radius.

electric charge The source of the electric and magnetic field. Particles can have positive, negative, or zero charge. All charges are integer multiples of the fundamental unit of charge, which is the charge on a proton. *See also* magnetic field.

electromagnetism The theory of the electric and magnetic fields, devised by Maxwell in the nineteenth century. A single theory unifying electricity and magnetism is required because moving electric charges create magnetic fields and oscillating magnetic fields create electric fields. *See also* electric charge, magnetic field.

electron neutrinos All leptons, including electrons, fall into three families. Each family contains the lepton, its anti-particle, and an associated neutrino and antineutrino. The numbers of leptons of each type must be conserved in a nuclear reaction. Thus, when an electron disappears, an electron neutrino must be created to take its place. *See also* lepton, anti-matter, neutrino.

electron The fundamental particle that carries the negative electric charge within atoms and which moves through solids to carry electric current. It is a lepton. It is affected by the electromagnetic forces because of its charge and by the weak interaction, but does not sense the strong interaction. *See also* lepton, strong interaction, weak interaction.

electroweak The theory that unifies the electromagnetic and weak interactions. It shows that the weak interaction, which is responsible for beta decay, is related to and of the same strength as the electromagnetic interaction at very high temperatures. Recent experiments have given very firm confirmation to this theory. *See also* electromagnetism, weak interaction.

electron volt A measure of energy, denoted by eV. One eV equals the energy an electron gains by falling through a voltage difference of one volt. This is 1.6022×10^{-19} kg m^2 s^{-2}. *See also* MeV.

elliptical galaxy A galaxy that appears smooth and elliptical in photographs. Some are probably oblate spheroids, but others may be genuinely tri-axial. Astronomers believe that their smooth appearance is caused by the extreme mixing that happens when two less regular galaxies collide and merge.

energy Physicists use the word energy to describe something very specific, not very closely related to the everyday uses of the word. The basic energy is kinetic, equal to $mv^2/2$ for a body of mass m and speed v. All other energies are defined in such a way that the sum of all energies is conserved, unchanging with time. In relativity, the energy includes the rest-mass of the object by the famous formula $E = mc^2$. For a system described by a classical theory of physics, the total energy of a system, including masses, is always positive, but in quantum systems negative energy is allowed, at least for short times. In general relativity, a sufficiently large curvature of time can make the total energy of a particle near a star or black hole negative, but even then the total energy of the particle plus the star must be positive. *See also* conservation of energy, classical, quantum theory, stationary limit.

entropy A measure of the disorder in a system. The larger the entropy, the less structure the system has. It is also a measure of *information*, since a more chaotic system contains more information: to reconstruct the system exactly would require a larger list of rules than to construct a well-ordered system. The second law of thermodynamics asserts that the total entropy of any closed system cannot decrease with time, and in any realistic system it will increase. Living systems manage to control their own entropy while they are alive, but to do so they must increase the entropy of their environments. Black holes are the objects with the highest known entropy.

epicycles *See* cycles.

ergosphere *See* stationary limit.

Euclidean geometry The geometry that follows the axioms of Euclid, which describe a flat space in which distances follow the usual Pythagorean theorem. A two-dimensional Euclidean geometry has the properties of the surface of a flat piece of paper. We normally assume that we live in a three-dimensional Euclidean geometry, but actually gravity makes tiny changes in the geometry that are not perceptible except over very large regions. *See also* Pythagorean theorem, curvature.

Euclidean plane A two-dimensional Euclidean space. *See also* Euclidean geometry.

events Points of spacetime, having a fixed location and time of occurrence. *See also* spacetime.

excited state In the quantum theory of atomic structure, electrons orbiting the nucleus normally have well-defined values of energy, and the electrons normally occupy all the lowest-energy states, one per state as required by the Pauli exclusion principle. In this configuration the atom is said to be its *ground state*. If an electron has a higher energy than required, so that a lower-energy state is empty, then the atom (as well as the electron) is said to be in an excited state. Left alone, the electron will normally rapidly drop into the lower-energy state, emitting a photon with a characteristic energy and frequency. *See also* quantum theory, atoms, photon, spectral lines.

expansion of the Universe Astronomers have discovered that objects that are more distant from us are systematically moving away from us with a faster speed, proportional to their distance. In such a circumstance, any astronomer anywhere else will also see the same thing, so that the Universe is expanding in a homogeneous way. This expansion can be measured for a variety of different objects independently. If one traces back the expansion in time, then all the observable Universe was compressed into a tiny volume about 14 billion years ago. The expansion at that time was explosive, and has been called the Big Bang. *See also* Big Bang, homogeneous.

experimenter In this book, a complete and careful system for gathering all possible information about events in spacetime. Experimenters carefully synchronize their clocks, they are not fooled into making errors because there is a delay in information reaching the experiment's headquarters from more distant information-gathering stations, they define things like distance and time in exactly the way one would expect, and they can make measurements with arbitrarily good accuracy. Such experimenters are the ones that will observe the unexpected effects of special relativity, such as the Lorentz–Fitzgerald length contraction or the time dilation. They are also called *observers*. *See also* special relativity, Lorentz–Fitzgerald contraction, time dilation.

exponential function The mathematical function describing quantities whose growth rate is proportional to their size. Denoted $\exp(x)$ or e^x, where $e = 2.7182818284$ to ten decimal places. Its inverse is the logarithm function. *See also* logarithmic scale.

fermion Each elementary particle has spin, an intrinsic angular momentum that, because of quantum effects, always is either an integer or half-integer multiple of $h/2\pi$, where h is Planck's constant. Particles that have integer spin are called bosons, those with half-integer are fermions. Fermions are unable to occupy the same quantum state, so they anti-bunch. All the standard elementary particles – electrons, protons, neutrons – are fermions. The fact that electrons cannot occupy the same state is essential for the structure of atoms as we know them, with diffuse clouds of electrons available to other atoms for bonding into chemical compounds. The fact that fermions cannot share a quantum state is called the *Pauli exclusion principle*. *See also* fermion, quantum theory.

finite differences The basis of computer calculations that approximate the continuous motion of something by a series of small but finite steps. The differences are used in place of the derivatives of calculus.calculus and finite-differences If the steps are small enough, the numerical approximation can be made as accurate as one wishes. *See also* calculus, derivatives.

flat space A space without curvature. This is not quite the same as a Euclidean space. A geometry can be flat without having a metric defined on it, and so without having a Pythagorean theorem. For example, an ordinary chart of, say, temperature in New York against time, is a two-dimensional flat space, but there is no meaning to the "distance" from one point on the curve to another. *See also* Euclidean geometry, curvature, metric tensor, Pythagorean theorem.

flux The term physicists use to denote how much of something is passing through a region in a certain time. The flux of energy is the amount of energy passing through a given surface, per unit area and per unit time. The flux of momentum would be the amount of momentum carried by particles passing through a surface, again per unit area and time. In particular, the energy flux measures what we normally mean by the (apparent) brightness of a source of light: multiplying the flux by the area of the pupil of my eye and by the (small) time it takes the eye to sense light, it tells me how much light I actually see from the source.

frame In relativity, the complete coordinate system that an experimenter (observer) constructs in order to locate events and measure distance and time relations among them. In special relativity, this frame can be a homogeneous coordinate system, with straight coordinate lines and uniform distances and time between them. Such a frame is called an *inertial frame*. In general relativity, the curvature of spacetime prevents large-scale homogeneous coordinate systems from being constructed, so the word *frame* is usually reserved for the local coordinates set up by a locally inertial observer to measure phenomena in a small region around a particular event. *See also* experimenter, homogeneous, inertia.

fundamental frequency The characteristic frequency that has the longest wavelength inside a vibrating body, therefore involving as much of the body as possible in a single coherent motion. *See also* characteristic frequencies, overtones.

galaxy A collection of stars well separated from other such collections and held together by its own gravitational attraction. Most galaxies contain 10^{10} stars or more, as does our own, the Milky Way. But some dwarf galaxies are much smaller, like the Magellanic Clouds. Globular clusters are smaller still, and are not referred to as galaxies, since they are normally part of true galaxies and not isolated on their own. Galaxies usually contain gas as well as stars, and they appear also to contain considerable dark matter, perhaps ten times as much as is luminous. The visible part of the Universe contains some 10^{12} galaxies. *See also* globular clusters.

galaxy cluster Galaxies are not distributed uniformly in space. Instead, they group into loose chains and more tightly bound clusters. The Virgo Cluster is the nearest large cluster, but our own galaxy belongs to a small cluster called the Local Group. Clusters can have anything from a handful to thousands of galaxies. They also contain dark matter, even more than in the individual galaxies. The statistical distribution of galaxies in clusters and chains provides important information about the way they were formed, and supports the cold-dark-matter model of galaxy formation. *See also* galaxy, dark matter, cold dark matter.

gamma-rays High-energy photons, above about 100 keV, with wavelengths shorter than about 10^{-11} m. This is the highest-energy (shortest-wavelength) section of the electromagnetic spectrum. *See also* infrared, microwave, ultraviolet radiation, X-ray, sub-millimeter.

gamma-ray bursts Every day the Earth receives two or more bursts of gamma-radiation from very distant sources. The bursts are likely produced by either the merger of two neutron stars or a neutron star and a black hole, or by a highly energetic form of supernova explosion called a *hypernova*. If they are mergers, then they will be accompanied by a strong burst of gravitational radiation from the orbital in-spiral that took place before the merger. *See also* gamma-rays, neutron star, black holes, supernova, hypernova, gravitational wave.

geodesic The mathematical name for a curve that follows a locally straight line through a curved space. Always going straight as determined by a locally flat observer, the line can nevertheless wander about because of the curvature of the space. They are therefore good tracers of curvature. On a sphere, geodesics are great circles. In a flat space they are normal straight lines. In spacetime, they are the world lines of freely-falling particles, unaffected by non-gravitational forces. *See also* curvature, flat space, spacetime, world line.

giant In astronomy, a star that has expanded to many times its normal size. This happens when the star exhausts its normal fuel of hydrogen, the central region contracts and heats up, and the star begins to process heavier nuclei into still heavier ones. The core of the star gets denser and its envelope thinner. If the star expands far enough, the outer layers become cool and the spectrum moves into the red. These are *red giants*. Stars that are not quite so big and so are at a higher surface temperature could be *blue giants*. *See also* cataclysmic variable, supergiant, nuclear reactions.

glitches A word that astronomers have adopted to describe rapid changes in the periods of pulsars. They typically show a sudden rapid speed-up followed by a longer slow-down. Their cause is not entirely clear, but it may have to do with the interaction between the superfluid liquid interior and the crust of the star. *See also* pulsar, superfluid, crust.

global Used by mathematicians and physicists to describe concepts that are valid everywhere in a large domain. Its opposite is *local*. As an example, one can say that an observer in special relativity can construct a global frame that is the same everywhere, while one in the curved spacetime of general relativity could construct a similar frame only locally. *See also* local, frame, spacetime, curvature.

globular clusters Roughly spherical, tightly bound star clusters containing hundreds of thousands of members or more. They have a distinctive shape and appearance. Since all their members were formed at the same time, they are good laboratories for learning about relative rates of evolution of different kinds of stars. Encounters among member stars can create binary systems of stars and black holes, and also can cause the cluster to evolve in various ways, and globular clusters may collide with one another or with other concentrations of stars in the Galaxy. They may have been much more plentiful in the past, and indeed perhaps they are among the building blocks of galaxies. *See also* star cluster, galaxy.

grand unified theory Unified theories of the strong interaction with the electroweak. *See also* electroweak, strong interaction, unified field theories.

gravitational collapse The inward fall of a self-gravitating body that can no longer produce enough pressure to resist the pull of gravity. This is the event that triggers supernovae of Type II. *See also* supernova of Type II.

gravitational lensing The action of gravity in bending the path along which light propagates in such a way that images of objects are distorted, duplicated, or changed in brightness.

gravitational slingshot When a small mass, like a spacecraft, encounters a large mass, like a planet, that is moving, the energy of the spacecraft is not the same before and after the encounter. It is possible to arrange the encounter so that the spacecraft gains energy and is slung into a different trajectory. This can enable the craft to reach parts of the Solar System that require more energy than the launch can give it. *See also* conservation of energy.

gravitational wave A ripple in the gravitational field that travels with the speed of light through space. It carries time-dependent tidal accelerations, which are the only ones measurable by a local experiment. The accelerations are transverse to the direction of motion of the wave, and they mimic the accelerations of the masses that produce the waves, as projected on a plane perpendicular to the direction of motion. Because gravity is a very weak force, the waves have only a tiny influence on matter that they pass through. This makes them hard to detect but also makes them good carriers of information, since they do not get distorted by intervening matter. *See also* tidal acceleration, bar detector, interferometer.

gravitoelectric field A term used in this book to describe the dominant part of the gravitational acceleration in general relativity, the part that is embodied in the curvature of time and is generated by the active gravitational mass. For weak fields, this is the ordinary Newtonian gravity. *See also* curvature, active gravitational mass.

gravitomagnetism The part of the gravitational acceleration in general relativity that is generated by momentum, and which acts only on bodies with momentum. It has some resemblance to ordinary magnetism, hence its name. Some physicists call it *magneto-gravity* instead. It is responsible for the dragging of inertial frames and the existence of stationary limits. *See also* stationary limit, frame, inertia, dragging of inertial frames, momentum.

graviton A quantum of a gravitational wave in the same sense that a photon is a quantum of light. However, while the photon is a well-defined concept grounded in the theory of quantum electrodynamics, the graviton is just a guess: there is no quantum theory of gravity yet. *See also* quantum theory, photon, quantum electrodynamics.

greenhouse effect The trapping of heat by using a material that is transparent to light but not to infrared radiation. Glass in a greenhouse allows sunlight to pass through, where it is absorbed by plants and the ground. The energy is re-radiated as infrared light, because that is where the peak of the black-body spectrum is for typical temperatures on the Earth. The glass is not as transparent in the infrared part of the spectrum as in the visible. Therefore, heat builds up in the greenhouse. This effect is a cause of warming on the Earth. Carbon dioxide, methane, and fluorocarbons are all greenhouse gases, which allow sunlight to reach the ground but which absorb some of the infrared light that is re-emitted back to space. This effect has existed for billions of years; without it the Earth would be much colder. The concern today is that human activity is raising the amount of greenhouse gas and hence the mean Earth temperature beyond safe levels. *See also* infrared, greenhouse gases.

greenhouse gases Gases that create a greenhouse effect by being transparent to visible light but at least partly absorbing at the longer infrared wavelengths. *See also* greenhouse effect.

half-life The time it takes for half of the members of a sample of radioactive particles to decay. Each particle decays at random and has no "memory" of how long it has been waiting to decay. Given a radioactive particle that was created a million years ago, the probability that it will decay in the next year is the same as the probability that it would have decayed in its first year. It follows that, if half the particles decay in a certain time, then half of the remaining ones will decay in the same period of time again. This is the half-life of the particles.

Hawking radiation The black-body radiation emitted by a black hole. This emission is a purely quantum effect: in classical general relativity, black holes cannot emit any radiation. *See also* black-body radiation, black holes, quantum theory, classical, Hawking temperature.

Hawking temperature The temperature of the black-body radiation emitted by a black hole. If the hole is spherical, this is inversely proportional to the mass M of the hole. *See also* Hawking radiation.

helioseismology The study of the characteristic frequencies of oscillation of the Sun. Thousands of frequencies have been measured, and this tightly constrains the solar model. In this way, astronomers "see" deep inside the Sun. This information has helped to point the way to solutions of the solar neutrino puzzle. *See also* characteristic frequencies.

Hertzsprung–Russell diagram A chart plotting the luminosity and temperature of a number of stars. Stars do not appear at random locations in this plot. Most fall in a narrow band called the *main sequence*. Others fall in regions called the giant branches. White dwarfs are located in another small region. These groups show that, for a star with a given mass and at a given stage of its evolution, the temperature and luminosity are related to one another. *See also* main sequence stars, giant, white dwarf.

homogeneity/isotropy problem The puzzle of why the Universe shows such uniformity in all directions and at all distances from us. In the standard Big Bang model, regions sufficiently far from us in different directions have not had time to make contact before emitting the light we see from them. They could not, therefore, have come to some kind of equilibrium, and yet they look very similar. *Inflation* solves this problem. *See also* homogeneous, isotropic, Big Bang, inflation.

homogeneous The same everywhere. The Universe is homogeneous at a given moment of cosmological time, provided one averages over volumes containing many clusters of galaxies. *See also* galaxy cluster.

horizon The outer boundary of a black hole or the limit of what we can see in cosmology. For black holes, the surface is technically known as the *event horizon*, and it is defined as the boundary between events that can send light rays to a very distant observer and those that cannot. For cosmology, the surface is technically known as the *particle horizon*, and it is defined as the boundary between events that could send light to us and those that could not, since the Big Bang. (We ignore scattering and absorption here, just asking whether a particle traveling at the speed of light could reach us.) The event horizon expands if matter falls into the black hole; the particle horizon expands all the time, since each moment we can see some regions that had, until then, been too far away. *See also* black holes, Big Bang.

hot dark matter Dark matter that is composed of particles whose masses are so small that, at the time and temperature of decoupling, the velocities of the particles were too large to allow fluctuations in their densities to grow fast enough to form the seeds of galaxy formation. This mass is about $10 \, \text{eV}$. Hot dark matter particles would today be distributed much more smoothly in the Universe than is the visible matter. *See also* decoupling, dark matter, cold dark matter.

Hubble constant The present relative rate of expansion of the Universe, that is the ratio of the speed of recession of a distant galaxy to the distance of the galaxy. This ratio is the same for all galaxies, no matter where the observer stands, in a perfectly homogeneous universe: hence the word "constant" in the name. The real Universe is not perfectly homogeneous, and so measuring the Hubble constant has not been easy. Only very distant galaxies give good values, since their random motions are small compared to the systematic expansion speed. But it is difficult to estimate the distances to such galaxies. Astronomers seek *standard candles*, objects whose luminosity is known so that their distance can be estimated from their apparent brightness. Recently Type Ia supernovae have become useful standard candles. *See also* cosmological scale-factor, homogeneous, standard candle, supernova of Type Ia.

Hubble time The time it would have taken the Universe to reach its present size if its expansion were constant in time. This is the reciprocal of the Hubble constant. *See also* Hubble constant.

hyperboloid A three-dimensional geometric figure obtained by rotating a hyperbola about its axis of symmetry.

hypernova An unusually powerful supernova resulting from gravitational collapse, as does a supernova of Type II. *See also* gravitational collapse, supernova of Type II.

hypotenuse The long side of a right-angled triangle, opposite the right angle.

indices Labels attached to components of vectors or tensors to indicate which directions the components are associated with. *See also* component, Pythagorean theorem.

inertia Essentially, the mass of a body. The word is often used to describe the property of a body that makes it resist acceleration and attempt to continue moving in a straight line. The concept is a little vague, and is made much more precise by Newton's laws of motion. Inertia is simply a property: something either has it or it does not. By contrast, mass is a quantity: it is measured in kilograms in the si system, and it is meaningful to say that one body has twice as much mass as another. In special relativity, the word is applied to the special coordinate system, or frame, that should be used by experimenters: an *inertial frame*. *See also* mass, frame.

inflation In cosmology, the postulated period of time during which the very early Universe expanded exponentially rapidly. Such a phase, if it occurred, would explain the homogeneity of the Universe and many other observed properties. *See also* negative pressure.

infrared Region of the electromagnetic spectrum extending from the red end of the visible spectrum to longer wavelengths, typically from 0.7 µm to about 1 mm. *See also* microwave, X-ray, gamma-rays, ultraviolet radiation, sub-millimeter.

inhomogeneity Non-uniformity, condition of not being homogeneous. *See also* homogeneous.

innermost stable circular orbit A unique feature of orbits around black holes and other ultra-compact objects, which is not present in Newtonian gravity, is that there is an inner limit to circular orbits; inside this limit, circular orbits exist but are unstable: any small disturbance or non-circularity will make the orbit diverge rapidly from the original circular form. These orbits set limits on how far inwards an accretion disk can extend. *See also* accretion disk.

interferometer An instrument designed to measure with great sensitivity changes in the difference between two lengths. The lengths are called the arms of the instrument, and the measurement technique is to split coherent light along the two arms, reflect it from the ends and look at the interference pattern

formed when the light re-combines. Changes in the difference between the arm-lengths change the pattern. Interferometers are used as gravitational wave detectors by making the ends of the two perpendicular arms free: when a gravitational wave comes along, it changes the lengths of the two arms in different ways, thereby creating a signal at the output. *See also* bar detector.

interstellar clouds Dense clumps of gas and dust in the Galaxy, where star formation occurs.

invariance Independence, the condition of being unchanged when something else changes. In physics, this is used to describe systems that have a symmetry. A system that is independent of time is time-invariant. If it is symmetrical under rotations, it is rotation-invariant. In dynamics, systems that are invariant have associated conserved quantities. The time-invariance of the spacetime of special relativity (Minkowski spacetime) insures that physical systems in special relativity have a conserved energy. The spherical symmetry of the non-rotating black hole (Schwarzschild solution) insures that particles orbiting the hole have a conserved angular momentum. *See also* conservation of angular momentum, conservation of energy, Minkowski spacetime, black holes.

invariant hyperbola The set of all events in the spacetime of special relativity (Minkowski spacetime) that have a fixed interval from the origin. This definition is independent of the observer, so the set is invariant under a change of observer. The set forms a hyperbola when drawn in just two dimensions, t and x, say, where the equation is $c^2t^2 - x^2 = k$ for some fixed k. *See also* Minkowski spacetime, spacetime-interval.

inverse-square Depending inversely on the square of a variable. In gravitation, the Newtonian gravitational force is proportional to $1/r^2$, so it is an inverse-square law in the distance r from the source of gravity.

ionized Having lost one or more electrons (said of atoms). An atom that has all of its electrons is charge-neutral. If one or more are removed, it has a positive charge and is called a positive ion.

Irregular galaxy A galaxy that is not classifiable as either spiral or elliptical. *See also* spiral galaxy, elliptical galaxy.

isothermal Literally, of uniform temperature. A gas that keeps its temperature constant when it expands or compresses is said to have an isothermal equation of state.

isotope Atoms of a given element must all have the same number of protons in their nuclei, and (unless ionized) the same number of electrons in orbit around the nucleus; but they do not have to have the same number of neutrons in the nucleus. The chemistry of the element depends on its charged particles; neutrons are important only because they help to hold the nucleus together by the strong interaction. Atoms of the same element that have different numbers of neutrons are said to be different *isotopes* of the element. *See also* ionized, chemical elements, nucleus, proton, electron, deuterium, strong interaction.

isotropic The same in all directions from a given point. A sphere is isotropic about its center, but not about other points. The Euclidean plane is isotropic about all points. If something is isotropic about all points it must also be homogeneous, and if something is homogeneous and also isotropic about one point, it must be isotropic about all points. *See also* anisotropy, homogeneous, Euclidean plane.

Jeans length The scale on which gravity overwhelms pressure in a homogenous gas. Small disturbances in a homogeneous gas on very short length-scales will bounce back and smooth out because of gas pressure, but on longer scales the attraction of gravity causes them to grow. This is the Jeans instability. It plays a role in star formation, and it may have been important in galaxy formation as well. The amount of gas enclosed in a sphere whose radius was the Jeans length at the time of decoupling was approximately the mass of a typical globular cluster today; this suggests that globular clusters formed as gas instabilities in the expanding universe. *See also* homogeneous, globular clusters.

jets Narrow linear plume of gas moving away from an astronomical object at high speed. A great many objects are seen to produce jets on different length-scales: newly forming stars, pulsars, black hole systems, radio galaxies, quasars, and more. *See also* pulsar, quasars.

joule si unit of energy, equal to the work done by a force of one newton moving through one meter, or twice the kinetic energy of a one-kilogram mass moving at $1 \, \text{m s}^{-1}$.

kelvin The si scale for absolute temperature. It has the same degree size as the standard celsius scale, but its zero is at absolute zero. *See also* celsius.

kiloparsec One thousand parsecs, or about 3.085678×10^{19} m. *See also* megaparsec, parsec.

Kuiper Belt The region outside the orbit of Neptune which seems to contain a large number of planetesimals, which is a reservoir from which large asteroids occasionally fall toward the Sun. It is thought to be a relatively thin ring, having been formed during planetary formation. *See also* planetesimal, Oort Cloud.

laser A source of coherent light, which is light whose photons have the same frequency and phase. Its creation depends on the fact that photons are bosons. A laser needs a "pump", which is the source of energy for the light and which arranges a large number of atoms in the laser to be in an excited state. When one of these atoms spontaneously decays to its ground state, emitting a photon, this photon will induce other atoms to decay and emit photons of exactly the same frequency and phase. This induced emission is a purely quantum effect. *See also* photon, boson, excited state, quantum theory.

latitude The usual measure of North–South position on the Earth, running from 90° South (or -90°) at the South Pole to 90° North (or 90°) at the North Pole. *See also* co-latitude, longitude.

law of sines The geometrical relation in a triangle in which the ratio of the sine of any angle to the length of the side opposite that angle is the same for all three angles.

laws of motion The three statements that Newton formulated as a sufficient set to determine the motions of bodies when forces are applied to them. The first law says that, if there is no net force on a body, it will move in a straight line. The second law says that the acceleration of a body is proportional to the net applied force divided by the body's mass. The third says that if one body exerts a force on another, then the second exerts and equal and opposite force on the first. These laws formed the basis of the study of mechanics until Einstein. *See also* mechanics.

lepton Literally, "light particle": particles that are affected by the weak interactions and (if charged) the electromagnetic force, but not by the strong interactions. Three kinds of leptons are known: electron-leptons, mu-leptons, and tau-leptons. Each is named for its "meson", and each family consists of the meson, an associated neutrino, and the anti-particles of these two. *See also* baryon, neutrino, anti-matter, weak interaction, strong interaction, lepton number, muons.

lepton number The net number of leptons in a system or reaction, with anti-particles counting negatively. There are actually three kinds of lepton numbers, for the three kinds of leptons: electron leptons, mu-leptons, and tau-leptons. *See also* lepton, weak interaction, muons.

light-cone In spacetime, the set of events that can be connected to a given event by a single null line, a line along which light could travel. *See also* spacetime.

lightlike In spacetime, a separation between two events is *lightlike* if the events can be connected by a line along which light can travel. *See also* light-cone, spacelike, timelike.

linear Described by a straight line. In mathematics, the relation between two variables y and x is *linear* if the equation relating them has the form $y = mx + b$, for constant m and b. *See also* non-linear.

local Used by mathematicians and physicists to describe concepts that are valid only sufficiently near a particular point. Its opposite is *global*. Smooth geometries are said to be *locally flat*. *See also* global, locally flat.

locally flat The property that smooth geometries have, that they can be approximated very well by a flat geometry in a sufficiently small region around any point, as is familiar by the fact that street-maps printed on flat sheets of paper work well within cities. They are said to be locally flat at that point. *See also* local.

logarithmic scale The scale on a graph in which the markings are separated by an amount proportional to the logarithm of the quantity being displayed. This implies that marks at uniform steps represent uniform factors in the increase of the number, not uniform steps in size. Typically these may be shown as factors of ten steps for each mark. Such a scale is useful for showing the structure of curves that change by large amounts over their range, and for showing exponential and power-law relationships between variables, since these plot as straight lines on graphs where one or both axes have logarithmic scales, respectively. *See also* power-law, exponential function.

longitude The usual measure of East–West position on the Earth, running from 180° West (or -180°) in the Pacific Ocean through 0° at Greenwich, England, to 180° East (or 180°) in the Pacific again. *See also* latitude.

longitudinal In the theory of waves, a *longitudinal* wave is one whose action is along the direction of its motion. Sound waves are longitudinal, whereas water waves are transverse. *See also* transverse.

loop In computer programs, a group of instructions that is executed repeatedly.

Lorentz–Fitzgerald contraction The change in length of a moving body in special relativity.

Lorentz–Fitzgerald transformation The rule for changing from the spacetime coordinate system (t, x, y, z) of one experimenter in special relativity to that of another. *See also* experimenter.

loss of simultaneity The fact that, in special and general relativity, there is no universally agreed notion of whether or not two different events occurred at the same time. One experimenter, measuring as carefully as possible, might assign the same time-of-occurrence to two events. A different experimenter, one who is moving along the direction separating the two events, and who is making the same set of measurements and using the same definition of simultaneity, will place the event that is toward the front in the direction of his motion at an earlier time than the other one.

luminosity The amount of energy radiated by an object per unit time. This is the intrinsic brightness. The apparent brightness depends on how far away the object is. Astronomers measure luminosity in absolute magnitudes. *See also* absolute magnitude, apparent magnitude, Einstein luminosity.

macroscopic The word physicists use to denote aspects of the world that are on a large enough scale to be perceived by the eye or other senses; opposite of microscopic.

magnetars Neutron stars with ultra-strong magnetic fields, seen as pulsars with very long pulse periods. *See also* magnetic field, neutron star, pulsar.

magnetic field Magnetism is the force created by moving charges that is not present if the charges are at rest. Magnetism acts only on moving charged particles, and the force is proportional to their charge and their speed of motion. The *magnetic field* is the term used for the magnetic force on a particle per unit charge and per unit speed, i.e. the part of the magnetic force that depends just on the particles that create it. The field extends through all space, but it only exerts a force wherever a moving charged particle may be. (This is what physicists mean by the word *field*.) The field has an energy spread out through space, and it can change with time, carrying waves: electromagnetic waves. Similar remarks apply to the electric field, which is created by charges and acts on charges, regardless of their state of motion. *See also* electric charge, electromagnetism.

magnetic monopoles Hypothetical particles with a magnetic charge, that is carrying just a North magnetic pole or a South magnetic pole. No such particles have been discovered, but there is reason in theories of high-energy physics to believe that they may have been abundant in the very early universe. They would behave like electric charges but with magnetic and electric fields interchanged: a static magnetic monopole would create a magnetic field, while a moving magnetic monopole would create an electric field. The theory of inflation explains why they are so rare now that they have not been seen. *See also* electric charge, magnetic field, inflation.

magnitude Normally used as shorthand for *apparent magnitude*. *See also* apparent magnitude.

main sequence stars Stars that are burning hydrogen in their cores to power their luminosity. The term refers to the narrow sequence of points in the Hertzsprung–Russell diagram occupied by such stars, forming a nearly one-dimensional sequence according to the mass of the star. When stars begin to burn heavier elements they become giants, and eventually evolve either to white dwarfs or to supernovae. *See also* Hertzsprung–Russell diagram, giant, white dwarf, supernova.

maser The radio-wave analog of a laser. Masers were invented in laboratories before lasers, and even earlier by Nature: many dense molecular clouds, stars, and accretion disks radiate masers. The radiation comes from transitions in molecules, which typically have much less energy and therefore longer wavelength than internal transitions in atoms. Masers have very small size and, since the wavelength of the radiation is known from measurements on molecular transitions in the laboratory, they allow astronomers to follow the motions of very precisely located regions of gas clouds. They have helped to measure the masses of some supermassive black holes. *See also* black holes, laser, molecules, accretion disk, excited state.

mass The substance of a body; its resistance to acceleration: the more mass an object has, the smaller will be its acceleration in response to an applied force. In relativity, energy has mass ($m = E/c^2$) and resists being accelerated. *See also* weight, inertia.

mass function A particular function of the masses of two stars in a circular binary orbit, and of the angle of inclination of their orbit, which is measurable if it is possible to follow the speed of one of the stars along the line-of-sight to the system. This information is all that is normally available for most binary systems observed optically. *See also* angle of inclination, spectroscopic binary.

mass-to-light ratio The ratio of the mass of an astronomical system to its luminosity, in units of the solar values: $(M/M_\odot)/(L/L_\odot)$. From studying many kinds of stars and galaxies, astronomers have a rough idea of what this ratio should be for a typical system. This allows them to estimate the mass of the system from a measurement of its luminosity. *See also* luminosity.

matrix A mathematical structure having the form of an array comprising rows and columns.

matter-dominated In cosmology, a cosmological model in which the dynamics is governed by matter rather than radiation or dark energy. In practice this means that pressure is negligible in determining the evolution of the cosmology. *See also* cosmology, dark energy.

mean Mathematical term for the average of a set of numbers.

mechanics The study of the motions of bodies in response to applied forces.

megaparsec One million parsecs, or about 3.085678×10^{22} m. *See also* kiloparsec, parsec.

metric tensor The mathematical structure that allows one to calculate distances in a curved space, or intervals in a curved spacetime. In any particular coordinate system it can be represented by a matrix of values, which may change from place to place. *See also* matrix, spacetime-interval.

MeV One million electron volts of energy, or 1.6022×10^{-13} kg m^2 s^{-2}. It is frequently used as a measure of mass; the energy equivalent of the mass of an electron is about 0.5 MeV. *See also* electron volt.

microlensing Gravitational lensing phenomenon in which the lens is a star rather than a galaxy. Since individual stars are very much smaller than galaxies and their random velocities are much higher, microlensing tends to be a short-lived phenomenon. It can give very useful information about the sizes of light-emitting regions in the object being lensed. *See also* gravitational lensing.

micron Another name for a micrometer, 10^{-6} of a meter.

microwave An electromagnetic wave with a wavelength longer than infrared; the beginning of the radio region of the spectrum. Typical wavelength range is 1 mm to 1 m. *See also* infrared, ultraviolet radiation, X-ray, gamma-rays, submillimeter.

millisecond pulsars Pulsars with periods shorter than 10 ms. *See also* pulsar.

Minkowski spacetime Spacetime with no gravitational effects; flat spacetime. *See also* spacetime, curvature.

missing mass Matter that is inferred to be present in galaxies, clusters, and the Universe as a whole by the fact that the dynamics of these systems cannot be explained by the masses of observed stars. The missing mass is some 100 times as much as the luminous mass. Some of it may be in ordinary gas, but most of it must be in some form of matter that has not yet been observed experimentally. *See also* cold dark matter, hot dark matter, galaxy cluster.

molecules Systems of atoms joined together by the mutual attractions of the electrons of the atoms for the nuclei of other atoms in the molecule. Molecules are the building blocks of chemicals, and the study of the combinations of atoms into molecules is the main subject of the science of chemistry. In astronomy molecules are formed in the cool outer regions of giant stars and in molecular clouds. *See also* atoms, nucleus, electron, giant.

momentum The product of the velocity of an object with its mass. Like angular momentum, this is conserved in Newtonian mechanics and special relativity, and in general relativity if the spacetime is invariant under translations in the direction of the momentum. *See also* invariance, spacetime.

muons One of the types of lepton mesons. Along with electrons and tau-mesons, muons form one of the three families of leptons. It has associated with it the mu-neutrino. *See also* lepton, lepton number, electron.

naked singularities Singularities in general relativity that are not hidden within a horizon but are visible to other parts of the universe. A naked singularity would represent the breakdown of the predictive power of general relativity, and would presumably mean that the theory had to be replaced. Serious singularities of this type are not known, and are postulated not to exist (cosmic censorship hypothesis). This has not yet been proved. *See also* singularity, cosmic censorship hypothesis.

nebula A diffuse cloud of gas around a star. Originally astronomers used this term for any diffuse clouds of light on the sky. As telescopes improved, some nebulae turned out to be star clusters in our Galaxy, others external galaxies. This obsolete usage is preserved in some traditional names, like the "Great Nebula in Andromeda" for the Andromeda Galaxy M31. *See also* star cluster, galaxy, planetary nebula.

negative pressure Normal gas pressure is positive, in that it pushes outwards on the walls of its container. Systems have negative pressure if they pull in on their containers. A stretched rubber band has negative pressure along its length; this is called *tension*. Since pressure contributes to the gravitational field through the active gravitational mass, a sufficiently large negative pressure can turn the gravitational field repulsive. This is the explanation for inflation. *See also* active gravitational mass, inflation.

neutrino Leptons of very small or zero mass that are produced in beta decay and many other nuclear reactions. There are three kinds of neutrinos, associated with each of the three kinds of leptonic mesons. The flux of neutrinos from the Sun is smaller than expected, and this has led to revisions in the theory of leptons and the determination that neutrinos of one kind seem to transform themselves into one another. *See also* lepton.

neutron The electrically neutral particle that, with the proton, is one of the building blocks of the atomic nucleus. Its mass is slightly larger than that of the proton, large enough that a single isolated neutron is can decay into a proton, an electron, and a neutrino. This process of beta decay can also happen inside a nucleus that has too many neutrons. Neutrons are stable in nuclei that don't contain too many of them, and they are stable within neutron stars, of which they are the main constituent. *See also* baryon, beta decay, neutron star, proton, strong interaction.

neutron star A star whose support against gravity comes from the pressure of a degenerate gas of neutrons. Its typical mass is about a solar mass, and its radius about 10 km. Some neutron stars are pulsars, but there are probably many more that are not detected through any emitted radiation. *See also* degenerate gas, neutron, pulsar, white dwarf.

non-linear A mathematical term for a relationship between two variables that is not linear, so it does not plot as a straight line. *See also* linear.

normal modes The patterns of vibration of an object that are associated with its characteristic frequencies. For each frequency there is a specific kind of motion at different points in the body, and this pattern is the normal mode associated with that frequency. *See also* characteristic frequencies.

nova A star that suddenly brightens up. The change in brightness is not as great as for a supernova. Where a supernova represents the destruction of a star, a nova is caused by changes in the rate of accretion onto a compact star, and therefore can recur in the same system. *See also* accretion, supernova.

nuclear reactions Reactions involving changes in the composition of nuclei, such as combining two protons and two neutrons to make a helium nucleus, or combining three helium nuclei to form a carbon nucleus. While chemical reactions involve only the electrons and leave the nuclei intact, nuclear reactions change the nuclei and therefore the element. They require higher temperatures, energies, and densities than chemical reactions. Reactions that release energy consitute the main source of the energy radiated by stars.

nucleon A neutron or proton: the constituents of the nucleus. Nucleons exert forces on one another via the strong interaction, which electrons and neutrinos do not feel. *See also* neutron, proton, strong interaction.

nucleus The positively charged center of an atom, containing protons and neutrons. The number of protons determines the element that the atom belongs to, and its chemical properties. *See also* chemical elements, neutron, proton.

observer *See* experimenter.

Occam's razor The principle that any hypothesis or theory devised to explain a new phenomenon should be as simple as possible and involve as few new assumptions and undetermined parameters as possible.

Olbers' Paradox The question: why is the sky dark at night? In an infinitely large and infinitely old universe filled uniformly with stars, the sky would be infinitely bright, and our Sun would not even be noticeable. Our Universe must be either of finite age or of finite size, or both.

Oort Cloud The roughly spherical cloud in which comets originate, far outside the orbit of Pluto and outside the Kuiper Belt. This is thought to have been left over from the formation of the Solar System; at such large distances, where light from the Sun is very weak, icy comets formed but never evolved into the planetesimals that inhabit the closer Kuiper Belt. *See also* Kuiper Belt, planetesimal.

overtones Characteristic frequencies that are higher than the fundamental frequency. For a simple stretched string, the overtones are at integer multiples of the fundamental, but in more complicated systems they are not. *See also* characteristic frequencies, fundamental frequency.

ozone The compound O_3, made of three oxygen atoms. It is only very weakly bound, and can be split up by the addition of a small amount of energy. It forms in the upper atmosphere of the Earth where collisions among molecules are infrequent enough that the molecule can have a long lifetime. It is a good absorber of ultraviolet light from the Sun, and protects living things from this damaging radiation. Man-made pollutants, particularly fluorocarbons, have reduced the concentration of ozone dramatically at some latitudes.

panspermia The idea that life could have originated somewhere else in the Universe and come to the Earth soon after it formed. The evolutionary record would not have been different, but the initial primitive living organisms would have come from somewhere else.

parallax The apparent change of position on the sky of an astronomical object, caused by the Earth's orbital motion around the Sun. Objects near the Earth seem to move back and forth relative to objects more distant. The amount of apparent motion of a near object measures its distance.

parsec The standard unit of distance in astronomy, about 3.085678×10^{16} m. It is defined as the distance to a star whose parallax is exactly one second of arc. It equals 3.26 light-years. *See also* megaparsec, kiloparsec.

particle horizon *See* horizon.

pattern matching The process of searching through a set of data to find something that matches a pre-determined pattern. The detection of gravitational waves relies on pattern matching, because the strength of the waves is not great enough to make them visible in the raw data output from a detector. Instead, scientists look for disturbances that, systematically over time, match a predicted waveform.

Penrose process A method of extracting energy from a black hole by making use of negative-energy orbits inside the stationary limit. If a positive-energy particle falls into the stationary limit and splits into two, one of which has negative energy, then when the other one escapes from the hole it will carry more energy than it began with. This energy comes from the rotation of the hole. *See also* stationary limit.

periastron *See* perihelion.

perihelion The point on an elliptical orbit around the Sun that is nearest to the Sun. If the central object is a star, the point is called the *periastron*; if the Earth, *perigee*. *See also* aphelion.

perpetual motion The idea that an isolated, realistic physical system could somehow execute a particular motion indefinitely. Since all real physical processes involve some kind of friction or dissipation, perpetual motion requires that the energy be replenished, and this conflicts with the principle of conservation of energy for an isolated system. No such systems have ever been found. *See also* conservation of energy.

photoelectric effect The ejection of an electron from a metal when light of a certain frequency falls on it. If the light has too low a frequency, nothing happens. The higher the frequency of light beyond the critical value at which electrons begin to be ejected, the greater the energy of the ejected electron. This was explained by Einstein to be a consequence of the relation between energy and frequency for photons. Once a photon has enough energy to tear the electron away from the metal, any extra energy (on account of the larger frequency) goes into the kinetic energy of the electron. *See also* photon.

photon A concept introduced by Einstein and now fully explained by the theory of quantum electrodynamics. According to this picture, light sometimes behaves like a particle whose energy E is proportional to its frequency f, $E = hf$, where h is Planck's constant. *See also* quantum electrodynamics.

photosphere The outer layer of the Sun from which comes the light that we see. Light scatters a huge number of times on making its way outwards from the Sun's central region, so the photosphere is the surface of last scattering.

physical cosmology The study of the physical processes in the early Universe, including the formation of elements, the first stars, and galaxies.

pixel The smallest area in an image that can be resolved, i.e. distinguished from neighboring areas. In a photograph, this would be the size of a grain of the emulsion, the smallest unit that is exposed by light. In a digital camera or video camera, this is the size of one element of the CCD that is used as the sensing device.

Planck length A number with the dimensions of length which is formed purely from the fundamental constants. Its value is $(hG/c^3)^{1/2} = 4 \times 10^{-35}$ m. *See also* Planck mass, Planck time.

Planck mass A number with the dimensions of mass which is formed purely from the fundamental constants. Its value is $(hc/G)^{1/2} = 5.5 \times 10^{-8}$ kg. *See also* Planck length, Planck time.

Planck time A number with the dimensions of time which is formed purely from the fundamental constants. Its value is $(hG/c^5)^{1/2} = 1.4 \times 10^{-43}$ s. *See also* Planck length, Planck mass.

plane of the sky The "sky" is the astronomers' word for the celestial sphere, which is a two-dimensional sphere around the Earth; all distant objects are projected onto this sphere to get their angular location on the "sky". Near any particular point, the celestial sphere may be approximated by a plane, because it is locally flat. This is the plane of the sky at that point. *See also* locally flat.

plane wave A wave that propagates with a planar wave-front. All waves that spread out from a localized source are effectively plane waves far away, when the distance over which the curvature of their wave-front is noticeable is much larger than an experimenter's apparatus.

planetary nebula The shell of gas expelled by a giant star during its transition to becoming a white dwarf. These are among the most beautiful objects in the Galaxy when photographed with sufficient resolution.

planetesimal A rocky fragment of the kind that accumulated into planets. Individual examples remaining today are asteroids. *See also* asteroids, Kuiper Belt.

plasma A gas consisting of free electrons and ions. It must be hot enough to prevent the ions and electrons from recombining into neutral atoms.

plate tectonics The process by which the continents have moved around the Earth.

point mass The idealization of a simple elementary particle as a point. This model cannot be realistic, since if the particle is charged the electric field would be infinitely large, and the energy required to assemble the particle would also be infinite. String theory attempts to remedy these problems by representing particles as two-dimensional loops. *See also* string theory.

polarization The direction, or set of directions, in which a wave acts. A longitudinal wave has only one polarization, along the direction of motion. But a transverse wave can be polarized in various ways in the transverse plane. An electromagnetic wave acts along a line, so there are two independent polarizations along the two perpendicular axes. A gravitational wave acts with ellipses, and these have two independent orientations in the transverse plane, rotated by 45° from one another. *See also* longitudinal, transverse.

polytrope In astronomy, a stellar model constructed using a power-law relation between pressure and density. *See also* power-law.

position vector The vector that locates the position of an object; it stretches from the origin to the location of the object. *See also* vector.

positron The anti-particle of the electron, with the same mass but a positive electric charge. *See also* anti-matter, electron.

post-Newtonian The name used for an approximation to general relativity which describes systems that have weak gravitational fields as basically Newtonian systems with corrections. The corrections are called post-Newtonian terms.

power The rate of doing work, or the rate of expending energy.

power-law A mathematical term for a relationship between two variables in which one is proportional to the other raised to a constant power.

primordial black holes Black holes formed in the very early Universe. Normal stellar evolution leads only to black holes larger than about a solar mass. The conditions to form smaller black holes do not exist today, because the required density is much larger than nuclear. However, in the early Universe, when the average density was larger than nuclear, density irregularities could conceivably have collapsed to black holes of smaller mass. These could contribute to the dark matter today, except that any holes smaller than about 10^{12} kg would have decayed by now due to the Hawking radiation. *See also* black holes, gravitational collapse.

principle of general covariance One of the ideas that guided Einstein's development of general relativity. It is the statement that the equations describing gravity and matter fields must take the same form in any coordinate system. No system is to be preferred, unlike the situation in special relativity, where inertial frames are singled out as the ones that should be used by experimenters. *See also* frame, inertia.

principle of mediocrity *See* Copernican principle.

principle of relativity The principle that the laws of physics should be independent of the motion of the experimenter who tests them. This was first enunciated in respect to gravity on the Earth by Galileo. Einstein made it a cornerstone of his special relativity. He later generalized it to the principle of general covariance when he created general relativity. *See also* principle of general covariance, experimenter.

proper distance The distance as measured by a local experimenter; independent of coordinate system.

proper time The time as measured by a local experimenter's clock; independent of coordinate system.

proton The fundamental positively charged particle, which is one of the building blocks of the nucleus of all atoms. *See also* atoms, baryon, electron, neutron, nucleus.

protostar The collapsing cloud of gas, on its way to forming a star, which is radiating light because of the release of gravitational energy, but within which nuclear reactions have not yet begun. *See also* nuclear reactions.

pulsar A spinning neutron star which emits a beam of radiation that sweeps the sky as the star turns. When observed from Earth, if the observer is in the beam, the radiation pulses on and off. Most pulsars emit radio waves, some are observed to pulse in optical light or even X-rays and gamma-rays. The beams appear to be formed at the poles of strong magnetic fields, but the mechanism is not understood. Most known pulsars spin several times per second; some spin several hundred times per second. *See also* neutron star, X-ray, gamma-ray, magnetars, millisecond pulsars.

Pythagorean theorem The theorem that gives the length of the hypotenuse c of a right triangle in terms of the other two sides a and b: $c^2 = a^2 + b^2$. The relation defines the metric of Euclidean space. *See also* hypotenuse, metric tensor.

quadratic equation In mathematics, an equation containing the square of an unknown variable, but no higher powers.

quadrupole formula The expression giving a first approximation to the gravitational radiation emitted by a system with weak internal gravitational fields.

quanta Discrete amounts of something.

quantum electrodynamics The quantum theory of the electromagnetic field. *See also* quantum theory, electromagnetism.

quantum fluctuations In quantum electrodynamics, the electromagnetic field undergoes fluctuations that would not be allowed in classical theory, where fields have perfectly well-defined values. The fluctuations lead to a number of effects, such as the Hawking radiation from black holes and a possible explanation for the cosmological constant. *See also* quantum electrodynamics, cosmological constant, Hawking radiation.

quantum gravity The hoped-for theory that will generalize general relativity to a quantum theory of the gravitational field. Most physicists expect that this will happen only through unifying gravity with other forces, as in string theory. Others hope to find a theory of the quantum gravitational field alone. *See also* quantum theory, string theory.

quantum theory A theory of physics that incorporates the characteristic features of quantum phenomena: uncertainty in measurements, radiation fields behaving sometimes like waves and other times like particles, predictions only of the probabilities of the outcomes of experiments rather than certainties.

quark matter Matter so dense that nucleons overlap and their constituents, the quarks, are the true particles of the gas. *See also* quarks.

quark soup *See* quark matter.

quarks Baryons are not the most elementary of particles. They are composed of three building blocks, called quarks. Quarks have a remarkable interaction among one another: it is impossible to pull a quark away from others and isolate it. *See also* baryon.

quasars The brightest continuous light sources in the Universe. They seem to be driven by accretion onto a supermassive black hole. Most are at great distances, which suggests that the ones in our neighborhood have died away. *See also* accretion.

quintessence A word used by some physicists for theories of dark energy that explain why we are seeing an acceleration in the expansion of the Universe today. *See also* dark energy.

radians The mathematical measure of angles that is more natural than degrees. The size of an angle in radians is the ratio of the length of an arc to its radius. This measure runs from 0 to 2π for a full circle.

radiation reaction The force on a system that is created by the radiation that the system emits. Any loss of energy must be reflected in a force that opposes the motion. This is a self-force, created by the particle's own field.

ram pressure The pressure exerted by a stream of gas when it encounters a wall; this depends on its speed and density.

redshift The lengthening of the wavelength of a wave. This can happen because of the motion of its source or its receiver, or because of gravity.

relaxation time The time it takes for a system to reach a form of equilibrium. For clusters of stars, this is the time to share out the energy of a perturbation among the stars, losing any memory of the original perturbation.

relaxed A cluster is relaxed if the distribution of its stars and their velocities bears no memory of the history of the cluster and the origin of its stars.

rest-mass The mass of a body when it is at rest, as inferred from Newton's second law. *See also* laws of motion, mass.

rotation curve The graph of the rotational speed of the stars and gas in a spiral galaxy against their distance from the center. From this it can be inferred how much mass the galaxy has and how it is distributed with radius. These curves have revealed that there is a great deal of mass outside the region of such galaxies which emits light. *See also* dark matter.

scalars The mathematical term for an ordinary quantity that is not associated with any direction. Temperature, density, and pressure are scalars. The position and velocity of an object are, on the other hand, vectors. *See also* vector.

scale-height A term used for a length that is typical for the change in some quantity. *See also* Jeans length.

Schwarzschild radius The radius of the horizon of a Schwarzschild black hole, equal to $2GM/c^2$, where M is the mass of the black hole. *See also* black holes.

selection effect A term astronomers use for an effect that systematically distorts a measurement because it is impossible to obtain a fair sample. For example, if one tried to estimate the mean distance to a collection of stars, the measurement would be distorted by the fact that the brighter stars can be seen at greater distances. The objects are not selected correctly for a fair estimate.

shot noise The random fluctuations in light that come from the fact that it is composed of discrete photons rather than a continuous wave of energy. *See also* photon.

singularity A place where the equations of general relativity fail to predict the future. If a particle encounters a singularity, it has no future. Most singularities in known solutions are places of infinite curvature, which would tear apart any particle, but in principle they could be weaker. Physicists regard singularities as unsatisfactory, indicating that general relativity is incomplete. Most hope that quantizing gravity will eliminate them. The most serious singularities in known solutions are inside black holes, hidden from view, unable to influence us. The cosmic censorship hypothesis expresses the hope that this is generally true. *See also* curvature, quantum gravity, cosmic censorship hypothesis.

solar constant The mean flux of light from the Sun on the Earth. *See also* flux.

Solar Neutrino Unit A measure of the flux of neutrinos from the Sun at the Earth. One SNU is defined to be the flux that would induce one nuclear reaction in every 10^{36} chlorine atoms in a neutrino detector per second. *See also* flux, neutrino.

spacelike The term used to describe spacetime-intervals that are positive, so that the two events could be simultaneous in some frame. *See also* spacetime-interval, event, frame.

spacetime The set of all events. *See also* events.

spacetime diagram The graphical representation of events in spacetime. It shows a vertical axis which is the time as measured by some particular experimenter, and one or more horizontal axes for the space coordinates of events. Points in the diagram are events in spacetime. *See also* spacetime, events.

spacetime foam The idea that, through quantum effects, spacetime on a very small length-scale (the Planck length) could consist of constant fluctuations producing Planck-mass black holes. *See also* spacetime, quantum theory, Planck length, Planck mass, black holes.

spacetime-interval The invariant measure of distance or time in spacetime. The interval between two events is independent of the experimenter who measures time and space separations. It is the generalization of the Pythagorean theorem to spacetime. It is calculated from the metric tensor of spacetime, and it encodes the curvature of the spacetime. *See also* invariance, timelike, spacelike, Pythagorean theorem, curvature, metric tensor.

spacetime metric The metric tensor of spacetime, carrying the information about the spacetime-interval. *See also* spacetime-interval, metric tensor.

special relativity Einstein's first theory of spacetime, which treats how measurements of length and time are made. It predicts the slowing of time and the shortening of distances with motion. *See also* spacetime, time dilation, Lorentz–Fitzgerald contraction.

spectral lines Narrow features in the spectrum of light from an object, which indicate the presence of particular elements in the object.

spectroscopic binary A binary star system in which only one object is observed, and whose orbital motion is inferred from the time-dependent Doppler shift it produces in spectral lines from the observed star. *See also* spectral lines.

spin In fundamental physics, the intrinsic angular momentum carried by a particle.

spiral galaxy A galaxy which presents a spiral pattern in a photograph. Normally the spiral has two arms and it is tightly wound. The spirals are density waves, locations of rapid star formation, bright because the bright massive stars live for much less than the time it takes for the wave to move around the galaxy. *See also* elliptical galaxy, density wave.

spontaneous symmetry breaking In fundamental physics, the idea that a unified theory of forces can produce very unsymmetrical physical effects depending on random details of how the symmetries behave when the gas cools. *See also* electroweak, unified field theories.

standard candle The term astronomers use for an object whose intrinsic luminosity is known, so that its distance can be inferred from its apparent brightness. Most distance measurements in astronomy are calibrated by the use of standard candles.

standard meter stick By analogy with a standard candle, an object whose physical size is known, so that its distance can be inferred from its angular diameter. *See also* standard candle.

star The basic producer of light in the Universe. The Sun is a star. An object that is too small to produce light is a planet.

star cluster A set of stars grouped together and presumably formed together. Some clusters are large spherical assemblies of millions of stars; these are called globular clusters. Other clusters have fewer members more loosely bound to one another, and are called open clusters. *See also* globular clusters.

stationary limit The surface that is the outer boundary of the region around a rotating black hole (or possibly a very compact rotating star) in which there are orbits that have negative total energy with respect to a distant experimenter. This is associated with the fact that, inside the stationary limit, all free particles must rotate with respect to the distant experimenter: to stand still requires going faster than light with respect to local inertial frames. The negative-energy orbits have negative angular momentum relative to the hole. Non-rotating, Schwarzschild black holes do not have stationary limit surfaces. Kerr black holes have stationary limits that are topological spheres, called *ergospheres*. *See also* dragging of inertial frames, energy.

statistical mechanics The branch of physics that derives the macroscopic properties of gases from statistical averages over the motions and interactions of huge numbers of atoms and molecules. *See also* macroscopic, atoms, molecules, thermodynamics.

Steady-State model of the Universe The model devised by Hoyle and collaborators to show how a cosmology can be expanding and yet the same for all time. This requires matter to be created from nothing all over the Universe, to fill in the gaps left by the expansion of previously created matter. The hypothetical C-field was invoked to power this creation. Although the model permitted the Copernican principle to be applied to cosmology in its fullest sense, many physicists did not like the *ad hoc* nature of the C-field, and the model never gained many adherents. It has been modified substantially in recent years in order to be compatible with the cosmic microwave background radiation and other cosmological observations. *See also* C-field, Copernican principle, cosmic microwave background radiation.

strain The relative stretching or expansion of a system. For one-dimensional systems like rubber bands, this is defined as the change in length divided by the original length. In more than one dimension one must take into account shear as well as stretching. *See also* stress.

strange matter The whimsical name given to baryons that include the "strange quark" in their composition. Strange baryons can be made abundantly in accelerators. Some scientists speculate that strange matter is the real ground state of matter, so that normal matter will, in the right circumstances, transform itself into strange matter spontaneously. In this case, neutron stars could turn out to be strange stars. *See also* baryon, strange stars.

strange stars Stars made of strange matter. If strange matter is more stable than normal neutron matter, then these neutron stars might transform themselves into strange stars, which would be even more compact. *See also* neutron stars, strange matter.

stress The pressure, ram pressure, and shear forces inside smooth matter. Stress is what maintains strain in systems. *See also* strain.

stress–energy tensor The source of gravity in general relativity. Einstein knew he could not use just the mass density as the source, so he used the tensor that contains the mass-density, momentum-density, and pressure: the stress-energy tensor. *See also* strain.

string theory A candidate for the ultimate theory of all the physical forces. It describes particles as small loops of string, which is easier to treat consistently than a point-particle model. *See also* point mass, quantum gravity.

strong interaction The nuclear force, which is attractive enough to beat the electric repulsion of the protons in a nucleus and bind them together. Stronger than electric forces over distances the size of a proton, the strong interaction is a short-range force, falling off with distance much more rapidly than the electric force. So over the large size of an atom, it exerts a negligible influence. *See also* electromagnetism, nucleus, nucleon.

sub-millimeter A range of wavelengths of the electromagnetic spectrum with wavelengths smaller than 1 mm. *See also* infrared, microwave, X-ray, gamma-rays, ultraviolet radiation.

sunspots Dark blotches on the face of the Sun. These are cooler regions where magnetic fields loop out of the Sun and back into it.

superclusters Clusters of clusters of galaxies. These are the largest organized structures observed in the Universe. *See also* galaxy cluster.

superconductor A material that conducts electricity without resistance. This can only happen because of quantum effects. *See also* quantum theory.

superfluid A fluid that moves without friction. This can only happen because of quantum effects. *See also* quantum theory.

supergiant A massive giant star, very likely to explode as a supernova and leave behind a black hole or neutron star. *See also* supernova, black holes, neutron star.

supernova An explosion in a star that results in the destruction or dramatic transformation of the star. *See also* supernova of Type Ia, supernova of Type II.

supernova of Type Ia A supernova explosion which originates in a white dwarf and results in the complete disintegration of the star through a gigantic nuclear chain reaction. These supernovae are thought to be standard candles, and measurements of very distant examples have revealed the acceleration of the expansion of the Universe. *See also* white dwarf, supernova, standard candle.

supernova of Type II A supernova explosion which originates in the gravitational collapse of the inner core of a giant star, leaving behind either a neutron star or a black hole. *See also* gravitational collapse, supernova, giant, supergiant, neutron star, black holes.

surface of last scattering In cosmology, the location where photons became able to move freely through the Universe. *See also* decoupling.

surface brightness The apparent brightness of an object, per unit angular area on an image, i.e. per square radian on the sky.

synchronous rotation Rotation or a body about its axis with the same period as it executes an orbital motion, so that it always presents the same face to the object about which it orbits. The Moon is in synchronous rotation about the Earth.

tachyon A hypothetical particle that travels faster than light. Relativity forbids the acceleration of a particle up to the speed of light, but it is not inconsistent with relativity alone for a particle to travel faster than light, provided it never slows down to the speed of light. However, tachyons are difficult to reconcile with causality, since they travel backwards in time with respect to some observer.

tension *See* negative pressure.

tensor The mathematical generalization of the idea of a vector to something that effectively involves several vectors at once. A vector's components have one index. The next-simplest tensor is a matrix, whose components have two indices. It is possible to define tensors that have three or more indices. The full (Riemann) curvature tensor, for example, has four. *See also* matrix, vector, curvature, indices.

terrestrial planets The planets with rocky surfaces: Mercury, Venus, Earth, and Mars.

thermodynamics The study of heat, its transfer from one body to another, and the work that it can do in engines.

tidal acceleration The difference in the acceleration of gravity across an object or between two nearby objects. This is their relative acceleration, and it exists even in a freely-falling frame where their common acceleration is not present. Einstein identified the tidal acceleration as the true, observer-independent, non-removable effect of gravity.

time dilation The slowing of time produced by the motion of a body, as explained by special relativity. The faster a body's speed, the slower time goes. For photons, time stands completely still.

time travel Moving backwards in time relative to other objects, while time moves forwards for oneself. Physicists have discovered that wormholes can be used for time travel if they can be kept open for long enough for objects to pass through them, but keeping them open may require negative energy. *See also* wormhole, energy.

timelike The term used to describe spacetime-intervals that are negative, so that the two events could be at the same location in some frame. *See also* spacetime-interval, events, frame.

transponder An amplifier that receives a signal from a distant transmitter and returns the signal, amplified so that it has enough power to be received by the original transmitter. This allows the transmitter to measure the distance to the transponder from the round-trip travel time of the signal. This is the main way in which spacecraft are tracked from Earth as they move through the Solar System. A transponder can be thought of as an *active mirror*.

transverse Across the line of motion. The action of a water wave is transverse: the wave moves along the surface of the water, but the water itself moves up and down. *See also* polarization, longitudinal.

ultraviolet radiation Region of the electromagnetic spectrum with wavelengths shorter than the violet end of the visible spectrum, running from about $0.4 \, \mu m$ to 10^{-8} m. *See also* infrared, microwave, X-ray, gamma-rays, sub-millimeter.

unified field theories Ever since Maxwell unified the electric and magnetic forces into the theory of electromagnetism, in the nineteenth century, physicists have followed a path of simplifying the laws of physics by finding ways in which different forces are related. The electromagnetic and weak interactions have now been unified into the electroweak theory, which is very successful at explaining and predicting phenomena in its domain. Many physicists now expect that the next step will be a unification of the strong interaction with the electroweak. This would be a grand unified theory. *See also* electroweak, electromagnetism, weak interaction, strong interaction, grand unified theories.

vector The mathematical term for a directed object with length. *See also* component, displacement, position vector, indices.

virial method A method astronomers use to estimate the mass of a system of stars or galaxies by measuring their velocities relative to one another, and assuming that the mass creates a strong enough gravitational field to hold the objects together with roughly the same distribution of speeds over long periods of time.

viscosity Friction in the motion of a fluid.

visual binaries A binary system in which both stars are visible as separated images. Only the binaries nearest the Earth can be resolved in this manner.

visual magnitude The magnitude (brightness) of a star in visible light. *See also* absolute magnitude, apparent magnitude.

vortices A vortex (plural form: vortices) is the center of a rotational fluid motion. In everyday use it can be the center of a whirlpool, hurricane, or other natural phenomenon. In the theory of superfluidity, it is a line about which the fluid can rotate without having a rotational flow. This apparent contradiction requires a careful understanding of rotational motion in fluids. A superfluid is irrotational because a little stick or flag embedded in the fluid will not spin as the fluid moves. Nevertheless, the fluid can flow past obstacles on curved paths, provided it always keeps locally irrotational. It can even rotate about a center if its angular velocity about the center is proportional to $1/r$, the distance from the center. This implies that the fluid description breaks down at the center, and so the vortex at the center is a kind of singular point, not a place where the fluid remains superfluid. Related phenomena allow magnetic fields to thread through superconductors, and also lead to cosmic strings in the expanding Universe. *See also* cosmic strings.

watt The unit used in the si system for power, abbreviated W: $1\,W = 1\,kg\,m^2\,s^{-2}$. *See also* power.

weak interaction The force between leptons. It is the force that is responsible for beta decay. For example, the decay of a neutron into a proton, electron, and electron neutrino is a beta decay, because it requires the creation of two leptons. *See also* beta decay, lepton, strong interaction.

weight The force of gravity on an object, proportional to its mass but also to the acceleration of gravity.

white dwarf A star whose support against gravity comes from the pressure of a degenerate gas of electrons. Its typical mass is about a solar mass, and its radius about that of the Earth. *See also* degenerate gas, neutron star.

white hole The time-reverse of a black hole, in which a compact object exists from the distant past, but nothing can fall into it; instead, it explodes and disappears. This behavior is possible according to general relativity, but would require very special initial conditions at the Big Bang to make the initial compact objects. There is no evidence for such objects in the real Universe. *See also* black holes.

work The product of force and distance, which represents the energy expended by the force in moving an object through the given distance. Work is measured in energy units, like joules. *See also* joule.

world line The set of events experienced by an object over time. This is a timelike line through spacetime.

wormhole A hypothetical connection between one region of space and another, or even between our Universe and another. They are relatives of black holes. Although no such connections have ever been observed, they are interesting theoretical objects because they allow physicists to explore the limits of what might be possible in general relativity. Even time travel may be possible with such objects. *See also* black holes, time travel.

X-ray Region of the electromagnetic spectrum with shorter wavelengths than ultraviolet. Typical wavelength range is 10^{-11} m to 10^{-8} m. For X-rays and gamma-rays it is typical to quote energies rather than wavelengths; this range runs from about 0.1 keV to 100 keV. *See also* infrared, ultraviolet radiation, microwave, gamma-rays, sub-millimeter.

X-ray binary A binary star system that emits X-rays because gas is flowing from one of the stars (usually a giant) onto its companion (usually a neutron star or black hole). The energy released as gas spirals down onto the companion heats the gas to temperatures where it emits X-rays. *See also* X-ray, giant, neutron star, black holes.

zero-point energy In quantum theory, nothing can be perfectly at rest, so when a material is cooled to absolute zero, its atoms retain a small motion, whose energy is the zero-point energy. *See also* quantum theory.

Index

Page numbers in *italics* refer to terms in figures and investigations.

SEVENTH EDITION

INTRODUCTION TO PERSONALITY
Toward an Integration

WALTER MISCHEL
Columbia University

YUICHI SHODA
University of Washington

RONALD E. SMITH
University of Washington

WILEY

JOHN WILEY & SONS, INC.

Acquisitions Editor   *Tim Vertovec*
Assistant Editor   *Lili DeGrasse*
Marketing Manager   *Kate Stewart*
Production Editor   *Sandra Dumas*
Designer/Design Director   *Madelyn Lesure*
Photo Editor   *Sara Wight*
Photo Researcher   *Elyse Reider*
Production Management Services   *Progressive Publishing Alternatives*

Cover: Pablo Picasso, *J'aime Eva*. Photo provided courtesy Columbus Museum of Art, Columbus Ohio, ©2003 Estate of Pablo Picasso/Artists Rights Society (ARS), New York.

This book was typeset in 10/12 New Caledonia by Progressive Information Technologies and printed and bound by R. R. Donnelley (Willard). The cover was printed by Lehigh Press.

The paper in this book was manufactured by a mill whose forest management programs include sustained yield harvesting of its timberlands. Sustained yield harvesting principles ensure that the number of trees cut each year does not exceed the amount of new growth.

This book is printed on acid-free paper.

Library of Congress Cataloging-in-Publication Data

Mischel, Walter.
  Introduction to personality / Walter Mischel. — 7th ed.
    p. cm.
  Includes bibliographical references and indexes.
  ISBN 0-471-27249-3
    I. Personality. I. Title.

BF698.M555 2003
155.2 — dc21

2003053487

Mischel, Walter; Shoda, Yuichi; Smith, Ronald, E.
Introduction to Personality: Toward An Integration, Seventh Edition

ISBN 0-471-27249-3
ISBN 0-471-45153-3 (WIE)

Printed in the United States of America.

10 9 8 7 6 5 4 3

PREFACE AND TEXT ORGANIZATION

This extensive revision by a new team of authors contains more than 50 percent fresh material. The book has been updated and reorganized to reflect the growth and transformation of personality psychology in recent years while also remaining true to and respectful of the field's rich history of theories and methods for understanding personality. We present personality as a growing, vibrant field that speaks directly to how students live their lives and how they think about themselves. We have rewritten the text to make it readily accessible to students, and to make mastery of the material more enjoyable as well as informative and stimulating.

First, a few words about the reasons for the substantive changes and for the new subtitle: *Toward an Integration*. Instructors familiar with earlier versions of this text, published in its first edition more than thirty years ago, may wonder what prompted these major changes in a well-established text. Personality psychology was founded as the area within psychology devoted to understanding how factors studied at different levels of psychological analysis combine and interact distinctively within the individual. The aim was to study the person—and personality—as a coherent and unique whole. The hope was for personality psychology to be the hub where all the levels of analysis become integrated within the functioning person.

For many years the field of personality—and particularly its texts, including earlier editions of this one—became divided into alternative approaches that competed against each other. The implication was that if a given approach at a particular level proved to be "right" and useful, the other approaches and levels were bound to be somehow "wrong" or less important. The questions usually asked were: "Which one is best? Which one is right?" In time, however, it has become increasingly clear that the different theoretical approaches ask different questions and address different phenomena at different levels of analysis. Consequently they usually deal only with selected aspects of personality versus the construct in its entirety. Historically, this was understandable, given the limits to what any one researcher or theorist can know and study, especially as the knowledge base in the area grew at an accelerated pace. But this approach also undermined the original ambitious mission of personality psychology to become the locus of integration.

The good news is that personality psychology is moving into a new stage (e.g., Duke, 1986; Carver, 1996; Cervone & Mischel, 2002; Morf, 2003). One gets a sense of this movement toward integration just from some of the titles of the articles, for example: *Rethinking and reclaiming the interdisciplinary role of personality psychology: The science of human nature should be the center of the social sciences and humanities.* (Baumeister & Tice, 1996). Fortunately, the explosion of research findings at different levels of analysis has been so great that the pieces of the puzzle are starting to fit together. The insights from different levels complement each other increasingly well and help to build a more integrated and cumulative view of the person as a whole.

This seventh edition of **Introduction to Personality** reflects these new developments in the field, as indicated in its subtitle. At the same time, we continue to provide

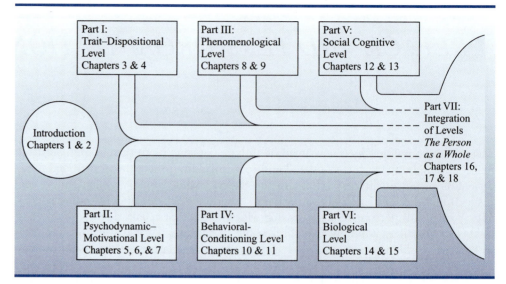

Organization of the text.

coverage of the essential features and contributions from this rich heritage. To do so, this edition first covers in a balanced manner the key ideas and pioneering work that shaped the field for many decades in the last century. But the focus is on distilling how they still speak to and inform each other, and how they add to the current state of the science.

A variety of features are used to help the student to see the interplay among the insights obtained at the six major levels of analysis pursued in personality psychology, namely the trait-dispositional, psychodynamic-motivational, phenomenological, biological, behavioral-conditioning, and social-cognitive levels. Throughout the book, we show how the discoveries made at each level enrich the understanding of the whole. We also show how each level has practical applications for benefiting personal adjustment, self-understanding, and effective coping. In a feature new to this edition, we highlight the "personal side of the science" by inviting students to ask specific questions about how this level applies to themselves. Consistent with the theme of integration, *part preludes* orient the reader to each major section of the text that follows by placing it in a broader conceptual framework. Review sections conclude each part with a segue into the next part to enhance the integration.

After providing a solid background in the six levels of analysis, and pointing out their interconnections, the final part of this text—*Integration of Levels: The Person as a Whole*—is the most notable innovation, found in no other current personality text. It consists of three completely fresh, integrative chapters that focus on the contemporary scene. These chapters demonstrate the complementary relations among all the levels for gaining a rich sense of the personality as a whole. They examine the person engaged in goal pursuit and self-regulation, functioning and adapting proactively within his or her context and culture. The organization of the entire text is depicted in the graphic above, showing the six levels and their flow into the integration presented in the last three chapters which illustrate how each contributes to the total picture of the person.

Much of the text was rewritten while retaining only its best, time-tested features. The extensive rewriting not only reflects the growth and transformation of personality psychology in recent years but also our goal to make the book more readily accessible to today's students and to make mastery of the material more enjoyable as well as informative and stimulating.

Based on previous teaching experience—cumulatively the authors have spent more than 60 years teaching the undergraduate course in personality—new features are added to facilitate mastery of the material. One important pedagogical addition consists of *focus questions,* which occur in the margin of the book adjacent to important concepts and facts. The focus questions are designed to facilitate active processing of content by the reader and to function as study guides and retrieval cues. Their inclusion was based on educational research in controlled studies that showed that questions like ours significantly enhanced retention of facts and concepts. With our own students over the years, this approach has proven so successful (increasing test scores appreciably in groups of randomly selected groups of students who used such questions compared with others who were not provided with them) that we have made it an integral learning tool in this text. In addition, to highlight important points more effectively, we have replaced the essay-style summaries of earlier editions with bullet point summaries at the end of each chapter.

We found the preparation of this edition a wonderful adventure in which we learned much about our own field. We hope that both instructors and students will share our excitement about the current state and future of personality psychology, and our appreciation for the richness of its heritage. Your comments on earlier editions have helped shape this text, and we trust that we have heard and been responsive to your suggestions.

Walter Mischel
Columbia University

Yuichi Shoda
University of Washington

Ronald E. Smith
University of Washington

May, 2003

ACKNOWLEDGEMENTS

The authors owe a great debt to more people than can be acknowledged here. Walter Mischel had the extraordinary benefit of studying and working directly with historical pioneers who helped build the field, and who influenced him profoundly. These notably included his mentors George Kelly and Jules Rotter at Ohio State University, and Gordon Allport, Henry Murray, and David C. McClelland, who were his senior colleagues at Harvard University, and he remains grateful to them all. In the preparation of this edition, Ozlem Ayduk, Rodolfo Mendoza-Denton, and Ethan Kross all played significant roles, commenting on many early drafts and contributing to them. A special debt is owed to Amy Blum who generously helped to prepare the manuscript, commenting and working diligently and constructively on it from its earliest versions into its final production.

Yuichi Shoda owes his beginning in psychology to the confluence of idealistic subculture at UC Santa Cruz in the early 1980s that encouraged asking big questions and such social and personality psychologists as Elliot Aronson, Tom Pettigrew, Brewster

Smith, and in particular, David Harrington. Through the years at Stanford and Columbia as a graduate student, he was fortunate to absorb by osmosis, the philosophies and approaches to science of such mentors and colleagues as Larry Aber, Ozlem Ayduk, Al Bandura, Geraldine Downey, Carol Dweck, Tory Higgins, Dave Krantz, Bob Krauss, Rudy Mendoza-Denton, Phil Peake, Ewart Thomas, Jack Wright, and above all, Dan Cervone, whose infectious love of good ideas and dedication to our field drew Yuichi to it and sustained him throughout. The University of Washington is now his intellectual home, and he gratefully acknowledges his colleagues and students, particularly Ron Smith, for their inspiration and support. Most important of all, however, he has had the privilege of pursuing a shared vision of personality as a dynamic, multifaceted yet coherent whole, with his mentor and friend, Walter Mischel, for over two decades. This textbook is an expression of this vision.

Ronald Smith would like to acknowledge the contributions made to his training by the clinical and personality faculty at Southern Illinois University, including Janet Rafferty and Donald Shoemaker, who were former students of Rotter and Kelly, respectively. Other important mentors at SIU were Loren Chapman, Peter Lewinsohn, Edward Lichtenstein, and Nathan Azrin. For the past 30 years, the University of Washington has offered a wonderful intellectual and collegial environment for the study of personality. Special thanks are due to Irwin and Barbara Sarason for being such an important part of this environment. Finally, he thanks the countless undergraduate and graduate students in his personality courses for stimulating his thinking about the field and how to present it as an instructor.

The editorial and production staff, and particularly Tim Vertovec, Lili DeGrasse, and Kimberly Martin, worked hard and intensively to make the project go well, and were a pleasure to deal with from start to finish: we thank them for going all the way. Last, but by no means least, we are indebted to the effort, care, and time of the many instructors and colleagues who helped in the review process, and thank them sincerely:

Bob Bornstein, Gettysburg College
Kelly A. Brennan, SUNY Brockport
Eva G. Clarke, Old Dominion University
Jean M. Edwards, Wright State University
Howard Ehrlichman, Queens College of CUNY
Jeannine Feldman, San Diego State University
Steve Funk, Northern Arizona University
David A. Haaga, American University
Heather Haas, Colby College
Christopher Layne, University of Toledo
Mark Lenzenweger, Harvard University
Rich Lewine, University of Louisville
Herbert Mirels, Ohio State University
Joseph A. Reilly, Drexel University
Carolin J. Showers, University of Oklahoma

Tables 1 and 2 present a suggested list of chapter assignments for a quarter system, or a semester system:

TABLE 1   Suggested schedule for a quarter system

Week 1	**INTRODUCTION**
	CHAPTER 1 *Orientation to Personality*
	CHAPTER 2 *Data, Methods, and Tools*
Week 2	**PART I THE TRAIT-DISPOSITIONAL LEVEL**
	CHAPTER 3 *Types and Traits*
	CHAPTER 4 *The Expressions of Dispositions*
Week 3	**PART II THE PSYCHODYNAMIC-MOTIVATIONAL LEVEL**
	CHAPTER 5 *Psychodynamic Theories: Freud's Conceptions*
	CHAPTER 6 *Post-Freudian Psychodynamics and Motives*
Week 4	CHAPTER 7 *Psychodynamic Processes*
	Exam 1
Week 5	**PART III THE PHENOMENOLOGICAL LEVEL**
	CHAPTER 8 *Phenomenological Conceptions*
	CHAPTER 9 *The Internal View*
Week 6	**PART IV THE BEHAVIORAL-CONDITIONING LEVEL**
	CHAPTER 10 *Behavioral Conceptions*
	CHAPTER 11 *Analyzing and Modifying Behavior*
Week 7	**PART V THE SOCIAL COGNITIVE LEVEL**
	CHAPTER 12 *Social Cognitive Conceptions*
	CHAPTER 13 *Social Cognitive Processes*
Week 8	Exam 2
	PART VI THE BIOLOGICAL LEVEL
	CHAPTER 14 *Heredity and Personality*
Week 9	CHAPTER 15 *Evolution, Brain, and Personality*
	PART VII INTEGRATION OF LEVELS: The Person as a Whole
	CHAPTER 16 *Personality in the Pursuit of Goals*
Week 10	CHAPTER 17 *Self-Regulation: From Goal Pursuit to Goal Attainment*
	CHAPTER 18 *Personality in Its Context and Culture*
Week 11	Final Exam

TABLE 2 **Suggested schedule for a semester course**

Week 1	**INTRODUCTION** CHAPTER 1 *Orientation to Personality*
Week 2	CHAPTER 2 *Data, Methods, and Tools* **Part I: THE TRAIT-DISPOSITIONAL LEVEL** CHAPTER 3 *Types and Traits*
Week 3	CHAPTER 4 *The Expressions of Dispositions* **Part II: THE PSYCHODYNAMIC-MOTIVATIONAL LEVEL** CHAPTER 5 *Psychodynamic Theories: Freud's Conceptions*
Week 4	CHAPTER 6 *Post-Freudian Psychodynamics and Motives* CHAPTER 7 *Psychodynamic Processes*
Week 5	Review and integration of Parts I and II Exam 1
Week 6	**Part III: THE PHENOMENOLOGICAL LEVEL** CHAPTER 8 *Phenomenological Conceptions* CHAPTER 9 *The Internal View*
Week 7	**Part IV: THE BEHAVIORAL-CONDITIONING LEVEL** CHAPTER 10 *Behavioral Conceptions* CHAPTER 11 *Analyzing and Modifying Behavior*
Week 8	Review and integration of Parts III and IV Exam 2
Week 9	**Part V: THE SOCIAL COGNITIVE LEVEL** CHAPTER 12 *Social Cognitive Conceptions* CHAPTER 13 *Social Cognitive Processes*
Week 10	**Part VI: THE BIOLOGICAL LEVEL** CHAPTER 14 *Heredity and Personality* CHAPTER 15 *Evolution, Brain, and Personality*
Week 11	Review and integration of Parts V and VI Exam 3
Week 12	**Part VII: INTEGRATION OF LEVELS: The Person as a Whole** CHAPTER 16 *Personality in the Pursuit of Goals* CHAPTER 17 *Self-Regulation: From Goal Pursuit to Goal Attainment*
Week 13	CHAPTER 18 *Personality in Its Context and Culture* Review and integration, Parts I to VII
Week 14	Final Exam

BRIEF CONTENTS

CONTENTS

CHAPTER 4   *THE EXPRESSIONS OF DISPOSITIONS*   **69**

PART III

THE PHENOMENOLOGICAL LEVEL **169**

CHAPTER 8 ## *PHENOMENOLOGICAL CONCEPTIONS* **171**

CHAPTER 9   *THE INTERNAL VIEW* **194**

PART IV

THE BEHAVIORAL-CONDITIONING LEVEL 217

PART V

THE SOCIAL COGNITIVE LEVEL **269**

PRELUDE TO PART V: The Social Cognitive Level **269**

CHAPTER 12 *SOCIAL COGNITIVE CONCEPTIONS* **271**

PART VI

THE BIOLOGICAL LEVEL **321**

PART VII

INTEGRATION OF LEVELS: THE PERSON AS A WHOLE **367**

CHAPTER 16 *PERSONALITY IN THE PURSUIT OF GOALS* **371**

CHAPTER

1

ORIENTATION TO PERSONALITY

▶ WHAT IS PERSONALITY PSYCHOLOGY?

Personality is a psychological concept that has many meanings, reflecting the richness and complexity of the phenomena to which the term refers. Here is an example of one aspect of the concept.

Stable, Coherent Individual Differences

Charles and Jane both are first year college students taking an introductory course in economics. Their instructor returns the midterm examination in class, and both receive a D grade. Right after class, Charles goes up to the instructor and seems distressed and upset: He sweats as he talks, his hands tremble slightly, he speaks slowly and softly, almost whispering. His face is flushed and he appears to be on the edge of tears. He apologizes for his "poor performance," accusing himself bitterly: "I really have no good excuse—it was so stupid of me—I just don't know how I could have done such a sloppy job." He spends most of the rest of the day alone in his dormitory, cuts his classes, and writes a long entry in his diary.

Jane, on the other hand, rushes out of the lecture room at the end of class and quickly starts to joke loudly with her friend about the economics course. She makes fun of the course, comments acidly about the instructor's lecture, and seems to pay little attention to her grade as she strides briskly to her next class. In that class, Jane participates more actively than usual and, surprising her teacher, makes a few excellent comments. This example illustrates a well-known fact: Different people respond differently

1

to similar events. One goal of personality psychology is to find and describe those *individual differences* between people that are psychologically meaningful and stable.

1.1 Which two aspects of individuality give rise to the concept of personality?

Though the concept of personality has to do with how an individual differs from others, it implies more. What if, on the next day, it was Jane, not Charles, who became very upset when they learned they didn't do well in another class? Does that suggest their behaviors on either day (or both days) aren't characteristic of their personality? If you answered yes, that suggests another key component of personality. That is, personality refers to qualities of individuals that are relatively *stable*. If a person's behavior changes from time to time, then it may not be indicative of personality.

Let's further suppose that you learned a little more about what happened the second day: The course in which they received a poor grade on the second day was English Composition. Not only did they receive a poor grade, they also learned that their classmates thought poorly of the essays they wrote and shared with their classmates. Now, does this additional information help make sense of their behaviors? If you answered yes, think about why.

1.2 How do situational factors (circumstances) contribute to perceived personality coherence?

One possibility is that with the new information about what happened on the second day, one can begin to see why their behaviors changed from the first day to the second day. One can begin to form a mental picture of the kind of person who doesn't seem upset by a bad grade in Economics but is devastated by a poor grade in English and/or her peer's unenthusiastic response to her essay. Similarly, one may form an impression of a person who is very upset by a poor grade in Economics but is unaffected by a bad grade in English composition or his peer's reactions. The information about the circumstances has transformed the change from one that is puzzling to one that makes sense. The change is potentially meaningful, because even though on the surface Charles and Jane's behaviors changed, there is coherence in the way they changed. That is, *coherence* in the pattern of change in an individual's behavior may be another key component of personality.

Describing, Predicting, Understanding

The term "personality" usually implies *continuity* or *consistency* in the individual. Personality psychologists therefore ask questions like: How consistent are the observed differences between people? How would Charles and Jane respond to a D in physical education? How would each respond if they were fired from their part-time jobs? What do the differences in the reactions of the two students to their grade suggest about their other characteristics? For example, how do they also differ in their academic goals and in their past achievements and failures?

1.3 Describe three goals of personality psychologists.

If the answers to such questions are yes, then the observed differences may be meaningful indicators of individual differences in the personality of these two students. Identifying consistent, stable individual differences is an important goal for personality psychologists — and for everyday life — because it makes it possible both to describe people and to try to predict their future behavior, and so to get to know what we can expect from them.

In addition to mapping out the differences between people in terms of their characteristic ways of behaving — that is, thinking, feeling, and acting — personality psychologists try to understand what it is that underlies these differences. They ask: Why did Jane and Charles react so differently to the same event? What within each person leads to his or her distinctive ways of behaving? What must we know about each person to understand — and perhaps sometimes even predict — what he or she will think and feel and do under particular conditions? Personality psychologists ask questions of this sort as they pursue the goal of trying to explain and understand the observed psychological differences between people.

Defining Personality

What *is* personality? The term *personality* has many definitions, but no single meaning is accepted universally. In popular usage, personality is often equated with social skill and effectiveness. In this usage, personality is the ability to elicit positive reactions from other people in one's typical dealings with them. For example, we may speak of someone as having "a lot of personality" or a "popular personality," and advertisements for self-help courses promise to give those who enroll "more personality."

Less superficially, personality may be taken to be an individual's most striking or dominant characteristic. In this sense, a person may be said to have a "shy personality" or a "neurotic personality," meaning that his or her dominant attribute appears to be shyness.

In personality psychology, the concept goes much beyond these meanings. The definition begins with the assumption that there are stable individual differences. It is further assumed that these differences reflect an underlying *organization*. In one classic and still influential working definition, the idea of organization is central to the definition. Personality psychology is:

> . . . the dynamic organization within the individual of those psychophysical systems that determine his characteristic behavior and thought (Allport, 1961, p. 28).

As the science matures, there is a growing consensus about the findings and concepts that have stood the test of time. Consequently, a unifying conception of personality and, more modestly, at least a broadly acceptable definition, is becoming possible. A good candidate for such a definition was offered by Pervin (1996, p. 414):

> Personality is the complex organization of cognitions, affects, and behaviors that gives direction and pattern (coherence) to the person's life. Like the body, personality consists of both structures and processes and reflects both nature (genes) and nurture (experience). In addition, personality includes the effects of the past, including memories of the past, as well as constructions of the present and future.

Consistent with that definition, David Funder (2001, p. 198) defines the mission of personality psychology as needing to "account for the individual's characteristic patterns of thoughts, emotion, and behavior together with the psychological mechanisms—hidden or not—behind those patterns."

As discussed above, individual differences are always a core part of the definition of this field, but they are not the whole of it. Thus the term personality psychology does not need to be limited to the study of differences between individuals in their consistent attributes. Rather, "personality psychology must also . . . study how people's [thoughts and actions] . . . interact with—and shape reciprocally—the conditions of their lives" (Mischel, 1980, p. 17).

This expanded view recognizes that human tendencies are a crucial part of personality. But it also recognizes the need to study the basic processes of adaptation through which people interact with the conditions of their lives. Personality thus includes in the person's unique patterns of coping with, and transforming, the psychological environment. This view of personality focuses not only on *personal tendencies* but also on *psychological processes* (such as learning, motivation, and thinking) that interact with *biological-genetic processes* to influence the individual's distinctive patterns of adaptation throughout the life span.

1.4 Cite five aspects of the construct of personality as currently conceived.

In summary, to capture the richness of human behavior, the personality construct has to encompass the following aspects:

- Personality shows continuity, stability, and coherence.
- Personality is expressed in many ways—from overt behavior through thoughts and feelings.
- Personality is organized. In fact, when it is fragmented or disorganized it is a sign of disturbance.
- Personality is a determinant that influences how the individual relates to the social world.
- Personality is a psychological concept but it also is assumed to link with the physical, biological characteristics of the person.

▶ THEORY AND LEVELS OF ANALYSIS IN PERSONALITY PSYCHOLOGY

Early "Big Picture" Theory

Personality psychology is a relatively young science but it has been practiced from the time that people began asking questions about human nature: Who should I fear? Who can I trust? Who do I select for a mate? Who am I?

In Western societies, since the time of the ancient Greeks, philosophers have long pondered questions about human nature and attempted classification schemes for making sense of the varieties of individual differences in important attributes and their causes. As early as 400 B.C., Hippocrates philosophized about the basic human temperaments (e.g., choleric, depressive), and their associated traits, guided by the biology of his time. For example, he thought physical qualities like yellow bile, or too much blood, might underlie the differences in temperament. He began a tradition—trait and type psychology—whose modern versions date to the start of the last century. It is a tradition that is still very much alive and well, although completely transformed in current scientific practice, drawing extensively both on modern measurement methods and on the biology of today.

Aristotle postulated the brain to be the seat of the rational mind, or the "conscious and intellectual soul that is peculiar to man" (Singer, 1941, p. 43). This view has become a foundation of the Western view of human mind. For example, in his dualistic view of the human being as consisting of mind and body, Descartes viewed the mind as what gives us the capacities for thought and consciousness, which sets us apart from the physical world of matter. The mind "decides" and the body carries out the decision.

In the early 1900s, Sigmund Freud, living and working in Vienna as a physician, upset the rational view of human nature that characterized his time with a powerful and comprehensive theory of personality. Freud's theory made reason secondary and instead made primary the unconscious and its often unacceptable, irrational motives and desires, thereby forever changing the view of human nature. The tradition he began also continues to influence contemporary personality psychology, but again in ways that are much changed, both by the work of his many followers and by developments in other areas of the science that made it possible to reinterpret much of his work and to revise it as needed.

From Grand Theories to Levels of Analysis

In the first half of the twentieth century, personality psychology was inspired by grand theories of personality that were being developed by several "big picture" innovators such as Sigmund Freud: Each proposed distinctive conceptions of the nature of personality, and tried to present a comprehensive view of all of personality in all of its diverse aspects. Like Freud, many of these theorists were working in Western Europe as therapists treating psychologically disturbed and distressed individuals. As practicing therapists, they used the cases of their patients as the basis for broader generalizations on the nature of personality. Consequently their ideas helped to shape clinical psychology and psychiatry as well as personality psychology. One hazard here was that because their work was based on their experience with emotionally disturbed patients, they may have focused more on the disturbed aspects of personality than on its healthier versions in less troubled people.

Broad theories like Freud's provide an orientation and perspective that stimulates different types of research within the field and different types of real-life applications, such as clinical practice with people experiencing psychological problems. Most notably, they lead to different lines of research and to different forms of therapy or intervention designed to modify or enhance personality constructively. They also lead to different approaches to assess personality and to think about persons, including oneself, and thus matter a great deal to the image one develops of personality and individuality, and indeed of oneself as a person. As such they are valuable.

1.5 What kinds of scientific contributions have been made by broad theories of personality? What are their scientific shortcomings?

In spite of the growth of personality psychology as a field of scientific research, however, most of the grand theories of personality, like Freud's, did not lend themselves to precise scientific testing that allowed them to be either supported or disconfirmed clearly on the basis of empirical studies (Meehl, 1990, 1997). Reasons for this range from the difficulty of specifying the theoretical premises in testable terms, to various types of experimental and statistical limitations in conducting and evaluating the test results. But even beyond these limitations, grand theories often function more like general guidelines or orientations for studying personality and interpreting the results from a particular perspective or framework. Thus it is difficult to firmly reject or support a given theory on the basis of empirical studies. As one pundit put it, many big theories in all areas of science generally are never really disconfirmed: They just die of loneliness as they gradually are replaced by approaches that seem more fruitful and lead to more informative new research that raises new questions and suggests new—always tentative—answers.

In the second half of the last century, after World War II, American personality psychology grew into a substantial field in its own right. It was influenced by European psychology but also developed in its own directions within the larger science of psychology. The influences in the United States came from traditions that sprang up in university psychology departments that were devoted to turning psychology away from philosophy and into science. Researchers working both with normal and disturbed populations developed and applied increasingly sophisticated scientific methods to address many central issues in personality psychology. In time, it became possible to examine important questions about personality with research evidence that accumulated at a rapid rate and pointed to exciting new directions as the science evolved.

From the 1960s to about the 1980s, the field of personality psychology was full of seemingly insoluble controversies among apparently irreconcilable broad theoretical approaches. The result was much debate and new research that helped to clarify important questions but that also created much divisiveness among alternative viewpoints. In contrast, in the current scene there are numerous encouraging signs of integration and constructive syntheses of the insights coming from theorists and researchers that are working at different **levels of analysis,** addressing different aspects of personality. It is increasingly seen that each level has its legitimacy and usefulness, and each requires distinctive methods and concepts. But the findings from different levels do not necessarily conflict. On the contrary, they usually add to the understanding and clarification of the whole. Each level of analysis yields many solid answers—as well as raising new questions—and each contributes to building a cumulative and coherent view of personality psychology.

▶ LEVELS OF ANALYSIS: ORGANIZATION OF THIS BOOK

1.6 At which six levels of analysis is personality being studied? What kinds of questions can we ask about ourselves and others at each level?

In this book you will learn some of the major theoretical approaches to personality that have guided thinking and research, and see how research and theory-building is done, at each level of analysis. We will survey some of the main concepts developed to describe and understand the important psychological differences among people, and we will consider the concepts and findings that are central to diverse views of human nature.

To capture the essentials, this text is organized into the six major levels of personality study from a century of work in psychology as a science and profession. Each part of the text presents the main concepts, methods, and findings associated with that level of analysis, and each focuses attention on distinctive aspects of personality. Each level adds to the appreciation of the richness and complexity of personality. Each level also led to discoveries that have important practical and personal applications that we will examine. In combination, the six levels provide an overview of the many complex and diverse aspects of human personality. The final part of the text shows how the levels interconnect and become integrated to give a more coherent view of the person as a whole. The organization of the text highlights how each level adds to the whole, and suggests their evolving integration. This organization can be seen at a glance in Figure 1.1.

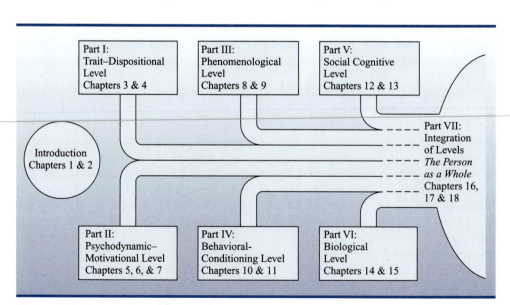

Figure 1.1 Organization of the text.

After an Introduction section that gives an overview of the data, methods, and tools of the science, Part I presents contributions from the Trait-Dispositional Level; Part II focuses on those coming from the Psychodynamic-Motivational Level; followed by Part III, the Phenomenological Level; Part IV, the Behavioral-Conditioning Level; Part V, the Social Cognitive Level; and Part VI, the Biological Level. Below we will look at an overview of each of the levels. As Figure 1.1 suggests, the contributions from each level are cumulative and come together in the final Part VII, Integration of Levels: The Person as a Whole, which explicitly shows their interconnections. Each level asks distinctive questions although both the questions and the levels overlap.

Almost everyone becomes interested in the science of personality because they want to understand people, and particularly themselves and those they care about, as fully as possible. In this sense, the science of personality also has a distinctly personal side: the questions that researchers ask at each level in formal scientific terms also have personal relevance. Often they are asked more informally by most people. Therefore, to make those connections explicit, the key questions pursued at each level of analysis are summarized in In Focus 1.1, and phrased in personal ways that invite you to ask them about yourself.

IN FOCUS 1.1

THE PERSONAL SIDE OF THE SCIENCE

Some Key Questions at Different Levels of Analysis Phrased as Questions That Can Be Asked About Oneself:

- **Trait-Dispositional Level:** What am I like as a person? How am I different from other people "on the whole"? In what general ways are people different from each other? Does what I usually do and think and feel depend mostly on myself or on the situation in which I find myself? When and how is my behavior influenced by the situation? How does my personality influence the situations I choose to be in? How does my personality influence the effects that different kinds of situations have on me?

- **Psychodynamic-Motivational Level:** Does what I do sometimes puzzle me? How and why? What are the real motives that drive or underlie my behavior? How can I explain irrational fears and anxieties? How do I try to protect myself psychologically against getting hurt? How much of what I do is unconscious or done without awareness? What might be some unconscious influences on my behavior? Do I have motives that make me uncomfortable? If yes, what do I try to do about that?

- **Phenomenological Level:** Who am I really? Who do I want to become? How do I see myself? How do I see my parents? What do I feel about myself when I don't meet my parents' expectations? How is my real self different from the self I would ideally like to be? What is my ideal self? How am I different from my mother but similar to my father? How can I think my way out of my childhood?

- **Behavioral-Conditioning Level:** How is what a person *does* linked to what happens to him or her when he does it? How are important behavior patterns, including emotions and fears, learned? How does what I do and feel depend on my earlier experiences? How can my behavior and feelings be modified by new learning experiences? Do aspects of my personality depend on the contexts in which I am? How am I different when with a good friend at school and when with my family at home for the holidays? Why?

- **Social Cognitive Level:** What is the role in personality of what people know, think, and feel? How does what I know, think, and feel about myself and the social world influence what I do and can become? What can I do to change how I think and feel? Will that change my personality and behavior? How much of who and what I am and do is "automatic"? How much is open to "willpower" and self-regulation? How do willpower and self-regulation work? How can I enhance my control over my life?

- **Biological Level:** What in my personality comes from my parents and the genes I inherited from them? How is my personality a reflection of my life experiences? How does my personality reflect my basic biological predispositions? Can my experiences change my biology; for instance, does my brain change when I'm depressed? How do the same experiences affect people with different genetic predispositions? Why is my personality so different (or similar) to my siblings? How does my biology influence my pursuit of life goals? How does evolutionary theory help me understand dating and social behavior today?

Now let's consider each level to get an overview.

The Trait-Dispositional Level

1.7 Describe the focus of the trait-dispositional level of analysis.

The trait-dispositional level seeks to identify the types of stable psychological qualities and behavioral dispositions that characterize different individuals and types consistently. In everyday life, people may ask themselves questions like those listed for this level in In Focus 1.1: "What am I like as a person? How am I different from other people 'on the whole'? In what general ways are people different from each other?" Using the natural language of trait terms, people often ask and answer such questions easily, not just about themselves but about other people: He or she seems friendly, assertive, aggressive, submissive, conscientious, and so on. Examples of such traits used in current research include broad characteristics such as agreeableness, conscientiousness, and open mindedness. At this level, one also studies the diverse but stable expressions over time and across situations of these types and traits to see how consistent they are.

This level of analysis has become one of the most vigorous and widely researched in recent years. This progress has been supported by the development of some straightforward, well-established self-report methods that are being used by researchers throughout the world. Consequently, the usefulness of trait level analyses is being extensively investigated and has led to findings showing the stability of personality over time. Work at this level is also yielding a broad taxonomy for classifying individuals with regard to major traits, providing a map on which people, groups, and even cultures can be compared.

Work at this level of analysis especially in recent years also has been getting answers to questions such as: Does what we do and think and feel characteristically depend mostly on the individual or on the situation? How do the two—the person and the situation—interact?

The Psychodynamic-Motivational Level

1.8 Which aspects of personality are addressed at the psychodynamic-motivational level of analysis?

The psychodynamic-motivational level probes the motivations, conflicts, and defenses, often without one's awareness, that can help explain complex consistencies and inconsistencies in personality. Questions you might ask yourself when thinking about this level of analysis, as In Focus 1.1 suggests, include: "Does what I do sometimes puzzle me? How and why? What are the real motives that drive or underlie my behavior? How can I explain irrational fears and anxieties? How do I try to protect myself psychologically against getting hurt?"

Work at this level is relevant for understanding many puzzles of personality, for example, when people turn out to be more complex and unpredictable than expected and seem to change as one knows them better. Here is an example:

Roberto was confusing his girlfriend. She felt she was getting more smoke than light—the longer she knew him, the less she felt she really understood him, and she was beginning to lose her trust. Before they moved in together, she thought she knew him—warm, friendly, fun and easy to be with. Later she began to see the sadness, the inside rage, the fears and the unpredictability. After a while, he stayed in bed most of the day. When asked what's going on he'd say, "I'm fine," denying any anxiety or depression, and he seemed to mean it. He was reassuring, yet she remained quite unconvinced.

Roberto's girlfriend intuitively understood that what Roberto said and seemed to honestly believe was not necessarily the whole story and that there were other reasons that he could not acknowledge even to himself. The kind of insight needed to understand Roberto's behavioral and emotional inconsistencies is at the heart of the psychodynamic motivational level of analysis.

Much of the work at this level has been done in clinical situations, beginning with Sigmund Freud a century ago. He worked, for example, on the case of Hans, a four-year-old child who had developed an irrational fear of going outdoors because horses might be there and he had become terrified of them although he had never been hurt by one. Freud created a theory that used the concept of the unconscious and the child's unacceptable sexual and aggressive wishes to explain how a fear like that could have developed.

An important key here was the discovery that certain impulses, such as a young child's aggressive impulses toward his father, are treated by socializing agents as taboo and punished, making the child anxious. Because the impulses still persist but create painful anxieties, the child may unconsciously redirect them at other objects, for example, by becoming afraid of horses that remind him of his father who might punish him.

The Phenomenological Level

Each person sees the world subjectively in his or her own personal ways. To understand this privately experienced side of personality, we must examine the nature of subjective experience; we have to try to see how people perceive their world. Workers at this level of analysis are genuinely interested in hearing and exploring fully the answers people give to questions (from In Focus 1.1) like these: "Who am I really? Who do I want to become? How do I see myself? How do I see my parents? What do I feel about myself when I don't meet my parents' expectations?

Here is a sample of self-reported personal feelings taken from a self-description by a college student about to take a final examination:

1.9 What is the major focus of the phenomenological level of analysis?

. . . When I think about the exam, I really feel sick . . . so much depends on it. I know I'm not prepared, at least not as much as I should be, but I keep hoping that I can sort of snow my way through it. . . . I keep trying to remember some of the things he said in class, but my mind keeps wandering. God, my folks—What will they think if I don't pass and can't graduate? Will they have a fit! Boy! I can see their faces. Worse yet, I can hear their voices: "And with all the money we spent on your education." Mom's going to be hurt. She'll let me know I let her down. She'll be a martyr . . . Oh hell! What about Anne [girlfriend]? She's counting on my graduating. We had plans. What will she think? . . . I've got to pass. I've just got to . . . What's going to happen to me? . . . The whole damn world is coming apart (extracted from Fischer, 1970, pp. 121–122).

Work at this level begins by listening closely and trying to understand the individual's experience as he or she perceives it. The focus is on subjective experience, feelings, the personal view of the world and the self. The focus also is on people's positive strivings and their tendencies toward growth and self-actualization.

These concerns require studying the internal or mental processes through which individuals interpret experience. A distinguished psychologist, Ulric Neisser, for example, put it this way (1967, p. 3): "Whether beautiful or ugly or just conveniently at hand, the world of experience is produced by [the person] who experiences it." That statement, of course, does not imply that there is no "real" world out there, but just that it is *the experienced world* that is basic for understanding phenomena like personality and the important differences between people. Ideally, researchers at this level would like to look at the world through the eyes of the persons they are studying, to stand in that person's shoes, to know emotionally as well as intellectually what it might be like to *be* that person.

Although for many years this level of analysis was treated with suspicion by the rest of the field, more recently there has been an explosion of solid research on the self and close relationships. This work is restoring the self as an important concept in current personality psychology. It is addressing the processes through which the self develops, and telling us much about the links between the self and personality adjustment, mental health, and positive functioning.

The Behavioral-Conditioning Level

1.10 What questions about personality are addressed at the behavioral-conditioning level?

As In Focus 1.1 indicates, work at this level of analysis has asked questions like: "How are important behavior patterns, including emotions and fears, learned? How does what I do and feel depend on my earlier experiences? How can my behavior and feelings be modified by new learning experiences?"

Consider this dilemma:

Jake was upset because he could not accept the management job he had been so eager to get when he learned that it was on the 80th floor: just the thought of riding up in the elevator terrified him—but he had never been afraid of elevators in the past.

If this description makes you think here come more examples for the psychodynamic-motivational level like Freud's case of little Hans, you are right. Work at the behavioral-conditioning level also tries to provide accounts of irrational behaviors that puzzle the people who are tortured by them, similar to many of the same basic personality phenomena that Freud discovered. But they get there through a different route and reach different conclusions that lead to important revisions in some of the earlier ideas for dealing with such problems.

The behavioral-conditioning level analyzes specific patterns of behavior that characterize individuals and the situations or conditions that seem to regulate their occurrence and strength. It studies the determinants of learning and applies learning principles to modify problematic patterns of behavior, including emotional reactions like fears. Behavioral analyses focus on an important behavior—such as the stutter of a person suffering from public speaking anxieties, or the inability to stay concentrated on one's studying before exams. Then they analyze the situations or conditions that seem to control that behavior, that is, the conditions in which the stutter or the studying becomes worse or improves. Finding the conditions under which the problem improves becomes the basis for designing treatments to modify the behavior to help reduce or eliminate the problem. Behavioral analyses have helped us understand the conditions through which behaviors relevant to personality—from stutters through poor self-concepts, and troublesome behavior in interpersonal relationships—are learned and can be modified.

The results have been applied to help people overcome a variety of serious personal difficulties, ranging from common but debilitating fears, to weight problems, to learning deficits and handicaps, to increasing personal assertiveness and self-esteem. Originally this level of analysis dealt mostly with learning and conditioning based on animal work. In new direction, behavioral levels of analysis have a second life because advances in brain imaging, for example, and in other areas of cognitive science—the modern study of mental processes—make it possible to analyze mental functions previously considered too mysterious for behavioral study with the objective methods of science.

The Social Cognitive Level

Personality research at this level focuses on the role of personal theories and beliefs, expectations, motivations, and emotions as determinants of individual differences in social behavior. The focus here includes the person's social knowledge of the world, and the goal is to understand how people make sense of other people and themselves and cope as they negotiate their interpersonal lives. Questions that you might ask about yourself at this level of analysis (In Focus 1.1) are: "How does what I know, think, and feel about myself and the social world influence what I do and can become? What can I do to change how I think and feel?"

1.11 How does the social cognitive level of analysis differ from the behavioral-conditioning approach?

This level examines individual differences in how social knowledge is used in dealing with the world, in the construction of the self, in self-regulation, and in self-control. The specific focus is on the individual's characteristic ways of thinking and processing information, both cognitively and emotionally, as determinants of his or her distinctive and meaningful patterns of experience and social behavior. For example:

Yolanda and Virginia, now college students, both lost their young mothers last year from breast cancer. Both women know that their total family histories put them at high risk for having a genetic predisposition for this type of cancer early in life. At the college health service the specialist they see points out the risks of their situation in great detail. He urges that they adhere strictly to monthly breast self-examination as an important part of their health-maintenance program.

When Yolanda tries self-examination, she remembers the risks the doctor described and she imagines them vividly. Her thoughts flow from anticipating finding the lump, to thinking "I'm going to die," to a flood of panic feelings. She thinks "what will be will be" and stops trying to self-examine. When Virginia begins, she also remembers the physician's words but she imagines that if there is a lump she will find it early, it can be removed, and she can be successfully treated. She thinks: Above all I will have at least some control over my fate.

An important challenge at this level is to understand such internal mental and emotional processes and their links to the characteristic behavior patterns that may enduringly distinguish different individuals and types of people. The goal is not only to "get inside the head" but to understand the stable mental-emotional processes and structures that generate the diverse individual differences that are observed. Most attention in recent years is given to studying the basic psychological processes through which individuals construct, interpret, and understand their social-personal world. Questions here include: How do individuals come to deal with their worlds in the stable cognitive, emotional, and behavior patterns that characterize them? What is the "self" and how do self-concepts and perceptions about the self influence what the person thinks, feels, and becomes?

The Biological Level

1.12 What kinds of biological factors are known to underlie individual differences in personality?

An important goal of personality study at the biological level is to try to specify the role of genetic determinants and of the social environment in shaping who and what we become. The focus in much of this work in the past has been to answer the age-old question: How much of personality reflects nature, and how much nurture—and above all, how do these two sources of influence interact in shaping our characteristics? When you learn what is now known at this level of analysis, you will be able to answer questions like: "To what extent does my personality come from my parents and the genes I inherited from them? To what extent is my personality a reflection of my life experiences? To what extent does my personality reflect my basic biological predispositions?"

This level of analysis also addresses the fact that humans are biological beings who evolved in adaptive ways that endowed the species with biological characteristics, constraints, and possibilities. These influence human nature and the way we fight, mate, socialize, and create. The goal at this level of analysis is to examine how aspects of personality may have evolved in response to the evolutionary pressures and history that shaped our species over time. Here is an example of the kinds of problems it studies and the questions this level asks:

> Consider two identical twin baby girls who were separated at birth and grew up in very different worlds. Jane was raised on a rural Iowa farm, an only child, with hard-working but unloving parents. Nahid's life unfolded in the capital of Iran, nurtured by loving parents in a large middle-class family. The identical twins started life with the same DNA and therefore with virtually identical brains. Suppose the twins were reunited at age 30 and tested extensively, how similar will their personalities turn out to be?

Levels of Analysis Applied to Understand Unexpected Aggression: The Texas Tower Killer

The different levels of analysis discussed throughout this text can complement each other constructively. Taken together, they increase the total understanding of personality as a whole. To illustrate, we next look at the types of questions that each level asks when confronted by the real-life puzzles of personality. In this example, you can see that the phenomena addressed by all levels interact concurrently within a personality.

The example here is the case of Charles Whitman, a University of Texas college student. Late one hot summer night, Charles Whitman killed his wife and mother. The next morning he went to a tower on the University of Texas campus and opened fire on the crowded campus below with a high-powered hunting rifle. In 90 horrifying minutes, he killed 14 people, wounded another 24, and even managed to hit an airplane before he was killed by police. After the Whitman incident, the first question asked was a familiar one: What caused this mild-mannered young man to explode into violence?

The night before the killing, Whitman wrote the following letter, reflecting his internal subjective experiences at the time:

> I don't really understand myself these days. I am supposed to be an average, reasonable, and intelligent young man. However, lately (I can't recall when it started) I have been the victim of many unusual and irrational thoughts. These thoughts constantly recur, and it requires a tremendous mental effort to concentrate on useful and progressive tasks. In March when my parents made a physical break I noticed a great deal of stress. I consulted a Dr. Cochrum at the University Health Center and asked him to recommend someone that I could consult with about some psychiatric disorders I felt I had. I talked with a doctor

Charles Whitman: The Texas Tower Killer

once for about two hours and tried to convey to him my fears that I felt overcome by over-whelming violent impulses. After one session I never saw the doctor again, and since then I have been fighting my mental turmoil alone, and seemingly to no avail. After my death I wish that an autopsy would be performed on me to see if there is any visible physical dis-order. I have had some tremendous headaches in the past and have consumed two large bottles of Excedrin in the past three months.

Applied to the Whitman case, at the **trait-dispositional level** of analysis, psychologists will test the person on measures like those described in the next chapter. The main questions will be: Would Whitman be likely to have a distinctive trait profile on such tests that shows, for example, high levels of angry hostility, impulsiveness, and neuroticism, with poor impulse control and little ability to handle stress? If he was not characterized by such a profile before the incident, did Whitman undergo personality change, at least as defined at the trait level? Or were there any subtle indications in his behavioral tendencies that might have allowed one to predict his actions? The profile resulting from such tests would provide a rich description of his characteristics. Many of these might not have been evident from his previous behaviors, but might now help one to make sense of his violent outburst and the character traits with which it was consistent.

At the **psychodynamic-motivational level,** to understand Whitman's violence one might first focus on the "unusual and irrational thoughts" to which he referred and to his "overwhelming violent impulses." Using the methods of psychodynamic theory (Chapters 5, 6, 7), one seeks to understand the unconscious conflicts and struggles that underlie them. According to psychoanalytic theory, for example, human aggression is an outgrowth of the continuous conflict between strong and often unconscious impulses and the defenses developed by the ego to keep them in check. Might

1.13 How do the six levels of analysis address different potential causes of Charles Whitman's killing spree?

Whitman's defenses against his own anger and hostility have become so rigid and extreme that he could not express his aggressive impulses even in indirect or disguised forms? A question asked at the psychodynamic level then is: Will the unreleased pressures build up to an explosion point? Work at this level assumes that the provocation that triggers unexpected destructive outburst is usually trivial. Instead, it searches to try to infer underlying conflicts and unconscious dynamics that might account for the unexpected and seemingly inexplicable change in his typical behavior.

At the **phenomenological level,** the focus would shift to trying to illuminate Whitman's own views of what he did and why, and of what he believed was happening to him, beginning with the letter he wrote, and extending in various other directions. The concern would be to understand how his perceptions and interpretations of what was happening to him misguided him to the actions that then erupted explosively. In the effort to unscramble his confusions and misperceptions, attention would be focused on his disturbed sense of self and his panic in trying to deal with the internal conflicts, loss of control, and feelings of fragmentation, despair, and helplessness that he was experiencing. In these efforts, work at this level and work at the psychodynamic level would become complementary, with researchers at both levels converging on some of the same questions albeit with somewhat different concepts and methods.

At the **behavioral-conditioning level,** the focus is on the ways in which the person's behavior reflects and is shaped by his or her learning history and present life conditions. Applied to Charles Whitman, at this level, one would seek the answer in Whitman's previous learning experiences and the culture he grew up in. A question at this level would be whether there was a history of fascination and rewarding experiences with guns, as well as exposure to role models that displayed violent behaviors. One may also ask whether there was an influence by the culture or subculture and the rewards it offered for aggression in diverse forms. Perhaps the environment in which he developed had primed him to solve his problems in a violent manner, particularly when he was overwhelmed by the recent life stresses that he described in his letter.

At the **social cognitive level,** how people perceive and interpret events, and the internal states and mental-emotional processes that these perceptions activate, determine how they behave. To understand what Whitman did requires understanding these mental and emotional processes that were activated in him at the time of his outburst, the specific situations he was exposed to, and his characteristic cognitive and affective dynamics that generated his extraordinary aggression. Perhaps his aggression was prompted by perceptions that he had been terribly wronged in ways that allowed him to justify his actions at least to himself. By blaming a person or group for real or imagined wrongs, people can create an image of a hated enemy fully deserving of whatever aggression is directed toward them. But why did Whitman do something so extreme that most people would not do under similar circumstances, and with equally good reasons? What was going on in Whitman's mind that was so different and that could plausibly account for the extremeness of his aggression? What kinds of skills, self-regulatory controls, and values did he lack that could allow such behavior? What were the beliefs and expectations that led him to his actions? These are the kinds of questions that drive work at this level of analysis. While there is no way to answer such questions in hindsight, the challenge is to be able to do so—at least sometimes—in advance. In pursuing the answers, work at this level and at several of the other levels converges and each again complements the others.

At the **biological level,** to understand Whitman, one would begin by considering the possible links between the brain and aggression that might underlie his ferocious outburst. To understand the Whitman case at this level, a postmortem examination was conducted to follow up on his reference to intense headaches. It revealed a highly malignant tumor in an area of the brain hypothesized to be involved in aggressive behavior. Some experts therefore suggested that Whitman's damaged brain might have predisposed him to violent behavior. On the other hand, although many efforts have been made to locate and study areas of the brain involved in aggressive behavior, even with modern methods these relations are still poorly understood, and emerging evidence suggests that the relationship between brain areas and behavior is not simple. Although certain areas of the brain may have coordinating functions in aggression, we also know that these regions are closely regulated by other areas of the brain that process information coming in from the environment. Certain kinds of brain damage or disorders can produce violent and unpredictable behavior in humans, too. In the majority of individuals who behave aggressively, however, there is no evidence of brain damage, although the aggression is being triggered by a variety of brain mechanisms.

Nevertheless, biological and genetic factors do appear to play a more general role in aggressive behavior (Baron & Richardson, 1994; Loehlin, 1992). Identical twins, who are genetic carbon copies of each other, are more similar in their aggressive and dominant behavior patterns than are fraternal twins who differ genetically from one another (Plomin & Rende, 1991). This is the case even if the identical twins are raised in different homes with presumably different social environments (Bouchard and others, 1990). But, behavior geneticists also remind us that genetic factors never operate in isolation; they always interact with environmental factors. In recent years, new discoveries about the brain, and new methods of studying its activities, have encouraged great interest and much research at this level. If Whitman were tested today, biological level measures (such as brain scans) would allow much fuller analyses of the links between his thoughts, actions, and brain processes. Most important, as you will see later in this text (Chapter 17), recent advances in the study of brain mechanisms are giving us an increasingly clear picture of the neural mechanisms in the brain that can go wrong when people have violent outbursts of aggression (Davidson, Putnam, & Larson, 2000). Recent work at this level, discussed in the final part of the text, also makes it clear that the biological and the psychological aspects of personality are in a continuous reciprocal influence process, each affecting the other. Consequently, contributions from the different levels of analysis again continuously enrich each other.

Integration of Levels: The Person as a Whole

In sum, the different levels all add their distinctive insights to understanding the total person "as a whole" and they inform each other and interact. In this text we will focus first on each level separately so that each can be studied in depth, and then consider their interconnections and integration in the final Part's three chapters. When taken together, the work done at these different levels addresses every conceivable cause for any behavior or mental event central for personality. At times, to be sure, work at the different levels can also produce critical findings that contradict each other and generate real conflicts, and that has happened often in the field's history, and will continue in the future as it does in every science. But those are some of the most exciting moments in science and often set the stage for dramatic progress. In fact, in recent years, personality psychologists working at different levels seem to be crossing more freely over what used to be rigid boundaries. As one reviewer of ongoing work within diverse research orientations put it:

1.14 What kinds of theoretical integrations are occurring as personality psychologists seek a unified understanding of the person?

Their research programs frequently inform one another. The complementary findings are beginning to portray a coherent (albeit incomplete) picture of personality structure and functioning. Personality psychologists have found common ground (Cervone, 1991, p. 371).

A more comprehensive view of the person is emerging that seeks to incorporate many of the insights and findings from each of the diverse levels within one broader, unifying framework (e.g., Baumeister & Tice, 1996; Carver, 1996; Cervone & Mischel, 2002; Mischel & Shoda, 1998). If this trend continues, it promises to be an exciting time for the field. It indicates that personality psychology is becoming a more cumulative science in which knowledge and insights add to each other, allowing each generation of researchers to revise earlier conclusions and to build progressively on each other's work. If so, major contributions provided by each stream of work will ultimately become more integrated, retaining those elements that stand the test of time and research as the science matures.

There also are indications that boundaries are being crossed productively between personality psychology and related fields, both at more molar, social-cultural levels of analysis (e.g., Nisbett, 1997) and at more molecular levels, particularly in cognitive neuroscience and in behavioral genetics (Plomin, DeFries, McClearn, & Rutter, 1997; Rothbart, Posner, & Gerardi, 1997). It has long been the hope of personality psychology that it could some day provide an integrated view of the person (e.g., Allport, 1937) that at least begins to capture the complexity and depth of its subject matter: optimists in the field are beginning to think that day might not be too far off (Cervone & Mischel, 2002, Chapter 1).

Practical Applications: Coping and Personal Adaptation

1.15 In what sense is personality an applied science?

To speak to why most people really want to study personality, we also will look at how the discoveries made already can allow a better understanding of oneself as a person and as at least a partial architect of one's own future. Personality theories are often *applied* to help improve the psychological qualities of one's lives. Even people whose problems are not severe enough to seek help from professionals still search for ways to live their lives more fully and satisfyingly. But what constitutes a fuller, more satisfying life? Given the diversity and complexity of human strengths and problems, it seems evident that simple notions of psychological adequacy in terms of "good adjustment" or "sound personality" are naive. More adequate definitions of "adaptation" and "abnormality," of "mental health" and "deviance," hinge on the personality theory that is used as a guide. The work discussed through the text offers distinctive notions about the nature of psychological adequacy and deviance. On closer examination, it will be seen that even these conceptions from different levels of analysis in fact have clear common themes. But each also adds to the strategies that can be chosen to try to change troublesome behaviors and to encourage better alternatives.

Many personality psychologists are searching for useful techniques to deal with the implications of personality for human problems, such as depression, anxiety, and poor health, and to foster more advantageous patterns of coping and growth. In addition to having enormous practical and social importance, attempts to understand and change behaviors provide one of the sharpest testing grounds for ideas about personality. These efforts include different forms of psychotherapy, drugs, and physical treatments, various special learning programs, and changes in the psychological environment to permit people to develop to their full potential. Research on these topics informs us about the usefulness and implications of different ideas about personality change in normal well-functioning people as well as in those who are distressed. The concepts, methods, and findings relevant to personality assessment, change, and growth will be discussed at many points as they apply to each of the major approaches and the levels of analysis that they guide.

SUMMARY

WHAT IS PERSONALITY?

- The term "personality" implies stable and coherent individual differences that can be described or predicted.
- In personality psychology, "personality" refers to the person's unique patterns of coping with and transforming the psychological environment.
- Personality psychologists study how personality dispositions and psychological and biological-genetic processes influence people's distinctive patterns of behavior.

THEORY AND LEVELS OF ANALYSIS

- In the first half of the last century, grand theories of personality (e.g., those of Freud) developed, introducing many lines of research and therapeutic practices.
- Work in personality psychology can now be grouped into six different major levels of analysis.

LEVELS OF ANALYSIS

- These six levels provide an overview of the many complex and diverse aspects of human personality.
- The trait-dispositional level tries to identify consistencies in the basic expressions of personality, conceptualized as stable personality characteristics.
- The psychodynamic-motivational level probes the motivations, conflicts, and defenses— often unconscious—that may underlie diverse aspects of personality.
- The phenomenological level focuses on the inner experiences of the person and his or her way of seeing and interpreting the world.
- The behavioral-conditioning level analyzes specific patterns of behavior that characterize individuals and identifies the conditions that regulate their occurrence.
- The social cognitive level focuses on the distinctive patterns of thoughts, expectations-beliefs, goals-values, emotional reactions, and self-regulatory efforts that characterize the person.
- The biological level explores the biological bases of personality, including the role of heredity, the brain, and evolution.
- The example of Charles Whitman shows how each level of analysis contributes to a fuller understanding of individual personality and behavior.

TOWARD AN INTEGRATIVE SCIENCE

- Work at each level provides basic concepts and strategies for seeking information about people and for constructively changing maladaptive behavior patterns.
- An increasingly comprehensive view of the person seems to be emerging that incorporates many of the insights and findings from each level of analysis.
- Boundaries are also being crossed between personality psychology and other related fields.

PRACTICAL APPLICATIONS

- The findings of personality psychologists address diverse human problems, such as depression, anxiety, impulse control, and poor health.

KEY TERMS

behavioral-conditioning
    level   14
biological level   14
levels of analysis   6

phenomenological
    level   14
psychodynamic-
    motivational level   13

social cognitive level   14
trait-dispositional level   13

2

DATA, METHODS, AND TOOLS

▶ WHY A SCIENCE OF PERSONALITY? BEYOND HINDSIGHT UNDERSTANDING

Much of our lives are spent trying to understand, after the fact, our own and others' behavior. In the words of the Danish philosopher Sören Kierkegaard, "Life is lived forwards, but understood backwards." And, as the saying goes, hindsight is 20/20. But that is also the problem with hindsight understanding. Past events can be explained in many ways, and there is no sure way to determine which, if any, of the alternative explanations is the right one.

Akira Kurosawa's classic film *Rashomon* tells the tale of the rape of a woman and the murder of a man, seen entirely in flashbacks from the perspectives of four narrators. Each story makes perfect sense when taken by itself, but together, they do not add up and the contradictions become evident. The characters are the same in all four versions of the story, as are many of the details. But much is different, as well. The film

2.1 Describe and compare two basic approaches to understanding behavior and its causes. What are the advantages of a scientific approach to understanding?

never tells what really happened, leaving the viewer with a feeling of ambiguity and an appreciation of how what we see depends on who we are. In real life, we don't have the moviegoer's advantage of viewing the same event through different people's eyes. Rather, like the characters in the film, we see only through our own eyes, and are convinced we see the truth, without realizing how different other accounts of the same event can be when seen through different eyes.

It is not that after-the-fact understanding is useless. Often, there may be no alternative and hindsight can help one at least try to make sense of events that otherwise cannot be understood. But that is very different from the essence of scientific inquiry. A science begins by creating a language to describe phenomena in a way that allows a single common understanding to emerge, in order to avoid multiple alternative accounts that vary with each observer. The goal is not just to find a common understanding, but to be able to use it to make accurate predictions and to test if they really are accurate, remaining ready to disconfirm and modify them if they are unsupported. Therefore, researchers try to arrange conditions under which they can test hypotheses about the various causal factors that might influence the occurrence of the behavior or event of interest. If one understands the causes of a given behavior, then it may be possible to predict when it will occur again, and when it will not, with reasonable accuracy. In time, this knowledge also makes it possible to build a theory to understand what is the process or mechanism that underlies the behavior being studied. That is why, at least in the view of most personality researchers, we need a science of personality. And that in turn requires attention to the methods and tools needed to pursue that goal.

Personality psychologists are committed to studying persons by means of scientific methods, but they are equally eager to avoid oversimplifying their subject matter and reducing the complexities of personality into a stereotypically "scientific" collection of formulae and variables. They recognize that "the most distinguishing feature of persons is that they construct meaning by reflecting on themselves, their past, and the future" (Cervone & Mischel, 2002, p. 5).

In that spirit, it makes sense to begin the study of persons by asking individuals how they see and understand themselves, and what they are like in their own eyes—while also recognizing that this personal "internal" vantage point will necessarily be limited and incomplete. So, as a first step in this text, we begin by considering a case study of Gary W. As you proceed through the text, you will learn about Gary's personality and the information made available about him, which is based on his clinical files but was modified sufficiently in order to protect confidentiality and to illustrate points in the text. Because researchers pay much attention to case material at some levels but not at others, information about Gary is more complete in some parts of the text and more limited in others.

In Focus 2.1 presents some of what Gary said when "asked to describe yourself as a person"—an exercise that you the reader may also want to try on yourself. He was asked to describe himself on his first visit to the university campus clinic in the psychology department of a large university.

The impressions Gary gives us about himself, while interesting, are of uncertain value: We know neither their accuracy nor their meaning. Then how can we find out more? To convert personality speculations about people into ideas that can be studied scientifically, we must be able to put them into *testable* terms using good measures. There is a wide range of these available to personality researchers, as considered next.

GARY W., THE TEXT'S CASE — GARY'S SELF-DESCRIPTION

I'm twenty-five years old, and a college graduate. I'm in business school working toward an MBA.

. . . I'm an introspective sort of person — not very outgoing. Not particularly good in social situations. Though I'm not a good leader and I wouldn't be a good politician, I'm shrewd enough that I'll be a good businessman. Right now I'm being considered for an important job that means a lot to me and I'm sweating it. I know the powers at the office have their doubts about me but I'm sure I could make it — I'm positive. I can think ahead and no one will take advantage of me. I know how to work toward a goal and stick with whatever I start to the end — bitter or not.

The only thing that really gets me is speaking in a large group. Talking in front of a lot of people. I don't know what it is, but sometimes I get so nervous and confused I literally can't talk! I feel my heart is going to thump itself to death. I guess I'm afraid that they're all criticizing me. Like they're almost hoping I'll get caught with my pants down. Maybe I shouldn't care so much what other people think of me — but it does get to me, and it hurts — and I wind up sweating buckets and with my foot in my mouth.

I'm pretty good with women, but I've never found one that I want to spend the rest of my life with. Meanwhile I'm enjoying life. I hope someday to find a woman who is both attractive and level-headed. Someone who is warm and good but not dominating and who'll be faithful but still lead a life of her own. Not depend on me for every little thing.

My childhood was fairly typical middle-class, uptight. I have an older brother. We used to fight with each other a lot, you know, the way kids do. Now we're not so competitive. We've grown up and made peace with each other — maybe it's just an armistice, but I think it may be a real peace — if peace ever really exists. I guess it was his accident that was the turning point. He got pretty smashed up in a car crash and I guess I thought, "There but for the grace of God. . . ." I count a lot on being physically up to par.

Dad wasn't around much when we were growing up. He was having business troubles and worried a lot. He and Mother seemed to get along in a low-key sort of way. But I guess there must have been some friction because they're splitting now — getting divorced. I guess it doesn't matter now — I mean my brother and I have been on our own for some time. Still, I feel sorry for my Dad — his life looks like a waste and he is a wreck.

My strengths are my persistence and my stamina and guts — you need them in this world. Shrewdness. My weaknesses are my feeling that when it comes to the crunch you can't really trust anybody or anything. You never know who's going to put you down or what accident of fate lies around the corner. You try and try — and in the end it's probably all in the cards. Well, I guess that's about it. I mean, is there anything else you want to know?

▶ THE RANGE OF PERSONALITY-RELEVANT MEASURES

Imagine going to a hospital for a comprehensive checkup. You may see a radiologist who would take X-rays and perform CAT scans. You may see an endocrinologist who analyzes the levels of various hormones in your blood. You may see a neurologist who performs a variety of diagnostic tests to study the functioning of your nervous system and taps your reflexes. You may see an immunologist who examines your immune system, and you may see a cardiologist who specializes in determining how well your heart is working. All these experts will offer a snapshot of a given aspect of your body. Yet they all reflect one thing, your body, which is a *system* of many, many interdependent components. The different levels of analyses of personality psychology work in a similar way. Each will provide data about one aspect of your total functioning personality, and an ultimate challenge is putting them all together for an understanding of how the personality as a system operates.

Psychologists approaching personality at different levels of analysis obtain information about people from many sources and through diverse strategies. The result is a collection of diverse observations about many aspects of persons. A central goal of

2.2 How do levels of analysis contribute to the diversity of personality measures available to personality psychologists?

personality psychology is to figure out how these diverse aspects about an individual relate to each other and help us to understand what is going on in the individual as a whole. But before we get there (which we will, in Chapter 16), we need to first see just what kinds of data are available; and the range is huge. This chapter is an introduction to the sources and types of data that are made available from each level, and to the concepts and tools used to make sense of those data.

Interviews

2.3 Describe some strengths and weaknesses of interviews.

A valuable source of information is the **interview**—a verbal exchange between the participant and the examiner, favored particularly by workers at the psychodynamic-motivational level and those at the phenomenological level. Some interviews are tightly structured and formal: the examiner follows a fixed, prescribed format. For example, in research to survey people's sexual activities, the interviewer might follow a standard series of questions, starting with questions about the person's earliest experiences and going on to potential problems in current relationships.

The interview is the oldest method for studying personality, and it remains the most favored for psychodynamic research and assessment (Watkins, Campbell, Nieberding, & Hallmark 1995). Its usefulness as an assessment tool depends on many considerations, including how the interview is guided and structured, and how the interviewee's responses are recorded, coded, and interpreted. Each of these steps requires attention to the same issues that apply to other methods that rely on the clinician's judgment, as was just discussed for projective techniques. In recent years, unobtrusive video and sound recording has made the interview a method that is more open to manageable scoring, coding, and data analysis. These procedures often can be made even more flexible by computerized programs. The interview therefore is being used with renewed interest in efforts to systematically improve psychodynamic assessment (Horowitz and associates, 1989; Perry & Cooper, 1989).

Interviews, while popular, tend to be expensive and time consuming to conduct, as well as to code or score, because it is not easy to have all interviews with different people conducted the same standard way so that they can be compared easily. Therefore many researchers seek short cuts by using various tests, often in the form of ratings and self-reports. These short cuts were used particularly early in the field's development as discussed in In Focus 2.2.

Tests and Self-Reports

2.4 Differentiate between performance and self-report tests.

A **test** is any standardized measure of behavior, including verbal behavior, as in self-reports. Some tests involve **performance measures.** For example, researchers interested in seeing how personality measures in childhood predict academic performance in later life might use measures like the SAT as an outcome assessment. Likewise, those interested in anxiety might use a measure in which the ability to repeat long strings of numbers under difficult, stressful conditions is measured. On this measure, the examiner verbalizes long lists of numbers and the respondent has to repeat them backwards, knowing that a poor score will be taken as an index of low intelligence.

Some tests are in the form of **self-reports**—a term that refers to any statements people make about themselves. Respondents are asked to react to sets of questions or items with one of a limited number of prescribed choices (e.g., "yes," "no," "strongly

IN FOCUS 2.2

EARLY PERSONALITY MEASUREMENT

Interest in self-description or self-report as a method of personality assessment was stimulated by an inventory devised during World War I (Watson, 1959). This was Woodworth's *Personal Data Sheet*, later known as the *Psychoneurotic Inventory*. It was aimed at detecting soldiers who would be likely to break down under wartime stress. Because it was impractical to give individual psychiatric interviews to recruits, Woodworth listed the kinds of symptoms psychiatrists would probably ask about in interviews. He then condensed them into a paper-and-pencil questionnaire of more than 100 items. Examples are: "Do you wet your bed at night?" and "Do you daydream frequently?" The respondent must answer "yes" or "no" to each question. Soldiers who gave many affirmative responses were followed up with individual interviews. This

method was valuable as a simplified and economic alternative to interviewing everyone individually. Often questionnaires are still employed as substitutes for interviews.

The Woodworth questionnaire was not used widely, but it was a forerunner of the many other self-report devices that flourished in the 1920s and 1930s, and new versions of similar measures are still used currently. These self-reports compared people usually with respect to a single summary score. This total score served as an index, for example, of their "overall level of adjustment," just as single scores or mental quotients were developed to describe the level of "general intelligence." In addition to efforts to assess adjustment, attempts to measure individuals on various personality dimensions soon became extremely popular.

agree," "frequently," "don't know"), not unlike multiple choices on tests in academic courses. Examples are shown in Figures 2.1 and 2.2.

Self-reports offer quick ways of getting information the person is willing and able to reveal. For example, Table 2.1 shows a self-report measure of anxiety with multiple items. The responses are scored and added to estimate the person's overall self-reported level of anxiety.

Projective Measures

Projective tests were developed more than 60 years ago and continue to be popular in clinical use. With these methods, assessors present the person with ambiguous stimuli and ask them ambiguous questions that have no right or wrong answers. For example, they ask "What could this remind you of?" [while showing an inkblot], or ask you to tell an imaginative story in response to a highly ambiguous scene shown on a card. Measures like these are of theoretical importance to much of the work at the psychodynamic-motivational level and therefore are described more fully in Part II, especially in Chapter 7.

2.5 What are projective tests? With what level of analysis are they most often associated?

Naturalistic Observation and Behavior Sampling

Just as astronomers cannot manipulate the actions of heavenly bodies, psychologists often cannot—or should not—manipulate certain aspects of human behavior. For example, one could not or would not create home environments in which children become delinquent or marital conflicts are provoked. Although such phenomena cannot be manipulated, often they can be observed closely and systematically. Ethical considerations often prevent psychologists from trying to create

2.6 Describe and compare naturalistic observation and remote behavior sampling methods.

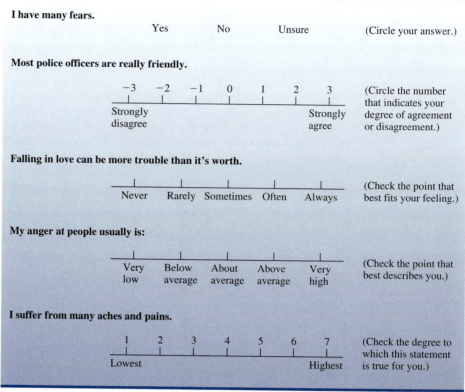

Figure 2.1 Examples of different types of structured test items.

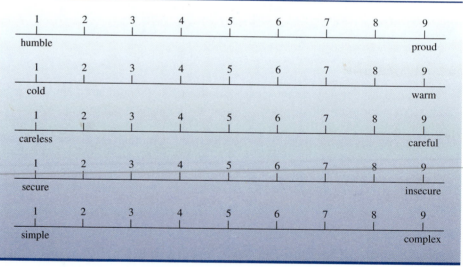

Figure 2.2 Examples of adjective scales. Rate yourself for the degree to which these terms describe you. (Check the point that describes you best on each scale.)

Source: Adapted as illustrations, selected from 40 adjective items used by McCrae & Costa to identify personality dimensions from ratings and questionnaires. McCrae, R. R., & Costa, P. T., Jr. Updating Norman's "adequacy taxonomy": Intelligence and personality in dimensions in natural language and in questionnaires. *Journal of Personality and Social Psychology, 49,* 710–721. © 1985 by the American Psychological Association. Reprinted with permission.

TABLE 2.1   Items Similar to Those on Anxiety Scales

Item	Responses Indicative of Anxiety
I cannot keep my mind focused on anything.	True
I am a worrier.	True
I am often afraid.	True
I feel safe most of the time.	False
I don't fret a lot.	False
I sleep well before exams.	False
Often I think I am nervous.	True

Note: The participant must respond "true" or "false."

powerful, lifelike experimental treatments in the laboratories (see Consent Form, Figure 2.5).

Even when some variables can be manipulated in experiments, the investigator often prefers to observe behavior as it naturally occurs, without any scientific interference. Some of the most informative work using this method, called *naturalistic observation*, comes from students of animal behavior, who unobtrusively observe the moment-by-moment lives of such animals as chimpanzees in their natural environment. Such methods have been adapted to study families interacting in their own homes (Patterson, 1990).

Gerald Patterson and coworkers at the Oregon Social Learning Center developed a behavioral coding system having 29 categories with very specific definitions. Both parents' and children's behavior could be coded, including specific types of aversive behaviors (e.g., yell, negativism, hit, whine, refused to comply). In one large project, trained observers came to the families' homes at dinnertime, when problem

Concealed video cameras and one-way mirrors allow unobtrusive observation by researchers who remain unseen by participants.

interactions most often occur. Every interaction of the child with another family member was coded, so that it was possible to study the entire sequence of interactions. The data indicated that in distressed families, the problem children's aversive behaviors continued in "chains" over longer periods of time, with an escalating pattern of hostile interchanges with family members. When the parents in the problem families reacted with punishment, it tended to prolong the escalation of aggression as the child reacted with defiance or resumed aversive behaviors shortly afterwards. These behaviors translated into poor social skills, noncompliance at school, poor school achievement, rejection by peers, and, in many cases, antisocial behavior as an adolescent (Patterson & Fisher, 2002).

In a somewhat similar fashion, but usually on a smaller scale, unseen observers may study children from behind a one-way mirror in such settings as a playroom or a preschool class (Mischel, Shoda, & Rodriguez, 1989). Of course, observation is a commonplace method in everyday life; through observation we form impressions and learn about events and people. The distinguishing feature of observation as a scientific tool is that it is conducted as precisely, objectively, and systematically as possible.

In clinical applications, direct observation may give both client and assessor an opportunity to assess life problems and to select treatment objectives. Direct observation of behavior samples also may be used to assess the relative efficacy of various treatment procedures.

Remote Behavior Sampling: Daily Life Experiences

It is not practical or possible for behavioral assessors to follow people around from situation to situation on a daily basis. In addition, assessors are frequently interested in unobservable events, such as emotional reactions and thinking patterns, that may shed considerable light on personality functioning. Through remote behavior sampling, researchers and clinicians can collect samples of behavior from respondents as they live their daily lives. A tiny computerized device carried by respondents pages them at randomly determined times of the day. When the "beeper" sounds, respondents record their current thoughts, feelings, or behaviors, depending on what the researcher or therapist is assessing (Csikszentmihalyi, 1990; Singer, 1988; Stone, Shiffman, & DeVries, 2000). Respondents may also report on the kind of situation they are in so that situation–behavior interactions can be examined. The data can either be stored in the computer or transmitted directly to the assessor.

Remote sampling procedures can be used over weeks or even months to collect a large behavior sample across many situations. This approach to personality assessment has great promise, for it enables researchers and clinicians to detect patterns of personal functioning that might not be revealed by other methods (Stone et al., 2000).

2.7 Describe methods that are used in remote behavior sampling research.

In recent years, many personality researchers have moved outside the lab to study people's daily experiences by obtaining the person's self-reported reactions to daily experiences that cannot be observed directly (Tennen, Suls, & Affleck, 1991). For example, researchers use daily mood measures on which participants indicate the degree to which they experienced various emotions (such as enjoyment/fun, pleased; depressed/blue) in each reporting period (Larsen & Kasimatis, 1991). Such reports can be linked to other aspects of experience, such as minor illnesses and psychological well-being (e.g., Emmons, 1991). Likewise, daily reports of everyday reactions to various stressors and hassles, such as interpersonal conflicts at home, can be related to other

TABLE 2.2 Illustrative Methods for Sampling Daily Life Experiences

Method	Examples	Source
Preprogrammed time samples	Digital watch alarm signals time for respondents to record their tasks, behavior, and perceptions at the moment	Cantor, Norem, Langston, Zirkel, Fleeson, & Cook-Flannagan, 1991
Systematic diaries	Self-reports of reactions to daily stressors (e.g., overload at work, family demands, arguments)	Bolger & Schilling, 1991
Sampling emotions, symptoms, and other internal states	Self-ratings of emotional states (e.g., pessimistic, optimistic, full – hungry), occurrence and duration of symptoms (e.g., backache, headache), reported personal strivings and well-being	Larsen & Kasimatis, 1991; Emmons, 1991; Diener et al., 1995; David et al., 1997

measures of personality (Bolger & Schilling, 1991; David, Green, Martin, & Suls, 1997). Experience samples also are used to study reactions to common life problems such as adjusting to college life in terms of such personal tasks as getting good grades and making friends (e.g., Cantor and associates, 1991). Examples of different behavior sampling methods are shown in Table 2.2.

Physiological Functioning and Brain Imaging

Personality researchers have long searched for practical methods to assess emotional reactions. One of the classic measures of physiological functioning is the **polygraph,** an apparatus that records the activities of the autonomic nervous system. Measures of bodily changes in response to stimulation also provide important information, especially when the stimuli are stressful or arousing. The polygraph apparatus contains a series of devices that translate indices of body changes into a visual record by deflecting a pen across a moving paper chart. A popular component of polygraphic measurement is the **electrocardiogram (EKG).** As the heart beats, its muscular contractions produce patterns of electrical activity that may be detected by electrodes placed near the heart on the body surface. Another component is the changes in blood volume that may be recorded by means of a **plethysmograph.** Other useful measures include changes in the electrical activity of the skin due to sweating (recorded by a galvanometer and called the **galvanic skin response** or **GSR**), changes in blood pressure, and changes in muscular activity (Cacioppo, Berntson, & Crites, 1996; Geen, 1997).

Intense emotional arousal is generally accompanied by high levels of "activation" in the brain (Malmo, 1959; Birbaumer & Ohman, 1993). The degree of activation in the cerebral cortex may be inferred from "brain waves" recorded by the **electroencephalograph (EEG),** as illustrated in the records shown in Figure 2.3. As the EEG patterns in this figure indicate, the frequency, amplitude, and other characteristics of brain waves vary according to the participant's arousal state, from deep sleep to great excitement.

The biological level of analysis has achieved increasing influence in the study of personality, thanks to technical advances in the measurement of physiological reactions, genetic functions, and brain processes, as you will see in later chapters. For example, new brain imaging procedures make it possible to examine relations between neural

2.8 What types of physiological measures are used to measure biological aspects of personality?

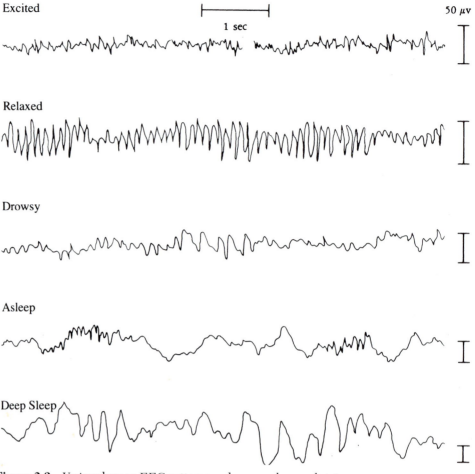

Figure 2.3 Various human EEG patterns under several arousal states.

Source: Jasper, H. (1941). In Penfield & Erickson (Eds.), *Epilepsy and cerebral localization.* Courtesy of Charles C. Thomas, Publisher, Springfield, IL and of Wilder Penfield Archive.

functions and behaviors. **Positron emission tomography (PET) scans** measure the amount of glucose (the brain's main fuel) being used in various parts of the brain and provide an index of activity as the brain performs a particular function.

Functional magnetic resonance imaging or fMRI measures the magnetic fields created by the functioning nerve cells in the brain and with the aid of computers depicts these activities as images. These pictures virtually "light up" the amount of activity in different areas as the person performs mental tasks and experiences different kinds of perceptions, images, thoughts, and emotions. They thus allow a much more precise and detailed analysis of the links between activity in the brain and the mental states we experience while responding to different types of stimuli and generating different thoughts and emotions. These can range, for example, from thoughts and images about what we fear and dread to those directed at that we crave the most. The result is a virtual revolution for work that uses the biological level of neural activity to address questions of core interest for personality psychology.

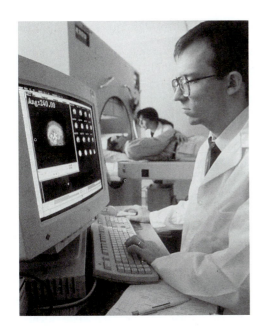

Advances in brain imaging offer a new way to study mental activity.

▶ CONCEPTUAL AND METHODOLOGICAL TOOLS

Constructs and Operational Definitions

To conduct scientific research (or even to carry on intelligent discourse), it is necessary to clearly identify and specify the phenomena that one wants to understand. The vocabulary of psychology is filled with terms like aggression, extraversion, intelligence, stress, learning, and motivation. All of these are simply words or concepts—scientists prefer the term **constructs**—that refer to classes of behaviors, thoughts, emotions, and situations. Every personality term that we will discuss in this book is a construct (including the term *personality*). These words represent nonmaterial ideas—concepts and not things—and they may have different meanings for different people. For example, the term *dependency* refers to a particular class of behaviors, but the specific types of behavior that are labeled "dependent" may differ from one person to another. Unless two people have a common definition of what dependent means, they can't be sure they're communicating effectively when they talk about "dependent" people. "What do you mean by that?" is a question psychologists must answer very precisely if they are to study a psychological phenomenon.

Operationalization translates these constructs into something observable and measurable. It refers to the specific procedures used to produce or measure it in a particular study. Sometimes, a construct is operationalized in terms of a condition to which someone is exposed. For example, the construct hunger could be operationally defined as "the number of hours that a person is deprived of food." At other times, a construct is operationalized in terms of some behavior of the participant. Thus, we could also define hunger in terms of people's ratings of how hungry they feel or in terms of how much effort they will make to obtain food. Regardless of the method used to operationalize a construct, it cannot be studied scientifically unless it can be tied to something observable.

2.9 What are psychological constructs? Why do they require operationalization?

An Example: Defining the Construct of Aggression

To illustrate these points, let us consider the construct of aggression. If we ask the question, "What is aggression?" we are trying to specify the meaning of the construct itself. That's no easy matter, for the term is used in many different ways in our daily discourse, and with different connotations. Thus, an athlete may be praised for being an "aggressive competitor," whereas an "aggressive schoolyard bully" may evoke disapproval.

2.10   How have psychologists defined aggression at a conceptual level?

Over the years, researchers have offered various construct definitions. One early one was "the delivery of noxious stimuli to another organism" (Buss, 1961). Other psychologists agreed that when we aggress, we deliver aversive stimulation of some kind to the target of the aggression, but they found the definition too broad. For example, if a person accidentally bumps into you, it may be aversive, but is it aggression? If a doctor gives a child a painful vaccination, she's delivering noxious stimulation, but is it an act of aggression? Over time, these concerns were addressed with more elaborate definitions. Today, a definition adopted by most psychologists who study aggression is the following: **aggression** is "any form of behavior directed toward the goal of harming or injuring another living being who is motivated to avoid such treatment" (Baron & Richardson, 1994, p. 7). Notice that this definition has a number of facets:

1. Aggression is behavior, not an emotion. The behaviors may take many forms, including physical and verbal acts. It may even involve not doing something, as when a person deprives another of something needed and thus harms that individual.

2. Aggression is motivated behavior, with the intent being to hurt another. Thus, unintentional acts that harm another would not be considered aggression.

3. The target of the behavior is a living being, not an inanimate object. However, if the purposeful destruction of an inanimate object, such as a prized vase, harms another person, the act would be considered aggression.

4. The recipient must be motivated to avoid the treatment. Thus, an assisted suicide would not be classified as an aggressive act if the deceased person wanted to die, even though it may legally be classified as manslaughter.

Given a working definition of the construct, we must find a way to operationally define it in terms of observable events if we are to study it scientifically. A variety of operational definitions, some quite ingenious, have been used to measure individual differences in aggression. Here are a few of the ways that aggression has been operationally defined and measured in various studies:

2.11   What kinds of measures have been used to provide operational definitions of aggression?

- Archival records, such as school suspensions, rates of violent behavior, or arrests for such acts.

- Verbal reports of aggressive behavior, obtained in interviews and on questionnaires.

- Scores on personality measures designed to measure aggressiveness.

- Ratings and reports from others on a person's aggressive behavior.

- Observations of aggressive behavior in natural and laboratory settings.

- Unnecessary honking of one's automobile horn at other motorists.

- Administration of electric shock or aversive noise to another person in a laboratory setting.

- Written or spoken insults during laboratory interactions with another person.
- Negative evaluations of another person who has provoked study participants.

▶ ESTABLISHING RELATIONSHIPS AMONG OBSERVATIONS

Once observations are made, the next step is to determine the relationships among them. First, how objective is the measure: do the measured results fluctuate across different occasions? Do multiple observers agree with each other? Are self-reports and observers' ratings consistent with direct observations of relevant behaviors? Second, how does the measure relate to measures of other constructs: is social anxiety related to loneliness? Do extraverts have happier lives than introverts? What personality factors distinguish happily married couples from those headed for divorce? Do firstborn versus later-born children differ in personality? Is the amount of television violence a child is exposed to related to antisocial aggression later in life? These and countless other psychological questions ask about *associations* between naturally occurring events or variables.

Correlation: What Goes with What?

Data that psychologists who study personality collect, regardless of their source, are conceptualized as variables. A **variable** is an attribute, quality, or characteristic that can be given two or more values. For example, a person's height is a variable, and a psychological characteristic such as attitude toward premarital sex can be quantified using a seven-point scale in which 0 is neutral, $+3$ is extremely positive, and -3 is extremely negative (Figure 2.1). Often, two or more variables seem to be associated—seem to "go together"—in such a way that when we know something about one variable, we can usually make a good guess about the other variables. For example, people who are taller generally tend to weigh more; when we know how tall someone is, we can roughly predict the person's weight. This "going together," "co-relationship" or joint relationship between variables, is what psychologists mean by the term **correlation.**

The degree of relationship or correlation may be expressed quantitatively by a number called a **correlation coefficient.** Correlation coefficients can range from -1.00 through .00 to $+1.00$. A coefficient of $+1.00$ means that there is a perfect positive relation between X and Y—that is, the person having the highest score on X also has the highest score on Y, the person having the second highest score on X has the second highest score on Y, and so on. A correlation of -1.00 signifies a perfect negative relation so that the higher a person's score is on X, the lower the score on Y. A correlation of .00 means that there is no relation at all between X and Y. The correlation coefficient thus indicates both the direction (positive or negative) and the strength of the statistical relation between the two measures.

Suppose we want to know how scores on a psychological test of anxiety are related to students' performance. We can obtain a score on each of the two measures for each student in our sample and graph these data in a scatter plot like those shown in Figure 2.4. Each point in the scatter plot represents the intersection of an individual participant's scores on the two variables of interest, which we'll call variable X (anxiety) and variable Y (average performance level).

The scatter plots in Figure 2.4 illustrate three kinds of correlational results: positive, in which high scores on anxiety are related to high performance averages; negative, in which high scores on anxiety are related to low performance averages; and a zero correlation, in which there is no relation between the two variables and the data points are

2.12 What is meant by correlation?

2.13 What is a correlation coefficient? In what two ways do correlation coefficients differ?

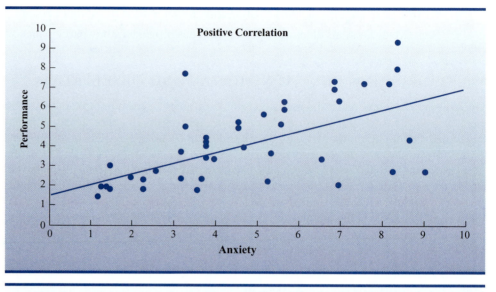

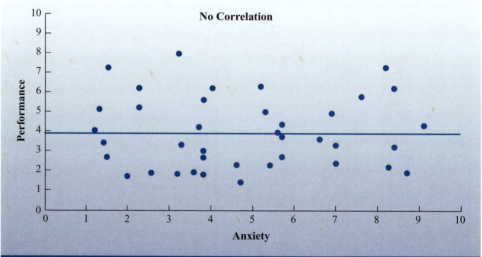

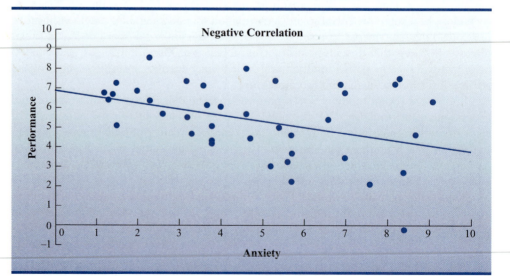

Figure 2.4 Examples of positive correlation, no correlation, and negative correlation between anxiety levels and performance levels.

scattered about in a random pattern. Correlations can be any value between $+1.00$ and -1.00. As mentioned above, a correlation of 0 (i.e., no relationship whatsoever) indicates the weakest relationship. The closer to $+1.00$ or -1.00 the correlation, the more strongly the two variables are related. Thus, a correlation of $-.59$ indicates a stronger association between X and Y than does a correlation of $+.37$. In case you're wondering about the previous examples, moderate negative correlations (typically around $-.30$) exist between anxiety and student performance. As a basis for comparison, a correlation of about $+.50$ exists between height and weight in the general population.

Interpreting Correlations

One useful feature of the correlation coefficient is that by squaring the coefficient, we arrive at an estimate of what percentage of the variation in one of the variables is accounted for by differences in the other. For example, squaring the correlation of $-.30$ between anxiety and grade point average tells us that 9 percent of the total group variation in grades can be accounted for (linked to) differences in anxiety ($-.30 \times -.30 = .09$). This means that there can be 10 other variables, each accounting for 9 percent of the total variance in grades, that are totally unrelated to anxiety and with each other (i.e., each can have a $+.30$ or $-.30$ correlation with grades, while the correlations among them and with anxiety are 0.0). Together, these 11 variables would account for 99 percent of the variance.

2.14 If we square a correlation coefficient, what does the product tell us?

Correlations that are even close to perfect are very rare in psychology, showing that although many psychological variables are, indeed, associated with each other, the association usually is not very strong. Correlations of about .30 to .50, either positive or negative, are fairly common in psychology. Such correlations may allow predictions that significantly exceed chance guesses, but they are still far from perfect. Statistical computations are used to evaluate the **statistical significance** of particular correlation coefficients reflecting how far a given association exceeds that which would be expected by chance.

Correlations are useful, but they do not indicate cause and effect. If variable X and Y are correlated with each other, there are actually three possibilities:

2.15 What kinds of causal interpretations can be made on the basis of a statistically significant correlation coefficient?

1. Variable X causes Variable Y.
2. Variable Y actually causes Variable X.
3. A third factor (Variable Z) causes both X and Y, so that there's no direct causal relation between X and Y.

Consider the positive correlation between people's shoe sizes and height. Does it mean larger shoes cause people to grow taller (the first possibility above), or does it mean that greater height causes larger shoes (the second possibility), or does it mean that a third factor was responsible for both (e.g., genes and nutrition that influence physical growth determined height and foot size, the latter of which is in turn reflected in shoe size)? The correlation would only alert you to the many things that might make the two tend to occur together.

Although correlations among measures do not allow us to establish causality, they are useful in prediction. If two measures are highly related, either positively or negatively, knowledge of the score on one measure allows us to predict (within certain limits) the score on the other. Thus, admissions officers use college entrance examination scores to predict probable college grades and success because these measures are positively correlated. Insurance premiums are likewise established on the basis of correlations among certain risk factors and medical/accident outcomes. In a sense, your insurance company is betting that you will not demolish your car, become seriously ill, or die before you are statistically "supposed to" based on the information about you, such as your driving record and whether or not you smoke. Because insurers' predictions are based on sound correlational data, the odds are solidly in their favor.

▶ RELIABILITY AND VALIDITY OF OBSERVATIONS AND MEASURES

In order to be scientifically useful, the observations and measures we use to operationally define the constructs we wish to study must have certain characteristics. First, they must be consistent, or reliable, in a number of ways. Secondly, they must be valid indicators of the constructs we are interested in. Much scientific activity is devoted to developing reliable and valid measures, because without good measurement tools, scientific activity cannot proceed.

Reliability: Are the Measurements Consistent?

2.16 What is meant by reliability of measurement?

A number of techniques are available for estimating the consistency or "reliability" of personality measures. When the same test is given to the same group of people on two occasions, a retest correlation or "coefficient of stability" is obtained. This measure provides an index of **temporal reliability.** Generally, for measures of constructs that are expected to remain stable, temporal reliability of a measure should be high.

Other reliability estimates are more concerned with the consistency with which different parts or alternate forms of a test measure behavior. The correlation between parts of a single form is called **internal consistency.** A special type of internal consistency is measured by the correlation between scores on alternate forms of a test administered to the same set of people. For example, a test of anxiety may be given before an intervening procedure, such as psychotherapy, and an "alternate form" of the same measure, using items that are not identical to the form used earlier is given after the intervention. By using two different forms, researchers can avoid the contaminating effects of administering the same form twice.

If subjective judgment enters into scoring decisions, a special kind of reliability check is needed. This check is called **interscorer agreement** or consistency. It is the degree to which different scorers or judges arrive at the same statements about the same test data. For example, if three judges try to infer personality traits from a person's interview behavior and dream reports, it would be necessary to establish the degree to which the three assessors reach the same conclusions. As noted before, interscorer agreement is easiest to achieve when scoring is objective, as on highly structured tests (for example, when all answers are given as either "yes" or "no").

Validity: What Is Being Measured?

2.17 What is meant by validity of measurement? Distinguish between content and criterion validity.

A woman's self-report on a 10-item questionnaire provides her *stated* reactions to the items under the specific testing conditions. Thus if a person reports that she is "very friendly," that is just what she *says* about herself on the test. To know more than that, one needs validity research to establish the meaning and implications of the test answers.

Content Validity. **Content validity** is the demonstration that the items on a test adequately represent a defined broader *class* of behavior. For example, judges would have to agree that the different items on a "friendliness" questionnaire all in fact seem to deal with the class or topic of friendliness. In practice, content validity often is assumed rather than demonstrated. Even if the content validity of the items is shown acceptably, it cannot be assumed that the answers provide an index of the individual's "true" trait position. We do not know whether or not the person who says she is friendly, for example, is really friendly. A self-description is a self-description, and a description by another person is another person's description, and the relationships between such data and other events or measures have to be determined.

Criterion Validity. To go beyond description of the sampled behavior, one has to determine the relationship between it and the score on other measures that serve as referents or standards, thus providing **criterion validity.** For example, psychiatrists' ratings about progress in therapy, teachers' ratings of school performance, the person's behavior on another test, or self-reported progress on another occasion, may be selected as criteria. Criterion validity may be established by a correlation among concurrently available data (such as current test score and present psychiatric diagnosis). This is called **concurrent validity.** Criterion validity also can be predictive if it comes from correlations between a measure and data collected at a later time, for example, pretherapy diagnosis and adjustment ratings after a year of psychotherapy. The term for that is **predictive validity.** Correlations may be looked for between data that seem to have a strong surface similarity in content, such as a child's arithmetic performance on an IQ test and his future success in an arithmetic course. Or they may be sought between measures whose contents appear quite dissimilar, such as a patient's drawings and a psychiatric diagnosis.

Tests based on criterion validity may be used for various practical purposes, depending on their specific validation procedures. Obviously a test may have criterion validity with regard to criteria measured roughly at the same time and still be unable to predict future behavior. Likewise, a test may have predictive validity without concurrent validity if, for example, it can predict suicide five years before a patient kills himself but relates to no other measure at the time of administration.

Construct Validity: Elaborating the Meaning of the Construct. Personality psychologists guided by trait theory usually want to infer and describe a person's dispositions from his or her test responses. **Construct validity** is the effort to elaborate the inferred traits determining test behavior (Campbell, 1960; West & Finch, 1997). Basically, it tries to answer the question: What does this test measure? The concept of "construct validity" was introduced for problems in which:

2.18 What is construct validity? How is it established?

> . . . *no existing measure as a definitive criterion of the quality with which he is concerned. Here the traits or qualities underlying test performance are of central importance (American Psychological Association, 1966, pp. 13–14).*

Traditionally, construct validity involves the following steps. The investigator begins with a hunch about a dimension on which individuals can be compared, for example, "submissiveness." The researchers might regard submissiveness as a "tendency to yield to the will and suggestions of others" (Sarason, 1966, p. 127). To study this tendency, they devise a measure of submissiveness. They have no one definite criterion, however, but instead the validity of the measure is indicated by the degree to which it predicts behaviors hypothesized to reflect the construct it measures (e.g., submissiveness). If the prediction turns out to be accurate, our confidence in the construct validity of the measure is increased. But what if the predicted behavior was not observed? Perhaps the measure does not reflect the construct. Or, perhaps the hypothesis is wrong. To resolve this dilemma, construct validity research will turn to test other hypotheses about behaviors believed to reflect submissiveness. If many of these behaviors are successfully predicted by the measure of submissiveness, our confidence in the measure is increased, and our confidence is decreased in the first hypothesis that failed to receive support. Construct validity therefore follows an ever-evolving spiral of increased precision in both the measures and the hypotheses about the construct (West & Finch, 1997).

▶ THE EXPERIMENTAL APPROACH

2.19   What is the advantage of experimental research over correlational research?

So far, we have looked at examples of research strategies that examine the relationships between naturally existing differences among people in various personal qualities as well as behavior. But how would one know if changes in one variable would cause changes in another? To answer such questions, psychologists employ an experimental method in which the experimenter varies one or more factors and then measures how another variable has been affected. The logic behind this approach is that if two or more groups of equivalent participants are treated identically in all respects but the variable that is intentionally varied, and if the behavior of the groups differs, then that difference in behavior is likely to have been caused by the factor that was varied.

Independent and Dependent Variables. In psychological experiments, the researcher is interested in relations between conditions that are manipulated and behaviors that are measured. The condition that is controlled or manipulated by the experimenter is called the **independent variable;** the resulting behavior that is measured is called the **dependent variable.** To look at it another way, the independent variable is the cause, or the stimulus, and the dependent variable is the effect, or the response. Both the independent and dependent variables reflect underlying constructs that have to be operationally defined.

2.20   Differentiate experimental and control groups.

Experimental and Control Groups.   Suppose that we want to test the hypothesis that frustration increases aggression. Frustration and aggression are both constructs that can be applied to a variety of circumstances and behaviors. To test the hypothesis experimentally, we need to operationally define both frustration and aggression. To operationally define frustration, we do the following: Participants are given a set of 15 items with strings of 7 random letters and told that the letters can be rearranged into words. Any person who can unscramble them within 10 minutes will win a monetary prize. However, half of the participants are given letter sets that are impossible to solve, which we assume will be frustrating to the participants. This group is termed the **experimental group.** To be sure that the frustration has an effect on subsequent aggression, a second group is given a set of easily solvable scrambled words. This group that is not subjected to the frustration is the **control group.**

After the participants are exposed to the frustrating or nonfrustrating condition, they participate in a second, supposedly unrelated, experiment where they are asked to assist the experimenter in a study of the effects of punishment on learning. They are seated before a box with 10 buttons that deliver increasingly intense levels of electric shock. Whenever the learner (seated in an adjoining room) makes a mistake, the participant selects the shock level and presses the button. In reality, the learner is not being shocked, but the participant does not know that. The average intensity of shocks administered by each participant is the operational definition of that person's aggression.

2.21   Why do experimenters employ (a) random assignment to conditions and (b) double-blind designs?

To make the experimental and control groups as similar as possible in all respects other than exposure to the frustration, we assign participants to the experimental and control groups on a random basis. To guarantee random assignment, we can assign each person a number and then either draw numbers out of a hat or use a table of randomly ordered numbers devised by statisticians for this purpose. Random assignment is intended to minimize group differences on other factors (such as intelligence or social class) that might affect the dependent variable.

Double-Blind Designs. Many studies have shown that if experimenters expect to obtain certain results, they are more likely to get them (Rosenthal & Rubin, 1978). In some of these studies, experimenters showed participants a standard set of facial photographs and obtained ratings of how successful the persons in the photos had been. Some of the experimenters were told that the people in the photos had been very successful and that most past participants had rated them as being successful. Other experimenters were told just the opposite. Actually, the people in the photos had been rated by earlier participants as being neither successful nor unsuccessful. Experimenters who expected "successful" ratings tended to get more of them, and those who expected low success ratings tended to receive such ratings, even though they apparently did nothing intentional to influence their participants.

Because of potential problems involving experimenter expectancy effects, experiments are usually set up so that an experimenter collecting data from participants is unaware of the experimental condition to which the participants have been assigned. When both the participant and the experimenter are blind to the independent variable manipulation, the design is referred to as a **double-blind experiment.**

Some types of behavior can be studied only in their natural settings where little or no control is possible; others can be studied under highly controlled laboratory conditions. The decision to study behavior in a natural setting as opposed to a laboratory involves some important trade-offs. On the one hand, identifying the true causes of behavior in a real-world setting poses problems because there is no way to rule out other possible causes by controlling them. On the other hand, when people are observed in their native habitat, the researcher can be more confident that the results can be applied to other similar real-life settings. In case studies and in observational research carried out in natural settings, it is also possible to observe the full complexity of person–environment relations. This very complexity, however, can make it difficult to identify the causal factors with complete confidence.

Because of the advantages and disadvantages of the various personality research methods, many personality psychologists stress the desirability of using all of the methods, moving from the real world to the laboratory and back again. When consistent results are found in both settings, we can have increased confidence in our observations. In many instances in personality research, the movement is from observations made in natural settings to the laboratory, where greater control of variables is possible.

▶ ETHICS IN PERSONALITY RESEARCH

Some of the studies we have reviewed in this chapter have involved exposing participants to frustrating conditions or using deception. Although the large majority of personality research does not involve such manipulations, some studies do. Sometimes, in order to study important problems, personality researchers walk an ethical tightrope, balancing the importance of the knowledge to be gained and the benefits that may result from its application against the use of deception or the exposure of participants to stressful conditions. When personality data are collected from people who can potentially be identified, invasion of privacy issues also arise.

The desire to protect the privacy and welfare of human participants has resulted in a set of explicit ethical guidelines. For example, according to the research guidelines of the American Psychological Association (APA), participants cannot be placed in either

2.22 What measures are taken by personality researchers to protect the privacy and well-being of participants in their studies?

physical or psychological jeopardy without their informed consent. A typical consent form is shown in Figure 2.5. As the form indicates, participants must be told about the procedures to be followed and warned about any risks that might be involved. If deception is necessary in order to carry out the research, then participants must be completely debriefed after the experiment and the entire procedure must be explained to them. Special measures must be taken to protect the confidentiality of data, and participants must be told that they are free to withdraw from a study at any time without penalty.

When children, seriously disturbed mental patients, or others who are not able to give true consent are involved, consent must be obtained from their parents or guardians. Strict guidelines also apply to research in prisons. Inmates cannot be forced to participate in research, nor can they be penalized for refusing to do so. In research dealing with rehabilitation programs, inmates must be permitted to share in decisions

CONSENT FORM

FOR PARTICIPATION IN AN EXPERIMENT IN _____
PSYCHOLOGY IN THE LABORATORY OF _____

1. In this experiment, you will be asked to
2. The benefit we hope to achieve from this work
3. The risks involved (if any)

CONSENT AGREEMENT

I have read the above statement and am consenting to
participate in the experiment of my own volition. I
understand that I am free to discontinue my
participation at any time without suffering
disadvantage. I understand that if I am dissatisfied
with any aspect of this program at any time, I may report
grievances anonymously to _____

Signed: _____
Date: _____

Figure 2.5 A typical consent form for participation in a psychological study. Ethical standards require that participation in research comes only after volunteers understand the task and freely consent.

concerning program goals. Researchers who violate the code of research ethics face serious legal and professional consequences.

The ethical and moral issues in psychology are not simple ones. They are quite similar to those that confront medical researchers. In some instances, the only way to discover important knowledge about behavior or to develop new techniques to enhance human welfare is to deceive participants or to expose them to potentially stressful situations. To help researchers balance the potential benefits against the risks involved and to ensure that the welfare of participants is protected, academic and research institutions have created scientific panels that review every research proposal. If a proposed study is considered ethically questionable, or if the rights, welfare, and personal privacy of participants are not sufficiently protected, the methods must be modified or the research cannot be conducted.

SUMMARY

STUDYING PERSONS: SOURCES OF INFORMATION

- Psychologists want to understand personality to predict future behavior and understand present and past behavior.
- Case studies are a method used to evaluate the individual intensively. They can be conducted at each level of analysis and over many occasions. The case of "Gary W." will be used as a case example throughout the text.
- Psychologists utilize various types of personality tests or structured interviews to assess personality in a quantifiable way.
- Naturalistic observation is especially useful when aspects of behavior cannot—or should not—be manipulated.
- Direct behavior measurement samples behavior in diverse situations. It includes both verbal and nonverbal behavior, as well as physiological measurements of emotional reactions.
- Through remote behavior sampling and daily diary studies, researchers can collect samples of behavior from respondents as they live their daily lives.
- To measure changes in the autonomic nervous system, researchers often utilize data from the EKG, plethysmograph, galvanic skin responses (GSR), and EEG.
- PET scans help to examine neural activity in the brain.

CONCEPTUAL AND METHODOLOGICAL TOOLS

- Constructs are concepts that refer to classes of behaviors and situations.
- Operationalization refers to the specific procedures used to produce or measure constructs in a particular study.

ESTABLISHING AND QUANTIFYING THE RELATIONSHIPS AMONG OBSERVATIONS

- The correlational approach utilizes statistical analysis to measure whether two phenomena or *variables* are related to one another.
- The degree to which two variables are related is mathematically represented by the correlation coefficient.

RELIABILITY AND VALIDITY

- Reliability is found when the results of a measure can be repeated both within the confines of the study and with different investigators.
- Validity refers to how well an assessment device actually measures what it claims to measure. It includes content validity, construct validity, and criterion validity, both concurrent and predictive.

THE EXPERIMENTAL APPROACH

- The experimental approach tries to demonstrate causal relations by manipulating one variable—the independent variable—and measuring the effects on a second variable—the dependent variable.
- In order to minimize experimenter expectancy effects, experiments may be set up to be double-blind.

ETHICS IN PERSONALITY RESEARCH

- To help researchers take into account the privacy and welfare of their participants, the American Psychological Association (APA) has set up guidelines to ensure that participants cannot be placed in either physical or psychological jeopardy without their informed consent.

KEY TERMS

aggression   30
concurrent validity   35
construct   29
construct validity   35
content validity   34
control group   36
correlation   31
correlation coefficient   31
criterion validity   35
dependent variable   36
double-blind
  experiment   37
electrocardiogram
  (EKG)   27

electroencephalograph
  (EEG)   27
experimental group   36
functional magnetic
  resonance imaging
  (fMRI)   28
galvanic skin response
  (GSR)   27
independent
  variable   36
internal consistency   34
interscorer
  agreement   34
interview   22

operationalization   29
performance
  measures   22
plethysmograph   27
polygraph   27
positron emission
  tomography (PET)
  scans   28
predictive validity   35
self-reports   22
statistical significance   33
temporal reliability   34
test   22
variable   31

I

THE TRAIT-DISPOSITIONAL LEVEL

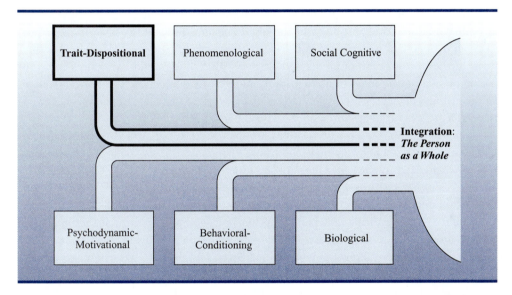

▶ PRELUDE TO PART I: THE TRAIT-DISPOSITIONAL LEVEL

"My roommate is a really closed-in person who just doesn't like to relate. Regardless of how nice I try to be to her the most I get is courtesy—and she seems like that with every-body—a loner. She's a total contrast from the fun-lover I was so lucky to have last year."

Individuals differ consistently from each other in their personality characteristics. These differences are readily perceived and with much agreement among observers, including in the self-perceptions of the people who are being characterized. Workers at this level of analysis search for these differences, and attempt to find those that are most important, consistent and perhaps even universal.

Two questions drive work at the trait-dispositional level:

• What are the basic psychological traits that characterize people?

• How can the consistent differences between people in these traits be best captured and described?

Theory and research at this level to answer these questions provides one of the most pro-ductive and enduring traditions for the study of personality. The answers have contributed greatly to the measurement, description, classification, and analysis of personality since the start of the field. Most work at this level of analysis has been driven by the assumption that the important qualities that characterize a person consist of a finite number of broad traits. Traits are defined as "dimensions of individual differences in tendencies to show consistent patterns of thoughts, feelings, and actions" (McCrae et al., 1998). It is assumed that these traits will be expressed consistently in behavior across many different situations and with much stability over time. It also is assumed that traits are quantifiable—some people are more sociable than others, and some are more modest than others, for example. Guided by these assumptions, the focus and goal has been to develop methods to quantify people's social–personal traits. The methods are used to create taxonomies for classifying and capturing the fundamental traits of personality so that people can be compared on the amounts of the different traits that they possess.

From the start of the field, work at this level has provided both concepts and methods for describing and comparing human individual differences on a vast array of attributes and qualities, and many of the highlights are discussed in Chapter 3. It provides an overview of major work in the early phases of this tradition. The chapter then turns to the current developments and advances that now provide measures for comparing people on the main traits that have emerged from a century of active research. In the last few decades of the twentieth century, work at this level of analysis has experienced a great resurgence and made significant advances.

Chapter 4 examines how dispositions are expressed in behavior. It shows that people are characterized not only by broad traits. People also display patterns of behaviors that are connected to particular types of situations in distinctive ways. For example, the phrase "Monica is warm and friendly when starting a relationship but she begins to withdraw when people get really close to her" may accurately capture that this person shows a distinctive "*if* . . . *then* . . ." pattern in her behavior: she is warm at the outset but cools as a relationship becomes close. Such patterns can also provide clues to characteristic underlying motivations and goals that may enrich the understanding of the individual's personality. This part of the text examines the various types of consistency that characterize personality, the attempts to measure them, and the theories that have been developed to try to account for them, at the trait-dispositional level of analysis.

The Personal Side of the Science

Some questions at the trait-dispositional level you might ask about yourself:

▶ What am I like as a person?

▶ How am I different from other people "on the whole"? In what general ways are people different from each other?

▶ Does what I usually do and think and feel depend mostly on myself or on the situation in which I find myself?

▶ When and how is my behavior influenced by the situation?

▶ How does my personality influence the situations I choose to be in?

▶ How does my personality influence the effects that different kinds of situations have on me?

CHAPTER

3

TYPES AND TRAITS

"My father is a really great guy. He's absolutely dependable; I can always count on him."

"Nancy's very quiet and withdrawn. She never says hello to anybody."

Descriptions like those above are examples of everyday trait psychology. We see analyses at the trait level whenever people describe and group the differences among themselves into slots or categories. We all tend to classify each other readily on many dimensions: sex, race, religion, occupation, friendliness, and competitiveness are a few examples. Good–bad, strong–weak, friend–enemy, winner–loser—the ways of sorting and classifying human qualities seem virtually infinite.

　　Most sciences classify and name things in their early efforts to find order. You see this, for example, in the classification system of biology, in which all life is sorted into genera and species. This effort to categorize also occurs in psychology, where, as the oldest and most enduring approach to individuality, it is known as the **trait approach.** Many psychologists working at this level of analysis try to label, measure, and classify people, often but not always using the trait terms of everyday language (for example, friendly, aggressive, honest) in order to describe and compare their psychological attributes and to make sense of them (John, 1990).

3.1　What are the general goals of the trait-dispositional approach to personality?

43

▶ TYPES AND TRAITS

3.2 Differentiate between traits and types.

Traditionally, analyses at this level have been guided by the assumption that behavior is primarily determined by stable generalized **traits**—basic qualities of the person that express themselves in many contexts. Many investigators have searched vigorously for these traits, trying to find the person's position on one or more trait dimensions (for example, intelligence, introversion, anxiety) by comparing the individual with others under similar uniform conditions. Guided by the belief that positions on these dimensions tend to be stable across situations and over time, the focus in the study of individuality at this level becomes the search to identify the person's basic stable and consistent traits or characteristics.

Types

Some categorizations sort individuals into discrete categories or *types* (Eysenck, 1991; Matthews, 1984). In the ancient theory of temperaments, for example, the Greek physician Hippocrates assigned persons to one of four types of temperament: *choleric* (irritable), *melancholic* (depressed), *sanguine* (optimistic), and *phlegmatic* (calm, listless). In accord with the biology of his time (about 400 B.C.), Hippocrates attributed each temperament to a predominance of one of the bodily humors: yellow bile, black bile, blood, and phlegm. A choleric temperament was caused by an excess of yellow bile; a depressive temperament reflected the predominance of black bile; the sanguine person had too much blood; and phlegmatic people suffered from an excess of phlegm.

Other typologies have searched for constitutional types, seeking associations between physique and indices of temperament. Such groupings in terms of body build have considerable popular appeal, as seen in the many stereotypes linking the body to the psyche: Fat people are "jolly" and "lazy," thin people are "morose" and "sensitive," and so on.

One of the important typologies used repeatedly by personality theorists has grouped all people into **introverts** or **extraverts.** According to this typology, the introvert withdraws into herself, especially when encountering stressful emotional conflict, prefers to be alone, tends to avoid others, and is shy. The extravert, in contrast, reacts to stress by trying to lose himself among people and social activity. He is drawn to an occupation that allows him to deal directly with many people, such as sales, and is apt to be conventional, sociable, and outgoing.

The very simplicity and breadth that makes such typologies appealing also reduces their value. Because each person's behaviors and psychological qualities are complex and variable, it is difficult to assign an individual to a single slot. Nevertheless, important typologies continue to be explored and are useful for many purposes. For example, the Type A pattern is particularly interesting because some of its ingredients, particularly chronic levels of hostility and anger, seem to have value in predicting a variety of dangerous health outcomes, most notably a proneness to premature coronary disease (see In Focus 3.1).

Traits: Individual Differences on Dimensions

3.3 What is meant by the scalability of a trait? How does this differ from a type conception?

Traits Defined.　While typologies assume discontinuous categories (like male or female), traits are continuous dimensions like "friendliness" (see Figure 3.1). On such dimensions, differences among individuals may be arranged quantitatively in terms of the degree of the quality the person has (like degrees of "conscientiousness"). Psycho-

AN EXAMPLE: TYPE A PERSONALITY

In collaborative research, psychologists and physicians have looked at the psychological variables in men at higher risk of coronary heart disease early in life. A coronary-prone behavior pattern was identified (Friedman & Roseman, 1974; Glass, 1977) and designated as *Type A*. This behavior pattern is characterized by:

1. ***Competitive Achievement Striving.*** Type As are likely to be involved in multiple activities, have numerous community and social commitments, and participate in competitive athletics. In laboratory studies, they are persistent and behave as though they believe that with sufficient effort they can overcome a variety of obstacles or frustrations.

2. ***Exaggerated Sense of Time Urgency.*** Type As show great impatience and irritation at delay (for example, in a traffic jam, on a waiting line, when someone is late for a meeting).

3. ***Aggressiveness and Hostility.*** Type As may not be generally more aggressive than other people, but they become more aggressive under circumstances which threaten their sense of task mastery, for example, when under criticism or high time pressure.

Individuals who manifest these behaviors to a great degree are called Type As. Those who show the opposite patterns of relaxation, serenity, and lack of time urgency are designated as Type B. The two types differ in many ways, including in their family environments (Woodall & Matthews, 1989).

A number of studies have suggested that Type A people may have at least twice the likelihood of coronary heart disease as Type B people. They also smoke more and have higher levels of cholesterol in their blood. Type A people also tend to describe themselves as more impulsive, self-confident, and higher in achievement and aggression. Both Type A men and women fail to report physical symptoms and fatigue (Carver *et al.*, 1976; Weidner & Matthews, 1978). This tendency to ignore symptoms may result in a Type A individual failing to rest or to seek medical care in the early phases of heart disease and may be one reason why these people push themselves into greater risk of premature death from coronary heart disease. Identifying individuals at high risk for heart disease and teaching them to pay more attention to physical symptoms may be an important part of programs aimed at reducing the toll of heart disease.

There may be a less strong relationship between the total pattern of Type A behavior and coronary disease than was suggested initially, especially among high-risk people (Matthews, 1984). Rather than looking at the relationship between the Type A pattern as a whole and coronary disease, it may be more useful to isolate such specific components of the pattern as anger and hostility. These components were found to be related to coronary disease even in the more recent studies (Miller and colleagues, 1996). In sum, it now seems that specific behaviors, rather than the more global typology, are linked to a higher risk of coronary disease.

---

logical measurements usually suggest a continuous dimension of individual differences in the degree of the measured quality: most people show intermediate amounts, and only a few are at each extreme, as Figure 3.1 shows. For example, on the introversion–extraversion typology, individuals differ in the extent to which they show either quality but usually do not belong totally to one category or the other. It is therefore better to think of a psychological continuum of individual differences for most qualities or traits.

Traits are assumed to be quantifiable and scalable:

> "By [scalability] we mean that a trait is a certain quality or attribute, and different individuals have different degrees of it. . . . If individuals differ in a trait by having higher or lower degrees of it, we can represent the trait by means of a single straight line. . . . Individual trait positions may be represented by points on the line" (Guilford, 1959, pp. 64–65). In its simplest meaning, the term "trait" refers to consistent differences between the behavior or characteristics of two or more people. Thus, a **trait** may be simply defined as ". . . any distinguishable, relatively enduring way in which one individual varies from another" (Guilford, 1959, p. 6).

3.4 Describe the characteristics of the Type A personality. What is its significance for physical health?

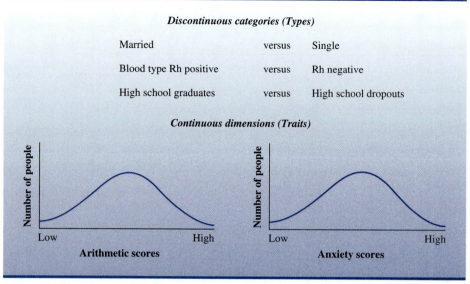

Figure 3.1 Examples of discontinuous categories (types) and continuous dimensions (traits).

The search for traits begins with the commonsense observation that individuals often differ greatly and consistently in their responses to the same psychological situation or stimulus. That is, when different people are confronted with the same event—the same social encounter, the same test question, the same frightening experience—each individual tends to react in a somewhat different way. The basic idea that no two people react identically to the same stimulus is shown schematically in Figure 3.2. Moreover, in everyday life most of us are impressed with the distinctive *consistency* of one individual's responses over a wide variety of stimulus situations: We expect an "aggressive" person to differ consistently from others in his or her responses to many stimuli.

Describing and Explaining.   In addition to using trait labels to describe individual differences, some theorists also see traits as explanations: In their view, the trait is the property within the person that accounts for his or her unique but relatively stable reactions to stimuli. Thus, the trait becomes a construct to explain behavior—a hypothesized reason for enduring individual differences. Before looking at formal trait theories,

3.5 What everyday observations stimulate the search for traits?

3.6 Describe the descriptive and explanatory uses of traits.

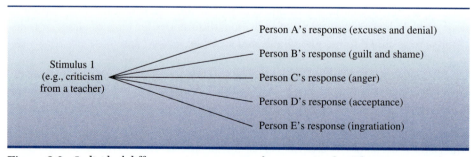

Figure 3.2 Individual differences in response to the same stimulus. The trait approach emphasizes consistent differences among people in their response to the same stimulus.

however, we should consider how traits are used informally by people in daily life. Indeed we are all trait theorists in the sense that we generate ideas about our own dispositions and the characteristics of other people.

Trait Attributions

When people describe each other in daily life, they spontaneously use trait terms. We all characterize each other (and ourselves) with such terms as aggressive, dependent, fearful, introverted, anxious, submissive—the list is almost endless. We see a person behaving in a particular way—for example, sitting at a desk for an hour yawning, and we attribute a trait, "unmotivated" or "lazy" or "bored" or "dull."

These simple trait attributions are often adequate to "explain" events for many everyday purposes in commonsense psychology (Heider, 1958; Kelley, 1973; Ross & Nisbett, 1991). In these commonsense explanations, traits are invoked not just as descriptions of what people do but also as the causes of their behavior. Thus in everyday practice, traits may be used first simply as adjectives describing behavior ("He behaves in a lazy way"), but the description is soon generalized from the behavior to the person ("He *is* lazy") and then abstracted to "He has a lazy disposition" or "He is unmotivated." These descriptions pose no problems as long as their basis is recalled—he is seen as behaving in a lazy way and no more. A hazard in trait attribution is that we easily forget that nothing is explained if the state *we* have attributed to the person from his behavior ("He has a trait of laziness") is now invoked as the *cause* of the behavior from which it was inferred. We then quickly emerge with the tautology, "He behaves in a lazy way because he has a lazy disposition," or because he is "unmotivated." The utility of trait terms therefore depends on their ability to make predictions about people's behaviors in a new situation based on their behaviors observed in the past in a different situation.

The trait approach to formal personality study begins with the commonsense conviction that personality can be described with trait terms. But it extends and refines those descriptions by arriving at them quantitatively and systematically. Efforts to explain individual differences by formal trait theories face some of the same problems that arise when traits are offered as causes by the layman. However, numerous safeguards have been developed to try to control some of these difficulties, as discussed in later sections.

Gordon Allport

One of the most outstanding trait psychologists was Gordon Allport, whose 1937 book *Personality: A Psychological Interpretation* launched the psychology of personality as a field and discipline. In this classic work and many later contributions, he made a convincing case that a distinctive field was needed, devoted to understanding the person as a coherent, consistent whole individual. His view of personality was broad and integrative, and he was sensitive and attentive to all its diverse aspects. Reacting against the tendency of researchers to study isolated part processes, such as learning and memory, in ways that failed to take account of individual differences, he wanted the field to pursue two goals. One was to understand the differences between people in personality; the other was to see how the different characteristics and processes (like learning, memory, and biological processes) that exist within an individual interact and function together in an integrated way. His vision underlies much of what is still the definition and main mission of personality psychology today, as discussed in the introduction of

3.7 In Allport's theory, what are the characteristics of traits, and how are they linked to basic psychological processes?

Gordon Allport (1897–1967)

this text. In particular, Allport's conception of traits continues to guide much of the work at the trait-dispositional level of analysis. In Allport's theory, traits have a very real existence: They are the ultimate realities of psychological organization. Allport favored a biophysical conception that

> *does not hold that every trait-name necessarily implies a trait; but rather that behind all confusion of terms, behind the disagreement of judges, and apart from errors and failures of empirical observation, there are none the less* bona fide *mental structures in each personality that account for the consistency of its behavior (1937, p. 289).*

According to Allport, traits are determining tendencies or predispositions to respond. In other words, a trait is

> *a generalized and focalized neuropsychic system (peculiar to the individual) with the capacity to render many stimuli functionally equivalent, and to initiate and guide consistent (equivalent) forms of adaptive and expressive behavior (1937, p. 295).*

Allport implied that traits are relatively general and enduring: they unite many responses to diverse stimuli, producing fairly broad consistencies in behavior. This relationship is seen in Figure 3.3.

Allport was convinced that some people have dispositions that influence most aspects of their behavior. He called these highly generalized dispositions **cardinal traits.** For example, if a person's whole life seems to be organized around goal achievement and the attainment of excellence, then achievement might be his or her cardinal trait. Less pervasive but still quite generalized dispositions are **central traits,** and

3.8 Differentiate between Allport's notions of cardinal and central traits. What is their relation to secondary dispositions?

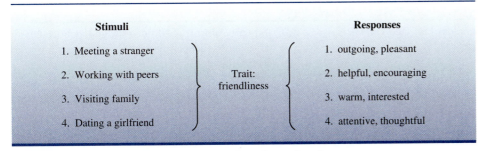

Figure 3.3   An example of a trait as the unifier of stimuli and responses.

Allport thought that many people are broadly influenced by central traits. Finally, more specific, narrow traits are called **secondary dispositions** or "attitudes."

Allport believed that one's pattern of dispositions or "personality structure" determines one's behavior. This emphasis on structure rather than environment or stimulus conditions is seen in his colorful phrase, "The same fire that melts the butter hardens the egg" (1937, p. 102). Allport was a pioneering spokesman for the importance of individual differences: No two people are completely alike, and hence no two people respond identically to the same event. Each person's behavior is determined by a particular **trait structure.**

Allport thought that traits never occur in any two people in exactly the same way: They operate in *unique* ways in each person. This conviction was consistent with his emphasis on the individuality and uniqueness of each personality. To the extent that any trait is unique within a person rather than common among many people, it cannot be studied by making comparisons among people. Consequently, Allport urged the thorough study of individuals through intensive and long-term case studies. He also believed, however, that because of shared experiences and common cultural influences, most persons tend to develop some *roughly* common kinds of traits: they can be compared on these common dispositions.

Many of Allport's theories were most relevant to the in-depth study of lives and experience rather than to the quantitative study of groups. He contributed to trait theory, but he was critical of many of the statistical methods and quantitative research strategies favored by other trait theorists. Nevertheless, his influence has been most important for the study of common "global" traits, for which he is still considered a model (Funder, 1991).

Raymond B. Cattell

Raymond B. Cattell (1950, 1965) is another important trait theorist. For Cattell, the trait is also the basic unit of study; it is a "mental structure," inferred from behavior, and a fundamental construct that accounts for behavioral regularity or consistency. Like Allport, Cattell distinguished between **common traits,** which are possessed by all people, and **unique traits,** which occur only in a particular person and cannot be found in another in exactly the same form.

Cattell also distinguished **surface traits** from **source traits** (see Table 3.1 for selected examples). Surface traits are clusters of overt or manifest trait elements (responses) that seem to go together. Source traits are the underlying variables that are the causal entities determining the surface manifestations. In research, trait elements (in the form of test responses or scores) are analyzed statistically until collections of

TABLE 3.1   Surface Traits and Source Traits Studied by Cattell

Examples of Surface Traits (Cattell, 1950)	Integrity, altruism—dishonesty, undependability Disciplined thoughtfulness—foolishness Thrift, tidiness, obstinacy—lability, curiosity, intuition
Examples of Source Traits (Cattell, 1965)	Ego strength—emotionality and neuroticism Dominance—submissiveness

Note: These are selected and abbreviated examples from much longer lists.

3.9   Describe Cattell's approach to identifying the structure of personality traits. What three types of data were used to discover source traits?

elements that correlate positively in all possible combinations are discovered. This procedure, according to Cattell, yields surface traits.

For Cattell, source traits can be found only by means of the mathematical technique of factor analysis (discussed below in this chapter). Using this technique, the investigator tries to estimate the factors or dimensions that appear to underlie surface variations in behavior. According to Cattell, the basic aim in research and assessment should be identification of source traits. In this view, these traits are divided between those that reflect environmental conditions (**environmental-mold traits**) and those that reflect constitutional factors (**constitutional traits**). Moreover, source traits may either be *general* (those affecting behavior in many different situations) or *specific*. Specific source traits are particularized sources of personality reaction that operate in one situation only, and Cattell pays little attention to them.

Cattell uses three kinds of data to discover general source traits: *life records,* in which everyday behavior situations are observed and rated; *self-ratings;* and *objective tests,* in which the person is observed in situations that are specifically designed to elicit

Raymond B. Cattell (1905–1998)

responses from which behavior in other situations can be predicted. The data from all three sources are subjected to factor analysis. In his own work, Cattell shows a preference for factor analysis of life-record data based on many behavior ratings for large samples of persons. Some 14 or 15 source traits have been reported from such investigations, but only six have been found repeatedly (Vernon, 1964).

In Cattell's system, traits may also be grouped into classes on the basis of how they are expressed. Those that are relevant to the individual's being "set into action" with respect to some goal are called **dynamic traits.** Those concerned with effectiveness in gaining the goal are **ability traits.** Traits concerned with energy or emotional reactivity are named **temperament traits.** Cattell has speculated extensively about the relationships between various traits and the development of personality (1965).

Hans J. Eysenck

The extensive researches of Hans Eysenck have complemented the work of the American trait theorists in many important ways. Eysenck (1961, 1991) has extended the search for personality dimensions to the area of abnormal behavior, studying such traits as *neuroticism–emotional stability.* He also has investigated *introversion–extraversion* as a dimensional trait (although Carl Jung originally proposed "introvert" and "extravert" as personality *types*). Eysenck and his associates have pursued an elaborate and sophisticated statistical methodology in their investigations of these personality dimensions. In addition to providing a set of descriptive dimensions, Eysenck and his colleagues have studied the associations between people's positions on these dimensions and their scores on a variety of other personality and intellectual measures, and developed an influential model of personality designed to account for the roots of these traits in ways that connect to the biological level of analysis, and will be considered in those sections of the text.

3.10 According to Eysenck, which two trait dimensions can be used to describe individual differences in personality?

Hans J. Eysenck (1916–1997)

Eysenck emphasized that his dimension of introversion–extraversion is based entirely on research and "must stand and fall by empirical confirmation" (Eysenck & Rachman, 1965, p. 19). In his words:

> *The typical extravert is sociable, likes parties, has many friends, needs to have people to talk to, and does not like reading or studying by himself. He craves excitement, takes chances, often sticks his neck out, acts on the spur of the moment, and is generally an impulsive individual. He is fond of practical jokes, always has a ready answer, and generally likes change; he is carefree, easygoing, optimistic, and 'likes to laugh and be merry.' He prefers to keep moving and doing things, tends to be aggressive and loses his temper quickly; altogether his feelings are not kept under tight control, and he is not always a reliable person.*
>
> *The typical introvert is a quiet, retiring sort of person, introspective, fond of books rather than people; he is reserved and distant except to intimate friends. He tends to plan ahead, 'looks before he leaps,' and mistrusts the impulse of the moment. He does not like excitement, takes matters of everyday life with proper seriousness, and likes a well-ordered mode of life. He keeps his feelings under close control, seldom behaves in an aggressive manner, and does not lose his temper easily. He is reliable, somewhat pessimistic and places great value on ethical standards.*

Eysenck and his colleagues recognized that these descriptions may sound almost like caricatures because they portray "perfect" extraverts and introverts, while in fact most people are mixtures who fall in the middle rather than at the extremes of the dimensions (see Figure 3.4). As Figure 3.4 shows, Eysenck suggested that the second

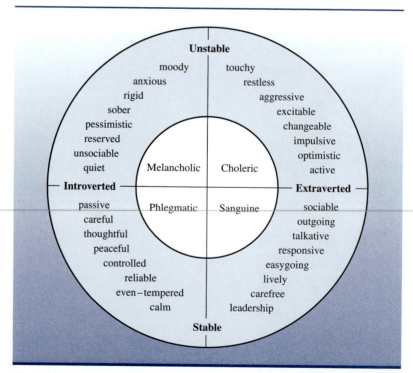

Figure 3.4 Dimensions of personality. The inner ring shows the "four temperaments" of Hippocrates; the outer ring shows the results of modern factor analytic studies of the intercorrelations between traits by Eysenck and others (Eysenck & Rachman, 1965).

TABLE 3.2 Sexual Activities Reported by Introverted (I) and Extraverted (E) Students[a]

Activity		Males		Females	
		I	E	I	E
Masturbation at present		86	72	47	39
Petting: at seventeen		16	40	15	24
at nineteen		3	56	30	47
at present		57	78	62	76
Coitus (intercourse): at seventeen		5	21	4	8
at nineteen		15	45	12	29
at present		47	77	42	71
Median frequency of coitus per month (sexually active students only)		3.0	5.5	3.1	7.5
Coitus partners in the last twelve months (unmarried students only)	One	75	46	72	60
	Two or more	18	30	25	23
	Four or more	7	25	4	17

[a] The numbers are the frequencies of endorsements by each group.

Source: Based on data from Giese, H. & Schmidt, S. (1968). *Studenten sexualitat.* Hamburg: Rowohlt; cited in Eysenck, H. J. (1973). Personality and the law of effect. In D. E. Berlyne & K. B. Madsen (Eds.), *Pleasure, reward, preference,* New York: Academic Press.

major dimension of personality is **emotional stability** or **neuroticism.** This dimension describes at one end people who tend to be moody, touchy, anxious, restless, and so on. At the other extreme are people who are characterized by such terms as stable, calm, carefree, even-tempered, and reliable. As Eysenck stressed, the ultimate value of these dimensions will depend on the research support they receive.

To clarify the meaning of both dimensions, Eysenck and his associates have studied the relations between people's positions on them and their scores on many other measures. An example of the results found is summarized in Table 3.2, which shows self-reported differences in the sexual activities of extraverts and introverts (reported in Eysenck, 1973). As expected, the extraverts generally reported earlier, more frequent, and more varied sexual experiences. While the groups differed on the average, there was still considerable overlap, making it difficult to predict any particular individual's behavior from her introversion–extraversion score alone. But the results of many studies of this type provide an increasingly comprehensive picture of Eysenck's dimensions. In addition, Eysenck's ideas are notable in stimulating a search for the biological foundations of dispositions.

▶ COMMON FEATURES OF TRAIT THEORIES

Now consider the principal common characteristics of trait approaches.

Inferring Traits from Behavioral Signs

In the search for dispositions, one always *infers* traits from behavior—for example, from what people say about themselves on a questionnaire. The person's responses or behaviors are taken as *indicators* of underlying traits. The trait approach to personality is a "sign" approach in the sense that there is no interest in the test behavior itself. That is, test responses are of value not in their own right but only as *signs* of the traits that underlie them; test behavior is always used as a sign of nontest behavior (Loevinger, 1957). Thus, the interest is not in answers on an inventory or IBM score sheet for their

3.11 From what kinds of observations are traits inferred?

own sake, but only in these responses to a test as cues or signs of dispositions. It is therefore essential to demonstrate the relation between test behaviors and the traits they supposedly represent.

Traits are inferred from questionnaires, ratings, and other reports about the person's dispositions. Usually, these self-reports are taken as direct signs of the relevant dispositions. For example, the more often you rate yourself as aggressive, the more you are assumed to have an aggressive disposition. So the relationship between what is sampled and the inferred trait is *direct* and *additive:* the more frequently a behavioral tendency is reported or described, the greater the amount of the underlying disposition.

Generality and Stability of Traits

3.12 What is the importance of behavioral consistency in trait theories?

Trait theorists often have disagreed about the specific content and structure of the basic traits needed to describe personality, but their general conceptions have much similarity and they remain popular (Funder, 1991). They all use the trait to account for consistencies in an individual's behavior and to explain why persons respond differently to the same stimulus. Most view traits as dispositions that determine such behaviors. Each differentiates between relatively superficial traits (e.g., Cattell's surface traits) and more basic, underlying traits (e.g., Cattell's source traits). Each recognizes that traits vary in breadth or generality. Allport puts the strongest emphasis on the relative generality of common traits across many situations. Each theorist also admits trait fluctuations, or changes in a person's position with respect to a disposition. At the same time, each is committed to a search for relatively broad, stable traits.

Traits and States Distinguished

3.13 How does an anger state differ from an anger trait?

It is one thing to be irascible, quite another thing to be angry, just as an anxious temper is different from feeling anxiety. Not all men who are sometimes anxious are of an anxious temperament, nor are those who have an anxious temperament always feeling anxious. In the same way there is a difference between intoxication and habitual drunkenness . . .

Cicero (45 B.C.)

This ancient wisdom (quoted in Chaplin, John, & Goldberg, 1988) is used by modern trait theorists to illustrate a distinction that is often made, both intuitively and by trait psychologists, between *traits* and *states* (Chaplin, John, & Goldberg, 1988; Eysenck, 1983). Both traits and states are terms that refer to the perceived attributes of people. Both refer to categories that have fuzzy boundaries, and both are based on prototypes or ideal exemplars (e.g., Cantor & Mischel, 1979). The difference between them is that prototypic traits are seen as enduring, stable qualities of the person over long time periods and as internally caused. In contrast, prototypic states refer to qualities that are only brief in duration and attributable to external causes, such as the momentary situation (Chaplin, John, & Goldberg, 1988). Examples of terms that people tend to classify as traits are gentle, domineering, and timid, while terms like infatuated, uninterested, and displeased tend to be seen as states.

Search for Basic Traits

Guided by the assumption that stable dispositions exist, trait psychologists try to identify the individual's position on one or more dimensions (such as neuroticism, extraversion). They do this by comparing people tested under standardized conditions. They believe that positions on these dimensions are relatively stable across testing situations and over long time periods. Their main emphasis in the study of personality is the development of instruments that can accurately tap the person's underlying traits. Less attention has been paid to the effects of environmental conditions on traits and behavior. The search for traits, and for useful ways to measure them, in modern personality psychology has a long history, and it began in response to urgent practical considerations at the time of the first world war, as was discussed in Chapter 2.

Quantification

A main feature of the trait approach has been its methodology, and you saw many examples in Chapter 2 of self-report questionnaires and ratings by self and others. This methodology is "psychometric" in the sense that it attempts to measure individual differences and to quantify them. Psychometricians study persons and groups on trait dimensions by comparing their scores on tests. To do this, they sample many people, compare large groups under uniform testing conditions, and devise statistical techniques to infer basic traits. Their methods over the years have become increasingly sophisticated and effective for meeting a wide range of measurement goals (e.g., Jackson & Paunonen, 1980; John, 1990).

Aggregating Across Situations to Increase Reliability

Although they acknowledge that the situation is important, many psychologists working at the trait level are convinced that past research has underestimated the personal constancies in behavior. They point out that if we want to test how well a disposition (trait) can be used to predict behavior, we have to sample adequately not only the disposition but also the behavior that we want to predict (Ajzen & Fishbein, 1977; Block, 1977; Epstein, 1979, 1983). Yet in the past, researchers often attempted to predict single acts (for example, physical aggression when insulted) from a dispositional measure (e.g., self-rated aggression on a personality scale). Generally, such attempts did not succeed. But while measures of traits may not be able to predict such single acts, they may do much better if one uses a *"multiple act criterion"*: a pooled combination of many behaviors that are relevant to the trait, and a pooled combination of many raters (e.g., McCrae & Costa, 1985, 1987).

The methods and results of this line of research are illustrated in a study in which undergraduate women were given the "dominance scale" from two personality inventories (Jaccard, 1974). The women also were asked whether or not they had performed a set of 40 dominance-related behaviors. For example, did they initiate a discussion in class, argue with a teacher, ask a male out on a date. The dominance scales from the personality inventories did not predict the individual behaviors well. But the researcher found that when the 40 behavioral items were summed into one pooled measure, they related substantially to the personality scores. Namely, women high on the dominance scales also tended to report performing more dominant behaviors, and the reverse was also true. Thus a longer, aggregated, and therefore more reliable behavioral measure revealed associations to other measures (the self-reports on the personality tests) that would not otherwise have been seen. Similar results were found when the behaviors were measured directly by observation (e.g., Weigel & Newman, 1976).

In the same direction, it has been shown that reliability will increase when the number of items in a test sample is increased and combined. Making this point, Epstein (1979) demonstrated that temporal stability (of, for example, self-reported

3.14 How did the Jaccard and Epstein studies illustrate the importance of aggregating behaviors to improve correlations between trait measures and behavior?

emotions and experiences recorded daily, and observer judgments) becomes much larger when it is based on averages over many days than when it is based on only single items on single days. His demonstrations also indicate that even when one cannot predict what an individual will do in a specific situation, it is often possible to predict the person's overall standing relative to other people when the behaviors are aggregated (combined) across many situations (Epstein, 1983).

▶ TAXONOMY OF HUMAN ATTRIBUTES

A widely shared goal is to find a universal taxonomy or classification system for sorting the vast array of human attributes into a relatively small set of fundamental dimensions or categories on which most individual differences can be described. From this perspective, researchers attempt to identify "the most important individual differences in mankind" (Goldberg, 1973, p. 1).

Psycholexical Approach

3.15 Describe the assumptions underlying the psycholexical approach to identifying the major traits that underlie personality.

Researchers in this approach assume that the most significant individual differences—those that are most important in daily human relationships—enter into the natural language of the culture as single-word trait terms. They use a variety of methods to identify basic trait terms in the language and to categorize them into smaller groupings. This is an enormous classification task, given that English includes thousands of trait terms (over 18,000 in one count of the dictionary). The researchers hope that an extensive, well-organized vocabulary for describing human attributes in trait terms will lead to better theories of personality and better methods of personality assessment. This research strategy is called the **psycholexical approach.** Its basic data are the words in the natural language that describe human qualities. In these studies, many people are asked to rate how well each of many trait terms describes or fits a particular person they know well. In some studies, this is a peer; in some studies, participants rate how well the words describe them. In each study, the results are then analyzed to see which sets of trait terms tend to cluster or "go together" when individuals are described. Using statistical procedures called factor analysis, the researchers try to specify a small number of factors or dimensions that seem to capture the common element among adjectives that are closely associated (e.g., Goldberg, 1990).

This approach is illustrated in a comprehensive taxonomy of the **domain of interpersonal behavior** (Wiggins, 1979, 1980) that yielded the dimensions shown in Figure 3.5. Note that each dimension is bipolar, that is, has two opposite ends or poles. The dimensions are structured in a circular pattern like a pie. Each pole is made up of a set of adjectives so that *Ambitious (Dominant),* for example, is defined with such terms as persevering, persistent, industrious. The opposite pole, *Lazy (Submissive),* includes such terms as unproductive, unthorough, unindustrious. Wiggins reports that these dimensions fit well the results of earlier descriptions of the interpersonal domain (Leary, 1957). They seem to be reasonably robust and useful when different samples of people are rated on them and continue to be revised and refined (Wiggins, Phillips, & Trapnell, 1989).

The "Big Five" Trait Dimensions

For many years in the long search for a "universal taxonomy" of traits, researchers disagreed actively as to which personality dimensions they should use to describe personality. Some proposed as many as 16; others, as few as two or three (Vernon, 1964).

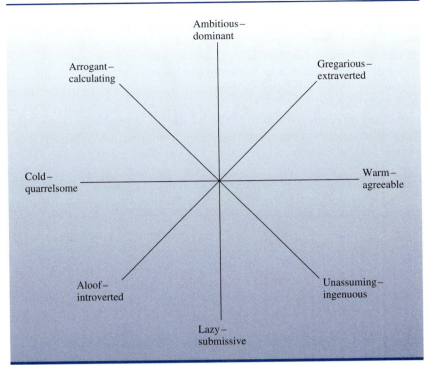

Figure 3.5 Wiggins' (1980) taxonomy of the interpersonal domain.
Source: Wiggins, J. S. A psychological taxonomy of trait-descriptive terms: The interpersonal domain. *Journal of Personality and Social Psychology, 37,* 399, fig. 2. © 1979 by the American Psychological Association. Reprinted with permission.

More recently, however, consensus has grown among many researchers to focus on five dimensions of personality (Goldberg, 1991; John, 1990; McCrae & Costa, 1985, 1987, 1999) that emerge from ratings using English-language trait adjectives. These dimensions are found by using the method of factor analysis, and we consider that method next and then turn to its findings.

Factor Analysis to Find Dimensions: The NEO-PI-R

Trait terms number in the thousands, as a look at the dictionary makes clear. But they have to be simplified and organized to become manageable units for describing people systematically. Consider, for example, the mass of data yielded by responses to a self-report measure with 550 items answered by 100 persons. To extract order from such a stack of facts, investigators searching for underlying traits try to group responses into more basic clusters. For this purpose, many trait psychologists turn to **factor analysis.** This is a mathematical procedure that helps to sort test responses into relatively homogeneous clusters of items that are highly correlated. Working in the psycholexical approach, a number of researchers have reached reasonable agreement about the five types of dimensions or factors on which English trait terms may be clustered, often called the "Big Five Structure" (e.g., Goldberg, 1991; John, 1990).

These dimensions emerged from mathematical analyses of the responses with the method of factor analysis. Factor analysis is a very useful tool for reducing a large set of correlated measures to fewer unrelated or independent dimensions. As such, it can be a powerful aid to psychological research by clarifying which response patterns go together. Suppose, for example, that 50 students have answered 10 personality questionnaires, each of which contains 100 questions. A factor analysis of this mass of information can show which parts of the test performances go together. It essentially finds and connects the items that tend to "go together" (covary) with each of the other items in the total set. The analogy would be a procedure that allows you to go to a stack of several hundred unsorted books in the library and that finds and groups together those that are alike on certain dimensions (e.g., size, color, length, language, content areas) but different in other respects. Factor analysis, however, does not necessarily reveal the basic traits of persons any more than it reveals the nature of the knowledge in the library sample you are sorting. The results depend on what is put into the procedures, such as the test items and the people selected by the researcher, and on the details of his or her procedures and decisions.

As has long been recognized (Overall, 1964), factor analysis cannot establish which characteristics of persons or things being measured are "real" or "primary." The factors obtained are simply names given to the correlations found among the particular measures. In other words, factor analysis yields a greatly simplified patterning of the data put into it. However, it cannot go beyond the limitations of the original tests or measures. And what comes out of the analysis depends on many decisions by the investigator (e.g., in the type of factor analysis he or she conducts). Consequently, while the factor analytic search for hypothesized underlying traits may yield mathematically pure factors, their psychological meaningfulness and relevance for the person's actual behavior has to be demonstrated.

The factor analytic approach to describing trait dimensions is illustrated in a series of pioneering factor analytic studies (Norman, 1961, 1963; Tupes & Christal, 1958, 1961). These studies investigated the factors obtained for diverse samples of people rated by their peers on rating scales. The scales themselves came from a condensed version of the thousands of trait names originally identified by Allport and Odbert's search many years earlier for trait names in the dictionary. After much research, 20 scales were selected and many judges were asked to rate other people on them. The results were carefully factor analyzed. The same set of five relatively independent factors appeared consistently across several studies and continues to form the basis of what has become the **Big Five Structure** (e.g., John, 1990; Goldberg, 1992). It consists of five factors measured with a personality inventory now called the **NEO-PI-R** (Costa & McCrae, 1997), as shown in Table 3.3.

The Big Five Factors include: Neuroticism (N), for example, worrying and insecurity; Extraversion (E) or surgency (positive emotionality), as reflected by terms like friendly, talkative; **Openness to experience (O); Agreeableness (A);** and **Conscientiousness (C).** See Table 3.3 for descriptions of each factor. In the NEO-PI-R, each factor is also more finely described in terms of a number of different facets through which the particular super trait may be expressed, illustrated in Table 3.4.

The Big Five resemble the dimensions initially proposed by Norman (1963) and found repeatedly in research, although sometimes given slightly different names. Each dimension includes a collection of bipolar rating scales, such as "calm–worrying" and "timid–bold," that refer to types of feelings or behaviors. For each dimension, Table 3.3 gives examples of the adjectives describing the two ends of some of the rating scales used.

To illustrate the type of research that underlies the development of the Big Five, in one study, 187 college students rated how well each of 1,710 trait terms described him or her (Goldberg, 1991). Statistical analysis showed that these terms clustered into five major factors or dimensions much like the Big Five (McCrae & Costa, 1987). Thus, when people are described with trait terms like those shown in Table 3.3, a reliable clustering occurs consisting of five large descriptive categories or "super traits." This Big Five Structure seems to characterize major dimensions of personality in natural

3.16 How was factor analysis used to identify the Big Five traits that are thought by some theorists to describe personality? Describe the five factors using the acronym OCEAN to capture their first letters.

TABLE 3.3   The Big Five Factors and Illustrative Components

Factor (Trait Dimension)	Adjective Items[a]
I. **Neuroticism (N)** (negative emotions, e.g., anxiety, depression)	Calm—worrying Unemotional—emotional Secure—insecure Not envious—jealous
II. **Extraversion (E)** (versus introversion)	Quiet—talkative Aloof—friendly Inhibited—spontaneous Timid—bold
III. **Openness to Experience** (versus closed-minded)	Conventional—original Unadventurous—daring Conforming—independent Unartistic—artistic
IV. **Agreeableness (A)** (versus antagonism)	Irritable—good-natured Uncooperative—helpful Suspicious—trusting Critical—lenient
V. **Conscientiousness (C)** (versus undirectedness)	Careless—careful Helpless—self-reliant Lax—scrupulous Weak-willed—goal-directed

[a] Illustrative adjectives describing the two ends of the scales that comprise the dimension.

Source: Adapted from McCrae, R. R., & Costa, P. T., Jr. (1987). Validation of the five-factor model of personality across instruments and observers. *Journal of Personality and Social Psychology, 52,* 81–90. Essentially similar results were found in John, O. P. (1990). The big-five factor taxonomy: Dimensions of personality in the natural language and questionnaires. In L. A. Pervin (Ed.), *Handbook of personality: Theory and research* (pp. 66–100). New York: Guilford Press; and in Normak, W. T. (1963). Toward an adequate taxonomy of personality attributes: Replicated factor structure in peer nomination personality ratings. *Journal of Abnormal and Social Psychology, 66,* 574–583.

English-language words. A number of personality trait questionnaires and personality ratings using these types of trait terms provide descriptions of persons that seem to fit the Big Five reasonably (Costa & McCrae, 1992a; Costa, McCrae, & Dye, 1991).

Considerable stability has been demonstrated on trait ratings and questionnaires related to the Big Five (e.g., McCrae & Costa, 1990) even for long time spans. Stability tends to be particularly high during the adult years (Costa & McCrae, 1997; McCrae & Costa, 1999). It is notable that in spite of the many changes that often occur in life structures during adulthood over long time periods—including the changes produced

3.17   How stable are the Big Five dimensions over time?

TABLE 3.4   Illustrative Facet Scales for the Big Five Factor

Neuroticism: anxiety, angry hostility, depression, impulsiveness, vulnerability

Extraversion: warmth, gregariousness, assertiveness, activity, excitement-seeking positive emotions

Openness to Experience: fantasy, aesthetics, feelings, actions, ideas, values

Agreeableness: trust, straightforwardness, altruism, compliance, modesty, tender-mindedness

Conscientiousness: competence, order, dutifulness, achievement, striving, self-discipline, deliberation

Source: Reproduced by special permission of the Publisher, Psychological Assessment Resources, Inc., 16204 North Florida Avenue, Lutz, Florida 33549, from *The NEO Personality Inventory-Revised,* by Paul Costa, and Robert McCrae, Copyright 1978, 1985, 1989, 1992 by PAR, Inc. Further reproduction is prohibited without permission of PAR, Inc.

TABLE 3.5   Stability of NEO-PI Scales (Ages 25–56)

NEO-PI Scale	Men	Women
N (Neuroticism)	.78	.85
E (Extraversion)	.84	.75
O (Open-minded)	.87	.84
A (Agreeable)	.64	.60
C (Conscientiousness)	.83	.84

Note: Retest interval is 6 years for N, E, and O scales, 3 years for short forms of A and C scales.

Source: Adapted from *Personality in Adulthood* (p. 88), by R. R. McCrae and P. T. Costa Jr., 1990, New York: Guilford.

by marriage, children, divorce, residential and occupational moves, and health issues— the status of most individuals on the Big Five dimensions tends to show high stability (see Table 3.5)

Supportive Evidence for the Five-Factor Model

3.18   Cite four lines of evidence for the generality of the Five Factor Model.

An explosion of research has provided extensive empirical documentation for the robustness of the Five Factor Model, or Big Five. The kinds of evidence that has accumulated is impressive (e.g., McCrae & Costa, 1997, 1999; McCrae, et al., 1998), and too large to summarize beyond the general conclusions to which it leads:

- The Big Five Factor Structure has often been replicated in research by diverse investigators using a variety of English-speaking samples.

- Especially the N, E, and A factors listed in Table 3.3 have been found to replicate well even when the languages, cultures, and item formats used differ. Replicability has been reported for diverse languages and language families that span Sino-Tibetan, Uraic, Hamito-Semitic, and Malayon-Polynesian.

- Overall, the results are impressive and broadly generalizable across diverse cultures (McCrae et al., 1998), although unsurprisingly some of the factors may take different forms in different samples and cultures.

- The factor structure of individuals as described by this model tends to be relatively stable in adults over long periods of time.

In sum, the total evidence is impressive for the broad robustness and potential universality of this taxonomy as a comprehensive descriptive system for describing people in trait terms at a broad, highly abstract "super trait" level. The value of having such a robust and comprehensive trait "map" for describing and comparing people is considerable and it is widely recognized (e,g., MacDonald, 1998).

The Big Five dimensions offer a useful replicable taxonomy of trait terms, potentially not only in the English language but across diverse languages (Costa & McCrae, 1997; DeRaad, Perugini, Hrebickova, & Szarota, 1998; McCrae et al., 1998). Its focus on language as the route for identifying what humans care about with regard to personality makes good sense. It is based on the idea that the languages people develop contain clusters of trait terms that go together and that reflect what they care about when they describe and evaluate other people and themselves (Saucier & Goldberg, 1996). Factor analysis has been used effectively to identify these clusters, and they provide a robust view of personality qualities that matter to people importantly in everyday life.

At the same time, the limitations of the model also are being expressed (e.g., McAdams, 1992). The deepest concern is the one stated by Block (1995, 2001; Pervin, 2003). While himself a major contributor to the trait-dispositional level of personality, Block calls attention to the fact that labeling a person's position (e.g., low on "conscientiousness") with regard to the Big Five, or on any other descriptive taxonomy, does not itself provide an understanding of the personality processes that underlie and account for the labeled phenomena, and that need to be clarified.

Stability of Traits over Time

Are a person's qualities stable? Do early characteristics predict later qualities? When parents say "Fred was always so friendly, even as a little baby," are their comments justified? Can we predict the six-year-old's behavior from responses in the first year of life? Is there much continuity in the qualities and behaviors of the child throughout childhood? Research has made it clear that although behavior depends on context, and the same person may be quite different in very different contexts, there also is significant trait stability over time, particularly after the first few years of life.

3.19 At what point does temporal stability begin to emerge, and which behaviors show stability?

Many important connections have been found over time in the life course. For example, lower sensitivity to touch on the skin in the newborn predicted more mature communications and coping at age two and one-half and at age seven and one-half. High touch sensitivity and high respiration rates at birth were related to low interest, low participation, lower assertiveness, and less communicativeness in later years (Bell, Weller, & Waldrop, 1971; Halverson, 1971a, 1971b). But these links between newborn and later behaviors were exceptions, and the associations that were found generally were not strong. So there are some connections between a newborn's qualities and characteristics later in life, but in the individual case, one could not predict confidently from responses at birth to later characteristics:

> To use an analogy, newborn behavior is more like a preface to a book than a table of its contents yet to be unfolded. Further, the preface is itself merely a rough draft undergoing rapid revision. There are some clues to the nature of the book in the preface, but these are in code form and taking them as literally prophetic is likely to lead to disappointment (Bell et al., 1971, p. 132).

As one progresses beyond the "preface of the book" to the first few chapters (to the early years of life), continuities in development do become increasingly evident (Block & Block, 1980; McCrae & Costa, 1996; Mischel et al., 1989). Children who are seen as more active, assertive, aggressive, competitive, outgoing, and so on at age three years are also more likely to be described as having more of those qualities later in development, for example. In sum, the specific links between qualities of the child, say in the fourth year and in the eighth year, may be complex and indirect. But a thorough analysis of the patterns indicates that significant threads of continuity do emerge over time (Caspi, 1987, 1998). Thus childhood characteristics may be connected coherently to later behavior and attributes to some degree (Arend, Gove, & Sroufe, 1979; Block, 1971; Mischel et al., 1989). Experiences in the early pages of the book affect what happens in the later pages, although these early experiences do not prevent the possibility of genuine changes later. The amount of stability or change over time varies for different types of characteristics and different types of experiences at different points in development (Caspi, 1998; Caspi & Bem, 1990).

To illustrate with one concrete example, studies of lives over many years indicate that "ill-tempered boys become ill-tempered men" (Caspi & Bem, 1990, p. 568). These continuities in development reflect the fact that stable qualities of the individual, whether adaptive (such as the ability and willingness to delay gratification), or aversive and maladaptive (such as inappropriate displays of temper), can have profound "chain effects" that quickly accelerate. Beginning early in life, they can trigger long sequences of interconnected events that impact on the person's subsequent opportunities and options, often greatly limiting them. As Caspi and Bem note, a child's ill temper rapidly produces trouble in school, which in turn makes school a negative experience, which provokes the school authorities and can lead to expulsion, which permanently limits occupational opportunities and constrains the future. The unhappy network of consequences is illustrated in Figure 3.6.

The network of outcomes associated with early ill-temperedness is broader than just school and career. For example, Bem and Caspi reported that almost half of the men (46 percent) with a history of ill-tempered behavior as children were divorced by age forty while only 22 percent of the other men studied were divorced at that age.

Taken collectively, studies of individual differences on trait dimensions after early childhood have produced many networks of meaningful correlations. These associations tend to be large and enduring when people rate themselves or others with broad trait terms (e.g., Block, 1971; Caspi & Bem, 1990; Costa & McCrae, 1988; E. L. Kelly, 1955). Such ratings suggest significant continuity and stability in how people are perceived over the years as well as in how they perceive themselves. For example, long-term stability was shown using rating methods in some early landmark studies that closely followed children into adulthood (Block, 1971), as summarized in Table 3.6. It is also high on many personality trait inventories, and ranges on average from median correlations of .34 to .77 (Costa & McCrae, 1997).

3.20 What is the temporal stability of results derived from self-descriptions and ratings?

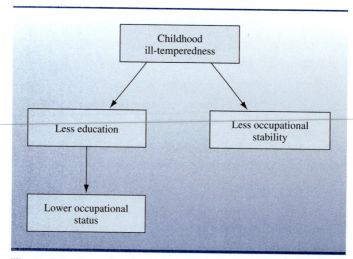

Figure 3.6 Some long-term outcomes associated with childhood ill-temperedness. Statistically significant correlations are shown. Data are for 45 men from middle-class origins.

Source: Figure and data adapted from "Moving against the world: Life-course patterns of explosive children," by A. Caspi, G. H. Elder, Jr. and D. J. Bem (1987). *Developmental Psychology, 23*, pp. 308–313.

TABLE 3.6   Examples of Significant Stability over Time in Ratings of Personality

	Correlations From		
	Junior to Senior H. S.	Senior H. S. to Adulthood	Item Rated
Males	.57	.59	Tends toward undercontrol of needs and impulses; unable to delay gratification
	.58	.53	Is genuinely dependable and responsible person
	.50	.42	Is self-defeating
Females	.50	.46	Basically submissive
	.48	.49	Tends to be rebellious, nonconforming
	.39	.43	Emphasizing being with others, gregarious

Source: Based on Block J. (1971). *Lives through time*. Berkeley, CA: Bancroft.

When ongoing behavior in specific situations is sampled objectively by different, independent measures, however, the association generally tends to be much more modest. Thus while people often show consistency on questionnaires and ratings, these data may not very accurately predict their actual behavior, as has been noted even by advocates of trait level analysis (Funder, 2001). Therefore, one has to be cautious about generalizing from a person's personality test behavior, including the robust and well-established measures of the Big Five, to his or her behavior outside the test.

Links Between Perceiver and Perceived: Valid Ratings

One enduring concern about the factors that have emerged from trait ratings was that they may reflect the social stereotypes and concepts of the judges rather than the trait organization of the rated persons. Mulaik (1964), for example, conducted three separate factor analytic studies, using many trait-rating scales, to determine the degree to which the method reveals the subject's personality factors as opposed to the rater's conceptual factors. The judges in one study rated real persons on the scales, including family members, close acquaintances, and themselves. In a second study, they rated stereotypes like "suburban housewife," "mental patient," and "Air Force general." The raters in the third study rated the "meaning" of 20 trait words. There was much similarity between the factors found for ratings of real persons and those found for ratings of stereotypes and words. Results like these led many investigators to conclude that personality factors that emerge from ratings may reflect the raters' conceptual categories rather than the traits of the subjects being judged (Mischel, 1968; Peterson, 1968).

It is quite possible that raters' stereotypes and preconceptions enter into these ratings, and it is also the case that different investigators may arrive at somewhat different views of trait organization. Nevertheless, a few basic trait dimensions have been found over and over again, as the evidence for the Big Five makes clear. And, most importantly, the characterizations of people on these dimensions, based on ratings by peers, reasonably agree with the self-ratings of the rated individuals themselves

3.21 What evidence exists that trait ratings may be influenced by stereotypes and preconceptions? Cite research evidence that trait ratings can nonetheless predict independent behavior ratings reflecting the same traits.

(e.g., McCrae & Costa, 1987). Thus, even if stereotypes and oversimplifications enter into these judgments, as they probably do, they are made reliably, shared widely, and seem significantly linked to qualities of the rated persons (Funder & Colvin, 1997).

In sum, there has been extensive debate about the limitations and usefulness of trait ratings (e.g., Block, 1995; Block, Weiss, & Thorne, 1979; Pervin, 1994; Romer & Revelle, 1984; Shweder, 1975). Nevertheless, it is clear that the descriptions of people obtained from different raters in different contexts often agree with each other. This basic conclusion was reached more than four decades ago and still stands. To illustrate, in one classic study, agreement was found between peer judgments on dimensions of aggressive and dependent behavior and separate behavior ratings of actual aggressive and dependent behavior (Winder & Wiggins, 1964). Participants were first classified into high, intermediate, and low groups on aggression and dependency on the basis of their peer reputations. Separate behavior ratings made later indicated that the three groups differed from each other significantly in the amount of aggression and dependency they displayed in an experimental situation. Similarly, college students were preselected as extremely high or low in aggressiveness (on the basis of ratings by their peers). They tended to be rated in similar ways by independent judges who observed their interaction (Gormly & Edelberg, 1974). Thus, there are linkages between a rater's trait constructs and the behavior of people he or she rates (e.g., Funder & Colvin, 1997; Jackson & Paunonen, 1980; McCrae & Costa, 1987; Mischel, 1984).

Personality Judgments to Predict Behavior

Do trait judgments reflect the constructs of perceivers rather than the behavior of the perceived? What are the links between the perceiver's judgments of what people "are like" and the behaviors that those people actually display? Some theorists argue that personality structure exists "all in the head" of the perceiver (Shweder, 1975). Others believe it is "all in the person" perceived (Epstein, 1977; Funder & Colvin, 1997). Still others favor the view that it depends both on the beliefs of observers and on the characteristics of the observed (e.g., Cantor & Mischel, 1979; Magnusson & Endler, 1977; Mischel, 1984). They believe that ratings may be influenced by the rater's expectations but have roots in the behavior of the perceived.

For many years, most researchers studied either the judgments of the perceiver or the behavior of the perceived; they rarely considered the fit between the two. One study tried to relate observers' overall judgments of children's aggressiveness (and other traits) to the children's independently coded actual behavior (Mischel, 1984; Wright & Mischel, 1987). Specifically, their trait-related acts were recorded during repeated observation periods on 15 separate occasions distributed across three camp situations during a summer. Examples of behaviors coded as aggressive are: "I'm gonna punch your face" or "Let's go beat up . . ." or lifting a dog by his collar and choking him.

The question was: Do raters' judgments of the child's overall aggression relate to the actual frequency of the child's aggressive acts as coded by other independent observers? The results clearly answered the question: Yes. Thus, children who are rated by independent observers as more aggressive, actually tend to be aggressive more frequently. For instance, they yell and provoke more. These differences, however, were seen primarily only in certain situations, namely in those stressful situations that greatly taxed or strained the competencies of the children. In sum, personality judgments seem to have links to the judged person's actual behavior, although that behavior may be visible only in certain types of situations in which the relevant individual differences emerge.

In sum, the findings discussed so far indicate that trait ratings involve judgments that are widely shared about characteristics of people that often are stable over time and that are expressed in behavior in different types of situations. Given that these judgments are widely made in everyday life and can have important consequences, it also becomes interesting to ask how such judgments of what someone (or something) "is like" are made, and that question is addressed in In Focus 3.2.

IN FOCUS 3.2

PROTOTYPES: "TYPICAL" PEOPLE

Psychologists concerned with the classification of human attributes also try to identify how we judge the **prototypicality** or "typicality" of different members of a category, whether it is a trait category, such as one of the Big Five, or an everyday, "natural" category, such as the category "birds." To demonstrate the idea of typicality to yourself, think of the most typical, representative, or "birdlike" bird. You probably will think of a bird that is something like a robin or sparrow—not a chicken or an ostrich. The point is simply that some members of a category (in this case, birds) are better or more typical examples of a category: Some reds are "redder" than others, some chairs are more chair-like than others. This point has been elegantly documented for everyday categories of natural objects like furniture (Rosch, Mervis, Gray, Johnson, & Boyce-Braem, 1976). Thus natural categories may be organized around *prototypical* examples (the best examples of the concept), with less prototypical or less good members forming a continuum away from the central one (Rosch, 1975; Tversky, 1977).

The important point here is that trait categories, like other categories, often are not well-defined, distinct, nonoverlapping categories in which each member of a category has all its defining features. (While many birds sing, not all do, and some nonbirds surely do.) Well-defined, nonoverlapping categories are built into artificial, logical systems, but they are rare in the real world. If we turn from the abstract world of logic and formal, artificial systems to common, everyday categories—to furniture, birds, and clothing—the categories become "fuzzy." As the philosopher Wittgenstein (1953) first pointed out, the members of common, everyday categories do not all share all of a set of single, essential features critical for category membership. If you closely examine a set of natural objects all labeled by one general term (like "birds") you will not find a single set of features that all members of the category share: Rather, a *pattern* of overlapping similarities—a *family resemblance* structure—seems to emerge.

"Typicality" and "family resemblance" and "fuzziness" also may characterize everyday judgments about categories of people (Cantor & Mischel, 1979; Cantor, Mischel, & Schwartz, 1982; Chaplin, John, & Goldberg, 1988; John, 1990). When making these judgments while forming our everyday impressions of people, we seem able to agree about who is a more or less typical or even "ideal" kind of personality, just as we agree that a robin is a more birdlike bird than a chicken. For instance, such qualities as "sociable" and "outgoing" are the characteristics of a "typical extravert."

The study of prototypicality rules and family resemblance principles in judgments of people helps in understanding how consistency and coherence are perceived in spite of variations in behavior (Kunda, 1999; Mischel, 1984). To illustrate, someone who really knows Rembrandt's work, who has seen dozens of his paintings, can easily identify whether a previously unseen painting is a real Rembrandt, an imitation, or a fake. The art expert seems able to extract a central distinctive gist from a wide range of variations. The same processes that permit such judgments must underlie how we identify personality coherence in the face of behavioral variability, and agree that someone is or is not a "real" introvert, or an anxious neurotic, or a sincere, friendly person. One attractive feature of this prototypicality approach to traits is that it seems highly compatible with attempts to quantify interpersonal behavior into dimensions of the sort discussed throughout this chapter (John, Hampson, & Goldberg, 1991).

Interaction of Traits and Situations

3.22 How can the study of interactions between traits and situations contribute to an understanding of personal individuality?

It has become increasingly recognized that a comprehensive approach to the study of traits must deal seriously with how the qualities of the person and the situation influence each other—that is, their "interaction" (Higgins, 1990; Magnusson, 1990; Mischel & Shoda, 1998). The behavioral expressions of a person's traits depend on his or her psychological situation at the moment. For example, rather than exhibit aggression widely, an individual may be highly aggressive but only under some set of relatively narrow circumstances, such as when psychological demands are very high (Shoda, Mischel & Wright, 1994; Wright & Mischel, 1987). Moreover, this aggressiveness may be expressed in some ways (such as verbally) but not in others (for example, physically). The implications of these specific interactions for contemporary personality theory will become evident when we consider the topic of interaction and the nature of consistency in the next chapter.

Are Traits Explanations or Summaries?

An enduring and difficult question at the trait level of analysis concerns the causal status of traits. The question is, do traits when defined and studied in the ways described in this chapter really serve to explain individual differences or do they simply provide useful ways to measure and describe those observed differences? The question is particularly important because the use of traits like "sociability" to explain individual differences in the behaviors that define sociability (warmth, friendliness) is readily criticized as circular. Sensitive to this concern, not all theorists working at this level of analysis view traits as causes.

3.23 How did Buss and Craik suggest a descriptive rather than causal interpretation of personality traits?

Buss and Craik (1983) are a notable case in point. They moved away from the traditional trait view of dispositions as underlying internal causes (or explanations) of cross-situational consistencies in behavior. Instead, they see dispositions like the Big Five as summary statements of **act trends** or behavioral trends, not explanations. In this view, dispositions are natural categories made up of various acts. The acts within a category differ in the degree to which they are prototypical or ideal members. Most importantly, this view of dispositions emphasizes that dispositions do not provide explanations of behavior; instead, they are summary statements of behavioral trends that must themselves be explained (Buss & Craik, 1983; Wiggins, 1980).

This revised view of dispositions reflects attempts to take account of the explanatory limitations of classical global trait theories like Allport's (1937), which have been widely criticized for several decades (Mischel, 1968; Ross & Nisbett, 1991; Wiggins, 1980). It is a revision, however, that is by no means completely shared: some voices still call for a return to the earlier approach to traits as global entities that provide adequate explanations (Funder, 1991).

In thinking about the meaning and uses of traits, keep in mind that even if traits like the Big Five don't explain why people behave as they do, they can be of great value for many goals. They are providing a common map for comparing individuals, groups, and even cultures on meaningful dimensions or factors of trait terms. Such terms are used by people in daily life to characterize and evaluate themselves and other persons. They provide a language and taxonomy for personality assessment that has long been needed for quantifying and comparing people in their broad characteristics. And they raise fascinating questions about the possible bases of these traits, their functions, and the ways they are expressed in the life course.

SUMMARY

TYPES AND TRAITS

- Personality types refer to discrete categories of people that have similar features of characteristics (physically, psychologically, or behaviorally).
- Carl Jung divided people into two types: introverts (withdrawn, shy) and extroverts (sociable, outgoing).
- Traits are basic, stable qualities of the person. Unlike types, traits are continuous dimensions on which individual differences may be arranged.
- Trait theorists conceptualize traits as underlying properties, qualities, or processes that exist in persons.
- Trait constructs have been used to account for observed behavioral consistencies within persons and for the behavioral differences among them.

GORDON ALLPORT

- For Allport, traits are the ultimate realities of psychological organization.
- In Allport's theory, traits are the general and enduring mental structures that account for consistency in behavior.
- They range in generality from highly generalized cardinal traits to secondary traits or more specific "attitudes."
- For Allport, "personality structure" is the individual's stable pattern of dispositions or traits.
- Allport urged the intensive study of the individual.

R. B. CATTELL

- Cattell distinguished between surface traits and source traits.
- Through factor analysis, Cattell tried to estimate the basic dimensions or factors underlying surface variations in behavior.
- Cattell differentiated between environmental-mold traits, which reflect environmental conditions, and constitutional traits, which reflect constitutional factors.

HANS J. EYSENCK

- Eysenck used empirical analyses to explore dimensional traits.
- He found that introverts are more reserved, quiet, and introspective than extraverts who are more sociable, active, and carefree.

COMMON FEATURES OF TRAIT LEVEL ANALYSES

- Trait theorists assume traits to be general underlying dispositions that account for consistencies in behavior.
- Some view traits as causes and explanations for behaviors; others interpret them as summaries of behavioral tendencies.
- Some traits are considered to be relatively superficial and specific; others are more basic and widely generalized.
- Traits are seen as enduring, stable qualities of the person over long time periods, while states refer to qualities that are brief in duration and attributable to external causes.
- This level of analysis tries to identify and measure the individual's position on one or more dispositional dimensions.

- Psychometric strategy is used to sample, quantify, and compare the responses of large groups of people to discover basic traits.
- To test the stability of a given trait, psychologists sample the individual's behavior over the course of multiple situations.

TAXONOMY OF HUMAN ATTRIBUTES

- The psycholexical approach to personality categorizes natural-language trait terms in order to describe and understand basic human qualities.
- The "Big Five" identified five primary dimensions of personality through factor analysis: neuroticism, extraversion, openness to new experience, agreeableness, and conscientiousness.
- Each of the five dimensions (or factors) includes specific personality characteristics represented as adjectives on a bipolar scale (for example, quiet–talkative, suspicious–trusting).
- The NEO-PI-R is a widely used personality inventory designed to rate individuals on these Big Five factors.
- Trait research focuses on correlations among trait ratings obtained from self-reports, paper-and-pencil inventories, and questionnaires. The results often show significant stability in traits over time and between independent raters.
- There is increasing recognition that the qualities of the person interact with those of the situation(s) in which he or she functions.

KEY TERMS

ability traits   51
act trends   66
agreeableness (A)   58
big five structure   58
cardinal traits   48
central traits   48
common traits   49
conscientiousness (C)   58
constitutional traits   50
domain of interpersonal
   behavior   56
dynamic traits   51

emotional stability 53
environmental-mold
   traits 50
extraversion (E)
   (extraverts)   44
factor analysis   57
introverts
   (introversion)   44
NEO-PI-R   58
neuroticism (N)   53
openness to
   experience (O)   58

prototypicality   65
psycholexical
   approach   56
secondary
   dispositions   49
source traits   49
surface traits   49
temperament traits   51
trait   44
trait approach   43
trait structure   49
unique traits   49

CHAPTER

4

THE EXPRESSIONS OF DISPOSITIONS

Carmen and Dolores are both moderately high on measures of extraversion and their total scores on that scale are similar. But those who know the two well also see a distinctive difference between their patterns of typical behavior. Carmen is more outgoing in big groups and you count on her to be the center of attention and the life of the party at campus events. When with just a few people having dinner, however, she tends to be more on the quiet side. She also seems to have many casual acquaintances but few intimate friends that she really relates to. Dolores, in contrast, is at her warmest in one-on-one relations—she is the close friend you are sure to have fun with when in a small group or just going out together for the evening.

Do differences like these tell us something about the nature of traits and dispositions? This chapter examines that question, and the answers have yielded some surprises. In Chapter 3 we saw that people differ reliably and stably from each other in their personality. Some people may be distinctly more benevolent, benign, congenial, cordial, generous, good-humored, gracious, likable, sociable, and welcoming than others. Observers tend to agree with each other about peoples' personal qualities. And these qualities are stable over time, sometimes over many decades. Is there also a distinctive *pattern* in the behaviors individuals engage in, and the thoughts and feelings they express characteristically? What are these patterns and what do they mean? In what ways are one person's behaviors distinct from others? This chapter explores the question of the regularities and distinctiveness that exist in the observable patterns of thoughts, feelings, and behaviors that characterize different individuals and types.

▶ TRAITS, SITUATIONS, AND THE PERSONALITY PARADOX

Individual Differences in Behavior Tendencies

Some people may be more likely to behave in a friendly manner; some tend to behave more aggressively than others; some often show disagreeable behavior—the list of individual differences in behavior tendencies is long. And for every type of behavior, one also can think of a corresponding personal quality that makes a person display that behavior more often, or more strongly, than other people. Think of adjectives that describe a type of behavior: helpful, kind, agreeable, conscientious, or aggressive. For every adjective like that there is a noun that refers to a personal quality: helpfulness, kindness, agreeableness, conscientiousness, and aggressiveness. In fact, for any given type of behavior, one can hypothesize a corresponding personal quality that leads people to enact it. The grammatical convention in English makes it easy to name these qualities: just attach -ness at the end of the adjective.

The Intuitive Assumption of Consistency

4.1 What was the surprising result of research on the consistency of behavior across situations? What did behavioral specificity suggest about personality dispositions?

As you saw in Chapter 3, people differ reliably in their personality characteristics, and many if not all of such characteristics do take the form of xxx-ness, where xxx refers to a type of behavior. It seems obvious that people differ reliably in their tendency to display any given type of behavior. So, if you see a person behave in a very friendly manner, it's reasonable to assume she has a friendly personality, and expect that next time you see her, she will behave in a friendly way also. That is, to the extent that people differ reliably in their tendency to display a behavior, we expect them to behave in a similar way in a variety of situations. Of course our behaviors do vary from one situation another; everyone is more somber when her softball team has suffered a season-ending loss in a close game, compared to winning a championship. But the person who is particularly excited after her team's success may be the one who manages to find something to laugh about in a crushing loss. In short if in one situation, a person displays more of behavior X than other people, then we expect him, in other situations as well, to display more of that behavior than other people. So imagine the surprise when in the 1960s and 1970s researchers examined people's behaviors and found that this is not necessarily the case. So although intuitive, it seems obvious that individuals do differ consistently in the kinds of behavioral tendencies that they exhibit in many different situations, the puzzle was that the results in study after study failed to support this assumption when researchers tried to predict actual social behavior from trait measures. This discrepancy between intuition and data became a challenge to the field that deeply influenced its future course.

The 1968 Challenge

Personality psychologists for many years had searched for evidence to document the breadth and power of global personality traits. In the late 1960s, a number of them began to note that these efforts consistently yielded disappointing results (Mischel, 1968;

Peterson, 1968; Vernon, 1964). Mischel (1968) in particular reviewed research that showed that what people do varies to a surprising degree with the particular situation and context. For example, the person who seems conscientious about work may show a very different pattern with family.

Thus, serious questions were raised about the use of broad trait scores for predicting what a given person will do in different kinds of situations and for explaining the seeming inconsistency or variability that was observed across those situations. It was routine for research articles to end with an apology that the tests used did not predict what people would actually do in particular situations. And generally these disappointing results were seen as due to poor methods, bad judges, and unreliable tests, not to any possible problems with the guiding assumptions about the nature of trait consistency.

Over the course of many years, a massive literature had developed on the low consistency of the individual's behaviors when measured in different situations. That implies that Jane, for example, might be highly conscientious in one type of situation but not necessarily in another type of situation. Reviews of this literature, beginning in the 1920s (e.g., Hartshorne & May, 1928; Newcomb, 1929), and reexamined later (e.g., Mischel, 1968; Mischel & Peake, 1982; Pervin, 1994; Shoda, 1990), repeatedly added to the bewilderment, noting that consistency across different types of situations (e.g., from home to school to work) was surprisingly low. Low consistency was found regularly in the behavioral referents for such traits as rigidity, social conformity, dependency, and aggression; for attitudes to authority; and for virtually any other personality dimension (Mischel, 1968; Peterson, 1968; Vernon, 1964). This raised a disturbing question: Might personality dispositions be less global—and more *situation-specific*—than had been assumed? If so, where was the consistency of personality that we all believe intuitively must be there?

The Paradox Defined

The Mischel (1968) book, *Personality and Assessment*, raised this question most forcefully and caused a crisis for the field by asking: "What if the problem is not just with bad methods and poor data but also with wrong assumptions?" The defiant part of this monograph was the proposal that the findings it reviewed might reflect not just imperfect methods and measurement noise, but also the state of nature. This created a challenge that upset the field of personality for many years and generated what Bem and Allen (1974) called the **personality paradox:** Namely, the data from research indicated lack of consistency. Nevertheless, human intuition, and a long tradition of Western thought dating to the ancient Greeks, and including much modern personality research and practice, led to the conviction that the opposite was true. Which one was right, the research or the intuition? A long debate resulted, as discussed in In Focus 4.1.

4.2 How did the "personality paradox" create conflict in the field of personality while promoting the position of situationism?

▶ INCORPORATING SITUATIONS INTO TRAITS

If we accept that people's behaviors vary widely across situations, how do we reconcile that with our intuitive conviction that each individual is characterized by stable and distinctive qualities? What remains consistent through the changing stream of thoughts, feelings, and behaviors? How might one capture what is constant? The finding of large variations in people's behaviors across situations challenged the ultimate goal of personality psychology, which is to identify the coherence and stability that underlie individuals' thoughts, feelings, and behaviors.

THE PERSON *VERSUS* SITUATION DEBATE

This debate consisted of a prolonged and heated controversy about personality dispositions, consistency, and the role of the situation that consumed much of the agenda of personality psychology in the 1970s and early 1980s.

Situationism. At the height of the debate, many social psychologists amassed evidence for the power of the situation as the main determinant of behavior. Some also argued that often the situation is so powerful that individual differences and personality don't make much difference. The emphasis on the power of situational variables, and the belief that personality was less important than the situation, was named **situationism.** This view also proposed that people tend to make erroneous explanations of the causes of behavior. The error is that they systematically neglect the role of the situation and instead invoke personality dispositions as favorite—but incorrect—explanations of social behavior (e.g., Nisbett & Ross, 1980). Called the **fundamental attribution error,** the tendency to focus on dispositions in causal explanations soon was seen as a mistake committed by laypersons in everyday life, as well as by the psychologists who study them (Ross, 1977).

Evidence of systematic judgmental errors in personality assessment and inferences, of course, had been noted often in the past. Before, however, the limitations of judgments about personality had been dismissed as merely due to unreliable, imperfect methods, open to correction by improving the quality of measurement. Now instead, they were read as reflecting human nature (e.g., Nisbett & Ross, 1980). In its most extreme form, some critics argued that personality was mostly a fictitious construction in the mind of the perceiver (e.g., Shweder, 1975).

Revival of the Traditional Paradigm. At the opposite pole, many personality psychologists renewed even more intensely their efforts to retain the traditional paradigm for the study of traits. In the early 1980s, a resurgence of the factor analytic approach occurred in the study of traits. It was founded on an agreement among many researchers to reach a consensus concerning the set of major traits or basic dispositions needed for a comprehensive taxonomy of personality using factor analyses based on traits ratings, in the form of the "Big Five" (e.g., McCrae & Costa, 1999). Many similar factor analytic studies and taxonomies had been done in earlier years, and their strengths and limitations had been duly noted (e.g., Fiske, 1994; Mischel, 1968; Peterson, 1968). The difference now was that they agreed about what those factors were and proposed that the Big Five at least approximated the basic structure of personality. It turned out that this consensus actually was a useful way to revitalize enthusiasm for the study of traits, by showing the power and breadth of the broad trait level of analysis, as you saw in the last chapter.

The Role of the Situation. To understand this debate and the passions it stirred requires a quick look at the history of personality psychology. Since the inception of the field a century ago, the trait level of analysis was mostly devoted to studying the person apart from the situation. In contrast, the field of social psychology was devoted to understanding the general effects and power of situations regardless of individual differences. In that framing, in social psychology the person became the **error variance**—the noise—that had to be removed. In contrast, in personality psychology it was believed that one had to remove the effects of situations—and treat them as noise or error—in order to glimpse the true situation-free personality that remained consistent.

Given the way that the mission of personality and social psychology as disciplines was defined then, many personality psychologists heard a strong argument for the importance of the situation as trying to undermine personality as a field and as a construct. In contrast, most social psychologists hailed it as supporting the importance and power of the situation and the relative insignificance of individual differences in personality. The mistaken belief that to the degree that the person was important, the situation was not, and vice versa, led to the unfortunate **person versus situation debate.** Its resolution is discussed throughout the remainder of this chapter.

4.3 What is the fundamental attribution error? How was it used to promote situationism among social psychologists?

Within the patterns of intraindividual variation, however, there may be a distinctive temporal order, a stable pattern over time that is unique to each individual. On the surface, the thoughts, feelings, and behaviors of an individual may vary considerably, and on the surface, this may appear to go against the central tenet of the construct of personality—that personality is relatively invariant or consistent over time and across situations. But when we look beneath the surface, and focus on how the variation occurs and on what external and internal situations it depends, there may be a regular pattern that is distinctive for each individual, and that goes beyond the surface level variation.

If . . . Then . . . Situation–Behavior Signatures

While still relatively small, the number of investigators focusing on, and finding, such **stable intraindividual patterning** is increasing (e.g., Eizenman, Nesselroade, Featherman, & Rowe, 1997; Fleeson, 2001; Rhodewalt & Morf, 1998; Zelenski & Larsen, 2000). For example, Larsen and Kasimatis (1990) found that some individuals' affective experiences clearly follow a seven-day weekly cycle, while others do not show such a pattern. Similarly, Brown and Moskowitz (1998) showed that some individuals have discernible daily cycles in their interpersonal behaviors such as dominance–submissiveness, and agreeableness–quarrelsomeness, whereas others do not. Rusting and Larsen (1998) found that an "evening-worse" pattern was associated with neuroticism, depression, and anxiety as well as a cognitive style indicative of hopelessness. Cote & Moskowitz (1998) found that in individuals who score high on agreeableness, the pleasant emotions they feel in a given interpersonal interaction and the agreeableness of their behavior were more strongly related.

Sometimes, however, regularities in the stream of behaviors contain even more information. Behaviors do not occur in a vacuum; they occur in specific situations. When the situation changes, so do the behaviors, but the relationship between the situations and behaviors may be stable, and may express an individual's distinctive underlying personality system. For instance, suppose two professors on the whole display the same average amount of sociability. However, the first one is extremely sociable and warm with students but unfriendly and cold with colleagues. In contrast, the second professor shows the opposite pattern, and is unfriendly and disinterested with students but very sociable with colleagues. Research indicates that many people are reliably characterized by such patterns. Such profiles seem to constitute a sort of **signature of personality** that does reflect some of the essence of personality coherence and promises to provide a route to glimpse the underlying system that generates them (Mischel & Shoda, 1995; Shoda, Mischel, & Wright, 1994).

To give another example, in situation A, people rarely initiate personal interactions while in situation B, they commonly do so. Suppose also that Mark tends to become irritated when he thinks he is being ignored (likely to happen in situation A), whereas April is happier when she is left alone and even becomes irritated when people tell her personal stories about themselves (which happens often in situation B). Then Mark will become irritated in situation A, but not in situation B, and April will show the opposite *if . . . then . . .* **pattern,** annoyed if B, but not if A. These feelings further activate other thoughts and feelings in each situation in a characteristic pattern for each individual. If so, then even if both people have similar overall levels of irritability, they will generate distinctive, predictable *if . . . then . . .* patterns.

The point is that an individual's personality may be seen not only in the overall average frequency of particular types of behavior shown but also in when and where that behavior occurs. The *if . . . then . . .* patterns of situation–behavior relationships that unfold—if they are stable—can then provide a key to the personality. That is an expression of personality coherence that is eliminated when the situation is removed by aggregating the behavior across different situations or by ignoring the situation altogether.

4.4 What are behavioral signatures? How does this "*if . . . then . . .*" concept preserve the coherence and consistency of personality?

Evidence for Signatures Indicative of Personality Types

Researchers have tested this proposition and found clear support for it (Shoda, Mischel, & Wright, 1989, 1993a, b, 1994). In these studies, children were systematically observed for more than 150 hours per child in a residential summer camp setting over the course of six weeks (e.g., Shoda et al., 1993a, 1994). As predicted, children tended to display stable, distinctive patterns of *if . . . then . . .* relationships. To illustrate (Fig. 4.1), some children were consistently more verbally aggressive than others when warned by an adult, but were much less aggressive than most when their peers approached them positively. In contrast, another group of children with a similar overall average level of aggression was distinguished by a striking and opposite *if . . . then . . .* pattern: they were more aggressive than any other children when peers approached them positively, but were exceptionally unaggressive when warned by an adult (Shoda et al., 1994). A child who regularly becomes aggressive when peers try to play with him is quite different from one who expresses aggression mostly to adults

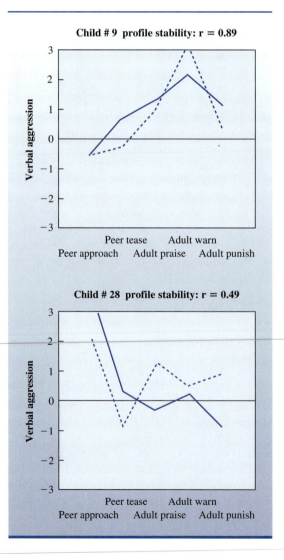

Figure 4.1   Individual *if . . . then . . .* situation-behavior signatures for two children. Their aggressive behavior was observed in five different situations many times. Half of the observations are shown as dotted lines; half as solid lines. Profile stability is the correlation between the two sets of observations.
Source: Shoda, Y., Mischel, W., & Wright, J. C. Intra-individual stability in the organization and patterning of behavior: Incorporating psychological situations into the idiographic analysis of personality. *Journal of Personality and Social Psychology, 67,* 674–687, fig. 1. © 1994 by the American Psychological Association. Adapted with permission.

who try to control him, even if both are equal in their overall aggressive behavior. In short, stable *if . . . then . . .* personality signatures, not just stable levels of average overall behavior, were found to characterize individuals, and their patterns seem to be meaningful reflections of the personality system.

To summarize, individuals differ stably in their distinctive *if . . . then . . .* strategies and behavior patterns. Consequently, they will behave in their characteristic ways within a given type of situation, but they will *vary* their behavior predictably when the "if" changes, thus producing behavioral variability across situations. Then, by aggregating behavior across situations in their search for broad behavioral dispositions, researchers risk eliminating the distinctive individuality and stability that they are trying to find! Like their handwritten counterpart, behavioral signatures can be seen as an expression of individuality that identifies the person. To illustrate this more concretely, Gary W's behavioral signatures are summarized next.

Gary W's Behavioral Signatures

Gary's behavior depends on the psychological situation as well as on the type of behavior involved. For example, while he may respond pessimistically and with defensive withdrawal in many interpersonal situations, when tasks call for quantitative skills he is self-confident and optimistic. And while on average, Gary is not a "generally aggressive person," he does have disruptive outbursts of anger, especially when he feels ignored or rejected. It is notable that, *if* Gary feels provoked or threatened in an intimate relationship with a woman, *then* he becomes vulnerable to outbursts of rage. On the other hand, *if* he feels secure, *then* he can be extremely caring. Such *if . . . then . . .* relationships provide a window for seeing Gary's unique but stable patterns, the "signature" of his personality. In short, this perspective emphasizes that both the person and the situation need to be considered when trying to predict and understand individuals and the important ways they differ from each other.

Consider the differences between Gary W (dotted line) and Charles W, his older brother (solid line). Figure 4.2 illustrates their self-reported (in daily diaries) emotional states (from extremely stressed, upset, and negative affect at the top of the scale to extremely pleased and positive at the bottom). As the figure shows, Gary and Charles are similar in becoming very upset in situation 1. This is when they feel "provoked or threatened" in close personal relations with women. They differ, however, in the other events that upset them most. In situation 2, public speaking (for example, in seminar presentations), Gary becomes distressed; Charles enjoys those occasions and looks forward to them. Likewise, while Gary becomes extremely unhappy and "mad" when he gets negative feedback about his work (situation 4), Charles readily dismisses complaints from unhappy patients or colleagues as "part of the grumbling you have to expect. . . . There are always people you can't please." Interestingly, the pattern reverses when the brothers must deal with quantitative work problems (situation 3), for example, when writing research reports in their technical specialties. Gary finds those tasks a "real high—a challenge"; Charles experiences them as anxiety-provoking, fearing that his inadequate grasp of the quantitative method he uses in his medical research will be revealed.

Behavioral Signatures of Different Personality Types

Are behavioral signatures meaningful, and can they help us generalize and predict behaviors in a different context? Can they also be used to predict future behaviors in new situations? One way to address these questions is by identifying patterns of *if . . . then . . .*

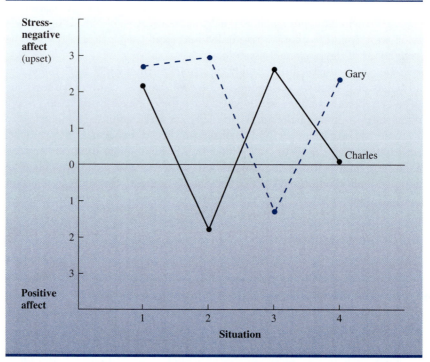

Figure 4.2 Illustrative *if . . . then . . .* signatures. Self-reported stress and emotion, for Gary and Charles, in four situations (sampled repeatedly in daily diaries for 3 months)

relations that characterize not just one person, but a group of people or a type. Once one knows that an individual has a certain type of *if . . . then . . .* signature, it would become possible to make predictions about her behavior based on what is already known about that type. That approach is being applied to important personality types, and we next look at two examples.

4.5 Summarize the behavioral signature of the narcissistic personality type.

The Narcissistic Signature. Narcissists are people whom most of us have met, and when we do, we rarely forget them; although when we get to know them, we often find them extremely difficult to deal with and wish we had never met them. Think of the vain would-love-to-be a movie star type, deeply in love with himself, and ever-eager to prove how wonderful he is—almost desperately trying to convince himself as well as the world that he is really the greatest. The most striking feature of the "signature" of narcissists (Morf & Rhodewalt, 2001a,b) is the narcissist's chronic vigilance for situational opportunities in which their grandiose self-concepts can be affirmed and bolstered (Morf, Ansara, and Shia, 2001; Morf, Weir, & Davidov, 2000; Rhodewalt & Eddings, 2002). Thus they are characterized by the pattern of *if* opportunities to promote self, *then* much effort toward providing evidence for self's superiority. Simultaneously, narcissists have cynical and unempathic views of others and seem insensitive to other's concerns and situational constraints. Thus, if in situations that call for modesty, as when others out-perform them, they nevertheless may derogate their competitors, belittling their achievements—even when the long-term costs are self-defeating and relationship-destructive. Furthermore, they provide post hoc interpretations of

Narcissists have grandiose self-concepts.

experiences that ingeniously amplify the positive feedback to them while discounting the negative (documented in Morf & Rhodewalt, 2001a,b)—in short, they walk around the world trying to convince themselves that everybody really loves them—although they also seem to have a hard time believing they are really lovable.

The Signature of Rejection Sensitivity.   It is widely assumed that dependency in adults is linked to a broad disposition of passivity–submissiveness, but fine-grain analyses of the actual behavior indicate that within some situations, dependency is in fact related to high levels of activity and assertiveness. Specifically, while dependent individuals who are focused on getting along with a peer will self-denigrate, they will self-promote when concerned mostly with pleasing an authority figure (Bornstein, Riggs, Hill, & Calabrese, 1996).

In particular, there has been much research on individuals who have intense anxieties about interpersonal rejection, called rejection sensitivity or RS (Downey & Feldman, 1996). Their rejection sensitivity is seen *if* in an intimate relationship they encounter what could be construed as uncaring behavior (e.g., partner is attentive to someone else), *then* becoming overly worried that "she doesn't love me." These thoughts in turn tend to trigger expectations of rejection, abandonment, feelings of anger and resentment, and anxiety and rage at the prospect of abandonment. Coercive and controlling behaviors then become activated, but typically are blamed on the partner's behavior. However, they are not overall more likely than others to express anger, disapproval, and coercive behaviors in certain types of intimate situations. For example, high RS women tend to be especially solicitous, putting the other person's needs first in their relationships (e.g., Purdie & Downey, 2000).

Distinctive situation–behavior signatures like these are interesting to many personality psychologists because they give clues for understanding the person's underlying

4.6 How is the trait of rejection sensitivity expressed as a personality signature?

The high RS person easily feels rejected when the partner withdraws attention.

personality. For example, in the narcissist signature described on page 76, narcissists treat interpersonal situations as challenges for eagerly showing off how good they are by out-doing and out-performing other people while trying to take center stage in the process. Their signatures hint at some of these underlying qualities. In contrast, rejection sensitive people view the same situations in order to anxiously scan them. They want to see how likely they are to be personally rejected and hurt, and are ready to magnify the slightest possible criticism and to overreact to it.

Two Types of Consistency

4.7 Differentiate between Type 1 and Type 2 consistency.

Taken collectively, the results in the search for consistency in the behavioral expressions of personality suggest that there are two different aspects of consistency. Both need to be considered in the study of individual differences and personality dispositions at the trait level of analysis, and each has its distinctive uses, advantages, and limitations.

- **Type 1: Average overall levels of behavior tendencies.** The first or "classic" form of consistency is in the overall average differences in the levels of typical behavior of different kinds (such as aggressiveness, sociability) that characterize the individual. The trait level of analysis captures these by aggregating indices of what the individual is like and does across many different types of situations, as when people are described with regard to their status on broad factors.

- **Type 2: If . . . then . . . (situation–behavior) signatures.** These consistencies are seen in patterns of stable links between types of situations and types of characteristic behavior, as illustrated in the signatures of the narcissistic and rejection sensitive types of persons.

Uses of the Two Types of Consistency

Is it more useful to try to infer broad traits or situation-specific signatures? The answer always depends on the particular purpose. Inferences about global traits may have limited value for the practical prediction of a person's specific future behavior in specific situations, or for the design of specific psychological treatment programs to help facilitate constructive change. But broad traits have many other uses. Indeed, they have value for everyday inferences about what other people seem like on the whole, as when you get first impressions about a stranger, or form a quick opinion about a new acquaintance (e.g., McAdams, 1999). With people we know well and who are important to us, however, we also want to know "what makes them tick." Therefore, we try to understand their goals, motivations, and feelings in order to make sense of the *if . . . then . . .* patterns—the personality signatures that characterize them as we get to know them better (Chen-Idson & Mischel, 2001).

The first type of consistency—average overall behavior tendencies on broad dimensions—can be enhanced by the method of aggregation introduced in the last chapter. In this method, behavior on a dimension is simply averaged across many different situations to estimate the person's overall true level of the characteristic. Bolstered by this method, the case can be made that global dispositions as traditionally conceptualized are "alive and well" if one routinely aggregated multiple observations and measures across different situations. That serves to eliminate the role of the situation by averaging it out. This strategy now acknowledges that specific behaviors across different types of situations could not be predicted by such a model (e.g., Epstein, 1979), and continues to treat the situation as a source of noise by removing it as before. On the positive side, by aggregating the individual's behavior across many different kinds of situations, it is in fact possible to increase greatly the strength of the overall average differences that will be found between individuals "on the whole" for diverse traits (e.g., Epstein, 1983).

Inferences about broad traits also have practical value for gross initial screening decisions (as in personnel selection), studying average differences between groups of individuals in personality research (Block & Block, 1980), or the layman's everyday perception of persons (e.g., Funder & Colvin, 1997; Schneider, 1973; Wright & Mischel, 1988). And as was seen in the last chapter, when measures are combined or aggregated over a variety of situations, one can demonstrate stable differences among individuals in their relative overall standing on many dimensions of social behavior.

In sum, overall average differences between individuals can be obtained easily, particularly with rating measures, and used to discriminate among them for many purposes (Funder & Colvin, 1997; Kenrick & Funder, 1988). For other goals, however, more specific levels of analysis that consider *if . . . then . . .* , situation–behavior signatures and that take the situation into account may be more useful. Knowing how your friend behaved before can help you predict how he or she probably will act again in similar situations. The impact of any situation or stimulus depends on the person who experiences it, and different people differ greatly in how they cope with most stimulus conditions. It is a truism that one person's favorite "stimulus" may be the stuff of another's nightmares and that in the same "stimulus situation" one individual may react with aggression, another with love, a third with indifference. Different people act differently with some consistency in particular classes of situations, but the particular classes of conditions tend to be narrower than traditional trait theories have assumed (e.g., Cantor, 1990; Cantor & Kihlstrom, 1987; Mischel, 1990). For purposes of important individual decision-making, one may need highly individualized assessments of what the specific situations mean to the person.

The significant consistency that exists is at least in part reflected in stable patterns of behavior *within* similar types of psychological situations (Mischel & Peake, 1982; Mischel & Shoda, 1995). As the similarity between situations decreases, so does the cross-situational consistency of the person's behavior (Krahe, 1990; Shoda, 1990; Shoda, Mischel, & Wright, 1994). Then it becomes difficult to predict from what individuals did

4.8 What are the practical advantages of the two methods of studying consistency? What is required in order to obtain Type 1 consistency?

4.9 In what situations are behavior signatures a more useful way to approach personality consistency? Under which circumstances will it be difficult to predict behavior using Type 2 consistency?

in one type of situation to what they will do in a very different type of situation. On the other hand, within particular types of situations, individuals do show characteristic stable patterns. These patterns are expressed as stable *if . . . then . . .* relationships, such that when the situation (the "if") remains stable (e.g., if "teased by peers about his glasses and appearance"), then so does the behavior (e.g., he becomes physically aggressive). So for the people we know well, we have a sense not only of what they are like in general (in trait terms) but also of what they are likely to do in particular types of situations (e.g., at the holiday dinner, in an argument about money, on a date).

To apply these abstract points more concretely, think again about Gary. It is certainly possible to form some generalizations about his seemingly major qualities, strengths, and problems. Such generalizations help us to differentiate Gary from other people, and to compare him with them. Suppose that we learned, for example, from trait tests that Gary tended to be low on extraversion, relatively high on neuroticism and anxiety, high on conscientiousness, and intermediate on both sociability and open-mindedness scales. Such characterizations may help one to gain a quick overall impression of Gary. But in order to predict what Gary will do in specific situations, or to make decisions about him (as in therapy or vocational counseling), it would be necessary to conduct a much more individually oriented study that considers the specific qualities of Gary as they relate to the specific situations of interest in his life.

Just when does Gary become more—or less—neurotic and anxious? Under what conditions does he *not* avoid close relations with people? When is he likely to be more and less sociable in ways that are distinctive and predictable? The analysis and prediction of specific behavior requires that we ask questions like these to link behavior to conditions, and to capture their distinctive pattern and meaning, in addition to painting personality portraits with more general characterizations that also have their many valuable uses.

▶ INTERACTIONISM IN PERSONALITY PSYCHOLOGY

Interactionism in personality is the idea that the individual's experience and action cannot be understood as the result of separate personal and situational factors. Rather, it is a product of dynamic interactions between aspects of personality and situations (Magnusson, 1999; Magnusson & Endler, 1977; Mischel, 1973). Interactionism focuses on how the expressions of the stable personality system are visible in the person's unique *patterns* of *if . . . then . . .* , situation–behavior relationships.

The Meaning of Person × Situation Interaction

4.10 What is meant by a person × situation interaction? What does it add to a knowledge of a person's traits?

In the interactionist view, knowledge of individual differences often tells us more if it is combined with information about the conditions and situational variables that influence the behavior of interest. Conversely, the effects of conditions depend on the individuals in them. Thus, the interaction of individual differences and particular conditions, and not just the individual or the context separately, is important (Bem & Funder, 1978; Magnusson, 1990; Shoda, Mischel, & Wright, 1994; Wright & Mischel, 1987). *If . . . then . . .* , behavioral signatures of individuals, and those of groups of individuals such as the "narcissists," illustrate the basic principles of interactionism.

An Example: Uncertainty Orientation. An example of **person × situation interaction** involving an individual difference dimension comes from research on **uncertainty orientation.** This personality dimension is defined at one end by individuals who are relatively comfortable dealing with uncertainty and strive to resolve it, and on the other end by those who are more uncomfortable with uncertainty and likely to avoid situations that increase their subjective sense of uncertainty (Sorrentino & Roney, 1986, 2000). Now consider this question: Would experiencing a situation in which one didn't have control over the outcome make people approach, rather than avoid, new information? Would students do better on a test if they are told that the test is diagnostic of an important ability? The answers to these questions depend on an individual's uncertainty orientation. For those who are high in uncertainty orientation (i.e., those who are comfortable with uncertainty and seek to master it), the answer is yes to both questions. But for those who are uncomfortable with uncertainty, the answer is very different (Huber, Sorrentino, Davidson, & Epplier, 1992). For example, they do better on a test if they are told it is not diagnostic of important abilities (Sorrentino & Roney, 1986). Further, when individuals low in uncertainty orientation experience uncontrollability, they avoid new information, especially if they are also mildly depressed (Walker & Sorrentino, 2000). So it all depends on the interaction between the type of person with the type of situation: the situation that suits one person may distress another.

4.11 How does research on uncertainty orientation illustrate the value of an interactionist approach?

Note that differences in personality are expressed here not in the overall average behavior tendencies but rather in the different effects of situations, be it the experience of uncontrollability or the perceived meaning of a test. A similar insight was captured in an example Gordon Allport (1937) used. He asked: What is the difference between butter and raw eggs? They differ in the *effect* of cooking heat: as heat increases one softens and melts, but the other one hardens. The question of which is harder *in general* is meaningless unless one specifies the temperature. It results in one answer if the food is in the refrigerator, but another if the food is in the frying pan. Depending on the situations in which participants in a study find themselves, one would make opposite predictions for their behavior. And that, put most simply, is the basic **principle of interactionism.**

Definition of Triple Typology. In a thoughtful analysis of the importance of interactions for personality psychology and for understanding personality dispositions, Daryl Bem (1983) summarized a mission of personality psychology, or any science for that matter, with great simplicity. Whether or not one is studying molecules, plants, or people, there are millions of specific instances of them. A goal of science is to produce a way to categorize them into a manageably small number of *meaningful* categories. Because objects can be categorized in many arbitrary ways, "meaningful" of course is the key concept. In what sense should it be meaningful?

4.12 What is the basic question posed in Daryl Bem's triple typology? How is it illustrated in the case of narcissism?

Bem's answer to that question was that a meaningful way to categorize people, behaviors, and situations was one that allows accurate statements in the form: *"this type of person will do these types of behaviors in these types of situations."* This is a **triple typology** in the sense that it classifies together three categories: types of people, types of behavior, and types of situations. A meaningful system of categorizing people, behaviors, and situations is one that allows the personality psychologist to spell out precise *if . . . then . . .* regularities in describing types of people. For example, "narcissists" is a more meaningful way of categorizing people than "people who shop at supermarkets." For narcissists, we can say, "*if* they encounter situations that can be seen as an opportunity to bolster their grandiose self-concept, *then* they will seek to demonstrate their superiority." It would be much more difficult to generate *if . . . then . . .* statements about the social and interpersonal behaviors of "people who shop at supermarkets" that apply to most members of that category.

Note that the *if . . . then . . .* statement describing narcissists already contains references to types of behaviors and types of situations. That is, "seeking to demonstrate their superiority" is a type of behavior, and "situations that can be seen as an opportunity to bolster their grandiose self-concept" is a type of situation. There are meaningful categories of behaviors and situations precisely because they allow a simple description of narcissists in one *if . . . then . . .* statement. The practical challenge of course is to find meaningful categories of people, behaviors, and situations.

In one promising direction, modern computer technology is beginning to provide hope that a triple typology is in fact attainable. For example, Van Mechelen and his colleagues have devised a computer algorithm that simultaneously tries out many different ways to categorize people, behaviors, and situations. It searches for the best system of typologies that allow the fewest number of *if . . . then . . .* statements to describe the data with the least amount of errors (Van Mechelen & Kiers, 1999; Vansteelandt & Van Mechelen, 1998). To see how this works, read In Focus 4.2.

4.13 Describe the approach taken to produce a triple typology of hostility. What psychological processes were found to underlie the tendency to keep anger in or let it out?

IN FOCUS 4.2

A TRIPLE TYPOLOGY FOR HOSTILITY

To test their methodology, the researchers asked study participants to describe what specific responses they would exhibit when faced with a specific type of situation. They chose these situations carefully to get a wide spectrum of potentially frustrating situations in order to obtain a variety of response patterns. Some of the situations they used were: you are in a restaurant and have been waiting a long time to be served (low frustration); you are trying to study and there is incessant noise (moderate frustration); your instructor unfairly accuses you of cheating on an examination (high frustration). The responses provided were also intended to reflect a wide variety of potential hostile responses, such as turning away; wanting to strike something or someone; hands tremble; heart beats faster; cursing; becoming tense; feeling irritated; or becoming enraged (taken from S-R Inventory of Hostility, Endler and Hunt, 1968).

When the data were analyzed by a computer program designed by Van Mechelen and colleagues, the result was a grouping of people, behaviors, and situations. In this classification, each group of people was succinctly and accurately characterized by a set of *if . . . then . . .* statements describing how this group of people tends to display these types of behaviors in these types of situations. For example, people of type 3 were likely to respond to a situation that is highly frustrating by becoming tense, having their heart beat faster, sweating, and having their hands tremble. People of type 2 would respond to the same type of situations by wanting to strike something, cursing, and becoming tense or irritated. In a situation that is not very frustrating, such as when you accidentally

bang your shins against a park bench, neither type 2 nor type 3 people would respond in a hostile manner. However, a person of type 7 would likely have an increased heart rate, grimace, become tense, and curse (Vansteelandt & Van Mechelen, 1998). Because these person types directly reflect *if . . . then . . .* behavioral signatures, this sort of typology can be used to predict the probability of a certain type of individual producing certain types of behavior when faced with a certain type of situation.

After Van Mechelen and his colleagues formulated these typologies, they were able to examine some of the psychological processes that linked the type of person to the type of situation–response patterns they produced. They designed a questionnaire that asked about the participant's interpretation of situations, expectations about the consequences of different types of behavior, and whether they kept their anger in or let it out. These questions were designed to engage thoughts and feelings that were related to hostile and aggressive behavior. For example, participants were asked to agree or disagree with the following statements: "When other people get me in trouble, I rapidly think that they do it on purpose" (interpretation); "I think that, in general, the expression of anger is not appreciated by others" (expectation); "I easily suppress feelings of frustration" (anger-in); "When I feel frustrated, I show it easily" (anger-out) (Vansteelandt & Van Mechelen, 1998). They also asked the participants to fill out a questionnaire designed to measure the participant's ability to tolerate frustration (Van Der Ploeg, Defares, & Spielberger, 1982). The researchers then looked at how the person types related to the presence or absence of each of the types of thoughts and feelings measured.

Their findings showed the feasibility of forming meaningful categories of people, behaviors, and situations simultaneously. They also indicated that the resultant typology provided glimpses into the internal psychological processes that underlie the observed *if . . . then . . .* behavior sig-

natures characteristic of various types. The different types were clearly different in the patterns they used for responding to different types of frustrating situations from very mild to the highly frustrating. The types also differed in the ways in which they typically expressed hostility, all the way from no hostile responses to cursing and readiness to strike out physically in different types of situations.

Interaction as a Rule in Science

Although the concept of interactionism has been controversial in personality psychology, the fact that organisms and environment interact to determine behavior and development is a general conclusion in most sciences dealing with living organisms. In biology, evidence of organism–environment interaction is commonplace. As Lewontin (2000) points out, genetic and environmental influences are intertwined, making development "contingent on the sequence of environments in which it occurs" (p. 20). In the example of plant growth, Lewontin notes that cloned samples of seven different individuals of a given plant species that were grown at different elevations showed little cross-environment consistency in the rank-ordering of their size of mature plants, because a particular plant that flourishes at one altitude may readily have below-average growth in another context. Consequently, one has to represent plant growth not in terms of the average growth tendencies of an individual organism but in graphs that take into account the environment in which the organism develops. To understand plant growth and predict the size of a given plant relative to others, one has to take into account the particular environmental conditions that are present—in other words, one has to identify and understand the *if . . . then . . .* contingencies.

Other examples abound in the biological sciences, including an analysis of the "plasticity" (modifiability) of neural systems (Edelman, 1992; Kolb & Whishaw, 1998) and the ways in which cultural and biological factors interact in human evolution (Durham, 1991). Ehrlich (2000) focusing on the interplay of genetic endowment and environmental experience obtained similar findings for the importance of interaction. He concluded that the psychologist's typical strategy of partitioning the determinants of behavioral characteristics into separate genetic versus environmental causes was no more sensible than asking which areas of a rectangle are mostly due to length and which mostly due to width. Yet curiously the implications of interactionism for understanding the multiple cognitive, behavioral, and physiological mechanisms through which personality functions have been extremely difficult to accept in traditional trait level analyses of personality (as discussed in Cervone & Mischel, 2001, Chapter 1).

4.14 How general is interactionism throughout the sciences? Give some examples.

Resolution of the Personality Paradox

If stable *if . . . then . . .* , situation–behavior patterns are meaningful reflections of the personality, they also should be linked to the person's self-perceptions about his or her own consistency. The relationship between the stability of the person–situation profile that characterizes an individual in a particular domain of behavior and the self-perception of consistency has been closely examined. The results directly speak to Bem's classic "personality paradox" which, as you saw earlier, has motivated much of

4.15 Are self-perceptions of behavioral consistency related to actual consistency? Consistency of what?

the research agenda in studies at the trait level. As Bem pointed out in the 1970s, while our intuitions convince us that people have broad behavioral dispositions that we believe are seen in extensive consistency in their behaviors across situations, the research results on cross-situational consistency in behavior persistently contradict our intuitions (Bem & Allen, 1974). Recall that to resolve this dilemma, and prove our intuitions are better than our research, Bem and Allen noted that traditional methodologies assume that all traits belong to all persons. But if a given trait is in fact irrelevant for some people, their inconsistency with regard to it will obscure the consistency of the subset of people for whom the trait is relevant. Therefore, Bem and Allen reasoned that a solution to the consistency problem requires first selecting only those persons who perceive themselves as consistent in the given disposition. We then should expect to find high cross-situational consistency in their behavior in that domain, but not in the behavior of those who see themselves as inconsistent with regard to it, or to whom it is irrelevant.

Initially, Bem and Allen (1974) obtained some encouraging support for this prediction. A few years later, in a more comprehensive test in a large field study at a college in the midwest of the United States, researchers observed behavior relevant to "college conscientiousness" and friendliness as it occurred over multiple situations and occasions (Mischel & Peake, 1982). Each of the 63 participating college students were observed repeatedly in various situations on campus relevant to their conscientiousness in the college setting. The specific behaviors and contexts selected as relevant were supplied by undergraduates themselves in pretesting at the college. Conscientiousness was sampled in various situations such as in the classroom, in the dormitory, in the library, and the assessments occurred over repeated occasions in the course of the semester.

These data were used to examine the links between the students' self-perceptions of consistency and their actual behavior (Mischel & Shoda, 1995). As the first set of two columns of Figure 4.3 shows, those who perceived themselves as consistent (the first light column) did not show greater overall cross-situational consistency than those who did not. In contrast, the second set of columns clearly supports the hypothesis of coherence in terms of pattern stability: For individuals who perceived themselves as consistent, the average *if . . . then . . .* , situation–behavior signature stability correlation was near .5, whereas it was trivial for those who saw themselves as inconsistent.

In short, the self-perception of consistency seems to be predictable from the stability in the situation–behavior signatures. That, in turn, indicates that the intuition of consistency is neither paradoxical nor illusory: it is linked to behavioral consistency, but not the sort for which the field was searching for so many years.

Summary: Expressions of Consistency in Traits – Dispositions

4.16 How does the concept of behavioral signatures help resolve the personality paradox? What does this resolution suggest about how we might view the coherence of personality?

This chapter started with the basic personality paradox as the problem to untangle. To recapitulate, the construct of personality rests on the assumption that individuals are characterized by distinctive qualities that are relatively invariant across situations and over a span of time. This assumption seems also intuitively true. The paradox was that in a century of personality research, evidence showed that individual differences in social behaviors tend to be surprisingly variable across different situations. Research that sought regularities in the patterns of variability has begun to provide a resolution of this paradox. Namely, the variability across situations seen as the person's behavior unfolds across different situations is not simply random fluctuations. Rather, it in part is a stable and meaningful characteristic of an individual that may provide a window into the underlying system that produced them. This type of behavioral signature of person-

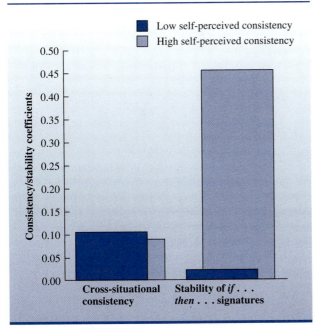

Figure 4.3   Cross-situational consistency and the stability of *if* . . . *then* . . . signatures for people high versus low in self-perceived consistency in conscientiousness. Self-perceived consistency is predicted by stability of the *if* . . . *then* signatures.

Source: Based on data from Mischel, W., & Shoda, Y. (1995). A cognitive-affective system theory of personality: Reconceptualizing situations, dispositions, dynamics, and invariance in personality structure. *Psychological Review, 102,* 246–268.

ality is lost if one treats the situation as error or noise and eliminates it by simply averaging behavior over diverse situations.

The intuition of consistency in the behavior of the people we know and in ourselves, then, is neither paradoxical nor illusory. It just turns out to be based on a different type of behavioral consistency than the one for which psychologists searched for so many years. Such variability seems to be an essential expression of the enduring underlying personality system itself and its organization. The person–situation debate is resolved not just by recognizing that the person and the situation both are important—a fact that has long been acknowledged. Rather, it dissolves if one thinks of personality in ways that make the predictable variability of behavior across situations an essential aspect of its behavioral expression in the form of meaningful and stable interactions between types of persons and types of situations. Furthermore, to the degree that people are characterized by stable and distinctive patterns of variations in their behavior across situations, their level of cross-situational consistency cannot be very high (Shoda, 1990) and there is no reason to expect it to be. It likewise follows that the "situation," then, is not necessarily a source of error to be removed in the search for personality. Rather, it is a locus in which personality is expressed and within which it can be more fully understood. The challenging question then becomes: what is the nature of that underlying system and organization that accounts for the two types of behavioral consistency that have been found to characterize people stably? That is the question we will continue to pursue in the rest of this text.

SUMMARY

TRAITS, SITUATIONS, AND THE PERSONALITY PARADOX

- Despite intuition and a long tradition of Western thought, researchers found it difficult to show that individuals display highly consistent types of behavior across a variety of different situations.

- In the person–situation debate, it seemed that psychologists had two choices: Either they could view situation information as "noise" that obscures the true, consistent personality, or they could recognize the power of situations.

INCORPORATING SITUATIONS INTO TRAITS

- Alternatively, they could take into account situational features when interpreting individual behavior patterns.

- Personality may be seen both in the overall average frequency of behaviors and in the link between the type of behavior and the type of situation in which it occurs.

- The *if . . . then . . .* , situation–behavior signatures that have been studied show significant stability. Two types of personality consistency emerge, a) the average of overall behavioral tendencies and b) distinctive *if . . . then . . .* , situation–behavior signatures.

- Measures of traits are often not able to predict single acts, but they may predict a pooled combination of behaviors across different situations.

- Knowledge of an individual's specific *if . . . then . . .* signature may help predict what that individual will do in similar types of situations.

INTERACTIONISM IN PERSONALITY

- *Interactionism* is the idea that the individual's behaviors are the product of dynamic interactions between personality and the psychological environment.

- Some personality psychologists try to categorize people, situations, and behaviors into triple typologies.

- Biological sciences treat the dynamic interaction between organisms and their environments as a given.

- The intuition of consistency is neither paradoxical nor illusory: Personality coherence is seen in the individual's meaningful, stable situation–behavior signatures.

KEY TERMS

error variance   72
fundamental attribution
   error   72
if . . . then . . . pattern
   (signature)   73
interactionism   80
person versus situation
   debate   72

person × situation
   interaction   80
personality
   paradox   71
principle of
   interactionism   81
signature of
   personality   73

situationism   72
stable intraindividual
   patterning   73
triple typology   81
uncertainty
   orientation   80

PART I

THE TRAIT-DISPOSITIONAL LEVEL

▶ OVERVIEW: FOCUS, CONCEPTS, METHODS

As we conclude Part One and the presentation of work at the trait-dispositional level of analysis, we pause to put it in perspective and consider the essence of its contributions. So much has been done within trait-dispositional approaches that it becomes easy to lose the essential characteristics, which are summarized in Table A. The summary reminds you that the main aim of the trait approach was to provide methods to infer and quantify people's social–personal traits. In classic trait theories of personality, these traits were assumed to be stable and broadly consistent dispositions that underlie a wide range of behaviors across a number of related situations. As summarized in Table A, traits are inferred from questionnaires, ratings, and other reports about the subject's dispositions. Usually, the person's self-reports are taken as direct signs of the relevant dispositions. For example, the more often you rate yourself as aggressive, the more you are assumed to have an aggressive disposition. The focus of research is on measurement to develop quantitative ways of finding and describing important stable individual differences. Traditionally, the trait approach has recognized that behavior varies with changes in the situation, but it has focused on individual differences in the overall response tendency averaged across many situations. Individual differences on any given dimension are more visible in some situations than in others, and not all dimensions are relevant for all individuals. Some psychologists within this perspective now view traits as "act trends"—summaries of behavior—rather than as explanations or determinants of behavior.

Increasingly, some theorists and researchers also try to take systematic account of the situation as it interacts with the individual's qualities and are incorporating the situation into the conception and assessment of personality dispositions. This alternative

TABLE A   Dispositions Expressed as Overall Behavioral Tendencies

Basic units:	Inferred trait dispositions
Causes of behavior:	Generalized (consistent, stable) dispositions
Behavioral manifestations of personality:	Direct indicators of traits
Favored data:	Test responses (e.g., on questionnaires); trait ratings, including self-ratings
Observed responses used as:	Direct signs (indicators) of dispositions
Research focus:	Measurement (test construction), description of individual differences and their patterning; taxonomy of traits
Approach to personality change:	Not much concerned with change: Search for consistent, stable characteristics
Role of situation:	Acknowledged but of secondary interest

TABLE B   Dispositions Expressed as *If . . . then . . .* Behavioral Signatures

Basic units:	*If . . . then . . .* behavioral signatures of personality
Causes of behavior:	Underlying personality processes/dynamics
Behavioral manifestations of personality:	Patterns of behavioral signatures; inconsistencies across different kinds of situations
Favored data:	Behavior sampling, situation-specific diaries, field studies of behavior in context as it unfolds naturally, *if . . . then . . .* self-reports, Person X Situation interactions
Observed responses used as:	Indicators of underlying personality processes (e.g., motives, goals)
Research focus:	Identifying stable personality signatures and understanding their meaning
Approach to personality change:	By understanding the meaning of the personality signatures and modifying them if desired (discussed in Part VII)
Role of situation:	Essential; needs to be incorporated into the conception and measurement of dispositions and personality

approach, summarized in Table B, includes the overall level of a given type of behavior in the study of individual differences. It focuses, however, on the situation-specific expression of traits in the form of stable *if . . . then . . .* , situation–behavior signatures of personality. These signatures are seen as a second type of behavioral consistency that needs to be captured in a comprehensive assessment of dispositions. They provide clues to the motives, goals, and other aspects of the personality system that may underlie them. As Table B indicates, that requires sampling behavior in relation to the situations in which it occurs, and leads to alternative methods for measuring and thinking about personality and its expressions.

▶ ENDURING CONTRIBUTIONS OF
THE TRAIT-DISPOSITIONAL LEVEL

Research and theory at this level of analysis represents one of the longest and richest research traditions in the study of personality. For a century, it has made essential contributions to the quantification and objective measurement of personality, for evaluating test results, and for measuring treatment effects. From the start of the field, it has provided both concepts and methods for describing and comparing human individual differences with regard to a vast array of attributes. Its yield has been great, identifying and quantifying the consistencies that characterize different individuals and types of people on trait dimensions. It now provides reliable taxonomies for describing people, identifying their stable positions on a map of trait factors. The Big Five provides such a map that is currently state-of-the-art, and there is considerable, although by no means complete, consensus that it usefully captures much or most of the trait domain. It is useful for many goals, enabling descriptions of the broad trait differences that characterize individuals and groups, and it has greatly stimulated research in the area. It helps psychologists to capture the general gist of what individuals or groups and even cultures "are like on the whole" and to obtain a broad impression that compresses much information.

Nevertheless, work at this level has limitations that even its advocates recognize. Basically, it is not that the work is wrong or technically flawed. The limitation, rather, is that it is only one piece—an extremely important piece—needed for putting together a conception of the whole of personality and all its diverse aspects, and therefore it is incomplete. It offers a first part or stage for an account of personality. Oliver John (1990), a major leader in research on the Big Five, also has been one of its most perceptive critics. He sees that the categories are at such a broad level of abstraction that "they are to personality what the categories 'plant' and 'animal' are to the world of biological objects—extremely useful for some initial rough distinctions, but of less value for predicting specific behaviors of a particular object" (John, 1990, p. 93). An ideal taxonomy, in John's view, needs to be built on causal and dynamic psychological principles and needs to be cast at different levels of abstraction from the broad to the more specific. The rest of this text discusses each of those other levels, and as noted in the Introduction, illustrates their integration in the final part.

The challenge for the trait level will be to connect its work to that coming from the other levels (e.g., Mischel & Shoda, 1998; John, 2001). That also will make it possible to link the individual's position on trait maps to the underlying psychological processes needed to help us understand more about the "why" of personality, that is, to what the mechanisms are that underlie and generate their observed consistencies. Even when we know that someone generally "is a sociable person," for example, the psychologist still needs to explain and understand *why* the person is that way and why and how the psychological and behavioral expressions of his or her sociability take the particular forms that they do.

Work at this level of analysis also has shown that the consistency that characterizes people is seen in their stable *if . . . then . . .* patterns, as well as in their overall average levels of different types of behaviors. Thus, within particular types of situations, individuals do show characteristic stable patterns, and these are reflections and expressions of their underlying personalities. Capturing these signatures, clarifying the underlying processes that generate them, and linking them to work at other levels of analysis will be an important piece of the agenda for the future. These signatures, as well as the overall average levels of behavior that characterize the person at the super trait level, will together provide useful windows for studying the underlying organization and nature of personality. In the history of personality psychology, a major step toward understanding such signatures was pioneered in the work described in the next part of the text, Part II: The Psychodynamic-Motivational Level.

PART

II

THE PSYCHODYNAMIC-MOTIVATIONAL LEVEL

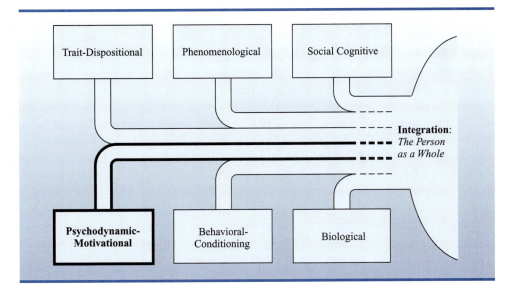

Integration:
The Person as a Whole

Trait-Dispositional

Phenomenological

Social Cognitive

Psychodynamic-Motivational

Behavioral-Conditioning

Biological

▶ PRELUDE TO PART II: THE PSYCHODYNAMIC-
MOTIVATIONAL LEVEL

The trait-dispositional level identified the consistencies and the variability in people's behavior, as you saw in Part I. A next question is: what accounts for and underlies these consistencies? And what accounts for the puzzling inconsistencies and changes in behavior that sometimes develop with distressing emotional consequences? These questions were first addressed in depth by Sigmund Freud, who was working with emotionally disturbed patients in Vienna at about the same time that the trait level of analysis began to develop in the United States.

A hundred years ago in Vienna, a young woman sought help from her physician, Dr. Freud, because she suddenly could no longer see. In a careful examination, Freud found nothing physically wrong with her visual system, and there had been no accident or plausible incident to make her lose her sight. Another patient, while waiting for her appointment, could not resist once again scrubbing her already immaculate hands; that morning she had already washed them dozens of times. In other respects she seemed normal, rational, and was functioning well except for these new symptoms. Sigmund Freud had to find a way to make sense of such problems and those of many of his other patients who needed help because their symptoms made no sense to them either and seemed to be in conflict with other aspects of their personalities. Work at the psychodynamic-motivational level of analysis was begun to meet these challenges.

Freud's approach to the question and his methods for answering it began with a hypothesis that he had formed early in his work with mental patients. Their behavior, he thought, simply could not be understood by looking only at the consistencies in what they did. The answers, he proposed, must be found by understanding their *inconsistencies.* For many years, he struggled to understand the ways in which those inconsistencies might reveal internal contradictions and motives of which the individual seemed to be unaware. If the causes were outside people's awareness, then the real meaning of their behavior could not be seen in what they said at the "surface level" of their self-descriptions.

The more deeply he delved, the more he became convinced that human behavior often is not rational and reflects unconscious motives, conflicts, and processes that are considered taboo by society and therefore condemned and avoided. To make sense of this, the theories that Freud and his followers proposed at first upset the world greatly, as revolutions often do. In time, however, his work set much of the agenda for personality psychology, and it is still relevant today, although often in extensively revised forms. Freud's theories were extremely broad. He tried to explain virtually everything about the human mind and its expressions in ways that he hoped would be universally true for all people across all cultures. His work has been debated and attacked, often with good reason, mostly for not being testable scientifically and for failing to be validated in many of the tests that have been attempted. Yet his contributions changed the view of human nature and gave the world new insights into the self-defensive and often self-defeating characteristics of the human personality. Freud addressed not just the development of abnormal problems but the ways in which essentially normal individuals may function to protect themselves from their own conflicts and anxieties beginning in early childhood.

Chapter 5 presents Freud's basic concepts. Chapter 6 highlights the work he directly influenced at this level for many years, including in the study of motives such as competence and achievement that go much beyond those that Freud emphasized. Chapter 7 examines basic mechanisms and processes central to these ideas, and the implications for personality assessment, change, and treatment.

The Personal Side of the Science

Some questions at the psychodynamic-motivational level you might ask about yourself:

▶ Does what I do sometimes puzzle me?

▶ How and why?

▶ What are the real motives that drive or underlie my behavior?

▶ How can I explain irrational fears and anxieties?

▶ How do I try to protect myself psychologically against getting hurt?

▶ How much of what I do is unconscious or done without awareness?

▶ What might be some unconscious influences on my behavior?

▶ Do I have motives that make me uncomfortable?

▶ If yes, what do I try to do about that?

PSYCHODYNAMIC THEORIES: FREUD'S CONCEPTIONS

Although his contribution to the conception of personality and of the human mind began a century ago, Freud's impact on society, as well as on the social sciences and humanities, still is almost universally hailed as profound; its significance is frequently compared with that of Darwin. Freud's monumental contributions set the stage for the psychodynamic level of analysis that focuses on the unconscious motivations and conflicts underlying much of our behavior. He revolutionized conceptions of the human psyche, provided strikingly powerful metaphors for the mind and human condition, opened the topic of childhood sexuality, and pioneered twentieth-century psychiatry with his approach to the treatment of psychological problems.

Sigmund Freud (1856–1939) started his career in medicine with research on cocaine. In 1884, the properties of this new drug intrigued him, partly because it helped him to get some relief from his own episodes of depression. When he became aware of cocaine's dangers, he soon gave it up, both for personal use and as a research problem. Studying for six months with the neurologist Jean Charcot in Paris in 1885, he became interested in the use of hypnosis to help patients deal with various nonorganic symptoms of "nervous disorders," particularly "hysteria."

Freud's breakthroughs began from his clinical observations of patients he was seeing in his private practice when he returned to Vienna. Imagine scenes like this one in his consulting room on a pleasant residential street in the Vienna of 1905. Freud is presented with a young girl who feels compelled to rinse out her wash basin over and over, dozens of times after each time she washes herself. Her habit becomes so intense and upsetting that her whole life revolves around it. Why? A young boy becomes terrified of horses although he himself was never hurt by one. Why? It was puzzles like these that intrigued the Viennese physician Sigmund Freud who invented psychoanalysis, reshaped the field of psychology, and influenced many later developments in all the social sciences and in Western concepts of human nature.

From these puzzles, Freud created a theory and a treatment method that changed our view of personality, health, and the mind itself. Working as a physician treating disturbed people in Vienna at the turn of the century, he formulated a theory that upset many cherished assumptions about human nature and startled the neo-Victorian world. Before Freud, people's behavior was believed to be under their conscious and rational control. Freud turned that conception upside down. Rather than seeing consciousness as the core of the mind, Freud compared personality to an iceberg: only the tip shows itself overtly, the rest lies below. Rather than viewing the person as a supremely rational being, he saw people as driven by impulses and striving to satisfy deep and lasting sexual

5.1 How did Freud's views conflict with the prevailing beliefs of his time?

Sigmund Freud (1856–1939)

and aggressive urgings. Rather than relying on people's reports about themselves as accurate self-representations, he interpreted what they said and did as highly indirect, disguised, symbolic representations of unconscious underlying forces.

In the course of more than 40 years of active writing and clinical research, Freud developed a theory of personality, a method of treatment for personality disturbances, and a wealth of clinical observations based on his therapeutic experiences as well as on his analyses of himself. Freud based both his theory and his psychoanalytic treatment on his extensive clinical observations of disturbed persons. He first noted certain **sensory anesthesias,** which are losses of sensory ability, as in blindness, deafness, or loss of feeling in a body part. He also found patients with motor paralyses that seemed to have no neurological origin. He proposed that these symptoms expressed a way of defending against unacceptable unconscious wishes. For example, a soldier who cannot admit his fear of facing battle develops a motor paralysis without a neurological basis. Or a young bride, unable to admit her hostility to her husband, becomes confined to her chair, although she shows no physical disease. All these examples illustrate **hysteria.** The fundamental feature of hysteria, according to Freud, is the presence of massive repression and the development of a symptom pattern that indirectly or symbolically expresses the repressed needs and wishes. On the basis of careful clinical observations, Freud gradually developed his theory of personality, continuously changing his ideas in the light of his growing clinical experiences and insights.

5.2 How did the study of hysteria influence Freud's views?

▶ BASIC ASSUMPTIONS: UNCONSCIOUS MENTAL DETERMINISM

Two key assumptions underlie much of Freud's conception:

5.3 What two major assumptions underlie Freud's ideas?

- First, his unique innovation was to propose that behavior is never accidental: it is psychologically determined by mental motivational causes. This is called the principle of **motivational determinism.**

- Second, these causes are outside of the person's complete consciousness or awareness.

The Unconscious

Freud the scientist wanted to try to explain the irrational behavior he witnessed in his patients. They seemed compelled to do things that they could not explain or sometimes even remember. Most puzzling, he could not attribute their symptoms to organic causes such as brain injuries or physical diseases: physically they were intact. They were consciously trying to stop their symptoms, desperate to relieve them, but they simply could not control them. Freud's insight was to propose that some unconscious irrational force was behind the symptom psychologically. The battles between the conscious will and the unconscious became the war of mental life in his theory.

Around the year 1900, Freud first divided mental processes into conscious, preconscious, and unconscious (Freud, 1905). We are instantly aware of our **conscious** thoughts. The immediately available level of consciousness refers to what is in one's attention at a given moment. The many events that we can bring into attention more or less easily, from the background music on the radio to memories of things experienced years ago, are **preconscious.** Thus even though we are not aware of preconscious thoughts at a given moment, we can bring them into awareness voluntarily and fairly easily. In contrast, outside this range of the potentially available lies the **unconscious.**

5.4 Define and give an example of each of Freud's three levels of consciousness.

This third zone is not responsive to our deliberate efforts at recall, and it is the layer that was Freud's core concern. Because their content is threatening, unconscious mental activities are kept beyond awareness by a mechanism of repression that works actively to keep them away from our awareness, so that we simply are unable to raise them into consciousness.

The Roads to the Unconscious

Freud was eager to find methods for his work that would make his ideas more than abstract claims or beliefs. It was through these methods that his conception achieved its richness and ultimately made its enormous impact. If the unconscious mind was so important for understanding psychological causes, then the challenge was how to get to it.

5.5 Why do dreams occupy a position of importance in psychoanalytic theory? How are they related to neurotic symptoms?

Dreams. Freud probed the unconscious most deeply through his explorations of dreams. He saw his 1899 book, *The Interpretation of Dreams,* like the adventures of a Columbus of the mind, a voyage into the darkest regions. The dream, Freud proposed, was the dreamer's unconscious effort to fulfill a wish that could not be expressed more directly. The analyst's task was to discover the hidden secrets underneath the surface content of the dream. To uncover those buried meanings, the analyst must overcome the dreamer's own resistance to facing himself or herself honestly, no matter how frightening or ugly the discoveries might prove to be.

In his voyage into the unconscious, Freud proceeded by scrutinizing his own dreams to try to face the motivations deep within his own personality. Unflattering self-revelation often resulted. For example, Freud is troubled by fears about plagiarizing, and he dreams about himself being treated as a thief stealing overcoats in lecture halls (Roazen, 1974, p. 99). Through self-analysis, Freud constructed his theory of the unconscious and the devious self-deceptions with which people try to disguise their own wishes from themselves.

In *The Interpretation of Dreams,* Freud built the case that in the dream we can find the hidden fulfillment of a desire that the person is trying to avoid experiencing consciously. Interestingly, this insight into the wish-fulfilling nature of dreams also led to the view that dreams, rather than disturbing sleep, actually function as the "guardians" of sleep. More than 50 years later, experiments discovered that in fact people need their dreams and become deeply troubled if their sleep is deprived of the phases in which dreaming naturally occurs (Roazen, 1974).

From his pioneering theory of dreams, Freud moved on to analyze the meaning of the disturbed behaviors and the anxieties his patients displayed and to develop a systematic theory about how to treat them. The main sources of anxiety, according to Freud, were the person's own unconscious sexual desires and aggressive impulses. He saw both sexual and aggressive urgings as basic human impulses or instincts, part of our heritage. He believed that sexuality does not begin with puberty but is visible early in childhood. It shows itself also in the young girl's affection for her father and the boy's infantile desires for his mother. Emotional attitudes, moreover, arise in these early relationships.

In the face of the objections and prudishness of the Victorian Age, he insisted that the route to self-acceptance was the honest recognition of one's instinctual sexuality and aggressiveness. Avoiding self-deception was the key. Making the unconscious impulses conscious was the road to health. The symptoms, Freud believed, were simply the indirect and sometimes symbolic expression of the unacceptable impulses that the person was unable to face consciously because of the anxiety they created.

Free Association.   Much of Freud's thinking was built on the analysis of dreams, but that was not the only method he favored. His second road to the unconscious, sometimes called the "royal road," became the therapeutic method of **free association.** In this method, the patient, reclining on a couch, is encouraged to simply say anything and everything that comes to mind, no matter what it is or how irrational it might seem, without censoring it. In Focus 5.1 discusses how free association is encouraged. In this way, the unconscious begins to become conscious. Although "resistance" to this process occurs often, it is gradually "worked through" until the unacceptable wishes can be faced. Then the patient is freed from having to manifest them indirectly through such symptoms as hysterical paralysis or other neurotic expressions.

IN FOCUS 5.1

ENCOURAGING FREE ASSOCIATION

Free association and the analysis of dreams are the methods of personality study that come most directly from Freud's work. Both methods are used in the context of the patient–therapist meetings during psychoanalysis.

In free association, you are instructed to give your thoughts complete freedom and to report daydreams, feelings, and images, no matter how incoherent, illogical, or meaningless they might seem. This technique may be employed either with a little prompting or by offering brief phrases ("my mother . . ."; "I often . . .") as a stimulus to encourage associations.

Freud believed that dreams were similar to the patient's free associations. He thought the dream was an expression of the most primitive workings of the mind. Dreams were interpreted as fulfilling a wish or discharging tension by inducing an image of the desired goal. Freud felt that through the interpretation of dreams he was penetrating into the unconscious.

The following passages illustrate some typical instructions and responses in the process of encouraging free association during psychoanalytic interviews. The patient complains that she does not have any thoughts at the moment.

Therapist: It may seem that way to you at first, but there are always some thoughts there. Just as your

heart is always beating, there's always some thought or other going through your mind.

Patient: Your mentioning the word "heart" reminds me that the doctor told my mother the other day she had a weak heart.

In a later interview the same woman became silent again and could not continue.

Therapist: Just say what comes to you.

Patient: Oh, odds and ends that aren't very important.

Therapist: Say them anyway.

Patient: I don't see how they could have much bearing. I was wondering what sort of books those are over there. But that hasn't anything to do with what I'm here for.

Therapist: One never can tell, and actually you're in no position to judge what has bearing and what hasn't. Let me decide that. You just report what comes into your mind regardless of whether you think it's important or not (Colby, 1951, p. 153).

▶ PSYCHIC STRUCTURE: ANATOMY OF THE MIND

To understand how we deal with unconscious wishes, Freud (1933) also developed an "anatomy" of the mind that occupied him in the early part of the 1920s. This led to the structural view of personality consisting of three "institutions" or mental "agencies": the id, ego, and superego. These institutions form in the course of early experience, with the superego, the last to emerge, crystallizing some time after the sixth year.

TABLE 5.1   The Freudian Conception of Mental Structure

Structure	Consciousness	Contents and Function
Id	Unconscious	Basic impulses (sex and aggression); seeks immediate gratification regardless of consequences; impervious to reason and logic; immediate, irrational, impulsive
Ego	Predominantly conscious	Executive mediating between id impulses and superego inhibitions; tests reality; seeks safety and survival; rational, logical, taking account of space and time
Superego	Both conscious and unconscious	Ideals and morals; strives for perfection; incorporated (internalized) from parents; observes, dictates, criticizes, and prohibits; imposes limitations on satisfactions; becomes the conscience of the individual

These three agencies are closely linked to the three layers of consciousness. The id is in the unconscious layer, characterized by mental processes outside one's awareness; the ego is predominantly conscious; and the superego includes a mix of conscious and unconscious processes. Although the three parts interact intimately, each has its own characteristics, which are summarized in Table 5.1 and discussed next.

The Id: At the Core

5.6 Describe the id and the process that governs its operation.

The **id** is the mental agency that contains everything inherited, especially the instincts. It is the basis of personality, the energy source for the whole system, and the foundation from which the ego and superego later develop. The id, according to Freud, is the innermost core of personality, and it is closely linked to biological processes.

The id's instincts, Freud thought, have their source biologically within the excitation states of the body. They act like drives, pressing for discharge (release). They are motivated, in the sense that their *aim* is to seek reduction, that is, to lower the state of excitation. Thus the tension that the build-up of the unexpressed drive creates has to be released, much like the build-up of steam in an overheating boiler has to be let out or the system explodes. Instinctual drives are biological and inborn, but the objects involved in attempts to reduce the drives depend on the individual's particular early experiences.

The Pleasure Principle.   Increases in energy from internal or external stimulation produce tension and discomfort that the id cannot tolerate. The id seeks immediate tension reduction, regardless of the consequences. This tendency toward immediate tension reduction Freud called the **pleasure principle.** The id obeys it, seeking immediate satisfaction of instinctual wishes and impulses, regardless of reason or logic or consequences.

Sexual and Aggressive Instincts.   Freud (1940) classified these impulses or instincts into the categories of life, or sexual instincts, and death, or aggressive instincts. The psychological representations of these instincts are wishes, and they often are irrational and unconscious.

Primary Process Thinking.   To discharge tension, the id forms an internal image or hallucination of the desired object. The hungry infant, for example, may conjure up an internal representation of the mother's breast. The resulting image is considered a wish fulfillment, similar to the attempted wish fulfillment that Freud believed characterized normal dreams and the hallucinations of psychotics. **Primary process thinking** was Freud's term for such direct, irrational, reality-ignoring attempts to satisfy needs. Because mental images by themselves cannot reduce tension, the ego develops.

The Ego: Tester of Reality

The **ego** is a direct outgrowth of the id. Freud described its origin this way:

> Under the influence of the real external world around us, one portion of the id has undergone a special development. From what was originally a cortical layer, equipped with the organs for receiving stimuli and with arrangements for acting as a protective shield against stimuli, a special organization has arisen which henceforward acts as an intermediary between the id and the external world. To this region of our mind we have given the name of ego (Freud, 1933, p. 2).

5.7 How does the ego derive from the id, and what are its functions and operating principles?

The ego is in direct contact with the external world. It is governed by considerations of safety, and its task is preservation of the organism. The ego wages its battle for survival against both the external world and the internal instinctual demands of the id. In this task, it has to continuously differentiate between the mental representations of wish-fulfilling images and the actual outer world of reality. In its search for food or sexual release, for example, it must find the appropriate tension-reducing objects in the environment so that tension reduction can actually occur. That is, it must go from image to object, and get satisfaction for id impulses while simultaneously preserving itself.

The Reality Principle. The ego's function is governed by the **reality principle,** which requires it to test reality and to delay discharge of tension until the appropriate object and environmental conditions are found. The ego operates by means of a "secondary process" that involves realistic, logical thinking and planning through the use of the higher or cognitive mental processes. That is, while the id seeks immediate tension reduction by such primary process means as wish-fulfilling imagery and direct gratification of sexual and aggressive impulses, the ego, like an executive, mediates between the id and the world, testing reality and making decisions about various courses of available action. For example, it delays impulses for immediate sexual gratification until the environmental conditions are appropriate.

Freud believed the ego was the only hope for the world, the part of the mind that would allow humans to emerge from the irrationality and primitivism of being driven wildly by their biological impulses. The ego was the way toward a life of reason: "Where id was," Freud wrote, "there shall ego be," and psychoanalysis was the road for that transformation from the person's domination by impulsivity to rationality and insightfulness.

The Superego: High Court in Pursuit of Perfection

Freud's third mental structure was the **superego.** He wrote:

5.8 How does the superego develop, and what are its functions?

> The long period of childhood, during which the growing human being lives in dependence on his parents, leaves behind it as a precipitate the formation in his ego of a special agency in which this parental influence is prolonged. It has received the name of superego. In so far as this superego is differentiated from the ego or is opposed to it, it constitutes a third power which the ego must take into account (Freud, 1933, p. 2).

Thus, the superego is the agency that internalizes the influence of the parents. It represents the morals and standards of society that have become part of the internal world of the individual in the course of the development of personality. The superego is the conscience, the judge of right and wrong, of good and bad, in accord with the internalized standards of the parents and thus, indirectly, of society. It represents the ideal. Whereas the id seeks pleasure and the ego tests reality, the superego seeks perfection. The superego, for Freud, involved the internalization of parental control in the form of self-control. For example, the individual with a well-developed superego resists "bad" or "evil" temptations, such as stealing when hungry or killing when angry, even when there are no external constraints (in the form of police or other people) to stop him.

In the theory, the superego develops around age five out of the human infant's long period of helplessness and extreme dependency on caregivers. The young child desperately fears the possible loss of this early love; the threat of parents withdrawing protection and gratification is terrifying. At first this fear is rooted in the objective anxiety of losing love and satisfaction due to the child's own actions (being "bad"). In time, an active *identification* occurs as the child incorporates the parental images and commands into itself psychologically. As the parental wishes become incorporated through this process, the conscience becomes an internal voice rather than an external control.

Once fully developed, the superego can become a compelling and even irrational force of its own, just as demanding as the id. Examples of this force are seen in severe depressions characterized by extreme self-hatred and self-destructiveness. The tyranny of the superego is thus added to the demands of the id. It is the burden of the ego to continuously try to compromise among these competing forces while testing the waters of "reality."

▶ CONFLICT, ANXIETY, AND PSYCHODYNAMICS

5.9 What three ego activities are involved in psychodynamic processes?

Psychodynamics are the processes through which personality works. In Freud's view, they concern three continuous tasks of the ego: (1) the control of unacceptable impulses from the id; (2) the avoidance of pain produced by internal conflict in the efforts to control and master those unacceptable impulses; and (3) the attainment of a harmonious integration among the diverse components of personality in conflict. Much of Freud's own energy was directed at understanding the **transformation of motives:** the basic impulses persist and press for discharge, but the objects at which they are directed and the manner in which they are expressed are transformed (1917).

Conflict

5.10 Describe the basic conflicts that characterize personality dynamics.

According to Freud (1915), the three parts of the psychic structure—id, ego, and superego—are always in dynamic conflict. The term *dynamics* refers to this continuous interaction and clash between id impulses seeking release and inhibitions or restraining forces against them—an interplay between driving forces and forces that inhibit them. These forces and counterforces propel personality.

The id's drive for immediate satisfaction of impulses reflects human nature: people are motivated to avoid pain and achieve immediate tension reduction. This drive for immediate satisfaction of instinctual demands leads to a clash between the individual and the environment. Conflict develops to the degree that the environment and its representatives in the form of other persons, notably the parents in childhood, and later the superego, punish or block immediate impulse expression.

Persons in time come to incorporate into their superegos the values by which they are raised, largely by internalizing parental characteristics and morals. In Freud's view, perpetual warfare and conflict exist between humans and environment. Insofar as societal values become internalized as part of the person, this warfare is waged internally between the id, ego, and superego, and it produces anxiety.

Anxiety is painful and therefore we automatically try to reduce the tension it produces. Freud first used defense as a psychoanalytic term (1899), but he focused primarily on one type of defense mechanism involving denial and repression. His daughter, Anna Freud, distinguished some of the other major defense mechanisms recognized today (A. Freud, 1936)—discussed in the next chapter.

Psychodynamic theorists emphasize that when a threat becomes especially serious, it may lead to intense inhibitions and defenses. In the psychodynamic view, such defensive inhibition is desperate and primitive. It is a massive, generalized, inhibitory reaction rather than a specific response to the particular danger. This **denial defense** occurs when the person can neither escape nor attack the threat. If the panic is sufficient, the only possible alternative may be to deny it. Outright denial may be possible for the young child, who is not yet upset by violating the demands of reality testing. When the child becomes too mature to deny objective facts in the interests of defense, denial becomes a less plausible alternative and repression may occur.

Defense: Denial and Repression

In psychodynamic theory, **repression** usually refers to a particular type of denial: ". . . the forgetting, or ejection from consciousness, of memories of threat, and especially the ejection from awareness of impulses in oneself that might have objectionable consequences" (White, 1964, p. 214).

Repression was one of the initial concepts in Freud's theory and became one of its cornerstones. Freud (1920) believed that the mechanisms of denial and repression were the most fundamental or primitive defenses and played a part in other defenses. Indeed, he thought that other defenses started with a massive inhibition of an impulse.

Freud based his ideas concerning repression and defense on his clinical observations of hysterical women at the turn of this century. Recall that he noted that some of these patients seemed to develop physical symptoms that did not make sense neurologically. For example, in a hysterical difficulty called "glove anesthesia," the patient showed an inability to feel in the hands—a symptom that is impossible neurologically. In their studies of hysteria, Freud and his associate Breuer hypnotized some of the patients and found, to their great surprise, that when the origins and meanings of hysterical symptoms were talked about under hypnosis, the symptoms tended to disappear. This finding proved beyond any doubt that the symptoms were not caused by organic damage or physical defects.

Partly to understand hysteria, Freud developed his theory of unconscious conflict and defense. In his view, such symptoms as hysterical blindness and **hysterical anesthesia (loss of sensation)** reflected defensive attempts to avoid painful thoughts and feelings by diversionary preoccupation with apparently physical symptoms. Freud thought that the key mechanism in this blocking was unconsciously motivated repression. Through repression, the basic impulses that are unacceptable to the person are rendered unconscious and thereby less frightening. Because such diversionary measures are inherently ineffective ways of dealing with anxiety-provoking impulses, these impulses persist. The impulses continue to press for release in disguised and distorted forms that are called "symptoms."

5.11 How did Freud use his concept of repression to explain hysterical disorders?

Libido

Freud thought that each person contained finite amount of energy, called libido, a term for the total energy that drives id impulses. **Libido is attached or fixed on aspects of the internal and external environment.** The energy available to the organism may be continuously transformed, fixed onto different "objects" (note that "objects" was a term that for Freud included people and not just inanimate things). However, the total amount of energy is conserved and stable. Freud's energy system thus was consistent with the hydraulic models of nineteenth-century physics.

The id was seen as a kind of dynamo, and the total mind (or psyche) was viewed as a closed system motivated to maintain equilibrium: any forces that were built up required discharge. The discharge could be indirect. Instinctual impulses could be displaced from one object to another, for instance, from one's parents to other authority figures or more remotely, from the genitals, for example, to phallic symbols.

▶ NEUROSIS

When Defenses Fail: Neurotic Anxiety and Conflict

Sometimes the defenses that disguise basic motives may become inadequate and denial and repression no longer work. Even under the usual circumstances of everyday life, the defenses are occasionally penetrated and the person betrays himself (Freud, 1901). Such betrayals of underlying motives are seen when defenses are relaxed, as in dream life or in jokes and slips of the tongue. The defense process involves **distortion** and **displacement,** which occur when private meanings develop as objects and events become transformed into symbols representing things quite different from themselves. It is believed that these meanings are partially revealed by behavioral "signs" or symptoms that may symbolize disguised wishes and unconscious conflicts. For example, phobias such as the fear of snakes may reflect basic sexual conflicts; in this case, the feared snake has symbolic meaning.

Development of Neurotic Anxiety. It is now possible to consider the Freudian conception of how neurotic anxiety and problems may develop. The sequence here (depicted in Fig. 5.1) begins with the child's aggressive or sexual impulses that seek direct release. These efforts at discharge may be strongly punished and blocked by dangers or threats (for example, intense parental punishment such as withdrawal of love). Hence, they lead to objective anxiety. The child may become especially afraid that these impulses will lead to loss of parental love and in time, therefore, may come to fear his or her own impulses. Because this state is painful, the child tries to repress

5.12 What is neurotic anxiety, and what is its role in neurotic symptoms?

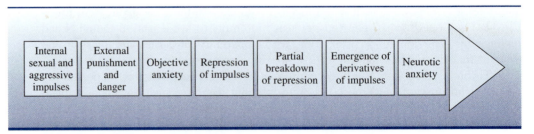

Figure 5.1   Sequence in Freudian conception of neurotic anxiety.

these impulses. If the ego is weak, the repression is only partly successful and the instinctual impulses persist. Unless expressed in some acceptable form, these impulses become increasingly "pent up," gradually building up to the point where they become hard to repress. Consequently, there may be a partial breakdown of repression, and some of the impulses may break through, producing some neurotic anxiety. Anxiety, in this view, functions as a danger signal, a warning to the individual that repressed impulses are starting to break through the defenses. Rather than emerging directly, however, the unacceptable impulses express themselves indirectly in disguised and symbolic ways.

The Meaning of Neurotic Acts. Freud felt that the symbolic meaning of behavior was clearest in neurotic acts. He cited the case of a girl who compulsively rinsed out her wash basin many times after washing. Freud thought that the significance of this ceremony was expressed in the proverb, "Don't throw away dirty water until you have clean." He interpreted the girl's action as a warning to her sister, to whom she was very close, "not to separate from her unsatisfactory husband until she had established a relationship with a better man" (Freud, 1959, vol. 2, p. 28).

Another patient was able to sit only in a particular chair and could leave it only with much difficulty. In Freud's analysis of her problem, the chair symbolized the husband to whom she remained faithful. Freud saw the symbolic meaning of her compulsion in her sentence "It is so hard to part from anything (chair or husband) in which one has once settled oneself" (Freud, 1959, vol. 2, p. 29). Thus, the important object of conflict—the husband—was transformed into a trivial one—the chair. Freud cited these and many similar cases as evidence for the view that neurotic behaviors express unconscious motives and ideas symbolically.

The clinician's task, then, is to decipher the unconscious meaning of the patient's behavior and to discover the conflicts and dynamics that might underlie seemingly irrational behavior patterns (see Table 5.2).

Origins of Neuroses

In Freud's view, serious problems, such as the neuroses, and the roots of the symptoms that characterize them, begin in early childhood:

> . . . It seems that neuroses are acquired only in early childhood (up to the age of six), even though their symptoms may not make their appearance till much later. The childhood neurosis may become manifest for a short time or may even be overlooked. In every case the later neurotic illness links up with the prelude in childhood.

5.13 At what developmental period do neuroses develop and what factors produce them?

TABLE 5.2   Possible Meanings of Some Behavioral Signs in Freudian Theory

Behavioral Sign	Possible Underlying Meaning
Fear of snakes	Sexual conflict regarding genitals
Compulsive cleanliness	Reaction against anal impulses
Obsessive thought: "My mother is drowning"	Imperfectly repressed hostility toward mother
Paranoid jealousy	Homosexual wishes
Preoccupation with money	Problems around toilet training

> . . . *The neuroses are, as we know, disorders of the ego; and it is not to be wondered at if the ego, so long as it is feeble, immature and incapable of resistance, fails to deal with tasks which it could cope with later on with the utmost ease. In these circumstances instinctual demands from within, no less than excitations from the external world, operate as 'traumas,' particularly if they are met halfway by certain innate dispositions (Freud, 1933, pp. 41–42).*

As these quotations indicate, neuroses were seen as the products of early childhood traumas plus innate dispositions. But even the behavior of less disturbed persons was believed to reflect expressions of underlying unconscious motives and conflicts. These manifestations could be seen in the "psychopathology of everyday life"—the occurrence of meaningful but common unconscious expressions, as discussed next.

The Psychopathology of Everyday Life: "Mistakes" That Betray

5.14 How do everyday "slips" reveal the unconscious? Provide some examples.

Some of Freud's most fascinating—and controversial—ideas involved the elaboration of possible hidden meanings that might underlie such common occurrences as slips of the tongue, errors in writing, jokes, and dreams. In Freud's (1901, 1920) view, "mistakes" may be unconsciously motivated by impulses that the individual is afraid to express directly or openly. To show that mistakes may really be motivated by underlying wishes, Freud pointed out many instances in which even the attempt to "correct" the error appears to betray a hidden, unacceptable meaning. In one case, for example, an official introduced a general as "this battle-scared veteran" and tried to "correct" his mistake by saying "bottle-scarred veteran." Other examples are summarized in Table 5.3. Some common Freudian dream symbols are shown in Table 5.4.

Motivational Determinism: Unconscious Causes

5.15 What is motivational determinism?

According to Freud's principle of motivational determinism, all behavior is motivated, and the causal chain that links wishes to actions can be complex and indirect.

TABLE 5.3 **Examples of Behaviors Motivated by Unconscious Wishes**

Behavior Involved	Unconscious Wish	Transformation
Slip of the tongue: May I "insort" (instead of escort) you?	To insult	Condensation: (insult + escort = "insort")
Slip of the tongue: "Gentlemen, I declare a quorum present and herewith declare the session *closed*."	To close the meeting	Association of opposites: (open = closed)
Dream of disappointment in quality of theater tickets, as a result of having gotten them too soon.	I married too soon; I could have gotten a better spouse by waiting.	Symbolism: (getting tickets too soon = marrying too soon)
Dream of breaking an arm	Desire to break marriage vows	Conversion into visual imagery: (breaking vows = breaking an arm)

Source: Freud, S. (1920). *A general introduction to psychoanalysis.* New York: Boni and Liveright.

TABLE 5.4   Some Freudian Dream Symbols and Their Meanings

Dream Symbol	Meaning
King, queen	Parents
Little animals, vermin	Siblings
Travel, journey	Dying
Clothes, uniforms	Nakedness
Flying	Sexual intercourse
Extraction of teeth	Castration

Source: Freud, S. (1920). *A general introduction to psychoanalysis.* New York: Boni and Liveright.

Does a dream of a broken arm imply a desire to break one's wedding vows?

Suppose, for example, a man fights with his wife about money, is personally fussy about his appearance, and becomes very upset when he loses his umbrella. These seemingly different bits of behavior might actually be motivated by a common cause. Much of psychoanalytic assessment and therapy is a search for such underlying causes. A psychodynamic explanation of behavior consists of finding the motives that produced it. The focus is not on behavior, but on the motivations that it serves and reflects. And usually these motives are disguised in their expressions and unconscious, rooted in early experiences.

▶ PERSONALITY DEVELOPMENT

Freud believed that every person normally progresses through five **psychosexual stages.** During the first five years of life, pleasure is successively focused on three zones of the body as the oral, anal, and phallic stages unfold. Then comes a quiet latency period of about five or six years. Finally, if progress through each stage has been successful, the person reaches the mature or genital stage after puberty. But special problems at any stage may retard or arrest (fixate) development and leave enduring marks on the person's character throughout life.

Stages of Development

5.16 Describe the five psychosexual stages and indicate how unresolved conflicts at the early stages create personality traits.

Oral. The **oral stage** occurs during the first year of life when pleasure is focused on the mouth and on the satisfactions of sucking, eating, and biting in the course of feeding (but see In Focus 5.2). The dependent, helpless person is said to be fixated at this stage, when the infant is totally dependent upon others for satisfaction of his or her needs.

According to Freud, the oral stage is divided into two periods: (1) sucking and (2) biting and chewing. Later character traits develop from these earliest modes of oral pleasure. More specifically, oral incorporation (as in sucking and taking in milk in the first oral period) becomes the prototype of such pleasures as those gained from the acquisition of knowledge or possessions. In his view, the gullible person (who is "easily taken in") is fixated at the oral, incorporative level of personality. The sarcastic, bitingly argumentative person is fixated at the second oral period—the sadistic level associated with biting and chewing.

Anal. In the second year of life, the **anal stage** is marked by a shift in body pleasure to the anus and by a concern with the retention and expulsion of feces. According to Freud, during toilet training, the child has his first experience with imposed control. The manner in which toilet training is handled may influence later personal qualities and conflicts. For example, extremely harsh, repressive training might produce a person characterized by obstinacy, stinginess, and a preoccupation with orderliness and cleanliness.

5.17 How does the *Oedipus complex* promote male sex role development?

Phallic. The **phallic stage** is the period in which the child observes the difference between male and female and experiences what Freud called **the Oedipus complex.** This complex, symbolized in the father–son conflicts of ancient Greek myths, occurs at about age five.

Freud thought that both boys and girls love their mother as the satisfier of their basic needs and resent their father as a rival for their mother's affections. In addition, the boy fears castration by the father as punishment for desiring his mother sexually. This **castration anxiety** is so terrifying that it results in the repression of the boy's sexual desire for his mother and hostility toward his father. To reduce the anxiety of possible castration by the father, the boy tries to become like him or to identify with him. In this identification, he gradually internalizes the father's standards and values as his own, becoming more like his father rather than battling him.

Identification with the father in turn helps the boy gain some indirect satisfaction of his sexual impulses toward his mother. In this last phase of the Oedipus complex of the male, the superego reaches its final development as the internalized standards of parents and society, and the opposition to incest and aggression becomes part of his own value system.

In the female, **penis envy,** resulting from the discovery that she lacks the male organ, is the impetus to exchange her original love object—the mother—for a new object—the father. Unlike the boy's Oedipus complex, which is repressed through fear, the girl—having nothing to lose—persists in her sexual desire for her father. This desire does, naturally, undergo some modification because of realistic barriers.

Latency. After the phallic stage, a **latency period** develops. Now there is less overt concern with sexuality; the child represses his or her memories of infantile sexuality and forbidden sexual activity by making them unconscious.

Genital. This is the final, mature stage of psychosexual development. Now the person is capable of genuine love for other people and can achieve adult sexual satisfactions. No longer characterized by the selfishness (narcissism) and mixed, conflicting

5.18 Does research support Freud's conception of the infant as primarily "oral"?

IN FOCUS 5.2

HOW ORAL IS THE INFANT?

Although the feeding situation is a critical phase of early development, it is only one part of the total relationship between the growing organism and the world. Thus, the baby is more than an "oral" creature. Babies respond to stimulation of the mouth, lips, and tongue, but in addition, they see, hear, and feel, obtaining stimulation visually, aurally, and from being handled.

Convinced that in spite of an abundance of theories personality psychology has much too little real data, Professor Burton L. White of Harvard University began by carefully observing infants as they lay in their cribs. (The participants were physically normal infants in an orphanage.) He and his colleagues recorded the quantity and quality of visual–motor activity to study the babies' attention. On the basis of these observations, they plotted the development of the infants' tendency to explore the visual surroundings, as depicted in Figure 5.2. The findings surprised the investigators:

> One important revelation for me which resulted from these weekly observations was that, contrary to my academically bred expectations, infants weren't really very oral during the first months of life. In fact, between two and six months, a far more appropriate

> description would be that they are visual-prehensory creatures. We observed subject after subject spend dozens of hours watching first his fists, then his fingers, and then the interactions between hands and fingers. Thumb-sucking and mouthing were rarely observed except for brief periods when the infant was either noticeably upset or unusually hungry (White, 1967, p. 207).

These observations point up how much more we need to know about the details of the infant's activities before we can reach conclusions about what events characterize early development. It may be, as this investigator's comments imply, that the "oral" infant will turn out to be much more attentive and active, and less oral and passive, than was believed in early formulations. Indeed, close observation of young children suggests a dramatic increase in the infant's competence by the second or third month. More wakefulness and alertness, greater receptivity to stimulation, more directed attention and less fussing begin to characterize the baby (Sroufe, 1977). Stimulation becomes less unsettling and may be sought out actively as the baby becomes more and more attentive, even to its own movements.

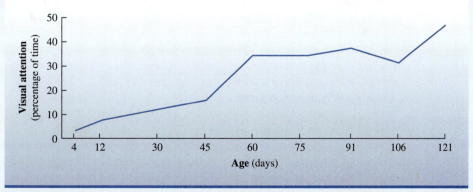

Figure 5.2 The development of the tendency to explore the surroundings.

Source: Adapted from White, B. L., & Held, R. (1966). Plasticity of sensorimotor development in the human infant. In J. F. Rosenblith and W. Allinsmith (Eds.), *The Causes of Behavior II*, pp. 60–70. Boston: Allyn and Bacon.

feelings that marked the earlier stages, he or she can relate to others in a mature, heterosexual fashion. But before he or she reaches the **genital stage,** excessive stress or overindulgence may cause the person to become fixated at earlier levels of psychosexual development.

Fixation and Regression

5.19   How do fixation, regression, and identification enter into personality development?

The concepts of **fixation** and **regression** are closely connected with Freud's conceptualization of psychosexual stages of development. Fixation means that a sexual impulse is arrested at an early stage. Regression is reversion to an earlier stage in the face of unmanageable stress. Fixation occurs when conflict at a particular stage of psychosexual development is too great. Severe deprivation or overindulgence at a particular stage, or inconsistent alterations between indulgence and deprivation, may lead to fixation.

In sum, personality is intimately related to the individual's mode of coping with problems at each stage of psychosexual development. The result is reflected in the nature of character formation, symptoms, and relations with other people. When individuals' resolution of problems at any stage of development is inadequate, later stress may cause them to regress to that earlier stage. They then display behavior typical of that less mature period.

Freud's Theory of Identification

Parts of Freud's theory of psychosexual stages have been modified and even rejected in recent years. Some of his closely related concepts regarding identification, however, have continued to be influential.

Early personality development occurs in the setting of the family. In that context, you saw Freud strongly emphasize the child's attachment to the mother and the rivalry between son and father for her attentions. This triangle of relations, called the Oedipal situation, is the basis for identification with the standards of the parent. This identification process Freud attributed to two mechanisms that operate during psychosexual development.

Anaclitic identification is based on the intense dependency of a child on the mother, beginning early in the course of the infant's development. Because of the helplessness of the infant, the dependency upon the caretaker is profound. Identification for girls is based mainly on this early love or dependency relation with the mother. In anaclitic identification, the child must first have developed a dependent love relationship with her caretaker (usually the mother). Later, when the mother begins to withdraw some of her nurturant attention, the child tries to recapture her by imitating and reproducing her in actions and fantasy.

For boys, dependency or anaclitic identification with the mother is followed later by **identification with the aggressor.** The "aggressor" is the father during the Oedipal phase of development. Identification with the aggressor is motivated by fear of harm and castration by the punitive father in retribution for the son's fantasies and his sexual wishes toward the mother. Freud described the situation vividly:

. . . *When a boy (from the age of two or three) has entered the phallic phase of his libidinal [sexual] development, is feeling pleasurable sensations in his sexual organ and has learnt to procure these at will by manual stimulation, he becomes his mother's lover. He wishes to possess her physically in such ways as he has divined from his observations and intuitions about sexual life, and he tries to seduce her by showing her the male organ which he is proud to own. In a word, his early awakened masculinity seeks to take his father's place with her; his father has hitherto in any case been an envied model to the boy, owing to the physical strength he perceives in him and the authority with which he finds him clothed. His father now becomes a rival who stands in his way and whom he would like to get rid of (Freud, 1933, p. 46).*

The hostile feelings that the boy experiences in the Oedipal situation create great anxiety in him; he desires the mother but fears castration from the father. To defend against the anxiety, he resolves the Oedipal conflict, repressing his aggressive wishes against his father and trying to become more like him. It is as though the boy believes that if he *"is"* the father, he cannot be hurt by him. Identification with the aggressor requires that the boy have a strong (but ambivalent) relation with the father. In this relationship, love for the father is mixed with hostility because the father possesses the mother and interferes with the son's urges. Freud thought that through identification with the aggressor, boys develop a stricter superego.

▶ IMPACT OF FREUD'S THEORIES

Image of the Person

Freud built a dramatic image of what a person might be, inventing a sweeping and novel theoretical system. Freud saw the person as struggling with himself and the world, blocked by anxieties, conflicted, and plagued by his own unacceptable wishes and unconscious secrets. This picture has captivated the imagination of many laymen as well as clinicians. Consequently, it has had an enormous impact on philosophical as well as psychological conceptions of human nature. In Freud's view, humans are not the unemotional, rational beings that Victorian society thought they were. Instead, people are torn by unconscious conflicts and wishes that push them in seemingly puzzling ways.

Freud's emphasis on unconscious impulses as the most basic determinants of behavior is seen in an analogy in which the relation of the id and the ego is likened to that between a horse and its rider:

> *. . . The horse provides the locomotive energy, and the rider has the prerogative of determining the goal and of guiding the movements of his powerful mount towards it. But all too often in the relations between the ego and the id we find a picture of the less ideal situation in which the rider is obliged to guide his horse in the direction in which it itself wants to go (Freud, 1933, p. 108).*

Thus, in Freud's psychology the id is stubborn and strong and often the ego cannot really control it effectively.

Freud believed that the environment is less important than inborn instincts in the dynamics of personality. He thought that external stimuli make fewer demands and, in any event, can always be avoided. In contrast, one's own impulses and needs cannot be escaped. Consequently, he made instinctual impulses the core of personality. Psychodynamic theories also have shaped ideas about adaptation, deviance, and personality change. And they have done this more than any other psychological approach.

5.20 Describe Freud's conception of the human being and relate it to your own beliefs.

The Healthy Personality

5.21 What is Freud's conception of healthy personality functioning?

For Freud, a healthy personality showed itself in the ability to love and work and required a harmony among id, ego, and superego. Referring to the goal of psychotherapy, Freud wrote, "Where id was, there shall ego be." He meant that, for the healthy personality, rational choice and control replace irrational, impulse-driven compulsion. A healthy personality also required mature (genital) psychosexual development.

From the psychodynamic perspective, adequate adaptation requires insight into one's unconscious motives. Persons who can cope adequately are the ones who can face their impulses and conflicts without having to be extremely defensive. Symptoms represent the return of unsuccessfully repressed materials, reemerging to torture the person in disguised forms. Breakdowns reflect the inadequacy of defenses to deal with unconscious conflicts. Symptoms and mental problems develop when the ego does not have the strength to cope with the conflicting demands of external reality and the internal pressures of id and superego. In Freud's words (1940, pp. 62–63), when:

> . . . the ego has been weakened by the internal conflict, we must come to its aid. The position is like a civil war which can only be decided by the help of an ally from without. The analytical physician and the weakened ego of the patient, basing themselves upon the real external world, are to combine against the enemies, the instinctual demands of the id, and the moral demands of the superego. We form a pact with each other. The patient's sick ego promises us the most candor, promises, that is, to put at our disposal all of the material which his self-perception provides; we, on the other hand, assure him of the strictest discretion and put at his service our experience in interpreting material that has been influenced by the unconscious. Our knowledge shall compensate for his ignorance and shall give his ego once more mastery over the lost provinces of his mental life. This pact constitutes the analytic situation.

The pact to which Freud refers above is also the basis for the close relationship between therapist and client that develops over the course of the analysis. It is called **transference.** The name reflects the view that the patient transfers onto the therapist many of the feelings experienced initially with the parents. In the course of the therapy, these feelings are examined closely and "worked through" until they become resolved.

Behaviors as Symptoms

The Freudian approach views an individual's problematic behavior as symptomatic (rather than of main interest in its own right). It searches for the possible causes of these symptoms by making inferences about the underlying personality dynamics. For example, an individual who has a bad stutter might be viewed as repressing hostility, one with asthma as suffering from dependency conflicts, and one with snake fears as victimized by unconscious sexual problems. This focus on the meaning of behavior as a symptom (sign) guides the psychodynamic strategy for understanding both normal and abnormal behavior. Thus, the psychodynamically oriented clinician seeks to infer unconscious conflicts, defense structure, problems in psychosexual development, and the symbolic meaning and functions of behavior.

SUMMARY

BASIC ASSUMPTIONS

- Freud's work was based on clinical observations of neurotic persons and self-analysis. This led him to posit the unconscious as a key component of personality.
- Freud used dreams and free association to tap into unconscious wishes.

PSYCHIC STRUCTURE

- The id, ego, and superego form the psychodynamic structure of the personality.
- The id is the primary, instinctual core, obeying the "pleasure principle."
- The ego mediates between the instinctual demands of the id and the outer world of reality, utilizing "secondary processes": logical thinking and rational planning.
- The superego represents the internalized moral standards of the society and parental control.

CONFLICT, ANXIETY, AND PSYCHODYNAMICS

- Personality dynamics involve a perpetual conflict between the id, ego, and superego.
- The desire for immediate gratification of sexual and aggressive instincts puts the person in conflict with the environment and ultimately the superego. This struggle produces anxiety.
- Defenses such as denial and repression may be used by the ego when it is unable to handle anxiety effectively. The person's unacceptable impulses and unconscious motives are transformed into "symptoms."
- The total mind (or psyche) is seen as a closed system motivated to maintain equilibrium: any forces that build up require discharge.

NEUROSIS

- When defenses fail, the conflicts may build up into neurotic anxiety, revealed indirectly through symbolic behavior.
- Neuroses are a product of early childhood trauma combined with innate predispositions.
- Small mistakes or slips of the tongue may reveal an unconscious need to express undesirable impulses.
- All behavior, even the seemingly insignificant or absurd, is motivated and significant.

PSYCHOSEXUAL STAGES AND PERSONALITY DEVELOPMENT

- Personality development includes a series of psychosexual stages: oral, anal, phallic, and genital.
- Later personality traits develop according to the individual's experience at each of these stages of maturation.
- In fixation, a sexual impulse is arrested at an early stage.
- Regression is a reversion to an earlier stage of development in the face of stress.
- Anaclitic identification and identification with the aggressor are two Freudian identification mechanisms.

IMPACT OF FREUD'S THEORIES

- Healthy individuals achieve a kind of truce within themselves by substituting rational choice for id impulse and arriving at the final stage of psychosexual development.
- The Freudian approach views an individual's problematic behavior as symptomatic and searches for the unconscious causes of these symptoms.

KEY TERMS

anaclitic
  identification   108
anal stage   106
castration anxiety   106
conscious   95
denial   101
distortion/displacement
  102
ego   99
fixation   108
free association   97
genital stage   108
hysteria   95
hysterical anesthesia   101

id   98
identification with the
  aggressor   108
latency period   106
libido   102
motivational
  determinism   95
Oedipus
  complex   106
oral stage   106
penis envy   106
phallic stage   106
pleasure principle   98
preconscious   95

primary process
  thinking   99
psychodynamics   100
psychosexual stages   105
reality principle   99
regression   108
repression   101
sensory anesthesia   95
superego   99
transference   110
transformation of
  motives   100
unconscious   95

6

POST-FREUDIAN PSYCHODYNAMICS AND MOTIVES

"Psychoanalysis was a well-guarded fortress, and most psychologists had little interest in scaling its walls . . . [in contrast] today . . . 'pluralism' characterizes contemporary psychoanalysis" (Westen, 1990, p. 21).

The psychodynamic level of analysis has expanded and diversified in the many years since Freud's original concepts were first expressed. This chapter presents some of the main ideas of theorists who were influenced by Freud and retained much of Freud's psychodynamic orientation, but transformed its focus and shape in crucial ways. These "neo-Freudians" or "post-Freudians" or "ego psychologists" represent a wide range of innovations, and have been given many different labels. They began with Freud's own followers at the start of the twentieth century and moved on to some radical departures from his ideas. Their unifying theme is that they retained a focus on the psychodynamic-motivational level of analysis. But they went on to call attention to the diversity of human motives and to mental and emotional processes in personality development after early childhood that had been neglected before.

▶ TOWARD EGO PSYCHOLOGY AND THE SELF

6.1   Describe four changes in emphasis from classical psychoanalysis that occurred among neo-Freudian theorists.

Although each neo-Freudian writer has made a distinctive contribution, certain common themes emerge, especially in recent years. These themes suggest a gradual shift in focus, summarized in Table 6.1. Less attention is paid to Freud's ideas about the basic sexual and aggressive instincts of the id, and the id itself is given a less dominant role. More attention is on the concept and functions of ego and "self," to the point where the newer theoretical trends have been named **ego psychology** and its practitioners often are called *ego psychologists.*

Ego psychologists recognize that the ego has crucial functions that may be relatively independent of underlying unconscious motivations. The ego deals often with "higher order" motives and goals, such as the striving for competence, for achievement, and for power, to name a few of the many motives that may drive human behavior. The ego psychologists see these motives are part of normal ego functions as people form and pursue their life goals. In short, with this growth of the role of ego, and expansion of the range of human motives, the person is viewed as a more competent, potentially creative problem solver, engaged in much more than the management of instincts that press for discharge.

The neo-Freudians, as Table 6.1 suggests, also saw human development as a more continuous process that extends throughout the life span. More than the product of early psychosexual experiences, personality began to be viewed as a lifelong development. Personality is rooted in social and interpersonal relations and in the context of culture and society. It is not isolated within the psyche of the individual and in the drama of the relations with the immediate family in the first few years of life.

Let us now consider some of the major relevant theorists at this level of analysis to illustrate the range and nature of their ideas about personality

Anna Freud and the Ego Defense Mechanisms

As noted in the last chapter, it was Freud who identified and focused on the mechanism of denial and repression. But it was another psychoanalyst, his own daughter, Anna Freud, who provided important analyses of the various forms of defense mechanisms that the ego uses, summarized in Table 6.2.

6.2   How are defense mechanisms involved in the energy exchanges that occur in psychodynamic processes?

Most analysts view the defense mechanisms as the core of psychodynamics. These are the processes through which the ego does much of its peace-keeping work. They are the mechanisms through which the ego tries to subordinate the impulses, test reality,

TABLE 6.1   Post-Freudian Developments: Some Characteristics of the Neo-Freudians

Less Attention to	More Attention to
Id and instincts	Ego and self; ego defenses; higher order motives
Purely intrapsychic causes and conflicts	Social, interpersonal causes; relationship issues
Earliest childhood	Later developments throughout the life span; adult functioning
Psychosexual stages	Social forces and positive strivings; the role of the culture and society

Anna Freud (1895–1982)

and accommodate the demands of the superego in the lifelong war within the psyche. The process of defense involves a continuous conflict between impulses seeking discharge and defenses designed to transform these wishes into an acceptable form for the ego. In the course of these transformations, energy is exchanged and directed toward different objects, mediated by the mechanisms of defense.

Transformation of Motives. For example, if sadistic aggressive impulses cannot be repressed but still are too threatening to self-acceptance, they might be transformed into a more socially sanctioned form, such as an interest in surgery (see Figure 6.1). Likewise, sexual wishes toward the mother might be displaced into a career of painting

TABLE 6.2 Definitions and Examples of Some Defense Mechanisms

Mechanisms	Definition	Example
Repression	Massive inhibition of a threatening impulse or event by rendering it unconscious (beyond awareness)	Guilt-producing sexual wishes are "forgotten"
Projection	Unacceptable aspects of oneself are attributed to someone else	Projecting one's own unacceptable sexual impulses by attributing them to one's boss
Reaction formation	Anxiety-producing impulse is replaced by its opposite in consciousness	Unacceptable feelings of hate are converted into "love"
Rationalization	Making something more acceptable by attributing it to more acceptable causes	Blaming an aggressive act on "being overworked" rather than on feeling angry
Sublimation	Expression of a socially unacceptable impulse in socially acceptable ways	Becoming a soldier to hurt others; becoming a plumber to indulge in anal desires

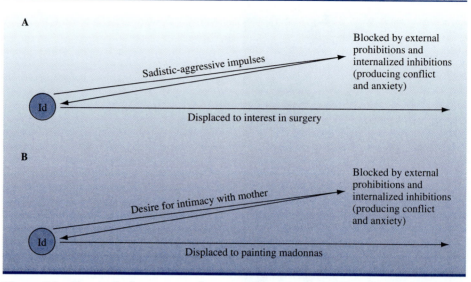

Figure 6.1 The psychodynamic transformation of motives: Examples of displacement in the form of sublimation.

madonnas, as some Freudians think happened among certain Renaissance painters. Freud (1909) himself suggested such dynamics in the case of Leonardo da Vinci.

6.3 What are the major signs that repression is occurring?

Freud in his original work was concerned with extreme examples of the mechanism of repression, but its workings also may be seen in many mild forms, as in the following illustration. Jim, a thirteen-year-old, has a girlfriend with whom he has had a few happy "dates," but she fails to show up at the movie after having promised to be there. Jim says that he feels no anger, is not annoyed, and just "doesn't care." Yet he explodes later at dinner with his family and gets into a squabble with his little sister. Is Jim using repression? If Jim privately knows he is just trying to cover his irritation and upset, then repression is not a relevant explanation for what he is doing. On the other hand, if his anger is evident to those who know him well, but truly hidden from his own awareness, then repression may be at work as a defense. This hypothesis is strengthened if Jim also shows resistance to efforts by his mother to suggest, for example, that he may be upset by the broken date. In genuine repression, if you push the person to face the underlying feelings that are being avoided unconsciously, it may only increase the defensive attempts to reject the interpretation and avoid the emotion. That tends to make the defense more elaborate and even irrational. In this type of example, when the person is no longer a young child and the threat is relatively mild, repression is unlikely to be very deep. The depth and desperation of the defenses tend to be much greater in infancy or early childhood, or when the threats are profoundly frightening and the individual is highly vulnerable.

6.4 How are unacceptable impulses masked by projection, reaction formation, rationalization, and sublimation?

Projection. In **projection,** the person's own unacceptable impulses are inhibited and the source of the anxiety is attributed to another person. For example, one's own angry feelings are attributed to one's innocent friend. Or a man who is attracted to his brother's wife sincerely believes she is trying to seduce him at a family gathering. Projection presumably gives relief because it reduces anxiety.

Reaction Formation. Another defense, termed **reaction formation,** occurs when a person transforms an anxiety-producing impulse into its opposite. For example, people frightened by their own sexual impulses and desires for unusual sexual adventures may become actively involved in a "ban the filth" vigilante group. They use their energy to vigorously censor books and movies they consider obscene, while secretly and unconsciously attracted to them. Through projection and reaction formation, the id impulses are expressed, but in a disguise that makes them acceptable to the ego: desire becomes disdain and hate.

A mother who was a leading psychoanalyst describes reaction formation as a mechanism shown by her own children in their early development (Monro, 1955). As a psychoanalyst, she was especially sensitive to the possible problems faced in the anal stage and the impulses activated then. She therefore allowed her children considerable freedom to express their infantile anal interests with few inhibitions. After experiencing that early phase of freedom, however, the children seemed to spontaneously develop an opposite pattern. Now they began to exhibit over-cleanliness to the point of finickiness, wanting everything to be "super clean," orderly, and neat. They found dishwater disgusting, for example, and insisted on refilling the sink with clean water repeatedly in spite of an extreme water shortage in the county. Likewise, they refused to clean up the puppy's "mistakes," and, as their insightful mother put it, "The reaction of finicking disgust, very genuinely experienced, was clearly related to the positive pleasures recently renounced. Housekeeping became much smoother as reaction formation involving an extreme if somewhat spotty orderliness also gave way to advancing maturity" (Monro, 1955, p. 252).

Rationalization. **Rationalization** is a defense that involves trying to deceive oneself by making rational excuses for unconscious impulses that are unacceptable. For example, a man who has unconscious, deeply hostile impulses toward his wife might invent elaborate excuses that serve to disrupt and even destroy their relationship without ever admitting his true feelings. He might invoke explanations such as "pressures at the office," "a hectic schedule," or "worrying about inflation and politics" as reasons for staying away from home. In doing so, he experiences little guilt over (and might even feel justified in) ignoring, avoiding, and frustrating his wife.

Sublimation. **Sublimation** is an ego defense that is particularly significant in the development of culture. It consists of a redirection of impulses from an object (or

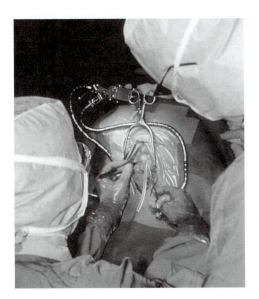

The concept of sublimation suggests that sexual and aggressive impulses may be redirected from their original objects and displaced to socially acceptable activities and careers.

target) that is sexual or aggressive to one that is social in character. Suppose that aggressive urges become too threatening to the young boy, for example the desire to kill the father while struggling with the Oedipus complex. He (or she) may sublimate (or transform) these impulses into socially acceptable forms that in the course of development evolve into the choice of surgery as a career.

Carl Jung

While Anna Freud remained basically loyal to the ideas proposed by her father, Carl Jung was a much more controversial figure who initiated a different form of the psychodynamic approach. Born in 1875, Carl Jung was raised in Basel, Switzerland, the son of a pastor in the Swiss Reformed Church. Upon earning his medical degree from the University of Basel, he began his career in psychology at the Psychiatric Institute in Zurich. Jung began as an admirer and associate of Freud but later became a dissenter and developed his own theory of psychoanalysis and his own method of psychotherapy. His approach became known as analytical psychology. Although it retains Freud's unconscious processes, it claims a **collective unconscious**—an inherited foundation of personality. The contents of the collective unconscious are **archetypes** or "primordial images." Unlike the personal unconscious, whose contents were once conscious but have been forgotten or repressed, the contents of the collective unconscious have never been in consciousness. Therefore the contents of the collective unconscious are not individually acquired; they are due to heredity. Examples of archetypes include God, the young potent hero, the wise old man, the Earth Mother, the Fairy Godmother, and the hostile brethren. They occur in myths, art, and dreams of all mankind.

In Jung's view, the psyche included not only a conscious side but also a covert or **shadow aspect** that is unconscious. Personal growth involves an unfolding of this shadow and its gradual integration with the rest of the personality into a meaningful, coherent life pattern. The unconscious of every female includes a masculine, assertive element

6.5 How did Jung's notion of the unconscious mind differ from Freud's conception?

Carl Gustav Jung (1875–1961)

TABLE 6.3   Examples of Jungian Concepts

The Collective Unconscious:	Found in everyone and said to contain inherited memories and ancestral behavior patterns
Archetypes:	Basic elements or primordial images forming the collective unconscious, manifested in dreams and myths (e.g., Earth Mother, the wise old man)
The Animus:	The masculine, assertive element in the unconscious of every woman
The Anima:	The feminine, passive element in the unconscious of every man
The Mandala:	Usually a circular shape, symbolizing the self, containing designs often divided into four parts

(the **animus**). The unconscious of every male includes a feminine, passive element (the **anima**). To be constructively masculine or feminine, individuals of each sex must recognize and integrate these opposite sex elements within themselves (see Table 6.3).

Jung described four basic ways of experiencing the world: *sensing, intuition, feeling,* and *thinking,* summarized in Table 6.4. According to Jung, people differ consistently in the degree to which they emphasize each way of experiencing. One person, for example, might typically prefer intuition to abstract thinking, choosing to become a psychoanalyst rather than a mathematician. Another might know the world mostly through his or her senses with little use of either intuition or reason. In addition, Jung was concerned with extraversion–introversion.

Like the four ways of experiencing, extraversion–introversion for Jung are divided: one is dominant in the conscious life while the other influences the unconscious side of the personality.

Jung broadened the concept of psychic energy. He did not exclude the sexual instinct of Freudian theory but thought it was only one among many instincts. For Jung, the meaning of behavior became fully intelligible only in terms of its end products or final effects; we need to understand humans not only in terms of their past but also in the light of their purposes and goal strivings.

Jung, like Freud, emphasized symbolic meanings. He believed, for example, that "abnormal behaviors" are expressions of the unconscious mind. Also like Freud, Jung thought that abnormal behaviors were merely one way in which the contents of the unconscious may reveal themselves. More often, he felt, they are expressed in dreams. Jung went beyond Freud, however, in his increasing fascination with dreams as unconscious expressions of great interest in their own right. This contrasts with their use merely as starting points for saying whatever comes to mind, that is, "free associations" (discussed in Chapter 5). As Jung put it: ". . . I came increasingly to disagree with free

6.6   Describe Jung's four basic ways of experiencing the world.

TABLE 6.4   Jung's Four Ways of Experiencing the World

Ways of Experiencing	Characteristics
Sensing	Knowing through sensory systems
Intuition	Quick guessing about what underlies sensory inputs
Feeling	Focus on the emotional aspects of beauty or ugliness, pleasantness or unpleasantness
Thinking	Abstract thought, reasoning

association as Freud first employed it; I wanted to keep as close as possible to the dream itself, and to exclude all the irrelevant ideas and associations that it might evoke" (Jung, 1964, p. 28).

In the same direction, Jung became intrigued by the unconscious for its own sake. He viewed the unconscious not just as the source of instincts. For him it was a vital, rich part of everyone's life, more significant than the conscious world, full of symbols communicated through dreams. The focus of Jungian psychology became the study of people's relations to their unconscious. Jung's method taught individuals to become more receptive to their own dreams and to let their unconscious serve as a guide for how to live.

6.7 How did Jung's conception and use of dreams differ from Freud's?

Jung's conception of personality is complex, more a set of fascinating observations than a coherent theory. His observations often dwelled on the multiple, contradictory forces in life: "I see in all that happens the play of opposites" (1963, p. 235). Yet he also was one of the first to conceptualize a *self* that actively strives for oneness and unity. Jung saw the self (the striving for wholeness) as an archetype that is expressed in many ways. The expressions of the striving for wholeness include the **mandala** (a magic circle archetype shown in Figure 6.2) and various religious and transcendental experiences. He devoted much of his life to the study of these expressions in primitive societies, alchemy, mythology, dreams, and symbols. To achieve unity and wholeness, the individual must become increasingly aware of the wisdom available in his or her personal and collective unconscious and must learn to live in harmony with it. His ideas continue to fascinate many psychologists and are being applied to topics that range from "feminist consciousness" (Lyons, 1997) to the role of the spiritual in healing

Figure 6.2 A mandala.

(Molina, 1996). However, his ideas remain difficult to study with the methods most psychologists favor.

Alfred Adler

Like Freud, Alfred Adler also was born in Austria, 14 years after Freud—in 1870. He earned his degree as doctor of medicine in 1895. After a brief period as an ophthalmologist, he practiced psychiatry, joining Freud's Vienna circle of associates at the turn of the century. A highly independent, even rebellious person, Adler broke from Freud after 10 years and began his own psychoanalytic movement, ultimately as a founder of the Society for Individual Psychology.

Adler's contributions have suffered an ironic fate. Much of what he said has become so widely accepted, and seems so plausible, that it has been incorporated into the everyday ideas and terms, the ordinary wisdom that we intuitively have about psychology. Some of these concepts are so common as to risk becoming cliches. Nevertheless, while the popularity of Adler's ideas makes them less distinctive, they remain important even in contemporary thinking about personality.

It is often said that every personality theory captures best the personality of the theorist who created it. Adler's own childhood was marked by chronic illness and hostile relations with his five siblings. Interestingly, both these themes—physical weakness, or **organ inferiority,** and **sibling rivalry**—became central concepts in his theory. Adler's theory begins with a recognition of the infant's profound *helplessness,* a state that makes him or her especially vulnerable to any biological *organ inferiority* or weakness. This biological vulnerability becomes the root for a psychological state that endures in the person and that has central importance in Adler's theory: feelings of inferiority.

It is the struggle to overcome these inferiority feelings that provides the underlying motivation for lifelong compensatory strivings. Throughout the life course, the person tries to make up for this perceived deficit by striving for perfection and superiority. The particular attitude the person adopts toward the inevitable state of inferiority, rather than the deficit itself, was most important for Adler; given a courageous attitude, a perceived deficit can become a positive asset. We all know dramatic examples of personal victories in overcoming biological deficits. Demosthenes was the ancient Greek who achieved fame as a great orator, overcoming a childhood stutter, and more than one great athlete reached the Olympics after long efforts to compensate for early concerns with physical weakness or illness. This **compensatory motivation** contrasts sharply with the id impulses, sexual and aggressive in nature, featured as the driving forces in Freud's theory. It is a much more social psychological view of motivation. It is rooted in the person's feelings about a biological deficit but goes much beyond that origin. The striving to compensate can have many constructive healthy outcomes when pursued with courage. If compensatory efforts fail, however, the person may develop an **inferiority complex,** continuing to feel extremely inadequate about the perceived inferiority and failing to grow beyond it.

Adler also showed a more social orientation in other parts of his theory in which he is alert to cultural influences and social, interpersonal situations. He saw the rivalry between siblings within the family as an important part of development. Thus he viewed the family as the context for significant relationships and conflicts beyond those captured in the Oedipal triangle of mother–father–child that was central for Freud. Indeed, Adler's ideas were notable as a major break from concern with inborn impulses and hereditary

6.8 How are Adler's own life experiences reflected in his theory?

In Adler's view of compensatory motivation, the striving to excel may be a way of overcoming early feelings of inferiority.

6.9 Which concepts in Adler's theory reflect his positive view of human nature?

causes to focus on the environmental forces and the social world as determinants of personality development. Although the individual functions with consistency and unity, the pattern that makes up the style of life can be modified. This happens when the person changes the goals toward which the whole pattern of striving is directed.

The striving for perfection plays a great role for Adler, but it is matched by his concern for the individual's social feeling or *social interest*. This focus on the positive, adaptive aspects of personality development is also seen in two other Adlerian concepts: courage and common sense. Taken together, social feeling, courage, and common sense constitute the set of characteristics that mark well-functioning, healthy persons.

Such persons cope with the realities of life, including their inevitable helplessness and inferiority, with confidence and constructive strivings, without excessive fear but also without unrealistic fantasies. In contrast, the unhealthy personality abandons appropriate effort and avoids facing realistic difficulties by a retreat into increasingly grandiose fantasies. These fantasies are maladaptive if they widen the gap to reality and provide an unrealistic avoidance of failure.

The positive qualities of social feeling, courage, and common sense are natural states: every person is capable of having them spontaneously unless they are blocked or frustrated in the course of development. For example, the excessively *pampered child* may develop a style of life characterized by being extremely demanding, while the severely *rejected child* may live life in a world seen as dangerous and hostile.

To help overcome this type of damage, the therapist in the Adlerian approach provides the encouragement and sympathetic understanding that allows the patient to face life more realistically and effectively. In this supportive atmosphere, the patient can abandon "mistaken" strivings for fantastic superiority and stop the retreat from reality to begin to face life with common sense, courage, and social feeling.

Erich Fromm (1900–1980)

Erich Fromm

Erich Fromm (1941, 1947) helped to expand Freudian concepts to take account of the important role of society in the development and expression of personality.

For Fromm, people are primarily social beings to be understood in terms of their relations to others. According to Fromm, individual psychology is fundamentally social psychology. People have psychological qualities, such as tendencies to grow, develop, and realize their potentialities. These basic tendencies lead them to freedom and to want to strive for justice. Thus, human nature has a force of its own that influences social processes.

Fromm's explanation of character traits illustrates the difference between Freud's biological orientation and Fromm's social orientation. Fromm criticized Freud's idea that fixation at certain pleasure-giving stages is the cause of later character traits. According to Fromm, character traits develop from experiences with others. Psychosexual problems are the body's expressions of an attitude toward the world that is socially conditioned. According to Freud, culture is the result of society's efforts to suppress instinctual drives. For Fromm, culture is molded by the structure and substance of a given society. Another major point of departure from Freud is Fromm's belief that ideals like truth, justice, and freedom can be genuine strivings and not simply rationalizations of biological motives. Freud's psychology is a psychology of instinctual drives that defines pleasure in terms of tension reduction. Fromm's psychology tries to make a place for positive attributes, such as tenderness and the human ability to love, and the desire for freedom as basic aspects of human nature.

6.10 Contrast Fromm's views on the origin of character traits and positive ideals with those of Freud.

Erik Erikson

The psychoanalyst Erik Erikson (1963) has proposed stages of development that call attention to problems of social adaptation (Table 6.5). As children grow up, they face

6.11 Describe Erikson's psychosocial stages and the psychosocial crisis expressed at each stage.

Erik H. Erikson (1902–1994)

a wider range of human relationships and new challenges. The solution of the specific problems at each of eight **psychosocial stages** (rather than psychosexual stages) determines how adequate they will become as adults. Erikson's focus on psychosocial development reflects the growing neo-Freudian emphasis on broad social and cultural forces, rather than instinctual drives alone.

TABLE 6.5 Erikson's Stages of Psychosocial Development

Stage and Age	Psychosocial Crisis	Optimal Outcome
I. Oral-sensory (1st year of life)	Trust vs. Mistrust	Basic trust and optimism
II. Muscular-anal (2nd year)	Autonomy vs. Shame	Sense of control over oneself and the environment
III. Locomotor-genital (3rd through 5th year)	Initiative vs. Guilt	Goal-directedness and purpose
IV. Latency (6th year to start of puberty)	Industry vs. Inferiority	Competence
V. Puberty and Adolescence	Identity vs. Role Confusion	Reintegration of past with present and future goals, fidelity
VI. Early Adulthood	Intimacy vs. Isolation	Commitment, sharing, closeness, and love
VII. Young and Middle Adult	Generativity vs. Self-absorption	Production and concern with the world and future generations
VIII. Mature Adult	Integrity vs. Despair	Perspective, satisfaction with one's past life, wisdom

Source: Adapted from Erikson, E. (1963). *Childhood and society.* New York: Norton.

At each stage of development, Erikson hypothesizes a **psychosocial crisis.** This crisis arises from the person's efforts to solve the problems at that stage. For example, in the first stage of life (the "oral-sensory" stage of the first year), the crisis involves "trust versus mistrust." Erikson hypothesizes that at this stage the child's relation to the mother forms basic attitudes about "getting" and "giving." If the crisis is properly resolved, the experiences at this stage lay the foundation for later trust, drive, and hope.

Erikson's stages extend beyond infancy to include crises of adolescence and adulthood. He sees development as a process that extends throughout life, rather than being entirely determined in the early years. In this developmental process, **ego identity** is central:

> The integration . . . of ego identity is . . . more than the sum of the childhood identifications. It is the accrued experience of the ego's ability to integrate all identifications with the vicissitudes of the libido, with the aptitudes developed out of endowment, and with the opportunities offered in social roles (Erikson, 1963, p. 261).

The underlying assumptions of his view of development are:

> (1) that the human personality in principle develops according to steps predetermined in the growing person's readiness to be driven toward, to be aware of, and to interact with, a widening social radius; and (2) that society, in principle, tends to be so constituted as to meet and invite this succession of potentialities for interaction and attempts to safeguard and to encourage the proper rate and the proper sequence of their enfolding (Erikson, 1963, p. 270).

Erikson's ideas have become popular in many parts of our culture. His thoughts concerning the "identity crises" of adolescence, for example, are discussed widely. Indeed the phrase **identity crisis** has become a part of everyday speech. Erikson's ideas have influenced concepts of human nature and the general intellectual culture. He believed that all young people must generate for themselves some central perspective that gives them a meaningful sense of unity and purpose. This perspective integrates the remnants of their childhood with the expectations and hopes of adulthood (Erikson, 1968). The sense of identity involves a synthesis of how individuals have come to see themselves and their awareness of what the important other people in their lives expect them to be. The stages of development that he discussed are summarized in Table 6.5.

▶ OBJECT RELATIONS THEORY AND THE SELF

Over many decades, psychodynamic theory and practice have undergone particularly important transformations. While many psychologists remain within an essentially psychoanalytic framework, many have moved far beyond Freud and his immediate followers. These innovators further changed how they think about personality, the roots of mental health, and ways to help troubled people.

The basic orientation of this approach has emerged clearly in an integrative review of this movement (Cashdan, 1988; Greenberg & Mitchell, 1983). There have been different variations in this shift. Leaders include such psychoanalysts as Melanie Klein (one of the earliest innovators) in England and, more recently, Otto Kernberg (1976, 1984) and Heinz Kohut (1971, 1977) in the United States. In this section, we will emphasize the common themes of change that seem to characterize this movement.

The developing self is defined from the start in relational, interpersonal terms in Kohut's theory.

6.12 What is the central emphasis within object relations theories?

The approach is called **object relations theory** and therapy (Cashdan, 1988), and the first point to note is that the "objects" in the language of this theory are simply other human beings. The term "objects" is a leftover from classic psychoanalysis, and the phrase "significant others" essentially could substitute for it. The important shift from classic psychoanalysis to object relations theory is that while the former focused on the instinctual drives, the latter focuses on the relationships to significant other people (i.e., object relations).

The most important object for the developing child generally, and unsurprisingly, is the mother. It is in the young child's relationship with the mother that the **relational self** begins to originate and emerge. Note that the self is defined from the start in relational or interpersonal terms.

"Good–Bad Splitting"

Within psychodynamic theory, one of the first to address the mother–child relationship in great depth was Melanie Klein, a British psychoanalyst who was Freud's contemporary. A theme that still persists from Klein's work is her clinical observation that the young child tends to divide the world into good and bad. Klein saw the core conflict throughout life as a struggle between positive feelings of love and negative feelings of hate. Her insight that in this conflict people tend to "split" the world into benevolent and malevolent components has been integrated into much current relational theorizing about personality structure and development.

Klein spoke of a nourishing good breast and an empty bad breast in the child's conflict-ridden representation of the mother. This notion has remained in a variety of contemporary psychodynamic ideas about "good" and "bad" self-representations, internal representations of the self and of other people. It is part of the belief that from infancy on there is a tendency to somehow "split" or partition experiences in good–bad, gratifying–frustrating terms (Cashdan, 1988), fragmenting rather than integrating them into a coherent whole. When these splits are severe, therapy seeks to help the person to integrate them.

The Development of Self

Briefly, development is seen as a process in which the newborn begins in a world that is experienced as split into good (gratifications) and bad (tensions) feelings. In this early world, other people, including the mother, are not yet differentiated. Emotional splitting of experiences and people (objects) in good–bad, positive–negative terms continues throughout later life.

6.13 How is splitting involved in the development of the self concept and in a person's level of self-esteem?

The most important object, the mother, soon begins to be represented by the young child internally as an image. With cognitive development and the growth of language skills, the child can start to internalize not only a maternal image but also maternal conversation in the form of an inner dialogue. Some of these early conversations are audible. You know this if you have ever heard the conversations youngsters sometimes have with themselves as they praise or scold their own performance, saying "good boy" or "no, no" aloud to themselves. This internal dialogue is especially evident during toilet training and other early exercises in the development of self-regulation.

In time, the internalizations of maternal images and conversations become the foundations of the developing self. You can see this development, for example, in the increasing use of "I" in the child's speech. The child's utterance changes from "Jane wants ice cream" to the personal pronoun in which "I" want it, "I" eat it, "I" am bad.

In this conception, emotional splitting continues as an aspect of the developing self: "Just as early splitting of the mother creates a split in the maternal presence, so the split in the inner maternal presence creates a split in the self. Early splits give birth to later splits" (Cashdan, 1988, p. 48). In time, individuals come to view themselves as good or bad depending on their earlier good–bad emotional experiences of splitting. The sense of self-esteem that ultimately emerges characterizes how persons feel about themselves. It is both the consequence of the earlier experiences and the determinant of much of what is experienced later in the course of life.

As the splitting process continues, a variety of identity splits occur. They yield such important categories as one's sexual identity, career identity, identity as a parent, and so on. Each is colored emotionally in good–bad terms. The emotional splitting represented by the enduring concern with goodness–badness never ends. When it is tilted toward a badness imbalance, it continues to corrode the person's relationships and is not adaptive. The therapeutic process, in turn, is viewed as the method for undoing the imbalance, recognizing and overcoming inner conflicts, and developing a more integrated and positive image of the self.

▶ ATTACHMENT: THE ROOTS OF OBJECT RELATIONS

Like Freud, object relations theorists focus on the importance of the early years. Unlike Freud's emphasis on how the instinctual drives are expressed and managed in the first few years, however, these theorists stress the type of relationship that develops with the early caretaker, usually the mother. This early relationship becomes the basic framework for the perception and experience of later relationships. The details of this developmental process have received increasing research attention by child psychologists. They study the quality and varieties of early attachment relations between mother and child (Ainsworth, Blehar, Waters, & Wall, 1978) and trace how these early relationships link to subsequent development (Sroufe & Fleeson, 1986). Much of this work was influenced by the attachment theory proposed by a British psychiatrist, John Bowlby, discussed next.

Attachment Theory

Bowlby was seeing a 3-year-old boy in psychotherapy. It was half a century ago, and he was being supervised by the psychoanalyst Melanie Klein, who refused to allow him to meet with the boy's mother. Frustrated by this experience, as well as by the unscientific nature of psychoanalytic theorizing, Dr. Bowlby developed his now-famous attachment theory (Holmes, 1993). This theory is consistent with object relations theory, giving center stage to the relationship between the young child and the primary caregiver (Bowlby, 1982). For Bowlby, however, the psychological characteristics of the "object"—particularly the mother—were crucial, and he emphasized the experienced relationship between child and the mother as the primary caregiver.

6.14 What is the role of working models in Bowlby's theory? How do these develop?

According to Bowlby, based on experiences in this relationship, the child develops **internal working models.** These are mental representations of others, of the self, or of relationships that guide subsequent experience and behavior. Children who have had positive, gratifying experiences with significant others in their environment will develop internal working models of others as responsive and giving, and of themselves as competent and worthy of affection; those who have painful or unsatisfying experiences develop internal models that reflect those troubled relationships. The basic message in attachment theory is that caregivers who are responsive to the infant and its needs provide a **secure base,** a safe haven of dependable comfort in the young child's life from which the world can then begin to be explored with trust, without fear of abandonment.

Early Attachment Relations: Secure/Insecure Attachment Patterns

Mary Ainsworth (1989) developed the "Strange Situation" to examine patterns of infant–parent attachment in everyday situations among 8 1/2 to 12-month-olds (e.g., Stayton and Ainsworth, 1973; Ainsworth et al., 1978). The **Strange Situation** assesses individual differences in the baby's relationship with the mother. Ainsworth chose to conduct her study in a "strange" setting, that is, one unfamiliar to the baby, because she found that most American children were accustomed to the frequent comings and goings of their mothers throughout the day. Placed in an unfamiliar setting, however, children were expected to exhibit their characteristic attachment behaviors in response to stress.

6.15 Describe the *Strange Situation* and what it has revealed about children's attachment patterns. How are these patterns related to childhood behavior?

In this situation, the toddler (about age 18 months) is introduced to a novel playroom environment with the mother and a stranger, Then the toddler is exposed to different levels of availability of the mother, from present and involved with the child, to present and mildly preoccupied, to absent. The child is separated from the mother twice during the Strange Situation; once left with the stranger and once left alone.

Three main patterns of behavior have been identified in this situation. Toddlers who avoided the mother throughout the paradigm, as well as on reunion, were considered **insecure–avoidant,** or A babies. Some were able to greet the mother positively upon reunion, and then return to play, and they attended to the mother and desired interaction with her throughout the procedure. These were termed **securely attached,** or B babies. Those whose reunion behavior combined contact-seeking and anger, and who were difficult to comfort upon reunion, were classified as the C or

Is early attachment the preface to adult romantic love?

insecure–ambivalent babies. Home visits revealed that the different types of babies experienced different patterns of maternal responsiveness. For example, mothers of infants rated as securely attached were most responsive toward their babies. Mothers of resistant babies were inconsistent in their responsiveness. The responsiveness of mothers of avoidant children varied with the context: they were unresponsive to bids for contact and comfort, but they were controlling and intrusive in reponse to their children's attempts at independent play. Likewise, children rated secure are more likely to remain confident and flexibly organized when faced with an insurmountable task as preschoolers (Arend et al., 1979) and suggest strategies for coping with the absence of a parent as 6-year-olds (Main, Kaplan, & Cassidy, 1985). Additionally, 5-year-olds with a history of a secure relationship with the mother are less likely to exhibit negative interactions with a peer (Youngblade & Belsky, 1992). Likewise, toddlers who at 18 months were able to cope effectively with separation from the mother in the strange situation, for example by distracting themselves and exploring the toys in the environment to reduce distress, were also more likely to become able to cope effectively with frustration in later childhood (Sethi and colleagues, 2000). Specifically, these toddlers also were able, at age 5, to delay immediate gratification longer for more valued rewards.

Attachment in Adult Relationships

Many other links have been found with infant attachment style and some of these may extend into adulthood. Studies of adult attachment suggest that individuals may carry specific attachment styles with them in relationships throughout life (e.g., Fraley & Shaver, 1997; Kobak & Sceery, 1988). For example, college students rated secure (by interview measures) were rated by their friends as warm and nurturant (Bartholomew & Horowitz, 1991).

Studies of adult attachment styles generally classify participants into one of three categories, based on their self-reports: secure, avoidant, and ambivalent. Participants also are asked about their childhood experiences within the family, and their past and current romantic relationship history and satisfaction. Studies try to link the self-reported attachment styles to the participants' reports about their personal relationships. Table 6.6 gives examples of the kinds of relationships found.

6.16 Describe the three adult attachment patterns found in self-report studies. How do these patterns relate to measures of adjustment?

TABLE 6.6   Some Correlates of Adult Relationship Style

Relationship Style	Associated with
Secure	Report they had supportive families; trusting, warm, happy parents; can tolerate separations from partners without high anxiety; have partners who tend to be satisfied with the relationship; able to give partners emotional support when they need it; generally construct positive romantic relationships; believe romantic love is real and can last.
Avoidant	Report they had aloof, emotionally distant parents, did not feel warmth, closeness or trust in family; tend to fear intimacy, find emotional commitment difficult; unable to be highly emotionally supportive of partner; cynical about romantic love and doubt that it lasts.
Ambivalent	Report they have many romantic relationships that don't last long; anxiously fearful of losing partners; ready and eager to change self to please partner; stressed by separations from partner. Believe falling in love is easy but does not last.

Source: Based on Hazan & Shaver, 1987, 1994.

As the table indicates, adults who see themselves as secure about relationships also report themselves to have had happier, more positive relationship histories, including in their families as children, and in their past and current relationships with romantic partners. Unsurprisingly, these differences are seen particularly when the secure adults are compared to those who describe themselves as avoidant in relationships. A study of adults at age 52 also used a more objective measure of relationship success, namely marriage and divorce statistics. It found the adults who were secure (self-described) also were almost all married (95 percent), whereas only 72 percent of the avoidant adults were married and half of them had been previously divorced (Klohnen & Bera, 1998). As the table also indicates, the three groups differ in their beliefs about romantic love, with the secure being believers, the avoidant more cynical, and the ambivalent thinking it comes easily but does not last.

What conclusions do these findings allow? On the one hand, people clearly differ in how secure they feel themselves to be in close relationships, both currently and in the past, and how they feel about those relationships. These feelings are associated also with what they expect to find in relationships, and what they are likely to experience within them, and how their partners will feel about them. Much less clear is the degree to which adult attachment patterns are directly linked to those observed in toddlers and young children. Although object relations theory generally assumes that the adult patterns are a direct outgrowth and continuation of those found in early life, the data are open to alternative interpretations and have been strongly questioned (e.g., Lewis, 1999, 2002).

6.17 How much temporal stability did Lewis find in attachment styles from childhood through adulthood? What are the implications of his findings?

Specifically, after an extensive and detailed review of the massive research on this topic Lewis (1999, p. 341) concludes that attachment patterns do not show continuity over long periods of time (e.g., from 1 year of age to age 18 years) in the course of development. However, at any given period of time, there are links between current attachment patterns and important indicators of functioning. In Lewis' view, this pattern—not stable forever but significant and meaningful at any given moment in life—points to the importance of the current social environment in which the individual is functioning. But it does not support the idea that the attachment styles of infancy become the model for all that follows.

Rather, individuals seem to acquire multiple mental representations of others, perhaps based on various significant people with whom they come in contact throughout their lives. These mental representations differ in the degree to which they are readily "accessible" or become easily activated in the person's mind. In that case, everything else being equal, the representation that has been activated most in the past is likely to guide their perception of others as well as their thoughts and behaviors in response to them. But everything is often not equal. In any given person's life, there are a variety of people who differ in their expression of love, availability, and supportiveness. Thus, different people could activate different attachment representations (Shaver & Mikulincer, 2002).

In addition, a person's current motives or mood can also affect the attachment representation that is most easily activated at a given moment (Shaver, Collins, & Clark, 1996). In short, the past experiences in earlier attachment relationships may or may not play a major role in the person's current relationships. To the many people in the world whose early childhood was not a picture of ideally experienced security, this is good news, even though it is disturbing to the classic theory of object relations and makes their assumptions of continuity tentative at best.

Kohut's Theory

. . . man can no more survive psychologically in a psychological milieu that does not respond empathically to him, than he can survive physically in an atmosphere that contains no oxygen (Kohut, 1977, p. 85).

The object relations theorists share several themes, as the last section showed. One leader in this movement, Heinz Kohut, is selected in this section for further attention, because his work is seen as especially influential for changing views of the healthy and the disturbed personality. Kohut, a psychiatrist who received his medical training at the University of Vienna, went on to psychoanalytic training and teaching in Chicago where he gained recognition as a theorist and clinician in the 1970s.

In Kohut's view, throughout the century profound changes in the family and culture have occurred: psychoanalysis and psychodynamic theory must be responsive

6.18 According to Kohut, what changes in family life have occurred since Freud's era, and how have they influenced the nature of psychological problems?

Heinz Kohut (1913–1981)

For Kohut, the modern child's life is characterized not by too much parenting and intimacy but by loneliness.

to them. An important change, he believes, is that Freud's patients typically came from a Western civilization in which life was concentrated in the home and family unit. Families tended to expose their children to **emotional overcloseness** (Kohut, 1977, p. 269), and these intense emotional relations in turn often produced neurotic problems involving internal conflicts such as those in the Oedipus complex. The developing child was likely to be trapped in too much intimacy, too much stimulation, too much intrusiveness.

In contrast, children now are more likely to see parents at most in leisure hours and to develop much less clear role definitions and models: "The environment which used to be experienced as threateningly close, is now experienced more and more as threateningly distant . . ." (Kohut, 1977, p. 271). While personal problems used to arise from being too stimulated emotionally by parents (Freud, 1963), now youngsters tend to be *under*-stimulated and may search for erotic sensations and other strong experiences to fill the emotional emptiness of their lives and to try to escape loneliness and depression.

Kohut's thinking has led the way for a new psychoanalytic interest in the self and for the treatment of problems such as disorders of the self. Rather than being driven by unconscious conflicts and impulses, Kohut sees patients today as often deprived of **empathic mirroring** and ideal "objects" (people) for identification. Because their parents were walled off from them emotionally, or too involved with their own narcissistic needs, they did not provide the necessary models for healthy development of the self and for the formation of meaningful, responsive relationships in adulthood.

People fear the destruction of the self when they don't feel the empathic human responses from the important others ("self objects") in their lives. Kohut compares this state to being deprived of **psychological oxygen.** The availability of empathic reactions from self objects is as vital to the survival of the self as the presence of oxygen is to the survival of the body:

> What leads to the human self's destruction is its exposure to the coldness, the indifference of the nonhuman, the nonempathical responding world (Kohut, 1984, p. 18).

What is feared most is not so much physical death, but a world in which our humanness would end forever (Kohut, 1980, 1984). In the same vein, Kohut does not see Freud's castration anxiety as the ultimate human anxiety: ". . . the little boy's manifest horror at the sight of the female genitals is not the deepest layer of this experience . . . behind it and covered by it lies a deeper and even more dreadful experience—the experience of the faceless mother, that is, the mother whose face does not light up at the sight of her child" (Kohut, 1984, p. 21).

This type of anxiety about being in a totally uncaring, unresponsive world is experienced in a dream that one of Kohut's patients had (Kohut, 1984, p. 19). In his "stainless steel world" dream, Mr. U was in an ice tunnel with walls from which large glistening strands of ice went down to the ground and up to the ceiling. It was like an enormous model of the human heart, large enough to be walked around in (like one in a museum the patient knew well). Walking within this icy heart, Mr. U felt the anxiety of an oncoming but unnamed danger to which he was exposed, all alone, except for a shadowy figure to whom he appealed but who was unresponsive. In a flash, he was pulled through a crack in the wall into a cityscape that was blindingly bright—a landscape that was utterly unreal, with busy but completely unapproachable people all around: a "stainless steel world" in the patient's own words, in a science fiction scene with no escape, no communication, trapped forever, unreachable in a world of cold-heartedness.

Kohut's theory also leads to a reinterpretation of such Freudian constructs as the "Oedipal period." In his view, during this period the boy fears confrontation from a mother who is nonempathic and sexually seductive rather than affectionate and accepting of him. He also fears confrontation from a father who is competitive and hostile with him rather than pleased and proud. In parallel fashion, during the same period the girl fears confrontation from a father who is nonempathic and sexually seductive instead of affectionate and accepting. At the same time, she also fears a mother who is competitive and hostile rather than reflecting that she is proud of the child and pleased by her.

In Kohut's theory, if parents fail to respond empathically and healthily to their child in this phase of development, it sets up a defect in the self. As a result, the child develops a tendency to experience sexual fantasies and the fragments of love rather than love. Likewise, the child with a defective self also tends to experience hostile fantasies and only the fragments of assertiveness rather than appropriate assertiveness. The individual's typical internal reaction to these experiences becomes great anxiety. These characteristics of the defective self contrast with those of a healthy, normal personality, which, instead of anxiety and fragmented experiences, feels the glow of appropriate sexual functioning and assertiveness, as summarized in Table 6.7.

TABLE 6.7   Kohut's Characteristics of the Self and Mental Health

Defective Self	Healthy Personality
Fragmented experience of love (sexual fantasies)	Feels glow of healthy pleasure in appropriate sexual functioning
Fragmented assertiveness (hostile fantasies)	Able to be self-confidently assertive in pursuit of goals

▶ MURRAY, THE HARVARD PERSONOLOGISTS, AND HIGHER MOTIVES

The psychodynamic approach stimulated many innovations in personality theory and assessment. One of the most extensive and imaginative efforts unfolded under the leadership of Henry A. Murray, Robert W. White, and their many colleagues at the Harvard Psychological Clinic in the 1940s and 1950s. This group, which became known as the **Harvard personologists,** provided a new model called **personology** for the intensive psychodynamic study of individual lives, and devoted itself to the portrayal of persons in depth. The Harvard personologists were influenced strongly by Freud and the post-Freudian ego psychologists. They also were influenced by "biosocial" organismic views that emphasized the wholeness, integration, and adaptiveness of personality. In turn, they trained and influenced a group of students who went on to study a variety of important human motives that are useful for understanding the goals that different people enduringly pursue with varying degrees of intensity.

Higher Order Motives

Perhaps most important for the advancement of personality theory, Henry Murray called attention to the finding that sex and aggression were by no means the only basic motives revealed by the in depth study of fantasies with methods that he helped to develop. His influential investigation, *Explorations in Personality* (1938), suggested that a comprehensive understanding of personality required taking account of the following points:

In-depth studies of lives revealed many human needs and motives. Although sex and aggression have retained an important place among the human motivations recognized by Freud's successors, Murray believed additional motives and needs have to be hypothesized to capture the complexity of personality. He and the Harvard personologists advanced the modern study of personality dynamics by identifying a wide range of such human motives that go far beyond the id impulses of classic Freudian theory.

The motives that the Murray group identified are called **higher order motives.** They used that name because unlike such basic biological needs as hunger and thirst, they do not involve specific physiological changes (such as increased salivation or stomach

6.19 How do Murray's higher order needs differ from biological drives? How do needs and presses conform to a person × situation perspective?

TABLE 6.8   Some Nonphysiological Human Needs Hypothesized by Henry Murray

Abasement (to comply and accept punishment)	Humiliation avoidance
Achievement (to strive and reach goals quickly and well)	Nurturance (to aid or protect the helpless)
Affiliation (to form friendships)	Order (to achieve order and cleanliness)
Aggression (to hurt another)	Play (to relax)
Autonomy (to strive for independence)	Rejection (to reject disliked others)
Counteraction (to overcome defeat)	Seclusion (to be distant from others)
Defendance (to defend and justify)	Sentience (to obtain sensual gratification)
Deference (to serve gladly)	Sex (to form an erotic relationship)
Dominance (to control or influence others)	Succorance (to ask for nourishment, love, aid)
Exhibition (to excite, shock, self-dramatize)	Superiority (to overcome obstacles)
Harm avoidance (to avoid pain and injury)	Understanding (to question and think)

Source: Based on Murray, H. A., Barrett, W. G., & Homburger, E. (1938). *Explorations in personality.* New York: Oxford University Press.

contraction). Instead, they are psychological desires (wishes) for particular goals or outcomes that the person values. Table 6.8 shows examples of the diverse needs inferred by Henry Murray and associates (1938) in their classical listing. Many of these motives have been investigated in detail (e.g., Emmons, 1997; Koestner & McClelland, 1990).

Motives in Murray's view do not operate regardless of the context or situation and its pressures, which he called **environmental presses.** Because Murray theorized that these situational presses influenced personality and its expressions, he and his colleagues developed intensive observational techniques that tried to assess them and take them into account.

The Harvard model for studying personality stimulated many research programs that helped clarify and specify the types of nonbiological higher needs that Murray's group identified. One good example was the motivation for competence—the need to be effective in its own right—described first by Robert White.

Competence Motivation

A variety of higher order motives such as curiosity, the need for stimulation, and the desire for play and adventure all may be seen as parts of a more basic motive: the desire for competence (White, 1959). According to Robert White, who also was a Harvard personologist working with Murray, everyday activities such as a child's exploring, playing, talking, and even crawling and walking reflect the desire for mastery and effective functioning; they are satisfying for their own sake (intrinsically) and create in the person a feeling of efficacy. White argues the point in these words:

6.20 What is the origin of competence motivation? How does it influence behavior?

> *If in our thinking about motives we do not include this overall tendency toward active dealing, we draw the picture of a creature that is helpless in the grip of its fears, drives, and passions; too helpless perhaps even to survive, certainly too helpless to have been the creator of civilization. It is in connection with strivings to attain competence that the activity inherent in living organisms is given its clearest representation—the power of initiative and exertion that we experience as a sense of being an agent in the living of our lives. This experience may be called a feeling of efficacy (White, 1972, p. 209).*

Competence motivation is a desire for mastery of a task for its own sake and may apply to such diverse tasks as running, piano playing, juggling, chess, or the mastery of a new surgical procedure. According to White, the desire for mastery arises independently of other biological drives (such as hunger and sex) and is not derived from them. Moreover, people engage in activities that satisfy competence needs for the sake of the activity, not for the sake of any external reward such as the praise, attention, or money to which it may lead. The concept of competence motivation is valuable in emphasizing the enormous range of creative activities that humans pursue and appear to enjoy in their own right, and it has played a major role in research on motivation and adaptive problem solving (e.g., Dweck, 1990). It is, however, only one of many motives that influence human behavior.

Diverse other higher order nonphysiological motives and needs have been identified in personality research. These include the need for achievement (McClelland et al., 1953; McClelland, 1985), the need for affiliation (McAdams & Constantion, 1983), the need for control (Glass, 1977), and the need to endow experience with meaning, called need for cognition (Cacioppo & Petty, 1982). In addition, researchers are examining individual differences in a large array of other goals, motives, and personal projects that motivate behavior (e.g., Emmons, 1997).

To study motives, the Harvard personologists wanted measures that would get beyond direct self-reports to tap underlying needs of which the participant might not be aware, or which might be uncomfortable to acknowledge. With this goal, they developed a new projective test that examined the person's fantasies as revealed by stories told to a set of pictures, described in In Focus 6.1.

Need for Achievement

David C. McClelland and his colleagues (1953), also working at Harvard University, used the TAT to explore the **need for achievement (n Ach)**. They defined this need as *competition with a standard of excellence.* To study it, they examined fantasies from stories told on the TAT by systematically scoring the occurrence of achievement imagery in the stories. They assumed that the more the stories told involved achievement themes and concerns, the higher the level of the achievement motive.

McClelland (1961) found intriguing relations between TAT achievement themes and many economic and social measures of achievement orientation in different cultures. He devised special TAT cards showing work-relevant themes (see Figure 6.4) to assess the achievement motive. If, for example, the person creates stories in which the hero is studying hard for a profession and strives to improve, to compete against standards of excellence, and to advance far in a career, the story gets high n Ach scores. This technique has become an important way of measuring the motive to achieve in an indirect or implicit way that avoids simple self-report and its potential biases. It has

---

IN FOCUS 6.1

THE THEMATIC APPERCEPTION TEST (TAT)

The **TAT** consists of a series of pictures and one blank card (see Figure 6.3 with TAT-like card). Scenes show, for example, a mother and daughter, a man and a woman in the bedroom, a father and son. The test was developed by Henry Murray in collaboration with Christina Morgan as part of the studies of lives and motivations conducted by the Harvard personologists (Murray et al., 1938). It was used to study a variety of motives.

If you take this test, you will be told that it is a story-telling test and that you are to make up a story for each picture: "Tell what has led up to the event shown in the picture, describe what is happening at the moment, what the characters are feeling and thinking, and then give the outcome." You are encouraged to give free reign to your imagination and to say whatever comes to mind. It is expected that people will interpret the ambiguous pictures presented according to their individual readiness to perceive in a certain way ("apperception"). The themes that recur in these imaginative productions are thought to reflect the person's underlying needs, as well as potential conflicts and problems. Various scoring manuals have been developed that make it possible to objectively code the stories into various categories in research on motives. Good examples of such coding systems are found in studies of the need for achievement described in the text.

Figure 6.3 Picture similar to those in the TAT.

Figure 6.4 A TAT card developed by David C. McClelland for measuring the "Need to Achieve".

been used often in research for many years beginning in the 1950s (Atkinson, 1958) and yielded a rich network of meaningful relationships (McClelland, 1985).

For example, the level of achievement motivation exhibited by unemployed blue-collar workers was related to how they dealt with their unemployment situation and tried to find work again. Those higher in measured achievement motivation used more strategies in their job search, and began to look for a job sooner after being laid off, than did those lower in this motive (Koestner & McClelland, 1990).

It is also noteworthy that achievement motivation, like most other motivations, is influenced by socialization experiences, especially those in childhood. Parents who set standards that challenge their children to compete and excel, who make their expectations clear, provide goals and standards that the child can meet, and who offer support along the work route toward attaining those goals, also tend to have children whose achievement motivation is higher.

Need for Power

Achievement motivation may be the most extensively researched motive, but it is only one of many that has received attention and yielded interesting findings. David Winter (1973), influenced by the work and methods of McClelland, focused on the **need for power,** defined in terms of the impact one wants to have on people. Individual differences are measured by scoring the power themes that appear in the stories people tell

6.21 What relations to behavior have been demonstrated in research on need for power?

to the TAT (e.g., expressed concern with the status of the characters and the strength and impact of their actions). An extensive network of personal qualities and preferences—from greater risk-taking and more leadership activity to more visible indicators of status—is associated with this motive. People high on power motivation express more concern with impact on others, status, and reputation in their stories. To get the image that engages and reflects the power motive, think here of glossy advertisements for luxury automobiles shown glistening in the driveways of elegant mansions. As the image suggests, the advertisement industry is fully aware of the power motive, and seeks to engage it effectively in those who are likely to be responsive to it.

Research on the power motive has found connections between power imagery in political communications, such as the speeches and writings of political leaders, and the warfare that follows at various points in history (Winter, 1993). Winter examined power images in the leaders' communication, for example between Britain and Germany at the time of the First World War. These images also were analyzed in leaders' communications between the United States and the Soviet Union during the Cuban missile crisis in the 1960s. These analyses showed that power images increase before military action and threat erupts. In contrast, power images tend to decrease preceding a decrease in actual threat levels and warfare. The analysis of these images promises to have practical value for predicting important political events.

Need for Intimacy

6.22 How does
McAdams's research
on need for intimacy
relate to Kohut's theory?

The **need for intimacy** was examined by Dan McAdams (1990) and revealed a common human desire: people are motivated to warmly and closely connect, share, and communicate with other people in their everyday life. Also a student of McClelland, and like him, using the TAT, McAdams found that individuals who are high in this motive make more eye contact, and laugh and smile more (McAdams, Jackson & Kirshnit, 1984). In social and group interactions, rather than dominating they tend to be oriented to the communal in their focus and goals (McAdams & Powers, 1981). They prefer close interactions with a few friends with whom they can communicate intimately, and they are not the typical extravert life-of-the party. Consequently, they are more likely to be seen as caring, loving, and sincere, rather than dominant (McAdams, 1990).

Implicit Methods for Assessing Motives

The needs discussed here—from competence and achievement to power and intimacy—merely illustrate the diverse array that continues to be studied in the search for important individual differences. A common theme runs through them all.

Namely, as Freud first noted, people often are unaware of all of their motivations, particularly when they are driven by motives that they do not like to see in themselves. That is why *indirect* and projective methods—like the TAT—are used to study motives rather than relying on simple self-reports on the traditional personality scales. Sometimes also called **implicit methods,** they provide a glimpse into human motives that bypasses the problems found on explicit measures that ask people about the self more directly. Namely, on explicit self-report measures and in face to face interviews, people try to make their responses socially desirable and appropriate, for example, by presenting themselves in a more positive light and hiding their vulnerabilities. Unsurprisingly, the two types of measures, implicit and explicit, often are not closely related so that responses on one do not predict responses on the other (e.g., Greenwald et al, 2002).

SUMMARY

COMMON THEMES: TOWARD EGO PSYCHOLOGY AND THE SELF

- The psychoanalytic followers of Freud deemphasized the role of instincts and psycho-sexual stages.
- They concerned themselves more with the social milieu and the ego.
- Anna Freud described defense mechanisms that may serve the ego in coping with anxiety and life tasks. These mechanisms include repression, projection, reaction formation, rationalization, and sublimation.
- Jung emphasized the collective unconscious and its symbolic and mystical expressions. He focused on dreams and on the need to achieve unity through awareness of the collective and personal unconscious.
- Adler saw individuals as struggling from birth to overcome profound feelings of helplessness and inferiority by striving for perfection.
- According to Adler, people are social beings who are influenced more by cultural influences and personal relations than by sexual instincts.
- Fromm likewise saw people primarily as social beings who can be understood best in relation to others. Culture does not exist to stifle instinctual drives; instead, it is a product of the people in the society.
- Erikson views social adaptation, not unconscious sexual urges, as the key force underlying development that takes place over an entire lifetime. Most critical during this development is the evolution of "ego identity."

OBJECT RELATIONS THEORY AND THE SELF

- In object relations theory, the developing self is defined in relation to other "objects" or human beings.
- Kernberg and Kohut emphasize the mental representation of the self and other persons that develop in the early relationship with the primary caregiver.

ATTACHMENT: THE ROOTS OF OBJECT RELATIONS

- In attachment theory, the early relationship between the individual and his or her caregiver becomes the basic framework for the perception and experience of later relationships.
- Ainsworth's "Strange Situation" study assessed individual differences in attachment relations (insecure–avoidant, securely attached, or insecure–resistant) by putting the toddler in an unfamiliar setting with different levels of exposure to his or her mother.
- Kohut theorized that in Freud's time, families tended to expose their children to excessive emotional closeness, whereas children in the 21st century are more likely to have less parental exposure and emotional support, hence they may lack "empathic mirroring."
- Psychotherapy in this framework tries to provide empathic care, to improve self-esteem and to help find suitable objects for identification.

MURRAY AND THE HARVARD PERSONOLOGISTS

- Henry Murray and the Harvard Personologists focused on the intensive psychodynamic study of individual lives.
- Murray proposed that there are higher order needs and motives, such as competence motivation, need for achievement, need for power, and need for intimacy, which are purely psychological and have no specific physiological basis.

KEY TERMS

anima   118
animus   118
archetypes   118
collective
   unconscious   118
compensatory
   motivation   121
competence
   motivation   135
ego identity   125
ego psychology   114
emotional
   overcloseness   132
empathic mirroring   132
environmental
   presses   135
Harvard
   personologists   134

higher order
   motives   134
identity crisis   125
implicit methods   138
inferiority complex   121
insecure–
   ambivalent   129
insecure–avoidant   128
internal working
   models   128
mandala   120
need for achievement
   (n Ach)   136
need for intimacy   138
need for power   137
object relations
   theory   126
organ inferiority   121

personology   134
projection   116
psychological
   oxygen   132
psychosocial
   crisis   125
psychosocial stages   124
rationalization   117
reaction formation   116
relational self   126
secure base   128
securely attached   128
shadow aspect   118
sibling rivalry   121
Strange Situation   128
sublimation   117
Thematic Apperception
   Test (TAT)   136

PSYCHODYNAMIC PROCESSES

In this chapter, we will focus on research relevant to the psychodynamic level of analysis. This research tells an important story for personality psychology because some of it has led to conclusions different from original hypotheses, and all of it has stimulated and challenged later developments throughout the field. Many of those ideas continue to influence how both professional psychologists and laypersons think about the nature of personality and the complexities of the mind.

▶ ANXIETY AND UNCONSCIOUS DEFENSES

The psychodynamic view originally saw **anxiety** as the emotional fear triggered when unacceptable impulses begin to push themselves into consciousness. More recently, other theorists have liberalized the definition and recognize that anxiety can be experienced for many different reasons. In spite of the many varieties of anxiety that people experience, the following three elements often are found (Maher, 1966):

7.1 What three elements typically occur in anxiety responses? How can internal processes produce them and defend against them?

- A conscious feeling of fear and danger, without the ability to identify immediate objective threats that could account for these feelings.

- A pattern of physiological arousal and bodily distress that may include miscellaneous physical changes and complaints (Cacioppo et al., 1996). Common examples include *cardiovascular* symptoms (heart palpitations, faintness, increased blood pressure, pulse changes); *respiratory* complaints (breathlessness, feeling of suffocation); and *gastrointestinal* symptoms (diarrhea, nausea, vomiting). If the anxiety persists, the prolonged physical reactions to it may have chronic effects on each of these bodily systems. In addition, the person's agitation may be reflected in sleeplessness, frequent urination, perspiration, muscular tension, fatigue, and other signs of upset and distress.

- A disruption or disorganization of effective problem-solving and cognitive (mental) control, including difficulty in thinking clearly and coping effectively with environmental demands.

An outstanding characteristic of human beings, emphasized in psychodynamic theories, is that they can create great anxiety in themselves even when they are not in any immediate external danger. A man may be seated comfortably in his favorite chair, adequately fed and luxuriously sheltered, seemingly safe from outside threats, and yet torture himself with anxiety-provoking memories of old events, with terrifying thoughts, or with expectations of imagined dangers in years to come. He also can eliminate such internally cued anxiety within his own mind without altering his external environment, simply by avoiding or changing his painful thoughts or memories.

Does such avoidance of anxiety also occur unconsciously, outside the range of one's awareness? A core assumption of the psychodynamic view is that the answer to the question is clearly affirmative. After years of research, however, that answer still remains controversial.

Defense mechanisms are attempts to cope mentally (cognitively) with internal anxiety-arousing cues. Usually it has also been assumed, in line with Freudian theory, that these efforts are at least partly unconscious—that is, they occur without the person's awareness. Because this assumption is so basic, research in the psychodynamic approach has tried to clarify this process. Especially important have been studies of unconscious processes and mechanisms of defense, which are a focus of this chapter. The greatest attention has been devoted to repression, probably because it is the most important one for Freud's theory.

The Concept of Unconscious Repression

Most people sometimes feel that they actively try to avoid painful memories and ideas and struggle to "put out of mind" thoughts that are aversive to them. Common examples are trying not to think about a forthcoming surgical operation and trying to prevent yourself from thinking about the unknown results of an important test. Psychologists often call such efforts to avoid painful thoughts "cognitive avoidance."

Cognitive avoidance obviously exists and everyone is familiar with it. No one doubts that thoughts may be inhibited. However, the mechanisms underlying cognitive avoidance have been controversial. The basic controversy is whether or not cognitive avoidance includes an unconscious defense mechanism of "repression" that forces unacceptable material into an unconscious region without the person's awareness.

Repression versus Suppression.   The psychoanalytic concept of **repression** as a defense mechanism is closely linked to the Freudian idea of an unconscious mind. The early Freudians saw the unconscious mind as a supersensitive entity whose perceptual alertness and memory exceeded the conscious mind (Blum, 1955). A chief function of the unconscious mind was to screen and monitor memories and the inputs to the senses. This screening served to inhibit the breakthrough of anxiety-arousing stimuli from the unconscious mind to the conscious, or from the outside world to consciousness. Just as the conscious mind was believed capable of deliberately (consciously) inhibiting events by **suppression,** so the unconscious was considered capable of inhibition or cognitive avoidance at the unconscious level by repression.

7.2 Differentiate between suppression and repression of anxiety-producing thoughts.

Suppression occurs when one voluntarily and consciously withholds a response or turns attention away from it deliberately. Unconscious repression, in contrast, may function as an automatic guardian against anxiety, a safety mechanism that prevents threatening material from entering consciousness. Psychoanalysts have offered clinical evidence for the existence of repression in the form of cases in which slips of the tongue, jokes, dreams, or free associations seemed to momentarily bypass the defenses and betray the person, revealing a brief glimpse of repressed unconscious impulses.

Studying Repression.   Repression has remained a cornerstone for most psychoanalysts (Erdelyi, 1985; Grunbaum, 1984), and it has been the subject of a great deal of research for many years. The early efforts to assess whether or not particular findings demonstrated the truth of Freud's concepts created more controversy than clarity. In more recent years, it has been recognized that well-designed experiments on the topic of cognitive avoidance can provide useful information about cognitive processes and personality regardless of their direct relevance to the Freudian theory of repression (Kihlstrom, 1999; Westen & Gabbard, 1999).

7.3 Summarize some of the challenges in studying repression experimentally and in attributing results to repression.

Early experiments on repression studied the recall of pleasant and unpleasant experiences (Jersild, 1931; Meltzer, 1930). These investigators assumed that repression showed itself in a tendency to selectively forget negative or unpleasant experiences rather than positive ones. Critics soon pointed out, however, that the Freudian theory of repression does not imply that experiences associated with unpleasant affective tone are repressed (Rosenzweig & Mason, 1934; Sears, 1936). Freudian repression, instead, depends on the presence of an "ego threat" (e.g., a basic threat to self-esteem) and not on mere unpleasantness.

Later researchers also recognized that to study repression adequately they should be able to demonstrate that when the cause of the repression (the ego threat) is removed, the repressed material is restored to consciousness (Zeller, 1950). This assumption was consistent with the psychoanalytic belief that when the cause of a repression is discovered by insight in psychotherapy, the repressed material rapidly emerges into the patient's consciousness. In other words, if the threat is eliminated, it becomes safe for the repressed material to return to awareness. Reports by psychoanalysts often have cited cases in which a sudden insight supposedly lifted a long-standing amnesia (memory loss).

Experiments to show repression effects have been inconclusive. For example, when college students were threatened by taking a test described as measuring their unacceptable unconscious sexual conflicts, they tended to recall anxiety-arousing words less well. When the threat was removed (by revealing that the test was really not an index for such conflicts), recall improved (D'Zurilla, 1965). Does this mean that repression occurred? After much debate, it has become clear that the answer is "not necessarily"; other interpretations are at least equally plausible.

Reduced recall for threat-provoking information may simply reflect that the person is upset. Therefore, other thoughts, produced by the anxiety as the person worries, may interfere with recall. If the recall improves later when the threat is removed, it may only mean that the competing, anxious thoughts now no longer interfere with recall. It does not necessarily mean either that unconscious repression occurred or that awareness returned when the threat was removed (Holmes & Schallow, 1969; Kihlstrom, 2003; Tudor & Holmes, 1973).

Perceptual Defense

If unconscious repression is a mechanism that keeps painful material out of consciousness, one might also expect it to screen and block threatening perceptual inputs to the eyes and ears. Indeed, clinical reports from psychoanalysts suggest that in some cases of hysteria, massive repression may prevent the individual from perceiving (consciously registering) threatening stimuli such as sexual scenes or symbols.

One very severe instance of this would be hysterical blindness. In these cases, the individual seems to lose his or her vision, although no physical damage to the eyes or to the perceptual system can be detected. Case reports have suggested that such psychological failures to see might be linked to traumatic sexual experiences with resulting repression of stimuli that might unleash anxiety. Although clinical case reports often provide suggestive evidence, they are never conclusive. To go beyond clinical impressions, researchers have tried to study possible anxiety-reducing distortions in perception experimentally. Because it was obviously both unfeasible and unethical to induce sexual traumas in human participants, considerable ingenuity was needed to find ways to study perceptual defense in experiments.

7.4 What methods were used in perceptual defense research, what results were obtained, and what alternative explanations were offered?

The Long History of Perceptual Defense. In the 1940s and 1950s researchers reasoned that persons who did not give sexual or aggressive responses to ambiguous stimuli must be defending against this type of ideation, especially if the same stimuli generally elicited many such responses from most normal people. Consequently, if a person fails to identify potentially threatening inputs to the senses, such as anxiety-arousing sexual words or threatening scenes, it might indicate perceptual inhibition or defense.

To study this process, researchers presented threatening perceptual stimuli in decreasing degrees of ambiguity. They began at a point at which participants could tell what the words were and could reasonably interpret them in many ways, to a point of definiteness that permitted only one clearly correct interpretation. A helpful device for this purpose was the **tachistoscope,** a machine through which potentially threatening words (e.g., "penis," "whore") and neutral words (e.g., "house," "flowers") could be flashed at varying speeds. These stimulus words were presented on a screen very rapidly at first and then gradually exposed for increasingly long durations. The length of time required before each person correctly recognized the stimulus served as the "defensiveness" score; the longer the time required to recognize threatening stimuli, the greater the individual's defensive avoidance tendencies were assumed to be.

In one classic study, college students viewed words presented tachistoscopically so rapidly that they could not perceive them consciously. The words were either emotional or neutral in meaning (McGinnies, 1949). Each student was asked what word had been seen after each exposure. If the answer was wrong, the same word was presented again at a slightly longer exposure time, and the participant again tried to recognize it. It was predicted that such "taboo" words as "penis" or "raped" would be anxiety laden and therefore more readily inhibited than neutral words such as "apple." The results confirmed this prediction, showing greater perceptual defense (longer recognition times) for taboo words than for neutral words.

But these results also can be interpreted quite differently. As was noted long ago (Howes & Solomon, 1951), the perceptual situation places the participant in an embarrassing predicament. In the typical procedure, an undergraduate was brought to the laboratory by a professor or an assistant and then exposed to brief and unclear stimulus presentations by the tachistoscope. The task is essentially a guessing game in which the partcipant tries to discern the correct word from fleeting fragments. On the first trial of a word, for example, something like an "r" and "p" may be seen and the participant may guess "rope." On the next trial, the participant may think "good grief, that looked something like 'rape'!" But rather than make this guess to a professor or an assistant in the academic atmosphere of a scientific laboratory, the participant may deliberately suppress the response. Instead of saying "rape," "rope" is offered again, and the taboo word is withheld until the participant is absolutely sure that this perception is correct.

In sum, a major problem in interpreting results from such studies is that it is extremely difficult to know whether individuals are slower to report some stimuli because they are unconsciously screening them from awareness. Even without such an unconscious mechanism, they may be inhibited about reporting such stimuli until they are absolutely sure, just to avoid embarrassment.

Limitations of Early Laboratory Studies. Psychoanalytically oriented critics have been quite skeptical of the relevance of many of these studies for their theory (Erdelyi, 1993; Erdelyi & Goldberg, 1979). They argue that it is confusing and misleading to study single processes (such as repression) in isolation, outside the context of the person's total psychic functioning. They see these experimental studies at best as suggestive, but clinically irrelevant. They doubt that long-term psychodynamic processes like unconscious reactions to traumatic events in early childhood can be studied under the artificial conditions of the typical laboratory experiment (e.g., Erdelyi, 1993; Madison, 1960). It is also risky to generalize from college sophomores in a laboratory to clinical populations and clinical problems. The mild anxiety induced by experimental threats to college students may have little relevance for understanding the traumas experienced by the young child trying to cope with Oedipal fantasies or the severely disturbed patient in the clinic. On the other hand, the clinician's judgments and intuitive procedures for inferring unconscious dynamics, for example, from projective tests, also have serious limitations (Kihlstrom, 2003; Rorer, 1990). There have been many studies of unconscious repression and unconscious perception, but the conclusions that may be drawn from them have long remained controversial.

> 7.5 Psychoanalysts have been critical of experimental studies of unconscious processes. What are their major criticisms?

▶ THE COGNITIVE UNCONSCIOUS

Although research on the unconscious remains controversial, some clear themes have emerged (e.g., Kihlstrom, 1999), and we examine them next.

Experimental Evidence for Unconscious Processes

Experiments that try to demonstrate repression and unconscious perception have been difficult to interpret (Erdelyi, 1985; D'Zurilla, 1965; Kihlstrom, 1999). Most of the early studies yielded unclear results. After more than 25 years of experiments, for example, one influential review of experimental research on repression led the reviewer to conclude that "there is no evidence to support the predictions generated by the theory of repression" (Holmes, 1974, p. 651). Two decades later, another review led him to the same conclusion (Holmes, 1992). But it is just as easy to conclude that all these studies were flawed.

The two main ways of looking at the research results are well summarized by Erdelyi (1985, pp. 104–105):

> One may begin to understand why experimental psychology, in contrast to dynamic clinical psychology, has taken such a different stance on phenomena such as repression. Experimental psychology, in the service of simplification and control, has studied manifest events, specific memory episodes, or percepts (as in perceptual defense). It is not typical, however, for a normal college subject to resort in the laboratory to such drastic defenses as to block out a clear memory episode or perceptual experience. Consequently, experimental psychology has had great difficulty in demonstrating repression in the laboratory and has understandably taken a skeptical attitude towards this phenomenon. Clinicians, on the other hand, deal continually with latent (hidden) contents, and in this realm the selective blocking or distortion of information through context manipulation is utterly commonplace. The clinician shakes his head in disbelief at the experimental psychologist, whose paltry methodology cannot even demonstrate a phenomenon as obvious and ubiquitous as repression. The experimentalist similarly shakes his head at the credulity of the psychodynamic clinician, who embraces notions unproven in the laboratory and which, moreover, rest on the presumed existence of unconscious, indeed, physically non-existent, latent contents.

In sum, most clinicians remain convinced that evidence for the unconscious is almost everywhere; in contrast, many experimental researchers have had trouble finding it anywhere. On the other hand, new data have come from experiments by social psychologists into the "automatic processing" of social information at levels that are at least partly outside our awareness. For example, information related to racial and sexual stereotypes can affect our judgments and actions even when we are not aware of ever having seen that information (Bargh, Chen, & Burrows, 1996; Kunda, 1999). After reviewing the extensive evidence rapidly accumulating on the topic of information processing outside awareness, Kunda (1999, pp. 287–288) concludes: ". . . there is considerable evidence that our judgments, feelings, and behaviors can be influenced by factors that we have never been aware of, by factors that we were aware of at one time but can no longer recall, and by factors that we can still recall but whose influence we are unaware of."

The Repressed Memory Debate

7.6 What evidence is there that long-forgotten memories of abuse that suddenly reappear may in some cases be "false" memories? What is the risk of concluding that all are false?

Based on the Freudian premise that the human mind purposefully but unconsciously hides frightening or potentially damaging memories, repression is seen by many clinicians and researchers as a major root of psychological dysfunction. But skeptics see it more as a fiction created by a biased theoretical perspective through which therapists may unwittingly damage their patients and their families (Loftus, 1993, 1994; Ofshe & Watters, 1993). While repressed memory continues to remain a controversial academic topic in psychology, it now also has become the center of a public debate that has spread into the media.

Return of the Repressed. The public debate began in recent years when many people claimed that they had been the victims of childhood abuse. They asserted that they had been unaware of these abuses until their repressed memories for the traumatic event were recovered, usually in the course of psychotherapy. Often they then tried to bring criminal charges against the alleged perpetrators of these abuses. The question raised then was: are these recovered memories accurate or might they be **false memories**?

The reported recovered memories of abuse have often produced great personal pain and been the bases for lawsuits and punishment of the alleged abusers. As the defense lawyers are quick to point out, however, the existence of repression itself remains controversial. On the one hand, a review of over 60 years of research led Holmes (1992) to still conclude that ". . . at the present time there is no controlled laboratory evidence supporting the concept of repression" (p. 95). Likewise, in a study of children (ages 5–10) who witnessed the murder of one of their parents, Malmquist (1986) reported that none of the children seemed to repress the memory: instead they frequently focused on it.

Nevertheless, most sophisticated and highly experienced clinicians still remain convinced that case studies compellingly document the existence of repression (Erdelyi, 1993). Supporting that view, some researchers make the case that memories for traumatic events from the early years of life, even after they have been forgotten, may reappear in adulthood under conditions of high stress (Jacobs & Nadel, 1985; Schooler, 1994, 1997; Schooler, Bendiksen, & Ambadar, 1997).

Did It Really Happen? But even assuming that a defensive mechanism of repression occurs and in some cases is activated as a way of coping with traumatic events such as sexual abuse, there is a major problem in determining accurately whether a particular memory for a traumatic event is accurate or instead invented or grossly distorted. Modern memory research makes it clear that memories are not stored like videotapes on a shelf and later replayed. Rather, they are reconstructions of the past, and they are subject to being influenced by suggestions and cues from the outside world and the inside of one's mind, including one's own fantasies and speculations (e.g., Loftus, 1993, 1994; Ofshe, 1992). In short, memories can be created in people by suggestion and self-suggestion, or by telling them that something occurred, often in ways that can even give the person confidence that they are based on reality.

The Power of Suggestion. One worry is that false memories may be unwittingly strengthened by therapists, particularly if they believe, as Freud did, that problems tend to stem from the abuse that the patient suffered as a child. Guided by that belief, some therapists may too easily encourage their patients to explore their unconscious minds, searching for repressed memories that may not exist. In this search, they may do what is sometimes called "memory work," using hypnosis, suggestive questioning, guided visualization, and dream interpretation (Loftus, 1994). Such exploration risks coming up with memories of traumatic events that may have been imagined or fantasized but that did not actually occur. In that case, instead of helping victims to reclaim lost pieces of their lives, they may hurt innocent people and create rather than reduce distress.

The Risk of Excessive Skepticism

The debate about repressed memory raised important questions but also may risk creating excessive skepticism leading people to doubt all reports about repressed memories that become accessible. The danger is that by making the topic so controversial, many victims of abuse and trauma in childhood and later life may be ignored or undermined in their efforts to find and express the truth. Relevant here is the discovery that when people have had traumatic experiences, it can greatly help them to discuss those memories candidly, rather than to try to hide or suppress them. The process of talking or writing about these traumatic experiences not only helps one feel better by "getting it off your chest," as folk wisdom has long suggested, but also can profoundly improve one's health. For example, by simply writing about their traumatic experiences during a period of four days, essentially healthy college students improved their health following the study, with fewer visits to the health center and improved blood pressure and immune system functioning (Pennebaker, 1993; Pennebaker, Kiecolt-Glaser, & Glaser, 1988). It would therefore be unfortunate if trauma victims failed to use opportunities to appropriately air rather than conceal their experience.

Optimistic Prospects for the Future of the Unconscious

In conclusion, what can be said about the role of the unconscious in modern personality psychology? This question was addressed in a special series of articles in which research experts in the area reviewed what is known about the nature of the unconscious (e.g., Bruner, 1992; Erdelyi, 1992; Greenwald, 1992). Although there was no consistent agreement about how to characterize unconscious processes in terms of their precise role and significance, there was ". . . absolute agreement that exciting times, both in research and theory, are ahead for the unconscious" (Loftus & Klinger, 1992, p. 764). This optimism seems justified because methods and concepts in the research on this topic are becoming more sophisticated (e.g., Lewicki, Hill, & Czyzewska, 1992). There is also growing attention to how motivational factors, such as the person's goals and needs, influence memory (e.g., Kihlstrom, 1999, 2003; Kunda, 1990, 1999; Singer & Salovey, 1993).

▶ PATTERNS OF DEFENSE: INDIVIDUAL DIFFERENCES IN COGNITIVE AVOIDANCE

Cognitive avoidance of anxiety-provoking information is not only a basic process, it also is a dimension of personality on which people differ substantially and interestingly. Some people react to stimuli that arouse anxiety by avoiding them cognitively, but other people do not. For many years, individual differences in "defensive" patterns have been found (Bruner & Postman, 1947; Lazarus, 1976; Miller, 1987; Paulhus, Fridhandler, & Hayes, 1997).

Repression–Sensitization

The dimension on which these differences seemed to fall was a continuum of behaviors ranging from avoiding the anxiety-arousing stimuli to approaching them more readily and being extra vigilant or super-sensitized to them. The former end of the continuum included behaviors similar to the defensive mechanisms that psycho-analysts called denial and repression; the latter pattern—vigilance or sensiti-zation to anxiety-provoking cues—seemed more like obsessive worrying. This dimension has become known now as the **repression–sensitization** continuum. Repression–sensitization became the focus of much research both as a dynamic process and as a personality dimension on which individuals might show consistent patterns.

7.7 What is meant by repression and sensitization? What research evidence exists that this is a meaningful personality dimension that relates to other behaviors?

In general, individuals show some consistency in their cognitive avoidance of anxiety-provoking cues such as threatening words (Eriksen, 1952; Eriksen & Kuethe, 1956). In one study, reaction time and other measures of avoidance were obtained in the auditory recognition of poorly audible sentences that had sexual and aggressive content. People who were slow to recognize such sentences also tended to avoid sexual and aggressive materials in a sentence completion test (Lazarus, Eriksen, & Fonda, 1951). People who more readily recalled stimuli associated with a painful shock also tended to recall their failures; those who forgot one were more likely to forget the other (Lazarus & Longo, 1953). A tendency for some consistency in cognitive avoidance may exist at least when ex-tremely high and low groups are selected (Eriksen, 1966). Correlations between cognitive avoidance on experimental tasks and various other measures of repression–sensitization also imply some consistency in these patterns (Byrne, 1964; McFarland & Buehler, 1997; Weinberger, 2002).

In short, in spite of disagreement about the concept and role of repression, there now is agreement that people do use a mechanism of mentally distancing themselves from awareness of threatening and negative thoughts or emotions. In modern research, the term repression currently often refers to that meaning. In Focus 7.1 illustrates that the expressions of individual differences in repressive tendencies depend on the interaction between the individual's disposition and the particular situation.

IN FOCUS 7.1

PERSON–SITUATION INTERACTION IN REPRESSIVE TENDENCIES

Current work on individual differences in the expressions of repressive tendencies tries to take into account both its dispositional determinants and its situational determi-nants in a Person × Situation interaction model. It thus addresses both overall individual differences in the ten-dency to distance oneself, and the particular types of situ-ations in which these differences will, and will not, be seen. To illustrate, Mendolia, Moore, and Tesser (1996) in one study assessed college students on self-report measures of anxiety and repressive tendencies (Wein-berger, 2002). They then observed their reactions in a series of experiments measuring their attentiveness to the emotional details of positive and negative situations—in this case, success or failure on a task claiming to identify their level of intelligence. After these experiences, partic-ipants were given a word familiarity task designed to measure how much emotion they were feeling. They were asked to recall emotion (e.g., angry, happy) vs. non-emotion (swift, interstate) words, and a variety of other measures.

Briefly, the researchers found that repressors are more attentive to the emotional details of both positive and neg-ative situations, in this case both when seeming to succeed or to fail on the supposed test of intelligence. Further, three factors differentiate when people will exhibit repres-sive behavior and when they will not: (1) they must possess an increased sensitivity to both positive and negative emo-tional events (i.e., be high in this disposition); (2) there must be situational information that poses a threat to their self-concept; and (3) the situation must provide them with the opportunity to distract or distance themselves from the self-threatening information. According to the researchers, without all of these factors, repressors and nonrepressors may respond in the same manner.

Selective Attention

Individual differences in repression–sensitization have often been measured on a self-report questionnaire (Byrne, 1964). On this scale, **repressors** are people who describe themselves as having few problems or difficulties. They do not report themselves as highly sensitive to everyday stress and anxieties, whereas those with an opposite pattern are called **sensitizers.** Individual differences on this scale can predict selective attention to important personal information about the self. In one study, college students were exposed to personal information about themselves supposedly based on their own performance on personality tests they had taken earlier (Mischel, Ebbesen, & Zeiss, 1973). The results were made available to students in individual sessions in which each was left alone to look at descriptions of his or her personal liabilities (in one computer file) and personal assets (in another file). For example, the personality assets included such feedback as, "Affiliative, capable of cooperating and reciprocating deeply in relations with others. . . ." In contrast, the personality liabilities information included such descriptions as, "Nonperseverative, procrastination, and distractibility . . . resultant failures lead to greater and greater apathy. . . ." (Actually, all students received the same feedback, but in the debriefing session in which they were told the truth, it became clear that they had been convinced that the test results really described them; in fact, they believed the information had captured much about their personalities.)

The question was: Would repressors and sensitizers (as measured on the self-report questionnaire) exhibit different attentional patterns for the positive, self-enhancing information versus the negative, threatening information about themselves? There were clear differences.

These results are summarized in Figure 7.1. Sensitizers attended more to their liabilities and spent little time on their assets; in sharp contrast, repressors attended as much to their assets as to their liabilities. Later research suggests that repressors may

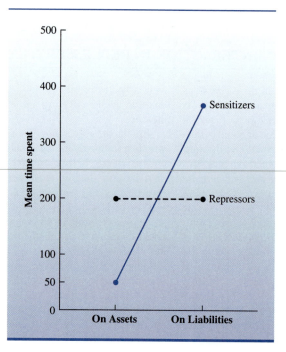

Figure 7.1 Attention deployment by sensitizers and repressors. Mean amount of time (in seconds) spent on assets and liabilities by sensitizers and by repressors.

Source: Based on data in the control condition from Mischel, W., Ebbesen, W. B., & Zeiss, A. R. (1973). Selective attention to the self: Situational and dispositional determinants. *Journal of Personality and Social Psychology, 27,* 129–142. (Figs. 1 and 2, p. 136.)

be people who are especially sensitive to criticism and threat, and use their defense to protect themselves against this vulnerability (Baumeister & Cairns, 1992).

In sum, the repressors and the sensitizers differed in how they dealt with positive and negative ego-relevant information. But these differences seem linked to their attentional strategies, not to unconscious distortions in their memory: both groups are able to recall threatening information about themselves with equal accuracy if they are made to attend equally to that information (Mischel, Ebbesen, & Zeiss, 1976). Normally, however, the sensitizers tended to focus on negative self-relevant information, whereas the repressors tended to avoid it and preferred to think about happier things. When the situation prevents repressors from simply ignoring the threat, they do begin to attend to it and worry anxiously about it (Baumeister & Cairns, 1992).

The two groups also differed in their self-descriptions: while repressors described themselves in positive, socially desirable terms, representing themselves consistently in a more favorable light, sensitizers painted a much more critical and negative self-portrait (Alicke, 1985; Joy, 1963). Most interesting, however, is that it is the repressor, not the sensitizer, who better fits the picture of the optimistic personality with a good mental and physical health prognosis.

This is surprising for psychodynamic theory, which has long assumed that accurate self-awareness and being in touch with one's personal limitations, anxieties, and flaws (i.e., sensitization) were important ingredients of the healthy personality. In contrast, repression and cognitive avoidance of negative information and threat were the hallmark of the brittle, vulnerable personality. It is, of course, likely that the massive emotional repression to which Freudians refer is quite different from the self-enhancing positive bias that seems to characterize repressors as identified on this scale. Likewise, the growth of self-awareness and accessibility of personal anxieties to which psychodynamic theory refers may also be different from "sensitization" measures on this scale. But as the next section also suggests (Miller, 1987), and as a great deal of work shows (Bonanno, 2001; Seligman, 1990; Taylor & Brown, 1988; see Chapter 13), under many circumstances an attitude of emotional blunting that deliberately avoids threatening information may be highly adaptive, a sign of mental health rather than of a fragile personality in need of insight.

Blunting versus Monitoring Styles

In a related direction, it has been shown that people differ considerably in their disposition to blunt and distract themselves or to monitor for (be alert to) danger signals (Miller, 1996). One promising scale tries to identify information avoiders and information seekers as two distinct coping styles. The **Miller Behavioral Style Scale (MBSS)** consists of four hypothetical stress-evoking scenes of an uncontrollable nature (Miller, 1981, 1987; Miller & Mangan, 1983). On this measure, people are asked to imagine vividly scenes like "you are afraid of the dentist and have to get some dental work done"; or "you are being held hostage by a group of armed terrorists in a public building"; or "you are on an airplane, 30 minutes from your destination, when the plane unexpectedly goes into a deep dive and then suddenly levels off. After a short time, the pilot announces that nothing is wrong, although the rest of the ride may be rough. You, however, are not convinced that all is well."

Each scene is followed by statements that represent ways of coping with the situation, either by **monitoring** or by **blunting.** Half of the statements for each scene are of a monitoring variety. For example, in the hostage situation: "If there was a radio present, I would stay near it and listen to the bulletins about what the police were doing";

7.8 Distinguish monitoring versus blunting responses to threatening events. How are MBSS scores related to information search under anxiety-arousing conditions?

Some people cope with unpleasant events by mentally "blunting" (distracting themselves).

or, in the airplane situation: "I would listen carefully to the engines for unusual noises and would watch the crew to see if their behavior was out of the ordinary." The other half of the statements are of the blunting type. For example, in the dental situation: "I would do mental puzzles in my head"; or, in the airplane situation: "I would watch the end of the movie, even if I had seen it before." The individual simply marks all the statements following each scene that might apply to him.

College students were threatened with a low probability shock and allowed to choose whether they wanted to monitor for information or distract themselves with music (Miller, 1981). As expected, the amount of time spent on information rather than on music was predicted reasonably well by an individual's MBSS score. In particular, the more blunting items endorsed on the scale, the less time the person spent listening to the information and the more time listening to the music. In sum, individuals differed in the extent to which they chose to monitor or distract themselves when faced with aversive events (possible electric shocks) in the experiment. These differences in coping style were related to their questionnaire scores, supporting the validity of the MBSS.

The Role of Control: When Don't You Want to Know?

7.9 How does the controllability of aversive stimuli influence the adaptiveness of approach–avoidance tendencies?

Whether or not persons react to negative stimuli by avoiding them "defensively" in their cognitions and perceptions may depend on whether or not they believe that they can somehow cope with them by problem solving and action. If adaptive action seems impossible, cognitive avoidance attempts become more likely. But if the painful cues can be controlled by the person's actions, then greater attention and vigilance to them may occur. This point is illustrated in the same blunting-monitoring study that gave individuals a choice between stress-relevant information or distraction just discussed (Miller, 1979). The information was a warning signal for when an electric shock would come; the distraction was listening to music. Half the participants were led to believe

TABLE 7.1   Number of Participants Who Seek Information (Warning Signal) or Distraction (Listen to Music) When Avoidance Is or Is Not Possible

	Information Seeking	Distraction Seeking
Avoidance Possible	24	10
Avoidance Not Possible	11	23

Source: Based on Miller, S, M. (1979). Coping with impending stress: Physiological and cognitive correlates of choice. *Psychopathology.* 16, 572–581.

the shock was potentially avoidable; the rest believed it was unavoidable. As Table 7.1 shows, when participants believed avoidance was possible, they preferred information; when they thought it was unavoidable, they preferred distraction. Other research has documented the many circumstances under which coping with problems is improved by the selective avoidance of threatening information (Janis, 1971; Lazarus, 1976, 1990; Lazarus & Folkman, 1984).

Matching the Medical Information to the Patient's Style

In a real-life application to a medical situation, Miller and Mangan (1983) gave the MBSS to gynecologic patients about to undergo colposcopy, a diagnostic procedure to check for the presence of abnormal (cancerous) cells in the uterus. Based on scale scores, patients were first divided into monitors or blunters. Half the women in each group were then given extensive information about the forthcoming procedure, and half were given (the usual) minimal information. Psychophysiological reactions (like heart rate), subjective reports, and observer ratings of arousal and discomfort were taken before, during, and after the procedures. The results revealed, again as expected, that monitors showed more arousal overall than blunters (e.g., see Fig. 7.2 for the physician's rating of patients' tension during the exam).

Most interesting, physiological arousal was reduced when the level of preparatory information was consistent with the patients' coping style. That is, physiological arousal was reduced for blunters who received minimal information and for monitors who received extensive information. The total results clearly show strong individual differences in informational preferences during the coping process when people are faced with threats. Most important, they showed that when monitors are told more, and blunters are told less, about an impending stress each type is likely to cope with it best: matching the information to the style can reduce the stress experienced and enhance resources.

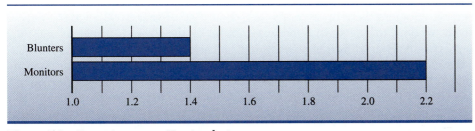

Figure 7.2   Doctor's report—Tension during exam.
Source: Miller, S. M., & Mangan, C. E. (1983). Cognition and psychopathology. In K. Dobson & P. C. Kendall (Eds.), *Cognition, stress, and health*. Reprinted with permission from Elsevier.

▶ PERSONALITY ASSESSMENT

Freudian psychology was especially exciting because it promised a way to understand and to treat each complex individual with the depth that he or she deserves. The last two chapters gave you a sense of some of the main concepts that underlie the psychodynamic approach to personality. Guided by these concepts, one tries to help the person to reveal unconscious motives, conflicts, and other dynamics. In this approach, the objective is to uncover disguises and defenses, to read the symbolic meanings of behaviors, and to find the unconscious motives that underlie action. In this way, the clinician tries to find the distinctive qualities that characterize the individual.

The Core Beneath the Mask

Psychodynamic theorists recognized that a person's overt actions across seemingly similar situations often seem inconsistent. They felt, however, that these inconsistencies in behavior were merely superficial, because beneath them were underlying motives that actually drove the person consistently over the years. The basic motives persist across diverse settings, but their overt expressions are disguised. Therefore, the task is to find the person's fundamental motives and dynamics under the defensive distortions of the overt behavior. The challenge is to discover the basic core hidden behind the mask, to find the truth beneath the surface. But how?

Relying on the Clinician

Psychodynamic interpretations depend more on intuitions than on tests. The rules for relating behavioral signs to unconscious meanings are not spelled out and require clinicians to form their own judgments based on clinical experience and the "feel" of the case. The merit of such assessments depends on two things. First, it depends on the evidence supporting the techniques upon which the psychologist relies. Second, it depends on the value of clinical judgment itself. Because psychodynamic theories rest on the belief that the core of personality is revealed by highly indirect behavioral signs, evidence for the value of these indirect signs of personality is most important. We next review some of the main clinical methods that have been studied in the search for valuable signs of personality. Probably the most important of these methods are the projective techniques.

Projective Methods

7.10 What are the major features of projective tests, and the rationale underlying their use? Why are they favored by psychodynamic clinicians?

The main characteristic of **projective methods** is the way in which the testing situation is usually structured so that the task is ambiguous. Typically, there are also attempts to disguise the purpose of the test (Bell, 1948; Exner, 1993), and the person is given freedom to respond in any way that he or she likes. In projective testing, assessors present you with ambiguous stimuli and ask ambiguous questions. For example, they ask, "What might this be?" "What could this remind you of?" [while showing an inkblot] or say, "Create the most imaginative story that you can [showing a picture], including what the people are thinking and feeling, what led up to this situation, and how it all comes out." Or they read words and ask you to "say the first thing that comes to mind."

Psychoanalytically oriented assessors favor projective techniques, because they assume that the "unconscious inner life" is at least partially "projected" and revealed in responses to the ambiguous projective test situation. The two most influential and popular projective techniques have been the Rorschach, and the Thematic Apperception Test (TAT) described below. Both the reliability and the validity of the Rorschach have been heavily disputed and generally found to be problematic. In a comprehensive review, Kihlstrom (2003) concludes that even in the few cases in which projective techniques had some validity, there was no evidence that the findings reflected the person's *unconscious* mental states.

Nevertheless, a 1995 survey with replies from over 400 clinical psychologists across a variety of settings indicated that both the Rorschach and TAT still remain among the most commonly used personality assessment procedures in everyday clinical practice (Watkins et al., 1995).

The Rorschach Test

Developed by the psychiatrist Hermann Rorschach in 1921, the **Rorschach test** consists of a series of inkblots on 10 separate cards (Figure 7.3). Some of the blots are black and white, and some colored. The individual is instructed to look at the inkblots one at a time and to say everything that the inkblot could resemble or look like. The examiner then generally conducts an inquiry into the details of the person's interpretation of each blot.

Responses may be scored for location (the place on the card that the response refers to) and such determinants as the physical aspects of the blot, such as shape, color, shading, or an expression of movement (Exner, 1993). The originality of the responses, the content, and other characteristics also may be scored and compared to

7.11 How are responses to the Rorschach and TAT analyzed and used to infer psychodynamics?

Figure 7.3 Taking a Rorschach test.

those of other people of similar age. The interpreter may try to relate these scores to aspects of personality, such as creative capacity, contact with reality, and anxiety.

Gary's Reactions to Two of the Rorschach Inkblot Cards

Response: This looks like two dogs, head-to-foot (laughs), licking each other. That's about it, that's all.

Inquiry answers (to the question "What about the inkblot made you think of two dogs?"): They're sort of fuzzy . . . kinda shapeless. It was the dark skin and the furry effect that made me think of it.

Response: Didn't we have this one already?
This could be an ogre laughing—his head thrown back and he's laughing, his eyes and mouth wide open.
These over here look like insects, tsetse flies in fact, with tiny, tiny legs, and small, delicate and rather beautiful wings.
That's it, that's enough.

Inquiry answers: It's the shaggyness and the hugeness, the massiveness of the shape. The wings over here, head here.

Assessment with the TAT

As noted in Chapter 6, the Thematic Apperception Test or TAT was developed in the Harvard Psychological Clinic research program during the 1930s, and used extensively to study motives like need for achievement. The TAT is a projective test in the sense that it is expected that people will interpret the ambiguous pictures presented according to their individual readiness to perceive in a certain way ("apperception"). The themes that recur in these imaginative productions are thought to reflect the person's underlying needs, as well as potential conflicts and problems. Special scoring keys have been designed for use with the TAT (Bellack & Abrams, 1997; McClelland et al., 1953; Mussen & Naylor, 1954). In clinical work, the stories usually are not scored formally, however, and instead the clinician interprets the themes intuitively in accord with his or her personality theory (Rossini & Moretti, 1997). On the other hand, formal scoring is often used in research applications. Two of Gary's answers to TAT cards are shown in In Focus 7.2.

IN FOCUS 7.2

GARY'S TAT STORIES

Card depicting two men: Two men have gone on a hunting trip. It is dawn now and the younger one is still sound asleep. The older one is watching over him. Thinking how much he reminds him of when he was young and could sleep no matter what. Also, seeing the boy sleeping there makes him long for the son he never had. He's raising his hand about to stroke him on the forehead. I think he'll be too embarrassed to go ahead with it. He'll start a fire and put on some coffee and wait for the younger man to wake up.

Card depicting young man and older woman: This depicts a mother–son relationship. The mother is a strong, stalwart person. Her son is hesitating at the doorway. He wants to ask her advice about something but isn't sure whether it's the right thing to do. Maybe he should make up his own mind. I think he'll just come in and have a chat with her. He won't ask her advice but will work things out for himself. Maybe it's a career choice, a girlfriend. I don't know what, but whatever it is, he'll decide himself. He'll make his own plans, figure out what the consequences will be, and work it out from there.

Studying Lives in Depth

Using the TAT and a large variety of other methods, Henry Murray and the Harvard Personologists developed a distinct research-based approach to personality and an assessment style that became widely respected by other psychologists and is still influential. As you saw in the last chapter, it opened the way to studying diverse human motives. It also provided a new research style: the intense study of lives in depth over time.

The Harvard group focused on intensive long-term studies of relatively small samples of people. In one classic project (Murray et al., 1938) called *Exploration in Personality,* researchers studied Harvard college undergraduates over a period of many years and gathered data on their personality development and maturation at many points in their lives. They administered tests of many kinds, obtained extensive autobiographical sketches, observed participants' behavior directly, and interviewed them in great detail. These methods probed thoroughly into many topics and most facets of their lives. They included, for example, inquiries into the individual's personal history and development, school and college, major life experiences, family relations, childhood memories, and sexual development (e.g., earliest recollections, first experiences, masturbation). They also assessed abilities and interests. A sampling of these topics is summarized in Table 7.2. The results often provided rich narrative accounts of life histories, as in Robert White's *Lives in Progress* (1952), which traced several lives over many years.

Assessment Strategy: Diagnostic Council

The researchers in the Harvard clinical studies were experienced psychologists who interpreted their data clinically. Usually a group of several assessors studied each individual participant. To share their insights, they pooled their overall impressions at a staff conference or "diagnostic council." These councils became a model for clinical practice. In them, a case conference was conducted in detail and in depth about each individual. On the basis of the council's discussions, inferences were generated about each participant's personality. They inferred basic needs, motives, conflicts, and dynamics; attitudes and values; main character strengths and liabilities. Each piece of information served as a sign of the individual's personality and was interpreted by the council of assessors.

7.12 How was the diagnostic council used by Harvard personologists in the in-depth study of individuals over time?

TABLE 7.2 Examples of Topics Included in the Study of Lives by the Harvard Personologists

Personal history (early development, school and college, major experiences)

Family relations and childhood memories (including school relations, reactions to authority)

Sexual development (earliest recollections, first experiences, masturbation)

Present dilemmas (discussion of current problems)

Abilities and interests (physical, mechanical, social, economic, erotic)

Aesthetic preferences (judgments, attitudes, tastes regarding art)

Level of aspiration (goal setting, reactions to success and failure)

Ethical standards (cheating to succeed, resistance to temptation)

Imaginal productivity (reactions to inkblots)

Musical reveries (report of images evoked by phonograph music)

Dramatic productions (constructing a dramatic scene with toys)

Source: Based on Murray, H. A., Barrett, W. G., & Homburger, E. (1938). *Explorations in Personality.* New York: Oxford University Press.

Selecting U.S. Spies: The OSS Assessment Project

7.13 How were situational tests used by diagnostic councils in the OSS Assessment Project?

This strategy is illustrated in one of the important applied projects of the personologists—their effort to select officers for the supersensitive Office of Strategic Services (OSS) during World War II. OSS officers in World War II had to perform critical and difficult secret intelligence assignments, often behind enemy lines and under great stress. The personologists obviously could not devote the same lengthy time to studying OSS candidates that they had given to Harvard undergraduates in the relaxed prewar days in Cambridge. Nevertheless, they attempted to use the same general strategy of global clinical assessment. For this purpose, teams of assessors studied small groups of OSS candidates intensively, usually for a few days or a weekend, in special secret retreats or "stations" located in various parts of the country. Many different measures were obtained on each candidate.

One of the most interesting innovations was the **situational test.** In this procedure, participants were required to perform stressful, lifelike tasks under extremely difficult conditions. For example, "The Bridge" task required building a wooden bridge under simulated dangerous field conditions and under high stress and anxiety. But such situational tests were not used to obtain a sample of their bridge-building skills. Instead, the clinicians made deep inferences, based on the behavior observed during the task, about each person's underlying personality. It was these inferences of unobserved attributes or dispositions, rather than the behavior actually observed in the sampled situation, that entered into the assessment report and became the bases for clinical predictions. In this fashion, behavior samples and situational tests were transformed into inferences about underlying dispositions and motives.

To illustrate, in the *Assessment of Men* by the OSS staff (1948), the bridge-building situation was used to answer questions like these (p. 326):

> *Who took the lead in finally crossing the chasm? And why did he do it? Was it to show his superiority? Why did each of the others fall back from the trip? Did they fear failure?*

It is obvious that the chief value of this situation was to raise questions about personality dynamics which required an explanation on the basis of the personality trends already explored. If these could not supply a reasonable explanation, then new information had to be sought, new deductions made.

In the situational test, just as on the projective test, behavior was interpreted as a clue revealing personality. Although behavior was sampled and observed, the observations served mainly as signs from which the researchers inferred the motives that prompted the behaviors.

From Situational Tests to Psychodynamic Inferences

In the global clinical assessment strategy, the assessors form their impressions of the person from performance on various projective and objective tests, the autobiography and total personal history, and reactions to thorough interviews. Several assessors study the same person and each generates his or her own clinical impressions. Later, at a conference, the assessors discuss and share their interpretations and pool their judgments. Gradually they integrate their impressions of each person's overall personality structure and dynamics. To predict the person's behavior in a new situation (for example, under attack behind enemy lines), they try to infer, from the personality they have hypothesized, how such an individual would probably react to the stresses of that situation. This global

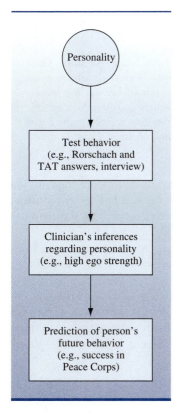

Figure 7.4 Global psychodynamic assessment. The clinician infers the participant's personality from his or her test behavior and predicts future behavior by judging how a person with that personality would probably react to specific future situations.

assessment model is schematized in Figure 7.4. It depicts the strategy of using test behavior to generate inferences about the underlying personality. The prediction of future behavior is based on judgments of how a person with that particular personality would probably act. The test behavior itself is used as a sign of the underlying personality.

▶ TREATMENT AND CHANGE

The psychodynamic approach to treatment has had an enormous influence on American psychiatry and clinical psychology. Its major version is **psychoanalysis** or psychoanalytic therapy, a form of psychotherapy originally developed by Freud and practiced by psychoanalysts.

Traditionally, in psychoanalysis several weekly meetings, each about an hour long, are held between the therapist and client (or "patient"), often for a period of many years. The treatment is based on the premise that neurotic conflict and anxiety are the result of repressed (unconscious) impulses. The aim is to remove the repression and resolve the conflict by helping patients achieve insight into their unconscious impulses.

Free Association and Dream Interpretation

To uncover unconscious material (or lift the repression), the techniques of **free association** and **dream interpretation** are used in traditional psychoanalysis. As was noted before, in free association the patient, usually reclining on a couch, is instructed to report whatever comes to mind without screening or censoring his or her thoughts

7.14 What is the basic goal in psychoanalytic treatment? What methods are used to achieve positive outcomes?

in any way. Here is a fragment of free association from a psychotherapy session as an example:

I wonder how my mother is getting along. You know how she and I don't get along. Once when I was about twelve, she and I were having an argument—I can't remember what it was about—argument 1001. Anyway, the phone rang and one of her darling friends offered her two tickets to the matinee performance of a ballet that day. What a day. She refused them to punish me. For a change! I don't think I even saw a ballet until I was grown up and married. Joe took me. I still get sad when I think about it. I could cry. All blue. It reminds me of all the times when I felt. . . .

Any difficulties or blocks in free association are considered as important as the material that is easily produced. These difficulties are interpreted as **resistance,** caused by unconscious defenses blocking access to material central to the patient's problems, and the person is encouraged to continue with the free association.

According to psychoanalytic theory, the ego's defense mechanisms are relaxed during sleep, making dreams an avenue to express repressed material. But the defenses still operate to distort the content of dreams, so interpretation is necessary to unravel their meaning. In treatment, the interpretations of blocks in association, and dreams, are done carefully so that the patient continues to relax the defenses. The therapeutic goal is to make the unconscious conscious as the patient gradually relaxes and overcomes resistances to facing unconscious conflicts and motives. For a more contemporary approach to some of Freud's key concepts, see In Focus 7.3.

7.15 In what ways do contemporary studies on trauma support Freud's views and suggest more effective methods of treatment?

IN FOCUS 7.3

TODAY'S VIEW OF FREUD'S THEORY OF TRAUMA

Freud's theory of trauma and neurosis illustrates how his ideas have remained influential, but only with substantial modifications. Freud's theory of the origin of neurosis emphasized the importance of childhood traumas—intense emotional experiences that the person cannot deal with and that "exert a disintegrating effect on the mind" (Spiegel, Koopman, & Classen, 1994, p. 11). Freud held that traumatic experiences can induce not only severe anxiety but also dissociative states of the sort he reported in his clinical observations of hysterical patients who developed physical symptoms (like "glove anesthesia") that made no sense neurologically.

Although these ideas continue to be controversial, they also have stimulated extensive contemporary research on reactions to acute stress, called **traumatic experiences.** These are experiences that abruptly and severely disrupt the person's normal daily routine, threatening him or her with physical injury or death, as happens under war conditions, in earthquakes, or to victims of violent crimes or terrorism. In contemporary research, it has been possible to document and study these experiences systematically. The results of these

new studies in part support many of Freud's insights, and in part serve to modify and improve the treatment procedures to help the victims deal with their traumas more effectively (e.g., Roth & Newman, 1990; Spiegel & Cardena, 1990, 1991).

In the contemporary view, consistent with Freud, when traumas profoundly endanger the victim's core beliefs about oneself and the world, dissociative reactions may occur to keep the threat outside full awareness, split off from the rest of one's experience. Consequently, treatment is aimed at helping the person to integrate the experience and to manage the painful emotions. Freud originally emphasized simply helping the person to reexperience and repeat the event and its emotions and to express them in the therapy relationship.

Current approaches take another step. Called **cognitive restructuring,** the focus is on reinterpreting the meaning of the event in a way that allows the person to deal with it. It involves helping the person to give up the sense of control over the event that often creates inappropriate guilt:

Traumatic experiences may be induced by such extreme events as wars that shatter normal life routines.

A soldier who survived a rocket attack may feel he traded his safety for that of a comrade who died. . . . Therapy is aimed at helping the victim acknowledge and bear the emotional distress which comes with traumatic memories, grieving the loss of control which occurred at the time and thereby admitting the uncomfortable sense of helplessness (Spiegel et al., 1994, p. 18).

In short, in contemporary applications of Freud's ideas about traumas and disassociation, the focus is on helping the person to: acknowledge the trauma itself; bear and reinterpret the memories of the traumatic events; and cast them into a more meaningful perspective with less self-blame. In these efforts, hypnosis, which Freud originally explored deeply but then abandoned as a technique, is often used. Support is given and encouraged, for example, by strengthening interpersonal relationships. These treatment features are characteristic of the "ego psychology" which grew out of, and built upon, Freud's work, and remain influential in current clinical practice.

The Transference Relationship and Working Through

The therapist in psychoanalysis is supposed to create an atmosphere of safety by remaining accepting and noncritical. Therapists deliberately reveal little about themselves and remain shadowy figures who often sit behind the patient and outside his view in order to facilitate a **transference** relationship. Transference is said to occur when the patient responds to the therapist as if he or she were the patient's father,

7.16 Why is the transference relationship so important in psychoanalytic treatment? What is involved in the working-through process?

mother, or some other important childhood figure. Feelings and problems initially experienced in childhood relations with these figures are transferred to the relationship with the therapist. Transference is regarded as inevitable and essential by most psychoanalysts. It is a concept that has been revitalized in current research on memory that shows how early feelings about a person can be reactivated later in life by other people who are perceived to be similar (Andersen & Chen, 2002) as discussed in Chapter 13. In the transference, the therapist demonstrates to patients the childhood origins of their conflicts and helps them to work through and face these problems. Here in the words of a distinguished psychoanalyst (Colby, 1951) is an example of how the transference is used and interpreted:

> The manner in which a patient acts and feels about his therapist is a bonanza of psychological information. In subtracting the inappropriate from appropriate responses the therapist has a first-hand, immediately observable illustration of the patient's psychodynamics in an interpersonal relationship. . . .
>
> A woman from an old Southern family broke away in late adolescence from family ties and values. She became a nomadic Bohemian vigorously opposed to all authority. She expressed her feelings by zealous work in Anarchist societies and other radical movements. In therapy she often told of fearlessly challenging policemen and openly sneering at successful businessmen.
>
> Yet her behavior toward the therapist was in marked contrast to this. She was very respectful, non-aggressive and acquiescent—all attitudes she faintly remembers having as a child toward her parents until adolescence. The therapist's concept was that the patient unconsciously saw him as a feared and loved parent who must not be antagonized. She really feared authority as a source of punishment (p. 113).

The insight to be achieved by the patient in psychoanalysis is not a detached, rational, intellectual understanding. People must work through their problems in the transference relationship. **Working through** involves repeated reexamination of basic problems in different contexts until one learns to handle them more appropriately and understands their emotional roots.

Relational Therapy and Restoration of the Self

Led by theorists like Kohut, **relational therapy** (also called "object relations therapy") has emerged as a coherent approach to treating personality problems. It is a psychodynamic approach in the sense that its roots are in earlier psychoanalytic theories. Like them, its focus is on the individual's often unconscious, long-standing conflicts and defenses. It is distinctive, however, in three ways. It sees the history of these problems in early relationships, especially with the mother. It sees their expression in current relationships. Finally, it treats them by focusing on the interpersonal relationship within the therapy context.

7.17 How does relational therapy differ from classical psychoanalytic treatment?

In this approach, the therapist actively and empathically "engages" the patient to build a close therapeutic relationship. Interpretation and confrontation of basic relational problems occur in this supportive context. Note that this focus on the carefully nurtured empathic relationship contrasts with the "blank screen" image of the traditional Freudian analyst. The Freudian patient free associates while reclining on a couch, with the therapist sitting behind the patient. In relational therapy, the two face each other and interact actively, as the therapist provides empathic support as well as gradual confrontation.

Alternative Psychodynamic Interpretations of Gary W.

One concrete implication of these results is that the psychodynamic meaning of the same information may be interpreted quite differently depending on the specific theory and biases of the particular psychologist. This is especially likely when broad inferences are made about a person's deep, unobservable psychodynamics. For example, if a disciple of Alfred Adler assesses Gary W., he may find evidence for an inferiority complex and sibling rivalry from the same observations that lead a Freudian to infer castration anxiety and an Oedipal complex. Likewise, in the same set of responses, Kohut's student may see Gary's fragmented self and his cry for empathic mirroring of his feelings.

Many psychologists now see such alternative views as a welcome development. It encourages the frank recognition that the same "facts" about a person are indeed open to multiple interpretations from different perspectives. Each view may contribute a somewhat different vision of the individual from a different angle, and no one view is necessarily exclusively correct or absolute. Critics of this approach, however, are understandably concerned that there are no clear standards for selecting among alternative interpretations. There also is no strong evidence to help decide how to select those that might be most useful for different goals. Fortunately, work at the levels of analyses that are presented in the subsequent parts of this text have helped to overcome these limitations.

SUMMARY

ANXIETY AND UNCONSCIOUS DEFENSES

- Conscious and unconscious defense mechanisms can be used to cope with anxiety-arousing stimuli.
- The nature and mechanisms of unconscious repression remain controversial.
- "Perceptual defense" may reduce anxiety by altering sensory perceptions, as in hysterical blindness.
- The length of time required to recognize threatening stimuli may be interpreted as a measure of defensive avoidance.
- Many psychoanalysts warn against generalizing from experiments with college students to clinical populations and problems.
- It is difficult to determine whether a particular memory for a traumatic event is accurate or invented or grossly distorted.

PATTERNS OF DEFENSE: INDIVIDUAL DIFFERENCES IN COGNITIVE AVOIDANCE

- Individuals tend to respond to stress in one of two ways: either avoidance or hypersensitization. This is the repression–sensitization personality dimension.
- These two types do not differ in their recall for positive or negative information.
- Repressors attend more to positive feedback, see themselves more favorably, and are more mentally and physically healthy, whereas sensitizers are preoccupied with negative feedback and are often self-critical.
- There are individual differences in the tendency toward blunting (self-distraction) versus monitoring (increased alertness) in stressful situations.
- If adaptive action in the face of a threat seems impossible, cognitive avoidance or blunting may be more likely. If the potentially painful events can be controlled by the person's actions, then greater attention and vigilance to them may be found.

PERSONALITY ASSESSMENT

- Psychodynamic assessment tries to find the person's fundamental motives and dynamics, and relies on the clinician.
- Projective methods such as the TAT (Thematic Apperception Test) or the Rorschach test are used often.
- In the Rorschach test, the individual is shown a series of inkblots and is asked to say what the inkblots resemble.
- The TAT consists of a series of ambiguous pictures for which the test-taker is asked to make up a story.
- Harvard personologists provided a model for the intensive clinical study of individuals.
- The OSS Assessment project analyzed individuals while they performed tasks under highly stressful situations to infer each person's personality.

TREATMENT AND CHANGE

- Psychoanalytic treatment aims to reduce unconscious defenses and conflicts by helping patients gain insight into their unconscious motivations, using methods like free association and dream interpretation.
- Patients "work through" their problems in the transference relationship with the therapist.
- In relational therapy, the therapist actively and empathically "engages" the patient to build a close working relationship.

KEY TERMS

anxiety   142
blunting   151
cognitive
   restructuring   160
dream interpretation   159
false memories   147
free association   159
Miller Behavioral Style
   Scale (MBSS)   151

monitoring   151
projective methods   154
psychoanalysis   159
relational therapy   162
repression   143
repression–
   sensitization   149
repressors   150
resistance   160

Rorschach test   155
sensitizers   150
situational test   158
suppression   143
tachistoscope   144
transference   161
traumatic
   experiences   160
working through   162

PART II

THE PSYCHODYNAMIC-MOTIVATIONAL LEVEL

▶ OVERVIEW: FOCUS, CONCEPTS, METHODS

Some of the essentials of work at the psychodynamic level are summarized in the Overview table. The table reminds you that unconscious motives and psychodynamics within persons are viewed as the basic causes of their behavior, including their feelings, conflicts, and problems. Clinicians working at this level of analysis try to infer and interpret these causes from their overt "symptom-like" expressions in the individual's behaviors. Dreams, free associations, and responses to unstructured situations such as projective stimuli (for example, the Rorschach inkblots), are especially favored sources of information. The responses from the person serve as indirect signs whose meaning and significance the clinician interprets. The role of the situation is deliberately minimized, guided by the belief that the more ambiguous the situation, the more likely it is that the individual's basic, underlying psychodynamics will be projected in how he or she interprets it and reacts.

The focus of research, like the focus of personality assessment and of psychotherapy at this level of analysis, is on the person's unconscious psychodynamics and defenses. Those defenses disguise the underlying motives and conflicts that must be revealed and confronted in order for the individual to function well. Personality change requires insight into the disguised (unacceptable to the person) unconscious motives and dynamics that underlie the behavioral symptoms, so that they can be made conscious, accepted, and managed more rationally. When that occurs, the symptoms should diminish. In more

Overview of Focus, Concepts, Methods: The Psychodynamic-Motivational Level

Basic units:	Inferred motives and psychodynamics
Causes of behavior:	Underlying stable motives and their unconscious transformations and conflicts
Behavioral manifestations of personality:	Symptoms and "irrational" patterns of behavior (including dreams, "mistakes," and fantasies)
Favored data:	Interpretations by expert judges (clinicians)
Observed responses used as:	Indirect signs
Research focus:	Personality dynamics and psychopathology; unconscious processes; defense mechanisms; the fragmented self
Approach to personality change:	By insight into the motives and conflicts underlying behavior, by making the unconscious conscious
Role of situation:	Deliberately minimized or ambiguous

recent work, however, increasing attention is being given to such concepts as the self, self-perception, and interpersonal relationship problems with significant others.

▶ ENDURING CONTRIBUTIONS OF THE PSYCHODYNAMIC-MOTIVATIONAL LEVEL

The impact of Freud's theory on society and on philosophy, as well as on the social sciences, is almost universally hailed as profound; its significance is frequently compared with that of Darwin. Freud's monumental contributions have been widely acknowledged. He opened the topic of childhood sexuality, revolutionized conceptions of the human psyche, provided strikingly powerful metaphors for the mind and human condition, and pioneered twentieth-century psychiatry with his approach to the treatment of psychological problems. The evidence relevant to Freud's theory as a scientific psychological system, however, has been questioned persistently (e.g., Grunbaum, 1984).

Although Freud attempted to create a general psychology, his main work was with conflict-ridden persons caught up in personal crises. Freud observed these tortured individuals only under extremely artificial conditions: lying on a couch during the psychotherapy hour in an environment deliberately made as nonsocial as possible. This drastically restricted observational base helped to foster a theory that originally was almost entirely a theory of anxiety and internal conflict. It paid little attention to the social environment and to the interpersonal context of behavior. We have already seen that many of Freud's own followers attempted to modify that initial emphasis and to devise a more ego-oriented and social approach. That trend is continuing.

This newer "ego psychology" and more "cognitive clinical psychology" is characterized by a greater focus on development beyond early childhood to include the entire life span. It attends more to the role of interpersonal relations and society, and to the nature and functions of the concepts of ego and self and how the person thinks rather than to the id and its impulses. Freud's ideas thus have been going through a continuing revolution that goes beyond his own writings to the extensions introduced by his many followers over the years. The object relations theories of analysts such as Kohut and Kernberg are especially important innovations that seem to be changing psychodynamic theories substantially. They emphasize the self (rather than the instincts and the unconscious defenses against their expression) and thus have some similarity to the self theories discussed in later parts of this text.

One of the main criticisms of Freud's theory is that it is hard to test. Unfortunately, this criticism also applies (although to a lesser degree) to most of the later psychodynamic thinkers. That is true in part because psychodynamic constructs tend to have both the richness and the ambiguity of metaphors; while they may seem compelling intuitively, they are hard to quantify. Rooted in clinical experience and clinical language, the terms often are loose and metaphoric and convey different meanings in different contexts. The theory also requires the user to have available a clinical background with much experience and training as the framework and language on which the constructs draw.

These criticisms notwithstanding, many of the key concepts from the theory are still central in different areas of contemporary psychology, psychiatry, and clinical practice, and some of these have been modified and extensively researched, leading to many new insights and discoveries (e.g., Andersen, Chen, & Miranda, 2002; Bargh, 1997; Kihlstrom, 1999; Miller, Shoda, & Hurley, 1996). Especially vigorous efforts have been made to deal with defense mechanisms and conflict, and to submit them to experimental study (e.g., Blum, 1953; Erdelyi, 1985; Holmes, 1974; Sears, 1943, 1944; Silverman, 1976; Westen & Gabbard, 1999). You already saw some of the research that psychoanalytic ideas have influenced, and you will see many more examples throughout the text.

Particularly exciting in recent years is the renewed interest in psychodynamic concepts such as the unconscious, in light of new discoveries about mental processes (e.g., attention, thinking, and memory) emerging from the study of cognitive psychology. There are many interesting parallels, and people certainly are not aware of all the things that happen in their minds. Modern cognitive psychology has understandably stimulated a sympathetic reexamination of the nature of unconscious processes (e.g., Kihlstrom, 1999).

There is evidence that unconscious processes and events influence us massively. For example, individuals with amnesia continue to be affected by previous experiences although they cannot remember them (Kihlstrom, Barnhardt, & Tataryn, 1992). All sorts of information that influences how different individuals encode (interpret) and evaluate their social environments is acquired unconsciously (e.g., Lewicki et al., 1992). And most of what we feel and believe and do may be elicited or triggered automatically without conscious control or awareness (e.g., Bargh, 1997). Thus, the reality of unconscious processes is now accepted as a fact. Its significance, however, continues to be debated (Loftus & Klinger, 1992), and firm conclusions cannot be reached until further research explores more directly the relevance of nonconscious information processing for Freud's concepts about the unconscious mind (e.g., Erdelyi, 1992).

In sum, when these contributions are put in perspective, they are most impressive. Especially notable is the continued recognition of the importance of motives and conflicts often operating at levels outside the person's full awareness. Likewise, work at this level called attention to the self-protective nature of the mind, and the diverse defense operations people use to reduce anxiety and to try to make themselves feel better about themselves, often with substantial long-tem costs to themselves and others. This list merely hints at the enormous contributions made by a genius and his colleagues, beginning more than a hundred years ago, and still being vigorously pursued in current thinking, theory, research, and clinical applications.

PART

III

THE PHENOMENOLOGICAL LEVEL

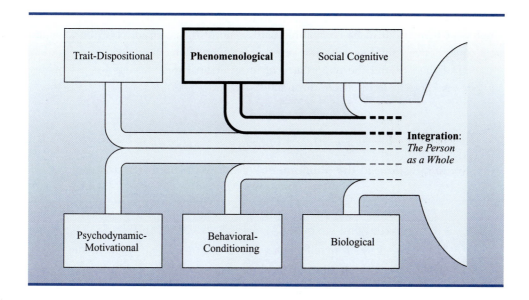

Integration:
The Person
as a Whole

▶ PRELUDE TO PART III: THE PHENOMENOLOGICAL LEVEL

The phenomenological level is devoted to understanding the individual's inner psychological experiences as perceived and understood by that person. Its focus is on how we see and experience ourselves, other people, and the social world. It is also concerned with the nature of the self, and the relationships between self, other people, and the social world. Work at this level began in the late 1940s and in the 1950s mostly as a protest against the psychoanalytic work of Freud and his followers which was dominant in clinical psychology and psychiatry at that time. Unlike the beginnings of the Freudian movement, there was no single leader. Instead, a number of different psychologists who cared about personality and psychotherapy began to raise questions that challenged basic assumptions that were accepted by many in the field at that time.

George Kelly, for example, proposed that what is most important about personality is that each person views or "construes" the world in distinctly different ways. It is these personal "construals" or appraisals that guide what people think, do, and become. They may even lead the person to develop maladaptive, self-defeating, and self-destructive problems like those shown by Freud's patients. Most important, Kelly believed that people are free to change how they construe and appraise themselves and their world, and they can do so in constructive ways that enhance their freedom. In other words, we do not have to be victims of the past, of our genes, and of our life situations. At the same time that Kelly was making his contributions and even in the same place—Ohio State University's Psychology Department—Carl Rogers, a former minister who also became a psychology professor, developed a new theory of personality that focused on the individual's potential for personal growth and genuineness and on the nature of the self.

The contributions at the phenomenological level are diverse and significant and remain vibrant in somewhat modified forms. For example, a great deal of current work is devoted to understanding the "self" and its role in personality (Part V), building directly on the contributions first made at the phenomenological level. In hindsight, although work at the phenomenological level began as a protest against earlier work, much of it turned out to be compatible with some of those earlier contributions and built constructively on them. But it evolved in new directions that added to the understanding of the self and the importance of interpersonal relationships for personality and personal growth, as discussed throughout this part.

The Personal Side of the Science

Some questions at the phenomenological level you might ask about yourself:

▶ Who do I want to become?

▶ How do I see myself? How do I see my parents?

▶ What do I feel about myself when I don't meet my parents' expectations?

▶ How is my real self different from the self I would ideally like to be?

▶ What is my ideal self?

▶ How am I different from my mother but similar to my father?

PHENOMENOLOGICAL CONCEPTIONS

Each person sees the world subjectively in his or her own personal way. To understand this privately experienced side of personality, we must examine the nature of subjective experience; we must see how people perceive their world. For example, we cannot understand anxiety as an aspect of personality fully without understanding how the individual experiences it. We begin with a sample of such personal experience in the form of a self-description by a college student about to take a final examination:

> *When I think about the exam, I really feel sick . . . so much depends on it. I know I'm not prepared, at least not as much as I should be, but I keep hoping that I can sort of snow my way through it. He [the professor] said we would get to choose two of three essay questions. I've heard about his questions . . . they sort of cover the whole course, but they're still pretty general. Maybe I'll be able to mention a few of the right names and places. He can't expect us to put down everything in two hours . . . I keep trying to remember some of the things he said in class, but my mind keeps wandering. God, my folks—What will they think if I don't pass and can't graduate? Will they have a fit! Boy! I can see their faces. Worse yet, I can hear their voices: "And with all the money we spent*

on your education." Mom's going to be hurt. She'll let me know I let her down. She'll be a martyr: "Well, Roger, didn't you realize how this reflects on us? Didn't you know how much we worked and saved so you could get an education? . . . You were probably too busy with other things. I don't know what I'm going to tell your aunt and uncle. They were planning to come to the graduation you know." Hell! What about me? What'll I do if I don't graduate? How about the plans I made? I had a good job lined up with that company. They really sounded like they wanted me, like I was going to be somebody. . . . And what about the car? I had it all planned out. I was going to pay for it and still have enough left for fun. I've got to pass. Oh hell! What about Anne [girlfriend]? She's counting on my graduating. We had plans. What will she think? She knows I'm no brain, but . . . hell, I won't be anybody. I've got to find some way to remember those names. If I can just get him to think that I really know that material, but don't have time to put it all down. If I can just . . . if . . . too goddamn many ifs. Poor dad. He'll really be hurt. All the plans we made—all the . . . I was going to be somebody. What did he say? "People will respect you. People respect a college graduate. You'll be something more than a store-keeper." What am I going to do. God, I can't think. You know, I might just luck out. I've done it before. He could ask just the right questions. What could he ask? Boy! I feel like I want to vomit. Do you think others are as scared as I am? They probably know it all or don't give a damn. I'll bet you most of them have parents who can set them up whether they have college degrees or not. God, it means so much to me. I've got to pass. I've just got to. Dammit, what are those names? What could he ask? I can't think . . . I can't. . . . Maybe if I had a beer I'd be able to relax a little. Is there anybody around who wants to get a beer? God, I don't want to go alone. Who wants to go to the show? What the hell am I thinking about? I've got to study. . . . I can't. What's going to happen to me? . . . The whole damn world is coming apart (Fischer, 1970, pp. 121–122).

Feelings and thoughts like those reported by this student are the raw materials of theories that deal with the self and with the person's subjective, internal experiences and personal concepts. These are the theories that traditionally have guided much of the work at this level of analysis. There are many complexities and variations in the orientation to personality presented in this chapter. In spite of these variations, however, a few fundamental themes emerge.

8.1 What is the focus of the phenomenological level of analysis?

We call the level of analysis in this chapter phenomenological, a term that refers to the individual's experience as he or she perceives it, because that is its most basic element. Some of the orientations guiding the work here have been given other labels also, such as "self" theories, "construct" theories, "humanistic" theories, "cognitive" theories, and "existential" theories. Most phenomenological theories are distinctive both in the concepts they reject and in the ones they emphasize. They tend to reject most of the dynamic and motivational concepts of the psychoanalytic level and also most of the assumptions of the trait level. Persons thus are viewed as experiencing beings rather than as the victims of their unconscious psychodynamic conflicts. Most of the approaches discussed in this chapter also stress people's positive strivings and their tendencies toward growth and self-actualization.

Most of the theories presented here are concerned broadly with how we know and understand the world and ourselves. They have tried to understand how the individual perceives, thinks, interprets, and experiences the world; that is, they have tried to grasp the individual's point of view. Their focus is on persons and events of life as seen by the perceiver. In sum, they are most interested in the person's experience as he or she perceives and categorizes it—the person's **phenomenology.** Ideally, they would like to look at the world through the "participant's" eyes and to stand in that person's shoes, to experience a bit of what it is to *be* that person. This phenomenological view is the main concern of the present chapter.

▶ SOURCES OF PHENOMENOLOGICAL PERSPECTIVES

The orientation presented in this chapter has numerous sources. Among the many early theorists who were fascinated with the self were William James, George H. Mead, and John Dewey. Another early theorist concerned with the self was Carl Jung. As early as the start of the century, Jung called attention to the organism's strivings for self-realization and integration. He believed in creative processes that go beyond the basic instincts of Freudian psychology. It is also noteworthy that in the psychodynamic tradition, Heinz Kohut and other "object relations" theorists (Chapter 7) also make the self a central concept.

Allport's Functional Autonomy

Gordon Allport (1937) was one of the first to emphasize the uniqueness of the individual and of the integrated patterns that distinguish each person. He also noted the *lack of* motivational continuity during the individual's life and criticized the Freudian emphasis on the enduring role of sexual and aggressive motives.

According to Allport, behavior is motivated originally by instincts, but later it may sustain itself indefinitely without providing any biological gratifications. Allport saw most normal adult motives as no longer having a functional relation to their historical roots. "Motives are contemporary. . . . Whatever drives must drive now. . . . The character of motives alters so radically from infancy to maturity that we may speak of adult motives as *supplanting* the motives of infancy" (1940, p. 545). This idea has been called **functional autonomy** to indicate that a habit, say practicing the violin at a certain hour each day, need not be tied to any earlier motive of infancy. The extent to which an individual's motives are autonomous is a measure of maturity, according to Allport.

Allport thus stressed the **contemporaneity of motives** (1961): motives are to be understood in terms of their role in the present regardless of their origins in the past. In his view, the past is not important unless it can be shown to be active in the present. He believed that historical facts about a person's past, while helping to reveal the total course of the individual's life, do not adequately explain the person's conduct today. In his words, "Past motives explain nothing unless they are also present motives" (1961, p. 220). In sum, Allport emphasized that later motives do not necessarily depend on earlier ones. Although the life of a plant is continuous with that of its seed, the seed no longer feeds and sustains the mature plant. A pianist may have been spurred to mastery of the piano through the need to overcome inferiority feelings, but the pianist's later love of music is functionally autonomous from its origins. Allport focused on the individual's currently perceived experiences, his or her phenomenological self and unique pattern of adaptation. He also favored a holistic view of the individual as an integrated, biosocial organism, rather than as a bundle of traits and motives. Table 8.1 summarizes some of Allport's main ideas about individuality.

Lewin's Life Space

Still another important post-Freudian influence came from field theories (Lewin, 1936). These theories saw behavior as determined by the person's psychological life space—by the events that exist in the total psychological situation at the moment—rather than by past events or enduring, situation-free dispositions. The most elegant formulation of this position was Kurt Lewin's **field theory.** In field theory, the way in which an object is perceived depends upon the total context or configuration of its surroundings. What is

8.2 What did Allport mean by functional autonomy? How does this concept relate to a phenomenological approach to behavior?

TABLE 8.1 Some Distinguishing Features of Individuality According to Allport (1961)

1. Motives become independent of their roots (*functional autonomy*).
2. A *proprium* or self develops, characterized by:

Bodily sense	Self-esteem
Self-identity	Rational thought
Self-image	

3. A *unique*, integrated pattern of adaptation marks the person as a whole.

perceived depends on the *relationships* among components of a perceptual field, rather than on the fixed characteristics of the individual components.

8.3 What are the major components of Lewin's life space? Describe his classic formula for the causes of behavior.

Lewin defined **life space** as the totality of facts that determine the behavior (B) of an individual at a certain moment. The life space includes the person (P) and the psychological environment (E), as shown in Figure 8.1. Behavior is a function of the person and the environment, as expressed in the formula

$$B = f(P, E)$$

Ordinary cause, based on the notion of causation in classical physics, assumes that something past is the cause of present events. Teleological theories assume that future events influence present events. Lewin's thesis is that neither past nor future, by definition, exists at the present moment and therefore neither can have an effect at the present. Past events have a position in the historical causal chains that create the present situation, but only those events that are functioning in the present situation need to be taken into account. Such events are, by definition, current or momentary. In other words, only present facts can cause present behavior.

To represent the life space, Lewin therefore took into account only that which is contemporary. He termed this the **principle of contemporaneity** (Lewin, 1936). This does not mean the field theorists are not interested in historical problems or in the effects of previous experience. As Lewin (1951) pointed out, field theorists have enlarged the psychological experiment to include situations that contain a history that is systematically created throughout hours or weeks. For example, college students in an experiment might be given repeated failure experiences (on a series of achievement tasks) during several sessions. The effects of these experiences on the students' subsequent aspirations and expectations for success might then be measured.

The boundaries between the person and the psychological environment and between the life space and the physical world are **permeable boundaries,** that is,

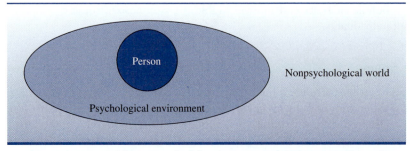

Figure 8.1 Lewin's life space. The life space contains the person in his or her psychological environment, which is delineated by a boundary (the ellipse) from the nonpsychological world.

they can be crossed easily. That makes prediction difficult because one cannot be sure beforehand when and what facts will permeate a boundary and influence a fact from another region. Lewin asserts that the psychologist might therefore concentrate on describing and explaining the concrete psychological situation in terms of field theory rather than trying to predict the future.

Lewin (1935) rejected the notion of constant, entity-like personality characteristics such as unchanging traits. As a result of dynamic forces, psychological reality is always changing. The environment of the individual does not serve merely to facilitate tendencies that are permanent in the person's nature (1936). Habits are not frozen associations, but rather the result of forces in the organism and its life space.

Lewin was similarly dissatisfied with the usual concept of needs. In descriptions of psychological reality, Lewin said, the needs that are producing effects in the momentary situation are the only ones that have to be represented. A need in Lewin's theory corresponds to a tension system. Lewin was also interested in reward and punishment. He saw rewards as devices for controlling behavior in momentary situations by causing changes in the psychological environment and in the tension systems of the person.

For Lewin, behavior and development are functions of the same structural and dynamic factors. Both are a function of the person and the psychological environment. In general, with increasing maturity there is greater differentiation of the person and the psychological environment.

Lewin's field theory had a major impact on experimental social psychology. His students extended his ideas and pursued them through ingenious experiments designed to alter the participant's life space—by altering perceptions about the self, about other people, about events. The effects of these alterations on attitudes, aspirations, and other indices were then examined carefully. Until recently Lewin's influence on personality psychology has been less extensive. There is now, however, an increasing recognition of the importance of the psychological situation in studies of traits and motives (e.g., Magnusson, 1999; Mischel & Shoda, 1999).

Kurt Lewin (1890–1947)

Phenomenology and Existentialism: The Here and Now

George A. Kelly and Carl Rogers developed views of personality in which private experiences, subjective perceptions, and the self all have an important part. Abraham Maslow put his emphasis on human growth motivation. In his view, growth motivation moves the individual through hierarchically ordered degrees of health to ultimate self-actualization. "Every person is, in part, his own project, and makes himself" (Maslow, 1965, p. 308). Behavior is goal directed, striving, purposeful, and motivated by higher needs to realize or **self-actualize** one's human potential rather than by primary biological drives alone.

The ideas of most of these theorists have much in common with the existential philosophical position developed by such European thinkers and writers as Kierkegaard, Sartre, and Camus and by the Swiss psychiatrists Binswanger and Boss. The key features of their orientation are expressed by Rollo May, an American proponent of existential psychology. Thinking about a patient of his in psychotherapy, May recognizes that he has available all sorts of information about her, such as hypotheses from her Rorschach and diagnoses from her neurologist. He then comments (1961, p. 26):

> But if, as I sit here, I am chiefly thinking of these whys and hows of the way the problem came about, I will have grasped everything except the most important thing of all, the existing person. Indeed, I will have grasped everything except the only real source of data I have, namely, this experiencing human being, this person now emerging, becoming, "building world," as the existential psychologists put it, immediately in this room with me.

8.4 What is the role of immediate experience in existential theories? Where are the true causes of behavior?

May's remarks point to the existentialist's focus on phenomenological experience, on the "here and now" rather than on distant historical causes in the person's early childhood. Furthermore, the existential orientation sees the human being as capable of choice and responsibility in the moment rather than as the victim of unconscious forces or of habits from the past.

The Swiss existential psychiatrist Binswanger commented that Freudian theory pictured human beings not yet as people in the full sense, but only as creatures buffeted about by life. Binswanger believes that for a person to be fully himself— that is, to be truly realized or actualized as a human being—he must "look fate in the face." In his view, the fact that human life is determined by forces and conditions is only one side of the truth. The other side is that we ourselves "determine these forces as our fate" (cited in May, 1961, p. 252). Thus, in the phenomenological and existential orientation, humans are seen as beings whose actualization requires much more than the fulfillment of biological needs and of sexual and aggressive instincts.

The existentialists propose that we are inevitably the builders of our own lives and, more specifically, that each person is:

1. a *choosing* agent, unable to avoid choices throughout the course of life
2. a *free* agent, who freely sets life goals
3. a *responsible* agent, accountable personally for his or her life choices

Our existence in life is given, but our essence is what we make of life, how meaningfully and responsibly we construct it. This is an often painful, isolated, agonizing enterprise. To find satisfying values, to guide our lives accordingly, to give life meaning—these goals are all part of the existential quest. They require the "courage to be"—the courage to break from blind conformity to the group and to strive instead for self-fulfillment by seeking greater self-definition and authenticity.

Finally, to grasp what it means to *be* also requires being in constant touch with the awareness of nonbeing, of alienation, of nothingness, and ultimately of the inevitability of death, everyone's unavoidable fate. The awareness of this inevitable fate and what that implies produces **existential anxiety.** The antidote for such anxiety is to face and live our lives responsibly, meaningfully, and with courage and awareness of our potential for continuous choice and growth.

8.5 What is meant by existential anxiety?

To understand some of the main features of the existential and phenomenological orientation more closely, we shall consider the ideas of one of its most articulate proponents in the next sections.

▶ GEORGE KELLY'S PSYCHOLOGY OF PERSONAL CONSTRUCTS

To the humanist every man is a scientist by disposition as well as by right, every subject is an incipient experimenter, and every person is by daily necessity a fellow psychologist (G. A. Kelly, 1966, in B. A. Maher, 1979, p. 205).

In addition to acting as motivated organisms, people also are perceivers and construers of behavior: they give meaning to their experience and interpret events. They generate abstractions about themselves and others, just like psychologists do. These hypotheses and constructions have long intrigued psychologists interested in subjective states, phenomenology, and the experience of the self.

The Person's Constructs

In the psychodynamic approach, the motive is the chief unit, unconscious conflicts are the processes of greatest interest, and the clinical judge is the favored instrument. Kelly's (1955) personal construct theory, in contrast, seeks to illuminate the person's own constructs rather than the hypotheses of the psychologist. Its main units are personal constructs—the ways we represent or view our own experiences. Rather than seeing people as victimized by their impulses and defenses, this position views the human being as an active, ever-changing creator of hypotheses.

8.6 What are personal constructs? How does Kelly's orientation and measurement approach differ from the trait approach? What did Kelly's approach reveal about Gary W.?

According to Kelly, trait psychology tries to find the person's place on the *theorist's* personality dimension. **Personal construct theory** instead tries to see how the person sees and aligns events on *his or her own* dimensions (Fransella, 1995). It is Kelly's hope to discover the nature of the personal construct dimensions rather than to locate the individual's position on the dimensions of the psychologist's theory. If next week's test is important to you, Kelly wants to explore how you see it, what it means to you, not what your score is on a scale of test-taking anxiety.

Suppose that you see a person quietly letting herself be abused by someone else. You might conclude that the person is "submissive." Yet the same behavior might be construed by other observers as sensitive, cautious, intelligent, tactful, or polite. Thus Kelly emphasized that different people may construe the same event differently and that every event can be construed in alternative ways.

The **personal construct** is the central unit of George Kelly's theory. Kelly's operation for measuring a construct is best seen in his **Role Construct Repertory Test** or **"Rep" Test.** On the Rep Test, you are asked to list many people or things that are important to you (for example, self, mother, brother). After these items are listed (or the assessor lists them), you are asked to consider them in groups of three. In each

George A. Kelly (1905–1967)

triad you have to indicate how two items are similar to each other and different from the third. In this way the subjective dimensions of similarity among events, and the subjective opposites of those dimensions, may be evoked systematically (Table 8.2). It is also possible to study the characteristics of the people's construct systems—for example, the number of different constructs they have in their construct repertory.

The Rep Test is a flexible instrument that can be adapted for many different purposes, and it provides a convenient and fairly simple way to begin the exploration of personal constructs. Some examples, taken from the study of Gary W., are shown in Table 8.2 (which illustrates the general type of procedure that may be used to elaborate personal constructs).

Research on the temporal stability of personal constructs from Kelly's Rep Test indicates a good deal of consistency over time (Bonarius, 1965). For example, a high

---

TABLE 8.2   Elaboration of Personal Constructs: Examples from Gary W.

1. List the three most important people in your life:

   Me, my brother, my father

   How are any two of these alike and different from the third?

   My brother and I both know how to be tough and succeed, no matter what—my father is soft, knocked out, defeated by life.

2. Think of yourself now, five years ago, and five years from now. How are any two of these alike and different from the third?

   Five years ago I was warmer, more open and responsive to others than I am now. Now I'm mostly a scheming brain. Five years from now I hope to have recaptured some of that feeling and to be more like I was five years ago.

---

retest correlation (.79) was found for constructs after a two-week interval (Landfield, Stern, &Fjeld, 1961), and factor analyses of the Rep Test suggest that its main factor is stable (Pedersen, 1958), and thus that an individual's main constructs may be relatively permanent.

According to Kelly, the individuals' personal constructs gradually become elaborated through their answers on the Rep Test and through their behaviors in the interview and on other tests. To illustrate some features of the assessment of personal constructs, here is an analysis of our case based on how Gary spontaneously elaborates and contrasts the constructs with which he views the world. What follows are excerpts from an attempt by an assessor to summarize some of Gary's main conceptions.

A Personal Construct Conceptualization of Gary W.

Rationality – Emotionality. This is a construct dimension that seems to be of considerable importance for Gary. He elaborates what this construct means to him most clearly when he is discussing his interpersonal relationships. He describes a sexual relationship with a woman in such terms as "spiritual," "instinctive," "sublime," and "beyond rationality." It is characterized by intense feeling and the primacy of emotions, and it is based on physical attraction. Real friendships, in contrast, are based upon rational grounds—such as interests and ways of thinking that are common to both parties.

The distinction between the rational and the emotional is echoed when Gary describes his worries in terms of those that are "rational" versus those that are "immediate and threatening." In discussing anger, he says that he has learned to cover up his feelings, but that his emotions sometimes "surface." He no longer gets violently angry, as he did when he was a child, but is "controlled," "stony," and "devious." He gives the most positive evaluation to reason, and contrasts what is reasonable with what is "worthless."

Transposed onto a time line, his distinction between reason and emotion forms part of the contrast between adults and children. After he was about 12 years old, Gary "psyched out" his father, so the latter was no longer his "enemy" but instead became his "friendly, rational adviser." He also describes shifts in his relations with his mother and with his brother that apparently involve handling his feelings toward them in a less explosive way.

Power and Control versus Dependence and Weakness. This seems to be a major dimension on which adults and children differ. Adults are the enemies of children. In his interview, descriptions of his childhood experiences, what parental figures require of a child, are typically the opposite of what the child wants. Gary describes life as a child as involving "denial, helplessness, nothing and nobody on my side." It was a time when he "couldn't control events," when he was being "manipulated" and "shamed." Gary contrasts foresight, and events that he can plan and control, with accidents, terror, and the unpredictable.

Defeat – Success. Defeat–Success is a closely related dimension around which a number of constructs are clustered. Defeat is defined in terms of lack of money, passivity, compliance, dependence, frustration, undermined masculinity, and physical pain. Success means money, activity, freedom, independence, control, and being a "real" man.

Security – Liberty. This is another major dimension for Gary. In describing jobs, acquaintances, and life styles, he talks in terms of "the ordinary 9 to 5 job complete with wife, kids, and mortgage," versus the "free and easy life." "Blind obedience" is contrasted to "judging the issues for oneself." Gary describes himself as being "uncertain," and contrasts being freewheeling with plodding determination. He sees himself as being currently without "acceptance" and "success" and he feels "cut off." His own "procrastination" hinders his "drive," but he hopes his "ambition will win out" and gain him both security and liberty.

As far as *role conceptions* are concerned, Gary now sees his father as "emasculated" and "knocked out," although once he saw him as "a giant" and as his "enemy." The father seems to have moved along the conceptual continuum from "power and control" to a point where he is seen as inadequate and as being competition no longer. He dislikes his father for the middle-class values that he feels he represents and for his passivity. There is also the implication that he resents his father for not comparing favorably with his mother. The turning point in Gary's feelings for his brother, whom he disliked for sharing many of their father's qualities, came when his brother was smashed up in a car crash. He now sees him as less conventional, more humorous and self-examining.

Gary sees his mother, and ideal women in general, as "independent partners" rather than "devouring" sources of affection. Instead of making their families central in importance, they achieve success and recognition in work outside their home. They keep the male "alive" by providing stimulation through their competence, which extends even to athletics, rather than being dependent and "clinging." Gary sees himself as similar to his mother and says he loves her best, next to himself. On the more negative side, he sees his mother as frigid and incapable of expressing affection. However, in view of his own evaluation of emotionality, this criticism is a highly qualified one. He sees his mother as having in many ways been the cause of his father's defeat, but constantly adds that she did not intend this result and feels bad about it, that it was a byproduct of other admirable qualities she possesses.

His relationship with his mother is characterized by control of expression of both anger and love. He sees her dominating tendencies as dangerous, as evidenced by his childhood conception of her as omniscient and omnipotent. This fear seems to have generalized to his grandmother and to other women, as evidenced by his TAT stories.

In his relationships with women, there seems to be a general distinction between sex objects and companions. In describing a sexual relationship that he felt had no potentialities for friendship, Gary says, "If we hadn't been able to 'make it' we would have stopped seeing each other." He generally prefers women who are stimulating and challenging, though he fears all forms of domination, through either authority or emotional ties.

In his relationships with men outside his family, Gary prefers distance and respect and finds that closeness leads to friction, as with his present roommate. At school he found two older men whom he could look up to: a teacher to whom he was grateful for not being "wishy-washy" and another person whom he describes as being a "real man."

Behavioral Referents for Personal Constructs

8.7 How are behavioral referents used to clarify personal constructs? How does this differ from a psychodynamic approach?

Some of Gary's main personal constructs emerged from his self-descriptions and verbalizations. When people start to express their constructs, they usually begin with very diffuse, oversimplified, global terms. For example, Gary called himself "shrewd," "too shy," "too sharp." He also said he wanted to "feel more real," to "adjust better," and to "be happier."

What can the construct assessor do with these verbalizations? As we have seen in earlier chapters, psychodynamically oriented clinicians rely chiefly on their intuitive inferences about the symbolic and dynamic meanings of verbal behavior. Personal construct assessors recognize that talk about private experiences and feeling tends to be ambiguous. For example, statements of the kind commonly presented in clinical contexts, like "I feel so lost," generally are not clear. Instead of inquiring into why the person feels "lost," personal construct assessments try to discover referents for just what the statement means. An adequate personal construct assessment of what people say involves the analysis of what they mean. For this purpose, the assessor's initial task is like the one faced when we want to understand a foreign language. A personal construct analysis of language tries to decipher the content of what is being conveyed and to discover its meaning for the person. Its aim is not to translate what is said into signs of underlying motives, of unconscious processes, or of personality dimensions. Often it is hard to find the words for personal constructs. Just as the psychologist interested in such concepts as extraversion, identity, or anxiety must find behavioral referents to help specify what he or she means, so must the client find such referents for his or her private concepts, difficulties, and aspirations.

In sum, Kelly urged a specific and elaborate inquiry into personal constructs by obtaining numerous behavioral examples as referents for them. Kelly (1955) has described in detail many techniques to explore the conditions under which the individual's particular constructions about emotional reactions may emerge and change. His ideas still influence current work to explore the meanings that underlie puzzling behavior patterns and seeming inconsistencies in the expressions of personality.

Exploring the Meaning Underlying Puzzling Behavior Patterns

Contemporary work that was inspired by Kelly also tries to understand the personal meaning that may underlie puzzling patterns of behavior. Consider Mary, a senior attorney at an international corporate law firm, who is concerned about the performance of one of her new staff attorneys, John. Before becoming a corporate lawyer, Mary had done graduate work in psychology many years ago at Ohio State University where she was influenced by one of her professors, George Kelly.

Mary has been puzzled by the variability in John's work and wants to find a way to improve his sense of well-being and productivity. Is he always distracted, or does it depend on situations? What are the situations in which he's less distracted and can concentrate? In short, Mary is trying to assess the *if . . . then . . .* patterns that constitute his behavioral signature (see Chapter 4) so that she can understand their meaning and implications. After she learns that John tends to be particularly distracted in meetings at the corporate headquarters but not when he travels to branch offices, she begins to wonder: What exactly is the important difference between the meetings at the headquarters and those at branch offices? In essence, Mary wants to know the personal meaning of the situations, seen from John's point of view. This is also one of the central questions for the science of personality. In particular, researchers have been asking: can personal meanings of situations be assessed in a systematic, objective fashion?

George Kelly anticipated this challenge, and in his personal construct theory offered a general strategy. Although his strategy was originally proposed for individual therapeutic settings, it is being extended to a method for systematically and objectively assessing the meaning of a situation to a given individual (e.g., Mischel & Shoda, 1995; Shoda, LeeTiernan & Mischel 2002). For example, meetings at company headquarters may be characterized by such features as the presence of the boss, the authority whose

8.8 How does selective attention to psychologically meaningful features of situations provide information about a person's constructs?

opinions directly influence promotion and salary decisions. Another feature might be the presence of highly competitive "career climbers" who focus only on advancement of their individual careers.

The subjective meaning of the situations may be understood by a person's tendency to selectively attend to particular types of features. While some people may not even notice the career-advance focus of their colleagues at headquarters, for others, like John, it is highly salient. Individuals also may differ reliably in the psychological impact of attending to these features. In this example, John's perception of his colleagues' focus on selfish career advancement reminds him that he feels himself losing sight of the sense of purpose that initially attracted him to his career. Now suppose that John has a stable tendency to attend to and react to the presence of these features in distinctive ways, for example by becoming upset and less motivated in his own work. If so, it is important to identify the features that are influencing him if one wants to understand their meaning for him and the ways they are affecting what he is thinking, feeling, and doing.

People as Scientists

8.9 In what sense is each individual a "personal scientist"? What are the "operational definitions" for personal constructs?

The psychology of personal constructs explores the maps that people generate in coping with the psychological terrain of their lives. Kelly emphasizes that, just like the scientist who studies them, human participants also construe behavior—categorizing, interpreting, labeling, and judging themselves and their world: Every person is a scientist.

The individuals assessed by psychologists are themselves assessors who evaluate and construe their own behavior; they even assess the personality psychologists who try to assess them. Constructions and hypotheses about behavior are formulated by all persons regardless of their formal degrees and credentials as scientists. According to Kelly, it is these constructions, and not merely simple physical responses, that must be studied in an adequate approach to personality. Categorizing behavior is equally evident when a psychotic patient describes his personal, private ideas in therapy, and when a scientist discusses her favorite constructs and theories at a professional meeting. Both people represent the environment internally and express their representations and private experiences in their psychological constructions. Personal constructions, and not objective behavior descriptions on clear dimensions, confront the personality psychologist.

Kelly noted that most psychological scientists view themselves as motivated to achieve cognitive clarity and to understand phenomena, including their own lives. Yet the "subjects" of their theories, unlike the theorists themselves, are seen as unaware victims of psychic forces and traits that they can neither understand nor control. Kelly tries to remove this discrepancy between the theorist and the participant and to treat all people as if they were scientists.

Just like the scientist, participants generate constructs and hypotheses with which they try to anticipate and control events in their lives. Therefore, to understand the participant, one has to understand his or her constructs or private personality theory. To study an individual's constructs, one has to find behavioral examples or "referents" for them. We cannot know what another person means when she says, "I have too much ego," or "I am not a friendly person," or "I may be falling in love," unless she gives us behavioral examples. Examples (referents) are required whether the construct is personal, for example the way a patient construes herself "as a woman," or theoretical, as when a psychologist talks about "introversion" or "ego defenses." Constructs can become known only through behavior.

Constructive Alternativism: Many Ways to See

If one adopts Kelly's approach to understanding people, then:

> *Instead of making our own sense out of what others did we would try to understand what sense they made out of what they did. Instead of putting together the events in their lives in the most scientifically parsimonious way, we would ask how they put things together, regardless of whether their schemes were parsimonious or not (Kelly, 1962, in B. A. Maher, 1979, p. 203).*

8.10 Define constructive alternativism. How do anticipations guide behavior?

The same events can be alternatively categorized. While people may not always be able to change events, they can always construe them differently (Fransella, 1995). That is what Kelly meant by **constructive alternativism.** To illustrate, consider this event: A boy drops his mother's favorite vase. What does it mean? The event is simply that the vase has been broken. Yet ask the child's psychoanalyst and he may point to the boy's unconscious hostility; ask the mother and she tells you how "mean" he is; his father says he is "spoiled"; the child's teacher may see the event as evidence of the child's "laziness" and chronic "clumsiness"; his grandmother calls it just an "accident"; and the child himself may construe the event as reflecting his "stupidity." While the event cannot be undone—the vase is broken—its interpretation is open to alternative constructions, and these may lead to different courses of action.

Kelly's theory began with this fundamental postulate: "A person's processes are psychologically channelized by the ways in which he anticipates events" (Kelly, 1955, p. 46). This means that a person's activities are guided by the constructs he or she uses to predict events. Like other phenomenological theories, this postulate emphasizes the person's subjective view, but it is more specific in its focus on how the individual predicts or anticipates events. Although the details of the theory need not concern us here, several of the main ideas require comment.

Kelly was concerned with the **convenience of constructs** rather than with their absolute truth. Rather than try to assess whether a particular construct is true, his approach attends to its convenience or utility for the construer. For example, rather than try to assess whether or not a client is "really a latent homosexual" or "really going crazy," one tries to discover the implications for the client's life of construing himself in that way. If the construction is not convenient, then the task is to find a better alternative—that is, one that predicts better and leads to better outcomes. Just as psychologists may get stuck with an inadequate theory, so their patients also may impale themselves on their constructions and construe themselves into a dilemma. Individuals may torture themselves into believing that "I am not worthy enough" or "I am not successful enough," as if these verdicts were matters of indisputable fact rather than constructions and hypotheses about behavior. The job of psychotherapy is to provide the conditions in which personal constructs can be elaborated, and tested for their implications. And, if they prove to be not helpful to the person, they then can be modified, just as a scientist can change a theory or idea that turns out not to work well. Like the scientist, every person needs the chance to test personal constructs and to validate or invalidate them, progressively modifying them in the light of new experience.

8.11 Why are constructs not evaluated in terms of truth or falsity? What did Kelly view as the goal for therapy?

Roles: Many Ways to Be

Rather than seeing humans as possessing fairly stable, broadly generalized traits, Kelly saw them as capable of enacting many different roles and of engaging in continuous

8.12 What is a role as defined by Kelly? How was role-taking used therapeutically?

change. A **role,** for Kelly, is an attempt to see another person through the other's glasses—that is, to look at a person through *his or her* constructs—and to structure one's actions in that light. To enact a role requires that behavior be guided by perception of the other person's viewpoint. Thus, to "role play" your mother, for example, you would have to try to see things (including yourself) as she does, "through her eyes," and to act in light of those perceptions. You would try to behave as if you really were your mother. Kelly used the technique of role playing extensively as a therapeutic procedure designed to help persons gain new perspectives and to generate more convenient ways of living.

People Are What They Make of Themselves: Self-Determination

Like other phenomenologists, Kelly rejected the idea of specific motives. His view of human nature focuses on how people construe themselves and on what they do in the light of those constructs (Fransella, 1995). Kelly (like Rogers) believed that no special concepts are required to understand why people are motivated and active: Every person is motivated "for no other reason than that he is alive" (Kelly, 1958, p. 49).

Like many existentialists, Kelly believed that the individual *is* what he *does* and comes to know his nature by seeing what he is doing. Starting from his clinical experiences with troubled college students in Fort Hays, Kansas, where he taught for many years, Kelly independently reached a position that overlaps remarkably with the views of such European existential philosophers as Sartre. In Sartre's (1956) existentialist conception, *"existence precedes essence"*: There is no human nature—man simply *is*, and he is nothing else but what he makes of himself.

▶ CARL ROGERS' SELF THEORY

Carl Rogers and George Kelly overlapped briefly when both held positions as professors at Ohio State University in the middle of the last century, and while their views shared many similarities, they developed them in quite different directions. Whereas Kelly as you saw earlier put the focus on personal constructs and alternative ways of finding more constructive ways of viewing the world, Rogers went on to develop a theory of the self and the conditions that allow its optimal growth and fulfillment.

Unique Experience: The Subjective World

At the start just like Kelly, Rogers' theory of personality emphasizes the unique, subjective experience of the person. He believed that the way you see and interpret the events in your life determines how you respond to them. Each person lives in a subjective world, and even the so-called objective world of the scientist is a product of subjective perceptions, purposes, and choices. Because no one else, no matter how hard he tries, can completely assume another person's "internal frame of reference," the person himself has the greatest potential for awareness of what reality is for him. In other words, each person potentially is the world's best expert on himself and has the best information about himself.

In Rogers' view, "behavior is typically the goal-directed attempt of the organism to satisfy its needs as experienced, in the field as perceived" (1951, p. 491). The emphasis is on the person's perceptions as the determinants of his or her actions: How one sees and interprets events determines how one reacts to them.

Carl Rogers (1902–1987)

Self-Actualization

Like most phenomenologists, Rogers wanted to abandon specific motivational constructs and viewed the organism as functioning as an **organized whole.** He maintained that "there is one central source of energy in the human organism; that it is a function of the whole organism rather than some portion of it; and that it is perhaps best conceptualized as a tendency toward fulfillment, toward actualization, toward the maintenance and enhancement of the organism" (1963, p. 6). Thus the inherent tendency of the organism is to actualize itself. "Motivation" then becomes not a special construct but an overall characteristic of simply being alive.

In line with his essentially positive view of human nature, Rogers' theory asserts that emotions are beneficial to adjustment: "emotion accompanies and in general facilitates . . . goal-directed behavior, . . . the intensity of the emotion being related to the perceived significance of the behavior for the maintenance and enhancement of the organism" (1951, p. 493).

In the course of actualizing itself, the organism engages in a valuing process. Experiences that are perceived as enhancing it are valued positively (and approached), while experiences that are perceived as negating enhancement or maintenance of the organism are valued negatively (and avoided). "The organism has one basic tendency and striving—to actualize, maintain, and enhance the experiencing organism" (Rogers, 1951, p. 487). The idea and implications of self-actualization are discussed more fully in In Focus 8.1.

The Self

The self is a central concept for most phenomenological theories, and it also is basic for Rogers—as it was for many of the post-Freudians working at the psychodynamic level. Indeed, his theory is often referred to as a self theory of personality. The self or

8.13 What is meant by self-actualization? What does it imply about the nature of the human being, and how does this conception differ from Freud's?

SELF-ACTUALIZATION AS A NEED (MASLOW)

Abraham Maslow (1968, 1971) was one of the most influential spokespersons for the importance of becoming "in touch" with one's true feelings. He considered this a core ingredient of well-being and self-fulfillment. Maslow also emphasized human beings' vast positive potential for growth and fulfillment, and believed that the striving toward actualization of this potential is a basic quality of being human:

> *Man demonstrates in his own nature a pressure toward fuller and fuller Being, more and more perfect actualization of his humanness in exactly the same naturalistic, scientific sense that an acorn may be said to be "pressing toward" being an oak tree, or that a tiger can be observed to "push toward" being tigerish, or a horse toward being equine . . . (Maslow, 1968, p. 160).*

Maslow's commitment was to study optimal man and to discover the qualities of those people who seemed to be closest to realizing all their potentialities. In his view one has **higher growth needs**—needs for self-actualization fulfillment—that emerge when more primitive needs—physiological needs, safety needs, needs for belongingness and esteem—are satisfied (see Fig. 8.2). Maslow wanted to focus on the qualities of feeling and experience that seem to distinguish self-actualizing, fully functioning people. Therefore, he searched for the attributes that seemed to mark such people as Beethoven, Einstein, Jefferson, Lincoln, Walt Whitman, as well as some of the individuals he knew personally and admired most. These positive qualities are an essential part of the humanistic view of the "healthy personality."

Maslow called the special state in which one experiences a moment of self-actualization a **peak experience.** It is a temporary experience of fulfillment and joy in which the person loses self-centeredness and (in varying degrees of intensity) feels a nonstriving happiness, a moment of perfection. Words that may be used to describe this state include "aliveness," "beauty," "ecstasy," "effortlessness," "uniqueness," and "wholeness." Such peak experiences have been reported in many contexts, including the aesthetic appreciation of nature and beauty, worship, intimate relationships with others, and creative activities.

8.14 How did Maslow study optimal human functioning? What were the most notable characteristics of highly functioning people?

The peak experience is characterized by a feeling of fulfillment and joy.

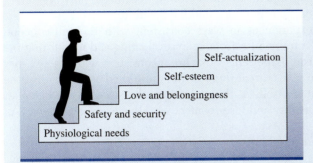

Figure 8.2 Maslow's hierarchy of needs. Maslow arranges motives in a hierarchy ascending from such basic physiological needs as hunger and thirst through safety and love needs to needs for esteem (e.g., feeling competent), and ultimately, self-actualization—the full realization of one's human potential, as in creativity. The lower needs are more powerful and demand satisfaction first. The higher needs have less influence on behavior but are more distinctly human. Generally, higher needs do not become a focus until lower ones have been at least partly satisfied.

In sum, in spite of its many different versions, the phenomenological-humanistic orientation tends to view "healthy people" as those who:

1. Become aware of themselves, their feelings, and their limits; accept themselves, their lives, and what they make of their lives as their own responsibility; have "the courage to be."

2. Experience the "here-and-now"; are not trapped to live in the past or to dwell in the future through anxious expectations and distorted defenses.

3. Realize their potentialities; have autonomy and are not trapped by their own self-concepts or the expectations of others and society.

To help achieve these ideals, several avenues for constructive personality change have been favored by advocates of the phenomenological approach, as discussed next.

self-concept (the two terms mean the same thing for Rogers) is an organized, consistent, whole. It consists of perceptions about oneself and one's relationships to others and to diverse aspects of life, and these all have values attached to them (Rogers, 1959, p. 200). As a result of interaction with the environment, a portion of the perceptual field gradually becomes differentiated into the self. This perceived self (self-concept) influences perception and behavior. That is, the interpretation of the self—as strong or weak, for example—affects how one perceives the rest of one's world.

The experiences of the self become invested with values. These values are the result of direct experience with the environment, or they may be taken over from others. For example, a young child finds it organismically enjoyable to relieve himself whenever he experiences physiological tension in the bowel or bladder. However, he may sometimes also experience parental words and actions indicating that such behavior is bad, and that he is not lovable when he does this. A conflict then develops that may result in distortion and denial of experience. That is, the parental attitudes may be experienced as if they were based on the evidence of the child's own experience. In this example, the satisfaction of defecating may start to be experienced as bad even though a more accurate symbolization would be that it is often experienced as organismically satisfying. Rogers goes on to suggest that in bowel training, denial or distortion of experience may be avoided if the parent is able genuinely to accept the child's feelings and at the same time accepts his or her own feelings.

8.15 What is the self? How does it develop, acquire values, and influence behavior?

Consistency and Positive Regard

Rogers proposes two systems: the **self** or **self-concept** and the organism. The two systems may be in opposition or in harmony. When these systems are in opposition or

8.16 How does the need for positive regard cause inconsistencies between self-concept and experience?

incongruence, the result is maladjustment, for then the self becomes rigidly organized, losing contact with actual organismic experience and filled with tensions. Perception is selective: we try to perceive experiences in ways consistent with the self-concept. The self-concept thus serves as a frame of reference for evaluating and monitoring the actual experiences of the organism. Experiences that are inconsistent with the self may be perceived as threats, and the more threat there is, the more rigid and defensive the self structure becomes to maintain itself. At the same time, the self-concept becomes less congruent with organismic reality and loses contact with the actual experiences of the organism.

Rogers (1959) assumed a universal need for positive regard. This need develops as the awareness of the self emerges and leads the person to desire acceptance and love from the important people in his life. Sometimes they may accept him conditionally (i.e., depending on his specific behavior), or they may accept him in his own right and give him unconditional regard. The person needs positive regard not only from others but also from his self. The need for self-regard develops out of self-experiences associated with the satisfaction or frustration of the need for positive regard. If a person experiences only unconditional positive regard, his self-regard also would be unconditional. In that case, the needs for positive regard and self-regard would never be at variance with **organismic evaluation.** Such a state would represent genuine psychological adjustment and full functioning.

Most people do not achieve such ideal adjustment. Often a self-experience is avoided or sought only because it is less (or more) worthy of self-regard. For example, a child may experience anger toward her mother but avoids accepting that feeling because she wants to be a "good girl." When that happens, Rogers speaks of the individual's having acquired a **condition of worth.** These are the conditions that other people, usually the parents, implicitly make a requirement for being loved and worthwhile. For example, a parent may lead a child to feel that to be worthy she has to get outstanding grades at school, or that the boy in the family has to be a robust athlete even if he is naturally inclined to become someone quite different. Experiences that are in accord with the individual's conditions of worth tend to be perceived accurately in awareness, but experiences that violate the conditions of worth may be denied to awareness and distorted grossly. When there is a significant amount of incongruence between the individual's self-concept and her evaluation of an experience, then defenses may become unable to work successfully. For example, if a young woman persistently experiences herself as painfully dissatisfied and "unhappy" in her efforts at schoolwork, but views herself as having to "succeed at college" in order to be an adequate person, she may experience great strain in her defensive efforts.

8.17 What role do conditions of worth play in anxiety and perception?

Rogers' theory, like Freud's, posits that accurate awareness of experiences may be threatening to the self. Anxiety in Rogers' theory might be interpreted as the tension exhibited by the organized concept of the self when it senses (without full awareness) that the recognition of certain experiences would be catastrophic to the self (1951). If a person's concept of the self has been built around his "masculinity," for example, experiences that might imply that he has some stereotypically feminine tendencies would threaten him severely. Anxiety thus involves a basic threat to the self, and defenses are erected to avoid it. Consistent with Rogers' theory, a great deal of research has demonstrated that people in fact engage in diverse strategies to protect their self-esteem when it is severely threatened. For example, they readily attribute important failures to chance rather than to themselves, but see success as due to their own abilities (Snyder & Uranowitz, 1978; Weiner, 1995).

Rogers assumed a need for unconditional positive regard not only from others but from the self.

Client-Centered Therapy

Rogers' ideals led to a new and still influential form of thinking about personality change and conducting therapy to facilitate positive change. **Client-centered, or person-centered (Rogerian), therapy** seeks to bring about the harmonious interaction of the self and the organism. The warm and unconditionally accepting attitude of the counselor hopefully enables the client to perceive and examine experiences that are inconsistent with the current self-structure. The client can then revise this self-structure to permit it to assimilate these inconsistent experiences. According to Rogers, the client gradually reorganizes the self-concept to bring it into line with the reality of organismic experience: "He will *be*, in more unified fashion, what he organismically *is*, and this seems to be the essence of therapy" (1955, p. 269).

In his therapy, Rogers rejected most of Freud's concepts regarding the nature of psychodynamics and psychosexual development. He also avoided all diagnostic terms, refusing to put his labels on the client. He maintained, however, the interview format for psychotherapy (using a face-to-face arrangement rather than the orthodox psychoanalyst's couch for the client). Rogers and his students focused on the client–clinician relationship. Usually they required many fewer sessions than did psychoanalytic therapy, and they dealt more with current than with historical concerns in the client's life.

For Rogers (1959), the therapist's main task is to provide an atmosphere in which the client can be more fully open to her own **organismic experience.** To achieve a growth-conducive atmosphere, the clinician must view the client as intrinsically good and capable of self-development. The clinician's function is to be nonevaluative and to convey a sense of unconditional acceptance and regard for the client (Brazier, 1993). To reach the client effectively the clinician must be "genuine" and "congruent"—an open, trustworthy, warm person without a facade (Lietaer, 1993). The **congruent therapist,** according to Rogers, feels free to be himself or herself and to accept the client fully and immediately in the therapeutic encounter, and conveys this openness to the client. When a genuinely accepting, unconditional relationship is established, the client will become less afraid to face and accept his or her own feelings and experiences. Becoming open to the experience of herself as she is, she can reorganize her

8.18 What is the objective in client (person)-centered therapy? Which therapist qualities contribute to a positive outcome?

self-structure. Now, it is hoped, she will accept experiences that she had previously denied or distorted (because they did not fit her self-concept) and thus achieve greater internal congruity and self-actualization.

Rogers thus sought an empathetic, interview-based relationship therapy. He renounced the Freudian focus on psychodynamics and transference. Instead, he wanted to provide the client an unconditionally accepting relationship—an atmosphere conducive to "growth" (self-actualization). In this relationship, the focus is on empathic understanding and acceptance of feelings rather than interpretation, although the latter is not excluded. The clinician is relatively "nondirective"; the objective is to let the client direct the interview while the clinician attempts to accurately reflect and clarify the feelings that emerge.

In client-centered therapy, now also called person-centered therapy, permissiveness and unqualified acceptance on the part of the therapist provide an atmosphere favorable to personal honesty. Psychologists are urged to abandon their "objective" measurement orientation and their concern with tests. Instead, they should try to learn from the client how he or she thinks, understands, and feels. "The best vantage point for understanding behavior is from the internal frame of reference of the individual himself" (Rogers, 1951, p. 494). Although their focus is on empathy, the Rogerians have not neglected objective research into the relationship, as was noted earlier in this chapter in the context of interview research. As a result, Rogerians have helped to illuminate some of the processes that occur during client-centered therapy and also have provided considerable evidence concerning its effectiveness (e.g., Truax & Mitchell, 1971).

8.19 Compare and contrast person-centered and psychoanalytic therapy.

Rogers' client-centered psychotherapy differs in many ways from Freudian psychotherapy. Indeed, when Rogers first proposed his techniques, they were considered revolutionary. Sometimes his approach to psychotherapy is even described as the polar opposite of Freud's. While there are major differences between Freudian and Rogerian approaches to psychotherapy, on closer inspection there also are some fundamental similarities. Both approaches retain a verbal, interview format for psychotherapy; both focus on the client–clinician relationship; both are primarily concerned with feelings; both emphasize the importance of unconscious processes (defense, repression); both consider increased awareness and acceptance of unconscious feelings to be major goals of psychotherapy.

To be sure, the two approaches differ in their focus. They differ in the specific content that they believe is repressed (for example, id impulses versus organismic experiences), in the motives they consider most important (such as sex and aggression versus self-realization), and in the specific insights they hope will be achieved in psychotherapy (the unconscious becomes conscious and conflict is resolved versus organismic experience is accepted and the self becomes congruent with it). But these differences should not obscure the fact that both approaches are forms of relationship treatment that emphasize awareness of hypothesized unconscious feelings and the need for the client to accept those feelings.

Rogers Reflects on His Own Work

Looking back at the almost 50 years of his contributions to psychology, Rogers (1974) tried to pinpoint the essence of his approach. In his view, his most fundamental idea was that:

> *the individual has within himself vast resources for self-understanding, for altering his self-concept, his attitudes, and his self-directed behavior—and that these resources can be tapped if only a definable climate of facilitative psychological attitudes can be provided* (Rogers, 1974, p. 116).

Such a climate for growth requires an atmosphere in which feelings can be confronted, expressed, and accepted fully and freely. His continued emphasis on the person's potential freedom, the hallmark of a humanistic orientation, remains unchanged:

> *My experience in therapy and in groups makes it impossible for me to deny the reality and significance of human choice. To me it is not an illusion that man is to some degree the architect of himself . . . for me the humanistic approach is the only possible one. It is for each person, however, to follow the pathway—behavioristic or humanistic—that he finds most congenial (Rogers, 1974, p. 119).*

In the same humanistic vein, he regretted modern technology and called for autonomy and self-exploration:

> *Our culture, increasingly based on the conquest of nature and the control of man, is in decline. Emerging through the ruins is the new person, highly aware, self-directing, an explorer of inner, perhaps more than outer, space, scornful of the conformity of institutions and the dogma of authority. He does not believe in being behaviorally shaped, or in shaping the behavior of others. He is most assuredly humanistic rather than technological. In my judgment he has a high probability of survival (Rogers, 1974, p. 119).*

In sum, Rogers' theory and the therapy he developed highlights many of the chief points of the phenomenological and humanistic approach to personality. It emphasizes the person's perceived reality, subjective experiences, organismic striving for actualization, the potential for growth and freedom (Rowan, 1992). It rejects or deemphasizes specific biological drives. It focuses on the experienced self rather than on historical causes or stable trait structures. A unique feature of Rogers' position is his emphasis on unconditional acceptance as a requisite for self-regard.

Other phenomenological theorists have emphasized different aspects in their formulations. For example, the **gestalt theory** (meaning completeness or fullness) of Fritz Perls focuses on awareness of one's own experience in dynamic interaction with the environment (e.g., Van De Riet, Korb, & Gorrell, 1989). Perls' theory continues to influence some psychotherapists.

▶ COMMON THEMES AND ISSUES

The conceptions surveyed in this chapter are quite diverse and far more complex than a brief summary suggests. In spite of their diversity, they share a focus on the self as experienced, on situations as perceived, on personal constructs, on feelings, and on the possibility of freedom. They search for data and methods that explore how people can reveal themselves more fully and honestly. They also see greater self-awareness, consistency, and self-acceptance as crucial aspects of personal growth and actualization (Spinelli, 1989).

8.20 What major assumptions and commonalities characterize the various phenomenological approaches to personality?

Potential for Growth and Change

The existential belief that "man is what he makes of himself" and what he conceives himself to be is extremely appealing to many people. It recognizes the human potential for growth and change, and for alternative ways of construing and dealing with life's challenges. It is optimistic in its belief that people do not have to be victimized by their biographies (as George Kelly put it). While emphasizing the potential for freedom and choice, it also is sensitive to the constraints and limitations of the human condition.

Psychologists who appreciate the attractiveness of these beliefs, but who are committed to a deterministic view of science, also have to ask, however: what are the causes that govern what individuals make of themselves and conceive themselves to be? And how do individuals come to make themselves and conceive themselves in particular ways? While philosophers may put the springs of action and cognition into the will (as Sartre does), the scientifically oriented psychologist seeks to understand the psychological mechanisms that underlie the will (Mischel, Shoda, & Rodriguez, 1989). Many psychologists accept the idea that individuals are what they make of themselves. However, as scientists they want to go further and search for the conditions and processes that make that possible. For example, they want to identify the influences that determine the person's self-conceptions and ability to choose. These challenges began to be addressed by work at the phenomenological level, as you will see in the next chapter, and it continues to be on the agenda of all levels in the rest of the text.

The existential idea that the person has potential control and responsibility for himself or herself, and that people are what they make of themselves, has profound implications for the study of personality. Instead of a search for where the individual stands with regard to the assessor's dimensions, the assessor's task is to help clarify what the individual is making of himself or herself, and the "projects" and goals and plans of that person (Cantor, 1990) as they unfold in the course of life (Emmons, 1997; Mischel, Cantor, & Feldman, 1996). The next chapter illustrates some of the main steps taken in this direction at the phenomenological level.

SUMMARY

SOURCES OF PHENOMENOLOGICAL PERSPECTIVES

- Phenomenological theories focus on the immediate perceived experience and concepts of individuals, and on their strivings toward growth and self-actualization.
- Allport stresses the functional autonomy of motives—current motives may be independent of their historical roots.
- Lewin's field theory introduces the notion of life space and the importance of the psychological environment. He stresses the immediate relationships between the person and the psychological environment.
- The existentialists focus on the "here and now." They emphasize that we build our own lives, and that each person is a choosing, free, responsible agent.

GEORGE KELLY'S PSYCHOLOGY OF PERSONAL CONSTRUCTS

- George Kelly focuses on how the individual views his or her own experiences. His Role Construct Repertory (Rep) Test is used to study personal constructs.
- Kelly emphasizes that all individuals think much like the scientists who study them.
- "Constructive alternativism" refers to the individual's ability to construe the same event in different ways, leading to different courses of action.
- Role play may help the person to select more satisfactory modes of construing the world.

CARL ROGERS' SELF THEORY

- Rogers' theory emphasizes the person's unique, subjective experience of reality. He proposes that the inherent tendency of the organism is to actualize itself.
- Maslow saw self-actualization as a basic human need.

- In Roger's theory, the *self* (self-concept) develops as the result of direct experience with the environment and may also incorporate the perceptions of others. The experienced self in turn influences perception and behavior.

- Maladjustment occurs when the self-concept and a person's experiences are in opposition.

- Client-centered therapy seeks to bring about the harmonious interaction of the self and the organism through unqualified acceptance by the therapist.

- Rogers emphasizes unconditional acceptance as a requisite for self-regard.

KEY TERMS

client-centered (person-centered) therapy   189
condition of worth   188
constructive alternativism   183
contemporaneity of motives   173
convenience of constructs   183
existential anxiety   177
field theory   173

functional autonomy   173
gestalt theory   191
higher growth needs   186
life space   174
organismic evaluation   188
organismic experience   189
organized whole   185
peak experience   186
permeable boundaries   174

personal construct   177
personal construct theory   177
phenomenology   172
principle of contemporaneity   174
role   184
Role Construct Repertory Test ("Rep" Test)   177
self-actualize (self-actualization)   176
self-concept (self)   187

THE INTERNAL VIEW

The phenomenological theories discussed in Chapter 8 deeply influenced personality theory, assessment, and change. To be useful, these theories require ways to access the person's perceived internal experience. In Kelly's own words, if a person's private domain is ignored, "it becomes necessary to explain him as an inert object wafted about in a public domain by external forces, or as a solitary datum sitting on its own continuum" (1955, p. 39). On the other hand, if individuals are to be understood within the framework of scientific rules, methods have to be found to reach those private experiences and to bring them into view. In this chapter, we will consider some of the methods used to examine the internal view of other people.

▶ EXPLORING INTERNAL EXPERIENCE

Phenomenologists like Rogers and Kelly wanted to go beyond introspection and anchor their theories to scientific methods. In Rogers' view, for example, the therapist enters the internal world of the client's perceptions not by introspection but by observation and inference (1947). The same concern with objective measurement of subjective experience characterized George Kelly's approach to assessment. This chapter considers some of the main efforts that have been made by psychologists working at this level to study experience objectively. Beyond finding such methods, they also have developed new strategies for personal growth and awareness. Although much of this work was begun more than 50 years ago, it has remained important in current work (e.g., Higgins, 1996; Lamiell, 1997; Leary & Tangney, 2003).

Why Self Matters: Consequences of Self-Discrepancies

The theories presented in Chapter 8 make a compelling case for the importance of subjective experience and of the self for personality. Going beyond theory to data, contemporary research is showing the powerful role of the individual's self-concepts, self-perceptions, and feelings about himself or herself (e.g., Hoyle, Kernis, Leary, & Baldwin, 1999; Leary & Tangney, 2003). The results show that the internal experiences on which the theorists in the last chapter focused change what people become and influence the types of problems and coping strategies they develop.

One of the most notable findings is that people experience different types of discrepancies between different aspects of the self, and these discrepancies influence their subsequent emotions and behaviors in predictable ways (see Table 9.1). It was Carl Rogers who first suggested that, beginning early in life, discrepancies develop between various mental representations of the self. For example, your **actual self,** that is, the representation of yourself as you are (e.g., a good basketball player) may be discrepant with your **ideal self,** the representation of who you would like to be (a great basketball player). Likewise, the actual self may be discrepant with the **ought self,** the representation of who you believe you should be (e.g., a doctor).

According to E. Tory Higgins (1987), such discrepancies may be experienced not only from one's own vantage point, but also from that of significant others, such as

9.1 Describe three types of self-discrepancies and the emotions to which they give rise according to Higgins.

E. Tory Higgins

TABLE 9.1   Types of Concepts about the Self

Self-Concept	Definition	Example
Actual	One's representation of oneself: the belief about the attributes one actually has	I am a caring and warm person, athletic and attractive
Ideal	One's representation of who one would hope, wish, or like to be: the beliefs about the attributes one would like to have ideally	I would love to be generous and giving, successful, popular, brilliant, and loved
Ought	One's representation of who one should be, or feels obligated to be: the beliefs about the attributes one is obligated to have, i.e., that are one's duty to possess	I should be more ambitious and tough, hardworking and disciplined

Note: In addition to one's own standpoint, each concept also can be represented from the viewpoint of a significant other. For example, your perception of who your father thinks you should be (e.g., strong-willed instead of caring) is an "ought/other" representation of the self.
Source: Based on Higgins, E. T. (1987). Self-discrepancy: A theory relating self and affect. *Psychological Review, 94,* 319–340.

a parent or an older sibling (see Table 9.2). An example of a discrepancy between the actual self and ought self would be a disagreement between the self you believe yourself to be and the "ought self" that you perceive your father thinks you should be. This could take a form like the thought: "I'm afraid my father will be angry with me because I didn't work as hard on my exams as he believes I should." That in turn can lead to feelings of agitation, such as fear and worry. In contrast, a perceived discrepancy between the actual self and ideal self makes an individual more vulnerable to feelings of dejection, such as disappointment and dissatisfaction. In short, Higgins

TABLE 9.2   Illustrative Self-Discrepancies

Types of Self-Discrepancies	Induced Feelings	Example
Actual/Own: Ideal/Own	Disappointment and dissatisfaction	I'm dejected because I'm not as attractive as I would like to be.
Actual/Own: Ideal/Other	Shame and embarrassment	I'm ashamed because I fail to be as kind a person as my parents wished me to be
Actual/Own: Ought/Own	Guilt and self-contempt	I hate myself because I should have more willpower
Actual/Own: Ought/Other	Fear or feeling threatened	I'm afraid my father will be angry with me because I didn't work as hard as he believes I should

Note: "Own" refers to the person's vantage point, "other" to the vantage point of another significant person (e.g., the father).
Source: Based on Higgins, E. T. (1987) Self-discrepancy: A theory relating self and affect. *Psychological Review, 94,* 319–340.

proposed and found evidence that various discrepancies like those shown in Table 9.2 have predictable emotional consequences.

Furthermore, different emotions lead to different patterns of coping with the perceived self-discrepancies (Higgins, 1987). Suppose for example that Gary experiences a discrepancy between the actual self and ought self and feels distressing negative emotions. If the discomfort associated with these negative feelings becomes too great, he may try to reduce or eliminate it in various ways. For example, to deal with a discrepancy between the perceived actual self and the own ideal self, one can reevaluate the negative interpretation of past painful events. For example, the high school student who feels rejected by others ("nobody likes me") because he was not elected class president then might think about the many close friends he has to reduce the negative feelings produced by the experienced actual-ought self-discrepancy.

Alternatively, to remove discrepancies people may change their actual behavior to match an important standard. Suppose an undergraduate studies very little for a midterm and receives a low grade on the exam. For the final, she may study very hard. In doing so, she relieves her guilt for not living up to what she herself perceives to be her responsibility to work diligently in college and receive exemplary grades. In this approach, regardless of the form the change actually takes, the motivation for the change arises from the conflicts each individual feels among his or her various representations of the self. These discrepancies, according to Higgins, cause specific types of emotional discomfort that individuals are motivated to reduce as best they can.

The key point from this is that the work on self-discrepancies makes clear that the self matters. It matters because the internal experiences that people have when they perceive such discrepancies are consequential for the emotions they experience and for the coping patterns they use to try to deal with them. For the scientist concerned with the experienced self and the internal view, the work by Higgins and colleagues also is interesting because it offers a way to study these experiences rigorously using the methods of the science. This work is also having practical applications for understanding serious personality problems, like eating disorders, as discussed in In Focus 9.1.

9.2 What kinds of methods are used to cope with the various self-discrepancies?

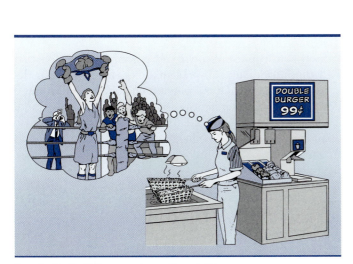

Actual–Ideal self discrepancies can be painful.

EFFECTS OF SELF-DISCREPANCY: ANOREXIA

Alarmed by the growing number of cases of eating disorders especially among adolescents and young adult women, researchers and clinical psychologists have been trying to understand this self-destructive, potentially fatal behavior. Explanations offered for eating disorders range from maladaptive interactions among family members to perceptual distortions in the way victims actually see their own bodies. Eating disorders also may reflect discrepancies among the individual's self-concepts, which impact his or her self-evaluation and can disturb self-regulation (Strauman et al., 1991). This explanation stems from self-discrepancy theory (Higgins, 1987) in which different kinds of negative feelings typically are associated with particular kinds of perceived self-discrepancies. As discussed in the text, according to the theory, **actual/ideal discrepancies** make one vulnerable to feelings of dejection. In contrast, **actual/ought discrepancies** make an individual susceptible to feelings of agitation. In Higgins' theory, the person's **self-evaluative standards** (self-guides) are represented by the ideal self and the ought self. When the actual self falls short of these self guides, the individual becomes prone to negative emotions and motivational states. These negative feelings, in turn, can produce distress and maladaptive behavior (Strauman and associates, 1991). Some of the negative effects may be seen in eating disorders.

Thus, **anorexic behavior** (self-starvation) has been linked to actual/ought discrepancies (Strauman et al., 1991). According to these researchers, anorexic behaviors tend to be more characteristic of individuals whose actual self-concepts are discrepant from their representations of how significant others believe they ought to be. The anorexic behaviors seem to be part of a pattern of self-punitive, self-critical efforts to meet what these individuals see as the demands and expectations of significant others. In contrast, bulimic eating problems, such as binging, tend to be more associated with discrepancies between people's actual self-concept and their own ideal self-concepts. Emotionally, those suffering from bulimia experience dejection and related feelings as a reflection of the discrepancy between the body types they perceive themselves to have and their ideals. Interestingly, this seems to be the case irrespective of the person's actual body mass.

Self-discrepancy theory also suggests another insight into eating disorders. The fact that eating disorders are much more prevalent among adolescent girls than boys has led to speculation that women are more commonly socialized to derive feelings of self-worth from their physical appearance. Because modern society mainly considers thin women to be beautiful, many women feel compelled to constantly monitor their body weight and thus are more prone to eating disorders. An alternative explanation, however, is that girls typically are more restricted and controlled than boys. Consequently, prior to adolescence they develop more rigid self-guides, that is, more clearly defined ideal and ought selves. The stronger the self-guides, the more vulnerable the individual is to experience self-discrepancy and negative feelings (Strauman et al., 1991, p. 947). According to the theory, then, women are more likely than men to develop disordered eating habits.

9.3 How has self-discrepancy theory contributed to our understanding of anorexia?

The View Through the Person's Eyes

Phenomenological study begins with the person's own viewpoint. To approach that viewpoint, one may begin with the individual's self-presentation, as expressed in the person's own self-description. Some of the raw data of phenomenology were illustrated in the self-description recorded by Gary W. when he was asked to describe himself as a person, which was presented in Chapter 2, In Focus 2.1. If you look back at it, you will see that it gives at least a preliminary sense of Gary looking at himself and his world through his own eyes, using his own phrases, beginning to show some of his perceptions, thoughts, beliefs, and feelings.

How can we begin to interpret Gary's self-portrait? It is possible to proceed in terms of one's favorite theory, construing Gary's statements as reflections of his traits, or as signs of his dynamics, or as indicative of his social learning history, or as clues to the social forces that are molding him. But can one also make Gary's comments a bridge for understanding his private viewpoint, for glimpsing his own personality theory and for seeing his self-conceptions?

Because each of us is intimately familiar with his own conscious, perceived reality, it may seem deceptively simple to reach out and see another person's subjective world. In fact, we of course cannot "crawl into another person's skin and peer out at the world through his eyes," but we can "start by making inferences based primarily upon what we see him doing rather than upon what we have seen other people doing" (Kelly, 1955, p. 42). That is, we can try to attend to him rather than to our stereotypes and theoretical constructs.

A most direct way to inquire about another person's experience is to ask him, just as Gary was asked, to depict himself. Virtually all approaches to personality have asked people for self-reports. Usually, these reports have served primarily as cues from which to infer the individual's underlying personality structure and dynamics. Perhaps because of the assumption that people engage in extensive unconscious distortion, the client's own reports generally have been used as a basis for the clinician (or the test) to generate inferences and predictions about her, rather than as a means of conveying the client's view of herself.

Uses of Self-Assessments

Can people be "experts" about themselves? Can their reports serve as reliable and valid indices of their behavior? For example, in his self-appraisal Gary predicts that he can succeed in the job for which he is being considered. Is this self-assessment accurate, or is it a defensive hope?

One way to address these questions is to examine whether people's self-assessments can predict their own future behavior. To establish the utility of a person's direct report about himself, you must compare it with the predictions about him that can be made from other data sources. For example, you may compare the individual's self-reports with the statements drawn from sophisticated psychometric tests or from well-trained clinical judges who use such techniques as the interview and the projective test to infer the individual's attributes.

9.4 What evidence exists that simple self-reports can have predictive validity that equals or exceeds that of more complex methods?

It has been a surprise for many psychologists to learn that simple self-reports may be as valid as, and sometimes better predictors than, more sophisticated, complex, and indirect tests designed to disclose underlying personality. In a pioneering study, researchers tried to predict future adjustment for schizophrenic patients (Marks, Stauffacher, & Lyle, 1963). They found that simple self-reports on attitude scales yielded better predictions than did psychometrically more sophisticated scales. Such attitude statements have also been one of the best predictors of success in the Peace Corps; they have been more accurate than far more costly personality inferences. Interviews and pooled global ratings from experts did not prove nearly as accurate as self-reports were (Mischel, 1965). In sum, useful information about people may be obtained most directly by simply asking them (e.g., Cantor & Kihlstrom, 1987; Emmons, 1997). These conclusions seem to hold for such diverse areas as college achievement, job and professional success, treatment outcomes in psychotherapy, rehospitalization for psychiatric patients, and parole violations for delinquent children (e.g., Emmons, 1997; Mischel, 1981; Rorer, 1990).

In sum, under some conditions, people may be able to report and predict their own behavior at least as accurately as experts. Of course, people do not always predict their own behavior accurately. Sometimes individuals lack either the information or the motivation to foretell their own behavior, or are motivated to not reveal it even if they know it. If a criminal plans to steal again, we cannot expect him to say so to the examining prosecutor at his trial. Moreover, many future behaviors may be determined by variables not in the person's control (for example, other people or accidents). The obtained findings do suggest that self-estimates and self-predictions are useful assessment tools.

The Q-Sort Technique

One problem with self-descriptions, however, is that different individuals may use different words, phrases, and expressions to describe the same experience, and therefore it becomes difficult to compare one person's self-description with another person's. To compare people, they need to use the same standard language to describe themselves. An especially useful technique for achieving this goal is the **Q-technique** or **Q-sort,** a tool that also has been used in work at other levels of analysis, including the trait-dispositional (Block, 1961, 1971). This technique consists of many cards, each with a printed statement such as "I am a submissive person," "I am likable," and "I am an impulsive person." Or the items might be "is a thoughtful person," "gets anxious easily," "works efficiently."

The Q-sort may be used for self-description, for describing how one would like to be (the ideal self), or even to describe a relationship. For a self-sort, clients would be instructed to sort the cards to describe themselves as they see themselves currently, placing cards in separate piles according to their applicability, ranging from those attributes that are least like them to those that are most like them. For example, the terms that Gary W. had indicated as most self-descriptive were: "haughty, determined, ambitious, critical, logical, moody, uncertain."Or people might be instructed to use the cards to describe the person they would most like to be—their ideal person. To describe a relationship, they would sort the cards into piles ranging from those that are most characteristic of the relationship to those least characteristic. As these examples indicate, the method asks one to sort the cards into a distribution along a continuum from items that are least characteristic (or descriptive) to those that are most characteristic of what one is describing.

The Q-sort is also used to describe the characteristics associated with successful performance in a given task. For example, one can find the profile of qualities "most characteristic" of people who succeed in a particular situation. One can then search for those individuals who best match that profile when trying to predict who will or will not do well in that type of situation (Bem & Funder, 1978). Likewise, Q-sorts are often used to characterize changes in development, for example by comparing people's Q-sort profiles at different stages (Mischel, Shoda, & Peake, 1988).

At this level of analysis, for many psychologists, self-statements from Q-sorts may be of interest in themselves. The phenomenologist simply wants to see the person's self-characterization for its own sake. Unlike other measures that compare individuals with each other along a given dimension to obtain *between-person* differences, the goal of Q-sort assessment is the pattern of the various characteristics *within each person*.

Because the Q-sort requires arranging the cards into a predetermined distribution (e.g., 5 cards in the "least characteristic," 5 cards in the "most characteristic," and 20 cards in the "neither characteristic nor uncharacteristic" piles), if one averages across all the characteristics, everyone will have the same score. This is like giving different painters tubes of red, blue, green, and yellow watercolor paint, and they are asked to paint while making sure they use up all the paint they are given. If you "average" all the colors on each canvas, they will be all the same. What is different is the spatial configuration of the colors on the canvas. Similarly, what is different in the Q-sort depiction of an individual is the *arrangement* among the set of characteristics. With this

9.5 Describe the Q-sort technique, and the various aspects of the self that it can assess.

focus on intraindividual arrangement or configuration, Q-sort has been used to determine how similar, or dissimilar, persons' descriptions are of themselves as they really view themselves ("real self") and as they would like to be ("ideal self"). For example, using a Q-sort that contains 100 cards, individuals can "paint" the picture of their real self. Using the same set of 100 cards, they also paint the picture of their ideal self. Now, the similarity between these two pictures of self can be indexed by the correlation between the two, because both real and ideal selves are described using the same 100 descriptors. As you'll see later, often the degree of similarity between these two descriptions is informative.

Interviews

Most modern phenomenologists have recognized that self-reports may not reveal everything important about behavior and may not give a complete picture of personality. Persons may be conscious of the reasons for their behavior but be unable or unwilling to report them, for example, if they are uncomfortable or ashamed about aspects of their own feelings, perceptions, and behaviors. Or they may not be conscious of all of their experiences, in which case, they cannot communicate them no matter how hard they try. Rather than considering that as a limitation, phenomenologists such as Rogers focus on the person's frame of reference as an important vantage point for understanding him or her. To illustrate this point, imagine two people who, objectively, are both good singers. But while the first thinks she is a great singer, the second thinks he is often out of tune and believes that others are only being polite when they say he sings well. Obviously these differences in their subjective views of themselves make them quite different people even though both sing equally well.

9.6 How is the interview used to assess personality characteristics?

The psychologist's task, in Rogers' view, is to provide conditions that are conducive to growth and that facilitate free exploration of feelings and self in a therapeutic context. This is because one cannot expect people to be honest about themselves when they fear that their statements may incriminate them or lead to negative decisions about their future. In order to become more aware of and articulate private feelings, one needs an unthreatening atmosphere that reduces anxieties and inhibitions, and fosters self-disclosure (Jourard, 1967; Lietaer, 1993). Phenomenologically oriented psychologists therefore try to create conditions of acceptance, warmth, and empathy in which the individual may feel more at ease for open self-exploration (Vanaerschot, 1993). These conditions of acceptance are illustrated vividly in "client-centered" (Rogerian) therapy, discussed later in this chapter.

Rogerians have not only tried to create conditions conducive to personal growth, they also have studied these conditions through methods such as the interview (e.g., Rogers, 1942; Rogers & Dymond, 1954). They use the interview to observe how the individual interprets himself and his experiences, regardless of the validity of the data he provides. This is done in an atmosphere conducive to genuine self-disclosure in which self-revelation and honest self-reports are actively encouraged (e.g., Cantor, 1990; Fodor, 1987; Jourard, 1974).

The Semantic Differential

The **semantic differential** is used to study what different stimuli, events, or other experiences mean to the individual—that is, their personal significance. If you take the semantic differential you are asked to indicate the meanings of diverse words, phrases, and concepts by rating them on many scales (Osgood, Suci, & Tannenbaum, 1957). You

9.7 Describe the semantic differential method. What three underlying factors consistently emerge from such ratings?

are given a stimulus word like "feather," or "me," or "my father," or a phrase like "my ideal self," and are asked to rate each stimulus on a seven-point, bipolar scale. Polar adjectives like "rough-smooth" or "fair-unfair" are the extremes of each scale, and you mark the point that most nearly indicates the meaning of the stimulus concept for you. For example, you might be asked to rate "my ideal self" on scales like those shown in Table 9.3. To see what that is like, you should try it for yourself, both for the concepts suggested and for any others you find interesting for yourself. The technique is both objective and flexible. It permits investigation of how people describe themselves and others, as well as how special experiences (for example, psychotherapy) affect them.

A great deal of research has repeatedly indicated that three main factors tend to emerge when the results are analyzed. A primary **evaluative (good-bad) factor** seems to be the most important (Kim & Rosenberg, 1980). In other words, evaluations in such terms as "good-bad" enter most extensively into how people characterize themselves, their experiences, and other people (Ross & Nisbett, 1991). The two other factors are **potency,** represented by scale items like hard-soft, masculine-feminine, strong-weak, and **activity,** tapped by scales like active-passive, excitable-calm, and hot-cold (Mulaik, 1964; Vernon, 1964).

Nonverbal Communication

9.8 What is the significance of nonverbal behavior in assessing inner experiences?

Techniques like the semantic differential and the Rep Test sample what people say—that is, their verbal behavior. But significant communication among people is often nonverbal—it can involve facial expressions, movements, and gestures. Nonverbal expressions have intrigued psychologists of many theoretical orientations who are interested in the individual's perceptions and inner states. Researchers explore the possible meanings and effects of such nonverbal expressions as eye contact and the stare.

For example, when an interviewer evaluates participants positively they increase eye contact with him but when he evaluates them negatively, they decrease eye contact

TABLE 9.3   Examples of Concepts and Rating Scales from the Semantic Differential

Concepts whose meanings are rated:
  My Actual Self
  My Ideal Self
  Masculinity
  Foreigner
  Mother

Scales for rating the meaning of each concept:

Strong	____:____:____:____:____:____:____	Weak
Pleasant	____:____:____:____:____:____:____	Unpleasant
Hard	____:____:____:____:____:____:____	Soft
Safe	____:____:____:____:____:____:____	Dangerous
Fair	____:____:____:____:____:____:____	Unfair
Active	____:____:____:____:____:____:____	Passive

Note: As an exercise, provide these ratings yourself for each concept.
Source: Based on Osgood, C. E., Suci, G. J., & Tannenbaum, P. H. (1957). *The measurement of meaning.* Urbana, IL: The University of Illinois Press.

with him (Exline & Winters, 1965). The effects of eye contact seem to interact with the verbal content conveyed in the relationship. One study varied whether an interviewer looked at the participant frequently or hardly at all, and whether the conversation was positive or threatening (Ellsworth & Carlsmith, 1968). When the verbal content was positive, more frequent eye contact on the part of the inteviewer produced more positive evaluations of the interviewer. In contrast, when the verbal content was negative, more frequent eye contact produced more negative evaluation.

Although much is still unknown about nonverbal communication, many results have been encouraging. It has been shown, for instance, that "when people look at the faces of other people, they can obtain information about happiness, surprise, fear, anger, disgust/contempt, interest, and sadness. Such information can be interpreted, without any special training, by those who see the face . . ." (Ekman, Friesen, & Ellsworth, 1972, pp. 176–177). In short, phenomenological, "inner" experiences may be visible in the form of nonverbal behaviors.

Studying Lives from the Inside

The Whole Person: Psychobiography.   The phenomenological approach has many of its deepest roots in the psychotherapies of theorists like Rogers and Kelly, but it also has been extended in other directions. Most notably, the approach has been adapted and combined with other methods to study lives in depth and over long periods of time (Runyan, 1997). Called **psychobiography,** the intensive study of individual lives has become a specialty area in its own right. These studies attempt to provide a comprehensive psychological understanding of one person, often selecting public figures like Adolf Hitler or Ghandi. As its advocates note, personality psychology has many sides, and does not have to be confined to quantitative comparisons between people or groups (Lamiell, 1997; Runyan, 1997). Instead, the study of lives focuses on one person at a time, and tries to cover the whole of his or her life in all its complexity over the life course. The methods employed borrow from biography, history, and other social sciences as well as from psychology and the phenomenological approach. As one of its most enthusiastic practitioners put it, there is a "softer human science end of psychology" whose advice to students is "Learn all you can about people and lives, including yourself" (Runyan, 1997, p. 61).

9.9 How have the methods of psychobiography and personal narrative been used to study people's lives?

Life Stories: Personal Narratives.   In a closely related vein, but with its own guiding philosophy and methodology, Dan McAdams (1999) has developed an approach to understanding persons and the ways in which they construct their lives. His focus is on the **personal narratives**—the stories—that people tell themselves as they try to make sense of their own lives and experiences. McAdams (1999, p. 488) proposes that "The I apprehends experience in narrative terms, casting the Me as one or more characters in an ongoing sequence of scenes." Essentially the I is a storyteller, and the story each person tells—both to himself or herself and to other people varies with time and circumstances. The stories you tell in a college application, a job interview, when building a relationship with a potential romantic partner, or in a private diary entry to yourself will vary. The contents and structure of the story will change, and as life is lived the stories of one's life are progressively edited, revised, rewritten, reconstructed. The core of this message is that regardless of their validity by any external standard, these stories people tell about themselves are worth hearing, and personality psychologists— as well as nonprofessionals—can benefit by listening closely.

▶ ENHANCING SELF-AWARENESS:
ACCESSING ONE'S EXPERIENCES

Given that people's subjective experiences of the self have far-reaching implications for their emotions, behavior, and their well-being, it should not be surprising that changing the subjective views of the self can have significant impact on individuals' well-being. Probably the most dramatic and controversial manifestation of the effort to alter the subjective experience of self on a massive scale was by means of psychedelic drugs, and in some ways it created at least a short-term cultural revolution in much of the western world. Initially, drugs such as psilocybin and LSD were advocated most energetically by Timothy Leary and Richard Alpert when they were psychologists at Harvard University in 1961 and 1962. In the 1960s, the "mind-expanding" movement through drug-induced "trips" or psychic "voyages" gained many enthusiastic participants throughout the western world. Such drugs as LSD undoubtedly produce major alterations in subjective experience, including the intensification of feelings (Leary, Litwin, & Metzner, 1963), but enthusiasm for them was soon tempered by the recognition that they entail serious risks. The trend became to search for greater awareness without the aid of any drugs by means of psychological experiences and changes within the self and in how life is lived.

Historically, routes to increasing awareness that rely on psychological experiences rather than on drugs include meditation (Ornstein, 1972), encounter groups, and "marathons" of the type developed originally at the Esalen Institute in Big Sur, California (Schutz, 1967). While meditative techniques were based mainly on Eastern religious sources (Ornstein & Naranjo, 1971), the encounter or "sensitivity training" movement drew on various role-play and psychodrama techniques, existential philosophy, and Freudian psychodynamic theory.

The resulting syntheses were seen in the ideas of the **Gestalt therapy** of Fritz Perls (1969), in the efforts to expand human awareness and to achieve "joy" and true communication (Schutz, 1967), and in the pursuit of "peak experiences" and "self-actualization" (Maslow, 1971). In its early versions, Gestalt therapy included confrontations and "encounters" that quickly challenged the person's self-reported experiences, sometimes interpreting them as superficial and defensive. Often the "leader" tried rapidly and directly to stimulate and probe the deeper feelings that might underlie what the person disclosed (Polster & Polster, 1993). In more recent versions, there is less rapid and dramatic confrontation and a slower, more empathic attempt to explore the person's internal experiences in his or her own terms. The aim is to focus awareness on what is being felt fully and honestly. The process of enhancing both self-awareness and interpersonal awareness becomes the center of the therapeutic relationship, encouraging the person to be more closely in touch with what is experienced and freer to experiment interpersonally (Fodor, 1987; Wheeler, 1991).

9.10 Indicate how gestalt therapy and T-groups are used to explore and alter internal experience? What anecdotal and scientific evidence is there for their effectiveness?

Group Experiences

As part of the search for growth and expanded awareness, in the 1960s and 1970s a variety of group experiences became popular. Encounter groups have had many different labels, such as **human-relations training group (T-group),** and **sensitivity training group,** but in this discussion the focus is on their common qualities. Schutz (1967) in his book *Joy* noted that encounter group methods involve doing something, not just talking. In this quest, he advocated a host of group methods that include body exercises, wordless meetings, group fantasy, and physical "games." Elliot Aronson (1972, p. 238), a pioneer of experimental social psychology and a major contributor to the science, described what is learned in group experiences this way: ". . . in a psychology course I learn how people behave; in a T-group I learn how I behave. But I learn much more than that: I also learn how others see me, how my behavior affects them, and how I am affected by other people." Referring to the process through which such learning occurs, Aronson (p. 239) emphasizes learning-by-doing; ". . . people learn by trying things out, by getting in touch with their feelings and by expressing those feelings to other people, either verbally or nonverbally." Such a process requires an atmosphere of trust so that members learn not how they are "supposed" to behave but rather what they really feel and how others view them.

At a theoretical level, the encounter group movement involved a complex synthesis of both Freudian and Rogerian concepts with a focus on nonverbal experiences and self-discovery. The psychodynamic motivational framework was largely retained and was used in many of the interpretations, but it was implemented by direct "acting-out" procedures for expressing feelings through action in the group, by body contact designed to increase awareness of body feelings, and by games to encourage the expression of affection and aggression. Thus, many of Freud's and Rogers' ideas were transferred from the consulting room to the group encounter, and from verbal expression to body awareness and physical expression. Indeed, Carl Rogers (1970) developed and extended many of his theoretical concepts to the encounter experience, and he became one of its leading advocates. Rather than talking about impulses, feelings, and fantasies, the individual is encouraged to act them out in the group. For example, rather than talk about repressed feelings of anger toward his father, the individual enacts his feelings, pummeling a pillow while screaming "I hate you, Dad, I hate you." Many people have reported positive changes as a result of group experiences. To illustrate, consider this testimonial cited by Rogers (1970, p. 129):

> I still can't believe the experience that I had during the workshop. I have come to see myself in a completely new perspective. Before I was "the handsome" but cold person insofar as personal relationships go. People wanted to approach me but I was afraid to let them come close as it might endanger me and I would have to give a little of myself. Since the institute I have not been afraid to be human. I express myself quite well and also am likeable and also can love. I go out now and use these emotions as part of me.

While such reports were encouraging, they are not firm evidence, and they were offset in part by reports of negative experiences (Lieberman et al., 1973). Some behavior changes did seem to emerge, but their interpretation is beset by many methodological difficulties (Campbell & Dunnette, 1968). When careful control groups were used, doubt was raised if the gains from encounter experience reflect more than the enthusiastic expectancies of the group members. For example, people in weekend encounter groups showed more rated improvement than did those who remained in an at-home control group, but improvement in the encounter groups did not differ from that found in an on-site control group whose participants believed they were in an encounter group although they only had recreational activities (McCardel & Murray, 1974).

Nevertheless, a number of experimental studies indicated specific changes that may occur in some types of groups. These changes include a decrease in ethnic prejudice (Rubin, 1967; Saley & Holdstock, 1993), an increase in empathy (Dunnette, 1969) and susceptibility to being hypnotized (Tart, 1970), and an increased belief by individuals that their behavior is under their own control (Diamond & Shapiro, 1973). This evidence is accompanied by a greater awareness on the part of therapeutic group enthusiasts that not all groups are for all people, that bad as well as good experiences may occur (Bates & Goodman, 1986), and that coerciveness in groups is a real hazard that needs to be avoided (Aronson, 1972).

Particularly important is the finding that **self-disclosure** and sharing of stressful, traumatic experiences, either in groups or in other forms (e.g., in diaries) can have dramatically beneficial effects on well-being and health (Pennebaker, 1993, 1997). An illustration of the value of sharing traumatic experiences comes from work with patients suffering from advanced breast cancer (Spiegel et al. 1989). The patients were divided into two conditions, both conditions receiving the usual medical care for cancer. However, people in the intervention received weekly supportive group therapy for a period of one year, sharing their experiences with others openly. The patients who had this experience survived almost twice as long (37 months) as those who did not.

Meditation

As Eastern cultures have long known, and as Western cultures have only recently learned, meditation can have powerful effects on subjective experience. The term "meditation" refers to a set of techniques that are the product of another type of psychology, one that aims at personal rather than intellectual knowledge. As such, the exercises are designed to produce an alteration in consciousness. It is a shift away from the active, outward-oriented mode and toward the receptive and quiescent mode. Usually it is a shift from an external focus of attention to an internal one (Ornstein, 1972, p. 107).

Students, businessmen, athletes, ministers, senators, and secretaries were among the more than 600,000 people in the United States alone who enthusiastically endorsed one version of meditation called **transcendental meditation (TM).** Introduced into the United States in 1959 by Maharishi Mahesh Yogi, a Hindu monk, TM has changed over the years in the public's mind from counterculture fad to mainstream respectability. TM is defined as a state of restful alertness from which one is said to emerge with added energy and greater mind-body coordination. It is practiced during two daily periods of 20 minutes each. The meditator sits comfortably with eyes closed and mentally repeats a Sanskrit word called a **mantra.**

Maharishi, his movement, and more recent followers maintain that the technique of TM can be learned only from specially trained TM teachers who charge a substantial fee. One of the pioneer researchers into the effects of TM, Herbert Benson, a Harvard cardiologist, disagreed (1975). He believed that the same kind of meditation can be self-taught with a one-page instruction sheet, achieving the same measurable results (see Table 9.4).

TABLE 9.4   The Mechanics of Meditation

1. Sit in a comfortable position in a quiet environment, eyes closed.
2. Deeply relax all muscles.
3. Concentrate on breathing in and out through your nose. As you breathe out, repeat a single syllable, sound, or word such as "one" silently to yourself.
4. Disregard other thoughts, adopt a passive attitude, and do not try to "force" anything to happen.
5. Practice twice daily for 20-minute periods at least two hours after any meal. (The digestive process seems to interfere with the elicitation of the expected changes.)

Source: Adapted from Benson, H. (1975). *The relaxation response.* New York: Morrow.

Scientific research indicates that there are direct physical responses to meditation (Alexander, Robinson, Orme-Johnson, Schneider, & Walton, 1994). These changes include decreased rate of metabolism (decreased oxygen consumption), and an increase in alpha waves (slow brain-wave patterns). Although meditation was initially introduced as a technique for expanding consciousness, the current emphasis is on reducing stress, lowering blood pressure, alleviating addictions (O'Connell & Alexander, 1994), and increasing energy and powers of concentration.

The research publicized by the TM movement is open to criticism. The fact that many of the researchers are dedicated meditators themselves makes it possible that their results and interpretations may be unintentionally biased. Perhaps most important, the characteristic brain-wave pattern that the TM movement claims to be a sign of the "alert reverie" produced by meditation does not appear to be unique to meditation (Pagano, Rose, Stivers, & Warrenburg, 1976). It can occur, for example, in hypnosis when a state of deep relaxation is suggested. And Benson's (1975) relaxation technique (see Table 9.4) produces the same reductions in oxygen consumption and respiration rate that are produced during transcendental meditation without the expense and the complex rituals of meditation training. In reply, those committed to the movement argue that TM produces a wide range of more fundamental changes than relaxation, including a tremendous improvement in the quality of life. Perhaps because the subjective experience produced by meditation may appear unique to the meditators, they often continue to claim that meditation is a distinct state of consciousness in spite of the contradictory physiological evidence.

9.11 What physiological and psychological effects occur from meditation?

The Person's Experience and the Unconscious

For many years, most psychologists did not consider subjective experiences as phenomena of interest in their own right. They preferred, instead, to infer what dispositions and motives might underlie the person's behavior. The historical neglect of the individual's viewpoint probably has many reasons. One was the belief that because of unconscious distortions and defenses, people's self-appraisals were biased and inaccurate. Some psychologists thus refrained from studying the person's perceptions, concepts, and intentions because they felt such data were not really scientific. But it is entirely legitimate philosophically and logically to take account of individuals' reported subjective perceptions—the rules they use and the reasons they give to explain their own actions (Lamiell, 1997; T. Mischel, 1964). It is not legitimate, however, to assume that these rules and reasons are useful bases for predicting what individuals will do—their behavior—in other situations. The links between persons' reported feelings and beliefs and their other behaviors have to be demonstrated empirically.

9.12 Why have some psychologists minimized the clinical and scientific value of subjective self-reports?

Although the phenomenological orientation focuses on the individual's viewpoint, in some of its variations it also seeks to infer his or her unconscious characteristics and conflicts. For example, Carl Rogers in some of his formulations (1963) has emphasized integration, unity, and achieving congruence with their "inner organismic processes." Rogers thought that these organismic processes were often unconscious. As a result of socialization procedures in our culture, persons often become dissociated, "consciously behaving in terms of static constructs and abstractions and unconsciously behaving in terms of the actualizing tendency" (Rogers, 1963, p. 20).

To the extent that psychologists accept the idea that unconscious processes are key determinants of personality, and rely on clinical judgment to infer them, they face all the challenges previously discussed in the context of psychodynamic theory. But methodological difficulties should not deter psychologists from listening more closely

Family therapy analyzes the transactions in a relationship, helping each partner to see the viewpoint of the other.

to what their "participants" can tell them; the caution applies to the interpretation process, not to the value of listening and empathizing. **Family therapies** have been influenced extensively by this perspective, and try to sensitize the members of a family to see each other's viewpoints and take them into account in their daily transactions within the family system.

Accessing Painful Emotions: Hypnotic Probing

9.13 How is hypnosis used to help people access dissociated experiences? What limitations of hypnotic probing have emerged?

In one direction, hypnosis is being used to help individuals access the painful feelings and memories that may become difficult to recall in the aftermath of traumatic experiences (Spiegel et al., 1994). Such mental avoidance may follow potentially life-threatening events and losses of the sort experienced by victims of disasters, violent crimes, or war. But they may also occur in the course of everyday life especially in early childhood when people have experiences and feelings that they may find traumatic and too difficult to deal with, as Freud stressed in his theory. The dissociated state itself is characterized by being out of touch with one's feelings, sometimes with amnesia, depression, a sense of numbing, detachment, and withdrawal.

In this type of therapy, hypnosis may be used within a highly supportive, structured setting to induce a trance state. While the person is in this state, memories of the painful events may be re-experienced in the safe therapeutic setting. Recall of the traumatic events is stimulated by the careful guidance and suggestions of the therapist, designed to elicit vivid images that make the memories more accessible (Spiegel, 1981, 1991). It may then become possible for the person to try to place the traumatic experience into perspective and to accept it emotionally. In this process, the person is encouraged to grieve for what has been lost rather than continue to "split off" the painful feelings and run away from them. According to this approach ". . . a loss not grieved leads to a life not lived— to a kind of numbing or withdrawal and depression . . ." (Spiegel, 1981, p. 35).

While this approach has received much attention, in recent years its limitations and hazards also have become clear. Hypnosis is not a magic route to the truth: what is re-experienced and reported is subject to all sorts of influences, including implicit suggestions from the therapist and expectations about what will be revealed. As a result, the reports that follow are not at all above suspicion (Loftus, 1994, 1993).

Peering into Consciousness: Brain Images of Subjective Experiences

Interestingly, it was Freud who initially thought that hypnosis would provide a window into the unconscious, but who soon decided to abandon it in favor of other techniques—a choice that now seems like a mistake to some of his critics. Especially exciting is that new functional brain imaging techniques are helping to specify the brain locations that become activated during hypnotic concentration and that underlie different types of mental states and events (e.g., Spiegel, 1991; Schachter, 1995). For example, these imaging techniques show that the practice of meditation activates distinctive brain structures involved in attention and control of the autonomic nervous system (e.g., Lazar and associates, 2000).

The Value of Self-Disclosure about Subjective Experiences

Over the course of many years, Jamie Pennebaker and his colleagues have been studying how writing about stressful and traumatic experiences influences people's mental and physical well-being (Pennebaker & Graybeal, 2001). In Pennebaker's typical method, volunteer participants, usually college undergraduates, are brought into the laboratory and asked to write about either traumatic experiences from their past or superficial topics (for comparison purposes) for 15 minutes a day for 3 to 4 consecutive days.

9.14 Summarize the research of Pennebaker on self-disclosure of painful events and its effects? How does Pennebaker account for the positive results?

As was noted in the discussion of psychodynamic process, these studies demonstrated that when people write about their traumatic and stressful experiences, they dramatically improve their well-being and health. Relative to people who are told to write about superficial topics, people who write about their emotional experiences subsequently make fewer visits to the doctor, at least in short-term follow-up studies (Smyth, 1998; Pennebaker, Colder, & Sharp, 1990; Pennebaker, Kiecolt-Glaser, & Glaser, 1988). They also show improved hormonal activity and immune function (Petrie et al. 1995). In addition, people who self-disclose their subjective experiences by writing about them tend to show a number of other behavioral improvements. College students who self-disclosed improved their grades (Pennebaker and associates, 1990), and unemployed workers got new jobs faster (Pennebaker, 1997).

Contrary to classical psychoanalytical theory, Pennebaker does not believe that the power of writing is a function of catharsis or the release of pent-up negative feelings. Instead, he and his collaborators have evidence that writing about emotional experiences influences people's health and behavior by changing the way they think about their emotions and themselves. Writing allows them to construct mental narratives—literally stories—of their experiences. The specific psychological processes that make writing about emotional experiences lead to such positive outcomes still needs more research and many studies are under way. Finally, note also that while it helps to write about one's negative feelings, brooding and ruminating about them can have the opposite effect, as discussed in In Focus, 9.2.

CAUTION: RUMINATION CAN INCREASE DEPRESSION

Being in touch with, and expressing, what you really think and feel, including with regard to painful emotions, may have considerable value, but thinking about the negative too much also has its hazards. Although many people believe that they should focus inward and attend in detail to their feelings, ruminating about the negative can become a pattern that makes depressive feelings even worse.

Going against the grain of popular wisdom, Susan Nolen-Hoeksema and colleagues have found that **rumination** tends to enhance angry and depressed moods rather than to foster improvements. People who ruminate are more likely to generate more negative memories, make more pessimistic predictions about the future, and interpret present situations more negatively than people who do not ruminate (Lyubomirsky & Nolen-Hoeksema, 1993, 1995). By ruminating about negative feeling it can

become more difficult to focus on and solve the problems that underlie one's distress and depression. In a study of people who lost a loved one to a terminal illness, those who ruminated more at the time of their loss were more depressed 18 months later than those who ruminated less (Nolen-Hoeksema, Parker, & Larson, 1994).

In moderation, the tendency to ruminate is a relatively common and stable coping style that is more prevalent among women than men (Butler & Nolen-Hoeksema, 1994; Nolen-Hoeksema et al., 1994). But chronic ruminators— people who reflexively ruminate in response to stress— experience more distressing thoughts and feelings compared to people who do not ruminate when under distress (Rusting & Nolen-Hoeksema, 1998). In fact, the tendency to ruminate may be an indicator that predicts future diagnoses of depression and anxiety (Nolen-Hoeksema, 2000).

9.15 How does rumination go beyond Pennebaker's approach? What negative effects result?

▶ CHANGE AND WELL-BEING

The Meaningful Life, the Healthy Personality

The phenomenological orientation implies a "humanistic" view of adaptation, personal health and "deviance." There are many variations, but in general personal genuineness, honesty about one's own feelings, self-awareness, and self-acceptance are positively valued. **Self-realization,** the ultimate in fulfillment in this perspective, involves a continuous quest to know oneself and to actualize one's potentialities for full awareness and growth as a human being. Denouncing "adjustment" to society and to other people's values as the road to dehumanization, the quest is to know oneself deeply and to be true to one's own feelings without disguise or self-deception. Human problems are seen as rooted in distortions of one's own perceptions and experiences to please the expectations of society, including the demands of one's own self-concept (Chapter 8).

9.16 What kinds of characteristics did Maslow identify in self-actualized people?

Maslow (1968) offered a description of the **healthy personality** from a humanistic viewpoint, and his view of the qualities of the **self-actualizing person** is summarized in Table 9.5.

In recent years, research has identified many of the key psychosocial ingredients that enhance well-being (e.g., Aspinwall & Staudinger, 2002), summarized in Table 9.6. The results are generally consistent with the expectations of the Phenomenological Approach. Namely, human resilience and strength—including physical well-being and health—in dealing with serious stressors is enhanced when the individual can find **meaning in life** and in the experience of living it. This is the case even when (or perhaps especially when) the experience is tragic, such as the loss of a life partner or the development of a life-threatening illness (O'Leary, 1997). A dramatic example is seen in how the actor Christopher Reeve, who had starred as Superman in the movies, dealt with his sudden paraplegic condition after an accident. He managed to construe this experience not as an occasion for self-pity but as an opportunity to make a contribution, becoming a dedicated spokesperson to increase research for spinal cord injury.

9.17 What psychosocial factors seem to contribute to physical and psychological well being? Cite relevant research results.

TABLE 9.5 Some Qualities of Maslow's "Self-Actualizing" People

1. Able to perceive reality accurately and efficiently.

2. Accepting of self, of others, and of the world.

3. Spontaneous and natural, particularly in thought and emotion.

4. Problem-centered: concerned with problems outside themselves and capable of retaining a broad perspective.

5. Need and desire solitude and privacy; can rely on their own potentialities and resources.

6. Autonomous: relatively independent of extrinsic satisfactions, for example, acceptance or popularity.

7. Capable of a continued freshness of appreciation of even the simplest, most commonplace experiences (for example, a sunset, a flower, or another person).

8. Experience "mystic" or "oceanic" feelings in which they feel out of time and place and at one with nature.

9. Have a sense of identification with humankind as a whole.

10. Form their deepest ties with relatively few others.

11. Truly democratic; unprejudiced and respectful of all others.

12. Ethical, able to discriminate between means and ends.

13. Thoughtful, philosophical, unhostile sense of humor; laugh at the human condition, not at a particular individual.

14. Creative and inventive, not necessarily possessing great talents, but a naive and unspoiled freshness of approach.

15. Capable of some detachment from the culture in which they live, recognizing the necessity for change and improvement.

Source: Based on Maslow, A. H. (1968). *Toward a psychology of being.* New York: Van Nostrand.

Beyond such vivid single cases, there is a great deal of evidence that the ingredients of well-being, summarized in Table 9.6, significantly improve the biological response to illness (Ickovics, 1997). To illustrate, Shelly Taylor found that HIV positive gay men who had lost a life partner but construed the experience as giving new meaning to their life maintained their level of immune functioning longer than did those who did not find meaning in the same experience (O'Leary 1997; Taylor, 1995).

A second ingredient of resilience is an **optimistic orientation** (as contrasted to a pessimistic, hopeless, helpless orientation), and a focus on the positive in oneself and in human nature (e.g., Aspinwall & Staudinger, 2002). Scheier and colleagues compared optimists and pessimists (identified through a self-report measure) on a well-researched measure of coping styles in dealing with stress (Scheier & Carver, 1992). Optimists characteristically tend to use more active coping and planning, seek emotional support

TABLE 9.6 Some Key Ingredients for Well-Being

• finding meaning in life

• optimism (versus helplessness/hopelessness)

• self-efficacy/agency (beliefs that one can do things effectively)

• social support (relatedness: groups or friends that share experience caringly)

The star of Superman became heroic also in real life.

more, use religion, and attempt to grow constructively with the adverse experience they are having. They also are less likely than pessimists to use negative kinds of coping strategies. For example, when dealing with breast cancer, optimists are less likely to deny that they had cancer, tend to disengage less, and use generally positive ways of coping with their disease. In turn, such active coping styles and realistic acceptance of the illness tend to be predictive of survival. As the table indicates, beliefs that one can do things effectively, called **self-efficacy,** and having relationships that provide support and connectedness also are important ingredients for effective coping and well-being.

SUMMARY

EXPLORING INTERNAL EXPERIENCE

- Discrepancies develop between various mental representations of the self and they have emotional consequences. People use various strategies to reduce these discrepancies.

- Research indicates that self-assessment sometimes can yield predictions as accurate as those from sophisticated personality tests.

- In the Q-sort procedure, participants are asked to sort attribute cards from most descriptive to least descriptive to describe, for example, the self, the ideal self, or a relationship.

- Phenomenologists use the interview to explore the person's feelings and self-concepts and to see the world from his or her framework and viewpoint.

- The "semantic differential" is a rating technique for the objective assessment of the meaning of the rater's words and concepts.

- Personal narratives and psychobiographies are other useful methods of exploring individuals at this level of analysis.

ENHANCING SELF-AWARENESS: ACCESSING ONE'S EXPERIENCES

- Methods to enhance self-awareness and interpersonal awareness include Gestalt therapy, meditation and encounter groups, and generally emphasize emotional aspects of experience and existence.

- Researchers have also examined the effects of self-disclosure (e.g., by writing about traumatic experiences) and of rumination on diverse positive and negative outcomes.

CHANGE AND WELL-BEING

- In this perspective, the healthy personality is characterized by personal genuineness, honesty about one's own feelings, self-awareness, self-acceptance, and self-realization, without destorting one's self-perceptions.

KEY TERMS

activity 202
actual self 195
actual/ideal discrepancies 198
actual/ought discrepancies 198
anorexic behavior 198
evaluative factor 202
family therapy 208
Gestalt therapy 204
healthy personality 210

human-relations training group (T-group)/ sensitivity training group 204
ideal self 195
mantra 206
meaning in life 210
optimistic orientation 211
ought self 195
personal narratives 203
potency 202
psychobiography 203
Q-sort (Q-technique) 200

rumination 210
self-actualizing person 210
self-disclosure 206
self-efficacy 212
self-evaluative standards (self-guides) 198
self-realization 210
semantic differential 201
systems therapy 215
transcendental meditation (TM) 206

PART III

THE PHENOMENOLOGICAL LEVEL

▶ OVERVIEW: FOCUS, CONCEPTS, METHODS

As this part concludes, we again take stock and seek an overview of this level of analysis. Many psychologists welcome the emphasis on the person's cognitions, feelings, and personal interpretations of experience stressed by the contributors at this level of analysis. This concern with how individuals construe events and see themselves and the world has been a most influential force, and it has generated a great deal of research. The resulting contributions are widely acknowledged. Some of the main features of phenomenological approaches are summarized in the next table, which shows some of their shared characteristics.

Overview of Focus, Concepts, Methods: The Phenomenological Level

Basic units:	The experienced self; personal constructs and self-concepts; subjective feeling and perceptions; self-discrepancies
Causes of behavior:	Self-concepts, personal construal (appraisal), feelings and conflicts, free choices
Manifestations of personality:	Private internal experiences, perceptions, and interpretations, self-actualization
Favored data:	Self-disclosure and personal constructs (about self and others); self-reports
Observed responses used as:	Signs (of the person's inner states, perceptions, or emotions)
Research focus:	Personal constructs; self-concepts, self-awareness and expression; human potential and self-actualization; emotion
Approach to personality change:	By increased awareness, personal honesty, internal consistency, and self-acceptance; by modifying constructs; by alternative construals (appraisals)
Role of situation:	As the context for experience and choice; focus on the situation as perceived

As the table indicates, the basic unit of personality is the experienced self and the personal concepts and feelings of the individual. These feelings and concepts and the conflicts experienced in relation to them are seen as basic causes of behavior, but the person has choice and responsibility for what he or she does. The favored data are people's reports and self-disclosures about their personal feelings and constructs. They provide the assessor with a glimpse of the person's inner states, perceptions, and emotions, and an empathic sense of what it is like to view the world through that person's eyes. The goal of research is to explore the individual's private feelings and concepts and examine their implications. It is assumed that by increasing self-awareness and "being in touch with" one's genuine experienced feelings, persons will enhance their perceived coherence, realize their potential for growth, and self-actualize. The effect of situations can be substantial but always depends on how they are perceived subjectively by the person.

▶ ENDURING CONTRIBUTIONS OF THE PHENOMENOLOGICAL LEVEL

Work at this level, begun by pioneers in the middle of the twentieth century, that focuses on how people construe or appraise themselves and their experiences is still timely today. There is a renewed research interest in personal constructs and construal, the self, and the nature of selfhood (e.g., Hoyle et al., 1999; Leary & Tanguey, 2003; Mischel & Morf, 2003), as discussed in Part V and in the final chapters on the integration of levels.

Work at the phenomenological level looked beyond the individual's previous history, or early life traumas and conflicts, or traits, to offer a more positive, optimistic view of the human condition and the prospects for personal growth and constructive change. This focus also has encouraged movement toward a more positive psychology directed at enhancing human potential (Aspinwall & Staudinger, 2002). It proposed that people do not forever have to be victims of their past and can move beyond the less constructive aspects of their personalities toward much more positive ways of fulfilling themselves.

One enduring practical application from work at this level came in the form of family therapy or **systems therapy** (e.g., Minuchin, Lee, & Simon, 1996). These therapies try to understand how the problems of an adolescent, for example, are interpersonal and reflect conflicts and issues in the family system within which he or she lives. Treatment therefore needs to take account of what is going on in the family as a whole, not just in the adolescent's mind. It examines for example, the role that was established for the "problem person" within the context of the larger system (e.g., "Anne's the one you can't count on"), and the ways in which that role feeds into the conflicts currently experienced. The goal is to help the family members to see each other more constructively. The hope is to reconstrue the "problem" to move beyond old stereotypes and essentially create new narratives that work better and fit the changing circumstances and the contexts of the family and all its members as they evolve and try to deal with each other. The same would be true if the case was one of an abused woman. Treatment would focus on understanding the abusive relationship and the ways in which it was maintained, not just on the personality of the abused person.

In sum, the applications of this approach have been broad, and have ranged from working to help overcome the psychological consequences of child abuse, battering, and incest to the trauma of HIV infection and AIDS (e.g., Walker, 1991; White & Epston, 1990) to modern family therapists (e.g., Minuchin, Lee, & Simon, 1996), to help for women in abusive relationships with violent men (Goldner et al., 1990). These applications have made clear that many key phenomenological concepts can help in understanding and treating some of the most difficult problems of contemporary life. They also illustrate how well the ideas of George Kelly and Carl Rogers, articulated half a century ago, have stood the test of time.

Theorists like Rogers and Kelly at first were lone voices calling for a new kind of personality psychology, and sketching some of its components and outlines. It is a tribute to the early theorists working at this level that many of the topics they championed are being built upon decades later with fresh and intense new interest. Stimulated by the explosion of interest in cognitive approaches to personality, the self and its perceptions and cognitions are receiving unprecedented attention in personality psychology, as you will see in later chapters. In fact they have become a centerpiece for current work at the social cognitive level (Part V). Before we get there however, we turn to a level of analysis that added a very different focus to the study of personality: the role of learning and conditioning as determinants of human action and personality.

PART

IV

THE BEHAVIORAL-CONDITIONING LEVEL

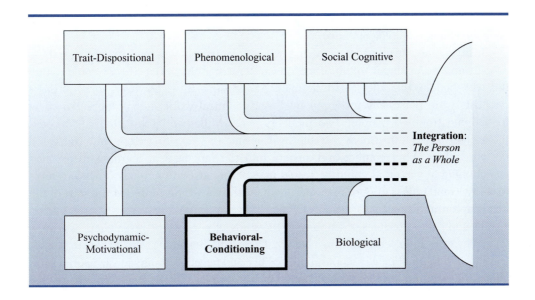

▶ PRELUDE TO PART IV: THE BEHAVIORAL-
CONDITIONING LEVEL

Mr. Z was institutionalized in a large midwestern state mental hospital for psychotic patients in the 1950s. Pacing in a day-room with broken chairs and filled with the screams of disturbed people, he was a portrait in distress and isolation. At the center of this huge bleak room was a glassed-in, walled nursing station, behind which the staff and occasional visiting physician took shelter. The ward was high security, and medications for helping to deal with psychotic outbursts and aggressive rages were still in their infancy. In this kind of bedlam, some psychologists formed a hypothesis. They thought that the patients' wild behavior might in part be a function of the fact that the only time staff emerged to have any contact with them was when they began to act up: might there be a relationship between the display of disturbed behavior and the conditions under which they received any kind of human attention? Might psychotic behavior actually be inadvertently strengthened by the hospital conditions under which they lived and the contingencies under which they received attention? Systematic studies over the years showed that even under less extreme conditions in human relationships, people reinforced and strengthened in each other the very behaviors that they were trying to control.

While Freud and his followers were capturing the attention of the clinical world with their fresh approaches to the treatment of personality problems in Europe, psychologists in the United States were committing themselves to the scientific, experimental study of behavior under well-controlled laboratory conditions. Often they used animals as subjects so that experimental manipulations were possible. Committed to building a science of personality, they wanted to only study phenomena that they could observe directly and objectively. Therefore, they deliberately avoided asking questions about anything that they felt they could not test experimentally. So they did not want to deal with questions about what people are like (Trait-Dispositional Level), or what their motives might be (Psychodynamic-Motivational Level), or what their "real" selves could be (Phenomenological Level).

Instead, they insisted on studying what organisms—whether people or rats— actually *do,* and linking it to the conditions under which they do it. If they could not do this with people, they did it with animals. At the same time, however, some of these scientists were personally intrigued by Freud's clinical work, and its potential value for helping troubled people. Two pioneers, Neal Miller and John Dollard, devoted themselves to make his ideas testable in experiments. As a first step, they translated the core ideas about conflict and anxiety into the language of learning theory and conditioning in ways that could be tested in experiments with animals. Their attempts to study Freud's ideas in rigorous experiments with rats are presented at the start of the first chapter in this part.

To apply the principles of learning to the study of personality, researchers also drew on earlier work in experimental psychology. That work, beginning with Pavlov in Russia, had shown how emotional reactions, both positive and negative, to previously neutral stimuli may be acquired through simple conditioning principles. The excitement here was that the analysis of learning promised to become a more scientific way of understanding some of the complex phenomena that had been discovered by Freud and his followers.

Learning, of course, occurs in various ways, and another breakthrough came in the contributions of the behaviorist B.F. Skinner and his many disciples. Not unlike Freud, Skinner also was often attacked, and often with good reason. Nevertheless, he provided a novel way for thinking about personality and above all for applying the principles of learning to improve the human condition, as described in this part of the text.

Because of the intense focus on behavior in the work done at this level, one may ask: The behavioral level seems to be all about *behavior,* but what about personality? For researchers at this level, the route to changing anything on the "inside"—feelings, cognitions, personality—is by understanding the person's behavior and the "conditions that control it," and then using that knowledge to help the individual to behave in more adaptive, functional ways (e.g., O'Donohue and associates, 2001). They assume that if problematic behaviors like stutters, tics, debilitating fears, or inappropriate patterns of impulsive or over-controlling behaviors are modified appropriately, then the internal states and stable qualities of the individual will essentially "catch up" and also change. A person who could not speak in public but now can is likely to feel and become less shy, for example.

Work at the behavioral-conditioning level originated in experimental psychology and remained isolated from mainstream work in personality research for many years. More recently, as part of the emerging integration among the various levels of analysis, these contributions are being connected to developments at the other levels of analyses, as you will see in subsequent parts of the text, particularly when we turn to the social cognitive level that builds directly on it.

The Personal Side of the Science

Some questions at the behavioral-conditioning level you might ask about yourself:

► How are important behavior patterns, including emotions and fears, learned?

► How does what I do and feel depend on my earlier experiences?

► How can my behavior and feelings be modified by new learning experiences?

► Do aspects of my personality depend on the contexts in which I find myself?

► How am I different when with a good friend at school and when with my family at home for the holidays? Why?

10

BEHAVIORAL CONCEPTIONS

Within the broad boundaries set by human genes, what people become is influenced importantly by learning. Through learning, things that attract one person may come to repel another, just as one individual's passions may become another's nightmares. Much is known about human learning, and it can be harnessed to influence people for good or for ill.

In this chapter and in the next, you will find some approaches to personality that focus on the level of learning through conditioning. These approaches are called behavioral theories, or learning-conditioning theories, and several different varieties have been formulated. In this chapter, we will consider some of the original classic concepts of this approach; in later chapters, we will examine its more recent developments and applications.

Work relevant to personality conducted at this level has been done primarily by psychologists dedicated to the rigorous, scientific study of psychology, who were heavily influenced by research on the nature of the learning process. In work that began more than 75 years ago, they devoted themselves to the development of an experimental methodology. Their goal was to conduct research that was experimental and yet relevant to understanding complex social behavior and individual differences. Guided by strategies in other natural sciences, they began with careful study of learning and performance in lower animals in highly controlled laboratory situations. While some were fascinated by the bold speculations about the mind that were coming at the same time from theorists like Sigmund Freud, most were skeptical about the use of informal clinical methods. Rather than probe the dreams and free associations of neurotic patients or theorize broadly about human nature and society, these researchers sought a system that would be objectively testable, preferably by laboratory techniques.

Psychologists working at this level studied the learning mechanisms through which certain events—"stimuli"—become associated with particular behaviors or responses. Like all scientific theorists, they wanted to understand causes—in this case, learning or the ways in which stimuli become associated with responses. The basic assumptions of the learning theories may appear at first glance quite different from those of psychodynamic theories. Yet they try to provide complementary accounts of many of the same basic personality phenomena. To illustrate this fundamental interrelatedness of these different levels of analyses, we begin this chapter with what may be the most systematic attempt to translate some of Freud's ideas into the concepts and language of learning.

10.1 What causal factors in personality are the focus of the behavioral level of analysis?

▶ THE BEHAVIORAL APPROACH TO PSYCHODYNAMICS: DOLLARD AND MILLER

In this section, we will concentrate on the theory developed at Yale University in the late 1940s by John Dollard and Neal Miller, who were fascinated by Freud's bold speculations about the mind. But they were also skeptical about his informal clinical methods and were motivated to really test his ideas. We will call their orientation **psychodynamic behavior theory** because it is the major effort to integrate some of the fundamental ideas of Freudian psychodynamic theory with the concepts, language, and methods of experimental laboratory research on behavior and learning. Although this work began as a translation of Freud's core concepts into the language and ideas of theory, it went much beyond translation by also opening the way to extensive experimental research into the basic processes involved.

Primary Needs and Learning

In this learning reinterpretation, the newborn infant begins life with a set of *innate* or **primary biological needs,** such as the need for food and water, oxygen, and warmth. Satisfaction of these needs to some minimal degree is essential for the organism's survival. But although these needs are innate, the behaviors required to satisfy them involve learning.

The most casual observation of other cultures quickly reveals that there are almost endless ways to fulfill even such primary needs as hunger and thirst. Through learning, great variability develops in the ways in which needs are fulfilled. Consider, for example, food preferences: The gourmet dishes of one culture may be the causes of nausea in another. The same learned variability seen in food preferences is also found in standards of shelter, clothing aesthetics, and values when one compares different cultures.

Most human behaviors involve goals and incentives whose relations to innate needs are extremely remote. People seem to strive for such exceedingly diverse goals as money, status, power, love, charity, competence, mastery, creativity, self-realization, and so on—as was discussed in earlier chapters. These and many more strivings have been characterized and classified as human motives. Neal Miller and John Dollard explored the learning processes through which such motives may evolve from primary needs.

10.2 Describe the four central concepts in the learning process, according to Dollard and Miller.

Starting with the basic assumption that behavior is learned, Dollard and Miller (1950) constructed a learning theory to explain the wide range of behavior involved in normal personality, neurosis, and psychotherapy. In their view, the four important factors in the learning process are drive (motivation), cue (stimulus), response (act or thought), and reinforcement (reward). In its simplest form, their idea is that

> *In order to learn one must want something, notice something, do something, and get something (Miller & Dollard, 1941, p. 2).*

These four events correspond respectively to "drive," "cue," "response," and "reward" ("reinforcement"). Learning, in their view, is the process through which a particular response and a cue stimulus become connected.

Think of an animal in the psychologist's laboratory. Motivated by the *drive* of hunger, the animal engages in diffuse activity. At one point, the animal happens to see a lever *(cue)*. Its *response*, at first accidental, is to press the lever, and this action releases food into his cup. The animal eats the food at once, thereby reducing tension of his hunger drive *(reward* or *reinforcement)*. Now in the future when the animal is hungry, it is more likely to press the lever again: The association between the cue stimulus (the lever) and the response (pressing it) has been strengthened. On subsequent trials, the hungry animal will press the lever sooner. Let us consider each of the four components separately.

Drive. For Dollard and Miller (1950), any strong stimuli (internal or external) may impel action and thus serve as drives. The stronger the stimulus, the greater its **drive** or motivating function. A mild stimulus (such as the faint sound of a distant horn) does not motivate behavior as much as a strong stimulus (the blare of the horn near one's ear). Examples of strong stimuli are hunger pangs and pain-inducing noise—they motivate behavior. While any stimulus may become strong enough to act as a drive, certain classes of stimuli (such as hunger, thirst, fatigue, pain, and sex) are the primary basis for most motivation. These stimuli are "primary" or innate drives. The strength of the primary drives varies with the **conditions of deprivation:** The greater the deprivation, the stronger the drive. Like Freud, Dollard and Miller's theory of drives was based on a "hydraulic" or steam boiler model. The build up of a drive, like the steam in a boiler as the temperature rises, increasingly presses for release of discharge. When release occurs, the drive is reduced. Drive reduction is satisfying or "reinforcing" (as discussed further in the concept of "reinforcement" below).

10.3 Differentiate between primary and secondary drives. In what sense is fear both a learned response and a learned drive that can result in reinforcement?

Often the operation of primary drives is not easy to observe directly. Society generally protects its members from the unpleasant force of strong primary drives by providing for their reduction before they become overwhelming. Moreover, social inhibitions—for example, in the area of sex—may further prevent the direct or complete public expression of primary drives. Consequently, much visible behavior is motivated by already altered "secondary" or **learned drives.** It is these transformed motives that are most evident under conditions of modern society and that are important in civilized human behavior.

According to Dollard and Miller, these learned drives are acquired on the basis of the primary (unlearned, innate) drives and are elaborations of them. The acquisition of fear as a learned drive has been studied carefully. (Some of the specific mechanisms of such learning are discussed in detail in the next chapter.) A fear is learned if it occurs in response to previously neutral cues (e.g., a white room). A learned fear is also a drive in the sense that it motivates behavior (e.g., escape from the room), and its reduction is reinforcing.

In one study, rats were exposed to electric shock in a white compartment and were permitted to escape to a black compartment where there was no shock (Miller, 1948). Eventually the rats responded with fear to the white compartment alone (that is, without the shock). Even when the shock (primary drive stimulus) was no longer present, the animals learned new responses, such as pressing a lever or turning a wheel, in order to escape from the harmless white compartment. In commonsense terms, they behaved as if they were afraid of an objectively harmless stimulus. The motivation for this new learning lies, according to Miller and Dollard, in the learned fear of the white compartment. Thus, fear is conceptualized as both a learned response and a learned drive, and its reduction is considered to be a reinforcement.

Dollard and Miller's emphasis on drives, both "primary" and "secondary," is reminiscent of the Freudian emphasis on motives and impulses as the forces underlying behavior. While Freud's conceptualization stresses instinctual wishes, however, Dollard and Miller's makes room for many learned motives, whose roots are in primary drives.

Cue. "The drive impels a person to respond. Cues determine when he will respond, where he will respond and which response he will make" (Dollard & Miller, 1950, p. 32). The lunch bell, for example, functions as a **cue** for hungry schoolchildren to put away their books and get their lunchboxes. Cues may be auditory, visual, olfactory, and so on. They may vary in intensity, and various combinations of stimuli may function as cues. Changes, differences, and the direction and size of differences may be more distinctive cues than is an isolated stimulus. For example, a person may not know the absolute length of an unmarked line but yet be able to tell which of two lines is longer.

Response. Before a **response** to a cue can be rewarded and learned, it must of course occur. Dollar and Miller suggest ranking the organism's responses according to their probability of occurrence. They call this order the "initial hierarchy." Learning changes the order of responses in the hierarchy. An initially weak response, if properly rewarded, may come to occupy the dominant position. The new hierarchy produced by learning is termed the "resultant hierarchy." With learning and development, the hierarchy of responses becomes linked to language, and is heavily influenced by the culture in which social learning has occurred.

Reinforcement. A **reinforcement** is a specific event that strengthens the tendency for a response to be repeated. For Miller and Dollard, reinforcement involved **drive reduction** or tension reduction. Guided by their hydraulic or boiler model of drives, they believed that drives that are not released continue to build up like steam in the boiler, creating tension and pressing for discharge, as noted above. It is the reduction of tension—drive-reduction—that is the organism's goal: when it happens, it is reinforcing or rewarding.

A reduction of a drive reinforces any immediately preceding response. The reduction or avoidance of painful stimulation, and of learned fears or anxieties associated with pain and punishment, also may function as a reinforcement (Miller, 1948). Reinforcement is essential to the maintenance of a habit as well as to its learning.

10.4 According to Dollard and Miller, how do reinforcement and extinction occur?

Extinction is the gradual elimination of a tendency to perform a response; it occurs when that response is repeated without reinforcement. The time required to extinguish a habit depends on the habit's initial strength and on the conditions of the extinction situation. According to Dollard and Miller, extinction merely inhibits the old habit; it does not destroy it. If new responses performed during extinction are rewarded, they may be strengthened to the point where they supersede the old habit. For example, if a child is praised and rewarded for independent, autonomous play but consistently unrewarded (extinguished) when he or she dependently seeks help, the independent pattern will become predominant over the dependent one.

Later, Miller (1963) speculated about possible alternatives to his drive-reduction concept of reinforcement learning. His hypothesis of an alternative to drive reduction includes the tentative assumption of "activating" or "go" mechanisms in the brain. These "go" mechanisms, Miller conjectured, could be activated in a variety of ways (such as by thinking), not merely by the reduction of drives or noxious stimuli.

Conflict. Individuals may experience **conflict** when they want to pursue two or more goals that are mutually exclusive. For example, a person may want to spend the evening with a friend but thinks he should prepare for an examination facing him the next morning; or she may want to express her anger at her parents but also does not want to hurt them. When an individual must choose among incompatible alternatives, he or she may undergo conflict.

Neal Miller's (1959) conceptualization of conflict, which is influenced by Lewin (1935), hypothesizes **approach** and **avoidance tendencies.** For example, in an **approach–approach conflict,** the person is torn, at least momentarily, between two desirable goals. Conversely, people often face **avoidance–avoidance conflicts** between two undesirable alternatives: to study tediously for a dull subject or flunk the examination, for example. The individual may wish to avoid both of these aversive events, but each time he starts to move away from his desk, he reminds himself how awful it would be to fail the test.

Some of the most difficult conflicts involve goals or incentives that are both positive and negative. These are the goals or incentives that are mixed feelings or ambivalent attitudes. For example, we may want the pleasure of gourmet treat but not the calories, or we may desire the fun of a vacation spree but not the expense, or we may love certain aspects of a parent but hate others.

Recall that approach–avoidance conflicts had a predominant place in Freud's hypotheses regarding intrapsychic clashes—for example, between id impulses and inhibitory anxieties. Just as conflict is central to Freud's conception of personality dynamics, so it is the core of Dollard and Miller's theory. But whereas Freud developed his ideas about conflict from inferences regarding id–ego–superego clashes in his neurotic patients, Dollard and Miller tested their ideas in controlled experiments with rats (e.g., Brown, 1942, 1948; Miller, 1959).

Briefly, the drive-conflict model proposes the simultaneous existence of drive-like forces (approach tendencies) and of inhibitory forces (avoidance tendencies). Predictions about behavior in an approach–avoidance conflict involve inferences about the strength of the approach tendencies and of the inhibiting forces, the resulting behavior being a function of their net effect. Within this general framework of drive-conflict theory, many formulations have been advanced that are similar to the intrapsychic conflicts between id impulses and ego defenses that are crucial in Freud's theory. Some of these ideas are still relevant in current work on approach and avoidance behaviors and will be discussed in that context (Chapter 17).

10.5 What is the basis for conflicts? What major forms do they take?

Neurotic Conflict: The Core

Freud conceptualized conflict and anxiety as fundamental for understanding personality, both normal and abnormal, and saw them as the core ingredients of neurotic behavior. In Freud's formulation, **neurotic conflict** involves a clash between id impulses seeking expression and internalized inhibitions that censor and restrain the expression of those impulses in accord with the culture's taboos. Dollard and Miller state the same basic ideas in the language of learning theory.

In their view of neurosis, strong fear (anxiety) is a learned drive that motivates a conflict concerning "goal responses" for other strong drives, such as sex or aggression—the impulses that also were basic for Freud. Specifically, when the neurotic person—or even the young child—begins to approach goals that might reduce such drives as sex or aggression, strong fear is elicited in him or her. Such fear may be elicited by thoughts relevant to the drive goals, as well as by any overt approach attempts. For example, sexual wishes or hostile feelings toward a parent may be frightening; hence a conflict ensues between the wishes and the fear triggered by their expression. These inhibitory fearful responses further prevent drive reduction, so that the blocked drives (such as sex and aggression) continue to "build up" to a higher level. The person is thus trapped in an unbearable neurotic conflict between frustrated, pent-up drives and the fear connected with approach responses relevant to their release.

The neurotic person in this dilemma may be stimulated simultaneously by the frustrated drives and by the fear that they evoke. The high drive state connected with this conflict produces "misery" and interferes with clear thinking, discrimination, and effective problem solving. The "symptoms" shown arise from the build up of the drives and of the fear that inhibits their release.

Anxiety and Repression. Like Freud, Dollard and Miller accept unconscious factors as critically important determinants of behavior, and, again like Freud, they give anxiety (or learned fear) a central place in dynamics. In their view, repression involves the learned response of **not-thinking** of something and is motivated by the secondary drive of fear. That is, due to past experiences, certain thoughts may have come to arouse fear as a result of their associations with pain or punishment. By not-thinking these thoughts, the fear stimuli are reduced and the response (of not-thinking) is further reinforced. Eventually, not-thinking (inhibiting, stopping, repressing) becomes anticipatory, in the sense that the individual avoids particular thoughts before they can lead to painful outcomes. This formulation is similar to Freud's idea that repression is the result of anxiety caused when unacceptable material starts to emerge from the unconscious to the conscious.

Dollard and Miller's account thus serves as a clear translation of the psychodynamic formulation of anxiety, repression, and defense into the terms of reinforcement learning theory. Defenses and symptoms (e.g., phobias, hysterical blindness) are reinforced by the immediate reduction of the fear drive. While the temporary effect of the symptom is drive reduction and momentary relief, its long-range effects may be debilitating. For example, a phobic symptom may prevent a person from working effectively and hence create new dilemmas, fear, guilt, and other conditions of high drive conflict.

Psychodynamic Behavior Theory. As was noted earlier, Freud constructed a theory of development without the benefit of learning concepts. He adopted a body of language and invented new terms that made it difficult to coordinate his theory with experimental psychology. Dollard and Miller demonstrated that this coordination could be achieved.

10.6 How did Dollard and Miller translate Freud's concept of neurotic conflict into learning theory terms?

10.7 How are anxiety and repression translated into learning theory terms? How does this approach explain repression?

10.8 Summarize the major contributions and criticisms of psychodynamic learning theory.

They drew on laboratory research with animals to devise a personality theory in learning terms that closely paralleled, and in many respects translated, Freudian theory. The psychodynamic emphasis on motives, on unconscious processes, and on internal conflicts and defenses, such as repression, remained largely unchanged. Many psychologists found Freud's basic ideas more congenial and easier to adopt when they were put into the language of learning and experimental psychology. Consequently, these concepts stimulated much research.

Other psychologists were troubled because the research by Miller and his colleagues was based mainly on animal studies. Indeed, this fact has earned it some of its greatest criticism over the years. Careful investigation carried out with a lower species, such as the rat, whose behavior is far removed from human problems, may not hold up when extrapolations are made to people. Many critics have objected that human social behavior is fundamentally different than the behavior of animals in the laboratory and therefore requires a different methodology. Some critics were repelled by the analogies between rat and person and believed that in the transition from the clinic to the laboratory, some of the most exciting features of Freud's view of people were lost. Of course the real test of a position is not its appeal to friends and critics, but the research and conceptual advances it produces.

We now turn to the basic research that provided the foundation for Miller and Dollard's theory, and that still is a basis for understanding how some important aspects of personality may develop, are maintained, and sometimes can be changed dramatically. It also offered a way of making irrational fears and some of the other phenomena Freud identified less mysterious and more open to scientific testing.

▶ CLASSICAL CONDITIONING: LEARNING EMOTIONAL ASSOCIATIONS

Strong human emotions, often seemingly irrational, including negative emotions such as intense fears, and positive feelings, as in attraction, love, and patriotism, may be acquired through the simple processes of classical conditioning. It is therefore important to understand the basic rules of conditioning because they help to take the mystery out of many of the complex emotions we all experience. Knowing these rules is especially useful because we often experience strong negative emotions without awareness of why and how they originated, or how we might be able to change our feelings when they create serious problems for us—like being unable to take a desired job because it happens to be on a high floor.

Classical conditioning or **conditioned-response learning** is a type of learning, first demonstrated by the Russian psychologist Ivan Pavlov, in which a neutral stimulus (for example, a bell) becomes conditioned by being paired or associated with an unconditioned stimulus (one that is naturally powerful).

How Classical Conditioning Works

A dog automatically salivates when food is in its mouth. The response of salivation is a **reflex** or **unconditioned response (UCR):** it is natural and does not have to be learned. Like most other reflexes, in humans and in animals alike, salivation helps the organism adjust or adapt: the saliva aids in digesting the food. Stimuli that elicit unconditioned responses are called **unconditioned stimuli (UCS).** The unconditioned stimulus (food in this example) can elicit behavior without any prior learning.

Any dog owner knows that a hungry dog may salivate at the mere sight of food, before it gets any in its mouth. The dog may even begin to salivate at the sight of the empty dish in which the food is usually served. Salivating at the sight of the empty dish that has been associated with food is an example of a learned or **conditioned response (CR)**. The stimulus that elicits a conditioned response is called a **conditioned** (learned) **stimulus (CS)**: its impact on behavior is not automatic but depends on learning.

Pavlov discovered some of the ways in which such neutral stimuli as lights and metronome clicks could become conditioned stimuli capable of eliciting responses like salivating. His pioneering experiments with dogs began with his repeatedly making a certain sound whenever he gave his dogs their food. After a while he found that the dogs salivated to the sound even when it was no longer followed by food: conditioning had occurred. This type of learning is what we now call classical conditioning.

To sum up, in classical conditioning the participant is repeatedly exposed to a neutral stimulus (that is, one that elicits no special response) together with an unconditioned stimulus that elicits an unconditioned response. When this association becomes strong enough, the neutral stimulus by itself may begin to elicit a response similar to the one produced by the unconditioned stimulus. (See Tables 10.1 and 10.2 for basic definitions and examples.)

> 10.9 Define UCS, UCR, CS, and CR. How are conditioned responses learned?

Higher-Order Conditioning

We have seen that when a previously neutral stimulus, such as a light, a bell, or a face, has become a conditioned stimulus through its association with an unconditioned stimulus, such as food or pain, it can in turn modify one's reactions to another neutral stimulus by being associated with it. This process is called **higher-order conditioning**. It was demonstrated when Pavlov found that after a metronome sound had

> 10.10 How does higher-order conditioning underlie many of our likes and dislikes?

Ivan Pavlov (1849–1936)

TABLE 10.1 The Language of Classical Conditioning

Term	Definition
Unconditioned stimulus (UCS)	A stimulus to which one automatically, naturally responds without learning to do so.
Unconditioned response (UCR)	The unlearned response one naturally makes to an unconditioned stimulus. The response may be positive or negative (e.g., salivating when food is placed in the mouth; jerking one's hand away from a hot stove).
Conditioned stimulus (CS)	A previously neutral stimulus to which one learns to respond after it has been paired or associated with an unconditioned stimulus.
Conditioned response (CR)	The learned response to a conditioned stimulus. This response was previously made only to an unconditioned stimulus, but now it is made to a conditioned stimulus as a result of the pairing of the two stimuli.

become a conditioned stimulus (by being paired with food), it could itself be paired with a neutral stimulus (such as a black triangle) and, as a result of that association, the neutral stimulus would also elicit the unconditioned response of salivation. In people, words and other complex symbols can be powerful conditioned stimuli capable of evoking emotional responses through higher-order conditioning.

A wide variety of stimuli, including activities, individuals, groups, and events, are valued according to their associations with positive or negative outcomes and even mere labels. For example, when neutral items are paired with words like "dirty" and "ugly," they take on negative valuations, but the same items become positively evaluated when they have been associated with words like "beautiful" and "happy" (Staats & Staats, 1957). Likewise, the names of countries and political parties and the sight of national flags, or the sounds of national anthems can come to arouse intense positive or negative feelings depending on their earlier associations.

TABLE 10.2 Examples of Possible Effects of Classical Conditioning

Before Conditioning	After Conditioning
Dog knocks child over (UCS)	Dog approaches (CS)
Child cries (UCR)	Child cries (CR)
Mother feeds and cuddles baby (UCS)	Baby smells mother's perfume (CS)
Baby relaxes (UCR)	Baby relaxes (CR)
Car accident injures woman (UCS)	Woman thinks about getting in car (CS)
Woman is afraid (UCR)	Woman is afraid (CR)
Man drives across swaying bridge (UCS)	Man approaches another bridge (CS)
Man is afraid (UCR)	Man is afraid and avoids bridge (CR)
Mother discovers her daughter masturbating, scolds her, slaps her hands (UCS)	Daughter looks at her nude body (CS)
Daughter is hurt and afraid (UCR)	Daughter feels anxious and negative about her body, particularly her genitals (CR)

Most experiments in classical conditioning are performed in the laboratory, but the knowledge they have generated may help us to understand many things that happen outside the laboratory, such as the development of affections and attractions (Byrne, 1969; Lott & Lott, 1968). For example, a liking for particular people and things may depend on the degree to which they have been associated with positive or pleasant experiences (Griffitt & Guay, 1969). If so, one's affection for a friend may be directly related to the degree to which he or she has been associated with gratifications for oneself.

Now consider the development of fear. How do initially neutral (or even positive) stimuli acquire the power to evoke fear? Suppose, for example, a person repeatedly sees a light and experiences an electric shock simultaneously. In time, the light by itself may come to evoke some of the emotional reaction produced by the shock. Neutral stimuli that are closely associated in time with any pain-producing stimulus then become conditioned stimuli that may elicit fear and avoidance reactions. Thus, the seemingly irrational fears that some people have may reflect a conditioned association between previously neutral stimuli and painful events.

Classical conditioning may influence development throughout a person's life. If, for example, sexual curiosity and fear-producing experiences (such as severe punishment) are closely associated for a child, fear may be generated by various aspects of the individual's sexual behavior even after there is no longer any danger of punishment. And conditioning can spread as well as persist: the child who is made to feel bad about touching the genitals may also become anxious about other forms of sexual expression and may even develop broader fears.

How much we like a person may depend on the degree to which he or she has been associated with gratification.

In a classic study, following a strategy that would not be tolerated now because of the ethical issues it raises, Watson and Rayner (1920) induced a severe fear of rats in a little boy named Albert, who had not been afraid of rats before. This was done by classical conditioning: just as Albert would reach for the rat, the experimenters would make a loud noise that frightened him. After he had experienced the rat and the aversive noise several times in close association, he developed a strong fear of the rat.

Albert's fear generalized so that later, when shown a variety of new furry stimuli such as cats, cotton, fur coats, human hair, and wool, he responded with obvious fear to them as well. His fear had spread to these new objects even though they had never been paired with the noise. This is a human example of the kind of learning found when rats who were shocked in a white compartment began to respond fearfully to the compartment itself even when the shock no longer occurred (Miller, 1948). The case of Little Albert has become one of the first bases for applications of classical conditioning to the analysis and treatment of human problems.

10.11 Compare Freud's analysis of Little Hans with the behavioral reinterpretation of the case.

As the In Focus 10.1 example illustrates, the behavioral view of neurosis is concerned with anxiety and avoidance no less than the psychodynamic view, but it tries to link them to external circumstances rather than to internal conflicts (Redd, 1995; Redd, Porterfield, & Anderson, 1978). Through direct or vicarious frightening experiences, people often develop anxiety in response to particular objects, persons, or

---

A BEHAVIORAL CHALLENGE TO THE PSYCHODYNAMIC THEORY OF NEUROSIS

Differences between theoretical approaches are seen most clearly when applied to the same case. Behavior theorists have strongly challenged Freud's theory of neurosis by reanalyzing in learning terms a case that he presented. Recall that Freud's view of how neuroses develop begins with the child's aggressive or sexual impulses, which seek direct, immediate release. Because expression of these impulses may be punished severely, the child may become anxious about his own impulses and try to repress them. But the impulses continue to seek release and become increasingly pent up. Eventually they may be impossible to repress, and components of them may break through, creating further anxiety. To reduce this anxiety, the person may attempt a variety of defense mechanisms. In neurosis, these defenses begin to break down: the unacceptable impulses start to express themselves indirectly and symbolically in various disguised forms, such as in phobias or obsessive–compulsive thoughts and actions. The roots of neurosis, in Freud's view, are always in childhood:

It seems that neuroses are acquired only in early childhood (up to the age of six), even though their symptoms may not make their appearance till much later. The childhood neurosis may become manifest for a short time or may even be overlooked. In every case the later neurotic illness
links up with the prelude in childhood (Freud, 1933, pp. 41–42).

An example from Freud's (1963) theory is his published case of Little Hans. Hans was a five-year-old boy who developed a horse phobia. He was afraid of being bitten by a horse and, after seeing a horse hitched to a wagon slip and fall on a street near his home, began to dread going out. Freud interpreted the phobia as an expression of the child's psychodynamic conflicts. These conflicts included his desires to seduce his mother and replace his father. These desires, in turn, made him fear castration by the father; symbolically, the horse came to represent the dreaded father.

A behavioral analysis of this case, however, explains Hans's phobia without invoking any internal conflicts or symbolism (Wolpe, 1997; Wolpe & Rachman, 1960). Namely, the scene Hans witnessed of a horse falling down and bleeding on the street was sufficiently frightening to the young child to produce fear. In turn, the fear generalized to all horses and resulted in Hans's avoidance behavior. Thus, a simple conditioning process might explain a phobia: since the horse was part of an intensely frightening experience, it became a conditioned stimulus for anxiety, and the anxiety generalized to other horses. The process is the same as in the case of Little Albert whose fear was induced by conditioning, as discussed in the text.

situations. Not only encountering these events in reality but even just thinking about them may be upsetting. These emotional reactions may generalize and take many forms. Common examples include muscle tensions, fatigue, and intense fear reactions to seemingly neutral stimuli.

Psychodynamic and behavioral theorists do agree that the neurotic individual may make all sorts of efforts to escape and avoid painful feelings. Many of his avoidance attempts may be maintained persistently because they serve to terminate the pain. For example, such elaborate avoidance defenses as obsessive–compulsive rituals, in the form of handwashing for many hours, may be maintained because they reduce the person's anxiety (Wolpe, 1963). In addition, attention and sympathy from relatives and friends for being sick or relief from pressures and obligations can also serve to maintain the anxious person's avoidance patterns by providing reinforcement for them.

From Trauma to Anxiety

Some of the clearest examples of anxiety reactions occur after the individual has experienced a threatening danger or trauma. A near-fatal automobile accident, an almost catastrophic combat experience, an airplane crash—such intense episodes of stress are often followed by anxiety. After the actual dangers have passed, stimuli that remind the individual of those dangers, or signs that lead him or her to expect new dangers, may reactivate anxiety and distort perceptions (e.g., Rachman & Cuk, 1992).

After severe trauma, the victim is more likely to respond anxiously to other stress stimuli that occur later in life (Archibald & Tuddenham, 1965; Milgram, 1993). Surviving victims of Nazi concentration camps, for example, sometimes continued for years to be hypersensitive to threat stimuli and to react to stress readily with anxiety and sleep disturbances (Chodoff, 1963). These observations support the idea that anxiety involves a learned fear reaction that is highly resistant to extinction and that may be evoked by diverse stimuli similar to those that originally were traumatic. That is, the fear evoked by the traumatic stimuli may be reactivated and also may *generalize* to stimuli associated with the traumatic episode. For example, after a child has been attacked and bitten by a dog, her fear reaction may generalize to other dogs, animals, fur, places similar to the one in which the attack occurred, and so on (Fig. 10.1). Moreover, if the generalization stimuli are very remote from the original traumatic stimulus, the person may be unable to see the connection between the two and the anxiety may appear (even to her) particularly irrational. Suppose, for example, that the child becomes afraid of the room in which the dog bit her and of similar rooms. If the connection between her new fear of rooms and the dog's attack is not recognized, the fear of rooms now may seem especially bizarre.

From a learning point of view, anxieties after traumas, like other learned fears, may be acquired through simple association or conditioning principles. If neutral stimuli have been associated with aversive events or outcomes, then they also may come to elicit anxiety in their own right. Such aversively conditioned emotional reactions may also generalize extensively to new stimuli (Fig. 10.1). Clinical examples of aversive arousal and avoidance include many phobic and anxious reactions to objects, people, and social and interpersonal situations. Not only external events, but also their symbolic representations in the form of words or of thoughts and fantasies, may create painful emotions. In our example of the child traumatized by the dog, even thinking about the incident, or the room in which it occurred or similar rooms may terrify the youngster.

Stimuli closer or more relevant to those associated with emotional arousal tend to elicit stronger reactions. In one study, novice sports parachutists and members of the control group took a specially constructed word association test (Epstein & Fenz, 1962;

10.12 Why are trauma victims more reactive to later life stresses?

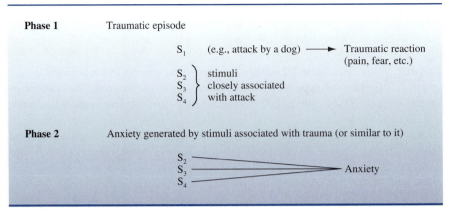

Figure 10.1   From trauma to anxiety.

Fenz, 1964). The words contained four levels of relevance to parachuting. Throughout the word association tests, participants' physiological reactions (*galvanic skin response*, or *GSR*, which as discussed earlier is a change in electrical activity of the skin due to sweating) were recorded to measure their emotional arousal in response to the various stimulus words. One testing occurred two weeks before the scheduled jump, another testing was done the day before the jump, and a final test was on the day of the parachute jump.

The results showed more arousal in parachutists for parachute-relevant words. The effect was greatest for the words most relevant to parachuting, and their arousal was highest when the testing time was closer to the emotion-arousing parachute jump itself.

▶ INSTRUMENTAL (OPERANT) CONDITIONING: LEARNING FROM RESPONSE CONSEQUENCES

The work of Pavlov and the many others who explored the implications of classical conditioning showed how emotional reactions, both positive and negative, to previously neutral stimuli may be acquired through the process of close association with unconditioned stimuli. But learning takes place in more than one form, and a second type, called instrumental or operant conditioning, provides another foundation for understanding the role of experience and social learning for the development of personality.

How Instrumental or Operant Conditioning Works

10.13 How does operant conditioning differ from classical conditioning?

Behavior is modified by its consequences: the outcome of any response (or pattern of responses, often called **operants** for how the organism "operates" on the environment) determines how likely it is that similar responses will be performed in the future. If a response has favorable (reinforcing) consequences, the organism is more likely to perform it again in similar situations. Contrary to some widespread misconceptions, reinforcers or favorable outcomes are not restricted to such primitive rewards as food pellets or sexual satisfactions. Almost all events may serve as reinforcers, including such cognitive gratifications as information (Jones, 1966) or the achievement of competence. Such learning, based on the consequences produced by responses, is called **operant conditioning** (or, in earlier usage, trial-and-error or **instrumental learning**).

TABLE 10.3   Summary of Two Types of Learning

Type	Arrangement	Effect	Example	Interpretation
Classical conditioning	A neutral stimulus (e.g., a bell) repeatedly and closely precedes a powerful unconditioned stimulus (e.g., food).	The originally neutral stimulus becomes a conditioned stimulus—that is, acquires some of the impact of the powerful unconditioned stimulus.	Bell begins to elicit a salivary response, even when not paired with food.	Organism learns that the conditioned stimulus (bell) signals (predicts) the occurrence of the unconditioned stimulus (food).
Instrumental (Operant) conditioning	A freely emitted response (operant) is repeatedly followed by a favorable outcome (reinforcement).	The operant increases in frequency.	If crying is followed by attention, its frequency is increased.	Organism learns that this response will produce that particular outcome.

When the consequences of a response pattern change, the probability of it and of similar response patterns occurring again also changes (Nemeroff & Karoly, 1991). If a little boy whines and clings to his mother and she drops everything in an attempt to appease him, the chances increase that he will behave in this way in the future. If she systematically ignores the behavior, however, and consistently fails to react to it, the chances decrease that the child will continue to behave this way.

Summary of Two Types of Learning

Table 10.3 summarizes the two types of learning discussed so far: classical conditioning and operant conditioning. There is much overlap between these types of learning, but each has some relatively distinct features, as indicated in the table, and each has a place within current social learning theory.

▶ B. F. SKINNER'S INFLUENCE ON PERSONALITY

The motivational-drive theory of Miller and Dollard provided an account of human personality that gave motivation an important role. It also allowed a learning reinterpretation of many of the motivational concepts that Freud invoked in his psychoanalytic theory. But while the motivational focus of Miller and Dollard spoke to the interests of many clinical psychologists, it was rejected within learning theory by the highly influential American behaviorist, B. F. Skinner. Skinner and his followers wanted to find a level of analysis that was entirely objective and required no inferences about underlying mental processes.

In Skinner's view of science, we can know people only by examining their behavior—the things they say and do. In a sense, all psychological approaches are based on the study of behavior, but they differ in how the behavior is used. For example, in analyses at the trait level, behaviors are used as signs for inferring the person's traits or attributes from the observable things the individual does. In the behavioral approach favored by B. F. Skinner and many of his students, however, the observed behavior is the basic unit, and the interest is in specifying the conditions and stimuli or situations that "control" it. The concept that the stimulus controls the response is called **stimulus control.**

In this tradition, behaviorists tried to sample the individual's behaviors directly but generally were reluctant to interpret them as signs (indicators) of the person's motives or other attributes. For example, from this perspective one might try to sample Gary's behaviors to find out just what he does before speeches, without drawing any inferences from them about his underlying anxiety, insecurity, or other personal qualities. To the degree that theorists limit themselves to behavior, their definition of personality itself becomes equated with the whole of an individual's behaviors: *the person "is" what the person "does."*

Importance of the Situation

10.14 What is meant by stimulus control of behavior?

Work at this level of analysis gives particular importance to the role of stimuli and situations in the regulation of behavior, including behaviors indicative of personality. Because the strategy used is often experimental, these stimuli are simply green or red lights or tones that have become associated with certain qualities through manipulated learning experiences. But evidence for the important role of the situation in social behavior and even as a determinant of how we perceive other people also comes from more naturalistic social experiments.

Psychologists guided by this approach point out that often behavior may be predicted simply from knowledge about relevant conditions. Consider, for example, studies that tried to predict the post-hospital adjustment of mental patients. Accurate predictions of post-hospital adjustment required knowledge of the environment in which the ex-patient would be living in the community—such as the availability of jobs and family support—rather than any measures of person variables or in-hospital behavior (e.g., Fairweather, 1967; Fairweather et al., 1969; Holahan & Moos, 1990).

Likewise, to predict intellectual achievement it helps to take into account the degree to which the child's environment supports (models and reinforces) intellectual development (Wolf, 1966). And to predict whether or not people respond to stress with illness, it helps to know the degree to which they have social supports (e.g., spouse, family) in their environments (Holahan & Moos, 1990; Nilson et al., 1981). Finally, when powerful treatments are developed, predictions about outcomes may be useful when based simply on knowing the treatment to which the individual is assigned (e.g., Bandura, Blanchard, & Ritter, 1969; Bandura, 1986).

In daily life, different people are exposed to extremely different kinds of situations, often in stable ways. The life of a college student contains sets of situations that may importantly influence not only what the students do but also in time, through their interactions with those situations, what they become (Cantor et al., 2002). Imagine a day in the life of a New York City taxi driver, or that of a monk, a professor, a company executive, or a hairdresser, and the differences in the kinds of situations they must encounter, day in and day out. Such differences are not limited to occupational and social role differences. A person with an irritating, defensive style of social interaction is likely to provoke different reactions from others compared to a person with a more agreeable style.

A dramatic illustration of the power of the situation and the social role it assigns to one is seen in a natural experiment that was considered shocking at the time (Rosenhan, 1973). Normal individuals (doctors, psychologists, and graduate students from Stanford University) admitted themselves to the local mental hospital by complaining of various psychiatric symptoms. Then, during their confinement, they proceeded to behave rationally, which they normally were. From the moment that they were admitted to the hospital, these individuals were consistently treated by the professional hospital staff (who did not know their true identity) as if they were insane, and labeled psychotic. Their rational, normal behavior not withstanding, they were still treated as insane by the staff, although some of the other patients suspected that they did not really belong there. Staff observations of the Stanford group included notes about their obsessive intellectualizing—frantically scribbling notes, and so on.

Rejection of Inferred Motives

To explain behavior, many earlier theorists hypothesized a wide range of human motives. Theories concerning motivation have helped to reveal the variety and complexity of human strivings, and also have contributed to the development of research about their causes. Investigators of motives originally were inspired by the model of experimental research on biological drives in animals. In animal studies of motivation, the hypothesized need of the animal (its hunger or sex drive, for example) has been linked clearly to observable conditions manipulated in the laboratory. For example, the strength of the hunger drive may be inferred in part from the amount of time that the animal has been deprived of food. When a dog has not been fed for two days, we may safely say that it has a high hunger drive. In such cases, references to drives and motives are straightforward. Likewise, some careful investigations of hypothesized higher-order motives in people have specified clearly the objective conditions that define the motive (e.g., Emmons, 1997; McClelland, 1992; McClelland et al., 1953).

Less rigorous applications of motivational theory to personality, however, may use motives loosely (e.g., as "wishes" or "desires"), and their value as explanations of behavior is open to question—and Skinner took the lead in raising those questions for many years. In his view, the tendency to invoke motives as explanations of why people behave as they do is understandable, because that is how we "explain" behavior in commonsense terms. To explain why a child spent an unusual amount of time cleaning and grooming himself neatly, we easily might say "because he had strong cleanliness needs" or "because he had a compulsive desire for order."

Such hypotheses about motives may sound like explanations, but they tell us little unless the motive is defined objectively and unless the causes of the motive itself are established. What makes the child have "cleanliness needs"? What determines his "compulsive desires"? Why does she "wish" to be clean? These are the kinds of questions raised by B. F. Skinner, who worked for many years at Harvard University.

A pioneer in the behavioral approach, Skinner criticized many concepts regarding human needs as being no more than motivational labels attached to human activities. Thus, orderly behavior may be attributed to a motive for orderliness, submissive behavior to submissiveness needs, exploratory behavior to the need to explore, and so on. To avoid such circular reasoning and to untangle explaining from naming, behaviorally oriented psychologists like Skinner prefer to analyze behaviors in terms of the observable events and conditions that seem to vary with them. Hence, they refuse to posit specific motivations for behavior. Rather, they try to discover the external events that strengthen its future likelihood and that maintain or change it. This approach leads to questions like: When does that child's cleaning activity increase, and when does it decrease in relation to observable changes in the environment? For example, how do the reactions of the parents influence the behavior?

For Skinner, psychology is the science of behavior: inferences about unobservable states and motives are not adequate explanations, and they add nothing to a scientific

10.15 Why did Skinner reject motives and inner conflicts as causes of behavior?

B. F. Skinner (1904–1990)

account of the conditions controlling behavior. "Motivation" is simply the result of depriving or satiating an organism of some substance such as water or food for a given period of time. Thus, a "drive" is just a convenient way of referring to the observable effects of such deprivation or satiation. Likewise, Skinner avoids any inferences about internal "conflicts," preferring an experimental analysis of the stimulus conditions that seem to control the particular behavior in the situation. In his words:

> *Man, we once believed, was free to express himself in art, music, and literature, to inquire into nature, to seek salvation in his own way. He could initiate action and make sponta-neous and capricious changes of course. . . . But science insists that action is initiated by forces impinging upon the individual, and that caprice is only another name for behavior for which we have not yet found a cause (Skinner, 1955, pp. 52–53).*

The essence of Skinner's behavioristic view is the belief that our behavior is shaped by the external environment, not by motives, dispositions, or "selves" that are "in" the person.

Identifying Stimuli (Situations) Controlling Behavior

10.16 What is involved in a functional analysis of behavior?

Skinner's work is based on the premise that a genuine science of human behavior is not only possible but desirable. In his view, science should try to predict and determine experimentally the behavior of the individual organism (Skinner, 1974).

Skinner proposed a **functional analysis** of the organism as a behaving system. Such an analysis tries to link the organism's behavior to the precise conditions that control or determine it. Skinner's approach therefore concentrates on the observable covariations between independent variables (stimulus events) and dependent variables (response patterns). The variables in a functional analysis, according to Skinner, must be external, observable, and described in physical and quantitative terms. It will not do to say that a child becomes concerned with cleanliness when she "fears her father's disapproval"; one must specify the exact ways that changes in the father's specific behavior (e.g., his praise) are related to specific changes in what the child does (e.g., how much she washes her hands per hour).

Skinner contends that the laboratory offers the best chance of obtaining a scientific analysis of behavior; in it, variables can be brought under the control of experimental manipulation. Furthermore, the experimental study of behavior has much to gain from dealing with the behavior of animals below the complex human level. Science, Skinner points out, advances from the simple to the complex and is constantly concerned with whether the processes and laws discovered at one stage are adequate for the next.

Skinner incorporated into his position many concepts regarding classical conditioning, but he concentrated on another kind of learning that is different in some ways from classical conditioning. He contends (1953) that most human social behavior consists of freely emitted response patterns, or operants. Even a little baby shows much spontaneous behavior: it reaches up to a mobile, turns its head, looks at objects, cries and gurgles, and moves its arms and legs. Through such operants, the organism operates on its environment, changing it and, in turn, being changed by it.

In research on operant conditioning, the typical experiment involves an animal or a person freely performing (emitting) operant responses. The experimenter has preselected a particular class of responses to reinforce (e.g., a young child successfully using a potty-training chair or an adult using personal pronouns in an interview). When the selected operant response is made (the child urinates in the potty or the adult says, "I," "you," "she," and so on), the reinforcement occurs: the child gets a small toy; the interviewer nods or murmurs "good." Figure 10.2 illustrates what happens in operant conditioning.

The outcomes a person obtains for a particular behavior influence his or her future behavior. A child's refusal to eat may gain attention from a father usually too busy to pay the child much attention. Since the child's behavior is reinforced by the attention, she may refuse to eat again. If she is offered special treat foods in an effort to tempt her,

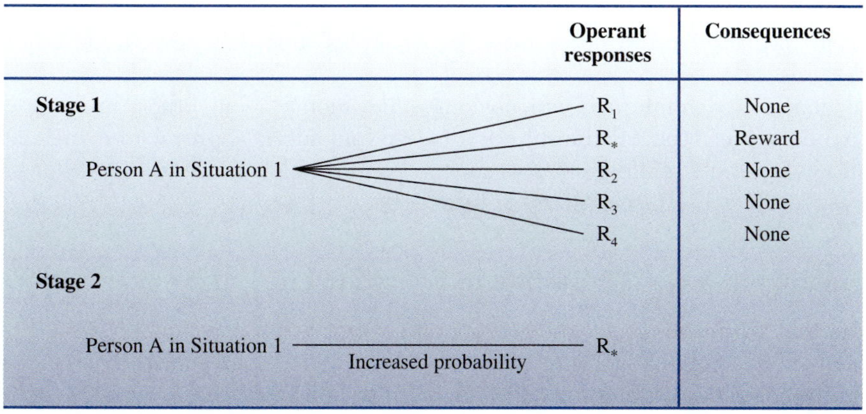

	Operant responses	Consequences
Stage 1	R_1	None
	R_*	Reward
Person A in Situation 1	R_2	None
	R_3	None
	R_4	None
Stage 2		
Person A in Situation 1 — Increased probability	R_*	

Figure 10.2 Operant conditioning: Skinner's view. A person performs (emits) many operant responses in any given situation. If one operant is followed by a favorable outcome (reward) in that situation, the person will be more likely to perform that operant again in a similar situation.

she may quickly turn into a finicky eater with limited food preferences. By changing the outcomes of responses, reinforcing previously unreinforced behaviors or discontinuing reinforcement for other behaviors, even behavior patterns that seem deeply ingrained may be changed.

10.17   How are operant techniques used to change behavior?

Influenced by Skinner, many psychologists have tried to modify maladaptive behavior by altering the consequences to which it leads. Working with people who have severe behavioral problems, they attempt to remove reinforcement for disadvantageous behavior and to make attention, praise, or other reinforcement contingent on the occurrence of more adaptive advantageous behavior. Learning programs of this type follow a set of definite steps. First, the problem behaviors are carefully defined and their frequency in a naturalistic context is measured. Next, one observes and records the reinforcing consequences that seem to maintain the behaviors. Guided by this analysis, the relearning program is designed and put into effect. Finally, the resulting changes are assessed over a period of time.

For example, in one case, parents sought help because their three-year-old daughter developed regressive behaviors and reverted to crawling rather than walking. This regression produced serious problems for the child and the family. An analysis of the girl's behavior suggested that her regressive, babyish actions were being encouraged and maintained unwittingly by the attention they brought her. Therefore, an effort was made to rearrange the response–reinforcement patterns so that crawling and infantile acts were not rewarded by the attention of worried adults. Instead, attention and other rewards were made contingent on more adaptive and age-appropriate behaviors, such as jumping, running, and walking, thereby increasing these desirable behaviors while the infantile ones decreased (Harris et al., 1964).

Conditioned Generalized Reinforcers

As noted before, neutral stimuli may acquire value and become conditioned reinforcers when they become associated with other stimuli that already have reinforcing properties. **Conditioned reinforcers** become generalized when they are paired with more than one primary reinforcer. A good example of a conditioned generalized reinforcer is money, because it can provide so many different primary gratifications (food, shelter, comfort, medical help, and alleviation of pain). Gradually, generalized reinforcers may become quite potent even when the primary reinforcers upon which they were initially based do not accompany them any more. Some people, for example, seem to learn to love money for its own sake and work to amass "paper profits" that they never trade in for primary rewards.

10.18   How are conditioned generalized reinforcers created?

Some generalized reinforcers are obvious—like money—but others are subtle and involve complex social relationships. Attention and social approval from people who are likely to supply reinforcement—such as parents, a loved one, or a teacher—often are especially strong **generalized reinforcers.**

Discrimination and Generalization in Everyday Life

10.19   Define general-ization and discrim-ination. Provide an example of a discriminative stimulus in your own life.

Discriminative stimuli indicate when an operant response will or will not have favorable consequences. Without such signals we would not know in advance the outcomes to which different behaviors are likely to lead and life would be chaotic. With the help of discriminative stimuli we learn to stop the car when coming to a railroad crossing; to eat certain foods with forks and spoons and to continue to eat others with our fingers; to shout and cheer at football games, but not at course examinations; to wear warmer clothes when the temperature starts to drop and to shed them when it becomes hot; to stop at red traffic lights and to go when they turn green.

When a particular response or pattern of responses is reinforced in the presence of one stimulus but not in the presence of others, discrimination occurs. It may be all right to belch in your own room when alone or with close friends but less acceptable to do so when talking to a faculty advisor in her office, and people soon get feedback that makes this clear to them. Discrimination results from the reinforcement or condoning of behavior in some situations but not in others. The individual is more likely to display the behavior in those situations in which it will probably be reinforced than in those in which it is unlikely to be reinforced.

If a response pattern is uniformly rewarded in many conditions or situations, **generalization** occurs. For example, a child is likely to develop generalized aggressive patterns if he is encouraged or allowed to behave aggressively with his parents and teachers as well as with his siblings and classmates both when he is at school and at home. Generalization depends on the similarity among stimulus situations. Stimuli that are physically similar or that have similar meanings result in the greatest generalization.

From the behavioral perspective, the socialization of children is based on discrimination training. For example, children must learn to control their bowel and bladder functions so that defecation and urination occur only in some situations and not in others. Active exploration of the toy box or the sandbox is permitted and encouraged, while forays into the medicine chest or mother's jewel box have quite different outcomes. As a result of such **discrimination training,** the child's behavior begins to depend on the specific conditions in which it unfolds.

When behavior yields similar consequences in a broad variety of settings, it can be expected to generalize from one situation to another. For example, if a little girl easily gets help in solving problems at home, at school, with parents, teachers, and siblings, she may develop widespread dependency. In contrast, when certain behaviors, such as

"Generalized reinforcers" often are subtle and involve complex human relationships.

curiosity, are punished in some situations but not in others, consistencies across the different situations should not be expected. A child becomes increasingly discriminating as the various roles of sibling, student, lover, and many more are learned. Each of these roles implies its own distinct set of appropriate behaviors in particular situations.

Shaping Behavior by Successive Approximations

10.20 How is shaping of behavior accomplished?

Before a response can be reinforced, it must occur. Extremely complex responses, such as saying new words in a foreign language, are unlikely ever to be performed spontaneously by the learner. If you do not know how to say "How do you do?" in Greek, you are unlikely ever to come out with the right phrase spontaneously, no matter how many sounds you utter. To try to overcome this problem, and to help an organism form new responses, Skinnerians often use a procedure called "shaping."

Shaping is a technique for producing successively closer approximations to a particular desired behavior. It consists of carefully observing and immediately rewarding any small variations of the behavior in the desired direction as they are spontaneously performed by the organism. At first, a large class of responses is reinforced; then gradually the class is narrowed, and reinforcement is given only for closer approximations to the final form of the desired behavior. For example, when teaching a pigeon to stand only in the center of a large bull's-eye target painted on the floor, one might reward the bird for standing increasingly close to the center.

The Patterning of Outcomes: Schedules of Reinforcement

The patterning, sequencing, or scheduling of reinforcement affects the future occurrence and strength of the reinforced behavior (Ferster & Skinner, 1957). Sometimes the scheduling of reinforcement may be even more important than the nature of the reinforcer (Morse & Kelleher, 1966). Continuous reinforcement usually increases the speed with which responses are learned. Intermittent reinforcement tends to produce more stable behavior that is more persistently maintained when reinforcement stops. For example, rewarding temper tantrums intermittently (by occasionally attending to them in an irregular pattern) may make them very durable. Since many potentially maladaptive behaviors, such as physical aggression and immature dependency, are rewarded intermittently, they can become very hard to eliminate (Plaud & Gaither, 1996).

10.21 Describe the effects of schedules of reinforcement on learning and extinction of behaviors.

Different schedules have different influences on operant responses. Operant strength is measured by the rate of responses: the more frequently a response is made in a given period of time, the greater its rate (and inferred strength).

Continuous reinforcement (CRF) is a schedule on which a behavior is reinforced every time it occurs. Responses are usually learned most quickly with continuous reinforcement. A child would become toilet trained more quickly if he were praised and rewarded for each successful attempt. While continuous reinforcement is easy to create in a laboratory, in life it is a rare experience; a partial reinforcement or intermittent schedule, in which a response is reinforced only some of the time, is much more common. We see partial reinforcement when a child's bids for attention succeed only occasionally in getting the parent to attend, or when the same sales pitch produces a sale once in a while, or when the gambler hits the jackpot but only in between many losing bets.

Behavior that has received **partial reinforcement** or **intermittent reinforcement** often becomes hard to eliminate even when reinforcement is withdrawn altogether. A mother who intermittently and irregularly gives in to her child's nighttime bids for attention (crying, calling for a drink of water, or for just one more story) may find the child's behavior very durable and unresponsive to her attempts to stop it by ignoring it. Many potentially maladaptive behaviors (facial tics, physical aggression, immature dependency) are hard to eliminate because they are rewarded intermittently.

The child with a speck of grit in her eye who successfully follows her father's instruction to blink to get it out may keep on blinking periodically long after the eye is clear of irritation. If her blinks are further reinforced by her parents' occasional attention (whether troubled concern or agitated pleas to "stop doing that!"), she may develop an unattractive facial tic that is extremely resistant to extinction. Likewise, as has often been noted, the gambler who hit a jackpot once may persist for a long time even when the payoff becomes zero. The persistence of behavior after partial reinforcement suggests that when one has experienced only occasional, irregular, and unpredictable reinforcement for a response, one continues to expect possible rewards for a long time after the rewards have totally stopped.

Superstitions: Getting Reinforced into Irrationality

The relationship between the occurrence of an operant response and the reinforcement that follows it is often causal. For example, turn the door knob and the door opens, the outcome reinforcing the action. Consequently, in the future, we are likely to turn door knobs to enter and leave rooms, and our behavior at the door seems rational. Often, however, the response–reinforcement relationship may be quite accidental, and then bizarre and seemingly superstitious behavior and false beliefs may be produced (Matute, 1994). For example, a primitive tribe may persist in offering human sacrifices to the gods to end severe droughts because occasionally a sacrifice has been followed by rain.

The development of superstition, according to Skinner, may be demonstrated by giving a pigeon a bit of food at regular intervals—say every 15 seconds—regardless of what he is doing. Skinner (1953, p. 85) describes the strange rituals that may be conditioned in this way:

> When food is first given, the pigeon will be behaving in some way—if only standing still—and conditioning will take place. It is then more probable that the same behavior will be in progress when food is given again. If this proves to be the case, the "operant" will be further strengthened. If not, some other behavior will be strengthened. Eventually a given bit of behavior reaches a frequency at which it is often reinforced. It then becomes a permanent part of the repertoire of the bird, even though the food has been given by a clock which is unrelated to the bird's behavior. Conspicuous responses which have been established in this way include turning sharply to one side, hopping from one foot to the other and back, bowing and scraping, turning around, strutting, and raising the head. The topography of the behavior may continue to drift with further reinforcements, since slight modifications in the form of response may coincide with the receipt of food.

Punishment

Skinner focused on the role of rewards, but punishment or **aversive stimulation** is also important. In laboratory studies of anxiety, the unconditioned stimulus is usually a painful electric shock and the stimulus to be conditioned is a discrete event such as a distinctive neutral tone or a buzzer. Generally, human life is not that simple and neat. Often "aversive stimuli" involve punishments that are administered in less obvious and less controlled

10.22 How are superstitious behaviors established? Why is it so difficult to abolish them?

10.23 Describe advantages and possible disadvantages of punishment involving aversive stimulation. What is the preferred approach to eliminating undesirable behavior?

ways. These punishments may be conveyed subtly, by facial expressions and words rather than by brute force, and in extremely complicated patterns, by the same individuals who also nurture the child, giving love and other positive reinforcement. Moreover, the events that are punished often involve more than specific responses; they sometimes entail long sequences of overt and covert behavior (Aronfreed, 1994).

The behaviors that are considered inappropriate and punishable depend on such variables as the child's age and sex as well as the situation. Obviously the helplessness and passivity that are acceptable in a young child may be maladaptive in an older one, and the traits valued in a girl may not be valued in a boy. While the mother may deliberately encourage her son's dependency and discourage his aggressiveness, his school peers may do the reverse, ridiculing dependency at school and modeling and rewarding aggression and self-assertion. Given this, the influence of punishment on personality development is, not surprisingly, both important and complex (Aronfreed, 1968, 1994).

A careful review of research on the effects of punishment upon children's behavior concluded, in part, that:

> . . . *aversive stimulation, if well timed, consistent, and sufficiently intense, may create conditions that accelerate the socialization process, provided that the socialization agents also provide information concerning alternative prosocial behavior and positively reinforce any such behavior that occurs* (Walters & Parke, 1967, p. 218).

The important point to remember here is that when punishment is speedy and specific it may suppress undesirable behavior, but it cannot teach the child desirable alternatives. Therefore, parents should use positive techniques to show and reinforce appropriate behavior that the child can employ in place of the unacceptable response that has to be suppressed (Walters & Parke, 1967). In that way, the learner will develop a new response that can be made without getting punished. Without such a positive alternative, the child faces a dilemma in which total avoidance may seem the only possible route. Punishment may have very unfortunate effects when the child believes there is no way in which he or she can prevent further punishment and cope (Linscheid & Meinhold, 1990). If you become convinced that no potentially successful actions are open to you, that you can do nothing right, depression, hopelessness, and negative thinking may follow (Nolen-Hoeksema, 1997; Seligman, 1975).

SUMMARY

THE BEHAVIORAL APPROACH TO PSYCHODYNAMICS: DOLLARD AND MILLER

- Dollard and Miller fused psychoanalytic concepts with the more objective language and methods of laboratory studies of animal learning.
- Their theory emphasizes drive, cue, response, and reinforcement as the basic components of learning. Events that reduce a drive serve as reinforcements.
- Conflict exists when two or more goals are mutually exclusive. This conflict can create anxiety and repression.

TWO TYPES OF CONDITIONING

- Classical conditioning principles have been extended to explain some complex social phenomena and neurotic or abnormal behaviors, such as irrational fears.
- Traumatic fear may generalize so that events and cognitions closely associated with the original traumatic experiences may later evoke anxiety even after the objective danger is gone.

- In operant conditioning, behavior patterns may be modified by changing the consequences (reinforcements) to which they lead. Information and attention, as well as food and sexual gratification, are among the many outcomes that can serve as reinforcers.

B. F. SKINNER'S INFLUENCE ON PERSONALITY

- Analysis of the stimulus conditions controlling behavior replaces inferences about internal conflicts and underlying motives in Skinner's conceptualization.

- Discrimination in learning is fundamental in the socialization process. When behavior yields similar consequences under many conditions, generalization occurs, and the individual may display similar behavior patterns across diverse settings.

- Behavior may be shaped by reinforcing successively closer approximations to a particular desired behavior.

- While continuous reward or reinforcement for behavior may result in faster learning, irregular or intermittent reinforcement often produces more stable behavior that persists even when reinforcement is withdrawn. Many potentially maladaptive behaviors are rewarded irregularly and may therefore become very resistant to change.

- Irrational behavior may be created by accidental/noncausal pairings of behavior and response.

- The influence of punishment is complex and depends on many conditions, such as its timing.

KEY TERMS

11

ANALYZING AND MODIFYING BEHAVIOR

Many psychologists recognized that the principles of classical and operant conditioning, discussed in the last chapter, might be useful for treatment of at least some human behavioral problems, such as specific fears. Most psychologists, however, were concerned that the early behavioral work was based largely on studies with animals constrained in artificial laboratory situations. Elegant experiments were done on the behavior of rats running in mazes and of pigeons pecking on levers as food pellets dropped down. But how could one meaningfully extend the results from these studies to personality and the complex lives of people?

At the same time (before the 1970s), most clinicians saw behavioral concepts as too superficial for understanding the complexities of personality and irrelevant for helping people with personality problems. Partially in response to these challenges, behaviorally oriented workers tried to apply their ideas and methods to people and to issues relevant to personality and personality change. For several decades, they began to treat some of the most difficult behavioral and personality problems that had resisted other forms of therapy (e.g., O'Donohue et al., 2001). For example, they were allowed to try to treat hospitalized people who were so severely disturbed that there was little to lose by attempting experimental innovations with them after other available methods had proved to be unsuccessful. This chapter gives you a sense of their main strategies for assessment and change that are still relevant in various forms. By learning about some of these procedures, you will also get a better picture of the underlying philosophy that guides work at this level.

▶ CHARACTERISTICS OF BEHAVIORAL ASSESSMENTS

Rather than trying to infer the person's broad traits and motives, behavioral approaches focus on the specific conditions and processes that might govern his or her behavior.

Keep in mind again that all scientific psychological approaches are based on behavioral observation: giving answers on personality inventories or saying what an inkblot looks like are behaviors just as much as crying or running or fighting. However, in most personality assessment, the observed behaviors serve as highly indirect *signs* of the traits or motives that might underlie them. In contrast, in behavior assessments, the observed behavior is treated as a *sample,* and interest is focused on how the specific sampled behavior is affected by alterations in conditions (Mischel, 1968). For example, if the interest is in a child's physical and verbal aggression when relating to peers at preschool, then that is the type of behavior and situation that will be observed and assessed directly, rather than relying on teacher's or parent's reports, or scores on personality tests (O'Donohue et al., 2001).

11.1 How does behavioral assessment differ from trait measures of personality?

Case Example: Conditions "Controlling" Gary's Anxiety

To illustrate the general strategy of behavior assessment, let's again consider the case of Gary W. An assessment of Gary at this level obviously would focus on his behavior in relation to stimulus conditions. But what behaviors, and in relation to which conditions? Rather than seek a portrait of Gary's personality and behavior "in general," or an estimate of his "average" or dominant attributes, the focus is much more specific at the behavioral level of analysis. The behavior patterns selected for study depend on the particular problem that requires investigation. In clinical situations, the client indicates the priorities; in research contexts, they are selected by the investigator.

During his first term of graduate school, Gary found himself troubled enough to seek help at the school's counseling center. As part of the behavioral assessment that followed, Gary was asked to list and rank in order of importance the three problems that he found most distressing in himself and that he wanted to change if possible. He listed "feeling anxious and losing my grip" as his greatest problem. To assess the behavioral referents for his felt "anxiety," Gary was asked to specify in more detail just what changes in himself indicated to him that he was or was not anxious and "losing his grip."

He indicated that when he became anxious he felt changes in his heart rate, became tense, perspired, and found it most difficult to speak coherently. Next, to explore the covariation between increases and decreases in this state and changes in stimulus conditions, Gary was asked to keep an hour-by-hour diary sampling most of the waking hours during the daytime for a period of two weeks and indicating the type of activity that occurred during each hour. Discussion with him of this record suggested that anxiety tended to occur primarily in connection with public speaking occasions—specifically, in classroom situations in which he was required to speak before a group. As indicated by the summary shown in Table 11.1, on only one occasion that was not close in time to public speaking did Gary find himself highly anxious. That occasion turned out to be one in which he was brooding in his room, thinking about his public speaking failures in the classroom.

TABLE 11.1   Occurrence of Gary's Self-Reported Anxiety Attacks in Relation to Public Speaking

Occurrence of Anxiety	Hours with Anxiety (10)	Hours without Anxiety (80)
Within 1 hour of public speaking	9 (90%)	0 (0%)
No public speaking within 1 hour	1 (10%)	80 (80%)

Having established a covariation between the occurrence of anxiety and public speaking in the social-evaluative conditions of the classroom, his assessors identified the specific components of the public speaking situation that led to relatively more and less anxiety. The purpose here was to establish a hierarchy of anxiety-evoking stimuli ranging from the mild to the exceedingly severe. This hierarchy then was used in a treatment designed to gradually desensitize Gary to these fear stimuli.

Note that this behavioral assessment of Gary is quite specific: It is not an effort to characterize his whole personality, to describe "what he is like," or to infer his motives and dynamics. Instead, the assessment restricts itself to some clearly described problems and tries to analyze them in objective terms without going beyond the observed relations. Moreover, the analysis focuses on the stimulus conditions in which Gary's behavior occurs and on the covariation between those conditions and his problem. Behavior assessment tends to be focused assessment, usually concentrating on those aspects of behavior that can be changed and that require change. Indeed, as you will see often in this chapter, behavior assessment and behavior change (treatment) are closely connected.

The assessment of Gary's anxiety illustrates one rather crude way to study stimulus–response covariations. Of course there are many different ways in which these covariations can be sampled. This chapter illustrates some of the main tactics developed for the direct measurement of human behavior at this level of analysis.

Direct Behavior Measurement

For many purposes in personality study, it is important to sample and observe behavior in carefully structured, lifelike situations. In clinical applications, direct observation may give both client and assessor an opportunity to assess life problems and to select treatment objectives. Direct observation of behavior samples also may be used to assess the relative efficacy of various treatment procedures. Finally, behavior sampling has an important part in experimental research on personality.

The types of data collected in the behavioral approach include situational samples of both nonverbal and verbal behavior, as well as physiological measurements of emotional reactions. In addition, a comprehensive assessment often includes an analysis of effective rewards or reinforcing stimuli in the person's life. Examples of all of these measures are given in the following sections.

11.2   Describe how behavioral assessment has been used to assess anxieties and psychotic behavior.

Situational Behavior Sampling.   You already saw examples of **behavior sampling** in Chapter 2; such sampling is basic for much of the work at this level. Given the important role of fears and avoidance behavior, much attention has been given to assessing in detail the strength of diverse avoidance behaviors reliably in clinical situations. This was done, for example, by exposing fearful individuals to a series of real or symbolic fear-inducing stimuli. For example, fear of heights was assessed by measuring the distance that the phobic person could climb on a metal fire escape (Lazarus, 1961).

The same people were assessed again after receiving therapy to reduce their fears. In this phase, they were invited to ascend eight stories by elevator to a roof garden and to count the passing cars below for two minutes. Claustrophobic behavior—fear of closed spaces—was measured by asking each person to sit in a cubicle containing large French windows opening onto a balcony. The assessor shut the windows and slowly moved a large screen nearer and nearer to the person, thus gradually constricting her space. Of course each person was free to open the windows, and thereby to terminate the procedure, whenever she wished, although she was instructed to persevere as long as possible. The measure of claustrophobia was the least distance at which the person could tolerate the screen. As another example, Table 11.2 shows a checklist for performance anxieties before making a public speech.

Direct behavior sampling has also been used extensively in the analysis of psychotic behavior. One study, for instance, employed a time-sampling technique. At regular, 30-minute intervals, psychiatric nurses sought out and observed each hospitalized patient for periods of one to three minutes, without directly interacting with him (Ayllon & Haughton, 1964). The behavior observed in each sample was classified based on the occurrence of three experimenter-defined behaviors (for example, psychotic talk), and the time-check recordings were used to compute the relative frequency of the various behaviors. This time-sampling technique was supplemented by recordings of all the interactions between patient and nurses (such as each time the patient entered the nursing office). The resulting data served as a basis for designing and evaluating a treatment

TABLE 11.2   Timed Behavioral Checklist for Performance Anxiety

Behavior Observed	Time Period 1 2 3 4 5 6 7 8
1. Paces	
2. Sways	
3. Shuffles feet	
4. Knees tremble	
5. Extraneous arm and hand movement (swings, scratches, toys, etc.)	
6. Arms rigid	
7. Hands restrained (in pockets, behind back, clasped)	
8. Hand tremors	
9. No eye contact	
10. Face muscles tense (drawn, tics, grimaces)	
11. Face "deadpan"	
12. Face pale	
13. Face flushed (blushes)	
14. Moistens lips	
15. Swallows	
16. Clears throat	
17. Breathes heavily	
18. Perspires (face, hands, armpits)	
19. Voice quivers	
20. Speech blocks or stammers	

Source: Paul, Gordon. *Insight vs. desensitization in psychotherapy.* © 1966 by The Board of Trustees of the LeLand Stanford Jr. University, renewed 1994. With the permission of Stanford University Press, www.sup.org.

program. Similar ways of sampling and recording family interactions have been developed by others (Patterson, 1976; Ramsey et al., 1990). They studied highly aggressive children in the course of everyday family life, for example, at dinner. The attempt was to analyze the exact conditions under which aggression increased or decreased.

Verbal Behavior. As illustrated in the assessment of Gary's public speaking anxieties, a daily record may provide another valuable first step in the identification of problem-producing stimuli. Many behaviorally oriented clinicians routinely ask their clients to keep specific records listing the exact conditions under which their anxieties and problems seem to increase or decrease (Wolpe & Lazarus, 1966). The person may be asked to prepare by himself lists of all the stimulus conditions or events that create discomfort, distress, or other painful emotional reactions.

Finding Effective Rewards

11.3 How can direct observation be used to identify reinforcers that control behavior? How have tokens been used to change behavior and to identify reinforcers?

So far, we have considered the direct measurement of various responses. Behavior assessments, however, analyze not just what people do (and say and feel), but also the conditions that regulate or determine what they do. For that reason, behavior assessments have to find the rewards or reinforcers that may be influencing a person's behavior. If discovered, these reinforcers also can serve as incentives in therapy programs to help modify behavior in more positive or advantageous directions. Psychologists who emphasize the role of reinforcement in human behavior have devoted much attention to discovering and measuring effective reinforcers. People's actual choices in lifelike situations, as well as their verbal preferences or ratings, reveal some of the potent reinforcers that influence them. The reinforcement value of particular stimuli also may be assessed directly by observing their effects on the individual's performance (Daniels, 1994; Weir, 1965).

Primary reinforcers such as food, and **generalized conditioned reinforcers** such as praise, social approval, and money, are effective for most people. For example, in one case study, researchers and teachers attempted to reduce the disruptive behaviors of a blind, learning-disabled boy (Heitzman & Alimena, 1991). His behaviors were problematic because they disrupted the class and also prevented the boy from achieving optimal academic success. Differential reinforcement was used to reduce the amount of inappropriate behaviors to a socially acceptable level. In this procedure, the boy could not exceed a certain amount of disruptive or inappropriate behaviors in one day if he wanted to be rewarded. Examples of rewards were listening to a favorite tape, free time to talk to friends, and sitting in the teacher's car (he liked the feel of velour seats). After 26 days, there was an 88 percent reduction in target behaviors.

Sometimes, however, it is difficult to find potent reinforcers that would be feasible to manipulate. With disturbed groups (such as hospitalized schizophrenic patients), for example, many of the usual reinforcers prove to be ineffective, especially with people who have spent many years living in the back wards of a mental hospital. Ayllon and Azrin (1965) have shown how effective reinforcers can be discovered even for seemingly unmotivated psychotic patients. These reinforcers then can serve to motivate the patients to engage in more adaptive behavior.

As a first step, the patients were observed directly in the ward to discover their most frequent behaviors in situations that permitted them freedom to do what they wished. Throughout the day, observers carefully recorded the things the patients did, or tried to do, without pressures from the staff. The frequency of these activities provided an index of their potential values as reinforcers.

Six categories of reinforcers were established on the basis of extensive observation. These categories were: privacy, leave from the ward, social interactions with the staff, devotional opportunities, recreational opportunities, and items from the hospital canteen. "Privacy," for example, included such freedoms as choice of bedroom or of eating group, and getting a personal cabinet, a room-divider screen, or other means of preserving autonomy. "Recreational opportunities" included exclusive use of a radio or television set, attending movies and dances, and similar entertainment.

The patients could obtain each of the reinforcers with a specific number of **tokens** which they earned by participating in such rehabilitative functions as self-care and job training. A sensitive index of the subjective reinforcement value of the available activities is obtained by considering the outcomes for which the patients later chose to exchange most of their tokens. Over 42 days, the mean tokens exchanged by eight patients for the available reinforcers are shown in Table 11.3. Note that chances to interact socially with the staff and opportunity for recreation and spiritual devotion are most unpopular. These results suggest that, with chronic hospitalized patients such as these, therapy programs that rely primarily on social motivations would not fare well. Instead, such reinforcers as privacy, autonomy, and freedom might be the most effective incentives.

Assessing Conditions Controlling Behavior

To assess behavior fully, behavior theorists believe that we have to identify the conditions that control it. But how do we know whether or not a response pattern is really controlled or caused by a particular set of conditions? Behaviorally oriented psychologists test the conditions by introducing a change and observing whether or not it produces the expected modification in behavior. They ask: Does a systematic change in stimulus conditions (a "treatment") in fact change the particular response pattern that it supposedly controls? If we hypothesize that a child's reading problem is caused by poor vision, we would expect appropriate treatment (such as corrective eye glasses or corrective surgery) to be followed by a change in the behavior (i.e., an improvement in reading). The same should be true for psychological causes. For example, if we believe that the child's reading difficulty is caused by anxiety about pressure to read from her mother, we should try to show that if the mother reduces her pressure, it will yield the expected improvement in reading. That is, to understand behavior fully, we need to know the conditions that cause it. We can be most confident that we understand those conditions when we can show that a change in them yields the predicted change in the response pattern.

TABLE 11.3   Mean Tokens Exchanged for Various Available Reinforcers (by 8 Patients During 42 Days)

Reinforcers	Mean Tokens Paid	Number of Patients Paying Any Tokens
Privacy	1352.25	8
Commissary items	969.62	8
Leave from ward	616.37	8
Social interaction with staff	3.75	3
Recreational opportunities	2.37	5
Devotional opportunities	.62	3

Source: Based on Ayllon, T., & Azrin, N. H. (1965). The measurement and reinforcement of behavior of psychotics. *Journal of the Experimental Analysis of Behavior, 8,* 357–383.

A rigid distinction between behavior assessment and treatment (i.e., behavior change) thus is neither meaningful nor possible. Indeed, some of the most important innovations in behavior assessment have grown out of therapeutic efforts to modify problematic behavior. A main characteristic of these assessment methods is that they are linked closely to behavior change and cannot really be separated from it.

The close connection between behavior assessment and behavior change is most evident in **functional analyses**—that is, analyses of the precise covariations between changes in stimulus conditions and changes in a selected behavior pattern. Such functional analyses are the foundations of behavior assessments, and they are illustrated most clearly in studies that try to change behavior systematically. The basic steps may be seen in a study that was designed to help a girl in nursery school.

Functional Analyses: Case Example

11.4 What was the usefulness of a reversal design in the functional analysis of Ann's problem behaviors?

Ann was a bright four-year-old from an upper-middle-class background who increasingly isolated herself from children in her nursery school (Allen et al., 1964). At the same time, she developed various ingenious techniques to gain prolonged attention from the adults around her. She successfully coerced attention from her teachers, who found her many mental and physical skills highly attractive. Gradually, however, her efforts to maintain adult attention led her to become extremely isolated from other children.

Soon Ann was isolating herself most of the time from other youngsters. This seemed to be happening because most of the attention that adults were giving her was contingent, quite unintentionally, upon behaviors that were incompatible with Ann's relating to other children. Precisely those activities that led Ann away from play with her own peers were being unwittingly reinforced by the attention that her teachers showered on her. The more distressing and problematic Ann's behavior became, the more it elicited interest and close attention from her deeply concerned teachers.

Ann was slipping into a vicious cycle that had to be interrupted. A therapeutic plan was formed where Ann no longer received adult attention for her withdrawal from peers and her attempts at solitary interactions with adults. At the same time, the adults gave her attention only when she played with other children. That is, attention from adults became contingent on her playing with her peers.

As part of the assessment, two observers continuously sampled and recorded Ann's proximity to and interactions with adults and children in school at regular 10-second intervals. The therapeutic plan was instituted after five days of baseline data had been recorded. Now, whenever Ann started to interact with children an adult quickly attended to her, rewarding her participation in the group's play activities. Even approximations to social play, such as standing or playing near another child, were followed promptly by attention from a teacher. This attention was designed to further encourage Ann's interactions with other children. For example: "You three girls have a cozy house. Here are some more cups, Ann, for your tea party." Whenever Ann began to leave the group or attempted to make solitary contacts with adults, the teachers stopped attending to her.

Figure 11.1 summarizes the effects of the change in the consequences to Ann for isolate behavior with her peers. Notice that in the baseline period before the new response–reinforcement contingencies were instituted, Ann was spending only about 10 percent of her school time interacting with other children and 40 percent with adults. For about half the time she was altogether solitary. As soon as the **contingencies of reinforcement** were changed and adults attended to Ann only when she was

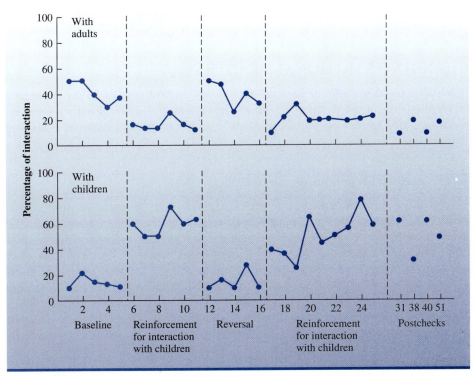

Figure 11.1 Percentages of time spent by Ann in social interaction during approximately two hours of each morning session.

Source: From Allen, E. K., Hart, B., Buell, J. S., Harris, F. R., & Wolf, M. M. (1964). Effects of social reinforcement on isolated behavior of a nursery school child. *Child Development, 35,* p. 515.

near children, her behavior changed quickly in accord with the new contingencies. When adult–child interactions were no longer followed by attention, they quickly diminished to less than 20 percent. On the first day of this new arrangement (day six), Ann spent almost 60 percent of her time with peers.

To assess the effects of reinforcement more precisely, the procedures were reversed on days 12 to 16. Adults again rewarded Ann with their attention for interacting with them and disregarded her interactions with children. Under these conditions (the "reversal" days in Figure 11.1), Ann's previous behavior reappeared immediately. In a final shift (beginning on day 17), in which attention from adults again became contingent upon Ann's interacting with children, her contact with peers increased to about 60 percent. After the end of the special reinforcement procedures (day 25), periodic postchecks indicated that Ann's increased play behavior with peers tended to remain fairly stable.

A complete analysis also must consider the total relations among stimulus conditions rather than focus on single aspects of reinforcement in isolation. These assessments showed, for example, that this child was highly discriminating in the very particular times and circumstances during which she became self-destructive. For example, massive withdrawal of attention—as when the experimenter withheld attention from an entire session—did not affect her self-destructive behavior. In contrast, the removal of smiles and attention *only* for previously reinforced responses changed her behavior (also see

Smith et al., 1992). Note that in this approach, assessment and behavior change become inextricably fused: the assessments guide the therapeutic program, and the efficacy of the treatment program is in turn continuously assessed (e.g., Frank & Hudson, 1990).

▶ CHANGING EMOTIONAL REACTIONS

Next we will consider some of the main techniques and findings of behavior therapy based on the concepts of the classic learning-conditioning theories. First we focus on methods designed to change previously learned disadvantageous emotional reactions, such as anxiety.

Desensitization: Overcoming Anxiety

11.5 What behavior change principle underlies systematic desensitization? Describe the three procedural steps in the treatment.

For many years, most therapists were afraid that symptom substitution would occur if they tried to remove the problematic behaviors. This concern stemmed from the Freudian assumption that attempts to change problematic behavior without first getting at the unconscious causes would lead to other problems that would be even worse than the original ones. This belief was based on the medical model of illness in which it is important to distinguish the observed behavior (e.g., pain in a leg) and its potential cause (e.g., a malignant tumor). Treating the symptoms with pain-killers rather than the cancer of course would soon be disastrous. Joseph Wolpe, a psychiatrist who became skeptical about psychoanalytic theory, took the risk of attempting

Systematic desensitization has been used effectively to treat anxiety about public speaking and many other fears.

direct behavior modification with many of his patients. In 1958 he published a book describing a method of **systematic desensitization** based on the principle of classical conditioning.

Wolpe was impressed by the work of such early learning theorists as Pavlov and believed that neurosis involves maladaptive learned habits, especially anxiety (fear) responses. In neurotic behavior, he hypothesized, anxiety has become the conditioned response to stimuli that are not anxiety-provoking for other people. He reasoned that therapy might help the neurotic individual to inhibit anxiety by counterconditioning him to make a competing (antagonistic) response to anxiety-eliciting stimuli. In his words, "If a response antagonistic to anxiety can be made to occur in the presence of anxiety-evoking stimuli so that it is accompanied by a complete or partial suppression of the anxiety responses, the bond between these stimuli and the anxiety response will be weakened" (Wolpe, 1958, p. 71). His attempt to desensitize the individual to anxiety-evoking stimuli includes three steps (summarized in Table 11.4):

1. *Establishing the anxiety stimulus hierarchy.* First the therapist helps the client identify the situations that evoke distressing emotional arousal and avoidance, usually in the course of interviews. Sometimes a person has many areas of anxiety, such as fear of failure, self-doubts, dating, guilt about sex, and so on. Regardless of how many areas or "themes" there are, each is treated separately.

For each theme, the person grades or ranks the component stimuli on a **hierarchy of severity of anxiety** ranging from the most to the least intensely anxiety-provoking events (see Table 11.5). For example, a person who is terrified of public speaking might consider "reading about speeches while alone in my room," a mildly anxiety-provoking stimulus, while "walking up before the audience to present the speech" might create severe anxiety in him (Paul, 1966). In Gary's case, "the minute before starting a formal speech" was the most anxiety-provoking, while "watching a friend practice a speech" and "taking notes in the library for a speech" were only moderately disturbing. As another example, a woman who sought treatment for sexual dysfunction indicated that "being kissed on cheeks and forehead" evoked merely mild anxiety but thinking about items like "having intercourse in the nude while sitting on husband's lap" produced the most intense anxiety in her (Lazarus, 1963).

TABLE 11.4   Three Basic Steps in the Desensitization of Anxiety

Step	Example
1. Establishing the anxiety stimulus hierarchy; anxiety-evoking situations ranked from least to most severe	*Low anxiety:* reading about speeches alone in your room
	Intermediate anxiety: getting dressed the morning on which you are to give a speech
	High anxiety: presenting a speech before an audience
2. Learning an incompatible response	Learning deep muscle relaxation by tensing and relaxing various muscle groups (head, shoulders, arms), deep breathing techniques, and similar methods
3. Counterconditioning: learning to make the incompatible response to items in the hierarchy	Practicing relaxation responses to the lowest item on the hierarchy and moving gradually to the higher items

Note: These items are examples from much longer hierarchies.

TABLE 11.5   Items of Different Severity from Four Anxiety Hierarchies°

Severity (Degree of Anxiety)	Anxiety Hierarchies (Themes)			
	1 Interpersonal Rejection	2 Guilt About Work	3 Test-Taking	4 Expressing Anger
Low	Thinking about calling Mary (a new girlfriend) tonight	Thinking "I still haven't answered all my mail"	Getting the reading list for the course	Watching strangers quarrel in street
Intermediate	Asking for a date on the telephone	Taking off an hour for lunch	Studying at my desk the night before the final	My brother shouting at his best friend
High	Trying a first kiss	Going to a movie instead of working	Sitting in the examination room waiting for the test to be handed out	Saying "No! I don't want to!" to mother

°These items are examples from much longer hierarchies.

2.   *Training the incompatible response (relaxation).* After identifying and grading the stimuli that evoke anxiety, the person needs to learn responses that can be used later to inhibit anxiety. Wolpe prefers to use **relaxation responses** because they can be taught easily and are always inherently incompatible with anxiety: no one can be relaxed and anxious simultaneously. The therapist helps the client to learn relaxation by elaborate instructions that teach first to tense and then to relax parts of the body (arms, shoulders, neck, head) until gradually an almost hypnotic state of total calm and deep muscle relaxation is achieved. Most people can learn how to relax within a few sessions. The critical problem is to learn to relax to anxiety-evoking stimuli, and that task is attempted in the next phase.

3.   *Associating anxiety stimuli and incompatible responses.* In the critical phase, **counterconditioning**, the client is helped to relax deeply, and then is presented with the least anxiety-arousing stimulus from the previously established hierarchy. Usually the stimulus event is described verbally or presented symbolically (in a picture) while the client is deeply relaxed and calm. As the therapist says the words for the item (e.g., "walking down a fire escape from the 6th floor") the client with a fear of heights tries to generate the most vivid image of it that his or her imagination can form. As soon as the client can concentrate on this item while remaining calm, the next, more severe item from the hierarchy is introduced until, step by step, the entire hierarchy is mastered (e.g.,"looking down from the roof of a skyscraper").

If at any point in the procedure the client begins to become anxious while presented with an anxiety stimulus, he or she signals the therapist. The client is promptly instructed to discontinue the image of the stimulus until calm again. Then a somewhat less severe item from the hierarchy is presented so that he or she can concentrate on it without anxiety. After that, the client is ready to advance to the next item in the anxiety hierarchy and the step-by-step progress up the list can be resumed.

In sum, the **desensitization** (counterconditioning) procedure attempts to have responses incompatible with anxiety (such as relaxation) occur in the presence of mildly anxiety-evoking stimuli. The incompatible response then will at least partially prevent the anxiety response. In that way, the association between the aversive stimulus and anxiety becomes reduced, while the association of the stimulus with the relaxation reaction becomes strengthened (Guthrie, 1935; Wolpe, 1958).

McClanahan (1995), for example, treated a woman who suffered from severe habitual nail-biting. McClanahan noted that the nail-biting was almost always precipitated by anxiety which included feeling overwhelmed, apprehensive, nervous, and worried. The desensitization techniques used were deep muscle relaxation and Transcendental Meditation. Systematic desensitization reduced anxiety and decreased the frequency and duration of nail-biting significantly. Clinical reports of successful desensitization may be encouraging, but more conclusive evidence comes from controlled experiments, and many studies indicate that desensitization is a valuable method for modifying phobias and reducing anxiety (Kazdin & Wilson, 1978; Wilson & O'Leary, 1980).

Conditioned Aversion: Making Stimuli Unattractive

While some people suffer because they have learned to react negatively to certain situations, others are plagued because they become pleasurably aroused by, or even addicted to, stimuli that most people in the culture find neutral or even aversive. One example of this problem is fetishistic behavior, in which the person may become sexually excited by such objects as undergarments. In these cases, things that are neutral or even disgusting for most people have acquired the power to produce pleasurable emotional arousal. Another example is drug addiction for which the arousal may have a significant and necessary physiological component to it. Such reactions can provide the person with some immediate reduction of pain, but they often have severely negative and destructive long-term consequences (Baker et al., 2003).

A positively valued stimulus may be neutralized by counterconditioning if it is presented with stimuli that evoke extremely unpleasant reactions. Gradually, as a result of repeated pairings, the previously positive stimulus acquires some of the aversive emotional properties evoked by the noxious events with which it has been associated.

11.6 What conditioning procedures are used in aversion therapy? What are some limitations of this approach?

An Example: Treating Cocaine Dependency

Chemical aversion therapy was used to treat cocaine addicts who volunteered to participate in a two-week study (Frawley & Smith, 1990). The stimulus used to represent cocaine was a mixture of chemicals which tasted and smelled like "street cocaine." In the treatment room, there were pictures of paraphernalia for cocaine use, as well as the utensils used to snort cocaine. The patients were instructed to "snort" the "cocaine" in their normal fashion. They then were given an injection of nausea-inducing drugs and were instructed to continue to "snort" the "cocaine." Afterwards, the patients were encouraged to focus on paraphernalia, pictures of cocaine, and to pair the cocaine use with negative consequences, all while experiencing nausea.

At 6 months posttreatment, more than half of the cocaine addicts had totally abstained from cocaine since the treatment. As predicted, the pairing of the nausea with the cocaine use produced a conditioned aversion for the recovering addicts and helped them break their habit. Systematic counterconditioning has also been attempted with other addictions such as alcoholism (Bandura, 1969, 1986).

Psychologists are reluctant to use **aversion therapies** like this one because they inflict aversive experiences on a troubled person. However, aversion therapies usually are attempted only after other forms of help (such as interview therapies) have been tried unsuccessfully. In some cases, aversion treatments have come as a last resort in lieu of more drastic treatments, such as long imprisonment or irreversible brain surgery (Rachman & Hodgeson, 1980; Raymond, 1956). And usually they are not imposed on the client: they are voluntary and the person submits to them with full knowledge and consent.

Indeed, it is this very dependence on the client's cooperation that limits the efficacy of the treatment. That is, after the initial counterconditioning trials, the client often may revert to his or her fetish without submitting voluntarily to further treatment. Since it becomes impractical to hospitalize him continuously or remove him from exposure to the problematic stimuli, he must learn to administer aversive stimulation to himself whenever necessary. For example, he may be taught to administer electric shock to himself from a small portable, battery-operated apparatus concealed in his clothing, or to induce aversive thoughts or imagery whenever he experiences the problematic urges. Thus, counterconditioning procedures ultimately provide the individual with a form of *self*-control. Whether or not he continues to practice and seek this self-control is up to him. And whether or not he practices self-control determines how effectively his new behavior will be maintained (Mischel et al., 1996).

▶ CHANGING BEHAVIOR

Many psychologists have tried to modify maladaptive behaviors by changing the consequences to which those behaviors lead. Guided to a large extent by B. F. Skinner's ideas about learning, they try to withdraw reinforcement for undesired behavior and to make attention, approval, or other reinforcement contingent on the occurrence of more appropriate, advantageous behavior (e.g., Haring & Breen, 1992). Their basic procedure is well illustrated in the work of Hawkins and his colleagues (1966).

Case Example: Hyperactivity

Hawkins' case was Peter, a young child of low intelligence. Peter was brought to a clinic by his mother because he was "hyperactive" and "unmanageable." Because the problems seemed to involve the relations between Peter and his mother, he was assessed and treated directly in his home. His mother served as a therapist under the guidance of the professional workers.

11.7 How have operant procedures been used to treat autistic children?

A first task was to specify the problematic behaviors. Direct observations of Peter in the home revealed the following problems to be among the most common and disturbing ones:

1. Biting his shirt or arm.
2. Sticking his tongue out.
3. Hitting and kicking himself, other people, or objects.
4. Using derogatory names.
5. Removing his clothing and threatening to remove it.

The frequency of these and similar behaviors was carefully recorded at 10-second intervals during one-hour observation sessions in the home. After the first assessments were completed, the researcher helped the mother to recognize the occurrence of Peter's nine most objectionable behaviors. Whenever these occurred during subsequent one-hour sessions at home, she was taught to respond to them with definite steps. These steps involved signaling to Peter when his behavior became disruptive and, if a verbal warning failed, isolating him briefly in a separated, locked "time out" room without toys and other attractions. Release from the room (and reinstatement of play, attention, and nurturance) was contingent on Peter's terminating the tantrum and showing more reasonable, less destructive behavior. This arrangement was opposite to the one the mother may have inadvertently used in the past, when she became increasingly concerned and attentive (even if distressed) as Peter became increasingly wild. Subsequent assessment revealed that the new regimen was effective in minimizing Peter's outbursts. While apparently helpful to Peter's development, reducing his tantrums may have been just one step toward the more extensive help he needed.

Using a combination of modeling and reinforcement procedures, Lovaas and his coworkers (Lovaas et al. 1966, 1991) modified the deficient speech and social behaviors of severely disturbed ("autistic") children who were unable to talk. First the therapist modeled the sounds himself. He rewarded the child only for vocalizing the modeled sounds within a specified time interval. As the child's proficiency increased, the therapist proceeded to utter more complicated verbal units. Gradually, training progressed from sounds to words and phrases. As the training continued, rewards from the therapist became contingent on the child's reproducing increasingly elaborate verbalizations more skillfully (i.e., more quickly and accurately). The combination of modeling and reinforcement procedures gradually helped the child to learn more complex meanings and complicated speech. Research like this shows the value of wisely used reinforcement; but rewards also may be hazardous, as discussed in In Focus 11.1.

11.8 How can overjustification effects be avoided in behavior change attempts involving the use of positive reinforcers?

IN FOCUS 11.1

REWARDS MAY BACKFIRE

Rewards are important for effective behavior, but they can be used unwisely. A major purpose of effective therapy (and socialization) is to wean the individual away from external controls and rewards so that his behavior becomes increasingly guided and supported by intrinsic gratifications—that is, satisfactions closely connected with the activity itself. Therefore, it is essential to use rewards or incentives only to the extent necessary to initiate and sustain prosocial (adaptive, desirable) behavior.

External incentives may be important in order to encourage a person to try activities that have not yet become attractive for him or her. When rewards are used to call attention to a good job, or to an individual's competence at an activity, they may actually bolster interest. They provide positive performance feedback and supply tangible evidence of excellence (Harackiewicz, Manderlink, & Sansone, 1984). Approval and praise from parents for trying to play a violin, for example, may be helpful first steps in encouraging the child's earliest musical interest. But when the youngster begins to experience activity-generated satisfactions (for example, from playing the music itself), it becomes important to avoid excessive

external rewards. Too much reward would be unnecessary and possibly harmful, leading the child to play for the wrong reasons and making him or her prone to lose interest easily when the external rewards are reduced or stopped altogether (Lepper, Greene, & Nisbett, 1973). Likewise, children need to be encouraged to develop sense of fairness, empathy, and helpful responsiveness to the needs of other people. To become a social being requires attention to the long-term consequences of one's behavior and not just to its immediate payoffs. While such sensitivities might initially be encouraged by external rewards, ultimately they need to be sustained by gratification from the activities themselves.

In sum, **overjustification** of an activity by excessive external reward may interfere with the satisfactions (intrinsic interests) that would otherwise be generated by the activity itself. Excessive external rewards may even have boomerang effects and lead the recipient to devalue and resist the rewarded activity. Children who get paid to do their homework may develop a feeling that school work is an aversive chore rather than a route to learning and self-development.

People often are judged to be maladjusted mainly because they have not learned how to perform the behavior patterns necessary to effectively meet the social or vocational demands they encounter. They cannot behave appropriately because they lack the skills required for successful functioning. For example, the socially deprived, economically underprivileged person may suffer because he or she never has acquired the response patterns and competencies needed to obtain success in vocational and interpersonal situations. Similarly, the high school dropout in our culture does indeed carry an enduring handicap. Such **behavioral deficits,** if widespread, may lead to many other problems, including severe emotional distress and avoidance patterns to escape the unhappy consequences of failure and incompetence. Many special learning programs have been designed to teach people a variety of problem-solving strategies and cognitive skills (Bijou, 1965), and to help them achieve many other positive changes in behavior (e.g., Kamps et al., 1992; Karoly, 1980).

Contingency Contracting

11.9   Describe the use and effectiveness of contingency contracting in self-control programs.

A move to enroll the person actively in his or her own behavior change program whenever possible is reflected in the use of **contingency contracting** (Rimm & Masters, 1974; Thoresen & Mahoney, 1974). An example was the treatment of "Miss X" for drug abuse as described by Boudin (1972). Miss X, a heavy user of amphetamines, made a contingency contract with her therapist. She gave him $500 (all of her money) in 10 signed checks of $50 each and committed him to send a check to the Ku Klux Klan (her least favorite organization) whenever she violated any step in a series of mutually agreed-upon specific actions for curbing her drug use. After applying the contract for three months, a follow-up for a two-year period indicated that Miss X did not return to amphetamine use. The principle of contingency contracting can be extended to a wide variety of commitments in which the client explicitly authorizes the therapist to use rewards to encourage more advantageous behaviors in ways formally agreed upon in advance.

Symptom Substitution?

11.10   What controversy gave rise to the concept of symptom substitution? What research evidence exists for symptom substitution?

Do behavior therapies neglect the causes of the person's problematic behavior and thus leave the "roots" unchanged while modifying only the "superficial" or "symptomatic" behaviors? It is often charged that behavior therapists ignore the basic or underlying causes of problems. Advocates of behavior therapy insist that they do seek causes but that they search for *observable* causes controlling the current problem, not its historically distant antecedents or its hypothesized but unobservable psychodynamic mechanisms. Traditional, insight-oriented approaches have looked, instead, for historical roots in the person's past and for theoretical mechanisms in the form of psychodynamics. The difference between these two approaches thus is not that one looks for causes while the other does not: Both approaches search for causes but they disagree about what those causes really are.

> *All analyses of behavior seek causes; the difference between social behavior and [psychodynamic] analyses is in whether current controlling causes or historically distant antecedents are invoked. Behavioral analyses seek the current variables and conditions controlling the behavior of interest. Traditional [psychodynamic] theories have looked, instead, for historical roots . . . (Mischel, 1968, p. 264).*

Traditional approaches ask about the patient, "Why did she become this kind of person?" Behavioral approaches ask, "What is now causing her to behave as she does and what would have to be modified to change her behavior?"

Does a neglect in treatment of the psychodynamics hypothesized by traditional therapies produce **symptom substitution**? In spite of many initial fears about possible symptom substitution, behavior change programs of the kind discussed in the preceding sections tend to be the most effective methods presently available. Researchers have found that the changed behaviors are not automatically replaced by other problematic ones (Bandura, 1986; Kazdin & Wilson, 1978; Lang & Lazovik, 1963; Paul, 1966; Rachman, 1967; Rachman & Wilson, 1980). On the contrary, when people are liberated from debilitating emotional reactions and defensive avoidance patterns, they generally tend to become able to function more effectively in other areas as well. As was noted years ago:

> Unfortunately, psychotherapists seem to have stressed the hypothetical dangers of only curing the symptoms, while ignoring the very real dangers of the harm that is done by not curing them (Grossberg, 1964, p. 83).

On the other hand, some enthusiastic proponents of behavior modification overlook the complexity of the client's problems. They may oversimplify the difficulties into one or two discrete phobias when in fact the client may have many other difficulties. In that case, it would not be surprising to find that even after removal of the initial problem, the individual still is beset with such other psychological troubles as self-doubts, feelings of worthlessness, and so on. Such a condition, of course, would imply that the person had an incomplete treatment rather than that symptom substitution had occurred. It would be extremely naive to think that reducing Gary's public speaking anxiety, for example, would make his life free of all other problems. Whatever other difficulties he might have would still require attention in their own right.

In sum, to avoid the emergence of disadvantageous behaviors, a comprehensive program must provide the person with more adaptive ways of dealing with life; such a program may have to go beyond merely reducing the most obvious problems. Behavior modification does not automatically produce generalized positive effects that remove all the person's troubles.

Evaluating the Consequences of Behavior, Not the Person

The behavioral approach avoids evaluating the health, adequacy, or abnormality of the person or personality as a whole. Instead, when judgments are made, they focus on evaluation of the individual's specific behaviors. Behaviors are evaluated on the basis of the kinds of consequences that they produce for the person who generates them and for other people who are around him or her. Evaluations about the positive or negative consequences of behavior are social and ethical judgments that depend on the values and standards of the community that makes them. Advantageous behaviors are those judged to have positive personal and interpersonal consequences (for example, helping people "feel good," or increasing constructive, creative outcomes) without any aversive impact on others. Behaviors that have negative, life-threatening, destructive consequences, or those that endanger the full potentialities of the person or other people (for example, debilitating fears, homicidal attempts), would be considered maladaptive.

The behavioral approach also implies a high value for the development of the individual's total competencies and skills so that he or she can maximize opportunities and options. Similarly, the person must be able to discern the important contingencies and rules of reinforcement in his or her life in order to maximize satisfactions and minimize aversive, disadvantageous outcomes. To be able to overcome unfavorable environments and life conditions, it is especially important to develop effective strategies for self-control.

Does Changing Behavior Change Personality?

11.11   Under what conditions can changes in specific behaviors foster personality change?

There is much controversy with regard to the depth and endurance of behavior change. Basically, the question is whether behavior change entails genuine, durable change or whether it is limited to relatively minor, specific behaviors that have little applicability to major life problems.

To facilitate transfer from treatment to life, workers at this level of analysis introduce into treatment stimulus conditions that are as similar as possible to the life situations in which the new behaviors will be used. For example, if the patient is afraid of going into elevators, the treatment may be conducted in elevators, not in the therapist's office.

The question remains: When fears and other emotional problems are reduced, and social competence improves through more adaptive, functional patterns of social behavior, does personality change? The answer depends on the definition of personality. There is considerable evidence that changing disadvantageous behaviors, such as a severe stutter or uncontrolled facial tics, and eliminating distressing, fear-producing emotional reactions, also improves how people feel about themselves, and their self-concepts (e.g., Bandura, 1997; Meichenbaum, 1995). Self-concepts, and self-esteem, tend to reflect at least in part the individual's actual competencies and how the person's behavior is seen by others (e.g., Leary & Downs, 1995). Our self-perceptions include the information that we get about the adequacy of our own behaviors, and if these self-perceptions are part of personality, then personality changes when behavior becomes more adaptive and gratifying. The individual who learns to perform more competently achieves more gratification and is also likely to develop a more positive attitude toward himself or herself. As a result of being able to overcome fears and anxieties, one should also become more confident. Reducing Gary's fears of public speaking would not be a cure-all, but it might certainly help him to feel more positively about himself and would open alternatives (for example, in his career opportunities) otherwise closed to him. Ultimately, if enough anxious behavior is brought under control, shyness may decrease, creating a cycle of improved behavior leading to improved confidence and positive expectations, leading to further personality change. At least, that is the rationale of much work at this level of analysis.

But while this may often be true, it does not always happen. Indeed, critics of behavior therapy note that people may suffer not because their behavior is inadequate but because they evaluate it improperly. That is, some people have problems with distorted self-concepts more than with performance. Often people label themselves and react to their own behaviors very differently than do the people around him and the rest of society. An esteemed financier, for example, may receive the rewards and praise of society while he is privately unhappy enough to commit suicide. Or a popular student who is the prom queen of her school might have secret

People's standards affect their reactions to performance outcomes. To some, an A− can feel like a failure.

doubts about her sexual adequacy and femininity and might be torturing herself with these fears.

Many personal problems involve inappropriate self-evaluations and self-reactions. In these cases, the difficulty often may be the person's appraisal of his or her performances and attributes rather than their actual quality and competence level. For example, a student may react self-punitively to his scholastic achievements even when their objective quality is high. The student who is badly upset with himself for an occasional grade less than an A may need help with self-assessment rather than with school work. Fortunately, work at the social cognitive level, reviewed in the next part of the text, goes on to address those issues.

Unexpected Similarities: Behavior Theory and Existentialism

Psychological theories sometimes have unexpected similarities. The behavioral approaches discussed in this part focus on what the person is *doing* in the here and now, rather than on reconstructions of personal history or inferences about hypothesized traits and motives. In this way, they have some unexpected similarities with work at the phenomenological level described in the previous part. The irony is that many leaders of the phenomenological approach were motivated originally to protest against behavioral-learning-conditioning approaches, just as the behavioral approaches were a protest against speculating about internal states and experiences that could not be directly observed. After the heat of the battles cools, however, one can see a surprising common theme. Thus, the behavioral focus on what the person is *doing*, rather than on attributes or motives, is compatible with the views of the famous existentialist thinker, Sartre, who declared: "existence precedes essence." He meant by that phrase that:

> . . . *man first of all exists, encounters himself, surges up in the world—and defines himself afterwards. If man as the existentialist sees him is not definable, it is because to begin with he is nothing. He will not be anything until later, . . . Thus, there is no human nature. . . . Man simply is (Sartre, 1965, p. 28).*

11.12 What similarities exist between behavioral and existential theories? What is their major area of disagreement?

A rejection of preconceptions about motives and traits was also true for George Kelly in his "personal construct theory" (Chapter 8), and it is what he meant when he said, "I *am* what I *do*," and urged that to know what one is, one must look at what one does. Skinner the arch behaviorist would have agreed with him on this point at least, and probably on a few others. The possible compatibility between aspects of behavior theory and of the existential-phenomenological movement hinge on several common qualities, summarized below.

The behavior level and the existential-phenomenological level are similar in that:

- Both share a focus on the here and now.
- Both emphasize what the person is doing rather than the constructs of the psychologist who studies the person.
- Both share a lack of interest in distant historical reconstruction in favor of a concern with new action possibilities for the individual.

These commonalities violate many common stereotypes about both approaches. They suggest that each is enriched by taking account constructively of the other's work. The similarities, although only partial, are impressive, the more so because philosophically it is hard to imagine two positions that on the surface would seem more incompatible.

However, the two positions may have one critical incompatibility. The existentialist takes the philosophical position that the individual is responsible and attributes to the person's choices the ultimate causes of behavior. In Sartre's phrase, a person "is what he wills to be" (1956, p. 291). A behavior theorist may share Sartre's desire to put "every man in possession of himself" rather than allow him to be possessed by unconscious psychic forces. But a behavioral analysis of causation cannot begin with the person's will as the fundamental cause of what he does, nor can it end with his constructs as a final explanation of his behavior.

George Kelly (personal communication, 1965) once emphasized his belief that personal constructs are the basic units and that it is personal constructs, rather than stimuli, that determine behavior. He recalled vividly from his Navy experience during World War II how very differently he related to the same officer on different occasions depending on how he construed him at the time. He remembered that the captain seemed different to him in an informal role, chatting with his jacket off, from the way he seemed when he wore his officer's coat. "You see," Kelly said, "it is not the stimulus—the captain—but how I construed him that [led] to my reactions to him."

But B. F. Skinner would find this story an excellent example, not of "construct control," but of "stimulus control": With his four stripes on, you see the captain one way; without his four stripes, you see him differently. To understand the change in the construct, in the behavioral view, you have to include in your understanding how those four stripes came to control it. The phenomenological and behavioral positions differ in their focus of attention. The phenomenologist wants to know and understand the person's experience; the behaviorally oriented psychologist wants to clarify the conditions that control the ultimate behavior.

The approaches to which we turn in the rest of the text seek to reconcile and integrate the two within a more comprehensive system that draws on contributions from each of the levels of analysis discussed throughout the text.

DEPRESSION: MORE THAN INSUFFICIENT REINFORCEMENT

According to one influential behavioral theory, depression may be understood as a result of a persistent lack of gratification or positive outcomes (reinforcement) for the person's own behavior (Lewinsohn, 1975; Lewinsohn, Clarke, Hops, & Andrews, 1990). That is, depressed persons feel bad and withdraw from life because their environments are consistently unresponsive to them and fail to provide enough positive consequences for what they are doing. The situation is analogous to an **extinction schedule** in which reinforcement for a behavior is withdrawn until gradually the behavior itself stops. In the case of the depressed person, the only reinforcement that does continue tends to be in the form of attention and sympathy from relatives and friends for the very behaviors that are maladaptive: weeping, complaining, talking about suicide. These depressive behaviors are so unpleasant that they soon alienate most people in the depressed person's environment, thus producing further isolation, lack of reinforcement, and unhappiness in a vicious cycle of increasing misery and increasing withdrawal.

Behavioral View of Depression. Depressed people, according to this theory, may suffer from three basic problems. First, they tend to find relatively few events and activities gratifying. Second, they tend to live in environments in which reinforcement is not readily available for their adaptive behaviors. For example, they may live highly isolated lives, as often happens with older people living alone, or with younger people in large universities that make them feel lost. Third, they lack the skills and behaviors they need to get positive reactions and feedback from other people. For example, they may be shy and socially awkward, making gratifying relationships with others very difficult. It follows that they in turn develop feelings such as "I'm not likable," or even "I'm no good." The essentials of the theory are shown schematically in Figure 11.2.

Although the theory has not been tested conclusively, a good deal of evidence is consistent with it. For example, depressed people do seem to elicit fewer behaviors from other people and thus presumably get less social reinforcement from them. Depressed people also tend to engage in fewer pleasant activities and enjoy such events less than individuals who are not depressed (Lewinsohn, 1975).

This concept of depression immediately suggests a treatment strategy (Lewinsohn et al., 1990): increase the rate of positive reinforcement for the depressed person's adaptive efforts. Note that such a plan would require increasing the rate of positive outcomes received by depressed people contingent on their own behavior; it does not mean simply giving more rewards regardless of what the individual does.

11.13 Describe the behavioral theory of depression and its implications for treatment.

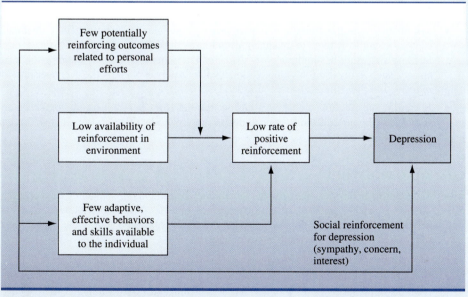

Figure 11.2 Schematic representation of Lewinsohn's theory of depression.
Source: Based on Lewinsohn, P. M. (1975). The behavioral study and treatment of depression. In M. Hersen (Ed.), *Progress in behavior modification*. Reprinted with permission from Elsevier.

SUMMARY

CHARACTERISTICS OF BEHAVIORAL ASSESSMENTS

- Behavioral approaches to personality all carefully measure behavior in relation to specific stimulus conditions. They treat observed behavior as a sample, and the focus is on how the specific sample is affected by variations in the stimulus conditions.

- The case of Gary W. illustrates the behavioral assessment of anxiety.

DIRECT BEHAVIOR MEASUREMENT

- Behavior may be measured directly by sampling it, both verbally and in performance, and by measuring physiological and emotional reactions.

- The reinforcing value of stimuli may be assessed from an individual's choices in lifelike situations, verbal preferences, and ratings, or the observed effects of various stimuli on actual behavior.

- In clinical work, it may be especially important to discover rewards that are effective for the individual to facilitate therapeutic progress.

ASSESSING CONDITIONS CONTROLLING BEHAVIOR

- Functional analyses, the foundations of behavior assessments, require careful observation of the behavior in question as it naturally occurs to specify the conditions that maintain this behavior.

- Then systematic changes are made in those conditions until the problem behavior no longer occurs and more satisfying behaviors are substituted.

- Behavior therapies attempt to modify disadvantageous behavior by planned relearning experiences and by rearranging stimulus conditions.

CHANGING EMOTIONAL REACTIONS

- Systematic desensitization can help people overcome fears or anxieties. The individual is exposed cognitively (e.g., in imagination) to increasingly severe samples of aversive or fear-arousing stimuli; simultaneously she is helped to make responses incompatible with anxiety, such as muscle relaxation. Gradually the anxiety evoked by the aversive stimulus is reduced and the stimulus is neutralized.

- In conditional aversion, a positive arousing stimulus (for example, cocaine for a cocaine addict) may be neutralized by repeatedly pairing it with one that is very aversive (such as chemically induced nausea).

CHANGING BEHAVIOR

- Maladaptive behaviors such as hyperactivity may be modified by changing the consequences to which they lead. Attention, approval, or other positive consequences are withdrawn from the maladaptive behavior and rewards become contingent instead on the occurrence of more advantageous behavior.

- The overuse of external rewards may backfire.

- In contingency contracting, the person makes a contract with another individual to self-reward for increasing positive behaviors (or self-punish for performing negative behaviors), as specified in the contract.

- To facilitate transfer from treatment to life, treatment samples the relevant situations and occurs in the same life setting in which improvement is desired. New behavior change methods encourage self-management so that individuals may gain relative independence and control of their own behavior as rapidly as possible.

• A promising theory of depression suggests that depressed people are caught in a vicious cycle of lack of reinforcement for their own efforts, leading to greater withdrawal from other people and, in turn, increased depression.

KEY TERMS

aversion therapy   255
behavior sampling   246
behavioral deficits   258
contingencies of
   reinforcement   250
contingency
   contracting   258
counterconditioning
   (desensitization)   254

extinction
   schedule   263
functional analysis   250
generalized conditioned
   reinforcers   248
hierarchy of anxiety
   (grading the
   stimuli)   253
overjustification   257

primary
   reinforcers   248
relaxation
   responses   254
symptom
   substitution   259
systematic
   desensitization   253
tokens   249

PART IV

THE BEHAVIORAL-CONDITONING LEVEL

▶ OVERVIEW: FOCUS, CONCEPTS, METHODS

The next table summarizes the main characteristics of work at the behavioral-conditioning level of analysis. As the table indicates, these approaches focus on the individual's behavior in its context. Prior learning and cues in the situation are seen as important determinants of behavior. Preferred data consist of direct observations of behavior as it changes in relation to changing situations. Responses are used as samples of behavior, not as indirect indicators of hypothesized inner states, motives, or traits. Research seeks to analyze the conditions that influence or control the behavior of interest, including the conditions that allow persons to enhance and control their own behavior, for example, by rearranging their environments. To change the person's behavior, therapy is directed at identifying and modifying the consequences behavior produces or the outcomes associated with it. The situation is treated as an integral aspect of behavior; it provides cues and outcomes that impact the maintenance and modification of behavior. Thus, the situation cannot be removed from the assessment of behavior or from attempts to change behavior.

Overview of Focus, Concepts, Methods: The Behavioral-Conditioning Level

Basic units:	Behavior-in-situation
Causes of behavior:	Prior learning and cues in situation (including behavior of others); reinforcement contingencies
Behavioral manifestations of personality:	Stable behavior equated with personality
Favored data:	Direct observations of behavior in the target situation; behavior change as a function of changing the situation (stimulus conditions)
Observed responses used as:	Behavior samples
Research focus:	Behavior change; analysis of conditions controlling behavior
Approach to personality change:	By changing conditions; by learning experiences that modify behavior
Role of situation:	Extremely important: regulates much behavior, elicits emotional reactions and behavior patterns

▶ ENDURING CONTRIBUTIONS OF THE BEHAVIORAL-CONDITIONING LEVEL

The contributions of behavioral approaches for applied purposes were widely applauded for identifying many effective, if incomplete, ways to help think about and improve diverse problems in everyday life (O'Donohue et al., 2001). Dollard and Miller translated many key psychodynamic concepts into learning terms in ways that encouraged experimental research. Work at the behavioral level helped take the mystery out of many of the "abnormal" phenomena that had long puzzled earlier workers. Notably, it showed how seemingly inexplicable and irrational behaviors—such as intense anxieties and bizarre behavior patterns—may be shaped by the circumstances of life and the ways in which learning and reinforcement occur. Over the years, much of this work became the basis for current "cognitive behavior therapies" which directly address issues in personality change and the coping processes of normal individuals (Meichenbaum, 1993). Gradually, these therapies have integrated discoveries from the behavioral level with those from more recent work on social cognition and mental processes, which are discussed in later sections.

Likewise, findings from research at the behavioral level on the responses of animals in laboratory conditions became the basis of later studies of complex social learning and personality at the social cognitive level (Part V). Recent breakthroughs at the biological level that identify brain centers involved in emotional reactions and those fundamental for other types of learning and memory functions also have given new life to work at the behavioral level. Brain processes and their links to behavior now can be studied experimentally, particularly in laboratory studies with animals. Consequently, the "black box" of the brain and the processes that occur within it, previously treated as "fictions" and off limits for a scientific psychology, have now become the arena for extensive research (e.g., LeDoux, 1996, 2001), at the vanguard of the science, as discussed in the final chapters of the text.

In the past, work at this level was faulted for overemphasizing the "stimulus" or "situation" and the role of rewards, and for missing much of the essentials of human personality, including the human potential for freedom and self-control: people are not just victims of their situations but also can actively create and change them, if not in the laboratory then certainly in real life. Critics saw behavioral work as reflecting a focus on the importance of the situation that minimizes the importance of dispositional or intrapsychic determinants such as traits and motives (Bowers, 1973; Carlson, 1971). A great deal of research in recent years has again demonstrated the power of stimuli to trigger automatically all sorts of cognitive, emotional, and behavioral reactions, leading practitioners to question the importance of the person's consciousness or internal thought processes (e.g., Bargh, 1996, 2001). The question of the degree to which the person can actually exert "agency" and intervene to change the ways in which stimuli impact on him or her remains one of the great open issues, actively pursued by a wide range of scientists and philosophers. We will revisit these big questions in the final sections of this text.

Historically, one of the persistent concerns about behavioral approaches is whether they really contribute to an understanding of the phenomena of interest to personality psychologists. An early worry about behavioral approaches was that they did not even try to study complex and distinctively human qualities, such as the emotions and internal states that seem so basic for understanding personality and appreciating its complexity. Indeed, some critics charged that behavioral psychologists only study what is easy to study with available methods. They accuse them of looking only under the light. Addressing this concern, researchers at the social cognitive level discussed in the next part of the text introduced some of the important missing pieces.

PART

V

THE SOCIAL COGNITIVE LEVEL

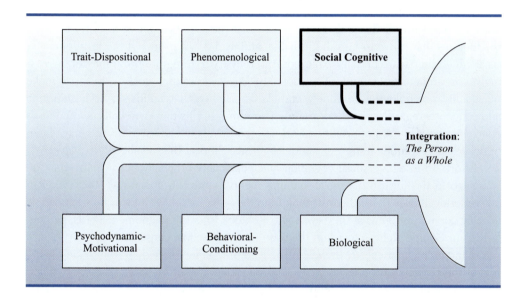

▶ PRELUDE TO PART V: THE SOCIAL COGNITIVE LEVEL

In the 1960s and 1970s, Albert Bandura and Walter Mischel were both young psychologists at Stanford University with offices across from each other. They both had been trained as researchers in approaches to personality that made learning and conditioning a cornerstone of personality development. But they both also had been trained as clinical psychologists interested in helping people deal with their life problems. As clinicians working with people in real-life situations, they saw the limitations of the simple forms of behavioral conditioning that were based mostly on experiments with rats or pigeons in the laboratory. Consequently, they tried to connect findings from the behavioral approach with new insights coming from research on social cognition and the extraordinary mental capacities of human beings. In time, their work and that of others (e.g., Rotter, 1954) led to a radically different social and cognitive approach to learning that had more direct relevance to personality. It also addresses the specific mechanisms that enable people to use both social learning and cognitive principles to enhance their own abilities to self-regulate and to pursue their chosen life goals as effectively and happily as possible.

Currently, researchers at the social cognitive level focus on the importance of social cognition for personality. *Social cognition* refers to the person's social knowledge of the world, to how we see and make sense of ourselves and other people, and how we use this knowledge in dealing with the world, with ourselves, with our relationships. These researchers are concerned with the links between what goes on in the mind of the person—their social cognitions, emotions, goals—and their social behaviors. Their aim is to understand both what people think and feel and want and what they actually do. With these broad goals, the social cognitive level tries to integrate contributions coming from the behavioral level, the phenomenological level, and from the study of social cognition, to form a more complete view of the person.

Thus, a distinctive feature of work at this level of analysis is that it is explicitly integrative: rather than building a single unique theory or approach with distinct boundaries, its edges are deliberately fuzzy. Workers at the social cognitive level try to distill the findings that prove to be valid, reliable, and useful, regardless of the level of analysis at which they were initially explored. And they try to integrate them into an increasingly comprehensive model of personality. An overarching goal is to help make personality psychology a genuinely cumulative science that builds on its best results and concepts, regardless of their origins. Therefore, in this part, you will see numerous concepts that also are important in work at other levels of analysis, and that have been enriched and extended in new directions by researchers at the social cognitive level.

Efforts to integrate social cognition and personality originally began more than 40 years ago in the form of a "social learning theory" approach (Bandura, 1969; Mischel, 1968; Rotter, 1954). These theorists took from the behavioral level a focus on what people actually do, not just on what they say. Some also took from the phenomenologists, particularly from George Kelly, a focus on the ways in which people construe or appraise themselves and their experience (Mischel, 1973). And they all increasingly paid attention to the role of the person's expectancies, goals, and values in how information is interpreted and processed by different types of people. Further advance came rapidly with the onset of the cognitive revolution in psychology in the 1970s, and was stimulated by progress in recent years in the understanding of brain functions coming from related disciplines.

The next chapter discusses the development and emergence of the social cognitive approach. The second chapter in this part then illustrates the social cognitive processes that are being studied and their implications for understanding personality.

The Personal Side of the Science

Some questions at the social cognitive level you might ask about yourself:

▶ What is the role in personality of what people know, think, and feel?

▶ How does what I know, think, and feel about myself and the social world influence what I do and can become?

▶ What can I do to change how I think and feel?

▶ Will that change my personality and behavior?

▶ How much of who and what I am and do is "automatic"?

▶ How much is open to "willpower" and self-regulation?

▶ How do willpower and self-regulation work?

▶ How can I enhance my control over my life?

12

SOCIAL COGNITIVE CONCEPTIONS

▶ DEVELOPMENT OF THE SOCIAL COGNITIVE PERSPECTIVE

For more than four decades, psychology has experienced a cognitive revolution that has made the mind and how it works the focus of the field. It made exciting discoveries about how people think, and how they selectively attend to, manipulate, transform, store, and generate information. The findings soon made it clear that cognition has a central role in personality (Bandura, 1986; Mischel, 1973, 1990; Pervin & John, 1999), clinical psychology (Arkowitz, 1989; Beck, 1976), and social psychology (e.g., Higgins & Kruglanski, 1996).

Historical Roots

The social cognitive approach to personality began in the late 1960s. It was conceived by many psychologists who were in rebellion, frustrated by the limitations of earlier theories. At that time, the field of personality was divided into three theoretical camps. In one were enthusiastic Freudians guarding Freud's original work against anyone who sought to revise or criticize it. In a second camp were students of individual differences searching for broad personality trait dimensions. In the third camp were radical behaviorists concerned with conditioning and the relations between stimuli and responses, not with the internal ways an organism mediates between them. This third camp opposed any constructs that invoked the mind or mental activity in any form that could not be directly or simply measured. Except for occasional exchanges to attack each others' work, there was little communication among the camps. It was a milieu unresponsive to any "constructive alternatives."

You have already seen one protest movement in response to this state of affairs in the humanistically oriented ideas developed by theorists at phenomenological level, such as Carl Rogers and George Kelly. In their focus on the concepts and constructs of the individual, as perceived and understood by that person, they were anticipating the cognitive revolution. But these voices were treated more as a sideline protest movement than as a route for redirecting mainstream theory and research. Major disturbances were needed in that mainstream before serious attention would be paid to possible alternatives, and they occurred in the late 1960s.

Cognitive Processes Underlying Behavior

In the 1960s, some personality psychologists were drawn to the behavioral perspective, attracted by its rigorous scientific emphasis (Bandura, 1969; Mischel, 1968; Rotter, 1954). In their theorizing, they drew heavily on principles of learning established originally in experimental work with animals, as in the operant and classical conditioning studies (reviewed in Chapter 10). Nevertheless, much as they tried to stretch existing behavioral concepts, these personality psychologists encountered more and more new research findings that forced a change in their view and required moving beyond behaviorism, as discussed next.

12.1 How did delay of gratification research illustrate the important role of cognitive transformations?

Traditional behavioral approaches asserted that stimuli control behavior. But in fact, the perceiver's mental representations and **cognitive transformations** of the stimuli can determine and even reverse their impact. Such transformations were illustrated in research on the factors that influence how long preschool children will actually sit still alone in a chair waiting for a desired but delayed outcome (e.g., tempting pretzels or marshmallows) that has been placed in front them (Mischel, Ebbesen, & Zeiss, 1972). The question was: how long will children voluntarily continue to delay and what makes it hard or easy? It turned out that the answer depends importantly on how the children mentally represent the rewards (Mischel, 1974; Mischel et al., 1996).

For example, if a young child is left during the waiting period with the actual desired objects—the pretzels, for example—in front of her, it becomes extremely difficult to wait for more than a few moments. But through self-instructions they can cognitively transform the objects in ways that permit them to wait for long periods of time. For example, if they think about the stick pretzels they want now as little logs, or think about the marshmallows as round white clouds or as cotton balls, they often can wait for the whole required time. In short, what is in the children's heads—not what is physically in front of them—determines their ability to delay. Through self-instructions about what to imagine during the delay period, it is possible to completely alter (indeed, to reverse) the effects of the physically present temptations in the situation.

This was important because at that time behaviorism was still the dominant view. These findings with the little children made clear that it's what's in the head that is influencing behavior. Therefore, how people think and represent the world has to be taken into account to understand what they are doing and why they are doing it. George Kelly and Carl Rogers were right.

▶ OBSERVATIONAL LEARNING (MODELING): ALBERT BANDURA

At about the same time, new discoveries in learning research revealed a third type of learning process independent both of classical conditioning and of reinforcement. It was recognized, of course, that reward or reinforcement has a powerful influence on behavior in a great variety of situations. But it also became clear that people learn by observing others, not merely by experiencing rewards for what they do themselves. Much social learning occurs through observation without any direct reinforcement administered to the learner. Classical and operant conditioning remain important types of learning. However, it is now also clear that some of the most important human learning occurs through observation.

Learning Through Observation

Observational learning, sometimes called **modeling,** is learning that occurs without the learner's receiving direct external reinforcement. Such learning occurs even without the person's ever performing the learned response at all. You can learn a lot about how to kill people, for example, just by watching how it's done on television.

12.2 What is required for observational learning to occur?

Observational learning occurs when people watch others or when they attend to their surroundings, to physical events, or to symbols such as words or pictures. Albert Bandura (1969, 1986) for the last four decades has led the way in the analysis of observational learning and its relevance for personality. Much human learning, from table manners, interpersonal relations, including cooperation and aggression, to school and work, depends on observation of this kind rather than on direct reinforcement for a particular action.

Albert Bandura

Children learn partly through observation and imitation.

Figure 12.1   Photos depicting children spontaneously imitating a previously viewed model's aggressiveness.
Source: Bandura, A. (1965). Vicarious processes: A case of no-trial learning. In L. Berkowitz (Ed.), *Advances in experimental social psychology* (Vol. 2, pp. 1–55). New York: Academic Press.

Observational learning is often indirect. We learn much from what others observe and then tell us about. The mass media, which are highly effective means of communicating experiences and observations, contribute heavily to the enormous amount we learn about the environment and the behavior of others.

The influence of observational learning through the mass media is seen in the contagion of aggression that has spread through the United States as a result of so many people's watching television violence. Although the television networks are reluctant to accept it, watching violence on television can have definite negative effects on viewers (Liebert, 1986; Murray, 1973). For example, after observing films of an aggressive model who punched, pummeled, and hurled a Bobodoll, children spontaneously imitated the model's aggressiveness when put in a similar situation (Bandura, 1965). (See Figure 12.1 for an illustration.) Similarly, after watching violent cartoons for some time, children became more assaultive toward their peers than did other youngsters who had viewed nonviolent cartoons for the same period of time (Steuer, Applefield, & Smith, 1971).

Completely new response patterns can be learned simply by observation of others performing them. Observation is especially important for learning a language. Bandura (1977) emphasizes the advantages of observation for language learning compared to direct reinforcement for uttering the right sounds. He notes that exposure to models who speak the language leads to relatively rapid acquisition, while shaping would take much longer.

Observing Other People's Outcomes: What Happens to Them Might Happen to You

People learn about the possible consequences of various behaviors from observing what happens to others when they engage in similar behaviors. Your expectations about the outcomes of a particular course of action depend not only on what has happened to you in the past, but also on what you have observed happening to others.

We are more likely to do something if we have observed another person (model) obtain positive consequences for a similar response. Seeing other children praised for cooperative play, for example, makes a child more likely to behave cooperatively in similar situations. If, on the other hand, models are punished for a particular pattern of behavior such as cooperation, observers are less likely to display similar behavior (Bandura, 1965).

Although laboratory studies offer clear demonstrations of the importance of expected consequences, examples from life are more dramatic. Consider, for instance, the role of modeling in airline hijackings (Bandura, 1973). Air piracy was unknown in the United States until 1961. At that time, some successful hijackings of Cuban airliners to Miami were followed by a wave of hijackings, reaching a crest of 87 airplane piracies in 1969 that intermittently continued and, in tragically altered form, were seen again in the disaster of the September 11, 2001 terrorist attack on the United States.

In sum, you do not have to perform particular actions yourself in order to learn about them and their consequences; the observed as well as the directly experienced consequences of performances influence subsequent behavior. You do not have to rob a bank to learn that it is punishable; you do not have to be arrested for hijacking to learn about its consequences; and you do not have to rescue a burning child from a fire or return found money to discover that such acts are considered good. Information that alters what people expect will happen as result of a behavior changes the probability that they will perform it. Models inform us of the probable consequences of particular behaviors and thus affect the likelihood that we will perform them.

Observation also influences the emotions we experience. By observing the emotional reactions of others to a stimulus, it is possible vicariously to learn an intense emotional response to that stimulus. Suppose you see someone wincing from a painful electric shock each time a red light goes on in the machine to which he is attached. Soon you will wince when the red light comes on. Thus, you may become "vicariously conditioned" when you observe repeatedly the close connection between a stimulus (red light on) and an emotional response (pain cues) exhibited by another person, and this happens without your receiving any direct aversive stimulation yourself.

This point was demonstrated in a study in which adults observed another person making fear responses in reaction to the sound of a buzzer supposedly associated with the onset of an electric shock. (Actually the person was a confederate of the experimenter and only feigned pain and fear without getting any shocks.) Gradually, after repeatedly watching the pairing of the buzzer and the responses made by the confederate, the observers themselves developed a measurable physiological fear response to the sound of the buzzer alone (Berger, 1962).

Rules and Symbolic Processes. Many studies showed that people did not seem to need trial by trial "shaping" but rather seemed to be helped most by the rules and self-instructions they use to link discrete bits of information meaningfully to learn and remember materials (Anderson & Bower, 1973). Likewise, studies with children indicated that it helps not only to reward appropriate behavior but also to specify the

12.3 How do observed consequences affect the likely occurrence of behaviors learned through modeling?

12.4 What evidence exists that emotional responses can be acquired through observation alone?

12.5 How are symbolic processes involved in learning?

relevant underlying rules and principles so that children can more readily learn the standards that they are supposed to adopt (Aronfreed, 1966). When children understand that particular performance patterns are considered good and that others are unsatisfactory, they adopt the appropriate standards more easily than when there are no clear verbal rules (Liebert & Allen, 1967). Beginning early in life, the young child is an active thinker and perceiver who forms theories about the world, not a passive learner shaped by external rewards (Bruner, 1957; Flavell & Ross, 1981).

Cognition plays a role even in classical conditioning. Suppose, for example, that a person has been conditioned in an experiment to fear a light because it is repeatedly paired with electric shock. Now if the experimenter simply informs her that the light (the conditioned stimulus) will not be connected again with the electric shock, her emotional reactions to it can quickly extinguish (Bandura, 1969). On later trials, she can see the light without becoming aroused. Findings like these forced both researchers and behavior therapists to pay more attention to the individual's mental processes. In the development of therapeutic methods to change behavior, they began to more directly engage the person's thought processes and social knowledge for therapeutic ends (Davison & Neale, 1990; Davison, Neale, & Kring, 2004).

New Directions. These sorts of findings suggested that a more cognitive approach to personality was required that takes into account how the individual characteristically deals mentally and emotionally with experiences. The search began for a theory of the cognitive-emotional-motivational processes that underlie the person's characteristic behavioral expressions and conflicts. As such, these efforts are in the tradition pioneered by Sigmund Freud, Henry Murray, and George Kelly, advanced over the century by many theorists reviewed in earlier chapters.

► SOCIAL COGNITIVE PERSON VARIABLES

12.6 Describe the five basic "person variables" in the social cognitive conception of personality.

Growing out of these contributions, a basic set of psychological **person variables** for describing individual differences in personality was proposed (Mischel, 1973), summarized in Table 12.1. These person variables try to capture the differences between people in the characteristic ways in which they interpret aspects of themselves and of their social worlds. It includes how they react to their own interpretations cognitively and emotionally, and in their behavior patterns.

TABLE 12.1   Types of Person Variables

- *Encodings (Construals, Appraisals):* Categories (constructs) for the self, people, events, and the situations (external and internal).
- *Expectancies and Beliefs:* About the social world, about outcomes for behavior in particular situations, about self-efficacy, about the self.
- *Affects:* Feelings, emotions, and affective responses (including physiological reactions).
- *Goals and Values:* Desirable outcomes and affective states; aversive outcomes and affective states; goals, values, and life projects.
- *Competencies and Self-Regulatory Plans:* Potential behaviors and scripts that one can do, and plans and strategies for organizing action and for affecting outcomes and one's own behavior and internal states.

Source: Mischel, W., & Shoda, Y. (1995). A cognitive-affective system theory of personality: Reconceptualizing situations, dispositions, dynamics, and invariance in personality structure. *Psychological Review, 102,* 253.

Walter Mischel

As the table shows, the person variables include encodings or construals, expectancies and beliefs, affects (feeling states and emotions), subjective values and goals, as well as competencies and self-regulatory strategies. These variables interact, but each provides distinctive information about the individual.

Encodings (Construals): How Do You See It?

People differ greatly in how they **encode** (represent, construe, appraise, interpret) themselves, other people and events, and their experiences. The same hot weather that upsets one person may be a joy for another who views it as a chance to go to the beach. The same stranger in the elevator who is perceived as dangerous by one person may be seen as interesting by another. Individuals differ stably in how they encode and categorize people and events, and these interpretations influence their subsequent reactions to them.

12.7 What is meant by encoding? How does encoding affect other thoughts, feelings, and behaviors?

Suppose Mark, a teenager, tends to encode (construe, perceive) peers in terms of their hostile threats, and is highly sensitive to their possible attempts to challenge, manipulate, and control him. If he cognitively represents his world in such terms, Mark will be vigilant to threats and primed to defend himself. He therefore may see an innocent accident, such as having someone push against him in the crowded staircase, as a deliberate affront or violation (Cantor & Mischel, 1979; Dodge, 1986).

Different persons group and encode the same events and behaviors in different ways (Argyle & Little, 1972) and they selectively attend to and seek out different kinds of information (Bower, 1981; Miller, 1987). For example, some individuals tend to encode ambiguous negative events as instances of personal rejection (e.g., Downey & Feldman, 1996) and keep looking for clues that they are being rejected, which then makes them depressed and withdrawn. Other people, however, may easily feel disrespected and become angry and hostile even when they hear a mumbled greeting (e.g., Dodge, 1993).

Different individuals may encode the same situation in different ways.

How people encode and appraise events and selectively attend to what they observe also greatly influences what they learn. For example, people who have poor social skills tend to encode situations in terms of the degree to which they might feel self-conscious versus self-confident in them. In contrast, more socially skilled individuals encode the same situations in terms of other dimensions such as how interesting or pleasant they might be (Forgas, 1983).

Expectancies and Beliefs: What Will Happen?

This analysis of personality does not stop with a description of what people know and how they interpret events. It also seeks to predict and understand actual performance in specific situations. Research on learning made it clear that the actions people perform depend on the consequences they expect (Bandura, 1986; Rotter, 1954). To predict specific behavior in a particular situation, one has to consider the individual's specific expectancies about the consequences of different behavioral possibilities in that situation (Mischel, 1973).

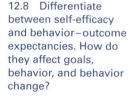

12.8 Differentiate between self-efficacy and behavior–outcome expectancies. How do they affect goals, behavior, and behavior change?

A particularly important type of expectancy consists of **self-efficacy expectations:** the person's belief that he or she *can* perform a particular behavior, like handling a

The development of self-efficacy takes many forms.

SELF-EFFICACY AND ITS IMPLICATIONS

Self-efficacy refers to the individual's belief that he or she can successfully execute the behaviors required by a particular situation. Perceptions of one's own efficacy importantly guide and direct one's behavior. The close connection between high self-efficacy expectations and effective performance is illustrated in studies of people who received various treatments to help reduce specific fears. A consistently high association was found between the degree to which persons improved from treatment (becoming able to handle snakes fearlessly) and their perceived self-efficacy (Bandura, 1977). If we assess perceived self-efficacy (asking people to specifically predict their ability to do a given act successfully), according to Bandura we can predict whether or not they will be able to perform it—for example, whether they now can pick up and play with the previously feared snake. (Bandura & Adams, 1977). Results of this kind suggest some clear links between self-perceptions of one's competence and the ability to actually behave competently.

Perceived self-efficacy influences the goals people set for themselves and the risks that they are willing to take: the greater their perceived self-efficacy, the higher the goals they choose and the stronger their commitment and perseverance in pursuing them (Bandura, 1997, 2001). Conversely, people who view themselves as lacking efficacy for coping with life tasks are vulnerable to anxiety and may develop avoidance patterns designed to reduce their fears. People who see themselves as lacking in essential efficacy also may become prone to depression. They may even show impairments in their immune system when coping with stressors that they believe they cannot control (Wiedenfeld et al., 1990).

The Role of Self-Efficacy.   Although many different therapy techniques may produce changes in behavior,

might they all work through the same basic mechanism? Bandura (1978, 1982) has proposed that there is such a common mechanism: people's expectations of self-efficacy. Behavior therapy works, according to Bandura, by increasing efficacy expectations and thus leading people to believe that they can cope with the difficult situations that threatened them before. In this view, individuals will try to overcome a particular fear to the extent that they expect they can do so successfully. Any methods that strengthen expectancies of personal efficacy will therefore help the person perform the relevant behavior. The best methods will be the ones that give the person the most direct, compelling success experiences in performing the particular behavior, thereby increasing efficacy expectancies most. For example, actually climbing a fire escape successfully is a better way to overcome a fear of heights than just thinking about it, because it provides a more complete success experience and a stronger expectation for future mastery.

In sum, diverse processes of behavior therapy in Bandura's view exert their beneficial effects mainly by enhancing the individual's sense of self-efficacy. Thus, he sees each of the major behavioral strategies for inducing change as sharing one crucial ingredient: They improve perceived self-efficacy, thereby freeing the individual to perform actions that previously were not possible.

Self-efficacy is Bandura's most basic construct for understanding human motivations and emotional reactions. People with a sense of self-efficacy about their intended actions are more able to carry them out. High efficacy expectations thus help the individual to persist in the pursuit of goals, even in the face of adversities that would derail or depress persons who are less sure of their relevant personal competencies.

---

snake or taking an exam. Measures of these expectations predict with considerable accuracy the person's actual ability to perform the relevant acts. Efficacy expectancies guide the person's selection of behaviors from among the many that he or she is capable of constructing within any situation (Bandura, 1986). (See In Focus 12.1 for a detailed discussion of self-efficacy.)

Another type of expectancy concerns **behavior–outcome** relations. These behavior–outcome expectancies (hypotheses, contingency rules) represent the expected *if . . . then . . . relations* between behavioral alternatives and expected probable outcomes in particular situations. In any given situation, we generate the response pattern that we expect is most likely to lead to the most subjectively valuable outcomes in that situation (Bandura, 1986; Mischel, 1973; Rotter, 1954). In the absence of new information about the behavior–outcome expectancies in any situation, behavior will depend on one's previous behavior–outcome expectancies in similar situations (Mischel & Staub, 1965). That is, if you do not know exactly what to expect in a new situation (a first job interview, for example), you are guided by your previous expectancies based on experiences in similar past situations.

For example, we generate behavior in light of our expectancies even when they are not in line with the objective conditions in the situation. If you expect to be attacked, you become vigilant even if your fears later turn out to have been unjustified. If you expect to succeed, you behave quite differently than if you are convinced you will fail. Indeed, we sometimes behave in ways that directly help to confirm our expectations, thus enacting self-fulfilling prophecies (Buss, 1987). The person who is easily suspicious, angry, and ready to aggress is likely to extract reciprocal hostility and defensiveness from others that will, in turn, confirm his beliefs about them.

Affects: Feelings and "Hot" Reactions

12.9   Why are "hot" cognitions" so named?

What we feel—our affects and emotions—profoundly influences other aspects of behavior (e.g., Contrada, Cather & O'Leary, 1999; Smith & Lazarus, 1990; Zajonc, 1980). They also impact on the person's efforts at self-regulation and the pursuit goals (e.g., Mischel et al., 1996). It has long been known that such cognitions as beliefs about the self and one's personal future are themselves **hot cognitions,** that is, thoughts that also activate strong emotion (Ayduk & Mischel 2002; Metcalfe & Mischel, 1999). As Smith and Lazarus (1990) noted, anything that implies important consequences, harmful or beneficial, for the individual can trigger an emotional reaction. For example, when one is feeling bad or sad—what is called a negative affective state—and gets negative feedback about performance (e.g., a disappointing test score), it is easy to become demoralized and to over-interpret the feedback, resulting in depression (Wright & Mischel, 1982).

Affective reactions to situation features (such as faces) may occur immediately and automatically (e.g., Murphy & Zajonc, 1993; Niedenthal, 1990), outside of awareness (Gollwitzer & Bargh, 1996). These emotional reactions in turn can trigger closely associated cognitions and behaviors (Chaiken & Bargh, 1993). The affective states and moods experienced are easily influenced by situational factors, even by such simple events as finding a coin on the street (e.g., Isen, Niedenthal, & Cantor, 1992; Schwarz, 1990). But what we feel also reflects stable individual differences which may be related to temperament and biological variables (Rothbart et al., 1994).

Goals and Values: What Do You Want? What Is It Worth?

12.10   Describe the organizational functions of goals and values. What is intrinsic motivation?

Goals and values drive and guide the long-term projects people pursue, the situations and outcomes they seek, and their reactions to them (e.g., Linville & Carlston, 1994; Martin & Tesser, 1989). They serve to organize and motivate the person's efforts, providing the direction and structure for the life tasks and projects they pursue (Grant & Dweck, 1999). Motives, such as achievement, power, intimacy, and others studied by the Harvard personologists, are also represented in the personality system by the person variable of goals.

Goals influence what is valued, and values also influence performance. For example, even if two individuals have similar expectancies, they may act differently if the outcomes they expect have different personal values for them (Rotter, 1954, 1972) or if they are pursuing different goals (e.g., Cantor, 1994). If everyone in a group expects that approval from a teacher depends on saying certain things the teacher wants to hear, there may be differences in how often they are said due to differences in the perceived value of obtaining the teacher's approval. Praise from the teacher may be important for a student striving for grades, but not for a rebellious adolescent who rejects school. What delights one person may repel his or her neighbor. That makes it necessary to consider the individual's goals and the subjective value of particular classes of events, that is, his or her preferences and aversions. These goals and values are particularly important because much human behavior is driven by **intrinsic motivation:** the gratification the individual receives from the activity or task itself (Cantor, 1990; Deci & Ryan, 1987). Such motivation is reflected in the life goals that the person pursues, as will be discussed in detail in later chapters (Chapters 16, 17).

What Can You Do? Overcoming Stimulus Control

People regulate their own behavior by self-imposed goals and self-produced consequences. Even in the absence of external constraints and social monitors, we set performance goals for ourselves. We react with self-criticism or self-satisfaction to our behavior depending on how well it matches our expectations and standards (Bandura, 1986; Higgins, 1990). The expert sprinter who falls below his past record may condemn himself bitterly; the same performance by a less experienced runner who has lower standards may produce self-congratulation and joy. To predict Mark's reaction to being pushed, for example, it helps to know the personal standards he uses to evaluate when and how to react aggressively. Will he react aggressively even if the peer who pushed him is much younger? Likewise, can he regulate his own response strategically, or will he react explosively and automatically?

    People also differ in the types of plans that guide their behavior in the absence of, and sometimes in spite of, immediate external situational pressures. Such plans specify

12.11 How do self-regulation processes counteract external stimulus control of behavior?

Self-observation plays an important role in the acquisition of skills and competencies.

the kinds of behavior appropriate under particular conditions, the performance levels (standards, goals) that the behavior must achieve, and the consequences of attaining or failing to reach those standards (Mischel et al., 1996). Plans also specify the sequence and organization of behavior patterns (Gollwitzer & Moskowitz, 1996; Schank & Abelson, 1977). Individuals may differ with respect to each of the components of self-regulation (e.g., Baumeister & Heatherton, 1996).

Self-regulation provides a route through which we can influence our interpersonal and social environment substantially. We can actively *select* many of the situations to which we expose ourselves, in a sense creating our own environment, choosing to enter some settings but not others (Buss, 1987; Ross & Nisbett, 1991). Such active choice, rather than automatic responding, may be facilitated by thinking and planning and by rearranging the environment itself to make it more favorable for one's goal (e.g., Gollwitzer & Moskowitz, 1996). Even when the environment cannot be changed physically (by rearranging it or by leaving it altogether and entering another setting), it may be possible to *transform* it psychologically, as discussed in Chapter 17.

In this section, we have summarized the person variables that are the basic units of the theory that we next describe. But first look at the long history of these person variables, as described in In Focus 12.2.

12.12 Which of the five person variables were especially influenced by the theories of Rotter and Kelly?

IN FOCUS 12.2

ON THE HISTORY OF PERSON VARIABLES

The development of person variables (Mischel, 1973; Mischel & Shoda, 1995) has a long history. These variables reflect an attempt to integrate contributions from two quite different theoretical perspectives that at first may seem incompatible. One was the personal construct theory of George Kelly (discussed in Chapter 8) and the other was the social learning theory developed by Julian B. Rotter (1954). Both Rotter and Kelly were professors in the graduate training program in clinical psychology at Ohio State University in the early 1950s, and both were important mentors for Walter Mischel. In the early 1950s, Mischel did his doctoral dissertation with Rotter, but most of his clinical psychology training was with Kelly. Although Professor Kelly's and Professor Rotter's offices were across the hall from each other, the two men were as different as their theories. They had little direct contact, and little influence on each other, although they both greatly influenced any students who were at Ohio State in those years. Their ideas were expressed in two extremely important books, Rotter's *Social Learning and Clinical Psychology*, published in 1954, and Kelly's two-volume *The Psychology of Personal Constructs* (1955). Together, these two pioneers contributed much to modern personality and clinical psychology.

The person variable of "encoding strategies" explicitly reflected George Kelly's core point—the importance of

how individuals construe their experience and themselves. This construal or appraisal characterizes the first phase of how people make sense of incoming social information and interpret its personal meaning. The conceptualization of personality in cognitive social terms, however, sought a personality theory that includes not only how people perceive and interpret but also what they actually choose and do (Mischel 1973). For that purpose, it explicitly drew on expectancy-value concepts which had been given a central place by Rotter (1954). In Rotter's early social learning theory, the probability that a particular pattern of behavior will occur was a joint function of the individual's outcome expectancies and the subjective value of those outcomes.

In the late 1940s and early 1950s, Rotter introduced the expectancy construct to personality psychology and made it a centerpiece of his version of social learning. He had argued convincingly for the importance of both expectancies and values as basic building blocks for a theory of social learning that he wanted to speak more directly to the assessment and treatment of clinical problems. While his theory was extremely elegant, it was, like his colleague George Kelly's theory of personal constructs, apparently ahead of its time: the impact of both these theorists seems to be felt more decades later in indirect forms than when Rotter and Kelly first advanced their ideas.

Julian B. Rotter

► THE COGNITIVE–AFFECTIVE PERSONALITY SYSTEM (CAPS)

Structure and Dynamics

The **cognitive–affective personality system,** abbreviated as **CAPS** (Mischel & Shoda, 1995, 1998; Shoda & Mischel, 1998), focuses on how the person variables are organized and function within the individual. The personality system consists of a distinctive stable organization of person variables whose interactions generate the patterns of behavior that characterize the person.

When a person experiences certain features of a situation, a characteristic pattern of cognitions and affects becomes activated. The personality structure—the person's stable organized system of interconnections among the person variables—guides their further activation. In other words, given the occurrence of particular situational features, a cascade of interpretations, expectations, feelings, goals, and possible behavior scripts becomes activated in a pattern that is distinctive and stable. You saw examples in the descriptions of the narcissistic and rejection sensitive personality types given in Chapter 4. For example, the rejection sensitive person is likely to perceive the partner's "need for space to devote more time to work" as a sign of potential rejection. In turn, fears of abandonment, feelings of anxiety, impulses of anger, and so on become activated internally and expressed in aggressive behavior in a distinctive sequence.

Personality processing dynamics in this theory consist of such stable distinctive sequences of thoughts, feelings, and behaviors that become activated by particular types of situations. These are the stable *if . . . then . . .* patterns, or personality signatures, discussed in Chapter 4. *If* Gary sees himself as being rejected, *then* he thinks about abandonment, feels panic, and erupts with anger, aggression, and insults against his partner. The pattern is self-defeating because his hostility in response to his own rejection fears is likely to lead to actual rejection even when there was none before, making his worst fears come true (Downey et al., 1997).

You can see the implications of the CAPS model and the social cognitive approach to personality more concretely if you look at how it is applied to understanding a particular life, rather than people in general or even different types. We therefore focus again on Gary W.

View of the Person: Gary W.

12.14 Describe how the CAPS model guided the personality assessment of Gary W.

To give a richer sense of how this level of analysis sees the person as a whole, we next apply it to Gary W.

Assessing Person Variables and Dynamics. The assessments of Gary on each of the person variables used many sources: his own self-reports, diary notes based on daily self-observations, and interviews. The focus was on each of the person variables and on the situations in which they were expressed. Diverse measures, including questionnaires, were devised to sample a wide range of information about Gary and his personal world.

These measures drew on recent research which has provided relevant strategies for such assessments of experiences and behaviors that take account of the situations in which they occur (Cervone & Williams, 1982; Mischel et al., 1996; Mischel & Shoda, 1998). The methods include time sampling of tasks, behavior, and perceptions (e.g., Cantor et al., 1991; Moskowitz, Suh, & Desaulniers, 1994), and self-reports of reactions to daily stressors (e.g., Bolger & Schilling, 1991). They also include sampling of physical symptoms and of emotional reactions to them (Larsen & Kasimatis, 1991), as well as records of personal strivings and well-being (Emmons, 1991). All these methods allow systematic analyses of the types of *ifs* and *thens* significant in the lives of individuals in which their characteristic behavioral patterns unfold.

The assessments of Gary were intended to obtain answers to questions such as the following:

1. **Encodings.** What are Gary's enduring views of himself (self-concepts) and of his relations to the significant people in his life that are most important for him? In which of these is he particularly vulnerable? How does he assign causal responsibility for the positive and negative events and outcomes experienced in different domains (e.g., work, interpersonal)?

2. **Expectancies and beliefs.** What are Gary's self-efficacy beliefs and expectations in the domains that matter most to him? What are the major anticipated consequences (outcomes) of different alternative courses of action that he is considering for dealing with his current life goals?

3. **Affects.** What are his characteristic "hot" reaction patterns? What triggers them?

4. **Goals and values.** What values seem to be guiding his behavior? What life goals motivate Gary and shape the direction of his efforts?

5. **Competencies and self-regulatory strategies.** What strategies and competencies are available and readily activated by him? How do they help him to sustain long-term effort and self-control in the pursuit of his goals? What standards does he use in self-evaluation?

Note that the assessment of Gary in this approach does not address each person variable in isolation. Rather, it seeks to clarify how they interact, that is, their organization. For example, how do Gary's expectations affect his mood, and how does his mood, in turn, influence the strategies that are activated? It's these interactions of person variables within the system that generate the distinctive, stable patterns of *if . . . then . . .* signatures that characterize the person.

Gary's Ways of Encoding Experience and Himself.   Gary tends to encode situations on the basis of their potential threat for him. He views himself as a person who needs and wants a warm and caring relationship with a woman but he also sees women as a source of threat and provocation who "can't really be trusted."

For example, he views his new girlfriend's bids for "more emotional space"—for example, when she says "I feel like you're strangling me"—as "cold" and "rejecting." He becomes quickly infuriated when she expresses her feelings of anger and frustration with his demands and expectations and feels it means "she doesn't really love me." He says he tries to control himself on such occasions but in fact he quickly loses his temper. He then becomes enraged, although he claims to have overcome his childhood tendency toward emotional outbursts. He justifies and dismisses this behavior as an inevitable result of his girlfriend's lack of understanding and sympathy for him. Thus, he sees his explosions of anger as not in his control, attributes them to her "cold and rejecting style" and his "temper"—"I got it from my dad"—and exempts himself from responsibility.

Gary sees himself as shrewd and analytic, able to persist tenaciously in pursuit of a work goal. He seems to be exceptionally sensitive to the reactions of other people. Although Gary often seems to devalue many of his work achievements, particularly those that are in a verbal form, he actually has high regard for his quantitative abilities and work (e.g., priding himself on an innovative financial software program he developed). In the quantitative domain, he takes appropriate credit for his achievements ("with numbers I have a mind like a sharp razor"). Here he accurately evaluates himself as exceptionally talented and sometimes even inflates the magnitude of his contribution (when compared to its assessment by his peers).

Affect and Distress Reactions.   When dealing with public speaking situations, Gary is prone to becoming swamped with debilitating anxiety, followed by some depressive feelings. Most important, Gary becomes easily upset by any feedback short of complete approval and applause, especially in close romantic relationships. Unless he gets rapt attention and praise from his girlfriend, he feels she is rejecting him. He then becomes angry and hostile. After only several months, his relationship with his new girlfriend is in danger of being soured by the same conflicts that ended his relationship with his former girlfriend.

Gary's Rejection Sensitivity Dynamics (Interpersonal Domain).   Figure 12.2 shows Gary's **rejection sensitivity (RS)** dynamics, which are typical of rejection sensitivity in the domain of interpersonal relations. These dynamics are shown in abstract terms in the figure. The solid arrows show the activation paths in the dynamics, the broken arrows indicate deactivation. Gary's rejection sensitivity dynamics become activated when he perceives and encodes rejection cues, which he does easily and for which he scans persistently. He anxiously expects and vigilantly looks for signs that he is being criticized, "violated," ignored, or rejected. He easily interprets even ambiguous events, like the partner's momentary wandering attention or preoccupation with other concerns, as if they were intended to hurt him deliberately. His anger and even rage become easily and automatically activated.

12.15 Describe how Gary's rejection sensitivity reflects the interaction of CAPS elements.

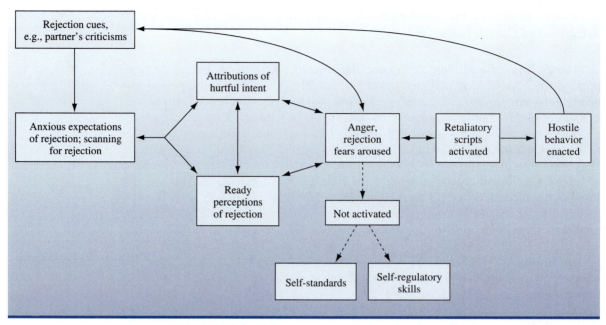

Figure 12.2 Gary's rejection sensitivity (RS) dynamics.

Note that the path to the activation of self-standards and values that might help him control his rage are not activated (broken arrows), nor are his self-regulatory skills accessed. Bypassing these, he becomes increasingly distressed, activating retaliatory hostile scripts and behaviors that he acts out. These lead to real rejection from his girl-friend, and the cycle is ready to be repeated again, spiraling into destruction of the relationship that he is so anxious to preserve. Although this dynamic is activated mostly in relations with the romantic partner, Gary's sensitivity to rejection sometimes is also visible in the domain of work and relations with authority. For example, he can suddenly become angry, impulsive, and hostile when he believes his work, especially quantitative, is not sufficiently respected and appreciated, once even starting a physical scuffle that later embarrassed him greatly.

Expectancies, Values, and Goals.   At first, Gary tries to present the self-confidence expected of an ambitious business student, but on closer examination he has many self-doubts. His genuine high self-efficacy expectations in the quantitative domain are an important exception to these self-doubts.

Most of Gary's goals, even in business school, seem to be directed at avoiding nega-tive reactions from other people. Rather than being characterized by learning goals and motivated by the intrinsic satisfaction of mastering a given activity itself, Gary seems driven by concerns about the evaluative reactions of other people, particularly male peers, and is oriented to preventing himself from getting hurt. He becomes especially anxious and even depressed at the prospect of intimate relations with women. This raises conflicts for him because he values and seeks such intimacy. Unfortunately, at present he does not seem to see clearly how his behavior, specifically his temper outbursts, impacts his closest relationships and therefore assumes no responsibility for it. As was noted in the discussion of his rejection sensitivity dynamics, when these become activated, he bypasses his own standards, values, and goals. Instead, automatic impulsive outbursts of anger become enacted without self-censorship, although they are followed by self-blame and self-criticism after they are enacted.

These excerpts from the case of Gary illustrate concretely how the CAPS personality system operates. To summarize, in CAPS theory, the personality system interacts continuously and dynamically with the social world in which it functions (Mischel et al., 1996; Shoda & Mischel, 1998). The interactions with the external world involve a two-way reciprocal interaction process. Gary is responding to his girlfriend, interpreting her expressed needs as rejections of him. His ways of reacting automatically with hurt, retaliation, and anger alienate her, and she in fact becomes increasingly rejecting. Ultimately, his expectations of abandonment are confirmed, but he was instrumental in shaping that outcome.

In the CAPS model, the behaviors that the personality system generates impact the social world, partly shaping and selecting the interpersonal situations the person subsequently faces. In turn, these situations influence the person (e.g.,Bandura, 1986; Buss, 1987). The personality system actively selects and changes situations. Thus it is *proactive* insofar as it does not just respond to the environment but generates, selects, modifies, and shapes situations in its reciprocal transactions with them (see Dodge, 1997a,b; Mischel & Shoda, 1995).

12.16 What is meant by the concept of reciprocal interactionism? How do we see it at work in Gary W.'s case?

▶ COMMON THEMES

The perspective that is illustrated by CAPS includes many different themes and variations. Some of the main shared features of this still-evolving approach are summarized below.

Work at this level of analysis has a social cognitive focus. It is social in its concern with patterns of social behavior and the interpersonal aspects of life. It is cognitive in its emphasis on mental processes and on understanding how different people process incoming information (Higgins & Kruglanski, 1996). It examines how people process and use information about themselves and their social worlds, and studies the cognitive strategies they develop for coping with important life challenges. Research tries to see how individuals differ in their distinctive encoding (appraisal) of different features of situations. It also examines how those encodings activate and interact with other cognitions and affects in the personality system to generate the person's distinctive patterns of behavior.

The situation is important in work at this level because the personality system is in continuous interaction with situations: the behaviors that are generated depend on the situation involved in the particular interaction. But this is a two-way relationship: the effect of the situation also depends on the characteristics of the personality system. The personality system in part creates the situations themselves through its own interpretations, thoughts, and actions (Bandura, 2001; Mischel & Shoda, 1998). Sometimes this two-way interaction process is called **reciprocal interactionism** (Bandura, 1977), to refer to the chain of causes. In this causal chain, the qualities of a person (including cognitive social person variables), the behaviors generated by the person, and the environment all interact with each other, and each influences the others.

Theory at this level is influenced by information processing models about how the mind works. Although originally based on simple computer systems, in recent years these models have gained more of the depth and complexity essential for a dynamic conception of personality. First, modern information processing models are more like biological and cognitive neuroscience models of how the brain works (e.g., Anderson,

1996; Churchland & Sejnowski, 1992; Shoda & Mischel, 1998). Their focus now is not just on how much of a particular quality (e.g., of self-efficacy expectations) a person has. Rather, it is on the way the person's internal mental representations—the specific cognitions and feelings that become activated—are related to each other and interconnected within the total network of the mental and emotional system. In this view, it is the person's distinctive network that makes up the structure of personality.

The metaphor for this network now tends to be the systems of connections within the human brain through which the neurons are interconnected and process information. This type of processing system operates rapidly and functions at multiple levels often outside conscious awareness (Kihlstrom, 1999, 2003). Theoretically, it is able to deal with the fact that personally important information processing is emotional and often automatic and unconscious. Models that deal with these processes now try to go beyond merely "cool" cognitions to "hot" affect-laden representations and feeling states (e.g., Metcalfe & Mischel, 1999; Smith & Lazarus, 1990; Kahneman & Snell, 1990).

Many researchers at this level of analysis also hope that their ideas and work will help people to enhance their options and decrease their personal vulnerabilities. Put most simply, they would like their work to ultimately help optimize human well-being, freedom, and choice. Thus, they also emphasize human potential and assume that people can change constructively and extensively under appropriate conditions. With that goal in mind, often they study such real-life problems as overcoming anxious, depressive, and health-threatening mental states (Aspinwall & Taylor, 1997; Cantor & Kihlstrom, 1987; Taylor & Armor, 1996). Much of this work tries to improve people's sense of mastery and competence so they can better fulfill their potential. The commitment is to enhancing the human potential for alternative, more constructive ways of coping, consistent with the phenomenological perspectives of such earlier theorists as George Kelly.

The emphasis on stimulus conditions in early behavioral approaches often implies a passive view of people—an image of organisms that are empty except for sets of automatic responses to external stimuli. In contrast, social cognitive approaches emphasize that people are active and agentic—that is they actively shape their worlds and social environments. Consequently much of the research at this level deals with issues of self-regulation and self-control and is designed to see how people can more effectively regulate their behavior to "take charge" of their lives (e.g., Bandura, 2001; Mischel et al., 1996).

Recall that trait theories emphasize differences *between* people in their response to the same situation or even regardless of the situation. Cognitive social theories also address these differences. However, they go on to try to understand the unique ways an individual's behavior may change in response to even slight changes in conditions, as indicated in Figure 12.3. They focus on these enduring within-individual differences (intra-individual stability) as an important aspect of the difference between individuals. Understanding the psychological impact or meaning of particular events and conditions in the person's life requires observing how changes in those conditions alter what he or she expects, thinks, and feels.

Work at this level also draws on the behavioral level of analysis to the degree that it asks: Why and when will individuals *do* what they do? That is, it goes from social cognition, feelings, and motivation to a focus on human action (e.g., Gollwitzer & Bargh, 1996). But although the approach includes a focus on behavior, its primary goal is to clarify the psychological processes that underlie the particular behavior of interest.

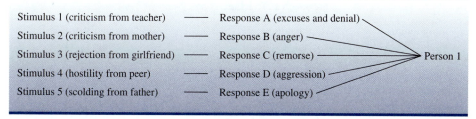

Figure 12.3 The same person behaves in distinctive but stable ways in response to various psychological situations.

SUMMARY

DEVELOPMENT OF THE SOCIAL COGNITIVE PERSPECTIVE

- The social cognitive level seeks to understand the cognitive, affective, and social processes that characterize the individual. It was stimulated by the finding that the perceiver's mental representations and cognitive transformations of stimuli can determine their impact on the individual.

OBSERVATIONAL LEARNING (MODELING): ALBERT BANDURA

- Albert Bandura called attention to the importance of observational learning for personality development and change.
- Observational learning can help create or remove fears and other strong emotional reactions.
- Human performance is dramatically improved by awareness of the rules or principles that influence the outcomes for behavior.

SOCIAL COGNITIVE PERSON VARIABLES

- In social cognitive theory, a set of *person variables* (cognitive–affective units or CAUs) is used to describe individual differences.
- They consist of *encodings, expectancies and beliefs, affects, goals and values, and self-regulatory strategies and competencies*.

THE COGNITIVE–AFFECTIVE PERSONALITY SYSTEM (CAPS)

- In CAPS theory, individual differences reflect (1) differences in the chronic accessibility of person variables, and (2) differences in their organization.
- In the CAPS system the person and the situation are in continuous interaction, influencing each other reciprocally.
- The system's structure is reflected and expressed in the person's stable patterns of *if . . . then . . .* relationships which form distinctive behavioral signatures.
- Gary's personality dynamics illustrate how CAPS interacts continuously and dynamically with the social world.

COMMON THEMES

- Social cognitive approaches focus on how individuals select, attend to, and process information about the self and the world and react to it.
- *Reciprocal interactionism* suggests that people's attributes and actions interact with the social environment continuously, each influencing the other.

KEY TERMS

✓ behavior–outcome
relations   279

✓ cognitive transformations
(of stimuli)   272

✓ cognitive–affective
personality system
(CAPS)   283

✓ encoding   277

✓ hot cognitions   280

✓ intrinsic motivation   281

✓ modeling   273

✓ observational
learning   273

✓ person variables   276

✓ personality processing
dynamics   283

✓ reciprocal
interactionism   287

✓ rejection sensitivity   285

self-efficacy
expectations   278

SOCIAL COGNITIVE PROCESSES

When Jana told her partner Rodolfo she had to spend the evening preparing for a work assignment due the next day he became upset, interpreting it as another sign of her increasingly uncaring attitude toward him. He felt rejected, and then quickly became irritated and angry, starting an argument. His interpretation of her behavior as rejecting him was of course only one way to encode it. Compare him with Mike who is less sensitive to rejection. When Mike heard a similar comment from his partner he took it at face value, respecting her need to concentrate on her work.

To explain individual differences in reactions to the same event, researchers at the social cognitive level of analysis begin by seeing how the different individuals construe and interpret its meaning and significance. These interpretations and the emotions and behavioral reactions that unfold are, in turn, influenced by the "self."

The importance of the self in psychology has had many dramatic ups and downs. For many years at the height of behaviorism it was almost totally down. There were, however, some notable exceptions, discussed earlier in this text. The self was important for the phenomenologists, and likewise in the clinical work of some of the post-Freudians and object relations theorists. But these remained lone voices for decades, muted by the dominance of behaviorists who judged the self to be a confusing fiction outside the realm of science. Currently, interest in the self is again extremely high (e.g., Leary & Tangney, 2003). After its long slumber, the self is attracting the attention of many theorists and researchers because new methods for studying social cognition have opened it to systematic study. The self now has become a core topic in personality at the social cognitive level of analysis. This chapter examines the self and self-relevant processes, focusing on concepts and research from the social cognitive level of analysis.

▶ THE SELF

At a common sense level, everyone knows what the self is: we have a sense of ourselves, refer to ourselves, talk about ourselves, evaluate ourselves—sometimes blaming ourselves, sometimes praising ourselves—and we often control ourselves—with varying degrees of success. The "self" has been central in personality psychology since the start of the field, reflected in concepts like "self-actualization," "self-evaluation," "self-regulation" and "self-control." For many researchers, the self has become a prefix for practically every important process and function in a personality system.

The "I" and the "Me"

13.1 In what two ways do we use the term "self"?

An important distinction between two different aspects of the self was originally made a century ago by America's pioneering psychologist William James (1890), and it still guides much of the work on the self.

The self as an "I" is an agent, actor, or doer, an executive who conducts basic psychological functions. The **"I" as an agent** engages in such functions as self-regulation (as in impulse control and planning for the future), and self-evaluation to assess progress along the way.

The self has a second important aspect that James also recognized: the "Me" that is observed and perceived. It is the Me that one sees when attention is focused on the self, the **"Me" as an object,** represented in self-concepts, in how we see ourselves. Each person develops a self-theory about his or her Me. This theory is a construction, a set of concepts about the self (Epstein, 1973, 1983; Harter, 1983; Wiley, 1979).

The Self as a Basic Schema

Intuitively the "I," "me," or "self" is a basic reference point around which experiences and the sense of personal identity itself seem to be organized. At the social cognitive level of analysis, the self is viewed not as a "thing" but as a schema or cognitive category that serves as a vital frame of reference for processing and evaluating experiences (e.g., Baumeister, 1997; Markus, 1977; Markus & Cross, 1990; Markus, Kitayama, & Heiman, 1996).

Schemas in social cognitive theories are important for understanding mental functioning because they are basic units for organizing information. You can think of them as knowledge structures made up of collections of attributes or features that have a "family resemblance" to each other. These structures often have clear exemplars or "best examples" (e.g., a sparrow is a better or more prototypic example of a bird than an ostrich). Schemas help one to make sense of new events by recognizing what they are like in terms of their similarity to the structures that already exist. That happens, for example, when you see a flying object of a type you don't recall having seen before, but nevertheless have no trouble deciding it must have been a bird—not a plane, or a stone, or a balloon. The "self" that develops also consists of interconnected knowledge structures of many different sorts.

In the course of development, the child acquires an increasingly rich concept of himself or herself as an active agent, an "I" separate from other people and objects. A sense also develops of a "Me" that has defining features and qualities reflected in multiple self-concepts.

Self-Schemas. Self-concepts or **self-schemas** (Markus, 1977) include generalizations about the self, such as "I am an independent person" or "I tend to lean on people." These cognitions arise from past experiences and, once formed, guide how we deal with new information related to the self. To illustrate, if you have strong self-schemata about being extremely dependent, passive, and conforming, you would process and remember information relevant to those schemata more quickly and effectively than do people for whom those schemata are not personally relevant (Markus, 1977). People have better recall for information about traits that they believe describe them than for traits that are not self-descriptive (Rogers, 1977; Rogers, Kuiper, & Kirker, 1977). Thus, we give information relevant to the self special cognitive treatment, for example, by being more attentive toward it.

Self-schemas are highly accessible personal constructs that a person is ready to use for encoding information. People differ stably in their self-schemas (Higgins, 1996; Higgins, King, & Mavin, 1982). These self-concepts also have motivational implications. For example, most people desire to maintain positive views of themselves, are motivated to pursue self-knowledge, and want to enhance and improve themselves (Baumeister, 1996).

Each person develops a **self-theory** about his or her Me. This theory is a construction, a set of concepts about the self. It is created by the child from experience, but it in turn affects future experience (Epstein, 1973, 1983; Harter, 1983, 1999; Wiley, 1979). The individual's self-concepts are not a simple mirror-like reflection of some absolute reality. Self-concepts, like impressions of other people and of the world, involve an integration and organization of a tremendous amount of information. Self-concepts are not a mirror of reality, but they are correlated with the reactions and outcomes that the person has experienced throughout the past and expects to obtain in the future (Leary & Tangney, 2003; Wiley, 1979). Although the concepts change over time, their foundations form early in life, and they influence what that future becomes (Markus & Cross, 1990; Markus et al., 1996).

The Relational Self and Transference

In recent years, the self has been reconceptualized in light of current theory and findings on social cognition and specifically on how memory works.

The Relational Self. Self-knowledge includes all the thoughts (cognitions) and feelings (affects) that develop about oneself. It consists of information that is stored and organized in memory as a cognitive–affective (mental and emotional) representation. This knowledge representation is closely connected in the memory system to knowledge

13.2 Describe the meaning and functions of the self-schema.

13.3 In what sense is the self inherently relational? What do social cognition researchers mean by transference, and how does this compare with Freud's use of the term?

representations about the significant other people in one's life (e.g., Andersen & Chen, 2002; Linville & Carlston, 1994), making these two types of information directly associated. Consequently, when the representation of a particular significant other is activated (e.g., you think about your mother), aspects of your own self-representation also become activated mentally. These close connections in memory make the self intrinsically *relational* and interpersonal: in a sense, the significant others to whom one is close become part of one's personal identity. Thus, the self evolves from, and is linked to, the important relationships in one's life: in this sense, the self that emerges is intrinsically entangled with significant others.

As the self develops in relation to a particular significant other person, expectations also develop about the most likely interactions that will occur with him or her. For example, an adolescent may come to expect that interactions with her mother about smoking are likely to play out along predictable lines (e.g., more hassles, more guilt, more avoidance). These scripts become reactivated in future interactions with the significant other or with other people who remind you of that person (Andersen, Reznik, & Chen, 1997).

Transference Reconsidered. The notion of the *relational self,* conceptualized in terms of memory representations that connect self-knowledge with representations of significant others, also allows a fresh view of psychodynamic concept of transference. Transference, as discussed in earlier chapters, plays an important role in psychoanalytic theory and psychotherapy beginning with Freud, and in the ideas of Kohut and the object relations theorists. The concept of transference has now been recast in social-cognitive terms, and this is not just a renaming: it informs the process in light of findings from memory research. In the social cognitive reinterpretation by Susan Andersen and colleagues (2002), when one develops relationships with new people, transference readily occurs. This happens to the degree that

The self is interpersonal.

Susan Anderson

representations of significant others in memory are activated by the newly encountered person.

Feeling attracted to—or repelled by—a newly encountered person, and easily making all sorts of inferences about his or her qualities (that may or may not turn out to be accurate), can be understood in terms of the particular significant other representation that is being triggered by, and applied to, the new person (Andersen et al. 1997; Andersen & Chen, 1998). If the psychiatrist's manner reminds the patient of her father, the cognitive–affective representation of the father, and of the patient's self in relation to him, also may become easily activated and brought to mind.

Note that this view of transference is compatible with Freud's (1912/1958) claim that the individual's mental representations of significant others, most notably the parents in early childhood, profoundly influence relationships to new people, including the psychoanalyst in the treatment process. It is different, however, because it sees this process in terms of social-cognitive information processing rather than as a reflection of the psychosexual drives and conflicts favored in psychodynamic theory.

Perceived Stability of Self and Potential for Change

Perhaps the most compelling quality of the self is its perceived continuity. The experience of **subjective continuity** in oneself—of basic oneness and durability in the self—seems to be a fundamental feature of personality. Indeed, the loss of a feeling of consistency and identity may be a chief characteristic of personality disorganization, as is seen in schizophrenic patients who sometimes report vividly experiencing two distinct selves, one of them disembodied (Laing, 1965).

13.4   What psychological mechanisms enable us to maintain consistency and continuity in the self-schema?

Each person normally manages to reconcile seemingly diverse behaviors into one self-consistent whole. A man may steal on one occasion, lie on another, donate generously to charity on a third, cheat on a fourth, and still readily construe himself as "basically honest and moral." People often seem to be able to transform their seemingly discrepant behaviors into a constructed continuity, making unified wholes out of diverse actions. How does this integration work?

Many complex factors are involved (Harter, 1983; Markus & Cross, 1990). One answer to this question may be that people tend to reduce cognitive inconsistencies and, in general, simplify and integrate information so that they can deal with it (Ross & Nisbett, 1991; Tversky & Kahneman, 1974). The human mind may function like an extraordinarily effective reducing valve that creates and maintains the perception of continuity even in the face of perpetual observed changes in actual behavior (Festinger, 1957; Mischel, 1973). The striving for self-consistency that the phenomenologists emphasized (Rogers, 1951) has received support in the work of other researchers working on related topics (Aronson & Mettee, 1968; Cooper & Fazio, 1984).

Another basis for perceived consistency is that people often know a good deal about their own characteristic *if . . . then . . .* , situation–behavior patterns. There is a strong relationship between the intra-individual stability of those patterns and the person's self-perception of consistency. This finding was discussed in Chapter 4. It was found that college students who perceived themselves as consistent with regard to their "conscientiousness" had much more stable patterns of situation–behavior relations in their conscientious behavior than did those who saw themselves as not highly consistent (Mischel & Shoda, 1995). For example, students who always conscientiously prepared for tests, but were not conscientious about keeping appointments on time would see themselves as consistent. Note that their consistency was in their stable *pattern*, rather than from one type of situation (test preparation) to another (punctuality).

In general, the self-concepts and personal constructs that people have seem to show a good deal of stability (Byrne, 1966; Gough, 1957; Mischel, 1968). Thus our personal theories about ourselves and the world tend to be relatively stable and resistant to change (Nisbett & Ross, 1980), as are our attitudes and values (E. L. Kelly, 1955). This perceived stability, however, coexists with the equally compelling fact that throughout the life course, people modify and transform their self-concepts as they envisage and construct "alternative future selves" (Cantor, 1990, p. 735) and strategically adapt to diverse life challenges (Mischel & Morf, 2003).

Multiple Self-Concepts: Possible Selves

The self traditionally has been viewed as unitary: a single self that is relatively consistent (Allport, 1955; Snygg & Combs, 1949). More recently, it is increasingly common to characterize the self as a multifaceted, dynamic set of concepts consisting of multiple selves. These different peceived selves reflect different aspects of an individual's total personality.

Consider a woman who seems hostile and fiercely independent some of the time but passive and dependent on other occasions. What is she really like? Which one of these two patterns reflects the woman that she really is? Must she be a really aggressive person with a facade of passivity—or is she a warm, passive-dependent woman with a surface defense of aggressiveness? In theory, at the social cognitive level, it is possible for her to be all of these: a hostile, fiercely independent, passive, dependent, aggressive, warm person all in one (Mischel, 1969). Indeed, each of these aspects may constitute a different **possible self** or potential way of being (Markus & Cross, 1990).

Hazel Markus

Of course, which of these selves she is at any particular moment would not be random; it would depend on who she is with, when, and why, and most importantly on how she construes and interprets the situation, that is, its meaning to her. But each of these aspects of her self may be a quite genuine and real aspect of her total being, part of her own unique but stable patterning of person variables (Mischel & Morf, 2003).

The self-concepts that encode different aspects of the person vary, depending on particular contexts, on the type of behaviors that are self-assessed, and on the culture. The beliefs and values of one's culture and group profoundly influence how individuals construe themselves and their future possibilities (e.g., Stigler, Shweder, & Herdt, 1990), and even the conception and definition of the "self" varies greatly across cultures. Thus, the self may be a more central concept for societies like ours that value self-enhancement and see the self as an entity, than for societies like the Japanese that value the relations between the person and other people rather than oneself alone (e.g., Markus et al., 1996; Markus & Kitayama, 1991).

The concepts of the self that the person can access easily comprise what has been called the **working self-concept** (Markus & Nurius, 1986), which derives from various self-conceptions that are present in thought and memory. According to Markus and her colleagues, this working self-concept includes ever-changing combinations of *past selves* and *current selves,* as well as the imagined *possible selves* that we hope to become or are afraid of becoming. These possible selves serve as guides for behavior and can have significant impact on one's emotional and motivational states. A schematic representation of some components of a working self-concept is shown in Figure 13.1.

Self-Esteem and Self-Evaluation

One of the most critical aspects of the self-concept is **self-esteem** (Harter, 1983). Self-esteem refers to the individual's personal judgment of his or her own worth (Coopersmith, 1967; Epstein, 1973, 1990). Self-esteem is such an important aspect of

13.5 How does self-esteem develop and influence behavior?

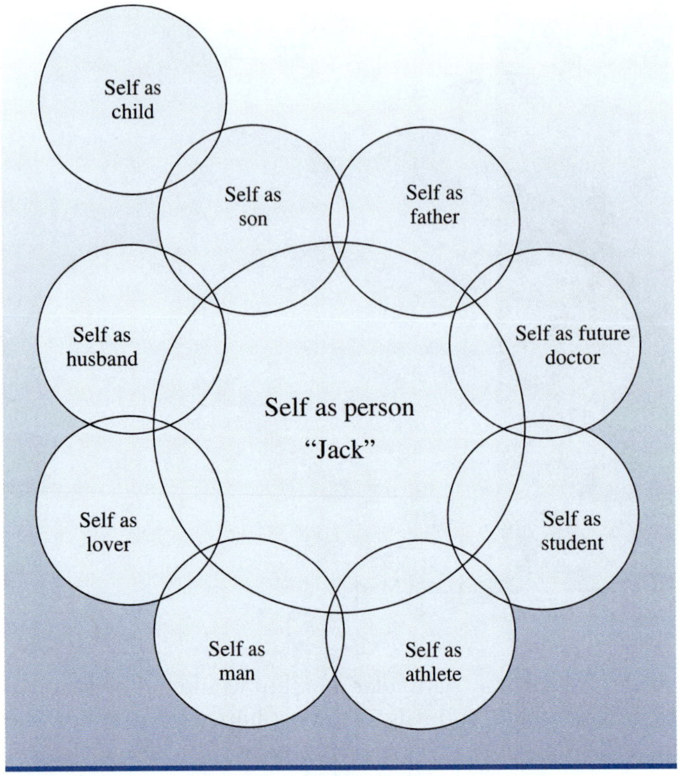

Figure 13.1 Schematic representation of a working self-concept. The working self-concept contains mental representations of diverse aspects of the self, from the present and the past, as well as imagined possible future selves.

the self-concept that the two terms are often used as if they were the same. Although self-esteem is sometimes discussed as if it were a single entity, persons may evaluate their functioning in different areas of life discriminatively (Crocker, 2002).

Self-esteem and self-evaluations are influenced by the feedback that people continuously get from the environment as they learn about themselves, beginning in the early course of development (Leary & Downs, 1995; Leary, 2002). In their self-appraisal, individuals are guided by their memories and interpretations of earlier experiences, and by the framework of self-concepts, self-standards, and self-perceptions through which they view and filter their experiences. People compare themselves to their own standards, as well as to their perceptions of the performance of relevant others (Bandura, 1986; Higgins, 1996; Norem & Cantor, 1986). For example, people who greatly value achievement and who are motivated to achieve tend to react quite differently to failure experiences than do those who are low in achievement striving (Heckhausen, 1969; Koestner & McClelland, 1990). However, the same outcome that is a discouraging "wipe out" for one may be a motivating challenge for the other.

People adopt many different strategies to cope with performance feedback relevant to self-esteem. For example, the impact of success and failure experiences depends on whether the person construes or "frames" the outcomes as reflecting on the self as a whole, or in terms that are more circumscribed and specific to the particular success or failure situation (Mendoza-Denton and associates, 2001). So if the experience of failure is framed in terms of being about the self as whole without any situational qualifiers (e.g., "I am a failure"), the emotional fall-out and consequences generalize much more broadly than when the experience is framed in situation-specific, *if . . . then . . .* terms (e.g., "I fail if I take this kind of test on this topic.")

Self-evaluation processes are basic for understanding how people see themselves and respond to their own experiences. They reflect each person's compromises between the need for accurate perception of his or her performance in the real world and the self-protective desire to maintain a favorable self-image. Personality theorists have long recognized that in this self-evaluation process, we generally manage to combine a mix of "thorough realism" and "protective maneuvering" (Cantor & Kihlstrom, 1987, p. 152).

Costs of Self-Esteem Pursuit. People with high as compared to low overall levels of self-esteem generally seem to function better in many aspects of their lives, and to feel better about themselves in many domains (Hoyle et al., 1999). However, while having genuinely high self-esteem may have its substantial benefits, its active pursuit can have high costs (e.g., Crocker, 2002). The risk in the pursuit of self-esteem is that it can lead people to focus too much on themselves while neglectful of the feelings and needs of others. This can encourage competitiveness rather than cooperation (e.g., Carver & Scheier, 1998). It can lead to preoccupation with "how am I doing?" rather than attention to "how are you?"—and a neglect of other people while busily trying to build up one's own self-esteem. The costs of pursuing self-esteem tend to be especially high for people who see high self-esteem as essential for feeling that they are worthy. For them, threats to self-esteem may trigger anxiety followed by negative defensive reactions to reduce the discomfort. These defensive reactions can range from hostility and aggression to withdrawing from relationships or creating emotional distance that undermines important connections with other people (e.g., Baumeister, 1998).

> 13.6 How can the pursuit of high self-esteem create maladjustment?

Features and Functions of the Self: Overview

The explosion of research on the self in the last few decades is leading to a consensus about its essential features (Hoyle et al., 1999; Mischel & Morf, 2003). In contemporary views on the self, the self is seen as having the following qualities:

> 13.7 Describe three important features of the self.

- *The self is essentially social and interpersonal—it arises out of social experiences with significant other people and is expressed in relation to them (e.g., Andersen & Chen, 2002).*

In earlier forms this social view of the self was seen in the discussion particularly of Carl Rogers and the later work of Tory Higgins. It is seen clearly in current work, for example, on the relational self and the reinterpretation of transference in social cognitive terms.

- *The self is important for understanding the processes of adaptation and coping—that is the doing or executive functions of personality (e.g., self-regulation, self-defense). In performing these functions it is a dynamic, motivated system (e.g., Baumeister, 2002; Derryberry, 2002; Leary, 2002).*

Many of these qualities and functions were emphasized in earlier forms by the ego psychologists, stemming originally from the psychodynamic Freudian tradition. It was seen, for example, in Anna Freud's focus on ego defenses and in the theories and clinical practice of Alfred Adler and Carl Jung (Chapter 6).

- *Closely related to the above is the view that the self has evaluative functions, and it is basic for the concept of identity (e.g., Crocker, 2002; Eisenberg, Spinrad, & Morris, 2002). It is motivated to protect self-esteem (Leary, 2002). These features were important in the theories of Erik Erikson, Heinz Kohut (1971), and Otto Kernberg (1976, 1984), which you already encountered in this text in Chapter 6. They are again being given new life in the work on self-regulation and goal pursuit discussed later in the text (e.g., Chapters 16 and 17).*

▶ CAUSAL ATTRIBUTIONS, HELPLESSNESS, AND MASTERY

Perceptions about one's adequacy and self-efficacy can have either self-enhancing or self-destructive consequences. In this section we consider how these positive and negative patterns develop and play out.

Causal Attributions Influence Emotions and Outcomes

13.8 What two kinds of causal attributions do people make? Define the self-enhancing attributional bias.

Did I do it? Was it my fault? Did I really deserve it? Was it just luck? Did he spill the milk on me on purpose? Is she mean or just upset today? The different ways in which individuals answer these kinds of questions about causation reflect some extremely interesting individual differences. *Causal attributions* are the explanations people make of the causes of events and they have predictable implications for how they feel about themselves and other people (Weiner, 1990). For example, we may see the same event—say, getting an A on an exam—as due to *internal causes* (such as high ability or hard work) or as due to *external causes* (such as the ease of the task or good luck). How we feel about the grade depends on whether we see it as due to internal or external causes (Phares, 1976; Rotter, 1966).

Pride and Shame.   Generally, "pride and shame are maximized when achievement outcomes are ascribed internally, and minimized when success and failure are attributed to external causes" (Weiner, 1974, p. 11). In other words, a success that is perceived to be the result of one's ability or effort (internal causes) produces more positive feelings about oneself than does the same success when it is viewed as merely reflecting luck or an easy task (external causes). Conversely, we feel worse (for example, experience "shame") when we perceive our failure as reflecting low ability or insufficient effort than when it is seen as due to bad luck or the difficulty of the particular task. For example, being fired from a job for one's incompetence has a different emotional impact than does being fired because the firm went bankrupt (see In Focus 13.1 for discussion).

As noted in the In Focus 13.1, in most people, perceptions of causes tend to be biased in self-enhancing ways (Greenwald, 1980; Harter, 1983; Leary, 2002). This bias can have beneficial effects, although in the extreme it can be self-deluding. But at the other extreme, to lack the self-enhancing, self-protective bias completely may risk developing a depressive orientation (Lewinsohn and colleagues, 1980).

Learned Helplessness and Apathy

13.9 How does learned helplessness develop and affect behavior?

When people believe that there is nothing they can do to control negative or painful outcomes, they may come to expect that they are helpless (Seligman, 1975) and encode themselves in helpless terms. They expect that the aversive outcomes are uncontrollable, that there is nothing they can do. In this state, called **learned**

THE PERCEPTION OF CONTROL AND MEANINGFULNESS

There is a strong bias to see events and behavior as meaningful, orderly, and controllable even when they are random. You have seen this attributional bias if you have ever watched behavior in a casino. If you observed closely, you probably noticed how often gamblers act as if they can control chance by shaking the dice just right, or waiting to approach the roulette wheel at the perfect moment, or pulling the slot machine levers with a special little ritual. Gamblers are not unique in believing they can control chance events.

Even when something is clearly the result of chance (like the cards one draws in a poker game), people may see it as potentially controllable and not just luck. There is a deep human tendency to see the world as predictable and fair. We expect a "just world" (Heider, 1958), a world in which the things that happen to people are deserved and caused by them—even things like whether or not they win a lottery, get cancer, or are raped and murdered. Much research (reviewed by Langer, 1977) suggests that

people often do not discriminate between objectively controllable and uncontrollable events. Instead they seem to have the "illusion of control," acting as if they can even control events that actually are pure chance.

In most people, this is part of a **self-enhancing bias**: we are more likely to see ourselves as causally responsible for our actions when they have positive rather than negative outcomes. When we do well or win, it is to our personal credit; when we do badly or lose, we could not help it (e.g., Fitch, 1970; Urban & Witt, 1990). Even when outcomes are negative, as in a tragic accident, we may find it hard to cope with events unless we somehow can see them as "just," meaningful, and orderly (e.g., Taylor & Armor, 1996; Taylor & Brown, 1988). Although this finding was first seen as evidence for a simple self-enhancing bias by perceivers, more recently it has been shown to have important, potentially beneficial and therapeutic effects for personality (e.g., Seligman, 1990; Seligman, Reivich, Jaycox, & Gillham, 1995; Taylor & Armor, 1996).

helplessness, they also may become apathetic and despondent and that state may generalize and persist.

The concept of learned helplessness originally was based on findings from some experiments with animals exposed to extreme frustration. Consider, for example, a dog placed in a situation where he can do nothing to end or escape an electric shock delivered to his feet through an electrified grid on the floor. Later he may sit passively and endure the shocks even though he can escape them now by jumping to a nearby compartment (Seligman, 1975, 1978). These findings on learned helplessness with animals also were consistent with less rigorous observations of humans forced to face extreme and persistent frustration. Withdrawal, listlessness, and seeming emotional indifference are often found among war prisoners and inmates of concentration camps, for example. The victims seem to have given up totally, presumably overwhelmed by their inability to do anything that will change their desperate lot.

Dramatic examples of learned helplessness may be found among the children of America's migrant families whose plight was described by Coles (1970). He notes that unlike typical children in the middle class, migrant children soon discover that their shouts and screams will not necessarily bring any relief from their pains and frustrations. Consider this description by a migrant mother of her own helpless feelings in the face of her children's suffering:

My children, they suffer, I know. They hurts, and I can't stop it. I just have to pray that they'll stay alive, somehow. They gets the colic, and I don't know what to do. One of them, he can't breathe right and his chest, it's in trouble. I can hear the noise inside when he takes his breaths. The worst thing, if you ask me, is the bites they get. It makes them unhappy, real unhappy. They itches and scratches and bleeds, and oh, it's the worst. They must want to tear all their skin off, but you can't do that. There'd still be mosquitoes and ants and rats and like that around and they'd be after your insides then, if the skin was all gone. That's what would happen then. But I say to myself it's life, the way living is, and there's not much to do but accept what happens. Do you have a choice but to accept? That's what I'd like to ask you, yes sir. Once, when I was little, I seem to recall asking my uncle if there wasn't something you could do, but he said no, there wasn't, and to hush up. So I did. Now I have to tell my kids the same, that you don't go around complaining—you just don't (Coles, 1970, pp. 9–10).

In sum, a good deal of research now suggests that when people believe they cannot control events and outcomes, they gradually develop a sense of helplessness (Wortman & Brehm, 1975; Seligman, 1990) and even severe depression. In the extreme, when people feel they cannot tolerate the frustrations in their lives, they may lose interest in all activities and virtually stop behaving, often spending a great deal of time in bed. They tend to become very sad and feel worthless, often suffering a variety of physical complaints.

13.10 What four classes of symptoms characterize depression?

There is general agreement that the behaviors listed in Table 13.1 characterize depressed people as a group. Each depressed individual displays his or her own combination of any of these behaviors, often including deep unhappiness, emotional numb-

---

TABLE 13.1   Some Indicators of Depression

Mood
   Feel sad and blue most of the time
   No longer enjoy things they used to
   General loss of feeling
   Fatigue, apathy, boredom
   Loss of interest in eating, sex, and other activities

Physical symptoms
   Headaches
   Difficulty sleeping
   Gastrointestinal symptoms (indigestion, constipation)
   Weight loss—loss of appetite
   Vague physical complaints

Behaviors
   Unsociable—often alone
   Unable to work, less sexual activity
   Complaining, worrying, weeping
   Neglect appearance
   Speak little (speech is slow, monotonous, soft)

Ideation
   Low self-esteem ("I'm no good")
   Pessimism ("Things will always be bad for me")
   Guilt, failure, self-blame, self-criticism
   Feel isolated, powerless, helpless
   Suicidal wishes ("I wish I were dead." "I want to kill myself.")

Source: Based on Lewinsohn, P.M. (1975). The behavioral study and treatment of depression. In M. Hersen (Ed.), *Progress in behavior modification* (pp. 19–64). New York: Academic Press.

ness and loss of interest, withdrawal from normal activities, and profoundly negative feelings about oneself and life. In contrast, when people believe that they can make an impact on their environment, that they can influence and control events, they become more alert, happier—and may even live longer (Seligman, 1990).

Pessimistic Explanatory Styles

Persons are most vulnerable to perceiving themselves as helpless when they see the bad things that happen in life as due to their own internal qualities rather than to more momentary, external, or situational considerations (Abramson, Seligman, & Teasdale, 1978). They explain why they failed the test, for example, by thinking "I'm incompetent" rather than as due to the flu they had and the fact that they could hardly see the page. People who do this consistently have a distinctive, essentially *pessimistic explanatory style.*

Pessimism is defined as an explanatory style that has three components: the person sees bad events as enduring, widespread, and due to the self. At an early age, this style can be a predictor of poor health for the future (Peterson, Seligman, & Vaillant, 1988). In one study, a group of healthy 25-year-old college graduates filled out questionnaires asking them to tell about difficult personal experiences in their lives. Researchers then analyzed the way the students explained the bad events that had occurred to determine whether or not their explanatory styles were generally pessimistic. Judges rated the explanations according to three criteria: the level of *stability* (seeing an event as having no end in sight; for example, "it won't ever be over for me"), *globality* (generalizing the event to many aspects of one's life), and *internality* (accepting one's self as causing or central to the problem rather than attributing it to some external factor or just plain circumstances). The more consistently stable, global, and internal the explanations, the higher the "pessimistic" style score.

Over a span of 35 years, the group was followed up with health examinations, and measures of illnesses were recorded. For about 20 years after college, there were no significant differences in the health of the participants. By the time they reached the ages of 45 through 60, however, those who at age 25 had been more pessimistic in their responses were more likely to be suffering from illness.

Researchers of optimistic–pessimistic explanatory styles also studied newspaper interviews with ballplayers from the Baseball Hall of Fame, published throughout the first half of the century. The interviews quoted the players' explanations of good and bad events as they discussed how and why they won or lost games. Players who attributed their losses to their own personal, stable qualities and thus saw them as their own fault, but saw their wins as due to momentary external causes (e.g., "the wind was right that afternoon") tended to live less long than those who used more optimistic explanatory styles to construe their good and bad experiences (Peterson & Seligman, 1987).

Learned Optimism

The opposite of the learned helplessness pattern is a pattern of experience and thinking called **learned optimism** (Seligman, 1990). To develop this style the person is helped to encode the daily hassles and setbacks in life by deliberately using self-enhancing explanations. Now negative experiences become viewed as momentary, specific, and not due to one's one failings. This optimistic, positive way of interpreting experience tends to be associated with a wide range of positive outcomes, as assessed by self-reports of feeling happier and evidence of more effective functioning and work success (also see In Focus 13.2).

13.11 Describe the three attributional elements of the pessimistic explanatory style. How does pessimism relate to physical well-being?

13.12 Contrast learned optimism with learned helplessness. How are optimism and self-enhancing illusions related to psychological adjustment?

THE ILLUSORY WARM GLOW OF OPTIMISM

Traditionally, psychologists have considered an accurate perception of self as essential for mental health (Jahoda, 1958). Researchers, however, have found that most psychologically healthy people have somewhat unrealistically positive illusory self-views, whereas those who perceive themselves more accurately tend to be the less mentally healthy (Taylor & Brown, 1988). For example, people with realistic self-perceptions are more likely to experience low self-esteem and depression, while more stable individuals tend to see positive personality traits as being most descriptive of themselves (Alicke, 1985; Brown, 1986).

The exaggeratedly positive self-perceptions that most people have became apparent in a study comparing depressed patients with nondepressed psychiatric and normal controls (Lewinsohn et al., 1980). Patients who had interacted in small group situations were asked to rate both themselves and each other with regard to personality characteristics. Nondepressed individuals' self-ratings were considerably more favorable than the ratings others had given them. Self-ratings of depressed individuals were consistent with the ratings given them by others,

suggesting that the nondepressed people had unrealistically positive self-views, seeing themselves through rose-colored glasses.

Individuals biased by such self-enhancing illusions also tend to feel they have an unrealistically large amount of control in pure chance situations, where in fact they cannot influence the outcome (Langer, 1975). In a study involving dice-throwing, for example, nondepressed participants felt that they would have greater control when throwing the dice themselves than when someone else did it for them (Fleming & Darley, 1986). The opposite state is found in depressed persons, who are likely to have a more realistic perception of the amount of control they have in chance situations—which in the case of dice-throwing is zero. Likewise, people in a depressed state tend to be reasonably accurate when predicting the future, while the nondepressed display an unrealistic optimism (Alloy & Ahrens, 1987). These results are clear, interesting, and consistent but, of course, they should not be misread as suggesting that gross distortions of reality characterize nondepressed individuals.

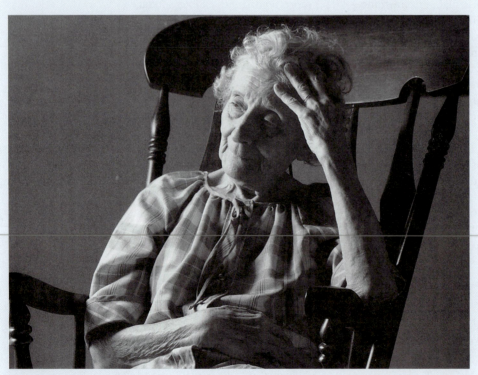

The depressed may perceive themselves *more* accurately than the non-depressed perceive themselves.

Similar results were found with different measures based on self-reported optimism in one's orientation to, and interpretation of, life events. For example, the degree of optimism in the person's orientation to life predicts recovery from coronary bypass surgery in patients; it also predicts fewer physical symptoms in college students (Scheier & Carver, 1987). People with an optimistic orientation to life seem to face stressful situations by thinking about them constructively to deal with them as effectively as possible, essentially trying to make the best of the situation even if it is an extremely difficult one. In contrast, the pessimistic orientation is associated with withdrawal, which automatically prevents problem solutions or constructive reinterpretation (Scheier, Weintraub, & Carver, 1986).

Helpless versus Mastery-Oriented Children

Following failure on a task, some individuals seem to fall apart and their performance deteriorates. But other people actually improve. What causes these two different types of responses to failure? The answer to this question again requires understanding how the person construes or interprets the reasons for the experience. Consistent with the work on learned helplessness, children who believe their failure is due to lack of ability (called *helpless children*) were found to perform more poorly after they experienced failure than did those who see their failure as due to lack of effort (called *mastery-oriented children*). Indeed, the **mastery-oriented** children often actually performed better after failure (Dweck, 1975).

When faced with failure, helpless children seem to have self-defeating thoughts that virtually guarantee further failure. This became clear when groups of helpless and mastery-oriented fifth-graders were instructed "to think out loud" while solving problems. When children in the two groups began to experience failure, they soon said very different things to themselves. The helpless children made statements reflecting their

13.13 Compare the attributional patterns of helpless and mastery-oriented children, and their effects on performance.

Carol Dweck

TABLE 13.2 Coping Strategies of Helpless and Mastery-Oriented Children

Helpless Children	Mastery-Oriented Children

Attributions for failure to self
  "I'm getting confused"
  "I'm not smart"
	Self-instructions to improve performance
	  "The harder it gets, the harder
	  I need to try"

Solution irrelevant statements
  "There's a talent show this
  weekend, and I am going
  to be Shirley Temple"
	Self-monitoring statements
	  "I'm really concentrating now"

Statements of negative affect "This isn't fun anymore"	*Statements of positive affect* "I love a challenge"

Source: Based on Diener, C. I., & Dweck, C. S. (1978). In analysis of learned help-lessness: Continuous changes in performance, strategy, and achievement cognitions following failure. *Journal of Personality and Social Psychology, 36,* 451–462.

loss or lack of ability, such as "I'm getting confused" and "I never did have a good memory" (Diener & Dweck, 1978, p. 458). None of the mastery-oriented children made lack-of-ability statements. Instead, these children seemed to search for a remedy rather than for a cause for their failure. They gave themselves instructions that could improve their performance, such as "I should slow down and try to figure this out" and "The harder it gets, the harder I need to try."

The helpless children made many statements that were irrelevant to the solution and that usually were ineffective strategies for problem-solving (see Table 13.2). For example, one helpless male repeatedly chose the brown-colored shape, saying "chocolate cake" in spite of the experimenter's repeated feedback of "wrong." Finally, the attitudes of the two groups toward the task differed markedly. Even after several failures, mastery-oriented children remained positive and optimistic about the possibility of success, while helpless children expressed negative feelings and resignation, declaring, for example, "I give up."

Incremental versus Entity Theories: Your Own Personality Theory Matters

13.14 Compare entity with incremental theories. How do they affect the goals people choose and their achievement outcomes?

Not just psychologists but everyone—including the young child—has theories. Often their theories are not spelled out and thus only implicit theories of personality, but they can have profound effects on daily life. Carol Dweck and colleagues have shown that people's implicit theories or beliefs about the malleability or fixedness of personality and character predict many of their reactions to others as well as to their own performance and feelings. This work began when Dweck and Leggett (1988) found that children who display the helpless pattern see their intelligence as a fixed trait or static entity that they cannot change or control. In contrast to such an entity interpretation, youngsters who are mastery-oriented tend to view their intelligence more flexibly as something they can increase and develop.

These different views or theories about intelligence also orient them toward different types of goals. The **entity theorists** seem to choose goals motivated by the desire to avoid unfavorable judgments and to gain approval about their competence.

The **incremental theorists** choose goals motivated by the desire to increase their competence, for example, seeking opportunities to learn new things and enhance their mastery.

These differences in children's theories about their intelligence also predict important real-life outcomes in their development. Children were tracked from sixth to seventh grade in their transition to junior high school (Henderson & Dweck, 1990). The most impressive gains in grades during this transition were found for the incremental theorists. In sharp contrast, children who saw their intelligence as fixed tended to remain low achievers if they had been low achievers before and, most distressing, among these entity theorists ". . . many of those who had been high achievers in sixth grade were now among the lowest achievers" (Dweck, 1990, p. 211). Clearly, our self-concepts and theories about our important qualities, our way of "encoding" or construing ourselves, impact our subsequent development (see In Focus 13.2).

Going further, Erdley and Dweck (1993) divided late grade school children into those who held entity versus incremental theories and showed them slides depicting some negative behaviors of a "new boy at school" For example, he lied about his family background, cheated from a classmate's paper, and stole a classmate's left-over art materials. They also received information about situational factors (the boy had moved in the middle of the school year to the new school) and about possible psychological mediators (the boy was nervous about making a good impression).

As predicted, entity theorists made significantly stronger judgments about the boy's global moral traits (e.g., bad, mean, nasty) than did incremental theorists. Furthermore, entity theorists did not revise their trait judgments of the boy when positive information was provided whereas incremental theorists responded to the inconsistent information. Entity theorists also expected their first impression of the boy to remain valid forever whereas incremental theorists predicted that the boy might act differently in the future. Once they have rendered a negative moral judgment, entity theorists generally recommend punishment for the transgressor (Chiu, Dweck, Tong, & Fu, 1997; Erdley & Dweck, 1993), whereas incremental theorists recommend education or rehabilitation, consistent with their belief in the possibility of personality change even in wrongdoers.

The implications of these two types of theories for other attitudes and behaviors also have been explored across different cultures. For example, although Hong Kong college students were significantly more "collectivistic" than U.S. students, the entity theorists in both cultures made stronger dispositional inferences than did incremental theorists, apparently assuming that what people do even in a single instance reflects their stable fixed traits. (Chiu, Hong, & Dweck, 1997). Likewise, entity theorists, in another study, endorsed both positive and negative stereotypes about ethnic groups (African–Americans, Asians, Latinos) and occupational groups (lawyers, politicians) more strongly than did incremental theorists in studies with college students (Levy, Stroessner, & Dweck, 1998). In short, one's implicit personality theory shows itself in many forms, influencing how one judges people and interprets behavior.

Is the entity theory or the incremental theory better? Although an entity view has so far been linked with more maladaptive outcomes, neither view necessarily reflects the truth. And each theory has advocates convinced of its greater value.

▶ PERSONALITY ASSESSMENT

The social cognitive level of analysis has made diverse contributions to personality assessment and personality change.

13.15 Describe three characteristics of social cognitive personality assessment. What kinds of measures are typically employed, and how do they differ from trait measures?

First, assessments at this level tend to be specific, focused on specific cognitions, feelings, and behavior in relation to particular types of situations, rather than in terms of situation-free broad trait terms. Recall, for example, that self-efficacy refers to individuals' beliefs that they can do what is required in a particular type of situation or task (such as if asking for a pay raise, or on a specific achievement task, or in approaching a dangerous object), not their overall efficacy expectations in general. This focus reflects the view that the expressions of the personality system and the person's goals and motivations are seen in *if . . . then . . .* personality signatures, not just in the overall levels of different types of behavior.

Second, assessments whenever possible also try to identify the implications for constructive personality change or treatment. They thus are aimed at identifying the psychological person variables which might be modified, for example by enhancing efficacy expectancies through exposure to relevant efficacy-building experiences. Such assessments, always closely linked to change or treatment programs, are being done effectively in areas that range from weight control in anorexic patients to recovery of sexual functioning after massive coronary problems, to overcoming debilitating fears.

Personality assessors in this approach also seek to identify the underlying person variables and processes that seem to account for the individual's stable behavior patterns. They tend to conceptualize these underlying variables in relatively specific (rather than global) terms.

Researchers and assessors therefore obtain self-reports, ratings, and other data to infer the particular person variable as directly and specifically as possible within these contexts. Some also try to sample and observe behavior as it occurs naturally. For example, they ask people to provide daily diary reports of what they actually did and experienced within specific situations (e.g., Ayduk and associates, 2003; Bolger & Schilling, 1991; Cantor et al., 1991). There have been many applications, and we illustrate these with the example of the measurement of self-efficacy expectations.

Measuring Self-Efficacy Expectancies

13.16 How is self-efficacy measured? How well do these measures predict behavior?

Given the importance of the self-efficacy construct in social cognitive theories it is also a person variable that is used extensively in personality assessment at this level of analysis (Bandura, 1978, 1986; Cervone, Shadel, & Jencius, 2001; Merluzzi, Glass & Genest, 1981). Self-efficacy is assessed by asking the person to indicate the degree of confidence that he or she can do a particular task. For example, Bandura et al. (1985) wanted to assess the recovery of patients who had suffered heart attacks. Many tasks were described to the patients. These included such potentially stressful things as driving a few blocks in the neighborhood, driving on a freeway, and driving on a narrow mountain road. They also included situations that would induce other kinds of emotional strain, as illustrated in Table 13.3. For each item the respondent indicates the confidence level for being able to do the task.

Self-efficacy measures are particularly useful because they tend to predict the relevant behaviors at high levels of accuracy. For example, a consistently strong association was found between rated self-efficacy and the degree to which people showed increased approach behavior toward previously feared objects after they had received treatment for their fears (Bandura, Adams, & Beyer, 1977). Note that a distinguishing feature of self-efficacy measurement is that it is always about a specific domain, such as the ability to approach a type of feared object, or to control your weight, or efficacy about college course work in your major. This specificity contrasts with global measures, for example when you are asked to rate your "over-all level of self-esteem."

TABLE 13.3 Measuring Self-Efficacy Expectancies

Listed below are situations that can arouse anxiety, annoyance, and anger. Imagine the feelings you might have in each situation, such as your heart beats faster and your muscles tense. Indicate whether you could tolerate now the emotional strain caused by each of the situations.

Under the column marked *Can Do,* check (✓) the tasks or activities you expect you could do *now*.

For the tasks you check under *Can Do,* indicate in the column marked *Confidence* how confident you are that you could do the tasks. Rate your degree of confidence using a number from 10 to 100 on the scale below.

10	20	30	40	50	60	70	80	90	100
Quite Uncertain				Moderately Certain			Certain		

	Can do	Confidence
Attend a social gathering at which there is no one you know.	_____	_____
At a social gathering, approach a group of strangers, introduce yourself, and join in the conversation.	_____	_____
At a social gathering, discuss a controversial topic (politics, religion, philosophy of life, etc.) with people whose views differ greatly from yours.	_____	_____
Be served by a salesperson, receptionist, or waiter whose behavior you find irritating.	_____	_____
Complain about poor service to an unsympathetic sales or repair person.	_____	_____
When complaining about bad service, insist on seeing the manager if you are not satisfied.	_____	_____
In a public place, ask a stranger to stop doing something that annoys you, such as cutting in line, talking in a movie, or smoking in a no-smoking area	_____	_____
Ask neighbors to correct a problem for which they are responsible, such as making noise at night, not controlling children or pets.	_____	_____
At work, reprimand an uncooperative subordinate.	_____	_____

Source: Examples reprinted from *American Journal of Cardiology, Vol 55,* Bandura, A., Taylor, C. B., Ewart, C. K., Miller, N. M., & Debusk, R. F., "Exercise testing to enhance wives' confidence in their husbands' cardiac capability soon after clinically uncomplicated acute myocardial infarction", 635–638 (1985), with permission from Excerpta Medica Inc.

The focus on the context, situation, or contingency (e.g., *if . . . then . . .*) is of course a key characteristic of assessment at this level in which person qualities are directly linked to contexts and situations. It is also well illustrated in the work discussed next.

Individual Differences in *If . . . then . . .* Signatures

Advances also are being made in assessments of individual differences that take close account of the situation. Iven Van Mechelen and his colleagues (Van Mechelen & Kiers, 1999; Vansteelandt & Van Mechelen, 1998) are using the concept of *if . . . then . . .* signatures to classify people into different types based on the kind of responses they give in specific types of situations, such as those that are especially frustrating, as discussed in Chapter 4. The results are encouraging for building a classification system for types of people that is both theory-guided and specific enough to predict behavior in different contexts. Recall that the ultimate goal of such typologies is to predict the probability of certain type of individuals exhibiting certain types of behavior when faced with certain types of situations. The hope would be that this could allow a level of precision not possible with typologies based on characterizations in broad trait terms like "disagreeable" or "unsociable." In their recent work, these researchers took the additional step of linking their typology to measures of social cognitive person variables. Using a questionnaire that asked about the participant's encodings, expectancies, and whether they kept their anger in or let it out, they demonstrated meaningful links between such person variables and types of *if . . . then . . . behavior signatures.*

The Implicit Association Test (IAT)

13.17 What kinds of cognitive processes are implicit measures designed to assess?

Behavioral measures have the advantage of being objective, in that most people would agree that a child hit another child when they see the behavior in front of them. But what about assessing thoughts and feelings, especially those that are not within the person's awareness? We could, of course, ask people to tell us what's on their mind. But then we will learn what they *think* are their thoughts or feelings. What if these are not their *true* thoughts and feelings? And what exactly would one mean by "true" thoughts and feelings? These are the issues that are central for researchers at the psychodynamic-motivational level of analysis, and they also are important for the social cognitive level of analysis. The CAPS system described in the last chapter, for example, assumes that many of the thoughts and feelings in the personality system may not be accessible to the person's awareness. In recent years there have been important advances in assessing such aspects of personality outside awareness. Consider the following scenario:

> *A boy goes fishing with his father, but slips on a slippery rock and suffers a serious head injury. When the ambulance takes him to the ER, the doctor there turns pale, and exclaims "Oh, my god, it's my son!"*

What's going on here? Is there something strange about this scenario? How do you explain this? Wasn't the boy's father with the boy when the accident occurred? In fact, nothing may be unusual with the scenario, and the most likely, and mundane, interpretation is that the doctor at the ER is the boy's mother. If you didn't think of that possibility right away, that may be because in your mind you may have associated doctors in emergency rooms more strongly with men than women, reflecting gender stereotypes in our society.

Such an association is *implicit* when people have no awareness of making it and being influenced by it as they process social information such as this scenario. It may be like a mental habit, not necessarily a motivated bias, which is applied automatically, without much awareness. Several ways to assess implicit associations have been developed (Greenwald et al., 2002). Here we will describe one, called the **Implicit Association Test (IAT),** for the assessment of **implicit self-esteem,** developed by Greenwald and Farnham (2000).

The IAT requires participants to make a series of judgments about each of the words presented to them briefly, one at a time, in quick succession.

Specifically, the words presented come from four distinct categories, and the participants' task is to indicate the category of the word presented, as quickly as possible, by pressing an appropriate computer key. One category of words refers to *self* such as "I," "me," and "mine." As soon as these words appear, participants press a particular computer key (e.g., the "A" key located on the left side of the keyboard). Another category represents *"not self,"* and so when they see words such as "they," "theirs," "it," they press a different computer key (e.g., the "5" key in the numerical keypad located on the right side of a keyboard). The remaining two categories refer to obviously good and obviously bad concepts. So when they see words such as *health, joy,* and *kindness,* they press one computer key, and when they see words such as *ugly, failure,* and *awful,* they press another.

If participants were given four separate keys to indicate each of the four categories, the task would be relatively simple. But the crucial innovation in the design of IAT is that there are only two keys available to press on the keyboard. Therefore two concepts must "share" a key. This is done in one of two ways. In the "self = bad" condition, participants press the same key (e.g., "A") to indicate that the word they see refers to either "self" or "bad," and they press another key (e.g., "5") when the word refers either to "not self" or "good." Most participants find this combination of keys very difficult to master, and their responses are slow, in order to prevent making mistakes. In contrast, in the "self = good" condition, in which the concepts are combined in a different way, they find it much easier to do this task. Namely, when they use the same key to indicate they saw a word that referred to either "self" or "good," and the other key to indicate they saw a word that referred to either "not self" or "bad," they are much faster in completing the task.

The difference in the impact of these two ways of combining the concepts on the difficulty of the task, reflected in the time needed to complete it, is taken to indicate the strength of association between "self" and "good" (and "not self" and "bad"). This is a measure of self-esteem, because a central element of the concept of self-esteem is how positively or negatively one regards oneself. And it is an implicit measure because the associations being measured do not rely on people's own awareness and report of how strongly they associate self with good, but rather based on an automatic response that has been shown to be difficult to control consciously. In fact, the implicit and explicit measures of self-esteem were largely unrelated to each other. In the Greenwald and Farnham (2000) study, those who had a high implicit self-esteem as measured by IAT were no more likely than those who had low implicit self-esteem to describe themselves as "good" when they were asked to rate themselves on a "good"–"bad" continuum.

Because the self-report measures used in this study were well-established in the field, one may be tempted to conclude that the implicit measure was not a good measure. However, the measure of implicit self-esteem also was able to predict a phenomenon that has been well-established as related to self-esteem (Brown & Dutton, 1995; Dodgson & Wood, 1998; Greenberg et al., 1992). Namely, people who were high in self-esteem, as determined by the implicit measure, were relatively unaffected by experience of failure, while those who were low in self-esteem by the implicit measure took the failure to heart (Greenwald & Fahnham, 2000, Study 3).

Since its introduction in 1998, the IAT has become widely used to measure not only implicit self-esteem, but also implicit attitudes, beliefs, and values in a number of domains. One very interesting finding is that IAT measures of anti-Black prejudice predicted the White participants' behaviors toward Black people on indicators such as body openness, eye contact, and friendly laughter (McConnell & Leibold, 2001).

▶ PERSONALITY CHANGE AND THERAPY

Work at the social cognitive level has had an important influence on current forms of therapy for a wide range of personality problems. These range broadly from fears and depression through difficulty in relationships, to troubles dealing with work challenges. Other examples include the treatment of impulsive and aggressive tendencies, and help

13.18 What are the objectives of social cognitive approaches to therapeutic behavior change?

for overcoming addictions and improving self-control in adhering to crucial health and medical regimens. Many psychologists who work at the social cognitive level also explore the links between personality and health (e.g., Miller, Shoda, & Hurley, 1996). In recent years they have been investigating individual differences in vulnerability to severe psychological disorders, such as depression, as well as to physical diseases, such as cancer and coronary–pulmonary disease. Social cognitive approaches are also being applied to understanding social and cultural differences (Mendoza-Denton, Shoda, Ayduk & Mischel, 1999).

Social Cognitive Influences

Therapists within this approach try to help clients to identify their disadvantageous ways of thinking about themselves, other people, and their problems. In the safety of the therapy situation, therapist and client interact to experiment and explore ways of reconstruing and modifying basic assumptions in thinking and the automatic emotional reactions that are not working well for the client. A fundamental aim is to increase the client's perceived and real freedom to change in desired directions. One route is to provide experiences that enhance efficacy expectations. Change becomes easier when clients begin to expect that they can cope more effectively and that they can face previously terrifying situations more calmly. Therapy includes a wide range of actual and imagined (symbolic) experiences designed to develop more effective strategies and plans for setting, and achieving, desired goals, for functioning more comfortably and effectively interpersonally, and for reducing anxiety, depression, and perceived stress.

The social cognitive approach to personality has influenced contemporary approaches to constructive educational and therapeutic change in many domains. They range from the treatment of all sorts of fears to the conduct of psychotherapy and the development of mass public education programs to improve health-enhancing behaviors, such as reduction of tobacco and other drug use, dieting, and reduction of high-risk sexual practices. These programs often use exposure to models who display the desired behaviors as part of their efforts to effect change (e.g., Bandura, 1986; Cervone et al., 2001).

13.19 How are modeling and role playing used in social skills training?

Modeling has been used, for example, to help people overcome shyness and assert themselves more effectively when they feel they should. Assertiveness skills may be sought by anyone who wishes to be more effective with other people, whether roommates, a boss, spouse, or parent. People who are unable to be assertive, who cannot stand up for their rights, may not only be exploited and deprived by others but may feel ineffective and incompetent and lack self-esteem. Thus, **assertiveness training** may have many positive effects on one's life. The procedure may include observation of models who assert themselves effectively. This step may be followed by role playing with the therapist and by rehearsing more assertive responses, first in safe situations and ultimately in real life when the assertive responses are needed.

Improvement in assertive behavior may be achieved by observation of models who display assertiveness. In **covert modeling,** this is done in imagination, as unassertive individuals visualize scenes in which a model performs assertively when it is appropriate. Here is a typical scene:

The person (model) is dining with friends in a restaurant. He (she) orders a steak, instructing the waiter that it be rare. The steak arrives and as the person cuts into it, it is apparent that something is wrong. He (she) signals the waiter, who comes to the table. The person says, "This steak is medium; I ordered it rare. Please take this back and bring me a rare one" (adapted from Kazdin, 1974, p. 242).

Cognitive Behavior Therapy

Because so many behavioral approaches in clinical psychology and therapy have become cognitive, their more accurate new name within the profession is **cognitive behavioral therapy** or CBT. Indeed they may more accurately be called "social cognitive behavioral therapy" except for the fact that it's much too cumbersome a label. Donald Meichenbaum (1995, p. 141), a CBT leader, describes how cognitive-behavioral therapists challenged the earlier behaviorists this way:

> *They [cognitive-behavioral therapists] questioned the tenets of classical learning theo-ries and psychoanalytic formulations, and caused the field to question how best to conceptualize the clients' thoughts and feelings. Moreover, they raised questions of how the clients' thoughts influence and, in turn, are influenced by their feelings, behaviors, resultant consequences, and physiological processes. They emphasized that indi-viduals not only respond to their environments but are also the architects of those environments.*

Within the last four decades, the cognitive-behavioral model has evolved into a promi-nent force in psychology (Dobson & Craig, 1996). CBT is currently used to treat con-ditions such as depression, anxiety, phobias, obsessional disorders, aggression, and hypochondriasis (Dunford, 2000; Kendall & Panichelli-Mindel, 1995; Rachman, 1996). Its growth reflects many changes and reasons, as summarized in Table 13.4. It also gives increasing attention to the client's personal feelings, to experiences as seen by the client, and to developing as well a balanced, internally harmonious life. In that sense it shows similarity to many of the therapeutic values and practices first emphasized in the phenomenological–humanistic approaches, for example in the work of Carl Rogers and Abraham Maslow.

Donald Meichenbaum

TABLE 13.4   Some Reasons for the Growth of Cognitive Behavior Therapy

Recognition of problems that go beyond specific behaviors (e.g., career conflict, depression)

Helps people interpret and construe experiences constructively

Address affect (feelings and emotions) as well as action

Deals with interactions between thoughts, feelings, and action

Uses and combines diverse methods and concepts that prove useful

Conceptually and historically, CBT is rooted deeply within behavioral approaches. But its growth also was influenced by other contributions, and it now integrates work from the behavioral and social cognitive levels with insights and methods from most of the other levels of analysis. At a theoretical level it has considerable similarity to George Kelly's work on personal constructs, particularly for the treatment of depression.

Beck's Cognitive Therapy

The social-cognitive trend in therapeutic practice is reflected in their labels; for example, Aaron T. Beck's **cognitive restructuring.** Working as a psychiatrist in Philadelphia, Beck developed one of the most influential and well-articulated versions of cognitive therapy. His approach is directed at changing how people encode or construe their experiences and themselves, and he has applied it systematically to the treatment of depression (Young, Weinberger, & Beck, 2001). He defines cognitive therapy this way:

13.20   What are the goals and procedures in Beck's cognitive therapy?

> *Cognitive therapy is an active, directive, time-limited, structured approach used to treat a variety of psychiatric disorders (for example, depression, anxiety, phobias, pain problems, etc.). It is based on an underlying theoretical rationale that an individual's affect and behavior are largely determined by the way in which he structures the world. His cognitions (verbal or pictorial "events" in his stream of consciousness) are based on attitudes or assumptions (schemas), developed from previous experiences. For example, if a person interprets all his experiences in terms of whether he is competent and adequate, his thinking may be dominated by the schema, "Unless I do everything perfectly, I'm a failure." Consequently, he reacts to situations in terms of adequacy even when they are unrelated to whether or not he is personally competent (Beck, Rush, Shaw, & Emery, 1979, p. 3).*

There are five basic steps in Beck's version of cognitive therapy:

1. Clients first learn to recognize and monitor their negative, automatic thoughts. These thoughts are "dysfunctional," that is, ineffective, and lead to serious dilemmas.

2. They are taught to recognize the connections between these negative thoughts (cognitions), the emotions they create, and their own actions. (See Figure 13.2 for examples of connections between thoughts and emotions.)

3. They learn to examine the evidence for and against their distorted automatic thoughts.

4. They substitute for these distorted negative thoughts more realistic interpretations.

5. They are taught to identify and change the inappropriate assumptions that predisposed them to distort their experiences. Examples of such assumptions are shown in Figure 13.3.

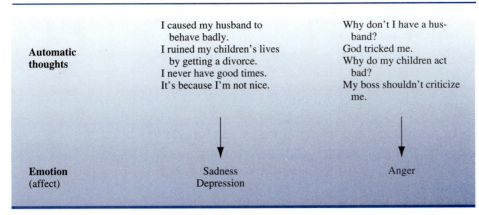

Figure 13.2 Examples of connections between negative automatic thoughts and emotion.
Source: Adapted from Beck, A. T., Rush, A. J., Shaw, B. F., & Emery, G. (1979). *Cognitive therapy of depression* (p. 250, fig. 3). New York: The Guilford Press.

A variety of ingenious techniques have been developed to encourage people to undertake these five basic steps and to use them effectively to alter their actions, thoughts, and feelings. Cognitive therapy of this sort appears to be a promising part of treatment for depression and related emotional and behavioral problems. Its value has been explored in a large number of studies with subjects ranging from those with mild problems to hospitalized, severely depressive patients. The results seem to be consistently encouraging (Beck et al., 1979; Wright & Beck, 1996).

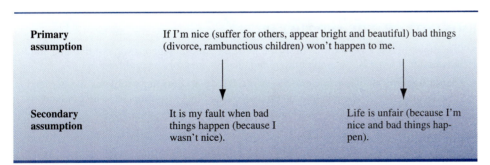

Figure 13.3 Examples of assumptions that encourage depression
Source: Adapted from Beck, A. T., Rush, A. J., Shaw, B. F., & Emery, G. (1979). *Cognitive therapy of depression* (p. 250, fig. 3). New York: The Guilford Press.

SUMMARY

THE SELF

- William James distinguished between two aspects of the self: the "I" and the "Me."
- Self-concepts or self-schemas are knowledge structures developed from the person's perceived life experiences. They influence how new information relevant to the self will be processed.

- The relational self is a representation of the self that is connected to representations of the significant other people in one's life. In new relationships, representations of significant others may be activated by the newly encountered person.
- The working self-concept comprises the most salient concepts of the self.
- Self-esteem refers to the individual's personal judgment of his or her own worth.
- Self-evaluation reflects the individual's performance, as well as earlier experiences and personal self-standards.

CAUSAL ATTRIBUTIONS, HELPLESSNESS, AND MASTERY

- Causal attributions influence emotions and behavioral reactions. Events that are attributed to *internal causes* are more likely to produce pride or shame than are results attributed to *external causes.*
- People with learned helplessness believe they have no control over negative outcomes in their lives and may become apathetic and depressed.
- Pessimism is an explanatory style in which bad life outcomes are attributed to one's own internal qualities. Learned optimism is an opposite, self-enhancing explanatory style that often has positive effects.
- *Helpless children,* who perceive their intelligence as unchanging, tend to perform poorly after failure. Conversely, *mastery-oriented children,* who feel they can improve their intelligence, cope better with failure.
- Entity theorists choose safe goals to avoid negative outcomes but incremental theorists prefer goals that enhance their competence.

PERSONALITY ASSESSMENT

- Assessments at the social-cognitive level of analysis tend to be specific and connected to treatment efforts, as exemplified in self-efficacy expectancy measures.
- The Implicit Association Test (IAT) uses a word-association task to evaluate participants' implicit level of self-esteem and diverse attitudes.

PERSONALITY CHANGE AND THERAPY

- Therapies at this level try to identify the person's disadvantageous ways of thinking and to encourage more adaptive ways of thinking, feeling, and solving problems.
- Aaron Beck's cognitive behavioral therapy (CBT) treats personal problems and depression by changing the way people encode themselves and their experiences to find more constructive ways of thinking and behaving.

KEY TERMS

PART V

THE SOCIAL COGNITIVE LEVEL

▶ OVERVIEW: FOCUS, CONCEPTS, METHODS

As Part Five, the Social Cognitive Level of Analysis, concludes, the next table summarizes its main characteristics and enduring contributions.

Overview of Focus, Concepts, Methods: The Social Cognitive Level

Basic units of personality:	Underlying person variables and processes: Encoding (construing, including self-concepts and explanatory styles), expectancies and beliefs, affects, values and goals, self-regulatory competencies
Causes of behavior:	Reciprocal interaction between person and situation, mediated by the person variables interacting within the personality system
Behavioral manifestations of personality:	Stable patterns of person–situation interactions; distinctive signatures of *if . . . then . . .* relationships (she does *X* when *Y*, but she does *A* when *B*)
Favored data:	Self-reports, diaries, ratings, and behavior samplings relevant to person variables; outcome information (such as symptoms, later school grades) within specific situations
Observed responses used to:	Infer underlying person variables and cognitive, emotional, motivational processes. Assess and predict behavior and outcomes (such as proneness to disease, well-being)
Research focus:	Refining theories about underlying processes, clinical implications (for health, for risk prevention in vulnerable individuals, for therapy, for enhancing self-efficacy)
Approach to personality change:	By changing person variables and mediating processes (dynamics), modifying expectations, developing effective self-regulatory strategies and plans for goal-attainment
Role of situation:	Important; provides psychological cues and interacts with person variables

As the overview table indicates, the social-cognitive level focuses on the psychological variables that underlie the differences between individuals in their cognitions, emotions, and actions. For that purpose, psychologists study the diverse person variables discussed in this section of the text. They see these variables as basic units of personality and as important—but not exclusive—determinants of the patterns of behavior that characterize individuals distinctively and enduringly. They try to understand 1) the processes through which the variables operate and exert their effects, and 2) the unique patterns through which different persons manifest these variables in their behavior. These behavioral expressions of personality are found in stable patterns of *if . . . then . . . ,* person–situation interaction.

▶ ENDURING CONTRIBUTIONS OF THE SOCIAL COGNITIVE LEVEL

Work on personality at the social cognitive level began in the early 1970s by building on the foundations provided by earlier behaviorally oriented theorists, particularly social learning theory (e.g., Bandura, 1969; Rotter, 1954), and George Kelly's focus on personal constructs as guides for behavior and social perception. Going beyond the "stimulus-response" concepts developed in simple learning experiments with animals, more complex social learning was studied, and social cognitive theories of personality were expanded to include the role of both cognitive and social variables (e.g., Bandura, 1969, 1986; Mischel, 1973, 1990; Rotter, 1954). Work at this level moved away from a focus on external stimuli and rewards and explicitly rejected the idea that situations are the only (or even the main) determinants of behavior. Instead, personality was conceptualized in terms of cognitive and social person variables. The reciprocal (mutual) interaction between the person and the situation was emphasized, recognizing that people select, create, and change situations actively and are not merely passively "shaped" by them (Bandura, 1977, 1986; Mischel, 1973, 1984).

In the last few decades, social cognitive theories still include principles of social learning as a centerpiece for understanding many aspects of personality and social behavior. But they now give an even greater role to cognition, emotion, and motivation, and address an increasingly wide range of personality phenomena. In one direction, work at this level of analysis is building on and revitalizing the contributions of phenomenological approaches and self-theorists, reopening many of the topics the self theorists pioneered more than thirty years ago. This renewal has already proved to be fruitful. Decades ago, phenomenological approaches were criticized because they relied heavily on people's perceptions, which are subject to all sorts of biases and, therefore, are potentially inaccurate sources of information. Researchers at the social cognitive level in recent years have turned this so-called problem into an exciting research topic. Instead of dismissing the individual's perceptions because of their possible biases, they have made those biases major topics in the study of social cognition, investigating how they influence social judgment and decisions.

The realization that how people perceive themselves and their experiences crucially influences their behavior, previously the distinctive hallmark of phenomenological approaches, is now also a central assumption within social cognitive approaches. Because researchers now use more sophisticated methods of measurement, difficult topics that previously resisted objective research are being opened. For example, the "self" in personality psychology has gone from an abstract concept about which theorists speculated to become an active research topic about which much has been learned. The same can be said for emotions, whose nature is becoming much clearer. Thus researchers at the social cognitive level are seeking ways to link social cognition and social perception to other aspects of personality and social behavior. We saw, for example, some of the determinants of perceptions of personal control, and that changes in those perceptions can influence behavior for good or ill. The results ultimately should help clarify the relations between what people perceive and think and what they feel and do.

The classic humanistic commitment to enhance personal growth and the human potential has long been a key feature of phenomenological approaches. Now it also seems to be absorbed increasingly into some forms of cognitive behavior therapy. Personality psychologists within this framework devote much attention to the therapeutic implications of their work for both psychological and physical health and well being (e.g., Contrada and associates, 1990).

Traditionally, most approaches to personality have proposed certain basic assumptions about the nature and causes of personality (e.g., in terms of unconscious psychodynamics and motives). Their proponents have devoted themselves to trying to prove the truth and importance of those assumptions, usually ignoring findings from other approaches. As a science progresses, however, it tends to develop a more cumulative strategy in which concepts are modified, deleted, and incorporated in light of new research findings, regardless of the theoretical orientation that guided the original researchers. The field then matures into one in which the contributions of individual theorists become moments in a larger history of continuous change and evolution. In time, the best of what proves useful and valid is retained; the rest is left behind.

The social cognitive level of analysis helps to provide a foundation for a cumulative science of individuals because it deliberately drew on earlier work that had proved promising rather than trying to emerge with a totally novel approach. A review of much of the work at the social cognitive level suggests the emerging "convergence and complementarity of theoretical conceptions and empirical findings" that may be seen as a basic indicator of the field's progress (Cervone, 1991, p. 371). The degree to which the various levels of analysis do in fact complement and add to each other will become especially clear in the final part of the text that focuses on the integration among all the levels.

VI

THE BIOLOGICAL LEVEL

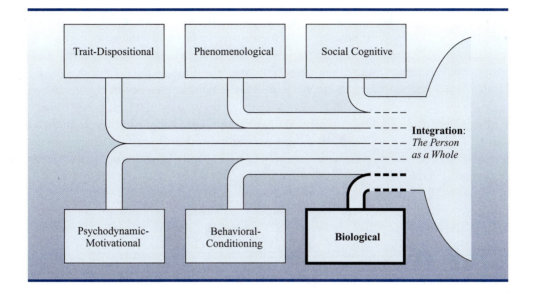

▶ PRELUDE TO PART VI: THE BIOLOGICAL LEVEL

Two identical twin girls were separated at birth, adopted and raised apart in different families from birth on. Nahid's life was spent in the capital of Iran while Jane grew up on a farm in rural Iowa. As identical twins, the two shared the same genetic heritage equally at the outset and consequently started their development with essentially identical brains and physical-biological structures. Nevertheless, the two girls grew up and experienced their lives in very different families, social worlds and cultures. One was reared by loving parents in a family that adored her, surrounded by six siblings, four older than she and two much younger. The other twin grew up as the only child with a neglectful, depressed mother and an alcoholic, abusive stepfather. If the twins are reunited at age 30 and tested extensively, how similar will their personalities turn out to be?

At the biological level of analysis researchers want to answer the age-old question: how much of personality reflects nature, and how much nurture—and how do these two sources of influence interact in shaping human characteristics?

A second question also springs from the recognition that we humans are biological beings. It asks: what are the consequences of having evolved as creatures within

321

the animal kingdom? Evolution endowed us with a host of biological characteristics, constraints, and possibilities that influence who and what we are and do and can become—from eating, drinking, and breathing, to fighting, mating, socializing, and creating. Work at this level shows that people are not simply a blank slate; we are born with a wide range of mechanisms that make it possible to perform an impressive array of complex feats of information processing that facilitated the evolutionary fitness of our species. Like other species, we are the products of the evolutionary processes that shaped us. Consequently, a major goal at this level of analysis is to examine the ways in which personality may have evolved in response to the evolutionary pressures and history that shaped our species over the span of time.

Work at the biological level for many years had little to say about the specific ways in which biology influences individuals' experiences and behaviors that interested personality psychologists. In just a few years that picture has changed dramatically: new ways have been found for examining the links between activity in the human brain and the thoughts, feelings, and behavioral tendencies of the person as they occur. The most dramatic advances are in brain imaging technology. These methods now allow studies of the relations between what happens at the brain level and all the psychological levels that have been discussed in the text. These relations are beginning to be traced as both the brain and the psychological processes are activated concurrently.

Chapter 14 discusses genes, heredity, and the interplay of genetic and social influences on personality. Chapter 15 considers the evolutionary processes that have shaped the genes over the course of the species' development, and examines the implications for understanding personality at that level of analysis. Then we turn to the specific biological processes and brain mechanisms that link each individual's genetic background to central aspects of personality and social behavior, such as emotion and motivation.

The Personal Side of the Science

Some questions at the biological level you might ask about yourself:

▶ What in my personality comes from my parents and the genes I inherited?

▶ How is my personality a reflection of my life experiences?

▶ How does my personality reflect my basic biological predispositions?

▶ Can my experiences change my biology; for instance, does my brain change when I'm depressed?

▶ How do the same experiences affect people with different genetic predispositions?

▶ Why is my personality so different from (or similar to) that of my siblings?

▶ How does my biology influence my pursuit of life goals?

CHAPTER

14

HEREDITY AND PERSONALITY

Our genes dictate whether we will be male or female, blue eyed or brown eyed, curly haired or straight haired. To some extent they also influence height and weight. But do the genes also underlie aspects of personality? If the answer is that they do, the next question is "how much?" Do they affect sexual preferences and choice of partners? Are they reflected in tendencies to develop severe mood disorders and mental illness?

The rapidly growing field of **behavior genetics** studies the role of genes in social behavior and personality. Current work on these topics is vigorous and moving in many new directions. It is propelled by better methods for studying the inheritance of personality characteristics and by rapid advances in genetics research. In this chapter, we examine the genetic and biochemical roots of personality, including temperament and attitudes, beliefs and behavior. We next consider the interplay between genetic determinants and social factors, such as the family and culture in which the child develops, as influences on personality.

▶ GENETIC BASES OF PERSONALITY

The Human Genome: The Genetic Heritage

As scientists race to map out the human genome, controversies sprout about the implications and the hazards of each new claim. The arguments become especially heated as they speak to the heritability of personality characteristics and behavioral tendencies—from thrill-seeking, alcoholism, and criminal tendencies to sexual attitudes and political conservatism. The conclusions reached influence how we judge other peoples' behavior and our own. If genes underlie antisocial and violent behavior, can one hold criminals responsible for their crimes? Can reform and change be possible? Is biology destiny? How you see the role of heredity in personality affects not just your view of human nature. It also can change your personal sense of your own possibilities and limitations in trying to build a life.

14.1 What is the human genome? How does it contribute to human similarities and differences?

The genetic heritage of each human being—the **human genome**—consists of roughly 30,000 to 40,000 genes. Each gene consists of DNA sequences contained in 23 pairs of chromosomes. In each chromosome pair, one set comes from the mother, the other from the father. Each of about a trillion cells in the human body houses within its nucleus this genetic inheritance.

The vast majority of human genes are the same for every human being, resulting in the enormous similarities that people share—10 toes, 10 fingers, 32 teeth, two arms, two legs, two eyes, two ears, one heart, and so on. But a relatively small minority of genes are different for different persons and these make the individual distinctive genetically. They are the ones that influence the diverse variations among people in such characteristics as eye color and height—and perhaps in aspects of personality and social behavior not anticipated until recently. The question to which we turn here is the role of these genetic differences between individuals as influences on personality.

Most of what is known about the genetic roots of personality comes from studies that compare the similarity in personality shown by individuals who vary in the degree to which they share the same genes and/or the same environments. Therefore, the "genetic" and "environmental" influences on personality discussed next do not refer to the effects of specific genes and specific environments. Rather, they refer to the overall effects of these two types of determinants on individual differences on average. A great deal has already been discovered in recent years about the effects of heredity and environment on personality and the results have profound implications for how one thinks about human nature.

To anticipate what you will see in this chapter, genetic research on personality is producing a huge and complex literature, but it has a clear message: Genes play a role in personality, and it appears to be a larger one than earlier research had suspected (e.g., Eaves, Eysenck, & Martin, 1989; Loehlin, 1992; Loehlin & Nichols, 1976; Plomin et al., 1997). For example, it is now widely believed that such dispositions as extraversion–introversion (Chapter 3) have a biological-genetic basis (Bouchard et al., 1990; Eysenck, 1973; Plomin & Caspi, 1999; Tellegen et al., 1988). Now let us look at some of the evidence that leads to that conclusion.

The Twin Method

14.2 How do twin studies shed light on genetic factors in personality?

Most often, these studies use the **twin method** to assess genetic influence. This method simply compares the degree of similarity on measures of personality obtained for genetically identical twins—those who are from the same egg or **monozygotic (MZ)**—as opposed to twins who are fraternal or **dizygotic (DZ),** that is with each from a different fertilized egg (Plomin et al., 1997). Identical twins are virtually clones, that is, they are genetically identical, but fraternal twins—like other siblings—are only 50 percent similar genetically. To the degree that genetic factors affect personality, it follows that identical twins must be more similar than fraternal twins with regard to that characteristic.

Studies comparing identical twins and fraternal twins help specify the role of genes in personality.

Results of Twin Studies

A pioneering study with nearly 800 pairs of adolescent twins and measuring dozens of personality traits reached a conclusion that has stood the test of time (Loehlin & Nichols, 1976): Identical twin pairs are much more alike than fraternal twin pairs. The resemblance within identical twin pairs tends to be strongest for general ability and is less strong for special abilities. The resemblance is somewhat lower for personality inventory scales and lowest for interests, goals, and self-concepts (see Table 14.1). For personality, twin correlations are about .50 for identical twins and .25 for fraternal twins. This same study also found that nearly all personality traits measured by self-report questionnaire show moderate genetic influence.

The Big Five. Genetic research on personality has focused on five broad dimensions of personality—the Big Five (Goldberg, 1990), discussed in Chapter 3. Extraversion and neuroticism have been studied most. Extraversion includes sociability, impulsiveness, and liveliness. Neuroticism (emotional instability) includes moodiness, anxiousness, and irritability. Table 14.2 summarizes results for extraversion and neuroticism (Loehlin, 1992). Results from five twin studies in five different countries, using a

14.3 In general, what do twin studies find regarding the degree of resemblance in personality as compared with other characteristics?

TABLE 14.1 Resemblance of Identical and Fraternal Twin Pairs: Typical Correlations within Pairs

Trait Area	Identical Twins	Fraternal Twins
General ability	.86	.62
Special abilities	.74	.52
Personality scales	.50	.28
Ideals, goals, interests	.37	.20

Source: Adapted from *Heredity, Environment, and Personality: A Study of 850 Sets of Twins* by John C. Loehlin and Robert C. Nichols, Copyright © 1976. By permission of the University of Texas Press.

TABLE 14.2   Resemblance of 24,000 Pairs of Reared-Together Twins in Five Countries and Identical Twins Reared Apart

	Correlations within Twin Pairs		
	Identical Twins Reared Together	Fraternal Twins Reared Together	Identical Twins Reared Apart
Extraversion	.51	.18	.38
Neuroticism	.46	.20	.38

Source: Adapted from Loehlin, J. C. *Genes and environment in personality development.* © 1992, reprinted by permission of Sage Publications, Inc.

total of 24,000 pairs of twins, consistently indicate moderate genetic influence. Correlations are about .50 for identical twins and about .20 for fraternal twins.

The role of heritability in extraversion and neuroticism has been studied extensively, but much less genetic research has been done for the other three Big Five traits, namely, agreeableness (likeability, friendliness), conscientiousness (conformity, will to achieve), and culture (openness to experience). These qualities have been investigated with diverse measures (rather than standardized tests) that also make it more difficult to compare results across different studies. Nevertheless, results of twin and adoption studies with measures related to these three traits also suggest genetic influence for agreeableness, conscientiousness, and culture at least to a moderate degree (Loehlin, 1992).

Temperaments.   The term **temperaments** refers to traits that are visible in early childhood (Buss & Plomin, 1984), and seem especially relevant to the individual's emotional life (Allport, 1961; Clark & Watson, 1999). Dispositions usually considered temperaments include the general level of emotionality, sociability, and activity. These temperaments are usually assessed through parental reports about their children on temperament rating scales (Rothbart et al., 1994). In adults, they typically are assessed by self-report measures (e.g., Buss & Plomin, 1984), with items like those shown in Table 14.3.

TABLE 14.3   Self-Report Items to Assess Temperament in Adults (Buss & Plomin, 1984)

EMOTIONALITY (easily aroused physiologically to experience negative emotions)
   Many things annoy me.
   I get emotionally upset easily.

SOCIABILITY (seeks social interaction)
   I prefer working with others rather than alone.
   I like to be with other people.

ACTIVITY (overall energy level, tempo/speed, intensity or vigor)
   My life is fast-paced.
   I usually seem to be in a hurry.

Note: Participants rate on five-point scales the degree to which items like these apply to them, from "not typical" of me to "very typical" of me. Based on Buss, A. H. & R. Plomin. (1984). *Temperament: Early developing personality traits* (Table 7.3). Hillsdale, NJ: Erlbaum. Copyright, 1984, reprinted by permission.

Emotionality is often defined as the tendency to become aroused easily physiologically (by ready activation of the autonomic nervous system) and especially to experience frequent and intense negative emotions such as anger, fear, and distress (Buss & Plomin, 1984). Not all researchers agree with this definition, however. They find that the intensity with which an individual experiences emotions is independent of how often he or she has such feelings (Larsen, Diener, & Emmons, 1986). That suggests that these two components of emotionality need to be considered separately, and that both need to be taken into account. For example, if Jane rarely experiences fear but becomes unbearably fearful in some situations, her emotional life would be quite different from someone who is often moderately fearful but never intensely afraid. Further, positive and negative emotions can function independently and they need to be measured separately. For example, people who often experience positive emotions may or may not also experience negative emotions often. In spite of these variations, there is agreement that emotionality is an important aspect of temperament.

Sociability refers to the degree to which the person seeks to interact with others and to be with people. (As such, it overlaps with the concept of extraversion versus introversion introduced in Chapter 3.)

Activity may be defined with regard both to the vigor or intensity of responses and their tempo or speed. It refers to individual differences on a dimension that ranges from hyperactivity to extreme inactivity.

In these dispositions genetic endowment seems to have a significant part (e.g., Thomas & Chess, 1977), and the evidence is increasingly strong (e.g., Plomin, 1990; Rowe, 1997; Clark & Watson, 1999). Figure 14.1 illustrates these types of results more concretely. It shows that on the dimension of emotionality, identical twins are rated as much more similar by their mothers than are fraternal twins.

Although the results are impressive, they are not easy to interpret. Some of the greater similarity found may reflect that the mothers themselves may treat the identical twins more similarly, as might other people in the environment. The mothers also may

14.4 In twin studies, how strongly do genetic factors seem to influence temperament? Why are these results difficult to interpret?

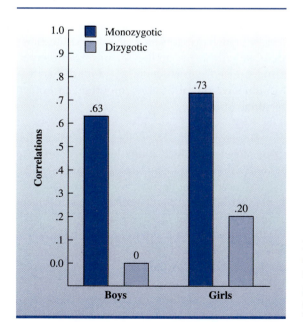

Figure 14.1 Similarity of emotionality: mother's ratings of monozygotic and dizygotic twin pairs.
Source: Correlation coefficients for degree of similarity from data in Buss, Plomin, & Willerman (1973).

have been influenced in their ratings not only by the twins' behavior but by their own expectations and preconceptions for identical versus fraternal twins. Nevertheless, results like these tend to be obtained so consistently that they suggest a significant genetic role in personality with regard to the temperaments of emotionality, activity, and sociability (Buss & Plomin, 1984; Plomin et al., 1997). A comprehensive review of this research concluded that: ". . . one-third to one-half of individual differences in temperamental traits can be attributed to genetic variation among children" (Rowe, 1997, p. 378).

Genetic researchers investigating differences between children in temperament could not use self-report questionnaires with their young participants and therefore used other measures, such as direct ratings of the children's behavior by observers (e.g., Cherny et al., 1994; Goldsmith & Campos, 1986; Saudino et al., 1996). With few exceptions (confined to the first few days of life), observational studies of young twins show genetic influence for diverse characteristics in both twin and adoption studies. These characteristics include the degree to which the child's behavior is inhibited—an aspect of fearfulness (e.g., Robinson et al., 1992); shyness both when observed at home and in the laboratory (Cherny et al., 1994); activity level (Saudino & Eaton, 1991); and empathy (Zahn-Waxler, Robinson, & Emde, 1992). Findings from these studies are notable because they supplement those that rely on self-reports—which are by far the most common.

14.5 How strongly are attitudes influenced by genetic factors? Which attitudes seem most strongly influenced?

Attitudes and Beliefs. Genetic influences also seem to play a role in individual differences in attitudes and beliefs. Results come from a number of twin studies (Eaves et al., 1989), including one of twins who were adopted and reared apart (Tellegen et al., 1988). To illustrate, a substantial genetic influence was found on *traditionalism,* a general orientation which taps conservative (as opposed to liberal) attitudes on diverse topics and many other attitudes also seem to show genetic influence (Eaves et al., 1989; see Plomin et al., 1997).

Attitudes that are more heritable may differ from those that are less heritable systematically, and examining such differences sheds some light on the nature of genetic influences. For example, in one study the researcher separated many specific attitudes into two sets (Tesser, 1993). One set contained those that twin studies had found were more heritable (such as attitudes about the death penalty and about jazz). The other set consisted of those that were less heritable (such as attitudes about coeducation, straightjackets, and the truth of the Bible). Then the researcher set up experimental situations designed to change these attitudes in college students. He found that the more heritable attitudes were harder to influence and also more important in determining the person's judgments of interpersonal attraction.

Aggressive and Altruistic Tendencies. Research with adult twins also points to the influence of genes on other aspects of social behavior. For example, self-reports on aggressiveness questionnaires were obtained from a large number of twin pairs in England (Rushton et al., 1986). The twins answered such questions as: "I try not to give people a hard time," and "Some people think I have a violent temper." The altruism questionnaire asked for the frequency of such behaviors as "I have donated blood," and "I have given directions to a stranger." Within the identical (monozygotic) twin pairs, the answers were more similar than would be expected by chance whereas between fraternal twins, the correlation was merely at a chance level. These results occurred both for the males and for the females. Using sophisticated statistical techniques, the researchers estimated that genetics accounted for approximately 50 percent of the individual differences in test answers.

Romantic Love.   Although genes seem to directly or indirectly influence individual differences on most measures of personality, as well as on social attitudes, including peoples' self-esteem (e.g., McGuire et al., 1994), one area that seems to be beyond DNA is romantic love. A behavior genetic twin family study focused on the genetic versus environmental influences on individual differences in adult romantic love styles. The participants were drawn from 890 adult twins and 172 spouses from the California Twin Registry and they had been married for an average of a dozen years (Waller & Shaver, 1994). Six different love styles were measured, ranging from one that values passion, excitement, intimacy, self-disclosure and "being in love from the start," to one that values a relationship that is affectionate and reliable and has companionship and friendship (with items like "It is hard for me to say exactly when our friendship turned into love.").

The findings showed that how people love is almost completely due to the environment and essentially unaffected by genetic influences. In fact, this is one domain in which it is the family environment that turned out to be particularly important (Waller & Shaver, 1994). As the researchers noted ". . . love styles may be learned during early familial or shared extra-familial interactions and subsequently played out in romantic relationships" (p. 272–273).

Twins Reared Apart

To try to separate the role of genetics and environment, it is especially informative to assess identical twins who have been reared apart in different families. Reports have come from two large-scale studies of twins reared apart in Minnesota (Bouchard et al., 1990; Tellegen et al., 1988) and in Sweden (Pedersen et al., 1988; Plomin et al., 1988). The results surprised even many of the researchers who have long been convinced that genes affect personality.

In this research, Bouchard, Tellegen, and their associates for more than a decade studied a sample of identical (monozygotic) twin pairs reared apart who were separated early in life (on average before the end of the second month). They grew up in different families, but mostly in English-speaking countries. As adults, their responses were assessed on many medical and psychological measures, including personality questionnaire scales and intelligence tests (as seen in Table 14.4). Most had not seen each other

14.6 What is the special value of comparing twins raised together and apart? What general result occurs in such studies that strengthens genetic explanations?

TABLE 14.4   Names of Personality Scales in Studies of Twins Reared Apart and Together

Well-being	Control
Social Potency	Harm Avoidance
Achievement	Traditionalism
Social Closeness	Absorption
Stress Reaction	Positive Emotionality
Alienation	Negative Emotionality
Aggression	Constraint

Note: On most scales, identical twins reared apart were as similar to each other as those reared together. The main exception was "social closeness" (on which those raised together were more similar).

Source: Tellegen, A., Lykken, D., Bouchard, T., Wilcox, K., Segal, N., & Rich, S. (1988). Personality similarity in twins reared apart. *Journal of Personality and Social Psychology, 54,* 1035 (Table 3).

for an average of about 30 years, although some had contact over the years. Comparisons were made with a larger sample of twin pairs who had been reared together and grew up in Minnesota.

There were instances of dramatic psychological similarities within the twin pairs, even for twins who grew up in radically different environments for 30 years or more in many cases. These twins seemed to share some quite distinctive mannerisms, postures, attitudes, and interests. For example, in some cases they posed alike for photos. Some turned out to have virtually the same height, the same weight, the same number of marriages and children, the same drinking and smoking habits, the same mannerisms, the same clothes, food, and jewelry preferences, similar physical symptoms—and similar scores on personality tests (e.g., Segal 1999). Many also quickly felt a close emotional connection with each other even though they had spent their entire lives apart.

There also were strong similarities in many of the test results of the identical twins. Especially interesting, the similarity was almost as high for the monozygotic twins who grew up in different homes as it was for those raised within the same family (as was seen in Table 14.2). Bouchard and colleagues (1990) attributed approximately 70 percent of the individual differences found in intelligence to heredity. They interpreted the effects of heredity on personality (as assessed on their questionnaires) to be approximately 50 percent and the effects of family environment to be trivial (Tellegen and associates, 1988). Likewise, twin studies of the Big Five factors suggest "The shared genes, not the shared experiences, mainly determine the family resemblance of 'blood relatives'" (Rowe, 1997, p. 380).

Beyond Self-Report Measures

14.7 What difficulties attend the use of self-report measures to assess heritability? How have researchers addressed these difficulties?

One of the most surprising findings from genetic research on self-report personality questionnaires is that of the many traits that have been studied virtually all show genetic influence. That may be because all these traits in fact reflect genetic predispositions. But at least some of the similarity may lie in the eyes of the beholder. For example, identical twins who have a higher level of frustration tolerance may *think* and honestly say that they experience less negative emotions than others, worry less, and rarely feel anxious, not because they experience these feelings less than others, but because they are less bothered by them. It is therefore particularly important to use measures of personality other than self-report questionnaires to investigate whether or not this result is somehow due to biases in the self-report measures themselves.

To address this, researchers gave a measure of the Big Five personality factors to almost a thousand pairs of twins (Riemann, Angleitner, & Strelau, 1997) in Germany and Poland. They also obtained ratings of each twin's personality by two different peers (who agreed reasonably with each other in their ratings about each twin's personality). On average, ratings by the peers correlated .55 with the twins' self-report ratings, providing moderate validity for those ratings. Table 14.5 shows twin correlations for the "Big Five" personality traits, which are similar to those summarized in Table 14.2, with average correlations of .52 for identical twins and .23 for fraternal twins. The fact that the peer ratings also indicated genetic influence helps support the conclusions reached earlier based on self-report questionnaires.

So far in this chapter we have reviewed some of the main findings that show significant effects of one's genetic heritage on the personality that develops. Most of these studies use the heritability index to estimate the size of these effects. Because the method and the conclusions it allows are easy to misunderstand they need to be examined carefully. In Focus 14.1 discusses these misunderstandings and clarifies the meaning of this index.

TABLE 14.5   Twin Study Using Self-Report and Peer Ratings of "Big Five" Personality Traits

	Correlations within Pairs			
	Self-Report Ratings		Peer Ratings	
	Identical Twins	Fraternal Twins	Identical Twins	Fraternal Twins
Extraversion	.56	.28	.40	.17
Neuroticism	.53	.13	.43	−.03
Agreeableness	.42	.19	.32	.18
Conscientiousness	.54	.18	.43	.18
Culture	.54	.35	.48	.31

Source: Riemann, R., Angleitner, A., & Strelau, J. (1997). Genetic and environmental influences on personality: A study of twins reared together using the self- and peer report NEO-FFI scales. *Journal of Personality, 65,* 449–476. With permission from Blackwell Publishing.

14.8   How are heritability coefficients computed? What do such estimates mean? What misconceptions and difficulties occur in interpreting them?

IN FOCUS 14.1

UNDERSTANDING HERITABILITY AND THE HERITABILITY INDEX

Attempts to understand the implications of heritability for personality have been undermined by widespread misconceptions. Let's address some of the most common.

The Heritability Index. Correlations like those shown in Table 14.5 are used by behavioral geneticists to estimate the percent of the variation in scores measuring individual differences that is attributable to genetic factors. The computation of this index is straightforward, but its interpretation is not. To compute it, pairs of same-sex fraternal and identical twins are compared on particular variables. For each type of twin, the researchers compute a correlation among pairs of identical twins to index their similarity, as well as a correlation among pairs of fraternal twins. Subtracting the latter from the former, and then multiplying the result by 2, results in the heritability index. It is an estimate of the role of heritability in accounting for individual differences on the variable studied. Results using these estimates to date suggest that heritability plays an important role in many aspects of personality.

The Meaning of High Heritability.   Twin studies are a valuable first step toward answering whether genetic factors contribute to individual differences in personality. But the heritability indexes they compute have to be interpreted cautiously for several reasons (e.g., Dickens & Flynn, 2001):

- First, **heritability estimates** always are limited to the specific population that was studied in the

research reported and for which they were computed. This is a crucial caution that behavioral geneticists generally emphasize but that many readers often fail to appreciate. (Goldsmith, 1991). Heritability estimates *do not* provide an absolute index of the degree to which a given characteristic is influenced by the genes. It is nonsense to think for example that Joe's aggressiveness is made up of 30 percent genetic influence and 70 percent environmental influences or that Susan's ballet skills are 20 percent genetic and 80 percent ballet school and practice. If all children took ballet lessons, then a greater portion of the differences among the children in ballet skills may be due to their variations in their genetic endowment. But if only some children took ballet lessons, then how well children can perform *pirouettes* may depend largely on the availability of ballet lessons. In general, the heritability index depends on the variations in the environment that exist within a given society in which the study is conducted. The more homogenous the environments are, the greater is the heritability index. Thus the heritability index reflects characteristics of a population, rather than of an individual. If the psychological environments to which children are exposed become more homogeneous across different cultures, for example due to global communication and media influences, the result would be an increase in the relative proportion of the genetic influences in the observed variations among people.

Heritability estimates *do* index the portion of the individual *variability* related to the genetic *variability* that exists in the population. If the characteristic is so important that any variation in the genes that affect it can reduce the evolutionary fitness of the individual, the heritability index would be close to 0. That happens because all individuals are likely to be identical with regard to that gene, and any variability observed in that characteristic is likely to come from environmental differences.

• Second, even when identical twins are reared apart, their similarities on personality measures are not necessarily due to their genes for personality itself. For example, similar interests and values in identical twins may in part reflect their similar physiques and appearance, constitutions, abilities, skills, and physical characteristics rather than any genes for personality. These physical qualities may lead other people to treat them similarly even when the twins live in different environments. A shared interest in becoming a fashion model, for example, may say more about the inheritance of faces than of personality. Their similar physical qualities also may lead each twin to see himself or herself in a somewhat similar way, for example, as highly attractive or unattractive, or as physically strong and skilled and competent or as weak and ineffective. This in turn could influence such aspects of personality as self-concepts and self-confidence and a host of related characteristics.

• Third, as mentioned earlier, many studies that compare the similarity of identical and fraternal twins who grow up together draw mostly on their answers on self-report personality questionnaires (the measure most often used). On these tests the identical twins might give more similar answers not just because they have the same genes but also because they might identify more closely with each other and emulate each other more, or might be treated more alike by parents and other people, and therefore become more alike in all sorts of ways. For example, they might often wear more similar clothes and share more activities and time

together. Just as greater similarity in dress among identical twins does not necessarily imply that clothing tastes are genetically determined, it is important also to be cautious before concluding that similarity in answers to self-report questionnaires about an attribute necessarily implies the specific genetic heritability of that attribute.

• Fourth, high heritability coefficients do *not* imply that the particular characteristic cannot be changed significantly. Even when genetics importantly influence individual differences in a trait, the trait may be modified by environmental influences. For example, substantial genetic influence for height means that height differences among the sampled individuals are largely due to genetic differences. But even when a trait is highly heritable it may be influenced importantly by the environment. The average height of Europeans has climbed more than 20 cm in the past 150 years (*TIME,* October 14, 1996 Volume 148, No. 16; see http://www.vwl.unimuenchen.de/ls_komlos/covereu.html). Thus, although height is highly heritable, such environmental interventions as improving children's nutrition and health can affect it. Indeed, environmental factors seem to account for the average increase in height across generations, although within each generation individual differences in height are highly heritable. As another example, consider the role of inheritance in intelligence. Research with twins raised in various environments (either together or apart) suggests that intelligence measured on standard tests tends to be increasingly similar to the degree that the individuals have an increasing proportion of genes in common (Cartwright, 1978; Plomin et al., 1997; Vandenberg, 1971). But whereas a person's genetic endowment may set an upper limit or ceiling on the degree to which his or her intelligence can be developed (Royce, 1973) it is the environment that may help or hinder achievement of that ceiling (e.g., Cantor & Kihlstrom, 1987).

• Finally, the heritability index itself does not address the mechanisms through which the genetic influences on personality operate and exert their effects on the individual.

Heredity versus Environment: Another False Dichotomy

Debates about heredity and environment have raged for years. Unfortunately, they readily turn into either/or competitions to see which one is more important. As a leader in research on the role of inheritance in behavior points out: ". . . evidence for significant genetic influence is often implicitly interpreted as if heritability were 100%, whereas heritabilities of behavior seldom exceed 50%" (Plomin, 1990,

p. 187). Plomin notes that this should make it clear that "non-genetic sources of variance" are also of self-evident importance. Indeed, Bouchard and colleagues in their studies of identical twins raised apart acknowledge that in their own results, "in individual cases environmental factors have been highly significant" (1990, p. 225), as with the case, for example, of a 29-point difference in the IQs of two identical twins.

In short, regardless of the exact percent used to estimate the influence of genetic factors, clearly their influence is considerable, especially given that few findings in personality account for as much as 20 percent of the variance. Just as noteworthy, however, is the other side of that finding: namely, the fact that the same data show that at least half of the variance of personality is *not* due to genetic factors and thus also attest to the importance of the environment for personality. The recurrent theme throughout this research is clear: unquestionably one's genetic endowment has extensive influences on one's life and personality development, and so also does the environment. The challenge will be to untangle the mechanisms through which both genes and experience interact throughout the life course to influence what we become. That will require specifying the mechanisms that produce these effects and clarifying the nature of the characteristics that are heritable and the specific aspects of the environment that are important.

Summary

Taken collectively, the findings support the view that genetic factors play a significant role in personality, in attitudes and values, as well as in self-esteem (e.g., McGuire et al., 1994). In some cases, genetic influences seem to account for as much as half of the individual differences observed. The magnitude of these effects may be exaggerated, however, particularly when the findings are based on self-reports. Plomin and colleagues (1990), critically reviewing twin studies of personality based on self-reports, point out certain method problems in these studies that may systematically overestimate the role of genetics. They believe the biases can be corrected if results from adoption studies are appropriately taken into account. When corrections are made for the erroneous inflation of the heritability estimates (see subsection, Stress Is Bad for Your Brain): "The true heritability estimate for self-reported personality is closer to the adoption study estimate of 20% than to the twin study estimate of 40%" (Plomin et al., 1990, p. 233). It is also clear, however, that even when one goes beyond self-reports the role of genetics in personality is still significant and has to be taken seriously. The task ahead is to understand the mechanisms through which these effects play out.

14.9 Compare heritability estimates derived from twin studies with those obtained in adoption studies.

▶ GENE–ENVIRONMENT INTERPLAY

Assuming that both nature (genetics) and nurture (environment) are important influences on personality, the next step is to understand the developmental interplay between them. This interplay is seen in the course of life as the personality of the person, partly influenced by genetics, interacts with and to some degree selects and shapes the situations in his or her psychological world. The causal relations go in both directions because those situations over time, in turn, exert their impact on what the person becomes. Let's examine this interplay and its implications for personality.

The Unique (Nonshared) Psychological Environment of Each Family Member

14.10 Differentiate between shared and unshared environmental influences within the family. Why is it difficult to interpret the causal influences between environmental and genetic factors?

Twin studies have inquired into the environmental influences that make genetically identical "clones"—identical twins—different from one another. For many years, it was widely thought that children growing up in the same family would be similar to one another in personality for environmental reasons because they shared the "same" family situation. Genetic researchers now believe that this assumption is false. They contend that family members resemble each other in personality largely because of genetic influences, and that the environment seems to make members of a family different (e.g., Plomin et al., 1997). Their reasoning is based on two findings. First, twins reared apart are only somewhat less similar in personality than twins who grew up in the same family. In addition, in studies of adoptive families that have more than one adopted child, the genetically unrelated children show little similarity in personality even though they had the same **shared environment,** that is, they were raised within the same family (Plomin, Chipuer, & Neiderhiser, 1994).

Note that the shared or **family environment** in the twin studies was treated as a global entity, as if families treat all the siblings in them the same way. The psychological environments experienced by each member of the family, however, may be quite different and each may receive distinctively different treatments within the same family. Each parent may relate differently to each child. In fact, findings of birth order effects suggest that children in the same family encounter quite different psychological environments (Sulloway, 1996). Moreover, much of the psychological environment experienced by individuals involves continuing significant encounters outside the family (with peers, school, spouse, and in the broader group and culture), the nature and effects of which may change with different phases of development.

Environmental Influences Within the Family. The **nonshared** or **unique environment** exerts its effects on each person through many routes, beginning in prenatal development and birth order effects. It includes biological events such as illness and nutrition, as well as psychosocial events such as interpersonal experiences that may range from parental reactions to peer influences, romantic partners, and mates (see Table 14.6). For example, earlier-born children are larger, more powerful, and more privileged than later-born children and that, in turn, may affect their personality development (Sulloway, 1996). From the start, each child is born into a slightly different family psychologically and structurally because parents treat siblings differently, and siblings treat each other differently.

It is understandable that most parents find it hard to admit that they treat their children differently, given the social norms that press for them to treat each one the

---

TABLE 14.6 Some Environmental Influences that Siblings in the Family Do Not Share Equally

Position in the family

Parental reactions

Accidents

Prenatal events, Illness

Peer group reactions and support

Other interpersonal experiences (e.g., mates)

Educational and occupational experiences

same. But children correctly believe that their parents do treat them differently—an impression confirmed by observational studies (Reiss and colleagues, 2000). So, for example, parents are likely to react differently to their first and their fifth child, perhaps hovering over their first baby and swamped by the time the fifth one arrives. On close examination, children growing up in the same family lead remarkably separate lives (Dunn & Plomin, 1990). Even such variables as parental divorce tend to affect children within the family differently, depending, for example, on the child's age and role in the family as well as many other variables (Hetherington & Clingempeel, 1992). In short, siblings growing up in the "same" family experience it differently so that psychologically it is not the same.

Research that traces how differences in such experiences lead to differences in outcomes is still just beginning, but some links have already been shown (Reiss et al., 2000). Most of these associations connect negative aspects of parenting (such as conflict) with such negative outcomes as antisocial behavior later in development. It is especially interesting that some associations (of modest strength) have been reported between differences in parental negativity toward identical twins and their adolescent maladjustment on such indices as depression and antisocial behavior (Reiss et al., 2000). Obviously such differences cannot be due to genetic factors and thus must be due to the nonshared environment (Pike et al., 1996). Generally weaker correlations are found between positive aspects of parenting (such as affection) and positive outcomes.

The correlations between measures of the children's nonshared environment and their personality also raise the old chicken and the egg question, that is, the direction of the effects: Does parental negativity cause negative outcomes in personality? Or do parents treat some of their children more negatively than others in reaction to the child's personality, responding more harshly, for instance, to siblings who are more difficult to deal with? If the latter holds, the differential treatment siblings receive may be due in part to each child's distinctive genetics (Plomin, 1994). In other words, differential parental treatment of siblings may be due to genetically influenced differences between the siblings, including in their personality.

Environmental Influences Outside the Family. The nonshared environment includes, but goes far beyond, differences in siblings' unique experiences within the family (see Table 14.6). Some of the most important aspects of the nonshared environment unfold in the experiences children have outside the family as they interact with their expanding worlds, most notably in relationships with peers and in school and play contexts. If you reflect on your own life, you will not find it difficult to come up with examples of relationships and experiences outside the immediate family that influenced your development. In these interactions even siblings living in the same family form quite different relations with peers, encounter different types of social support, build different lives with different educational, occupational, and interpersonal experiences and events along the route (Plomin, 1994). Further, such factors as accidents and illnesses, as well as chance encounters and experiences, may initiate significant differences between siblings (Dunn & Plomin, 1990). While such events initially may be relatively minor they can snowball and become compounded over time into large outcome differences years later.

To recapitulate, we saw that environmental influences that affect personality development do not seem to operate on a family-by-family "on the whole" basis, but rather on an individual-by-individual basis. That suggests that environmental effects on personality are specific to each child rather than general for all children in a given family.

It makes the unit of analysis each child as he or she distinctively interacts with relevant specific environmental situations, including with particular family members. Environmental influences on personality development seem to operate mostly in a nonshared manner, making children growing up in the same family different from one another. As was discussed, attempts to identify specific sources of nonshared environment indicate that many sibling experiences differ substantially. Even such seemingly shared variables as parental attitudes about childrearing and parents' marital relationship might not be experienced the same way and might, in fact, be subtly different for each sibling.

Interactions Among Nature-Nurture Influences

14.11 What research results suggest genetic effects on experience within the family? Describe some ways in which genetic factors influence nonshared aspects of the environment.

There are continuous complex interactions between the expressions of genetic influences and the situations and events that the person experiences (e.g., Rutter and colleagues, 1997). These interactions make it difficult to isolate the contribution of genetics versus environmental influences because their interplay becomes virtually indivisible.

For example, adult twins reared apart rated the family environments in which they grew up more similarly than did fraternal twins reared apart (Rowe, 1981, 1983). Presumably this happened for at least two possible reasons. First, identical twins even reared apart may be more alike in how they perceive and interpret their experiences. Second but equally important, they may have been treated more similarly due to their genetically-influenced shared characteristics such as their more similar physical appearance, abilities, skills, and temperaments, for example (Plomin et al., 1988). Likewise, beyond the family environment, genetic similarity may lead to greater similarity in the experienced environments in peer groups (e.g., Manke et al., 1995), in the classroom (Jang, 1993) and in work environments (Hershberger, Lichteinstein, & Knox, 1994). It may also influence such other life events as proneness to childhood accidents or illness (Phillips & Mathews, 1995), exposure to trauma (Lyons et al., 1993), or to drugs (Tsuang et al., 1992).

Perhaps the clearest evidence for genetic effects on experience comes from observational studies that also show such genetic effects (although often of lower magnitude), making it plain that they do not just depend on self-reports and questionnaires. For example, the Home Observation for Measurement of the Environment or HOME is a widely used measure of the home environment that combines observations and interviews (Caldwell & Bradley, 1978). It assesses aspects of the home environment such as parental responsiveness and encouraging developmental advance. In an adoption study with this measure, the home environment of each sibling was assessed when the child was one year old and again when each child was two years old (Braungart, Fulker, & Plomin, 1992). HOME correlations for genetically unrelated children adopted into the same home (adoptive siblings) were compared to those for genetically related siblings in nonadoptive homes (nonadoptive siblings). HOME scores were more similar for nonadoptive siblings than for adoptive siblings at both one and two years, suggesting genetic influence on this measure (see Table 14.7).

In another observational study, O'Connor and associates (1995) obtained videotaped observations of adolescents' interactions with their mothers or their fathers in 10-minute discussions of problems and conflicts within each parent–adolescent dyad. The adolescent participants included six groups of siblings: identical twins, fraternal twins, full siblings in nondivorced families, and full siblings, half siblings, and unrelated siblings in step-families. Using sophisticated estimates of heritability the researchers found some significant genetic influences both on the positive and negative interaction with both the mothers and fathers.

TABLE 14.7 HOME Score Correlations for Nonadoptive and Adoptive Siblings at Ages 1 and 2 Years

Environmental Measure	Sibling Correlations	
	Nonadoptive	Adoptive
1 year	.58	.35
2 years	.57	.40

Note: Adoptive = genetically unrelated, adopted into same home; *Nonadoptive* = genetically related in nonadoptive homes

Source: Braungart, J. M., Fulker, D. W., & Plomin, R. (1992). Genetic influence of the home environment during infancy: A sibling adoption study of the HOME. *Developmental Psychology, 28*, 1048–1055.

Genes Also Influence Environments

Genetic factors may contribute to the experienced environment in several ways (Plomin et al., 1997; Rutter and colleagues, 1997). First, people encounter the environments that their genetic relatives, in part, make for them. Take activity level, for example, which seems to have a heritable component and is thus shared to some degree between the child and the parents (Saudino & Plomin, 1996). From the start, parents construct aspects of their child's early environment, and tend to make it more (or less) stimulating and activity-filled in a way that is consistent with both their own and their child's genetic propensities. Thus highly active children are likely to have active parents who model and reward high activity and who also provide them with both genes and an environment conducive to the development of high activity.

Second, the individual's genetically influenced characteristics affect how other people will react to him or her. For example, highly active children might receive more positive reactions from their peers or, in the opposite direction, more negative reactions from their schoolteachers.

Third, and most important, individuals actively seek and create situations and social environments in ways congruent with their genetically influenced dispositions and qualities. Whereas extremely active children are likely to create a high-energy environment by actively selecting highly active friends and activities, less active children are apt to make their environment less energy demanding. This self-directed process of selecting and creating one's own situations is the most central for personality; it is literally the seat of the sort of dynamic Person × Situation interactions through which dispositions and the environment reciprocally influence each other.

Given the multiple paths through which genetic influences impact on the environment, it is understandable that genetic factors often contribute substantially to measures of the environment in research. But while it is clear that genetic factors influence the environments we experience and select, the effects are complex and the direct genetic influence is only one part of the variance, and does not account for most of it.

These findings suggest that researchers need to move away from passive models of how the environment or the genes separately affect individuals and turn to models of person-environment interaction. These interaction models recognize the active role that persons play in selecting, modifying, and creating their own environments. In this process there is a continuous interaction between dispositions (partly influenced by genetics, partly by environmental influences) and situations as the individual deals with his or her world in the course of development. Such interaction implies that some of the most important questions in genetic research will involve the environment and some of the most important questions for environment research will involve genetics (Rutter et al., 1997).

The interplay of biological and psychological processes is evident at every level of analysis. It is apparent even at the molecular level: the synapses in the brain change physically when new learning occurs. In this interaction, genes are switched on to make new proteins that are crucial in long-term memory, wherein the person's history resides. Furthermore, even relatively small heritable differences in qualities of temperament, such as activity and energy levels, emotionality, and sociability, which appear to be visible in early childhood (Buss & Plomin, 1984), can be biological foundations for diverse enduring behavioral tendencies that may develop from these roots (Kagan, 2003).

For example, temperamentally more active, energetic children tend to explore and interact more vigorously and forcefully with their environments, rapidly encountering its challenges and gratifications as well as its dangers and frustrations. In time they also are likely to become more aggressive than children who are temperamentally inhibited from exploring the unfamiliar and thus inclined toward shyness (e.g., Daniels et al., 1985). Heritable variations in arousal thresholds in certain loci of the brain also could influence such behaviors as shyness (e.g., Kagan, Reznick, & Snidman, 1988). Heritable differences in sensitivity and physiological reactivity in response to sensory stimulation partly predispose people to become introverted rather than extroverted (Stelmack, 1990). In turn, introverts may be more disposed to avoid the types of social stimulation that extroverts desire and actively select for themselves (Plomin, Manke, & Pike, 1996).

Search for Specific Gene–Behavior Connections

Given the evidence for genetic influence on personality traits that has been found in recent years, researchers in the vanguard of the field are now trying to become more precise and to identify specific genes that might connect to specific characteristics. They are beginning to go much beyond demonstrations that genetic influences are important for personality.

In earlier research, specific genetic defects have been linked to various abnormalities such as mental deficiencies. **PKU** is a case in point (see Table 14.8). **PKU** (*phenylketonuria*) disease is an inherited disorder in which a genetic abnormality results in the lack of an enzyme necessary for normal metabolism. Because of this enzyme deficiency, a toxic chemical accumulates in the body and results in central nervous system damage and mental retardation. Diagnosis of PKU disease is now possible immediately after birth, and highly successful treatment has been devised. The child is

---

TABLE 14.8 Effects of Some Genetic Abnormalities: Two Examples

Name of Disorder	Description	Cause
Down Syndrome	Severe mental retardation Physical appearance: small skull, sparse hair, flat nose, fissured tongue, a fold over the eyelids, short neck	A third, extra chromosome in the twenty-first chromosome pair Appears to be associated with advanced age in the mother
PKU (phenylketonuria)	Results in mental retardation if not treated soon after birth	A gene that produces a critical enzyme is missing

placed on a special diet that prevents the toxic substance from building up in the bloodstream. When the biological mechanisms underlying other forms of mental deficiency are known, equally effective cures may be possible.

Some psychological characteristics are determined by an individual's specific genetic structure. For example, when the twenty-first chromosome in the body cell of an individual has a third member instead of occurring as a pair, the individual will have **Down Syndrome,** a form of mental retardation. A technique for drawing amniotic fluid from the uterus of the pregnant mother enables doctors and prospective parents to know in advance if the developing fetus has this chromosome abnormality. This procedure is performed routinely for pregnant women 35 years of age or older because women in this age group are more likely to give birth to children with Down Syndrome.

To link genes to personality, promising examples include the report that a gene for a dopamine receptor was correlated with novelty-seeking in two studies (Benjamin et al., 1996; Ebstein et al., 1996), with other researchers finding that the same gene was related to hyperactivity (LaHoste et al., 1996). While such findings are exciting for genetic researchers, they must be interpreted with much caution and still be treated as tentative and suggestive rather than conclusive.

These reservations are necessary because findings of associations between personality and specific genes in the past have failed to replicate. That happened, for example, for the reported correlation between neuroticism and a gene important in the functioning of the chemical neurotransmitter serotonin (Lesch et al., 1996), which two other studies could not reproduce (Ball et al., 1997; Ebstein et al., 1997). Perhaps even more important, personality traits involve extremely complex patterns of characteristics and behaviors such that an association with any specific gene probably will have only a small effect (Plomin, Owen, & McGuffin, 1994). That is, genetic influence on personality involves the action of many genes, each of which has a small effect rather than the direct effect of any single major gene.

14.12 Have researchers been able to link specific genes to personality variables?

Given this great but unsurprising complexity, it can be valuable to use the sorts of promising methods for identifying genes in lower animals that cannot be employed with people (Plomin & Saudino, 1994). For example, in such animal studies it is possible to use powerful techniques that alter specific genes, called knock-outs, to test just how they influence behavior (Capecchi, 1994). In mice, for instance, several genes that, if disrupted, would result in fearfulness have been located (Flint et al., 1995). As predicted, mice display greater aggression when the researchers knock out the genes for an important neurotransmitter (Saudou et al., 1994) or enzyme (Nelson et al., 1995). The limitation of course is that it is difficult to generalize results from studies with mice to humans. But there are exciting prospects for discoveries in molecular genetics that do speak to the human condition and will prove generalizable to personality (Hamer & Copeland, 1998). However, that is the story that is not yet written but that should unfold in the next few years.

Causal Mechanisms: The Role of Neurotransmitter Systems

As discussed above, genetic and environmental influences are always in close interaction. The question then becomes: What are the causal mechanisms that underlie this interaction? To address that question at the molecular biological level of analysis, researchers now try to examine how genetic variability underlying neurotransmitter systems can be linked to variability in personality traits and behavioral patterns (e.g. Grigorenko, 2002).

Neurotransmitter systems are the physiological pathways that communicate and carry out the functions of signal detection and response via chemical receptors (neurotransmitters). The variability of each of these neurotransmitters is dictated by the variants of its corresponding gene. There are nine neurotransmitter systems, but based on their biochemical functions, some are more relevant to the carrying out of psychological functions than others. In general, personality psychologists have focused on three primary systems involved in behavioral functions (the dopaminergic, seritonergic, and GABAergic systems).

Researchers have been analyzing how certain behavioral patterns correlate with specific genetic variants. Originally, researchers aimed to show direct, one-to-one correlations between specific traits or behavioral attributes and specific genetic mutations. So far these studies have provided mixed results. Many of the experiments that seemed to exhibit this kind of direct correlation could not be repeated. Though it is clear that neurotransmitter systems are linked to the expression of behavioral patterns, it is also evident that these neurotransmitter systems do not necessarily act alone in producing their affect on behavioral attributes. In fact, recent studies have revealed that the various neurotransmitter systems work together very closely, and many of the behavioral functions originally attributed to specific neurotransmitter systems are actually a function of the interaction of more than one of these systems.

As we begin to understand more about these neurotransmitter systems and the genetic variations that effect their functioning, it is clear that the task of correlating genes to behavior is not as simple as was originally believed. The interaction between genetic expression and the phenotypic expression of personality is complex. Specifically, researchers must take into account that the genetic variability that underlies human behavior is influenced not only by the additive and interactive factors on the genetic level, but also by characteristics in the surrounding environment.

14.13 What were the results of Grigorenko's review of genetics and personality? What conclusions did she reach?

Let us look in detail at a comprehensive review of research on genes and personality (Grigorenko, 2002). The results challenge some of the basic, commonly accepted assumptions of recent years. Take the claim that genetic factors contribute extensively to the similarity of MZ twins, whereas the family–environment plays no serious role, discussed earlier in this chapter. Elena Grigorenko's review and research makes it clear that *it really all depends*. She reports a recent behavior–genetic analysis conducted with more than 700 Russian families in which shared family–environment factors contributed significantly to variation in the majority of the traits studied. Note that this contrasts with the fact that in a large subsample of the population (218 families in which at least one parent had a criminal record), genetic factors did *not* contribute significantly to variation in 13 of 15 traits. Why?

There are good biological reasons for these puzzling results. The main point here is that neurotransmitter systems interact with one another. Consequently, personality characteristics will not reflect the action of a single neurotransmitter system. The complexities get even greater because gene expression and surface-level behavioral characteristics are not stable, and instead tend to fluctuate across development in interaction with environmental conditions.

The fundamental message from all this for the student of personality is that the links between our biochemistry and our personality characteristics reflect interactions both within and between systems. These are not simple one-way causal influences—a message very much like the one that emerged in our review of the "person versus situation" debate at the trait level of analysis (Chapter 4). One specific implication is that it is unlikely that there will be either simple or direct relationships between particular genetic factors and very broad categories of personality traits like those represented by the Big Five. Thus, any expectation that a few genes might account for qualities like "open-mindedness" is probably unjustified.

Genetic and Environmental Influences on Person × Situation Interactions

The interplay of genes as well as the environment also is reflected in the kinds of **Person × Situation interaction patterns** and *if . . . then . . .* personality signatures discussed in earlier chapters. These signatures may reflect both genetic and environmental influences, depending on the specific type of behavior involved. An early study using an adult twin sample showed, for example, that the person × situation interaction patterns for anxiety were influenced significantly by genetics whereas for dominance shared sibling-environmental influences were found (Dworkin, 1979). The results supported the general conclusion that an individual's behavior shows characteristic, meaningful patterns across situations that partly reflect genetic influences.

More recent studies with larger samples and other methods provide further and even stronger evidence for the same basic point (Cherney et al., 1994; Plomin et al., 1997, p. 202). An important implication is that genetic influences may be expressed not simply in how much of a given trait a person "has." Genetics also may influence the characteristic pattern in which that trait-relevant behavior is typically expressed in relation to particular types of situations (Wright & Mischel, 1987, 1988; Mischel & Shoda, 1995).

14.14 How might genetic factors influence Person × Situation interactions?

Social Environments Change the Expression of Genes, the Brain, and Personality

Researchers pursuing the genetic approach to personality readily acknowledge that even highly heritable traits can be constrained or limited in their full expression. We see that, for example, when the person's growth and ultimate height is affected by nutrition or disease in development. But although they refer to the interplay of genes and environment often they do *not* see this interplay as a two-way reciprocal or mutual influence process. Obviously social environments and the experiences in the world (barring extreme radiation or other biochemical effects on the genes directly) cannot affect the structure of your DNA. So in that sense the interplay between genes and environment in these analyses refers to a one-way influence process from genes to environment, in which genetic influences impact through various routes on the environments experienced with no modification of the genetic structure itself.

But it is also the case that social-psychological environmental influences can and do affect the expression of the genes: just by reading this paragraph you increase DNA transcription rates of certain neurotransmitters. And environmental influences also change the hardwiring of the brain—the neuronal structures themselves—and thus produce stable changes within the person at an organic level, even though they do not alter the structure of the DNA. This is evident for example in the finding that stress actually shrinks the size of the hippocampus—a brain structure basic for higher-order mental functions (e.g., Sapolsky, 1996).

14.15 Can the environment change genes? Cite results from stress research.

Stress Is Bad for Your Brain

Sapolsky reviewed studies showing that sustained stress increases glucocorticoids (GCs), a chemical substance that at high rates can have negative effects on health.

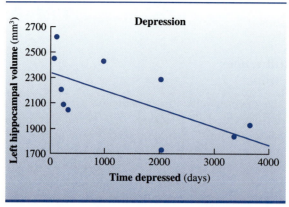

Figure 14.2 Relation between Hippocampal volume and duration of depression among individuals with a history of major depression.

Reported in: Sapolsky, R. M. (1996). Why stress is bad for your brain. *Science, 273,* 749–750. Figure is referenced as adapted from Sheline, Y., Wang, P. W., Gado, M. H., Csernansky, J. G., & Vannier, M. W. (1996). Hippocampal atrophy in recurrent major depression. Proceedings of the National Academy of Sciences of the United States of America. 93(9):3908–13.

Consistent with this finding, Sapolsky (1996) reports research with rodents showing that exposure to excessive amounts of GCs can impair the brain, with particularly unfortunate effects on the hippocampus, a brain structure crucial for learning and memory. Studies with depressed patients (as summarized in Fig. 14.2) also showed that the volume of this brain structure was significantly reduced, and the longer the depression the greater the amount of brain atrophy (Sapolsky, 1996). Furthermore, combat veterans suffering from post-traumatic stress reactions (e.g., after terrifying war experiences) also displayed not only greater exposure to GCs but also substantial reduction in their hippocampi on both sides of the brain.

In short, although the social environment does not influence the structure of the genes it can influence their expression, the brain, and the person's personality. As other sections of the text discuss, situations and environments importantly influence what people experience and do in stable relationship to those contexts. When the situations remain stable, so does their characteristic pattern of social behavior; when the situations change, the behavior pattern also does so predictably (e.g., Mischel & Shoda, 1995, 1998). And as was just noted, such environmental events as stress levels not only impact dramatically on behavior and experience, but also change the hardwiring, i.e., the structures in the brain. Person–environment interactions are two-way interactions in which the person's characteristics show some change over time, just as the characteristics also in part change the environment (Rutter et al., 1997).

In sum, environmental, genetic, and brain influences are in continuous interaction. These interactions affect what we feel and do, which, in turn, produces further changes. The challenge in future research on the interplay of genes, brain, and environment will be to clarify the complex and dynamic processes that lie between genes, brain, and behavior.

SUMMARY

GENETIC BASES OF PERSONALITY

- Twin studies examine the separate role of genetics and environment by comparing the similarities between identical and fraternal twins raised together and raised apart.

- In these studies, answers on self-report personality questionnaires typically yield correlations of about .50 for the similarity of identical twins reared together and about .25 for fraternal twins reared together. For identical twins reared apart, correlations are only slightly lower. About 40–50 percent of the self-reported personality differences among individuals may be accounted for by their genetic differences.

- The genetic contribution to personality is considerable, particularly to such temperaments as emotionality, activity level, and sociability.

- Most twin data on personality come from self-report personality questionnaires and are open to criticism. However, in recent twin studies such measures as ratings by peers or observational measures also indicate a strong genetic influence.

- The meaning of heritability estimates and their uses and misuses were discussed.

GENE–ENVIRONMENT INTERPLAY

- Nonshared environmental influences on personality development are substantial and children growing up in the same family experience it differently.

- Individual differences in what is experienced and in the environments one encounters are partly influenced by genetic factors that exert their impact through several routes. One's personality, itself influenced by genetics as well as environment, also affects the situations that one selects, influences, and creates in the course of development.

- Even highly heritable traits can be constrained or limited by aspects of their social environment. Although the social environment does not influence the structure of the genes it can change the brain and thus produce stable changes within the person at an organic level.

- Stable Person × Situation interaction patterns also reflect both genetic and environmental influences.

KEY TERMS

activity   327
behavior genetics   323
dizygotic twins   324
Down Syndrome   339
emotionality   327
heritability estimates
   (index)   331
human genome   324

monozygotic twins   324
neurotransmitter
   systems   340
nonshared (unique)
   environment   334
person × situation
   interaction patterns
   341

PKU (phenylketonuria)
   338
shared environment
   (or family environment)
   334
sociability   327
temperaments   326
twin method   324

15

EVOLUTION, BRAIN, AND PERSONALITY

▶ EVOLUTIONARY THEORY AND PERSONALITY

15.1 How does the evolutionary approach differ from the genetics approach?

Both evolutionary theory and the genetics approach try to link personality to its biological foundations, but there is a distinctive difference. The genetics approach focuses on the genetic influences that shape the specific biological processes and brain mechanisms underlying personality and social behavior. The evolutionary approach focuses on the processes that have shaped the genes over the long course of the species' development.

The Evolutionary Approach

The evolutionary approach is based on Charles Darwin's well-established theory of the evolution of the species (e.g., Buss, 1991, 1997, 1999, 2001; Cosmides & Tooby, 1989). In this view, the important personality differences between people reflect the process of natural selection through which change takes place in all organic forms over the course of evolution. The processes of adaptation and selection help to

344

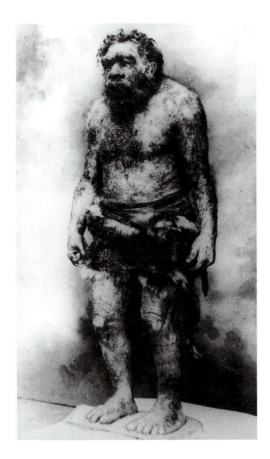

What personality attributes allowed our
ancestors to survive and reproduce?

explain the evolution of organisms and to understand *why* they become what they
are (e.g., why kidneys, larynxes, feet, and keen sight develop). These attributes are
in us now because, in the evolutionary view, they allowed our ancestors to survive
and reproduce.

The focus on reproduction and on passing forward one's own genes, rather than
those of genetic competitors, has many implications for social life and personality. It
suggests, for example, that such traits as dominance, conscientiousness, emotional stability, and sociability would have an especially significant role because they facilitate
reproduction and survival (e.g., Kenrick, Sadalla, Groth, & Trost, 1990). Through the
evolutionary process, people also should develop mechanisms that predispose them to
be particularly fine-tuned to potential mates' indicators of health and various types of
abilities, because they impact on their survival and reproductive success (McCrae &
Costa, 1989, 1997; Wiggins, 1979, 1997). These mechanisms become stable and robust
because of their evolutionary value (Buss, 1997).

In the evolutionary view, it is not surprising if, as language developed in the relatively recent evolutionary past, many personality trait terms have become established
that describe the variations of the most important features of the social terrain—the
landscape—to which human beings have to adapt (Buss, 1989). In fact, the lexicon
contains thousands of trait terms that describe these features, summarized in taxonomies like the Big Five. People use these terms to try to answer questions about the
status of others in the social hierarchy, their value for providing or sharing needed
resources, their potential harmfulness or helpfulness, and so on.

15.2 What are the
roles of directional
selection and adaptation
in evolutionary theory?

These functions of trait terms make it understandable that the dimensions captured in trait taxonomies like the Big Five are both highly evaluative and interpersonal. They seem intuitively to summarize the types of judgments people tend to make as they try to know others within their "adaptive landscape." Looked at from this perspective, these dimensions provide a map of potential resources. They identify, for example, who to go to for advice and help (those with high intellect, the open-minded), or who to avoid (the unstable, the unconscientious, the hostile), or who to choose as a mate or partner.

In evolutionary theory, the focus is not on the survival of the individual but on the fate of the gene pool that is distributed in groups across a population. The groups that survive, thrive and rapidly reproduce will pass their genes forward into future generations. The characteristics that are passed on include many variations on which individuals differ—from height and weight through diverse abilities and social-personal traits. Over the course of many generations, the variations in characteristics that are transmitted are influenced by **directional selection.** Through directional selection, versions of the characteristics that enhance survival and reproduction gradually become increasingly represented, while those that handicap survival and reproduction tend to fade out. Such selection has a directional effect, leading ultimately to more adaptive qualities in future generations.

If that were the whole story, in time, individual differences would disappear and everyone in the surviving group would have the ideal qualities for survival. The script becomes complicated by the fact that the characteristics that may be highly adaptive in some contexts and environments may be dysfunctional and even catastrophic in others—and the species has to deal with all kinds of changing environments and challenges. Hence the maintenance of genetic variability and intermediate values of certain characteristics is often adaptive for functioning under ever-changing life conditions. When it is adaptive to have characteristics that are not at the extreme ends of dimensions, genetic variability may be maintained through a mechanism of **stabilizing selection** (Plomin, 1981).

Implications of Evolution for Personality

15.3 How does the major goal of life explain such behaviors as male jealousy, standards for mate selection, and altruism?

Evolutionary theory has potentially great implications for personality (e.g., MacDonald, 1998). It provides provocative but often speculative insights and reinterpretations of everyday phenomena like human courtship and aggression among males. Consistently, its contributions are guided by the view that passing your genes or your group's genes—rather than those of your genetic competitor—forward to the next generation is much of what life is about.

Mate Selection. Evolutionary theory has a lot to say particularly about mate selection and the competition for mates in humans. The theory predicts that in mate selection there will be important sex differences in the desired characteristics. To the degree that some genes predispose males to search for fertile partners, those genes are more likely to be passed on over the generations. If cues for fertility are youth and beauty, males may evolve to seek such features. In contrast, genes will be passed forward if they predispose females to seek partners who will have the resources, status, dominance, and power that provide for and protect their offspring.

Sexual Jealousy. Consider next the evolution of male sexual jealousy. From this perspective, male sexual jealousy evolved throughout the evolutionary past of the species because those who had such feelings and behaviors prevented competing males from sexual contact with their mates. This mechanism prevented their competitors from interfering with their reproductive success (Buss, 1997) and so it survived.

Likewise, the competition for mating and reproduction among males also may play a role in the violence and killing not infrequently seen among young males, and only very rarely among females. Male aggression is often provoked by disputes about respect, status, and dominance, and it may not be coincidental that the age at which the killing is highest is also the age at which the mating is highest (Wilson & Daly, 1996).

Explanations Are Not Justifications.   At a theoretical level, these explanations may help one understand all sorts of behavior but they do not justify it. Critics of the approach therefore are quick to note that today's abusive, violent, criminal behaviors, or hurtful macho attitudes and biases, cannot be excused by pointing to their possible origins in evolutionary struggles at the dawn of human life. Attitudes and behaviors that may have served prehistoric ancestors can be inappropriate and dysfunctional for many aspects of modern social life. Fortunately, human beings also have the capacity for self-regulation that enables them to modify their behavior adaptively without waiting for the species to evolve further over the eons of time (Bandura, 2001).

With these hedges in mind, the evolutionary approach nevertheless offers insights into social behavior today. Look through the lenses of evolutionary theory at human courtship on a weekend evening at a crowded urban singles bar. The males try to strut and impress with talk and demonstrations about their ambitions, virility, accomplishments, and possessions. In the same scene, the females display their distinctively different strategies for enhancing their appeal. And the motivations that drive the sexes in the courtship dance may also be quite different. The male is eager to mate and reproduce while the female is more motivated to delay reproduction until the resources and opportunities for long-term offspring protection and care are favorable.

Altruism.   Evolutionary theory also speaks to human qualities that at first glance do not seem consistent with evolution, most notably altruism. Why should humans behave altruistically, taking risks that can kill them, as firemen and police officers do daily? Over the course of evolution, shouldn't such risk-taking and unselfishness fade out since it does nothing to increase the chance of passing your genes forward? The answer is again put in terms of the survival of the group, and not of the individual. To the degree that altruistic acts toward members of one's group facilitate their survival and reproduction, the genes from your gene pool will continue to be passed forward, even if you don't survive as an individual. Consistent with the theory, the most altruistic behaviors are often confined to one's own kin and kind, reflected in the phrase "charity begins at home." But the theory even sees an evolutionary basis for altruism toward strangers as a tendency acquired over the course of time as our ancestors discovered **reciprocal altruism.** This is the recognition that if we help others they are likely to reciprocate in kind, in turn, enhancing the potential for survival and reproduction (e.g., Trivers, 1971).

Evolutionary Theory and Inborn Constraints on Learning

Because evolution seems to prepare organisms to learn some associations much more easily than others, evolutionary theory also has implications that need to be considered when thinking about the relations between learning and personality. As you saw in work at the behavioral-conditioning level (Part IV), for many years, psychologists

15.4 How might innate factors affect organisms' learning capabilities? Why might humans differ from other animals in preparedness?

searched for general laws of learning which they assumed would hold for all species and for all types of responses and stimuli. This assumption has been seriously challenged (Seligman, 1971) and the evidence against it has mounted (Marks, 1987; Marks & Nesse, 1994; Pinker, 1997). By now it will not surprise you that evolutionary theory proposes that the differences found between species in the types of associations they readily learn reflect differences in what it was necessary to learn in the evolutionary struggle for survival (Buss, 1996, 1997; Seligman, 1971; Seligman & Hager, 1972).

Biological Preparedness. Let's look at evidence that, when organisms learn, not all associations between stimuli are formed with equal ease. Rats fed a distinctively flavored new food and then made sick to their stomachs will avoid that food even if their illness does not occur until 12 hours later. If the same food is followed by electric shock, however, the rat does not learn to avoid eating it, even when the shock is delivered immediately (Garcia, McGowan, & Green, 1972). A rat can be taught in one trial to avoid shock from a grid floor if it can escape to a compartment with a smooth black floor. It takes the rat close to 10 trials to learn to avoid the shock when the escape compartment has a grid floor continuous with that of a shock compartment (Testa, 1974). Little Albert was conditioned by Watson and Rayner to fear rats by pairing the appearance of the animal with a loud noise, as discussed in Part IV. But a student of Watson's was unable to produce fears when either a block of wood or a dark curtain was paired with the noise. These variations in the ease of associating different types of events suggest selective **biological preparedness.**

Evolution of Fears. Evolutionary theory also argues that people are disposed biologically to fear things that have threatened human survival throughout evolution (Buss, 1997; Seligman, 1971). For example, there are only a small number of common human fears, and they seem to be virtually universal. Fears of snakes, spiders, blood, storms, heights, darkness, and of strangers are typical, and they share a common theme: they endangered our evolutionary ancestors, and now we seem to be prewired to fear them. As Pinker (1997, p. 387) puts it: "Children are nervous about rats, and rats are nervous about bright rooms, before any conditioning begins, and they easily associate them with danger." These findings are only small examples of a wide range of data that in recent years increasingly point to prewired dispositions in the brain that arose in the course of evolution. And these dispositions seem to make humans distinctively prepared not only for some fears rather than for others, but also for all sorts of high-level mental activities, from language acquisition to mathematical skills to music appreciation to space perception (Pinker, 1997).

On the other hand, evidence that learning in animals, especially lower animals, is constrained by their biological capacities does not necessarily mean that humans are innately programmed in favor of specific associations. Such preprogramming in humans might be very disadvantageous in evolution because of our need to adapt to more complex and rapidly changing circumstances—circumstances that we ourselves often create. The human ability to symbolize experience allows enormous potential for learning. The extraordinary variety of human behaviors is well documented both by formal evidence and by casual observation. While chickens and pigs may have serious biological learning constraints, people are generally not so prewired, according to the counterargument (Bandura, 1986, 2001).

Specificity of Psychological Mechanisms

Some of the most important challenges that required adaptive solutions and that created problems for human survival came not from the physical challenges of nature but from other hostile humans. As Buss (1997) points out, this recognition comes from studies on topics that range from group warfare to the evolution of language and higher mental functions (e.g., Pinker, 1997).

15.5 What is domain specificity? How is it explained by evolutionary theory?

To illustrate, consider the formation of dyadic alliances, such as lifelong mating relationships. These relationships pose problems that demand adaptive social solutions for reproduction and survival. To enable this, we need to accurately assess the resources and characteristics of potential friends and enemies as they initiate efforts to develop friendships and connections to enhance their adaptation and survival. Relevant to this function, it has been suggested that humans have developed a **cheater detector**—a mechanism to detect cheaters who will seek the benefits of social exchange but refuse to reciprocate appropriately (Cosmides, 1989; Cosmides & Tooby, 1989).

The cheater detector mechanism also illustrates another key point of current evolutionary thinking: the concept of **domain specificity.** Because humans face a host of different types of social problems, each of which requires somewhat different strategies and solutions, evolutionary theory proposes that the mechanisms that emerge will be highly domain specific. Thus the psychological mechanisms that evolve were targeted to solve quite specific evolutionary problems like mate choice and mate retention (Buss, 1997). We saw that specificity in mate choice takes a somewhat different form for males and females. In turn, the psychological problem-solving mechanisms in mate choice are different from those needed in other social exchanges, for example, when dealing with potential enemies. Specificity appears to be a necessary requirement for achieving the enormous discriminativeness and flexibility in behavior that has been observed in most domains. In fact, specificity appears to be the rule, as seen in domains that range from the diversity of human motives to the specificity of learning, to the specificity of fears (Buss, 1997), to most patterns of interpersonal behavior (Cantor & Kihlstrom, 1987; Mischel, 1968, 1973; Mischel & Shoda, 1998).

Addressing the question of domain specificity, David Buss puts it this way:

> A carpenter's flexibility comes not from having a single domain-general 'all purpose tool' that is used to cut, saw, screw, twist, wrench, plane, balance, and hammer, but rather from having many, more specialized tools each designed to perform a particular function (Buss, 1997, p. 325).

Evolutionary theory emphasizes that it is the "specificity, complexity, and numerosity of evolved psychological mechanisms" (Buss, 1999, p. 41) that enable the behavioral flexibility crucial for adaptation. The availability of many specific tools, not any single highly elastic tool, gives the carpenter—and the human being—the flexibility required for the diverse challenges encountered.

Taking the analogy of the carpenter and his diverse highly specialized tool kit a bit further, different carpenters are particularly skilled with some tools and less skilled with others and each has tools that vary in quality. Similarly, people differ in the domains in which they are likely to be especially vulnerable and those in which they excel: the pattern of each person is unique and domain-specific in its behavioral expressions. These patterns are the stable *if . . . then . . .* , behavioral signatures of personality discussed in Chapter 4.

The Value of Discriminativeness in Coping with Stress

15.6 How do coping flexibility and discriminative facility relate to successful adaptation?

To illustrate domain specificity and the discriminative behavior and flexibility it demands, let's look at how people cope with stress effectively. An individual's ability to cope with daily stressors in the psychological and physical environment is fundamental to survival. Psychologists have distinguished two methods of coping with physical and psychological stress: problem-focused and emotion-focused coping (Chan, 1994; Lazarus & Folkman, 1984; Parker & Endler, 1996). The method that works best depends on the nature of the situation. Consequently, adaptive behavior ideally requires people to be flexible in the methods they use, and to discriminate among different types of situations so that they apply the method that best fits the particular context. For example, adaptive coping has to take into account whether or not the situation is controllable (Aldwin, 1994; Miller 1992). If a situation is controllable, then it is more adaptive to respond in a problem-focused manner. In contrast, in an uncontrollable situation, problem-focused coping is ineffectual, and emotion-focused coping is more adaptive.

One of the important factors that influence an individual's ability to successfully cope with stressful situations is **discriminative facility** (Chiu, Hong, Mischel, & Shoda, 1995; Cheng, Hui, & Lam, 2000). Discriminative facility is the ability to appraise situations as they present themselves and to respond appropriately. Cecilia Cheng and her associates found that individuals who had more discriminative facility used more flexible coping strategies in an experimental setting, and they also showed decreased levels of anxiety (Cheng et al., 2000; Cheng, Chiu, Hong, & Cheung, 2001, Study 1; Roussi, Miller, & Shoda, 2000). In short, to cope well, one needs to discriminate carefully and identify the type of situation faced so that one can then use the problem-solving approach that fits it best. Like the carpenter, in addition to having the right specific tools available, people need to select the tool that fits the task.

In short, evolutionary theory emphasizes the adaptive value of flexibly allowing one's behavior to take into account the requirements of the particular context, and carefully discriminate among situations so that one can identify the specific requirements and act accordingly. Knowing when to fight, when to flee, who to trust and rely upon, and who to fear are of long-term value for survival and reproduction (MacDonald, 1998). To illustrate further, in that view, people who are highly prone to intimacy may under some conditions also be highly prone to aggression and hostility. Consequently their relationships become 'compartmentalized.' They may be highly intimate with family and friends, while hostile, aggressive, and exploitative with out-group members or competitors.

▶ BRAIN–PERSONALITY LINKS

Over millions of years, our species has accumulated genes that have proven useful for individual survival and reproductive success, passing them from generation to generation. The result is the set of what is currently estimated to be about 30,000 genes that make up the human genome. Some of these genes are so important to an individual's basic functions that their DNA sequence is believed to be identical for all individuals. Any deviations from the optimal sequence perfected through evolution are likely to reduce adaptability. For other genes, there are variations. For example, there are variations in the genes that influence eye color, and individuals differ in the particular variations of those genes they have. Heritable differences in personality of the kind described in Chapter 14 reflect such variations.

AN EARLY EFFORT: PHYSIQUE AND PERSONALITY?

Attempts to connect the physical-biological aspects of individuals to their personality are hardly new. Long before genetics was a word, people noticed the dramatic differences in the kinds of bodies that different personalities seem to live in, and asked questions about the possible connections between physical make up and personality. In ancient Greece, Hippocrates described four types of personality. In turn, Galen's theory linked each type to the amounts of different types of bodily humors present, with excessive amounts of different humors interpreted as the biological factors underlying the different personality types.

In early versions of personality work at the biological level, formal classifications of the possible connections between mental disorders and body type were developed by the German psychiatrist Kretschmer a century ago. In 1942, an American physician, William H. Sheldon, suggested three dimensions of physique and their corresponding temperaments. These are summarized in Figure 15.1 and for many years received much attention. As

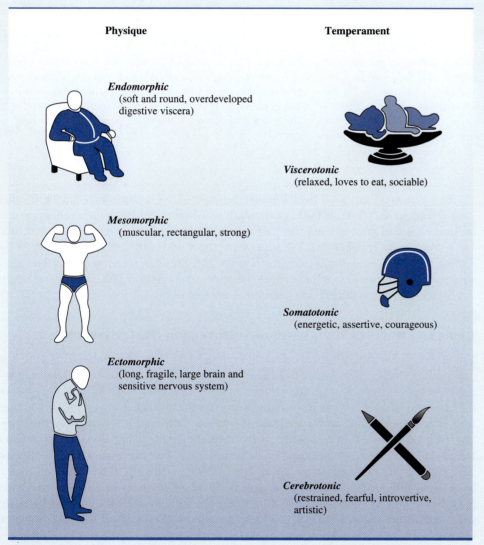

Physique

Endomorphic
(soft and round, overdeveloped digestive viscera)

Mesomorphic
(muscular, rectangular, strong)

Ectomorphic
(long, fragile, large brain and sensitive nervous system)

Temperament

Viscerotonic
(relaxed, loves to eat, sociable)

Somatotonic
(energetic, assertive, courageous)

Cerebrotonic
(restrained, fearful, introvertive, artistic)

Figure 15.1 Sheldon's physique dimensions and their associated temperaments.

Figure 15.1 suggests, according to Sheldon the **endomorph** is obese, the **mesomorph** has an athletic build, and the **ectomorph** is tall, thin, and stoop-shouldered. Rather than dividing people into three distinct types, Sheldon considered every individual's status on each dimension. He developed a seven-point rating system for measuring body types. For example, a 7-3-1 would be high on endomorphy, moderate on mesomorphy, and low on ectomorphy, presumably with corresponding levels of the associated temperaments. Sheldon's typology thus was quite sophisticated, especially by comparison with earlier attempts.

Sheldon's ideas about the association between body build and temperament get some support when untrained people rate the personality characteristics of others. In part these findings may reflect the fact that stereotyped ideas about the characteristics of fat, athletic, and skinny people are shared by the raters. For example, if raters think most fat people are jolly and thin people are sensitive, they may base their judgments of the individuals they rate on these stereotypes rather than on observed behavior. Thus they may rate a fat person as jolly, no matter how he or she behaves. Studies of behavior that avoid such stereotypes generally provide less evidence for the value

of this system (Tyler, 1956) and leave Sheldon's theory largely unsupported (Herman, 1992).

In addition to such effects from the stereotypes of others, people perceive and interpret their own body types. The thoughts and feelings we develop about our bodies may influence our behavior and who we become, increasing the connections between body types and personality. Thus, physical appearance and physical characteristics certainly affect the ways in which others perceive us, and ultimately even what we feel and experience about ourselves.

Physical characteristics such as strength, height, and muscularity also affect what we can and cannot do easily and thus also influence the situations we select, the work and avocations we pursue, and the interests and values that develop. In short, many different indirect causes may underlie the relations between physique and particular personality characteristics. The links between the two turned out to be much more complex and indirect than the early typologies suggested, and they are difficult to unscramble. Consequently, interest in these connections has gone down, particularly as better routes for exploring the relations between biology and personality have been opened.

15.7 What links did Sheldon propose between physique and personality? What factors could affect such relations?

But genes do not directly influence behavior. They influence behavior through their effects in shaping our body, particularly our brain, and the large number of specialized chemical components that regulate their functioning, such as enzymes, hormones, and neurotransmitters. We experience the world with distinctive brains. In the rest of this chapter, we will explore theories and findings about how our biological structures and processes influence our thoughts, feelings, and behavior, giving rise to the distinctive personalities that characterize each of us.

Biological Bases of Extraversion–Introversion (H. J. Eysenck)

15.8 According to Eysenck, what biological differences underlie extraversion and introversion?

You already learned about the work at the trait level by Hans Eysenck (1916–1997), an outstanding English personality theorist (Chapter 3). Eysenck also was a pioneer in trying to connect psychological dispositions to their biological foundations, focusing on the characteristics of extraverts versus introverts. Recall that extraverts are characterized as active, outgoing, and venturesome, whereas introverts are defined as the opposite: withdrawn, quiet, introspective, with a preference for being alone or with a good friend or two, but an aversion for large groups. According to Eysenck's theory (Eysenck, 1990; Eysenck & Eysenck, 1985, 1995), extraverts differ from introverts because of differences in their physiological **level of arousal** or **LOA** in the brain. Specifically, he proposed that these differences are influenced by the **ascending reticular activation system (ARAS)** of the brain, the system believed to regulate overall arousal in the cortex.

When Eysenck (1967) first developed his theory in the 1960s, the ARAS was viewed as a gateway that controls the level of stimulation into the cortex. In Eysenck's theory, introverts need only small amounts of stimulation to overstimulate them physiologically, which then leads them to become distressed and withdrawn in their behavior. In extraverts, in contrast, according to the theory, the ARAS is not easily stimulated, which leads them to seek activities that will increase the level of stimulation, for example, by socializing more actively and seeking activities—such as parties and adventures—more than introverts.

To test this theory, Eysenck and colleagues conducted studies with extraverts and introverts and measured their brain wave activity and cardiovascular activity. Researchers found, for example, that introverts (compared to extraverts) show greater changes in their brain wave activity in response to low frequency tones, indicating their lower threshold for stimulation to the CNS (central nervous system), as the theory predicted (Stelmack & Michaud-Achorn, 1985). Overall, a good deal of research indicates that while extraverts and introverts don't differ in their level of brain activity at resting levels, for example while sleeping, they do differ as predicted by the theory in their physiological reactivity to stimulation. Many different types of studies suggest this conclusion (e.g., Eysenck, 1983).

15.9 How did Eysenck test his theory? What physiological differences exist between introverts and extraverts?

The differences between extraverts and introverts in their reactivity to stimulation also should influence their performance under different arousal levels because different activities require different arousal levels to do them well. In earlier work, Hebb (1955) defined the **optimal level of arousal** or **OLA** as the arousal level that is most appropriate for doing a given task effectively, and showed that too much or too little arousal undermines performance. This is intuitively familiar: as you know from your own experience, prepping for an important examination, for example, requires a higher level of arousal and alertness than you need when trying to relax or to go to sleep. The point is that to be able to perform effectively requires a good match between one's arousal level and the optimal arousal level needed for the particular task. Consequently, the differences between extraverts and introverts in their ease of arousal should have important implications for how well they do on many life tasks: they should do well on tasks that require arousal levels that match their own, but not on those where there is a major mismatch.

An extravert faced with monotonous work tasks such as closely monitoring control panels that remain generally stable, or proofreading, is likely to soon become bored and have trouble maintaining attention, becoming underaroused, and apt to tune out or even fall asleep. In this situation, an introvert will probably remain attentive and perform more effectively. Suppose, however, the job is that of a fireman or emergency room worker, and all the alarms and sirens suddenly are ringing, the emergency lights are flashing, the scene is chaotic. Now the situation is reversed in terms of the amount of arousal activated, and the introvert who is easily overaroused has the disadvantage, prone to becoming distressed and ineffective, whereas the extravert may be in his ideal element and functioning at his best.

This point was made in a classic, well-controlled experiment by Geen (1984) whose results supported the expectations from Eysenck's theory. Geen allowed extraverts and introverts to choose their preferred noise level of background stimulation while working on a potentially boring but difficult task (learning paired associations between words). As expected, extraverts chose higher levels of stimulation than did introverts. Under these conditions, according to the theory, both groups will be working under their preferred level of arousal and therefore should perform equally well—and they did. It also follows that when extraverts are given the lower levels of arousal preferred by introverts during the task, they should become underaroused and bored, and therefore do less well. Likewise, when introverts are given the higher arousal levels preferred by introverts, they should become overaroused and upset, and their performance should deteriorate. To test these expectations, Geen in fact reversed the preferred stimulation conditions on the work task, giving the introverts the noise levels chosen by the extraverts and vice versa. Again the results were as predicted: performance was poorer when extraverts worked under the lower stimulation conditions preferred by the introverts, and vice versa.

The Behavioral Inhibition and Activation Systems

15.10 Describe the BIS and BAS, as well as their functions. How do they relate to introversion and extraversion?

Two neurological systems in the brain seem to be closely related to the distinction between the extraversion–introversion personality types. One is the **behavioral inhibition system (BIS),** which causes individuals to withdraw from certain undesirable stimuli or punishments and thus inhibits behavior. You can think of it as a withdrawal–avoidance system that allows one to pause and then contemplate alternatives before taking action. Shakespeare's Hamlet probably would have scored high on measures of the strength of this system. The other is the **behavioral activation system (BAS),** which directs individuals toward certain desirable goals or incentives and thus activates approach behavior. You can think of it as an energizing system that activates and facilitates rapid action.

According to some researchers, the BIS is the more influential, active system in introverts, and the BAS is believed to be the more active system in extraverts (Gray, 1991). The theory suggests, for example, that because BAS is the more active system in extraverts, they tend to focus on rewards and seek situations in which they could find them to actively pursue their desires. This is in contrast to the more passive, avoidant behavior of introverts who are, in the theory, physiologically more sensitized to punishment and threats (e.g., Bartussek, Diedrich, Naumann, & Collet, 1993). Gray's theory and related work also indicates that the BIS is linked to anxiety as a personality dimension, whereas the BAS is related to individual differences in impulsivity or the inability to inhibit and control one's immediate responses and urges.

Although there still is considerable disagreement among theorists about the nature of the exact mechanisms and links to personality (e.g., Cloninger, 1988; Davidson, 1995; Depue & Collins, 1999) several tentative conclusions seem to be emerging from a great deal of work at this level of analysis. Individuals who are extremely reactive in the BIS tend to be especially sensitive to potential threat, punishment, and danger. They therefore may focus more on the potential negative outcomes rather than on the possible gains and rewards when considering alternative action possibilities. For example, they are likely to attend more to how much money they could lose from a business decision than to how much they might gain. They are more prone to anxiety and in the extreme even to panic, more tuned to negative outcomes and punishment, less sensitive to possible rewards, incentives, and gratifications. On the other hand—or more literally, on the other side of the brain—people who are highly reactive in the BAS are likely to show the opposite pattern: they are more responsive to incentives and possible positive outcomes, more ready to experience positive emotions such as eagerness, hope, and excitement. These differences in the nature and meaning of BIS and BAS reactivity are becoming especially clear from studies on brain asymmetry, considered next.

Brain Asymmetry and Emotional Reactivity

In recent years, the implications of the brain's activity for personality also are being illuminated by investigations of the electrical activity in the frontal lobe areas of the cortex that are involved in such positive emotions as eagerness and hope (Davidson & Sutton, 1995), or in negative emotions such as disgust and fear (Davidson et al., 1990). This work begins with the finding that people differ consistently in the degree to which the right versus the left sides of their brains are activated, and that difference is called **brain asymmetry.** In these studies, the brain's electrical activity is assessed by the electroencephalograph (EEG). (This measure was more fully described in Chapter 2.)

To calculate differences in the brain's asymmetry, researchers subtract the level of EEG brain wave activity on the left side of the brain from the level of activity on the right side of the brain. A positive asymmetry measure indicates a relatively higher degree of activity on the right side of the brain; a negative measure indicates a relatively higher degree of activity on the left side of the brain (Sutton, 2002, p. 136). In these studies, stable individual differences have been found repeatedly in the brain's asymmetry in anterior (frontal) brain regions (e.g., Sutton & Davidson, 2000).

You can get a sense of these individual differences and their implications for personality from some experiments in which participants were exposed to videotapes depicting funny or aversive, fear-inducing, and disgusting themes. While they were watching these videos, their EEGs were recorded, their facial and behavioral reactions were filmed, and self-reports of their emotional reactions also were obtained. Individuals who had higher resting levels of activity in their left frontal area of the brain gave more positive responses to happy stimuli (Wheeler, Davidson, & Tomarken, 1993). They also showed higher levels of self-reported BAS sensitivity on questionnaires (Harmon-Jones & Allen, 1997; Sutton & Davidson, 1997). In contrast, higher resting levels of activity in the right frontal area of the brain were indicative of more negative feelings in response to aversive stimuli (Davidson and associates, 1990), as well as higher levels of self-reported BIS sensitivity on the self-report measures of reactivity to threats.

Evidence for the stability of these individual differences in affective styles also comes from studies of very young children. For example, 10-month-old children were exposed to a brief separation from the mother while she left the playroom in an experimental situation. While some infants became greatly distressed, others remained much calmer. Those who cried and were more upset also tended to be the ones who characteristically have more brain wave activity on the right than on the left side, even when they were measured during the resting stage. Thus individual differences in this type of emotional reactivity seem to be a stable quality already seen in the young child (Fox, Bell, & Jones, 1992). Likewise in adults, brain asymmetry is sufficiently stable over time to suggest an enduring disposition that relates to how one deals with different kinds of emotion-inducing events and experiences (Davidson, 1993).

The overall implication from this line of research for personality is clear. Individuals who are dominantly left-side active in their brains experience pleasant, positive emotions more easily when positive stimuli are encountered. People who are dominantly right-side active are more ready to respond with negative emotional reactions and distress when they face unpleasant or threatening events. Some people are more ready to feel and be happy than others; some people are more ready to feel and be distressed and to experience negative emotions than others.

Summary and Implications

In sum, findings on the BAS and BIS and on brain asymmetry suggest that these are specific and distinct incentive and threat systems in the brain that are responsive, respectively, to environmental signals of reward and threat, and that are related, respectively, to positive and negative mood (e.g., Gable, Reis, & Elliot, 2000). This is

15.11 How are hemispheric activation differences related to types of emotions?

important for understanding individual differences in personality because people differ in stable ways in their levels of reactivity in these two brain systems. These differences link to their behavioral dispositions with regard to active positive approach tendencies versus fearful behavioral inhibition and avoidance tendencies. These differences are reflected on measures of the relevant personality dispositions (extraversion, impulsivity for the BAS; inhibition, withdrawal, anxiety, or neuroticism for the BIS). Studies of brain asymmetry likewise suggest stable individual differences in the level of reactivity in different brain areas that are directly correlated with individual differences in positive and negative emotional and behavioral reactions to different types of stimuli. In short, there are clear and meaningful specific links between peoples' characteristic brain activities and their characteristic emotional and behavioral reaction patterns to different types of positive and negative stimulation.

15.12   How is the action of biological systems influenced by situational factors?

It is also noteworthy that brain systems such as BIS and BAS are linked to personality in context-specific ways that again reflect the importance of taking account of the relationship between the expressions of personality and the characteristics of the stimulus situation. You see this in the finding that high BIS reactivity relates to more sensitivity and attention to possible punishments, less to possible rewards, and is essentially diagnosed by the difference between the two. The distinctive dispositional pattern of people high in BIS activity takes the form: *if* threats-punishments are encountered, *then* inhibition, high anxiety-distress, negative emotions. The opposite *if . . . then . . .* pattern seems to characterize those high in BAS. For them, *if* rewards, incentives, *then* approach, eagerness, positive emotions. While the two types also link to broad dispositions like extraversion, their behavioral expressions are especially evident when the relevant "ifs" are present. Once again, we find that the expressions of personality are seen in person-situation interactions as well as in overall behavioral tendencies.

New insights about personality dispositions are coming from studies that probe the relationships that do—and that don't—exist between individual differences with regard to various types of relatively broad trait categories like neuroticism on the one hand, and biological processes on the other. For example, Heller and colleagues (2002) find that anxious emotional arousal and worry may reflect two separate neural networks in the brain. Specifically, these networks are in the left versus right hemispheres, and in the frontal versus posterior regions, as measured by EEG and blood flow indices of cortical activity. Their research suggests that psychological conditions that have long been grouped under the term "neuroticism" may actually consist of two distinct mechanisms that need to be understood separately. These two systems thus may involve different mechanisms that lead to different experiences and life problems, and that require different treatment.

What are the implications? New discoveries from multiple levels of analysis, in this case, the biological level utilizing modern brain assessment techniques, are allowing a greater level of precision in the analysis of broad dispositions like neuroticism. The ability to distinguish the different phenomena covered by a broad term like neuroticism makes it possible to examine their different causes. Then one also can develop distinctive therapies targeted to treat these different disorders more effectively.

Sensation Seeking: A Trait with a Specific Biological Basis?

15.13   How do high and low scorers on the Sensation Seeking Scale differ behaviorally?

Why do some people go parachuting and drive fast, whereas others prefer TV and rarely exceed the speed limit? According to one researcher, Marvin Zuckerman, the answer may lie in a trait called **sensation seeking.** Of particular interest is that this trait also seems to have clear connections with findings from the biological level that point both to the role of brain chemicals and of genetics as determinants. First, let us consider the nature of individual differences in this characteristic. These differences are measured with a **Sensation Seeking Scale (SSS)** which taps into four different aspects: Thrill and Adventure Seeking (engaging in risky sports and fast driving, for example), Experience Seeking (seeking novelty), Disinhibition (seeking sensation through social stimulation and activities), and Boredom Susceptibility (lack of tolerance for repetitive events).

This scale has helped to predict how people cope with situations in which there is a lack of stimulation: in this context, the sensation seekers get more restless and upset. It also predicts reactions to the opposite experience, as when one is placed in close confinement with another person. In that situation, those low in sensation seeking become more stressed, both in their own reports of the experience and on biochemical measures (e.g., Zuckerman, and associates 1968). Sensation seeking now has been related to a wide range of diverse behaviors (Zuckerman, 1979, 1983, 1984, 1994), as the examples in Table 15.1 summarize.

How does this personality characteristic connect to the biological level? As with extraversion, individual differences in sensation seeking also may arise in part from physiological differences in physical arousal, such as cortical activity in reaction to novel and familiar stimuli (Zuckerman, 1990). Hebb's (1955) ideas about optimal level of arousal (OLA) again are relevant here just as they were in our discussion of extraverts and introverts and their very different reactions to different types of situations that require different arousal level for optimal performance. Consistent with Hebb, if individuals differ in their preferred optimal level of arousal, then those with a low optimal level of arousal will attempt to maintain a low level of arousal which requires them to sometimes work toward reducing the stimulation in their environment. Those with a high OLA—the sensation seekers—will work toward increasing the stimulation they get in order to achieve and maintain their optimal level of arousal, and therefore they keep looking for change and searching for novel and complex sensations and experiences. Individuals who are high versus low on Zuckerman's measures of sensation seeking tend to choose occupations and display life patterns that are consistent with these expectations. High compared to low sensation seekers are more likely to be skydivers,

15.14 What is the hypothesized biological basis for differences in sensation seeking?

TABLE 15.1   Characteristics of High Sensation Seekers

Risk-Taking Behaviors:
  More varied sexual experience
  Greater use and variety of illegal drugs
  Risky driving habits
  More risky sports, take risks within sports

Intellectual Preferences:
  Prefer complexity
  Have a high tolerance for ambiguity
  Are more original and creative
  Rich imagery and dreams

Interests and Attitudes:
  Liberal, permissive, and nonconforming attitudes and choice
  Prefer high stimulation vocations (e.g., aircraft controllers; high-risk security officers, war-zone
    journalists, emergency room doctors)
  May view love more as a game and show less commitment to their relationships

or police officers who prefer riot duty, to take greater risks in gambling situations, to self-report having more diverse sexual experiences and partners (e.g., Zuckerman, 1978, 1984, 1991)—in other words, to behave in ways consistent with their needs for higher arousal states to obtain their "optimal" levels as Table 15.1 indicated.

Sensation seeking also may be linked to impulsiveness. Impulsiveness is a quality that in extreme and unsocialized form characterizes people who are considered antisocial personalities, sometimes also called sociopaths or psychopaths. These are people who frequently engage in criminal and antisocial behavior, seem unable to control their own impulses or to show the type of conscientiousness and adherence to conventions generally expected in social interactions and relationships within the society and culture. Zuckerman (1993) suggests that these may be unsocialized, impulsive sensation seekers—individuals who have difficulty inhibiting their impulses for the sake of social adaptation. They seek sensation even when the long-term costs to them and society are high. Unsurprisingly, such individuals also tend to be low on measures of conscientiousness, and high on measures of aggression.

To find the biological bases for individual differences in sensation seeking, Zuckerman has paid particular attention to the role of the chemicals in the nerve cells of the brain that are crucial for transmitting nerve impulses from cell to cell across the slight space—the synapses—that separates them. The **neurotransmitters** are the chemicals that enable the nerve impulses to jump across the synapses. When their levels are appropriate, the transmission across cells continues, allowing signals to get to their destinations. As with most processes in nature, there is a system of checks and balances, of activation and inhibition. The proper level of neurotransmitters is maintained by **monoamine oxidase (MAO)**, enzymes that break down the neurotransmitters after they have passed along the route. The theory here is that high sensation seekers are the individuals who are low in their MAO levels, and therefore lack the chemical inhibitors required to keep their neurotransmitter levels down.

15.15 How might testosterone interact with socialization factors to influence antisocial personality development?

IN FOCUS 15.2

TESTOSTERONE AND THE ANTISOCIAL PERSONALITY

As an example of the multiple biological factors that may interact and contribute jointly to personality, let us again consider antisocial personalities. You already saw that the MAO inhibitors may play a role here, as does the characteristic of impulsive sensation seeking. Research suggests that hormones, and specifically testosterone levels, may be another biological contributor to such antisocial behaviors. High levels of the sex hormone testosterone relate to, and may possibly promote, such behaviors as assaulting others, going AWOL while on military duty, abusing drugs, having multiple sex partners, or encountering other social problems throughout development and adulthood (Dabbs & Morris, 1990). Also, the higher the levels of this hormone, the lower the level of education attained, and consequently the lower also the ultimate socioeconomic level (Dabbs, 1992). Therefore, the causal chain here may be very complicated, influenced perhaps by a variety of socialization factors such as lower education rather than directly by testosterone.

The point here is like the one Zuckerman (1994) makes in his discussion of the biology of antisocial impulsivity: the interactions between neurotransmitters, enzymes, and hormones that underlie the behavioral and psychological manifestations of personality will probably turn out to be multiple and complex—and these biological factors are likely to also interact with a host of social and psychological factors. Thus, to adequately understand complex patterns of social behavior, whether antisocial or prosocial, one needs to consider determinants from every level of analysis discussed throughout this text.

Some support for these ideas comes from moderate or low correlations found between sensation seeking on the sensation-seeking scale and lower MAO levels in the bloodstream (Zuckerman, 1993). Although the evidence is still limited, this work continues to stimulate exciting developments at the interface of personality and biology that seeks to connect personality characteristics to specific neurotransmitters. Some of these studies are using either pharmacological or psychophysiological techniques to examine the physiological correlates of social behaviors.

The hope of finding such simple and direct links is understandably appealing, but Zuckerman (1994), himself is cautious about the prospects, noting that there are likely to be extremely complex interactions between neurotransmitters, enzymes, and hormones that underlie the behavioral and psychological manifestations of personality (also see In Focus 15.2). Nevertheless, and beyond sensation seeking, there is promising pharmacological work, for example, indicating that the neurotransmitter serotonin and the neurohormone oxytocin may mediate social bonding and affiliation, as well as social dominance and aggression (e.g., Carter, 1998; Taylor et al., 2000). Likewise, there is interesting work suggesting correlations between testosterone levels and antisocial personality (see In Focus 15.2).

Recent research on sensation seeking is itself producing something of a sensation with reports of links to a specific gene. In two studies, associations have been reported between a DNA marker (for a particular neuroreceptor gene) and measures of novelty seeking, an aspect of the broader pattern of sensation seeking (Benjamin et al., 1996; Ebstein et al., 1995). Findings like these await much further research before conclusions can be reached, but they again point to the potentially exciting connections between aspects of behavior and their biological roots which may well be found in the future. They make sensation seeking a particularly promising disposition for linking the biological and the dispositional-psychological levels of analysis.

▶ BIOLOGICAL ASSESSMENT AND CHANGE

Throughout this chapter and the last, you have seen that the interrelationships between biological processes and psychological processes central for personality, such as our emotions and motivations, are becoming increasingly visible. It has become possible to explore the relations between brain, evolution, and personality through the development of assessment methods that now enable researchers at the biological level to investigate these links systematically. In recent years, a revolution has occurred in such measurement that opens new routes for seeing brain–behavior connections.

New Windows on the Brain

Technological advances in brain imaging now allow researchers to use methods that capture even subtle nerve activity in the brain. As noted in Chapter 2, the technique of functional magnetic resonance imaging or fMRI measures the magnetic fields created by the functioning nerve cells in the brain and with the aid of computers depicts these activities as images. These images enable researchers to see the brain areas that are most active as the person performs different kinds of mental tasks and experiences different kinds of perceptions, images, thoughts, and emotions, from fears to anticipated

15.16 Describe fMRI and PET techniques for measuring brain processes.

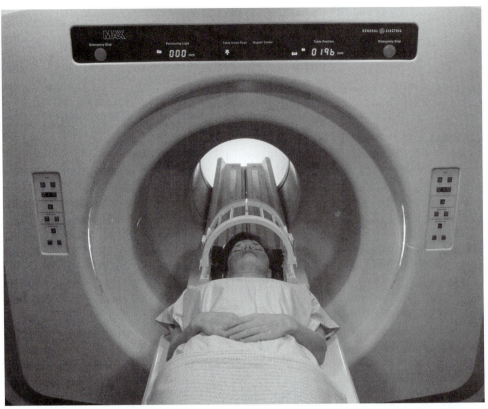

Woman entering an fMRI brain scanner.

gratifications. Brain imaging makes it possible to trace the relations between activity at the biological level and activity that involves different aspects of the person's mental and emotional representations as studied at the social cognitive, phenomenological, and psychodynamic-motivational levels. In short, the new windows on the brain allow a much more precise and detailed analysis of the links between activity in the brain and at the psychological levels reviewed throughout this text.

A second technique to link brain and behavior is the PET procedure or positron emission tomography method, which also was briefly introduced in Chapter 2. This technique creates images and maps of brain functioning by assessing metabolic activity in different areas. It records the radioactivity in the brain that occurs after the participant has been given a nontoxic but radioactively labeled form of glucose, which is the energy source for the brain's activity. These methods are being used by cognitive neuroscientists, increasingly working in collaboration with psychologists from the areas of social-personality and clinical psychology (e.g., Kosslyn et al., 2002; Ochsner & Lieberman, 2001).

Particularly exciting is the recognition that in these studies individual differences in personality are not "noise" that needs to be removed to decipher the mechanisms involved. On the contrary, attention to personality and the differences between individuals in responsiveness on these brain measures allows a much deeper and more complete understanding. For example, in studies of mental imagery using the amount of blood flow measurable in different areas of the brain, it was found that:

. . . monitoring individual differences in blood flow in relevant areas . . . provides enormous power in predicting behavior . . . individual differences not only can be used to establish that a particular type of representation is used during the task . . . but also can help identify the neural underpinnings of such processing (Kosslyn et al., 2002).

Biological Therapies

As we saw, some psychological dispositions have biological roots. If problematic behaviors, such as severe depression, in part reflect biological dispositions and biochemical problems, might they also respond to biological treatment? Researchers have actively pursued this question.

Biological treatments attempt to change an individual's mood or behavior by direct intervention in bodily processes. Pharmacotherapy, or treatment with drugs, has so far proved to be the most promising biological therapy for psychological problems. The types of drugs used for specific purposes are summarized in Table 15.2.

Antidepressants.   The antidepressants, or psychic energizers, are used to elevate the mood of depressed individuals. Two of the largest categories are the cyclics and the monoamine oxidase inhibitors, or MAOIs (Lader, 1980). Fluoxetine (trade name Prozac) is currently one of the most favored and widely used cyclic antidepressants, which increases the chemical neurotransmitter serotonin (Kramer, 1993). Excessively low serotonin levels seem to be related to such feelings as chronic pessimism, rejection sensitivity, and obsessive worry. Antidepressants have been shown to have varying degrees of efficacy and to produce various side effects (e.g., Davis, Klerman, & Schildkraut, 1967; Klein et al., 1980; Levine, 1991). The MAOIs, for example, appear highly effective in certain types of depressed individuals, although their side effects can include a dangerous rise in blood pressure when foods high in tyramine, such as red wine and cheese, are consumed (Howland, 1991; Kayser et al., 1988; Potter, Rudorfer,

15.17 Describe the drugs commonly used for therapeutic purposes.

TABLE 15.2   Some Drugs Used in Pharmacotherapy

Type of Drug	Application	Therapeutic Effects
Antidepressants (cyclics, monoamine oxidase inhibitors)	Depression	Appear effective for elevating mood in some people
Antipsychotics (phenothiazines)	Schizophrenia	Well substantiated: gets patients out of hospitals, have practically eliminated need for restraints in hospitals
Minor tranquilizers (benzodiazepines)	Anxiety, tension, milder forms of depression	Seem to slow down transmission of nerve impulses in the brain
Lithium	Manic behavior	Reduces mood swings
Methadone (a synthetic narcotic)	Heroin addiction	May eliminate craving for heroin; blocks "highs"

Note: These drugs are classified according to their effects on behavior not according to their chemical composition. Chemically dissimilar drugs can produce similar effects.

& Manji, 1991). Lithium, an alkali metal in a drug category of its own, is used to stabilize mood swings and occasionally to treat severe depression (Grilly, 1989).

Antipsychotics. The phenothiazines (most notably chlorpromazine) have proved to be so useful in managing schizophrenic patients that they are referred to as antipsychotic drugs. Their use in mental hospitals has been widespread since the 1950s and has changed the character of many hospitals, eliminating the need for locked wards and straitjackets. Discharged patients are often on maintenance dosages of these drugs and must occasionally return to the hospital for dose level adjustments.

The major tranquilizers have potentially serious side effects that may include motor disturbances, low blood pressure, and jaundice. There are also unpleasant subjective effects such as fatigue, blurred vision, and mouth dryness, which may explain why patients on their own may simply stop taking these drugs and often have to return to the hospital for an extended stay.

Tranquilizers. The barbiturates were the first widely used so-called "minor tranquilizers"—drugs that relieve relatively mild anxiety. However, these drugs were replaced by the benzodiazepines, which proved more effective with fewer side effects (Lader, 1980). For many years, the most widely used of these drugs was a synthetic chemical known by its trade name, Valium. It became the medicine most frequently prescribed in the United States for several years as Americans spent almost half a billion dollars a year on it.

Valium, like its predecessors the barbiturates, acts on the limbic system of the brain and is useful in the treatment of anxiety, panic disorder, and some convulsive disorders (Gitlin, 1990). Although Valium seemed less harmful than the barbiturates, it also proved to have side effects and to be potentially addictive. It can endanger a developing fetus and can have adverse effects of confusion and agitation, especially in the elderly. Now other benzodiazepines such as Xanax, Klonodin, and Ativan are popular alternatives, although their potentially severe side effects also require extremely cautious monitoring. The use of antidepressant medication in treatment of extreme anxiety or panic attacks also may be effective (Gitlin, 1990).

Other Common Drugs. Other widely used drugs include the psychostimulants, such as Ritalin, which are used in treating impulse disorders and severe attention deficits. Methadone, a drug which blocks the craving for heroin and prevents heroin "highs," is often used to overcome heroin addiction, either by weaning the person off it or by maintaining him or her on a fixed dose; however, methadone itself is addictive, and the process can take a long time (Lawson & Cooperrider, 1988).

As this brief survey suggests, some drugs appear to be positive contributors to a treatment program for some disorders. However, no drug by itself constitutes an adequate complete treatment for psychological problems. To the extent that the person's difficulties reflect problems of living, it would be naive to think that drugs can substitute completely for learning and practicing more effective ways to cope with the continuous challenges of life. And the fact that most drugs have negative side effects (Maricle, Kinzie, & Lewinsohn, 1988) makes it all the more important to seek psychological treatment for psychological problems whenever possible, often in conjunction with a medically supervised form of pharmacotherapy.

In spite of the problems encountered by efforts to treat psychological problems chemically, there is much exciting progress, and the new field of neuropharmacology is thriving (e.g., Cooper, Bloom, & Roth, 1986). For example, biological responses in panic disorders are becoming better understood and, in turn, the effects of various chemicals (such as sodium-lactate) on panic states are becoming known (e.g., Hollander et al., 1989). As a result, treatments for a wide range of anxiety and mood disorders are taking into account both biochemical and psychological processes (Barlow, 1988; Klein & Klein, 1989; Simons & Thase, 1992).

SUMMARY

EVOLUTIONARY THEORY AND PERSONALITY

- At the evolutionary level of analysis, personality traits and individual differences reflect the processes of natural selection and adaptation.

- For example, such traits as dominance, emotional stability, and sociability are seen as particularly robust because they have an especially significant role in mate selection and retention.

- The evolutionary perspective provides a fresh view of behaviors from mate selection and jealousy to altruism, aggression, and the development of fears.

- Evolutionary theory emphasizes the adaptive value of flexibly allowing one's behavior to take into account the requirements of the particular context. This requires carefully discriminating among situations.

BRAIN–PERSONALITY LINKS

- William H. Sheldon was an American physician who classified individuals into three different body types and their corresponding temperaments.

- More recently, Hans J. Eysenck measured brain wave and cardiovascular activity in introverts and extraverts to further understand the physiological differences that underlie these types.

- The behavioral inhibition system (BIS) is the neurological system that may dispose individuals to withdraw from negative stimuli. The behavioral activation system (BAS) is the neurological system that may lead individuals to seek out positive stimuli or rewards.

- BAS and BIS research suggests that these are specific and distinct incentive and threat systems in the brain with important links to personality.

- Sensation-seeking behavior has been linked to several biological factors including the level of the MAO enzyme in the brain and the presence of a specific DNA marker.

- The interactions between neurotransmitters, enzymes, and hormones that underlie the behavioral and psychological manifestations of personality seem to be multiple and complex and also interact with a host of social and psychological factors.

BIOLOGICAL ASSESSMENT AND CHANGE

- Personality psychologists utilize magnetic resonance imaging (MRI) and positron emission tomography (PET) scans to examine brain activity during various psychological experiences.

- Biological treatments for psychological disorders favor pharmacotherapy, which makes use of a variety of drugs including antidepressants, antipsychotics, and minor tranquilizers.

KEY TERMS

PART VI

THE BIOLOGICAL LEVEL OF ANALYSIS

▶ OVERVIEW: FOCUS, CONCEPTS, METHODS

As summarized in the Overview table, work at the biological level in the last few decades is changing the conception of personality in many crucial ways. It has demonstrated the significant role of heredity in personality, particularly with regard to such qualities as emotionality, activity, sensation seeking, extraversion, and a host of psychosocial traits. It thus has called attention to the important role of biology and the genes in personality and social behavior as well as in human abilities. It has opened the way to studying the activity in the brain, and to linking biological processes to psychological processes—including the person's thoughts, feeling, memories. The breakthroughs that are occurring rapidly in the tools available for studying these links are exciting, and they promise to pave the way for tracing the relationships between what goes on in the body and in the mind with increasing depth and precision.

Overview of Biological Level: Focus, Concepts, Methods

Basic units:	Biological variables: genetics, brain and other biological systems; brain–behavior links; role of biology in social behavior
Causes of behavior:	Biological and evolutionary processes; interactions between biological and environmental-social variables
Behavioral manifestations of personality:	Social behavior, personality traits, mental-emotional states
Favored data:	Physiological measures: PET Scans, EEG, MRI; test responses; trait ratings, adaptive and maladaptive behavior patterns; brain–behavior links, interactions of genetic and environmental factors
Observed responses used as:	Data for linking brain–behavior relations; behavior analyzed in terms of evolutionary theory
Research focus:	Heritability of personality; analyzing brain–behavior relations; effects of biological variables on personality
Approach to personality change:	Personality change is linked to brain/chemical changes; biological treatments for disorders (e.g., medications)
Role of the situation:	Important for evolutionary analyses of social behavior

▶ ENDURING CONTRIBUTIONS OF THE BIOLOGICAL LEVEL

The biological level of analysis is justifiably creating great interest at present on many fronts. As you saw in the discussion in Chapter 14 on studies of identical twins, many qualities of personality, especially as assessed on standardized trait questionnaires, are substantially influenced by genetic-biological determinants. Humans are biological creatures, and personality is importantly influenced by biological processes, beginning with the genes that are our heritage. The findings that genes influence personality have become sufficiently clear and strong to avoid debates about whether or not they are important: they are very important—but so is the environment. Research now is going beyond demonstrations that genes have a substantial impact on personality to identify the specific biochemical and psychosocial processes and their interactions that underlie these effects. Fortunately, such studies are now being pursued vigorously and promise to yield increasingly precise answers.

By discovering new methods for viewing the activity in the brain, work at the biological level of analysis also has opened a new route for personality research. It makes it possible to study the relationship between the individual's psychological activities, such as thoughts, memories, and intense feelings experienced by the person, for example, when exposed to a feared stimulus, and the activation of specific brain areas and centers in the brain. Studies of brain systems like the BIS and BAS, and of brain asymmetry, are already yielding new insights into the biological processes that link to crucial aspects of personality. These features of personality are fundamental, influencing the approach and avoidance dispositions that characterize each person distinctively. They affect one's typical reactions to rewards and incentives versus punishments and threats, and the positive and negative emotions and moods that influence how we experience the world and relate to it.

The challenge ahead for this line of research will be to examine more fully how these biological system relate to, and interact with, relevant aspects of personality assessed at the other levels of analysis, for example, in studies of human motivation, planning, and self-regulation. These new directions are already being pursued and are the topics of the final chapters in the text.

Questions for work at this level in the future will include the ways in which learning and therapeutic and educational experiences might be used to influence the level of activity in different brain areas in constructive directions. Such research ultimately may help liberate individuals from biological burdens of which they may not have to be the victims. The long-term yield from such research may help in the diagnosis and treatment of an array of human problems not yet imaginable in biochemical terms. If so, the approach also may make a great contribution to treatment and improvement of the human condition, but that possibility lies mostly in the years ahead. Concurrently, work on social evolution is providing new models, concepts, and methods for thinking in fresh ways about diverse aspects of social behavior—from mate selection and courtship to aggression, altruism, and human competencies. In the future, it will be important to try to link evolutionary processes to personality at the theoretical level in ways that allow empirical tests of the implications with increasing precision.

PART

VII

INTEGRATION OF LEVELS: THE PERSON AS A WHOLE

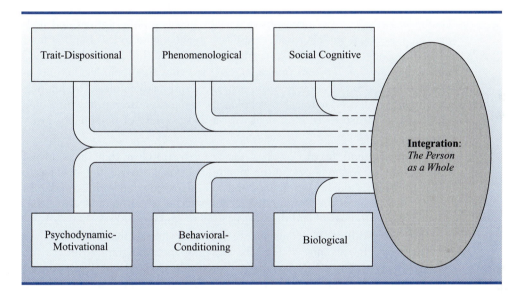

▶ PRELUDE TO PART VII: THE INTEGRATION OF LEVELS

In this final part of the text we look at the ways in which the levels come together when the person is studied as a whole, drawing on contributions from all levels. As a first step and preview, we begin with a review of the most basic contributions from each level. Even the short summary of these contributions below suggests that each level adds something significant to the others. Thus, far from being incompatible alternatives, the six levels provide insights on different aspects of the whole person. To reiterate a key theme in this text: when the levels are seen together and integrated, they enrich rather than fragment the view of the whole person.

Overview of Contributions from Each Level

1. **Trait-dispositional level.** Work at this level asks: "What units shall we use to describe the important stable, consistent individual differences in personality?" It has answered with the Five Factor Model (or Big Five) of traits, which pro-

vides a useful map of broad individual differences. Research at this level also has demonstrated that traits may be expressed as consistencies in behavior that take two complementary forms. Type I consistency is seen in overall average differences between people in their levels of different behavioral dispositions. Type II consistency is seen in differences in people's stable *if . . . then . . .* , situation–behavior patterns or personality signatures. The consistencies in personality identified by this level open the way for exploring how these patterns develop.

2. **Psychodynamic-motivational level.** This level identifies some of the possible causes of the consistencies and inconsistencies that characterize the individual. It discovered that people have motives and feelings that are threatening (anxiety-provoking) to them and to other people (e.g., parents) on whom they are dependent. Consequently, self-protective mechanisms and defenses develop to reduce the anxiety and conflicts that these motives and feelings create. As a result, the expressions of socially unacceptable motives and feelings, such as sexual and aggressive impulses, may become indirect and at least partially outside awareness. The contributions from this level have helped build a comprehensive view of personality. It is a view that includes the unconscious and self-protective aspects of personality and its dynamics. The findings profoundly influenced clinical psychologists and other therapists, providing methods and measures to help deal with a wide variety of personality problems and disorders.

3. **Phenomenological level.** Work at this level made clear that the study of what goes on inside the person at the mental-emotional experiential level has to be included in personality psychology and can be investigated with its methods. It called attention to the importance of the ways in which people view (construe, appraise) the world and themselves as determinants of how they feel and how they cope with life challenges and stresses. It showed that individuals are not necessarily the victims of their biology and their biography but can do much to change both their life course and their internal subjective experiences. It also innovated methods both for understanding and influencing the internal states of the person. And it gave center stage to the experience of the self as both an object that "knows itself" and as an active agent that can act to influence the life course. In all these ways, the pioneers at this level were years ahead of the field. Their contributions have been foundation stones for some of the most interesting developments in the evolution of personality psychology as a cumulative science, for example, in current work at the social cognitive level.

4. **Behavioral-conditioning level.** Work at this level began with discoveries about how simple processes of learning and conditioning can help make scientific sense of such phenomena as seemingly irrational fears and impulsive emotional reactions. It provided methods for examining experimentally the relationship between changes in conditions and changes in social behavior, opening the way to research into basic processes of learning. It demonstrated the importance of the environment in the development and modification of even extremely complex patterns of behavior. It insisted that the study of personality has to include what people *do.* It has to examine actual social behavior as it unfolds within particular contexts and interactions, not just what people *say* they do or what they say they are like. Thus, this level provided methods for closely observing and studying behavior and its links to the situations that preceded it and the consequences that

followed. It opened the way for a more rigorous approach that went beyond self-reports and ratings to understand what people really do.

For years, workers at this level, however, were reluctant to even consider the internal mental and emotional processes within the person because they lacked the methods for investigating them objectively. Consequently, their contributions to understanding personality, while important, were limited in scope. This was true especially when behaviorism was at its height in the middle of the last century. In more recent years, advances in ways of studying brain and hormonal processes and emotional reactions are bringing fresh life to work at this level. Currently, it again provides a valuable component for the evolution of an increasingly cumulative science of persons.

5. **Social cognitive level.** Work at the social cognitive level attempts to combine contributions from several of the other levels. First, it expanded the understanding of the role of social learning in personality development, noting the crucial importance of cognition and observation in the learning processes most relevant for personality. It showed that humans learn through what they see and observe and hear and read, and do so often even without making any external response or receiving any external rewards. It went on to integrate these insights with findings on work from other levels, including studies of the self, personal constructs, expectancies, values, and goals. It then applied these integrations to investigate the ways in which individuals can self-regulate and modify their own behavior in light of their long-term goals and values. Work at the social cognitive level also has provided a model for combining the contributions of multiple levels. The goal is to help personality psychology to evolve into an increasingly cumulative, integrated science for understanding the person as a functioning whole, not just piece by piece.

6. **Biological level.** This level links personality to its biological bases. It is exploring genetic-brain-behavior links and analyzing their interactions as they play out in the genesis of individual differences. It also shows the value of applying concepts and methods from the study of human evolution and adaptation to understand a wide range of social phenomena, from mating practices to altruism to coping with threat and stress. Insights, findings, and methods from this level open the way to connecting biological processes with work at all the other levels. For example, the feelings, thoughts, and personal constructs of interest to the phenomenologists now can be studied using fMRI to trace how what we think and understand relates to our brain activities. It is a sign of how far the field has moved that mental phenomena considered beyond the pale of scientific inquiry are now the central subject matter for brain research. Few would have anticipated just a few years ago that the self and consciousness would be the topics for a major conference of leading brain researchers and psychologists in New York City in 2002. Research now illustrates in detail that what people think, feel, and do depends on their genes and their brains, neural networks, and biochemistry. At the same time, the reverse is equally true: what people think, feel, and do changes their genes, brains, neural networks, and biochemistry. The implications are profound, and will become clear as you read the three chapters in this part.

With all the levels in mind as background, this final part of the text illustrates steps toward building a cumulative science of personality. It is based on promising ideas, findings, and methods from all levels of analysis and integrates them to help understand personality as completely as the state of the science allows. With that purpose, we now turn to the question of how people pursue their life goals, and answer it by drawing on contributions from diverse levels and focusing on their interconnections.

Chapter 16 applies the integrated view to understand how people with different personality structures and organizations characteristically pursue their life goals. It is an integration that is built on the field's century-long past but guided in new directions by research findings in the last few decades.

Chapter 17 examines work on personality and the mechanisms that enable self-regulation. Chapter 18 considers the person in context. It examines the diverse aspects of the personality and the multiple influences—from the biological to the cultural—that affect the life trajectories that people construct and the options they have for influencing their own future.

Will the applications from these findings allow people to become more self-directed, effective, fulfilled, and happier? The hope is that what is known about the ways in which the personality system functions will help individuals to "take charge" of their own lives with greater freedom to influence their future in the directions they choose for themselves.

The Personal Side of the Science

Some questions on the integration of levels you might ask about yourself:

▶ How does my gender influence my life goals and the ways in which I pursue them?

▶ How does the culture in which I live affect my answer to the first question?

▶ How do my most important characteristics reflect the interactions between my biological inheritance and the social environment in which I grew up?

▶ What are the various factors from all levels that influence my ability to exert self-discipline, e.g., to study for exams when tempted to head for a movie?

▶ How is my personality more than the sum of its parts?

16

PERSONALITY IN THE PURSUIT OF GOALS

Veronica and Martha are close college friends but could hardly be more different. Veronica "goes for the gold"—she is open to possibilities for new challenges and gratifications every day. This is true for her, whether in the sports world in which she excels already, or in taking on academic challenges, or in her readiness for new weekend adventures. Her roommate Martha is at the opposite pole—she always "plays it safe." When Veronica says "Let's get pizza," Martha says "But it's so full of calories!" Ever ready to worry about the possible downside, Martha sees opportunities more as threats to be avoided than challenges to be taken on. She typically tries out several classes before staying in any, and drops those that look like they could be too much for her—and that is often many of them. Nevertheless, she's an excellent student even in difficult courses and on average her grades are better than Veronica's.

► APPROACH–AVOIDANCE DISPOSITIONS IN GOAL PURSUIT

This chapter illustrates how contributions from each different level of analysis supplement and enrich the understanding of the functioning of the whole personality. When taken together, they can help to explain in depth the differences described in the Veronica–Martha example. In this chapter, the focus is on how work from the different levels illuminates the differences between people in how they pursue their goals. The questions here are: how and why do different personalities think, feel, and behave in distinctively different ways in their approach and avoidance patterns?

Individual differences in goal pursuit are seen in the differences among people, for example, in how they make and pursue their New Year's resolutions and try to deal with the frustrations and temptations that get in the way. You see these differences in the goals they set (to diet, to stop smoking, to save more money, to get a new job, to build a better relationship, to have more fun, to get a medical check-up), and in the life projects they pursue (to go to college, to get into business school, to get to know that person, to be a more caring, worthy human being). Dramatic differences are especially evident in the strategies that different individuals use to pursue their goals in the face of the obstacles that arise in daily life. Individuals differ in how they deal with competing and often conflicting life goals, motives, and tasks, and in how they approach or avoid different alternatives in goal pursuit. These differences are a central aspect of personality, and they are this chapter's concern.

The Biological Level: BIS and BAS

At the biological level, there are two distinct neural systems in the brain, one associated with approach, and the other with avoidance behavior (as discussed in Chapter 15). We reintroduce these biological systems here to show their links to the other levels of analysis. One is the behavioral inhibition system (BIS), and its activity leads individuals to withdraw from certain undesirable stimuli or punishments and thus inhibits their behavior. This withdrawal–avoidance motivational system is the one that enhances attention to threats and risks and possible punishment. From an evolutionary perspective, it is adaptive to humans because it is useful for avoiding danger and for surviving. But excessive BIS activity also has its costs. It can potentially deprive one of all sorts of rewarding and life-affirming approach activities in goal pursuit.

In contrast to the BIS, the behavioral activation system (BAS) is the energizing system. In Chapter 15, you saw that activity in this appetitive, positive-approach motivational system enhances attention to reward cues. It facilitates approach behaviors in the search for possible gratifications and positive outcomes in the pursuit of goals. Its downside is that it does not sensitize the person to the potential long-term risks and negative consequences of tempting approach behaviors that can be dangerous.

The Trait-Dispositional Level

16.1 How does BIS or BAS prominence influence people's goals and their strategies for goal attainment?

Some people have more activity in the BIS, whereas some are more active in the BAS. In turn, the degree of activity in these two systems also is linked to individual differences at the broad dispositional-trait level, as assessed for example by the NEO-PI-R and the Five Factor (Big Five) model.

Greater activity in the BAS seems to be more characteristic of extraverts, and greater activity in the BIS is theoretically and empirically linked to introversion and anxiety, and to related measures of temperament beginning early in life (Derryberry &

Rothbart, 1997; Gray, 1991). The BAS also seems to be associated with dominance/sensation seeking and impulsivity, and the BIS with conscientiousness/behavioral inhibition (MacDonald, 1998).

Theoretically, because of their greater reactivity in the BAS, extraverts and those high in dominance/sensation seeking will focus more on rewards, especially immediate rewards, and look eagerly for situations in which they could get them. Think here of a person like Veronica in the chapter introduction who loves adventures, gets easily involved, seizes opportunities, and often can be impulsive and indifferent to the long-term costs. In contrast, the greater reactivity in the BIS of introverts and anxiety-prone persons is related to their focus on threats and potential negative outcomes. The anxious introvert's tendency to be more passive and avoidant may reflect greater physiological sensitivity and alertness to all the possible things that could go wrong and produce trouble. Think here of the person who, like Martha in the chapter opening, foregoes strong temptations and pleasures (from pizzas to potentially dangerous thrills and adventures) out of the fear of the delayed consequences (from upset stomachs to unwanted pregnancies and HIV infections).

The Phenomenological Level: Positive–Negative Moods

Some researchers probe the links of the biological level to the different psychological levels without direct biological measurement. That became possible because reliable and valid self-report measures of BIS and BAS reactivity (Carver & White, 1994) were developed that correlate well with the brain measures (e.g., Gable, Reis, & Elliot, 2000). Table 16.1 lists items from the BIS/BAS self-report scales.

For example, people who had higher resting levels of activity in their left frontal area of the brain (indicative of BAS activity) also self-reported higher BAS sensitivity on items like those shown in Table 16.1. In contrast, higher resting levels of activity in the right frontal area of the brain (indicative of BIS activity) were related to higher self-reported BIS sensitivity (Harmon-Jones & Allen, 1997; Sutton & Davidson, 1997).

Using these self-report measures, some researchers examined the impact of the two systems on college students' emotional experiences in everyday life. They explored the joint effects of dispositional differences in BIS and BAS sensitivities to cues of rewards and punishments on participants' daily experiences. These experiences included the emotions of undergraduate students, specifically, their positive affect (PA) and negative affect (NA), and the positive and negative events that they encountered in their daily lives at college (Gable, Reis, & Elliot, 2000). According to BIS/BAS theory (e.g., Gray, 1991), individuals with high BIS dispositions will seek out and select negative cues and events in their environments. Those high in BAS dispositions will attend to and pursue positive events and possibilities for gratifications. Veronica and Martha again are examples you might have in mind here: one woman anticipates the pleasure of late night pizza and a beer with gusto and goes for it, the other thinks of the calories and even the prospect of an upset stomach and decides to call it a night.

Dr. Gable and her research group wanted to test if people's BIS/BAS dispositions in fact influence their experiences and their daily emotional states (Gable et al., 2000). They hypothesized that because BAS increases sensitivity to positive events, individuals high in BAS will have stronger reactions to them. The reverse should be true for individuals high in BIS: people high in this disposition over time should have magnified reactions to negative events.

16.2 How did Gable study emotional correlates of the BIS/BAS scales? What did she find?

TABLE 16.1   Some Items from the BIS/BAS Scale (Carver & White, 1994)

	1	2	3	4	
	Strongly Agree			Strongly Disagree	

BIS
1. If I think something unpleasant is going to happen, I usually get pretty "worked up."
2. I worry about making mistakes.
3. I feel pretty worried or upset when I think or know somebody is angry at me.
4. Even if something bad is about to happen to me, I rarely experience fear or nervousness.

BAS Reward Responsiveness
1. When I get something I want, I feel excited and energized.
2. When I'm doing well at something, I love to keep at it.
3. When good things happen to me, it affects me strongly.
4. When I see an opportunity for something I like, I get excited right away.

BAS Drive
1. When I want something, I usually go all out to get it.
2. I go out of my way to get things I want.
3. If I see a chance to get something I want, I move on it right away.
4. When I go after something, I use a "no holds barred" approach.

BAS Fun Seeking
1. I will often do things for no other reason than that they might be fun.
2. I crave excitement and new sensations.
3. I'm always willing to try something new if I think it will be fun.
4. I often act on the spur of the moment.

Source: Carver, C. S., & White, T. L. Behavioral inhibition, behavioral activation, and affective responses to impending reward and punishment: The BIS/BAS scales. *Journal of Personality and Social Psychology, 67,* 319–333. © 1994, American Psychological Association.

To test these ideas, participants were divided into BIS and BAS groups based upon their scores on the BIS–BAS Scale (Carver & White, 1994). In a series of studies, they were then asked for daily reports of their positive or negative affect on a 20-item scale to rate how much "you feel from day to day" emotions described by words like happy, distressed, nervous, enthusiastic (Watson, 1988). They also reported the occurrence of positive and negative daily events on a questionnaire consisting of 16 positive events (social and achievement) and 19 negative events (social and achievement), and indicated their frequency and importance for each.

These results show the important role of BIS/BAS as personality dispositions. Dispositional BAS sensitivity magnified peoples' reactions to positive events, and BIS sensitivity magnified their reactions to negative events. Furthermore, in their daily lives, on average, higher BAS people experienced more positive affect, more strongly.

Gable and colleagues also found that, as predicted, the number of positive events that occurred during the day on average was strongly related to participants' positive affect, and their negative affect was strongly related to the number of negative events. However, consistent with the BIS/BAS theory, on the whole, the two types of emotional reactions, positive and negative affect, seem to be functionally independent like the BAS and BIS systems themselves. This finding also fits with other data showing the independence of positive and negative affect (Cacioppo & Gardner, 1999; Watson, 1988).

The Social Cognitive Level: Promotion-Prevention Focus

So far, we have considered dispositions at a level of broad super traits (e.g., overall level of approach versus avoidance tendencies; Big Five dispositions such as extraversion). The dispositional differences illustrated above, however, also are seen in different characteristic self-systems and motivational patterns identified at the social cognitive level of analysis (e.g., Cantor & Mischel, 1977, 1979; Higgins & Kruglanski, 1996; Kunda, 1999) in which personality is expressed in daily life. At that level, one of the most important aspects of individual differences is in the characteristic, distinctive ways in which the person responds to new goals or tasks.

Consistent with work on BIS and BAS, and its related personality dispositions, some people are excited by the prospect of achieving a new goal, while others are more cautious and fear the possibility of failure. At the social cognitive level, a key question is: How are these different achievement attitudes shaped by the individual's personal experiences and cognitive structures? And how do the attitudes that develop in turn influence the person's goal pursuit strategies, motivations, and emotional experiences?

In the mid-twentieth century, researchers working at the motivational level discovered that people differ in their levels of achievement motivation (Chapter 6), and showed that how they view their achievements also influences their emotional experiences. Those who believe that they have generally achieved most of their goals have pride in their achievements—**"achievement pride"**—and are therefore more eager to pursue new goals, while those who believe that they have generally failed to achieve most of their goals are ashamed of their lack of achievement and tend to avoid pursuing new goals (Atkinson, 1964; McClelland, 1951, 1961).

Recently, psychologists working at the social cognitive level have reexamined these findings and reinterpreted them in ways that directly link to discoveries about the BIS and BAS. **Regulatory focus theory** (Higgins 1997, 1998) proposes that all goal-directed behavior is governed by two motivational systems—the **prevention system** and the **promotion system,** and these systems conceptually relate to the BAS and BIS. The promotion system (like the BAS) is focused on obtaining positive outcomes such as accomplishments or rewards. In the promotion system, the presence of positive rewards provides pleasure, and the absence of positive rewards brings pain. In contrast, the prevention system, like the BIS, attends to negative outcomes such as punishment or failure. In this system, the presence of negative outcomes brings pain, and the absence of negative outcomes provides pleasure. So the experience of people characterized either by more BIS sensitivity or by more BAS sensitivity would include both pleasure and pain, but they are experienced in different ways and in different circumstances.

If goal-directed behavior is regulated by these two motivational systems, how do they affect the strategies people use to achieve their goals at the social cognitive level? And, at the level of social learning, how are these motivational systems influenced by people's history of success and failure? The social cognitive level focuses on how people interpret events and their experience. At this level, it is therefore of particular interest to understand how the attributions they make about the causes of those experiences influence their strategies for goal pursuit.

Higgins and colleagues proposed that there are two types of achievement pride—promotion pride and prevention pride (Higgins, Friedman, Harlow, Chen-Idson, Ayduk, & Taylor, 2001). **Promotion pride** would motivate people to eagerly pursue their goals, while **prevention pride** would motivate people to vigilantly avoid factors that may hinder them from achieving their goals. The researchers further proposed that these two types of pride are influenced by the strategies an individual has successfully used to attain his or her goals in the past (thus connecting to the level of learning and past experience).

16.3 Which concepts in regulatory focus theory correspond to the BIS/BAS systems?

16.4 Describe two variants of achievement pride, and the factors that lead to their development.

Above all, the emotions experienced and the strategies preferred for goal pursuit will depend on how the person interprets those experiences. In the examples in the chapter opener, if Martha believes that she has often succeeded because she pays strict attention to the rules and avoids potentially goal-threatening situations, then she will be more likely to exhibit prevention pride and use vigilant means to achieve her goals. If Veronica sees successes as reflecting courageous risk-taking and willingness to take big chances for big prospects, promotion pride and a promotion-focused approach strategy—consistent with dispositional BAS tendencies—should emerge.

16.5 What kinds of items are used to measure the two types of achievement pride? How do scores on the RFQ subscales relate to one another and to achievement behaviors?

Higgins and colleagues examined these ideas in a series of studies. Their findings again underline the close links between the different levels of analysis, and the fact that when they are considered together the level of understanding is enriched. The studies are described in In Focus 16.1.

IN FOCUS 16.1

STUDYING PROMOTION VERSUS PREVENTION PRIDE

To test the propositions discussed in the text, Higgins and colleagues (Higgins et al., 2001)[1] conducted several studies in order to fully explore the relationship between a person's history of achievement and his or her strategy for pursuing goals in the future. The first study included the preparation of a scale to measure promotion versus prevention pride.

They developed the Regulatory Focus Questionnaire (RFQ) which measures how often participants feel that they have experienced success (or failure) through promotion or prevention self-regulation (see Table 16.2). They tested this new questionnaire by examining how well scores on this scale predicted participants' responses on a decision-making task.

TABLE 16.2   Sample Items from the Regulatory Focus Questionnaire

This set of questions asks you about specific events in your life. Please indicate your answer to each question by circling the appropriate number below it.

1. Compared to most people, are you typically unable to get what you want out of life?

1	2	3	4	5
(never or seldom)		sometimes		very often

2. How often have you accomplished things that got you "psyched" to work even harder?

1	2	3	4	5
never or seldom		sometimes		(very often)

3. Growing up, would you ever "cross the line" by doing things that you parents would not tolerate?

1	2	3	4	5
(never or seldom)		sometimes		very often

4. Not being careful has gotten me into trouble at times.

1	2	3	4	5
(never or seldom)		sometimes		very often

Note: Items 1 and 2 measure promotion pride, 3 and 4 prevention pride. Answers in parentheses are scored as high pride.

Source and Permission: Higgins, E. T., Friedman, R. S., Harlow, R. E., Chen-Idson, L., Ayduk, O. N., & Taylor, A. Achievement orientations from subjective histories of success: Promotion pride versus prevention pride. *European Journal of Social Psychology, 31,* 3–23. © 2001, John Wiley & Sons Limited. Reproduced with permission.

There were two decision-making tasks—one in which the correct answer required the participant to take advantage of an opportunity, and the other in which the correct answer required the participant to be more cautious in the approach. Participants also were asked to list their goals, the relative importance of each goal, and up to seven activities they could do to attain each goal, and answered a number of other questionnaires. In still another study, participants were asked to complete the RFQ and the Self-Control Strategies Scale (SCSS; Ayduk, 1999) which measures the self-control tactics a person uses to maintain a diet. The researchers also measured how often participants experienced the motivational states of eagerness vs. apathy and vigilance vs. carelessness. Finally, they examined how often participants felt they had experienced success or failure by using different kinds of strategies in their goal pursuit. Responses on these and other measures were coded and their relationships were analyzed statistically. Higgins and colleagues found that:

- Promotion pride was unrelated to prevention pride—a finding that is consistent with the view of BAS and BIS as independent systems.
- Individuals with high prevention pride experienced vigilance—anxiety and concern with making mistakes—more often in their activities, whereas those with high promotion pride experienced eagerness more often.

- People with a history of promotion success (promotion pride) responded to challenges with approach-oriented strategies. For example, someone with promotion pride will err on the side of trying too many options or taking full advantage of an opportunity. In contrast, people with a history of prevention success (prevention pride) are more vigilant and cautious in their goal-directed behavior. They err on the side of caution and avoid more risky avenues to success.

- When people feel that they have had success in the past due to promotion-related behavior, they practice more eager, approach-oriented behavior to attain their goals. If they feel that they have succeeded because of prevention-related behavior (e.g., their cautiousness), then they practice vigilant, more cautious strategies to attain their goals.

Taken collectively, the studies by Higgins and colleagues indicate that people can succeed in one of two ways. One may succeed by being cautious and avoiding failure at all costs, or one may succeed by eagerly pursuing one's desires. Though these are both positive experiences, they are distinct in their strategic orientation and the way that they influence future behavior and emotions.

The studies of promotion and prevention pride and orientation discussed above illustrate an *if . . . then . . .* analysis of the interaction of persons and situations in approach–avoidance patterns. The findings show the close links between broad personality dispositions and their more specific personality signatures (Chapter 4).

Thus, while people with promotion pride thrive in situations that offer opportunities and challenges, those with prevention pride do best when the tasks require caution and vigilance. Each orientation has its advantages and disadvantages. The individual who is able to flexibly access the orientation that best suits the situation is likely to be able to deal well with the widest possible range of encounters that might be faced in life. Since promotion and prevention are independent dimensions, it is possible for the same individual to be high (or low) in both types of pride.

Convergence with Evolutionary Theory

Evolutionary theory also calls attention to the need for organisms to take the context into careful account in their behavioral responses to the environment. Theoretically, the evolutionary level of analysis, like the social cognitive level, thus emphasizes interactions in the ways in which approach and avoidance tendencies and their associated dispositions are expressed and contextualized in goal pursuit. As noted in the discussion of evolutionary theory (Chapter 15), it is adaptive to flexibly suit one's actions to the requirements of the particular situation—essentially knowing when to approach, and when to avoid, to cope and survive. Consequently, in evolutionary theory (MacDonald, 1998, p. 128), people tend to compartmentalize. As seen in the last chapter, they may seek intimacy and approach under some conditions (e.g., with mates and offspring) but are also ready to flee or to be aggressive and hostile when necessary (e.g., with competitors or perceived enemies). How the patterns are played out in the process of adaptation depends in part on the overall level of broad dispositions like BIS/BAS. But it also depends on the particular domain and on the types of conditions and situations that are being encountered.

Linking Dispositions to Situations

In traditional approaches to dispositions, the "situation" was often treated as a source of measurement noise to be removed in personality assessment. More recently, the need to include situations in the analysis of personality is being recognized even by advocates of the Big Five broad dispositional approaches (e.g., John, 2001). The situation activates the processing dynamics or mechanisms through which a disposition functions and is expressed and in that sense, it becomes part of the disposition itself and its *if . . . then . . .* behavioral signature (Mischel & Shoda, 1998).

In time, individuals who are particularly high or low on factors like the Big Five are likely to actively influence the environments they typically encounter and experience. Just as you saw in Chapter 14, people's genetics in part influence the social environments in which they develop, so do their dispositional preferences in part shape the worlds that they experience. Those who are predominantly reactive in the BAS, for example, may change their personal environment by maximizing opportunities for challenges and rewards. In contrast, those high in BIS will tend to avoid risks and potential danger and threat, and subsequently try to shield themselves from them. In turn, the social world will soon reciprocate and react in distinctively different ways to these different types of individuals, further changing the environments they typically encounter.

You do not invite friends to try new rock-climbing adventures when you know they are risk-averse and will shy away from them. Likewise, high BAS people (think again of extraverts and people like Veronica) are likely to get and accept more party invitations than those high in BIS. Thus, both the brain and the external world interact and contribute jointly to the motivational qualities and emotional dispositions that characterize different individuals and types, which then actively influences the environments they experience and encounter, which further strengthens the dispositional tendencies. Again we see the dynamic interactions between the biological level, the trait-dispositional level, and the social environment. We next illustrate the links between multiple levels by looking at relations between the psychodynamic and behavioral levels when dealing with approach–avoidance conflicts.

Links to the Psychodynamic-Motivational and Behavioral Levels

16.6 How do current approach–avoidance models relate to Dollard and Miller's learning-based model of psychodynamic conflict?

The study of approach–avoidance tendencies and the dilemmas that they can produce has been a core topic of personality psychology ever since Sigmund Freud made conflict the center of his theory a hundred years ago. Because these conflicts create anxiety, the individual, according to Freud, engages in massive unconscious efforts to reduce it, which in turn can produce all sorts of difficulties and symptoms. The problem for scientists early in the last century was that such conflicts were very difficult to study experimentally. To make approach–avoidance conflicts open to experimental research, recall that John Dollard and Neal Miller (1950) working at the behavioral level of analysis years later attempted to rethink psychodynamic conflicts in the language of learning. As noted in earlier chapters, these were pioneering attempts to integrate the levels of analysis in the study of goal pursuit. Given the renewed interest in the approach and avoidance systems as key aspects of personality, the classic work of these pioneers has renewed relevance.

To resummarize, these researchers hypothesized *approach* and *avoidance* tendencies, and analyzed them in careful experiments with animals (Chapter 10). In such conflicts, the organism, at least momentarily, is torn between two desirable goals. Just as conflict was central to Freud's conception of personality dynamics, so it was at the core of Dollard and Miller's theory. But whereas Freud developed his ideas about conflict from inferences regarding id–ego–superego clashes in his neurotic patients, Dollard and Miller tested their ideas in controlled experiments with rats. Their findings shed light on understanding some of the problems of Freud's patients in more objective terms.

Their original theory of conflict was based on a number of animal experiments (e.g., Brown, 1942, 1948; Miller, 1959). In one study, for example, hungry rats learned how to run down an alley to get food at a distinctive point in the maze. To generate **ambivalence** (approach–avoidance tendencies), the rats were given a quick electric shock while they were eating. To test the resulting conflict, the rats were later placed again at the start of the alley. The hungry rat started toward the food but halted and hesitated before reaching it.

Dollard and Miller applied the concept of **goal gradients** to analyze these conflicts. Goal gradients are changes in response strengths as a function of distance from the goal object. To assess the strength of approach and avoidance tendencies at different points from the goal, a harness apparatus was devised to measure a rat's pull toward a positive reinforcement (food) or away from a negative reinforcement or punishment (shock). The harness enabled the experimenter to restrain the rat for a moment along the route to the goal and measure (in grams) the strength of the animal's pull on the harness at each test point (Brown, 1948). In this type of situation, the rats could be given different experiences that led the same response from them to become associated both with approach motivation (the food) and with avoidance motivation (the electric shock). First the hungry rats learned to run the maze to get the food that was waiting for them at the end, thus developing approach tendencies. Then sometimes they also received the electric shock at the end of the maze, which led their maze-running response to also become associated with punishment, analogous to the dilemmas of Freud's patients whose socially unacceptable desires also had become associated with fear of punishment.

When a hungry but now also fearful rat with such a history next gets into this maze he faces an approach–avoidance conflict. This conflict is shown in Figure 16-1. Running toward the goal is associated with hunger reduction and induces approach tendencies. But it is also associated with getting electric shock, which makes running in the opposite direction a way to reduce the fear drive. As a result, the rat would run in both directions, forward toward the food, and then away. Interestingly, the researchers found that when far from the goal, the approach tendency was stronger than the avoidance tendency, hence producing approach behavior. However, as the goal got nearer, the strength of the avoidance tendency increased sharply, leading the rat to run in the opposite direction. The animal stopped at the point where the approach and avoidance pulls became equalized,

16.7 Describe the role of goal gradients in explaining approach–avoidance conflicts. How were the gradients measured in the laboratory?

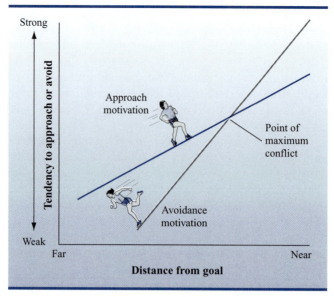

Figure 16.1 According to Neal Miller's analysis of approach-avoidance conflicts, both the tendency to approach and the tendency to avoid grow stronger as one moves closer to the goal. However, the tendency to avoid increases faster than the tendency to approach. Maximum conflict is experienced when the two gradients cross, because at this point the opposing motives are equal in strength.

Source: Smith, R.E. (1993). *Psychology.* St. Paul, MN: West. Reprinted with permission.

essentially torn equally in both directions. And at least by analogy that is also the neurotic dilemma of persons afflicted with intense approach–avoidance conflicts in which they feel psychologically paralyzed.

The work at the various levels of analysis discussed in this section points to some of the many ways in which individuals facing such conflicts are likely to behave. For example, those with much higher BAS reactivity and promotion focus might approach more readily and persistently compared to those with higher BIS reactivity and prevention focus.

▶ THE GOAL PURSUIT PROCESS

Life Tasks, Personal Goals, and Projects

The goals an individual pursues are organized and coherent. They reflect the activity of the entire personality system as the person interacts with the situations within his or her world at any given point in development. The goals of a college student, and the situations, challenges, and threats faced, are obviously different from those experienced in early childhood, or those that will be salient at mid-life and beyond. In this section, we will focus on how personality plays out and influences the mental and emotional processes that unfold, as individuals pursue their important life goals (e.g., Cantor & Kihlstrom, 1987; Cantor, 1990; Zirkel, 1992).

Current life tasks are defined as those central projects to which individuals perceive themselves as being committed during particular periods in their lives, such as in the transition from high school to college (Cantor & Kihlstrom, 1987, p. 168). These self-created tasks help give meaning to the individual's life and provide organization and direction for many more specific activities and goal pursuits that are in their service.

Examples of common life tasks include "getting promoted," "finding the right person and getting married," "making myself more fit," and "taking off a couple of years to do something meaningful for needier people." Life tasks thus constitute significant long-term goals that are meaningful for the individual at some point in his or her life. While they are typically experienced as personally urgent, they are also often ill-defined and loosely formulated.

Certain types of life tasks are shared by many people in the same phase of life. For example, first-year students in an honors college at the University of Michigan often had life tasks like the following ones in common: "Being on my own," "making friends," "establishing an identity," "getting good grades," and "establishing a future direction" (Cantor & Kihlstrom, 1987, p. 172). When the students were encouraged to elaborate their plans, feelings, and specific strategies, some common themes emerged concerning achievement, intimacy, and gaining better self-control. Each student, however, also uniquely construed his or her tasks and focused on somewhat different aspects of the same common themes. For example, while "being on my own" meant learning to cope with personal failures without the help of parental hugs for one student, for another it meant working on how to manage money responsibly.

Investigations of this type (e.g., Cantor and associates, 2002; Harlow & Cantor, 1996; Snyder & Cantor, 1998) help to provide systematic methods for accessing these personally significant, often unique aspects of experience. They draw on the person's own expertise to articulate his or her personal constructs, plans, and problem-solving strategies for dealing with everyday real-life challenges. In addition, they also help identify certain problem-solving strategies that different individuals prefer to guide and monitor their own progress as they pursue their chosen life tasks.

In this section we will illustrate the personality processing dynamics—the flow of interacting thoughts and feelings and potential behaviors—that become activated in goal pursuit. These dynamics are activated in ways that are distinctive for the particular individual's personality organization and structure, as people choose and pursue their life tasks and goals. The goals themselves are organized in their degree of importance for the person in **goal hierarchies** with some life goals at higher-order levels than others, as discussed in the section below on goals and standards.

Table 16.3 shows aspects of the mental and emotional dynamics in the personality system activated in goal pursuit. Goal pursuit in a situation begins with the ways in which the individual encodes—that is, interprets or appraises it. Then expectations, beliefs, feelings, and goals relevant to the appraisal are activated. These all interact within the system in ways that are guided and constrained by its unique organization. It is this distinctive personality organization that is relatively stable. The organization reflects the multiple influences of the person's unique total biological and psychosocial history.

Illustrative Processing Dynamics: Thinking about a Career

To illustrate this more concretely, we take another look at Veronica and Martha, the two women introduced in the chapter opener. In this simple example, the situation is inside the head: it happens when each woman is thinking about her future and a career in medicine.

16.8 How do CAPS dynamics help us explain Veronica and Martha's career goals?

TABLE 16.3 Overview of Personality Dynamics in the Goal Pursuit Process

Encoding (Construal/ Appraisal)	Expectancies/ Beliefs	Affects	Goals/ Values	Regulatory Competencies (Effortful Control)
Self-relevance (personal meaningfulness) appraisal	Perceived regulatory control/ helplessness	Emotional arousal / Fears and desires	Goal hierarchy / Standards	Self-control / Strategies (distraction/ "cooling")
Automaticity (trigger reactions)	Self-efficacy/ expected outcomes	Feelings states/"hot" (impulsive) automatic responses		Self-monitoring Planning / Self-instruction

Source: Based in part on material in Mischel, W., Cantor, N., & Feldman, S. (1996). Principles of self-regulation: The nature of willpower and self-control. In E. T. Higgins & A. W. Kruglanski (Eds.), *Social Psychology: Handbook of Basic Principles* (pp. 329–360). New York: Guilford.

What might they think and feel as they consider their futures? Assume that both are equally qualified academically for medical school, and have long had it as a life goal. One plausible script that seems to fit Veronica is in Figure 16.2. It shows that given her highly active BAS, and extraverted dispositional tendencies, she is likely to frame her future in medicine as full of opportunities. Theoretically, this activates her promotion focus and the eager, enthusiastic emotions that this entails. She could even imagine herself discovering or finding something exciting, maybe even a cure for cancer.

Martha's processing dynamic in goal pursuit, shown in Figure 16.3, is a sharp contrast. Given her generally high BIS and readiness to worry, her prevention focus snaps into action and she is anticipating the potential threats. She imagines herself failing,

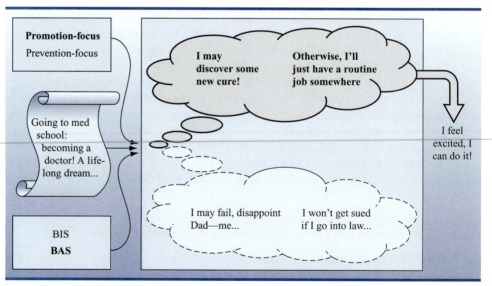

Figure 16.2 Veronica's processing dynamics in goal pursuit. Thinking about medical school activates promotion focus and BAS.

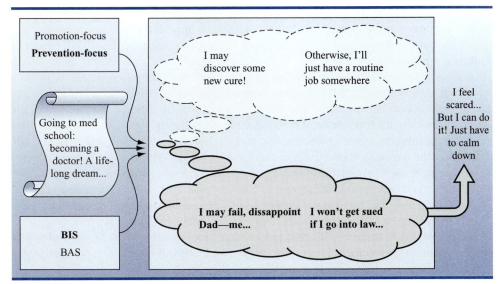

Figure 16.3 Martha's processing dynamics in goal pursuit. Thinking about medical school activates prevention focus and BIS.

disappointing her father and herself, perhaps even inadvertently hurting a patient and being sued for malpractice. She is strongly motivated to prevent these negative outcomes, and eager to not disappoint her father and to not get into trouble. But she also has high efficacy expectancies about her own ability to get into med school and have a successful career in medicine, in spite of her anxieties. And she places high value in having a life in medicine. Therefore, she decides to go forward with med school. Thus in spite of the large differences between Veronica and Martha in personality, both women reach the same decision—to apply to med school. But they do so with very different mental and emotional experiences and processing dynamics that affect how they feel and behave during the long process of trying to create a career in medicine.

We next consider the steps in goal pursuit shown in Table 16.3. This discussion is without sustained reference to the two women, but you might want to speculate about what went on in their heads as they moved forward in their goal pursuit.

Encoding/Appraisal of the Situation

Self-Relevance. The motivation to try to pursue a goal tends to increase to the extent that the activity or situation is encoded or appraised as personally meaningful (Bandura, 1986; Cantor, 1990). Consider, for example, how new mothers try to cope with the often exhausting and sometimes anxiety-provoking or boring chores of parenting. Research showed that they cope better when they view those tasks as fulfilling important self-obligations rather than as taking time away from other modes of self-fulfillment, such as a career (Alexander & Higgins, 1993).

The social interactions of daily life contain such episodes as an argument with parents, a movie with friends, a visit to the doctor, or a shopping trip to the mall. The differences among people in how they encode these events are related to their differences on other measures of personality, perceived social competence, and skills (Forgas, 1983).

Emotional Meanings. To appraise situations, you have to interpret their emotional meanings. Is your father really angry or teasing? Is your roommate being friendly

16.9 How do customary encodings and appraisals influence goal activation? Which encoding factors are particularly important?

or just polite? Usually these interpretations are made automatically, depending on previous experience and expectations. But the automatic appraisal *can* be inaccurate.

This point becomes most evident from anthropological reports of differences in cultural expressions of emotions. Visitors to a strange culture would make many mistakes if they tried to assess the meaning of emotional expressions in terms of their own past experience. A Masai warrior honors a young man who looks promising by spitting in his face; an Andaman Islander greets a visitor by sitting down on his lap and sobbing his salutation tearfully; a scolded Chinese schoolboy takes a reprimand with cheerful grinning as a sign of his respect; and to show anger, Navajo and Apache Indians lower the voice instead of raising it (Opler, 1967). Within each culture, people reach an agreement about the meaning of emotional cues. But certain facial expressions also may have universal meaning. The emotions of anger, happiness, disgust, sadness, fear, and surprise, shown in photographs of faces, tended to be correctly identified in a variety of extremely different cultures (Ekman, 1982).

On visits to extremely different cultures, people usually try to be careful not to misinterpret the emotional signals they observe and to avoid reacting inappropriately. A similar caution may be useful in carefully attending to the emotional signals that come from people in one's own personal life. Focused attention, rather than automatic responding, may allow one to see more clearly what is really going on.

Biological and Trait-Dispositional Levels. As you saw before, at the biological level, the appraisal process reflects the reactivity in the BIS/BAS. As they appraise situations, people with high chronic BIS reactivity are likely to focus on threats and to anticipate and experience them with increased negative emotions and anxiety. At a physiological level, this will be visible in bodily changes. Those who are extremely high in their BAS reactivity are likely to appraise even potentially threatening and stressful situations in ways that focus on what can be gained or maximized from the experience. They are more likely to minimize the downside and to focus on potential benefits.

For example, suppose both Veronica and Martha love cross-country skiing, but fall badly, sustaining serious fractures that greatly interfere with their plans and schedules. Martha is apt to focus on all the things she will not be able to do while in casts, and the negative consequences (e.g., incompletes in courses, disappointing others and herself, missed appointments). Veronica is likely to focus on how she can maximize or create some gains from her mishap. She imagines that soon she will have friends visit, be able catch up with sleep and see some good videos, treating herself to a bit of an unscheduled vacation. These differences between the two women also are reflected at the trait-dispositional level in their scores on measures like the Big Five and the NEO-PI-R: Veronica is likely to be high in extraversion and impulsivity, Martha is apt to be high in anxiety and neuroticism, for example. They are seen at the social cognitive level in their different promotion-focused versus prevention focused styles and behavioral signatures. Again, data from each level enrich the view of the whole person.

Expectancies and Beliefs

16.10 Describe how self-efficacy, perceived control, and behavior-outcome expectancies relate to goal striving and success.

The appraisal of the situation activates various relevant expectations and beliefs. Especially relevant for goal pursuit are efficacy, outcome expectations, and control beliefs.

Self-Efficacy Expectancies. Expectancies and beliefs that one will be able to exert control and successfully execute necessary actions are essential for effective goal pursuit. They support one's efforts and guide whether, where, when, and how one should pursue relevant goals (Mischel et al., 1996). To even try to "take charge" in self-guided goal pursuit requires a representation of the self "as a causal agent of the intended action" (Kuhl, 1984, p. 127). As discussed in Chapter 12, self-efficacy—the belief that "I can do it"—is a foundation for the successful pursuit of a difficult goal, or for changing and improving one's situation or oneself. Its psychological opposite, perceived helplessness, is the route to giving up, apathy, and depression (Chapter 13). Both Veronica and Martha may have equally high self-efficacy expectations about being able to pass the entry tests for getting into medical school—indeed Martha's expectancies may be higher than Veronica's. These expectancies may be good predictors of how well they actually will do on those tests, in spite of the very different thoughts and feelings they have when they think about themselves as becoming physicians (Figs. 16.2 & 16.3).

Perceived Control and Predictability. Even when the particular task is something aversive that has to be endured and cannot be controlled—like a painful dental procedure—the belief that one can predict or control the event or the stress is an important ingredient for adaptive behavior. For example, in a classic study, college women were exposed to an aversive noise while they worked on a task. The noise either occurred at a predictable time or unpredictably (Glass et al., 1969). Their tolerance for frustration and the quality of their performance on the task were impaired only when the noise was unpredictable. Equally interesting, the negative effects were reduced considerably if during the stress period the participants believed they could do something to control the end of the stress. Generally, most people tend to become less upset when they think they can predict and control stressful or painful events (Staub, Tursky, & Schwartz, 1971), even if the perception is illusory (Taylor & Brown, 1988).

Outcome Expectancies. Goal striving is also bolstered—or undermined—by outcome expectancies about the likelihood that the time and energy spent on the task will actually result (or not result) in the desired outcome (Brunstein, 1993; Carver & Scheier, 1982; Klinger, 1975). When, and where, and what we choose to wait or work for depends in part on the anticipated consequences. To the extent that the individual believes that self-regulation will lead to desirable consequences, he or she will be more likely to choose to perform it (Feather, 1990; Rotter, 1954).

People base these outcome expectations both on information in the current situation and on information from previous similar situations. Such expectancies strongly influence whether people will try to perform a task that is a prerequisite for attaining a larger reward or instead settle for a smaller reward that is available noncontingently and immediately. Those who expect to succeed on the task are more willing to try to perform it (e.g., Mischel & Staub, 1965). In sum, people are more likely to choose to perform an action that requires effort if they believe that they can perform the action (i.e., have high self-efficacy expectancies) and expect that it will lead to favorable consequences (e.g., Bandura, 1986; Mischel, 1973). A good example comes from studies designed to increase commitment to condom use among inner-city African–American adolescents (e.g., Jemmott and colleagues, 1992). A key component for achieving this goal required raising the participants' self-efficacy beliefs and their outcome expectancies about the positive consequences of using condoms.

Enhancing Self-Efficacy in Goal Pursuit. People who expect failure are likely to fulfill their prophecy. But if they are led to think they can do better, will their performance actually improve? One study of academically borderline college students examined just this question (Meichenbaum & Smart, 1971) and concluded that students who were led to expect success in fact became more successful in their schoolwork.

The power of positive thinking is shown even more dramatically through "mental practice." In one experiment, participants were instructed to imagine throwing darts at a target (Powell, 1973). Half of them were asked to imagine that their darts were hitting near the target's center; the other half imagined that their darts were striking outside the target area. Dart throwing improved significantly for people who had imagined successful performances during mental practice, but not for those who had imagined poor performances. Numerous studies to increase self-efficacy have shown similar effects (Bandura, 1986). The moral is plain: Think success!

Affects/Emotions

"Hot" Emotions and Impulsive Reactions. The personality system is not a cool information-crunching machine in which calm thoughts about possible outcomes are contemplated thoughtfully and rationally from a detached perspective. In the century since Sigmund Freud and his colleagues began their work at the psychodynamic-motivational level, it has become clear that humans are also emotion-driven creatures who are prey to their emotions, from which they can either benefit or suffer.

Even those who reject many of Freud's ideas recognize that people are intrinsically self-protective and sometimes even severely self-deceiving in how they deal with feelings and conflicts. Consider, for example, the student with public-speaking anxiety who panics at the thought of having to make a presentation to a group. Or take the dilemma of the addict who is trying to quit but is tempted with heroin, or the starving dieter faced with the ultimate chocolate cake. These kinds of hot situations and stimuli tend to automatically trigger hot reactions, rapidly generating the associated feelings of fear or desire and the urge to respond impulsively, bypassing self-regulatory controls just when they are most important.

In situations in which "hot" emotional reactions such as intense fear or cravings are elicited, the person may be subject to what the behaviorists call "stimulus control." If, for example, intense anger is aroused or strong temptations encountered that distract from one's commitments, an impulsive automatic response may follow. These also are the emotional conditions in which self-protective mechanisms, such as extreme avoidance, are automatically triggered, consistent with findings at the psychodynamic-motivation level.

Consider again the Veronica–Martha examples in Figures 16.1 and 16.2. When they imagine their medical futures, Veronica's focus on such positive possible outcomes as finding a new cure, and Martha's worry about possible malpractice risks, illustrate the very different internal states these women are likely to experience in their goal pursuit. The risk for Martha is that her sensitivity to potential problems may create so much anxiety that she ultimately drops out, or becomes extreme in her defenses to cope with those negative feelings. For Veronica, the risk may be that her insensitivity to the potential difficulties and stresses of a medical career may leave her inadequately prepared to cope with them when they do arise. In the next chapter, we will discuss the processes of self-regulation that can allow, or undermine, continued goal pursuit under difficult and emotion-arousing conditions.

16.11 How does anxiety affect goal striving?

Tuning in to the Wrong Thoughts: Anxiety. Effective goal pursuit is severely sabotaged by anxious, self-preoccupying thoughts (e.g., "I'm no good at this—I'll never be able to do it"). These anxious thoughts compete and interfere with task-relevant thoughts (such as, "Now I have to recheck my answers"). The result is that performance (as well as the person) suffers (Sarason, 1979). Self-preoccupying thoughts interfere most when the task to be done is complex and requires many competing responses. One just cannot be full of anxious, negative thoughts about oneself and simultaneously concentrate effectively on difficult work, like preparation for a chemistry final exam. Likewise, as the motivation to do well increases (as when success on the task is especially important), the highly anxious person may become particularly self-defeating. That happens because under such highly motivating conditions, test-anxious people tend to catastrophize, imagining all the terrible things that could happen. They then become even more negatively self-preoccupied, dwelling on how poorly they are doing. In contrast, if one can become less anxious, then attention can be devoted to the task and the person can concentrate on how to master it effectively.

Goals/Values, Motivations

Almost by definition, goals are pursued because the person is motivated to pursue them: the motivation for goal pursuit depends on the importance of the goal within the personality system.

Goal Hierarchies. A person's goals are organized hierarchically, with some more important or "super-ordinate" than others (e.g., Carver & Scheier, 1998; Vallacher & Wegner, 1987). The goal to be and feel "safe and secure" is a higher-level goal, than "finding a partner I can trust" or "getting a job I can count on." The latter goals are at a lower level than the super-ordinate one, but they are much higher than "making the apartment safer." Making the apartment safe is executed at the even more subordinate level of "going to the hardware store" or cleaning out potential fire hazards. Likewise, the goal of being "a worthy person" is higher in the organization of the system than a goal such as being "kind and caring." Being kind and caring connects to even lower-level goals like "helping the needy person cross the street."

When goal attainment at a given level (e.g., getting a job) is blocked and frustrated, people may continue to strive toward the higher-level goal (being safe and secure) toward which the lower-level activity was directed. They may try, for example, to focus more on building the relationship with the partner now and postponing the job search, as long as it still serves the same higher-level goal (Martin, Tesser, & McIntosh, 1993). Goals are central for understanding the personal and life goals that people pursue, but the processing dynamics of how they pursue them depend on their own standards and self-evaluations as they monitor their progress.

Standards and Self-Evaluation. In goal pursuit, people evaluate their own behavior and perceived progress, and they reward and punish themselves accordingly, thereby further influencing how they progress—or fail to do so (e.g., Bandura, 1989; Carver & Scheier, 1990). They congratulate themselves for what they see as positive, and they feel good or bad or uncertain about their own achievements. They self-administer psychological, social, and material rewards and punishments. In short, people assess themselves and become their own internal judges and reward–punishment system, using the standards that they have developed for themselves (Bandura, 1986, 2001; Higgins, 1990). Self-praise and censure, self-imposed treats and punishments, self-indulgence and self-laceration are signs of this pervasive human tendency to congratulate and condemn oneself.

16.12 Explain the concept of goal hierarchies and the factors that influence which goals are pursued.

16.13 How do self-evaluative standards influence striving and emotional responses? What kinds of conditions negate their influence?

You see it when people celebrate their good test scores and their acceptance letters, or kick themselves for what they read as disappointments and setbacks. A critical aspect of goal pursuit thus stems from the fact that beginning early in development, people learn to set their own performance standards and to make their own self-reward contingent upon their achieving these self-prescribed criteria. It's common to make a treat, like going to the movies, contingent on first finishing an assignment.

In this self-evaluation process, people compare their current state of performance with those standards. If they perceive a discrepancy, they tend to be motivated to reduce it or to reset their standards to a lower level (e.g., Carver & Scheier, 1981, 1990). As Bandura explained: "When people make self-satisfaction or tangible gratifications conditional upon certain accomplishments, they motivate themselves to expend the effort needed to attain the requisite performance" (Bandura, 1986, p. 350). By generating consequences for their own actions, individuals can bring them into line with their higher-order values and goals.

16.14 How are modeling and social comparison involved in self-standard development? How does self-awareness activate self-standards?

Acquisition of Standards. The standards and goals people observe in the behavior of the models they value guide their self-regulatory efforts. In a pioneering series of studies on the acquisition of self-regulatory standards, Bandura and Kupers (1964) examined young children's patterns of self-reward during a rigged bowling game. They found that children who observed adults reward themselves for low levels of performance (i.e., low-standard models) subsequently treated themselves to available rewards after achieving low or moderate scores. In contrast, those who had been exposed to high-standard models adopted a stricter criterion for self-reward, making it contingent on high levels of performance. Social comparisons with peers also influence adult standards and self-regulation. For example, the levels of drinking that college students consider acceptable depend on the standards and values of the groups to which they belong (Prentice & Miller, 1993).

Bypassing Self-Standards. Self-standards may be bypassed easily under various pressures of the moment (Carver & Scheier, 1978, 1998; Higgins, 1990; Wright & Mischel, 1987). Just as looking at yourself in the mirror makes it harder to bypass your "conscience" (that is, the personal standards and values to which you strive to adhere), listening to the "leader," the "authority," or the "group" can make it easy to forget those standards. Films about lynch mobs, gang rapes, wartime atrocities, the Holocaust, and urban riots document what can happen when personal responsibility is forsaken. Although we become desensitized easily by viewing them repeatedly in the media, they illustrate how often self-standards and careful self-examination are forgotten, with terrible consequences both to the victim and to the perpetrator. Research also has shown how easily people can be urged by authority figures into acts of aggression, failing to access their own self-standards and readily doing to others what they would consider unthinkable for themselves (Milgram, 1974). On the other hand, when self-regulatory scripts and plans for self-control are activated, the person may be able to overcome such pressures.

If you had to predict how easily Veronica and Martha would bypass their self-standards, and whose standards would be more stringent, what would you hypothesize, and why? Who would be more likely to make it through medical training? How sure are you, and on what basis?

Activation of Self-Standards: From Mindlessness to Self-Focus. Much of what we do as we pursue goals is **mindless** (Langer, 1978), guided by more or less automatic

If the bypasser sees herself in the store window she is more likely to help the sick old man.

behavioral routines or scripts that are familiar and require little active attention or self-regulatory effort (Abelson, 1976). In a fast-food restaurant or at a football game, for example, we tend to act out more or less automatic routines. Under such conditions, active, conscious self-regulation plays little role and we do not experience self-awareness; that is, we do not focus on ourselves and our self-standards, but behave more or less as if on automatic pilot.

Self-standards are more likely to be accessed and brought into play when attention is focused on the self. That happens, for example, when one is observed, or sees oneself in a mirror. Under such self-focus, people become more conscious of their behavior and how it compares to their standards. In turn, awareness of any discrepancies motivates them to modify their actions in order to bring them more into line with their standards and ideals. Thus, a focus of attention on the self motivates the person to try to reduce perceived discrepancies between what he or she is doing and the relevant standards for comparison (Carver & Scheier, 1981; Duval, Duval, & Mulilis 1992; Scheier & Carver, 1988). For example, a self-perceived "caring" person who sees herself reflected in a store window about to pass a sick old man sprawled on a city sidewalk is more likely to bring herself in line with her own standards and pause to try to help.

Alternative Routes in Effective Goal Pursuit

Before ending this discussion, let's look again at the finding by Higgins and colleagues (2001) that people can succeed through either prevention or promotion focus. They concluded that people may succeed in their goal pursuit by being cautious and avoiding failure at all costs, as exemplified by the Martha illustration. Or they can also succeed by eagerly pursuing their desires, like Veronica. Both are potential routes to success, although they are distinctively different strategic orientations. They are experienced very differently in the thoughts and feelings of the individuals, and they influence their future behavior, mental activity, and emotions in quite different ways.

Thus, the two women in our example have extremely different personality systems. These differences are seen at multiple levels of analysis. They differ in their BIS/BAS reactivity and in their Big Five factor scores and profiles especially with regard to extraversion, impulsivity, neuroticism, and anxiety. At the social cognitive level they experience their worlds in dramatically different ways. They contrast in typical moods and feeling states, and in how they mentally and emotionally represent their experiences and their possible futures. This is seen even in how they imagine their possible careers in medicine. They use very different strategies in their pursuit of goals—one "psyches" herself up, imagining the exciting possibilities; the other scares herself about the possible negatives and stressors along the route. Nevertheless, both might be able to persist and succeed in their goal pursuit. What are the probable psychological determinants of those ultimate outcomes?

In Table 16.3 on p. 382, summarizing the processes and person variables that influence goal pursuit, the last column lists regulatory competencies. The nature of those competencies, and their role in personality are discussed in the next chapter. There we will examine how the different levels of analysis discussed throughout the text help clarify the mechanisms of self-regulation and effortful control. These are mechanisms through which goal pursuit, whether motivated for rewards and opportunities or to avoid punishments and reduce fears, can be maintained effectively. Specifically, the next chapter details how individuals can self-regulate their own behavior and internal reactions in ways that allow them to achieve increasing mastery—and to feel better, more in control, and more effective—as they go from pursuing their goals to attaining them.

SUMMARY

APPROACH–AVOIDANCE DISPOSITIONS IN GOAL PURSUIT

- At the biological level, the behavioral inhibition system (BIS) regulates avoidance behavior; the behavioral activation system (BAS) regulates approach behavior.
- At the trait-dispositional level, increased activity in the BAS is linked to extraverted personality characteristics; activity in the BIS is linked to introversion and anxiety.
- At the internal-experiential level, BAS and BIS activity are related respectively to sensitivity to positive (reward) cues and to negative cues (threats, punishment).
- In regulatory focus theory, goal-directed behavior is governed by two motivational systems—the prevention system and the promotion system.
- Evolutionary theory emphasizes the adaptive value of situation-specific behaviors.
- Genetics and dispositional preferences influence the social environments in which people develop.

THE GOAL PURSUIT PROCESS

- During goal pursuit, mental and emotional dynamics are activated. These include encodings, expectancies/beliefs, feelings and emotions, personal goals/values, and regulatory competencies.
- The BIS and BAS biological systems influence how individuals appraise and respond to situations.
- In "hot," emotional situations, the person may be subject to "stimulus control," triggering impulsive, automatic responses.
- Anxious, self-preoccupying thoughts can sabotage goal pursuit.

- When a lower-level goal (e.g., getting a job) is blocked and frustrated, people often continue to strive toward the higher-level goal (being safe and secure).
- In goal pursuit, people influence their own progress by rewarding and punishing themselves.
- Self-standards may be bypassed easily under various pressures of the moment.
- Self-standards are more likely to be accessed when attention is focused on the self.
- People may succeed in their goal pursuit by being cautious and avoiding failure (prevention focus) and/or by eagerly pursuing their desires (promotion focus).

KEY TERMS

achievement pride   375
ambivalence   379
current life tasks   380
goal gradients   379

goal hierarchies   381
mindless   388
prevention pride   375
prevention system   375

promotion pride   375
promotion system   375
regulatory focus
　theory   375

17

SELF-REGULATION: FROM GOAL PURSUIT TO GOAL ATTAINMENT

Veronica and Martha are now in their third year in medical school. They both have successfully finished the grueling coursework in biochemistry, and today, on the first day of class in anatomy, they dissect a cadaver. With their shiny new dissecting kits, dissecting manuals, and scrub suits, each begins dissecting a real human body for the first time. Two classmates leave the lab 15 minutes into dissection, visibly pale and distressed. This semester, they will be spending many hours next to their cadaver learning the seven ins and outs of the pterygopalatine fossa and other structures in the nasal cavity. That evening, Veronica thinks: "Wow, finally, real medical education. I'm a bit uneasy about this right now, but like everyone says, I'm sure I'll get used to it, and who knows, I might be good at this—maybe I'll choose surgery as my specialty." Martha, on the other hand, is thinking: "Ok, this is the real thing. If I can't do this, no matter how many A's I get in biochemistry, I won't make it as a doctor. But the truth is, I almost fainted today. How am I going to get through this semester when we have this class every week?"

In this chapter, we will consider findings that can help Martha answer her question: what can she do if she wants to continue to pursue her goals in medicine to overcome the stress she is experiencing? She is committed to not quitting and does not want to shift to other routes for pursuing her life goals. Given her choice, her personal challenge becomes learning how to self-regulate her own thoughts and emotions to cope more adaptively with the stresses that are sure to arise along the route to her goals. For Martha, who focuses on threats and punishments, this is likely to be very difficult. This chapter examines the cumulative contributions from different levels of analysis in personality theory and research that speak to this fundamental human dilemma: *How can people protect themselves against their own dispositional vulnerabilities when they interfere with their life goals and well-being?*

▶ SELF-REGULATORY PROCESSES

In the last chapter we looked at individual differences in types of approach–avoidance dispositions at many different levels of analysis, from the biological, to the trait-dispositional, to the phenomenological, to the social cognitive, to the psychodynamic. We then turned to the mental and emotional processes that become activated in the personality system in goal pursuit. This chapter goes beyond such descriptions. Now we will examine the specific mechanisms that make self-regulation possible as individuals engage in their characteristic approach and avoidance patterns. Again we will draw on findings from multiple levels of analysis—from the biological to the social cognitive to the experiential, motivational, and emotional.

When BIS/BAS, approach–avoidance dispositions, or promotion–prevention goals and strategies are activated, people become motivated, pursuing their goals with their preferred strategies. Who perseveres and reaches those goals even when the inevitable obstacles arise along the road? Who fails? And why? The answers show that the differences between people in this regard are large, and the implications for their lives can be profound.

Why Self-Regulate?

Automaticity. Most of what people do in their goal pursuit is virtually automatic. Automatic behavior is seen in routine activities from attending classes and going to work or interacting with friends and family, to dealing with stress and coping with temptation and frustration. Most of the time, these processes appear essentially to run

17.1 How does self-regulation relate to the concept of automaticity?

on their own as if the person were on automatic pilot (e.g., Bargh, 1997, 2001). In this sense, the reactions of both Veronica and Martha to their first experience dissecting a cadaver are their first automatic reactions. They reflect the reactivity of their respective personality systems to the new events encountered. Some of the details of those mental and emotional processes were illustrated in the last section of the previous chapter, and were summarized in Table 16.3.

These automatic mental–emotional processes activated in goal pursuit are adaptive for most life functions. But sometimes what runs off automatically may be exactly what the person is eager to *not* do—for example, to yield to the temptation of the dessert when trying to diet, or to head for the movies soon after deciding to spend the evening working.

Examples of the difficulty of overcoming the usual automatic reactive tendencies are easy to find in daily life. Every dieter knows the quick failure of good intentions when the waiter flashes the chocolate pastry on the dessert tray. Strong motivation is not enough when the temptations are intense and "in your face."

Beyond Automaticity? Self-regulation becomes important—in fact crucial—when the person experiences a distressing conflict between strong approach tendencies and strong avoidance tendencies in goal pursuit. The dieter on the first day of her new health plan experiences such a conflict when she sees her friend consuming the tempting double-chocolate cake. The medical students eager to start becoming real doctors experience such a conflict when facing their first human cadaver.

Approach (Appetitive) and Avoidance (Aversive) Dilemmas

Approach–avoidance conflicts produce dilemmas of two types:

17.2 Distinguish between appetitive and aversive dilemmas in terms of approach and avoidance motives.

1. Approach or **appetitive dilemmas** occur when facing a strong approach temptation (take the cake, grab the money, be selfish in a relationship dispute). The conflict occurs because pursuing these temptations also threatens to interfere with another even more important, higher-order goal (to have self-respect, to be a worthy person). These conflicts are common and difficult to resolve because often the higher-order goal is more abstract and temporally delayed (to stay slim, to stay healthy in the future, to avoid shame or embarrassment later, to be moral, to be caring to your partner or parent). The human paradox is that although the delayed goal is ultimately more important to the self, it also is more abstract and far off. It therefore has less immediate emotional impact and "hot stimulus pull" in the situation compared to the temptation that is staring you in the face (e.g., Trope & Lieberman, 2003; Mischel & Ayduk, 2002).

2. Avoidance or **aversive, anxiety-producing dilemmas** occur when the immediate impulse is to flee because aversive emotions like fear, distress, and anxiety are experienced. The conflict arises when to serve a higher-order goal one needs to remain in the situation and behave rationally within it. The medical students staring at the human cadaver's decomposing face impulsively feel driven to rush away, but they are tortured because they don't want to humiliate themselves, and they do want to become good doctors in the future. Again the conflict is between the impulses and pressures within the immediate situation and the more distant, higher-order outcomes later. If you rely on what comes naturally, often the immediate pull of the situation, not you, becomes the winner.

Willpower. The impressive power of the situation or "stimulus control" is seen in work at the behavioral level, and in more recent studies in social cognition (e.g., Bargh, 1997, 2001). John Bargh and his colleagues have shown that most of what people do runs off automatically without conscious intervention. It is elicited by the particular stimulus conditions in the situation, often without the person's awareness. But these findings coexist with the intuitive conviction that human beings have the capacity to take control and exert willpower at least some of the time (e.g., Ayduk & Mischel, 2002; Derryberry, 2002). Often people do overcome obstacles and temptations along the way to achieving their valued long-term goals, and manage to resist the pull of even strong situational pressures. How do they do it? What makes willpower possible?

Motivation and Competence.   Effective self-regulatory behavior and self-control in goal pursuit depends both on the person's *motivation* and on his or her *competencies*. First let's consider the role of motivation. Take the motivation to refuse a slice of double-chocolate fudge cake. The motivational strength of this may depend, for example, on whether the individual construes the cake as "unhealthy and fattening" or as "a great treat." It also depends on how much the person values the long-term super-ordinate goals that are served by eating healthy and being fit. If these self-regulatory behaviors are serving a higher goal central to the self, like being a worthy self-respecting person, their motivational significance will be high and they will be mentally more accessible.

It is important, however, to distinguish between regulatory motivation and regulatory competence. Often people have one of these but not the other: a good example is a recent president of the United States. His adroit handling of political and foreign affairs showed that he had such skills, yet he either was unable or insufficiently motivated to apply them to himself when it came to his personal affairs, and he was disgraced and almost impeached as a result.

Even when motivation is high, self-control in the face of temptations and frustrations requires more than good intentions. As William James noted more than a hundred years ago, the gap from "desiring and wanting" to "willing" cannot be bridged unless "certain preliminaries" are met (1890). Even in the presence of high regulatory motivation, goal attainment depends critically on the availability and accessibility of effective self-regulatory competencies. **Regulatory competencies** refers to the cognitive and attentional mechanisms that help us execute goal-directed behavior. In the remainder of this chapter, we will examine the processes that underlie these competencies.

17.3 What besides regulatory motivation is needed in order to manifest willpower?

The Biological Level: Executive Functions

First, we consider the nature of these regulatory competencies and mechanisms at the biological-brain level. In the last few decades, much has been learned about the neural mechanisms that underlie and enable **executive functions** in self-regulation. A century earlier, Freud and his followers had used the same term "executive functions" to refer to the ways in which the ego enables delay of gratification and impulse control. Findings from cognitive neuroscience now indicate that at the biological level many of these functions involve executive systems in the frontal cortex.

17.4 Define executive functions. Which brain regions are involved?

Brain Mechanisms in Effortful Control.   Michael Posner and colleagues have described an "anterior attentional system" that regulates the pathways involved in executive functions throughout the cortex (Derryberry, 2002; Posner & Rothbart, 1991, 1998). These brain systems enable **effortful control,** or in lay language, willpower, in goal pursuit. They do so by allowing people to regulate their attention. This attention regulation process includes the ability to focus attention in perception, to

The same person can have high self-control in some situations but not in others.

switch attention between tasks, and to control thoughts flexibly (Derryberry & Reed, 2002). These processes and brain mechanisms are crucial for effective goal pursuit and self-regulation.

Self-Report Measures.   Self-report measures also have been developed to identify individual differences in the executive attention control mechanisms that are basic for effortful control. These scales (see Table 17.1 for sample items) measure individual differences in attention control ability and capacity for effortful control at the psychological level that are correlated with the brain measures (Derryberry, 2002).

The Trait-Dispositional Level

At the broad dispositional level, attention control scores are related positively to extraversion and negatively to anxiety and impulsivity (Derryberry, 2002). These findings closely parallel those from work on the BAS/BIS, as should be expected theoretically. Individual differences in self-regulation and the ability to exert effortful control are dramatic. While some people adhere to stringent diets or give up cigarettes after years of smoking them addictively, others fail in spite of affirming the same initial good intentions. The ancient Greeks considered a weakness of the will a character trait and called it *akrasia*—the deficiency of the will. To this day, this type of trait remains important in personality theories. It is seen in the factor of "conscientiousness" within the Big Five (Chapter 3) and in the dispositions of "ego resilience" and "ego control" (e.g., Block & Block, 1980; Eisenberg, Spinrad, & Morris, 2002), two personality characteristics discussed next.

Ego Control and Ego Resilience.   **Ego control** refers to the degree of impulse control in such functions as inhibition of aggression and the ability to plan. A related construct, **ego resilience**, refers to the individual's ability to adapt to environmental demands by appropriately modifying his or her habitual level of ego control. Ego resiliency allows functioning with some "elasticity" and "permeability" (Block, 2001, p. 123). Together, these two constructs represent core qualities for adaptive functioning from an influential trait-dispositional perspective (Eisenberg et al., 2002).

TABLE 17.1   Items from the Attention Control Scale

Please read each item below and then indicate how often you experience it, using the following response scale:

1	2	3	4
almost never	sometimes	often	always

____ 1.   It's very hard for me to concentrate on a difficult task when there are noises around.

____ 2.   When I need to concentrate and solve a problem, I have trouble focusing my attention.

____ 3.   When I am working hard on something, I still get distracted by events around me.

____ 4.   My concentration is good even if there is music in the room around me.

____ 5.   When concentrating, I can focus my attention so that I become unaware of what's going on in the room around me.

____ 6.   When I am reading or studying, I am easily distracted if there are people talking in the same room.

____ 7.   When trying to focus my attention on something, I have difficulty blocking out distracting thoughts.

____ 8.   I have a hard time concentrating when I'm excited about something.

____ 9.   When concentrating, I ignore feelings of hunger or thirst.

____ 10. I can quickly switch from one task to another.

Source: Derryberry, D., & Reed, M. A. (2002). Anxiety-related attentional biases and their regulation by attentional control. *Journal of Abnormal Psychology, 111,* 225–236.

As Jack Block (2001, pp. 123–124) puts it: "The resilient person anticipates wisely when to stop something unfruitful (like repetitively hitting a large boulder with a tiny hammer) or to continue something that may ultimately prove fruitful

Jack Block

(like shattering a small boulder with a repetitive sledge hammer). In its adaptiveness, resiliency well serves evolution." Note that this illustration also highlights the importance of the ability to discriminate, and the flexibility that this allows—an emphasis now found in the evolutionary approach, in the social cognitive approach, and in findings from the biological level of analysis (e.g., Grigorenko, 2002).

17.5 How have ego-control and ego resilience been measured in children? How are they related to adaptive behavior?

A number of tasks are used to measure the construct of ego control and ego resilience. For example, in one study researchers rated children's tendency to inhibit impulses to infer their level of ego control and also observed their delay behavior in experimental situations (Block & Martin, 1955). They exposed the children to a frustration in which a barrier separated the child from desired and expected toys. The **undercontrolling** children (those who had been rated as not inhibiting their impulses) reacted more violently to the frustrating barrier than did **overcontrolling,** inhibited children. The undercontrolling youngsters also became less constructive in their play.

Individuals who are high (rather than low) on indices of ego control tend to be somewhat more able to control and inhibit their motor activity. For example, they may be able to sit still longer or draw a line more slowly without lifting their pencil. These are only a few examples from much larger networks of correlations that support the constructs of ego resiliency and ego control (Block & Block, 1980; Mischel, 1984; Shoda, Mischel, & Peake, 1990).

In studies of the ego-resiliency construct, toddlers were evaluated for the degree to which they seemed secure and competent (in a problem-solving task). The toddlers who were secure and competent also scored higher on measures of ego resiliency when they reached the age of four to five years (Gove, Arend, & Sroufe, 1979; Matas, Arend, & Sroufe, 1978). As another example, ego-resilient children at age three are also viewed as more popular, interesting, and attractive at later ages (Block & Block, 1980). The resilience concept also is related to delay of immediate gratification for the sake of more valued but delayed outcomes (Shoda, Mischel, & Peake, 1990), discussed below. In short, both the concepts of ego control and ego resiliency offer useful characterizations of important individual differences (Eisenberg et al., 2002). These differences remain stable over many years in the course of development (e.g., Caspi, 1987; Caspi & Bem, 1990; Mischel, Shoda, & Rodriguez, 1989).

The Social Cognitive and Phenomenological Levels

In sum, work at the biological level is identifying the brain centers that enable executive skills and self-regulation. Work at the dispositional level on characteristics like ego resilience provides useful descriptions of what a person with self-regulatory competence is like. What about the social cognitive and phenomenological levels? Work at these levels currently tries to understand the mental mechanisms—the thoughts and strategies—that enable self-regulatory competence.

17.6 What questions about self-regulation are explored at the social cognitive level of analysis?

Cognitions, Emotions, and Attention in Self-Regulation. Specifically, the question asked is:

> *How does what you think, or where you focus your attention, enable or undermine your ability to self-regulate in emotion-arousing dilemmas?*

Applied to the dieter who has resolved to forego the double-chocolate cake tonight, how will what he thinks and how he deploys his attention make it possible for him to stick to his goal? Applied to Martha's dilemma as she tries her first dissection, how will what she says to herself and thinks make it possible for her to appropriately dissect the cadaver?

To answer such questions, researchers at this level have examined how mental activities—cognition and attention—during goal pursuit influence the ability to persist and reach the goal when there are strong approach–avoidance conflicts (e.g., Mischel & Ayduk, 2002). To orient you to those findings, let's anticipate their gist before they are discussed next in more detail. Briefly, they show that how the person construes and interprets the situation, and deploys attention while attempting effortful control, has effects on effortful control ability that directly parallel and complement those identified at the brain level in the work of Posner and his colleagues.

▶ EMOTION REGULATION IN APPROACH (APPETITIVE) DILEMMAS

In this section, we will examine how certain mental processes enable—or under-mine—self-regulation and goal pursuit in approach or appetitive dilemmas, as in the dieter's struggles with dessert temptations. Then in the section that follows, we will do the same with regard to avoidance and aversive dilemmas like the one Martha faced in her anatomy exercise. In both types of dilemmas people need willpower, or rather, the mental processes that make willpower possible. One of the clearest examples of the need for willpower, or in scientific terms, effortful control, is seen when people try to defer immediate gratification for the sake of important but delayed consequences.

Delay of Gratification as a Basic Human Task

The ability to voluntarily delay immediate gratification, to tolerate self-imposed delays of reward, is at the core of most concepts of willpower, ego strength, and ego resilience. It is hard to imagine civilization without such self-imposed delays, and it was Freud who called attention to the great importance of this ability as a requirement for ego development a century ago. Learning to wait for desired outcomes and to behave in light of expected future consequences is essential for the successful achievement of long-term, distant goals.

17.7 Why is delay of gratification an essential self-regulation skill? Describe the methods used to study this ability in children.

Beginning with toilet training, every person must learn to defer impulses and needs and to express them only under special conditions of time and place. To achieve educational and personal life goals, such as a college degree or a professional or busi-ness career, the route requires a continuous series of delays of gratification. This is seen in the progression from one grade to the next, and from one barrier to another in the long course from occupational choice to occupational success. In social relationships, the culture also requires delays, for example, the expectation that people should post-pone sexual relations, marriage, and children until they are "ready for them." The importance of self-control patterns that require delay of gratification has been widely recognized by theorists from Freud to the present. It is personally familiar to every student who has tried to persist with school work due the next day when greater temptations were available at the moment. The challenge for self-regulation research is to answer the question: Given that one has chosen to wait or work for a larger deferred goal, how can the delay period be managed and the impulse to take the immediate reward be resisted?

The Goal-Directed Delay Situation. A good method for the study of willpower in such "hot" approach–appetitive situations has proved to be the preschool delay of gratification paradigm (see Figure 17.1). It is a method that has been used extensively to study self-regulatory processes at early periods in development when they begin to emerge in the young child (Mischel and associates, 1989; Mischel & Ayduk, 2002).

In this method, the experimenter deliberately creates a dilemma for the young child. Imagine that a four-year-old is shown some desired treats, for example, marshmallows or little pretzel sticks. The child faces the following conflict: wait until the experimenter returns and get two of the desired treats, or, ring a bell, and the experimenter will come back immediately—but then the child gets only one treat. After the child chooses to wait for the larger outcome, the delay soon becomes very difficult and the frustration grows quickly. This situation has become a prototype for studying the conflict between an immediate smaller temptation and a higher-order but delayed larger goal, the bigger treat that will come later (when the experimenter returns). In this type of situation, more than 500 preschoolers in the Stanford University community were studied systematically through observation and experiments (e.g., Ayduk & Mischel, 2002; Mischel et al., 1989).

One of the most striking delay strategies used by some youngsters was as simple as it was effective. These children seemed to manage to wait for the preferred reward for long periods apparently by converting the aversive waiting situation into a more pleasant, nonwaiting one. They seemed to do this by elaborate self-distraction techniques through which they spent their time psychologically doing something (almost anything) other than waiting. Rather than focusing prolonged attention on the objects for which they were waiting, they avoided looking at them. Some of these children covered their eyes with their hands, rested their heads on their arms, or found other similar techniques for averting their eyes from the reward objects. Others seemed to try to reduce the frustration of delay of reward by generating their own diversions: they talked to themselves, sang little songs, invented games with their hands and feet, and when all

17.8 What methods were used by children able to delay gratification? How was this strategy studied experimentally, and with what results?

Figure 17.1 Waiting for delayed gratification.
Source: Based on Mischel, W., Ebbsen, E. B., & Zeiss, A. R. (1972). Cognitive and attentional mechanisms in delay of gratification. *Journal of Personality and Social Psychology, 21,* 204–219.

other distractions seemed exhausted, even tried to fall asleep during the waiting situation—as one child successfully did.

These observations suggested that diverting yourself from attention to the delayed reward stimulus (while maintaining behavior directed toward its ultimate attainment) may be a key step for effective delay of reward. That is, learning *not* to attend to or think about what you are awaiting may enhance effective delay of gratification much more than focusing attention on the outcomes. When hungry, for example, it seems easier to wait for supper if one is not confronted with the sight and smell of food. A series of experiments summarized next tested these hypotheses.

Cooling Strategies: It's How You Think That Counts

Strategic Self-Distraction. Directing attention away from the rewards turned out to be an excellent strategy. For example, Figure 17.2 shows results of an experiment that manipulated the extent to which children could attend to the reward objects while they were waiting (Mischel, Ebbesen, & Zeiss, 1972). As shown in the figure, in one condition, the children waited with both the immediate (less preferred) and the delayed (more preferred) reward facing them in the experimental room so that they could attend to both outcomes. In another group, neither reward was available for the child's attention, both having been removed from sight. In the remaining two groups, either the delayed reward only or the immediate reward only was available for attention while the child waited. The measure was the length of time before each child voluntarily stopped waiting. As the graph shows, the child waited much longer when the rewards were not available for attention, supporting the observations and hypotheses discussed above.

Many experiments have demonstrated that voluntary delay of gratification can be aided by activities that serve as distractions from the rewards and thus from the

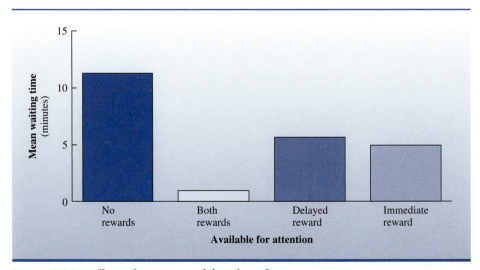

Figure 17.2 Effects of attention on delay of gratification.

Source: Based on data from Mischel, W., & Ebbsen, E. B. (1970). Attention in delay of gratification. *Journal of Personality and Social Psychology, 16,* 239–337.

aversiveness of wanting them but not having them. Through such distraction, it is possible to convert the frustrating delay-of-reward situation into a psychologically less aversive condition. For example, children waited much longer for a preferred reward when they were distracted cognitively from the goal objects than when they directly attended to them (Mischel, Ebbesen, & Zeiss, 1972; Mischel, Shoda, & Rodriguez, 1989). In contrast, distracting oneself from the rewards, overtly or cognitively (e.g., by thinking of "fun things") made it possible to wait longer for them. These results showing the value of self-distraction for avoiding temptation are consistent with the old saying "Satan, get thee behind me."

Rather than trying to maintain aversive activities such as delay of reward through "acts of will" and focused attention, effective self-control is helped by mentally *transforming* the difficult into the easy, the aversive into the pleasant, and the boring into the interesting, while still maintaining the activity on which the ultimate reward depends. Rather than "willing" oneself to heroic bravery, one needs to enact the necessary "difficult" response while engaging in another one cognitively (Mischel et al., 1996; Peake, Hebl, & Mischel, 2002). To do this effectively, when the task is complex, the individual may need to extensively rehearse and plan to implement the necessary actions (Gollwitzer & Moskowitz, 1996).

17.9 How do hot and cool construals of rewards differ? How can they be used to maintain motivation and resist immediate gratification?

Hot and Cool Construal. The outcomes or rewards in this type of situation may be construed in terms of their "hot," consummatory properties, or in terms of their "cool," informative properties. A hot representation of a reward such as a desired treat activates the behaviors associated with experiencing or consuming it. When young children look at the actual rewards, or think about them, they may focus spontaneously on these hot, arousing qualities. If they do this, they increase their own frustration and arousal, making it more difficult to continue to wait (e.g., Mischel & Ebbesen, 1970; Mischel, et al., 1972; Toner, 1981). In contrast, a psychologically distant or cool representation of the rewards has the opposite effect on self-regulation, enhancing goal-directed waiting (see Figure 17.3). Thus, presenting children with a picture of the rewards (e.g., in the form of life-size slide-projected images) or instructing them to cognitively transform them (e.g., by turning the pretzels into thin, brown logs in their imagination) actually facilitates waiting (Mischel, Shoda, & Rodriguez, 1989). In short, just how one thinks about the rewards (hot or cool) in the contingency crucially

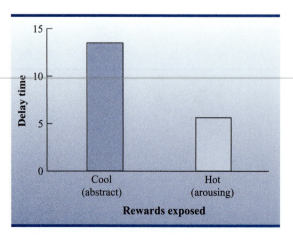

Figure 17.3 Effects of hot–cool thinking on delay time.

Source: Based on studies reported in Mischel, W., Shoda, Y., & Rodriguez, M. L. (1989). Delay of gratification in children. *Science, 244*, 933–938.

influences how long one can wait or work to attain them (Metcalfe & Mischel, 1999; Peake et al., 2002).

The children's behavior during the delay period (Mischel & Ebbesen, 1970) also suggested how they support their efforts at self-control. At points throughout the waiting period, most of the children's spontaneous attempts at self-distraction wavered. Those children who persisted often used strategies such as verbalizing self-instructions to remind themselves of the contingency (e.g., "If I wait I get [the preferred reward] but if I ring the bell I get [the less preferred reward]") and of their goal. In short, in effective delay of gratification the child tunes out the hot properties of the reward. Instead, to sustain her waiting behavior she focuses on its cool properties and the contingency.

Flexible Attention. Adaptive self-regulation involves more than application of cooling strategies. Successful delay and goal pursuit often require shifting attention flexibly rather than unconditional use of cooling strategies. Experiments confirm that some of the children who delay best seem to focus briefly on the hot features in the situation to sustain motivation but then quickly switch back to the cool features and self-distraction to avoid excessive arousal and frustration (Peake et al., 2002).

Effective self-regulation requires sensitivity to the demands, constraints, and opportunities of the particular situation. Such **discriminative facility**—taking into account characteristics of each situation and responding accordingly—may play a central role in coping, self-regulation, and social-emotional competence in general (see Cheng, 2003; Chiu, Hong, Mischel, & Shoda, 1995). Distraction, or cooling, then, is likely to be adaptive when applied to coping in aversive or frustrating situations that must be tolerated for goal attainment, but not in many other types of situations. The key is to know when to cool, when to stay warm, and to be able to do so when you need to. The mechanisms described underlie the kind of flexible, discriminative behaviors that Jack Block (2001) also saw as the essence of the resilient person.

In sum, in this section we have focused on voluntary delay of gratification when trying to control impulses in the process of waiting or working for a positive delayed outcome or goal. The findings also fit those from many other areas of self-regulation (Harter, 1983; Kopp, 1982; Kuhl, 1985), as illustrated next.

▶ EMOTIONAL REGULATION IN AVOIDANCE (AVERSIVE) DILEMMAS

The strategies that help people deal with the control of appetitive impulses as in the delay situation also apply to emotional self-regulation for dealing with aversive stimulation, fears, and stress-inducing situations. An example to keep in mind is the anxiety-producing situation in the anatomy room in which Veronica and Martha faced the dissection of their first human cadaver.

In a study that directly speaks to the self-regulatory problem these students had at the anatomy table, researchers interviewed medical students witnessing a medical autopsy for the first time (Lief & Fox, 1963). They found that many aspects of the autopsy procedure appear to have been designed as an institutionalized ritual to help students cope with their distress by making it easier for them to detach themselves

17.10 Describe some advantages and undesirable consequences of cognitive detachment in aversive situations.

emotionally. The autopsy room itself is arranged to provide a sterile, clinical, impersonal atmosphere. The face and genitalia of the corpse are kept covered unless they are being examined, and so on. But beyond the physical arrangements, it is a detached, emotionally distanced scientific stance that further helps to keep the whole procedure abstract and impersonal, helping to reduce emotional arousal. But as is always the case, adopting such a stance is easier for some people than for others. For Martha, with her readiness to scan for potential troubles down the road and to become anxious at the prospects of all the things that can go wrong, both in the lesson and in her career, dealing with this was particularly challenging.

Experimental research confirms that an attitude of detachment helps people react more calmly when exposed to gory scenes portraying bloody accidents and death (Koriat et al., 1972) or when expecting electric shock (Holmes & Houston, 1974). These results are consistent with reports showing that soldiers may immunize themselves against emotion by distancing themselves psychologically from their victims, for example, by calling them "gooks" and labeling them as subhumans. While highly effective for reducing feeling, a detachment strategy to reduce emotionality can be misused, producing callous, insensitive attitudes and cold-bloodedness toward others, in war, in medical practice, and in much of social life. On the other hand, in many life situations, detachment through cognitive reappraisal to attain emotional cooling provides a basic route for effective emotion regulation, as discussed in In Focus. 17.1.

Cognitive Transformations to Deal with Stress

Not surprisingly, a good deal of related research leads to the conclusion that self-distraction, when possible, can be an excellent way to manage unavoidable stresses like unpleasant medical examinations (Miller, 1987; Miller & Green, 1985) and coping with severe life crises (Taylor & Brown, 1988). Self-distraction (e.g., watching travel slides or recalling pleasant memories) increases tolerance of experimentally induced physical pain (e.g., Berntzen, 1987; Chaves & Barber, 1974). Similarly, distracting and relaxation-inducing activities such as listening to music reduce anxiety in the face of uncontrollable shocks (Miller, 1979, 1996), helps people cope with the daily pain of rheumatoid arthritis (Affleck et al., 1992) and even with severe life crises (e.g., Taylor & Brown, 1988). "Cooling" strategies generally can help one to transform potentially stressful situations to make them less aversive. For example, if surgical patients are encouraged to reconstrue their hospital stay as a vacation to relax a while from the stresses of daily life, they show better postoperative adjustment (Langer, Janis, & Wolfer, 1975), just as chronically ill patients who reinterpret their conditions more positively also show better adjustment (Carver et al., 1993). In sum, when stress and pain are inevitable, the adage to look for the silver lining and to "accentuate the positive" seems wise.

The type of cognitive strategy that helps one to deal best with stress also depends on the individual (Miller, 1987). Recall, for example, individual differences in the tendency to use distraction or "blunt" rather than to sensitize or "monitor," discussed in the chapter on repression as a mechanism of defense (Chapter 6). The point to remember is that such preferences in the type of information sought for dealing with stress affect how the person copes best with stress; while one strategy may help some people, the opposite strategy may help others. The task is to get the right match.

COOLING UNWANTED EMOTIONS: COGNITIVE REAPPRAISAL IS BETTER THAN DENYING NEGATIVE FEELINGS

When people are trying to hide their secret thoughts, or reduce the pain and fear of viewing a horrible event, what processes allow them to regulate their negative emotions? Stanford University psychologist James Gross and his colleagues have focused on two specific emotion regulation strategies. One is **cognitive reappraisal,** the other is **suppression.**

In a typical study, Gross brings participants into the laboratory and informs them that they will be watching a movie. The film they will see shows detailed, close-up views of severe burn victims or of an arm amputation. Participants then are divided into one of three different groups and given different instructions prior to viewing the film. In the cognitive reappraisal condition, they are asked to use a cooling strategy. Specifically, they are asked to try to think about the movie in a detached unemotional way, objectively, and to focus attention on the technical details of the event, not feeling anything personally (e.g., pretend that you're a teacher in medical school). In the suppression condition, participants are asked to try to hide their emotional reactions to the film as they watch it so that anyone seeing them would not know that they were feeling anything at all. In the control condition, participants are simply asked to watch the movie.

This and similar studies find that cognitive reappraisal is an overwhelmingly more adaptive way to regulate negative emotions. It is much better than suppression in which people hide the feelings they are experiencing. These differences are seen in how the two strategies influence the intensity of people's experiences as well as their effect on the person's level of physiological autonomic nervous system arousal and distress. Specifically, people who are told to think about the movie in a way that cools the emotional content (reappraisal condition) experience fewer feelings of disgust and less physiological activation (evidenced by less blood vessel constriction) when compared to people who are asked to try to hide and suppress their emotional responses to the film so that no one could see their real feelings on their faces (Gross, 1998).

Cognitive reappraisal and suppression produce differences at the cognitive level of analysis as well. In a variation of the study described above, participants saw a slide show instead of a movie and were later asked how much of the information from the slide show they remembered. People who suppressed during the experiment performed more poorly on the memory task relative to people who engaged in cognitive reappraisal (Richards & Gross, 2000). In addition, people who have a tendency to suppress perform more poorly on memory tasks than those who don't tend to suppress (Gross & John, 2003). Specifically, people who score high on the suppression scale of the Emotion Regulation Questionnaire perform more poorly than those who score low on the suppression scale.

Finally, note that the results from this work on emotion regulation are consistent with a psychodynamic level view of the relative ineffectiveness of emotional denial: trying to hide one's feelings is a poor defense mechanism (Chapters 6 and 7).

▶ INTERACTION AMONG LEVELS: HOT AND COOL SYSTEMS IN SELF-REGULATION

17.11 How did James Gross study the relative effectiveness of emotional suppression and cognitive restructuring? What were the results?

So far, we have considered the findings on self-regulation from each level of analysis, one at a time. This can create a somewhat misleading impression that these are independent sources of influence. In fact, the biological brain system and the social cognitive system of processing dynamics activated in the particular situation are in a continuous dynamic two-way interaction process. What Martha thinks when dealing with the dissection, and where she focuses her attention (on the ghastly tortured expression on the face, on the intricate ways in which the capillary system is organized) influence what becomes activated in the brain, which influences what she next feels and thinks, on and on. To reiterate, the psychological system and the biological system interact: to understand them requires attention to how they play out jointly.

Specifically, with regard to both approach and avoidance dilemmas, the person's thoughts, cognitive appraisals, and attention deployment are related to, and influence, what happens in the relevant brain systems. That process, in turn, influences what is experienced emotionally and cognitively in continuous interaction. At the social cognitive level, cooling operations make it easier to activate long-term goals and reduce the hot pull of the immediate emotion-producing triggers in the situation. Consequently, what becomes activated in the brain and at the neural level is modified as well. To understand this more deeply, we have to examine the relevant interactions.

The Emotional (Hot) Brain/The Rational (Cool) Brain

To reiterate, the situations in which people most need and want to control their impulses often are those in which it is most difficult for them to do so. These are the situations that elicit "hot" emotional reactions such as intense fear and anxiety, or strong appetites and cravings. These were seen in miniature form in the studies of little children trying so hard to wait for bigger treats but unable to hold out as the temptation to "get it now" becomes too strong. These kinds of hot situations tend to automatically trigger hot reactions—to eat the dessert, take the drug, grab the hundred-dollar bill dropped in your path. Such situations rapidly generate the associated feelings of fear or desire and the urge to respond impulsively. They lead people to bypass self-regulatory controls and self-standards just when they need them most. Why does that happen so easily?

Bodily Changes: Emotion in Stress. When an event triggers automatic fear reactions or is perceived as stressful, physiological reactions in the body are triggered at the biological level. Such feelings as "makes my heart race," and "choked up with rage" or "ready to vomit" are experienced psychologically in emotional states that involve bodily changes. It is the activation of the sympathetic division of the autonomic nervous system that produces many of the physiological changes that are experienced in reaction to perceived stress. They include a cascade of events such as channeling of the blood supply to the muscles and brain, slowing down of stomach and intestinal activity, increase in heart rate and blood pressure, and the pupils in the eyes widen. There is an increase in rates of metabolism and respiration, increase in sugar content of the blood, decrease in electrical resistance of the skin, increase in speed of the blood's ability to clot, and even "goose pimples" resulting from the erection of hairs on the skin.

17.12 How can stress responses have evolutionary value but be maladaptive in today's world?

Fight or Flight Reactions. These changes are referred to as parts of the automatic **fight or flight reaction,** because they are considered to be the body's emergency reaction system to threat and danger. At an evolutionary level, this automatic reaction pattern reflects an adaptive system developed over the course of the species' history for getting ready to cope with environmental challenges. These biological reactions enable the organism to cope rapidly and effectively with diverse environmental dangers, in part by providing the alertness and energy necessary for survival in the face of attack. This mechanism was crucial for survival in earlier ages in the species' history when physical dangers lurked everywhere.

Activation of this system can save your life when you face a real emergency—for example, when about to be assaulted in a dark alley, or in the midst of a sudden fire.

However, in contemporary life, it is often elicited in response to situations that do not objectively endanger survival. When you face a French examination, or a cadaver you have to dissect in an anatomy class, or walk into the boss's office to ask for a raise, it may feel high risk; but adaptive behavior in dealing with such challenges requires more fine-grain reactions—like thinking and planning—rather than fight or flight. In these situations, automatic hot emotional reactions, particularly intense fears, become debilitating rather than adaptive (e.g., Le Doux, 1996, 2001). And when individuals have these experiences persistently over long periods of time, they can become dangerous to physical health and well-being. The Type A pattern used to illustrate a dispositional typology in Chapter 3 is an example.

In sum, hot reflexive reactions may be part of the overall arousal state that helps initiate quick adaptive action, as in an emergency response to sudden dangers that mobilize the body's resources like a fire alarm. In the course of human evolution they are sure to have had important adaptive advantages for survival. However, this alarm system can make reflection and self-regulation most difficult (Metcalfe & Mischel, 1999).

The Hot Amygdala. A small almond-shaped region in the forebrain, buried under the prefrontal cortex called the **amygdala** (which means "almond" in Latin), is crucially important in emotional reactions, particularly fear (LeDoux, 1996). The central nucleus of this brain structure reacts almost instantly to signals that warn of danger, immediately sending out behavioral, physiological (autonomic), and endocrine responses. It mobilizes the body for action, readying it to flee or flight. As noted above, presumably this reflexive emergency reaction was useful for adaptation. From an evolutionary perspective, it had survival value to be able to react automatically to the snake in the grass without having to think about it, or to fight the opponent who is ready to strike when flight is not possible. The amygdala also seems to play an important role in strong appetitive behaviors, leading to impulsive approach behaviors, as when the sizzling pizza becomes irresistible to the hungry dieter, or the preschooler can't resist the tempting immediately available treat in the delay experiments.

17.13 What brain areas underlie the hot and cool behavior regulation systems, and how do they interact with one another?

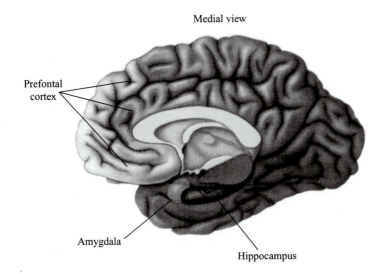

Medial view

Prefontal cortex

Amygdala

Hippocampus

The amygdala is important in fear and emotional reactions.

The Rational Cool Brain.   But these automatic reactions are only a "quick fix" that actually doesn't work well for most situations people face in modern life (LeDoux, 1996, p. 176). Unlike lower animals on the evolutionary ladder, human beings have the capacity to take control with higher-level brain centers (the prefrontal cortex). This makes it possible for the person to start cool rational thinking to try to solve the problem that the amygdala has already begun to respond to automatically and emotionally. How you think—hot or cool—can change the attention control centers activated, which in turn makes self-regulatory efforts either more or less difficult (Derryberry, 2002; Posner & Rothbart, 1998). In short, humans have an emotional brain. But they also have other areas of the brain that are crucially important for the more rational higher-level processes that make the species human. While part of the brain is emotional, there also is a thoughtful part that can be rational and used to solve problems effectively. Both areas of the brain interact in dealing with emotion-arousing dilemmas. The Hot System/Cool System Interaction model of self-regulation considered next was designed to understand these interactions.

Biosocial Model of Self-Regulation: Hot System/Cool System Interaction

To understand how the person's thoughts and feelings interact to enable or prevent self-control, a two-system biosocial model of willpower was developed. In this model, self-regulation is a function of a balance between two processing systems: the **emotional, hot system** and the **cognitive, cool system** (Metcalfe & Mischel, 1999). The theory proposes that the emotional, hot system generates automatic approach–avoidance or fight-or-flight responses. Biologically it is based primarily in the amygdala. In contrast, the cognitive cool system, based in the **hippocampus** and frontel lobe, is the locus of cognitive thought processes, generating thoughtful reflective reactions and plans. At relatively low levels of stress, the two systems work in concert, but as stress levels and aversive arousal (e.g., frustration) increase, hot-system processing begins to dominate the cool system. Thus, effective self-regulation hinges on being able to access cooling mechanisms to reduce negative arousal and suppress hot system activation when needed (Metcalfe & Mischel, 1999).

The *cool cognitive system* is specialized for complex mental and spatial representation and thought. It is cognitive, emotionally neutral, contemplative, flexible, integrated, coherent, slow, strategic—the seat of self-regulation and self-control. The *hot system* is specialized for quick emotional processing and automatic responding to "hot" trigger stimuli, as when rejection cues elicit abusive behavior from highly rejection sensitive people. It is quick and automatic, and is the basis of emotionality. It activates fears as well as passions, and it is impulsive and reflexive. This system is fundamental for emotional (classical) conditioning, but its activation undermines efforts at self-control, reflective thought, and planfulness.

The balance between the hot and cool systems is influenced both by stress and by the person's level of development. Stress—both chronic and situational—enhances the hot system and reduces activity in the cool system. With maturation beginning at about age 4 years, the cool system becomes more developed and active, and the hot system becomes more manageable (see In Focus 17.2 for discussion of the brain mechanisms involved as maturation normally occurs).

Attention Control.   How does all this translate into what people have to do for effective goal pursuit and resistance to temptations when dealing with the kinds

NEURAL MECHANISMS IN IMPULSIVE VIOLENCE

Recent advances in brain imaging technology have allowed researchers to understand how impulsive "hot" behavioral tendencies, seen for example in outbursts of violence and impulsive aggression, play out in underlying neural processes at the brain level of analysis. Among the many scientists that are active in this area of research, University of Wisconsin psychologist Richard Davidson and his colleagues have been pinpointing specific patterns of brain activation that underlie the self-regulation of impulsive behavior (Davidson et al., 2000).

Drawing on both animal and human brain research, Davidson and his colleagues have found that emotion is normally regulated in the human brain by a complex circuit of brain structures involving several interconnected regions. Brain imaging studies suggest that in a typical sequence of events, threatening behaviors (staring eyes, shouting, etc.) are conveyed to the amygdala, which then sends information to other higher regions of the brain that are responsible for initiating behavioral reactions.

In normal people, however, this pattern of events (threatening stimuli → amygdala activation → response) does not always lead to impulsive emotional responses. A number of prefrontal brain regions connect to the amygdala and interact to constrain the impulsive expression of emotional behavior. These connections provide people with the ability to self-regulate their emotional responses. It is the prefrontal brain regions that seem to constitute the core structures underlying the process of emotion regulation and are thus essential for achieving emotional restraint at the biological level. Consequently, functional or structural abnormalities in one or more of these regions, or in the interconnections among them, are believed to increase people's impulsive tendencies. Indeed, Davidson and his colleagues' research suggest that individuals who are predisposed to aggression and violence many suffer from an abnormality in this prefrontal circuitry.

of dilemmas discussed throughout this chapter? Extrapolating from the delay of gratification research and other studies on self-regulation (Mischel et al., 1996; Derryberry, 2002), effective self-regulation requires guiding the automatic hot emotional responses. When the person thinks of the rewards that can be obtained right away in a "hot" way, they become irresistible. But through mental transformations and strategic self-distractions, one can effectively continue to delay by "cooling" the hot pull of the immediate gratifications (Mischel et al., 1989). In Martha's case, she is struggling to reduce the aversive arousal created in the dissection situation, and to control herself from automatically dashing out of the room. For that goal, she has to shift attention away from her distressing feelings and focus instead on the complex technical details of the brain structures to prevent her from messing up her dissection.

Cool cognitions focused on the technical details of the procedures allow her to regulate her feelings adaptively, and shift them away from her distress and empathy with a dead human on to the medical task that needs doing. But that can change in a millisecond if suddenly something in the face reminds her that it looks a little like her father when he died.

When people manage to cool their impulsive hot reactions by attention control, they can make use of the vast cognitive resources that give humans their evolutionary advantage (Metcalfe & Jacobs, 1998; Metcalfe & Mischel, 1999). In other words, the trick is to think cool when your impulse is to act hot, and reduce hot ideation so that automatic hot responses are not made. Then problems can be dealt with more rationally, allowing long-term goals to be pursued effectively even in the face of strong momentary temptations to forget them. The ability to use attention control effectively in this manner is related closely to mechanisms that are activated in the brain. The first

17.14 How can voluntary attentional control guide hot and cool systems in an adaptive fashion?

chapter of this text described the case of Charles Whitman, the Texas Tower killer who suddenly massacred people on the college campus below him. Many years later it now is becoming possible to understand what may go wrong in such cases at the biological level, as discussed in In Focus 17.2.

Social Emotions Enable Self-Regulation: Links to Evolution

17.15 Why are social emotions an important evolutionary mechanism for human life? What is their relation to Freud's superego?

The discussion so far might easily create the impression that self-regulation always requires cooling one's emotions so that rational, problem-focused thinking can proceed effectively. In fact, it all depends on which emotions are involved. There are many social emotions that activate feelings like loyalty to the group or a partner. Such emotions can support self-regulation by calling attention to the long-term consequences of behaving, for example, in selfish or disloyal ways. Observations of life in diverse cultures throughout the world lead anthropologists and other social scientists to that conclusion as well (e.g., Fiske, 2002; Nesse, 2001). Their analysis again draws on evolutionary theory.

For example, the anthropologist-psychologist Alan Fiske (2002) proposes that from an evolutionary perspective, certain emotions have developed and are universally experienced because they are useful for self-regulation and make social life and long-term relationships possible in human society everywhere. The emotions that are experienced, however, also are shaped by the local demands and structures of the particular culture. Fiske argues that people need to have emotions that represent their important relationships, reflected in such feelings as empathy, affection, and loyalty.

Further, because humans are naturally impulsive and self-serving creatures driven to immediate gratification, they need to experience these essentially social and moral emotions, including guilt and shame, that allow them to take account of the long-term consequences of their actions. Such emotions help people to curb their appetites and prevent themselves from acting in impulsive, relationship-destructive and community-destructive ways. They are experienced as quick internal signals that help activate rapidly the self-control necessary for cooperation, for resisting temptation, and for maintaining duties and obligations. In short, this evolutionary and anthropological view casts what Freud called the superego in a new light. It sees the social-moral emotions as reflecting adaptive evolutionary processes. The basic point is that humans have intense social and moral emotions because they motivate them to "curb their non-social appetites in the interest of the relationships that are so crucial to their survival, reproduction, and welfare" (Fiske, 2002, pp. 173–174). These ideas are still in need of much research before they allow for firm conclusions. However, they help make clear that self-regulation reflects a basic human process shaped by a wide range of interacting influences, and is central for understanding the nature of personality.

▶ SELF-REGULATORY COMPETENCE

So far, we examined the processes that make self-regulation possible and that underlie individual differences in this ability. Much research also documents that self-regulatory competence is a person variable that has substantial stability and significantly predicts many aspect of the person's future.

A Stable Person Variable

Long-Term Stability of Self-Regulatory Competence. Self-regulatory competence as assessed in the delay of gratification paradigm in early childhood turns out to be a reliable index of a stable competence basic for many aspects of personality. First of all, it predicts important outcomes in development, adolescence, and adulthood. Studies with children tested in preschool showed that the number of seconds the child was able to delay gratification as a preschooler predicted positive social and cognitive outcomes many years later, including in adulthood.

17.16 What is the evidence that self-regulatory competence is a stable personality factor that aids adjustment?

To illustrate, those who waited longer in this situation as preschoolers were described by their parents as more socially and cognitively competent teenagers. They were perceived as more able to manage stress and exert effective self-control in diverse frustrating situations in adolescence (Mischel et al., 1989). They also had substantially higher SAT scores than the children who could not wait (Shoda et al., 1990). In a follow-up conducted when these individuals reached their early thirties, correlations remained significant between their preschool delay behavior and adult-relevant measures of social-cognitive competence, goal-setting, planning, and self-regulatory abilities (Ayduk et al., 2000).

Long-Term Protective Effects. More exciting than these kinds of predictions is the finding that this type of self-regulatory competence can have significant long-term protective effects. Self-regulatory competence helps to protect people from experiencing many of the negative consequences associated with their dispositional vulnerabilities. Consider, for example, the tendency to be highly rejection sensitive (RS) in interpersonal relationships. Often, highly rejection sensitive people in time develop lower self-esteem and become either more aggressive or more depressed which, in turn, undermines the quality of their lives. But that is not necessarily their fate. In an adult follow-up of the preschoolers who had participated in the original delay of gratification studies 20 years earlier, preschool delay ability predicted adult resiliency against the potentially destructive effects of RS (Ayduk et al., 2000, Study 1). Specifically, high RS people who were able to delay gratification longer in preschool were buffered in adulthood against low self-esteem and self-worth, were better able to cope with stress, and had greater ego resiliency. High RS people who were unable to delay gratification in preschool had lower academic achievement and more frequent cocaine/crack use than low RS people. As Fig. 17.1 shows, high RS people who had high preschool delay ability were buffered against such negative outcomes.

A parallel study, conducted among low-income, urban, minority middle-school children who are at higher demographic risk for maladjustment, replicated these findings with population-appropriate measures (Ayduk et al., 2000, Study 2). Again, among children high in RS, delay of gratification ability was associated with lower aggression against peers, greater interpersonal acceptance, and higher levels of self-worth.

Individual differences in attention control are visible already in the behavior of toddlers, and they predict self-regulatory competencies years later in development. In one study, mother–toddler interactions were observed as they unfolded at 19 months of age (Sethi and colleagues, 2000). Some toddlers used effective attention deployment strategies, such as self-distraction, to cope with and "cool" distress while briefly separated from their mothers in a playroom experiment. These tended to become the preschoolers who also significantly used more effective attention strategies to help themselves to delay gratification when tested in preschool 3 1/2 years later. When they were toddlers, they had actively directed attention away from the mother's absence, distracting themselves by exploring the room and playing with toys. In contrast, other toddlers were unable to activate such cooling strategies at 19 months. Instead, they stayed focused on the absence of the mother (e.g., clinging to the door) and became distressed. These tended to be the youngsters who also were less able to effectively

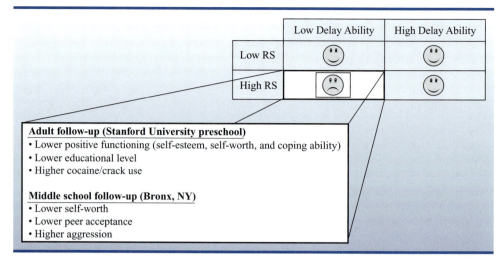

Figure 17.4 Chronic vulnerability (rejection sensitivity) is linked to negative outcomes (☹) when delay ability is low.

Note: Smiley faces ☺ show positive outcomes, frowning faces ☹ , negative outcomes. High delay ability protected even high rejection sensitive people from the negative outcomes experienced by equally rejection sensitive people who also were low in delay ability. Only the combination of High RS, Low Delay predicted significantly negative life outcomes.

Source: Based on studies in Ayduk, O., Mendoza-Denton, R., Mischel, W., Downey, G., Peake, P. K., & Rodriguez, M. (2000). Regulating the interpersonal self: Strategic self-regulation for coping with rejection sensitivity. *Journal of Personality and Social Psychology, 79,* 776–792.

self-distract during preschool delay and were unable to wait to get the bigger treat that they wanted.

In short, individual differences in the use of effective attention strategies for reducing stress and making self-regulation easier are seen early in life and seem to endure. Stable individual differences in attention processes such as eye-gaze aversion, flexible attention shifting and focusing, also have been documented as early as in infancy. Further, effective attention control is related to having lower impulsivity and less negative affect later in life, as seen both in children and in adults (Derryberry, 2002; Derryberry & Rothbart, 1997). Thus, attention control is part of a generalized self-regulatory competency that enables effective cooling of arousal and self-control of impulsive behaviors. These skills have adaptive long-term benefits and can protect against dispositional vulnerabilities.

17.17 In what ways can excessive self-control be maladaptive?

The Down-Side of Self-Regulation. On the one hand, cooling operations and effortful attention control seem to have significant value for the adaptive pursuit of life goals and effective functioning over time. On the other hand, these competencies may come with a cost, making some of these "high delay" individuals susceptible to other kinds of difficulties such as tendencies to become avoidant and withdrawn (Eisenberg et al., 2002). Likewise, whether or not one *should* or *should not* delay gratification or "exercise the will" in any particular choice is often everything but self-evident. Lionel Trilling, a distinguished humanist with much to say about psychology, captured the risks of too much self-control in a few words. After reminding us of the place of passion, he said "the will is not everything," and spoke of the "panic and emptiness which make their onset when the will is tired from its own excess" (Trilling, 1943,

p. 139). An excess of will can certainly be as self-defeating as its absence. Postponing gratification can be an unwise—and even stifling—joyless choice. But unless people develop the competencies to sustain delay and continue to exercise their will when they want and need to do so, the choice itself is lost.

For a clinical description of Gary W's Self-Regulatory System, see In Focus 17.3. It illustrates the multiple factors that influence the self-regulatory system, and the complex ways it plays out in the life of an individual, sometimes in self-defeating ways that undermine attainment of the most important goals.

Conclusions

Putting it together, "willpower" requires a combination of self-regulatory competencies and regulatory motivation: both desire (goals and motivations) and regulatory competencies (e.g., flexible attention deployment strategies) are needed. Both are necessary if one is to go from good but difficult intentions to their realization under frustrating, temptation-filled conditions. Even when regulatory competency is available, it will not be activated in the absence of regulatory motivation. Case studies and dramatic examples abound of individuals, including a recent U.S. president who exerted remarkable self-regulation in many areas of life—but did not do so in other areas, although the consequences were destructive. Presumably these individuals had the ability to control themselves but not sufficient motivation. Conversely, regulatory motivation is unlikely to sustain difficult long-term goal pursuit unless regulatory competency is available and accessible to smooth the journey.

The ingredients of willpower and particularly the processing dynamics that enable regulatory competence have long been mysterious, but some of the essentials now seem clear. Individual differences in self-regulatory ability are visible early in life, and a good deal is also known about the basic attention control mechanisms that enable it. These mechanisms help to demystify willpower and identify some of its key ingredients. Attention control is important in dealing both with fear and with the regulation of appetitive-approach behaviors. The implications of regulatory ability for the self are straightforward. This ability influences self-concepts and self-esteem, interpersonal strategies (e.g., aggression), coping, and the ability to buffer or protect the self against the maladaptive consequences of chronic personal vulnerabilities such as rejection sensitivity.

Potential for Self-Directed Change. Perhaps the most important question remains: can self-regulation be taught to empower the self and improve resilience? We do know that attention control strategies can be briefly modified in experiments (Ayduk, Mischel, & Downey, 2002; Mischel et al., 1989). Moreover, when effective control strategies are modeled, they may be readily adopted and generalize to self-control behavior outside the lab for at least a few months (Bandura & Mischel, 1965). What is not yet known is whether effective attention control to reduce stress and frustration during goal pursuit can be increased long-term via socialization, education, and therapy.

This chapter opened with Martha asking herself what can she do to overcome the stress she experiences when faced with tasks like her first dissection of a human cadaver, and we have tried to address her question. In a case like Martha's, it would not be surprising if on her own she began to more readily access the same self-regulatory competencies that allowed her to progress to this point in her impressive academic career. One can imagine her learning spontaneously to desensitize herself to the aversive emotions in the situation. With more exposure gradually she can detach herself

17.18 How does the social cognitive conception differ from the notion of willpower as an internal force that people either have or don't have?

GARY W'S SELF-REGULATORY SYSTEM

Although Gary is strongly motivated by concerns about how other people—particularly romantic partners—will see him and treat him, he does not seem to acknowledge approval or praise from others. He undermines his own hopes and potential with a pessimistic explanatory style. Privately, he tends to dismiss and trivialize his successes and magnify his perceived shortcomings, on which he focuses much of his attention, ruminating about them obsessively. But he can also flip and present a glowing picture of his abilities in his self-presentation in social situations.

At work or in school, he drives himself harshly. He holds himself to rigid goals for perfectionist performance, setting standards so high they virtually guarantee self-perceived failure. He then criticizes himself severely and ruminates about his shortcomings, depressing himself with his own thoughts. A notable exception to this pattern of stringent self-evaluation is in the domain of aggressive behavior. When he feels highly frustrated and stressed (which is not infrequent), he becomes capable of angry outbursts of temper and rage, particularly in his relations with his current girlfriend. Sometimes this also happens when he is frustrated in work situations with people he sees as less competent, especially when they are his seniors in authority. His hot emotional outbursts undermine his relationships and reflect his poor self-regulatory "cooling" skills for dealing effectively with impulsive angry reactions.

Gary lacks understanding of how his behavior undermines his ability to achieve the caring interpersonal relations he desires. He seems especially immature and vulnerable in this domain and flounders between bursts of anger and a sense of uncontrolled helplessness. He feels unable to influence his relationship or himself in directions that allow effective communication with a partner, a feeling that only furthers his frustration and anger and erodes his relationships.

Gary has excellent academic and cognitive skills. However, there is an obvious and urgent need to assess his social, interpersonal, and self-regulatory competencies, as well as his personal theories about relationships and himself. Many of his problems may reflect disadvantageous styles of interpreting and explaining his experiences. He also lacks interpersonal skills. He would benefit from role-play and practice with alternative ways of construing and solving conflicts when he feels threatened. Although he has self-regulatory competencies, as evident in his schoolwork and achievements, he has to learn to access those skills when he becomes too "hot" and frustrated to cope rationally in interpersonal situations.

Self-improvement efforts for Gary need to be directed at identifying and controlling social and internal cues that signal eruptions of his anger. The focus should be on ways in which he can more effectively regulate those emotions, and overcome the anger and hostility that jeopardize relationships he cares about most. He needs to develop less rigid and obsessive control patterns for coping with anxiety, particularly in public speaking contexts. At the same time, Gary needs to reexamine and rethink his interpersonal assumptions and personal theories. It will be especially important for him to probe alternative ways of construing criticism in order to find more effective ways of coping.

Thus, Gary needs to identify the interpersonal situations in which he feels provoked. He then can explore how his perceptions of rejection in those situations make him vulnerable to eruptions and automatic reactions that ultimately destroy the relations and intimacy that he is trying to build. Likewise he can be helped to see the consequences of his assumptions and habitual reactions in his close relations with women. Alternative ways of thinking likely to generate better outcomes then can be explored and practiced (Meichenbaum, 1995).

After recognizing and understanding the situations he finds most provoking and in which he feels violated, he then can try to reframe them in alternative, more constructive ways that make them less threatening and/or modify his own reactions to them. For example, by understanding the interpersonal situations in which he feels especially vulnerable he can reappraise them and activate his own self-regulatory controls. He can then begin to respond with appropriate assertiveness rather than with massive anxiety or uncontrolled rage. He also can learn to communicate more effectively about his feelings and needs. Developing more adaptive ways of thinking constructively about his relationships can make it easier to stop being destructive with the women he cares about and allow him to get what he wants more effectively.

17.19 How do self-regulation principles help us understand Gary W's adaptive skills and his problems in living?

and begin to cool her anxieties about anatomy, concentrating on the details of the dissecting method and the anatomical structures. In time, one can see her working carefully and patiently with great concentration. She begins to use some self-calming cooling strategies whenever anxiety begins to set in, and is sensitive enough to the cues of their onset to rapidly refocus her attention on where it will serve her best.

In fact, the person on whom the example of Martha is based did well in medical training, excelled working with oncology patients suffering from advanced cancers, and went on to become an expert in pathology, specializing in the diagnosis of brain tumors. This physician is still far from immune or insensitive to the many anxieties that are part of the daily routine, making potential life and death decisions based on reading the slides of brain tissues rushed to the pathology lab from patients still in the midst of surgery, and winning the admiration of colleagues and grateful patients. With intense focus on the many things that can go wrong in the diagnosis, "Martha" carefully scans the slides and tries hard to prevent most of them from happening.

SUMMARY

SELF-REGULATORY PROCESSES

- Approach or appetitive dilemmas occur when facing a strong approach temptation that interferes with one's more important, higher-order goal. Avoidance or aversive dilemmas occur when one is tempted to flee a situation that will ultimately help one to attain a high-order goal.

- Effective self-regulatory behavior and self-control in goal pursuit depends both on motivation and on regulatory competencies. The latter require cognitive and attention control mechanisms that help execute goal-directed behavior.

- Ego control refers to individual differences in the degree of impulse control in such functions as inhibition of impulses.

- Ego resilience is a construct that refers to the individual's stable ability to adapt flexibly to environmental demands by appropriately modifying his or her habitual level of ego control.

EMOTION REGULATION IN APPROACH (APPETITIVE) DILEMMAS

- The ability to voluntarily delay immediate gratification is at the core of most concepts of willpower and ego resilience.

- Successful delay of gratification in goal pursuit involves the ability to prevent oneself from focusing attention on the frustrating, emotion-arousing aspects of difficult situations, and instead strategically shifting attention elsewhere.

EMOTIONAL REGULATION IN AVOIDANCE (AVERSIVE) DILEMMAS

- To cope with aversive situations, individuals who practice cognitive reappraisal (cognitive "cooling" strategies) experience less negative feelings and physiological arousal than individuals who try to suppress their feelings.

- It also helps to cognitively transform or redefine stressful situations in more positive ways.

INTERACTION AMONG LEVELS: HOT AND COOL SYSTEMS IN SELF-REGULATION

- The emotional, hot system, centered in the amygdala, underlies automatic approach–avoidance or fight-or-flight responses.

- The cognitive, cool processing system, based in higher brain centers, particularly the prefrontal cortex, enables thoughtful, reflective thinking and planning.

- Anxious and self-preoccupied thoughts compete and interfere with task-relevant thoughts and undermine effective self-regulation.
- The balance between the hot and cool systems is influenced both by stress and by the individual's level of development. To effectively self-regulate, the individual must prevent or postpone automatic, hot, emotional, impulsive responses and activate cooling strategies by shifting attention and cognitions away from the hot features of the reward or the frustrating features of the conflict.
- "Hot," automatic responses are necessary in some situations. For example, strong social emotions can help people to resist temptation and avoid shirking responsibilities.

SELF-REGULATORY COMPETENCE

- The attention control mechanisms that underlie delay of gratification can help shield individuals against the negative consequences of their chronic personal vulnerabilities.
- Although always postponing gratification creates problems, unless people develop the competencies to delay when they want and need to do so, they lose the choice itself.

KEY TERMS

amygdala   407
approach (appetitive)
   dilemma   394
avoidance (aversive)
   dilemma   394
cognitive cool system   408
cognitive reappraisal   405
discriminative facility   403

effortful control   395
ego control   397
ego resilience   398
emotional
   hot system   408
executive functions   395
fight or flight
   reaction   406

hippocampus   408
overcontrolling
   children   398
regulatory competencies
   395
suppression   405
undercontrolling
   children   398

CHAPTER

18

PERSONALITY IN ITS CONTEXT AND CULTURE

This chapter gives a broader perspective on the person in his or her context and culture, integrating findings from all levels of analysis. First, we sketch a picture of personality as a whole. Then, we will look at the major influences that affect the diverse aspects of personality and that interact in the functioning person.

▶ PERSONALITY SYSTEM IN ACTION

An integrative view that draws on all the levels of analysis sees personality as having many diverse aspects: cultural, social, psychological, and biological. This section provides an overview of these aspects and how they interact in everyday experience. A simple representation of these diverse aspects and their interactions is given in Figure 18.1 that shows the personality system in its context at one moment in time for a given person. To make the example specific, keep what has been learned about Gary in mind.

Personality Dispositions and Dynamics

Figure 18.1 provides a glimpse of the multiple processes and interactions operating in the personality system. This system contains a host of dispositions, internal characteristics, and person variables that have been identified and studied by researchers working at all levels of analysis. You saw these throughout the text, and they were applied concretely to the case of Gary in many earlier chapters. They include his biological dispositions (e.g., BIS/BAS), psychological traits, appraisals, expectancies, self-concepts, goals, self-protective defenses, and self-regulatory processes.

An example of such a personality system and its internal processing dynamics was illustrated in the CAPS model in Chapter 12. Although discussed at the social cognitive level of analysis, this type of model is not specific to that level. Instead, it is a more general framework for examining personality processes and the personality system in ways that build on contributions from all the levels.

18.1 What two assumptions are made about individual differences in the personality system?

The main point here is that in current views of the personality system, two assumptions are commonly made about individual differences (e.g., Cervone & Shoda, 1999). The first is that people differ in the ease of accessibility of different mental representations. For example, we know that for Gary, anxious expectations of rejection are easily accessible, that is, they are readily activated in many situations.

The second assumption is that the personality system has a coherent organization of interconnections—essentially a coherent but complex network of associations—among the mental and emotional representations within it. Thus, the dispositional tendencies, expectations, goals, affects, and so on that are accessible become activated in a predictable set of interconnections, and these make up the "personality dynamics." These dynamics are activated under specific conditions, and then play out, for example,

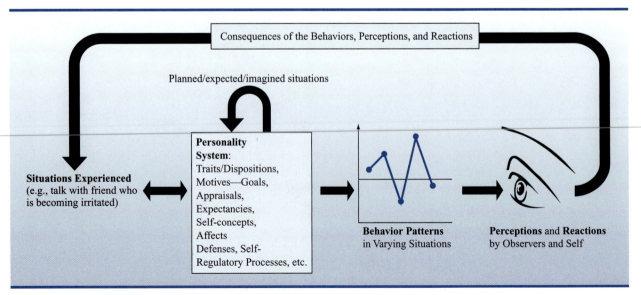

Figure 18.1   The personality system in context.

in the form of Gary's distinctive cognitive and emotional processes and the behavior patterns that they generate.

Sources of System Activation

The personality system is in a continuous state of activation, and the activations come from two sources. One source is external activation that comes from the situations encountered at a given moment in time, for example, the conversation Gary is having with his friend in which the friend is becoming irritated. The **active ingredients** are the features of those situations to which Gary is particularly attuned, such as potential rejection cues in an interpersonal context—in this case, the signs of the friend's irritation. These are the features for which Gary scans attentively and to which his personality system reacts distinctively and characteristically in stable patterns, that is, in terms of his personality dynamics. The two-way arrow from the situations to his personality system indicates that the two are in **reciprocal interaction,** each influencing the other. Gary's personality system selects and focuses on those active ingredients in the situations experienced, and it is also influenced by them, for example, leading him to an outburst of temper. This was illustrated earlier in the text in the discussion of Gary's rejection sensitivity dynamics.

The second source is internal activation, in the form of Gary's own thoughts and feelings and plans, for example, as he ruminates about himself and his relationships or thinks and plans what to do tomorrow. These internal experiences and anticipations impact the events that are encountered or generated next. Thus, as the last two chapters showed, the personality system is not merely reactive: it is also *proactive,* that is, future-oriented, as seen in the active pursuit of life goals.

18.2 What two classes of stimuli activate the personality system? What role does the system play in such activation?

Expressions of the System—and Their Consequences

At the behavioral level, the behaviors Gary generates are represented in the figure as an *if . . . then . . .* profile or behavioral signature to indicate that they occur in particular distinctive situations (e.g., his temper outbursts occur *if* rejection is perceived) and play out over time in stable ways. This profile also has an overall level in terms of the total amount of the characteristic behavior (e.g., hostility, temper-anger) he tends to display on average (see Chapter 4 to review those distinctions).

Gary's behaviors influence the situations he encounters in the future. The eye that is seeing his behavior in Figure 18.1 indicates that what he does is also observed and experienced by perceivers. One of these perceivers is Gary himself: we not only generate behavior, we also perceive and interpret it. Obviously, how he sees and interprets his outburst may be quite different from how others, for example, his partner, will view it and in turn react. Those perceptions affect what happens next by changing how the people who observe Gary respond to his behavior. This is indicated in the figure by the large curved arrow that shows those perceptions and their behavioral consequences influencing the subsequent situations that Gary is likely to experience. Over the course of time, the characteristics of the personality system change the stable psychological environment the individual tends to experience. Thus, people not only impose their own meanings on situations, but through their own behavior, they also select and create many of the situations they experience (e.g., Buss, 1996; Kenrick & Funder, 1988; Swann, 1983).

Gary's case illustrates this process of how one can shape one's own situations. Because he dreads social-evaluative situations, he avoids them whenever he can, preferring to isolate himself. This further strengthens his inclination to withdraw and be "a loner" and isolates him even more. A side effect is that he never learns to

18.3 As typified by Gary W., how do behavioral signatures influence the situations one encounters?

confront and cope more effectively with the social encounters he fears most. This is painfully evident in public speaking situations that he simply cannot avoid, and in which he easily panics, especially when he feels himself being evaluated by respected male peers in business contexts. His panic response further elicits the disrespect and "snickers" of his peers, thereby intensifying his dread and avoidance pattern in a vicious cycle.

Gary's preference for isolation and withdrawal has by now become a well-established pattern. He even announces to acquaintances that he prefers his own company to meeting other people, which they perceive as a rejection by him or an expression of aloofness. Then they reciprocate by avoiding him, which further isolates him even more and fuels his sense of alienation and distrust. In this way, personality variables influence the choice of the types of situations to which people expose themselves (e.g., Bolger & Schilling, 1991; Buss, 1987). For example, individuals who cooperate tend to elicit reciprocal cooperation from their partners on a task, which in turn confirms and strengthens their choice of a cooperative rather than a competitive strategy (Kelley & Stahelski, 1970).

The analysis of complex social interactions shows how each of us continuously selects, changes, and generates conditions just as much as we are affected by them (e.g., Patterson, 1976, 1990). In classic studies, husband–wife interactions were observed as the couples coped with such conflicts as how to celebrate their first wedding anniversary when each had made different plans (Raush et al., 1974). For example, Bob has arranged and paid in advance for dinner at a restaurant, but Sue has spent half the day preparing for a special dinner at home. As the couple realize their conflict and try to resolve it, their interactions continuously reveal that each antecedent act (what Sue has just said to Bob) constrains each consequent act (how Bob responds).

The interpersonal strategy in such interactions depends both on the type of person and the type of situation. For example, in the context of getting along with a peer, a more dependent person is likely to use a self-denigrating strategy. On the other hand, when the situation focuses on pleasing an authority, the dependent person is apt to use a self-promoting strategy (Bornstein et al., 1996). Once more, we see the effects of person × situation interactions.

▶ INTERACTING INFLUENCES OF BIOLOGY AND CULTURE

Where does personality come from? What shapes the various aspects of personality outlined above? What influences the accessibility of various thoughts and feelings, and the network of connections among them? What influences how an individual's behavior is seen by others? Throughout this text, each level of analysis added partial answers to these questions, and it is time now to look at the whole picture that emerges.

The individual's personality system develops from a foundation that is both biochemical and psychosocial; it is guided by the genes in interaction with the environment, as well as reflecting the influence of social learning, and the culture. As indicated in Figure 18.2, these processes interact in the course of development, influencing such person variables as how the person construes—and shapes—the situations encountered (Mischel & Shoda, 1998; Saudino & Plomin, 1996).

Interacting Influences on Personality Development

In the life of each person, biological and social-environmental-cultural influences interact in the development of the personality system, as was discussed in the interplay of genes and social influences (Chapter 14). If individuals are genetically and temperamentally disposed toward greater aggressiveness, for example, they soon become more likely to

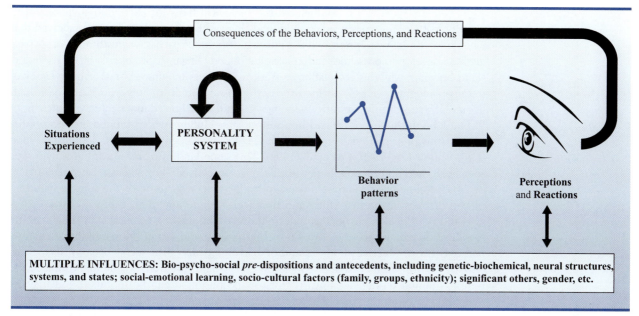

Figure 18.2 Influences on the personality system.

interpret the motives of other people as aggressive and to anticipate aggression from others (Dodge, 1986). They also may confirm these expectations and beliefs by behaving more aggressively, thereby evoking more aggressive reactions. Likewise, as aggressiveness becomes an established pattern, it also may be incorporated into the values and self-standards and goals of the individual. Each time the aggressive pattern is activated, it becomes more easily accessible in the system as a dominant response tendency, triggered even by minor frustrations and stress. Increasingly the person also may seek out peers and groups likely to further support this type of behavior and to strengthen it as a source both for self-esteem and peer approval. A common example is when aggressively inclined adolescents are drawn to like-minded peers in aggressive groups and gangs that then collectively and mutually further support their aggressive tendencies (e.g., Bandura & Walters, 1959).

In sum, as the example suggests, the early behavioral tendencies in the person interact with the psychological environment over the course of time as personality develops and becomes expressed in relatively stable ways (e.g., Contrada, Leventhal, & O'Leary, 1990). The above example shows how a dispositional tendency like aggression may be reinforced early in life by the social environment and then becomes well established. But your dispositional tendencies are not necessarily determinants of your fate in life. For example, the studies described next show that socialization experiences can help infants who are temperamentally shy to develop into people who are not shy: because social experiences can modify temperament, biology does not have to be destiny (Schmidt & Fox, 2002).

Biology-Trait-Socialization Interactions: Shyness

Shyness is a problem that plagues many people. Researchers have been examining how shyness as a trait in children and young adults manifests itself, and how biological and social factors might jointly influence what happens ultimately to the development of temperamentally shy young children (Schmidt & Fox, 2002; Schmidt & Fox, 1999). Their question was: Must the temperamentally shy infant and young child become a shy person in the course of development?

18.4 How has research on shyness illuminated biological–environmental interactions in personality development?

Their first goal was to understand the biological basis for childhood shyness and to identify features that may predict childhood shyness in infancy. Initial studies showed that when new stimuli were presented to infants, a distressed reaction was indicative of shyness likely to develop in preschool and in later childhood (Kagan & Snidman, 1991). To understand how and why this correlation existed, researchers monitored brain activity via electroencephalograph (EEG) during exposure to new stimuli. Infants who exhibited a high degree of stress during exposure to the new stimuli also showed more activity in their right frontal lobe during exposure and in a resting state. Infants who exhibited little to no distress at the presentation of new stimuli showed more activity in their left frontal lobe (Davidson & Fox, 1989; Fox et al., 1992). This same kind of pattern of brain activity was seen in shy and nonshy preschool age children, respectively (Fox and associates, 1996).

In a follow-up, researchers tried to understand why some of the infants who displayed distress in the first study and thus were temperamentally disposed toward shyness did not develop shyness as older children. The answer, in part, is in their specific socialization experiences. Many of the children who did not develop childhood shyness had less overprotective mothers and/or had been placed in social situations (such as daycare) more often (Fox and associates, 2001). The researchers concluded that initial temperamental tendencies toward shyness in childhood can be modified and may be overcome by parenting practices and social experiences.

Furthermore, shyness, like many broad dispositions such as neuroticism, comes in different forms and types. Studies with college students as well as with children suggest that shyness and sociability are not part of a single dimension (Schmidt & Fox, 2002). They are distinguishable on measures of electrical activity in the cortex as well as on indices such as EEG, and heart rate measured both when at rest and when experiencing a social stressor. So the individual who is highly shy but also highly sociable will be characterized by both biological and psychological-behavioral features that are quite different from an equally shy person who is low on the sociability dimension. Again it is the interaction of multiple variables and processes that determine how temperament plays out in personality development.

▶ GENDER AND SEX DIFFERENCES

18.5 Differentiate between sex and gender.

Established at birth, biological sex soon begins to direct much of one's psychological and social development. It influences self-concepts and identity, as well as goals and values, and continues to have a dominant influence throughout life.

Gender is a concept that refers to the social meaning of being a male or a female. Almost universally, gender is one of the most powerful determinants of how people view themselves and how other people treat them. Gender is probably one of the most important psychological categorizations that people are subjected to, and one that has become increasingly controversial. The controversy is based on many challenges to conventional, rigid **gender stereotypes** about what it means to be a man or a woman within a particular culture or society (e.g., Deaux & Major, 1987).

Overview and Issues

18.6 What behavioral sex differences exist in infants?

Neonatal Sex Differences. Some researchers have concentrated on sex differences in newborns (neonates) in order to study possible innate sex differences before socialization practices exert their huge effects. Human neonatal behavior indicates some sex differences in activity level and in reactivity to a variety of stimuli. For example, sex differences occur in infants' responses to facial stimuli during the first year of life (Lewis, 1969, 1990). Girls vocalized and smiled more and showed greater differential expression to the facial stimuli, although boys looked longer. At age two to three months, girls are more sensitive to skin exposure than boys (Wolff, 1966). New-born females seem to react more to the removal of a covering blanket, and show lower thresholds to air-jet stimulation of the abdomen (Bell & Costello, 1964). Newborn boys raise their heads higher than newborn girls do (Bell & Darling, 1965), and there are also sex differences in infant play behavior (Goldberg & Lewis, 1969). The interpretation sometimes drawn from these early sex differences in response to stimulation is that they are innate. But as in other domains of personality, nature and nurture—heredity and environment—tend to be deeply and often inextricably entwined.

Gender Concepts. **Gender concepts**—beliefs about the types of behavior more appropriate for one sex than for the other—are widely shared within a particular culture and to some extent across cultures (D'Andrade, 1966). Individuals differ in the sanctioned sex-role standards that they adopt or reject. Some modern western cultures or subcultures have been shifting toward greater flexibility in the range that tends to be explored by individuals and tolerated by others, although signs in opposite directions also can be seen in other cultures throughout the world. The concept of **sex role identity** refers to the "degree to which an individual regards himself as masculine or feminine" (Kagan, 1964, p. 144). The degree of match or mismatch between the sex-role standards of the culture and the individuals' assessment of his or her own attributes can have far-reaching implications on the personality that each develops and the social world in which he or she functions.

Expression of Gender-Relevant Behavior. As Deaux and Major (1987) note, sex and gender are constructs that deeply influence social relations. Gender-linked behaviors are expressed by males and females often at automatic levels and play out as almost reflexive scripts. Both sexes respond to the gender-relevant cues in their social interactions in ways that activate their conceptions about themselves as males and females. These gender-relevant conceptions include a host of expectations and values that guide the behavior they think appropriate to the context (see In Focus 18.1).

As an example of the influence of gender roles in the automatic processing of social information, consider a study conducted by Sadalla, Kenrick, and Vershure (1987). Students were asked to watch films of men in high or low dominance conditions. To create a condition of low dominance, an actor engaged in conversation with another male, repeatedly nodding his head, looking down often, and speaking little. To establish the reverse condition, the actor was more relaxed, nodded less, and made more assertive gestures. Students of both sexes rated those in the higher-dominance condition as more attractive. When shown films of women in similar conditions, however, no relationship was found. Thus the person's sex determined whether assertiveness-dominance led to perceived attractiveness—behaviors that made the male more attractive did not increase the perceived attractiveness of the female in this situation. Results like these lead many researchers to the conclusion that gender "is something we enact . . . a pattern of social organization that structures the relations, especially the power relations, between women and men" (Riger, 1992, p. 737).

Interactions of Biology, Sex, and Culture in Response to Threat

When thinking about the influence of biological and social-cultural influences on personality, it is easy to fall into "either–or" ways of casting the question, and to want to know, "bottom line, which one is more important?" We saw this clearly in the discussion of the

ADULT SEX DIFFERENCES AND THEIR IMPLICATIONS

For half a century, the nature and extent of sex differences in personality have been topics of extensive research and passionate debates (e.g., Gilligan, 1982; Maccoby & Jacklin, 1974; Mischel & Mischel, 1973). In adults, a variety of sex differences have been reported over the years. For example, the physical fight response to stress and threat shows higher levels for males than for females in humans, primates, and rats (e.g. Archer, 1990; Eagly & Steffen, 1986). Distinct sex differences also are reported in nonverbal behavior, such as facial expressiveness and body expressiveness (reviewed by Eagly, 1987) and greater rough and tumble play and physical aggressive tendencies in boys than girls (Collaer & Hines, 1995). Possible sex differences also have been found in some aspects of cognitive functioning but taken as a whole, the largest differences, unsurprisingly, have been shown for physical abilities (e.g., Eagly & Crowley, 1986; Eagly & Steffen, 1986).

Differences in mental and social abilities and in behaviors are considerably more modest, although a few minor differences between the sexes "are sprinkled across all domains" (Ashmore, 1990, p. 500). The exception may be physical aggression, in which fairly large differences tend to be found even when diverse measures are used. In most other domains, however, some argue that the more attention that is paid to avoiding sex bias in the research, the fewer the gender differences that are actually found (e.g., Riger, 1992).

Nevertheless, there is evidence for a number of nontrivial sex differences although their interpretation has been extensively debated (Eagly, 1995). For example, sex differences have been reported on measures of the Big Five Factor model that use trait ratings (e.g., Hoyenga & Hoyenga, 1993). Examples include moderately greater rated assertiveness in males, and greater trust, tender-mindedness, and anxiety in women. These data may reflect widely shared social perceptions and possible gender stereotypes, but they may or may not speak to differences in the actual social behavior of the sexes when it is studied by direct observation.

What is the conclusion one can reasonably draw when all the findings are looked at together? To some critics, the most impressive finding about sex differences in psychological characteristics is how unimpressive they generally seem to be and how much smaller they are than had been assumed in earlier years. They note that it is important to keep in mind that the variability *within* each sex on these measures is often more impressive than the differences found on average *between* the sexes. To others, it seems that the differences found are large enough to be taken seriously and that the reasons they are minimized are mostly political. In that view, sex differences are significant but they are being trivialized because of a "feminist commitment to gender similarity as a route to political equality" (Eagly, 1995, p. 155). Both interpretations have validity. Compared to within-sex differences, the differences between the sexes on most variables are probably not significant enough for practical purposes. Nevertheless, they may be reliable enough to provide clues as to how biological differences that exist at birth interact with culture to result in sex differences that are observed in adults. How the data are read depends on one's philosophy and politics on the topic, as well as on the methods used to estimate the size and meaning of the statistical differences that are reported.

18.7 In which areas do adult sex differences exist? Why is it difficult to interpret the cause of these differences?

influence of nature and nurture—genetics and social experience—on personality (Chapter 14). The same issues and the same answers apply in the present discussion of genetics, heredity, and social-cultural influences. To make sense of the whole story of sex differences in social behavior and personality, for example, the answer has to take into account the importance *both* of social factors, as discussed above, *and* of biological-evolutionary factors. Above all, one has to look at how the two influences *interact* so that what emerges depends critically on both. This point is made convincingly in a theory developed by University of California Los Angeles psychologist Shelley E. Taylor and her colleagues. They reached a radical conclusion that challenged much of the work on how humans, and particularly females, respond to threat and stress (Taylor and associates, 2000).

Recall that Chapter 17 discussed response to stress in terms of the fight-or-flight patterns that have long been assumed to be the universal human and animal reaction when severely threatened, for example, by animal predators or natural disasters. Taylor

Shelley Taylor

and colleagues note that the widely accepted belief (beginning with Cannon, 1932) that humans respond to high stress and threat with automatic fight or flight responses is based on many thousands of studies, but almost all were conducted with males, especially male rats. For females, the biobehavioral response to threat may be very different than it is for males, and for many good reasons.

Men Fight or Flee: Women Tend and Befriend. The theory Taylor and colleagues propose builds on recently discovered biological differences between human males and females, particularly in their neuroendocrine responses to threat. At the biological level, females are less likely to make a physical fight response to stress because they lack **androgens,** the hormones that in many species "act to develop the male brain for aggression . . . and then activate aggressive behavior in specific threatening contexts (such as responses to territorial establishment and defense)" (Taylor et al., 2000, p. 413). The role of testosterone, the male sex hormone, in aggression is controversial, but its level in humans increases with acute stress and is associated with hostility. In contrast, according to Taylor, in female humans, aggressive responses are not organized by testosterone or androgens and therefore do not involve the same automatic reaction patterns in response to stress.

Taylor integrates these findings with insights from the evolutionary level of analysis. According to her theory, the aggressive response to threat and stress from an evolutionary perspective may have been adaptive for men when dealing with danger (e.g., when on the hunt in the wilderness) but not for the challenges that females faced as the mothers of offspring. Females of most species, including humans until relatively recently, were pregnant, nursing, or caring for their young offspring, during their fertile lives. For these functions, a female fight-or-flight response to stress may have been strikingly nonadaptive for survival and hence became inhibited. The theory suggests that instead, females developed a **tending response to stress** that involved caretaking and grew out of the attachment system in humans. Many studies support this argument

18.8 What difference in behavioral responses to stress does Taylor propose, and what biological differences are thought to underlie this difference?

by showing that nurturing behaviors, such as touching the infant and carrying it close on the chest, benefit both mothers and offspring under high stress conditions (Taylor et al., 2000). A similar argument, based both on biology and evolution, is made for the value of **befriending,** for example, through group living and activities, as a highly adaptive way for females to cope with stress.

For the student of personality, and for the purposes of this concluding chapter, a clear message comes from the analysis by Shelley Taylor and colleagues:

> *Rather than viewing social roles and biology as alternative accounts of human behavior, a more productive theoretical and empirical strategy will be to recognize how biology and social roles are inextricably interwoven to account for the remarkable flexibility of human behavior (Taylor et al., 2000, p. 423).*

Interactions in the Genesis of Gender Roles

Cross-cultural studies of gender roles by anthropologists show the wide range of gender roles across different cultures (Gilmore, 1990). This diversity in sex roles spans "—from marked stereotypic male and female patterns to extremely muted, even quite difficult-to-detect gender differences—" (Nisbett, 1990, p. 258). For example, a highly stereotypic pattern of machismo, characterized by pride in male sexual prowess, courage in the face of physical danger, and economic dominance, is extremely common throughout the world, but even in this pattern there are striking exceptions. Analyses of these **macho patterns** by Gilmore and other anthropologists indicate that they reflect an adaptation process to extreme risks. These adaptations seem to be commonly connected with the economic role of males, for example, as hunters in the wild, or as protectors of vulnerable animal herds.

18.9 What cultural differences in male "macho" behavior did Gilmore find? What conclusion did he draw?

Gilmore's (1990) comparisons of Truk and Tahiti, two tropical "paradise" islands in the South Pacific, make the case convincingly. Trukese males became violent fighters who personified macho qualities—competitive and eager to show their physical prowess, accumulating many love affairs beginning early in life, while the women of their culture were expected to personify submissiveness. In contrast, Tahitian males did not compete with each other for material possessions. They were expected to be submissive, and allowed their women free and open sexual activity, for example, with Westerners, almost as soon as they landed. Why? As Nisbett (1990, p. 260) notes in his review of this work:

> *The Tahitians fish in a lagoon at no risk to themselves and food is quite plentiful. The Trukese must obtain their fish in the open sea. When a Trukese male leaves for a day of fishing, there is a genuine possibility he will not return. Thus the fearless, aggressive macho style is an adaptation to danger. Males are taught to fear unmanliness more than death because this is the only way to encourage men to produce when there is great danger . . . a muting of gender differences and surcease from the crushing requirements of assertive masculinity are a privilege of those societies that can put food on the table without great risk.*

The fact that at least in some societies men as a whole are gentle (the Tahitians) and even timid (the Semai people of central Malaysia) has serious implications for understanding both gender roles and culture. According to Gilmore (1990, p. 230), these exceptions to the norm for macho aggressive masculinity suggest that ". . . manliness is a symbolic script, a cultural construct, endlessly variable and not always necessary."

In the course of evolution, when there was dangerous work to be done, men were more likely to do it because of their anatomy and greater physical strength. But now that more of the "high risk–high glory" jobs do not depend on physical strength, one may well ask Gilmore's (p. 231) searching question: "Why should only males be permitted to be 'real men' and earn the glory of a risk successfully taken?"

▶ CULTURE AND PERSONALITY

The Link Between the Cultural and the Personal Meaning Systems

As part of the mix of biological and social-environmental influences on personality, the culture in which a person is raised and becomes socialized also importantly influences the development and expressions of personality. Language provides an analogy. Humans are uniquely endowed with the biological apparatus that makes language possible. Without this apparatus, it would simply not be possible to process language, as seen in how extremely difficult it has proven to be to teach language to primates (e.g., using sign rather than spoken languages). But the biological readiness alone will not result in an adult who can use a language: the specific language one speaks is determined by the human environment in which one grows up. Likewise, humans are endowed with a biological apparatus that makes the processing of social information possible, such as categorizing others' behaviors, inferring their intentions, as well as generating thoughts, emotions, and behaviors in response to them. Each individual develops a distinctive unique personality system that carries out such processing. But the shapes that one's personality system takes and the forms in which it is expressed, including the development of gender roles, are greatly influenced by the culture, as was illustrated above. Thus, how one categorizes many social situations is influenced by the concepts, or the vocabulary, of the culture in which one grows up. Furthermore, these concepts are not isolated, but rather are related to each other to form a meaning system unique to each culture.

The **cultural meaning system** includes distinctive views about what the world is like and what different situations mean and call for (e.g., Markus & Kitayama, 1998). Much like the personality system, it contains concepts and feelings about the nature of the self and its ideals, and the relationships between self and other people. It includes basic goals-values, emotions, standards, self-regulatory rules, and scripts for appropriate and inappropriate interpersonal behavior and control strategies in goal pursuit.

The significant people in the life of the child model and communicate the meaning system of their culture just as they communicate the vocabulary and grammar of the language. In time, each individual develops a **personal meaning system** that incorporates some components from the shared cultural meaning system while other components reflect unique life experiences.

Attempts to explore the links between culture and personality have a long history, and here we only look at some of the highlights. One of the first researchers to systematically examine the relationship between culture and personality was the anthropologist Ruth Benedict. In her 1934 book *Patterns of Culture*, Benedict described two personality types of the Southwest Pueblos of Mexico: the Appolonian and the Dionysian, based on a distinction used by the philosopher Nietzsche in his studies of Greek tragedy. Benedict characterized the Dionysian Pueblos as having a generalized love for excess, and the Appolonian Pueblos as sober and mistrustful of excessive behavior—qualities which were evident across a wide range of domains. Following Benedict, much subsequent research on culture and personality focused on identifying

18.10 How are cultural and personal meaning systems related?

broad behavior trends in the national character or "dominant personality" of cultural groups (e.g., Kardiner, 1945; Linton, 1945).

Despite the intuitive appeal of the culture and personality approach, by the 1970s it faced serious challenges. The most difficult one was how to account for mounting evidence that cultural group members sometimes behaved according to their dominant personality—and sometimes did not. The Appolonian Pueblos, for example, were documented to at times engage in typically Dionysian behavior, as well as vice versa (Barnouw, 1973). Faced with growing criticism that research in culture and personality did little else than promote cultural stereotypes, by the 1980s this line of inquiry lost favor—and credibility—among researchers. Nevertheless, the challenge remained of how to account for both the homogeneity and heterogeneity in a cultural group's behavior. The question is *how* does culture influence the expressions of personality?

Starting in the 1990s, a new approach, influenced by work at the social cognitive level, has given new impetus to understanding how cultural forces shape individual behavior (Markus & Kitayama, 1998). This approach has provided a natural bridge—and a renewed interest—in culture through its focus on the processes that underlie and motivate behavior. The psychological "units" that have been identified in this work are similar to the social-cognitive person variables discussed in Chapter 12, such as a person's beliefs, values, goals, and ways of construing the world. These are like the **units of culture** proposed by leading researchers as being transmitted and shared within a particular community or culture (Triandis, 1997; Betancourt & Lopez, 1993). At the same time, these units are only activated and guide behavior in relation to specific situations (Mischel & Shoda, 1995). Consequently, while people who share a culture and its meaning units may converge in their experiences and behavior in some situations, such convergence will not occur in situations that do not activate culturally-shared cognitions (Mendoza-Denton et al., 1999).

Culturally Specific Personality Dispositions. Current research is beginning to illustrate how culturally specific personality dispositions can arise from differences in the socialization and learning experiences that are specific to members of a cultural group. Mendoza-Denton et al. (2002), for example, have proposed that because African Americans in the United States are especially likely to be socially excluded, marginalized, or mistreated on the basis of their race, members of this group in particular may develop race-based rejection sensitivity. This processing dynamic is similar to rejection sensitivity in personal relationships (Downey & Feldman, 1996). However, its antecedents (prior experiences of discrimination), situational trigger features (situations where racism is possible), and consequences (mistrust in historically White institutions) make it a processing dynamic that is more likely to occur among African Americans than European Americans.

18.11 Describe research on race-based rejection sensitivity and the situations that evoke it.

Importantly, this processing dynamic, while stable over time, is relevant and activated only in situations where discrimination on the basis of race is applicable, and does not necessarily characterize the individual's responses and behavior in other contexts. As such, the approach allows for an analysis of how shared cultural experience can lead to similar behavior by members of cultural groups when they are in certain situations. At the same time, it still allows for diversity in group members' behavior in many other contexts. This approach takes into account individual differences as well as commonality among members of cultural groups. It also avoids having to rely on global cultural stereotypes that limited earlier efforts to understand the impact of culture on personality.

Cross-Cultural and Intracultural Differences

Cultures can differ in many ways. For example, some cultures are more complex than others, seen in the contrast between hunters/gatherers and large urban societies living in a modern computer-dependent environment. Likewise, cultures differ in the degree to which deviations from norms are tolerated. Of greater interest for personality psychology is a typology that contrasts **individualism** and **collectivism** as two meaning systems.

18.12 Distinguish between individualist and collectivist cultures.

Individualism Versus Collectivism.   Individualism involves a sense of the self as autonomous, devoted to the pursuit of personal goals, and relatively independent and self-focused. In contrast, collectivism is defined by closer connections to groups and family. People at this pole of the typology attempt to take extensive account of the goals of the family and groups with which they are connected, attentive to fulfilling duties and obligations. Even as adults, they remain interdependent in their social relationships, values, and goal pursuits. Likewise, in collectivist culture, the focus is less on self and self-enhancement and more on relationships, social obligations, and roles in comparison with an individualistic culture (e.g., Markus & Kitayama, 1998).

As the discussion above emphasizes, when cultures are described in broad trait terms like individualism versus collectivism, the fit tends to be imperfect. This is true because the terms capture something only about the average tendencies and characteristics of the majority of people within the culture. For example, rough estimates suggest that these descriptions may fit about 60 percent of the people in a culture (Triandis & Suh, 2002). That means that in a culture of individualism, 40 percent of the people don't show individualistic characteristics, and in a collectivist culture, 40 percent of the people cannot be characterized as collectivistic.

Throughout this text, you have seen that the variability *within* a person or type is often as impressive as the differences between persons and types. Likewise, the variability within a culture is often as impressive as the differences between cultures. Nevertheless, attention to the broad overall average differences can be valuable. It can give at least a rough map of what to expect in a given culture, and some of the outlines of its meaning system. It is certainly useful to know that in many situations in Asia, tactics that might be appropriate in New York City would be embarrassing and vice versa, and there is good reason that visitors come fortified with guides on what to expect and when and where to expect it. It is also useful to know that at a very broad level of trait description, in spite of cultural differences, there are important commonalities in the ways in which people characterize each other. This is seen in the finding that at least three or four of the factors in the Big Five factor structure can be found in most cultures, albeit not without exceptions. Perhaps the most important reservation to keep in mind here, however, is that even when the same factors appear in diverse cultures the particular meanings and usage may be quite different (Triandis & Suh, 2002).

Summary: Interacting Influences of Biology–Culture on Personality.   In sum, culture impacts on each aspect of personality, as depicted in Figure 18.1 and discussed at the start of this chapter. It may influence the situations that people tend to experience (the left-hand side of the figure), the ways they interpret these situations, and thus how they respond to them. This happens both within the personality system—that is, internally in the person's mental and emotional experience—and in the characteristic behavioral patterns that are generated. Culture also influences personality by affecting the ways in which other people, as well as the self, react to those behaviors, in turn changing the subsequent situations the person is likely to encounter. This interplay of personality and culture over time involves continuous interactions between the inside and the outside, the individual and the social world.

▶ WHAT DEVELOPS? THE EVOLVING SELF

In this chapter we have considered the multiple aspects of personality and the diverse influences that help answer the question: Where does personality come from? But what is it that emerges from all these influences? As Carlson (1971) put it years ago: Where is the *person* in personality research? Unsurprisingly, this big question has more than one answer.

Let's look again at the diverse aspects of personality summarized in Figure 18.1. It shows a dynamic personality system with its stable organization of mental and emotional representations interacting with its social world. The system generates patterns of behavior that are perceived and reacted to in the social world in ways that change the situations subsequently experienced. Looked at from the outside, as in this figure, the person can be studied and understood at many levels of analysis, and when the levels are combined the view is enriched. That has been a main message of this concluding part of the text.

But there also is another answer to the questions "where is the person?" and "what emerges?" Namely, throughout the life course it is the self or self-system that evolves and continues to construct itself (e.g., Hoyle et al., 1999; Leary & Tangney, 2003; Mischel & Morf, 2003). Then what is this "self" about which so many different claims have been made? So far in this text, although the self was discussed as a concept with many different meanings at each level of analysis, there is also a common theme that does emerge. As one modern summary puts it:

> The self is a dynamic psychological system, a tapestry of thoughts, feelings, and motives that define and direct—even destroy—us. Minus the self, there is little more to human beings than meets the eye: frail, relatively hairless creatures ruled by instinct and circumstance (Hoyle et al., 1999, p. 1).

18.13 What is meant by "selfhood"?

The term used in current work to capture the essence of the self is **selfhood.** First introduced in a presidential address by Brewster Smith to the American Psychological Association in 1978, the term was used to talk about what it means to be human. In its scientific use it refers to the *"thoughts, feelings and behaviors that arise from the awareness of self as an object and agent"* (Hoyle et al., 1999, p. 2).

The point here is that each person has the experience of self-awareness, of consciousness of himself or herself as a coherent entity, a distinct whole person. This is the sense of **self as "object."** And each person tries to pursue these actively in the process of self-construction to continue to build a self and a life within the interpersonal world (Tesser, 2002). This is the sense of **self as "agent."**

Taking Charge: Human Agency

The Self-Construction Process.   This text has attempted a scientific account of personality and the personality system in its context. Such an account can easily sound as if the personality is a machine-like system that simply reacts almost reflexively to the many influences that formed it and that now push and pull it around. But that interpretation misses one of the most essential features of personality and of selfhood: namely, the individual has the potential to behave in ways that are forward-looking and to actively change the situations of his or her life, rather than just reacting to them passively. In short, selfhood involves a process of self-construction (e.g., Hoyle et al. 1999; Mischel & Morf, 2003; Morf & Rhodewalt, 2001). It is a process in which individuals actively pursue life goals central for the self as they try to build their lives. To reiterate, you saw this concretely in the discussion in the previous two chapters of how individuals pursue life goals. Most importantly, in Chapter 17 you saw that the process of self-regulation allows people to influence their own life course, to transform and change at least some of the situations they experience, and to actively take charge of their life trajectory.

The same chapter also noted that toddlers already show individual differences in their ability to regulate their attention, and that these differences predict how well they can delay gratification which in turn predicts many outcomes in their lives as adults. These outcomes ranged from their SAT scores to their educational attainments, to their long-term goal-setting, to their sense of self-worth and ability to cope with life stresses. So is the toddler's future more or less sealed by the time the diaper stage of life is over? The answer of course is emphatically "no."

The Self as an Active Agent

Self-Direction/Agency.   First it's important to remember that when findings "predict" outcomes, they are based on correlations. When these correlations are statistically significant, as they are in most of the research discussed here, they still always leave much of the variance in the outcome unaccounted for. It is highly unusual for any single variable to account for more than 25 percent of the variance. Even when multiple variables are considered together, research in most areas indicates that at least 50 percent of the "variance" remains unexplained, open, and unpredicted.

A key question for each person then becomes: "what can I do to influence those outcomes—and myself—in ways that I would like my life and myself to develop?" In psychological terms, this is the question of **human agency** (e.g., Bandura, 2001), and the role of the self as a potentially active agent. To the degree that people have agency—and perceive themselves to have such agency—they may influence proactively the course of their personal experience, their views of the world, and their life paths (e.g., Mischel & Morf, 2003).

The Relational Self.   Discussions about agency and self-direction can risk sounding as if one is advocating a self-centered person seeking to advance in the world as an autonomous agent indifferent to others. But on closer examination, the self, as we have seen, is relational in nature (Chapter 13). People are not self-sufficient agents; they need each other, and to thrive requires connection as much as self-direction and agency. This recognition is seen especially in cultures characterized by collectivism, but it is not absent in cultures that focus on individualism.

For example, effective coping with long-term stress generally depends not just on what the person does inside the head but also on the psychological support environment. People deal better with stress, and are less likely to respond to it with illness, when they have social ties and supports, such as spouses, relatives, close friends, and groups to which they belong (Antonovsky, 1979; Holahan & Moos, 1990). Coping is also better when people can share their stressful experience with others (Nilson et al., 1981). When people are members of a group to which they "belong," they can receive emotional support, help with problems, and even a boost

18.14 Need human agency be self-centered?

The agentic self: People can influence who and what they become.

in self-esteem (Cantor et al., 2002; Cobb, 1976; Cohen & McKay, 1984). When they have a spouse to whom they feel they can talk, they tend to thrive, but having a spouse to whom one feels unable to talk makes the risk of depression considerably greater (Coyne & Downey, 1991). The self is relational and thrives with social support and connectedness, not in isolation (Leary & Tangney, 2003).

What Do People Need to Thrive? The View from Multiple Levels

18.15 Summarize what is required for psychological health from trait, psychodynamic, and phenomenological levels of analysis. What five "bottom line" requisites emerge?

If people have the capacity to take charge of what they become and their own life course, one has to also ask questions about the psychological qualities of the life one wants to live. While, of course, people differ in the specific goals they pursue in their lives, it may be reasonable to assume that they strive to have mental health, experience personal growth, and in general, *thrive*. But what does it mean to thrive? Work at several of the levels has had something to say about the essentials for personal growth, mental health, and thriving as a human being. Let's review some of the main examples to see the view of well-being that emerges when they are considered together. Table 18.1 summarizes some of the key ingredients identified.

At the *trait-dispositional level*, it is easy to identify the negative traits, for example, emotional instability or neuroticism, excessive impulsivity and anxiety, disagreeableness-hostility, lack of conscientiousness. Likewise, the positive qualities are clear since they are at the opposite ends of each of these dimensions. While it is informative to have descriptions of positive and negative qualities, this level focuses on describing the current and ideal functioning, and does not really address change possibilities in psychological terms.

The good news is that other levels of analysis do directly address the possibilities for constructive change. At the *psychodynamic-motivational level,* Freud's answer to the question was stated simply in his hope for the outcome of successful therapy: *Where id was there ego shall be,* was the phrase he used. Philosophically, for Freud that meant having to recognize and accept one's unconscious, animal-like biological urges, but to live life without enacting them. In his view, psychoanalysis helped people gain deep emotional insight into unconscious wishes and conflicts, and ultimately accept them without either having to act on them or deny them. The healthy person for Freud was one who becomes able to *love,* which is seen in the ability to have unambivalent close relationships, and to *work.* For Freud's followers, the answers shifted toward the

TABLE 18.1   Overview of Ingredients for Psychological Thriving, Well-Being, Mental Health

Level of Analysis	Key Ingredients
Trait-Dispositional	Positive traits (emotional stability, conscientiousness, agreeableness)
Psychodynamic-Motivational (Freud)	Insight into unconscious; able to love (unambivalent relationships) and work
Phenomenological (Humanistic, Existential)	Self-acceptance; responsible for self and choices: agentic
Social Cognitive	Competence/self-efficacy, relatedness (interpersonal connectedness); able to self-regulate: agentic

need for a healthy ego and self. In Kohut's words (Chapter 6), each person needs, above all, a self that feels empathic human responses from significant others. The healthy self feels the glow of pleasure in appropriate mature sexual functioning, as well as self-confident assertiveness in the pursuit of goals.

At the *phenomenological level*, the focus was not on recognizing and accepting that people are animals but rather that they are humans and as such have distinctive qualities. Most notably, a person has higher-order needs that go beyond the biological, and a self that has its own distinctive characteristics. For Carl Rogers, the essential point for thriving was to allow yourself to be genuine and to grow as you really are. That requires, in his view, that one listens to and accept one's organismic true self without distorting it to reduce anxiety or to please other people. For the existentialists and humanists also working at the phenomenological level, attention was on the combination of freedom and responsibility in making choices in life (e.g., May, 1961). The commonly used terms for these ingredients currently are autonomy or agency—namely a sense that one is at least partly causing one's own behavior, thus responsible and in charge of oneself or **agentic.**

Closely related to these ideas are those used at the *social cognitive level* to describe the essentials for well-being. Here the focus is on self-perceived **competence,** as reflected in Bandura's concept of self-efficacy, and in Dweck's concern with mastery and a belief that one is able to change and improve. Researchers who work on the self, selfhood, and self-determination (Deci & Ryan, 1995, 2000) also see competence as essential for thriving but they include an additional important component: **relatedness.** Because the self is intrinsically relational or interpersonal, people require the support and connection with others with whom they can feel close and mutually supported—they need relatedness as well as a sense of agency, competence, and efficacy, as was also noted earlier in this text. Finally, a common denominator that is seen at most levels is the importance of appropriate self-regulation, since that is an essential component for attaining the goals that one is pursuing (as discussed in detail in the previous chapter), and enables the person to be agentic.

Taken together, there is a clear consensus in the ingredients for well-being identified by these very different levels over the course of a century of work. To function well psychologically, people need to:

- Understand and accept themselves
- Have a sense of competence

- Feel agentic and responsible for their behavior and choices
- Be connected to other people
- Self-regulate appropriately

Potential for Change

18.16 To what extent do biological factors limit capacity for self-change?

Given what is known so far about personality, how great is the individual's potential for change and self-direction?

The Role of Genetics. On the one hand, the findings of behavior geneticists underline the role of the DNA and the person's genetic heritage (Wright, 1998). Recall, for example, that identical twins reared apart often display amazing similarities in what they do and become. But as a counter example, mate choice, for instance, although it is extensively influenced by the evolutionary history of the species, seems exempt from genetic influences (Chapter 14), and surely these choices are important expressions of the self. As one writer noted: "Are we not defined in a profound sense by the relationships we forge with others . . . ?" (Angier, 1998, p. 9). And again you have to remind yourself that even the strongest correlations in the heritability studies usually leave more than half the variance unexplained. Most of that unexplained variance is attributable to the environment—and that includes most importantly what the person does within that environment in the course of life.

The Role of the Brain. The same question now also is often raised about the role of the brain: doesn't the brain we inherit pretty much dictate the rest of the script that unfolds? Isn't biology destiny? A clear answer comes from the work of Eric Kandel, a recent Nobel prize-winning scientist who studied the well-mapped nervous system of Aplysia, a lowly sea slug.

In elegant studies, Dr. Kandel showed that learning leads the organism to produce new proteins that, in turn, remodel the neurons in the nervous system. The sea slugs that were put into controlled learning conditions that led to long-term memory also developed two times as many neuronal connections as the control group of slugs who were untrained. The implication for humans was clear: learning changes the structures and functions of the brain. And if the simple learning that slugs can do profoundly changes their nervous system, given the human capacities for learning, we have the potential to alter what happens in our brains dramatically.

In this text, you have seen much evidence about how individuals can construe and reinterpret themselves and explain events in ways that help to empower them and make their lives better and more constructive. Learning about the personality processes that allow change and emotional growth, as well as those that provide stability and coherence, also provides a potentially powerful route for self-direction. It can allow one perhaps to have more agency and to become thoughtful and wise about one's future. As George Kelly argued (Chapter 8), every person is a scientist, and in that sense can use what is known to form and test hypotheses about what might be possible.

The exciting possibilities for freedom and growth that each person has are not limitless. People may constructively rethink and reappraise their possible selves and expand their efficacy to a great degree, but their DNA influences the tools with which they do so. In addition to biology, cultural and social forces in part influence and limit both the events that people can control and their perceptions of their own possibilities (e.g., Kunda, 1999; Nisbett, 2001; Stigler et al., 1990). Within these boundaries, people have the potential to

Eric Kandel

gain substantial control over their own lives, shaping their futures in ways whose limits remain to be seen.

A few hundred years ago the French philosopher Descartes wrote his famous phrase "cogito ergo sum"—I think: therefore I am—and opened the way to modern psychology. With what is becoming known about personality, we can modify his assertion to say: "I think: therefore I can change what I am." Because by changing how I think, I can change what I feel, do, and become.

SUMMARY

PERSONALITY SYSTEM IN ACTION

- People differ in the ease of accessibility of different mental representations.
- The personality system is activated both through external situations and internally through the individual's own thoughts and feelings.
- The behavioral expressions of the personality system influence the situations the individual encounters and how others react, in turn, affecting the personality system and future behaviors.

INTERACTING INFLUENCES OF BIOLOGY AND CULTURE

- The multiple influences on the personality system include the interactive effects of biological determinants, learning, and social and cultural influences.
- Gender influences the views the individual develops about himself or herself as well as how other people treat the person.
- There are significant psychological differences both between the sexes and within the same sex.

- Females may respond to stress by tending and befriending instead of by fight or flight.
- There are distinct cultural differences in the way that gender roles are interpreted.

CULTURE AND PERSONALITY

- The cultural meaning system contains basic goals-values, emotions, standards, self-regulatory rules, interpersonal scripts, and control strategies that are activated only in certain situations.
- Cultures may differ in the typology of individualism versus collectivism, but this typology represents an average cultural tendency that may not fit many people within the culture.
- Culture influences the situations that people tend to experience, and the ways they interpret these experiences and respond.

WHAT DEVELOPS? THE EVOLVING SELF

- In self-construction, individuals actively pursue life goals central for the self as they build their lives.
- The self is relational: it thrives on social support and connectedness, not in isolation.
- People have the potential to learn and thus alter the structures and functions of their own brains.
- Ideal qualities of a well-functioning personality include: self-understanding and self-acceptance, a sense of competence, responsibility, agency, connection to other people, and the ability to self-regulate appropriately.

KEY TERMS

active ingredients   419
agentic   433
androgens   425
befriending (response to
   stress)   426
collectivism   429
competence   433
cultural meaning
   system   427

gender   422
gender concepts   423
gender stereotypes   422
human agency   431
individualism   429
macho patterns   426
personal meaning
   system   427
reciprocal interaction   419

relatedness   433
self as "agent"   430
self as "object"   430
selfhood   430
sex role
   identity   423
tending (response to
   stress)   425
units of culture   428

PART VII

THE INTEGRATION OF LEVELS

Personality psychology is evolving toward becoming an increasingly cumulative science that is based on promising ideas, findings, and methods from all areas of the field and all levels of analysis. The last three chapters in this text illustrated steps toward a more integrative approach to personality. It builds on concepts and findings from diverse levels of analysis that historically were considered incompatible but that, in fact, can enrich each other.

▶ PROSPECTS FOR PERSONALITY PSYCHOLOGY

Of course, integration for its own sake is not useful if the result is a conglomeration of concepts and findings with no clear criteria for inclusion or exclusion. Genuine differences in theory and research are often a necessary step in scientific progress. Not everyone can win, and not every idea will survive. The challenge is to gain breadth and perspective, but without indiscriminate inclusion, and with rigorous testing of the ideas, fine-tuning them progressively as the state of knowledge changes. By including useful concepts and discarding those that are not useful from all areas and levels of analysis, a solid integration begins to emerge.

▶ PERSONOLOGY REVISITED

More than 50 years ago, Henry Murray and his colleagues at Harvard University's Social Relations Department wanted to create a science of persons as individuals. To describe their ambitious work, they created a new term, "personology," for the study of persons and lives in depth (Chapter 6). While this term is no longer widely used, the goal that these pioneers pursued remains, and it is more timely now than ever before. Innovative as Murray and his colleagues were, they still lacked the methods to do their work as effectively as they wished. Fortunately, half a century later, advances in the science, for example, in biological approaches to personality and in the study of mental and emotional processes, are making it feasible to study individuals in increasing depth as Murray and his colleagues had hoped to do. For example, and as noted in the text, it is now possible to study what happens in the brain of an individual at the same time that his or her thoughts and feelings are studied. And it is possible to link these levels of analysis to intensive direct observations of how the person's behavior patterns interact with diverse situations sampled over extensive periods of time, either directly recorded or sampled through diaries. Valid, reliable, and useful tests and methods are available at each level of analysis, and all can be brought together to study personality with the depth and detail that this formidable task requires, and that makes personality an exciting field for now and for the future.

Personality psychology always has had two goals—description of individual differences and analysis of the processes that underlie them. By studying both together it becomes possible to build an increasingly cumulative and precise science of personality and human nature. The pursuit of both goals together provides an increasingly rich and clear view not just of personality constructs (e.g., neuroticism, self-regulatory competence) but also of the persons described with those terms and their distinctive characteristics. When the beautiful details begin to be put together, the coherent whole person emerges.

GLOSSARY

Ability traits   Cattell's term for traits that are concerned with effectiveness in gaining the goal.

Achievement pride   A type of pride that individuals feel when they have achieved some goals that they have set for themselves and are therefore eager to pursue new goals.

Act trends   General behavioral trends; Buss and Craik view traits as descriptions of individual differences in *act trends* instead of underlying internal causes.

Active ingredient   Features of situations that activate certain psychological processes in the perceiver.

Activity   Temperament that effects the vigor or intensity of responses; also refers to individual differences on a dimension that ranges from hyperactivity to extreme inactivity.

Actual self   The representation of oneself as one is.

Actual/ideal discrepancies   The differences between the self one perceives oneself to be and the self one would like to be.

Actual/ought discrepancies   The differences between the self one perceives oneself to be and the self one thinks one should be.

Affect   Feelings and emotions.

Agentic   To be at least partly responsible and in charge of oneself and one's behavior.

Aggression   Any behavior intended to harm or injure another living being who is motivated to avoid such treatment.

Agreeableness (A)   Good-natured, helpful, trusting, lenient; opposite of *antagonism.*

Ambivalence   Simultaneous attraction and aversion toward someone or something.

Amygdala   A small, almond-shaped region in the forebrain that is crucially important in emotional reactions, particularly fear.

Anaclitic identification   Hypothesized by Freud as the earliest form of identification based on the infant's intense dependency on the mother.

Anal stage   The second of Freud's psychosexual stages, it occurs during the child's second year; pleasure is focused on the anus and on the retention and expulsion of feces.

Analytical psychology   Carl Jung's theory of personality; humans are viewed as purposive and striving toward self-actualization; the unconscious includes a collective as well as a personal unconscious and is a healthy force.

Androgens   Hormones related to aggression, found in the male brain in many species.

Anima   In Jung's theory, the feminine, passive element in the unconscious of every male.

Animus   In Jung's theory, the masculine, assertive element in the unconscious of every female.

Anorexic behavior   Self-starvation; has been linked to *actual/ought discrepancies.*

Anorexia nervosa   A disorder in which the individual (usually an adolescent female) refuses to eat, without any apparent cause, sometimes starving to death.

Antidepressants   Drugs used to elevate the mood of depressed individuals.

Antipsychotic drugs   Drugs used in the treatment of major psychosis, notably schizophrenia.

Anxiety   A state of emotional arousal which may be experienced as a diffuse fear; in Freud's theory, the result of a struggle between an impulse and an inhibition.

Approach   Behavior that is directed toward actively achieving a goal or positive reward; commonly associated with the *behavioral activation system (BAS).*

Approach (appetitive) dilemma   When an individual faces an approach temptation that threatens to interfere with another more important, higher-order goal.

Approach tendencies   Individual's tendency to approach desired stimuli.

Approach-approach conflict   Conflict that occurs when a person must choose one of several desirable alternatives.

Approach-avoidance conflict   A conflict that occurs when a person confronts an object or situation that has both positive and negative elements.

Archetypes   Jung's term for the contents of the collective unconscious—images or symbols expressing the inherited patterns for the organization of experience (e.g., mother archetype).

Ascending reticular activation system (ARAS)   The system in the brain that is believed to regulate overall arousal in the cortex.

Assertiveness training   Training usually involving modeling and practice in developing and using effective assertive skills; a type of behavior therapy.

Automaticity Responses made automatically with little or no control, thought, or awareness.

Availability heuristic A cognitive rule or principle suggesting that the more easily we can think of something, the more likely we are to believe it to be in reality.

Aversion therapy Procedures that pair attractive, arousing, but problem-producing stimuli with another stimulus that evokes extremely negative reactions; the positive stimulus comes to evoke some of the negative or aversive reactions or is at least neutralized; for example, alcohol may be combined with nausea-producing drugs.

Aversive stimulation Punishment for a given response.

Avoidance Impulse to flee from a person or situation.

Avoidance (aversive) dilemma When an impulse to flee threatens to interfere with the pursuit of a higher-order goal.

Avoidance tendencies Individual's tendency to avoid undesirable stimuli.

Avoidance-avoidance conflict A conflict that occurs when a person must choose one of several undesirable alternatives.

Befriending (response to stress) Responding to stressful situations by making friends and offering assistance.

Behavior genetics Study of the role of genes in social behavior and personality.

Behavior modification Techniques used in behavior therapy which are derived from learning principles and intended to change behavior predictably for therapeutic goals.

Behavior sampling Method of selecting representative behaviors from an individual's daily life; can be done through self-reports, diary studies, and behavioral observation.

Behavioral activation system (BAS) Neurological system in the brain that activates approach behavior.

Behavioral deficits Lack of skills required for successful functioning.

Behavioral inhibition system (BIS) Neurological system in the brain that activates withdrawal from certain stimuli and inhibits behavior.

Behavioral medicine An interdisciplinary field that focuses on the influence of social and psychological factors on health.

Behavioral-conditioning level Approach to psychology emphasizing observable, objectively measurable behaviors and the relationships between these behaviors and specific events or stimuli in the environment.

Behavior-outcome relations The relationship between possible behavior patterns and expected outcomes in particular situations; an aspect of the expectancies that

constitute one of the person variables in the cognitive social level of analysis.

Benzodiazepines Minor tranquilizers that replaced the barbiturates as primary pharmacologic treatment for anxiety; an example is Valium.

Big Five Structure A popular taxonomy for characterizing individuals in terms of five major traits based on factor analysis of bipolar trait ratings and questionnaires.

Biofeedback Use of equipment to provide immediate feedback about the activities of the autonomic and somatic systems; for example, giving information about heart rate or brain waves to the person in whom they occur, at the time they occur.

Biological level Approach to personality that deals with genetic, hereditary, and evolutionary influences on personality.

Biological preparedness Biological predisposition to learn some associations more readily than others.

Bipolar representations According to object relations theorists, the combination of an individual's perception of self, his or her perception of another individual significant to him or her, and the emotions produced in him or her as a result of interaction between them; they become the templates through which the individual perceives ensuing relationships.

Blunting Ignoring anxiety-arousing stimuli as a means of coping with them; a style of information processing designed for dealing with stress.

Brain asymmetry The degree to which the right versus the left sides of the brain are activated.

Cardinal traits Allport's term for highly generalized dispositions or characteristics that influence most aspects of an individual's behavior throughout life.

Castration anxiety A male's fear of losing his penis; Freud believed this anxiety was central in the resolution of the Oedipus complex and the boy's identification with the father.

Catharsis The belief that the verbal or fantasy expression of an impulse leads to its reduction; a key concept in psychoanalytic theory.

Causal attribution Perception (judgment) of the causes of behavior, either to internal or external causes.

Central traits Allport's term for traits that are less important and pervasive than cardinal traits but that still influence much of a person's behavior.

Cheater detector A mechanism that humans have developed to detect individuals who seek the benefits of social exchange without reciprocating appropriately.

Chronic accessibility The ease with which particular cognitive-affective units or internal mental representations become activated or "come to mind."

Classical conditioning (conditioned-response learning) A type of learning, emphasized by Pavlov, in which the response to an unconditioned stimulus (e.g., food) becomes conditioned to a neutral stimulus (e.g., a bell) by being paired or associated with it.

Client-centered (person-centered) therapy Approach to therapy developed by Carl Rogers; emphasizes a nonevaluative, accepting atmosphere conducive to honesty and concentrates on present relationships and feelings.

Cognition Thought; mental activity.

Cognitive Pertaining to thoughts; mental.

Cognitive behavioral therapy (CBT) Therapy aimed at changing problematic behavior by thinking about one's problems and oneself more constructively and less irrationally (e.g., by modifying one's assumptions).

Cognitive cool system Processing system that generates cognitive thought processes and thoughtful, reflective reactions and plans; based in the hippocampus and frontal lobe.

Cognitive learning See *modeling* and *observational learning*.

Cognitive reappraisal Changing the way that one thinks about an experience in order to maintain self-regulation.

Cognitive restructuring Therapeutic technique aimed at learning to think about one's problems more constructively and less irrationally; Albert Ellis's *rational emotive therapy* is a form of cognitive restructuring.

Cognitive social competence (social and cognitive competencies) Person variable referring to an individual's abilities to cognitively process and use social information.

Cognitive social theories Approaches to personality that focus on the cognitive processes and structures underlying personal differences.

Cognitive transformations (of stimuli) Cognitively changing the mental representation of a stimulus by focusing on selected aspects of it or imagining it differently.

Cognitive-Affective Personality System (CAPS) The personality system as conceptualized within the social cognitive framework (see Mischel & Shoda, 1995).

Cognitive-affective units Mental-emotional representations—the cognitions and affects or feelings—that are available to the person.

Cognitive-physiological theory Theory of emotion developed by Stanley Schachter, stating that our experience of an emotion depends on the cognitive interpretation of physiological arousal; the same state of physiological arousal may be labeled differently under different circumstances; also known as the *two-factor theory of emotions.*

Collective unconscious Inherited portion of the unconscious, as postulated by Jung; consists of ancestral memories and archetypes that are part of each person's unconscious.

Common traits According to Allport, traits that are shared in different degrees by many people.

Community psychology Treating individuals by practicing behavior modification within their environments, not a mental hospital or therapeutic setting.

Compensatory motivation Motivation for the individual to compensate for early concerns with physical weakness or illness.

Competence Ability to perform a given task or types of tasks; related to self-efficacy.

Competence motivation Desire to acquire mastery of a task for its own sake.

Competency demand hypothesis The theory that an individual's characteristic coping behavior will tend to be manifested in situations that put greater demand on the individual's competencies.

Concordance rate (in schizophrenia) Percentage of twins in which the second twin is diagnosed as schizophrenic if the first twin has been diagnosed as schizophrenic; higher for monozygotic twins.

Concurrent validity Degree of correlation between a measure and another behavior/measure recorded at about the same point in time.

Condition of worth Conditions that other people, usually parents, implicitly make as a requirement for being loved and worthwhile.

Conditioned reinforcers Neutral stimuli that have acquired value by becoming associated with other stimuli that already have reinforcing power

Conditioned response (CR) A learned response to a conditioned stimulus; a response previously made to an unconditioned stimulus is now made to a conditioned stimulus as the result of pairing the two stimuli.

Conditioned stimulus (CS) A previously neutral stimulus to which one begins to respond distinctly after it has been paired with an unconditioned stimulus.

Conditioning A basic form of learning (see also *classical conditioning*; *operant conditioning*).

Conditions of deprivation Length of time an organism has been deprived of relief from an innate drive (e.g., hunger, thirst, sex).

Conflict Occurs when motivated to pursue two or more goals that are mutually exclusive.

Congruent therapist Therapist who feels free to "be himself" and to accept himself and the client fully and immediately in the therapeutic encounter.

Conscientiousness (C) Careful, self-reliant, goal-directed, scrupulous; opposite of *undirectedness.*

Conscious   Within awareness.

Constitutional traits   Those source traits that reflect constitutional factors.

Construct   Any class of behavior, thoughts, emotions, and situations.

Construct validity   The process of establishing that the theory about what accounts for behaviors on a particular test is valid; it involves validation of both the test and the theory that underlies it.

Constructive alternativism   Recategorization of individuals or events to facilitate problem solving; a concept in George Kelly's theory.

Constructivist view of science   Theory of science that there is no single, absolute Truth awaiting discovery, merely different ways of viewing and conceptualizing phenomena that are more or less useful and valid for particular purposes and contexts.

Contemporaneity of motives   Motives are understood in terms of their role in the present regardless of their origins in the past.

Content validity   The demonstration that the items on a test adequately represent a defined broader class of behavior.

Contingencies of reinforcement   Conditions that must be present in order for a response to be rewarded.

Contingency contracting   Self-imposed regulations that guide a person's behavior in the absence of immediate external pressures.

Continuous reinforcement   Schedule of reinforcement in which a response is reinforced every time it occurs.

Control group   The group of participants that does not receive the experimental treatment but is otherwise comparable to the experimental group; responses by this group can be compared with those by the experimental group to measure any differences.

Convenience of constructs   According to Kelly, the utility or convenience of a construct for the individual.

Correlation   The relationship between two variables or sets of measures; may be either positive or negative, and is expressed in a correlation coefficient.

Correlation coefficient   Quantitative expression of a correlation; ranges from 0 to +1 or −1.

Counterconditioning (desensitization)   Replacement of a response to a stimulus by a new response in behavior therapy.

Covert modeling   A type of behavior therapy in which the individual imagines a model performing the desirable behavior in an appropriate situation (see also *modeling*).

Criterion validity   Correlation between scores on a given test and the scores on other measures that serve as referents or standards.

Cue   Stimulus that directs behavior, determining when, where, and how the response (behavior) will occur.

Cultural meaning system   Culturally distinctive views about what the world is like; contains concepts and feelings about the self and its relation to others.

Cumulative science   A science that builds over time on its best findings and concepts and integrates them into a unifying general framework.

Current life tasks   Long-term projects to which individuals commit themselves during designated periods of their lives.

Defense mechanisms   According to Freud, ways in which the ego unconsciously tries to cope with unacceptable, anxiety-producing id impulses or feelings and events, as in repression, projection, reaction formation, sublimation, or rationalization.

Denial   In Freudian theory, a primitive defense mechanism in which a person denies a threatening impulse or event even though reality confirms it; the basis for development of repression.

Dependent variable   Aspect of the person's behavior that is measured after the independent variable has been manipulated.

Desensitization   Method of eliminating anxiety or fear responses to a stimulus in which the individual learns to make an incompatible response, such as relaxation, to a series of anxiety-evoking stimuli.

Directional selection   Selection process in which versions of characteristics that enhance survival and reproduction gradually become increasingly represented, while those that handicap survival and reproduction tend to fade out.

Discrimination training   Conditioning that involves reinforcement in the presence of one stimulus but not in the presence of others.

Discriminative facility   The ability to appraise situations as they present themselves and to respond accordingly.

Discriminative stimuli   Stimuli that indicate when a response will or will not have favorable consequences.

Distortion/Displacement   Occurs when private meanings develop as objects and events become symbols representing things quite different from themselves.

Dizygotic twins   Fraternal twins; two organisms that develop in the uterus at the same time but from two egg and two sperm cells; not genetically identical.

Domain of interpersonal behavior   According to Wiggins, a comprehensive taxonomy of personality dimensions discovered through factor analysis.

Domain specific knowledge and expertise   The group of competencies essential for everyday problem-solving and coping behaviors.

Domain specificity   Specific psychological problem-solving strategies that facilitated reproduction, adaptation, and survival in the course of evolution.

Double-blind experiment   Experimental procedure in which neither participants nor experimenters know whether participants are in experimental or control conditions.

Down syndrome   Genetic abnormality consisting of a third chromosome in the 21st chromosome pair; causes severe mental retardation and a distinctive appearance.

Dream interpretation   A method used in psychoanalysis to better understand the unconscious fears and desires of the patient through an analysis of the patient's dreams.

Drive   Any strong stimuli (internal or external) that impels action.

Drive reduction   Reduction of tension caused by the fulfillment of a drive.

Dynamic traits   Cattell's term for traits that are relevant to the individuals being "set into action" with respect to some goal.

Ectomorph   According to Sheldon, an individual who has a tall, thin, and stoop-shouldered figure and has an artistic, restrained, and introvertive temperament.

Effortful control   Willpower in goal pursuit.

Ego   In Freudian theory, the conscious part of the personality that mediates between the demands of the id and the demands of the world; operates on the reality principle.

Ego identity   The ego's ability to integrate changes in the libido with developmental aptitudes and social opportunities.

Ego psychology   A variety of psychoanalytic theory that stresses ego functions and de-emphasizes instinctual drives.

Ego control   Degree of impulse control; important in delay of gratification, planfulness, and aggression inhibition (see also *ego resilience*).

Ego resilience   Refers to the individual's ability to adapt to environmental demands by appropriately modifying his or her habitual level of ego control, thus functioning with some flexibility.

Electrocardiogram (EKG)   A polygraph that measures the patterns of electricity produced by muscular contractions of the heart.

Electroencephalograph (EEG)   Records the degree of activation in the cerebral cortex through "brain waves."

Emotional hot system   Processing system in the brain that generates automatic approach-avoidance, or fight-or-flight responses; based in the *amygdala*.

Emotional overcloseness   In Kohut's view, a state in which the family exposes the child to too much intimacy, stimulation, and intrusiveness.

Emotional stability   The opposite of neuroticism, according to Eysenck, who viewed emotional stability-neuroticism as an important trait dimension.

Emotionality   Tendency to become physiologically aroused easily and experience frequent and intense negative emotions such as anger, fear, and distress.

Empathic mirroring   Learning emotions and behaviors from the examples of others.

Encoding   A social cognitive person variable that includes the individual's personal constructs and units for categorizing people, events, and experiences.

Endomorph   According to Sheldon, an individual who is obese and correspondingly tends to be relaxed, sociable, and likes to eat.

Entity theorists   Those who see abilities and traits as fixed, unchangeable characteristics; they tend to choose goals that will assure them favorable judgment and approval.

Environmental presses   According to Murray, contextual or situational pressures that influence personality and its expression.

Environmental psychology   A field of study that applies psychology to understanding human behavior in relation to the person's environment.

Environmental-mold traits   Those source traits (according to Cattell) that reflect environmental conditions.

Eros   One of two sides of the personality, as seen by early psychoanalysts, that represented sexuality and love (see also *Thanatos*).

Error variance   The "noise" or the non-essential variance to be discounted from statistical analysis.

Evaluative factor   Evaluations such as "good" or "bad."

Executive functions   Self-regulatory processes, such as planning, self-control, and willpower.

Existential anxiety   An awareness of the nothingness and death that necessarily accompanies being and that must be resolved by enhanced attention to how we choose to lead our lives (in existential theories).

Expectancies   A person variable that includes behavior-outcome and stimulus-outcome expectancies and guides an individual's choices.

Experiment   An attempt to manipulate a variable of interest while controlling all other conditions so that their influence can be discounted and the effects of the variable measured.

Experimental group   The group in an experiment into which the independent variable is introduced; in order to determine the effects of the independent variable, the results of this group are compared with those in the control group.

Explanatory styles   The way in which people interpret and construe the reasons for different events and outcomes.

External control   When an individual believes that positive and/or negative events are a result of factors outside of one's control.

Extinction   The decrease in the frequency of a response that follows the repetition of the response (or in classical conditioning, the repetition of the conditioned stimulus) in the absence of the unconditioned stimulus.

Extinction schedule   Reinforcement for a behavior is withdrawn until gradually the behavior itself stops.

Extraversion (E) (Extraverts)   According to Jung, an individual who is conventional, sociable, and outgoing and who reacts to stress by trying to lose himself or herself among people.

Extraversion-introversion   According to Eysenck, a basic dimension of personality along which all individuals can be placed at some point.

Extrinsic motivation   Motivations inferred when a person engages in a task for the sake of the expected outcome, not out of enjoyment of the task itself.

Factor analysis   A mathematical procedure for sorting trait terms or test responses into clusters or factors; used in the development of tests designed to discover basic personality traits; it identifies items that are homogeneous or internally consistent and independent of others.

False memories   Memories of events that never actually occurred.

Family therapy   A therapeutic approach based on the premise that the roots of problems lie within the family system and therefore must be treated by improving family dynamics and relations.

Field theory   Position that construes behavior as determined by the person's psychological life space—by the events in his or her total psychological situation at the moment—rather than by past events or by enduring situation-free dispositions.

Fight or flight reaction   Automatic physical and emotional reaction to perceived threat or danger, leading to attack or withdrawal; involves increase in heart rate, metabolic rate, and blood pressure, and a decrease in stomach and intestinal activity.

Fixation   A psychodynamic term referring to a process by which a person remains arrested at an early stage of psychosexual development, or moves on but later regresses to that stage.

Free association   A technique used in psychoanalytic therapy in which the patient is instructed to report whatever comes to mind, no matter how irrational it may seem.

Functional analysis   A system of analysis proposed by Skinner to link the organism's behavior to the precise conditions that control it.

Functional autonomy   The idea that adult motives replace the motives of infancy.

Functional magnetic resonance imaging (fMRI)   Method of capturing and creating an image of the activity in the brain through measurement of the magnetic fields created by the functioning nerve cells in the brain.

Fundamental attribution error   The tendency to focus on dispositions as causal explanations of behavior.

Galvanic skin response (GSR)   Changes in electrical activity of the skin due to sweating, as recorded by a galvanometer, and used as an index of emotional state, for example, in a lie detector test.

Gender   The social meaning of being a male or a female.

Gender concepts   Beliefs about the types of behavior more appropriate for one sex than for the other.

Gender stereotypes   Rigid beliefs about what it means to be a man or a woman within a particular culture or society.

Generalization   Responding in the same way to similar stimuli, for example, when a child who has been bitten by a dog becomes afraid of all dogs.

Generalized conditioned reinforcers   Conditioned reinforcers that have been paired with more than one primary reinforcer.

Generalized reinforcers   See *generalized conditioned reinforcers*.

Genital stage   The last of Freud's psychosexual stages, in which the individual becomes capable of love and adult sexual satisfaction.

Gestalt theory   The completeness or fullness theory of Fritz Perls; focuses on the awareness of one's own experience in dynamic interaction with the environment.

Gestalt therapy   An approach developed by F. Perls that aims at expanding the awareness of self and putting the person in touch with his or her own feelings and creative potential; often practiced in groups, use is made of body exercises and the venting of emotions.

Globality   Generalizing an event to pertain to many aspects of one's life.

Goal gradients   Changes in response strengths as a function of distance from the goal object.

Goal hierarchies   Organization of goals in order of their degree of importance to the individual.

Harvard personologists   A group of psychologists in the 1940s and 1950s whose study of personality was strongly influenced by the work of Freud and by biosocial, organismic theory stressing the integrated, whole aspect of personality.

Healthy personality   According to Maslow, a self-actualized person.

Heritability estimates (index)   A measure used in behavior genetic research to try to assess the degree to which a trait or attribute is due to inheritance.

Hermaphrodite   A person who is born with moth male and female sex organs.

Hierarchy of anxiety (grading the stimuli)   A hierarchy of events that range from the most to the least intensely anxiety-provoking.

Higher growth needs   Needs for self-actualization fulfillment.

Higher-order conditioning   Process that occurs when a conditioned stimulus modifies the response to a neutral stimulus with which it has been associated.

Higher-order motives   A hypothesized motive that, unlike thirst or hunger, does not involve specific physiological changes.

Hippocampus   A horseshoe-shaped region in the temporal lobe of the brain that regulates emotional responses and complex mental and spatial thought.

Hostile attribution bias   Tendency to perceive others who are implicated in, though not necessarily responsible for, negative events being driven by hostile intentions.

Hot cognitions   Thoughts that activate strong emotion.

Human agency   The individual's ability to proactively influence his or her personal experience.

Human genome   The human genetic code consisting of roughly 30,000 to 40,000 individual genes contained in 23 chromosome pairs.

Human-relations training group (T-group)/Sensitivity training group   Encounter groups that focus on non-verbal experience and self-discovery.

Hysteria   A neurotic condition consisting of two subcategories: conversion reaction (physical symptoms such as paralysis or loss of sensation without organic cause) and dissociative reaction (disruption of a consistent unitary sense of self that may include amnesia, fugue, and/or multiple personalities).

Hysterical anesthesia   Loss of sensation in a part of the body without physiological impairment, reflecting a defensive attempt to avoid painful thoughts and feelings, according to Freud.

"I" as an agent   The self as an executive who conducts basic psychological functions; engages in self-regulation and planning for the future.

Id   In Freudian theory, the foundation of the personality and a basic component of the psyche, consisting of unconscious instincts and inherited biological drives; it operates on the pleasure principle.

Ideal self   The representation of who you would like to be.

Identification with the aggressor   Identification with the father or the "aggressor" during the Oedipal stage of

development; motivated by fear of harm and castration by the father.

Identity crisis   According to Erikson, a point in psychological development when the adolescent or young defines his or her identity.

I-E Scale   Trait dimension that reflects whether a person perceived locus of control is internal (I) or external (E).

If . . . then . . . Pattern (Signature)   Distinctive patterns of situation-behavior interactions that characterize an individual (e.g., *if* situation A, *then* behavior B).

Implicit Association Test (IAT)   A word-association task that is used to measure implicit self-esteem; may also be used to evaluate implicit attitudes, beliefs, and values as well.

Implicit methods   Indirect and projective methods of personality measurement.

Implicit self-esteem   How positively or negatively one regards oneself on indirect measures taken without one's conscious awareness.

Incremental theorists   Those who view abilities and traits as open to change; they tend to choose goals that will enhance their competence, though not necessarily guarantee success.

Independent variable   Stimulus or condition that the investigator systematically varies in an experiment.

Individualism versus collectivism   Two contrasting cultural meaning systems: one focuses on the importance of the individual, the other on the importance of the group or community.

Inferiority complex   According to Adler, feelings of inferiority in the individual that stem from the experience of helplessness and organ inferiority in infancy; results from a failure to compensate for early weakness through mastery in life tasks.

Insecure-ambivalent   Attitude presented by infants in the Strange Situation whose reunion behavior seemed to be a combination of contact-seeking and anger.

Insecure-avoidant   Attitude presented by infants in the Strange Situation who avoided their mother throughout the paradigm, even upon reunion.

Interactional analysis   An analysis of the ways in which the individual's behavior varies predictably across different situations (e.g., she does A when X but B when Y).

Interactionism   Idea that the individual's experiences and actions are the product of dynamic interactions between aspects of personality and situations.

Interjudge reliability   Degree of consistency among different judges scoring or interpreting the same information; if scoring decisions are subjective, then it becomes especially necessary to demonstrate interscorer agreement.

Internal consistency A high correlation between different parts of a single measure.

Internal control An individual's beliefs that positive and/or negative events are a result of his or her own behavior.

Internal working models Mental representations of the others or self, or of relationships, that guide subsequent experiences and behavior.

Internality Tendency to perceive oneself, and not external circumstances, as responsible for a problem.

Internalized Aspects of the personality that were acquired from external sources (e.g., parents) but become part of oneself.

Interscorer agreement The degree to which different scorers or judges arrive at the same statements about the same test data.

Interview A verbal method in which a person interacts directly with an interviewer in a one-on-one situation (e.g., to study personality, to survey beliefs).

Intrinsic motivation Motivation (e.g., curiosity, achievement, affiliation, identity, stimulation, and social approval) that does not depend on the reduction of primary drives (such as hunger or sex) and does not have specific physiological correlates.

Introversion (Introverts) Tendency to withdraw into oneself, especially when encountering stressful emotional conflict; according to Jung, the introvert is shy and withdrawn, and prefers to be alone (see also *extravert*).

IQ (Intelligence Quotient) Concept formulated by Binet to summarize an individual's mental level based on test scores; IQ means "mental age" divided by chronological age × 100; the average IQ at any given chronological age is set at 100.

Latency period In Freud's theory of psychosexual stages, the period between the phallic stage and the mature, genital stage, during which the child represses memories of infant sexuality.

Learned drive A motivation that has been transformed from a primary drive by social learning.

Learned helplessness A condition in animals and humans that results from exposure to inescapable painful experiences in which passive endurance persists even when escape becomes possible; it can lead to hopelessness and depression.

Learned optimism Believing that one can induce positive outcomes and is not responsible for negative events.

Learning goals Goals that are directed toward gaining new skills or competencies.

Learning theories Theories that seek to identify how and when new behaviors are acquired, performed, and modified.

Level of arousal (LOA) Level of stimulation in the brain.

Levels of analysis Major theoretical approaches to the study of personality that have guided thinking and research.

Libido In Freudian theory, psychic energy that may be attached to different objects (e.g., to the mouth in the oral stage of psychosexual development).

Life space Lewin's term for the determinates of an individual's behavior at a certain moment; it includes the person and his or her psychological environment.

"Me" as an object The self as an object that one sees when one focuses attention on the self; represented in self-concepts.

Macho patterns (machismo) Pride in male physical and sexual prowess.

Mandala One of Jung's archetypes, a circle symbolizing the self's search for wholeness and containing designs often divided into four parts.

Mantra A Sanskrit word used in transcendental meditation.

Mastery-oriented People who believe that their failure is due to lack of effort, not lack of ability.

Meaning in life An individual's sense of purposefulness in his or her life.

Mesomorph An individual who has an athletic build and has a very energetic, assertive, and courageous temperament according to Sheldon.

Methadone A drug that appears promising in the treatment of heroin addiction; blocks the craving for heroin and prevents "highs" if heroin is taken.

Miller Behavioral Style Scale (MBSS) Measure of monitoring-blunting tendencies.

Mindless Guided by automatic behavioral routines or scripts that are familiar and require little active attention or self-regulatory effort.

MMPI (Minnesota Multiphasic Personality Inventory) Most popular and influential personality questionnaire, consisting of more than 500 statements to which the person responds "true," "false," or "cannot say"; items cover a wide range of topics and have been grouped into nine clinical and three control scales; scores are summarized in a profile.

Modeling A technique used in behavior therapy in which the client observes the successful performance of the desirable behaviors by a live or symbolic model; effective in teaching complex, novel responses in short time periods and in overcoming fears (see *observational learning*).

Monitoring A cognitive coping mechanism or style of information processing in which people attend to anxiety-arousing stimuli, often in the hope of controlling them.

Monoamine oxidase (MAO) Enzymes that break down the neurotransmitters after they have passed along the route, maintaining the proper level of neurotransmitters.

Monozygotic twins   Identical twins; two organisms that develop from a single fertilized egg cell and share identical genes.

Moral anxiety   In Freud's theory, guilt about one's unacceptable feelings, thoughts, or deeds (see also *neurotic anxiety* and *reality anxiety*).

Motivational determinism   Freud's belief that everything a person does may be determined by his or her pervasive, but unconscious, motives.

Multiple act criterion   A criterion measure consisting of a combination of many acts or behaviors that are expected to be interrelated; combining these components increases the reliability of the measure.

Nature-nurture   Phrase delineating the long-standing controversy in psychology over inheritance versus environment as significant determinants of individual differences in personality; nature *and* nurture are important, and their interaction may be of greatest interest.

Need for achievement (n Ach)   Need for achievement (in theory of achievement motivation).

Need for intimacy   The motivation to warmly and closely connect, share, and communicate with other people in one's everyday life.

Need for power   Individual's desire to have an impact on other people.

Neo-Freudians   Post-Freudian innovators in the field of psychology who expanded on Freud's original work, putting less emphasis on the significance of the id and paying greater attention to the ego and the self.

NEO-PI-R   A personality inventory that measures individuals along the Big Five personality trait dimensions.

Neural networks   Interconnection of neurons in the brain that are activated during information processing and mental activities.

Neuropharmacology   The use of chemicals in treating problems with psychological symptoms and disorders of the nervous system.

Neurotic anxiety   In Freud's theory, fear that one's own impulses will go out of control and lead to punishment (see also *moral anxiety* and *reality anxiety*).

Neurotic conflict   Clash between id impulses seeking expression and internalized inhibitions

Neuroticism (N)   The opposite of emotional stability, according to Eysenck, who viewed neuroticism-emotional stability as an important trait dimension.

Neurotransmitter systems   Physiological pathways that communicate and carry out the functions of signal detection and response via chemical receptors (neurotransmitters).

Neurotransmitters   Chemicals that enable the nerve impulses to jump across nerve synapses from one nerve to the next.

Nominal situations   The routine activities and places that exist within a given environment.

Nonshared environment   All of the aspects of the individual's environment that are not shared with other members of the family (e.g., birth order effects, illness, and peer influences).

Not-thinking   According to Dollard and Miller, the learned response to fear, producing the phenomenon of repression in psychodynamic theory.

Object relations theory   An approach to psychoanalysis that stresses study of the interactions between individuals, especially in childhood.

Observational learning   The process of learning through observation of a live or symbolic model; requires no direct reinforcement.

Oedipus complex   According to Freud, the love for the opposite-sex parent during the phallic stage of development, particularly the son's love for the mother and hostility toward the father.

Openness to experience (O)   Willingness to try new things, original, independent; opposite of closed-mindedness.

Operant conditioning (instrumental learning)   The increase in frequency of an operant response after it has been followed by a favorable outcome (reinforced).

Operants   Freely emitted response patterns that operate on the environment; their future strength depends on their consequences.

Operationalization   Translation of a construct into something observable and measurable.

Optimal level of arousal (OLA)   Arousal level that is most appropriate for performing a given task effectively.

Optimistic orientation   A focus on the positive aspects of oneself and in human nature.

Oral stage   First of Freud's psychosexual stages, when pleasure is focused on the mouth and on the satisfactions of sucking and eating, as during the first year of life.

Organ inferiority   Alfred Adler's term for physical weakness associated with the helplessness of infancy.

Organismic evaluation   The self's evaluation of the experiences of the organism (in Rogers' theory).

Organismic experience   Experience of the self as a whole organism (in Rogers' theory).

Organized whole   Rogers' desired state of being, where the individual functions as an integrated unit.

Ought self   The representation of who you believe you should be.

Outcome expectancies   Belief that a particular behavior will lead to the anticipated outcome (e.g., that waiting for the promised dessert will actually lead to getting it).

Overcontrolling children   Children who tend to extensively inhibit their behavioral impulses

Overjustification   Excessive external reward for a given response; may interfere with the development of intrinsic rewards.

Parapraxes   In psychoanalytic theory, slips of the tongue that express unconscious thoughts.

Partial (intermittent) reinforcement   Reinforcement in which a response is sometimes reinforced, sometimes not reinforced.

Peak experience   According to Maslow, a temporary experience of fulfillment and joy in which the person loses self-centeredness and feels a nonstriving happiness.

Penis envy   Envy of the male sex organ; believed by Freud to be universal in women, to be responsible for women's castration complex, and to be central to the psychology of women.

Perceived self-efficacy   The belief that one is capable of performing or achieving the relevant task or goal (e.g., "If I jump into that pool, I can swim to the other side.").

Perceptual defense   Unconscious repressive mechanisms that screen and block threatening visual and auditory inputs.

Performance goals   Goals that are directed toward earning acclaim for one's actions and skills.

Performance measures   Measures that directly test the individual's ability to perform a task.

Permeable boundaries   Boundaries between the person and the psychological environment that can be crossed easily.

Person variables   Relatively stable social, cognitive, and emotional variables on which individuals differ (e.g., expectations, goals, values, and competencies); sometimes these are called cognitive social person variables.

Person versus situation debate   The belief that to the degree that the person was important the situation was not and vice versa.

Person × situation interaction   The idea that individual differences in behavior are reflected in the way each person responds to a particular situation and that the way a particular situation will affect behavior depends on the individual.

Person × situation interaction patterns   Stable patterns that emerge when the individual's behavior is measured in relation to its situational context.

Personal construct   The subjective dimensions through which the individual experiences the world and the self; central unit of Kelly's theory.

Personal construct theory   George Kelly's theoretical approach to personality that tries to see how the person sees or aligns events on his or her own dimensions.

Personal meaning system   The individual's system of concepts and feelings that guides and constrains how he or she interprets the self and the world.

Personal narratives   The stories people tell themselves to try to make sense of their own lives and experiences.

Personality dispositions   Behavioral tendencies or patterns that characterize individuals or types distinctively in reliable ways.

Personality paradox   Conflict between intuition of intra-individual consistency and the research results which show a lack of intra-individual consistency.

Personality processing dynamics   Stable, distinctive sequences of thoughts, feelings, and behaviors that become activated by particular types of situations in *if . . . then . . .* signatures.

Personality signature   Distinctive *if . . . then . . . ,* situation-behavior profiles that characterize individuals.

Personologists   Psychologists who study personality; phrase used by the "Harvard personologists" for their style of personality research.

Personology   Intensive psychodynamic study of individual lives as integrated, organized units, conceived by the "Harvard personologists."

Pessimism   An explanatory style in which the individual sees negative events as being widespread and largely a result of his or her own doing, while failing to take credit for positive events.

Phallic stage   The third of Freud's psychosexual stages (at about age five), when pleasure is focused on the genitals and both males and females experience the "Oedipus complex."

Pharmacotherapy   The use of drugs to treat psychological disorders or psychological symptoms.

Phenomenological   Term that refers to the individual's experience as he or she perceives it.

Phenomenological level   Theory that emphasizes the person's experience as he or she perceives it.

Phenomenology   The study of an individual's experience as he or she perceives and categorizes it, with emphasis on the self and interactions with other people and the environment.

Phenothiazines   Major tranquilizer drug useful in controlling schizophrenia (also called *antipsychotic drugs*).

Physiological measures   Measures of how a person responds physiologically (e.g., change in heart rate, degree of arousal) to different events.

PKU (phenylketonuria)   A genetic abnormality in which the gene that produces a critical enzyme is missing; it results in mental retardation if not treated soon after birth.

Placebo   An inert substance administered to someone who believes it is an active drug.

Pleasure principle In Freud's theory, the basis for id functioning; irrational, seeks immediate satisfaction of instinctual impulses.

Plethysmograph An instrument that measures changes in blood volume.

Polygraph An apparatus that records the activities of the autonomic nervous system.

Positron emission tomography (PET scans) Measures the amount of glucose being used in various parts of the brain, providing an index of activity as the brain performs a particular function.

Possible selves An individual's potential ways of being.

Potency Factor in semantic differential measures represented by scale items like hard-soft, masculine-feminine, strong-weak.

Practical intelligence The competencies, skills, and knowledge (expertise) essential for everyday problem-solving and coping behaviors.

Preconscious Thoughts, experiences, and memories not in a person's immediate attention but that can be called into awareness at any moment.

Predictive validity Correlation between current measure and future behaviors/measures.

Prevention pride Type of achievement pride that motivates people to vigilantly avoid factors that may hinder them from achieving their goals.

Prevention system System that attends to negative outcomes such as punishment or failure.

Primary biological needs Innate set of needs required for the organism's survival, such as food, water, oxygen, and warmth.

Primary process thinking Freud's term for the id's direct, reality-ignoring attempts to satisfy needs irrationally.

Primary reinforcers Innate reinforcers that cause automatic responses without the need for conditioning (e.g., food).

Principle of contemporaneity According to Lewin, the psychological life space that takes into account only what is happening and experienced at a given point in time.

Principle of interactionism In personality psychology, states that important personality differences may be seen in the ways in which an individual's behavior is effected by and interacts with different situations.

Processing (approaches) Theoretical approaches that focus on mental and emotional (psychological) processes or determinants that underlie behavior.

Processing dynamics Characteristic pattern of thoughts, feelings, and behavioral reactions activated in relation to situations.

Projection A defense mechanism by which one attributes one's own unacceptable aspects or impulses to someone else.

Projective methods Tests (such as the Rorschach or TAT) that present the individual with materials open to a wide variety of interpretations based on the belief that responses reveal important aspects of the respondent's personality; central in psychodynamic assessment.

Promotion pride Type of achievement pride that motivates people to eagerly pursue their goals.

Promotion system System focused on obtaining positive outcomes and gains, such as accomplishments or rewards.

Proprium Allport's term for the region of personality that contains the root of the consistency that characterizes attitudes, goals, and values; not innate, it develops in time.

Prosocial aggression Verbal threats and evaluative statements about the goodness or badness of behavior.

Prototypicality The degree to which the member of a category is representative of that category or exemplifies it.

Psychoanalysis A form of psychotherapy developed by Freud that aims at relieving neurotic conflict and anxiety by airing repressed, unconscious impulses over the course of regular meetings between patient and analyst.

Psychobiography The intensive study of individual lives using narrative methods.

Psychodynamic Behavior Theory Developed by John Dollard and Neal Miller in the late 1940s to integrate some of the fundamental ideas of psychoanalytic theory with the concepts and methods of experimental research on behavior and learning.

Psychodynamic-motivational level Approaches, beginning with Freud's work, that probe the motivations, conflicts, and defenses that help explain complex consistencies and inconsistencies in personality.

Psychodynamics In psychoanalytic theory, the processes through which personality is regulated; it is predicated on the concept of repressed, unconscious impulses and the significance of early childhood experience.

Psycholexical approach A research strategy that seeks to classify people into different trait groups by identifying differences among individuals on the basis of ratings with natural language terms (adjectives) that are factor analyzed.

Psychological features The aspects or ingredients of situations that activate the person's characteristic reaction patterns (e.g., being rejected by peers in a social situation).

Psychological oxygen According to Kohut, the psychological deprivation of empathic human responses in important others is analogous to the deprivation of oxygen.

Psychological situation   The circumstances and events within a nominal situation that affect behavior.

Psychometric trait approach   Approach that emphasizes quantitative measurement of psychological qualities, comparing the responses of large groups of people under standard conditions, often by means of paper-and-pencil tests.

Psychosexual stages   According to Freudian theory, development occurs in a series of psychosexual stages; in each stage (oral, anal, phallic, and genital) pleasure is focused on a different part of the body.

Psychosocial crisis   According to Erikson's theory, the person's efforts to solve the problems that occur at a given stage of psychosocial development.

Psychosocial stages   Erikson's eight stages of development; extending throughout life, each stage centers around a "crisis" or set of problems and the individual's attempts to solve it.

Q-sort (Q-technique)   A method of obtaining trait ratings; consists of many cards, on each of which is printed a trait description; the rater groups the cards in a series of piles ranging from those that are least characteristic to those that are most characteristic of the rated person.

Randomization   The assignment of research participants to different conditions on the basis of chance; if many participants are used, differences should average out except for the effects produced by the experiment itself.

Rational emotive therapy   Albert Ellis's approach, based on the idea that if people learn to think more rationally, their behavior will become more rational and their emotional problems will be reduced.

Rationalization   A defense mechanism that occurs when one makes something more acceptable by attributing it to more acceptable causes.

Reaction formation   A defense mechanism that occurs when an anxiety-provoking impulse is replaced in consciousness by its opposite.

Reality anxiety   In Freud's theory, the fear of real dangers in the external world (see also *moral anxiety* and *neurotic anxiety*).

Reality principle   In Freud's theory, the basis for ego functioning; rational; dictates delay in the discharge of tension until environmental conditions are appropriate.

Reciprocal altruism   The recognition that if we help others they are likely to reciprocate in kind.

Reciprocal interaction   Interaction between the person and the situation in which both variables influence one another reciprocally.

Reflex   An instinctive, unlearned response to a particular stimulus (see also *unconditioned response*).

Regression   In psychodynamic theory, reversion to an earlier stage; the return of the libido to its former halting places in development.

Regulatory competencies   The cognitive and attentional mechanisms that help execute goal-directed behavior.

Regulatory focus theory   Higgins' theory that all goal-directed behavior is governed by two motivational systems—the prevention system and the promotion system.

Reinforcement (Reinforcer)   Any consequence that increases the likelihood that a response will be repeated or strengthened.

Rejection sensitivity   A disposition to anxiously expect and easily perceive potential rejection even in ambiguous events.

Relatedness   Support and connection with others.

Relational self   The self perceived not as a single entity but as an object in relation to other objects, as in Kohut's object relations theory.

Relational therapy   A therapeutic process that emphasizes the role of early, current, and analyst-patient relationships in the development and resolution of personality problems.

Relaxation responses   Wolpe's preferred conterconditioning method for anxiety; involves elaborate instructions to teach individuals how to consciously relax themselves.

Repression   According to psychoanalytic theory, an unconscious defense mechanism through which unacceptable (ego-threatening) material is kept from awareness; the repressed motives, ideas, conflicts, memories, etc. continue to influence behavior.

Repression-sensitization   A dimension of differences in defensive patterns of perception, ranging from avoiding the anxiety-arousing stimuli to approaching them more readily and being extra vigilant or supersensitized.

Repressors   People who describe themselves as having few problems or difficulties and who do not report themselves as highly sensitive to everyday stress and anxieties.

Resistance   Difficulties in achieving progress in psychotherapy due to unconscious defenses as anxiety-producing material emerges during the treatment.

Response   Any observable, identifiable activity of an organism.

Response freedom   The condition in a situation that does not produce a similar reaction in all individuals but that allows individual differences in the behaviors of each person to become visible.

Role   An attempt to see another person through his or her constructs.

Role Construct Repertory Test ("Rep" Test)   A technique for measuring personal constructs developed by Kelly.

Rorschach test Projective test consisting of 10 symmetrical inkblots to which the person describes his or her reactions, stating what each blot looks like or might be.

Rumination Dwelling on particular types of cognitions and emotions, usually negative; in a ruminative style of dealing with depression, the person focuses on the fact that he or she is depressed, on the symptoms (like fatigue and disinterest) that are experienced, and on their negative consequences (e.g., "I might lose my job").

Schemas (schemata) Knowledge structures made-up of collections of attributes or features that have "family resemblance" to each other.

Schizophrenia Most common form of psychosis; prominent symptoms may include thought disorder, delusions, highly inappropriate or bizarre emotions, and hallucinations, all without known organic cause.

Scholastic Aptitude Test (SAT) A standardized measure of verbal and quantitative skills given routinely to high school seniors before entrance into college.

Secondary dispositions The most specific, defined traits or "attitudes" that influence an individual's behavior.

Secure base Safe haven of dependable comfort in a young child's life from which the world can be explored with trust.

Securely attached Attitude presented by infants in Strange Situation who greeted mother positively upon reunion and then returned to play.

Self actualize (self-actualization) The realization of one's human potential.

Self as "agent" The active process of self-construction in the course of building a self and a life within the interpersonal world.

Self as "object" The consciousness of the self as a coherent entity.

Self-actualizing person For Maslow, a person who has the personal qualities listed in Table 9.5.

Self-concept (self) System that serves as a frame of reference for evaluating and monitoring the actual experiences of the organism.

Self-consciousness Attention directed toward the self.

Self-disclosure Sharing of stressful, traumatic experiences either in groups or other forms (e.g., diaries).

Self-efficacy Belief that one can do things effectively.

Self-efficacy expectations The person's confidence that he or she can perform a particular behavior, like handling a snake or making a public speech.

Self-enhancing bias Increased likelihood of seeing oneself as causally responsible for actions when they have positive rather than negative outcomes.

Self-enhancing illusions Unrealistically positive view of the self and one's ability to control pure chance situations.

Self-esteem Refers to the individual's personal judgment of his or her own worth.

Self-evaluative standards (self-guides) Standards for assessing oneself.

Selfhood Term used to refer to the awareness of the self as "agent" and "object."

Self-instruction Talking to oneself to control one's behavior; an aspect of some types of control training.

Self-observation Systematic observation of one's own behavior; an important step in some types of behavior therapy.

Self-realization A continuous quest to know oneself and to actualize one's potentialities for full awareness and growth as a human being.

Self-regulation The monitoring and evaluation of one's own behavior, modifying one's own behavior and influencing his/her environment.

Self-regulatory systems A person variable that includes the individual's rules, plans, and self-reactions for performance and for the organization of complex behavior (see *person variables*).

Self-reinforcement The process of providing positive consequences to oneself contingent upon enacting desired behaviors or achieving specific performance criteria (e.g., you get to go to a movie if you get an A on the exam).

Self-relevant Meaningful to the self.

Self-reports Statements that the individual makes about himself or herself.

Self-schemas Cognitions about the self that arise from past experience and guide the processing of new information.

Self-theory A construction or set of concepts about the self that affect future experience.

Semantic differential Measure that tests individuals on their perceptions of the meanings of diverse words, phrases, and concepts by having them rate each item on a bipolar scale.

Sensate focus A method for overcoming sexual-performance fears in which the couple concentrates on sensual pleasures without engaging in sexual intercourse.

Sensation seeking Trait that represents an individual's level of desire to experience new things and take risks.

Sensation seeking scale (SSS) Measure that taps into four different aspects of sensation seeking behavior: thrill and adventure seeking, experience seeking, disinhibition, and boredom susceptibility.

Sense of helplessness Feeling that one's efforts and actions are not effective.

Sensitizers Individuals who are highly sensitive to everyday stress and anxieties.

Sensory anesthesia   Loss of sensory ability, such as blindness, deafness, or loss of feeling in a body part.

Serotonin   A neurotransmitter associated with depression when available in excess amounts in the body.

Sex role identity   The degree to which an individual identifies himself or herself as masculine or feminine.

Sex-typing   The process whereby the individual comes to acquire, to value, and to practice (perform) sex-typed behaviors.

Shadow aspect   According to Jung, the unconscious part of the psyche that must be absorbed into the personality to achieve full emotional growth.

Shaping   Technique for producing successively better approximations of behavior by reinforcing small variations in behavior in the desired direction and by reinforcing only increasingly close approximations to the desired behavior.

Shared environment (or family environment)   Individuals raised in the same family/household.

Sibling rivalry   Competition between the siblings of a family that, according to Adler, plays a major role in development.

Signature of personality   Stable patterns or profiles that show intraindividual behavior variation as it relates to specific types of situations.

Single-blind method   An experimental procedure in which participants do not know whether they are in experimental or control conditions.

Situational test   Procedure in which participants are observed performing a task within a lifelike situation; the Harvard personologists used stressful, lifelike tasks under extremely difficult situations to assess OSS candidates and used their performance to make clinical inferences about each person's underlying personality.

Situationism   Theory that dispositional consistency is a myth and individual behavior is entirely determined by situational attributes.

Sociability   The degree to which the person seeks interpersonal interaction.

Social cognitive level   A level of analysis in personality psychology that focuses on the social and cognitive meanings of events or situations for the individual, and the processes through which these meanings lead to social behavior.

Social intelligence   The competencies, skills, and knowledge needed to generate the person's cognitions and behavior patterns.

Sociobiology   An approach to explain social behavior in terms of evolutionary theories.

Source traits   In R. B. Cattell's theory, the traits that constitute personality structure and determine surface traits and behavior.

Stability   The durability of aspects of personality over time.

Stabilizing selection   Mechanism that weeds out characteristics at both extremities of a given dimension.

Stable intraindividual patterning   The idea that the temporal intraindividual variation in behavior is meaningful and stable.

State anxiety   A person's momentary or situational anxiety; it varies in intensity over time and across settings.

Statistical significance   A statistical computation that reflects how far a given association exceeds that which would be expected by chance.

Stimulus control   Behavior that is expressed stably but only under specific, predictable conditions.

Stimulus-outcome relations   Stable links between stimuli and other events that allow one to predict the outcomes from the stimuli.

Stimulus-response covariation   Stable link between stimulus and response that allow predictions of the response based on the stimulus.

Strange situation   An experimental study that puts a young child in an unfamiliar setting to assess individual differences in attachment relations.

Subjective continuity   Sense of basic oneness and durability in the self; feeling of consistency and identity.

Subjective values and goals   The particular outcomes and goals to which an individual assigns greatest import; a cognitive social person variable.

Sublimation   A process through which socially unacceptable impulses are expressed in socially acceptable ways.

Subliminally (subliminal)   Occurring outside of a person's consciousness or awareness.

Superego   In Freud's theory, the conscience, made up of the internalized values of the parents; strives for self-control and perfection; it is both unconscious and conscious.

Suppression   Occurs when one voluntarily and consciously withholds a response or turns attention away from something (see also *repression*).

Surface traits   R. B. Cattell's term for clusters of overt or manifest trait elements (responses) that seem to go together; the manifestations of source traits.

Symptom substitution   The controversial psychoanalytic belief that new symptoms will automatically replace problematic behaviors that are removed directly (e.g., by behavior therapy) unless their underlying unconscious emotional causes also have been removed.

Systematic desensitization   A behavior therapy procedure designed to reduce incapacitating anxiety; an incompatible response (usually relaxation) is paired with progressively more anxiety arousing situations until the individual is able to imagine or be in these situations without becoming anxious.

Systems theory   The analysis of units like the family as a system of relationships.

Systems therapy   A type of therapy that seeks to understand the individual within the context of the family or system.

Tachistoscope   A machine (used in studies of perceptual defense) that projects words onto a screen at different speeds.

Temperament traits   R. B. Cattell's term for traits that determine emotional reactivity.

Temperaments   Characteristic individual differences relevant to emotional expression, often visible early in life.

Temporal reliability   A high test-retest correlation when the same test is given on multiple occasions to the same group of people.

Tending (response to stress)   Responding to stressful situations with nurturance and care-taking behavior (rather than fight or flight reactions).

Test   A means of obtaining information about a person through standardized measures of behavior and personal qualities.

Thanatos   One of two aspects of the personality considered by early psychoanalysts to represent destruction and aggression; the darker side of human nature (see also *Eros*).

Thematic Apperception Test (TAT)   Projective test consisting of a set of ambiguous pictures about which the person being tested is asked to make up an interesting story.

Tokens   Items used to barter for rewards, used to assess the effectiveness of different rewards in operant conditioning.

Trait   A persistent (enduring) characteristic or dimension of individual differences; defined by Allport as a generalized "neuropsychic system," distinctive to each person, that serves to unify many different stimuli by leading the person to generate consistent responses to them.

Trait anxiety   A person's stable, characteristic overall level of anxiety.

Trait approach   An approach to personality that categorizes individuals in terms of traits.

Trait structure   Unique structure of a trait as it exists in a particular individual.

Trait theorists   Psychologists who study personality in terms of the different trait dimensions that characterize each individual.

Trait-dispositional level   Approach to personality that categorizes individuals in terms of traits or dispositions.

Transactional analysis   See *systems theory* and *family therapy*.

Transcendental Meditation (TM)   A form of deep meditation in which the meditator sits comfortably with eyes closed and repeats a special Sanskrit word called a "mantra."

Transference   In psychoanalysis, the patient's response to the therapist as though the therapist were a parent or some other important figure from childhood; considered and essential aspect of psychoanalytic therapy.

Transformation of motives   Defense mechanism in which basic impulses persist but the objects at which they are directed and the manner in which they are expressed are transformed.

Traumatic experiences   Experiences that abruptly and severely disrupt a person's life.

Triple typology   A classification system that brings together three categories in a meaningful way; in the case of personality, these three categories are types of people, types of behavior, and types of situations.

Twin method   Method of assessing genetic influence by comparing the degree of similarity on trait measures for genetically identical twins versus dizygotic twins.

Uncertainty orientation   Personality dimension that is defined at one end by individuals who are comfortable dealing with uncertainty and strive to resolve it, and at the other end by those who are uncomfortable with uncertainty and likely to avoid situations that increase their sense of uncertainty.

Unconditioned response (UCR)   The unlearned response one naturally makes to an unconditioned stimulus (e.g., withdrawing the hand from a hot object).

Unconditioned stimuli (UCS)   Stimuli to which one automatically, naturally responds without learning to do so (e.g., food, electric shock).

Unconscious   In psychoanalytic theory, the part of the personality of which the ego is unaware but that profoundly effects actions and behaviors.

Undercontrolling children   Children who do not adequately inhibit their behavioral impulses.

Unique traits   In Allport's theory, traits that only exist in one individual and cannot be found in another in exactly the same form.

Units of culture   Beliefs, values, goals, and ways of construing the world that are transmitted and shared within a particular community or culture.

Value   The subjective importance of an outcome or event for an individual.

Variable   An attribute, quality, or characteristic that may be given two or more values and measured or systematically varied.

Vicarious conditioning   Conditioning of a response to a stimulus through observation.

Visceral responses   Internal bodily responses to external events that occur without our willing them or even thinking about them (e.g., increased heart rate, changes in gland secretion).

Willpower The ability to voluntarily self-control for desired but difficult goals, for example, delay of gratification in anticipation of achieving a more far-reaching but distant goal.

Working self-concept The prominent concepts of the self that are foremost in an individual's thought and memory and that can be easily accessed.

Working through Process that occurs in psychoanalytic therapy when the patient, in the context of the transference relationship, re-examines his or her basic problems until their emotional roots are understood and learns to handle them more appropriately.

REFERENCES

Abelson, R. P. (1976). A script theory of understanding, attitude, and behavior. In J. Carroll & T. Payne (Eds.), *Cognition and social behavior*. Hillsdale, NJ: Erlbaum.

Abramson, L. Y., Seligman, M. E. P., & Teasdale, J. D. (1978). Learned helplessness in humans: Critique and reformation. *Journal of Abnormal Psychology, 87,* 49–74.

Affleck, G., Urrows, S., Tennen, H., & Higgins, P. (1992). Daily coping with pain from rheumatoid arthritis: Patterns and correlates. *Pain, 51,* 221–229.

Ainsworth, M. D. S. (1989). Attachments beyond infancy. *American Psychologist, 44,* 709–716.

Ainsworth, M. S., Blehar, M. C., Waters, E. C., & Wall, S. (1978). *Patterns of Attachment.* Hillsdale, NJ: Lawrence Erlbaum Associates.

Ajzen, I., & Fishbein, M. (1977). Attitude-behavior relations: A theoretical analysis and review of empirical research. *Psychological Bulletin, 84,* 888–918.

Aldwin, C. (1994). *Stress, coping, and development.* New York: Guilford Press.

Alexander, C. N., Robinson, P., Orme-Johnson, D. W., Schneider, R., & Walton, K. (1994). Effects of Transcendental Meditation compared to relaxation in promoting health and reducing mortality in the elderly. *Homeostasis, 35,* 4–5.

Alexander, M. J., & Higgins, E. T. (1993). Emotional trade-offs of becoming a parent: How social roles influence self-discrepancy effects. *Journal of Personality and Social Psychology, 65,* 1259–1269.

Alicke, M. D. (1985). Global self-evaluation as determined by the desirability and controllability of trait adjectives. *Journal of Personality and Social Psychology, 49,* 1621–1630.

Allen, E. K., Hart, B. M., Buell, J. S., Harris, F. R., & Wolf, M. M. (1964). Effects of social reinforcement on isolate behavior of a nursery school child. *Child Development, 35,* 511–518.

Alloy, L. B., & Ahrens, A. H. (1987). Depression and pessimism for the future: Biased use of statistically relevant information in predictions for self versus others. *Journal of Personality and Social Psychology, 52,* 366–378.

Allport, G. W. (1937). *Personality: A psychological interpretation.* New York: Holt, Rinehart and Winston.

Allport, G. W. (1940). Motivation in personality: Reply to Mr. Bertocci. *Psychological Review, 47,* 533–554.

Allport, G. W. (1955). *Becoming.* New Haven, CT: Yale University Press.

Allport, G. W. (1961). *Pattern and growth in personality.* New York: Holt, Rinehart and Winston.

American Psychological Association. (1966). *Standards for educational and psychological tests and manuals.* Washington, DC: APA.

Andersen, S. M., & Chen, S. (1998). Measuring transference in everyday social relations: Theory and evidence using an experimental social-cognitive paradigm. In H. Kurtzman (Ed.), *Cognition and Pychodynamics.* New York: Oxford University Press.

Andersen, S. M., & Chen, S. (2002). The relational self: An interpersonal social-cognitive theory. *Psychological Review, 109,* 619–645.

Andersen, S. M., Chen, S., & Miranda, R. (2002). Significant others and the self. *Self and Identity, 1,* 159–168.

Andersen, S. M., Reznik, I., & Chen, S. (1997). The self in relation to others: Cognitive and motivational underpinnings. In J. G. Snodgrass & R. L. Thompson (Eds.), *The self across psychology: Self-recognition, self-awareness, and the self-concept* (pp. 233–275). New York: New York Academy of Science.

Anderson, J. R. (1996). ACT: A simple theory of complex cognition. *American Psychologist, 51,* 355–365.

Anderson, J. R., & Bower, G. H. (1973). *Human associative memory.* New York: Wiley.

Angier, N. (1998). Separated by birth? *New York Times Book Review.* Feb. 8, p. 9.

Antonovsky, A. (1979). *Health, stress and coping.* San Francisco: Jossey-Bass.

Archer, J. (1990). The influence of testosterone on human aggression. *British Journal of Psychology, 82,* 1–28.

Archibald, H. C., & Tuddenham, R. D. (1965). Persistent stress reaction after combat. *Archives of General Psychiatry, 12,* 475–481.

Arend, R., Gove, F. L., & Sroufe, L. A. (1979). Continuity of individual adaptation from infancy to kindergarten: A predictive study of ego-resiliency and curiosity in preschoolers. *Child Development, 50,* 950–959.

Argyle, M., & Little, B. R. (1972). Do personality traits apply to social behavior? *Journal of Theory of Social Behavior (Great Britain), 2,* 1–35.

Arkowitz, H. (1989). From behavior change to insight. *Journal of Integrative and Eclectic Psychotherapy, 8,* 222–232.

Aronfreed, J. (1966). The internalization of social control through punishment: Experimental studies of the role of conditioning and the second signal system in the development of conscience. *Proceedings of the XVIIIth International Congress of Psychology*. Moscow, USSR, August, 35, 219–230.

Aronfreed, J. (1968). *Conduct and conscience: The socialization of internalized control over behavior*. New York: Academic Press.

Aronfreed, J. (1994). Moral development from the standpoint of a general psychological theory. In B. Puka (Ed.), *Defining perspectives in moral development. Moral development: A compendium* (Vol. 1, pp. 170–85). New York: Garland Publishing, Inc.

Aronson, E. (1972). *The social animal*. San Francisco: Freeman.

Aronson, E., & Mettee, D. (1968). Dishonest behavior as a function of differential levels of induced self-esteem. *Journal of Personality and Social Psychology, 9*, 121–127.

Ashmore, R. D. (1990). Sex, gender, and the individual. In L. A. Pervin (Ed.), *Handbook of personality: Theory and research* (pp. 486–526). New York: Guilford Press.

Aspinwall, L. G., & Staudinger, U. M. (2002). A psychology of human strengths: Fundamental questions and future directions for a positive psychology. Washington, DC: American Psychological Association.

Aspinwall, L. G., & Taylor, S. E. (1997) A stitch in time: Self-regulation and proactive coping. *Psychological Bulletin, 121*, 417–436.

Atkinson, J. W. (1964). *An introduction to motivation*. Princeton, NJ: Van Nostrand.

Atkinson, J. W. (Ed.). (1958). *Motives in fantasy, action and society*. Princeton, NJ: Van Nostrand.

Ayduk, O. N. (1999). Impact of self-control strategies on the link between rejection sensitivity and hostility: Risk negotiation through strategic control. *Dissertation Abstracts International, 60 (1-B)*, 0401.

Ayduk, O. N., Downey, G., Testa, S., Yen, Y., & Shoda, Y. (1998). Does rejection elicit hostility in rejection sensitive women? *Social Cognition, 17*, 245–271.

Ayduk, O., May, D., Downey, G., & Higgins, T. (2003). Tactical differences in coping with rejection sensitivity: The role of prevention pride. *Personality and Social Psychology Bulletin, 29*, 435–448.

Ayduk, O., Mendoza-Denton, R., Mischel, W., Downey, G., Peake, P. K., & Rodriguez, M. (2000). Regulating the interpersonal self: Strategic self-regulation for coping with rejection sensitivity. *Journal of Personality and Social Psychology, 79*, 776–792.

Ayduk, O., & Mischel, W. (2002). When smart people behave stupidly: Inconsistencies in social and emotional intelligence. In R. J. Sternberg (Ed.), *Why smart people can be so stupid* (pp. 86–105). New Haven, CT: Yale University Press.

Ayduk, O., Mischel, W., & Downey, G. (2002). Attentional mechanisms linking rejection to hostile reactivity: The role of "hot" vs. "cool" focus. *Psychological Science, 13*, 443–448.

Ayllon, T., & Azrin, N. H. (1965). The measurement and reinforcement of behavior of psychotics. *Journal of the Experimental Analysis of Behavior, 8*, 357–383.

Ayllon, T., & Haughton, E. (1964). Modification of symptomatic verbal behaviour of mental patients. *Behaviour Research and Therapy, 2*, 87–97.

Baker, T. B., Piper, M. E., McCarthy, D. E., Majeskie, M. R., & Fiore, M. C. (2003). Addiction motivation reformulated: An affective processing model of negative reinforcement. *Psychological Review*.

Ball, D., Hill, L., Freeman, B., Eley, T. C., Strelau, J., Riemann, R., Sinath, F. M., Angleitner, A., & Plomin, R. (1997). The serotonin transporter gene and peer-rated neuroticism. *NeuroReport, 8*, 1301–1304.

Bandura, A. (1965). Vicarious processes: A case of no-trial learning. In L. Berkowitz (Ed.), *Advances in experimental social psychology* (Vol. 2, pp. 1–55). New York: Academic Press.

Bandura, A. (1969). *Principles of behavior modification*. New York: Holt, Rinehart and Winston.

Bandura, A. (1973). *Aggression: A social learning analysis*. Englewood Cliffs, NJ: Prentice-Hall.

Bandura, A. (1977). *Social learning theory*. Englewood Cliffs, NJ: Prentice-Hall.

Bandura, A. (1978). Reflections on self-efficacy. In S. Rachman (Ed.), *Advances in behavior research and therapy* (Vol. 1). Elmsford, NJ: Pergamon.

Bandura, A. (1982). Self-efficacy mechanisms in human agency. *American Psychologist, 37*, 122–147.

Bandura, A. (1986). *Social foundations of thought and action: A social cognitive theory*. Englewood Cliffs, NJ: Prentice-Hall.

Bandura, A. (1989). Human agency in social cognitive theory. *American Psychologist, 44*, 1175–1184.

Bandura, A. (1997). *Self-efficacy: The exercise of control*. New York: Freeman.

Bandura, A. (2001). Social cognitive theory: An agentic perspective. *Annual Review, 52*, 1–26.

Bandura, A., & Adams, N. E. (1977). Analysis of self-efficacy theory of behavioral change. *Cognitive Therapy and Research, 1*, 287–310.

Bandura, A., Adams, N. E., & Beyer, J. (1977). Cognitive processes mediating behavioral change. *Journal of Personality and Social Psychology, 35*, 125–139.

Bandura, A., Blanchard, E. B., & Ritter, B. (1969). Relative efficacy of desensitization and modeling approaches for inducing behavioral, affective, and attitudinal changes. *Journal of Personality and Social Psychology, 13,* 173–199.

Bandura, A., & Kupers, C. J. (1964). Transmission of patterns of self-reinforcement through modeling. *Journal of Abnormal and Social Psychology, 69,* 1–9.

Bandura, A., & Mischel, W. (1965). Modification of self-imposed delay of reward through exposure to live and symbolic models. *Journal of Personality and Social Psychology, 2,* 698–705.

Bandura, A., Taylor, C. B., Ewart, C. K., Miller, N. M., & Debusk, R. F. (1985). Exercise testing to enhance wives' confidence in their husbands' cardiac capability soon after clinically uncomplicated acute myocardial infarction. *American Journal of Cardiology, 55,* 635–638.

Bandura, A., & Walters, R. H. (1959). *Adolescent aggression.* New York: Ronald Press.

Bargh, J. A. (1996). Automaticity in social psychology. In E. T. Higgins & A. W. Kruglanski (Eds.), *Social psychology: Handbook of basic principles* (pp. 169–183). New York: Guilford.

Bargh, J. A. (1997). The automaticity of everyday life. In R. S. Wyer Jr. (Ed.), *The automaticity of everyday life: Advances in social cognition* (Vol. 10, pp. 1–61). Mahwah, NJ: Erlbaum.

Bargh, J. A. (2001). Caution: Automatic social cognition may not be habit forming. *Polish Psychological Bulletin, 32,* 1–8.

Bargh, J. A., Chen, M., & Burrows, L. (1996). Automaticity of social behavior: Direct effects of trait construct and stereotype activation on action. *Journal of Personality and Social Psychology, 71,* 230–244.

Barlow, D. H. (1988). *Anxiety and its disorders: The nature and treatment of anxiety and panic.* New York: Guilford.

Barnouw. (1973). *Culture and personality.* Homewood, IL: Dorsey Press.

Baron, R. A., & Richardson, D. R. (1994). *Human aggression* (2nd ed.). New York: Plenum.

Bartholomew, K., & Horowitz, L. (1991). Attachment styles among young adults: A test of a four-category model. *Journal of Personality and Social Psychology, 61,* 226–244.

Bartussek, D., Diedrich, O., Naumann, E., & Collet, W. (1993). Introversion–extraversion and event-related potential (ERP): A test of J. A. Gray's theory. *Personality and Individual Differences, 14,* 565–574.

Bates, B., & Goodman, A. (1986). The effectiveness of encounter groups: Implications of research for counseling practice. *British Journal of Guidance and Counseling, 14,* 240–251.

Bateson, G. (1979). *Mind and nature: A necessary unit.* New York: Dutton.

Baumeister, R. F. (1996). Self-regulation and ego threat: Motivated cognition, self deception, and destructive goal setting. In P. M. Gollwitzer & J. A. Bargh (Eds.), *The psychology of action: Linking cognition and motivation to behavior* (pp. 27–47). New York: Guilford.

Baumeister, R. F. (1997). Identity, self-concept, and self-esteem: The self lost and found. In R. Hogan, J. Johnson, & S. Briggs (Eds.), *Handbook of Personality Psychology* (pp. 681–710). San Diego, CA: Academic Press.

Baumeister, R. F. (1998). The self. In D. T. Glibert, S. T. Fiske, & G. Lindzey (Eds.), *The handbook of social psychology* (Vol. 1, 4th ed., pp. 680–740). New York: Oxford University Press.

Baumeister, R. F. (2002). Ego depletion and self-control failure: An energy model of the self's executive function. *Self and Identity, 1,* 129–136.

Baumeister, R. F., & Cairns, K. H. (1992). Repression and self-presentation: When audiences interfere with self-deceptive strategies. *Journal of Personality and Social Psychology, 62,* 851–862.

Baumeister, R. F., & Heatherton, T. F. (1996). Self-regulation failure: An overview. *Psychological Inquiry, 7,* 1–15.

Baumeister, R. F., & Tice, D. M. (1996). Rethinking and reclaiming the interdisciplinary role of personality psychology: The science of human nature should be the center of the social sciences and humanities. *Journal of Research in Personality. 30,* 363–373.

Beck, A. T. (1976). *Cognitive therapy and the emotional disorders.* New York: International Universities Press.

Beck, A. T., Rush, A. J., Shaw, B. F., & Emery, G. (1979). *Cognitive therapy of depression.* New York: Guilford.

Bell, J. E. (1948). *Projective techniques.* New York: Longmans, Green.

Bell, R. Q., & Costello, N. (1964). Three tests for sex differences in tactile sensitivity in the newborn. *Biologia Neonatorum, 7,* 335–347.

Bell, R. Q., & Darling, J. (1965). The prone head reaction in the human newborn: Relationship with sex and tactile sensitivity. *Child Development, 36,* 943–949.

Bell, R. Q., Weller, G., & Waldrop, M. (1971). Newborn and preschooler: Organization of behavior and relations between periods. *Monographs of the Society for Research in Child Development, 36,* (Nos. 1 & 2).

Bellak, L., & Abrams, D. M. (1997). *The Thematic Apperception Test, the Children's Apperception Test, and the Senior Apperception Technique in clinical use (6th ed.).* Boston, MA: Allyn & Bacon.

Bem, D. J. (1972). Self-perception theory. In L. Berkowitz (Ed.), *Advances in experimental social psychology* (Vol. 6, pp. 1–62). New York: Academic Press.

Bem, D. J. (1983). Constructing a theory of the triple typology: Some (second) thoughts on nomothetic and idiographic approaches to personality. *Journal of Personality, 51,* 566–577.

Bem, D. J., & Allen, A. (1974). On predicting some of the people some of the time: The search for cross-situational consistencies in behavior. *Psychological Review, 81,* 506–520.

Bem, D. J., & Funder, D. C. (1978). Predicting more of the people more of the time: Assessing the personality of situations. *Psychological Review, 85,* 485–501.

Benedict, R. (1934). *Patterns of culture.* New York: Mentor.

Benjamin, J., Li, L., Patterson, C., Greenberg, B. D., Murphy, D. L., & Hamer, D. H. (1996). Population and familial association between the D4 dopamine receptor gene and measures of novelty seeking. *Nature Genetics, 12,* 81–84.

Benson, H. (1975). *The relaxation response.* New York: Morrow.

Berger, S. M. (1962). Conditioning through vicarious instigation. *Psychological Review, 69,* 450–466.

Berntzen, D. (1987). Effects of multiple cognitive coping strategies on laboratory pain. *Cognitive Therapy and Research, 11,* 613–623.

Betancourt, H., & Lopez, S. R. (1993). The study of culture, ethnicity, and race in American psychology. *American Psychologist, 48,* 629–637.

Bijou, S. W. (1965). Experimental studies of child behavior, normal and deviant. In L. Krasner & L. P. Ullmann (Eds.), *Research in behavior modification* (pp. 56–81). New York: Holt, Rinehart and Winston.

Birbaumer, N., & Ohman, A. (1993). *The structure of emotion: Psychophysiological, cognitive, and clinical aspects.* Seattle, WA: Hogrefe & Huber.

Block, J. (1961). *The Q-sort method in personality assessment and psychiatric research.* Springfield, IL: Charles C. Thomas.

Block, J. (1971). *Lives through time.* Berkeley, CA: Bancroft.

Block, J. (1977). Advancing the psychology of personality: Paradigmatic shift or improving the quality of research. In D. Magnusson & N. S. Endler (Eds.), *Personality at the crossroads: Current issues in interactional psychology.* Hillsdale, NJ: Erlbaum.

Block, J. (1995). A contrarian view of the five-factor approach to personality description. *Psychological Bulletin, 117,* 187–215.

Block, J. (2001). Millennial contrarianism: The five-factor approach to personality description 5 years later. *Journal of Research in Personality, 35,* 98–107.

Block, J., & Block, J. H. (1980). The role of ego-control and ego resiliency in the organization of behavior. In W. A. Collins (Ed.), *The Minnesota symposium on child psychology* (Vol. 13). Hillsdale, NJ: Erlbaum.

Block, J., Weiss, D. S., & Thorne, A. (1979). How relevant is a semantic similarity interpretation of personality ratings? *Journal of Personality and Social Psychology, 37,* 1055–1074.

Block, J. H., & Martin, B. (1955). Predicting the behavior of children under frustration. *Journal of Abnormal and Social Psychology, 51,* 281–285.

Blum, G. S. (1953). *Psychoanalytic theories of personality.* New York: McGraw-Hill.

Blum, G. S. (1955). Perceptual defense revisited. *Journal of Abnormal and Social Psychology, 51,* 24–29.

Bolger, N., & Schilling, E. A. (1991). Personality and the problems of everyday life: The role of neuroticism in exposure and reactivity to daily stressors. *Journal of Personality, 59,* 355–386.

Bolger, N., & Zuckerman, A. (1995). A framework for studying personality in the stress process. *Journal of Personality and Social Psychology, 69,* 890–902.

Bonanno, G. A. (2001). Grief and emotion: A social–functional perspective. In M. S. Stroebe & R. O. Hansson (Eds.), *Handbook of bereavement research: Consequences, coping, and care* (pp. 493–515). Washington, DC: American Psychological Association.

Bonarius, J. C. J. (1965). Research in the personal construct theory of George A. Kelly: Role construct repertory test and basic theory. In B. A. Maher (Ed.), *Progress in experimental personality research* (pp. 1–46). New York: Academic Press.

Bornstein, R. F., Riggs, J. M., Hill, E. L., & Calabrese, C. (1996). Activity, passivity, self-denigration, and self-promotion: Toward an interactionist model of interpersonal dependency. *Journal of Personality, 64,* 637–673.

Bouchard, T. J., Lykken, D. T., McGue, M., Segal, N. L., & Tellegen, A. (1990). Sources of human psychological differences: The Minnesota study of twins reared apart. *Science, 250,* 223–228.

Boudin, H. M. (1972). Contingency contracting as a therapeutic tool in the deceleration of amphetamine use. *Behavior Therapy, 3,* 604–608.

Bower, G. H. (1981). Mood and memory. *American Psychologist, 36,* 129–148.

Bowers, K. (1973). Situationism in psychology: An analysis and a critique. *Psychological Review, 80,* 307–336.

Bowlby, J. (1982) Attachment and loss: Retrospect and prospect. *American Journal of Orthopsychiatry, 52,* 664–678.

Braungart, J. M., Fulker, D. W., & Plomin, R. (1992). Genetic influence of the home environment during infancy: A sibling adoption study of the HOME. *Developmental Psychology, 28,* 1048–1055.

Brazier, D. (1993). The necessary condition is love: Going beyond self in the person-centered approach. In D. Brazier (Ed.), *Beyond Carl Rogers* (pp. 72–91). London: Constable.

Brockner, J. (1983). The roles of self-esteem and self-consciousness in the Wortman-Brehm model of reactance and learned helplessness. *Journal of Personality & Social Psychology, 45,* 199–209.

Brown, J. D. (1986). Evaluations of self and others: Self-enhancement biases in social judgment. *Social Cognition, 4,* 353–376.

Brown, J. D., & Dutton, K. A. (1995). The thrill of victory, the complexity of defeat: Self-esteem and people's emotional reactions to success and failure. *Journal of Personality and Social Psychology, 68,* 712–722.

Brown, J. S. (1942). The generalization of approach responses as a function of stimulus intensity and strength of motivation. *Journal of Comparative Psychology, 33,* 209–226.

Brown, J. S. (1948). Gradients of approach and avoidance responses and their relation to level of motivation. *Journal of Comparative and Physiological Psychology, 41,* 450–465.

Brown, K. W., & Moskowitz, D. S. (1998). Dynamic stability of behavior: The rhythms of our interpersonal lives. *Journal of Personality, 66,* 105–134.

Bruner, J. (1992). Another look at New Look 1. *American Psychologist, 47,* 780–783.

Bruner, J. S. (1957). Going beyond the information given. In H. Gruber et al. (Eds.), *Contemporary approaches to cognition.* Cambridge, MA: Harvard University Press.

Bruner, J. S., & Postman, L. (1947). Emotional selectivity in perception and reaction. *Journal of Personality, 16,* 69–77.

Brunstein, J. (1993). Personal goals and subjective well-being: A longitudinal study. *Journal of Personality and Social Psychology, 65,* 1061–1070.

Buss, A. H. (1961). *The psychology of aggression.* New York: Wiley.

Buss, A. H. (1989). Personality as traits. *American Psychologist, 44,* 1378–1388.

Buss, A. H., & Plomin, R. (1984). *Temperament: Early developing personality traits.* Hillsdale, NJ: Erlbaum.

Buss, A. H., Plomin, R., & Willerman, L. (1973). The inheritance of temperaments. *Journal of Personality, 41,* 513–524.

Buss, D. M. (1987). Selection, evocation, and manipulation. *Journal of Personality and Social Psychology, 53,* 1214–1221.

Buss, D. M. (1991). Evolutionary personality psychology. *Annual Review of Psychology, 42,* 459–491.

Buss, D. M. (1994). Personality evoked: The evolutionary psychology of stability and change. In T. F. Heatherton & J. L. Weinberger (Eds.), *Can personality change?* (pp. 41–57). Washington, DC: American Psychological Association.

Buss, D. M. (1996). The evolutionary psychology of human social strategies. In E. T. Higgins & A. W. Kruglanski (Eds.), *Social psychology: Handbook of basic principles* (pp. 3–38). New York: Guilford.

Buss, D. M. (1997) Evolutionary foundations of personality. In R. Hogan, J. A. Johnson, & S. R. Briggs (Eds.), *Handbook of personality psychology* (pp. 317–344). San Diego: Academic Press.

Buss, D. M. (1999). Human nature and individual differences: The evolution of human personality. In L. A. Pervin & O. P. John (Eds.), *Handbook of personality: Theory and research* (2nd ed., pp. 31–56). New York: Guilford.

Buss, D. M. (2001). Human nature and culture: An evolutionary psychological perspective. *Journal of Personality, 69,* 955–978.

Buss, D. M., & Craik, K. H. (1983). The act frequency approach to personality. *Psychological Review, 90,* 105–126.

Butler, L. D., & Nolen-Hoeksema, S. (1994). Gender differences in response to depressed mood in college sample. *Sex Roles, 30,* 330–346.

Byrne, D. (1964). Repression–sensitization as a dimension of personality. In B. A. Maher (Ed.), *Progress in experimental personality research* (Vol. 1). New York: Academic Press.

Byrne, D. (1966). *An introduction to personality.* Englewood Cliffs, NJ: Prentice-Hall.

Byrne, D. (1969). Attitudes and attraction. In L. Berkowitz (Ed.), *Advances in experimental social psychology* (Vol. 1). New York: Academic Press.

Cacioppo, J. T., & Gardner, W. L. (1999). Emotion. *Annual Review of Psychology, 50,* 191–214.

Cacioppo, J. T., & Petty, R. E. (1982). The need for cognition. *Journal of Personality and Social Psychology, 42,* 116–131.

Cacioppo, J. T., Berntson, G. G., & Crites Jr., S. L. (1996). Social neuroscience: Principles of psychophysiological arousal and response. In E. T. Higgins & A. W. Kruglanski (Eds.), *Social Psychology: Handbook of Basic Principles* (pp. 72–101). New York: Guilford.

Caldwell, B. M., & Bradley, R. H. (1978). *Home Observation for Measurement of the Environment.* Little Rock: University of Arkansas Press.

Campbell, D. T. (1960). Recommendations for APA Test Standards regarding construct, trait, or discriminant validity. *American Psychologist, 15,* 546–553.

Campbell, J., & Dunnette, M. (1968). Effectiveness of T-group experiences in managerial training and development. *Psychological Bulletin, 70,* 73–104.

Cannon, W. B. (1932). *The wisdom of the body*. New York: Norton.

Cantor, N. (1990). From thought to behavior: "Having" and "doing" in the study of personality and cognition. *American Psychologist, 45,* 735–750.

Cantor, N. (1994). Life task problem-solving: Situational affordances and personal needs. Presidential address of the society for personality and social psychology (Division 8 of the American Psychological Association, 1993, Toronto, Canada), *Personality and Social Psychology Bulletin, 20,* 235–243.

Cantor, N., Kemmelmeier, M., Basten, J., & Prentice, D. A. (2002). Life task pursuit in social groups: Balancing self-exploration and social integration. *Self and Identity, 1,* 177–184.

Cantor, N., & Kihlstrom, J. F. (1987). *Personality and social intelligence*. Englewood Cliffs, NJ: Erlbaum.

Cantor, N., & Mischel, W. (1977). Traits as prototypes: Effects on recognition memory. *Journal of Personality and Social Psychology, 35,* 38–48.

Cantor, N., & Mischel, W. (1979). Prototypes in person perception. In L. Berkowitz (Ed.), *Advances in experimental social psychology* (Vol. 12). New York: Academic Press.

Cantor, N., Mischel, W., & Schwartz, J. (1982). A prototype analysis of psychological situations. *Cognitive Psychology, 14,* 45–77.

Cantor, N., Norem, J., Langston, C., Zirkel, S., Fleeson W., & Cook-Flannagan, C. (1991). Life tasks and daily life experience. *Journal of Personality, 59,* 425–451.

Capecchi, M. R. (1994). Targeted gene replacement. *Scientific American, 270,* 52–59.

Carlson, R. (1971). Where is the personality research? *Psychological Bulletin, 75,* 203–219.

Carter, C. S. (1998). Neuroendocrine perspectives on social attachment and love. *Psychoneuroendocrinology, 23,* 779–818.

Carter, C. S., Lederhendler, I. I., & Kirkpatrick, B. (1997). *The integrative neurobiology of affiliation*. New York: New York Academy of Sciences.

Cartwright, D. S. (1978). *Introduction to personality*. Chicago: Rand McNally.

Carver, C. S. (1996). Emergent integration in contemporary personality psychology. *Journal of Research in Personality, 30,* 319–334.

Carver, C. S., Coleman, A. E., & Glass, D. C. (1976). The coronary-prone behavior pattern and the suppression of fatigue on a treadmill test. *Journal of Personality and Social Psychology, 33,* 460–466.

Carver, C. S., Pozo, C., Harris, S. D., Noriega, V., Scheier, M. F., Robinson, D. S., Ketchem, A. S., Moffat, F. L., Jr., & Clark, K. C. (1993). How coping mediates the effects of optimism on stress: A study of women with early stage breast cancer. *Journal of Personality and Social Psychology, 65,* 375–391.

Carver, C. S., & Scheier, M. F. (1978). Self-focusing effects of dispositional self-consciousness, mirror presence, and audience presence. *Journal of Personality and Social Psychology, 36,* 322–324.

Carver, C. S., & Scheier, M. F. (1981). *Attention and self-regulation: A control theory approach to human behavior*. New York: Springer-Verlag.

Carver, C. S., & Scheier, M. F. (1982). Control theory: A useful conceptual framework for personality-social, clinical, and health psychology. *Psychological Bulletin, 92,* 111–135.

Carver, C. S., & Scheier, M. F. (1990). Principles of self-regulation: Action and emotion. In E. T. Higgins & R. M. Sorrentino (Eds.), *Handbook of motivation and cognition* (Vol. 2, pp. 3–52). New York: Guilford.

Carver, C. S., & Scheier, M. F. (1992). Confidence, doubt, and coping with anxiety. In D. G. Forgays, T. Sosnowski, & K. Wrzesniewski (Eds.), *Anxiety: Recent developments in cognitive, psychophysiological, and health research* (pp. 13–22). Washington, DC: Hemisphere Publishing Corp.

Carver, C. S., & Scheier, M. F. (1998). *On the self-regulation of behavior*. New York: Cambridge University Press.

Carver, C. S., & White, T. L. (1994). Behavioral inhibition, behavioral activation, and affective responses to impending reward and punishment: The BIS/BAS scales. *Journal of Personality and Social Psychology, 67,* 319–333.

Cashdan, S. (1988). *Object relations theory: Using the relationship*. New York: Norton and Co.

Caspi, A. (1987). Personality in the life course. *Journal of Personality and Social Psychology, 53,* 1203–1213.

Caspi, A. (1998). Personality development across the life course. In W. Damon (Ser. Ed.) and N. Eisenberg (Vol. Ed.), *Handbook of Child Psychology, Vol. 3: Social, emotional, and personality development* (pp. 311–388). New York: Wiley.

Caspi, A., & Bem, D. J. (1990). Personality continuity and change across the life course. In L. A. Pervin (Ed.), *Handbook of personality: Theory and research* (pp. 549–575). New York: Guilford.

Cattell, R. B. (1950). *A systematic theoretical and factual study*. New York: McGraw-Hill.

Cattell, R. B. (1965). *The scientific analysis of personality*. Baltimore: Penguin Books.

Cattell, R. B. (1982). *The inheritance of personality and ability*. New York: Academic Press.

Cervone, D. (1991). The two disciplines of personality psychology [Review of *Handbook of personality: Theory and research*]. *Psychological Science, 2,* 371–376.

Cervone, D., & Mischel, W. (2002). Personality Science. In D. Cervone & W. Mischel (Eds.) *Advances in Personality Science* (pp. 1–26). New York: Guilford.

Cervone, D., & Mischel, W. (Eds.) (2002). *Advances in Personality Science.* New York: Guilford.

Cervone, D., Shadel, W. G., & Jencius, S. (2001). Social-cognitive theory of personality assessment. *Personality and Social Psychology Review, 5,* 33–51.

Cervone, D., & Shoda, Y. (1999). Social cognitive theories and the coherence of personality. In D. Cervone and Y. Shoda (eds.), *The coherence of personality: Social-cognitive bases of consistency, variability, and organization.* New York: Guilford.

Cervone, D., & Williams, S. L. (1982). Social cognitive theory and personality. In G. V. Caprara & G. L. Van Heck (Eds.), *Modern personality psychology: Critical reviews and new directions* (pp. 200–252). New York: Harvester Wheatsheaf/Simon & Schuster.

Chaiken, S., & Bargh, J. A. (1993). Occurrence versus moderation of the automatic attitude activation effect: Reply to Fazio. *Journal of Personality and Social Psychology, 64,* 759–765.

Chan, D. W. (1994). The Chinese Ways of Coping Questionnaire: Assessing coping in secondary school teachers and students in Hong Kong. *Psychological Assessment, 6,* 108–116.

Chaplin, W. F., John, O. P., & Goldberg, L. R. (1988). Conceptions of states and traits: Dimensional attributes with ideals as prototypes. *Journal of Personality and Social Psychology, 54,* 541–557.

Chaves, J. F., & Barber, T. X. (1974). Acupuncture analgesia: A six-factor theory. *Psychoenergetic Systems, 1,* 11–20.

Chen-Idson, L., & Mischel, W. (2001). The personality of familiar and significant people: The lay perceiver as a social cognitive theorist. *Journal of Personality and Social Psychology, 80,* 585–596.

Cheng, C. (2001). Assessing coping flexibility in real-life and laboratory settings: A multimethod approach. *Journal of Personality and Social Psychology, 80,* 814–833.

Cheng, C. (2003). Cognitive and motivational processes underlying coping flexibility: A dual-process model. *Journal of Personality and Social Psychology, 84,* 425–438.

Cheng, C., Chiu, C., Hong, Y., & Cheung, J. S. (2001). Discriminative facility and its role in the quality of interactional experiences. *Journal of Personality, 69,* 765–786.

Cheng, C., Hui, W., & Lam, S. (2000). Perceptual style and behavioral pattern of individuals with functional gastrointestinal disorders. *Health Psychology, 19,* 146–154.

Cherney, S. S., Fulker, D. W., Emde, R. N., Robinson, J., Corley, R. P., Reznick, J. S., Plomin, R., & Defries, J. C. (1994). Continuity and change in infant shyness from 14 to 20 months. *Behavior Genetics, 24,* 365–379.

Chiu, C., Dweck, C. S., Tong, J. Y., & Fu, J. H. (1997). Implicit theories and conceptions of morality. *Journal of Personality and Social Psychology, 73,* 923–940.

Chiu, C., Hong, Y., & Dweck, C. S. (1997). Lay dispositionism and implicit theories of personality. *Journal of Personality and Social Psychology, 73,* 19–30.

Chiu, C., Hong, Y., Mischel, W., & Shoda, Y. (1995). Discriminative facility in social competence: Conditional versus dispositional encoding and monitoring-blunting of information. *Social Cognition, 13,* 49–70.

Chodoff, P. (1963). Late effects of concentration camp syndrome. *Archives of General Psychiatry, 8,* 323–333.

Churchland, P. S., & Sejnowski, T. J. (1992). *The computational brain.* Cambridge, MA: MIT Press.

Clark, L. A., & Watson, D. (1999). Temperament: A new paradigm for trait psychology. In L. A. Pervin & O. P. John (Eds.), *Handbook of personality: Theory and research* (2nd ed., pp. 399–423). New York: Guilford.

Cloninger, C. R. (1988). A unified biosocial theory of personality and its role in the development of anxiety states: A reply to commentaries. *Psychiatric Developments, 2,* 83–120.

Cobb, S. (1976). Social support as moderator of life stress. *Psychosomatic Medicine, 38,* 300–314.

Cohen, S., & McKay, G. (1984). Social support, stress, and the buffering hypothesis: A theoretical analysis. In A. Baum, J. E. Singer, & S. E. Taylor (Eds.), *Handbook of psychology and health, Vol. 4. Social psychological aspects of health.* Hillsdale, NJ: Erlbaum.

Colby, K. M. (1951). *A primer for psychotherapists.* New York: Ronald.

Coles, R. (1970). *Uprooted children.* New York: Harper & Row.

Collaer, M. L., & Hines, M. (1995). Human behavioral sex differences: A role for gonadal hormones during early development? *Psychological Bulletin, 118,* 55–107.

Contrada, R. J., Cather, C., & O'Leary, A. (1999). Personality and health: Dispositions and processes in disease susceptability and adaptation to illness. In L. A. Pervin & O. P. John (Eds.), *Handbook of personality: Theory and research* (2nd ed., pp. 576–604). New York: Guilford.

Contrada, R. J., Leventhal, H., & O'Leary, A. (1990). Personality and health. In L. A. Pervin (Ed.), *Handbook of personality: Theory and research* (pp. 638–669). New York: Guilford.

Cooper, J., & Fazio, R. H. (1984). A new look at dissonance theory. In L. Berkowitz (Ed.), *Advances in experimental social psychology, Vol. 17. Theorizing in social psychology: Special topics* (pp. 229–262). New York: Academic Press.

Cooper, J. R., Bloom, F. E., & Roth, R. H. (1996). *The biochemical basis of neuropharmacology (7th ed.).* New York: Oxford University Press.

Coopersmith, S. (1967). *The antecedents of self-esteem*. San Francisco: Freeman.

Cosmides, L. (1989). The logic of social exchange: Has natural selection shaped how humans reason? Studies with the Wason selection task. *Cognition, 31,* 187–276.

Cosmides, L., & Tooby, J. (1989). Evolutionary psychology and the generation of culture: II. Case study: A computational theory of social exchange. *Ethology and Sociobiology, 10,* 51–97.

Costa, P. T., Jr., & McCrae, R. R. (1988). Personality in adulthood: A six-year longitudinal study of self-reports and spouse ratings on the NEO personality inventory. *Journal of Personality and Social Psychology, 54,* 853–863.

Costa, P. T., Jr., & McCrae, R. R. (1992a). Normal personality assessment in clinical practice: The NEO personality inventory. *Psychological Assessment, 4,* 5–13.

Costa, P. T., Jr., & McCrae, R. R. (1992b). *Revised NEO Personality Inventory (NEO-PI-R) and NEO Five Factor (NEO-FFI): Professional manual*. Odessa, FL: Psychological Assessment Resources.

Costa, P. T., Jr., & McCrae, R. R. (1997). Longitudinal stability of adult personality. In R. Hogan, J. Johnson, & S. Briggs (Eds.), *Handbook of personality psychology* (pp. 269–291). San Diego, CA: Academic Press.

Costa, P. T., Jr., McCrae, R. R., & Dye, D. A. (1991). Facet scales for agreeableness and conscientiousness: A revision of the neo-personality inventory. *Personality and Individual Differences, 12,* 887–898.

Costa, P. T., Jr., McCrae, R. R., & Siegler, I. C. (1999). Continuity and change over the adult life cycle: Personality and personality disorders. In C.R. Cloninger (Ed.), *Personality and psychopathology* (pp. 129–153). Washington, DC: American Psychiatric Press.

Cote, S., & Moskowitz, D. S. (1998). On the dynamic covariation between interpersonal behavior and affect: Prediction from neuroticism, extraversion, and agreeableness. *Journal of Personality and Social Psychology, 75,* 1032–1046.

Coyne, J. C., & Downey, G. (1991). Social factors and psychopathology: Stress, social support, and coping process. *Annual Review of Psychology, 42,* 401–425.

Crocker, J. (2002). Contingencies of self-worth: Implications for self-regulation and psychological vulnerability. *Self and Identity, 1,* 143–149.

Csikszentmihalyi, M. (1990). The domain of creativity. In M. Runco & R. S. Albert (Eds.) *Theories of creativity. Sage focus editions* (Vol. 115, pp. 190–212). Thousand Oaks, CA: Sage Publications.

Dabbs, J. M., Jr., & Morris, R. (1990). Testosterone, social class, and antisocial behavior in a sample of 4,462 men. *Psychological Science, 1,* 209–211.

Dabbs, J. M., Jr. (1992). Testosterone and occupational achievement. *Social Forces, 70,* 813–824.

D'Andrade, R. G. (1966). Sex differences and cultural institutions. In E. E. Maccoby (Ed.), *The development of sex differences* (pp. 174–204). Stanford: Stanford University Press.

Daniels, A. C. (1994). *Bringing out the best in people: How to apply the astonishing power of reinforcement*. New York: McGraw-Hill.

Daniels, D., Dunn, J. F., Furstenberg, F. F., Jr., & Plomin, R. (1985). Environmental differences within the family and adjustment differences within pairs of adolescent siblings. *Child Development, 56,* 764–774.

David, J. P., Green, P. J., Martin, R., & Suls, J. (1997). Differential roles of neuroticism, extraversion, and event desirability for mood in daily life: An integrative model of top-down and bottom-up influences. *Journal of Personality and Social Psychology, 73,* 149–159.

Davidson, R. J. (1993). The neuropsychology of emotion and affective style. In M. Lewis & J. M. Haviland (Eds.), *Handbook of emotions* (pp. 143–154). New York: Guildford.

Davidson, R. J. (1995). Cerebral asymmetry, emotion, and affective style. In R. J. Davidson & K. Hugdahl (Eds.), *Brain asymmetry* (pp. 361–387). Cambridge, MA: MIT Press.

Davidson, R. J., & Fox, N. A. (1989). The relation between tonic EEG asymmetry and ten-month-olds' emotional response to separation. *Journal of Abnormal Psychology, 98,* 127–131.

Davidson, R. J., & Sutton, S. K. (1995). Affective neuroscience: The emergence of a discipline. *Current Opinion in Neurobiology, 5,* 217–224.

Davidson, R. J., Ekman, P., Saron, C. D., Senulis, J. A., & Friesen, W. V. (1990). Approach-withdrawal and cerebral asymmetry: Emotional expression and brain physiology I. *Journal of Personality and Social Psychology, 58,* 330–341.

Davidson, R. J., Putnam, K. M., & Larson, C. L. (2000). Dysfunction in the neural circuitry of emotion regulation—A possible prelude to violence. *Science, 289,* 591–594.

Davis, J. M., Klerman, G., & Schildkraut, J. (1967). Drugs used in the treatment of depression. In L. Efron, J. O. Cole, D. Levine, & J. R. Wittenborn (Eds.), *Psychopharmacology: A review of progress*. Washington, DC: U.S. Clearing-House of Mental Health Information.

Davison, G. C., & Neale, J. M. (1990). *Abnormal psychology: An experimental clinical approach* (5th ed.). New York: Wiley.

Davison, G. C., Neale, J. M., & Kring, A. M. (2004). *Abnormal psychology* (9th ed.). Hoboken, NJ: Wiley.

Deaux, K., & Major, B. (1987). Putting gender into context: An interactive model of gender-related behavior. *Psychological Review, 94,* 369–389.

Deci, E. L., & Ryan, R. M. (1987). The support of autonomy and the control of behavior. *Journal of Personality and Social Psychology, 53,* 1024–1037.

Deci, E. L., & Ryan, R. M. (1995). Human autonomy: The basis for true self-esteem. In M. Kernis (Ed.), *Efficacy, agency, and self-esteem* (pp. 31–49). New York: Plenum.

Deci, E. L., & Ryan, R. M. (2000). The support of autonomy and the control of behavior. In E. T. Higgins & A. W. Kruglanski (Eds.), *Motivational science: Social and personality perspectives* (pp. 128–145). New York: Psychology Press.

Depue, R. A., & Collins, P. F. (1999). Neurobiology of the structure of personality: Dopamine, facilitation, of incentive motivation, and extraversion. *Behavioral and Brain Sciences, 22,* 491–569.

DeRaad, B., Perugini, M., Hrebickova, M., & Szarota, P. (1998). Lingua franca of personality: Taxonomies and structures based on the psycholexical approach. *Journal of Cross Cultural Psychology, 29,* 212–232.

Derryberry, D. (2002). Attention and voluntary self-control. *Self and Identity, 1,* 105–111.

Derryberry, D., & Reed, M. A. (2002). Anxiety-related attentional biases and their regulation by attentional control. *Journal of Abnormal Psychology, 111,* 225–236.

Derryberry, D., & Rothbart, M. K. (1997). Reactive and effortful processes in the organization of temperament. *Development and Psychopathology, 9,* 633–652.

Diamond, M. J., & Shapiro, J. L. (1973). Changes in locus of control as a function of encounter group experiences: A study and replication. *Journal of Abnormal Psychology, 82,* 514–518.

Dickens, W. T., & Flynn, J. R. (2001). Heritability estimates versus large environmental effects: The IQ paradox resolved. *Psychological Review, 108,* 346–369.

Diener, C. I., & Dweck, C. S. (1978). An analysis of learned helplessness: Continuous changes in performance, strategy, and achievement cognitions following failure. *Journal of Personality and Social Psychology, 36,* 451–462.

Dobson, K. S., & Craig, K. D. (1996). *Advances in cognitive-behavioral therapy.* Thousand Oaks, CA: Sage Publications.

Dodge, K. A. (1986). A social information processing model of social competence in children. *Cognitive perspectives on children's social behavioral development. The Minnesota symposium on child psychology, 18,* 77–125.

Dodge, K. A. (1993). New wrinkles in the person-versus-situation debate. *Psychological Inquiry, 4,* 284–286.

Dodge, K. A. (1997a, April). *Testing developmental theory through prevention trials.* Presented at Biennial Meeting for the Society for Research in Child Development, Washington, DC.

Dodge, K. A. (1997b, April). *Early peer social rejection and acquired autonomic sensitivity to peer conflicts: Conduct problems in adolescence.* Presented at Biennial Meeting for the Society for Research in Child Development, Washington, DC.

Dodgson, P., & Wood, J. V. (1998). Self-esteem and the cognitive accessibility of strengths and weaknesses after failure. *Journal of Personality and Social Psychology, 75,* 178–197.

Dollard, J., & Miller, N. E. (1950). *Personality and psychotherapy: An analysis in terms of learning, thinking, and culture.* New York: McGraw-Hill.

Downey, G., & Feldman, S. I. (1996). Implications of rejection sensitivity for intimate relationships. *Journal of Personality and Social Psychology, 70,* 1327–1343.

Downey, G., Freitas, A., Michaelis, B., & Khouri, H. (1998). The self-fullfilling prophecy in close relationships. Do rejection sensitive women get rejected by their partners? *Journal of Personality and Social Psychology, 75,* 545–560.

Duke, M. P. (1986). Personality science: A proposal. *Journal of Personality and Social Psychology, 50,* 382–385.

Dunford, F. W. (2000). The San Diego Navy experiment: An assessment of interventions for men who assault their wives. *Journal of Consulting and Clinical Psychology, 68,* 468–476.

Dunn, J., & Plomin, R. (1990). *Separate lives: Why siblings are so different.* New York: Basic Books.

Dunnette, M. D. (1969). People feeling: Joy, more joy, and the "slough of despond." *Journal of Applied Behavioral Science, 5,* 25–44.

Durham, W. H. (1991). *Coevolution.* Stanford, CA: Stanford University Press.

Duval, T. S., Duval, V. H., & Mulilis, J. (1992). Effects of self-focus, discrepancy between self and standard, and outcome expectancy favorability on the tendency to match self to standard or to withdraw. *Journal of Personality and Social Psychology, 62,* 340–348.

Dweck, C. S. (1975). The role of expectations and attributions in the alleviation of learned helplessness. *Journal of Personality and Social Psychology, 31,* 674–685.

Dweck, C. S. (1990). Self-theories and goals: Their role in motivation, personality, and development. In R. A. Dienstbier (Ed.), *Nebraska symposium on motivation* (Vol. 38, pp. 199–235). Lincoln, NE: University of Nebraska Press.

Dweck, C. S., & Leggett, E. L. (1988). A social-cognitive approach to personality and motivation. *Psychological Review, 95,* 256–273.

Dworkin, R. H. (1979). Genetic and environmental influences on person-situation interactions. *Journal of Research in Personality, 13,* 279–293.

D'Zurilla, T. (1965). Recall efficiency and mediating cognitive events in "experimental repression." *Journal of Personality and Social Psychology, 1,* 253–257.

Eagly, A. H. (1987). *Sex differences in social behavior: A social-role interpretation.* Hillsdale, NJ: Erlbaum.

Eagly, A. H. (1995). The science and politics of comparing women and men. *American Psychologist, 50,* 145–158.

Eagly, A. H., & Crowley, M. (1986). Gender and helping behavior: A meta-analytic review of the social psychological literature. *Psychological Bulletin, 100,* 283–308.

Eagly, A. H., & Steffen, V. J. (1986). Gender and aggressive behavior: A meta-analytic review of the social psychological literature. *Psychological Bulletin, 100,* 309–330.

Eaves, L. J., Eysenck, H. J., & Martin, N. G. (1989). *Genes, culture, and personality: An empirical approach.* London: Academic Press.

Ebstein, R. P., Gritsenko, I., Nemanov, L., Frisch, A., Osher, Y., & Belmaker, R. H. (1997). No association between the serotonin transporter gene regulatory region polymorphism and the Tridimensional Personality Questionnaire (TPQ) temperament of harm avoidance. *Molecular Psychiatry, 2,* 224–226.

Ebstein, R. P., Novick, O., Umansky, R., Priel, B., Osher, Y., Blaine, D., Bennett, E. R., Nemanov, L., Katz, M., & Belmaker, R. H. (1996). Dopamine D4 receptor (D4DR) exon III polymorphism associated with human personality trait of novelty seeking. *Nature Genetics, 12,* 78–80.

Edelman, G. M. (1992). *Bright air, brilliant fire: On the matter of the mind.* New York: Basic Books.

Ehrlich, P. R. (2000). *Human natures: Genes, cultures, and the human prospect.* Washington, DC: Island Press.

Eisenberg, N., Spinrad, T. L., & Morris, A. S. (2002). Regulation, resiliency, and quality of social functioning. *Self and Identity, 1,* 121–128.

Eizenman, D. R., Nesselroade, J. R., Featherman, D. L., & Rowe, J. W. (1997). Intraindividual variability in perceived control in an older sample: The MacArthur successful aging studies. *Psychology and Aging, 12,* 489–502.

Ekman, P. (Ed.). (1982). *Emotion in the human face* (2nd ed.). New York: Cambridge University Press.

Ekman, P., Friesen, W. V., & Ellsworth, P. (1972). *Emotion in the human face.* Elmsford, NY: Pergamon Press.

Ellsworth, P. C., & Carlsmith, J. M. (1968). Effects of eye contact and verbal content on affective response to a dyadic interaction. *Journal of Personality and Social Psychology, 10,* 15–20.

Emmons, R. A. (1991). Personal strivings, daily life events, and psychological and physical well-being. *Journal of Personality, 59,* 453–472.

Emmons, R. A. (1997). Motives and goals. In R. Hogan, J. A. Johnson, & S. R. Briggs (Eds.), *Handbook of personality psychology* (pp. 485–512). San Diego, CA: Academic Press.

Endler, N. S., & Hunt, J. M. (1968). S-R inventories of hostility and comparisons of the proportions of variance from persons, responses, and situations for hostility and anxiousness. *Journal of Personality and Social Psychology, 9,* 309–315.

Epstein, S. (1973). The self-concept revisited or a theory of a theory. *American Psychologist, 28,* 405–416.

Epstein, S. (1977). Traits are alive and well. In D. Magnusson and N. Endler (Eds.), *Personality at the crossroads: Current issues in interactional psychology* (pp. 83–98). Hillsdale, NJ: Erlbaum.

Epstein, S. (1979). The stability of behavior: I. On predicting most of the people much of the time. *Journal of Personality and Social Psychology, 37,* 1097–1126.

Epstein, S. (1983). Aggregation and beyond: Some basic issues on the prediction of behavior. *Journal of Personality, 51,* 360–392.

Epstein, S. (1990). Cognitive-experimental self-theory. In L. A. Pervin (Ed.), *Handbook of personality: Theory and research* (pp. 165–192). New York: Guilford.

Epstein, S., & Fenz, W. D. (1962). Theory and experiment on the measurement of approach-avoidance conflict. *Journal of Abnormal and Social Psychology, 64,* 97–112.

Erdelyi, M. H. (1985). *Psychoanalysis: Freud's cognitive psychology.* New York: W. H. Freeman & Company.

Erdelyi, M. H. (1992). Psychodynamics and the unconscious. *American Psychologist, 47,* 784–787.

Erdelyi, M. H. (1993). Repression: The mechanism and the defense. In D. M. Wegner & J. W. Pennebaker (Eds.), *Handbook of mental control. Century psychology series* (pp. 126–148). Englewood Cliffs, NJ: Prentice-Hall.

Erdelyi, M. H., & Goldberg, B. (1979). Let's not sweep repression under the rug: Towards a cognitive psychology of repression. In J. F. Kihlstrom & F. J. Evans (Eds.), *Functional disorders of memory.* Hillsdale, NJ: Erlbaum.

Erdley, C. A., & Dweck, C. S. (1993). Children's implicit personality theories as predictors of their social judgements. *Child Development, 64,* 863–878.

Eriksen, C. W. (1952). Individual differences in defensive forgetting. *Journal of Experimental Psychology, 44,* 442–446.

Eriksen, C. W. (1966). Cognitive responses to internally cued anxiety. In C. D. Spielberger (Ed.), *Anxiety and behavior* (pp. 327–360). New York: Academic Press.

Eriksen, C. W., & Kuethe, J. L. (1956). Avoidance conditioning of verbal behavior without awareness: A paradigm of repression. *Journal of Abnormal and Social Psychology, 53*, 203–209.

Erikson, E. (1963). *Childhood and society*. New York: Norton.

Erikson, E. (1968). *Identity: Youth and crisis*. New York: Norton.

Exline, R., & Winters, L. C. (1965). Affective relations and mutual glances in dyads. In S. Tomkins & C. Izard (Eds.), *Affect, cognition, and personality*. New York: Springer.

Exner, J. E. (1993). *The Rorschach: A comprehensive system, Vol. 1: Basic foundations* (3rd ed.). New York: Wiley.

Eysenck, H. (1983). Cicero and the state-trait theory of anxiety: Another case of delayed recognition. *American Psychologist, 38*, 114–115.

Eysenck, H. J. (1961). The effects of psychotherapy. In H. J. Eysenck (Ed.), *Handbook of abnormal psychology: An experimental approach* (pp. 697–725). New York: Basic Books.

Eysenck, H. J. (1967). *The biological basis of personality*. Springfield, IL: Charles C. Thomas.

Eysenck, H. J. (1973). Personality and the law of effect. In D. E. Berlyne & K. B. Madsen (Eds.), *Pleasure, reward, preference*. New York: Academic Press.

Eysenck, H. J. (1990). Biological dimensions of personality. In L. A. Pervin (Ed.), *Handbook of personality: Theory and research* (pp. 244–276). New York: Guilford.

Eysenck, H. J. (1991). Personality, stress, and disease: An interactionist perspective. *Psychological Inquiry, 2*, 221–232.

Eysenck, H. J., & Eysenck, M.W. (1985). *Personality and individual differences: A natural science approach*. New York: Plenum.

Eysenck, H. J., & Eysenck, M. W. (1995). *Mindwatching: Why we behave the way we do*. London: Prion Books.

Eysenck, H. J., & Rachman, S. (1965). *The causes and cures of neurosis: An introduction to modern behavior therapy based on learning theory and the principles of conditioning*. San Diego, CA: Knapp.

Fairweather, G. W. (1967). *Methods in experimental social innovation*. New York: Wiley.

Fairweather, G. W., Sanders, D. H., Cressler, D. L., & Maynard, H. (1969). *Community life for the mentally ill: An alternative to institutional care*. Chicago: Aldine.

Feather, N. T. (1990). Bridging the gap between values and actions: Recent applications of the expectancy-value model. In E. T. Higgins & R. M. Sorrentino (Eds.), *Handbook on motivation and cognition: Foundations of social behavior* (Vol. 3, pp. 151–191). New York: Guilford.

Fenz, W. D. (1964). Conflict and stress as related to physiological activation and sensory, perceptual and cognitive functioning. *Psychological Monographs, 78*, No. 8 (Whole No. 585).

Ferster, C. B., & Skinner, B. F. (1957). *Schedules of reinforcement*. New York: Appleton.

Festinger, L. (1957). *A theory of cognitive dissonance*. Stanford, CA: Stanford University Press.

Fischer, W. F. (1970). *Theories of anxiety*. New York: Harper.

Fiske, A. P. (2002). Using individualism and collectivism to compare cultures—A critique of the validity and measurement of the constructs comment on Oyserman et al. (2002). *Psychological Bulletin, 128*, 78–88.

Fiske, D. W. (1994). Two cheers for the Big Five. *Psychological Inquiry, 5*, 123–124.

Fitch, G. (1970). Effects of self-esteem, perceived performance, and choice on causal attribution. *Journal of Personality and Social Psychology, 16*, 311–315.

Flavell, J. H., & Ross, L. (Eds.). (1981). *Social cognitive development: Frontiers and possible futures*. New York: Cambridge University Press.

Fleeson, W. (2001). Toward a structure- and process-integrated view of personality: Traits as density distributions of states. *Journal of Personality and Social Psychology, 80*, 1011–1027.

Fleming, J., & Darley, J. M. (1986). *Perceiving intention in constrained behavior: The role of purposeful and constrained action cues in correspondence bias effects*. Unpublished manuscript, Princeton University, Princeton, NJ.

Flint, J., Corley, R., DeFries, J. C., Fulker, D. W., Gary, J. A., Miller, S., & Collins, A. C. (1995). A simple genetic basis for a complex psychological trait in laboratory mice. *Science, 269*, 1432–1435.

Fodor, I. (1987). Moving beyond cognitive-behavior therapy: Integrating Gestalt therapy to facilitate personal and interpersonal awareness. In N. S. Jacobson (Ed.), *Psychotherapists in clinical practice: Cognitive and behavioral perspectives* (pp. 190–231). New York: Guilford.

Forgas, J. P. (1983). Episode cognition and personality: A multidimensional analysis. *Journal of Personality, 51*, 34–48.

Foucault, M. (1980). *Power/knowledge: Selected interviews and other writings, 1972–1977* (C. Gordon, Ed. and Trans.). Brighton, England: Harvester.

Fox, N. A., Bell, M. A., & Jones, N. A. (1992). Individual differences in response to stress and cerebral asymmetry. *Developmental Neuropsychology, 8*, 161–184.

Fox, N. A., Henderson, H. A., Rubin, K. H., Calkins, S. D., & Schmidt, L. A. (2001). Continuity and discontinuity of behavioral inhibition and exuberance: Psychophysiological and behavioral influences across the first four years of life. *Child Development, 72*, 1–21.

Fox, N. A., Schmidt, L. A., Calkins, S. D., Rubin, K. H., & Copan, R. J. (1996). The role of frontal activation in the regulation and dysregulation of social behavior during the preschool years. *Development and Psychopathology, 8,* 89–102.

Fraley, R. C., & Shaver, P. R. (1997). Adult attachment and the suppression of unwanted thoughts. *Journal of Personality and Social Psychology, 73,* 1080–1091.

Frank, K. G., & Hudson, S. M. (1990). Behavior management of infant sleep disturbance. *Journal of Applied Behavior Analysis, 23,* 91–98.

Fransella, F. (1995). *George Kelly*. London: Sage Publications.

Frawley, P. J., & Smith, J. W. (1990). Chemical aversion therapy in the treatment of cocaine dependence as part of a multimodal treatment program: Treatment outcome. *Journal of Substance Abuse Treatment, 7,* 21–29.

Freud, A. (1936). *The ego and the mechanisms of defense.* New York: International Universities Press.

Freud, S. (1899). The interpretation of dreams. *Standard edition, Vol. 4.* London: Hogarth, 1955.

Freud, S. (1901). Psychopathology of everyday life. *Standard edition, Vol. 6.* London: Hogarth, 1960.

Freud, S. (1905). Fragments of an analysis of a case of hysteria. *Standard edition, Vol. 7.* London: Hogarth, 1953.

Freud, S. (1909). Leonardo da Vinci: A study in psychosexuality. *Standard edition, Vol. 2.* London: Hogarth, 1957.

Freud, S. (1911). Formulations regarding the two principles of mental functioning. *Collected papers, Vol. IV.* New York: Basic Books, 1959.

Freud, S. (1915). Instincts and their vicissitudes. *Standard edition, Vol. 14.* London: Hogarth, 1957.

Freud, S. (1917). On transformations of instinct as exemplified in anal eroticism. *Standard edition, Vol. 18.* London: Hogarth, 1955.

Freud, S. (1920). *A general introduction to psychoanalysis.* New York: Boni and Liveright, 1924.

Freud, S. (1933). *New introductory lectures on psychoanalysis* (W. J. H. Sproutt, Trans.). New York: Norton.

Freud, S. (1940). An outline of psychoanalysis. *International Journal of Psychoanalysis, 21,* 27–84.

Freud, S. (1958). A note on the unconscious in psychoanalysis. In J. Strachey (Ed. and Trans.), *The standard edition of the complete psychological works of Sigmund Freud* (Vol. 12, pp. 255–266). London: Hogarth Press. (Original work published 1912).

Freud, S. (1959). *Collected papers, Vols. I–V.* New York: Basic Books.

Freud, S. (1963). *The sexual enlightenment of children.* New York: Macmillan.

Friedman, M., & Roseman, R. H. (1974). *Type A behavior and your heart.* New York: Knopf.

Fromm, E. (1941). *Escape from freedom.* New York: Holt, Rinehart and Winston.

Fromm, E. (1947). *Man for himself.* New York: Holt, Rinehart and Winston.

Funder, D. C. (1991). Global traits: A neo-Allportian approach to personality. *Psychological Science, 2,* 31–39.

Funder, D. C. (2001). Personality. *Annual Review of Psychology, 52,* 197–221.

Funder, D. C., & Colvin, C. R. (1997). Congruence of others' and self-judgments of personality. In R. Hogan, J. Johnson, & S. Briggs (Eds.), *Handbook of Personality Psychology* (pp. 617–647). San Diego, CA: Academic Press.

Gable, S. L., Reis, H. T., & Elliot, A. J. (2000). Behavioral activation and inhibition in everyday life. *Journal of Personality and Social Psychology, 78,* 1135–1149.

Garcia, J., McGowan, B. K., & Green, K. F. (1972). Biological constraints on conditioning. In A. H. Black & W. F. Prokasy (Eds.), *Classical conditioning II: Current research and theory.* New York: Appleton-Century-Crofts.

Geen, R. G. (1984). Preferred stimulation levels in introverts and extraverts: Effects on arousal and performance. *Journal of Personality and Social Psychology, 46,* 1303–1312.

Geen, R. G. (1997). Psychophysiological approaches to personality. In R. Hogan, J. A. Johnson, & S. R. Briggs (Eds.), *Handbook of personality psychology* (pp. 387–416). San Diego: Academic Press.

Geertz, C. (1986). Making experiences, authoring selves. In V. Turner & E. Bruner (Eds.), *The anthropology of experience.* Chicago: University of Illinois Press.

Giese, H., & Schmidt, S. (1968). *Studenten sexualitat.* Hamburg: Rowohlt.

Gilligan, C. (1982). *In a different voice: Psychological theory and women's development.* Cambridge, MA: Harvard University Press.

Gilmore, D. D. (1990). *Manhood in the making.* New Haven, CT: Yale Press.

Gitlin, M. J. (1990). *The psychotherapist's guide to psychopharmacology.* New York: Free Press.

Glass, D. C. (1977). *Behavior patterns, stress, and coronary disease.* Hillsdale, NJ: Erlbaum.

Glass, D. C., Singer, J. E., & Friedman, L. N. (1969). Psychic costs of adaptation to an environmental stressor. *Journal of Personality and Social Psychology, 12,* 200–210.

Goldberg, L. R. (1973). *The exploitation of the English Language for the development of a descriptive personality taxonomy.* Paper delivered at the 81st Annual Convention of the American Psychological Association, Montreal, Canada.

Goldberg, L. R. (1990). An alternative "description of personality": The Big-Five factor structure. *Journal of Personality and Social Psychology, 59,* 1216–1229.

Goldberg, L. R. (1992). The development of markers for the big-five factor structure. *Psychological Assessment, 4,* 26–42.

Goldberg, L. R., & Lewis, M. (1969). Play behavior in the year old infant: Early sex differences. *Child Development, 40,* 21–31.

Goldner, V., Penn, P., Sheinberg, M., & Walker, G. (1990). Love and violence: Gender paradoxes in volatile attachments. *Family Process, 29,* 343–364.

Goldsmith, H. H. (1991). A zygosity questionnaire for young twins: A research note. *Behavior Genetics, 21,* 257–269.

Goldsmith, H. H., & Campos, J. J. (1986). Fundamental issues in the study of early temperament: The Denver twin temperament study. In M. E. Lamb, A. L. Brown, & B. Rogoff (Eds.), *Advances in developmental psychology* (pp. 231–283). Hillsdale, NJ.

Gollwitzer, P. M., & Bargh, J. A. (Eds.). (1996). *The psychology of action: Linking cognition and motivation to behavior.* New York: Guilford.

Gollwitzer, P. M., & Moskowitz, G. B. (1996). Goal effects on action and cognition. In E. T. Higgins & A. W. Kruglanski (Eds.), *Social Psychology: Handbook of Basic Principles* (pp. 361–399). New York: Guilford.

Gormly, J., & Edelberg, W. (1974). Validation in personality trait attribution. *American Psychologist, 29,* 189–193.

Gottlieb, G. (1998). Normally occurring environmental and behavioral influences on gene activity: From central dogma to probabilistic epigenesis. *Psychological Review, 105,* 792–802.

Gough, H. G. (1957). *Manual, California Psychological Inventory.* Palo Alto, CA: Consulting Psychologists Press.

Gove, F. L., Arend, R. A., & Sroufe, L. A. (1979). *Competence in preschool and kindergarten predicted from infancy.* Paper presented at the Meeting of the Society for Research in Child Development, San Francisco, CA.

Grant, H., & Dweck, C. S. (1999). A goal analysis of personality and personality coherence. In D. Cervone & Y. Shoda (Eds.), *Social-cognitive approaches to personality coherence.* New York: Guilford Press.

Gray, J. A. (1991). The neuropsychology of temperament. In J. Strelau & A. Angleitner (Eds.), *Explorations in temperament: International perspective on theory and measurement.* New York: Plenum Press.

Greenberg, J. R., & Mitchell, S. (1983). *Object relations in psychoanalytic theory.* Cambridge: Harvard University Press.

Greenberg, J., Solomon, S., Pyszczynski, T., & Rosenblatt, A. (1992). Why do people need self-esteem? Converging evidence that self-esteem serves an anxiety-buffering function. *Journal of Personality and Social Psychology, 63,* 913–922.

Greenwald, A. G. (1980). The totalitarian ego: Fabrication and revision of personal history. *American Psychologist, 7,* 603–618.

Greenwald, A. G. (1992). New look 3: Unconscious cognition reclaimed. *American Psychologist, 47,* 766–779.

Greenwald, A. G., Banaji, M. R., Rudman, L. A., Farnham, S. D., Nosek, B. A., & Mellott, D. S. (2002). A unified theory of implicit attitudes, stereotypes, self-esteem, and self-concept. *Psychological Review, 109,* 3–25.

Greenwald, A. G., & Farnham, S. D. (2000). Using the Implicit Association Test to measure self-esteem and self-concept. *Journal of Personality and Social Psychology, 79,* 1022–1038.

Griffit, W., & Guay, P. (1969). "Object" evaluation and conditioned affect. *Journal of Experimental Research in Personality, 4,* 1–8.

Grigorenko, E. L. (2002). In search of the genetic engram of personality. In D. Cervone & W. Mischel (Eds.), *Advances in Personality Science* (pp. 29–82). New York: Guilford.

Grilly, D. M. (1989). *Drugs and human behavior.* Boston: Allyn and Bacon.

Gross, J. J. (1998). Antecedent- and response-focused emotion regulation: Divergent consequences for experience, expression, and physiology. *Journal of Personality and Social Psychology, 74,* 224–237.

Gross, J. J., & John, O. P. (2003). Individual differences in two emotion regulation processes: Implications for affect, relationships, and well-being. *Journal of Personality and Social Psychology.*

Grossberg, J. M. (1964). Behavior therapy: A review. *Psychological Bulletin, 62,* 73–88.

Grunbaum, A. (1984). *The foundations of psychoanalysis.* Berkeley, CA: University of California Press.

Guilford, J. P. (1959). *Personality.* New York: McGraw-Hill.

Guthrie, E. R. (1935). *The psychology of learning.* New York: Harper & Brothers.

Halverson, C. (1971a). Longitudinal relations between newborn tactile threshold, preschool barrier behaviors, and early school-age imagination and verbal development. *Newborn and preschooler: Organization of behavior and relations between period.* SRCD Monograph, Vol. 36.

Halverson, C. (1971b). *Relation of preschool verbal communication to later verbal intelligence, social maturity, and distribution of play bouts.* Paper presented at the meeting of the American Psychological Association.

Hamer, D., & Copeland, P. (1998). *Living with our genes: Why they matter more than you think.* New York: Doubleday.

Harackiewicz, J. M., Manderlink, G., & Sansone, C. (1984). Rewarding pinball wizardry: Effects of evaluation and cue on intrinsic motivation. *Journal of Personality and Social Psychology, 47*, 287–300.

Haring, T. G., & Breen, C. J. (1992). A peer-mediated social network intervention to enhance the social integration of persons with moderate and severe disabilities. *Journal of Applied Behavior Analysis, 25*, 319–333.

Harlow, R. E., & Cantor, N. (1996). Still participating after all these years: A study of life task participation in later life. *Journal of Personality and Social Psychology, 71*, 1235–1249.

Harmon-Jones, E., & Allen, J. J. (1997). Behavioral activation sensitivity and resting frontal EEG asymmetry: Covariation of putative indicators related to risk for mood disorders. *Journal of Abnormal Psychology, 106*, 159–163.

Harris, F. R., Johnston, M. K., Kelley, S. C., & Wolf, M. M. (1964). Effects of positive social reinforcement on regressed crawling of a nursery school child. *Journal of Educational Psychology, 55*, 35–41.

Harter, S. (1983). Developmental perspectives on the self-system. In P. H. Mussen (Ed.), *Handbook of child psychology* (Vol. 4, E. M. Hetherington, Ed.). New York: Wiley.

Harter, S. (1999). *The construction of the self: A developmental perspective*. New York: Guilford.

Hartshorne, H., & May, A. (1928). *Studies in the nature of character, Vol. 1. Studies in deceit*. New York: Macmillan.

Hawkins, R. P., Peterson, R. F., Schweid, E., & Bijou, S. W. (1966). Behavior therapy in the home: Amelioration of problem parent-child relations with the parent in a therapeutic role. *Journal of Experimental Child Psychology, 4*, 99–107.

Hazan, C., & Shaver, P. (1987). Romantic love conceptualized as an attachment process. *Journal of Personality and Social Psychology, 52*, 511–524.

Hazan, C., & Shaver, P. R. (1994). Attachment as an organizational framework for research on close relationships. *Psychological Inquiry, 5*, 1–22.

Hebb, D. O. (1955). Drives and the CNS (conceptual nervous system). *Psychological Review, 62*, 243–259.

Heckhausen, H. (1969). Achievement motive research: Current problems and some contributions towards a general theory of motivation. In W. J. Arnold (Ed.), *Nebraska Symposium on Motivation* (pp. 103–174). Lincoln: Nebraska University Press.

Heider, F. (1958). *The psychology of interpersonal relations*. New York: Wiley.

Heitzman, A. J., & Alimena, M. J. (1991). Differential reinforcement to reduce disruptive behaviors in a blind boy with a learning disability. *Journal of Visual Impairment and Blindness, 85*, 176–177.

Heller, W., Schmidtke, J. I., Nitschke, J. B., Koven, N. S., & Miller, G. A. (2002). States, traits, and symptoms: Investigating the neural correlates of emotion, personality, and psychopathology. In D. Cervone & W. Mischel (Eds.) *Advances in Personality Science* (pp. 106–126). New York: Guilford.

Henderson, V., & Dweck, C. S. (1990). Adolescence and achievement. In S. Feldman & G. Elliot (Eds.), *At the threshold: Adolescent development*. Cambridge, MA: Harvard University Press.

Herman, C. P. (1992). Review of W. H. Sheldon "Varieties of Temperament." *Contemporary Psychology, 37*, 525–528.

Hershberger, S. L., Lichtenstein, P., & Knox, S. S. (1994). Genetic and environmental influences on perceptions or organizational climate. *Journal of Applied Psychology, 79*, 24–33.

Hetherington, E. M., & Clingempeel, W. G. (1992). Coping with marital transitions: A family systems perspective. *Monographs of the Society for Research in Child Development*, Nos. 2–3, Serial No. 277.

Higgins, E. T. (1987). Self-discrepancy: A theory relating self and affect. *Psychological Review, 94*, 319–340.

Higgins, E. T. (1990). Personality, social psychology, and person-situation relations: Standards and knowledge activation as a common language. In L. A. Pervin (Ed.), *Handbook of personality: Theory and research* (pp. 301–338). New York: Guilford Press.

Higgins, E. T. (1996). Ideals, oughts, & regulatory focus: Affect and motivation from distinct pains and pleasures. In P. M. Gollwitzer & J. A. Bargh (Eds.), *The psychology of action: Linking cognition and motivation to behavior* (pp. 91–114). New York: Guilford.

Higgins, E. T. (1997). Beyond pleasure and pain. *American Psychologist, 52*, 1280–1300.

Higgins, E. T. (1998). Promotion and prevention: Regulatory focus as a motivation principle. In M. P. Zanna (Ed.), *Advances in Experimental Social Psychology* (Vol. 30, pp. 1–46). New York: Academic Press.

Higgins, E. T., & Kruglanski, A. W. (Eds.). (1996). *Social Psychology: Handbook of Basic Principles*. New York: Guilford.

Higgins, E. T., & Sorrentino, R. M. (Eds.) (1990). *Handbook of motivation and cognition: Foundations of social behavior* (Vol. 2). New York: Guilford.

Higgins, E. T., Friedman, R. S., Harlow, R. E., Chen-Idson, L., Ayduk, O. N., & Taylor, A. (2001). Achievement orientations from subjective histories of success: Promotion pride versus prevention pride. *European Journal of Social Psychology, 31*, 3–23.

Higgins, E. T., King, G. A., & Mavin, G. H. (1982). Individual construct accessibility and subjective impressions and recall. *Journal of Personality and Social Psychology, 43,* 35–47.

Holahan, C. J., & Moos, R. N. (1990). Life stressors, resistance factors, and improved psychological functioning: An extension of the stress resistance paradigm. *Journal of Personality and Social Psychology, 58,* 909–917.

Hollander, E., Liebowitz, M. R., Gorman, J. M., Cohen, B., Fyer, A., & Klein, D. F. (1989). Cortisol and sodium lactate-induced panic. *Archives of General Psychiatry, 46,* 135–140.

Holmes, D. S. (1974). Investigations of repression: Differential recall of material experimentally or naturally associated with ego threat. *Psychological Bulletin, 81,* 632–653.

Holmes, D. S. (1992). The evidence for repression: An examination of sixty years of research. In J. L. Singer (Ed.), *Repression and dissociation* (pp. 85–102). Chicago: University of Chicago Press.

Holmes, D. S., & Houston, K. B. (1974). Effectiveness of situation redefinition and affective isolation in coping with stress. *Journal of Personality and Social Psychology, 29,* 212–218.

Holmes, D. S., & Schallow, J. R. (1969). Reduced recall after ego threat: Repression or response competition? *Journal of Personality and Social Psychology, 13,* 145–152.

Holmes, J. (1993). *John Bowlby and Attachment Theory.* New York: Routledge.

Horowitz, L. M., Rosenberg, S. E., Ureno, G., Kalehzan, B. M., & O'Halloran, P. (1989). Psychodynamic formulation, consensual response method, and interpersonal problems. *Journal of Consulting and Clinical Psychology, 57,* 599–606.

Howes, D. H., & Solomon, R. L. (1951). Visual duration threshold as a function of word-probability. *Journal of Experimental Psychology, 41,* 401–410.

Howland, R. H. (1991). Pharmacotherapy of dysthymia: A review. *Journal of Clinical Psychopharmacology, 11,* 83–92.

Hoyenga, K. B., & Hoyenga, K. T. (1993). *Gender-related differences: Origins and outcomes.* Boston, MA: Allyn & Bacon.

Hoyle, R. H., Kernis, M. H., Leary, M. R., & Baldwin, M. (1999). *Selfhood: Identity, esteem, regulation.* Boulder, CO: Academic Press.

Huber, G. L., Sorrentino, R. M., Davidson, M. A., & Epplier, R. (1992). Uncertainty orientation and cooperative learning: Individual differences within and across cultures. *Learning and Individual Differences, 4,* 1–24.

Ickovics, J. (1997, August). Smithsonian seminar on health and well-being sponsored by Society for the Psychological Study of Social Issues and American Psychological Society conducted at the Ninth Annual Conference of the American Psychological Society, Washington, DC.

Insel, T. R., & Winslow, J. T. (1998). Serotonin and neuropeptides in affiliative behaviors. *Biological Psychiatry, 44,* 207–219.

Isen, A. M., Niedenthal, P. M., & Cantor, N. (1992). An influence of positive affect on social categorization. *Motivation and Emotion, 16,* 65–78.

Jaccard, J. J. (1974). Predicting social behavior from personality traits. *Journal of Research in Personality, 7,* 358–367.

Jackson, D. N., & Paunonen, S. V. (1980). Personality structure and assessment. In M. R. Rosenzweig & L. W. Porter (Eds.), *Annual review of psychology* (Vol. 31). Palo Alto, CA: Annual Reviews, Inc.

Jacobs, W, J., & Nadel, L. (1985). Stress-induced recovery of fears and phobias. *Psychological Review, 92,* 512–531.

Jahoda, M. (1958). *Current concepts of positive mental health.* New York: Basic Books.

James, W. (1890). *The principles of psychology* (Vols. 1 and 2). New York: Holt.

Jang, K. L. (1993). *A behavioral genetic analysis of personality, personality disorder, the environment, and the search for sources of nonshared environmental influences.* Unpublished doctoral dissertation, University of Western Ontario, London, Ontario.

Janis, I. L. (1971). *Stress and frustration.* New York: Harcourt.

Jasper, H. (1941). Electroencephalography. In W. Penfield & T. Erickson (Eds.), *Epilepsy and cerebral localization.* Springfield, IL: Charles C Thomas.

Jemmott, J. B., Jemmott, L. S., Spears, H., Hewitt, N., & Cruz-Collins, M. (1992). Self-efficacy, hedonistic expectancies, and condom-use intentions among inner-city Black adolescent women: A social cognitive approach to AIDS risk behavior. *Journal of Adolescent Health, 13,* 512–519.

Jersild, A. (1931). Memory for the pleasant as compared with the unpleasant. *Journal of Experimental Psychology, 14,* 284–288.

John, O. P. (1990). The big-five factor taxonomy: Dimensions of personality in the natural language and questionnaires. In L. A. Pervin (Ed.), *Handbook of personality: Theory and research* (pp. 66–100). New York: Guilford.

John, O. P. (2001, February). What is so big about the Big Five, anyway? Invited address presented at the 2nd Annual Meeting of the Society of Personality and Social Psychology, San Antonio, TX.

John, O. P., Hampson, S. E., & Goldberg, L. R. (1991). The basic level in personality-trait hierarchies: Studies of traits use and accessibility in different contexts. *Journal of Personality and Social Psychology, 60,* 348–361.

Jones, A. (1966). Information deprivation in humans. In B. A. Maher (Ed.), *Progress in experimental personality research* (Vol. 3, pp. 24–307). New York: Academic Press.

Jourard, S. M. (1967). Experimenter-subject dialogue: A paradigm for a humanistic science of psychology. In J. Bugental (Ed.), *Challenges of humanistic psychology* (pp. 109–116). New York: McGraw-Hill.

Jourard, S. M. (1974). *Healthy personality: An approach from the viewpoint of humanistic psychology.* New York: Macmillan.

Joy, V. L. (1963, August). *Repression-sensitization and interpersonal behavior.* Paper presented at the meeting of the American Psychological Association, Philadelphia.

Jung, C. G. (1963). *Memories, dreams, reflections.* New York: Pantheon.

Jung, C. G. (1964). *Man and his symbols.* Garden City, NY: Doubleday.

Kagan, J. (1964). The acquisition and significance of sex typing and sex role identity. In M. Hoffman & L. Hoffman (Eds.), *Review of child development research* (Vol. 1). New York: Russell Sage.

Kagan, J. (2003). Biology, context, and developmental inquiry. *Annual Review of Psychology, 54,* 1–23.

Kagan, J., Reznick, J. S., & Snidman, N. (1988). Biological bases of childhood shyness. *Science, 240,* 167–171.

Kagan, J., & Snidman, N. (1991). Infant predictors of inhibited and uninhibited profiles. *Psychological Science, 2,* 40–44.

Kahneman, D., & Snell, J. (1990). Predicting utility. In R. M. Hogarth (Ed.), *Handbook of personality: Theory and research* (pp. 66–100). New York: Guilford.

Kamps, D. M., Leonard, B. R., Vernon, S., Dugan, E. P., Delquadri, C., Gershon, B., Wade, L., & Folk, L. (1992). Teaching social skills to students with autism to increase peer interactions in an integrated first-grade classroom. *Journal of Applied Behavior Analysis, 25,* 281–288.

Kardiner, A. (1945). *Psychological frontiers of society.* New York: Columbia University Press.

Karoly, P. (1980). Operant methods. In F. H. Kanfer & A. P. Goldstein (Eds.), *Helping people change: A textbook of methods* (2nd ed.). Elmsford, NY: Pergamon Press.

Kayser A., Robinson, D. S., Yingling, K., Howard, D. B., Corcella, J., & Laux, D. (1988). The influence of panic attacks on response to phenelzine and amitriptyline in depressed outpatients. *Journal of Clinical Psychopharmacology, 8,* 246–253.

Kazdin, A. E. (1974). Effects of covert modeling and model reinforcement on assertive behavior. *Journal of Abnormal Psychology, 83,* 240–252.

Kazdin, A. E., & Wilson, G. T. (1978). *Evaluation of behavior therapy: Issues, evidence and research strategies.* Cambridge, MA: Ballinger.

Kelley, H. H. (1973). The processes of casual attribution. *American Psychologist, 28,* 107–128.

Kelley, H. H., & Stahelski, A. J. (1970). The social interaction basis of cooperators' and competitors' beliefs about others. *Journal of Personality and Social Psychology, 16,* 66–91.

Kelly, E. L. (1955). Consistency of the adult personality. *American Psychologist, 10,* 659–681.

Kelly, G. A. (1955). *The psychology of personal constructs* (Vols. 1 and 2). New York: Norton.

Kelly, G. A. (1958). Man's construction of his alternatives. In G. Lindzey (Ed.), *Assessment of human motives* (pp. 33–64). New York: Holt, Rinehart and Winston.

Kelly, G. A. (1962). Quoted in B. A. Maher (Ed.), (1979), *Clinical psychology and personality: The selected papers of George Kelly.* Huntington, NY: Kreiger.

Kelly, G. A. (1966). Quoted in B. A. Maher (Ed.), (1979), *Clinical psychology and personality: The selected papers of George Kelly.* Huntington, NY: Kreiger.

Kendall, P. C., & Panichelli-Mindel, S. M. (1995). Cognitive-behavioral treatments. *Journal of Abnormal Child Psychology, 26,* 107–124.

Kenrick, D. T., & Funder, D. C. (1988). Profiting from controversy: Lessons from the person-situation debate. *American Psychologist, 43,* 23–34.

Kenrick, D. T., Sadalla, E. K., Groth, G., & Trost, M. R. (1990). Evolution, traits, and the stages of human courtship: Qualifying the parental investment model. *Journal of Personality, 58,* 97–116.

Kernberg, O. (1976). *Object relations theory and clinical psychoanalysis.* New York: Jason Aronson.

Kernberg, O. (1984). *Severe personality disorders.* New Haven: Yale University Press.

Kihlstrom, J. F. (1990). The psychological unconscious. In L.A. Pervin (Ed.), *Handbook of personality: Theory and research* (pp. 445–464). New York: Guilford.

Kihlstrom, J. F. (1999). The psychological unconscious. In L. A. Pervin & O. P. John (Eds.), *Handbook of personality: Theory and research* (2nd ed., pp. 424–442). New York: Guilford.

Kihlstrom, J. F. (2003). Implicit methods in Social Psychology. In C. Sansone, C. C. Morf, & A. Panter (Eds.), *Handbook of methods in social psychology.* Thousand Oaks, CA: Sage.

Kihlstrom, J. F., Barnhardt, T. M., & Tataryn, D. J. (1992). The psychological unconscious: Found, lost, and regained. *American Psychologist, 47,* 788–791.

Kim, M. P., & Rosenberg, S. (1980). Comparison of two-structured models of implicit personality theory. *Journal of Personality and Social Psychology, 38,* 375–389.

Klein, D. F., Gittelman, R., Quitkin, F., & Rifkin, A. (1980). *Diagnosis and drug treatment of psychiatric disorders: Adults and children* (2nd ed.). Baltimore, MD: Williams & Wilkins.

Klein, D. F., & Klein, H. M. (1989). The definition and psychopharmacology of spontaneous panic and phobia: A critical review I. In P. J. Tyrer (Ed.), *Psychopharmacology of anxiety.* New York: Oxford University Press.

Klinger, E. (1975). Consequences of commitment to and disengagement from incentives. *Psychological Review, 82,* 1–25.

Klohnen, E. C., & Bera, S. (1998). Behavioral and experiential patterns of avoidantly and securely attached women across adulthood: A 31-year longitudinal perspective. *Journal of Personality and Social Psychology, 74,* 211–223.

Kobak, R. R., & Sceery, A. (1988). Attachment in late adolescence: Working models, affect regulation, and representations of self and others. *Child Development, 59,* 135–146.

Koestner, R., & McClelland, D. C. (1990). Perspectives on competence motivation. In L. A. Pervin (Ed.), *Handbook of personality: Theory and research* (pp. 549–575). New York: Guilford.

Kohut, H. (1971). *The analysis of the self.* New York: International Universities Press.

Kohut, H. (1977). *The restoration of the self.* New York: International Universities Press.

Kohut, H. (1980). *Advances in self psychology.* New York: International Universities Press.

Kohut, H. (1984). *How does analysis cure?* Chicago, IL: University of Chicago Press.

Kolb, B., & Whishaw, I. Q. (1998). Brain plasticity and behavior. *Annual Review of Psychology, 49,* 43–64.

Kopp, C. G. (1982). Antecedents of self-regulation: A developmental perspective. *Developmental Psychology, 18,* 199–214.

Koriat, A., Melkman, R., Averill, J. R., & Lazarus, R. S. (1972). The self-control of emotional reactions to a stressful film. *Journal of Personality, 40,* 601–619.

Kosslyn, S. M, Cacioppo, J. T, Davidson, R. J, Hugdahl, K., Lovallo, W. R, Spiegel, D., & Rose, R. (2002). Bridging psychology and biology: The analysis of individuals in groups. *American Psychologist, 57,* 341–351.

Krahe, B. (1990). *Situation cognition and coherence in personality: An individual-centered approach.* Cambridge, England: Cambridge University Press.

Kramer, P. D. (1993). *Listening to Prozac.* New York: Viking.

Kuhl, J. (1984). Volitional aspects of achievement motivation and learned helplessness: Toward a comprehensive theory of action control. In B. A. Maher (Ed.), *Progress in experimental personality research* (Vol. 13, pp. 91–171). New York: Academic Press.

Kuhl, J. (1985). From cognition to behavior: Perspectives for future research on action control. In J. Kuhl (Ed.), *Action control from cognition to behavior.* New York: Springer-Verlag.

Kunda, Z. (1990). The case for motivated reasoning. *Psychological Bulletin, 108,* 480–498.

Kunda, Z. (1999). *Social cognition: Making sense of people.* Cambridge, MA: The MIT Press.

Lader, M. (1980). *Introduction to psychopharmacology.* Kalamazoo, MI: Upjohn.

LaHoste, G. J., Swanson, J. M., Wigal, S. S., Glabe, C., Wigal, T., King, N., & Kennedy, J. L. (1996). Dopamine D4 receptor gene polymorphism is associated with attention deficit hyperactivity disorder. *Molecular Psychiatry, 1,* 128–131.

Laing, R. D. (1965). *The divided self: An existential study in sanity and madness.* New York: Penguin.

Lamiell, J. T. (1997). Individuals and the differences between them. In R. Hogan, J. Johnson, & S. Briggs (Eds.), *Handbook of personality psychology* (pp. 117–141). San Diego, CA: Academic Press.

Landfield, A. W., Stern, M., & Fjeld, S. (1961). Social conceptual processes and change in students undergoing psychotherapy. *Psychological Reports, 8,* 63–68.

Lang, P. J., & Lazovik, A. D. (1963). Experimental desensitization of a phobia. *Journal of Abnormal and Social Psychology, 66,* 519–525.

Langer, E. J. (1975). The illusion of control. *Journal of Personality and Social Psychology, 32,* 311–328.

Langer, E. J. (1977). The psychology of chance. *Journal for the Theory of Social Behavior, 7,* 185–207.

Langer, E. J. (1978). Rethinking the role of thought in social interaction. In J. H. Harvey, W. J. Ickes, & R. F. Kidd (Eds.), *New directions in attribution research* (Vol. 2). Hillsdale, NJ: Erlbaum.

Langer, E. J., Janis, I. L., & Wolfer, J.A. (1975). Reduction of psychological stress in surgical patients. Unpublished manuscript, Yale University.

Larsen, R. J., & Kasimatis, M. (1990). Individual differences in entrainment of mood to the weekly calendar. *Journal of Personality and Social Psychology, 58,* 164–171.

Larsen, R. J., & Kasimatis, M. (1991). Day-to-day physical symptoms: Individual differences in the occurrence, duration, and emotional concomitants of minor daily illnesses. *Journal of Personality, 59,* 387–424.

Larsen, R. J., Diener, E., & Emmons, R. A. (1986). Affect intensity and reactions to daily life. *Journal of Personality and Social Psychology, 51,* 803–814.

Lawson, G. W., & Cooperrider, C. A. (1988). *Clinical psychopharmacology: A practical reference for nonmedical psychotherapists.* Rockville, MD: Aspen Publishers.

Lazar, S., Bush, G., Gollub, R. L., Fricchione, G. L., Khalsa, G., & Benson, H. (2000). Functional brain mapping of the relaxation response and meditation. *NeuroReport, 11,* 1581–1585.

Lazarus, A. A. (1961). Group therapy of phobic disorders by systematic desensitization. *Journal of Abnormal & Social Psychology, 63,* 504–510.

Lazarus, A. A. (1963). The treatment of chronic frigidity by systematic desensitization. *Journal of Nervous and Mental Diseases, 136,* 272–278.

Lazarus, R. S. (1976). *Patterns of adjustment.* New York: McGraw-Hill.

Lazarus, R. S. (1990). Theory-based stress measurement. *Psychological Inquiry, 1,* pp. 3–13.

Lazarus, R. S., Eriksen, C. W., & Fonda, C. P. (1951). Personality dynamics and auditory perceptual recognition. *Journal of Personality, 58,* 113–122.

Lazarus, R. S., & Folkman, S. (1984). *Stress, appraisal, and coping.* New York: Springer.

Lazarus, R. S., & Longo, N. (1953). The consistency of psychological defense against threat. *Journal of Abnormal and Social Psychology, 48,* 495–499.

Leary, M. R, & Downs, D. L. (1995). Interpersonal functions of the self-esteem motive: The self-esteem system as a sociometer. In M. H. Kernis (Ed.), *Efficacy, agency, and self-esteem* (pp. 123–144). New York: Kluwer Academic/Plenum Publishers.

Leary, M. R. (2002). The self as a source of relational difficulties. *Self and Identity, 1,* 137–142.

Leary, M. R., & Tangney, J. P. (Eds.) (2003). *Handbook of self and identity.* New York: Guilford.

Leary, T. (1957). *Interpersonal diagnosis of personality.* New York: Ronald Press.

Leary, T., Litwin, G. H., & Metzner, R. (1963). Reactions to psilocybin administered in a supportive environment. *Journal of Nervous and Mental Diseases, 137,* 561–573.

LeDoux, J. (1996). *The emotional brain.* New York: Simon & Schuster.

LeDoux, J. E. (2001). *Synaptic self.* New York: Viking.

Lepper, M. R., Greene, D., & Nisbett, R. E. (1973). Undermining children's intrinsic interest with extrinsic reward: A test of the "overjustification" hypothesis. *Journal of Personality and Social Psychology, 28,* 129–137.

Lesch, K., Bengel, D., Heils, A., & Sabol, S. Z. (1996). Association of anxiety-related traits with a polymorphism in the serotonin transporter gene regulatory region. *Science, 274,* 1527–1531.

Levine, R. (1991). *New developments in pharmacology* (Vol. 8, pp. 1–3). New York: Gracie Square Hospitals Publication.

Levy, S. R., Stroessner, S. J., & Dweck, C. S. (1998). Stereotype formation and endorsement: The role of implicit theories. *Journal of Personality and Social Psychology, 74,* 1421–1436.

Lewicki, P., Hill, T., & Czyzewska, M. (1992). Nonconscious acquisition of information. *American Psychologist, 47,* 796–801.

Lewin, K. (1935). *A dynamic theory of personality.* New York: McGraw-Hill.

Lewin, K. (1936). *Principles of topological psychology.* New York: McGraw-Hill.

Lewin, K. (1951). *Field theory in social science; selected theoretical papers.* D. Cartwright (Ed.). New York: Harper & Row.

Lewinsohn, P. M. (1975). The behavioral study and treatment of depression. In M. Hersen (Ed.), *Progress in behavior modification* (pp. 19–63). New York: Academic Press.

Lewinsohn, P. M., Clarke, G. N., Hops, H., & Andrews, A. (1990). Cognitive-behavioral treatment for depressed adolescents. *Behavior Therapy, 21,* 385–401.

Lewinsohn, P. M., Mischel, W., Chaplin, W., & Barton, R. (1980). Social competence and depression: The role of illusory self-perceptions. *Journal of Abnormal Psychology, 89,* 203–212.

Lewis, M. (1969). Infants' responses to facial stimuli during the first year of life. *Developmental Psychology, 1,* 75–86.

Lewis, M. (1990). Self-knowledge and social development in early life. In L. A. Pervin (Ed.), *Handbook of personality: Theory and research* (pp. 486–526). New York: Guilford.

Lewis, M. (1999). On the development of personality. In L. A. Pervin & O. P. John (Eds.), *Handbook of personality: Theory and research* (2nd ed., pp. 327–346). New York: Guilford.

Lewis, M. (2002). Models of development. In D. Cervone & W. Mischel (Eds.), *Advances in personality science* (pp. 153–176). New York: Guildford.

Lewontin, R. (2000). *The triple helix: Gene, organism, and environment.* Cambridge, MA: Harvard University Press.

Lieberman, M. A., Yalom, I. D., & Miles, M. B. (1973). *Encounter groups: First facts.* New York: Basic Books.

Liebert, R. M. (1986). Effects of television on children and adolescents. *Journal of Developmental and Behavioral Pediatrics, 7,* 43–48.

Liebert, R. M., & Allen, K. M. (1967). The effects of rule structure and reward magnitude on the acquisition and adoption of self-reward criteria. Unpublished manuscript, Vanderbilt University, Nashville, TN.

Liebert, R. M., & Baron, R. A. (1972). Some immediate effects of televised violence on children's behavior. *Developmental Psychology, 6,* 469–475.

Lief, H. I., & Fox, R. S. (1963). Training for "detached concern" in medical students. In H. I. Lief, V. F. Lief, &

N. R. Lief (Eds.), *The psychological basis of medical practice* (pp. 12–35). New York: Harper.

Lietaer, G. (1993). Authenticity, congruence, and transparency. In D. Brazier (Ed.), *Beyond Carl Rogers* (pp. 17–46). London: Constable.

Linscheid, T. R., & Meinhold, P. (1990). The controversy over aversives: Basic operant research and the side effects of punishment. In A. C. Repp & N. N. Singh (Eds.), *Perspectives on the use of nonaversive and aversive interventions for persons with developmental disabilities* (pp. 435–450). Sycamore, IL: Sycamore Publishing Company.

Linton, R. (1945). *The cultural background of personality.* New York: Appleton.

Linville, P. W., & Carlston, D. E. (1994). Social cognition of the self. In P. G. Devine, D. C. Hamilton, & T. M. Ostrom (Eds.), *Social cognition: Impact on social psychology* (pp. 143–193). New York: Academic Press.

Loehlin, J. C. (1992). *Genes and environment in personality development.* Newbury Park, CA: Sage.

Loehlin, J. C., & Nichols, R. C. (1976). *Heredity, environment, and personality: A study of 850 sets of twins.* Austin, TX: University of Texas Press.

Loevinger, J. (1957). Objective tests as instruments of psychological theory. *Psychological Reports Monographs, No. 9.* Southern University Press.

Loftus, E. F., & Klinger, M. R. (1992). Is the unconscious smart or dumb? *American Psychologist, 47,* 761–765.

Loftus, E. F. (1993). The reality of repressed memories. *American Psychologist, 48,* 518–537.

Loftus, E. F. (1994). The repressed memory controversy. *American Psychologist, 49,* 443–445.

Lott, A. J., & Lott, B. E. (1968). A learning theory approach to interpersonal attitudes. In A. G. Greenwald, T. C. Brock, & T. M. Ostrom (Eds.), *Psychological foundations of attitudes.* New York: Academic Press.

Lovaas, O. I., Berberich, J. P., Perloff, B. F., & Schaeffer, B. (1966). Acquisition of imitative speech by schizophrenic children. *Science, 151,* 705–707.

Lovaas, O. I., Berberich, J. P., Perloff, B. F., & Schaeffer, B. (1991). Acquisition of imitative speech by schizophrenic children. *Focus on Autistic Behavior, 6,* 1–5.

Lyons, D. (1997). The feminine in the foundations of organizational psychology. *Journal of Applied Behavioral Science, 33,* 7–26.

Lyons, M. J., Goldberg, J., Eisen, S. A., True, W., Tsuang, M. T., Meyer, J. M., & Henderson, W. G. (1993). Do genes influence exposure to trauma: A twin study of combat. *American Journal of Medical Genetics (Neuropsychiatric Genetics), 48,* 22–27.

Lyubomirsky, S., & Nolen-Hoeksema, S. (1993). Self-perpetuating properties of dysphoric rumination. *Journal of Personality and Social Psychology, 65,* 339–349.

Lyubomirsky, S., & Nolen-Hoeksema, S. (1995). Effects of self-focused rumination on negative thinking and interpersonal problem solving. *Journal of Personality and Social Psychology, 69,* 176–190.

Maccoby, E. E., & Jacklin, C. N. (1974). *The psychology of sex differences.* Stanford, CA: Stanford University Press.

MacDonald, K. (1998). Evolution, culture, and the Five-Factor Model. *Journal of Cross-Cultural Psychology, 29,* 119–149.

Madison, P. (1960). *Freud's concept of repression and defense: Its theoretical and observational language.* Minneapolis, MN: University of Minnesota Press.

Magnusson, D. (Ed.). (1980). *The situation: An interactional perspective.* Hillsdale, NJ: Erlbaum.

Magnusson, D. (1990). Personality development from an interactional perspective. In L. A. Pervin (Ed.), *Handbook of personality: Theory and research* (pp. 193–224). New York: Guilford.

Magnusson, D. (1999). Holistic interactionism: A perspective for research on personality development. In L. A. Pervin & O. P. John (Eds.), *Handbook of personality: Theory and research* (2nd ed., pp. 219–247). New York: Guilford.

Magnusson, D., & Endler, N. S. (1977). Interactional psychology: Present status and future prospects. In D. Magnusson & N. S. Endler (Eds.), *Personality at the crossroads: Current issues in interactional psychology.* Hillsdale, NJ: Erlbaum.

Maher, B. A. (1966). *Principles of psychotherapy: An experimental approach.* New York: McGraw-Hill.

Maher, B. A. (1979). *Clinical psychology and personality: The selected papers of George Kelly.* Huntington, NY: Wiley.

Main, M., Kaplan, N., & Cassidy, J. (1985). Security in infancy, childhood, and adulthood: A move to the level of representation. In I. Bretherton & E. Waters (Eds.), *Growing Points in Attachment Theory and Research. Monographs of the Society for Research in Child Development, No. 209, Vol. 50,* 66–104.

Malmo, R. B. (1959). Activation: A neuropsychological dimension. *Psychological Review, 66,* 367–386.

Malmquist, C. P. (1986). Children who witness parental murder: Post traumatic aspects. *Journal of the American Academy of Child Psychiatry, 25,* 320–325.

Manke, B., McGuire, S., Reiss, D., Hetherington, E. M., & Plomin, R. (1995). Genetic contributions to children's extrafamilial social interactions: Teachers, friends, and peers. *Social Development, 4,* 238–256.

Maricle, R. A., Kinzie, J. D., & Lewinsohn, P. (1988). Medication-associated depression: A two and one-half year follow-up of a community sample. *International Journal of Psychiatry in Medicine, 18,* 283–292.

Marks, I. M. (1987). *Fears, phobias, and rituals.* New York: Oxford Press University.

Marks, I. M., & Nesse, R. M. (1994). Fear and fitness: An evolutionary analysis of anxiety disorders. *Ethology and Sociobiology, 15,* 247–261.

Marks, J., Stauffacher, J. C., & Lyle, C. (1963). Predicting outcome in schizophrenia. *Journal of Abnormal and Social Psychology, 66,* 117–127.

Markus, H. (1977). Self-schemata and processing information about the self. *Journal of Personality and Social Psychology, 35,* 63–78.

Markus, H., & Cross, S. (1990). The interpersonal self. In L. A. Pervin (Ed.), *Handbook of personality: Theory and research* (pp. 576–608). New York: Guilford.

Markus, H. R., & Kitayama, S. (1991). Culture and the self: Implications for cognition, emotion, and motivation. *Psychological Review, 98,* 224–253.

Markus, H. R., & Kitayama, S. (1998). The cultural psychology of personality. *Journal of Cross-Cultural Psychology, 29,* 63–87.

Markus, H. R., Kitayama, S., & Heiman, R. J. (1996). Culture and basic psychological principles. In E. T. Higgins & A. W. Kruglanski (Eds.), *Social psychology: Handbook of basic principles.* New York: Guilford.

Markus, H., & Nurius, P. (1986). Possible selves. *American Psychologist, 41,* 954–969.

Martin, L. L., & Tesser, A. (1989). Toward a motivational and structural theory of ruminative thought. In J. S. Uleman & J. A. Bargh (Eds.), *Unintended thought* (pp. 306–323). New York: Guilford.

Martin, L. L., Tesser, A., & McIntosh, W. D. (1993). Wanting but not having: The effects of unattained goals on thoughts and feelings. In D. M. Wegner & J. Pennebaker (Eds.), *Handbook of mental control. Century psychology series* (pp. 552–572). Englewood Cliffs, NJ: Prentice-Hall.

Maslow, A. H. (1965). Some basic propositions of a growth and self-actualization psychology. In G. Lyndzey & C. Hall (Eds.), *Theories of personality: Primary sources and research* (pp. 307–316). New York: Wiley.

Maslow, A. H. (1968). *Toward a psychology of being* (2nd ed.). New York: Van Nostrand.

Maslow, A. H. (1971). *The farther reaches of human nature.* New York: Viking.

Matas, W. H., Arend, R. A., & Sroufe, L. A. (1978). Continuity of adaptation in the second year: The relationship between quality of attachment and later competence. *Child Development, 49,* 547–556.

Matthews, K. A. (1984). Assessment of type A, anger, and hostility in epidemiological studies of cardiovascular disease. In A. Ostfeld & E. Eaker (Eds.), *Measuring psychosocial variables in epidemiological studies of cardiovascular disease.* Bethesda, MD: National Institute of Health.

Matute, H. (1994). Learned helplessness and superstitious behavior as opposite effects of uncontrollable reinforcement in humans. *Learning and Motivation, 25,* 216–232.

May, R. (1961). Existential psychology. In R. May (Ed.), *Existential psychology* (pp. 11–51). New York: Random House.

McAdams, D. P. (1990). Motives. In V. Derlega, B. Winstead, & W. Jones (Eds.), *Contemporary research in personality* (pp. 175–204). Chicago, IL: Nelson Hall.

McAdams, D. P. (1992). The five-factor model in personality: A critical appraisal. *Journal of Personality, 60,* 329–361.

McAdams, D. P. (1999). Personal narratives and the life story. In L. A. Pervin & O. P. John (Eds.), *Handbook of personality: Theory and research* (2nd ed., pp. 478–500). New York: Guilford.

McAdams, D. P., & Constantian, C. A. (1983). Intimacy and affiliation motives in daily living: An experience sampling analysis. *Journal of Personality and Social Psychology, 45,* 851–861.

McAdams, D. P., Jackson, R. J., & Kirshnit, C. (1984). Looking, laughing, and smiling in dyads as a function of intimacy, motivation, and reciprocity. *Journal of Personality, 52,* 261–273.

McAdams, D. P., & Powers, J. (1981). Themes of intimacy in behavior and thought. *Journal of Personality and Social Psychology, 40,* 573–587.

McCardel, J. B., & Murray, E. J. (1974). Nonspecific factors in weekend encounter groups. *Journal of Consulting and Clinical Psychology, 42,* 337–345.

McClanahan, T. M. (1995). Operant learning (R-S) principles applied to nail-biting. *Psychological Report, 77,* 507–514.

McClelland, D. C. (1951). *Personality.* New York: Holt, Rinehart and Winston.

McClelland, D. C. (1961). *The achieving society.* New York: Van Nostrand.

McClelland, D. C. (1985). How motives, skills and values determine what people do. *American Psychologist, 40,* 812–825.

McClelland, D. C. (1992). Motivational configurations. In C. P. Smith, J. W. Atkinson, D. C. McClelland, & J. Veroff (Eds.), *Motivation and personality: Handbook of thematic content analysis* (pp. 87–99). New York: Cambridge University Press.

McClelland, D. C., Atkinson, J. W., Clark, R. A., & Lowell, E. L. (1953). *The achievement motive.* New York: Appleton.

McConnell, A. R., & Leibold, J. M. (2001). Relations among the Implicit Association Test, discriminatory behavior, and explicit measures of racial attitudes. *Journal of Experimental Social Psychology, 37,* 435–442.

McCrae, R. R., & Costa, P. T. (1989). The structure of personality traits: Wiggins' circumplex and the five-factor model. *Journal of Personality and Social Psychology, 56,* 586–595.

McCrae, R. R., & Costa, P. T., Jr. (1985). Updating Norman's "adequacy taxonomy": Intelligence and personality in dimensions in natural language and in questionnaires. *Journal of Personality and Social Psychology, 49,* 710–721.

McCrae, R. R., & Costa, P. T., Jr. (1987). Validation of the Five-Factor model of personality across instruments and observers. *Journal of Personality and Social Psychology, 52,* 81–90.

McCrae, R. R., & Costa, P. T., Jr. (1990). *Personality in adulthood.* New York: Guilford.

McCrae, R. R., & Costa, P. T., Jr. (1996). Toward a new generation of personality theories: Theoretical contexts for the five-factor model. In J. S. Wiggins (Ed.), *The five-factor model of personality: Theoretical perspectives* (pp. 51–87). New York: Guilford.

McCrae, R. R, & Costa, P. T. (1997). Conceptions and correlates of openness and to experience. In R. Hogan, J. Johnson, & S. Briggs (Eds.), *Handbook of personality psychology* (pp. 825–847). San Diego, CA: Academic Press.

McCrae, R. R, & Costa, P. T., Jr. (1999). A Five-Factor theory of personality. In L. A. Pervin & O. P. John (Eds.), *Handbook of personality: Theory and research* (2nd ed., pp. 139–153). New York: Guilford.

McCrae, R. R, Costa, P. T. Jr, Del Pilar, G. H, Rolland, J., & Parker, W. D. (1998). Cross-cultural assessment of the five-factor model: The Revised NEO Personality Inventory. *Journal of Cross-Cultural Psychology, 29,* 171–188.

McFarland, C., & Buehler, R. (1997). Negative affective states and the motivated retrieval of positive life events: The role of affect acknowledgment. *Journal of Personality and Social Psychology, 73,* 200–214.

McGinnies, E. (1949). Emotionality and perceptual defense. *Psychological Review, 56,* 244–251.

McGuire, S., Neiderheiser, J. M., Reiss, D., Hetherington, E. M., & Plomin, R. (1994). Genetic and environmental influences on perceptions of self-worth and competence in adolescence: A study of twins, full siblings, and step siblings. *Child Development, 65,* 785–799.

Meehl, P. E. (1990). Why summaries of research on psychological theories are often uninterpretable. *Psychological Reports, 66,* 195–244.

Meehl, P. E. (1997). Credentialed persons, credentialed knowledge. *Clinical Psychology-Science and Practice, 4,* 91–98.

Meichenbaum, D. (1993). Changing conceptions of cognitive behavior modification: Retrospect and prospect. *Journal of Consulting and Clinical Psychology, 61,* 202–204.

Meichenbaum, D. H. (1995). Cognitive-behavioral therapy in historical perspective. In B. Bongar & L. E. Beutler (Eds.), *Comprehensive textbook of psychotherapy* (pp. 140–158). New York: Oxford University Press.

Meichenbaum, D. H., & Smart, I. (1971). Use of direct expectancy to modify academic performance and attitudes of college students. *Journal of Counseling Psychology, 18,* 531–535.

Meltzer, H. (1930). The present status of experimental studies of the relation of feeling to memory. *Psychological Review, 37,* 124–139.

Mendolia, M., Moore, J., & Tesser, A. (1996). Dispositional and ituational determinants of repression. *Journal of Personality and Social Psychology, 70,* 856–867.

Mendoza-Denton, R., Ayduk, O., Mischel, W., Shoda, Y., & Testa, A. (2001). Person × Situation interactionism in self-encoding (*I am . . . when . . .*): Implications for affect regulation and social information processing. *Journal of Personality and Social Psychology, 80,* 533–544.

Mendoza-Denton, R., Downey, G., Purdie, V. J., Davis, A., & Pietrzak, J. (2002). Sensitivity to status-based rejection: Implications for African-American students' college experience. *Journal of Personality and Social Psychology, 83,* 896–918.

Mendoza-Denton, R., Shoda, Y., Ayduk, O., & Mischel, W. (1999). Applying CAPS theory to cultural differences in social behavior. In W. J. Lonner, D. L. Dinnel, D. K. Forgays, & S. A. Hayes (Eds.), *Merging past, present, and future: Selected papers from the 14th international congress of the international association for cross-cultural psychology* (pp. 205–217). Lisse, The Netherlands: Swets & Zeitlinger.

Merluzzi, T. V., Glass, C. R., & Genest, M. (Eds.). (1981). *Cognitive assessment.* New York: Guilford.

Metcalfe, J., & Jacobs, W. J. (1998). Emotional memory: The effects of stress on "cool" and "hot" memory systems. In D. L. Medin (Ed.), *The psychology of learning and motivation: Advances in research and theory* (Vol. 38, pp. 187–222). San Diego, CA: Academic Press.

Metcalfe, J., & Mischel, W. (1999). A hot/cool-system analysis of delay of gratification: Dynamics of willpower. *Psychological Review, 106,* 3–19.

Milgram, N. (1993). War-related trauma and victimization: Principles of traumatic stress prevention in Israel. In J. P. Wilson & B. Raphael (Eds.), *International handbook of traumatic stress syndromes. The Plenum series on stress and coping* (pp. 811–820). New York: Plenum Press.

Milgram, S. (1974). *Obedience to authority.* New York: Harper & Row.

Miller, N. E. (1948). Theory and experiment relating psychoanalytic displacement to stimulus response generalization. *Journal of Abnormal and Social Psychology, 43,* 155–178.

Miller, N. E. (1959). Liberalization of basic S-R concepts: Extensions to conflict behavior, motivation, and social learning. In S. Koch (Ed.), *Psychology: A study of a science* (Vol. 2, pp. 196–292). New York: McGraw-Hill.

Miller, N. E. (1963). Some reflections on the law of effect produce a new alternative to drive reduction. In M. R. Jones (Ed.), *Nebraska Symposium on Motivation* (Vol. 11, pp. 65–112). Lincoln: University of Nebraska Press.

Miller, N. E., & Dollard, J. (1941). *Social learning and imitation.* New Haven: Yale University Press.

Miller, S. M. (1979). Coping with impending stress: Physiological and cognitive correlates of choice. *Psychophysiology, 16,* 572–581.

Miller, S. M. (1981). Predictability and human stress: Towards a clarification of evidence and theory. In L. Berkowitz (Ed.), *Advances in experimental social psychology* (Vol. 14, pp. 203–256). New York: Academic Press.

Miller, S. M. (1987). Monitoring and blunting: Validation of a questionnaire to assess styles of information seeking under threat. *Journal of Personality and Social Psychology, 52,* 345–353.

Miller, S. M. (1992). Individual differences in the coping process: What to know and when to know it. In B. N. Carpenter (Ed.), *Personal coping: Theory, research, application* (pp. 77–91). Westport, CT: Praeger.

Miller, S. M. (1996). Monitoring and blunting of threatening information: Cognitive interference and facilitation in the coping process. In I. G. Sarason, G. R. Pierce, & B. R. Sarason (Eds.), *Cognitive interference: Theories, methods, and findings. The LEA series in personality and clinical psychology* (pp. 175–190). Mahwah, NJ: Erlbaum.

Miller, S. M., & Green, M. L. (1985). Coping with threat and frustration: Origins, nature, and development. In M. Lewis & C. Soarni (Eds.), *Socialization of Emotions* (Vol. 5). New York: Plenum Press.

Miller, S. M., & Mangan, C. E. (1983). The interacting effects of information and coping style in adapting to gynecologic stress: Should the doctor tell all? *Journal of Personality and Social Psychology, 45,* 223–236.

Miller, S. M., Shoda, Y., & Hurley, K. (1996). Applying cognitive-social theory to health-protective behavior: Breast self-examination in cancer screening. *Psychological Bulletin, 119,* 70–94.

Miller, T. Q., Smith, T. W., Turner, C. W., Guijarro, M. L. & Hallett, A. J. (1996). A meta-analytic review of research on hostility and physical health. *Psychological Bulletin, 119,* 322–348.

Minuchin, S., Lee, W., & Simon, G. M. (1996). *Mastering family therapy: Journeys of growth and transformation.* New York: Wiley.

Mischel, H. N., & Mischel, W. (1973). *Readings in personality.* New York: Holt, Rinehart & Winston.

Mischel, T. (1964). Personal constructs, rules, and the logic of clinical activity. *Psychological Review, 71,* 180–192.

Mischel, W. (1965). Predicting the success of Peace Corps Volunteers in Nigeria. *Journal of Personality and Social Psychology, 1,* 510.

Mischel, W. (1968). *Personality and assessment.* New York: Wiley.

Mischel, W. (1969). Continuity and change in personality. *American Psychologist, 24,* 1012–1018.

Mischel, W. (1973). Toward a cognitive social learning reconceptualization of personality. *Psychological Review, 80,* 252–283.

Mischel, W. (1974). Processes in delay of gratification. In L. Berkowitz (Ed.), *Advances in experimental social psychology* (Vol. 7). New York: Academic Press.

Mischel, W. (1980). Personality and cognition: Something borrowed, something new? In N. Cantor & J. Kihlstrom (Eds.), *Personality, cognition, and social interaction.* Hillsdale, NJ: Erlbaum.

Mischel, W. (1981). Metacognition and the rules of delay. In J. H. Flavell & L. Ross (Eds.), *Social cognitive development: Frontiers and possible futures.* New York: Cambridge University Press.

Mischel, W. (1984). Convergences and challenges in the search for consistency. *American Psychologist, 39,* 351–364.

Mischel, W. (1990). Personality dispositions revisited and revised: A view after three decades. In L. A. Pervin (Ed.), *Handbook of personality: Theory and research* (pp. 111–134). New York: Guilford Press.

Mischel, W. (1993). *Introduction to personality* (Fifth Edition). Fort Worth, TX: Harcourt, Brace, Jovanovich.

Mischel, W., & Ayduk, O. (2002). Self-regulation in a cognitive-affective personality system: Attentional control in the service of the self. *Self and Identity, 1,* 113–120.

Mischel, W., Cantor, N., & Feldman, S. (1996). Principles of self-regulation: The nature of willpower and self-control. In E. T. Higgins & A. W. Kruglanski (Eds.), *Social psychology: Handbook of basic principles* (pp. 329–360). New York: Guilford.

Mischel, W., & Ebbesen, E. B. (1970). Attention in delay of gratification. *Journal of Personality and Social Psychology, 16,* 239–337.

Mischel, W., Ebbesen, E. B., & Zeiss, A. R. (1972). Cognitive and attentional mechanisms in delay of gratification. *Journal of Personality and Social Psychology, 21,* 204–218.

Mischel, W., Ebbesen, E. B., & Zeiss, A. R. (1973). Selective attention to the self: Situational and dispositional determinants. *Journal of Personality and Social Psychology, 27,* 129–142.

Mischel, W., Ebbesen, E. B., & Zeiss, A. R. (1976). Determinants of selective memory about the self. *Journal of Consulting and Clinical Psychology, 44,* 92–103.

Mischel, W., & Morf, C.C. (2003). The self as a psycho-social dynamic processing system: A meta-perspective on a century of the self in psychology. In M. Leary & J. Tangney (Eds.), *Handbook of self and identity* (pp. 15–43). New York: Guilford.

Mischel, W., & Peake, P. K. (1982). In search of consistency: Measure for measure. In M. P. Zanna, E. T. Higgins, & C. P. Herman (Eds.), *Consistency in social behavior: The Ontario symposium* (Vol. 2, pp. 187–207). Hillsdale, NJ: Erlbaum.

Mischel, W., & Shoda, Y. (1995). A cognitive-affective system theory of personality: Reconceptualizing situations, dispositions, dynamics, and invariance in personality structure. *Psychological Review, 102,* 246–268.

Mischel, W., & Shoda, Y. (1998). Reconciling processing dynamics and personality dispositions. *Annual Review of Psychology, 49,* 229–258.

Mischel, W., & Shoda, Y. (1999). Integrating dispositions and processing dynamics within a unified theory of personality: The cognitive affective personality system (CAPS). In L. Pervin & O. John (Eds.), *Handbook of personality: Theory and research* (2nd ed., pp. 197– 218). New York: Guilford.

Mischel, W., Shoda, Y., & Peake, P. K. (1988). The nature of adolescent competencies predicted by preschool delay of gratification. *Journal of Personality and Social Psychology, 54,* 687–696.

Mischel, W., Shoda, Y., & Rodriguez, M. L. (1989). Delay of gratification in children. *Science, 244,* 933–938.

Mischel, W., & Staub, E. (1965). Effects of expectancy on working and waiting for larger rewards. *Journal of Personality and Social Psychology, 2,* 625–633.

Molina, M. A. N. (1996). Archetypes and spirits: A Jungian analysis of Puerto Rican Espiritismo. *Journal of Analytical Psychology, 41,* 227–244.

Monro, R. (1955). *Schools of psychoanalytic thought.* New York: Holt, Rinehart and Winston.

Morf, C. C., (2002). Personality at the hub: Extending the conception of personality psychology. *Journal of Research in Personality, 36,* 649–660.

Morf, C. C., Ansara, D., & Shia, T. (2001). *The effects of audience characteristics on narcissistic self-presentation.* Manuscript in preparation, University of Toronto.

Morf, C. C., & Rhodewalt, F. (2001a). Expanding the dynamic self-regulatory processing model of narcissism; Research directions for the future. *Psychological Inquiry, 12,* 243–251.

Morf, C. C., & Rhodewalt, F. (2001b). Unraveling the paradoxes of narcissism: A dynamic self-regulatory processing model. *Psychological Inquiry, 12,* 177–196.

Morf, C. C., Weir, C. R., & Davidov, M. (2000). Narcissism and intrinsic motivation: The role of goal congruence. *Journal of Experimental Social Psychology, 36,* 424–438.

Morse, W. H., & Kelleher, R. T. (1966). Schedules using noxious stimuli I. Multiple fixed-ratio and fixed-interval termination of schedule complexes. *Journal of the Experimental Analysis of Behavior, 9,* 267–290.

Moskowitz, D. S., Suh, E. J., & Desaulniers, J. (1994). Situational influences on gender differences in agency and communion. *Journal of Personality and Social Psychology, 66,* 753–761.

Mulaik, S. A. (1964). Are personality factors raters' conceptual factors? *Journal of Consulting Psychology, 28,* 506–511.

Murphy, S. T., & Zajonc, R. B. (1993). Affect, cognition, & awareness: Affective priming with optimal and suboptimal stimulus exposures. *Journal of Personality and Social Psychology, 64,* 723–739.

Murray, H. A., Barrett, W. G., & Homburger, E. (1938). *Explorations in personality.* New York: Oxford University Press.

Murray, J. (1973). Television and violence: Implications of the surgeon general's research program. *American Psychologist, 28,* 472–478.

Mussen, P. H., & Naylor, H. K. (1954). The relationship between overt and fantasy aggression. *Journal of Abnormal and Social Psychology, 49,* 235–240.

Neisser, U. (1967). *Cognitive psychology.* New York: Appleton.

Nelson, R. J., Demas, G. E., Huang, P. L., Fishman, M. C., Dawson, V. L., Dawson, T. M., & Snyder, S. H. (1995). Behavioural abnormalities in male mice lacking neuronal nitric synthase. *Nature, 378*(6555), 383–386.

Nemeroff, C. J., & Karoly, P. (1991). Operant methods. In F. H. Kanfer & A. P. Goldstein (Eds.), *Helping people change: A textbook of methods* (4th ed.). Pergamon general psychology series, Vol. 52. New York: Pergamon Press.

Nesse, R. M. (2001). *Evolution and the capacity for commitment.* New York: Russell Sage.

Newcomb, T. M. (1929). *Consistency of certain extrovert-introvert behavior patterns in 51 problem boys.* New York: Columbia University, Teachers College, Bureau of Publications.

Niedenthal, P. M. (1990). Implicit perception of affective information. *Journal of Experimental Social Psychology, 25,* 505–527.

Nilson, D. C., Nilson, L. B., Olson, R. S., & McAllister, B. H. (1981). *The planning environment report for the Southern California Earthquake Safety Advisory Board.* Redlands, CA: The Social Research Advisory and Policy Research Center.

Nisbett, R. (1990). Evolutionary psychology, biology, and cultural evolution. *Motivation and Emotion, 14,* 255–263.

Nisbett, R. (1997, May). Cultures of honor: Economics, history, and the tradition of violence. Address given at the Ninth Annual Convention of the American Psychological Society, Washington, DC.

Nisbett, R. E., Peng, K., Choi, I., & Norenzayan, A. (2001). Culture and systems of thought: Holistic versus analytic cognition. *Psychological Review, 108,* 291–310.

Nisbett, R. E., & Ross, L. D. (1980). *Human inference: Strategies and shortcomings of social judgment.* Century Psychology Series. Englewood Cliffs, NJ: Prentice-Hall.

Nolen-Hoeksema, S. (1997, May). Emotion regulation and depression. Closing plenary session at the Ninth Annual American Psychological Society Convention, Washington, DC.

Nolen-Hoeksema, S. (2000). The role of rumination in depressive disorders and mixed anxiety/depressive symptoms. *Journal of Abnormal Psychology, 109,* 504–511.

Nolen-Hoeksema, S., Parker, L. E, & Larson, J. (1994). Ruminative coping with depressed mood following loss. *Journal of Personality and Social Psychology, 67,* 92–104.

Norem, J. K., & Cantor, N. (1986). Anticipatory and post hoc cushioning strategies: Optimism and defensive pessimism in "risky" situations. *Cognitive Therapy and Research, 10,* 347–362.

Norman, W. T. (1961). Development of self-report tests to measure personality factors identified from peer nominations. *USAF ASK Technical Note,* No. 61–44.

Norman, W. T. (1963). Toward an adequate taxonomy of personality attributes: Replicated factor structure in peer nomination personality ratings. *Journal of Abnormal and Social Psychology, 66,* 574–583.

Ochsner, K. N., & Lieberman, M. D. (2001). The emergence of social cognitive neuroscience. *American Psychologist, 56,* 717–734.

O'Connell, D. F., & Alexander, C. N. (1994). Recovery from addictions using transcendental meditation and Maharishi Ayur-Veda. In D. F. O'Connell & C. N. Alexander (Eds.), *Self-recovery: Treating addictions using transcendental meditation and Maharishi Ayur-Veda* (pp. 1–12). New York: Haworth Press.

O'Connor, T. G., Hetherington, E. M., Reiss, D., & Plomin, R. (1995). A twin–sibling study of observed parent-adolescent interactions. *Child Development, 66,* 812–829.

O'Donohue, W., Henderson, D., Hayes, S., Fisher, J., & Hayes, L. (Eds.) (2001). *The history of the behavioral therapies: Founders personal histories.* Reno, NV: Context Press.

Office of Strategic Services Administration. (1948). *Assessment of men.* New York: Holt, Rinehart and Winston.

Ofshe, R. J. (1992). Inadvertent hypnosis during interrogation: False confession due to dissociative state, misidentified multiple personality and the satanic cult hypothesis. *International Journal of Clinical and Experimental Hypnosis, 40,* 125–156.

Ofshe, R. J., & Watters, E. (1993). Making monsters. *Society, 1,* 4–16.

O' Leary, V. (1997, August). Smithsonian seminar on health and well-being sponsored by Society for the Psychological Study of Social Issues and American Psychological Society conducted at the Ninth Annual Conference of the American Psychological Society, Washington, DC.

Opler, M. K. (1967). Cultural induction of stress. In M. H. Appley & R. Trumbull (Eds.), *Psychological stress* (pp. 209–241). New York: Appleton.

Ornstein, R. E. (1972). *The psychology of consciousness.* San Francisco: Freeman.

Ornstein, R. E., & Naranjo, C. (1971). *On the psychology of meditation.* New York: Viking.

Osgood, C. E., Suci, G. J., & Tannenbaum, P. H. (1957). *The measurement of meaning.* Urbana, IL: The University of Illinois Press.

Overall, J. (1964). Note on the scientific status of factors. *Psychological Bulletin, 61,* 270–276.

Pagano, R. R., Rose, R. M., Stivers, R. M., & Warrenburg, S. (1976). Sleep during transcendental meditation. *Science, 191,* 308–309.

Parker, J. D. A., & Endler, N. S. (1996). Coping and defense: A historical overview. In M. Zeidner & N. S. Endler (Eds.), *Handbook of coping: Theory, research, applications* (pp. 3–23). New York: Wiley.

Patterson, G. R. (1976). The aggressive child: Victim and architect of a coercive system. In L. A. Hamerlynck, L. C. Handy, & E. J. Mash (Eds.), *Behavior modification and families, Vol. 1. Theory and research.* New York: Brunner/Mazel.

Patterson, G. R. (Ed.). (1990). *Depression and aggression in family interaction.* Hillsdale, NJ: Lawrence Erlbaum Associates.

Patterson, G. R., & Fisher, P. A. (2002). Recent developments in our understanding of parenting: Bidirectional effects, causal models, and the search for parsimony. In M. H. Bornstein (Ed.), *Handbook of parenting: Vol. 5: Practical issues in parenting* (2nd ed., pp. 59–88). Mahwah, NJ: Lawrence Erlbaum Associates.

Paul, G. L. (1966). *Insight vs. desensitization in psychotherapy.* Stanford, CA: Stanford University Press.

Paulhus, D. L., Fridhandler, B., & Hayes, S. (1997). Psychological defense: Contemporary theory and research. In R. Hogan, J. A. Johnson, & S. R. Briggs (Eds.), *Handbook of personality psychology* (pp. 543–579). San Diego, CA: Academic Press.

Peake, P., Hebl, M., & Mischel, W. (2002). Strategic attention deployment in waiting and working situations. *Developmental Psychology, 38,* 313–326.

Pedersen, N. L., Plomin, R., McClearn, G. E., & Friberg, L. (1988). Neuroticism, extraversion, and related traits in adult twins reared apart and reared together. *Journal of Personality and Social Psychology, 55,* 950–957.

Pederson, F. A. (1958). Consistency data on the role construct repertory test. Unpublished manuscript, Ohio State University, Columbus.

Pennebaker, J. W. (1993). Social mechanisms of constraint. In D. M. Wegener & J. W. Pennebaker (Eds.), *Handbook of mental control* (pp. 200–219). Englewoood Cliff, NJ: Prentice-Hall.

Pennebaker, J. W. (1997). Writing about emotional experiences as a therapeutic process. *Psychological Science, 8,* 162–166.

Pennebaker, J. W., Colder, M., & Sharp, L. K. (1990). Accelerating the coping process. *Journal of Personality and Social Psychology, 58,* 528–537.

Pennebaker, J. W., & Graybeal, A. (2001). Patterns of natural language use: Disclosure, personality, and social integration. *Current Directions in Psychological Science, 10,* 90–93.

Pennebaker, J. W., Kiecolt-Glaser, J. K., & Glaser, R. (1988). Disclosure of traumas and immune function: Health implications for psychotherapy. *Journal of Consulting and Clinical Psychology, 56,* 239–245.

Perls, F. S. (1969). *Gestalt therapy verbatim.* Lafayette, CA: Real People Press.

Perry, J. C., & Cooper, S. H. (1989). An empirical study of defense mechanisms, I. Clinical interviews and life vignette ratings. *Archives of General Psychiatry, 46,* 444–452.

Pervin, L. A. (1994). A critical analysis of trait theory. *Psychological Inquiry, 5,* 103–113.

Pervin, L. A. (1996). *The science of personality.* New York: Wiley.

Pervin, L. A. (2003). *The science of personality* (2nd ed.). New York: Oxford University Press.

Pervin, L. A., & John, O. P. (Eds.) (1999). *Handbook of personality theory and research* (2nd ed.). New York: Guilford.

Peterson, C., & Seligman, M. E. P. (1987). Explanatory style and illness. *Journal of Personality, 55,* 237–265.

Peterson, C., Seligman, M. E. P., & Vaillant, G. E. (1988). Pessimistic explanatory style is a risk factor of physical illness: A thirty-five-year longitudinal study. *Journal of Personality and Social Psychology, 55,* 23–27.

Peterson, D. R. (1968). *The clinical study of social behavior.* New York: Appleton.

Petrie, K. J., Booth, R. J., Pennebaker, J. W., Davison, K. P., & Thomas, M. G. (1995). Disclosure of trauma and immune response to a hepatitis vaccination program. *Journal of Consulting and Clinical Psychology, 63,* 787–792.

Phares, E. J. (1976). *Locus of control in personality.* Morristown, NJ: General Learning Press.

Phillips, K., & Mathews, A. P., Jr. (1995). Quantitative genetic analysis of injury liability in infants and toddlers. *American Journal of Medical Genetics (Neuropsychiatric Genetics), 60,* 64–71.

Pike, A., Reiss, D., Hetherington E. M., & Plomin, R. (1996). Using MZ differences in the search for nonshared environmental effects. *Journal of Child Psychology and Psychiatry, 37,* 695–704.

Pinker, S. (1997). *How the mind works.* New York: W. W. Norton & Company.

Plaud, J. J., & Gaither, G. A. (1996). Human behavioral momentum: Implications for applied behavior analysis and therapy. *Journal of Behavior Therapy and Experimental Psychiatry, 27,* 139–148.

Plomin, R. (1981). Ethnological behavioral genetics and development. In K. Immelmann, G. W. Barlow, L. Petrinovich, & M. Main (Eds.), *Behavioral development: The Bielefeld interdisciplinary project.* Cambridge, England: Cambridge University Press.

Plomin, R. (1990). The role of inheritance in behavior. *Science, 248,* 183–188.

Plomin, R. (1994). The Emanuel Miller Memorial Lecture 1993: Genetic research and identification of environmental influences. *Journal of Child Psychology and Psychiatry, 35,* 817–834.

Plomin, R., & Caspi, A. (1999). Behavioral genetics and personality. In L. A. Pervin & O. P. John (Eds.), *Handbook of personality theory and research* (2nd ed., pp. 251–276). New York: Guilford.

Plomin, R., Chipuer, H. M., & Loehlin, J. C. (1990). Behavioral genetics and personality. In L. A. Pervin (Ed.), *Handbook of personality: Theory and research* (pp. 225–243). New York: Guilford.

Plomin, R., Chipuer, H. M., & Neiderhiser, J. M. (1994). Behavioral genetic evidence for the importance of nonshared environment. In E. M. Hetherington, D. Reiss, & R. Plomin (Eds.), *Separate social worlds of siblings: Impact of nonshared environment on development* (pp. 1–31). Hillsdale, NJ: Erlbaum.

Plomin, R., DeFries, J. C., McClearn, G. E., & Rutter, M. (1997). *Behavioral Genetics* (3rd Ed.). New York: W. H. Freeman and Company.

Plomin, R., Manke, B., & Pike, A. (1996). Siblings, behavioral genetics, and competence. In G. H. Brody (Ed.), *Sibling relationships: Their causes and consequences* (pp. 75–104). Norwood, NJ: Ablex Publishing Corp.

Plomin, R., McClearn, G. E., Pedersen, N. L., Nesselrode, J. R., & Bergeman, C. S. (1988). Genetic influence on childhood family environment perceived retrospectively from the last half of the life span. *Developmental Psychology, 24*, 738–745.

Plomin, R., Owen, M. J., & McGuffin, P. (1994). The genetic basis of complex human behaviors. *Science, 264*, 1733–1739.

Plomin, R., & Rende, R. (1991). Human behavioral genetics. *Annual Review of Psychology, 42*, 161–190.

Plomin, R., & Saudino, K. J. (1994). Quantitative genetics and molecular genetics. In J. E. Bates & T. D. Watts (Eds.), *Temperament: Individual differences at the interface of biology and behavior* (pp. 143–171). Washington, DC: American Psychological Association.

Polster, E., & Polster, M. (1993). Frederick Perls: Legacy and invitation. *Gestalt Journal, 16*, 23–25.

Posner, M. I., & Rothbart, M. K. (1991). Attentional mechanisms and conscious experience. In M. Rugg & A. D. Milner (Eds.), *The neuropsychology of consciousness* (pp. 91–112). San Diego, CA: Academic Press.

Posner, M. I., & Rothbart, M. K. (1998). Attention, self-regulation, and consciousness. *Philosophical Transactions of the Royal Society of London B, 353*, 1915–1927.

Potter, W. Z., Rudorfer, M. V., & Manji, H. K. (1991). The pharmacologic treatment of depression. *New England Journal of Medicine, 325*, 633–642.

Powell, G. E. (1973). Negative and positive mental practice in motor skill acquisition. *Perceptual and Motor Skills, 37*, 312–313.

Prentice, D. A., & Miller, C. T. (1993). Pluralistic ignorance and alcohol use on campus: Some consequences of misperceiving the social norm. *Journal of Personality and Social Psychology, 64*, 243–256.

Purdie, V., & Downey, G. (2000). Rejection sensitivity and adolescent girls' vulnerability to relationship-centered difficulties. *Child Maltreatment, 5*, 338–349.

Rachman, S. (1967). Systematic desensitization. *Psychological Bulletin, 67*, 93–103.

Rachman, S., & Cuk, M. (1992). Fearful distortions. *Behaviour Research and Therapy, 30*, 583–589.

Rachman, S., & Hodgeson, R. J. (1980). *Obsessions and compulsions.* Englewood Cliffs, NJ: Prentice-Hall.

Rachman, S., & Wilson, G. T. (1980). *The effects of psychological therapy.* Oxford, England: Pergamon Press.

Rachman, S. J. (1996). Trends in cognitive and behavioural therapies. In P. M. Salkovskis (Ed.) *Trends in cognitive and behavioural therapies* (pp. 1–23). New York: Wiley.

Ramsey, E., Patterson, G. R., & Walker, H. M. (1990). Generalization of the antisocial trait from home to school settings. *Journal of Applied Developmental Psychology, 11*, 209–223.

Raush, H. L., Barry, W. A., Hertel, R. K., & Swain, M. A. (1974). *Communication conflict and marriage.* San Francisco: Jossey-Bass.

Raymond, M. S. (1956). Case of fetishism treated by aversion therapy. *British Medical Journal, 2*, 854–857.

Redd, W. H. (1995). Behavioral research in cancer as a model for health psychology. *Health Psychology, 14*, 99–100.

Redd, W. H., Porterfield, A. L., & Anderson, B. L. (1978). *Behavior modification: Behavioral approaches to human problems.* New York: Random House.

Reiss, D., Neiderhiser, J. M., Hetherington, E. M., & Plomin, R. (2000). *The relationship code: Deciphering genetic and social influences on adolescent development.* Cambridge, MA: Harvard University Press.

Rhodewalt F., & Eddings, S. K. (2002). Narcissus reflects: Memory distortion in response to ego-relevant feedback among high- and low-narcissistic men. *Journal of Research in Personality, 36*, 97–116.

Rhodewalt, F., & Morf, C. (1998). On self-aggrandizement and anger: A temporal analysis of narcissism and affective reactions to success and failure. *Journal of Personality and Social Psychology, 74*, 672–685.

Richards, J. M., & Gross, J. J. (2000). Emotion regulation and memory: The cognitive costs of keeping one's cool. *Journal of Personality and Social Psychology, 79*, 410–424.

Riemann, R., Angleitner, A., & Strelau, J. (1997). Genetic and environmental influences on personality: A study of twins reared together using the self- and peer report NEO-FFI scales. *Journal of Personality, 65*, 449–476.

Riger, S. (1992). Epistemological debates, feminist voices: Science, social values, and the study of women. *American Psychologist, 47*, 730–740.

Rimm, D. C., & Masters, J. C. (1974). *Behavior therapy: Techniques and empirical findings.* New York: Academic Press.

Roazen, P. (1974). *Freud and his followers.* New York: Meridian.

Robinson, J. L., Kagan, J., Reznick, J. S., & Corley, R. (1992). The heritability of inhibited and uninhibited behavior: A twin study. *Developmental Psychology, 28*, 1030–1037.

Rogers, C. R. (1942). *Counseling and psychotherapy: Newer concepts in practice.* Boston: Houghton Mifflin.

Rogers, C. R. (1947). Some observations on the organization of personality. *American Psychologist, 2*, 358–368.

Rogers, C. R. (1951). *Client-centered therapy: Its current practice, implications and theory.* Boston: Houghton Mifflin.

Rogers, C. R. (1955). Persons or science? A philosophical question. *American Psychologist, 10,* 267–278.

Rogers, C. R. (1959). A theory of therapy, personality and interpersonal relationships, as developed in the client-centered framework. In S. Koch (Ed.), *Psychology: A study of a science* (Vol. 3, pp. 184–526). New York: McGraw-Hill.

Rogers, C. R. (1963). The actualizing tendency in relation to "motives" and to consciousness. In M. R. Jones (Ed.), *Nebraska symposium on motivation* (pp. 1–24). Lincoln, NE: University of Nebraska Press.

Rogers, C. R. (1970). *Carl Rogers on encounter groups.* New York: Harper & Row.

Rogers, C. R. (1974). In retrospect: Forty-six years. *American Psychologist, 29,* 115–123.

Rogers, C. R., & Dymond, R. F. (Eds.). (1954). *Psychotherapy and personality change, co-ordinated studies in the client-centered approach.* Chicago: University of Chicago Press.

Rogers, T. B. (1977). Self-reference in memory: Recognition of personality items. *Journal of Research in Personality, 11,* 295–305.

Rogers, T. B., Kuiper, N. A., & Kirker, W. S. (1977). Self-reference and the encoding of personal information. *Journal of Personality and Social Psychology, 35,* 677–688.

Romer, D., & Revelle, W. (1984). Personality traits: Fact or fiction? A critique of the Shweder and D'Andrade systematic distortion hypothesis. *Journal of Personality and Social Psychology, 47,* 1028–1042.

Rorer, L. G. (1990). Personality assessment: A conceptual survey. In L.A. Pervin (Ed.), *Handbook of personality: Theory and research* (pp. 693–720). New York: Guilford.

Rosch, E. (1975). Cognitive reference points. *Cognitive Psychology, 1,* 532–547.

Rosch, E., Mervis, C., Gray, W., Johnson, D., & Boyce-Braem, P. (1976). Basic objects in natural categories. *Cognitive Psychology, 8,* 382–439.

Rosenhan, D. L. (1973). On being sane in insane places. *Science, 179,* 250–258.

Rosenthal, R., & Rubin, D. (1978). Interpersonal expectancy effects: The first 345 studies. *Behavioral and Brain Sciences, 3,* 377–415.

Rosenzweig, S., & Mason, G. (1934). An experimental study of memory in relation to the theory of repression. *British Journal of Psychology, 24,* 247–265.

Ross, L. & Nisbett, R. E. (1991). *The person and the situation: Perspectives of social psychology.* New York: McGraw-Hill.

Ross, L. D. (1977). The intuitive psychologist and his shortcomings: Distortions in the attribution process. In L. Berkowitz (Ed.), *Advances in experimental social psychology* (Vol. 10). New York: Academic Press.

Rossini, E. D., & Moretti, R. J. (1997). Thematic Apperception Test (TAT) interpretation: Practice recommendations from a survey of clinical psychology doctoral programs accredited by the American Psychological Association. *Professional Psychology Research and Practice, 28,* 393–398.

Roth, S., & Newman, E. (1990). The process of coping with sexual trauma. *Journal of Traumatic Stress, 4,* 279–297.

Rothbart, M. K., Derryberry, D., & Posner, M. I. (1994). A psychobiological approach to the development of temperament. In J. E. Bates & T. D. Wachs (Eds.), *Temperament: Individual differences at the interface of biology and behavior* (pp. 83–116). Washington, D.C.: American Psychological Association.

Rothbart, M. K., Posner, M. I., & Gerardi, G. M. (1997). *Effortful control and the development of temperament.* Symposium presented at the 1997 Biennial Meeting of the Society for Research in Child Development. (Washington, DC: April 4, 1997).

Rotter, J. B. (1954). *Social learning and clinical psychology.* Englewood Cliffs, NJ: Prentice-Hall.

Rotter, J. B. (1966). Generalized expectancies for internal versus external control of reinforcement. *Psychological Monographs, 80,* 1–28.

Rotter, J. B. (1972). Beliefs, social attitudes, and behavior: A social learning analysis. In J. B. Rotter, J. E. Chance, & E. J. Phares (Eds.), *Applications of a social learning theory of personality.* New York: Holt, Rinehart and Winston.

Roussi, P., Miller, S. M., & Shoda, Y. (2000). Discriminative facility in the face of threat: Relationship to psychological distress. *Psychology and Health, 15,* 21–33.

Rowan, J. (1992). What is humanistic psychotherapy? *British Journal of Psychotherapy, 9,* 74–83.

Rowe, D. C. (1981). Environmental and genetic influences on dimensions of perceived parenting: A twin study. *Developmental Psychology, 17,* 203–208.

Rowe, D. C. (1983). A biometrical analysis of perceptions of family environment: A study of twin and singleton sibling kinships. *Child Development, 54,* 416–423.

Rowe, D. C. (1997). Genetics, temperament, and personality. In R. Hogan, J. Johnson, & S. Briggs (Eds.), *Handbook of personality psychology* (pp. 367–386). San Diego, CA: Academic Press.

Royce, J. E. (1973). Does person or self imply dualism? *American Psychologist, 28,* 833–866.

Rubin, I. J. (1967). The reduction of prejudice through laboratory training. *Journal of Applied Behavioral Science, 3,* 29–50.

Runyan, W. M. (1997). Studying lives: Psychobiography and the conceptual structure of personality psychology. In R. Hogan, J. Johnson, & S. Briggs (Eds.), *Handbook of Personality Psychology* (pp. 41–69). San Diego, CA: Academic Press.

Rushton, J. P., Fulker, D. W., Neale, M. C., Nias, D. K. B., & Eysenck, H. J. (1986). Altruism and aggression: The heritability of individual differences. *Journal of Personality and Social Psychology, 50,* 1192–1198.

Rusting, C. L., & Larsen, R. J. (1998). Diurnal patterns of unpleasant mood: Associations with neuroticism, depression, and anxiety. *Journal of Personality, 66,* 85–103.

Rusting, C. L., & Nolen-Hoeksema, S. (1998). Regulating responses to anger: Effects on rumination and distraction on angry mood. *Journal of Personality and Social Psychology, 74,* 790–803.

Rutter, M., Dunn, J., Plomin, R., Simonoff, E., Pickles, A., Maughan, B., Ormel, J., Meyer, J., & Eaves, L. (1997). Integrating nature and nurture: Implications of person-environment correlations and interactions for developmental psychopathology. *Development and Psychopathology, 9,* 335–364.

Sadalla, E. K., Kenrick, D. T., & Vershure, B. (1987). Dominance and heterosexual attraction. *Journal of Personality and Social Psychology, 52,* 730–738.

Saley, E., & Holdstock, L. (1993). Encounter group experiences of black and white South Africans in exile. In D. Brazier (Ed.), *Beyond Carl Rogers* (pp. 201–216). London: Constable.

Sapolsky, R. M. (1996). Why stress is bad for your brain. *Science, 273,* 749–750.

Sarason, I. G. (1966). *Personality: An objective approach.* New York: Wiley.

Sarason, I. G. (1979). Life stress, self-preoccupation, and social supports. Presidential address, Western Psychological Association.

Sartre, J. P. (1956). Existentialism. In W. Kaufman (Ed.), *Existentialism from Dostoyevsky to Sartre* (pp. 222–311). New York: Meridian.

Sartre, J. P. (1965). *Existentialism and humanism* (Mairet, Trans.). London: Methuen.

Saucier, G., & Goldberg, L. R. (1996). Evidence for the Big Five in analyses of familiar English personality adjectives. *European Journal of Personality, 10,* 61–77.

Saudino, K. J., & Eaton, W. O. (1991). Infant temperament and genetics: An objective twin study of motor activity level. *Child Development, 62,* 1167–1174.

Saudino, K. J., & Plomin, R. (1996). Personality and behavioral genetics: Where have we been and where are we going? *Journal of Research in Personality, 30,* 335–347.

Saudino, K. J., Plomin, R., & DeFries, J. C. (1996). Testerrated temperament at 14, 20, and 24 months: Environmental change and genetic continuity. *British Journal of Developmental Psychology, 14,* 129–144.

Saudou, F., Amara, D. A., Dierich, A., LeMur, M., Ramboz, S., Segu, L., Buhot, M. C., & Hen, R. (1994). Enhanced aggressive behavior in mice lacking 5-HT_{1B} receptor. *Science, 265,* 1875–1878.

Schachter, D. (1995). *Searching for memory.* New York: Basic Books.

Schank, R., & Abelson, R. P. (1977). *Scripts, plans, goals, and understanding.* Hillsdale, NJ: Erlbaum.

Scheier, M. F., & Carver, C. S. (1987). Dispositional optimism and physical well-being: The influence of generalized outcome expectancies on health. *Journal of Personality, 55,* 169–210.

Scheier, M. F., & Carver, C. S. (1988). A model of behavioral self-regulation: Translating intention into action. In L. Berkowitz (Ed.), *Advances in experimental social psychology* (Vol. 21, pp. 322–343). San Diego, CA: Academic Press.

Scheier, M. F., & Carver, C. S. (1992). Effects of optimism on psychological and physical well-being: Theoretical overview and empirical update. *Cognitive Therapy and Research 16,* 201–228.

Scheier, M. F., Weintraub, J. K., & Carver, C. S. (1986). Coping with stress: Divergent strategies of optimists and pessimists. *Journal of Personality and Social Psychology, 51,* 1257–1264.

Schmidt, L. A., & Fox, N. A. (1999). Conceptual, biological, and behavioral distinctions among different categories of shy children. In L. A. Schmidt & J. Schulkin (Eds.), *Extreme fear, shyness, and social phobia: Origins, biological mechanisms, and clinical outcomes* (pp. 47–66). New York: Oxford University Press.

Schmidt, L. A., & Fox, N. A. (2002). Individual differences in childhood shyness: Origins, malleability, and developmental course. In D. Cervone & W. Mischel (Eds.), *Advances in personality science* (pp. 83–105). New York: Guilford.

Schneider, D. J. (1973). Implicit personality theory: A review. *Psychological Bulletin, 73,* 294–309.

Schooler, J. W. (1994). Seeking the core: The issues and evidence surrounding recovered accounts of sexual trauma. *Consciousness and Cognition, 3,* 452–469.

Schooler, J. W. (1997). Reflections on a memory discovery. *Child Maltreatment, 2,* 126–133.

Schooler, J. W., Bendiksen, M., & Ambadar, Z. (1997). Taking the middle line: Can we accommodate both fabricated and recovered memories of sexual abuse? In M. Conway (Ed.), *False and recovered memories* (pp. 251–292). Oxford, England: Oxford University Press

Schutz, W. C. (1967). *Joy: Expanding human awareness.* New York: Grove.

Schwarz, N. (1990). Feelings and information: Informational and motivational functions of affective states. In R. M. Sorrentino & E. T. Higgins (Eds.), *Handbook of motivation and cognition: Foundations of social behavior* (Vol. 2, pp. 527–561). New York: Guilford.

Sears, R. R. (1936). Functional abnormalities of memory with special reference to amnesia. *Psychological Bulletin, 33,* 229–274.

Sears, R. R. (1943). *Survey of objectives studies of psychoanalytic concepts* (Bulletin 51). New York: Social Sciences Research Council.

Sears, R. R. (1944). Experimental analysis of psychoanalytic phenomena. In J. McV. Hunt (Ed.), *Personality and the behavior disorders* (pp. 306–332). New York: Ronald Press.

Segal, N. L. (1999). *Entwined lives: Twins and what they tell us about human behavior.* New York: Plume.

Seligman, M. E. P. (1971). Phobias and preparedness. *Behavior Therapy, 2,* 307–320.

Seligman, M. E. P. (1975). *Helplessness — On depression, development, and death.* San Francisco: Freeman.

Seligman, M. E. P. (1978). Comment and integration. *Journal of Abnormal Psychology, 87,* 165–179.

Seligman, M. E. P. (1990). *Learned optimism.* New York: A. A. Knopf.

Seligman, M. E. P., Reivich, K., Jaycox, L., & Gillham, J. (1995). *The optimistic child.* Boston: Houghton Mifflin Co.

Seligman, M. E., & Hager, J. L. (1972). *Biological boundaries of learning.* New York: Appleton-Century-Crofts.

Sethi, A., Mischel, W., Aber, L., Shoda, Y., & Rodriguez, M. (2000). The role of strategic attention deployment of self-regulation: Prediction preschoolers' delay of gratification from mother-toddler interactions. *Developmental Psychology, 36,* 767–777.

Shaver, P. R., Collins, N. L., & Clark, C. L. (1996). Attachment styles and internal working models of self and relationship partners. In G. J. O. Fletcher & J. Fitness (Eds.), *Knowledge structures in close relationships: A social psychological approach* (pp. 25–61). Mahwah, NJ: Erlbaum.

Shaver, P. R., & Mikulincer, M. (2002). Attachment-related psychodynamics. *Attachment and Human Development, 4,* 133–161.

Sheline, Y., Wang, P. W., Gado, M. H., Csernansky, J. G., & Vannier, M. W. (1996). Hippocampal atrophy in recurrent major depression. *Proceedings of the National Academy of Science of the United States of America, 93,* 3908–3913.

Shoda, Y. (1990). Conditional analyses of personality coherence and dispositions. Unpublished doctoral dissertation, Columbia University, New York.

Shoda, Y., LeeTiernan, S., & Mischel, W. (2002). Personality as a dynamical system: Emergence of stability and consistency from intra- and inter-personal interactions. *Personality and Social Psychology Review, 6,* 316–325.

Shoda, Y., & Mischel, W. (1998). Reconciling processing dynamics and personality dispositions. *Annual Review of Psychology, 49,* 229–258.

Shoda, Y., Mischel, W., & Peake, P. K. (1990). Predicting adolescent cognitive and self-regulatory competencies from preschool delay of gratification: Identifying diagnostic conditions. *Developmental Psychology, 26,* 978–986.

Shoda, Y., Mischel, W., & Wright, J. C. (1994). Intra-individual stability in the organization and patterning of behavior: Incorporating psychological situations into the idiographic analysis of personality. *Journal of Personality and Social Psychology, 67,* 674–687.

Shoda, Y., Mischel, W., & Wright, J. C. (1989). Intuitive interactionism in person perception: Effects of situation-behavior relations on dispositional judgements. *Journal of Personality and Social Psychology, 56,* 41–53.

Shoda, Y., Mischel, W., & Wright, J. C. (1993a). The role of situational demands and cognitive competencies in behavior organization and personality coherence. *Journal of Personality and Social Psychology, 56,* 41–53.

Shoda, Y., Mischel, W., & Wright, J. C. (1993b). Links between personality judgments and contextualized behavior patterns: Situation-behavior profiles of personality prototypes. *Social Cognition, 4,* 399–429.

Shweder, R. A. (1975). How relevant is an individual difference theory of personality? *Journal of Personality, 43,* 455–485.

Silverman, L. H. (1976). Psychoanalytic theory: The reports of my death are greatly exaggerated. *American Psychologist, 31,* 621–637.

Simons, A. D., & Thase, M. E. (1992). Biological markers, treatment outcome, and 1-year follow-up in endogenous depression: Electroencephalographic sleep studies and response to cognitive therapy. *Journal of Consulting and Clinical Psychology, 60,* 392–401.

Singer, C. J. (1941). *A short history of science to the nineteenth century.* Oxford: Clarendon Press.

Singer, J. A., & Salovey, P. (1993). *The remembered self: Emotion and memory in personality.* New York: Free Press.

Singer, J. L. (1988). Sampling ongoing consciousness and emotional experience: Implications for health. In M. J. Horowitz (Ed.). *Psychodynamics and cognition* (pp. 297–346). Chicago, IL: University of Chicago Press.

Skinner, B. F. (1953). *Science and human behavior.* New York: Macmillan.

Skinner, B. F. (1955). Freedom and the control of men. *American Scholar, 25,* 47–65.

Skinner, B. F. (1974). *About behaviorism.* New York: Knopf.

Smith, C. A., & Lazarus, R. S. (1990). Emotion and adaptation. In L. A. Pervin (Ed.), *Handbook of personality: Theory and research* (pp. 609–637). New York: Guilford.

Smith, R. E. (1993). *Psychology.* St. Paul, MN: West.

Smith, R. G., Iwata, B. A., Vollmer, R., & Pace, G. M. (1992). On the relationship between self-injurious behavior and self-restraint. *Journal of Applied Behavior Analysis, 25,* 433–445.

Smyth, J. M. (1998). Written emotional expression: Effect sizes, outcome types, and moderating variables. *Journal of Consulting and Clinical Psychology, 66,* 174–184.

Snyder, M., & Cantor, N. (1998). Understanding personality and social behavior: A functionalist strategy. In D. T. Gilbert, S. T. Fiske, & G. Lindzey (Eds.), *The handbook of social psychology* (4th ed., Vol. 1, pp. 635–679). New York: McGraw-Hill.

Snyder, M., & Uranowitz, S. (1978). Reconstructing the past: Some cognitive consequences of person perception. *Journal of Personality and Social Psychology, 36,* 941–950.

Snygg, D., & Combs, A. W. (1949). *Individual behavior.* New York: Harper & Row.

Sorrentino, R. M., & Roney, C. J. (1986). Uncertainty orientation, achievement-related motivation, and task diagnosticity as determinants of task performance. *Social Cognition, 4,* 420–436.

Sorrentino, R. M., & Roney, C. J. R. (2000). *The uncertain mind: Individual differences in facing the unknown.* Philadelphia, PA: Psychology Press/Taylor & Francis.

Spiegel, D. (1981). Vietnam grief work using hypnosis. *The American Journal of Clinical Hypnosis, 24,* 33–40.

Spiegel, D. (1991). Neurophysiological correlates of hypnosis and dissociation. *Journal of Neuropsychiatry, 3,* 440–445.

Spiegel, D. (1994). Cancer and depression. *Verhaltenstherapie, 4,* 81–88.

Spiegel, D., & Cardena, E. (1990). New uses of hypnosis in the treatment of posttraumatic stress disorder. *Journal of Clinical Psychiatry, 51* (10 Suppl.), 39–43.

Spiegel, D., & Cardena, E. (1991). Disintegrated experience: The dissociative disorders revisited. *Journal of Abnormal Psychology, 100,* 366–378.

Spiegel, D., Koopman, C., & Classen, C. (1994). Acute distress disorder and dissociation. *Australian Journal of Clinical and Experimental Hypnosis, 22,* 11–23.

Spiegel, D., Kraemer, H. C., Bloom, J. R., & Gottheil, E. (1989). Effect of psychosocial treatment on survival of patients with metastatic breast cancer. *The Lancet,* 888–891.

Spinelli, E. (1989). *The interpreted world: An introduction to phenomenological psychology.* London: Sage.

Sroufe, L. A. (1977). *Knowing and enjoying your baby.* Englewood Cliffs, NJ: Prentice-Hall.

Sroufe, L. A., & Fleeson, J. (1986). Attachment and the construction of relationships. In W. Harte & Z. Rubin (Eds.), *The nature of relationships.* Hillsdale, NJ: Erlbaum.

Staats, C. K., & Staats, A. W. (1957). Meaning established by classical conditioning. *Journal of Experimental Psychology, 54,* 74–80.

Staub, E., Tursky, B., & Schwartz, G. E. (1971). Self-control and predictability: Their effects on reactions to aversive stimulation. *Journal of Personality and Social Psychology, 18,* 157–162.

Stayton, D. J., & Ainsworth, M. D. S. (1973). Individual differences in infant responses to brief, everyday separations as related to infant and maternal behaviors. *Developmental Psychology, 9,* 226–235.

Stelmack, R. M. (1990). Biological bases of extraversion: Psychophysiological evidence. *Journal of Personality, 58 ,* 293–311.

Stelmack, R. M., & Michaud-Achorn, A. (1985). Extraversion, attention, and habituation of the auditory evoked response. *Journal of Research in Personality, 19,* 416–428.

Steuer, F. B., Applefield, J. M., & Smith, R. (1971). Televised aggression and the interpersonal aggression of preschool children. *Journal of Experimental Child Psychology, 11,* 442–447.

Stigler, J. W., Shweder, R. A., & Herdt, G. (1990). *Cultural psychology: Essays on comparative human development.* New York: Cambridge.

Stone, A. A., Schiffman, S. S., & DeVries, M. (2000). Rethinking our self-report assessment methodologies: An argument for collecting ecologically valid, momentary measurements. In D. Kahneman, E. Diener, & N. Schwartz (Eds.), *Understanding quality of life: Scientific perspectives on enjoyment and suffering.* New York: Russell Sage.

Strauman, T. J., Vookles, J., Berenstein, V., Chaiken, S., & Higgins, E. T. (1991). Self-discrepancies and vulnerability to body dissatisfaction and disordered eating. *Journal of Personality and Social Psychology, 61,* 946–956.

Sulloway, F. J. (1996). *Born to rebel: Birth order, family dynamics, and creative lives.* London: Little, Brown and Company.

Sutton, S. K., (2002). Incentive and threat reactivity: Relations with anterior cortical activity. In D. Cervone & W. Mischel (Eds.), *Advances in Personality Science* (pp. 127–150). New York: Guilford Press.

Sutton, S. K., & Davidson, R. J. (1997). Prefrontal brain asymmetry: A biological substrate of the behavioral approach and inhibition systems. *Psychological Science, 8,* 204–210.

Sutton, S. K., & Davidson, R. J. (2000). Resting anterior brain activity predicts the evaluation of affective stimuli. *Neuropsychologia, 38,* 1723–1733.

Swann, W. B., Jr. (1983). Self-verification: Bringing social reality into harmony with the self. In J. Suls & A. G. Greenwald (Eds.), *Social psychology perspectives* (Vol. 2, pp. 33–66). Hillsdale, NJ: Erlbaum.

Tart, C. (1970). Increases in hypnotizability resulting from a prolonged program for enhancing personal growth. *Journal of Abnormal Psychology, 75,* 260–266.

Taylor, S. E., (1995). *Health psychology* (3rd ed.). New York: McGraw-Hill.

Taylor, S. E, & Armor, D. A. (1996). Positive illusions and coping with adversity. *Journal of Personality, 64,* 873–898.

Taylor, S. E., & Brown, J. D. (1988). Illusion and well-being: A social psychological perspective on mental health. *Psychological Bulletin, 103,* 193–210.

Taylor, S. E., Klein, L. C., Lewis, B. P., Gruenewald, T. L., Gurung, R. A. R., & Updegraff, J. A. (2000). Biobehavioral responses to stress in females: Tend-and-befriend, not fight-or-flight. *Psychological Review, 107,* 411–429.

Taylor, S. E., & Schneider, S. (1989). Coping and the simulation of events. *Social Cognition, 7,* 174–194.

Tellegen, A., Lykken, D., Bouchard, T., Wilcox, K., Segal, N., & Rich, S. (1988). Personality similarity in twins reared apart. *Journal of Personality and Social Psychology, 54,* 1031–1039.

Tennen, H., Suls, J., & Affleck, G. (1991). Personality and daily experience: The promise and the challenge. *Journal of Personality, 59,* 313–338.

Tesser, A. (1993). The importance of heritability in psychological research: The case of attitudes. *Psychological Review, 100,* 129–142.

Tesser, A. (2002). Constructing a niche for the self: A biosocial, PDP approach to understanding lives. *Self and Identity, 1,* 185–190.

Testa, T. J. (1974). Causal relationships and the acquisition of avoidance responses. *Psychological Review, 81,* 491–505.

Thomas, A., & Chess, S. (1977). *Temperament and development.* New York: Brunner/Mazel.

Thoresen, C., & Mahoney, M. J. (1974). *Self-control.* New York: Holt, Rinehart and Winston.

Toner, I. J. (1981). Role involvement and delay maintenance behavior in preschool children. *The Journal of Genetic Psychology, 138,* 245–251.

Triandis, H. C. (1997). Cross-cultural perspectives on personality. In R. Hogan, J. Johnson, & S. Briggs (Eds.), *Handbook of personality* (pp. 440–459). San Diego, CA: Academic Press.

Triandis, H. C., & Suh, E. M. (2002). Cultural influences on personality. *Annual Review of Psychology, 53,* 133–160.

Trilling, L. (1943). *E. M. Forster.* Norfolk, CT: New Directions Books.

Trivers, R. L. (1971). The evolution of reciprocal altruism. *Quarterly Review of Biology, 46,* 35–57.

Trope, Y., & Liberman, N. (2003). Temporal construal. *Psychological Review.*

Truax, C. B., & Mitchell, K. M. (1971). Research on certain therapist interpersonal skills in relation to process and outcome. In A. E. Bergin & S. I. Garfield (Eds.), *Handbook of psychotherapy and behavior change* (pp. 299–344). New York: Wiley.

Tsuang, M. T., Lyons, M. J., Eisen, S. A., True, W. T., Goldberg, J., & Henderson, W. (1992). A twin study of drug exposure and initiation of use. *Behavior Genetics, 22,* 756 (abstract).

Tudor, T. G., & Holmes, D. S. (1973). Differential recall of successes and failures: Its relationship to defensiveness, achievement motivation, and anxiety. *Journal of Research in Personality, 7,* 208–224.

Tupes, C., & Christal, R. E. (1958). Stability of personality trait rating factors obtained under diverse conditions. *USAF WADC Technical Note, No.* 58–61.

Tupes, E. C., & Christal, R. E. (1961). Recurrent personality factors based on trait ratings. *USAF ASD Technical Report, No.* 61–67.

Tversky, A. (1977). Features of similarity. *Psychological Review, 84,* 327–352.

Tversky, A., & Kahneman, D. (1974). Judgment under uncertainty: Heuristics and biases. *Science, 185,* 1124–1131.

Tyler, L. E. (1956). *The psychology of human differences.* New York: Appleton.

Urban, M. S., & Witt, L. A. (1990). Self-serving bias in group member attributions of success and failure. *Journal of Social Psychology, 130,* 417–418.

Uvnas-Moberg, K. (1998). Oxytocin may mediate the benefits of positive social interaction and emotions. *Psychoneuroendocrinology, 23,* 819–835.

Vallacher, R. R., & Wegner, D. M. (1987). Action indentification theory: The representation and control of behavior. *Psychological Review, 94,* 3–15.

Van De Reit, V., Korb, M., & Gorrell, J. (1989). *Gestalt therapy: An introduction* (3rd ed.). New York: Pergamon Press.

Van Der Ploeg, H. M., Defares, P. B., & Spielberger, C. D. (1982). *Zelf-Analyse Vragenlijst (ZAV) [State-trait anger scale (STAS)].* Lisse, The Netherlands: Swets & Zeitlinger.

Van Mechelen, I., & Kiers, H. A. L. (1999). Individual differences in anxiety responses to stressful situations: A three-mode component analysis model. *European Journal of Personality, 13,* 409–428.

Vanaerschot, G. (1993). Empathy as releasing several microprocesses in the client. In D. Brazier (Ed.), *Beyond Carl Rogers* (pp. 47–71). London: Constable.

Vandenberg, S. G. (1971). What do we know today about the inheritance of intelligence and how do we know it? In R. Cancro (Ed.), *Intelligence: Genetic and environmental influences* (pp. 182–218). New York: Grune & Stratton.

Vansteelandt, K., & Van Mechelen, I. (1998). Individual differences in situation-behavior profiles: A triple typology model. *Journal of Personality and Social Psychology, 75*, 751–765.

Vernon, P. E. (1964). *Personality assessment: A critical survey.* New York: Wiley.

Walker, A. M., & Sorrentino, R. M. (2000). Control motivation and uncertainty: Information processing or avoidance in moderate depressives and nondepressives. *Personality and Social Psychology Bulletin, 26*, 436–451.

Walker, G. (1991). *In the midst of winter: Systematic therapy with families, couples, and individuals with AIDS.* New York: Norton.

Waller, N. G., & Shaver, P. R. (1994). The importance of non-genetic influences on romantic love styles: A twin-family study. *Psychological Science, 5*, 268–274.

Walters, R. H., & Parke, R. D. (1967). The influence of punishment and related disciplinary techniques on the social behavior of children: Theory and empirical findings. In B. A. Maher (Ed.), *Progress in experimental personality research* (Vol. 4, pp. 179–228). New York: Academic Press.

Watkins, C. E., Campbell, V. L., Nieberding, R., & Hallmark, R. (1995). Comtemporary practice of psycholgical assessment by clinical psychologists. *Professional Psychology, Research & Practice 26*, 54–60.

Watson, D. (1988). The vicissitudes of mood measurements: Effects of varying descriptors, time frames, and response formats on measures of positive and negative affect. *Journal of Personality and Social Psychology, 55*, 128–141.

Watson, J. B., & Rayner, R. (1920). Conditioned emotional reaction. *Journal of Experimental Psychology, 3*, 1–14.

Watson, R. I. (1959). Historical review of objective personality testing: The search for objectivity. In B. M. Bass & I. A. Berg (Eds.), *Objective approaches to personality assessment* (pp. 1–23). Princeton: Van Nostrand.

Weidner, G., & Matthews, K. A. (1978). Reported physical symptoms elicited by unpredictable events and the type A coronary-prone behavior pattern. *Journal of Personality and Social Psychology, 36*, 1213–1220.

Weigel, R. H., & Newman, S. L. (1976). Increasing attitude-behavior correspondence by broadening the scope of the behavioral measure. *Journal of Personality and Social Psychology, 33*, 793–802.

Weinberger, J. (2002). *Unconscious processes.* New York: Guilford.

Weiner, B. (1974). An attributional interpretation of expectancy value theory. Paper presented at the AAAS Meetings, San Francisco.

Weiner, B. (1990). Attribution in personality psychology. In L. A. Pervin (Ed.), *Handbook of personality: Theory and research* (pp. 465–484). New York: Guilford.

Weiner, B. (1995). *Judgments of responsibility: A foundation for a theory of social conduct.* New York: Guilford.

Weir, M. W. (1965). Children's behavior in a two-choice task as a function of patterned reinforcement following forced-choice trials. *Journal of Experimental Child Psychology, 2*, 85–91.

West, S. G., & Finch, J. F. (1997). Personality measurement: Reliability and validity issues. In R. Hogan, J. Johnson, & S. Briggs (Eds.), *Handbook of personality psychology.* (pp. 143–165). San Diego, CA: Academic Press.

Westen, D. (1990). Psychoanalytic approaches to personality. In L. A. Pervin (Ed.), *Handbook of personality: Theory and research* (pp. 21–65). New York: Guilford.

Westen, D., & Gabbard, G. O. (1999). Psychoanalytic approaches to personality. In L. A. Pervin & O. P. John (Eds.), *Handbook of personality: Theory and research* (2nd ed., pp. 57–101). New York: Guilford.

Wheeler, G. (1991). Gestalt reconsidered: A new approach to contact and resistance. New York: Gardner Press.

Wheeler, R. E., Davidson, R. J., & Tomarken, A. J. (1993). Frontal brain asymmetry and emotional reactivity: A biological substrate of affective style. *Psychophysiology, 30*, 82–89.

White, B. L. (1967). An experimental approach to the effects of experience on early human behavior. In J. P. Hill (Ed.), *Minnesota symposia on child psychology* (Vol. 1, pp. 201–226). Minneapolis: University of Minnesota Press.

White, B. L. & Held, R. (1966). Plasticity of sensorimotor development in the human infant. In J. F. Rosenblith & W. Allinsmith (Eds.), *The causes of behavior II* (pp. 60–70). Boston: Allyn & Bacon.

White, M., & Epston, D. (1990). *Narrative means to therapeutic ends.* New York: Norton & Co.

White, R. W. (1952). *Lives in progress.* New York: Dryden.

White, R. W. (1959). Motivation reconsidered: The concept of competence. *Psychological Review, 66*, 297–333.

White, R. W. (1964). *The abnormal personality.* New York: Ronald.

White, R. W. (1972). *The enterprise of living.* New York: Holt.

Wiedenfield, S. A., Bandura, A., Levine, S., O'Leary, A., Brown, S., & Raska, K. (1990). Impact of perceived self-efficacy in coping with stressors on components of the immune system. *Journal of Personality and Social Psychology, 59*, 1082–1094.

Wiggins, J. S. (1979). A psychological taxonomy of trait-descriptive terms: The interpersonal domain. *Journal of Personality and Social Psychology, 37*, 395–412.

Wiggins, J. S. (1980). Circumplex models of interpersonal behavior in personality and social psychology. In L. Wheeler (Ed.), *Review of Personality and Social Psychology* (pp. 265–294).

Wiggins, J. S. (1997). In defense of traits. In R. Hogan, J. Johnson, & S. Briggs (Eds.), *Handbook of Personality Psychology* (pp. 95–115). San Diego, CA: Academic Press.

Wiggins, J. S., Phillips, N., & Trapnell, P. (1989). Circular reasoning about interpersonal behavior: Evidence concerning some untested assumptions underlying diagnostic classification. *Journal of Personality and Social Psychology, 56*, 296–305.

Wiley, R. C. (1979). *The self concept, Vol. 2. Theory and research on selected topics.* Lincoln: University of Nebraska Press.

Wilson, G. T., & O'Leary, K. D. (1980). *Principles of behavior therapy.* Englewood Cliffs, NJ: Prentice-Hall.

Wilson, M. I., & Daly, M. (1996). Male sexual propietariness and violence against wives. *Current Directions in Psychological Science, 5*, 2–7.

Winder, C. L., & Wiggins, J. S. (1964). Social reputation and social behavior: A further validation of the peer nomination inventory. *Journal of Abnormal and Social Psychology, 68*, 681–685.

Winter, D. G. (1973). *The power motive.* New York: Free Press.

Winter, D. G. (1993). Power, affiliation, and war: Three tests of a motivational model. *Journal of Personality and Social Psychology, 65*, 532–545.

Wittgenstein, L. (1953). *Philosophical investigations.* New York: Macmillan.

Wolf, R. (1966). The measurement of environments. In A. Anastasi (Ed.), *Testing problems in perspective* (pp. 491–503). Washington, D.C.: American Council on Education.

Wolpe, J. (1958). *Psychotherapy by reciprocal inhibition.* Stanford, CA: Stanford University Press.

Wolpe, J. (1963). Behavior therapy in complex neurotic states. *British Journal of Psychiatry, 110*, 28–34.

Wolpe, J. (1997). From psychoanalytic to behavioral methods in anxiety disorders: A continuing evolution. In J. K. Zeig (Ed.), *The evolution of psychotherapy: The third conference* (pp. 107–116). New York: Brunner/Mazel, Inc.

Wolpe, J., & Lazarus, A. A. (1966). *Behavior therapy techniques: A guide to the treatment of neuroses.* Elmsford, NY: Pergamon Press.

Wolpe, J., & Rachman, S. (1960). Psychoanalytic evidence: A critique based on Freud's case of Little Hans. *Journal of Nervous and Mental Diseases, 31*, 134–147.

Woodall, K. L., & Matthews, K. A. (1989). Familial environments associated with Type A behaviors and psychophysio-

logical responses to stress in children. *Health Psychology, 8*, 403–426.

Wortman, C. B., & Brehm, J. W. (1975). Responses to uncontrollable outcomes. In L. Berkowitz (Ed.), *Advances in experimental social psychology* (Vol. 8, pp. 278–336). New York: Academic Press.

Wright, J. C., & Mischel, W. (1982). The influence of affect on cognitive social learning person variables. *Journal of Personality and Social Psychology, 43*, 901–914.

Wright, J. C., & Mischel, W. (1987). A conditional approach to dispositional constructs: The local predictability of social behavior. *Journal of Personality and Social Psychology, 53*, 1159–1177.

Wright, J. C., & Mischel, W. (1988). Conditional hedges and the intuitive psychology of traits. *Journal of Personality and Social Psychology, 55*, 454–469.

Wright, J. H., & Beck, A. T. (1996). Cognitive therapy. In R. E. Hales & S. C. Yudofsky (Eds.), *The American Psychiatric Press synopsis of psychiatry* (pp. 1011–1038). Arlington, VA: American Psychiatric Publishers.

Wright, L. (1998). *Twins and what they tell us about who we are.* New York: Wiley.

Young, J. E., Weinberger, A. D., & Beck, A. T. (2001). Cognitive therapy for depression. In D. H. Barlow (Ed.), *Clinical handbook of psychological disorders: A step-by-step treatment manual* (3rd ed., pp. 264–308). New York: Guildford.

Youngblade, L. M., & Belsky, J. (1992). Parent-child antecedents of 5-year-olds' close friendships: A longitudinal analysis. *Developmental Psychology, 28*, 700–713.

Zahn-Waxler, C., Robinson, J. L., & Emde, R. N. (1992). The development of empathy in twins. *Developmental Psychology, 28*, 1038–1047.

Zajonc, R. B. (1980). Feeling and thinking: Preferences need no inferences. *American Psychologist, 35*, 151–175.

Zelenski, J. M., & Larsen, R. J. (2000). The distribution of emotions in everyday life: A state and trait perspective from experience sampling data. *Journal of Research in Personality, 34*, 178–197.

Zeller, A. (1950). An experimental analogue of repression, I. Historical summary. *Psychological Bulletin, 47*, 39–51.

Zirkel, S. (1992). Developing independence in a life transition: Investing the self in the concerns of the day. *Journal of Personality and Social Psychology, 62*, 506–521.

Zuckerman, M. (1978). Sensation seeking. In H. London & J. E. Exner (Eds.), *Dimensions of personality* (pp. 487–559). New York: Wiley Interscience.

Zuckerman, M. (1979). Attribution of success and failure revisited: Or the motivational bias is alive and well in attribution theory. *Journal of Personality, 47*, 245–287.

Zuckerman, M. (1983). A rejoinder to Notarius. *Journal of Personality & Social Psychology, 45*, 1165–1166.

Zuckerman, M. (1984). Sensation seeking: A comparative approach to a human trait. *Behavioral and Brain Sciences, 7,* 413–471.

Zuckerman, M. (1990). The psychophysiology of sensation seeking. *Journal of Personality, 58,* 313–345.

Zuckerman, M. (1991). *The psychobiology of personality.* Cambridge, NY: Cambridge University Press.

Zuckerman, M. (1993). P-impulsive sensation seeking and its behavioral, psychophysiological, and biochemical correlates. *Neuropsychology, 28,* 30–36.

Zuckerman, M. (1994). *Behavioral expressions and biosocial bases of sensation seeking.* New York: Cambridge University Press.

Zuckerman, M., Persky, H., Link, K. E., & Basu, G. K. (1968). Responses to confinement: An investigation of sensory deprivation, social isolation, restriction of movement and set factors. *Perceptual and Motor Skills, 27,* 319–334.

PHOTO CREDITS

Chapter 1   Page 13: © AP/Wide World Photos.

Chapter 2   Page 25: Marcia Weinstein. Page 29: Roger Tully/Stone/Getty Images.

Chapter 3   Page 48: Courtesy Harvard University News Office. Page 50: Times Newspapers Ltd., London. Page 51: Courtesy Hans E. Eysenck.

Chapter 4   Page 77: Image State.

Chapter 5   Page 94: Courtesy National Library of Medicine.

Chapter 6   Pages 115, 118, 123 and 124: Bettmann/Corbis Images. Page 117: Gregory G. Dimijian/Photo Researchers. Page 122: © AP/Wide World Photos. Page 126: J. Nourok/Photo Researchers. Page 129 (left): Joe Gemignani/Corbis Images. Page 129 (right): Corbis Images. Page 131: From "The Restoration of Self" by Heinz Kohut, International Edit. Page 132: Jerome Tisne/Stone/Getty Images.

Chapter 7   Page 152: © AP/Wide World Photos. Page 155: Palmer and Brilliant/Index Stock. Page 161: Bud Gray/Stock, Boston.

Chapter 8   Page 175: © AP/Wide World Photos. Page 178: Brandeis University. Page 185: Douglas A. Land, La Jolla, California. Page 186: David Young-Wolff/PhotoEdit. Page 189: Spencer Grant/PhotoEdit.

Chapter 9   Page 208: Bruce Ayres/Stone/Getty Images.

Chapter 10   Pages 227 and 236: Bettmann/Corbis Images. Page 229: Milton Steinberg/Stock, Boston. Page 239: Bob Daemmrich Photography.

Chapter 11   Page 252: Bob Daemmrich/Stock, Boston.

Chapter 12   Page 273 (left) and 274: Courtesy Albert Bandura. Page 273 (right): F. Martinez/PhotoEdit. Page 277: Courtesy Walter Mischel. Page 278 (top): Sean Arbabi/Stone/Getty Images. Pages 278 (bottom) and 281: SUPERSTOCK. Page 283: Courtesy Julian B. Rotter.

Chapter 13   Page 294: Michael Newman/PhotoEdit. Page 295: Courtesy Susan Anderson. Page 297: Courtesy Hazel Markus. Page 304: Robin Sachs. Page 305: Courtesy Carol Dweck. Page 313: Courtesy University of Waterloo. Graphics Photo/Imaging.

Chapter 14   Page 325 (left): Mary Kate Denny/PhotoEdit. Page 325 (right): Renee Lynn/Stone/Getty Images.

Chapter 15   Page 345: Mansell/Time Pix. Page 360: Stone/Getty Images.

Chapter 16   Page 389: Mark Richards/PhotoEdit.

Chapter 17   Page 396 (top): Reuters New Media/Corbis Images. Page 396 (bottom): © AP/Wide World Photos. Page 398: Courtesy Jack Block. Page 400: Based on Mischel W., Ebbesen, E.B., & Zeiss, A.R., "Cognitive and Attentional Mechanisms in Delay of Gratification," *Journal of Personality and Social Psychology* **21**, 204–218 (1972). Page 407: Courtesy Barry Connors. From *Neuroscience: Exploring the Brain*, Williams & Wilkins, Philadelphia, PA, p. 444.

Chapter 18   Page 425: Courtesy Shelley Taylor. Page 435: AFP/Corbis Images.

NAME INDEX

SUBJECT INDEX